A Dictionary of Genetics

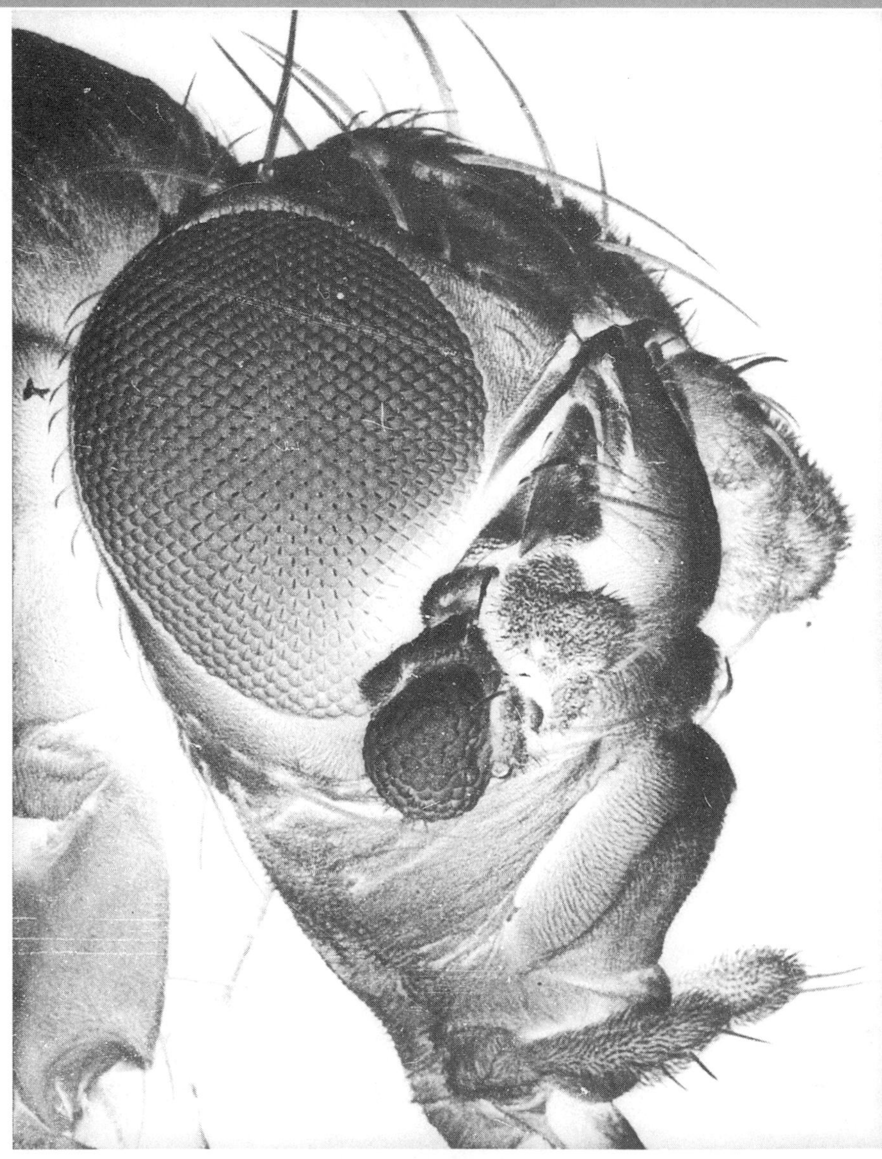

The head of a fruitfly, *Drosophila melanogaster*, viewed by scanning electron microscopy. Targeted expression of the *eyeless* gene has induced the formation of a cluster of eye facets on the distal segment of the antenna, which lies in front of the compound eye. For further details, consult the *eyeless* entry. (Reprinted with permission from Walter Gehring and from *Science*, Vol. 267, No. 5205, 24 March 1995. Photo by Andreas Hefti and George Halder. © 1995, American Association for the Advancement of Science.)

A Dictionary of GENETICS

Seventh Edition

ROBERT C. KING
Emeritus Professor, Northwestern University

WILLIAM D. STANSFIELD
Emeritus Professor, California Polytechnic State University

PAMELA K. MULLIGAN

OXFORD
UNIVERSITY PRESS

2006

OXFORD
UNIVERSITY PRESS

Oxford New York
Auckland Bangkok Buenos Aires Cape Town Chennai
Dar es Salaam Delhi Hong Kong Istanbul Karachi Kolkata
Kuala Lumpur Melbourne Mexico City Mumbai Nairobi
São Paulo Shanghai Singapore Taipei Tokyo Toronto

and an associated company in Berlin

Published by Oxford University Press, Inc.
198 Madison Avenue, New York, New York 10016
www.oup.com

Library of Congress Cataloging-in-Publication Data
King, Robert C., 1928–
A dictionary of genetics / by Robert C. King, William D. Stansfield, Pamela K. Mulligan.— 7th ed.
p. cm.
Includes bibliographical references (p.).
ISBN-13 978-0-19-530762-7
ISBN 0-19-530762-3
ISBN-13 978-0-19-530761-0 (pbk)
ISBN 0-19-530761-5 (pbk)
1. Genetics—Dictionaries. I. Stansfield, William D., 1930– . II. Mulligan, Pamela Khipple, 1953– III. Title.
QH427.K55 2006
576.503—dc22 2005045610

9 8 7 6 5 4 3 2 1

Printed in the United States of America
on acid-free paper

Preface

The field of genetics continues to advance at an astounding pace, marked by numerous extraordinary achievements in recent years. In just the past ten years, the genomic sequence of a multitude of organisms, from archaebacteria to large eukaryotes, has been determined and in many cases, comparatively analyzed in remarkable detail. Expressed sequence tags are being used for the detection of new genes and for genome annotation. DNA microarray technology has taken the study of gene expression and genetic variation to a global, genome-wide scale. Hundreds of new genes and microbial species have been identified by reconstructing the DNA sequences of entire communities of microorganisms collected in environmental samples. A wide variety of new regulatory functions have been assigned to RNA, and RNA interference has become an effective tool for creating loss of function phenotypes.

Such momentous advances in genetics have been accompanied by a deluge of new experimental techniques, computational technologies, databases and internet sites, periodicals and books, and, of course, concepts and terms. Furthermore, as new terminology emerges, many old terms inevitably recede from use or require revision. All this is reflected in the changing content of *A Dictionary of Genetics*, from the publication of its first edition to this seventh edition, 37 years later. This new edition has undergone an extensive overhaul, involving one or more changes (additions, deletions, or modifications of entries) on 95% of the pages of the previous one. The seventh edition contains nearly 7,000 definitions, of which 20% are revised or new, and nearly 1,100 Chronology entries, of which 30% are revised or new. Three hundred of the definitions are accompanied by illustrations or tables, and 16 of these are new. In addition, dozens of recent research papers, books, periodicals, and internet sites of genetic importance have been added to the appropriate Appendices of the current edition.

The year 2006 marks the 100th anniversary of the introduction of the term *genetics* by the British biologist William Bateson. In this seventh edition of *A Dictionary of Genetics*, the term *genetics* itself has been updated, reflecting progress in understanding and technique over the years, and necessitated by the convergence of classical and molecular genetics. Genetics today is no longer simply the study of heredity in the old sense, i.e., the study of inheritance and of variation of biological traits, but also the study of the basic units of heredity, i.e., genes. Geneticists of the post-genomics era identify genetic elements using forward or reverse genetics and decipher the molecular nature of genes, how they function, and how genetic variation, whether introduced in the lab or present in natural populations, affects the phenotype of the cell or the organism. The study of genes is increasingly at the core of genetic research, whether it is aimed at understanding the basis of Alzheimer disease in humans, flower development in *Arabidopsis*, shell pattern variation in *Cepaea* colonies, or speciation in *Drosophila*. Today's genetics thus also unifies the biological sciences, medical sciences, and evolutionary studies.

As a broad-based reference work, *A Dictionary of Genetics* defines terms that fall under this expansive genetics umbrella and includes not only strictly genetic terms, but also genetics-related words encountered in the scientific literature. These include terms referring to biological and synthetic molecules (e.g., *DNA polymerase, Morpholinos,* and *streptavidin*); cellular structures (e.g., *solenoid structure, spectrosome,* and *sponge body*); medical conditions (e.g., *Leber hereditary optic neuropathy [LHON], Marfan syndrome,* and *Tay-Sachs disease*); experimental techniques (e.g., *P element transformation, community genome sequencing,* and *yeast two-hybrid system*); drugs, reagents, and media (e.g., *ethyl methane sulfonate, Denhardt solution,* and *HAT medium*); rules, hypotheses, and laws (e.g., *Haldane rule, wobble hypothesis,* and *Hardy-Weinberg law*); and acronyms (e.g., *BACs, METRO,* and *STS*). Included also are pertinent terms from such fields as geology, physics, and statistics (e.g., *hot spot archipelago, roentgen,* and *chi-square test*).

As in previous editions, the definitions are cross-referenced and comparisons made whenever possible. For example, the *maternal effect gene* entry is cross-referenced to *bicoid, cytoplasmic determinants, cytoplasmic localization, grandchildless genes,* and *maternal polarity mutants,* and the reader is directed to compare it with *paternal effect gene* and *zygotic gene* entries.

In this edition of the *Dictionary* we have made every effort to identify the sources of the more than 120 eponyms appearing among the definitions, and following the example of Victor A. McKusick (distinguished editor of *Mendelian Inheritance in Man*), we have eliminated the possessive form, i.e., apostrophes, in most of the eponyms. Thus, the *Creutzfeld-Jakob disease* entry traces the names of the physicians who first described this syndrome in their patients and the time period when this occurred, and the *Balbiani body* definition identifies the biologist who first described these cellular structures and the time period during which he lived. This additional information under each eponym adds a personal, geographical, and historical perspective to the definitions and is one of the distinguishing features of this dictionary.

The Appendices

A Dictionary of Genetics is unique in that only 80% of the pages contain definitions. The final fifth of the *Dictionary* is devoted to six Appendices, which supply a wealth of useful resource material.

Appendix A, Classification, provides an evolutionary classification of the five kingdoms of living organisms. This list contains 400 words in parentheses, many of which are common names for easy identification (e.g., cellular slime molds, marine worms, and ginkgos). The italicized words in parentheses are genera which contain species notable for their economic importance (e.g., *Bos taurus, Gossypium hirsutum,* and *Oryza sativa*), for causing human diseases (e.g., *Plasmodium falciparum, Staphylococcus aureus,* and *Trypanosoma brucei*), or for being useful laboratory species (e.g., *Arabidopsis thaliana, Neurospora crassa,* and *Xenopus laevis*).

Appendix B, Domesticated Species, lists the common and scientific names of approximately 200 domesticated animal and plant species not found elsewhere in the *Dictionary*.

Appendix C, Chronology, is one of the most distinctive elements of the *Dictionary*, containing a list of notable discoveries, events, and publications, which have contributed to the advancement of genetics. The majority of entries in the Chronology report discoveries (e.g., 1865–66, Mendel's discovery of the existence of hereditary factors; 1970, the finding of RNA-dependent DNA polymerase; 1989, the identification of the cystic fibrosis gene). In addition, there are entries that present unifying concepts and theories (e.g., 1912, the concept of continental drift; 1961, the operon hypothesis; 1974, the proposition that chromatin is organized into nucleosomes). The Chronology also includes important technological advances and techniques that have revolutionized genetic research (e.g., 1923, the building of the first ultracentrifuge; 1975, the development of Southern blotting; 1985, the development of polymerase chain reaction; 1986, the production of the first automated DNA sequencer). There are also entries that contain announcements of new terms that have become part of every geneticist's vocabulary (e.g., 1909, gene; 1971, C value paradox; 1978, intron and exon).

Developments in evolutionary genetics figure prominently in the Chronology. Included in this category are important evolutionary breakthroughs (e.g., 1868, Huxley's description of *Archaeopteryx*; 1977, the discovery of the Archaea by Woese and Fox; 2004, the proposal by Rice and colleagues that viruses evolved from a common ancestor prior to the formation of the three domains of life), and publication of books which have profoundly affected evolutionary thought (e.g., 1859, C. Darwin's *On the Origin of Species*; 1963, E. Mayr's *Animal Species and Evolution*; 1981, L. Margulis's *Symbiosis in Cell Evolution*).

Relatively recent additions to the Chronology are entries for sequencing and analysis of the genomes of species of interest (e.g., 1996, *Saccharomyces cerevisiae*; 1997, *Escherichia coli*; 2002, *Mus musculus*). Finally, the Chronology lists 59 Nobel Prizes awarded to scientists for discoveries that have had a bearing on the progress of genetics (e.g., 1965, to F. Jacob, J. Monod, and A. Lwoff for their contributions to microbial genetics; 1983, to B. McClintock for her discovery of mobile genetic elements in maize; 1993, to R. J. Roberts and P. A. Sharp for discovering split genes). We hope that these and other Chronology entries, spanning the years 1590–2005, provide students, researchers, educators, and historians alike with an understanding of the historical framework within which genetics has developed.

The Chronology in Appendix C is followed by an alphabetical List of the Scientists cited in it, together with the dates of these citations. This list includes Francis Crick, Edward Lewis, Maurice Wilkins, and Hampton Carson (who all died late in 2004), and Ernst Mayr (who died early in 2005), and it provides the dates of milestones in their scientific careers. Finally, Appendix C includes a Bibliography of 170 titles, and among the most recent books are four that give accounts of the lives of David Baltimore, George Beadle, Sidney Brenner, and Rosalind Franklin. Also listed is a video collection (*Conversations in Genetics*) of interviews with prominent geneticists.

Appendix D, Periodicals, lists the titles and addresses of 500 periodicals related to genetics, cell biology, and evolutionary studies, from *Acta Virologica* to *Zygote*.

Appendix E, Internet Sites, contains 132 prominent web site addresses to facilitate retrieval of the wealth of information in the public domain that can be accessed through the World Wide Web. These include addresses for "master" sites (e.g., National Center for Biotechnology Information [NCBI], National Library of Medicine, National Institutes of Health), for individual databases (e.g., GenBank, Single Nucleotide Polymorphisms [SNPs], and Protein Data Bank [PDB]), and for species web sites (e.g., *Agrobacterium tumefaciens*, *Chlamydomonas reinhardii*, and *Gossypium* species).

Appendix F, Genome Sizes and Gene Numbers, tabulates the genome sizes and gene numbers for 49 representative organisms, viruses, or cell organelles that appear in the *Dictionary*. These are listed in order of complexity. The smallest genome listed is that of the MS2 virus, with 3.6×10^3 base pairs encoding just 4 proteins, and the largest listed is that of man, consisting of 3.2×10^9 base pairs of DNA encoding 31,000 genes. Between these entries appear the genome sizes and gene numbers of other viruses, organelles, and a diverse range of organisms representing all five kingdoms.

This is but a small representation of the larger and increasingly complex collections of genomic data which are being generated at an exponential rate and transforming the way we look at relationships between organisms that inhabit this planet. A quick glance at Appendix F raises some intriguing questions. For example, why does *Streptomyces*, a prokaryote, have more genes than *Saccharomyces*, a eukaryote, whose genome size is 28% larger? And why do the genomes of the puffer fish, *Takifugu rubripes*, and man encode roughly the same number of protein-coding genes, even though the puffer fish genome is nearly 88% smaller than the human? Such questions and others are at the forefront of current whole-genome research, as the massive sequence data are evaluated and the information encoded within them extracted. Comparative genomic analyses promise new insights into the evolutionary forces that shape the size and structure of genomes. Furthermore, the intertwining of genetics, genomics, and bioinformatics makes for a strong force for identifying new genetic elements and for unraveling the mysteries of cellular processes in the most minute detail.

Appendix Cross-References. Whenever possible, cross references to the Appendices appear under the appropriate definition. The cross references provide information which complements that in the definition. For example, *nucleolus* is cross-referenced to entries in Appendix C, which indicate that this structure was first observed in the nucleus in 1838, that it was first shown to be divisible into subunits in 1934, that in 1965 the sex chromosomes of *Drosophila melanogaster* were found to contain multiple rRNA genes in their nucleolus organizers, and that in 1967 amplified rDNA was isolated from *Xenopus* oocytes. Furthermore, nucleolar Miller trees were discovered in 1969, in 1976 ribosomal proteins were found to attach to precursor rRNAs in the nucleolus, and in 1989 the cDNA for human nucleolin was isolated. Another example is *Streptomyces*, which is cross-referenced to Appendices A, E, and F. In this case, the material in the Appendices indicates that this organism is a prokaryote belonging to the phylum Actinobacteria, that there is web-based information pertaining to *S. coelicolor* at http://www.sanger.ac.uk, and that the genome of this species has 12.07×10^6 base pairs and contains 7,825 predicted genes. The cross-referenced information in the Appendices thus greatly broadens the reader's perspective on a particular term or concept.

Genetics has clearly entered an exciting new era of exploration and expansion. It is our sincere hope that *A Dictionary of Genetics* will become a helpful companion for those participating in this marvelous adventure.

Rules Regarding the Arrangement of Entries

The arrangement of entries in the current edition has not changed since the publication of the previous edition. Each term appears in boldface and is placed in alphabetical order using the letter-by-letter method, ignoring spaces between words. Thus, *Homo sapiens* is placed between *homopolymer tails* and *homosequential species*, and *H-Y antigen* appears between *hyaluronidase* and *hybrid*. In the case of identical alphabetical listings, lowercase letters precede uppercase letters. Thus, the *p* entry is found before the *P* entry. In entries beginning with a Greek letter, the letter is spelled out. Therefore, β *galactosidase* appears as *beta galactosidase*. When a number is found at the beginning of an entry, the number is ignored in the alphabetical placement. Therefore, *M5 technique* is treated as *M technique* and *T24 oncogene* as *T oncogene*. However, numbers are used to determine the order in the series. For example, *P1 phage* appears before *P22 phage*. For two- or three-word terms, the definition sometimes appears under the second or third word, rather than the first. For example, definitions for *embryonic stem cells* and *germ line transformation* occur under *stem cells* and *transformation*, respectively.

Acknowledgments

We owe the greatest debt to Ellen Rasch, whose critical advice at various stages during the evolution of the dictionary provided us with wisdom and encouragement. We also benefited by following wide-ranging suggestions made by Lloyd Davidson, Joseph Gall, Natalia Shiltsev, Igor Zhimulev, and the late Hampton Carson. Rodney Adam, Bruce Baldwin, Frank Butterworth, Susanne Gollin, Jon Moulton, and Patrick Storto suggested changes that improved the quality of many definitions. Atsuo Nakata kindly brought to our attention many typographical errors that we had missed.

We are grateful to the many scientists, illustrators, and publishers who kindly provided their illustrations to accompany various entries. Robert S. King, who took over secretarial functions from his mother, Suja, and elder brother Tom, worked cheerfully and tirelessly throughout the project. Vikram K. Mulligan suggested various terms and modified others, and Rob and Vikram's drawings illustrate eight of the entries.

Robert C. King
William D. Stansfield
Pamela K. Mulligan

Contents

A Dictionary of Genetics

A

A 1. mass number of an atom; 2. haploid set of autosomes; 3. ampere; 4. adenine or adenosine.

Å Angstrom unit (*q.v.*).

A_2 *See* hemoglobin.

A 23187 *See* ionophore.

AA-AMP amino acid adenylate.

A, B antigens mucopolysaccharides responsible for the ABO blood group system. The A and B antigens reside on the surface of erythrocytes, and differ only in the sugar attached to the penultimate monosaccharide unit of the carbohydrate chain. This minor chemical difference makes the macromolecule differentially active antigenically. The I^A, I^B, and *i* are alleles of a gene residing on the long arm of chromosome 9 between bands 34.1 and 34.2. The I^A and I^B alleles encode A and B glycotransferases, and the difference in their specificities is due to differences in their amino acid sequences at only four positions. These in turn result from different missense mutations in the two alleles. The A and B transferases add *N*-acetyl galactosamine or galactose, respectively, to the oligosaccharide terminus. The *i* allele encodes a defective enzyme, so no additional monosaccharide is added to the chain. Glycoproteins with properties antigenically identical to the A, B antigens are ubiquitous, having been isolated from bacteria and plants. Every human being more than 6 months old possesses those antibodies of the A, B system that are not directed against its own blood-group antigens. These "preexisting natural" antibodies probably result from immunization by the ubiquitous antigens mentioned above. The A and B antigens also occur on the surfaces of epithelial cells, and here they may mask receptors that serve as binding sites for certain pathogenic bacteria. *See* Appendix C, 1901, Landsteiner; 1925, Bernstein; 1990, Yamomoto *et al.*; blood group, *Helicobacter pylori*, H substance, Lewis blood group, MN blood group, null allele, oligosaccharide, P blood group, *Secretor* gene.

ABC model *See* floral identity mutations.

ABC transporters a family of proteins that span the plasma membranes of cells and function to transport specific molecules into or out of the cell. The name is an abbreviation of *A*TP-*B*inding *C*assette. ABC transporters all contain an ATP binding domain, and they utilize the energy of ATP to pump substrates across the membrane against a concentration gradient. The substrates may be amino acids, sugars, polypeptides, or inorganic ions. The product of the cystic fibrosis gene is an ABC transporter. *See* Bacillus, cystic fibrosis (CF), *Escherichia coli*.

Abelson murine leukemia virus an oncogenic virus identified in 1969 by Dr. H. T. Abelson. The transforming gene *v-abl* has a cellular homolog *c-abl*. This is actively transcribed in embryos at all stages and during postnatal development. A homolog of *c-abl* occurs in the human genome at 9q34, and it encodes a protein kinase (*q.v.*). It is this gene which is damaged during the reciprocal interchange that occurs between chromosome 9 at q34 and chromosome 22 at q11, resulting in myeloid leukemia. *See* Philadelphia (Ph^1) chromosome, myeloproliferative disease.

aberrations *See* chromosomal aberration, radiation-induced chromosomal aberration.

ABM paper aminobenzyloxy methyl cellulose paper, which when chemically activated, reacts covalently with single-stranded nucleic acids.

ABO blood group system system of alleles residing on human chromosome 9 that specifies certain red cell antigens. *See* AB antigens, blood groups, Bombay blood group.

abortion 1. The expulsion of a human fetus from the womb by natural causes, before it is able to survive independently; this is sometimes called a miscarriage (*q.v.*). 2. The deliberate termination of a human pregnancy, most often performed during the first 28 weeks of pregnancy. 3. The termination of development of an organ, such as a seed or fruit.

abortive transduction failure of a transducing exogenote to become integrated into the host chromosome, but rather existing as a nonreplicating particle in only one cell of a clone. *See* transduction.

abortus a dead fetus born prematurely, whether the abortion was artificially induced or spontaneous. Over 20% of human spontaneous abortions show chromosomal abnormalities. *See* Appendix C, 1965, Carr.

abscisic acid a plant hormone synthesized by chloroplasts. High levels of abscisic acid result in the abscission of leaves, flowers, and fruits. The hormone also causes the closing of stomata in response to dehydration.

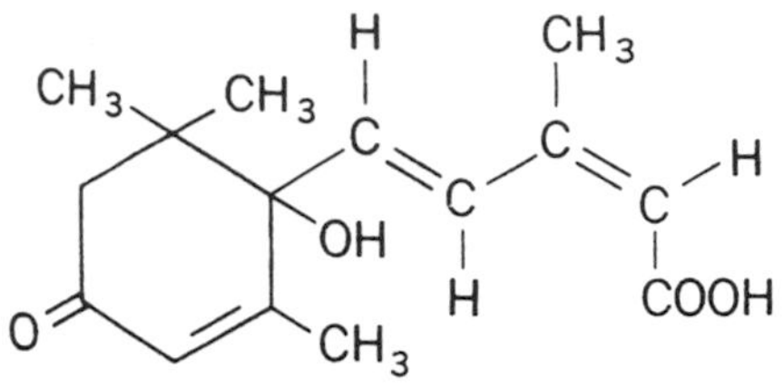

abscission the process whereby a plant sheds one of its parts, such as leaves, flowers, seeds, or fruits.

absolute plating efficiency the percentage of individual cells that give rise to colonies when inoculated into culture vessels. *See* relative plating efficiency.

absorbance (*also* absorbancy) a measure of the loss of intensity of radiation passing through an absorbing medium. It is defined in spectrophotometry by the relation log (I_o/I), where I_o = the intensity of the radiation entering the medium and I = the intensity after traversing the medium. *See* Beer-Lambert law, OD_{260} unit.

abundance in molecular biology, the average number of molecules of a specific mRNA in a given cell, also termed *representation.* The abundance, $A = NRf/M$, where N = Avogadro's number, R = the RNA content of the cell in grams, f = the fraction the specific RNA represents of the total RNA, and M = the molecular weight of the specific RNA in daltons.

abzymes catalytic antibodies. A class of monoclonal antibodies that bind to and stabilize molecules in the transition state through which they must pass to form products. *See* enzyme.

acatalasemia the hereditary absence of catalase (*q.v.*) in humans. Mutations in the structural gene on chromosome 11 at p13 result in the production of an unstable form of the enzyme. The gene is 34 kb in length and contains 13 exons.

acatalasia synonym for acatalasemia (*q.v.*).

acceleration *See* heterochrony.

accelerator an apparatus that imparts kinetic energy to charged subatomic particles to produce a high-energy particle stream for analyzing the atomic nucleus.

acceptor stem the double-stranded branch of a tRNA molecule to which an amino acid is attached (at the 3′, CCA terminus) by a specific aminoacyl-tRNA synthetase. *See* transfer RNA.

accessory chromosomes *See* B chromosomes.

accessory nuclei bodies resembling small nuclei that occur in the oocytes of most Hymenoptera and those of some Hemiptera, Coleoptera, Lepidoptera, and Diptera. Accessory nuclei are covered by a double membrane possessing annulate pores. They are originally derived from the oocyte nucleus, but they subsequently form by the amitotic division of other accessory nuclei.

***Ac, Ds* system** *Activator–Dissociation* system (*q.v.*).

ace *See* symbols used in human cytogenetics.

acentric designating a chromatid or a chromosome that lacks a centromere. *See* chromosome bridge.

Acer the genus of maple trees. *A. rubrum*, the red maple, and *A. saccharum*, the sugar maple, are studied genetically because of their commercial importance.

Acetabularia a genus of large, unicellular green algae. Each organism consists of a base, a stalk, and a cap. The base, which contains the nucleus, anchors the alga to the supporting rocks. The stalk, which may be 5 cm long, joins the base and the cap. The cap carries out photosynthesis and has a species-specific shape. For example, the disc-shaped cap of *A. mediterranea* is smooth, whereas the cap of *A. crenulata* is indented. Hammerling cut the base and cap off a *crenulata* alga and then grafted the stalk on a *mediterranea* base. The cap that regenerated was smooth, characteristic of the species that provided the nucleus. Heterografts like these provided some of the earliest evidence that the nucleus could send messages that directed developmental programs at distant regions of the cell. *See* Appendix A, Protoctista, Chlorophyta; Appendix C, 1943, Hammerling; graft.

Acetobacter a genus of aerobic bacilli which secure energy by oxidizing alcohol to acetic acid.

aceto-orcein a fluid consisting of 1% orcein (*q.v.*) dissolved in 45% acetic acid, used in making squash preparations of chromosomes. *See* salivary gland squash preparation.

acetylcholine a biogenic amine that plays an important role in the transmission of nerve impulses across synapses and from nerve endings to the muscles innervated. Here it changes the permeability of the sarcolemma and causes contraction. Acetylcholine is evidently a very ancient hormone, since it is present even in protists.

$$H_3C-\overset{\overset{O}{\|}}{C}-O-CH_2-CH_2-\overset{+}{N}(CH_3)_3$$

acetylcholinesterase the enzyme that catalyses the hydrolysis of acetylcholine (*q.v.*) into choline and acetate. Also called *cholinesterase*.

acetyl-coenzyme A *See* coenzyme A.

acetyl serine *See* *N*-acetyl serine.

achaete-scute complex a complex locus in *Drosophila* first identified by mutations that affected the development of adult bristles. Lack of the entire complex results in the failure of neurogenesis during the embryo stage. The complex contains four ORFs that encode DNA-binding proteins that contain helix-turn-helix motifs (*q.v.*).

achiasmate referring to meiosis without chiasmata. In those species in which crossing over is limited to one sex, the achiasmate meiosis generally occurs in the heterogametic sex.

Achilles' heel cleavage (AHC) a technique that allows a DNA molecule to be cut at a specified site. The name comes from the legend in Greek mythology where Achilles' mother dipped him in the river Styx. The waters made him invulnerable, except for the heel by which she held him. In the AHC procedure a sequence-specific DNA-binding molecule is complexed with the DNA under study. A methyltransferase is then added to methylate all CpG sequences except those hidden under the sequence-specific DNA-binding molecule. Next, this molecule and the methyltransferases are removed, and a restriction endonuclease is added. This will cut the DNA only in the region where methylation was blocked, i.e., the "Achilles' heel."

achondroplasia a form of hereditary dwarfism due to retarded growth of the long bones. It is the most common form of dwarfism in humans (1 in 15,000 live births) and is inherited as an autosomal dominant trait. Homozygotes die at an early age. The gene responsible has been mapped to chromosome 4p16.3. The *ACH* gene has been renamed *FGFR3*, since it encodes the Fibroblast Growth Factor Receptor 3, a protein containing 806 amino acids. The gene contains 14,975 bp of DNA and produces two alternative transcripts. Homologous genes have been identified in rat, mouse, *Xenopus*, and zebrafish. The genes are expressed in the chondrocytes of developing bones. *See* bovine achondroplasia, *de novo* mutation, fowl achrondroplasia, positional candidate approach.

achromatic figure the mitotic apparatus (*q.v.*).

A chromosomes *See* B chromosomes.

acid fuchsin an acidic dye used in cytochemistry.

acidic amino acid an amino acid (*q.v.*) having a net negative charge at neutral pH. Those universally found in proteins are aspartic acid and glutamic acid, which bear negatively charged side chains in the pH range generally found in living systems.

acidic dye an organic anion that binds to and stains positively charged macromolecules.

Acinonyx jubatus the cheetah, a carnivore that has the distinction of being the world's fastest land animal. Cheetahs are of genetic interest because, while most other species of cats show heterozygosity levels of 10–20%, cheetahs have levels close to zero. This high degree of homozygosity is correlated with low fecundity, high mortality of cubs, and low disease resistance.

Acoelomata a subdivision of the Protostomia-containing species in which the space between the epidermis and the digestive tube is occupied by a cellular parenchyma. *See* classification.

acquired characteristics, inheritance of inheritance by offspring of characteristics that arose in their parents as responses to environmental influences and are not the result of gene action. *See* Lamarckism.

acquired immunodeficiency syndrome *See* AIDS, HIV.

Acraniata a subphylum of Chordata containing animals without a true skull. *See* Appendix A.

acrasin a chemotactic agent produced by *Dictyostelium discoideum* that is responsible for the aggregation of the cells. Acrasin has been shown to be cyclic AMP (*q.v.*).

Acrasiomycota the phylum containing the cellular slime molds. These are protoctists that pass through a unicellular stage of amoebas that feed on bacteria. Subsequently, these amoebas aggregate to form a fruiting structure that produces spores. The two most extensively studied species from this phylum are *Dictyostelium discoideum* and *Polysphondylium pallidum*.

acridine dyes heterocyclic compounds that include acridine (shown below) and its derivatives. These molecules bind to double-stranded DNAs as intercalating agents. Examples of acridine dyes are acridine organe, acriflavin, proflavin, and quinicrine (*all of which see*).

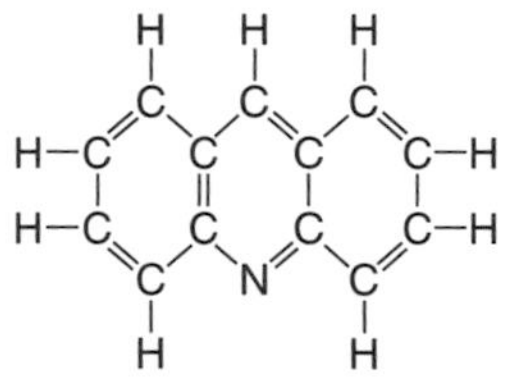

acridine orange an acridine dye that functions both as a fluorochrome and a mutagen.

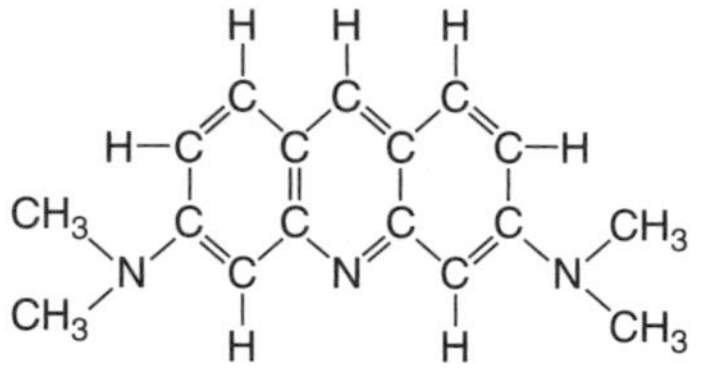

acriflavin an acridine dye that produces reading frame shifts (*q.v.*).

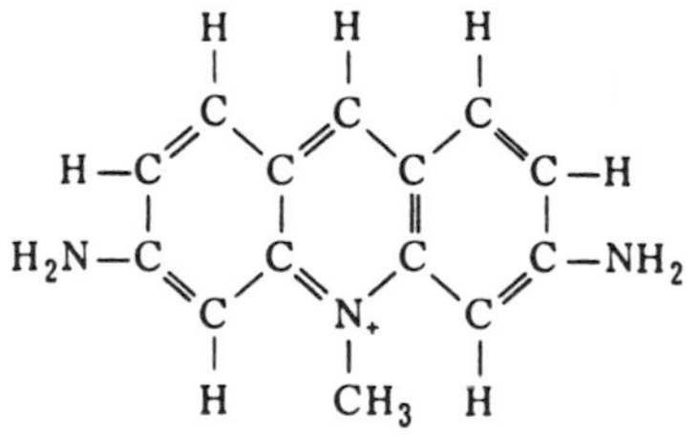

acritarchs spherical bodies thought to represent the earliest eukaryotic cells, estimated to begin in the fossil record about 1.6 billion years ago. Most acritarchs were probably thick-walled, cyst-forming protists. *See* Proterozoic.

acrocentric designating a chromosome or chromatid with a nearly terminal centromere. *See* telocentric chromosome.

acromycin *See* tetracycline.

acron the anterior nonsegmented portion of the embryonic arthropod that produces eyes and antennae. *See* maternal polarity mutants.

acrosome an apical organelle in the sperm head that is secreted by the Golgi material and that digests the egg coatings to permit fertilization.

acrostical hairs one or more rows of small bristles along the dorsal surface of the thorax of *Drosophila*.

acrosyndesis telomeric pairing by homologs during meiosis.

acrotrophic *See* meroistic.

acrylamide *See* polyacrylamide gel.

ACTH adrenocorticotropic hormone (*q.v.*).

actidione cycloheximide.

actin a protein that is the major constituent of the 7-nanometer-wide microfilaments of cells. Actin microfilaments (F actin) are polymers of a globular subunit (G actin) of Mr 42,000. Each G actin molecule has a defined polarity, and during polymerization the subunits align "head to tail," so that all G actins point in the same direction. F actin grows by the addition of G actin to its ends, and cytochalasin B (*q.v.*) inhibits this process. All the actins that have been studied, from sources as diverse as slime molds, fruit flies, and vertebrate muscle cells, are similar in size and amino acid sequence, suggesting that they evolved from a single ancestral gene. In mammals and birds, there are four different muscle actins. α_1 is unique to skeletal muscle; α_2, to cardiac muscle; α_3, to smooth vascular muscle; and α_4, to smooth enteric muscle. Two other actins (β and γ) are found in the cytoplasm of both muscle and nonmuscle cells. *See* alternative splicing, contractile ring, fibronectin, *hu-li tai shuo (hts)*, isoform, *kelch*, myosin, ring canals, spectrin, stress fibers, tropomyosin, vinculin.

actin-binding proteins a large family of proteins that form complexes with actin. Such proteins include certain heat-shock proteins, dystrophin, myosin, spectrin, and tropomyosin (*all of which see*).

actin genes genes encoding the various isoforms of actin. In *Drosophila*, for example, actin genes have been localized at six different chromosomal sites. Two genes encode cytoplasmic actins, while the other four encode muscle actins. The amino acid–encoding segments of the different actin genes have very similar compositions, but the segments specifying the trailers (*q.v.*) differ considerably in nucleotide sequences.

actinomycete any prokaryote placed in the phylum actinobacteria (*see* Appendix A). Actinomycetes belonging to the genus *Streptomyces* produce a large number of the antibiotics, of which actinomycin D (*q.v.*) is an example.

actinomycin D an antibiotic produced by *Streptomyces chrysomallus* that prevents the transcription of messenger RNA. *See* RNA polymerase.

activated macrophage a macrophage that has been stimulated (usually by a lymphokine) to enlarge, to increase its enzymatic content, and to increase its nonspecific phagocytic activity.

activating enzyme an enzyme that catalyzes a reaction involving ATP and a specific amino acid. The product is an activated complex that subsequently reacts with a specific transfer RNA.

activation analysis a method of extremely sensitive analysis based on the detection of characteristic radionuclides produced by neutron activation.

activation energy the energy required for a chemical reaction to proceed. Enzymes (*q.v.*) combine transiently with a reactant to produce a new complex that has a lower activation energy. Under these circumstances the reaction can take place at the prevailing temperature of the biological system. Once the product is formed, the enzyme is released unchanged.

activator a molecule that converts a repressor into a stimulator of operon transcription; e.g., the repressor of a bacterial arabinose operon becomes an activator when combined with the substrate.

***Activator-Dissociation* system** a pair of interacting genetic elements in maize discovered and analyzed by Barbara McClintock. *Ac* is an *autonomous* element that is inherently unstable. It has the ability to excise itself from one chromosomal site and to transpose to another. *Ac* is detected by its activation of *Ds*. *Ds* is *nonautonomous* and is not capable of excision or transposition by itself. *Ac* need not be adjacent to *Ds* or even on the same chromosome in order to activate *Ds*. When *Ds* is so activated, it can alter the level of expression of neighboring genes, the structure of the gene product, or the time of development when the gene expresses itself, as a consequence of nucleotide changes inside or outside of a given cistron. An activated *Ds* can also cause chromosome breakage, which may yield deletions or generate a breakage-fusion-bridge cycle (*q.v.*). It is now known that *Ac* is a 4,500 bp segment of DNA that encodes a transposable element (*q.v.*) which contains within it the locus of a functional transposase (*q.v.*). The transposase gives *Ac* the ability to detach

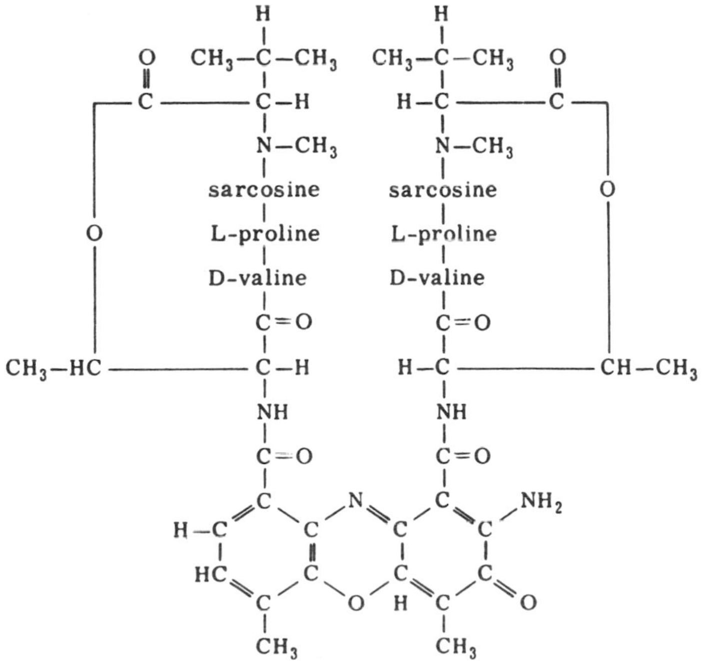

Actinomycin D

from one chromosome and then insert into another. The excision of *Ac* may cause a break in the chromosome, and this is what generated the breakage-fusion-bridge cycles that McClintock observed. *Ds* is a defective transpon that contains a deletion in its transposase locus. Therefore the *Ds* transposon can move from chromosome to chromosome only if *Ac* is also in the nucleus to supply its transposase. *Ac* and *Ds* were originally classified as mutator genes, since they would sometimes insert into structural genes and modify their functioning. *See* Appendix C, 1950, McClintock; 1984, Pohlman *et al.*; *Dotted*, genomic instability, mutator gene, terminal inverted repeats (TIRs), transposon tagging.

active center in the case of enzymes, a flexible portion of the protein that binds to the substrate and converts it into the reaction product. In the case of carrier and receptor proteins, the active center is the portion of the molecule that interacts with the specific target compounds.

active immunity immunity conferred on an organism by its own exposure and response to antigen. In the case of immunity to disease-causing agents, the antigenic pathogens may be administered in a dead or attenuated form. *See also* passive immunity.

active site that portion(s) of a protein that must be maintained in a specific shape and amino acid content to be functional. Examples: **1.** in an enzyme, the substrate-binding region; **2.** in histones or repressors, the parts that bind to DNA; **3.** in an antibody, the part that binds antigen; **4.** in a hormone, the portion that recognizes the cell receptor.

active transport the movement of an ion or molecule across a cell membrane against a concentration or electrochemical gradient. The process requires specific enzymes and energy supplied by ATP.

activin a protein first isolated from the culture fluid of *Xenopus* cell lines. Activin is a member of the transforming growth factor-β (*q.v.*) family of intercellular signaling molecules. It acts as a diffusible morphogen for mesodermal structures, and the type of differentiation is determined by the concentration of actin (i.e., high concentrations produce head structures, low concentrations tail structures).

actomyosin *See* myosin.

***acute myeloid leukemia 1* gene (*AML1*)** a gene that maps to 21q22.3 and is one of the most frequent targets of chromosome translocations associated with leukemia. The involvement of *AML1* with the oncogenic transformation of blood cells is worth noting, since acute myeloid leukemia is hundreds of times more common in children with trisomy 21 than in other children. *See* Down syndrome, *lozenge*, myeloproliferative disease.

acute transfection infection of cells with DNA for a short period of time.

acylated tRNA a transfer RNA molecule to which an amino acid is covalently attached. Also referred to as an activated tRNA, a charged tRNA, or a loaded tRNA.

adaptation **1.** the process by which organisms undergo modification so as to function more perfectly in a given environment. **2.** any developmental, behavioral, anatomical, or physiological characteristic of an organism that, in its environment, improves its chances for survival and of leaving descendants.

adaptive enzyme an enzyme that is formed by an organism in response to an outside stimulus. The term has been replaced by the term *inducible enzyme*. The discovery of adaptive enzymes led eventually to the elucidation of the mechanisms that switch gene transcription on and off. *See* Appendix C, 1937, Karström; regulator gene.

adaptive immunity the immunity that develops in response to an antigens (*q.v.*), as opposed to innate or natural immunity. *Contrast with* innate immunity.

adaptive landscape a three-dimensional graph that shows the frequencies of two genes, each present in two allelic forms (aA and bB in the illustration) plotted against average fitness for a given set of environmental conditions, or a comparable conceptual plot in multidimensional space to accommodate more than two loci.

adaptive melanism hereditary changes in melanin production that cause the darkening in color of populations of animals in darkened surroundings. By improving their camouflage, this makes them less conspicuous to predators. For example, desert mice are preyed upon by owls, hawks, and foxes. The mice that live among sand and light-colored rocks are tan and blend in well with their surroundings. However, the fur from populations of the same species that live among outcrops of dark, ancient lava flows is much darker. *See Chaetodipus intermedius.*

adaptive norm the array of genotypes (compatible with the demands of the environment) possessed by a given population of a species.

adaptive peak a high point (perhaps one of several) on an adaptive landscape (*q.v.*), from which movement in any planar direction (changed gene frequencies) results in lower average fitness.

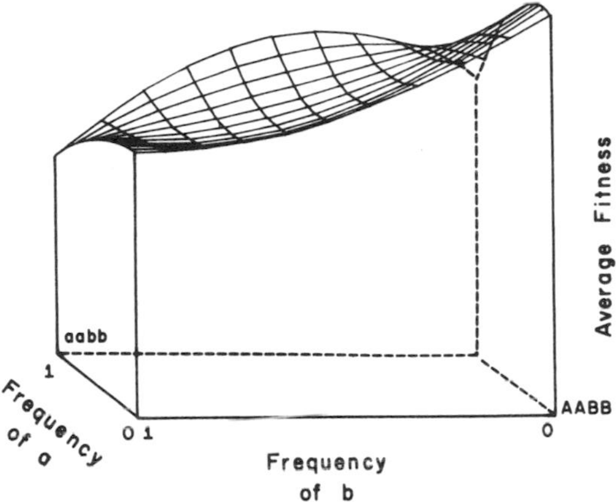

Adaptive landscape

adaptive radiation the evolution of specialized species, each of which shows adaptations to a distinctive mode of life, from a generalized ancestral species. Darwin observed the adaptive radiation of finch species on the Galapagos islands. The Hawaiian archipelago shows perhaps the most spectacular examples of adaptive radiations. *See* Darwin's finches, Hawaiian Drosophilidae, silversword alliance.

adaptive surface, adaptive topography synonyms for adaptive landscape (*q.v.*).

adaptive value the property of a given genotype when compared with other genotypes that confers fitness (*q.v.*) to an organism in a given environment.

adaptor a short, synthetic DNA segment containing a restriction site that is coupled to both ends of a blunt-ended restriction fragment. The adaptor is used to join one molecule with blunt ends to a second molecule with cohesive ends. The restriction site of the adaptor is made identical to that of the other molecule so that when cleaved by the same restriction enzyme both DNAs will contain mutually complementary cohesive ends.

adaptor hypothesis the proposal that polynucleotide adaptor molecules exist that can recognize specific amino acids and also the regions of the RNA templates that specify the placement of amino acids in a newly forming polypeptide. *See* Appendix C, 1958, Crick; transfer RNA.

ADCC antibody-dependent cellular cytotoxicity; also known as antibody-dependent cell-mediated cytotoxicity. Cell-mediated cytotoxicity requires prior binding of antibody to target cells for killing to occur. It does not involve the complement cascade. *See* K cells.

additive factor one of a group of nonallelic genes affecting the same phenotypic characteristics and each enhancing the effect of the other in the phenotype. *See* quantitative inheritance.

additive gene action 1. a form of allelic interaction in which dominance is absent; the heterozygote is intermediate in phenotype between homozygotes for the alternative alleles. 2. the cumulative contribution made by all loci (of the kind described above) to a polygenic trait.

additive genetic variance genetic variance attributed to the average effects of substituting one allele for another at a given locus, or at the multiple loci governing a polygenic trait. It is this component of variance that allows prediction of the rate of response for selection of quantitative traits. *See* quantitative inheritance.

adducin a ubiquitously expressed protein found in the membranes of animal cells. Mammalian adducin is a heterodimeric protein whose subunits share sequence similarities and contain protease-resistant N-terminal and protease-sensitive C-terminal domains. Adducin has a high affinity for Ca^{2+}/calmodulin and is a substrate for protein kinases. *In vitro* it causes actin filaments to form bundles and promotes spectrin-actin associations in regions where cells contact one another. In *Drosophila*, a homolog of mammalian adducin is encoded by the *hts* gene. *See* calmodulin, fusome, *hu-li tai shao (hts)*, heterodimer, protein kinase, spectrosome.

adduct the product of a chemical reaction that results in the addition of a small chemical group to a relatively large recipient molecule. Thus the alkylating agent ethyl methane sulfonate (*q.v.*) can add ethyl groups to the guanine molecules of DNA. These ethylated guanines would be examples of DNA adducts.

adenine *See* bases of nucleic acids.

adenine deoxyriboside *See* nucleoside.

adenohypophysis the anterior, intermediate, and tuberal portions of the hypophysis, which originate from the buccal lining in the embryo.

adenohypophysis hormone *See* growth hormone.

adenosine *See* nucleoside.

adenosine deaminase deficiency a rare immune deficiency disease due to mutations in a gene located on the long arm of human chromosome 20. The normal gene encodes an enzyme that controls the metabolism of purines, and ADA deficiency impairs the functioning of white blood cells. The division of T cells is depressed, and antibody production by B cells is reduced. As a result, ADA-deficient children die from viral, bacterial, and fungal infections. ADA deficiency is the first hereditary disease to be successfully treated by gene therapy. *See* Appendix C, 1990, Anderson; immune response.

adenosine phosphate any of three compounds in which the nucleoside adenosine is attached through its ribose group to one, two, or three phosphoric acid molecules, as illustrated here. AMP, ADP, and ATP are interconvertible. ATP upon hydrolysis yields the energy used to drive a multitude of biological processes (muscle contraction, photosynthesis, bioluminescence, and the biosynthesis of proteins, nucleic acids, polysaccharides and lipids). The most important process in human nutrition is the synthesis of ATP. Every day human beings synthesize, breakdown, and resynthesize an amount of ATP equaling their body weight. *See* Appendix C, 1929; Lohmann; ATPase, ATP synthase, cellular respiration, citric acid cycle, cytochromes, electron transport chain, glycolysis, mitochondria, oxidative phosphorylation, mitochondrial proton transport.

adenovirus any of a group of spherical DNA viruses characterized by a shell containing 252 capsomeres. Adenoviruses infect a number of mammalian species including humans. *See* human adenovirus 2 (HAdV-2), virus.

adenylcyclase the enzyme that catalyzes the conversion of ATP into cyclic AMP (*q.v.*). Also called adenylate cyclase. *See* adenosine phosphate.

adenylic acid *See* nucleotide.

ADH the abbreviation for alcohol dehydrogenase (*q.v.*).

adhesion plaques *See* vincullin.

adhesive molecules any pair of complementary cell-surface molecules that bind specifically to one another, thereby causing cells to adhere to one another, as do carbohydrates and protein lectins (*q.v.*). Phenomena dependent on adhesive molecules include invasion of host cells by bacteria and viruses, species-specific union of sperms and eggs, and aggregation of specific cell types during embryological development. *See* cell affinity, hemagglutinins, P blood group, selectins.

adjacent disjunction, adjacent segregation *See* translocation heterozygote.

adjuvant a mixture injected together with an antigen that serves to intensify unspecifically the immune response. *See* Freund's adjuvant.

adoptive immunity the transfer of an immune function from one organism to another through the transfer of immunologically active or competent cells. Also called *adoptive transfer*.

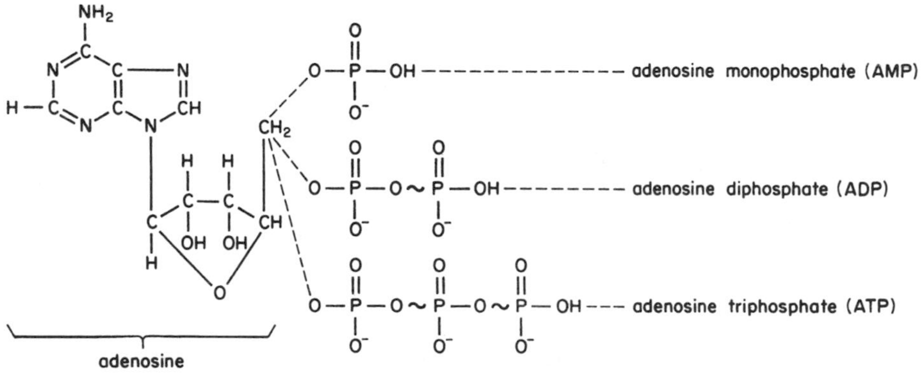

Adenosine phosphate

ADP adenosine diphosphate. *See* adenosine phosphate.

adrenal corticosteroid a family of steroid hormones formed in the adrenal cortex. There are more than 30 of these hormones, and all are synthesized from cholesterol by cortical cells that have been stimulated by the adrenocorticotropic hormone (*q.v.*).

adrenocorticotropic hormone a single-chain peptide hormone (39 amino acids long) stimulating secretion by the adrenal cortex. It is produced by the adenohypophysis of vertebrates. Abbreviated ACTH. Also called *corticotropin*.

Adriamycin an antibiotic produced by *Streptomyces peucetius* that interacts with topoisomerase. DNA isolated from Adriamycin-poisoned cells contains single- and double-strand breaks. *See* gyrase, mitotic poison.

adult tissue stem cells *See* stem cells

advanced in systematics, the later or derived stages or conditions within a lineage that exhibits an evolutionary advance; the opposite of primitive.

adventitious embryony the production by mitotic divisions of an embryonic sporophyte from the tissues of another sporophyte without a gametophytic generation intervening.

Aedes a genus of mosquitoes containing over 700 species, several of which transmit important human diseases. *A. aegypti*, the vector of yellow fever, has a diploid chromosome number 6, and about 60 mutations have been mapped among its three linkage groups. Among these are genes conferring resistance to insecticides such as DDT and pyrethrins (*both of which see*).

Aegilops a genus of grasses including several species of genetic interest, especially *A. umbellulata*, a wild Mediterranean species resistant to leaf rust. A gene for rust resistance has been transferred from *A. umbellulata* to *Triticum vulgare* (wheat).

aerobe an organism that requires molecular oxygen and lives in an environment in contact with air.

aestivate to pass through a hot, dry season in a torpid condition. *See also* hibernate.

afferent leading toward the organ or cell involved. In immunology, the events or stages involved in activating the immune system. *Compare with* efferent.

affinity in immunology, the innate binding power of an antibody combining site with a single antigen binding site. *Compare with* avidity.

affinity chromatography a technique for separating molecules by their affinity to bind to ligands (e.g., antibodies) attached to an insoluble matrix (e.g., Sepharose). The bound molecules can subsequently be eluted in a relatively pure state.

afibrinogenemia an inherited disorder of the human blood-clotting system characterized by the inability to synthesize fibrinogen; inherited as an autosomal recessive.

aflatoxins a family of toxic compounds synthesized by *Aspergillus flavus* and other fungi belonging to the same genus. Aflatoxins bind to purines, making base pairing impossible, and they inhibit both DNA replication and RNA transcription. These mycotoxins are highly toxic and carcinogenic, and they often are contaminants of grains and oilseed products that are stored under damp conditions. The structure of aflatoxin G_1 is shown.

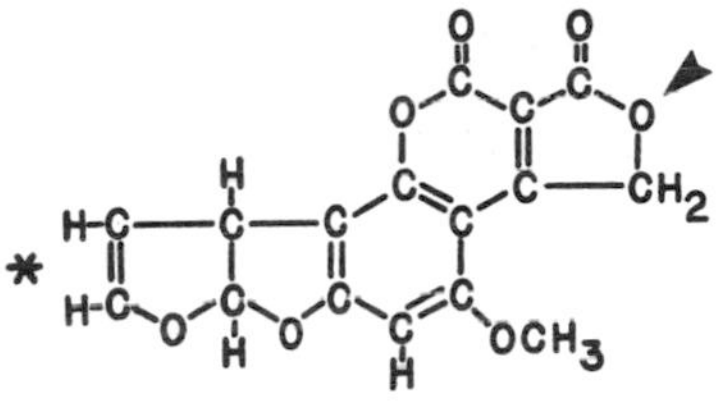

Aflatoxin B_1 has a CH_2 substituted for the O at the position marked by the arrow. Aflatoxin B_2 and G_2 are identical to B_1 and G_1, except that the ring labeled with an asterisk lacks a double bond.

African bees *Apis mellifera scutellata*, a race of bees, originally from South Africa, that was accidentally introduced into Brazil in 1957 and has spread as far as the southern United States. African bees are poor honey producers and tend to sting much more often than European bees. Because of daily differences in flight times of African queens and European drones, hybridization is rare. *See Apis mellifera.*

African Eve *See* mitochondrial DNA lineages.

African green monkey *See Cercopithecus aethiops.*

agamete a haploid, asexual reproductive cell resulting from meiosis in an agamont. Agametes disperse and grow into gamonts (*q.v.*).

agammaglobulinemia the inability in humans to synthesize certain immunoglobulins. The most common form is inherited as an X-linked recessive trait, which is symbolized XLA (X-linked agammaglobulinemia) in the early literature. When O. C. Bruton described the condition in 1952, it was the first hereditary immune disease to be reported. The disease is now known to be caused by mutations in a gene at Xq21.3–q22. The gene is 36,740 bp long, and it encodes a protein containing 659 amino acids. The protein is a tyrosine kinase that has been named in Bruton's honor, and the gene is now symbolized *BTK*, for its product, the Bruton tyrosine kinase. The enzyme is a key regulator in the development of B lymphocytes. Boys with XLA lack circulating B cells. The bone marrow contains pre-B cells, but they are unable to mature. *See* antibody.

agamogony the series of cellular or nuclear divisions that generates agamonts.

agamont the diploid adult form of a protoctist that also has a haploid adult phase in its life cycle. An agamont undergoes meiosis and produces agametes. *See* gamont.

agamospermy the formation of seeds without fertilization. The male gametes, if present, serve only to stimulate division of the zygote. *See* apomixis.

agamous *See* floral identity mutations.

Agapornis a genus of small parrots. The nest building of various species and their hybrids has provided information on the genetic control of behavior patterns.

agar a polysaccharide extract of certain seaweeds used as a solidifying agent in culture media.

agarose a linear polymer of alternating D-galactose and 3,6-anhydrogalactose molecules. The polymer, fractionated from agar, is often used in gel electrophoresis because few molecules bind to it, and therefore it does not interfere with electrophoretic movement of molecules through it.

agar plate count the number of bacterial colonies that develop on an agar-containing medium in a petri dish seeded with a known amount of inoculum. From the count, the concentration of bacteria per unit volume of inoculum can be determined.

age-dependent selection selection in which the values for relative fitness of different genotypes vary with the age of the individual.

agglutination the clumping of viruses or cellular components in the presence of a specific immune serum.

agglutinin any antibody capable of causing clumping of erythrocytes, or more rarely other types of cells.

agglutinogen an antigen that stimulates the production of agglutinins.

aggregation chimera a mammalian chimera made through the mingling of cells of two embryos. The resulting composite embryo is then transferred into the uterus of a surrogate mother where it comes to term. *See* allophenic mice.

aging growing old, a process that has a genetic component. Hereditary diseases are known in humans that cause premature aging, and mutations that speed up or delay aging have been isolated in *Saccharomyces*, *Caenorhabditis*, and *Drosophila*. *See* Appendix C, 1994, Orr and Sohal; 1995, Feng *et al.*; antioxidant enzymes, apoptosis, *daf-2*, free radical hypothesis of aging, *Indy*, *methuselah*, *Podospora anserina*, progeria, SGSI, telomerase, senescence, Werner syndrome.

agonistic behavior any social interaction between members of the same species that involves aggression or threat and conciliation or retreat.

agouti the grizzled color of the fur of mammals resulting from alternating bands of yellow (phaeomelanin) and black (eumelanin) pigments in individual hairs. The name is also given to the genes that control the hair color patterns. In the mouse more than 20 alleles have been described at the *agouti* locus on chromosome 2. The gene encodes a cysteine-rich, 131 amino acid protein that instructs the melanocytes in the hair follicle when to switch from making black to yellow pigment. The protein is translated by nearby follicle cells rather than in the melanocytes themselves. Therefore, the agouti protein acts as a paracrine-signaling molecule. *See* Appendix C, 1905, Cuénot; autocrine, *MC1R* gene, melanin.

agranular reticulum *See* smooth endoplasmic reticulum (SER).

agranulocytes white blood cells whose cytoplasm contains few or no granules and that possess an unlobed nucleus; mononuclear leucocytes including lymphocytes and monocytes.

agriculturally important species *See* Appendix B.

Agrobacterium tumefaciens the bacterium responsible for crown gall disease (*q.v.*) in a wide range of dicotyledonous plants. The bacterium enters only dead, broken plant cells and then may transmit a tumor-inducing plasmid into adjacent living plant cells. This infective process is a natural form of genetic engineering, since the bacterium transfers part of its DNA to the infected plant. This is integrated into the plant genome, and here it induces changes in metabolism and tumor formation. The genome of strain 58 of *A. tumefaciens* was sequenced in 2001 by a group led by E. W. Nestor. Its genome contained 5.67 mbp of DNA distributed among four replicons: a circular chromosome (CC), a linear chromosome (LC), and two plasmids (pAt and pTi). The plasmids are DNA circles, and most of the genes responsible for crown galls are on pTi. The Table lists the general features of the four replicons.

Replicon	Size (kbp)	ORFs	tRNAs	rRNAs
CC	2,841	2,789	40	2
LC	2,076	1,882	13	2
pAt	543	550	0	0
pTi	214	198	0	0

Strains of *A. tumefaciens* carrying the plasmid may be artificially genetically engineered to introduce foreign genes of choice into plant cells, and then by growing the cells in tissue culture, whole plants can be regenerated, every cell of which contains the foreign gene. *See* Appendix A, Bacteria, Proteobacteria; Appendix C, 1907, Smith and Townsend; 1981, Kemp and Hall; 2001, Wood *et al.*; Appendix E; Ti plasmid.

Agropyron elongatum a weed related to crabgrass noted for its resistance to stem rust. Genes conferring rust resistance have been transferred from this species to *Triticum aestivum* (wheat).

AHC Achilles' heel cleavage (*q.v.*).

AHF antihemophilic factor. *See* blood clotting.

AI, AID, AIH *See* artificial insemination.

AIA anti-immunoglobulin antibodies, produced in response to foreign antibodies introduced into an experimental animal.

AIDS the *a*cquired *i*mmuno*d*eficiency *s*yndrome, a disease caused by the human immunodeficiency virus (HIV). This virus attacks lymphocytes of helper T subclass and macrophages. The depletion of these cells makes the patient susceptible to pathogens that would easily be controlled by a healthy immune system. The infection is transmitted by sexual intercourse, by direct contamination of the blood (as when virus-contaminated drug paraphernalia is shared), or by passage of the virus from an infected mother to her fetus or to a suckling baby. AIDS was first identified as a new infectious disease by the U.S. Centers for Disease Control and Prevention in 1981. *See* Appendix C, 1983, Montagnier and Gallo. HIV, lymphocyte, retroviruses.

akinetic acentric (*q.v.*).

ala alanine. *See* amino acid.

albinism 1. deficiency of chromoplasts in plants. 2. the inability to form melanin (*q.v.*) in the eyes, skin, and hair, due to a tyrosinase deficiency. In humans the condition is inherited as an autosomal recessive. Tyrosinase (TYR) is an essential enzyme for melanin synthesis, and some mutations in the tyrosinase gene (*tyr*) result in oculocutaneous albinism (OCA). The *TYR* gene is located in 11q14-21; it contains five exons, and its mRNA is 2,384 nucleotides long. More than 90 mutations have been identified, most of the missense type. One such mutation in codon 422 results in the substitution of glutamine for arginine. The changed enzyme is heat-sensitive and so mimics the temperature-sensitive enzymes known for the Himalayan strains of mice, rabbits, and other species. *See* Himalayan mutant, ocular albinism, temperature-sensitive mutation, tyrosinase.

albino 1. a plant lacking chromoplasts. 2. an animal lacking pigmentation. *See* melanin.

albomaculatus referring to a variegation consisting of irregularly distributed white and green regions on plants resulting from the mitotic segregation of genes or plastids.

albumin a water-soluble 70-kilodalton protein that represents 40–50% of the plasma protein in adult mammals. It is important both as an osmotic and as a pH buffer and also functions in the transport of metal ions and various small organic molecules. Albumin is synthesized and secreted by the liver. In the mouse the albumin gene resides on chromosome 5, separated from the alpha fetoprotein gene by a DNA segment about 13.5 kilobases long. In humans, these two genes are in the long arm of chromosome 4. *See* Appendix C, 1967, Sarich and Wilson; alpha fetoprotein.

alcaptonuria alkaptonuria (*q.v.*).

alcohol any hydrocarbon that carries one or more hydroxyl groups. The term is often used to refer specifically to ethyl alcohol, the product of yeast-based fermentations. Hereditary differences in alcohol preference are known to exist in mice. *See* Appendix C, 1962, Rodgers and McClearn.

alcohol dehydrogenase (ADH) a zinc-containing enzyme found in bacteria, yeasts, plants, and animals that reversibly oxidizes primary and secondary alcohols to the corresponding aldehydes and ketones. In the case of yeast, ADH functions as the last enzyme in alcoholic fermentation. In *Drosophila melanogas-*

ter, ADH is a dimeric protein. By suitable crosses between null activity mutants it is possible to generate heteroallelic individuals that exhibit partial restoration of enzyme activity. This is often due to the production of a heterodimer with improved functional activity. The gene is of interest to developmental geneticist because its expression is controlled by *two* promotors. The proximal promotor lies adjacent to the initiation codon and switches the gene on during the larval stage. The distal promotor lies 700 base pairs upstream and controls the production of ADH in the adult. *See* allelic complementation, promotor.

aldehyde any of a class of organic compounds having the general formula $C_nH_{2n}O$ and containing a terminal $-C\begin{smallmatrix}\diagup H\\ \diagdown\!\!\diagdown O\end{smallmatrix}$ group.

aldosterone an adrenal corticosteroid hormone that controls the sodium and potassium balance in the vertebrates.

aleurone the outer layer of the endosperm of a seed. Genes controlling the inheritance of aleurone color in maize provided early examples of epistasis (*q.v.*) and parental imprinting (*q.v.*). *See* kernel.

aleurone grain a granule of protein occurring in the aleurone.

Aleutian mink an autosomal recessive mutation in *Mustela vison* producing diluted pigmentation of the fur and eyes. The homozygotes show a lysosomal defect similar in humans to the Chédiak-Higashi syndrome (*q.v.*).

alga (*plural* **algae**) any of a large group of aquatic, chlorophyll-bearing organisms ranging from single cells to giant seaweeds. *See* Appendix A: Cyanobacteria, Dinoflagellata, Euglenophyta, Xanthophyta, Chrysophyta, Bacillariophyta, Phaecophyta, Rhodophyta, Gamophyta, Chlorophyta.

algorithm a set of simple mathematical procedures that are followed in a specified order to solve a problem in a finite period of time. Computers are instructed to perform tasks with programs containing one or more algorithms.

alien addition monosomic a genome that contains a single chromosome from another species in addition to the normal complement of chromosomes.

alien substitution replacement of one or more chromosomes of a species by those from a different species.

aliphatic designating molecules made up of linear chains of carbon atoms.

aliquot a part, such as a representative sample, that divides the whole without a remainder. Two is an aliquot of six because it is contained exactly three times. Loosely used for any fraction or portion.

alkali metal any of five elements in Group IA of the periodic table: lithium (Li), sodium (Na), potassium (K), rubidium (Rb), and cesium (Cs).

alkaline earth any element of Group IIA of the periodic table: beryllium (Be), magnesium (Mg), calcium (Ca), strontium (Sr), barium (Ba), and radium (Ra).

alkaline phosphatase an enzyme that removes 5′-P termini of DNA and leaves 5′-OH groups. The alkaline phosphatase of *E. coli* is a dimer made up of identical protein subunits encoded by a single structural gene. *In vitro* complementation was demonstrated using this enzyme. *See* allelic complementation.

alkaloid any member of a group of over 3,000 cyclic, nitrogenous, organic compounds, many of which have pharmacological properties. They occur mainly in plants, but are also synthesized by some fungi, amphibians, and arthropods. They include caffeine, cocaine, quinine, morphine, nicotine, reserpine, strychnine, and theobromine.

alkapton 2,5-dihydroxyphenylacetic acid. *See* homogentisic acid.

alkaptonuria (*also* **alcaptonuria**) a relatively benign hereditary disease in humans due to a recessive gene located on the long arm of chromosome 3. Alkaptonurics cannot make the liver enzyme homogentisic acid oxidase. Therefore, homogentisic acid (*q.v.*) is not broken down to simpler compounds but is excreted in the urine. Since the colorless homogentisic acid is readily oxidized to a black pigment, the urine of alkaptonurics darkens when exposed to air. This disease enjoys the historic distinction of being the first metabolic disease studied. *See* Appendix C, 1909, Garrod.

alkylating agent a compound causing the substitution of an alkyl group (usually methyl or ethyl) for an active hydrogen atom in an organic compound. According to the number of reactive groups they contain, alkylating agents are classified as mono-, bi-, or polyfunctional. Many chemical mutagens are alkylating agents. *See* busulfan, chlorambucil, cyclophosphamide, epoxide, ethylmethane sulfonate, melphalan, Myleran, nitrogen mustard, sulfur mustard, TEM, Thio-tepa, triethylenethiophosphoramide.

alkyl group a univalent radical having the general formula C_nH_{2n+1} derived from a saturated aliphatic hydrocarbon by removal of one atom of hydrogen. Named by replacing the ending *-ane* of the hydrocarbon with *-yl* (e.g., meth*ane* becomes meth*yl*).

allantois a saclike outgrowth of the ventral side of the hindgut present in the embryos of reptiles, birds, and mammals. The allantois represents a large and precocious development of the urinary bladder.

allatum hormones hormones synthesized by the insect corpus allatum. The titer of allatum hormones influences the qualitative properties of each molt in holometabolous insects. At high concentrations, larval development ensues; at lower levels, the insect undergoes pupal metamorphosis, and in the absence of the allatum hormones adult differentiation takes place. The allatum hormones thus have a juvenilizing action and for this reason have also been called juvenile hormones (JHs). The structural formulas for three of the juvenile hormones are illustrated on page 14. In adult females, the allatum hormone is required for vitellogenesis. The JH analog, ZR515 (*q.v.*), is often used as a substitute for natural JHs in *Drosophila* experiments. *See* Appendix C, 1966, Röller *et al.*; ring gland, *status quo* hormones.

allele a shorthand form of **allelomorph,** one of a series of possible alternative forms of a given gene (cistron, *q.v.*), differing in DNA sequence, and affecting the functioning of a single product (RNA and/or protein). If more than two alleles have been identified in a population, the locus is said to show *multiple allelism. See* heteroallele, homoallele, isoallele, null allele, silent allele.

allele-specific oligonucleotide testing a technique used to identify a specific mutation in a collection of DNA fragments isolated from a mutant organism. An oligonucleotide is synthesized that has a base sequence complementary to the segment under study, and it is used as a probe. All segments binding to the probe are then collected and analyzed.

allelic complementation the production of nearly normal phenotype in an organism carrying two different mutant alleles in *trans* configuration. Such complementation is sometimes caused by the reconstruction in the cytoplasm of a functional protein from the inactive products of the two alleles. When such a phenomenon can be demonstrated by mixing extracts from individuals homozygous for each allele, the term *in vitro complementation* is used. Synonymous with intra-allelic complementation. *See* Appendix C, 1963, Schlesinger and Levinthal; alcohol dehydrogenase, alkaline phosphatase, transvection.

allelic exclusion the situation in a diploid nucleus where either the parental or the maternal allele, but not both, is expressed, even though both parental alleles are capable of being transcribed and may even be identical. This situation is seen during recombina-

```
             CH3                                      CH3    O
             |                                        |      ||
JH 1  CH3—CH2—C—CH—CH2—CH2—C=CH—CH2—CH2—C=CH—C—O—CH3
              \/               |
              O                CH2
                               |
                               CH3

             CH3                                      CH3    O
             |                                        |      ||
JH 2  CH3—CH2—C—CH—CH2—CH2—C=CH—CH2—CH2—C=CH—C—O—CH3
              \/               |
              O                CH3

             CH3                                      CH3    O
             |                                        |      ||
JH 3      CH3—C—CH—CH2—CH2—C=CH—CH2—CH2—C=CH—C—O—CH3
              \/               |
              O                CH3

             CH3                                    CH3    O      CH3
             |                                      |      ||    /
ZR 515    CH3—C—CH2—CH2—CH2—CH—CH2—CH=CH—C=CH—C—O—CH
              \                |                                 \
               O—CH3           CH3                                CH3
```

Allatum hormones

tion within the segmented *Ig* genes of immature lymphocytes. In any one B lymphocyte (*q.v.*), a light chain or heavy chain can be synthesized from a maternal or paternal homolog, not both. *See* immunoglobulin genes, isotype exclusion, somatic recombination.

allelic frequency the percentage of all alleles at a given locus in a population gene pool represented by a particular allele. For example, in a population containing 20 *AA*, 10 *Aa*, and 5 *aa*, the frequency of the *A* allele is $[2(20) + 1(10)]/2(35) = 5/7 = 0.714$. *See* gene frequency.

allelism test complementation test (*q.v.*).

allelomorph commonly shortened to allele (*q.v.*). *See* Appendix C, 1900, Bateson.

allelopathy an interaction involving two different species in which chemicals introduced into the environment by one suppress the growth or reproduction of the other.

allelotype the frequency of alleles in a breeding population.

allergen a substance inducing hypersensitivity.

allergy an immune hypersensitivity response to an agent that is nonantigenic to most of the individuals in a population.

allesthetic trait any individual characteristic that has an adaptive function only via the nervous systems of other organisms, for example, odors, display of color patterns, mating calls, etc., which are important components of courtship in various species. *See* courtship ritual, pheromone.

Allium the genus that includes *A. cepa*, the onion; *A. porrum*, the leek; *A. sativum*, the garlic; and *A. schoenoprasum*, the chive—all classic subjects for cytological studies of mitotic chromosomes.

alloantigen an antigen (*q.v.*) that elicits an immune response (*q.v.*) when introduced into a genetically different individual of the same species. Antibodies produced in response to alloantigens are called *alloantibodies*. *See* histocompatibility molecules.

allochromacy the formation of other coloring agents from a given dye that is unstable in solution. Nile blue (*q.v.*) exhibits allochromacy.

allocycly a term referring to differences in the coiling behavior shown by chromosomal segments or whole chromosomes. Allocyclic behavior characterizes the pericentric heterochromatin, the nucleolus organizer, and in some species entire sex chromosomes. If a chromosome or chromosomal segment is tightly condensed in comparison with the rest of the chromosomal complement, the chromosome or chromosomal segment is said to show *positive heteropycnosis* (*q.v.*). Allocycly is also used to describe asynchronous separation of bivalents during the first anaphase in meiosis. In man, for example, the X and Y chromosomes segregate ahead of the autosomes and are said to show *positive allocycly*.

allogeneic disease *See* graft-versus-host reaction.

allogeneic graft a graft of tissue between genetically different members of the same species, especially with regard to alloantigens (*q.v.*). *See* allograft, heterograft. *Compare with* xenograft.

allograft a graft of tissue from a donor of one genotype to a host of a different genotype but of the same species.

allolactose *See* lactose.

allometry the relation between the rate of growth of a part of an individual and the growth rate of the whole or of another part. In the case of isometry, the relative proportions of the body parts remain constant as the individual grows; in all other cases, the relative proportions change as total body size increases. *See* heterauxesis.

allomone any chemical secreted by an organism that influences behavior in a member of another species, benefiting only the producer. If both species benefit, it is a *synamone*. If only the receiver benefits, it is a *kairomone*.

alloparapatric speciation a mode of gradual speciation in which new species originate through populations that are initially allopatric, but later become parapatric before completely effective reproductive isolation has evolved. Natural selection may enhance incipient reproductive isolating mechanisms in the zone of contact by character displacement (*q.v.*), and other mechanisms. *Compare with* parapatric speciation.

allopatric speciation the development of distinct species through differentiation of populations in geographic isolation. Such populations are called allopatric.

allopatry referring to species living in different geographic locations and separated by distance alone or by some barrier to migration such as a mountain range, river, or desert. *Compare with* sympatry.

allophene a phenotype not due to the mutant genetic constitution of the cells of the tissue in question. Such a tissue will develop a normal phenotype if transplanted to a wild-type host. *See* autophene.

allophenic mice chimeric mice produced by removing cleaving eggs from mice of different genotypes, fusing the blastomeres *in vitro,* and reimplanting the fused embryos into the uterus of another mouse to permit embryogenesis to continue. Viable mice containing cells derived from two or more embryos have been obtained and used in cell lineage studies. *See* Appendix C, 1967, Mintz.

alloplasmic referring to organisms or cells bearing chromosomes of one species and cytoplasm of a different species; for example, bread wheat (*Triticum aestivum*) chromosomes and rye (*Secale cereale*) cytoplasm. *Compare with* heteroplasmic, heteroplastidy.

allopolyploid (*also* alloploid) a polyploid organism arising from the combination of genetically distinct chromosome sets. *See* isosyndetic alloploid, segmental alloploid.

alloprocoptic selection a mode of selection in which association of opposites increases the fitness of the associates. An example involves the loci governing alcohol dehydrogenase in *Drosophila melanogaster.* The fertility is greater than expected when two mating individuals are homozygous for different alleles and smaller than expected when they are homozygous for the same allele.

allostery the reversible interaction of a small molecule with a protein molecule, which leads to changes in the shape of the protein and a consequent alteration of the interaction of that protein with a third molecule.

allosteric effectors small molecules that reversibly bind to allosteric proteins at a site different from the active site, causing an allosteric effect.

allosteric enzyme a regulatory enzyme whose catalytic activity is modified by the noncovalent attachment of a specific metabolite to a site on the enzyme other than the catalytic site.

allosteric protein a protein showing allosteric effects.

allosteric site a region on a protein other than its active site (*q.v.*), to which a specific effector molecule may bind and influence (either positively or negatively) the functional activity of the protein. For example, in the lactose system of *E. coli,* the *lac* repressor becomes inactive (cannot bind to the *lac* operator) when allolactose is bound to the allosteric site of the repressor molecule. *See lac* operon.

allosyndesis the pairing of homoeologous chromosomes (*q.v.*) in an allopolyploid (*q.v.*). Thus if the genetic composition of an alloploid is given by AABB, where AA represent the chromosomes derived from one parent species and BB the chromosomes derived from the other parent species, then during meiotic prophase, A undergoes allosyndetic pairing with B. Such pairing indicates that the A and B chromosomes have some segments that are homologous, presumably because the two parent species have a common ancestry. In the case of *autosyndesis,* A pairs only with A, and B with B. *Segmental alloploids* form both bivalents and multivalents during meiosis because of allosyndesis.

allotetraploid an organism that is diploid for two genomes, each from a different species; synonymous with amphidiploid (*q.v.*).

allotypes proteins that are products of different alleles of the same gene. The term is often used to refer to serologically detectable variants of immunoglobins and other serum proteins.

allotype suppression the systematic and long-term suppression of the expression of an immunoglobulin allotype in an animal induced by treatment with antibodies against the allotype.

allotypic differentiation *See in vivo* culturing of imaginal discs.

allozygote an individual homozygous at a given locus, whose two homologous genes are of independent origin, as far as can be determined from pedigree information. *See* autozygote.

allozymes allelic forms of an enzyme that can be distinguished by electrophoresis, as opposed to the more general term *isozyme* (*q.v.*). *See* Appendix C, 1966, Lewontin and Hubby.

alpha amanitin *See* amatoxins.

alpha chain one of the two polypeptides found in adult and fetal hemoglobin (*q.v.*).

alpha fetoprotein the major plasma protein of fetal mammals. AFP is a 70-kilodalton glycoprotein that is synthesized and secreted by the liver and the yolk sac. The genes encoding AFP and serum albumen arose in evolution as the result of a duplication of an ancestral gene $(3–5) \times 10^8$ years ago. *See* albumen.

alpha galactosidase an enzyme that catalyzes the hydrolysis of substrates that contain α-galactosidic residues, including glycosphingolipids and glycoproteins. In humans, α-galactosidase exists in two forms, A and B. The A form is encoded by a gene on the X chromosome. Fabry disease (*q.v.*) is caused by mutations at this locus. The B form is encoded by a gene on chromosome 22.

alpha helix one of two common, regularly repeating structures seen in proteins (*compare with* beta pleated sheet). The alpha helix is a compact spiral with the side chains of the amino acids in the polypeptide extending outward from the helix. The helix is stabilized by hydrogen bonds that form between the CO group of each amino acid and the NH group of the amino acid, which lies four residues ahead in the sequence. All main-chain CO and NH groups are hydrogen-bonded according to this pattern. One turn of the helix occurs for each 3.6 amino acid residues. Alpha helices are built from a continuous sequence that contains as few as 4 to as many as 50 amino acids. *See* Appendix C, 1951, Pauling and Corey; 1958, Kendrew *et al.*; protein structure.

alpha particle a helium nucleus consisting of two protons and two neutrons, and having a double positive charge.

alpha tocopherol vitamin E (*q.v.*).

alphoid sequences a complex family of repetitive DNA sequences found in the centromeric heterochromatin of human chromosomes. The alphoid family is composed of tandem arrays of 170 base pair segments. The segments isolated from different chromosomes show a consensus sequence, but also differences with respect to individual bases, so that the 170 base pair units may vary in sequence by as much as 40%. The repeats are organized in turn into groups containing several units in tandem, and these groups are further organized into larger sequences 1 to 6 kilobases in length. These large segments are then repeated to generate segments 0.5 to 10 megabase pairs in size. Such larger, or "macro," DNA repeats are chromosome-specific. Since alphoid sequences are not transcribed, they play an as yet undefined structural role in the chromosome cycle. The variation in the sequences within the alphoid DNA results in a high frequency of RFLPs. These are inherited and can be used to characterize the DNAs of specific individuals and their relatives. *See* DNA fingerprint technique, restriction fragment length polymorphisms.

alteration enzyme a protein of phage T4 that is injected into a host bacterium along with the phage DNA; this protein modifies host RNA polymerase by linking it to ADP-ribose. RNA polymerase modified in this way renders it incapable of binding to sigma factor and thus unable to initiate transcription at host promoters. *See* RNA polymerase.

alternate disjunction, alternate segregation *See* translocation heterozygote.

alternation of generations reproductive cycles in which a haploid phase alternates with a diploid phase. In mosses and vascular plants, the haploid phase is the gametophyte, the diploid the sporophyte.

alternative splicing a mechanism for generating multiple protein isoforms from a single gene that involves the splicing together of nonconsecutive exons during the processing of some, but not all, transcripts of the gene. This is illustrated in the diagram, where a gene is made up of five exons joined by introns i^1–i^4. The exons may be spliced by the upper pathway shown by the dotted lines to generate a mature transcript containing all five exons. This type of splicing is termed *constitutive*. The alternative mode of splicing shown generates a mature transcript that lacks exon 4. If each exon encodes 20 amino acids, the constitutive splicing path would result in a polypeptide made up of 100 amino acids. The alternative path would produce a polypeptide only 80 amino acids long. If the amino acid sequences of the two proteins were determined, the first 60 and the last 20 would be identical. The premessenger RNAs (*q.v.*) of at least 40% of all human genes undergo alternative splicing. This removes the intron RNAs and joins the adjacent exon RNAs by phosphodiester linkages. The splicing takes place in spliceosomes (*q.v.*) that reside within the nucleus. Therefore the number of proteins encoded by the human genome is many times larger than the number of structural genes it contains. *See* Appendix C, 1977, Weber *et al.*; adenovirus, DSCAM, fibronectin, Human Genome Project, isoforms, posttranscriptional processing, myosin genes, RNA splicing, tropomyosin.

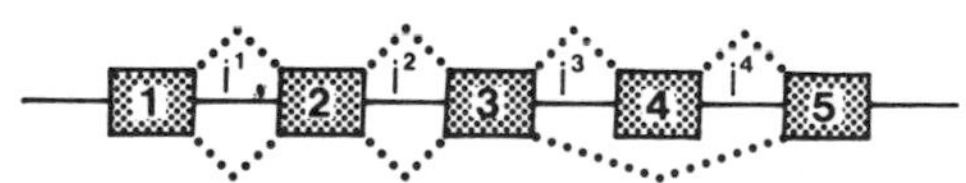

altricial referring to the type of ontogeny seen in vertebrate species characterized by large litters, short gestations, and the birth of relatively undeveloped, helpless young. *Compare with* precocial.

altruism behavior of an individual that benefits others. To the extent that the "others" are related to the altruist (the one exhibiting altruistic behavior), such actions may actually be an expression of fitness. *See* inclusive fitness.

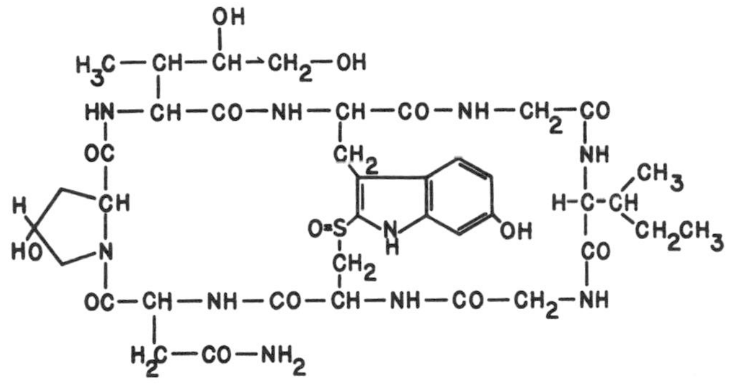

Alpha amanitin

Alu family the most common dispersed, repeated DNA sequence in the human genome. There are at least 750,000 Alu elements, each consisting of about 300 base pairs, accounting for 11% of human DNA. Each element is made up of two 130 base pair sequences joined head to tail with a 32 base pair insert in the right-hand monomer. Alu sequences are targeted by cohesins (*q.v.*). The family name is derived from the fact that these sequences are cleaved by restriction endonuclease *Alu I*. *See* human gene maps, repetitious DNA.

Alzheimer disease (AD) a multifactorial syndrome that causes a devastating decline in mental ability and is accompanied by the appearance of amyloid plaques in the cerebral cortex. These deposits were first observed in 1906 by a German physician, Alois Alzheimer, in brain tissue from a woman who had died of an unusual mental illness. Amyloid plaques contain aggregates of amyloid-beta-peptides (AβPs), and these are derived from an amyloid beta precursor protein (AβPP), which is encoded by a gene on human chromosome 21. Patients with trisomy 21 (Down syndrome, *q.v.*) commonly develop AD by age 40. Familial, early onset AD is often associated with mutations of genes that encode presenilins (PS1 and PS2). The genes for PS1 and PS2 are located at 14q24.3 and 1q42.1, respectively. The proteins they encode are made of 467 and 448 amino acids, respectively, and they contain seven to nine transmembrane domains. Both proteins are bound to membranes and play a role in cutting AβPP into AβPs, some of which are toxic. A gene homologous to the PS1 gene has been isolated from nematodes. *See* Appendix C, 1995, Sherrington, St. George-Hyslop *et al.*, Schellenberg *et al.*; *Caenorhabiditis elegans*, neuregulins (NRGs).

Amanita phalloides a poisonous mushroom which is the source of amatoxins and phallotoxins (*both of which see*). *See* Appendix A, Fungi, Basidiomycota.

amastigote *See* undulipodium.

amatoxins a group of bicyclic octapeptides that are among the poisons produced by *Amanita phalloides* (*q.v.*). These poisons inhibit transcription in eukaryotic cells because of their interaction with RNA polymerase II. However, they do not affect the RNA polymerases of mitochondria or chloroplasts. Alpha amanitin (formula, above) is an amatoxin most commonly used experimentally to inhibit transcription. *See* phallotoxins, RNA polymerase.

amaurosis blindness occurring without an obvious lesion in the eye, as from a disease of the optic nerve or brain. The term is sometimes found in the early descriptions of hereditary diseases leading to blindness (e.g., Leber congenital amaurosis). *See* Leber hereditary optic neuropathy (LHON).

amber codon the mRNA triplet UAG that causes termination of protein translation, one of three "stop" codons. The terms *amber* and *ochre* (*q.v.*) originated from a private laboratory joke and have nothing to do with colors.

Amberlite trade name for a family of ion-exchange resins.

amber mutation a mutation in which a polypeptide chain is terminated prematurely. Amber mutations are the result of a base substitution that converts a codon specifying an amino acid into UAG, which signals chain termination. In certain strains of *E. coli* amber mutations are suppressed. These strains contain a tRNA with an AUC anticodon, which inserts an amino acid at the UAG site and hence permits translation to continue. *See* ochre mutation, nonsense mutation.

amber suppressor any mutant gene coding for a tRNA whose anticodon can respond to the UAG stop codon by the insertion of an amino acid that renders the gene product at least partially functional.

For example, a mutant tyrosine-tRNA anticodon 3′AUC would recognize 5′UAG, tyrosine would be inserted, and chain growth would continue.

Ambystoma mexicanum the Mexican axolotl, a widely used laboratory species. The urodele for which the most genetic information is available. It was in the nuclei of oocytes of this species that the giant lampbrush chromosomes (*q.v.*) were first observed. *Ambystoma* has 14 pairs of chromosomes, and a genome size of about 35 gbp of DNA. It is one of the few vertebrates able to regenerate entire body structures. *See* Appendix A, Animalia, Chordata, Vertebrata, Amphibia, Urodela; Appendix C, 1882, Flemming; neoteny, regeneration.

amelogenins highly conserved proteins that are secreted by ameloblasts and constitute 90% of the organic matrix in the enamel of teeth. The amelogenins of humans come in a number of isoforms; the most common one contains 192 amino acids. The genes that encode amelogenins reside on the X and Y chromosomes. *AMELX* is at Xp22.22 and *AMELY* is at Yp11.2, and both are transcribed in the tooth buds of males. The X-linked gene contains 7,348 bases and the Y-linked gene 8,109. During forensic analyses, *amelogenin* primers from human X-Y DNA are often used in gender determination.

amelanogenesis imperfecta defects in the mineralization of teeth that affect the enamel layer. The teeth are small, pitted, and show yellow to brown discolorations. The condition often results from mutations in the genes that encode amelanogenins (*q.v.*).

amensalism a species interaction in which one is adversely affected and the other is unaffected.

Ames test a bioassay for detecting mutagenic and possibly carcinogenic compounds, developed by Bruce N. Ames in 1974. Reverse mutants to histidine independence are scored by growing *his⁻ Salmonella typhimurium* on plates deficient in histidine in the presence of the chemical (test) and in its absence (control).

amethopterin methotrexate (*q.v.*).

amino acid activation a coupled reaction catalyzed by a specific aminoacyl synthetase that attaches a specific amino acid (AA) to a specific transfer RNA (tRNA) in preparation for translation (*q.v.*).

$$AA + ATP \rightarrow AA\text{-}AMP + 2P$$
$$AA\text{-}AMP + tRNA \rightarrow AA\text{-}tRNA + AMP$$

amino acid attachment site the 3′ end of a tRNA molecule to which an amino acid is covalently attached by an aminoacyl bond. *See* amino acid activation, aminoacyl-tRNA synthetases, transfer RNA.

amino acids aminocarboxylic acids that are components of proteins and peptides. They also occur in their free form or attached to transfer RNAs (*q.v.*). There are 20 different amino acids for which at least one specific codon exists in the DNA genetic code. These universal amino acids are illustrated on page 21. Their abbreviations and messenger RNA code designations are on page 22. Amino acids are joined together to form polypeptides. Polymers containing 50 or more amino acids are called proteins. All amino acids contain a central carbon atom (designated alpha) to which an amino group, a carboxyl group, and a hydrogen atom are attached. There is also a side chain or residue (R), and this gives each amino acid its characteristic properties. Note that proline is unique in that the alpha C and its amino group are incorporated into the side chain, which is in the form of a five-atom ring. At pH 7 the side chains of lysine, arginine, and histidine are positively charged and the side chains of aspartic acid and glutamic acid are negatively charged. Therefore, the net charge born by a protein is determined by the relative proportions of these five amino acids in it. Other amino acids control the shape of proteins. Amino acids like isoleucine, leucine, phenylalanine, and valine are repelled by water molecules and therefore tend to be found buried within the interior of the protein structure. *See* genetic code, peptide bond, translation.

amino acid sequence the linear order of the amino acids in a peptide or protein. *See* protein structure.

amino acid side chain a group attached to an amino acid, represented by R in the general formula for an amino acid:

$$\begin{array}{c} NH_2-CH-COOH \\ | \\ R \end{array}$$

aminoaciduria the presence of one or more amino acids in the urine in abnormal quantities because of a metabolic defect.

aminoacyl adenylate the activated compound that is an intermediate in the formation of a covalent bond between an amino acid and its specific transfer RNA; abbreviated AA-AMP. *See* AMP, transfer RNA.

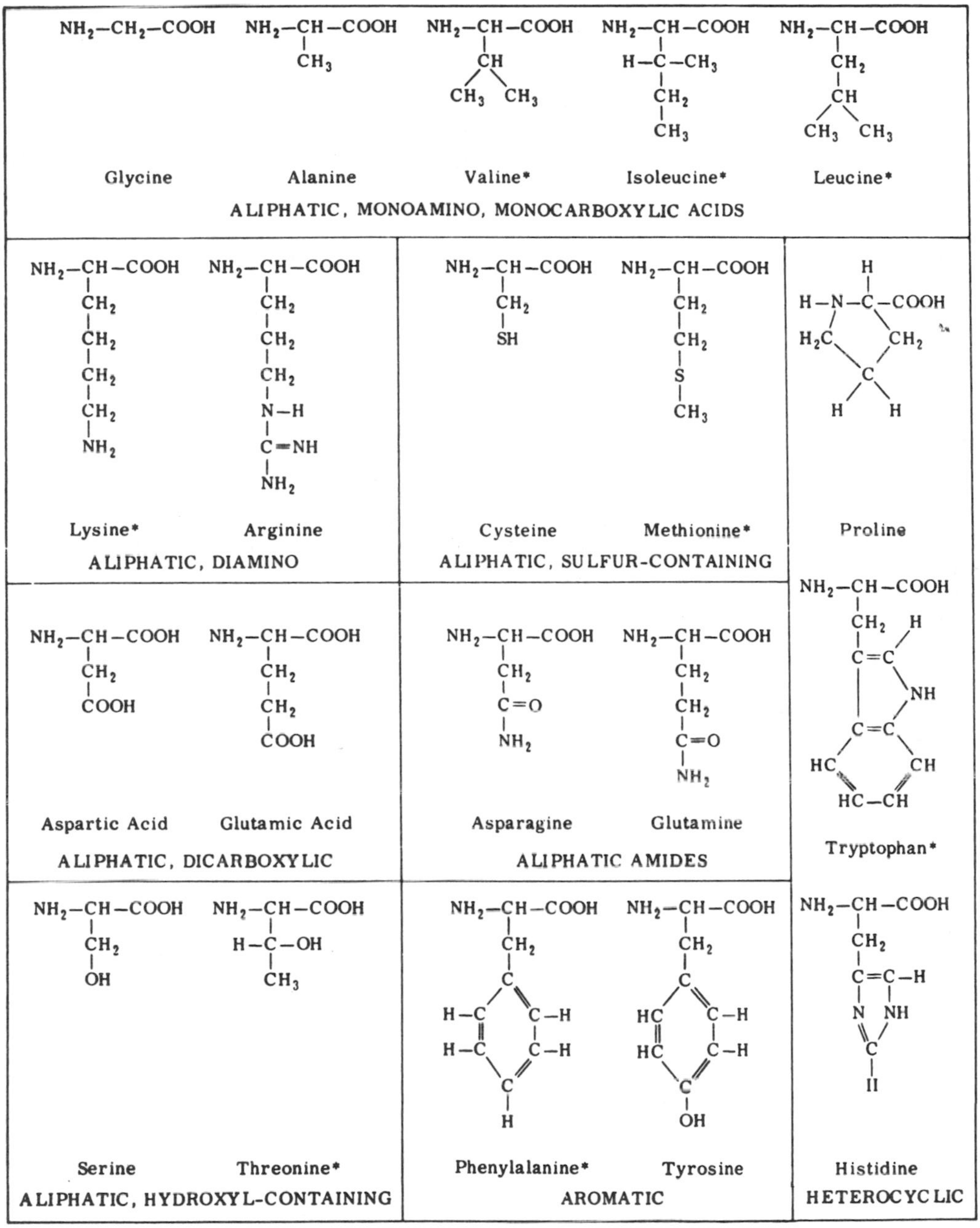

Glycine — Alanine — Valine* — Isoleucine* — Leucine*

ALIPHATIC, MONOAMINO, MONOCARBOXYLIC ACIDS

Lysine* — Arginine

ALIPHATIC, DIAMINO

Cysteine — Methionine*

ALIPHATIC, SULFUR-CONTAINING

Proline

Aspartic Acid — Glutamic Acid

ALIPHATIC, DICARBOXYLIC

Asparagine — Glutamine

ALIPHATIC AMIDES

Tryptophan*

Serine — Threonine*

ALIPHATIC, HYDROXYL-CONTAINING

Phenylalanine* — Tyrosine

AROMATIC

Histidine

HETEROCYCLIC

*required in the diet of mammals.

Structural formulas of the universal amino acids

aminoacyl site one of two binding sites for tRNA molecules on a ribosome; commonly called the *A site*. *See* translation.

aminoacyl-tRNA an aminoacyl ester of a transfer RNA molecule.

aminoacyl-tRNA binding site *See* translation.

aminoacyl-tRNA synthetases enzymes that activate amino acids and attach each activated amino acid to its own species of tRNA. These enzymes catalyze: (1) the reaction of a specific amino acid (AA) with adenosine triphosphate (ATP) to form AA-AMP, and (2) the transfer of the AA-AMP complex to a specific transfer RNA, forming AA-tRNA and

Amino Acids

AMINO ACID	ONE-LETTER SYMBOL	THREE-LETTER SYMBOL	mRNA CODE DESIGNATION
alanine	A	ala	GCU, GCC, GCA, GCG
arginine	R	arg	CGU, CGC, CGA, CGG, AGA, AGG
asparagine	N	asn	AAU, AAC
aspartic acid	D	asp	GAU, GAC
cysteine	C	cys	UGU, UGC
glutamic acid	E	glu	GAA, GAG
glutamine	Q	gln	CAA, CAG
glycine	G	gly	GGU, GGC, GGA, GGG
histidine	H	his	CAU, CAC
isoleucine	I	ile	AUU, AUC, AUA
leucine	L	leu	UUA, UUG, CUU,CUC, CUA, CUG
lysine	K	lys	AAA, AAG
methionine	M	met	AUG
phenylalanine	F	phe	UUU, UUC
proline	P	pro	CCU, CCC, CCA, CCG
serine	S	ser	UCU, UCC, UCA, UCG, AGU, AGC
threonine	T	thr	ACU, ACC, ACA, ACG
tryptophan	W	trp	UGG
tyrosine	Y	tyr	UAU, UAC
valine	V	val	GUU, GUC, GUA, GUG

free AMP (adenosine monophosphate). *See* adenosine phosphate.

amino group a chemical group ($-NH_2$) which with the addition of a proton can form $-NH_3^+$.

p amino benzoic acid a component of folic acid (*q.v.*).

aminopeptidase an enzyme (in both prokaryotes and eukaryotes) that removes the formylated methionine (fMet) or methionine from the NH_2 terminus of growing or completed polypeptide chains.

aminopterin *See* folic acid.

aminopurine 2-aminopurine (2-AP) is a fluorescent analog of adenine (6-aminopurine). The fluorescence of 2-AP ia quenched once it is incorporated into the base-stacked structure of dsDNA. However, if the base stacking or base pairing of DNA is locally perturbed, the fluorescence of 2-AP is enhanced. Therefore the intensity of fluorescence emissions by 2-AP molecules can be used to monitor perturbations in DNA structure caused by interactions between DNA and proteins, such as DNA polymerases, helicases, repair enzymes, and methyl transferases (*all of which see*).

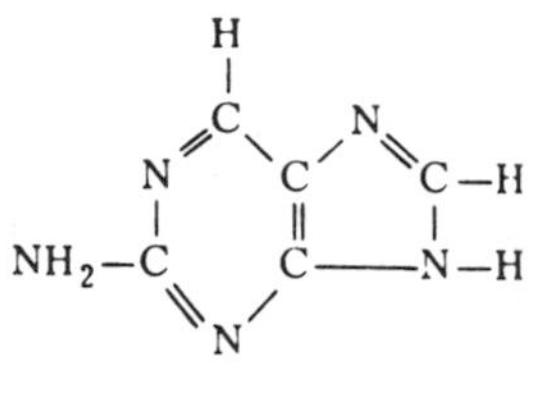

```
          H
          |
          C      N
       //   \  /   \\
      N      C       C—H
      |      ||      |
NH2—C       C———N—H
       \\   /
          N
```

amino terminal end the end of a polypeptide chain that has a free amino group.

Amish a human population descended from a limited number of founders who emigrated from southwestern Germany to the United States during the eighteenth century. The population is highly inbred, since marriage is allowed only within the community. Beneficial collaboration between geneticists and religious leaders has led to discoveries concerning certain genetic diseases that occur at unprecedentedly high frequencies among the Amish. *See* cartilage-hair hypoplasia (CHH), consanguinity, Ellis-van Creveld syndrome, inbreeding.

Amitochondriates a subkingdom of protoctists that includes the Archaeprotista and the Microspora (*see* Appendix A). These phyla contain anaerobic microorganisms that lack mitochondria and presumably were without them from the outset of their evolution.

amitosis the division of a nucleus into two parts by constriction without the participation of a mitotic apparatus. Accessory nuclei (*q.v.*) grow by amitosis.

amixis a reproductive cycle lacking meiosis and fertilization. Asexual reproduction. *Contrast with* amphimixis, apomixis.

***AML1* gene** *See acute myeloid leukemia 1* gene.

amniocentesis sampling of amniotic fluid for the prenatal diagnosis of fetal disorders. During the procedure, a hollow needle is inserted through the skin and muscle of the mother's abdomen, through the uterus, and into the amniotic sac that surrounds the fetus. Cells that have sloughed from the fetus are suspended in the fluid. Cells in the sample are cultured for about three weeks to raise their numbers to the point where chromosomal and biochemical analyses can be made. Amniocentesis cannot be done until about 16 weeks from the last menstrual period, since the sac containing the embryo is not large enough to permit safe withdrawal of the fluid until this time. *See* Appendix C, 1967, Jacobson and Barter; chorionic villi sampling, informed consent, genetic counseling, prenatal genetic testing.

amniocytes cells obtained by amniocentesis (*q.v.*).

amnion A fluid-filled sac within which the embryos of reptiles, birds, and mammals develop. The wall of this sac has a two-layered epithelium. The inner epithelium of the wall is the amnion, although the term is sometimes applied to the whole sac. The outer epithelium is usually called the chorion. Amniotic fluid within the sac provides a liquid environment for the embryo.

amniote a land-living vertebrate (reptile, bird, or mammal) whose embryos have an amnion and allantois.

Amoeba proteus a common species of rhizopod; a giant protozoan used for microsurgical nuclear transplantations. *See* Appendix C, 1967, Goldstein and Prescott.

amoeboid movement cellular motility involving cytoplasmic streaming into cellular extensions called *pseudopodia*.

amorphic mutation a mutation in which the altered gene product fails in its molecular function. Also called a *loss of function mutation* or a *null mutation*.

AMP adenosine monophosphate. *See* adenosine phosphate.

amphidiploid an organism that is diploid for two genomes, each from a different species; synonymous with allotetraploid. *See* Appendix C, 1925, Goodspeed and Clausen.

amphimixis sexual reproduction resulting in an individual having two parents: synonymous with mixis. *Contrast with* amixis, apomixis. The adjective form is *amphimictic*.

Amphioxus *See Branchiostoma.*

amphipathic descriptive of a molecule that has distinct polar and nonpolar segments (e.g., membrane phospholipids).

amphoteric compound (*also* ampholyte) a substance that can act both as an acid and a base. Thus a protein is amphoteric because it tends to lose protons on the more alkaline side of its isoelectric point and to gain protons on the acid side of its isoelectric point.

ampicillin *See* penicillin.

***amp*^{R}** a selectable gene which encodes the enzyme β-lactamase, which inactivates ampicillin (*q.v.*). Cells containing a plasmid vector (*q.v.*) which expresses *amp*^{R} can be selected from those that do not by growth in an ampicillin-containing medium. *See* penicillin, R (resistance) plasmid.

amplicon a segment of the genome that forms multiple linear copies after exposure of the organism to a compound that inhibits the functioning of a gene in the segment. For example, in mammals the enzyme dihydrofolate reductase (*q.v.*) is inhibited by methotrexate (*q.v.*). Exposure to this inhibitor causes amplification of the DHFR gene. More generally, the term *amplicons* is used for DNA fragments that have been generated in experiments utilizing the polymerase chain reaction (*q.v.*).

amplification *See* gene amplification, polymerase chain reaction, RNA amplification.

amplified RNA *See* RNA amplification.

amylase an enzyme that hydrolyzes glucosidic bonds in polyglucosans such as glycogen.

amyloid-beta-precursor protein (AβPP), amyloid-beta-peptides (AβPs), amyloid plaques. *See* Alzheimer disease (AD).

amyloplast a starch-rich plastid.

amyotrophic lateral sclerosis (ALS) a disease in humans resulting from the degeneration of motor neurons in the lateral columns of the spinal cord. The disease begins with an asymmetric weakness in the limbs and progresses to complete paralysis and death. ALS is sometimes called Lou Gehrig disease after the famous American baseball player who suffered from ALS. In familial cases of ALS, the condition was first shown to be due to mutations in a gene (*ALS1*) located at 21q22.1 which encoded the enzyme superoxide dismutase (SOD) (*q.v.*). Next *ALS2* was mapped to 2q33, and it encoded a protein (alsin) thought to be a GTPase regulator protein. Familial ALS due to *ALS1* is an adult-onset disease, and it shows dominant inheritance; whereas FALS due to *ALS2* causes a disease with juvenile onset which shows recessive inheritance. Two other forms of FALS occur: one caused by mutations in *ALS3* at 18q21 and the other by mutations in *ALS4* at 9q34. However, the cause of 90 percent of ALS cases (sporadic or non-familial) is unknown. *See* Appendix C, 1993, Rosen, Siddique *et al.*

anabolism the metabolic synthesis of complex molecules from simpler precursors, usually requiring the expenditure of energy and specific anabolic enzymes. *Contrast with* catabolism.

anaerobe a cell that can live without molecular oxygen. A strict anaerobe cannot live in the presence of oxygen. *See* Appendix C, 1861, Pasteur.

anagenesis phyletic evolution within a single lineage without subdivision or splitting; the opposite of cladogenesis.

analog a compound related to, but slightly different structurally from a biologically significant molecule, such as an amino acid (*see* azaserine), a pyrimidine or purine (*see* base analogs), or a hormone (*see* ZR515).

analogous referring to structures or processes that have evolved convergently, as opposed to the term *homologous* (*q.v.*). Analogous structures have similar functions but are different in evolutionary origin: e.g., the wing of a butterfly and of a bat. *See* homoplasy.

analysis of variance a statistical technique that allows the partitioning of the total variation observed in an experiment among several statistically independent possible causes of the variation. Among such causes are treatment effects, grouping effects, and experimental errors. Checking the absence of an effect due to the treatment is often the purpose of the inquiry. The statistical test of the hypothesis that the treatment had no effect is the F test, or variance-ratio test. If the ratio of the mean square for treatments to the mean square for error exceeds a certain constant that depends on the respective degrees of freedom of the two mean squares at a chosen significance level, then the treatments are inferred to have been effective. Analysis of variance is particularly useful in judging which sources of uncontrolled variation in an experiment need to be allowed for in testing treatment effects.

anamnestic response *See* immune response, immunological memory.

anaphase *See* mitosis.

anaphase lag delay in the movement of one or more chromosomes from the metaphase plate during anaphase, often resulting in chromosome loss (*q.v.*).

anaphylaxis a systemic allergic or hypersensitivity response leading to immediate respiratory and/or vascular difficulties.

Anas platyrhyncha the mallard duck, ancestor to the domestic or Pekin duck, *A.p. domestica*.

anastomosis the joining of two or more cell processes or tubular vessels to form a branching system.

anastral mitosis the type of mitosis characteristically found in plants. A spindle forms, but no centrioles or asters are observed.

anautogenous insect an adult female insect that must feed for egg maturation. *See* autogenous insect.

anchorage-dependent cells cells (or *in vitro* cell cultures) that will grow, survive, or maintain function only when attached to an inert surface such as glass or plastic; also known as *substrate-dependent cells*. The only normal animal cells that are designed to survive without attachment and spreading are cells that circulate in the blood. Some tumor cells acquire this ability to be anchorage-independent and leave their original tissue sites to form metastases. *See* microcarriers, suspension culture.

Anderson disease *See* Fabry disease.

androdioecy a sexual dimorphism in plants having bisexual and separate male individuals.

androecious referring to plants having only male flowers.

androecium the aggregate of the stamens in a flower.

androgen any compound with male sex hormone activity. In mammals, the most active androgens are synthesized by the interstitial cells of the testis. *See* testosterone.

androgenesis **1.** development from a fertilized egg followed by disintegration of the maternal nucleus prior to syngamy. The resulting individual possesses only paternal chromosomes and is haploid. **2.** production of an embryo having a diploid set of paternal chromosomes by nuclear transfer (*q.v.*). *Compare with* gynogenesis.

androgenic gland a gland found in most crustaceans belonging to the subclass Malacostraca. When implanted into maturing females, the gland brings about masculinization of primary and secondary sex characters.

androgen insensitivity syndrome a condition in which XY individuals develop as normal-appearing, but sterile, females. Spermatogenesis does not occur in the testes, which are generally located inside the abdomen. Also called *testicular feminization. See* Appendix C, 1988, Brown *et al.*; androgen receptor, androgen receptor gene.

androgenote a cell or embryo produced by androgenesis. *Compare with* gynogenote.

androgen receptor (AR) a protein belonging to a subfamily of steroid hormone receptors within a larger family of DNA-binding proteins. The human AR is made up of 919 amino acids and is subdivided into three domains. The N-terminal domain has a regulatory function, and the C-terminal domain binds dihydrotestosterone. The central domain, which binds to DNA, contains zinc fingers. The receptor binds to DNA as a homodimer. *See* androgen, testosterone, vitamin D receptor, zinc finger protein.

androgen receptor gene a gene symbolized *AR* that is located on the X chromosome at q12. It contains eight exons and is 180,245 bp long. *AR* encodes the androgen receptor (*q.v.*). Homologous genes occur in the rat and mouse. Mutations in critical portions of *AR* cause the loss of receptor activity, and this results in abnormalities in sexual phenotype referred to as androgen insensitivity syndrome (*q.v.*) or testicular feminization. See Appendix C, 1988, Brown *et al.*

androgynous 1. being neither distinguishably masculine nor feminine in appearance or behavior. 2. bearing staminate and pistillate flowers on distinct parts of the same inflorescence. *See* flower.

andromonecy a sexual condition in which plants develop both staminate flowers (that do not develop fruit) and hermaphroditic flowers.

androphages "male-specific" bacteriophages that absorb on the surface of F pili. Examples are MS2, R17, and Qβ (*all of which see*). *See* F factor (fertility factor).

anemia a disorder characterized by a decrease in hemoglobin per unit volume of blood. In the case of *hemolytic* anemia, there is a destruction of red blood cells. In the case of *hypochromic* anemia, there is a reduction in the hemoglobin content of the erythrocyte.

anemophily pollination by the wind.

anergy the lack of an expected immune response.

aneucentric referring to an aberration generating a chromosome with more than one centromere.

aneuploidy the condition in which the chromosome number of the cells of an individual is not an exact multiple of the typical haploid set for that species. The nomenclature employs the suffix *somic,* as the following examples illustrate. Down syndrome (*q.v.*) and Turner syndrome (*q.v.*) are examples of a human trisomic and monosomic, respectively. Nullisomics result from the loss (2N – 2) and tetrasomics from the gain (2N + 2) of a chromosome pair. If more than one different chromosome is lost or gained, the condition is described as doubly monosomic (2N – 1 – 1) or doubly trisomic (2N + 1 + 1). Early studies of aneuploids led to the conclusion that genes carried by specific chromosomes controlled morphological traits. For example, in *Datura stramonium* (*q.v.*) extra doses of chromosome G broaden and reduce the seed capsule and increase the size of the spines (see page 26). *See* Appendix C, 1934, Blakeslee; hyperploid, hypoploid, symbols used in human cytogenetics, polyploid.

aneurin vitamin B_1; more commonly known as *thiamine.*

aneusomy the condition in which an organism is made up of cells that contain different numbers of chromosomes. Aneusomy is widespread in flowering plants possessing B chromosomes (*q.v.*). In animals, the term generally refers to a diploid organism with subpopulations of aneuploid, somatic cells. The term aneusomy has been misused in the recent literature of human cytogenetics to refer to a genetic imbalance within a chromosome pair. For example, an individual heterozygous for a deficiency including one or more genes is hemizygous for those genes on the normal homolog. To call such an individual a *segmental aneusomic* is confusing, since aneusomy traditionally implies mosaicism. *See* aneuploidy.

Angelman syndrome (AS) children with this condition are hyperactive and are unable to develop normal speech. Because they show impaired motor control and tend to laugh excessively, the condition is sometimes called "happy puppet syndrome." The British pediatrician Harry Angelman gave the first description of children with the disease in 1965. Later the condition was found to be the result of a deletion in the long arm of chromosome 15. *See* Prader-Willi syndrome (PWS).

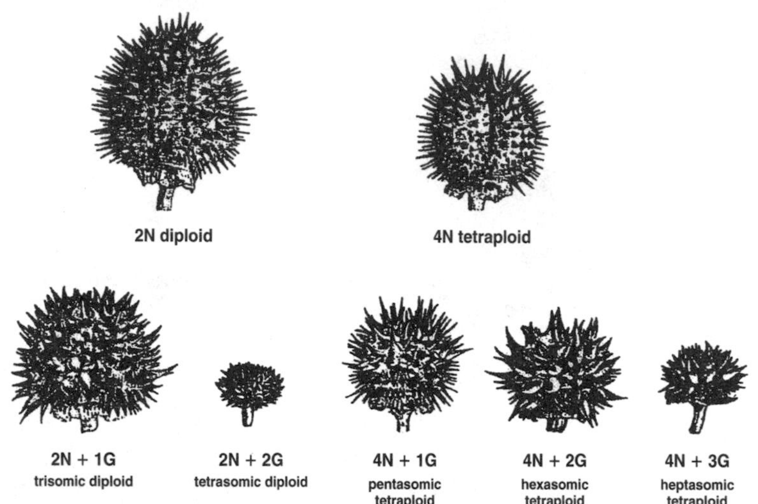

Aneuploidy in the Jimson weed, *Datura stramonium.* Extra doses of G chromosomes make the seed capsules smaller and broader and their spines larger.

angiosperm a flowering plant. Any species in the Superclass Angiospermae (*see* Appendix A, Kingdom Plantae) characterized by having seeds enclosed in an ovary. Almost all agriculturally important plants (apart from conifers) belong to the Angiospermae.

Angstrom unit a unit of length equal to one ten-thousandth of a micron (10^{-4} micron; a micron being 10^{-6} meter); convenient for describing atomic dimensions; also equivalent to 10^{-1} nanometers (nm) or 10^{-10} meter. Abbreviated A, A°, Å, Å.U., or A.U. Named in honor of the Swedish physicist Anders Jonas Ångstrom.

Animalia the kingdom containing animals (heterotrophic organisms developing from a blastula). *See* Appendix A, Kingdom 4; opisthokonta.

animal pole that pole of an egg which contains the most cytoplasm and the least yolk.

anion a negatively charged ion. *Contrast with* cation.

Aniridia a dominant mutation of the *Pax-6* gene located at 11p13, which causes defects in the iris, lens, cornea, and retina of humans. The *Pax-6* gene of humans, the *Sey* gene of mice and rats, and the *ey* gene of *Drosophila* are homologous. *See eyeless.*

anisogamy that mode of sexual reproduction in which one of the sex cells, the egg, is large and nonmobile, whereas the other (the sperm) is small and motile. In most anisogamous eukaryotes, the centriole is paternally inherited. *See* isogamy, parthenogenesis.

anisotropy a directional property of crystals and fibers having a high degree of molecular orientation. Anisotropic substances have different physical properties when tested in different directions. When a ray of plane polarized light passes through anisotropic material, it is split into two rays polarized in mutually perpendicular planes. This property of anisotropic material is called *birefringence.* Muscle fibers and the metaphase spindle are examples of living materials exhibiting birefringence. Materials showing no birefringence are said to be *isotropic. See* polarization microscope.

ankylosing spondylitis an arthritic disease resulting in a stiffening and bending of the spine; inherited as an autosomal dominant with reduced penetrance. Over 90% of patients with this disease carry the B27 HLA antigen. *See* histocompatibility.

ankyrin a protein that binds to β-spectrin as well as to the cytoplasmic domains of a variety of integral membrane proteins, and that is thought to interconnect the spectrin-based membrane cytoskeleton and the overlying lipid bilayer. Based on sequence similarity, ankyrins have been identified in various tissues and cell types from a variety of organisms. These proteins generally have three structural domains: a conserved, N-terminal region containing membrane-binding properties, a highly conserved spectrin-binding region, and a variable, regulatory C-terminal domain. In humans, three forms of ankyrin have been characterized. Ankyrin-1 (also called ankyrin-R) is encoded by the *ANK1* gene at chromosomal map position 8p11 and expressed in erythrocytes and the brain. Ankyrin-2 (also called ankyrin-B) is encoded by the *ANK2* gene at position 4q25-q27 and expressed primarily in the brain. A third protein, ankyrin-3 (also known as ankyrin-G), is encoded by the *ANK3* gene mapping to 10q21, and alternatively spliced isoforms are expressed in nervous tissue, muscle, and other tissues. Mutations in the erythrocytic ankyrin gene, *ANK1*, are associated with hereditary spherocytosis (*q.v.*). *See* fusome, integral protein, spectrin, spectrosome.

anlage the embryonic primordium from which a specific part of the organism develops.

anneal to subject first to heating then to cooling. In molecular genetics experiments, annealing is used to produce hybrid nucleic acid molecules containing paired strands, each from a different source. Heating results in the separation of the individual strands of any double-stranded, nucleic-acid helix, and cooling leads to the pairing of any molecules that have segments with complementary base pairs.

annidation the phenomenon where a mutant is maintained in a population because it can flourish in an available ecological niche that the parent organisms cannot utilize. A wingless mutant insect, for example, might be poorly adapted in its ancestral habitat but able to live in tunnels and crevices that a winged form could not occupy.

annotation *See* genome annotation.

annulate lamellae paired membranes arranged in stacks and possessing annuli resembling those of the nuclear membrane. Annulate lamellae may serve to transfer nuclear material to the cytoplasm by the replication of the nuclear envelope and may be a mechanism for storing gene-derived information to be used for cytoplasmic differentiation during early embryogenesis. During insect oogenesis, annulate lamellae occur alongside nurse cell nuclei and germinal vesicles, and they are abundant in the ooplasm.

annulus a ring. Applied to any of a number of ring-shaped parts of animals and plants. Used in cytology to refer to the ring-shaped nuclear pores.

anode the positive electrode; the electrode to which negative ions are attracted. *Contrast with* cathode.

anodontia the congenital absence of teeth. *Hypodontia* is currently the preferred term.

anonymous DNA a segment of DNA of unknown gene content that has been localized to a specific chromosome.

Anopheles a genus containing about 150 species of mosquitoes, many of which are of medical importance. Africa's principal malaria vector is *A. gambiae*. Other vector species are *A. funestus*, *A. quadrimaculatus*, *A. atroparvus*, *A. nili*, *A. moucheti*, and *A. pharoensis*. Polytene chromosomes occur in both larval salivary gland cells and adult ovarian nurse cells. Sibling species can often be separated by differences in the banding patterns of their polytene chromosomes. The genome of *A. gambiae* has been shown to contain 278 mbp of DNA and about 13,700 genes. *See* Appendix A, Arthropoda, Insecta, Diptera; Appendix C, 1898, Ross; 1899, Grassi; 2002, Holt *et al.*; Appendix E; intron dynamics, malaria, mariner elements, Minos element, shotgun sequencing.

anosmia the absence or loss of the sense of smell. Anosmia may be caused by injury to or loss of olfactory receptor neurons (*q.v.*) or by injury to region(s) in the brain or elsewhere where olfactory signals are processed. Anosmia may also result due to defects in any element of the olfaction signaling pathway. For example, mice and nematodes with defects in G proteins (*q.v.*) found in olfactory receptor neurons exhibit olfactory and chemosensory defects, respectively.

Anser anser the Gray Lag goose, a favorite experimental organism for students of animal behavior and its hereditary components.

antagonist a molecule that bears sufficient structural similarity to a second molecule to compete with that molecule for binding sites on a third molecule. *See* competition.

antagonistic pleiotropy a phenomenon in which alleles (that are detrimental late in life) improve fitness earlier in life.

antenatal before birth; during pregnancy.

antennae the first paired appendages on the head of arthropods.

Antennapedia a gene residing at 47.9 on the genetic map and within segment 84B of the salivary map of *Drosophila melanogaster*. The *Antp* gene is one of a cluster of three genes that specify the type of differentiation that cells in the segments from the

Arg Lys Arg Gly Arg Gln Thr Tyr Thr Arg Tyr Gln Thr Leu Glu Leu Glu Lys Glu Phe His Phe Asn Arg Tyr Leu Thr Arg Arg Arg
1 30

Helix Turn Recognition helix

Arg Ile Glu Ile Ala His Ala Leu Cys Leu Thr Glu Arg Gln Ile Lys Ile Trp Phe Gln Asn Arg Arg Met Lys Trp Lys Lys Glu Asn
31 60

Antennapedia homeobox

head to the anterior portion of the second thoracic segment will undergo. Mutations in the *Antp* gene cause the transformation of the segment that normally produces the antenna into one that produces a middle leg. The gene encodes a protein characterized by a homeobox (*q.v.*). This is a segment of 60 amino acids that lies close to the C terminus of the *Antp* protein. The amino acid sequence of this segment is shown above. It binds to target DNA sequences by its helix-turn-helix motif (*q.v.*). The complete three-dimensional structures of the *Antp* homeodomain as well as of the homeodomain-target DNA complex have been determined using NMR spectroscopy (*q.v.*) and x-ray crystallography (*q.v.*). *See* Appendix C, 1983, Scott *et al.*; 1989, Qian *et al.*; 1990, Maliki, Schughart, and McGinnis; *bithorax*, homeotic mutants, *Hox* genes, *Polycomb*, *proboscipedia*, segment identity genes.

anther the terminal portion of a stamen bearing pollen sacs.

anther culture a technique that utilizes anthers or pollen cells to generate haploid tissue cultures or even plants. *See* Appendix C, 1973, Debergh and Nitsch, haploid sporophytes.

antheridium the male gametangium of algae, fungi, bryophytes, and pteridophytes. *Contrast with* oogonium.

anthesis the time of flowering.

anthocyanins the red, violet, or blue glycosidic pigments that give color to flowers, fruits, seeds, stems, and leaves of plants. The common structural unit is a 15-carbon flavone skeleton to which sugars are attached. An example is pelargonidin, a scarlet pigment produced by geraniums. Unlike the carotenoids and chlorophylls, which are lipid-soluble pigments of plastids, anthocyanins are water-soluble and are found dissolved in the vacuoles (*q.v.*) of plant cells. A primary function of anthocyanins is to attract insect pollinators to plants. *See* kernel, pelargonidin monoglucoside, *R* genes of maize.

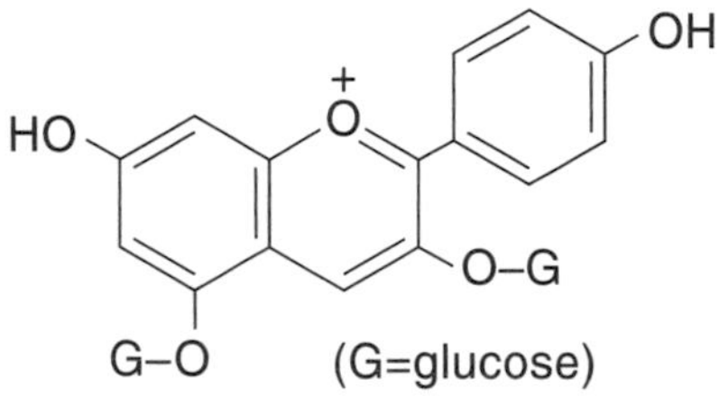

HO, O+, OH, O–G, G–O (G=glucose)

Pelargonidin, an anthocyanin

anthrax a disease caused by *Bacillus anthracis*, a spore-forming bacterium. The condition is seen in cows, pigs, goats, horses, and sheep. In humans the condition was first called *wool-sorter's disease*, and it resulted from inhalation of dust that contained spores. The first successful artificially produced vaccine (*q.v.*) was against anthrax. *See* Appendix A, Bacteria, Endospora; Appendix C, 1881, Pasteur.

anthropocentrism also called *anthropomorphism*. **1.** explanation of natural phenomena or processes in terms of human values. **2.** assuming humans to be of central importance in the universe or ultimate end of creation. **3.** ascribing human characteristics to a non-human organism.

anthropoid designating the great apes of the family Pongidae, including the gibbons, orangutans, gorillas, and chimpanzees.

anthropometry the science that deals with the measurements of the human body and its parts.

antiauxin a molecule that competes with an auxin (*q.v.*) for auxin receptor sites. A well-known antiauxin is 2,6-dichlorophenoxyacetic acid.

antibiotic a bacteriocidal or bacteriostatic substance produced by certain microorganisms, especially species of the genera *Penicillium*, *Cephalosporium*, and *Streptomyces*. *See* actinomycin D, ampicillin, chloramphenicol, cyclohexamide, erythromycin, kanamycin, neomycin, novobiocin, penicillin, puromycin, semisynthetic antibiotic, streptomycin, tetracycline.

antibiotic resistance the acquisition of unresponsiveness to a specific antibiotic by a microorganism that was previously adversely affected by the drug. Such resistance generally results from a mutation or the acquisition of R plasmids (*q.v.*) by the microorganism. *Compare with* antibiotic tolerance. *See* efflux pump.

antibiotic tolerance a bacterial trait characterized by the cessation of growth when exposed to penicillin, vancomycin, and other antibiotics but not by cell death. Antibiotic tolerance is more widespread than complete resistance to antibiotics. *See* antibiotic resistance.

antibody a protein produced by lymphoid cells (plasma cells) in response to foreign substances (antigens) and capable of coupling specifically with its homologous antigen (the one that stimulated the immune response) or with substances that are chemically very similar to that same antigen. *See* Appendix C, 1890, von Behring; 1900, Ehrlich; 1939, Tiselius and Kabat; abzymes, immunoglobulin.

antibody-antigen reaction *See* antigen-antibody reaction.

anticipation *See* genetic anticipation.

anticlinal referring to a layer of cells cutting the circumference of a cylindrical plant organ at right angles. *Compare* periclinal.

anticoding strand *See* strand terminologies.

anticodon the triplet of nucleotides in a transfer RNA molecule which associates by complementary base pairing with a specific triplet (codon) in the messenger RNA molecule during its translation in the ribosome.

antidiuretic hormone *See* vasopressin.

antigen 1. a foreign substance that, upon introduction into a vertebrate animal, stimulates the production of homologous antibodies; a complete antigen or immunogen. A complex antigenic molecule may carry several antigenically distinct sites (determinants). 2. a substance that is chemically similar to certain parts of an immunogen and can react specifically with its homologous antibody, but is too small to stimulate antibody synthesis by itself; an incomplete antigen or hapten (*q.v.*).

antigen-antibody reaction the formation of an insoluble complex between an antigen and its specific antibody. In the case of soluble antigens, the complex precipitates, while cells that carry surface antigens are agglutinated. *See* Appendix C, 1900, Ehrlich; 1986, Amit *et al.*

antigenic conversion 1. appearance of a specific antigen(s) on cells as a consequence of virus infection. 2. antibody-induced shift in certain protozoans or parasites to a new cell-surface antigen and cessation of expression of another antigen; resulting from switches in gene activity; synonymous in this sense with *serotype transformation; compare with* antigenic modulation.

antigenic determinant a small chemical complex (relative to the size of the macromolecule or cell of which it is a part) that determines the specificity of an antigen-antibody interaction. That portion of an antigen that actually makes contact with a particular antibody or T cell receptor. An epitope. *Contrast with* paratope.

antigenic mimicry acquisition or production of host antigens by a parasite, enabling it to escape detection by the host's immune system (as occurs in *Schistosoma*). *See* schistosomiasis.

antigenic modulation suppression of cell-surface antigens in the presence of homologous antibodies.

antigen variation the sequential expression of a series of variable surface glycoproteins (VSGs) by trypanosomes while in the bloodstream of a mammalian host. The production of VSGs allows the parasite to evade the immune defenses of the host by keeping one step ahead of the antibodies the host raises against them. Trypanosomes contain hundreds of different genes for the individual VSGs, but in a single trypanosome only one of these genes is expressed at a given time. The switch from one VSG to another is accompanied by rearrangements of the DNA that generates additional copies of the genes being expressed. The glycoprotein being synthesized forms a macromolecular coating about 15 nanometers thick over the body and flagellum of the parasite. This coating functions only during the mammalian stage of the life cycle, since it is shed when the trypanosome enters the tsetse fly vector. *See Glossina, Trypanosoma.*

antihemophilic factor a protein (also called factor 8) that participates in the cascade of reactions that results in blood clotting (*q.v.*). A deficiency in AHF results in classic hemophilia (*q.v.*). AHF contains 2,351 amino acids. The first 19 code for a leader sequence peptide (*q.v.*), and the mature clotting factor contains 2,332 amino acids. Three domains in factor 8 are also found in ceruloplasmin (*q.v.*). *See* Appendix C, 1984, Gitschier *et al.*; crossreacting material, von Willebrand disease.

antimetabolite in general, a molecule that functions as an antagonist or metabolic poison.

antimitotic agent any compound that suppresses the mitotic activity of a population of cells.

antimongolism a syndrome of congenital defects accompanying hypoploidy for chromosome 21. Such children carry a normal chromosome 21 and one that contains a large deficiency. From the standpoint of chromosome balance, mongolism and antimongolism represent reciprocal phenomena.

antimorphic mutation a mutation in which the altered gene product acts antagonistically to the wild type gene product. Antimorphic mutations behave as dominant or semidominant alleles. *See* Huntington disease (HD).

antimutagen a compound (generally a purine nucleoside) that antagonizes the action of mutagenic agents (generally purines or purine derivatives) on bacteria. Some antimutagens reduce the spontaneous mutation rate.

anti-oncogenes a class of genes (also called tumor suppressor genes) that are involved in the negative regulation of normal growth. The loss of these genes or their products leads to malignant growth. The *Rb* gene of humans is an example of an anti-oncogene. *See* Appendix C, 1969, Harris *et al.*; 1971, Knudson; 1990, Baker *et al.*; breast cancer susceptibility genes, retinoblastoma, *TP53*, Wilms tumor.

antioxidant enzymes enzymes that prevent the buildup of reactive oxygen molecules in cells. Superoxide dismutase (*q.v.*) and catalase (*q.v.*) are the primary enzymes involved in this process. Superoxide dismutase converts the superoxide radical (O_2^-) to H_2O_2 and catalase breaks down H_2O_2 into water and O_2. Transgenic *Drosophila* containing additional copies of the genes encoding superoxide dismutase and catalase have increased longevity. *See* Appendix C, 1994, Orr and Sohol; free radical theory of aging.

antiparallel 1. the opposite strand orientations with which all nucleic acid duplexes (DNA-DNA, DNA-RNA, or RNA-RNA) associate; if one strand is oriented left to right 5′ to 3′, the complementary strand is oriented left to right 3′ to 5′ antiparallel to it. 2. two segments of a polypeptide chain may lie antiparallel with one end *N*-terminal to C-terminal and the other end C-terminal to *N*-terminal, respectively. *See* Appendix C, 1952, Crick; 1961, Josse, Kaiser, and Kornberg; deoxyribonucleic acid, strand terminologies.

antipodal one of a group of three haploid nuclei formed during megasporogenesis. In corn, these divide mitotically and eventually form a group of 20 to 40 antipodal cells, which aid in nourishing the young embryo. *See* double fertilization.

antirepressor *See cro* repressor.

Antirrhinum a genus of decorative flowers commonly called snapdragons. The scientific name was given by Linnaeus in the first (1753) edition of the *Species Plantarum*. *A. majus* has 8 chromosomes, and it was a classical species for the study of multiple alleles. *Antirrhinum* research has made important contributions to the understanding of the biosynthesis of floral pigments and of photosynthetic pathways and their regulation. Studies of alleles of pigment genes that produced variegated color patterns led to the discovery of transposable elements (*q.v.*). One of these, Tam1, was one of the first plant transposons to have its nucleotide sequence determined. Transposon tagging (*q.v.*) has been used to identify genes that control floral organ identity. The base sequences are known for *Antirrhinum* genes that encode RNases, phytochromes, rRNAs, the large rubisco subunit, chloroplast proteins, and ubiquitin. *See* Appendix A, Plantae, Angiospermae, Dicotyledonae, Scrophulariales; anthocyanins, floral organ identity mutations.

antisense RNA (asRNA) a single-stranded RNA molecule with a nucleotide sequence complementary to a sense strand RNA, i.e., messenger RNA. The combining of sense and antisense RNA in a cell is a powerful silencer of gene expression. asRNAs occur naturally in many organisms (e.g., as small interfering RNAs (*q.v.*) or small temporal RNAs (*q.v.*)). These natural asRNAs silence genes at the transcriptional, post-transcriptional, or translational level. asRNAs can also be experimentally introduced into a cell or organism by such means as transformation (*q.v.*) with an artificial gene that transcribes asRNA or by injecting double-stranded RNA (*q.v.*) into cells or embryos. This is a powerful means of creating loss of function mutations in distinct genes. *See* Morpholinos, RNA interference (RNAi), strand terminologies.

antisense strand *See* strand terminologies.

antiserum a serum containing antibodies.

antisigma factor a protein that is synthesized during infection of *E. coli* by T4 phage and prevents recognition of initiation sites by the sigma factor of RNA polymerase.

antitermination factor a protein that allows an RNA polymerase to ignore signals to stop translation at particular sites in DNA molecules.

antitoxin an antibody that neutralizes a specific toxin. *See* *Corynebacterium diphtheriae.*

anucleate without a nucleus.

anucleolate without a nucleolus.

anucleolate mutation a mutation arising from the loss of the nucleolus organizer. Such mutations have been observed in *Drosophila, Chironomus,* and *Xenopus.* The study of such mutations led to the understanding of their role in the transcription of rRNA. *See* Appendix C, 1966, Wallace and Birnstiel; *bobbed,* ribosome.

AP endonuclease any enzyme that cleaves DNA on the 5′ side of a site in a DNA molecule that lacks a purine or pyrimidine. An *AP site* is a "hole" in a double-stranded DNA left by removal of a base, but leaving the sugar phosphate backbone intact. AP is the abbreviation for *ap*urinic or *ap*yrimidinic, referring to the loss of purines or pyrimidines. AP sites result when DNA glycosylases (*q.v.*) remove chemically altered bases from the polymer. An AP endonuclease will then break the phosphodiester backbone, and this allows excision of the damaged region. *See* cut-and-patch repair.

apetala *See* floral identity mutations.

aphasic lethal a lethal mutation in which death may occur randomly throughout development.

apical meristem *See* meristem.

Apicomplexa a phylum of the Protoctista (*see* Appendix A) that contains a group of obligate endoparasites, including the species that causes malaria (*q.v.*). All apicomplexans contain a vestigial plastid, the *apicoplast.* This chloroplast-derived organelle is the remnant of a secondary endosymbiosis between an ancestral apicomplexan and a chlorophyte (*q.v.*).

apicoplast an apical complex of organelles for which the phylum Apicomplexa is named. They function during the attachment and entry of the parasite into the host cell. The apicoplast is thought to have arisen from an algal chloroplast by symbiogenesis (*q.v.*). The *Plasmodium* apicoplast contains a circular DNA molecule made up of 35 kbp in which are found genes that encode ribosomal RNAs, a complete set of tRNAs, ribosomal proteins, RNA polymerase subunits, and various proteases. Certain apicomplexans have secondarily lost their apicoplasts. *See* Appendix A, Protoctista; *Cryptosporidium.*

Apis mellifera the domesticated honeybee. There are about 25 races recognized, but the ones most commonly used are the Italian, Caucasian, Carniolan, and German races. These are generally referred to as European bees, to distinguish them from African bees (*q.v.*). Bees reproduce by arrhenotokous parthenogenesis (*q.v.*). An alternative spelling, *A. mellifica,* also occurs in the literature. The honeybee is of interest because of its neural and behavioral plasticity with regard to social behavior, learning, and memory. Recently DNA microarray technology (*q.v.*) has been used to identify expression profiles of the genes that characterize worker/queen caste differentiation and to detect the subset of genes that is transcribed specifically in the brains of adult workers. *See* Appendix A, Animalia, Arthropoda, Hexapoda, Pterygota, Neoptera, Holometabola, Hymenoptera; Appendix C, 1999, Evans and Wheeler.

aplasia the failure of development of an organ.

Aplysia the genus of gastropods which includes the sea hares: large, sluglike, marine molluscs with rudimentary, internal shells. *Aplysia californica* has been much used in the study of the molecular basis of memory. *See* Appendix C, 1982, Kandel and Schwartz; 2003, Si *et al.*; CPEB protein.

apoenzyme the protein portion of a holoenzyme (*q.v.*), which requires a specific coenzyme for its function.

apoferritin *See* ferritin.

apoinducer a protein that, when bound to DNA, activates transcription by RNA polymerase.

apomict an organism produced by apomixis.

apomixis reproduction in which meiosis and fertilization are altered, so that only one parent contributes genes to the offspring (e.g., agamospermy or thelytoky). *Contrast with* amixis, amphimixis.

apomorphic an adjective applied to those derived characters of species that have evolved only within the taxonomic group in question. Plesiomorphic characters, on the other hand, are shared with other taxonomic groups as a consequence of their common ancestry. Thus, in mammals, the possession of hair would be an apomorphic character, whereas the possession of a backbone would be a pleisiomorphic character. *See* cladogram.

apoptosis (pronounced "apo-tosis") the programmed death of cells in various tissues at specific times during embryogenesis and metamorphosis or during cell turnover in adult tissues. For example, 12% of the cells formed during the development of an adult hermaphroditic *Caenorhabditis elegans* are destined to die because of a genetically controlled suicide program. If genes functioning in this system are inactivated by mutation, cells that normally die will survive. Recently the term apoptosis has been broadened to include all forms of cell death controlled by caspases (*q.v.*). Apoptosis as a part of normal development (programmed cell death), is now called *physiologic* apoptosis; whereas that occurring in diseased tissues is called *aberrant* apoptosis. *See* Appendix C, 1986, Ellis and Horvitz.

aporepressor a regulatory protein that, when bound to another molecule (corepressor, *q.v.*), undergoes an allosteric transformation that allows it to combine with an operator locus and inhibit transcription of genes in an operon. For example, in the histidine system of *E. coli*, excess histidine functions as *corepressor* by binding to an aporepressor to form a functional repressor (holorepressor); the holorepressor binds to the histidine operator and inhibits transcription of ten genes of this operon.

aposematic coloration warning coloration (*q.v.*).

apospory the development in plants of a diploid embryo sac by the somatic division of a nucellus or integument cell; a form of agamospermy (*q.v.*).

apostatic selection frequency-dependent selection for the rarer morph (e.g., in Batesian mimicry, the rarer the mimic, the greater its selective advantage). *See* mimicry.

a posteriori descriptive of biometric or statistical tests in which the comparisons of interest are unplanned and become evident only after experimental results are obtained.

apposition a process of increasing volume by adding new material to old at the periphery.

a priori descriptive of biometric or statistical tests in which the comparisons of interest are determined on theoretical grounds in advance of experimentation.

aptamers RNAs and DNAs derived from *in vitro* selection experiments that, starting from random sequence libraries, optimize the nucleic acids for high affinity binding to given ligands. Many applications of antibodies can be realized using aptamers, which possess even higher ligand affinities. Some aptamers find applications as molecular sensors and switches.

aptation any character currently subject to selection whether its origin can be ascribed to selective processes (adaptation) or to processes other than selection or selection for a different function (exaptation).

apterygote any of the primitive insects belonging to the cohort Apterygota, which contains the orders Thysanura (silverfish) and Diplura (campodeids and japigids). The species are wingless and possess abdominal styli. *See* Appendix A.

aptitude in microbial genetics, the specific physiological state of a lysogenic bacterium during which it can produce infectious bacteriophages when exposed to an inducing agent.

apyrene sperm *See* sperm polymorphism.

aquaporins a family of proteins that control the passage of water molecules through cell membranes. The first aquaporin studied was AQP1 which exists as a homotetramer and was originally classified as a human blood group antigen (Colten blood group) (*q.v.*). The AQP1 protein is present in many tissues besides erythrocytes (kidney tubules, lens epithelia, hepatic bile ducts, and lung endothelial cells are examples). The *AQP1* gene was mapped to 7p14. Mutations in the *AQP2* gene located at 12q13 are the cause of the hereditary disease diabetes insipidus (*q.v.*) inherited as an autosomal recessive. The AQP2 protein is translated in the collecting tubules of the kidney. The AQP4 protein occurs in the brain and retina and is encoded by the *AQP4* gene residing at 18q11.2–q12.1. The AQP7 protein is found in the plasmalemma of spermatids and is encoded by the *AQP7* gene at 9p13. Aquaporins have been isolated from many other animals, plants, and prokaryotes. *See* Appendix C, 1991, Preston and Agre.

Arabidopsis thaliana thale cress, a monoecious herb that belongs to the mustard family (Cruciferae). Its small size, small chromosome number ($N = 5$), short life cycle, and prolific production of seeds makes it ideal for genetic studies. *Arabidopsis* has a smaller genome size (1.2×10^8 bp) than agriculturally important species such as rice (4.2×10^8 bp) or wheat (1.6×10^{10} bp). About 25,500 genes have been identified in *Arabidopsis*, and 70% of these are present in duplicate copies. For this reason it seems likely that *Arabidopsis* had a tetraploid ancestor, and the actual number of *different* genes may be 15,000 or less. Also, 17% of all genes are arranged in tandem arrays, and these may have arisen by unequal crossing over (*q.v.*). Transposable elements (*q.v.*) account for at least 10% of the genome, and they are clustered near centromeres. Most structural genes are in the euchromatin. They are compact, about 2 kb long and about 4.6 kb apart. The average gene contains five exons, each about 250 bp long, whereas the average intron is only 170 bp long. About 100 *Arabidopsis* genes are homologous to human genes, which, in mutant form, cause hereditary diseases. There are over 800 genes that are involved in photosynthetic activities unique to plants. Many of these are orthologs of genes found in *Synechocystis* (*q.v.*). This suggests that these genes were acquired from a cyanobacterial symbiont that evolved into the progenitor of the chloroplasts. *See* Appendix A, Plantae, Tracheophyta, Angiospermae, Dicotyledonae, Cruciales; Appendix C, 2000, *Arabidopsis* Genome Initiative; Appendix E; centromere, ethylene, evolutionary mechanisms, floral identity mutations, sorting signals, transcription factors.

Arachnida a class of Arthropoda containing the spiders, scorpions, opilionids, and mites. *See* Appendix A.

arachnodactyly a synonym for Marfan syndrome (*q.v.*) found in the early literature.

Araenida the subclass of the arachnida containing the spiders.

Araldite trademark for a plastic commonly used for embedding tissues for electron microscopy.

arbovirus a virus that replicates in both an arthropod and a vertebrate host.

Archaea *See* next entry.

Archaebacteria a subkingdom of the Prokaryotae (*see* Appendix A). The archaebacteria are placed in a group separate from the rest of the bacteria (the eubacteria) on the basis of a variety of biochemical characteristics (distinctive compounds in their cell walls and membranes, differences in rare bases found in their tRNAs, and distinctive structures of RNA polymerase subunits). The archaebacteria are thought to have been dominant organisms in the primeval biosphere, since its atmosphere was rich in carbon dioxide and included hydrogen but virtually no oxygen. The magnitude of the molecular differences between these microorganisms and other bacteria supports the proposal that the taxonomic name be changed from Archaebacteria to simply Archaea. *See* Appendix C, 1977, Woese and Fox; extremophiles, flagellin, halophiles, methanogens, sulfur-dependent thermophiles, TATA box-binding protein (TBP).

Archaeoglobus fulgidus an anaerobic, sulfate-reducing archaeon found in hyperthermal environments. Its genome is a DNA circle that contains 2,178,400 base pairs. There are 2,436 ORFs, with an average size of 822 base pairs. Of 78 genes assigned to amino-acid biosynthetic pathways, over 90% have homologs in *Methanococcus jannashii* (*q.v.*). Over half of *A. fulgidus* ORFs are URFs. No inteins (*q.v.*) occur in this species, whereas *M. jannashii* has 18. *See* Appendix A, Archaea, Euryarchaeota; Appendix C, 1997, Klenk *et al.*; Appendix E; Archaebacteria, hyperthermophiles.

archaeon a bacterium that belongs to the subkingdom Archaea. The spelling *archaean* also occurs frequently. *See* Appendix A, Prokaryotes.

Archaeopteryx the genus name given to a group of ancient birds. The first fossil was discovered in 1861 and provided Darwin and his converts with ammunition to support the theory of evolution. Here was a bird whose reptilian characters made it a transitional form between these two classes of vertebrates. So far, six specimens have been recovered from the Jurassic Solnhofen formation in southern Germany. The most perfect specimen, generally considered to be the most beautiful and important fossil in the world, was discovered in 1877 and is housed in the Berlin Humboldt Museum. The fossil is birdlike in having feathers, a wishbone, and a big toe that rotated backward, opposite the other toes, as an adaptation to gripping branches. Its reptilian characteristics include claws on three of the five fingers, a long flexible tail, and teeth. The limb bones lack the air sacs found in birds. *See* Appendix C, 1868, Huxley.

Archaeprotista a phylum of protoctists (see Appendix A) that lack conventional mitochondria. Archaeprotists are anaerobes, and many are parasites. Some parasitic species, such as *Giardia intestinales* (*q.v.*), may once have had mitochondria, but these organelles underwent regressive evolution to become mitosomes (*q.v.*).

Archean the earlier of the two eras making up the Precambrian eon. Life began during the Archean, and the prokaryotes evolved. *See* Appendix C, 1977, Knoll and Barghoom; geologic time divisions.

archegonium the female sex organ of liverworts, mosses, ferns, and most gymnosperms.

archenteron the primitive digestive cavity of any metazoan embryo. It is formed by gastrulation.

area code hypothesis a theory proposing that as a cell passes through successive stages of differentiation, new patterns of molecules are displayed on its surface as a kind of area code that designates to which tissue or organ the cell should ultimately belong.

arena-breeding birds species of birds whose males gather for competitive courtship displays in open areas called arenas or leks (*q.v.*). Classic examples are species of peacock-pheasants, prairie chickens, and turkeys. *See* Appendix A, Aves, Neognathae, Galliformes; mate choice, sexual selection.

arg arginine. *See* amino acid.

arginine-urea cycle *See* ornithine cycle.

arithmetic mean an average; the number found by dividing the sum of a series by the number of items in the series.

arithmetic progression a series of numbers in which each number is larger (or smaller) than the number that precedes it by a constant value.

aRNA amplified RNA. *See* RNA amplification.

aromatic designating a chemical compound that contains a closed ring of carbon atoms, as found in benzene (*q.v.*). The aromatic amino acids phenylalanine and tyrosine serve as examples. *See* amino acid.

arrhenotokous parthenogenesis the phenomenon by which unfertilized eggs produce haploid males and fertilized eggs produce diploid females.

arrhenotoky arrhenotokous parthenogenesis (*q.v.*).

artemisinin a compound extracted from the leaves of *Artemisia annua L.*, a weed known in the United States as *annual wormwood.* Artemisinin is activated by iron to produce free radicals. The malaria parasite infects erythrocytes and causes iron to be released from infected cells. When artemisinin is administered to a patient with malaria, free radicals from it attack the cell membranes of the parasites in this iron-rich environment and destroy them. Malaria parasites such as *Plasmodium falciparum* (*q.v.*), which have developed resistance to older drugs like quinine, primaquine, or chloroquine (*q.v.*), are killed by artemisinin.

arthritis an autoimmune disease of the joints involving an attack by the body's immune system upon the synovial membranes.

Arthropoda the largest phylum in the animal kingdom in number of species. It contains the Chelicerata and Mandibulata. *See* Appendix A; Eumetazoa, Bilateria, Protostomia, Coelomata, Articulata.

Articulata a division of the invertebrates containing segmented, coelomate protostomes. *See* Appendix A.

artifact any structure that is not typical of the actual specimen, but that results from cytological processing, postmortem changes, etc.

artificial chromosomes *See* yeast artificial chromosomes.

artificial insemination (AI) the placement of sperm into a female reproductive tract or the mixing of male and female gametes by other than natural means. In AID, the sperm are provided by a donor other than the woman's husband; in AIH, the sperm are those of her husband. *See* Appendix C, 1769 (1780), Spallanzani.

artificial parthenogenesis induction of the development of an unfertilized egg by chemical or physical stimulation. *See* Appendix C, 1900, Loeb.

artificial selection the choosing by humans of the genotypes contributing to the gene pool of succeeding generations of a given organism.

Ascarididae a family of roundworms (*See* Appendix A, Animalia, Eumetazoa, Pseudocoelomata, Nematoda). Economically important species in this family are intestinal parasites of humans (*Ascaris lumbricoides* L.), horses (*Parascaris equorum*), and pigs (*Ascaris suum*). All show chromatin diminution (*q.v.*) early during embryogenesis.

Ascaris megalocephala species of roundworms in which chromatin diminution (*q.v.*) was first described. The preferred scientific name is *Parascaris equorum* (*q.v.*). *See* Appendix C, 1887, Boveri.

ascertainment bias *See* proband method.

Aschelminthes a phylum containing the pseudocoelomate animals with an anterior mouth, posterior anus, and straight digestive tube. *See* Appendix A.

ascidian a name given to a group of marine filter feeders, each enclosed in a tough sac. *See* *Ciona intestinalis*.

Ascobolus immersus an ascomycete fungus that is convenient for tetrad analysis (*q.v.*). *See* Appendix A.

ascogenous hypha a hypha that develops from the surface of an ascogonium (*q.v.*) after plasmogamy (*q.v.*). The hypha therefore contains nuclei of both mating types. After crozier formation (*q.v.*) such hyphae produce asci.

ascogonium in fungi, a female cell that receives haploid nuclei from an antheridium (*q.v.*).

ascomycete a fungus of class Ascomycetes that produce ascospores. *See* Appendix A.

ascorbic acid vitamin C (*q.v.*). Scurvy is a deficiency disease resulting from inadequate ascorbic acid in the diet.

ascospore a meiospore (*q.v.*) contained within an ascus (*q.v.*).

ascus a sac containing ascospores. The fact that all the products of a meiotic division are contained in an ascus in some fungi makes tetrad analysis (*q.v.*) possible.

asexual reproduction reproduction without sexual processes; vegetative propagation. *See* binary fission, budding, cloning, conidium, mitosis, monozygotic twins, parthenogenesis.

Ashkenazi (*plural* Ashkenazim) referring to Jews of eastern European origin. Most migrated from ancient Israel during the Middle Ages and established themselves first in Germany and France. In Hebrew, Ashkenazi means "Germans." The use of the Yiddish (Judeo-German) language distinguishes Ashkenazi Jews from the Sephardic Jews, who speak Ladino (Judeo-Spanish). Both languages use the Hebrew alphabet. Certain hereditary lysosomal diseases are found in relatively high frequencies among Ashkenazi Jews. It is assumed that the mutations that caused these diseases when homozygous conferred some as yet unknown advantage on heterozygotes in early ghetto environments. *See* diaspora, Gaucher disease, heterozygote advantage, Jews, lysosomal storage diseases, Mendelian Inheritance in Man (MIM), Nieman-Pick disease, Tay-Sachs disease.

Asilomar Conference a meeting where the potential hazards of recombinant DNA research were discussed for the first time. The Asilomar Conference Center is located on California's Monterey Peninsula. *See* Appendix C, 1975, NIH Recombinant DNA Committee; containment.

A-site-P-site model *See* translation.

asn asparagine. *See* amino acid.

asp aspartate; aspartic acid. *See* amino acid.

asparagine *See* amino acid.

Asparagus officinalis the asparagus.

aspartic acid *See* amino acid.

Aspergillus a genus of filamentous fungi belonging to the phylum Deuteromycota. *See* Appendix A. *A. flavus* is the source of aflotoxin (*q.v.*). *A. nidulans* is the species in which Pontecorvo and Roper discovered parasexuality (*q.v.*). *See* Appendix C. *A. nidulans* has a haploid chromosome number of eight, and there are eight well-mapped linkage groups. Its estimated genome size is 25,400 kilobase pairs. Mitochondrial genes are also characterized.

asRNA antisense RNA (*q.v.*).

association the joint occurrence of two genetically determined characteristics in a population at a frequency that is greater than expected according to the product of their independent frequencies.

associative overdominance linkage of a neutral locus to a selectively maintained polymorphism that increases heterozygosity at the neutral locus. *Compare with* hitchhiking.

associative recognition requirement for the initiation of an immune response of the simultaneous recognition by T lymphocytes of the antigen in association with another structure, normally a cell-surface alloantigen encoded within the major histocompatibility complex.

assortative mating sexual reproduction in which the pairing of male and female is not random, but involves a tendency for males of a particular kind to breed with females of a particular kind. If the two parents of each pair tend to be more (less) alike than is to be expected by chance, then positive (negative) assortative mating is occurring.

assortment the random distribution to the gametes of different combinations of chromosomes. Each 2N individual has a paternal and a maternal set of chromosomes forming N homologous pairs. At anaphase of the first meiotic division, one member of each chromosome pair passes to each pole, and thus the gametes will contain one chromosome of each type, but this chromosome may be of either paternal or maternal origin.

aster *See* mitotic apparatus.

asynapsis the failure of homologous chromosomes to pair during meiosis. *Contrast with* desynapsis.

atavism the reappearance of a character after several generations, the character being caused by a recessive gene or by complementary genes. The aberrant individual is sometimes termed a "throwback."

ataxia-telangiectasia mutated (ATM) referring to a gene that has been mapped at 11q22.3 and is responsible for a rare neurological disorder. *Ataxia* refers to the unsteady gate of patients due to the death of certain of their brain cells. Dilated blood vessels called *telangiectases* develop on the surfaces of their eyes and facial skin. The functional allele encodes a proetin that is a member of a family of phosphatidylinositol-3-kinases, enzymes that respond to DNA damage by phosphorylating key substrates involved in DNA repair. *See* ATM kinase, DNA damage checkpoint, *RAD*.

Atebrin a trade name for quinacrine (*q.v.*). Also spelled Atabrine.

ateliosis retarded growth resulting in a human of greatly reduced size but normal proportions. Such midgets generally show a marked deficiency in pituitary growth hormone. *See* pituitary dwarfism.

(A + T)/(G + C) ratio the ratio between the number of adenine-thymine pairs and the number of guanine-cytosine pairs in a given DNA sample.

athenosphere *See* plate tectonics.

ATM kinase a human DNA repair protein encoded by the *ataxia-telangiectasia mutated* (*ATM*) gene (*q.v.*). This enzyme plays a role in the DNA damage checkpoint (*q.v.*). *See* RAD.

atom the smallest particle of an element that is capable of undergoing a chemical reaction. *See* chemical element.

atomic mass the mass of a neutral atom of a nuclide, usually expressed in terms of atomic mass units.

atomic mass unit one-twelfth the weight of a ^{12}C atom; equivalent to 1.67×10^{-24} g.

atomic number the number of protons in the nucleus, or the number of positive charges on the nucleus: symbolized by Z. It also represents the number of orbital electrons surrounding the nucleus of a neutral atom.

atomic weight the weighted mean of the masses of the neutral atoms of an element expressed in atomic weight units.

ATP adenosine triphosphate. *See* adenosine phosphate.

ATPase any enzyme capable of hydrolyzing a phosphate bond in ATP. The energy derived from the hydrolysis is often used directly for transport of ions through membranes or for mechanical work. *See* adenosine phosphate, Wilson disease (WD).

ATP7B gene a gene located on the long arm of human chromosome 13 in region 14.2-21. It encodes an enzyme called *copper transporting ATPase 2*. This molecule belongs to a family of enzymes that transport metals in and out of cells using ATP as an energy source. *See* adenosine phosphate, ATPase, Wilson disease (WD).

ATP synthase a multimeric protein complex that plays a key role in the energy metabolism of all organisms. The inner membranes of the mitochondria of eukaryotes are covered with these organelles. They appear as particles 8–9 nm in diameter. Each particle has a spherical head piece (the F1 domain) that projects into the matrix of the mitochondrion and contains the catalytic sites for ATP synthesis. Imbedded in the lipid bilayer that forms the inner membrane of the mitochondrion is a cylindrical hydrophobic tailpiece (the Fo domain) that contains a channel for a stream of protons. This powers a rotation generator which is made up of a ring of 10–12 subunits that rotate 50–100 times a second. This rotation force is mechanically coupled to the catalytic sites in the F1 domain. These undergo a cycle of conformational changes which first loosely bind ATP and inorganic phosphate molecules, then rigidly bind the nascent ATP, and finally release it. Together the components of the different domains of ATP synthase contain at least 16 different proteins, and many mutations have been detected in the genes in the nuclei which encode these. Such loss of function mutations are generally lethal when hemizygous or homozygous. There are also two genes located in mtDNA, and individuals with mutations in their mtDNA can survive, provided wild-type mitochondria are also present in their cells. The ATP synthases of bacteria and chloroplasts resemble those of mitochondria with respect to the proteins that occupy the Fo and F1 domains. In bacteria, eight of the genes that encode these proteins are clustered in a single operon. *See* Appendix C, 1973, Boyer; 1981, Walker; adenosine phosphate, chemiosmotic theory, heteroplasmy, Leigh syndrome.

atresia congenital absence of a normal passageway. Absence of a normally open lumen.

atrichia hairlessness. In the domestic dog, the condition is inherited as an autosomal dominant. Homozygotes are stillborn.

attached X chromosome monocentric elements containing two doses of the X chromosome. *Drosophila* females carrying attached X chromosomes usually also have a Y chromosome. They produce XX and Y eggs and therefore generate patroclinous sons (which inherit their X from their father and their Y from their mother) and matroclinous daughters (which inherit their X from their mother and their Y from their father). The terms *double X* and *compound X* are also used for aberrant chromosomes of this types. *See* Appendix C, 1922, Morgan; detached X.

attachment efficiency *See* seeding efficiency.

attachment point (ap) a hypothetical analog of the centromere on chloroplast DNA in *Chlamydomonas*. Chloroplast genes may be mapped with respect to the attachment point.

attenuation 1. in physics, the loss in energy of an electromagnetic radiation as it passes through matter. 2. in microbiology, the loss in virulence of a pathogenic organism as it is repeatedly subcultured or is let multiply on unnatural hosts. 3. in immunology, the reduction in virulence of a substance to be used as an immunogen. Attenuation may result from aging, heating, drying, or chemically modifying the immunogen. 4. in molecular genetics, a mechanism for regulating the expression of bacterial operons that encode enzymes involved in amino acid biosynthesis. *See* attenuator.

attenuator a nucleotide sequence that is located upstream of those bacterial operons which encode the enzymes that are involved in the synthesis of amino acids. The expression of such operons is switched on and off by controlling the transcription of the messages for these operons. The leader sequence of the tryptophan operon of *E. coli* illustrates how attenuators function. In drawing A, an RNA polymerase is moving along the coding strand of a bacterial DNA molecule. The RNA molecule transcribed by the polymerase dangles behind it. Near its 5′ end is a site for binding ribosomes. One has attached and is moving along the RNA, forming a polypeptide as it goes. How fast the ribosome moves is determined by the availability of amino acid–charged tRNAs. The detailed structure of the RNA transcript is illustrated here in drawing B. The blocks designated A and B can pair because their base sequences are complementary and so can C and D. However, B can also pair with C, so there are three hairpin loops that can form, A/B, B/C, C/D. However, only the A segment contains tryptophan codons (symbolized by Xs). When tryptophan is abundant, the ribosome moves without pause past A to B. As the polymerase transcribes C and D, these pair to form a termination hairpin (*q.v.*), and the RNA polymerase together with its transcript detach from the DNA strand. Therefore, the trp operon is silenced. However, when there is no tryptophan in the environment, the ribosome pauses at the trp codons. Since A is covered by the ribosome, B pairs with C, and now D cannot form a termination hairpin. Therefore, the polymerase continues to the operon and transcribes it. The transcript is later translated, and the enzymes that result catalyze the formation of tryptophan. Thus, enzymes required for making an essential amino acid are synthesized only when the amino acid is scarce. *See* Appendix C, 1977, Lee and Yanofsky; leader sequence.

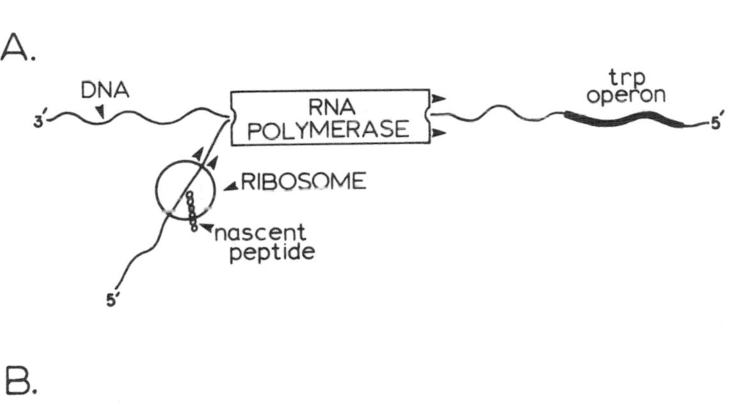

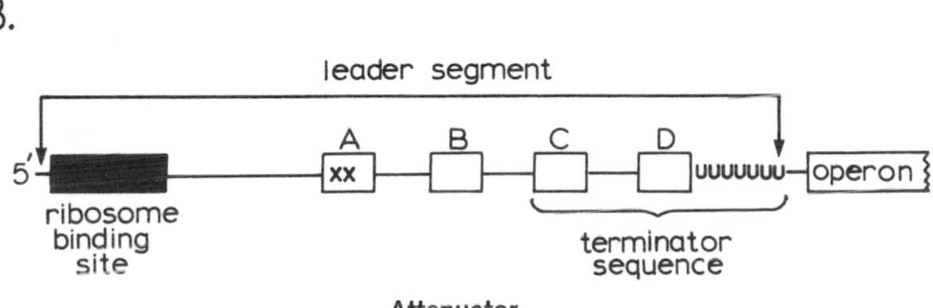

Attenuator

att sites loci on a phage and the chromosome of its bacterial host where recombination integrates the phage into or excises it from the bacterial chromosome.

audiogenic seizure convulsions induced by sound. Certain strains of mice, rats, and rabbits are especially prone to such seizures.

aureomycin *See* tetracycline.

Australian one of the six primary biogeographic realms (*q.v.*), comprising Australia, the Celebes, New Guinea, Tasmania, New Zealand, and the oceanic islands of the South Pacific.

Australopithecine referring to early hominids, fossils of which have been found in Africa. According to some classifications, the genus *Australopithecus* contains four species: *A. afarensis*, *A. africanus*, *A. robustus*, and *A. boisei*, all of which lived between 4 and 1 million years ago. The famous fossil "Lucy" is a 40% complete skeleton found in 1977. She lived about 3 million years ago and belonged to *A. afarensis*.

autapomorphic character 1. a derived character evolved from a plesiomorphic character state in the immediate ancestor of a single species 2. uniquely derived characters shared by several synapomorphous taxa. For example, hair was an autapomorphy of the first mammalian species and is also a synapomorphy of all mammals.

autarchic genes in mosaic organisms, genes that are not prevented from manifesting their phenotypic effects by gene products diffusing from genetically different neighboring tissues, whereas *hyparchic* genes are so inhibited.

autism a psychiatric disorder that usually is manifested by 3 years of age. The child shows an impaired ability to interact socially by eye-to-eye gaze, facial expression, body posture, and gestures. Spoken language either never develops or is delayed, and the child exhibits repetitive patterns of behavioral interests and activities and an abnormal preoccupation with routines. The prevalence in the U.S. population is 4 in 10,000, with a male-to-female ratio of 4:1. Susceptibility to autism is caused by genes at a variety of sites on autosomes 2, 6, 7, 13, 15, and the X. The most common chromosomal aberration causing autism is a maternally inherited duplication in the region 15q11–13.

autocatalysis the promotion of a reaction by its product. *See* protein splicing, ribozyme.

autochthonous pertaining to a species that has evolved within the region where it is native.

autocrine descriptive of a substance (e.g., a cytokine, *q.v.*) that binds to a surface receptor of the same cell that secreted it. *Paracrine* action is the binding of a secreted substance to a receptor on a nearby cell. *Endocrine* action is the binding of a secreted substance to a receptor on a distant cell. *See* agouti.

autofertilization *See* thelytoky.

autogamy that mode of reproduction in which the zygote is formed by fusion of two haploid nuclei from the same individual. In *Paramecium*, a process of self-fertilization resulting in homozygosis. In single individuals, the two micronuclei each undergo meiosis and seven of the eight resulting nuclei degenerate. The remaining haploid nucleus divides mitotically and the two identical nuclei fuse. The fusion nucleus gives rise to the micro- and macronuclei of the paramecium and its progeny. Autogamy is therefore a mode of nuclear reorganization that constitutes an extreme form of inbreeding. In *P. aurelia*, autogamy occurs spontaneously at regular intervals.

autogenous insect species in which females can produce eggs without first feeding. *See* anautogenous insect.

autogenous control regulation of gene expression by its own product either inhibiting (negative autogenous control) or enhancing (positive autogenous control) its activity. For example, in *E. coli*, AraC protein (the regulatory factor for the arabinose operon) controls its own synthesis by binding to the AraC promoter.

autograft the transplantation of a living piece of tissue from one site to another on the same animal.

autoimmune disease any pathological condition resulting from an individual's immune response to its own cells or tissues. *See* AIDS, arthritis, lupus erythematosus.

autologous referring to a graft from one region to another on the same animal.

automimic a palatable individual that is an automatic mimic of members of the same species that are unpalatable to predators. *See* automimicry.

automimicry the phenomenon, seen for example in the Monarch butterfly (*q.v.*), in which the species has a polymorphism in terms of its palatability to predators. The polymorphism arises from the different food plants chosen by the ovipositing female.

Most members of the species are rendered unpalatable because they feed as larvae upon plant species rich in substances toxic to birds. Those insects that feed on plants containing no toxins are palatable. However, such insects mimic perfectly the more abundant, unpalatable members of the species. *See* mimicry.

automixis fusion of nuclei or cells derived from the same parent to yield homozygous offspring. An example would be autogamy in paramecia or automictic parthenogenesis, as seen in certain species of Lepidoptera. *See* autogamy, thelytoky.

automutagen any mutagenic chemical formed as a metabolic product by an organism.

autonomous controlling element a controlling element (*q.v.*), apparently having both receptor and regulator functions, that enters a gene and makes it an unstable mutant.

autonomously replicating sequences eukaryotic DNA sequences that function as origins of replication, even though all of them may not be used in every cell cycle. ARSs have been isolated in yeast, and their functioning requires a 14 base pair core region. This contains an 11 base pair consensus sequence that consists almost entirely of A-T base pairs. *See* replication origin, replicon.

autophagic vacuole an enlarged lysosome containing mitochondria and other cellular organelles in the process of being digested.

autophene a phenotype due to the genetic constitution of the cells showing it. Transplantation of such mutant cells to a wild-type host does not modify their mutant phenotype.

autopoiesis the ability of an organism to maintain itself through its own metabolic processes at the expense of carbon and energy sources. Cells are autopoietic; viruses and plasmids are not.

autopolyploid a polyploid that originates by the multiplication of one basic set of chromosomes. *See* autotetraploid.

autoradiograph a photographic picture showing the position of radioactive substances in tissues, obtained by coating a squash preparation or a section with a photographic emulsion in the dark, and subsequently developing the latent image produced by the decay radiations. In the case of colony hybridization (*q.v.*), a filter containing the radioactive chimeric vectors is taken to a dark room, placed in an x-ray film holder, and covered with a sheet of x-ray film. The film is then left to expose for several hours or a few days before it is processed. The position of the silver grains on the film marks the location of the colonies of interest. *See also* DNA fiber autoradiography.

autoradiographic efficiency the number of activated silver grains (produced in a photographic emulsion coating a section) per 100 radioactive disintegrations occurring within the tissue section during the exposure interval.

autoradiography a technique for localizing radioactively labeled molecules by applying a photosensitive emulsion to the surface of a radioactive specimen. *See* autoradiograph.

autoregulation regulation of the synthesis of a gene product by the product itself. In the simplist autoregulated systems, excess gene product behaves as a repressor and binds to the operator locus of its own structural gene. *Contrast with* end product inhibition.

autoselection the process whereby a genetic element tends to increase in frequency by virtue of the nature of its transmission, even though it has no effect on the viability, fecundity, or fertility of the individual that bears it.

autosexing the use of sex-linked genes with obvious phenotypic effects to permit the identification by external inspection of the sex of immature organisms (larval silkworms or chicks, for example) before sexual dimorphic traits become obvious.

autosome any chromosome other than a sex chromosome. Each diploid organism will have a set of autosomes contributed by the male and a set contributed by the female parent. Therefore, there will be a maternal and a paternal representative of autosomes 1, 2, . . . *n*. The genes residing on autosomes follow the mode of distribution of these chromosomes to the gametes during meiosis. This pattern (autosomal inheritance) differs from that of genes on the X or Y chromosomes, which show the sex-linked mode of inheritance. *See* chromosome set, human pseudoautosomal region, sex linkage.

autosyndesis *See* allosyndesis.

autotetraploid an autopolyploid with four similar genomes. If a given gene exists in two allelic forms *A* and *a*, then five genotypic classes can be formed in an autotetraploid: *AAAA* (quadruplex), *AAAa* (triplex), *AAaa* (duplex), *Aaaa* (simplex), and *aaaa* (nulliplex).

autotrophs organisms that can build their own macromolecules from very simple, inorganic molecules, such as ammonia and carbon dioxide. Autotrophs include photosynthetic bacteria, protoctists, and plants that can convert visible light into chemical energy. In addition, chemoautotrophic bacteria can produce organic molecules from CO_2 in the absence of light. They use as energy sources for biosynthesis the oxidation of inorganic compounds such as molecular hydrogen, ammonia, and hydrogen sulfide. *Contrast with* heterotrophs.

autozygote an individual homozygous at a given locus whose two homologous genes are identical by descent, in that both are derived from the same gene in a common ancestor. *See* allozygote.

auxesis growth in size by increase in cell volume without cell division.

auxins a family of plant hormones that promote longitudinal growth and cell division. Natural auxins are indole derivatives biosynthesized from tryptophan. The most common auxin is indole acetic acid (*q.v.*), which is synthesized in all plants. *See* antiauxin.

auxocyte a cell whose nucleus is destined to enter meiotic prophase; a primary oocyte, primary spermatocyte, megasporocyte, or microsporocyte.

auxotroph a mutant microorganism that can be grown only upon minimal medium that has been supplemented with growth factors not required by wild-type strains.

Avena the genus to which the various species of oats belong. The most commonly cultivated oat is *A. sativa*.

Avena test a technique using the curvature of *Avena* coleoptiles as a bioassay for auxins.

average life in nuclear physics, the average of the individual lives of all the atoms of a particular radioactive substance. It is 1.443 times the radioactive half-life (*q.v.*).

avian leukosis *See* leukemia.

avian myeloblastosis virus an oncogenic RNA virus.

avidity the total combining power of an antibody with an antigen. It involves both the affinity of each binding site and the number of binding sites per antibody and antigen molecule. *Compare with* affinity.

Avogadro's number the number of atoms (6.025×10^{23}) in one gram atomic weight of an element; also the number of molecules in the gram molecular weight of a compound.

awn a stiff, bristlelike appendage occurring on the flowering glumes of grasses and cereals.

axenic growth of organisms of a given species in the complete absence of members of any other species.

axolotl a salamander, *Ambystoma mexicanum*, that exhibits neoteny (*q.v.*). It does not metamorphose, but mates and reproduces while a "juvenile," and never leaves the water.

axon the long process of a nerve cell, normally conducting impulses away from the nerve cell body.

axoneme a shaft of microtubules extending the length of a cilium, flagellum, or pseudopod of a eukaryotic cell. Axonemes from all cilia and flagella (including sperm tails) contain the same "9 + 2" arrangement of microtubules. In the center of each axoneme are two singlet microtubules that run the length of the shaft. The central tubules are surrounded by a circle of doublet microtubules, each consisting of an A and B subfiber. Each A subfiber has longitudinally repeating pairs of armlike projections that contain dynein (*q.v.*). *See Chlamydomonas reinhardi*, flagellum, tektin, Y chromosome.

axoplasm the cytoplasm contained in axons.

axopodia rigid, linear cellular projections composed mostly of microtubules found in species belonging to the Actinopoda. *See* classification, Protoctista.

5-azacytidine an analog of cytidine in which a nitrogen atom is substituted for a carbon in the number 5 position of cytosine (*q.v.*). The analog is incorporated into newly synthesized DNA, and such DNA is undermethylated. Since a reduction in the number of methyl groups attached to genes is associated with an increase in their transcriptional activities, 5-azacytidine can switch on certain genes. For example, patients given the drug may start making fetal hemoglobin, which implies that their gamma genes have been switched on. *See* hemoglobin, hemoglobin genes.

azaguanine a purine antagonist first synthesized in the laboratory and later shown to be identical to an antibiotic synthesized by *Streptomyces spectabalis*.

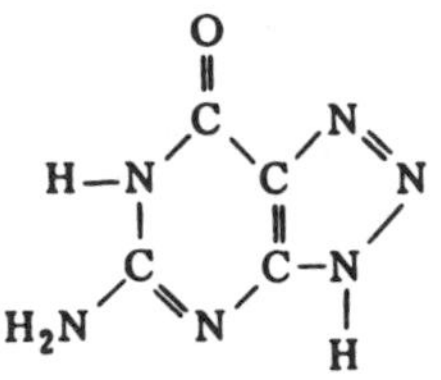

```
          O
          ‖
          C     N
        /   \  / \\
   H—N        C     N
     |        ‖    /
      C       C—N
    /  \\    /   |
 H₂N      N      H
```

Azaguanine is incorporated into mRNA and causes errors in the translation.

azaserine a glutamine analog synthesized by various species of *Streptomyces*. Azaserine inhibits purine biosynthesis and produces chromosome aberrations. It is mutagenic and has anti-tumorigenic activity.

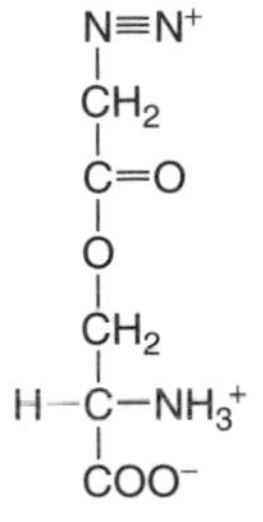

Azaserine

azoospermia absence of motile sperm in the ejaculate.

Azotobacter a genus of free-living, rod-shaped, soil bacteria capable of nitrogen fixation (*q.v.*). Polynucleotide phosphorylase (*q.v.*) was isolated from *Azotobacter vinelandii*. *See* Appendix A, Bacteria, Proteobacteria; Appendix C, 1955, Grungerg-Manago and Ochoa.

azure B a basic dye used in cytochemistry. *See* metachromasy.

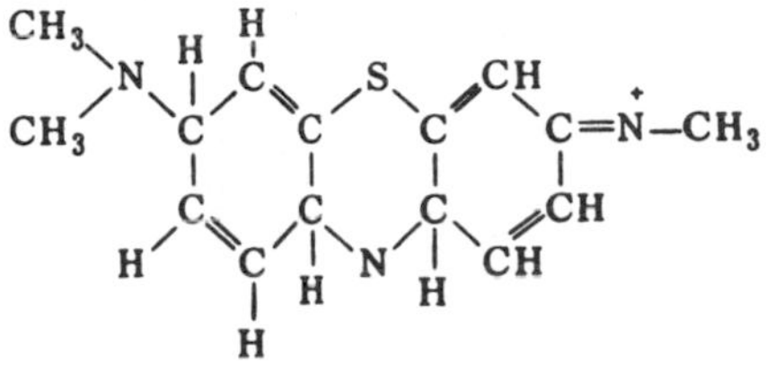

B

B_1, B_2, B_3, etc. the first, second, third, etc., backcross generations. The first backcross is made by mating an individual with one of its parents or with an individual of that identical genotype. The offspring produced belong to the B_1 generation. The second backcross is made by crossing B_1 individuals again with individuals of genotype identical to the parent referred to in the first backcross, etc.

BACs bacterial artificial chromosomes (*q.v.*).

Bacillus a genus of rod-shaped bacteria. *B. subtilis* is a Gram-positive, spore-forming, soil bacillus that grows readily in a chemically defined medium and undergoes genetic exchange by transformation and transduction (*q.v.*). Its genome contains 4,214,810 base pairs. There are about 4,100 protein-coding genes, and 53% of these are present as single copies. However, 25% of the genome is represented by families of duplicated genes. One family contains 77 genes for ATP-binding transport proteins. The genome is also home for at least 10 prophages, and this suggests that bacteriophage infection may have allowed horizontal gene transfer during evolution. *Bacillus megaterium* is the species in which the lysogenic cycle (*q.v.*) was deciphered. *B. thuringiensis* during sporulation produces an insecticidal crystalline deposit that is innocuous to vertebrates. The genes encoding the toxic proteins in these crystals have been identified and found to be carried on extrachromosomal plasmids. Some successful attempts have been made to splice these "insecticide genes" into the chromosomes of agriculturally important plant species. *See* Appendix A, Bacteria, Endospora; Appendix C, 1950, Lwoff and Gutman; 1997, Kunst *et al.*; ABC transporters, Bt designer plants, horizontal transmission, prophage.

Bacillus anthracis *See* anthrax.

backbone in biochemistry, the supporting structure of atoms in a polymer from which the side chains project. In a polynucleotide strand, alternating sugar-phosphate molecules form such a backbone.

backcross a cross between an offspring and one of its parents or an individual genetically identical to one of its parents.

backcross parent that parent of a hybrid with which it is again crossed or with which it is repeatedly crossed. A backcross may involve individuals of genotype identical to the parent rather than the parent itself.

background constitutive synthesis the occasional transcription of genes in a repressed operon due to a momentary dissociation of the repressor that allows a molecule of RNA polymerase to bind to its promoter and initiate transcription. Sometimes called "sneak synthesis."

background genotype the genotype of the organism in addition to the genetic loci primarily responsible for the phenotype under study.

background radiation ionizing radiation arising from sources other than that under study. Back ground radiation due to cosmic rays and natural radioactivity is always present, and there may also be background radiation due to man-made contaminating radiation.

back mutation reverse mutation (*q.v.*).

bacteria in the broader sense, all prokaryotes; more specifically, organisms belonging to the subkingdom Bacteria. *See* Appendix A, superkingdom Prokaryotes; Appendix C, 1677, van Leeuwenhoek; Appendix F, Genome sizes and gene numbers (of bacteria); Eubacteria.

bacterial artificial chromosomes (BACs) cloning vectors derived from the naturally occuring F factor (*q.v.*) of *Escherichia coli* (*q.v.*) and designed to accept large inserts (i.e., those in the size range of 80–350 kilobases). Insert-containing BACs are introduced into *E. coli* cells by electroporation (*q.v.*), where they can be maintained as circular plasmids. In contrast to yeast artificial chromosomes (YACs) (*q.v.*), which can show structural instability of inserts, DNA fragments closed in BACs remain structurally intact. This is because BACs contain F factor regulatory genes that control their replication and maintain their copy number to one or two per cell. The absence of multiple artificial chromosomes in a single cell minimizes sequence rearrangement in inserts by reducing the likelihood of recombination between the inserted fragments. BACs are useful for cloning DNA from large genomes, chromosome walking (*q.v.*), physical mapping, and shotgun sequencing

(*q.v.*) of complex genomes. *See* DNA vector, genome, genomic library, kilobase, physical map, plasmid cloning vector, P1 artificial chromosomes (PACs), regulator gene.

bacterial cell wall a structure forming a layer external to the plasma membrane. It controls the shape of the bacterium and serves as a permeability barrier. In some species a capsule may be formed external to the cell wall. The walls of Gram-positive bacteria are 30–100 nm thick and appear homogeneous. They are composed mainly of peptidoglycans. Gram-negative cell walls are thinner (20–30 nm) and are layered. The inner layer (adjacent to the plasma membrane) contains peptidoglycans. The outer layer contains peptidoglycans cross-linked to lipoproteins. Certain bacteria lack cell walls. *See* Gram-staining procedure, Mycoplasma, peptidoglycan.

bacterial transformation *See* transformation.

bacteriochlorophyll *See* chlorophyll.

bacteriocins proteins synthesized by various bacterial species that are toxic when absorbed by bacteria belonging to sensitive strains. Resistance to and the ability to synthesize bacteriocins are controlled by plasmids. *Escherichia coli* strains produce bacteriocins called colicins (*q.v.*). Bacteriocins from *Pseudomonas aeruginosa* are called pycocins.

bacteriophage a virus whose host is a bacterium; commonly called *phage*. Below are listed some common bacteria and their viral parasites:

Escherichia coli	P_1, P_2, P_4, $Q\beta$, λ, MS2, Mu 1, N_4, $\phi X174$, R_{17}, T_1 through T_7.
Salmonella typhimurium	P_{22}
Corynebacterium diphtheriae	β
Bacillus subtilis	SP_{82}, $\phi 29$
Shigella dysenteriae	P_1, P_2, P_4

Bacterial viruses show extreme variations in complexity. For example, the RNA phage R_{17} has a genome size of 1.1×10^6 daltons, whereas the DNA of the T4 phage weighs 130×10^6 daltons. *See* Appendix C, 1915, Twort; 1917, d'Herelle; 1934, Schlesinger; 1934, Ellis and Delbruck; 1942, Luria and Anderson; 1945, Luria; 1949, Hershey and Rotman; 1952, Hershey and Chase; 1953, Visconti and Delbrück; 1966, Edgar and Wood; 1970, Alberts and Frey; 1973, Fiers *et al.*; Appendix F, bacteriophages; filamentous phage, lambda (λ) bacteriophage, MS2, phi X174, plaque, P1 phage, P22 phage, Q beta phage, temperate phage, T phages, transduction, virulent phage.

bacteroids intracellular, nitrogen-fixing symbionts found in the root nodules of leguminous plants. Bacteroids are derived from free-living species of *Rhizobium* (*q.v.*). *See* leghemoglobin.

bacteriophage packaging insertion of recombinant bacteriophage lambda DNA into *E. coli* for replication and encapsidation into plaque-forming bacteriophage particles.

bacteriostatic agent a substance that prevents the growth of bacteria without killing them.

baculoviruses a group of viruses that infect arthropods, especially insects. Baculoviruses utilize the synthetic machinery of the insect host cell to synthesize polyhedrin, a protein that coats the virus particle. The gene for polyhedrin has a very strong transcriptional promoter to which foreign genes can be spliced to enhance their expression. These baculovirus expression vectors (BEVs) have been used in basic research and by commercial biotechnology enterprises aimed at the production of vaccines, therapeutics, and diagnostic reagents. By appropriate gene-splicing techniques baculoviruses have been engineered to synthesize foreign proteins, including the envelope protein of HIV (*q.v.*).

bag of marbles (bam) an autosomal gene in *Drosophila melanogaster* that is thought to regulate early germ cell development in both sexes. In the female, *bam* mutations produce ovarian tumors characterized by mitotically active stem cells which fail to produce cystoblast cells and their progeny, the interconnected cystocytes. In the male, *bam* mutations produce germ line cysts containing excessive numbers of primary and secondary spermatogonial cells which fail to differentiate. The *bam* gene encodes a novel protein which is found in the spectrosome (*q.v.*), the fusome (*q.v.*), and cytoplasm of cystocytes and spermatogonial cells. Fusomes in *bam* mutant females are structurally abnormal, leading to the suggestion that normal fusome biogenesis is essential for a switch from stem cell-like to cystoblast-like cell division. In males, *bam* gene product is thought to restrict the overproliferation of spermatogonial cells undergoing incomplete cytokinesis. *See* cystocyte divisions, spermatogonia.

balanced lethal system a strain of organisms bearing nonallelic recessive lethal genes, each in a different homologous chromosome. When interbred such organisms appear to breed true, because one-half of the progeny are homozygous for one or the other lethal gene and die prior to their detection. The surviving progeny, like their parents, are heterozygous for the lethal genes. *See* Appendix C, 1918, Muller; 1930, Cleland and Blakeslee.

balanced polymorphism genetic polymorphism maintained in a population because the heterozygotes for the alleles under consideration have a higher adaptive value than either homozygote. *See* Appendix C, 1954, Allison.

balanced selection selection favoring heterozygotes that produces a balanced polymorphism (*q.v.*)

balanced stock a genetic stock that, though heterozygous, can be maintained generation after generation without selection. Such stocks may contain balanced lethal genes, or a recessive lethal gene, which kills hemizygous males, combined with a nonallelic recessive gene, which confers sterility on homozygous females. *See* M5 technique.

balanced translocation synonymous with *reciprocal translocation*. *See* translocation.

Balbiani body a transitory, membrane-less structure consisting of cell organelles (e.g., mitochondria, centrioles, Golgi bodies, and endoplasmic reticulum) and macromolecules (e.g., RNAs, proteins, lipids, and ribonucleoproteins) found in the early ooplasm (*q.v.*) of a wide variety of vertebrate and invertebrate species. A Balbiani body generally originates near the nuclear envelope (*q.v.*) and differentiates into a well-defined mass often surrounded by mitochondria, before breaking down to release its constitutents into the ooplasm. It is thought to function in the organization and transport of RNAs and organelles that later become localized in specific regions of the ooplasm. Balbiani bodies show heterogeneity in number, morphology, and composition among different species and are named after the French biologist, Edouard-Gérard Balbiani (1823–1899), who was among the first to describe them. Also called *Balbiani vitelline body*. *See* mitochondrial cloud, sponge body, cytoplasmic localization.

Balbiani chromosome a polytene chromosome. So called because such banded chromosomes were first discovered by E. G. Balbiani in *Chironomus* larvae in 1881.

Balbiani ring a giant RNA puff present on a polytene chromosome of a salivary gland cell during a significant portion of larval development. The largest and most extensively studied Balbiani rings are on chromosome 4 of *Chironomus tentans* (see illustration). The transcription product of one of these puffs (BR 2) is a 75S RNA, which encodes the message for a giant polypeptide of the saliva. These secretory polypeptides are used by the larvae to build the tubes in which they reside. Balbiani rings contain thousands of DNA loops upon which mRNAs are being transcribed. These combine with proteins to form RNP granules (Balbiani ring granules) that eventually pass into the cytoplasm through nuclear pores. *See* Appendix C, 1881, Balbiani; 1952, Beermann; 1972, Daneholt; *Chironomus*, chromosomal puff.

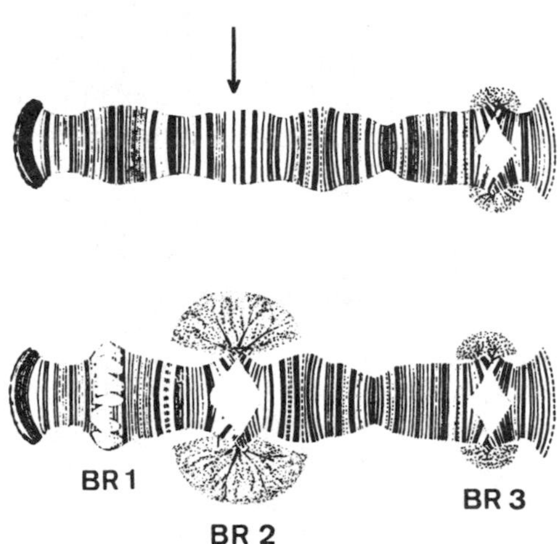

Chromosome 4 in the salivary glands of *Chironomus tentans* with the BR 2 band in an unpuffed (*above*) and in a puffed stage (*below*). The position of the BR 2 band is indicated by an arrow.

Bal 31 exonuclease a nuclease that digests linear, double-stranded DNA fragments from both ends. The enzyme is used *in vitro* to shorten restriction fragments. The shortened segment can then be religated with a DNA ligase (*q.v.*) to generate deletion mutants. *See* restriction endonucleases.

Baltimore classification of viruses an organizational scheme proposed by David Baltimore in 1971 that classified viruses according to the relationship between the viral genome and the messenger RNA used for viral protein synthesis. *Class I viruses* have double-stranded DNA genomes, and their mRNA is encoded by the template (negative) strand of the DNA. *Class II viruses* have a ssDNA genome and use a dsDNA intermediate during mRNA synthesis. *Class III viruses* have a dsRNA genome and one of the strands is equivalent to its mRNA. *Class IV viruses* are positive-strand RNA viruses. Therefore the genomic RNA is equivalent to mRNA. *Class V viruses* are negative-strand RNA viruses. Here the negative strand can serve as a template for mRNA synthesis. *Class VI viruses* are positive-strand RNA viruses that undergo reverse transcription (*q.v.*). *See* messenger RNA, virus.

Balzer freeze-fracture apparatus *See* freeze etching.

bam *See bag of marbles.*

***Bam*H1** *See* restriction endonuclease.

bananas perennial giant herbs of the tropics. The diploid fertile species *Musa acuminata* and *M. balbisiana* have genomes that may be symbolized *AA* and *BB*. Most cultivated bananas are sterile triploids that are vegetatively propagated. The triploid (AAA, ABB, or AAB) cannot form pollen, but the unpollinated ovary grows into a seedless fruit with edible pulp. The stimulus for parthenocarpy depends on the presence of at least three dominant complementary genes in the triploid. *See* Musaceae, parthenocarpy.

band **1.** in an electropherogram (*q.v.*), a region of the gel that contains clustered molecules of a particular size class visualized by staining, autoradiography, immunofluorescence, and so forth. **2.** in chromosome studies, a vertical stripe on a polytene chromosome that results from the specific association of a large number of homologous chromomeres at the same level in the somatically paired bundle of chromosomes. *See* Balbiani ring, deficiency loop, *Drosophila* salivary gland chromosomes, human chromosome band designations.

BAP 6-benzylaminopurine (*q.v.*).

Bar a sex-linked dominant mutation in *Drosophila melanogaster* which results in a reduction in the number of facets in the compound eye. This mutation, symbolized by *B*, is commonly used as a marker when constructing balancer X chromosomes. The original mutation contained a tandem duplication of chromosome segment 16A. Unequal crossing over (*q.v.*) can cause this mutation to revert to wild type. The 16A duplication is now thought to be a transposon-induced rearrangement. Analysis of the Bar phenomenon led to the discovery of position effects (*q.v.*). *See* Appendix C, 1925, Sturtevant; M5 technique, transposable elements.

barley *See Hordeum vulgare.*

Barr body the condensed single X chromosome seen in the nuclei of somatic cells of female mammals. *See* Appendix C, 1949, Barr and Bertram; dosage compensation, drumstick, late replicating X chromosome, Lyon hypothesis, sex chromatin.

basal body (granule) a structure generally composed of a ring of nine triplet microtubules surrounding a central cavity, found at the base of cilia. *See* axoneme, centriole, flagellum, kinetosome.

***Basc* chromosome** *See* M5 technique.

base analog a purine or pyrimidine base that differs slightly in structure from the normal base. Some analogs may be incorporated into nucleic acids in place of the normal constituent. *See* aminopurine, azaguanine, mercaptopurine. Analogs of nucleosides behave similarly. *See* 5-bromodeoxyuridine.

basement membrane a delicate acellular membrane that underlies most animal epithelia. *See* epidermolysis bullosa, laminin.

base pair a pair of hydrogen-bonded nitrogenous bases (one purine and one pyrimidine) that join the component strands of the DNA double helix. *Abbreviated* bp. *See* deoxyribonucleic acid, genome size, gigabase, Hoogsteen base pairs, hydrogen bond, kilobase, megabase, nucleotide pair.

base-pairing rules the rule that adenine forms a base pair with thymine (or uracil) and guanine with cytosine in a double-stranded nucleic acid molecule.

base-pair ratio *See* (A + T)/(G + C) ratio.

base-pair substitution a type of lesion in a DNA molecule that results in a mutation. There are two subtypes. In the case of *transitions*, one purine is substituted by the other or one pyrimidine by the other, and so the purine-pyrimidine axis is preserved. In the case of *transversions*, a purine is substituted by a pyrimidine or vice versa, and the purine-pyrimidine axis is reversed. *See* Appendix C, 1959, Freese.

bases of nucleic acids the organic bases universally found in DNA and RNA (see page 46). In a nucleotide sequence, a purine is often symbolized R, while a pyrimidine is symbolized by Y. The purines adenine and guanine occur in both DNA and RNA. The pyrimidine cytosine also occurs in both classes of nucleic acid. Thymine is found only in DNA, and uracil occurs only in RNA. *See* rare bases.

base stacking the orientation of adjacent base pairs with their planes parallel and with their surfaces nearly in contact, as occurs in double-stranded

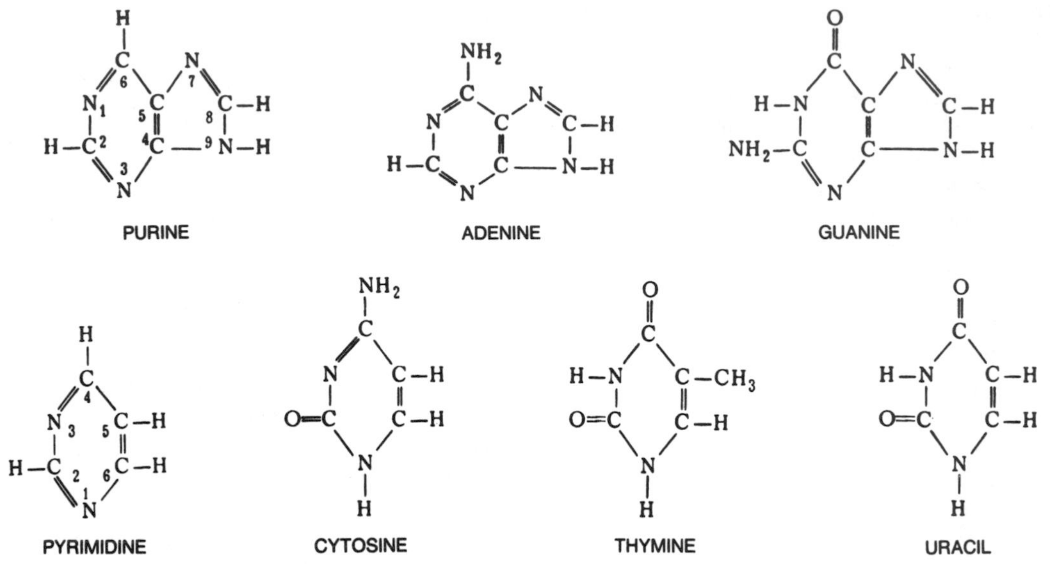

PURINE ADENINE GUANINE

PYRIMIDINE CYTOSINE THYMINE URACIL

Bases of nucleic acids

DNA molecules. Base stacking is caused by hydrophobic interactions between purine and pyrimidine bases, and results in maximum hydrogen bonding between complementary base pairs.

basic amino acids amino acids that have a net positive charge at neutral pH. Lysine and arginine bear positively charged side chains under most conditions.

basic dye any organic cation that binds to and stains negatively charged macromolecules, such as nucleic acids. *See* azure B.

basic number the lowest haploid chromosome number in a polyploid series (symbolized by x). The monoploid number.

basikaryotype the karyotype corresponding to the monoploid.

basophilic designating an acidic compound that readily stains with basic dyes.

Basques a human population living in the western part of the Pyrenees mountains. About 900,000 live in Spain and another 80,000 in France. Their language is unrelated to any Indo-European language, and they are believed to be the direct descendents of tribes dating back to paleolithic times. *See* cystic fibrosis, Rh factor.

Batesian mimicry a form of mimicry (*q.v.*) described by the British naturalist Henry Walter Bates in 1862.

B cell B lymphocyte. *See* lymphocyte.

B chromosomes supernumerary chromosomes that are not duplicates of any of the members of the basic complement of the normal or "A" chromosomes. B chromosomes are devoid of structural genes. During meiosis, B chromosomes never pair with A chromosomes, and Bs show an irregular and non-Mendelian pattern of inheritance. B chromosomes are extremely widespread among flowering plants and have been extensively studied in rye and maize. B chromosomes are believed to perpetuate and spread themselves in populations because they replicate faster than A chromosomes. *See* Appendix C, 1928, Randolph.

beads on a string the concept that genes in a chromosome resembled beads on a string and that recombination involved the string and not the beads. S. Benzer disproved this idea by showing that recombination did occur within genes. In fact it could occur between any adjacent nucleotides within a gene. *See rII*, T phages.

Becker muscular dystrophy (BMD) a less severe form of muscular dystrophy (*q.v.*). It was described in 1955 by Peter Emil Becker and shown to develop later in life and to progress more slowly than the common form. BMD constitutes about 10 percent of all cases of X-linked muscular dystrophy.

becquerel a unit of activity of a radioactive substance. 1 Bq = 1 disintegration per second.

bee dances circling and waggling movements *(Rundtanzen* and *Schwanzeltanzen)* performed by worker bees to give their hive mates information as to the location of a new source of food.

beef thymus the gland from which DNA was isolated for early structural studies. *See* Appendix C, 1951, Wilkins and Gosling; 1952, Franklin and Gosling; nucleic acid.

bees *See Apis mellifera*, African bees.

Beer-Lambert law the law that the absorption of light by a solution is a function of the concentration of solute. It yields a relation commonly used in photometry: $E = \log_{10} I_0/I = kcb$, where E = optical density; I_0 = the intensity of the incident monochromatic light; I = the intensity of the transmitted light; k = a constant determined by the solvent, wavelength, and temperature; c = the concentration of absorbing material in moles per liter; and b = the thickness in centimeters of the solution traversed by the light. The Beer-Lambert law is based on discoveries by the German mathematician, chemist, and physicist, August Beer (1825–1863) and the French mathematician, physicist, and astronomer, Johann Heinrich Lambert (1728–1777).

behavioral isolation a prezygotic (premating) isolating mechanism in which two allopatric species refuse to mate because of differences in courtship behavior; ethological isolation.

behavior genetics a branch of genetics that concerns the inheritance of forms of behavior such as courtship displays, nest building, etc., in lower animals and intelligence and personality traits in humans. Many traits that are of interest to behavioral scientists are quantitative characters (*q.v.*).

Bence-Jones proteins proteins identified by the English physician Henry Bence-Jones in 1847. B-J proteins are excreted in the urine of patients suffering from malignancies of antibody-secreting cells (plasmacytomas or myelomas). B-J proteins consist of dimers of immunoglobulin chains. The proteins are synthesized by clones of identical cells, and therefore each patient produces identical peptides in sufficient quantities for the amino acid sequences to be determined. B-J proteins played a critical role in investigations of the chemical structure of the immunoglobulins (*q.v.*).

benign *See* neoplasm.

benign subtertian malaria *See* malaria.

benzene the simplest of the aromatic organic compounds.

beta carotene *See* carotenoids, retinene.

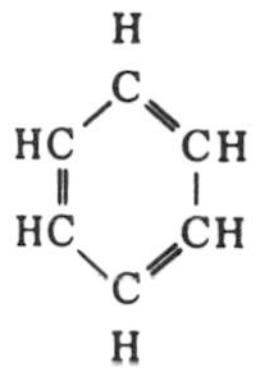

H
C
HC CH
HC CH
C
H

beta chain one of the two polypeptides found in adult hemoglobin.

beta galactosidase an enzyme that breaks lactose into glucose and galactose. In *E. coli*, the enzyme is a tetramer of about 500,000 daltons encoded by the *lac Z* gene. *See* homomeric protein, operon, *lac* operon, lactose.

beta lactamase *See* penicillinases.

beta-2 microglobulin *See* MHC molecules.

beta particle a high-energy electron emitted from an atomic nucleus undergoing radioactive decay.

beta pleated sheet one of two common, regularly repeating structures seen in proteins (*compare with* alpha helix). Each of the component polypeptide chains in the beta pleated sheet is fully extended, and the sheets are stabilized by hydrogen bonds between the NH and CO groups of the same or different polypeptide chains. Adjacent chains can run in the same or in opposite directions (parallel versus antiparallel beta sheets). For example, silk fibroin is composed of antiparallel beta pleated sheets. *See* Appendix C, 1951, Pauling and Corey; silk.

BEV the acronym for baculovirus expression vector. *See* baculoviruses.

bicoid (*bcd*) a gene that is essential for normal axis formation in *Drosophila*. The transcription of *bcd* takes place in the nurse cells (*q.v.*), and these are of maternal genotype. This mRNA is transported in a cytoplasmic stream that is pumped by the nurse cells (*q.v.*) into the oocyte, and here it is localized at the anterior pole. The sequence required for this localization resides in the trailer portion of the mRNA. This *bcd* mRNA is not translated until after the egg is fertilized. The product is a homeodomain protein that functions as a morphogen for anterior structures. Mothers who lack a functional allele of *bcd* produce embryos with aberrant or missing head and thoracic structures and anterior abdominal segments. Transplantation of anterior polar cytoplasm from wild type embryos or injection of purified *bcd* mRNA into *bcd*-deficient embryos can rescue (*q.v.*) the mutant phenotype. *bcd* protein is distributed in an anterioposterior concentration gradient in the zygote (*q.v.*) and controls the expression of several zygotic genes, depending on its local concentration. The *bcd* gene is a member of the anterior class of maternal polarity genes, and it lies within the *Antennapedia* (*q.v.*) complex. *See* Appendix C, 1988, Macdonald and Struhl; 1988, 1989, Driever and Nüsslein-Volhard; cytoplasmic localization, cytoplasmic determinants, *goosecoid*, homeobox, *hunchback*, maternal polarity mutants, *nanos* (*nos*), trailer sequence, zygotic gene.

bidirectional genes a pair of open reading frames (*q.v.*), one on the plus strand and the other on the minus strand of the same DNA double helix and overlapping to a certain degree. *Compare with* overlapping genes.

bidirectional replication a mechanism of DNA replication involving two replication forks moving in opposite directions away from the same origin.

biennial designating a plant that requires two years to complete its life cycle, from seed germination to seed production and death. The plant develops vegetatively during the first growing season and flowers in the second.

bifunctional vector *See* shuttle vector.

bilateral symmetry a form of symmetry in which the body can be divided by a longitudinal plane into two parts that are mirror images of each other.

Bilateria animals with bilateral symmetry. *See* classification.

bilharziasis *See* schistosomiasis.

bilirubin an orange pigment formed as a breakdown product of the heme component of hemeproteins, especially the hemoglobin released during the normal destruction of erythrocytes by the reticuloendothelial system. Bilirubin released into the circulation by the reticuloendothelial system is taken up by the liver and excreted into the bile. The accumulation of bilirubin in plasma and tissues results in jaundice. *See* Crigler-Najjar syndrome.

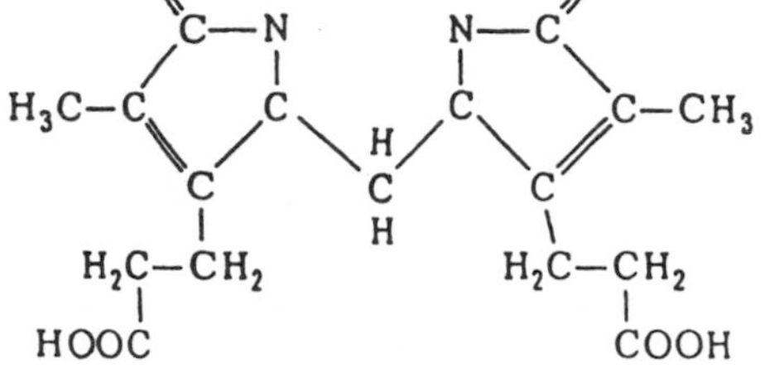

bimodal population a population in which the measurements of a given character are clustered around two values.

binary fission an amitotic, asexual division process by which a parent prokaryote cell splits transversely into daughter cells of approximately equal size. *Compare with* septal fission.

Binet-Simon classification *See* intelligence quotient classification.

binomial distribution a probability function so named because the probabilities that an event will or will not occur, n, $n-1$, $n-2$, . . . , 0 times are given by the successive coefficients in the binomial expansion $(a+b)^n$. Since a and b are the probabilities of occurrence and non-occurrence, respectively, their sum equals 1. The coefficients in a given binomial expansion can be found by referring to Pascal's pyramid (see illustration on page 49). Here each horizontal row consists of the coefficients in question for consecutive values of n. The expansions for n equaling 1, 2, 3, 4, or 5 are shown below:

$(a+b)^1 = 1a + 1b$
$(a+b)^2 = 1a^2 + 2ab + 1b^2$
$(a+b)^3 = 1a^3 + 3a^2b + 3ab^2 + 1b^3$
$(a+b)^4 = 1a^4 + 4a^3b + 6a^2b^2 + 4ab^3 + 1b^4$
$(a+b)^5 = 1a^5 + 5a^4b + 10a^3b^2 + 10a^2b^3 + 5ab^4 + 1b^5$

Note that each term of the triangle is obtained by adding together the numbers to the immediate left and right on the line above. Such a binomial distribution can be used for calculating the frequency of families in which a certain proportion of individuals show a given phenotype. If we ask, for example, what will be the distribution of girls and boys in families numbering four children and let the frequency of boys = a, and that of girls = b; then using the formula $(a+b)^4$, we conclude that the distribution would be 1/16 all boys; 4/16 3 boys and 1 girl; 6/16 2 boys and 2 girls; 4/16 1 boy and 3 girls; and 1/16 all girls. Pascal's pyramid is named after its inventor, the French mathematician Blaise Pascal (1623–1662).

binomial nomenclature *See* Linnean system of binomial nomenclature.

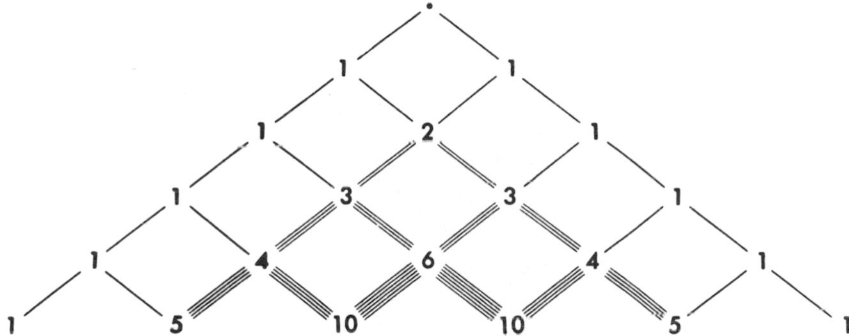

Binomial distribution (Pascal's pyramid)

bioassay determination of the relative potency or effectiveness of a substance (e.g., a drug or a hormone) by comparing its effect on a group of test organisms or cells (e.g., tissue culture) using appropriate controls.

biochemical genetics that branch of genetics that seeks to elucidate the chemical nature of hereditary determinants and their mechanisms of action during the life cycles of the organisms and their viruses. *See* Appendix C, 1909, Garrod; 1935, Beadle *et al.*

biocoenosis a group of plant and animal species living together as a community in a particular habitat.

biogenesis the production of a living cell from a parent cell. *Compare with* biopoesis.

biogenetic law *See* recapitulation.

biogeographic realms divisions of the land masses of the world according to their distinctive plant and animal inhabitants. The six realms shown in the map on page 50 are those first presented in 1876 by A. R. Wallace in his *Geographical Distribution of Animals*. Note the dashed line that separates Borner (B) and Java (J) from Sulawesi (S) and the smaller islands of Indonesia. This line corresponds to the Wallace line (*q.v.*).

biogeography the study of distributions of organisms over the earth and of the principles that govern these distributions. This field originated with studies of Alfred Russel Wallace. *See* Appendix C, 1855, 1869, 1876, Wallace.

bioinformatics a new field in which computer hardware and software technologies are developed and used to gather, store, analyze, and disseminate biological data, images, and other information. Wide-scale, collaborative works in genomics (*q.v.*) and proteomics (*q.v.*) rely heavily on bioinformatics. *See* genome annotation.

biological clock 1. any mechanism that allows expression of specific genes at periodic intervals. 2. any physiological factor that regulates body rhythms. *See* clock mutants.

biological evolution *See* evolution.

biological species groups of naturally interbreeding populations that are reproductively isolated from other such species.

bioluminescence the emission of light by living organisms.

biomass the total weight of organic material in a particular sample, region, trophic level, etc. The dry biomass of the earth is estimated to be 3×10^{15} kg. Prokaryotes (*q.v.*) represent the bottom of the food chain and the major source of C, N, and P.

biome a grouping of ecosystems (*q.v.*) into a larger group occupying a major terrestrial region (e.g., tropical rainforest biome, mixed conifer and deciduous forest biome).

biometry the application of statistics to biological problems. *See* Appendix C, 1889, Galton.

Biomphalaria glabrata *See* schistosomiasis.

biopoesis spontaneous generation (*q.v.*).

bioremediation the use of living organisms (e.g., bacteria, fungi, and other microorganisms) to reclaim contaminated environmental sites, such as soil, groundwater, or lakes. This is achieved by enhancing the natural ability of microbes already present at the site to break down contaminating molecules. Alternatively, such microbes are added to the affected site. *See Pseudomonas, Streptomyces.*

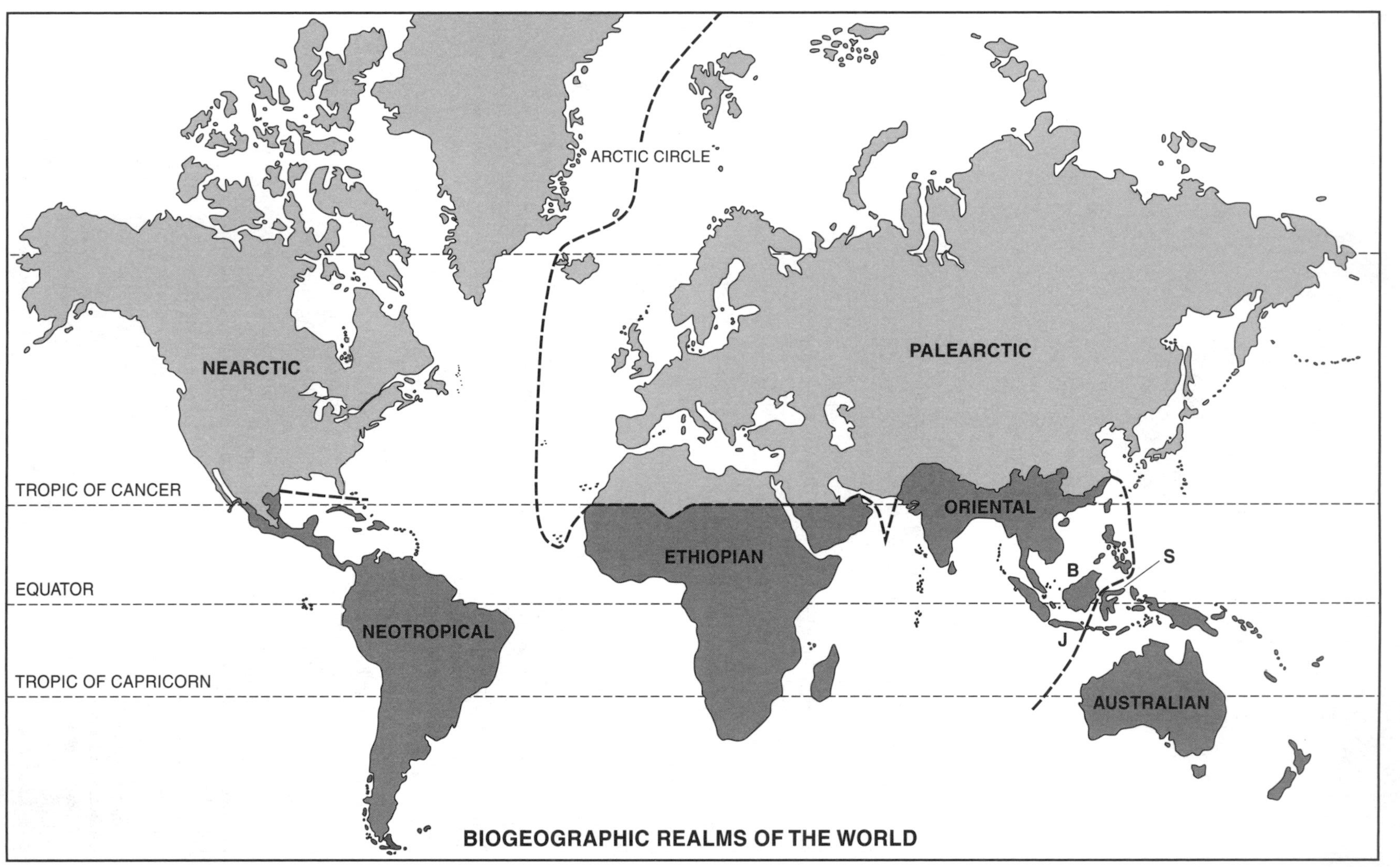

BIOGEOGRAPHIC REALMS OF THE WORLD

biorhythm a recurring cycle in the physiology or functioning of an organism, such as the daily cycle of sleeping and waking; a cyclic pattern of physical, emotional, or mental activity said to occur in the life of a person. *See* circadian rhythm.

biosphere the surface of the earth where life resides.

biosynthesis the production of a chemical compound by a living organism.

biota a collective term to include all the organisms living in a given region.

biotechnology the collection of industrial processes that involve the use of biological systems. For some industries, these processes involve the use of genetically engineered microorganisms.

biotic potential *See* reproductive potential.

biotin one of the water-soluble B vitamins. It functions as a cofactor in enzymes that catalyze carboxylation reactions. Biotin is often used as a chemical tag on nucleic acid probes because biotin-scavenging proteins like streptavidin (*q.v.*) bind it with high affinity. Biotin-binding proteins coupled with fluorescent dyes allow their detection on cytological preparations. *See* biotinylated DNA.

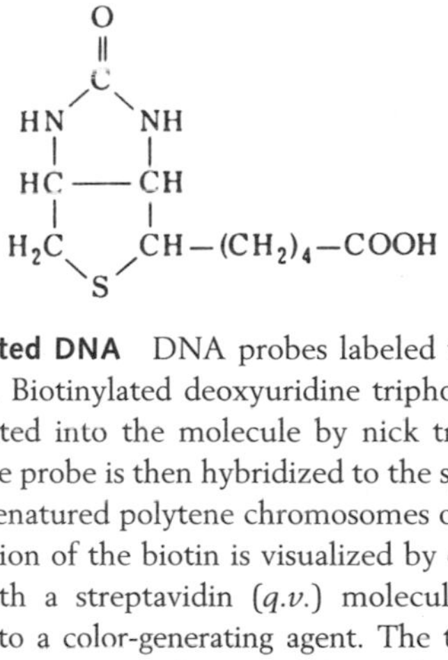

```
        O
        ‖
        C
      /   \
   HN       NH
    |        |
   HC ------ CH
    |        |
   H2C       CH—(CH2)4—COOH
      \    /
        S
```

biotinylated DNA DNA probes labeled with biotin (*q.v.*). Biotinylated deoxyuridine triphosphate is incorporated into the molecule by nick translation (*q.v.*). The probe is then hybridized to the specimen, such as denatured polytene chromosomes on a slide. The location of the biotin is visualized by complexing it with a streptavidin (*q.v.*) molecule that is attached to a color-generating agent. The technique is less time consuming than autoradiography and gives greater resolution. *See* Appendix C, 1981, Langer *et al.*

biotron a group of rooms designed for the control of environmental factors, singly and in combinations. Biotrons are used for producing uniform experimental organisms, and for providing controlled conditions for experiments.

biotype a physiologically distinct race within a species. If the biotype allows the race to occupy a particular environment, it is equivalent to an ecotype (*q.v.*).

biparental zygote 1. the common state for nuclear genes in diploid zygotes to contain equal genetic contributions from male and female parents. 2. the rare state for cytoplasmic genes in diploid zygotes to contain DNA from both parents (e.g., chloroplast DNA in *Chlamydomonas*).

biparous producing two individuals at one birth.

Bipolaris maydis a fungus responsible for disasterous epidemics of corn leaf blight that have caused losses of billions of dollars to the U.S. corn crops. Formerly called *Helminthosporium maydis*. *See* cytoplasmic male sterility.

birefringence *See* anisotropy.

birth defect 1. any morphological abnormality present at birth (congenital); such abnormalities may have a genetic basis or they may be environmentally induced (*see* phenocopy). 2. any biochemical or physiological abnormality present at birth; such abnormalities usually have a genetic basis and have been called "inborn errors of metabolism." *See* Appendix C, 1909, Garrod.

bisexual 1. pertaining to a species made up of individuals of both sexes. 2. pertaining to an animal having both ovaries and testes, or to a flower having both stamens and pistils.

Bison a genus including the American bison, *B. bison*, and the European bison, *B. bonasus*, the latter of which has been used in studies of the effects of inbreeding.

Biston betularia the peppered moth, the species used as the classic example of industrial melanism (*q.v.*). A black form of the species, called *carbonaria*, was first observed in 1848, and it spread through the industrial areas of England and soon became the most common form. This melanic form is due to a dominant gene, and moths of this phenotype appeared to be less conspicuous than the ancestral form in environments polluted by soot (see diagram on page 52). Therefore it was assumed that the melanics were selected because they avoided bird predation. During the 1950's experiments were done which seemed to show that birds preyed selectively on the more conspicuous moths. However, these early experiments were discredited by later work, and so the cause of the phenomenon has not been resolved. The declines in atmospheric pollutants following clean air legislation have been accompanied by reduction in the frequency of melanic peppered moths. For example, in the area around Liverpool the frequency of carbonaria fell from 90% to 10% over a 40-year period. *See* Appendix C, 1891, Tutt; 1958, Kettlewell; Bibliography, 2003, Hooper.

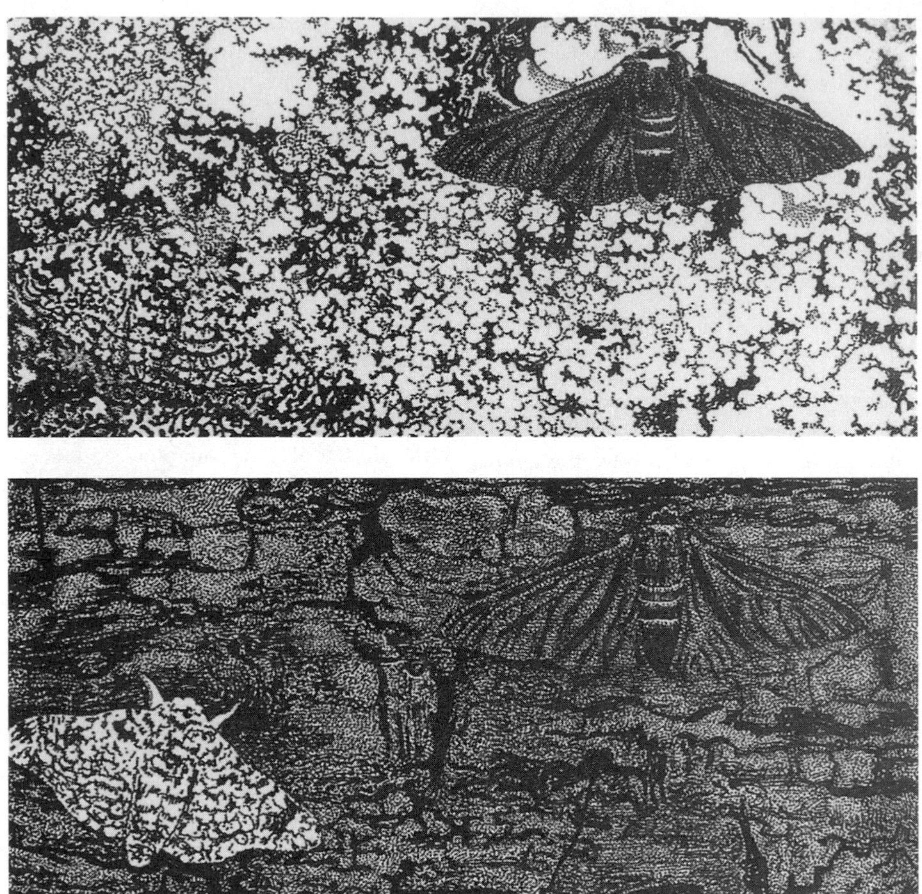

The ancestral (*left*) and melanic (*right*) form of the peppered moth. The backgrounds are non-polluted bark (*upper drawing*) and polluted bark (*lower drawing*).

Biston betularia

bithorax a gene residing at 58.8 on the genetic map and within segment 89E of the salivary map of *Drosophila melanogaster*. The *bx* gene is one of a cluster of three genes that specify the type of differentiation that cells in the segments starting at the posterior portion of the second thoracic segment through the eighth abdominal segment will undergo. Illustration D on page 210 shows a *bx* mutant with four wings. Therefore the normal function of this gene is to specify that the hind pair of imaginal discs (*q.v.*) form halteres (*q.v.*) not wings during metamorphosis. The three genes of the *bithorax* complex encode DNA-binding proteins with homeodomains. *See* Appendix C, 1978, Lewis; 1983, Bender *et al.*; floral identity mutations, homeotic mutations, *Hox* genes, metamerism, *Polycomb*, *proboscipedia*, segment identity genes, *spineless-aristapedia*.

Bittner mouse milk virus *See* mammary tumor agent.

bivalent a pair of homologous, synapsed chromosomes. *See* meiosis.

Bkm sequences a satellite DNA containing repeats of the tetranucleotide sequences GATA and GACA that was first isolated from the banded krait. In this and many other snakes, the sequences are concentrated in the W chromosome. Bkm sequences also occur in the W chromosomes of birds. *See* W, Z chromosomes.

blackwater fever a name for malaria, with reference to the urinary excretion of heme. The malaria parasite uses only the globin fraction of hemoglobin for its metabolism and discards the heme, which is excreted, darkening the urine.

BLAST Basic Local Alignment Search Tool. This algorithm is widely employed for determining similarity between nucleic acid or protein sequences, using sequences present in data bases.

blast cell transformation the differentiation, when antigenically stimulated, of a T lymphocyte to a larger, cytoplasm-rich lymphoblast.

blastema a small protuberance composed of competent cells from which an animal organ or appendage begins its regeneration.

blastocyst the mammalian embryo at the time of its implantation into the uterine wall.

blastoderm the layer of cells in an insect embryo that completely surrounds an internal yolk mass. The cellular blastoderm develops from a syncytial blastoderm by the partitioning of the cleavage nuclei with membranes derived from infoldings of the oolemma.

blastodisc a disc-shaped superficial layer of cells formed by the cleavage of a large yolky egg such as that of a bird or reptile. Mitosis within the blastodisc produces the embryo.

blastokinin *See* uteroglobin.

blastomere one of the cells into which the egg divides during cleavage. When blastomeres differ in size, the terms *macromere* and *micromere* are often used.

blastoporal lip the dorsal rim of the amphibian blastopore, which functions as the organizer inducing the formation of the neural tube. *See* chordamesoderm, Spemann-Mangold organizer.

blastopore the single external opening of the primitive digestive tract in the gastrula of most animals. The fate of the blastopore differs in those species belonging to the two subdivisions of the Bilateria. *See* Appendix A, Deuterostomia, Protostomia.

blastula an early embryonic stage in animals consisting of a hollow sphere of cells.

Blatella germanica the German cockroach, a species found throughout the world in association with humans. It is the hemimetabolous insect for which the most genetic information is available. *See* Appendix A, Arthropoda, Insecta, Dictyoptera.

blending inheritance 1. an obsolete theory of heredity proposing that certain traits of an offspring are an average of those of its parents because of the blending of their fluidlike germinal influences; hereditary characters transmitted in this way would not segregate in later generations. 2. a term incorrectly applied to codominant traits, to genes lacking dominance, or to additive gene action.

blepharoplast the basal granule of flagellates.

blocked reading frame *See* reading frame.

blood clotting a cascade of enzymatic reactions in blood plasma that produces strands of fibrin to stop bleeding. Fibrinogen, a protein found in the blood plasma, is acted upon by the enzyme thrombin. As a result a negatively charged peptide is split off the fibrinogen molecule, leaving monomeric fibrin, which is capable of rapid polymerization to produce a clot. Active thrombin is formed from an inactive precursor prothrombin, also found in blood plasma. The conversion of prothrombin into thrombin is a very complex process that requires a number of factors, including a lipoprotein factor liberated from rupturing blood platelets, plasma thromboplastin component, a complex of antihemophilic factor and von Willebrand factors, calcium, ions, and others. *See* hemophilia, von Willebrand disease.

blood coagulation blood clotting (*q.v.*).

blood group a type in a system of classification of blood, based on the occurrence of agglutination of the red blood cells when bloods from incompatible groups are mixed. The classical human blood groupings were A, B, AB, and O. However, a multitude of more recently identified groups exists. There are 33 blood group genes that have been localized to specific chromosomes. Seventeen of these reside on three chromosomes: the X has 5 genes, and autosomes 1 and 19 have 6 genes each. Nine other autosomes each have between 1 and 3 genes. *See* Appendix C, 1900, Landsteiner; 1925, Bernstein; 1951, Stormont *et al.*; A, B antigens, Bombay blood group, Colton blood group, Duffy blood group gene, H substance, Kell-Cellano antibodies, Kidd blood group, Lewis blood group, Lutheran blood group, MN blood group, P blood group, Rh factor, Secretor gene, XG.

blood group chimerism the phenomenon in which dizygotic twins exchange hematopoietic stem cells while *in utero* and continue to form blood cells of both types after birth. *See also* radiation chimera.

bloodline in domesticated animals, a line of direct ancestors.

blood plasma the straw-color fluid remaining when the suspended corpuscles have been removed from blood. *See* plasma lipoproteins, plasma thromboplastin component, plasma transferrins, plasmin, serum.

blood typing determination of antigens on red blood cells, usually for the purpose of matching donor and recipient for blood transfusion. Conventionally, only antigens of the ABO and Rh systems are typed for this purpose.

Bloom syndrome (BS) children suffering from this rare heriditary disease have short stature, sun-sensitive facial erythema, and a high rate of bacterial infections due to defects in their immune systems. The condition was first described in 1954 by the American dermatologist D. Bloom. Somatic cells from BS patients are hypermutable and show high frequencies of microscopically visible chromatid gaps, breaks, and rearrangements. BS is inherited as an autosomal recessive, and homozygotes are prone to develop a wide variety of cancers. The syndrome is caused by mutations in a gene on the long arm of chromosome 15 at band 26.1. The gene product (BLM) is a helicase (*q.v.*). Men with BS produce no spermatozoa and are sterile. Antibodies against BLM have been shown to localize near synaptomemal complexes (*q.v.*) in mouse spermatocytes at the zygotene and early pachytene stages of meiosis (*q.v.*).

blotting the general name given to methods by which electrophoretically or chromatographically resolved RNAs, DNAs, or proteins can be transferred from the support medium (e.g., gels) to an immobilizing paper or membrane matrix. Blotting can be performed by two major methods: (1) capillary blotting involves transfer of molecules by capillary action (e.g., Southern blotting, northern blotting, *both of which see*), and (2) electroblotting, which involves transfer of molecules by electrophoresis.

blue-green algae, blue-green bacteria *See* Cyanobacteria.

blunt end ligation the use of a DNA ligase (*q.v.*) to join blunt-ended restriction fragments. *Compare with* cohesive end ligation.

blunt ends *See* restriction endonuclease.

B lymphocyte a cell belonging to the class of lymphocytes that synthesize immunoglobulins. B lymphocytes mature within a microenvironment of bone marrow (in mammals) or within the bursa of Fabricius (in birds). At this time, the immunoglobulins synthesized by B lymphocytes are transferred to the cell surface. After the binding of an antigen molecule to a B lymphocyte, it goes through a cycle of mitotic divisions during which the immunoglobulins disappear from the cell surface. The plasma cells that result synthesize immunoglobulins and secrete them into the blood. However, some B lymphocytes do not differentiate into plasma cells, but retain membrane-bound immunoglobulins. These "memory" B lymphocytes function to respond to any subsequent encounter with the same antigen *See* lymphocyte, V(D)J recombination.

bobbed a gene (*bb*) in *Drosophila melanogaster* producing a small bristle phenotype. The locus of *bb* is very near the centromere, and *bb* is the only gene known to have alleles on both the X and Y chromosomes. The plus allele of *bb* is the nucleolus organizer, and the various hypomorphic alleles may represent partial deletions of ribosomal DNA. *See* Appendix C, 1966, Ritossa, Atwood, and Spiegelman.

Bolwig organs a pair of eyes first described in 1946 by N. Bolwig in larvae of *Musca domestica*. In *Drosophila*, Bolwig organs consist of two groups each containing 12 photoreceptors juxtaposed to the larval mouth hooks. *See Dscam*.

Bombay blood group a rare human variant of the ABO blood group system (first discovered in Bombay, India) that does not have A, B, or O antigens. Individuals homozygous for an autosomal recessive allele (*h/h*) cannot make the precursor H substance (*q.v.*) from which the A and B antigens are formed. This is a classical case of recessive epistasis in human genetics, because without the product of allele *H*, the products of the ABO locus cannot be formed. Bombay bloods appear to be group O when routinely tested by antibodies against the A or B antigens, but an individual with the Bombay phenotype may be carrying unexpressed genes for the A and/or B antigens. However, they make anti-H that is not found in individuals of groups A, B, or O. Therefore it is possible for a child of group A or B to be produced from parents that appear to be group O, if one of them is a Bombay phenotype and carries the genes for antigens A or B or both. *See* A, B antigens.

Bombyx mori the commercial silkmoth, which was domesticated in China from its wild progenitor *B. mandarina* about 5,000 years ago. *Bombyx* is estimated to contain about 18,510 genes in its genome of 429 mbp of DNA. There are about 400 visible phenotypes, and ~200 of these have been assigned to linkage groups. The diploid chromosome number is 28, and the chromosomes are holocentric (*q.v.*). The female is a female-heterogametic species (ZZ in male, ZW in female). Sex is determined by a dominant feminizing gene on the W chromosome. *See* Appendix A, Arthropoda, Insecta, Lepidoptera; Appendix C, 1913, Tanaka; 1933, Hashimoto; Appendix F; endopolyploidy, silk.

bond energy the energy required to break a given chemical bond. For example, 58.6 kilogram calories per mol are required to break a carbon to carbon (C–C) bond.

bonobo the pigmy chimpanzee. *See Pan.*

border cells in *Drosophila* oogenesis, a group of anterior cells that detach from the follicular epithelium and migrate between the nurse cells in a posterior direction until they reach the anterior surface of the oocyte. Here they later participate in the formation of the micropylar apparatus which allows the sperm to enter the egg. Mutations in the sex-linked genes *domeless, hopscotch,* or *unpaired* disturb the migration of the border cells.

Borrelia burgdorferi the spirochaete, transmitted by ticks, that causes Lyme disease in humans. Genome sequencing has revealed that the bacterium contains a 910,725 bp megachromosome and 17 different plasmids with a combined size of 533,000 bp. The main chromosome contains about 850 genes, and there are at least 430 genes on the plasmids. Unlike the chromosomes of most bacteria which are circular, the main chromosome and some of the plasmids are linear. The DNA in the telomeres (*q.v.*) forms covalently closed hairpin structures. *See* Appendix A, Prokaryotes, Bacteria, Spirochaetae; Appendix C, 1997, Fraser *et al.*

Bos the genus that includes the domestic cow, *B. taurus*, the Brahman, *B. indicus*, and the yak, *B. grunniens*. The haploid chromosome number for the domestic cow is 30, and about 500 genes have been mapped. *See* cattle for a listing of domestic breeds, beef thymus; Appendix E.

bottleneck effect fluctuations in gene frequencies occurring when a large population passes through a contracted stage and then expands again with an altered gene pool (usually one with reduced variability) as a consequence of genetic drift (*q.v.*).

botulism poisoning by an exotoxin (*q.v.*) synthesized by *Clostridium botulinum* (*q.v.*). The poison is called the *botulin toxin* (*botox*), and when eaten it blocks nerve impulses and causes muscle paralysis.

bouquet configuration a polarized arrangement of chromosome ends at the periphery while the remaining chromatin fills the volume of the nucleus. This is the result of telomeres (*q.v.*) moving along the inner surface of the nuclear envelope during leptonema and eventually bunching together at the bouquet site. The tethering of telomeres to the nuclear periphery requires a specific meiotic telomere protein. In *Saccharomyces cerevisiae* this protein is encoded by a gene called *Ndj1* (*nondisjunction 1*). Deletion of *Ndj1* prevents bouquet formation and causes a delay in the pairing of homologues. *See* meiosis.

Boveri theory of cancer causation the proposal that a malignant tumor arises by the proliferation of a single cell which has acquired an excess or deficiency of chromosomes due to errors in the number of chromosomes it received during mitosis. *See* Appendix C, 1914, Boveri.

bovine referring to members of the cattle family, especially to those of the domestic cattle species *Bos taurus*.

bovine achondroplasia hereditary chondrodystrophy seen in "bull-dog" calves of the Dexter breed. The condition is inherited as an autosomal recessive. *See* achondroplasia.

bp abbreviation for "base pairs."

Bq becquerel (*q.v.*).

brachydactyly abnormal shortness of fingers or toes or both.

brachyury a short-tailed mutant phenotype in the mouse governed by a gene on chromosome 17. It was through this mutant that the T complex (*q.v.*) was discovered. *See* T box genes.

Bracon hebetor *See Microbracon hebetor* (also called *Habrobracon juglandis*).

bradyauxesis *See* heterauxesis.

bradytelic used to refer to a lower-than-average rate of evolution. *See* evolutionary rate.

Brahman a breed of humped domestic cattle (*Bos indicus*).

brain hormone prothoracicotropic hormone (*q.v.*).

Branchiostoma a genus of lancelets, commonly called *Amphioxus. Branchiostoma lanceolatum* is the sole living representative of the Cephalochordata. *See* Appendix A, Animalia, Chordata, *Hox* genes.

branch migration *See* Holliday model.

branch site *See* lariat.

BRCA1, BRCA2 *See* breast cancer susceptibility genes.

Brdu 5-bromodeoxyuridine (*q.v.*).

breakage and reunion the classical and generally accepted model of crossing over by physical breakage and crossways reunion of broken chromatids during meiosis. *See* Holliday model.

breakage-fusion-bridge cycle a cycle that begins with a dicentric chromosome forming a bridge as it is pulled toward both poles at once during anaphase.

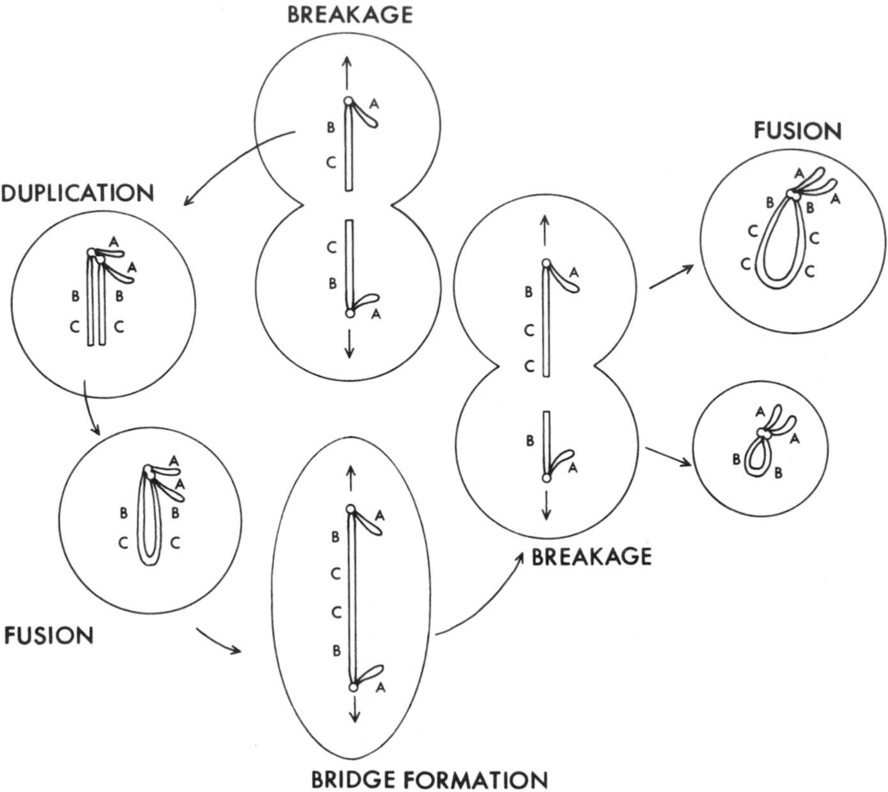

Breakage-fusion-bridge cycle

Such dicentric chromosomes may arise from an exchange within a paracentric inversion or may be radiation-induced. Once the dicentric breaks, the broken ends remain sticky, and these fuse subsequent to duplication. The result is another dicentric that breaks at anaphase, and so the cycle continues, with the chromosomes being broken anew at every mitosis (see illustration). Since each subsequent break is likely to be at a different place than the previous ones, there will be a repeated regrouping of the genetic loci to produce duplications and deficiencies. *See* Appendix C, 1938, McClintock; chromosome bridge, telomere.

breakage-reunion enzymes enzymes that use continuous stretches of DNA molecules, rather than preexisting termini, as substrates. The DNA duplex is broken and rejoined. The energy released by breakage is stored in a covalent enzyme-DNA intermediate and utilized in rejoining the molecules.

breakthrough an individual that escapes the deleterious action of its genotype. In a population of individuals homozygous for a given recessive lethal gene, almost all will die at a defined developmental stage. Those that develop past this stage are called "breakthroughs," or "escapers."

breast cancer susceptibility genes genes that when mutated greatly increase the susceptibility of heterozygous women to breast cancer. The first gene, *BRCA1*, resides at 17q21. It was cloned in 1994 and shown to encode a protein containing 1,863 amino acids. *BRCA2* resides at 13q12–13. It was cloned in 1995 and is now known to encode a protein containing 3,418 amino acids. Together, *BRCA1* and *BRCA2* are responsible for most cases of hereditary breast cancer. *BRCA1* also increases the risk of ovarian cancer, but *BRCA2* does not. The *BRCA1* protein contains two zinc finger domains and is therefore believed to function as a transcription factor. *See* Appendix C, 1994, Miki *et al.*; 1995, Wooster *et al.*; anti-oncogene, Mendelian Inheritance in Man (MIM), zinc finger protein.

breathing in molecular genetics, the periodic, localized openings of a DNA duplex molecule to produce single-stranded "bubbles."

breed an artificial mating group derived from a common ancestor for genetic study and domestication.

breeding the controlled propagation of plants and animals.

breeding size the number of individuals in a population that are actually involved in reproduction during that generation.

breeding true to produce offspring of phenotype identical to the parents; said of homozygotes.

bridge migration synonymous with branch migration. *See* Holliday model.

bridging cross a mating made to transfer one or more genes between two reproductively isolated species by first transferring them to an intermediate species that is sexually compatible with the other two species.

bristle organ each insect bristle is an organ consisting of four cells: the cell that secretes the bristle, the socket cell that secretes the ring that encloses the bristle, a sensory nerve cell whose process ends near the base of the bristle, and the sheath cell that surrounds the nerve axon. *See* trichogen cell.

broad bean *Vicia faba* (*q.v.*). This is the European plant to which the term *bean* was originally applied.

broad heritability the proportion of the total phenotypic variance (for a polygenic trait in a given population) that is attributed to the total genetic variance (including additive, dominance, epistatic, and other types of gene action); symbolized H^2. *See* heritability.

5-bromodeoxyuridine a thymidine analog that can be incorporated into DNA during its replication. This substitution profoundly affects that structure of the DNA. When both strands are substituted with BUDR, a chromatid stains less intensely than when only one strand is so substituted. Thus when cells are grown in the presence of BUDR for two replication cycles, the two sister chromatids stain differentially and therefore are called harlequin chromosomes. Consequently, the BUDR labeling method can be used to detect sister chromatid exchanges. BUDR causes breakage in chromosomal regions rich in heterochromatin. Additional acronyms are *Budr* and *Brdu*. *See* Appendix C, 1972, Zakharov, and Egolina.

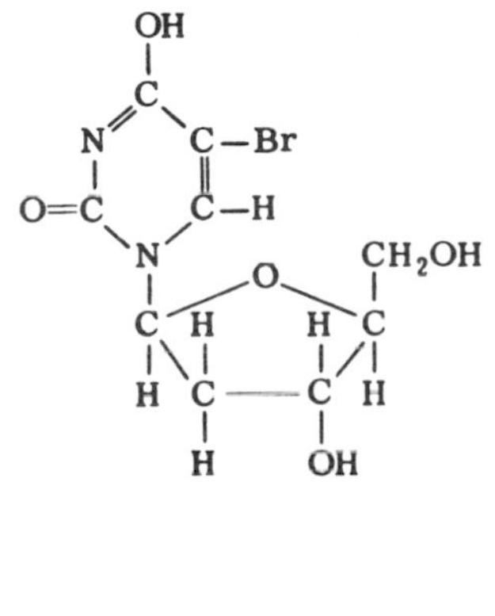

OH, C, N, C—Br, O=C, C—H, N, CH₂OH, O, C, H, H, C, H, C, C, H, H, OH

5-bromouracil a mutagenically active pyrimidine analog.

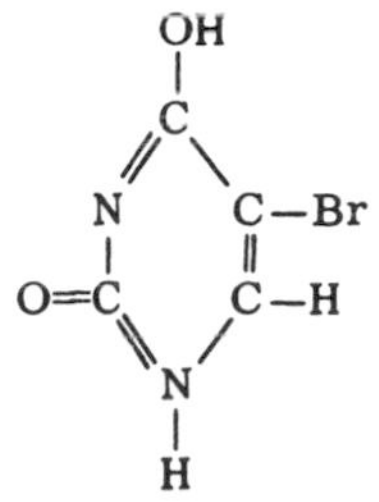

OH, C, N, C—Br, O=C, C—H, N, H

brood the offspring from a single birth or from a single clutch of eggs.

broodiness the tendency of female birds to incubate eggs.

Bruton tyrosine kinase *See* agammaglobulinemia.

Bryophyta the plant phylum containing mosses, liverworts, and hornworts. Bryophytes lack a vascular system. *See* Appendix A.

Bt designer plants transgenic plants (*q.v.*) with the gene for the toxin produced by *Bacillus thuringiensis* spliced into their genomes. Bt corn is an example of such a genetically engineered crop, and the Bt toxin kills off its major enemy, the European corn borer. One third of the corn sold in the U.S. is Bt corn. Farmers plant even more Bt soybeans and Bt cotton. *See Bacillus*, GMO.

bubble a structure formed in a duplex DNA at the site of initiation and strand separation during replication.

bud 1. A sibling cell produced during the division of a budding yeast, like *Saccharomyces cerevisiae*. The daughter cell develops as a protrusion of the cell wall of the mother cell. The mother retains its old cell wall components while the bud gets newly synthesized wall material (contrast with *septal fission*). The nucleus migrates to the neck of the bud. Here mitosis occurs with the nuclear envelope remaining intact, and a set of telophase chromosomes is delivered to each cell. 2. an underdeveloped plant shoot, consisting of a short stem bearing crowded, overlapping, immature leaves.

budding 1. in bacteria, yeast, and plants, the process by which a bud (*q.v.*) is produced. 2. in enveloped viruses such as influenza virus and Sindbis virus, a mode of release from the host cell in which a portion of the cell membrane forms an envelope around the nucleocapsid. The envelope contains viral proteins, but no cellular proteins.

BUDR 5-bromodeoxyuridine (*q.v.*).

buffer a compound that, in solution, tends to prevent or resist rapid changes in pH upon the addition of small quantities of acid or base.

buffering the resistance of a system to change by outside forces.

Bufo a genus of toads. Wild populations of species of this genus have been extensively studied by population geneticists.

bulb a modified shoot consisting of a very much shortened underground stem enclosed by fleshy scalelike leaves. It serves as an organ of vegetative reproduction. The onion, daffodil, tulip, and hyacinth produce bulbs.

bull the adult male of various animals including domesticated cattle, elephants, moose, and elk.

bull-dog calf *See* bovine achondroplasia.

buoyant density the equilibrium density at which a molecule under study comes to rest within a density gradient. *See* centrifugation separation.

Burkitt lymphoma a monoclonal malignant proliferation of B lymphocytes primarily affecting the jaw and associated facial bones. The cancer is named after Denis Burkitt, who first described it in central African children in 1958. Most Burkitt tumors occurring in Africans contain Epstein-Barr virus (EBV) (*q.v.*), and this virus is believed to be mosquito borne. Burkitt lymphomas from United States and European patients lack EBV. Burkitt lymphoma cells always contain a reciprocal translocation involving the long arm of chromosome 8 and chromosome 14, or less frequently 22 or 2. The break point on chromosome 8 is always near the *myc* oncogene (*q.v.*). The break point on the other chromosome is always near an immunoglobulin gene, namely, 14 (heavy chains), 22 (lambda light chains), or 2 (kappa light chains). In its translocated state *myc* is activated and the cancer ensues. *See* immunoglobulin chains, Philadelphia (Ph^1) chromosome.

bursa of Fabricius a saclike structure connected to the posterior alimentary canal in birds. The bursa is the major site where B lymphocytes become mature immunoglobulin (antibody)-secreting plasma cells. The equivalent organ in mammals has not been definitely identified. Most evidence suggests the bone marrow. The organ bears the Latinized name of the Italian anatomist Girolamo Fabrizio (1578–1657) who first described it.

bursicon an insect hormone that appears in the blood after molting and is required for the tanning and hardening of new cuticle.

burst size the average number of bacteriophages released from a lysed host. *See* Appendix C, 1939, Ellis and Delbrück.

busulfan a mutagenic, alkylating agent.

$$CH_3-\overset{\overset{\displaystyle O}{\|}}{\underset{\underset{\displaystyle O}{\|}}{S}}-O-(CH_2-CH_2)_2-O-\overset{\overset{\displaystyle O}{\|}}{\underset{\underset{\displaystyle O}{\|}}{S}}-CH_3$$

C

C **1.** Celsius (also Centigrade). **2.** carbon. **3.** the haploid amount of DNA. *See* C value, C paradox. **4.** cytosine or cytidine.

^{14}C a radioactive isotope of normal carbon (^{12}C) emitting a weak beta particle. The half-life of ^{14}C is 5,700 years. This radioisotope is extensively used as a tracer in molecular biology.

CAAT box part of a conserved DNA sequence of about 75 base pairs upstream from the initiator for eukaryotic transcription; possibly involved in binding RNA polymerase II. *See* Hogness box.

cadastral genes genes that restrict the action of other genes to specific regions of the organism. An example of such a boundary-setting gene is SUPERMAN in *Arabidopsis thaliana* (*q.v.*). Flowers that contain inactive alleles of SUPERMAN have stamens in whorl 4. Since stamens require activities B and C (*see* floral identity mutations), these abnormal flowers suggest that genes capable of producing B are normally inhibited in whorl 4 by cadastral genes like SUPERMAN. *See* floral identity mutations.

cadherins glycoproteins composed of 700–750 amino acids that function as cell–cell adhesion molecules. The *N*-terminal end of the molecule projects from the membrane surface and contains Ca^{2+} binding sites. The C-terminal tail binds to the actin of the cytoskeleton. In between is a segment that functions as an integral part of the cell membrane. E-cadherins are the best characterized of the cadherins. They are present in many types of epithelial cells and are usually concentrated in the adhesion belts that hold the cells together. *See* cell–cell adhesion molecules (CAMs).

Caenobacter taenospiralis *See* killer paramecia.

***Caenorhabditis* databases** *See* Appendix E.

Caenorhabditis elegans a small nematode whose developmental genetics has been extensively investigated. The worm is about 1 mm in length, and its life cycle, when reared at 20°C, is 3.5 days. Its transparent cuticle allows the visualization of every cell. The adult has 816 somatic cells, of which 302 are neurons. The complete lineage history and fate of every cell is known. C. *elegans* normally reproduces as a self-fertilizing hermaphrodite, which has two X chromosomes per cell, plus five pairs of autosomes. Loss of an X by meiotic nondisjunction leads to the production of males. These arise spontaneously among the progeny of hermaphrodites at a frequency of about 0.2%. The mating of hermaphrodites with males made genetic analysis possible. The C. *elegans* genome contains 100 million base pairs and about 19,100 protein-coding genes. Exons and introns each make up about 24% of the genome, and each gene has an average of 5 introns. About a quarter of the genes are organized into operons (*q.v.*). The ribosomal and the 5S RNA genes occur in tandom arrays in autosomes I and V, respectively. The *sel-12* gene of *Caenorhabditis* is homologous to a gene in humans that confers susceptibility to Alzheimer disease (*q.v.*). *See* Appendix A, Animalia, Pseudocoelomata, Nematoda; Appendix C, 1974, Brenner; 1977, Sulston, and Horvitz; 1981, Chalfe and Sulston; 1983, Greenwald *et al.*; 1998, C. *elegans* Sequencing Consortium; 2000, Rubin *et al.*, Fraser *et al.*; Appendix E, Individual Databases; apoptosis, cell lineage mutants, *daf-2*, helitron, *Hox* genes, *Pangrellus redivius, Turbatrix aceti,* zinc finger proteins. RNA interference (RNAi), trans-splicing,

caffeine a stimulant found in coffee and tea. The usual portion of these beverages contains about 100 mg of caffeine, making it the most common drug taken regularly by human beings. Caffeine is a purine analog that is mutagenically active in microbial systems. *See* alkaloid, bases of nucleic acids, theobromine.

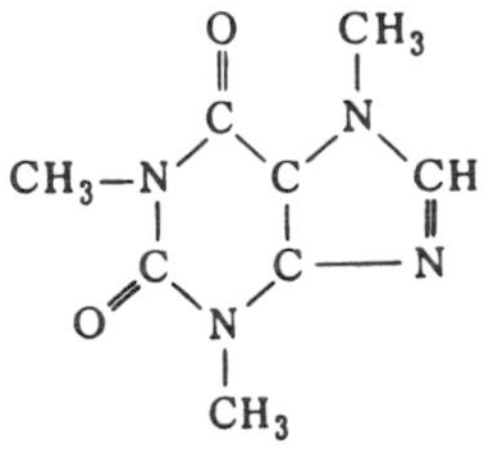

```
          O     CH3
          ‖     |
          C     N
        /   \ /   \
CH3—N       C     CH
    |       |     ‖
    C       C ——— N
  //  \   /
 O      N
        |
       CH3
```

Cairns molecule *See* theta replication.

Cajal body a nuclear organelle first identified in 1903 by the Spanish neurobiologist Santiago Ramon y Cajal in mammalian neurons and called by him the *accessory body*. In 1969 A. Monneron and W. Bernhard rediscovered these organelles within the interphase nuclei of mammalian liver cells and named them *coiled bodies* on the basis of their appearance in electron micrographs. Cajal bodies are now generally identified by immunofluorescence with specific antibodies against the protein *coilin* (*q.v.*), which is concentrated in them. The giant nucleus of amphibian oocytes (the germinal vesicle) contains 50 to 100 large Cajal bodies. All three eukaryotic RNA polymerases are found in oocyte Cajal bodies, along with many factors involved in transcription and processing of all types of RNA (pre-mRNA, pre-rRNA, tRNA, etc.). Based on studies of oocytes, Gall et al. have suggested that Cajal bodies are sites for preassembly of the transcription machinery of the nucleus, much as nucleoli are sites for preassembly of the translation machinery (ribosomes). *See* Appendix C, 1999, Gall *et al.*; nucleolus, posttranscriptional processing, snurposomes, transcriptosomes.

calciferol vitamin D (*q.v.*).

calcium an element universally found in small amounts in tissues. Atomic number 20; atomic weight 40.08; valence 2 +; most abundant isotope ^{40}Ca; radioisotope ^{45}Ca, half-life 164d, radiation emitted–beta particles. Extracellular calcium plays a role in blood clotting and maintaining the integrity of biological membranes. For example, calcium chloride treatment of bacteria makes them permeable to plasmids. Internally, calcium activates a variety of enzymes, especially protein kinases (*q.v.*). *See* Appendix C, 1972, Cohen *et al.*

calico cat *See* tortoiseshell cat.

Calliphora erythrocephala a large fly in which polytene chromosomes occur in the ovarian nurse cells in certain inbred lines. The banding pattern of these giant chromosomes has been compared with those of pupal trichogen cells (*q.v.*).

callus the cluster of plant cells that results from tissue culturing a single plant cell.

calmodulin an intracellular calcium receptor protein that regulates a wide spectrum of enzymes and cellular functions, including the metabolism of cyclic nucleotides and glycogen. It also plays a role in fertilization and in the regulation of cell movement and cytoskeletal control, as well as in the synthesis and release of neurotransmitters and hormones. Calmodulin is a heat- and acid-stable, acidic protein with four calcium-binding sites. It is found in all eukaryotic cells and has a molecular weight of 16,700. It appears to be the commonest translator of the intracellular calcium message. *See* second messenger.

calnexin an integral membrane protein of the endoplasmic reticulum (*q.v.*). Calnexin is a chaperone (*q.v.*) that binds misfolded glycoproteins and targets them for subsequent degradation by a proteasome (*q.v.*). The product of the delta F508 allele of cystic fibrosis gene is an example of a glycoprotein that interacts with this chaperone. *See* cystic fibrosis (CF).

calyx the sterile, outer whorl of floral parts composed of sepals.

cambium the lateral meristem of vascular plants.

Cambrian the earliest period in the Paleozoic era. Representatives of most animal phyla are present in Cambrian rocks. Algae, sponges, and trilobites (*q.v.*) were abundant. The Cambrian ended with a mass extinction. Seventy-five percent of all trilobite families and 50% of all sponge families died off. *See* geologic time divisions.

Camelus the genus of camels including C. *bactrianus*, the two-humped camel; and C. *dromedarius*, the one-humped camel, also called the dromedary.

cAMP *See* cyclic AMP.

Campbell model of lambda integration a hypothesis that explains the mechanism of integration of phage lamda into the *E. coli* host chromosome. According to the model, linear lambda DNA is first circularized. Then prophage integration occurs as a physical breakage and reunion of phage and host DNA molecules precisely between the bacterial DNA site for phage attachment and a corresponding site in the phage DNA. *See* Appendix C, 1962, Campbell.

canalization the existence of developmental pathways that lead to a standard phenotype in spite of genetic or environmental disturbances.

canalized character a trait whose variability is restricted within narrow boundaries even when the organisms are subjected to disturbing environments or mutations.

canalizing selection elimination of genotypes that render developing individuals sensitive to environmental fluctuations.

cancer a class of diseases of animals characterized by uncontrolled cellular growth. *See* Appendix E; anti-oncogenes, Burkitt lymphoma, carcinoma, immunological surveillance theory, leukemia, lymphoma, malignancy, melanoma, metastasis, myeloma, neoplasm, oncogene, oncogenic virus, p53, papilloma, sarcoma, teratocarcinoma, teratoma.

Canis familiaris the dog, the first animal domesticated by man and his companion for at least 15,000 years. The dog is a close relative of the gray wolf, *Canis lupus*. They both have a chromosome number of 39, and species hybrids are fertile. At least 400 different genetic diseases have been identified in dogs, and most of these are homologous to the human conditions (Duchenne and Becker types of muscular dystrophy, Niemann-Pick disease, von Willebrand disease, hemophilia A and B, and testicular feminization are examples). The dog genome contains 2.4 Gbp of DNA. *See* Appendix A, Mammalia, Eutheria, Carnivora; Appendix E; dog breeds, wolf.

canonical sequence an archetypical sequence (also known as a consensus sequence) to which all variants are compared. A sequence that describes the nucleotides most often present in a DNA segment of interest. For example, in the Pribnow box and the Hogness box, the canonical sequences are TATAAT and TATAAAA, respectively. The 14 nucleotide consensus sequence CCGTNTGYAARTGT has 11 nucleotides that are constant throughout the populations sampled. However, at position 5 any nucleotide (N) can be present, at the position 8 either pyrimidine (Y) can occur, and at position 11 either purine (R) can occur. *See* promoter.

cap *See* methylated cap.

CAP catabolite activator protein (*q.v.*).

capacitation a process of physiological alterations whereby a sperm becomes capable of penetrating an egg as a consequence of exposure to one or more factors normally present in the female reproductive tract. It is theorized that a substance coating the sperm head must be removed by these female factors before the sperm can become fully functional for fertilization.

capon a castrated domestic fowl.

capped 5′ ends the 5′ ends of eukaryotic mRNAs containing methylated caps (*q.v.*).

capping **1.** addition of a cap (*q.v.*) to mRNA molecules. **2.** redistribution of cell surface structures to one region of the cell, usually mediated by cross-linkage of antigen-antibody complexes.

Capsicum a genus that includes red peppers and pimentos, C. *annum*, and green pepper, C. *frutescens*.

capsid the protein coat of a virus particle.

capsomere one of the subunits from which a virus shell is constructed. Capsomeres may contain several different polypeptide chains. The virus shell is formed by assembling capsomeres about the nucleic acid core in a precise geometrical pattern. *See* icosahedron, Q beta (Qβ) phage, tobacco mosaic virus.

Carassius auratus the aquarium goldfish. A member of the carp family first described in China 2,300 years ago and bred for ornament since that time. *See* Appendix A, Chordata, Osteichthyes, Neopterygii, Cypriniformes.

carbohydrate a compound, having the general formula $C_xH_{2x}O_x$. Common examples of carbohydrates are glucose, cellulose, glycogen, and starches (*q.v.*).

carbon the third most abundant of the biologically important elements. Atomic number 6; atomic weight 12.01115; valence 4; most abundant isotope ^{12}C; radioisotope ^{14}C (*q.v.*).

3′ carbon atom end nucleic acids are conventionally written with the 3′ carbon of the pentose to the right. Transcription or translation from a nucleic acid proceeds from 5′ to 3′ carbon.

5′ carbon atom end nucleic acids are conventionally written with the end of the pentose containing the 5′ carbon to the left. *See* deoxyribonucleic acid.

carbon dioxide sensitivity *See* sigma virus.

Carboniferous the Paleozoic period that generated the great coal deposits. At this time the land was covered by extensive forests. Seed-bearing ferns and conifers appeared for the first time. Amphibians diversified, and the winged insects and reptiles arose. Cartilagenous fishes were the dominant marine vertebrates. In North America, where the stratigraphic record allows Carboniferous strata to be conveniently subdivided into upper and lower segments, the Carboniferous is replaced by the Pennsylvanian and Mississippian periods. *See* geologic time divisions.

carbonyl group a doubly bonded carbon-oxygen group (C=O). The secondary structure of a polypeptide chain involves hydrogen bonds between the carbonyl group of one residue (amino acid) and the imino (NH) group of the fourth residue down the chain. *See* alpha helix.

carboxyl group a chemical group (COOH) that is acidic because it can become negatively charged ($-\underset{\|}{\overset{}{C}}-O^-$ with =O below C) if a proton dissociates from its hydroxyl group.

carboxyl terminal C-terminus (*q.v.*).

carboxypeptidases two pancreatic enzymes (A and B) that hydrolyze protein chains beginning at

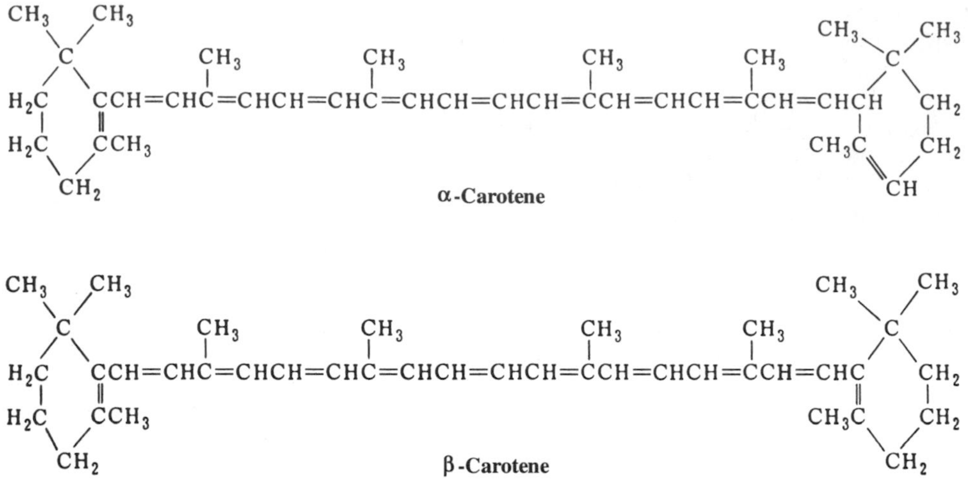

α-Carotene

β-Carotene

Carotenoids

the carboxyl terminal end of the chain and liberating amino acids one at a time. These enzymes are useful for amino acid sequence studies.

carboxysomes *See* cyanobacteria.

carcinogen a physical or chemical agent that induces cancer. A carginogen is usually mutagenic, and it either damages nucleic acids directly or indirectly, or it causes a genetic imbalance by inducing a chromosomal aberration (*q.v.*). *See* alkylating agent, antioncogenes, Boveri theory of cancer causation, ionizing radiation, oncogene, oncogenic virus, proto-oncogene, ultraviolet radiation.

carcinoma a cancer of epithelial tissues (e.g., skin cancer); *adenocarcinoma* is a cancer of gland epithelia.

carcinostasis inhibition of cancerous growth.

carnivore a meat-eating animal. Also applied to a few insectivorous plants. In classification, a member of the mammalian order Carnivora which contains cats, mongooses, dogs, bears, raccoons, pandas, otters, etc.

carotenoids lipid-soluble pigments ranging in color from yellow to red. The carotenes whose structures appear in the illustration are plant carotenoids. Beta carotene can be enzymatically hydrolyzed into two molecules of vitamin A (*q.v.*) and is therefore an important provitamin. *See* anthocyanins.

carpel the meristematic whorl of cells that produces the female reproductive organs in angiosperms. At maturity the *carpel* refers to the part of the flower that encloses the ovules and extends upward to form the pistil.

carrier 1. an individual heterozygous for a single recessive gene. 2. a stable isotope of an element mixed with a radioisotope of that element to give a total quantity sufficient to allow chemical operations. 3. an immunogenic molecule (e.g., a foreign protein) to which a hapten (*q.v.*) is coupled, thus rendering the hapten capable of inducing an immune response.

carrier-free radioisotope a radioisotope essentially undiluted with a stable contaminating isotope.

carrying capacity the size or density of a population that can be supported in stable equilibrium with the other biota of a community; symbolized K.

cartilage a skeletal connective tissue formed by groups of cells that secrete into the intercellular space a ground substance containing a protein, collagen (*q.v.*), and a polysaccharide, chondroitin sulfuric acid.

cartilage-hair hypoplasia (CHH) a disease inherited as an autosomal recessive. Homozygous children have short limbs because of arrested cartilage growth, and their hair is sparse and light colored. The disease is the first one shown to be caused by mutations in an untranslated gene. The gene, *RMRP* (*q.v.*), transcribes an RNA that is used directly as a subunit of a mitochondrial enzyme. CHH was first observed among the Amish (*q.v.*), where its frequency is about 1.5 per 1,000 live births. *See* Appendix C, 2001, Ridanpaa *et al.*

Carya a genus that includes C. *ovata*, the shagbark hickory, and C. *pecan*, the pecan.

caryonide a lineage of paramecia that derive their macronuclei from a single macronuclear primordium. Such paramecia are generally immediate descendants of the exconjugants.

caryopsis a dry indehiscent multiple-seeded fruit derived from a compound ovary. The corn ear is an example.

caspases a specific group of proteases that function during apoptosis (*q.v.*). Caspase is an abbreviation for *c*ysteine-dependent, *asp*artate-specific prote*ase*. Such proteins are initially secreted as inactive precursors. Upon receiving a chemical signal, such procaspases break down into subunits, which are then reassembled into heterotetrameric caspases. There are two classes of caspases. The first, called upstream initiators, serve to transduce signals from the cell surface. These initiators then interact with caspases of the second class, the downstream effectors which begin to destoy key cellular substrates. The death process enters its final phase when caspases activate the breakdown of DNA. *See* cellular signal transduction, separase, tumor necrosis factor.

cassette mutagenesis a technique that involves removing from a gene a stretch of DNA flanked on either end by a restriction site (*q.v.*) and then inserting in its place a new DNA segment. This cassette can contain base substitutions or deletions at specific sites, and the phenotypic effects that result give insight into relative importance of specific subsegments of the region to the functioning of the gene or its product.

cassettes loci containing functionally related nucleotide sequences that lie in tandem and can be substituted for one another. The mating-type reversals observed in yeast result from removing one cassette and replacing it by another containing a different nucleotide sequence. Mating-type loci in yeast contain homeoboxes (*q.v.*).

caste a class of structurally and functionally specialized individuals within a colony of social insects.

cat any of a number of domesticated breeds of the species *Felis catus*. Popular breeds include SHORT-HAIRED BREEDS: Domestic Shorthair, Siamese, Burmese, Abyssinian, Russian Blue, Havana Brown, Manx, and Rex; LONG-HAIRED BREEDS: Persian, Angora, and Himalayan.

catabolism metabolic breakdown of complex molecules to simpler products, often requiring catabolic enzymes and accompanied by the release of energy.

catabolite a compound generated by the breakdown of food molecules.

catabolite activating protein (CAP) a constitutively produced, dimeric, positive regulator protein in bacteria that, when bound to a promoter region and cAMP, facilitates transcription by RNA polymerase of certain catabolite-sensitive adjacent genes in inducible and glucose-sensitive operons (such as the *lac* operon of *E. coli*). Also known as cyclic AMP receptor protein (CRP) or catabolite gene activator (CGA) protein.

catabolite repression the reduction or cessation of synthesis of enzymes involved in catabolism of sugars such as lactose, arabinose, etc., when bacteria are grown in the presence of glucose. The enzyme adenyl cyclase is inhibited by glucose from converting ATP to cyclic adenosine monophosphate (cAMP); cAMP must complex with catabolite activator protein (CAP) in order for RNA polymerase to bind to promoters of genes responsible for enzymes capable of catabolizing sugars other than glucose. Therefore, in the presence of glucose, less CAP protein is available to facilitate the transcription of mRNAs for these enzymes.

catalase an enzyme that catalyzes the reaction of $H_2O_2 \rightarrow H_2O + \frac{1}{2}O_2$. Catalase is especially abundant in the liver, where it is contained in peroxisomes (*q.v.*). *See* acatalasemia, antioxidant enzymes, superoxide dismutase.

catalyst a substance that increases the rate of a chemical reaction without being consumed. Enzymes are biological catalysts.

catarrhine referring to primates of the infraorder Catarrhini that includes the Old World (African and Asian) monkeys, great apes, and humans. These primates are characterized by nostrils that are close-set and directed forward or downward, and they do not have prehensile tails. *Compare with* platyrrhine. *See Cercopithecus ethiops, Hylobates, Macacca mulatta, Pan.*

catastrophism a geological theory proposing that the earth has been shaped by violent events of great magnitude (e.g., worldwide floods, collisions with asteroids, etc.); the opposite of uniformitarianism (*q.v.*).

cat cry syndrome a syndrome of multiple congenital malformations in humans with a deficiency in the short arm of chromosome 5. Infants with this condition produce a peculiar cry that sounds like a cat mewing. Also known as the *cri du chat* syndrome.

category a rank in a taxonomic hierarchy to which one or more taxa may be assigned: e.g., phylum, class, order, family, genus, species.

catenane a structure made up of two or more interlocking rings.

catenate to convert two or more rings into a system of interlocking rings.

catenins a family of intracellular proteins that are a component of the junctional complexes which mediate adhesion between cells and signal contact inhibition (*q.v.*). In humans, alpha and beta catenins are encoded by genes at 5q31 and 3p22, respectively. The sequences of the two genes show no similarity. The alpha and beta proteins form 1 : 1 heterodimers, and they attach the inward-reaching carboxyl ends of molecules of cadherins (*q.v.*) to actin (*q.v.*) filaments within the cell. In *Xenopus*, beta catenins provide the first signal of dorsal ventral polarity in the embryo.

cathepsin any of certain proteolytic enzymes thought to reside in lysosomes (*q.v.*). Such enzymes are abundant, for example, in metamorphosing tadpoles during the resorption of the tail.

cathode the negative electrode to which positive ions are attracted. *Contrast with* anode.

cation a positively charged ion so named because it is attracted to the negatively charged cathode. *Contrast with* anion.

Cattanach translocation a translocation in the mouse discovered by B. M. Cattanach. The aberration involves an X chromosome into which a segment of autosome 7 has been translocated. The insertion carries the wild-type alleles of three autosomal genes that control the color of the fur. Studies on mice heterozygous for the Cattanach translocation have shown that during X-chromosome inactivation in somatic cells, the genes in the inserted autosomal segment are turned off sequentially in order of their distances from the X chromosomal element. Thus, the X inactivation spreads into the attached autosomal segment, but does not travel unabated to the end of the segment.

cattle any of a number of domesticated breeds of the species *Bos taurus*. Popular breeds include BEEF CATTLE: Hereford, Shorthorn, Aberdeen-Angus, and Santa Gertrudis; DAIRY CATTLE: Holstein-Friesian, Jersey, Guernsey, Ayrshire, and Brown Swiss. *See* ruminant mammals.

caudal (cad) a gene in *Drosophila* (located at 2-54) which produces a transcript that is localized at the posterior pole of the embryo. The *cad* gene is essential for the development of the hindgut. It encodes a protein 472 amino acids long that contains a homeodomain (*q.v.*). This protein (CAD) activates the transcription of various target genes, including *fushi tarazu* (*q.v.*).

caveolae flask-shaped invaginations 50–100 nm in diameter that are observed in the plasma membranes of mammalian cells such as adipocytes, endothelial cells, and muscle cells. Caveolae are involved in cholesterol transport and cellular signal transduction (*q.v.*) through the binding of immune and growth factor receptors. Endocytosis involving caveolae does not feed into the lysosome pathway, and therefore macromolecules that are internalized in caveolar vesicles avoid being degraded. Bacteria that can express FimH (*q.v.*) use caveolae to invade phagocytes, and since the phagosomes do not fuse with lysosomes, the bacteria remain viable.

caveolins principal protein components of caveolae (*q.v.*). Caveolin 3 is a muscle-specific form of caveolin encoded by a human gene at 3p25. Null mutations of this gene cause an autosomal dominant form of muscular dystrophy.

Cavia porcellus the guinea pig or cavy, a rodent living wild in the Andean region of South America, but domesticated and used as a laboratory animal. Numerous mutants are known, affecting hair color and texture. Immune response genes were discovered in this species. *See* Appendix A, Chordata, Mammalia, Rodentia; Appendix C, 1963, Levine, Ojia, and Benacerraf.

C banding a method for producing stained regions around centromeres. *See* chromosome banding techniques.

cc cubic centimeter. *See* milliliter.

C^{13}/C^{12} ratio the ratio between the heavy, stable isotope of carbon and the normal isotope in a sample of interest. Since organisms take up C^{12} in preference to C^{13}, the ratio is used to determine whether or not the carbon in the specimen is of biological origin.

cccDNA covalently closed, circular DNA.

$CD4^+$ cells, $CD8^+$ cells *See* T lymphocyte.

CD4, CD8 receptors proteins on the surface of T lymphocytes (*q.v.*) that determine their responses to antigens. Lymphocytes with the CD8 proteins on their surfaces function as killer T lymphocytes (T_k cells). Lymphocytes with the CD4 proteins on their surfaces function as helper T lymphocytes (T_h cells). These secrete interleukins (*q.v.*), which activate T_k cells and B lymphocytes. *See* immunoglobulin domain superfamily.

CD99 the protein encoded by the human gene *MIC2* (*q.v.*).

Cdc 14, *cdc* genes *See cell division cycle* genes.

cdc kinases cell division cycle kinases. *See* cyclins.

cdks cyclin-dependent kinases. *See* cyclins.

cDNA (copy DNA) single-stranded, complementary DNA produced from an RNA template by the action of RNA-dependent, DNA polymerase (reverse transcriptase) *in vitro*. If the RNA template has been processed to remove the introns, the cDNA will be much shorter than the gene from which the RNA was transcribed. The single-stranded, cDNA molecule may subsequently serve as a template for a DNA polymerase. The symbol cDNA is sometimes also applied to the double-stranded DNA molecule that results. *See* posttranscriptional processing.

cDNA clone a duplex DNA sequence complementary to an RNA molecule of interest, carried in a cloning vector.

cDNA library a collection of cDNA (*q.v.*) molecules, representative of all the various mRNA molecules produced by a specific type of cell of a given species, spliced into a corresponding collection of cloning vectors such as plasmids or lambda phages. Since not all genes are active in every cell, a cDNA library is usually much smaller than a gene library (*q.v.*). If it is known which type of cell makes the desired protein (e.g., only pancreatic cells make insulin), screening the cDNA library from such cells for the gene of interest is a much easier task than screening a gene library.

CD3 proteins *See* T lymphocyte.

Ceboidea the superfamily containing the monkeys of Central and South America.

cecidogen a gall-forming substance.

Celera Genomics a company founded by J. Craig Venter, who served as its president from 1998 to 2002. Its initial task was to complete the sequence and assembly of the human genome. A factory was set up in Rockville, MD, where 300 automated DNA sequencing machines were kept in continuous operation along with advanced computer systems for assembly of the sequenced fragments. As a test of its capabilities, Celera collaborated with the Berkeley *Drosophila* Genome Project to sequence and assemble the *Drosophila* genome. This task was completed during 1999 in only four months. The first rough draft of the human genome was completed in 2000, and the event was announced in a ceremony held at the White House. At first, Celera made money by charging subscription fees for the genomic data it uncovered. However, in May of 2005, Celera closed its subscription service and released all its genomic data to the public. *See* Appendix C, 2000, Adams *et al.*; 2001, Collins and Venter *et al.*; Appendix E, Individual databases; DNA sequencers, Human Genome Project, *Mus musculus*, TIGR.

cell the smallest, membrane-bound protoplasmic body capable of independent reproduction. *See* Appendix C, 1665, Hooke.

cell affinity a property of eukaryotic cells of the same type to adhere to one another but not to those of a different type; this property is lost when the cell transforms to the cancerous state.

cell–cell adhesion molecules (CAMs) molecules that are responsible for the selective adhesion of cells to form specific tissues during the early embryogenesis of vertebrates. The cadherins (*q.v.*) are an example of CAMs that require Ca^{2+} for their functioning.

cell culture a term used to denote the growing cells *in vitro*, including the culturing of single cells. In cell cultures the cells are not organized into tissues. *See* Appendix C, 1940, Earle; 1956, Puck *et al.*

cell cycle the sequence of events between one mitotic division and another in a eukaryotic cell. Mitosis (M phase) is followed by a growth (G_1) phase, then by DNA synthesis (S phase), then by another growth (G_2) phase, and finally by another mitosis. In HeLa cells (*q.v.*), for example, the G_1, S, G_2, and M phases take 8.2, 6.2, 4.6, and 0.6 hours, respectively. The period between mitoses ($G_1 + S + G_2$) is called interphase. Cells may have different doubling times, depending on their developmental stage or tissue type. The variation in doubling times is usually a function of the time spent in G_1. When a cell differentiates, it leaves the cycle and enters a phase designated G_0. Such "resting" cells are mitotically quiescent, but metabolically active. *See* Appendix C, 1953, Howard and Pelc; centriole, checkpoint, cyclins, maturation promoting factor (MPF).

cell determination an event in embryogenesis that specifies the developmental pathway that a cell will follow.

cell differentiation the process whereby descendants of a single cell achieve and maintain specializations of structure and function. Differentiation presumably is the result of differential transcriptions.

cell division the process (binary fission in prokaryotes, mitosis in eukaryotes) by which two daughter cells are produced from one parent cell. *See* Appendix C, 1875, Strasburger.

***cell division cycle* genes** genes first isolated from yeast which encode proteins that control critical steps in the cell division cycle. An example of such a protein is Cdc 14, a phosphatase that is localized in the nucleolus (*q.v.*). When the contents of the nucleolus are dispersed in late anaphase, this enzyme digests mitotic cyclins. *See* Appendix C, 1973, Hartwell *et al.*; cyclins.

cell division cycle kinases *See* cyclins.

cell-driven viral transformation a method for creating immortalized human antibody-producing cells *in vitro* without forming a hybridoma (*q.v.*). Normal B lymphocytes from an immunized donor are mixed with other cells infected with the Epstein-Barr virus (*q.v.*). The virus enters the B lymphocytes. The cells originally infected with the virus are experimentally destroyed, and the virally transformed cells producing the antibody of interest are isolated. In cell-driven viral transformation, about 1 in 50 B lymphocytes is transformed, whereas with the cell hybridization technique only about 1 human cell in 10 million is transformed.

cell fate the developmental destiny of a cell in terms of the differentiated structure(s) that it will inevitably give rise to during normal development.

cell fractionation the separation of the various components of cells after homogenization of a tissue and differential centrifugation. Four fractions are generally obtained: (1) the nuclear fraction, (2) the mitochondrial fraction, (3) the microsomal fraction, and (4) the soluble fraction or cytosol. *See* Appendix C, 1946, Claude.

cell-free extract a fluid obtained by rupturing cells and removing the particulate material, membranes, and remaining intact cells. The extract contains most of the soluble molecules of the cell. The preparation of cell-free extracts in which proteins and nucleic acids are synthesized represent milestones in biochemical research. *See* Appendix C, 1955, Hoagland; 1961, Nirenberg and Matthaei; 1973, Roberts and Preston.

cell fusion the experimental formation of a single hybrid cell with nuclei and cytoplasm from different somatic cells. The cells that are fused may come from tissue cultures derived from different species. Such fusions are facilitated by the adsorption of certain viruses by the cells. *See* polyethylene glycol, Sendai virus, Zimmermann cell fusion.

cell hybridization the production of viable hybrid somatic cells following experimentally induced cell fusion (*q.v.*). In the case of interspecific hybrids, there is a selective elimination of chromosomes belonging to one species during subsequent mitoses. Eventually, cell lines can be produced containing a complete set of chromosomes from one species and a single chromosome from the other. By studying the new gene products synthesized by the hybrid cell line, genes residing in the single chromosome can be identified. *See* Appendix C, 1960, Barski *et al.*, HAT medium, hybridoma, syntenic genes.

cell interaction genes a term sometimes used to refer to some genes in the I region of the mouse H2 complex that influence the ability of various cellular components of the immune system to cooperate effectively in an immune response.

cell line a heterogeneous group of cells derived from a primary culture (*q.v.*) at the time of the first transfer. *See* isologous cell line.

cell lineage a pedigree of the cells produced from an ancestral cell by binary fission in prokaryotes or mitotic division in eukaryotes. *Caenorhabditis elegans* (*q.v.*) is the only multicellular eukaryote for which the complete pattern of cell divisions from single-celled zygote to mature adult has been elucidated. Cell lineage diagrams are available that detail each cell or nuclear division and the fate of each cell produced by a terminal division.

cell lineage mutants mutations that affect the division of cells or the fates of their progeny cells. Cell lineage mutants generally fall into two broad classes. The first contains mutations that affect general cellular processes, such as cell division or DNA replication. Mutants perturbing the cell division cycle have been analyzed most extensively in *Saccharomyces cerevisiae*. The second class of mutations shows a striking specificity in their effects. For example, cell lineage mutants are known in *Caenorhabditis elegans* where particular cells are transformed to generate lineages or to adopt differentiated fates characteristic of cells normally found in different positions, at different times, or in the opposite sex. Some of these mutants result from transformations in cell fates. For example, a particular cell "A" will adopt the fate of another cell "B," and this results in the loss of the cells normally generated by A and the duplication of cells normally generated by B. Such transformations resemble the homeotic mutations (*q.v.*) of *Drosophila*. In *Caenorhabditis*, mutations of this type are generally symbolized by *lin*. *See* Appendix C, 1983, Greenwald *et al.*; developmental control genes, heterochronic mutations, selector genes.

cell lysis disruption of the cell membrane, allowing the dissolution of the cell and exposure of its contents to the environment. Examples: bacteria undergo *bacteriolysis*, red blood cells experience *hemolysis*.

cell-mediated immunity immune responses produced by T lymphocytes rather than by immunoglobulins (humoral- or antibody-mediated immunity); abbreviated CMI.

cell-mediated lympholysis the killing of "target" cells by activated T lymphocytes through direct cell–cell contact. Often used as an *in vitro* test of cell-mediated immunity.

cell plate a semisolid structure formed by the coalescence of droplets that are laid down between the daughter nuclei following mitosis in plants. The cell plate is the precursor of the cell walls, and it is synthesized by the phragmoplast (*q.v.*).

cell strain cells derived from a primary culture or cell line by the selection and cloning of cells having specific properties or markers. The properties or markers must persist during subsequent cultivation. *See in vitro* marker, *in vivo* marker.

cell-surface receptors transmembrane proteins on the surface of target cells. When they bind to appropriate extracellular signaling molecules, they are activated and generate a cascade of intracellular signals that alter the behavior of the target cells. Cell-surface receptors are grouped into three classes: (1) receptors that are linked to ion channels, (2) receptors linked to G proteins (*q.v.*), and (3) receptors linked to enzymes. These enzymes are generally protein kinases (*q.v.*). *See* ABC transporters, cellular signal transduction, receptor-mediated endocytosis.

cell theory the theory that all animals and plants are made up of cells, and that growth and reproduction are due to division of cells. *See* Appendix C, 1838, Schleiden and Schwann; 1855, Virchow.

cellular immunity immune responses carried out by active cells rather than by antibodies. *See* Appendix C, 1901, Mechnikov.

cellular signal transduction the pathways through which cells receive external signals and transmit, amplify, and direct them internally. The pathway begins with cell-surface receptors (*q.v.*) and may end in the cell nucleus with DNA-binding proteins that suppress or activate replication or transcription. Signaling pathways require intercommunicating chains of proteins that transmit the signal in a stepwise fashion. Protein kinases (*q.v.*) often participate in this cascade of reactions, since many signal transductions involve receiving an extracellular, chemical signal, which triggers the phosphorylation of cytoplasmic proteins to amplify the signal. *See* ABC transporter, cyclic AMP, gene-for-gene hypothesis, G proteins, polycystic kidney disease, transforming growth factor-β (TGF-β), *Wnt*.

cellular transformation *See* transformation.

cellulase an enzyme that degrades cellulose to glucose.

cellulifugal moving away from the center of the cell.

cellulose a complex structural polysaccharide that makes up the greater part of the walls of plant cells. As illustrated, cellulose is composed of a linear array of beta-D-glucose molecules.

cell wall a rigid structure secreted external to the plasma membrane. In plants it contains cellulose and lignin; in fungi it contains chitin; and in bacteria it contains peptidoglycans.

cen *See* symbols used in human cytogenetics.

cenospecies a group of species that, when intercrossed, produce partially fertile hybrids.

Cenozoic the most recent geologic era, occupying the last 65 million years and often called the age of mammals. *See* geologic time divisions.

CENP-A centromeric protein A, a histone variant that replaces H3 in centromeric nucleosomes. CENP-A confers a unique structural rigidity to the nucleosomes into which it assembles.

center of origin an area from which a given taxonomic group of organisms has originated and spread.

center of origin hypothesis the generalization that the genetic variability is greatest in the territory where a species arose. Conversely, marginal populations are likely to show a limited number of adaptations. Therefore, the regions where various

glucose

Cellulose

agriculturally important plant species arose can sometimes be identified by determining the amounts of genetic polymorphism in different geographic races. *See* Appendix C, 1926, Vavilov.

centimorgan *See* Morgan unit.

central dogma the concept describing the functional interrelations between DNA, RNA, and protein; that is, DNA serves as a template for its own replication and for the transcription of RNA which, in turn, is translated into protein. Thus, the direction of the transmission of genetic information is DNA → RNA → protein. Retroviruses (*q.v.*) violate this central dogma during their reproduction.

centric fusion breakage in the very short arms of two acrocentric chromosomes, followed by fusion of the long parts into a single chromosome; the two small fragments are usually lost; also termed a *Robertsonian translocation* or *whole arm fusion*. Centric fusions are seen in newborn infants with a frequency of 1 in 10,000. There is a marked excess of 21/21, 13/14, and 14/21 translocations. Centric fusions are an important cause of uniparental disomy. *See* Appendix C, 1911, Robertson; 1960, Polani *et al.*; disomy, telomeric fusion site.

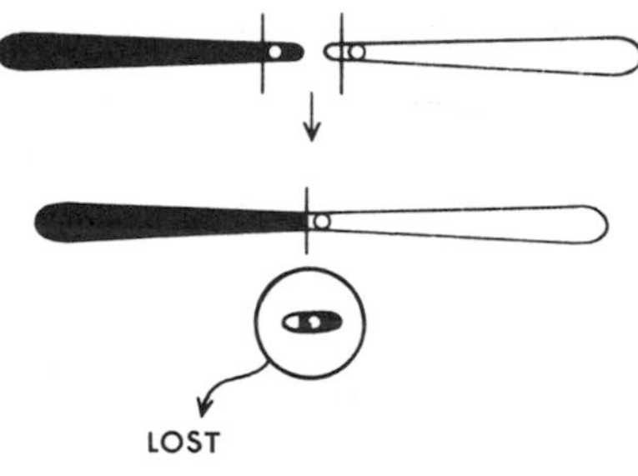

centrifugal acting in a direction away from the center.

centrifugal selection *See* disruptive selection.

centrifugation separation any of various methods of separation dispersions by the application of centrifugal force. In the case of *density gradient equilibrium centrifugation*, a gradient of densities is established in a centrifuge tube by adding a high molecular weight salt such as cesium chloride. The mixture of molecules to be studied is layered in the surface of the gradient and then centrifuged until each molecule reaches the layer in the gradient with a buoyant density equal to its own. In the case of *density gradient zonal centrifugation*, the macromolecules are characterized by their velocities of sedimentation through a preformed sucrose gradient. In this case the sedimentation velocity is determined by molecular size and shape.

centrifuge an apparatus used for the separation of substances by the application of centrifugal force generated by whirling at a high rate of rotation a vessel containing a fluid in which the substances are suspended. *See also* ultracentrifuge.

centriole a self-reproducing cellular organelle generally consisting of a short cylinder containing nine groups of peripheral microtubules (each group composed of three fused microtubules) disposed about a central cavity. Like DNA, centrioles replicate once during the cell division cycle, but they do so conservatively by forming a completely new centriole. This "daughter" centriole always lies at right angles to the "mother," and it grows outward until it reaches its mature size. Centrioles are capable of movement and always come to lie at the polar regions of the spindle apparatus in dividing animal cells. The behavior of the centrioles is illustrated in the *meiosis* entry. During anaphase the mother and daughter centrioles separate, move apart, and go on to form partner centrioles. Centrioles are required for animal somatic cells to progress through G1 and into the S phase of mitosis. The organelle that is ultrastructurally identical to the centriole forms the basal body of a cilium. Centrioles do not occur in the cells of higher plants. *See* Appendix C, 1888, Boveri; cell cycle, centrosome, kinetosome, microtubule organizing centers, p34 (CDC2).

centripetal acting in a direction toward the center.

centripetal selection *See* stabilizing selection.

centrolecithal egg one having centrally placed yolk. *See* isolecithal egg, telolecithal egg.

centromere a region of a chromosome to which spindle traction fibers attach during mitosis and meiosis. The position of the centromere determines whether the chromosome will appear as a rod, a J, or a V during its poleward migration at anaphase. In a very few species the traction fibers seem to attach along the length of the chromosome. Such chromosomes are said to be *polycentric* or to have a *diffuse centromere*. A replicated chromosome consists of two chromatids joined at the centromere region. Late in prophase, kinetochores develop on the two faces of the centromere that point toward the spindle poles. The microtubules of the traction fiber attach to the kinetochores, as illustrated on page 69. In the older literature, the terms *centromere* and *kinetochore* were used synonymously. However, the kinetochore is now defined as a complex structure, known to con-

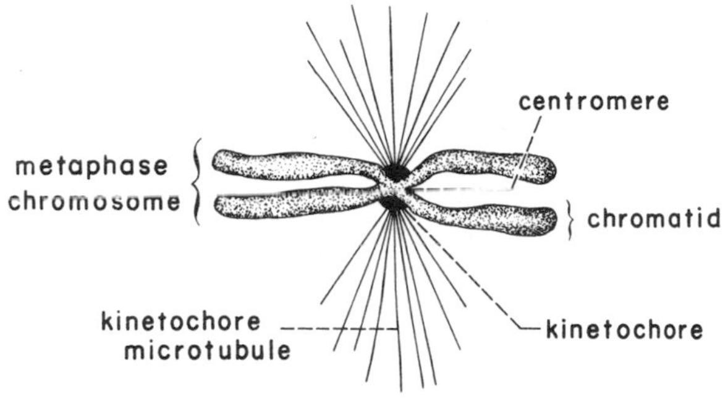

Centromere

tain several proteins, that binds to centromeric DNA and captures the microtubules that come from one of the two spindle poles. The centromere of metaphase chromosomes is narrower than the regions distal to it, and therefore it is called the *primary chromosomal constriction*. The centromere is generally bordered by heterochromatin that contains repetitious DNA (*q.v.*) and it is late to replicate. From a structural standpoint, centromeres are of two major types, those that occupy a very small region (~200 bp) of the chromosomal DNA and those that occupy large regions (40 kb to 5 mb). *Saccharomyces cerevisiae* has centromeres of the first type (*point* centromeres). Such small centromeres are expected, since yeast chromosomes are 100 times smaller than those from higher eukaryotes. The centromeres are not chromosome-specific and function normally in inverted sequence or when swapped between chromosomes. The minimal functional centromere is only about 112 base pairs and occupies about 40 nm of the B-form DNA. It is made up of three elements. The central one, containing 88 kb, is about 93% AT. The lateral elements have conserved sequences that contain about 80% AT. Specific proteins bind to the lateral elements and form a complex, which attaches the chromosome to a single spindle microtubule. Larger chromosomes have *regional* centromeres, which bind 30 to 40 microtubules simultaneously. For example, *Drosophila* centromeres contain 420 kb of DNA and are made up of simple repetitive DNA segments that are required for the special chromosomal organization at the centromere. There are also AT-rich segments, which may function in microtubule binding. Four transcriptionally active genes have been mapped within the centromere of chromosome 2 of *D. melanogaster*. In *Arabidopsis thaliana* (*q.v.*) the centromeres vary in length from 1.4 to 1.9 megabases and make up about 7% of the chromosomal DNA. Within the centromere regions are sequences of 180 base pairs that are repeated hundreds of times on all five chromosomes. Sequences that resemble retroposons (*q.v.*) are abundant in centromeric regions. Crossing over is dramatically suppressed within centromeres. There are about 200 genes contained in *Arabidopsis* centromeres, but many of these may be inactivated. However, at least 50 are transcriptionally active. *See* Appendix C, 1903, Waldeyer; 1980, Clark and Carbon; 1999, Copenhaver *et al.*; CENP-A, lamins, *Luzula*, MAD mutations, meiosis, microtubule, mitosis, yeast artificial chromosomes (YACs).

centromere interference the inhibitory effect of the centromere upon crossing over in adjacent chromosomal regions.

centromere misdivision *See* isochromosome.

centromeric coupling the forming of paired centromeres early in diplonema. At first a protein encoded by *Zip 1* (*q.v.*) holds the centromeres of homologous and nonhomologous chromosomes together indiscriminately. But as time passes, the number of homologous pairs increases, even though the total number of coupled centromeres remains the same. This observation suggests that the Zip 1 protein holds the centromeres together while chromosome homology is assessed. Then when correct pairing is achieved, synaptonemal complexes are constructed between the homologs. *See Gowen crossover suppressor.*

centromeric index the percentage of the total length of a chromosome encompassed by its shorter arm. For example, in human somatic cells during metaphase, chromosomes 1 and 13 have centromeric indexes of 48 and 17, respectively. Therefore, chromosome 1 is metacentric with its short arm occupying 48% of the total length of the chromosome, and chromosome 13 is acrocentric with a short arm that only occupies 17% of the total length.

centrosome a cytoplasmic region surrounding a pair of centrioles (*q.v.*), but devoid of a limiting membrane. The dense material surrounding the paired centrioles is called pericentriolar material or the centrosome matrix. All eukaryotes possess centrosomes, and during cell division centrosomes at opposite poles of the cell initiate the growth of the microtubules of the spindle apparatus. In the mitotic cells of the higher plants, centrioles are absent from the centrosomes, but the centrosomal matrix contains the necessary microtubule organizing activities for spindle formation. *See* cyclin-dependent kinease 2 (Cdk2), microtubule organizing centers (MTOCs), parthenogenesis, spindle pole body, tubulin.

Cepaea a genus of land snails belonging to the family Helicidae. C. *hortenses* and C. *nemoralis* exhibit extensive variation in color and ornamentation of the shell with longitudinal bands. These species have been extensively studied in the field and in laboratory colonies by population geneticists.

cephalic designating the head or the anterior end of an animal.

cephalosporin an antibiotic with structural similarities to penicillin (*q.v.*). It has the advantage of not causing allergic reactions in patients that are allergic to penicillin and of being inert to penicillinases (*q.v.*). *See* Appendix C, 1964, Hodgkin.

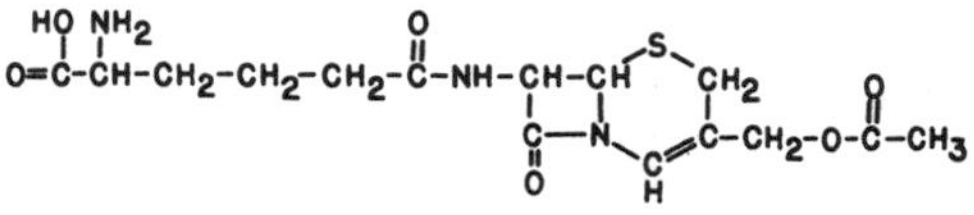

```
 HO NH2               O
  |  |                ‖        S
O=C-CH-CH2-CH2-CH2-C-NH-CH-CH   CH2     O
                        |   |    |      ‖
                        C---N   C-CH2-O-C-CH3
                        ‖     C
                        O     H
```

Cephalosporium a genus of molds of importance because of the cephalosporin antibiotics they produce.

Cercopithecoidea a superfamily of primates containing the Old World (African and Asian) monkeys, baboons, macaques, colobines, etc. A sister group to the Hominoidea (*q.v.*). The divergence of the Cercopithecoidea and Hominoidea took place about 30 million years ago.

Cercopithecus aethiops the African green monkey. A catarrhine primate with a haploid chromosome number of 30. About 20 genes have been assigned to nine different linkage groups. Monolayers of cultured African green monkey kidney cells are often used for growing viruses and mycoplasmas.

cereal a cultivated grass whose seeds are used as food; for example, wheat, oats, barley, rye, maize, etc.

cerebroside a molecule composed of sphingosine, a fatty acid, and a sugar; abundant in the myelin sheaths of nerve cells.

certation competition for fertilization among elongating pollen tubes.

ceruloplasmin a blue, copper protein present among the α_2 globulins of the plasma. Approximately 95% of the circulating copper of human beings is bound to ceruloplasmin. Ceruloplasmin is made up of eight subunits, each of molecular weight 18,000. *See* antihemophilic factor, Wilson disease.

cesium-137 a radioisotope of cesium with a half life of about 30 years. Generated during the explosion of certain nuclear weapons, it is one of the major sources of radiation contamination from fallout.

cesium chloride gradient centrifugation *See* centrifugation separation.

CFTR cystic fibrosis transmembrane-conductance regulator. *See* cystic fibrosis.

C(3)G *Gowen crossover suppressor* (*q.v.*).

C genes genes that code for the constant region of immunoglobulin protein chains. *See* immunoglobulin.

chaeta a bristle, especially of an insect.

Chaetodipus intermedius the rock pocket mouse, a species of rodents living in rocky habitats in adjacent deserts of Arizona, New Mexico, and Mexico. The genetic basis for adaptive melanism (*q.v.*) was first elucidated in this species. The color of the rodents matches their natural substrates, and so provides camouflage. The color of the dorsal fur is controlled by the MC1R gene (*q.v.*), and mutations at this locus determine the relative amounts of black *vs.* yellow melanin present in the hair. As predation eliminated mice with coat colors that failed to match their surroundings, genotypes were selected that provided the appropriate crypsis. *See* Appendix A, Chordata, Mammalia, Rodentia; Appendix C, 2003, Nachman, Hoekstra, and D'Agostino; melanin.

chaetotaxy the taxonomic study of the bristle pattern of insects.

Chagas disease a disease in humans caused by the parasite *Trypanosoma cruzi*. It is transmitted by bloodsucking bugs in the genera *Rhodnius* and *Triatoma* and by infected blood transfusion. The symptoms include swelling at the site of the vector's bite, fatigue, and fever during the acute stages, to cardiac, liver, and gastrointestinal problems, and eventually, death. This disease is estimated to affect 16–18 million people and is a major problem in Central America, South America, and Mexico. Darwin is thought to have contracted Chagas disease in South America, and as a result, spent the remainder of his life as a semi-invalid. The disease is named after Carlos Chagas, a Brazilian doctor, who first described it in 1909 and who later determined the life cycle of the parasite and identified the insects that transmit it. *See Glossina, Trypanosoma*; Appendix C, 2005, El-Sayed *et al.*

chain reaction a biological, molecular, or atomic process in which some of the products of the process, or energies released by the process, are instrumental in the continuation or magnification of the process.

chain termination codon *See* stop codon.

chain terminator a molecule that stops the extension of a DNA chain during replication. *See 2′,3′*-dideoxyribonucleoside triphosphates.

chalcones a group of pigments biogenetically related to anthocyans (*q.v.*). Chalcones give yellow to orange colors to the flowers of composites (*q.v.*).

chaperones eukaryotic proteins that help some nascent polypeptide chains fold correctly into their tertiary shapes, stabilizing and protecting them in the process, and/or preventing them from making premature or nonproductive intermolecular associations. Note that a chaperone forms a complex with a second protein to facilitate its folding, but chaperones are not part of the mature structure. Some of these molecular chaperones are heat-shock proteins (*q.v.*). Some chaperones may bind to nascent polypeptide chains while they are being synthesized on ribosomes, and they may also help the polypeptide move out of the tunnel of 60S ribosomal subunit. Other chaperones may keep the polypeptide in an unfolded conformation as it is being translated. This facilitates subsequent passage across membranes, as when protein enters the endoplasmic reticulum or a mitochondrion. Also called *chaperonins* or *molecular chaperones*. *See* prions.

character any detectable phenotypic property of an organism; synonymous with phenotype, trait.

character displacement the exaggeration of species markers (visual clues, scents, mating calls, courtship rituals, etc.) or adaptations (anatomical, physiological, or behavioral) in sympatric populations relative to allopatric populations of related species. This phenomenon is attributed to the direct effects of natural selection intensifying allesthetic traits useful for species discrimination or for utilizing different parts of an ecological niche (thereby avoiding direct competition). *See* Appendix C, 1956, Brown and Wilson.

character states a suite of different expressions of a character in different organisms. These different states are said to be homologs. A character may have a minimum of two states (present/absent or primitive/derived) or have many states.

Chargaff rule for the DNA of any species, the number of adenine residues equals the number of thymine residues; likewise, the number of guanines equals the number of cytosines; the number of purines (A + G) equals the number of pyrimidines (T + C). *See* Appendix C, 1950, Chargaff.

charged tRNA a transfer RNA molecule to which an amino acid is attached; also termed *aminoacylated tRNA*.

charon phages a set of 16 derivatives of bacteriophage lambda that are designed as cloning vectors. They were named by their originators (F. R. Blattner and 11 colleagues) after the old ferryman of Greek mythology who conveyed the spirits of the dead across the River Styx.

chase *See* pulse-chase experiment.

chasmogamous designating a plant in which fertilization takes place after the opening of the flower. *See* cleistogamous.

chDNA chloroplast DNA.

cheating genes any genetic elements that tend to increase in a population by meiotic drive (*q.v.*) even if they confer no selective advantage or perhaps even if they are harmful to the organisms in which they are present. *Compare with* selfish DNA. *See* segregation distortion.

checkpoint any one of several points in the cell cycle at which the progression of the cell to the next stage can be halted until more suitable conditions prevail. One major checkpoint is in G_1, just before the start of the S phase; the other is in G_2, just before the entry into mitosis. *See* Appendix C, 1989, Hartwell and Weinert; cell cycle, cyclins, DNA damage checkpoint, *MAD* mutations, maturation promoting factor (MPF), spindle checkpoint, *RAD9*.

Chédiak-Higashi syndrome (CHS) a hereditary disease of humans causing decreased pigmentation of the hair and eyes and the production of defective lysosomes in leukocytes and melanocytes. CHS is caused by mutations in the lysosomal trafficking regulator gene (*LYST*) (*q.v.*) which is located at 1q 42.1–2. A similar syndrome occurs in mice, mink, and cattle. The syndrome is named after the Cuban physician M. Chédiak and the Japanese pediatrician O. Higashi who described the condition in 1952 and 1954, respectively. *See* Aleutian mink.

cheetah *See Acinonyx jubatus.*

chelating agent a compound made up of heterocyclic rings that forms a chelate with metal ions. Heme (*q.v.*) is an example of an iron chelate. The porphyrin ring in chlorophyll (*q.v.*) forms a magnesium chelate.

chelation the holding of a metal ion by two or more atoms of a chelating agent.

Chelicerata a subphylum of arthropods containing the species that have no antennae and possess pincerlike chelicerae as the first pair of appendages. *See* classification.

chemical bonds *See* disulfide linkage, electrostatic bond, glycosidic bonds, high-energy bond, hydrogen bond, hydrophobic bonding, ionic bond, peptide bond, phosphodiester, salt linkage, van der Waals forces.

chemical elements listed alphabetically by their symbols. The biologically important elements are shown in boldface type. *See* periodic table.

Ac	Actinium
Ag	Silver
Al	Aluminum
Am	Americium
Ar	Argon
As	Arsenic
At	Astatine
Au	Gold
B	Boron
Ba	Barium
Be	Beryllium
Bi	Bismuth
Bk	Berkelium
Br	Bromine
C	**Carbon**
Ca	**Calcium**
Cd	Cadmium
Ce	Cerium
Cf	Californium
Cl	**Chlorine**
Cm	Curium
Co	**Cobalt**
Cr	Chromium
Cs	Cesium
Cu	**Copper**
Dy	Dysprosium
Er	Erbium
Es	Einsteinium
Eu	Europium
F	Fluorine
Fe	**Iron**
Fm	Fermium
Fr	Francium
Ga	Gallium
Gd	Gadolinium
Ge	Germanium
H	**Hydrogen**
He	Helium
Hf	Hafnium
Hg	Mercury
Ho	Holmium
I	**Iodine**
In	Indium
Ir	Iridium
K	**Potassium**
Kr	Krypton
La	Lanthanum
Li	Lithium
Lr	Lawrencium
Lu	Lutetium
Md	Mendelevium
Mg	**Magnesium**
Mn	**Manganese**
Mo	**Molybdenum**
N	**Nitrogen**
Na	**Sodium**
Nb	Niobium
Nd	Neodymium
Ne	Neon
Ni	**Nickel**
No	Nobelium
Np	Neptunium
O	**Oxygen**
Os	Osmium
P	**Phosphorus**
Pa	Protactinium
Pb	Lead
Pd	Palladium
Pm	Promethium
Po	Polonium
Pr	Praseodymium
Pt	Platinum
Pu	Plutonium
Ra	Radium
Rb	Rubidium
Re	Rhenium
Rh	Rhodium
Rn	Radon
Ru	Ruthenium
S	Sulfur
Sb	Antimony
Sc	Scandium
Se	Selenium
Si	Silicon
Sm	Samarium
Sn	Tin
Sr	Strontium
Ta	Tantalum
Tb	Terbium
Tc	Technetium
Te	Tellurium
Th	Thorium
Ti	Titanium
Tl	Thallium
Tm	Thulium
U	Uranium
V	Vanadium
W	Tungsten
Xe	Xenon
Y	Yttrium
Yb	Ytterbium
Zn	Zinc
Zr	Zirconium

chemiosmotic theory the concept that hydrogen ions are pumped across the inner mitochondrial membrane, or across the thylakoid membrane of chloroplasts, as a result of electrons passing through the electron transport chain (*q.v.*). The electrochemical gradient that results is the proton motive force (pmf). ATP synthase harnesses the pmf to make ATP *See* Appendix C, 1961, Mitchell; adenosine phosphate, mitochondrial proton transport.

chemoautotrophy *See* autotroph, methanogens.

chemokines a large family of structurally homologous cytokines (*q.v.*), 8 to 10 kilodaltons (kDa) in size. The name "chemokine" is a contraction of "*che*motactic cyto*kine*." Chemokines share the ability to stimulate leukocytic movement (chemokinesis) and directed movement (chemotaxis), especially of inflammatory cells to damaged or infected sites. Examples of chemokines include interleukin-8 (IL-8) that attracts neutrophils, basophils, and eosinophils, and monocyte chemotactic protein-1 (MCP-1) that acts specifically only on monocytes. Chemokines are produced by several types of cells, including activated mononuclear phagocytes, tissue cells (endothelium, fibroblasts), and megakaryocytes (which give rise to platelets that contain stored chemokine).

chemolithoautotroph an autotroph that gets its energy from oxidation of inorganic substances in the absence of light. Many hyperthermophiles use inorganic electron donors and acceptors in their energy metabolism and obtain their carbon from CO_2. *See* lithotroph, *Methanococcus jannaschii.*

chemostat an apparatus allowing the continuous cultivation of bacterial populations in a constant, competitive environment. Bacteria compete for a limiting nutrient in the medium. The medium is slowly added to the culture, and used medium plus bacteria are siphoned off at the same rate. The concentration of the limiting nutrient in the fresh medium determines the density of the steady-state population, and the rate at which the medium is pumped into the chemostat determines the bacterial growth rate. In chemostat experiments, environmental variables can be changed, one by one, to ascertain how these affect natural selection, or the environment can be held constant and the differential fitness of two mutations can be evaluated.

chemotaxis the attraction or repulsion of cells or organisms toward or away from a diffusing substance. Also known as *chemotropism.*

chemotherapy the treatment of a disease with drugs of known chemical composition that are specifically toxic to the etiological microorganisms and do not harm the host. The term was coined by Paul Ehrlich, who also gave such drugs the nickname *magic bullets. See* Salvarsan.

chemotrophs organisms whose energy is the result of endogenous, light-independent chemical reactions. A chemotroph that obtains its energy by metabolizing inorganic substrates is called a *chemolithotroph,* whereas one that metabolizes organic substrates is called a *chemoorganotroph. Contrast with* prototrophs. *See* autotrophs.

chiasma (*plural* chiasmata) the cytological manifestation of crossing over; the cross-shaped points of junction between nonsister chromatids first seen in diplotene tetrads. *See* Appendix C, 1909, Janssens; 1929, Darlington; crossing over, meiosis, recombination nodules.

chiasma interference the more frequent (in the case of negative chiasma interference) or less frequent (in the case of positive interference) occurrence of more than one chiasma in a bivalent segment than expected by chance.

chiasmata *See* chiasma.

chiasmatype theory the theory that crossing over between nonsister chromatids results in chiasma formation.

chicken *See Gallus domesticus.*

chimera an individual composed of a mixture of genetically different cells. In plant chimeras, the mixture may involve cells of identical nuclear genotypes, but containing different plastid types. In more recent definitions, chimeras are distinguished from mosaics (*q.v.*) by requiring that the genetically different cells of chimeras be derived from genetically different zygotes. *See also* aggregation chimera, heterologous chimera, mericlinal chimera, periclinal chimera, radiation chimera.

chimpanzee *See Pan.*

Chinchilla lanigera a rodent native to the Andes mountains of South America. It is bred on commercial ranches for its pelt, and many coat color mutants are available. Its haploid chromosome number is 32.

CHIP-28 an abbreviation for *CH*annel-forming *I*ntegral *P*rotein of *28* kDa relative molecular mass. CHIP-28 was purified from the plasma membranes of human erythrocytes and later shown to form channels permeable to water. It was renamed aquaporin1 (AQP1).

chiral descriptive of any molecules that exist in two mirror-image versions (enantiomers, *q.v.*).

Chironomus a genus of delicate, primitive, gnat-like flies that spend their larval stage in ponds and slow streams. Nuclei from various larval tissues contain giant polytene chromosomes. The salivary gland chromosomes of *C. thummi* and *C. tentans* have been mapped, and the transcription processes going on in certain Balbiani rings (*q.v.*) have been studied extensively. *See* Appendix C, 1881, Balbiani; 1952, Beermann; 1960, Clever and Karlson; chromosomal puff.

chi sequence an octomeric sequence in *E. coli* DNA, occurring about once every 10 kilobases, acting as a "hotspot" for RecA-mediated genetic recombination.

chi-square (χ^2) test a statistical procedure that enables the investigator to determine how closely an experimentally obtained set of values fits a given theoretical expectation. The relation between χ^2 and probability is presented graphically on page 74. *See* Appendix C, 1900, Pearson; degrees of freedom.

chi structure a structure resembling the Greek letter χ, formed by cleaving a dimeric circle with a restriction endonuclease that cuts each DNA circle only once. The parental monomeric duplex DNA molecules remain connected by a region of heteroduplex DNA at the point where crossing over oc-

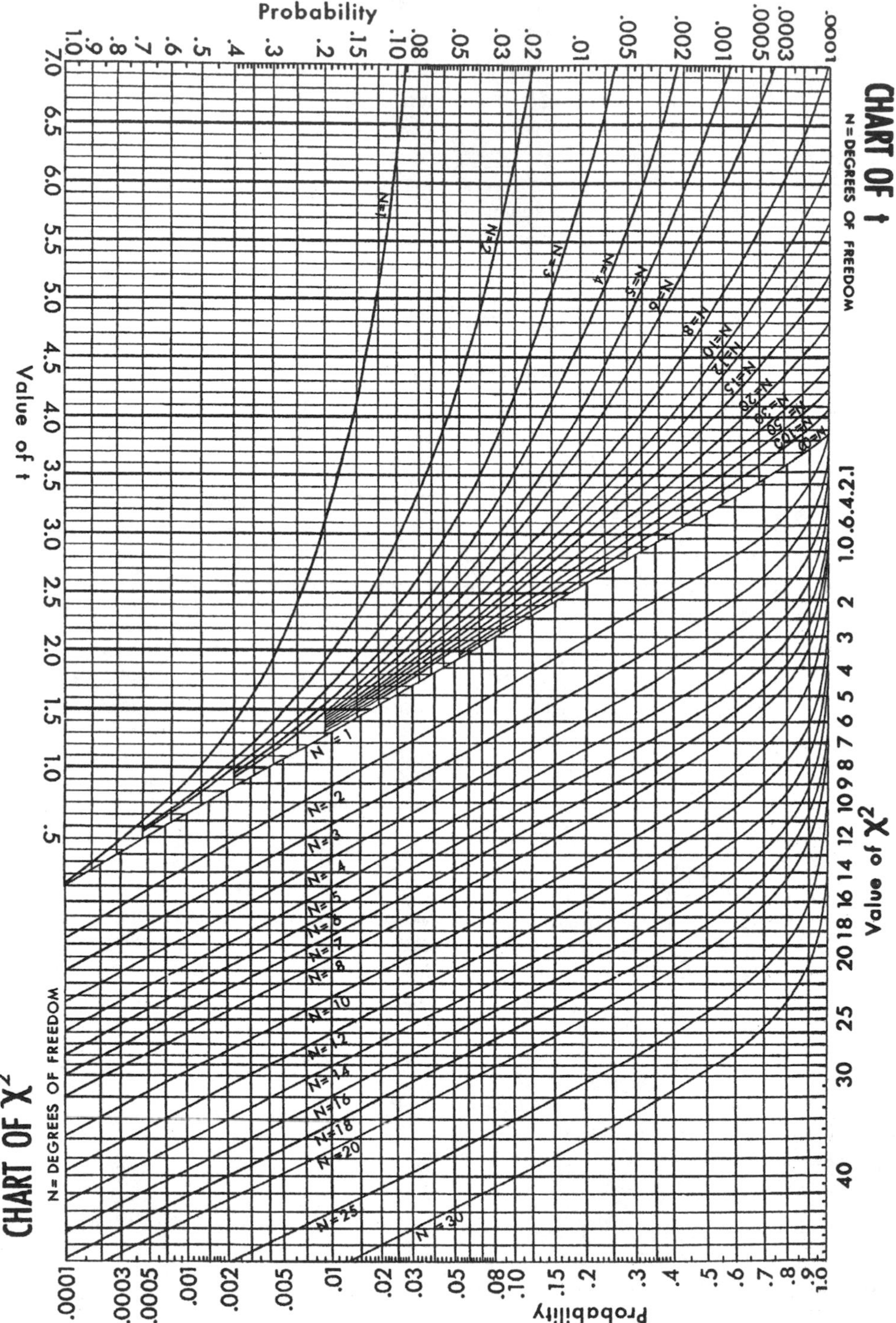

Chi square (χ^2) test and Student *t* test probability chart

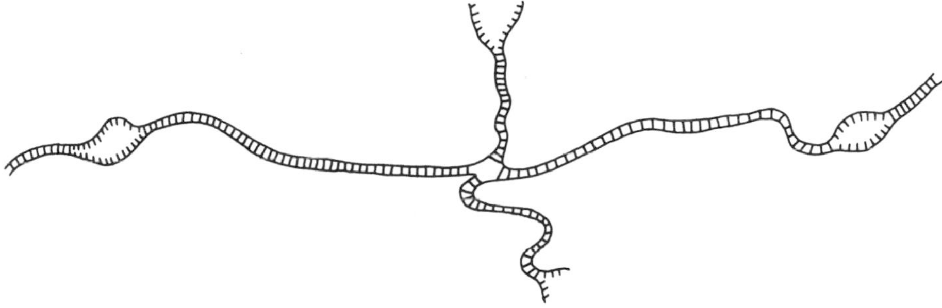

Chi structure

curred. Thus, the identification of such chi structures provides evidence for cross-over events taking place between circular DNA molecules.

chitin a polymer of high molecular weight composed of *N*-acetylglucosamine residues joined together by beta glycosidic linkages between carbon atoms 1 and 4. Chitin is a component of the exoskeletons of arthropods.

chlamydia obligate intracellular parasites or symbionts found throughout the animal kingdom. Some chlamydia have existed as endosymbionts of free-living amoebae since Precambrian times. The most common sexually transmitted disease in the U.S. is caused by *Chlamydia trachomatis*. Its genome consists of a 1,042,519 base pair chromosome and a 7,493 bp plasmid. *See* Appendix A, Prokaryotes, Bacteria, Pirellulae; Appendix C, 1998, Stephens *et al.*; Appendix F.

Chlamydomonas reinhardi a species of green algae in which the interaction of nuclear and cytoplasmic genes has been extensively studied. The nuclear gene loci have been distributed among 17 linkage groups. Nuclear genes are transmitted to the offspring in a Mendelian fashion, but chloroplast and mitochondrial DNAs are transmitted uniparentally. The mating-type "+" parents transmit chloroplasts, while the "–" parents transmit mitochondria. The chloroplast DNA contains nearly the same set of genes that encode rRNAs, tRNAs, ribosomal proteins and photosynthetic proteins as are found in the chloroplast DNAs of higher plants. The genetic analysis of the flagellar apparatus of *Chlamydomonas* has identified more than 80 different mutations that affect its assembly and function. The axoneme contains more than 200 proteins, most of which are unique to this structure. An alternative spelling, *C. reinhardtii*, occurs in the literature. *See* Appendix A, Protoctista, Chlorophyta; Appendix C, 1963, Sager and Ishida; 1970, Sager and Ramis; Appendix E.

chlorambucil a drug that binds DNA strands so that the double helix cannot unzip and replicate. Therefore it suppresses cell division and is used in cancer chemotherapy. *See* alkylating agent.

chloramphenicol an antibiotic produced by *Streptomyces venezuelae*. Chloramphenicol is a potent inhibitor of protein synthesis on the 70S ribosomes of prokaryotes. It attaches to the 50S ribosomal subunit and prevents the addition of an amino acid to the growing polypeptide chain. Chloroamphenicol does not bind to the 80S ribosomes of eukaryotes, but it does bind to the smaller ribosomes of the mitochondria present in eukaryotic cells. This is one of

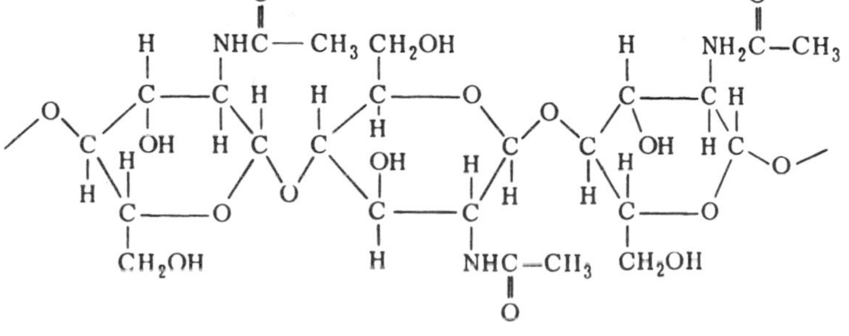

Chitin

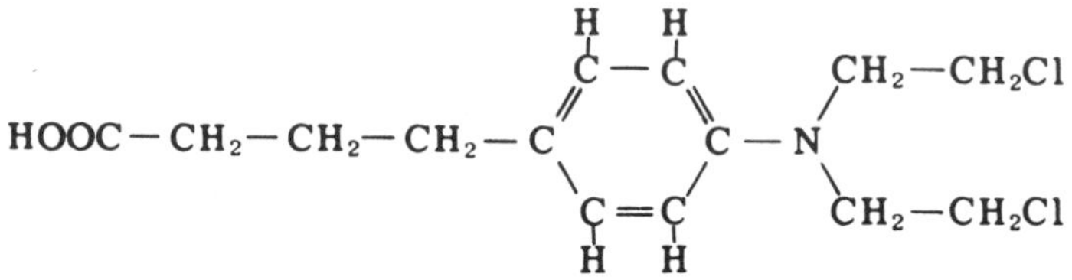

Chlorambucil

the lines of evidence that symbiotic prokaryotes were the ancestors of eukaryotic ribosomes. *See* cyclohexamide, endosymbiont theory, ribosome, ribosomes of organelles, serial symbiosis theory, translation.

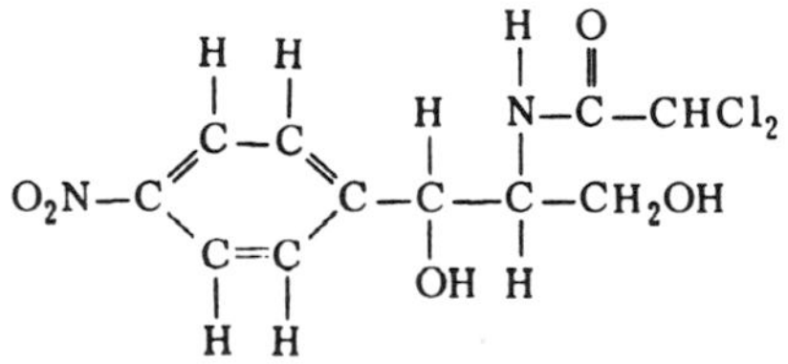

Chlorella vulgaris a species of green algae extensively used in studies of photosynthesis and its genetic control. *See* Appendix A, Protoctista, Chlorophyta.

chlorenchyma tissue possessing chloroplasts.

chlorine an element universally found in small amounts in tissues. Atomic number 17; atomic weight 35.453; valence 1^-; most abundant isotope ^{35}Cl; radioisotopes ^{33}Cl, half-life 37 minutes, ^{39}Cl, half-life 55 minutes, radiation emitted—beta particles.

chlorolabe *See* color blindness.

chloromycetin chloramphenicol (*q.v.*).

chlorophyll a group of pigments that mediate photosynthesis. These include chlorophyll a and b, the green pigments found in the chloroplasts of plants. The structural formula for the chlorophyll a molecule and the appropriate dimensions of the porphyrin ring and phytol chain are illustrated below. The chlorophyll b molecule differs from chlorophyll a in that the CH_3 indicated by the arrow is replaced by a CHO group. Other chlorophylls include chlorophyll c (found in the brown algae and some red algae), chlorophyll d (found in the red algae), and the bacteriochlorophylls (found in the green sulfur bacteria). *See* anthocyanins, Cyanobacteria.

chlorophyte any protoctist that belongs to the Chlorophyta. This phylum contains green algae that have undulipodia at some stage of their life cycle. Green land plants presumably arose from a chlorophyte lineage. *See Mesostigma viride*, undulipodium.

chloroplast the chlorophyll-containing, photosynthesizing organelle of plants. Chloroplasts are thought to be the descendants of endosymbiotic cyanobacteria. A typical chloroplast is illustrated on page 67. Each is surrounded by a double membrane and contains a system of internal thylakoid membranes. These form stacks of flattened discs called grana in which chlorophyll molecules are embedded. Chloroplasts contain DNA and can multiply. Replication of chloroplast DNA occurs throughout the cell cycle. Chloroplasts contain 70S ribosomes and in this respect resemble bacterial ribosomes rather than those of the plant cytoplasm (*see* ribosome). The repro-

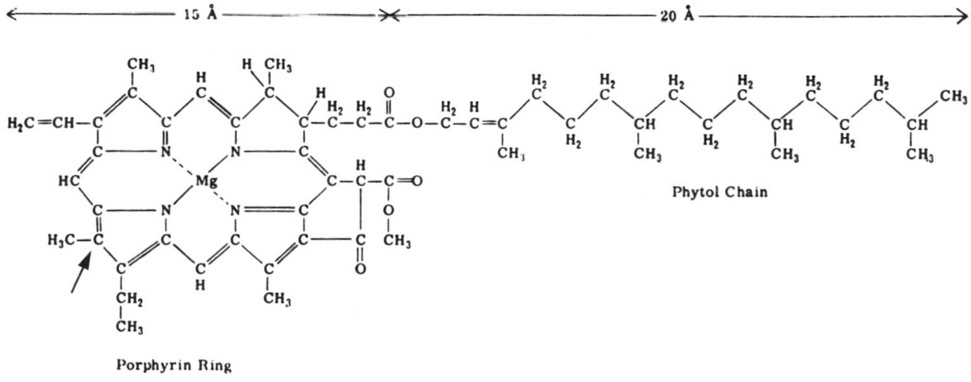

15 Å — 20 Å

Porphyrin Ring

Phytol Chain

Chlorophyll a

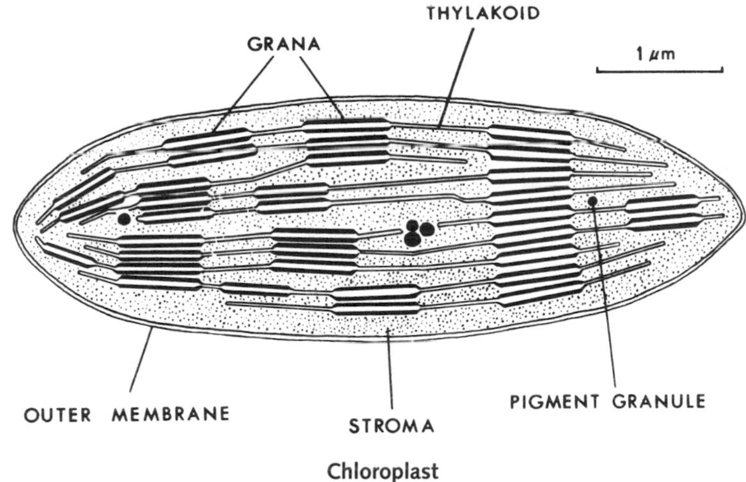

Chloroplast

duction and functioning of chloroplasts is under the control of both nuclear genes and those of the organelle. Chloroplasts develop from protoplastids. These are small organelles surrounded by a double membrane. The inner one gives rise to a sparse internal membrane system from which the thylakoids develop. Chloroplasts are generally inherited uniparentally. Most angiosperms show maternal inheritance, whereas most gymnosperms show paternal inheritance of chloroplasts. Some chloroplast genes confer an advantage to the plants that possess them by encoding proteins that immobilize herbicides. *See* Appendix C, 1837, von Mohl; 1883, Schimper; 1909, Correns and Bauer; 1951, Chiba; 1953, Finean *et al.*; 1962, Ris and Plaut; 1981, Steinbeck *et al.*; Appendix F; *Chlamydomonas reinhardi*, chloroplast DNA, chloroplast er, cyanelles, *Prochloron*, ribulose–1, 5-bisphosphate carboxylase-oxygenase, protein sorting, serial symbiosis theory, 5S RNA, *Synechocystis.*

chloroplast DNA (chDNA) chloroplast DNAs are circular, like those of mitochondria, but many times longer. There are 40–80 DNA molecules per organelle. The DNA molecules form clusters in the stroma and are thought to attach to the inner membrane. The DNAs are devoid of histones. The chloroplast genome encodes all the rRNA and tRNA molecules needed for translation, plus about 50 proteins. These include RNA polymerase, ribosomal proteins, components of the thylakoid membranes, and a family of proteins involved in oxidation-reduction reactions. Some chloroplast genes contain introns. Chloroplast genomes do not contain transposons, and since chDNAs are usually transmitted uniparentally, there is no opportunity for recombination. These facts may account for the observation that the protein-coding genes of chloroplasts evolve at a rate about five times slower than that of plant nuclear genes. For this reason chDNA variation has been extensively used in reconstructing plant phylogenies. *See* Appendix C, 1971, Manning *et al.*; 1972, Pigott and Carr; 1986, Ohyama *et al.*, Shinozaki *et al.*; 1987, Wolf, Li and Sharp; 1993, Hallick *et al.*; 2000, Lemieux, Otis and Turmel; 2004, Matsuzaki *et al.*

chloroplast ER *See* Chromista, Chrysomonads.

chloroquine an anti-malarial drug that accumulates in the food vacuole of the merozoite (*q.v.*) and interferes with the detoxification of the heme (*q.v.*) released during the digestion of hemoglobin (*q.v.*). *See* artemisinin, hemozoin, malaria.

chlorosis failure of chlorophyll development.

CHO cell line A somatic cell line derived from Chinese hamster ovaries. The cells have a near diploid number, but over one-half of the chromosomes contain deletions, translocations, and other aberrations that have occurred during the evolution of the cell line. *See Cricetulus griseus.*

cholecystokinen a hormone secreted by the duodenum that causes gallbladder contraction.

cholera an epidemic disease caused by the bacterium *Vibrio cholerae* (*q.v.*). Each is a curved rod about 0.5 by 3 micrometers that possesses a polar flagellum. Vibrios are highly motile and spread in water contaminated with feces. If sufficient vibrios are ingested, some will reach the intestine and multiply. Virulent strains produce a toxin encoded by a chromosomal gene. The cholera toxin is made up of multiple polypeptide chains, and these are organized into A and B subunits. The B subunits bind to gan-

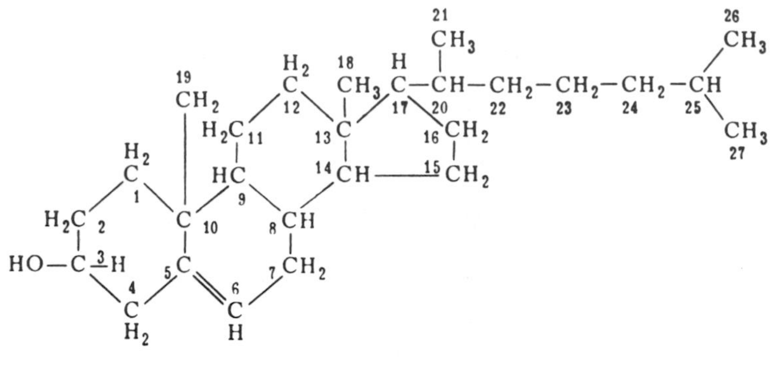

Cholesterol

glioside receptors on the microvilli of the cells lining the intestine. B subunits, in groups of five, form a hydrophilic channel through which the A subunit enters the cell. It interacts with a specific G protein, converting it into a permanently active form. This catalyzes reactions that stimulate secretion into the gut lumen of fluids rich in electrolytes. Therefore, the primary symptom of cholera is a profuse diarrhea. The watery discharge may amount to 30 liters per day, and without treatment the victim will die in a few days of dehydration and the loss of ions, especially potassium and bicarbonate. Mutations causing cystic fibrosis (*q.v.*) may have been retained in human populations because of the resistance shown by heterozygotes to cholera. The CF gene encodes a protein that functions as an ion channel. Mutations at this locus therefore impair fluid transport and reduce the fluid loss transport and reduce the fluid loss induced by the cholera toxin.

cholesterol the quantitatively predominant steroid of humans. A 27-carbon compound made up of a fused ring system. Cholesterol is a component of biomembranes and of the myelin sheaths that surround nerve axons. In insects, cholesterol serves as a precursor for ecdysone (*q.v.*). *See* Appendix C, 1964, Hodgkin; adrenal corticosteroid, familial hypercholesterolemia, vitamin D.

cholinesterase *See* acetylcholinesterase.

chondriome all the mitochondria of the cell referred to collectively.

chondriosome a rarely used term, superceded by mitochondrion.

chondroitin sulfuric acid a mucopolysaccharide commonly found in cartilage.

chordamesoderm a layer of cells derived from the blastoporal lip that later in embryogenesis forms the mesoderm and notochord of the vertebrate embryo. The chordamesoderm acts as an organizer upon the overlying ectoderm to induce differentiation of neural structures. *See* Spemann-Mangold organizer.

Chordata the phylum of animals with a notochord, a hollow dorsal nerve cord, and gill slits at some developmental stage. *See* Appendix A.

chorea a nervous disorder, characterized by irregular and involuntary actions of the muscles of the extremities and face. *See* Huntington disease.

chorioallantoic grafting the grafting of pieces of avian or mammalian embryos upon the allantois of the chick embryo under the inner shell-membrane. The implant becomes vascularized from the allantoic circulation and continues development.

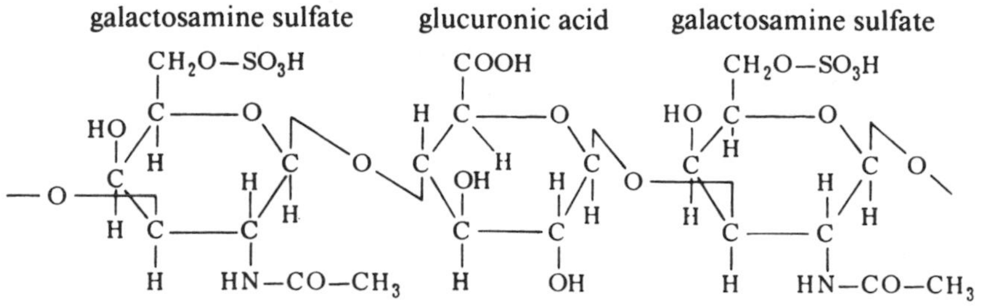

galactosamine sulfate — glucuronic acid — galactosamine sulfate

Chondroitin sulfuric acid

chorion 1. the insect egg shell. 2. *see* amnion.

chorionic appendages the anterior, dorsal projections of the *Drosophila* egg shell that serve as breathing tubes when the egg is submerged.

chorionic gonadotropin (CG) a hormone produced by the placenta that continues to stimulate the production of progesterone by the corpus luteum and thus maintains the uterine wall in a glandular condition. This hormone is the one for which most pregnancy tests assay. For humans, it is abbreviated HCG.

chorionic somatomammotropin *See* human growth hormone.

chorionic villi sampling the harvesting of chorionic cells by introducing a catheter through the vagina and into the uterus until it touches the chorion, which surrounds the human embryo. The chorionic cells withdrawn can be used to detect enzymatic and karyotypic defects, since they have the same genotype as embryonic cells. The sampling of chorionic villi can be performed between 10 and 11 weeks following the last menstrual period, so this technique has the advantage over amniocentesis (*q.v.*) of allowing fetal disorders to be detected earlier. Thus, when necessary, the pregnancy can be terminated at an earlier stage with less risk to the mother. *See* informed consent, genetic counseling, prenatal genetic testing.

Christmas disease the B form of hemophilia (*q.v.*), named after a British patient, Stephen Christmas. In 1952, when only 5 years old, he was shown to be suffering from a type of hemophilia different from the classic form. He could not produce clotting factor 9. This is sometimes called the *Christmas factor* in his honor.

chromatid conversion a form of gene conversion (*q.v.*) rendered evident by identical sister-spore pairs in a fungal octad that exhibits a non-Mendelian ratio. For example, if an ordered octad of ascospores from a cross of + × m = (++) (++) (++) (mm), one chromatid of the m parental chromosome appears to have been converted to +. In *half-chromatid conversion*, an octad of (++) (++) (+m) (mm) indicates that one chromatid of the m parental chromosome was "half-converted." *See* ordered tetrad.

chromatid interference a deviation from the expected 1 : 2 : 1 ratio for the frequencies of 2-, 3-, and 4-strand double crossovers signaling a nonrandom participation of the chromatids of a tetrad in successive crossovers.

chromatids the two daughter strands of a duplicated chromosome that are joined by a single centromere. Upon the division of the centromere, the sister chromatids become separate chromosomes. *See* Appendix C, Flemming; meiosis, mitosis.

chromatin the complex of nucleic acids (DNA and RNA) and proteins (histones and nonhistones) comprising eukaryotic chromosomes. *See* Appendix C, 1879, Flemming; 1976, Finch and Klug.

chromatin diminution the elimination during embryogenesis of certain chromosomes or chromosomal segments from the cells that form the somatic tissues. Germ cells, however, retain these chromosomes. The process occurs in some ciliates, nematodes, copepods, and insects. The discarded DNA often contains highly repetitive sequences and sometimes genes encoding ribosomal proteins. *See* Appendix C, 1887, Boveri; Ascarididae, C value paradox, genomic equivalence, *Parascaris equorum*.

chromatin fibers fibers of 30 nm diameter seen when interphase nuclei are lysed and their contents viewed under an electron microscope. The fibers are composed of solenoids of nucleosomes held together by hexamers of histone 1 molecules. *See* solenoid structures.

chromatin immunoprecipitation a technique for identifying a specific DNA sequence in the genome (*q.v.*) to which a particular protein binds *in vivo* (*q.v.*). In this procedure, the protein in question is cross-linked to DNA with formaldehyde *in vivo*. The DNA is then extracted from cells and sheared into small fragments. Antibodies against the bound protein are then used to isolate the protein-DNA complex, the protein is released, and the polymerase chain reaction (PCR) (*q.v.*) is used to amplify the DNA sequence to which the protein was bound. The amplified DNA can then be identified by sequence analysis.

chromatin-negative an individual (normally a male) whose cell nuclei lack sex chromatin.

chromatin-positive an individual (normally a female) whose cell nuclei contain sex chromatin. *See* Barr body, sex chromatin.

chromatograph the record produced by chromatography (*q.v.*). Generally applied to a filter paper sheet containing a grouping of spots that represent the compounds separated.

chromatography a technique used for separating and identifying the components from mixtures of molecules having similar chemical and physical properties. The population of different molecules is dissolved in an organic solvent miscible in water, and the solution is allowed to migrate through a stationary phase. Since the molecules migrate at slightly different rates, they are eventually separated. In *paper chromatography*, filter paper serves as the stationary phase. In *column chromatography* (*q.v.*), the stationary phase is packed into a cylinder. In *thin layer chromatography*, the stationary phase is a thin layer of absorbent silica gel or alumina spread on a flat glass plate. *See also* affinity chromatography, counteracting chromatographic electrophoresis, gas chromatography, ion exchange column. *See* Appendix C, 1941, Martin and Synge.

chromatophore submicroscopic particles isolated from photosynthetic bacteria that contain the photosynthetic pigments. *Compare with* chromophore.

chromatosome a DNA protein complex consisting of a nucleosome (*q.v.*), the linker DNA segment, and its H1 histone. *See* histones, nucleosome.

chromatrope *See* metachromasy.

Chromista the name proposed by T. Cavalier-Smith for a kingdom to contain those eukaryotic species that show the following specific ultrastructural similarities. They all posses, at some time during their life cycles, cells containing specialized undulopodia that have regularly spaced tufts of fibers extending from them. Such an undulopodium is often called a *hairy* or *tinsel* flagellum. The lateral fibers are called *mastigonemes*, and they are tubules that are rooted in the flagellar axonemes. The other organelle that characterizes the Chromista is the *chloroplast ER*. This refers to a chloroplast that is surrounded by four membranes, the last of which is continuous with the endoplasmic reticulum (*q.v.*). Species belonging to the phyla Chrysophyta, Cryptophyta, Xanthophyta, Eustimatophyta, Bacillanophyta, Phaeophyta, Labyrinthulomycota, Hypochytridiomycota, and Oomycota fall into this group. *See* Appendix A, Protoctista; Chrysophyta.

chromocenter a central amorphous mass found in the nuclei of larval salivary gland cells of *Drosophila*. Since the chromocenter results from the fusion of the heterochromatic elements that surround the centromeres of all the chromosomes in each nucleus, the distal, banded, euchromatic arms of the somatically paired, polytene chromosomes radiate from the chromocenter. In mitotic cells the pericentromeric, heterochromatic elements are rich in highly repetitive, DNA sequences. These fail to replicate during polytenization (*q.v.*) and are therefore underrepresented in the chromocenter. *See Drosophila* salivary gland chromosomes.

chromomere one of the serially aligned beads or granules of a eukaryotic chromosome, resulting from local coiling of a continuous DNA thread; best seen when most of the rest of the chromosome is relatively uncoiled as in the leptotene and zygotene stages of meiosis (*q.v.*). In polytene chromosomes (*q.v.*), the chromomeres lie in register and give the chromosome its banded appearance.

chromonema (*plural* **chromonemata**) the chromosome thread.

chromoneme the DNA thread of bacteria and their viruses.

chromophore 1. the part of a dye that gives the molecule its color. 2. any receptor molecule that absorbs light (thereby producing a color) and is usually complexed with a protein. For example, phytochromes (*q.v.*) consist of a linear tetrapyrole molecule that functions as a chromophore and a protein portion that anchors the chromophore to the inside surface of the plasma membrane and communicates a signal to one or more signal pathways. The same is true for retinal (*q.v.*) and its opsin (*q.v.*). *Compare with* chromatophore.

chromoplast a carotenoid-containing plastid that colors ripe fruits and flowers.

chromosomal aberration an abnormal chromosomal complement resulting from the loss, duplication, or rearrangement of genetic material. *Intrachromosomal* or *homosomal* aberrations involve changes that occur in but one chromosome. Such aberrations include *deficiencies* and *duplications* that result in a reduction or increase in the number of loci borne by the chromosome. *Inversions* and *shifts* involve changes in the arrangement of the loci, but not in their number. In the case of an inversion a chromosomal segment has been deleted, turned through 180°, and reinserted at the same position on a chromosome, with the result that the gene sequence for the segment is reversed with respect to that of the rest of the chromosome. In the case of shift a chromosomal segment has been removed from its normal position and inserted (in the normal or reversed sequence) into another region of the same chromosome. *Interchromosomal* or *heterosomal* aberrations arise from situations in which nonhomologous chromosomes are broken, and interchange occurs between the resulting fragments, producing a *translocation*. *See* radiation-induced chromosomal aberrations.

chromosomal mutation *See* chromosomal aberration.

chromosomal polymorphism the existence within a population of two or more different structural arrangements of chromosomal material (e.g., inversions, translocations, duplications, etc.).

chromosomal puff a localized swelling of a specific region of a polytene chromosome due to localized synthesis of DNA or RNA. Extremely large RNA puffs are called *Balbiani rings* (*q.v.*). *See* Appendix C, 1952, Beerman; 1959, Pelling; 1960, Clever and Karlson; 1961, Beerman; 1980, Gronemeyer and Pongs; heat-shock puffs, insulator DNAs, *Rhynchosciara*, *Sciara*.

chromosomal RNA ribonucleic acid molecules associated with chromosomes during either division (e.g., primer, *q.v.*) or interphase (e.g., incomplete transcripts).

chromosomal sterility sterility from the lack of homology between the parental chromosomes in a hybrid.

chromosomal tubules microtubules of the spindle apparatus that originate at kinetochores of centromeres. The chromosomal tubules interpenetrate on the spindle with polar tubules (*q.v.*) and are hypothesized to slide by one another during anaphase as a consequence of making and breaking of cross bridges between them.

chromosome 1. in prokaryotes, the circular DNA molecule containing the entire set of genetic instructions essential for life of the cell. *See* genophore. 2. in the eukaryotic nucleus, one of the threadlike structures consisting of chromatin (*q.v.*) and carrying genetic information arranged in a linear sequence. *See* Appendix C, 1883, Roux; 1888, Waldeyer.

chromosome arms the two major segments of a chromosome, whose length is determined by the position of the centromere. *See* acrocentric, metacentric, submetacentric, telocentric.

chromosome banding techniques there are four popular methods for staining human chromosomes. To produce *G banding*, chromosomes are usually treated with trypsin and then stained with Giemsa. Most euchromatin stains lightly, and most heterochromatin stains darkly under these conditions. *C bands* are produced by treating chromosomes with alkali and controlling the hydrolysis in a buffered salt solution. C banding is particularly useful for staining and highlighting centromeres and polymorphic bands (especially those of meiotic chromosomes). With *Q banding*, chromosomes are stained with a fluorochrome dye, usually quinacrine mustard or quinacrine dihydrochloride, and are viewed under ultraviolet light. The bright bands correspond to the dark G bands (with the exception of some of the polymorphic bands). Q banding is especially useful for identifying the Y chromosome and polymorphisms that are not easily demonstrated by the G-banding procedure. *R bands* are produced by treating chromosomes with heat in a phosphate buffer. They can then be stained with Giemsa to produce a pattern that is the reverse (hence the R in the term) of G bands, thereby allowing the evaluation of terminal bands that are light after G banding. Alternatively, chromosomes can be heated in buffer and then stained with acridine orange. When viewed under ultraviolet light, the bands appear in shades of red, orange, yellow, and green. They can also be photographed in color, but printed in black and white to reveal more distinctive R bands. Q-dark, R-positive bands have a low AT:CG ratio and are rich in SINE repeats and Alu sequences. Q-bright, R-negative bands have a high AT:CG ratio and are rich in LINE repeats. *See* Appendix C, 1970, Caspersson *et al.*; 1971, O'Riordan *et al.*; repetitious DNA.

chromosome bridge a bridge formed between the separating groups of anaphase chromosomes because the two centromeres of a dicentric chromosome are being drawn to opposite poles (see illustration on page 82). Such bridges may form as the result of single- or three-strand double-exchanges within the reverse loop of a paracentric inversion heterozygote (*see* inversion). They may also arise as radiation-induced chromosome aberrations (*q.v.*). Such chromosome bridges are always accompanied by acentric chromosome fragments. *See* breakage-fusion-bridge cycle.

chromosome condensation the process whereby eukaryotic chromosomes become shorter and thicker during prophase as a consequence of coiling and supercoiling of chromatic strands. *See* Appendix C, 1970, Johnson and Rao; nucleosome, solenoid structure.

chromosome congression the movement of chromosomes to the spindle equator during mitosis.

chromosome diminution or elimination *See* chromatin diminution.

chromosome jumping *See* chromosome walking.

chromosome loss failure of a chromosome to be included in a daughter nucleus during cell division. *See* anaphase lag.

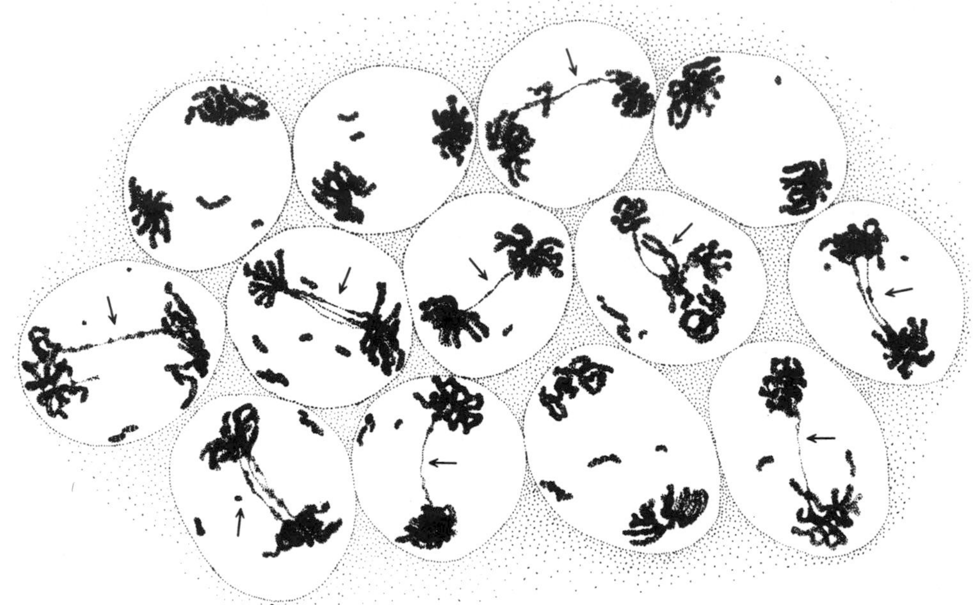

Chromosome bridges (identified by arrows)

chromosome map *See* cytogenetic map, genetic map.

chromosome painting a variation of fluorescence *in situ* hybridization (FISH) (*q.v.*) technique using fluorescent dye-tagged DNA segments that can hybridize to numerous sites along each human chromosome. Because there are too few known fluorescent dyes to distinctively mark each human chromosome, a combinatorial approach is taken. Since the total combinations given by a number of dyes (N) is $2^N - 1$, as few as five dyes can give enough combinations to produce probes that will individually label each chromosome. The fluorochrome colors cannot be distinguished by the unaided eye but can be detected by using a series of filters or an interferometer linked with computers to analyze the various dye combinations assigned to each chromosome. Chromosome painting is also known as multicolor fluorescence *in situ* hybridization (MFISH). Cross species-chromosomal painting and digital imaging are now commonly used in studies of the evolution of mammalian karyotypes. *See* Appendix C, 1997, Yang *et al.*

chromosome polymorphism the presence in the same interbreeding population of one or more chromosomes in two or more alternative structural forms.

chromosome puff *See* chromosomal puff.

chromosome rearrangement a chromosomal aberration involving new juxtapositions of chromosomal segments; e.g., inversions, translocations.

chromosome scaffold when histones are removed from isolated metaphase chromosomes and these are centrifuged onto electron microscope grids, extremely long loops of DNA can be seen to project from an irregular mass (the scaffold) whose dimensions are similar to the original intact chromosome.

chromosome set a group of chromosomes representing a genome (*q.v.*), consisting of one representative from each of the pairs characteristic of the somatic cells in a diploid species. *See* polyploidy.

chromosome sorting *See* flow cytometry.

chromosome substitution replacement by a suitable crossing program of one or more chromosomes by homologous or homoeologous chromosomes from another source. This may be a different strain of the same species or a related species that allows hybridization. *See* homoeologous chromosomes, homologous chromosomes.

chromosome theory of heredity the theory put forth by W. S. Sutton in 1902 that chromosomes are the carriers of genes and that their meiotic behavior is the basis for Mendel's laws (*q.v.*).

chromosome walking the sequential isolation of clones carrying overlapping restriction fragments to span a segment of chromosome that is larger than can be carried in a phage or a cosmic vector. The technique is generally needed to isolate a locus of interest for which no probe is available but that is known to be linked to a gene that has been identified and cloned. This probe is used to screen a genome library. As a result, all fragments containing the marker gene can be selected and sequenced. The fragments are then aligned, and those segments farthest from the marker gene in both directions are subcloned for the next step. These probes are used to rescreen the genome library to select new collections of overlapping sequences. As the process is repeated, the nucleotide sequences of areas farther and farther away from the marker gene are identified, and eventually the locus of interest will be encountered. If a chromosomal aberration is available that shifts a particular gene that can serve as a molecular marker to another position on the chromosome or to another chromosome, then the chromosome walk can be shifted to another position in the genome. The use of chromosome aberrations in experiments of this type is referred to as *chromosome jumping*. *See* Appendix C, 1978, Bender, Spierer, and Hogness.

chromotrope a substance capable of altering the color of a metachromatic dye (*q.v.*).

chronic exposure radiation exposure of long duration. Applied to experimental conditions in which the organism is given either a continuous low level exposure or a fractionated dose.

chronocline in paleontology, a character gradient in the time dimension.

chronospecies a species that can be studied from its fossil remains through a defined period of time.

chrysalis the pupa of a lepidopteran that makes no cocoon (*q.v.*).

chrysomonads fresh water members of the Chromista (*q.v.*). They are called golden algae because their cytoplasm contains yellow plastids. An example is any species belonging to the genus *Ochromonas*. Each cell has two undulipodia attached to its anterior pole. One is long and hairy, and the one behind it is shorter with fewer lateral fibers. *Ochromonas* also contains a chloroplast ER.

chymotrypsin a proteolytic enzyme from the pancreas that hydrolyzes peptide chains internally at peptide bonds on the carboxyl side of various amino acids, especially phenylalanine, tyrosine, and tryptophan.

Ci abbreviation for curie (*q.v.*).

cichlid fishes cichlids (pronounced 'sick-lid') are a family of freshwater fishes distributed throughout the tropics and subtropics. There are about 1,300 species described so far, but almost 1,000 are found in the rift lakes of East Africa (Lakes Victoria, Malawi, and Tanganyika). Speciation in the Rift Lake cichlids has been extremely rapid and has resulted in part from sexual selection (*q.v.*). *See* Appendix A, Chordata, Osteichythes, Neopterygii, Perciformes.

cilia (*singular*, **cilium**) populations of thin, motile processes found covering the surface of ciliates or the free surface of the cells making up a ciliated epithelium. Each cilium arises from a basal granule in the superficial layer of cytoplasm. The movement of cilia propels ciliates through the liquid in which they live. The movement of cilia on a ciliated epithelium serves to propel a surface layer of mucus or fluid. *See* axoneme.

ciliate a protozoan belonging to the phylum Ciliophora. *See* Appendix A, Protoctista.

Ciona intestinalis a species of ascidians, commonly called sea squirts. As adults they are hermaphrodites that live as filter feeders in shallow seas. They are sometimes called tunicates because each is enclosed in a tough tunic made up of tunicin, a cellulose-like fiber. Fertilized eggs develop into free-swimming larvae, and each tadpole has a prominent notochord and dorsal nerve tube. For this reason, they are classified as Urochordates, a primitive branch of the Acraniata (*q.v.*). *Ciona* has 14 chromosomes and a genome size of about 156 mbp. Its 15,900 ORFs are more closely packed than those of most vertebrates. About 1/6 of the *Ciona* genes have vertebrate homologs. The *Ciona* homologs are generally present in only a few copies compared to the multiple copies found in vertebrate gene families (i.e., actin genes or myosin genes). *See* Appendix A, Eumetazoa, Deuterostomia, Chordata, Acraniata, Urochordata; Appendix C, 2002, Dehal *et al.*; isoforms, tunicin.

circadian rhythm an oscillation in the biochemistry, physiology, or behavior of an organism that has a natural period of exactly 24 hours. If the organism is placed under constant conditions, the rhythm will continue with a period that is close to, but not exactly, 24 hours. For example, in *Neurospora* the number of hours between peaks of conidiation is 21.5 hours under constant conditions. This endogenous periodicity can be reset to 24 hours by a change in light level (e.g., a shift from constant darkness to a 12-hour light: 12-hour dark cycle). *See* clock mutants, entrainment, *frequency*.

circular dichroism the property of molecules to show differences in absorption between the clockwise and counterclockwise component vectors of a beam of circularly polarized light. Since helical molecules in solution often exhibit these properties, circular dichroism spectra have been used to study coiling changes of physiological significance, which various proteins can undergo. In the cases of chromatin fragments oriented in electric fields and exposed to circularly polarized ultraviolet light, measurements of circular dichroism allow conclusions to be drawn as to the way nucleosomes (*q.v.*) are stacked in 30-nanometer fibers.

circular linkage map the linkage map characteristic of *Escherichia coli*. In preparation for genetic transfer during conjugation, the ring-shaped chromosome of an *Hfr* bacterium breaks in such a way that when the ring opens the *F* factor is left attached to the region of the chromosome destined to enter the F^- cell last. Circular linkage maps have been constructed for several other bacteria and for certain viruses. *See also* linkage map.

circularly permuted sequences *See* cyclically permuted sequences.

circular overlap the phenomenon in which a chain of continuous and intergrading populations of one species curves back until the terminal links overlap each other. Individuals from the terminal populations are then found to be reproductively isolated from each other; that is, they behave as if they belonged to separate species. A ring of races so formed is referred to as a *Rassenkreis*.

circumsporozoite protein *See* sporozoite.

***cis*-acting locus** a genetic region affecting the activity of genes on that same DNA molecule. *Cis*-acting loci generally do not encode proteins, but rather serve as attachment sites for DNA-binding proteins. Enhancers, operators, and promoters are examples of *cis*-acting loci. *Contrast with trans*-acting locus.

***cis* dominance** the ability of a genetic locus to influence the expression of one or more adjacent loci in the same chromosome, as occurs in *lac* operator mutants of *E. coli*.

cis face *See* Golgi apparatus.

cisplatin one of the most widely used antitumor drugs, especially effective for the management of testicular and ovarian cancers. When cisplatin binds to DNA, it loses two chloride ions and forms two platinum-nitrogen bonds with the N7 atoms of adjacent guanines on the same strand. This localized disruption of the double helix inhibits replication.

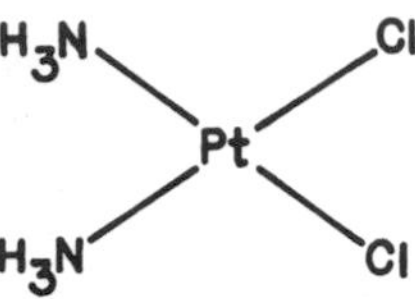

cis-splicing the splicing together of the exons within the same gene that occurs during the modification of the primary RNA transcript in the nucleus. *Compare with* trans-splicing and *see* posttranscriptional processing.

cisterna a flattened, fluid-filled reservoir enclosed by a membrane. *See* endoplasmic reticulum.

***cis, trans* configurations** terminology that is currently used in the description of pseudoallelism. In the *cis* configuration both mutant recons are on one homolog and both wild-type recons are on the other ($a^1a^2/{+}{+}$). The phenotype observed is wild type. In the *trans* configuration each homolog has a mutant and a nonmutant recon ($a^1{+}/{+}a^2$), and the mutant phenotype is observed. In the case of pseudoallelic genes, the terms *cis* and *trans* configurations correspond to the *coupling* and *repulsion* terminology used to refer to nonallelic genes. *See* transvection effect.

***cis-trans* test** a test used to determine whether two mutations of independent origin affecting the same character lie within the same or different cistrons. If the two mutants in the *trans* position yield the mutant phenotype, they are alleles. If they yield the wild phenotype, they represent mutations of different cistrons. However, different mutated alleles may represent cistrons with mutations at different sites. If these mutons are separable by crossing over, it is possible to construct a double mutant with the mutant sites in the *cis* configuration ($m^1m^2/{+}{+}$). Individuals of this genotype show the wild phenotype. *See* pseudoalleles.

cistron originally the term referred to the DNA segment that specified the formation of a specific polypeptide chain (*see* Appendix C, 1955, Benzer). The definition was subsequently expanded to include the transcriptional start and stop signals. In cases where an mRNA encodes two or more proteins, it is referred to as *polycistronic*. The proteins specified by a polycistron are often enzymes that function in the same metabolic pathway. *See* gene, transcription unit.

citrate cycle a synonym for citric acid cycle (*q.v.*).

citric acid cycle a cycle of enzyme-controlled reactions by which the acetyl-coenzyme A produced by the catabolism of fats, proteins, and carbohydrates is oxidized, and the energy released used to form ATP from ADP. The condensation of acetyl-coenzyme A with a four-carbon compound (oxaloacetic acid) yields the six-carbon compound citric acid for which the cycle is named. Through a further series of oxidative decarboxylations, citric acid is broken down to oxaloacetic acid, and so the cycle is completed. As a result, 1 molecule of activated acetate is converted to 2 CO_2 molecules, 8 H atoms are taken up and oxidized to water, and 12 molecules of ATP are formed concurrently. The enzymes are

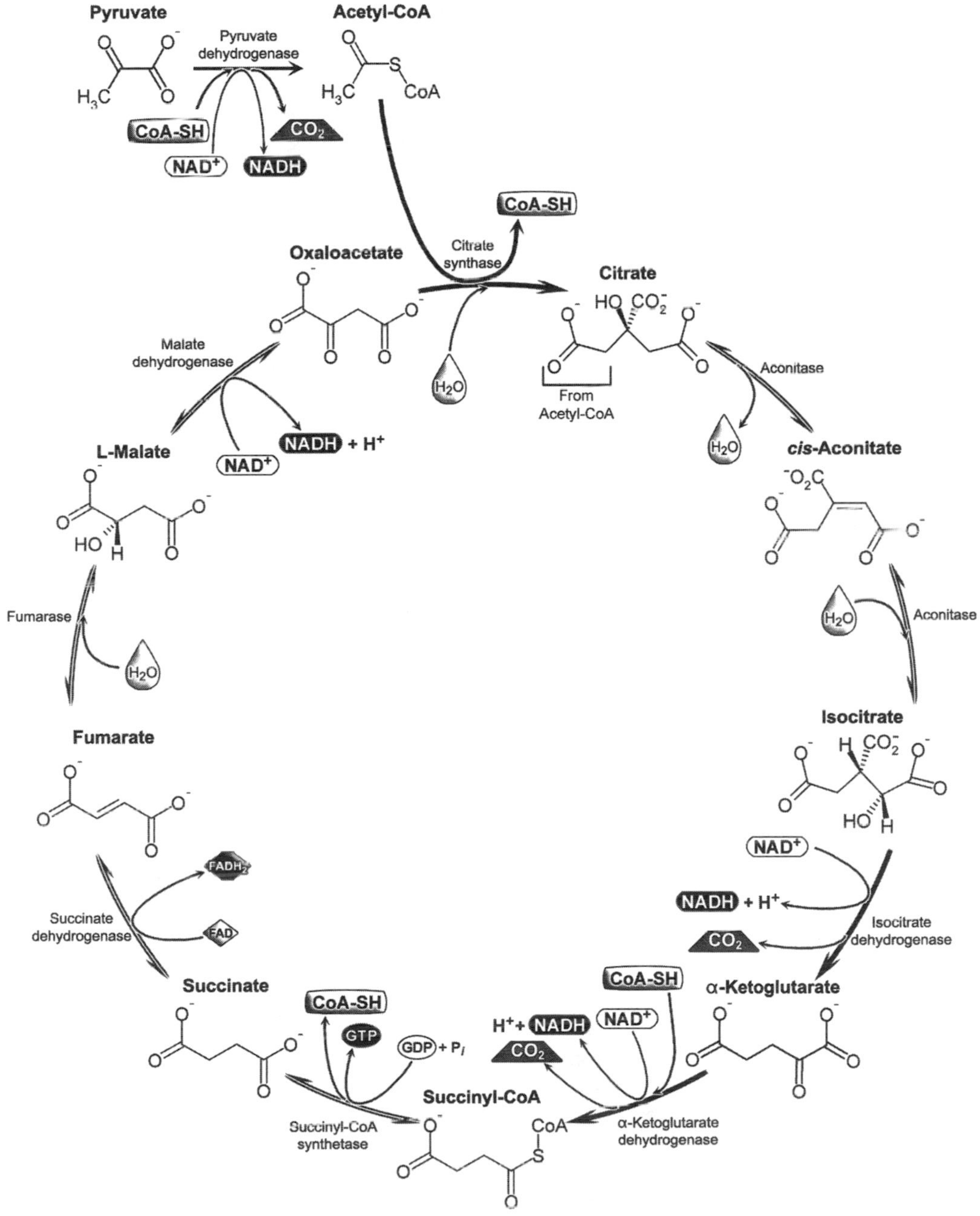

Citric acid cycle

localized in mitochondria (*q.v.*). The cycle was invented by prokaryotes about 2.5 billion years ago, and today virtually all organisms that use oxygen employ the citric acid cycle. The cycle is illustrated in the diagram on page 85. The sulfur atom in coenzyme A (*q.v.*) is the site where chemical groups are exchanged at various stages in the cycle. This is the reason for adding an S to the CoA-SH abbreviation. *See* Appendix C, 1937, Krebs; adenosine phosphate, ATP synthase, cytochrome system, glycolysis, glyoxylate cycle, nicotine-adenine dinucleotide (NAD).

citrullinuria a hereditary disease in humans arising from a deficiency of the enzyme arginosuccinate synthetase. It is caused by a recessive gene on autosome 9. *See* ornithine cycle.

Citrus a genus of fruit-bearing trees including C. *aurantium*, the Seville orange; C. *aurantifolia*, the lime; C. *paradisi*, the grapefruit; C. *limon*, the lemon; C. *reticulata*, the tangerine; C. *sinensis*, the sweet orange. The navel orange is a cultivar (*q.v.*) of the last-named species. *See* Appendix A, Plantae, Angiospermae, Dicotyledoneae, Sapindales.

clade 1. in classification, any group of organisms that is defined by characters exclusive to all its members and that distinguish the group from all others. 2. in evolutionary studies, a taxon or other group consisting of a single species and its descendents; a holophyletic group; a set of species representing a distinct branch on a phylogenetic tree. Graphically a clade includes the species represented by the node and all branches that spring from it. *See* cladogram, PhyloCode.

cladistic evolution splitting of a line of descent into two species. *Contrast with* phyletic evolution.

cladistics a method of classification that attempts to uncover the genealogic relationships between any two species by stressing only the advanced characteristics that they share. The graphic representation of the results of such an analysis is the cladogram (*q.v.*). *See* Appendix C, 1950, Hennig; apomorphic.

cladogenesis branching evolution; the splitting of a lineage into two or more lineages. *Contrast with* orthogenesis.

cladogram a branching diagram that displays the relationship between taxa in terms of their shared character states and attempts to represent the true evolutionary branchings of the lineage during its evolution from the ancestral taxon. In the accompanying cladogram, we start with ancestral species A, which is now extinct. It is characterized by six phenotypic characters (a–f) of taxonomic importance. The lineage undergoes speciation events (denoted by numbers 1 to 4) to produce five extant species, B–F. Each character (except f) undergoes a single change from the ancestral (plesiomorphic) state to a derived (apomorphic) state. The time and place of the change in the lineage is marked by letters carrying primes. Note that species E and F, which have the most recent common ancestor, share more apomorphic characters than species with more remote common ancestors. An ancestral character shared by several species is called symplesiomorphic, and a derived character shared by two or more species is called synapomorphic. Thus, character f is symplesiomorphic for the entire group, and character c is synapomorphic for species E and F. *See* node.

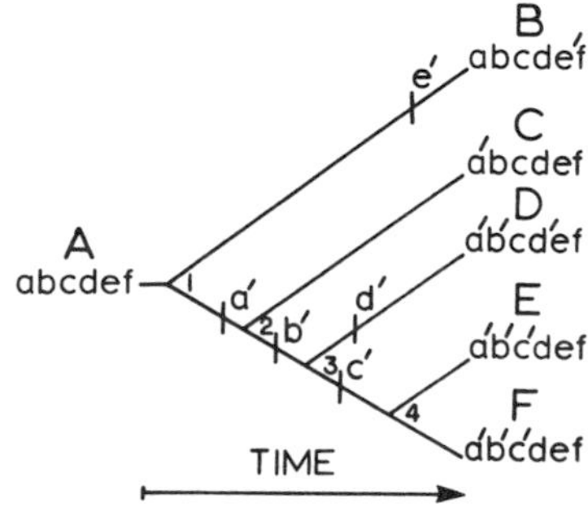

class a grouping used in the classification of organisms. It is a subdivision of a phylum, and it is in turn subdivided into orders.

classification 1. the process of grouping organisms on the basis of features they have in common, or on the basis of their ancestry, or both. 2. the resulting arrangement of living things into groups as a consequence of the above process. The traditional division of living organisms into two kingdoms (Plant and Animal) dates back to Linnaeus. This was supplanted by a system of five kingdoms (Bacteria, Protists, Fungi, Plants, and Animals) proposed by Whittaker. An updated variant of this system is used in Appendix A. More recently, a seven-kingdom system has been proposed by T. Cavalier-Smith. This system retains four of the five kingdoms but splits the protists into three separate kingdoms: Archaeozoa, Protozoa, and Chromista (*all of which see*).

class switching *See* heavy chain class switching.

clathrin *See* receptor-mediated endocytosis.

***ClB* technique** a technique invented by H. J. Muller and used for the detection of sex-linked lethal and viable mutations in *Drosophila melanogaster*. The name is derived from the X chromosome used.

It contains a Crossover suppressor (an inversion), a *l*ethal, and the dominant marker *B*ar eye.

cleavage the processes by which a dividing egg cell gives rise to all the cells of the organism. Such cleavages involve rapid cell divisions without intervening periods of cell growth. Therefore the embryo remains the same size, but the component cells get smaller and smaller. Cleavage in some species follows a definite pattern, where it is said to be determinate (permitting the tracing of *cell lineages*); in other species the pattern is lost after the first few cell divisions. *See* contractile ring, discoidal cleavage, radial cleavage, spiral cleavage.

cleavage furrow *See* contractile ring.

cleidoic egg an egg enclosed within a shell that is permeable only to gases.

cleistogamous designating a plant in which fertilization occurs within closed flowers, and therefore where self-pollination is obligatory. *See* chasmogamous.

cline a gradient of phenotypic and/or gene frequency change along a geographical transect of the population's range. *See* isophene.

clock mutants mutants that disrupt the normal 24-hour circadian cycle of an organism. The gene *period* (*q.v.*) in *Drosophila* and *frequency* (*q.v.*) in *Neurospora* are the best-studied clock mutants.

clomiphene any chemical that stimulates ovaries to release eggs.

clonal analysis the use of mosaics generated by genetic methods or surgical operations to investigate cell-autonomous and cell-nonautonomous developmental processes. Cell-autonomous genetic markers are used to label cells at an early stage in development, and the subsequent morphological fates of the offspring of these cells can then be followed. *See* compartmentalization.

clonal selection theory a hypothesis for explaining the specific nature of the immune response in which the diversity among various cells for the recognition of specific antigens exists prior to their exposure to the antigen. Subsequent exposure to a particular antigen causes the appropriate cells to undergo a clonal proliferation. *See* Appendix C, 1955, Jerne; 1959, Burnet; B lymphocyte.

clone 1. a group of genetically identical cells or organisms all descended from a single common ancestral cell or organism by mitosis in eukaryotes or by binary fission in prokaryotes. 2. genetically engineered replicas of DNA sequences.

cloned DNA any DNA fragment that passively replicates in the host organism after it has been joined to a cloning vector. Also called *passenger DNA*. *See* recombinant DNA technology.

cloned library a collection of cloned DNA sequences representative of the genome of the organism under study.

cloning in its most common usage, creating an organism genetically identical to another organism. This occurs naturally in the production of monozygotic twins. The nine-banded armadillo regularly produces a clone of four genetically identical offspring at each birth cycle. This type of cloning can also be done artifically by embryo splitting. For example, the eight-cell embryo of many animals can be divided in halves or quarters, and each of these portions can develop into identical twins or quadruplets. Mitotically produced cells are (barring mutations) genetically identical, and thus all the somatic cells of an individual are technically a clone. Likewise, bacterial cells produced by binary fission from a single parent cell also form a clone (or produce a colony on solid medium). A single individual or cell may be referred to as a clone, but only with reference to at least one other genetically identical individual or cell. Most plant cells are totipotent and usually are easier to clone than animal cells through asexual propagation (grafting, budding, cuttings, etc.). In mammals, sheep, cows, goats, horses, mules, pigs, rabbits, rats, mice, cats, and dogs have been cloned experimentally. However, the rate of success in such experiments has been low. *See* Appendix C, 1958, Steward, Mapes, and Mears; 1997, Wilmut *et al.*; *Dasypus*, Dolly, gene cloning, mosaic, nuclear reprogramming, nuclear transfer, sheep, therapeutic cloning, totipotency, twins.

cloning vector, cloning vehicle *See* bacterial artificial chromosomes, cosmid, DNA vector, lambda cloning vector, P1 artificial chromosomes, pBR322, plasmid cloning vector, yeast artificial chromosome.

clonogenicity the ability of cells from multicellular organisms (e.g., malignant cells or stem cells) to proliferate and form colonies (or clones), each consisting of one or more cell types. A cell with this ability is called a *clonogenic cell.*

clonotype the phenotype or homogenous product of a clone of cells.

close pollination the pollination of a flower with pollen of another flower on the same plant.

Clostridium botulinum an anaerobic, spore-forming, Gram-positive bacterium that is often present in decaying food and can cause a severe type of food poisoning called *botulism* (*q.v.*). *See* Appendix A, Bacteria, Endospora.

club wheat *Triticum compactum* (N = 21). *See* wheat.

clutch a nest of eggs.

cM centimorgan. *See* Morgan unit.

C-meiosis colchicine-blocked meiosis. *See* metaphase arrest.

C-metaphase colchicine-blocked metaphase. *See* metaphase arrest.

C-mitosis colchicine-blocked mitosis. *See* metaphase arrest.

c-myc *See myc.*

Cnidosporidia *See* microsporidia.

CNS abbreviation for central nervous system (the brain and spinal cord).

coadaptation the selection process that tends to accumulate harmoniously interacting genes in the gene pool of a population.

coated pit, vesicle *See* receptor-mediated endocytosis.

coat protein the structural protein making up the external covering of a virus.

cobalamin vitamin B_{12}, a coenzyme for a variety of enzymes. It is essential for deoxyribose synthesis. The vitamin is characterized by a tetrapyrole ring system similar to that of heme (*q.v.*). The molecule contains a central cobalt atom linked to four pyrrole nitrogens. The vitamin is unique in being synthesized only by microorganisms. *See* Appendix C, 1964, Hodgkin.

cobalt a biological trace element. Atomic number 27; atomic weight 58.9332; valence 2, 3^+; most abundant isotope Co^{59}; radioisotope Co^{60}, half-life 5.2 years, radiation emitted—beta and gamma rays (used extensively as a gamma-ray source).

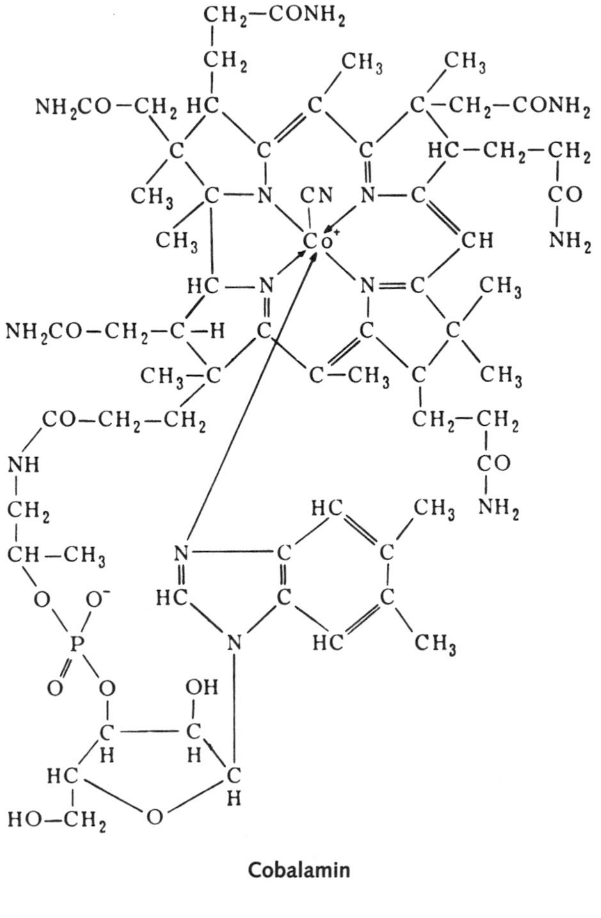

Cobalamin

coconversion the concurrent correction of two sites during gene conversion.

cocoon 1. a covering of silky strands secreted by the larva of many insects in which it develops as a pupa. Commercial silk is derived from the cocoon of *Bombyx*. 2. a protective covering of mucus secreted about the eggs of some species.

code for the genetic code dictionary, *see* amino acid, start codon, stop codon.

code degeneracy *See* degenerate code.

coding strand *See* strand terminologies.

coding triplet codon. *See* amino acid.

CODIS the acronym for the Combined DNA Index System operated by the Federal Bureau of Investigation. This is a national DNA databank which utilizes computer and DNA technologies to match DNA profiles developed from crime scene evidence against the profiles of individuals convicted of felony sex offenses and other violent crimes. The profiles stored in CODIS are generated using STR analysis (*q.v.*). All 50 states have laws requiring that DNA profiles of violent criminals be sent to CODIS, where they are stored in its Convicted Offender Index.

codominant designating genes when both alleles of a pair are fully expressed in the heterozygote. For example, the human being of AB blood group is showing the phenotypic effect of both I^A and I^B codominant genes. *See* semidominance.

codon (*also* coding triplet) the nucleotide triplet in messenger RNA that specifies the amino acid to be inserted in a specific position in the forming polypeptide during translation. A complementary codon resides in the structural gene specifying the mRNA in question. The codon designations are listed in the amino acid and genetic code entries. *See* start codon; synonymous codons, termination codon.

codon bias the nonrandom use of synonymous codons. Such biases have been observed in genes from both prokaryotes and eukaryotes. For example, 90% of the leucines in the outer membrane protein of *E. coli* are encoded by one of the six synonymous leucine codons. There seems to be a positive correlation between the relative frequencies of the synonymous codons in the genes of a given species and the relative abundances of their cognate tRNAs. So codon usage may be determined by translational efficiency. *See* genetic code, transfer RNA.

coefficient of coincidence *See* coincidence.

coefficient of consanguinity the probability that two homologous genes drawn at random, one from each of the two parents, will be identical and therefore be homozygous in an offspring. The inbreeding coefficient of an individual is the same as the coefficient of consanguinity of its parents. *See* Wright's inbreeding coefficient.

coefficient of inbreeding Wright's inbreeding coefficient.

coefficient of kinship coefficient of consanguinity (*q.v.*).

coefficient of parentage coefficient of consanguinity (*q.v.*).

coefficient of relationship (r) the proportion of alleles in any two individuals that are replicas inherited from a common ancestor.

coefficient of selection (s) *See* selection coefficient.

coelacanth *See* living fossil.

coelenterate referring to a marine organism characterized by radial symmetry, a simple diploblastic body, a gastrovascular cavity, and usually showing metagenesis (*q.v.*). *See* Appendix A, Animalia, Eumetazoa, Radiata, Cnidaria.

coelomate referring to animals having a coelom or body cavity formed in and surrounded by mesoderm. *See* Appendix A, Animalia, Eumatozoa, Bilateria, Protostomia, Coelomata.

coenobium a colony of unicellular eukaryotes surrounded by a common membrane.

coenzyme an organic molecule that must be loosely associated with a given enzyme in order for it to function. Generally, the coenzyme acts as a donor or acceptor of groups of atoms that have been added or removed from the substrate. Well-known coenzymes include NAD, NADP, ATP, FAD, FMN, and coenzyme A. Some coenzymes are derivatives of vitamins.

coenzyme A a coenzyme made up of adenosine diphosphate, pantothenic acid, and mercaptoethylamine. The acetylated form of the coenzyme plays a central role in the citric acid cycle (*q.v.*). *See* formula illustrated on page 90.

coenzyme Q a molecule (illustrated on page 90) that functions as a hydrogen acceptor and donor in the electron transport chain (*q.v.*). *See* cytochrome system (Q and QH_2)

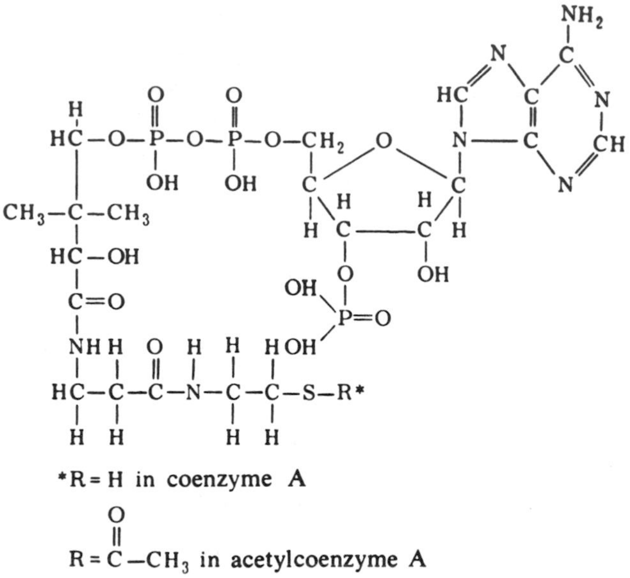

*R = H in coenzyme A

$R = \overset{O}{\overset{\|}{C}}-CH_3$ in acetylcoenzyme A

Coenzyme A

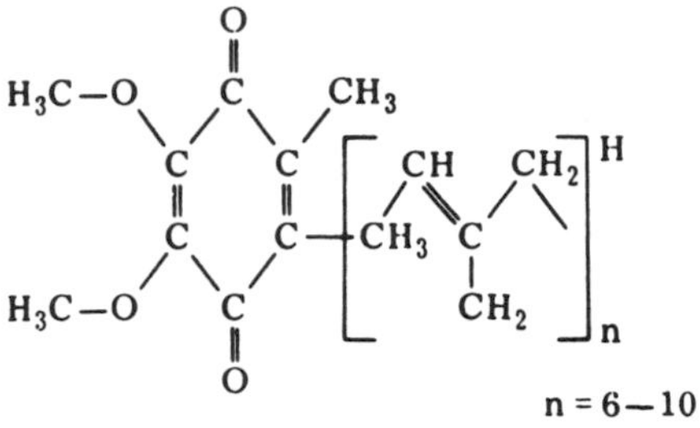

n = 6—10

Coenzyme Q

coevolution the evolution of one or more species in synchrony with another species as a consequence of their interdependence. Such a reciprocal adaptive evolution determines the patterns of host-plant utilization by insects. Also, parasites often evolve and speciate in harmony with their hosts. Even parasitic DNA molecules—that is, lysogenic viruses and transposable elements (*q.v.*)—coevolve with their hosts to prevent serious disruptions in the gene activities of the host that could reduce its fitness. *See* Fahrenholz's rule, gene-for-gene hypothesis.

cofactor a factor such as a coenzyme or a metallic ion required in addition to a protein enzyme for a given reaction to take place.

cognate tRNAs transfer RNA molecules that can be recognized by aminoacyl-tRNA synthetase enzymes.

cohesin a multi-subunit complex first identified in *Saccharomyces cerevisiae* that is essential for holding sister chromatids together from the time of DNA replication until the onset of anaphase. Cohesin is assembled from four subunits (Smc1, Smc3, Scc1, and Scc3). Loss of function mutations in any of the genes that encode these proteins results in precocious dissociation of sister chromatids. There are high concentrations of cohesin in the pericentric regions of chromosomes. The size of the pericentric cohesin-rich domains is the same for all chromosomes. Cohesin is also found in intergenic regions of the arms of yeast chromosomes at intervals of about 11 kb. The chromosomal regions to which cohesin complexes bind contain DNA sequences that belong to the Alu family (*q.v.*). During the anaphase stage of mitosis, the cohesin complexes joining the chromatids are cleaved by separase (*q.v.*). Therefore the chromatids become independent chromosomes and can move to the opposite poles of the dividing cell. During anaphase I of meiosis, the separase enzymes cleave the cohesins in the arms of the chromatids, so they are free to separate. However, the cohesins loaded on the centromeric DNA are protected by Sgo, and so the sister chromatids remain tethered at their centromeres. Sgo is degraded during telophase I, and therefore the centromeric cohesins are cleaved during anaphase II. Now the sister chromatids are independent chromosomes and are drawn to opposite poles of the spindle. *See* meiosis, *Sgo*.

cohesive-end ligation the use of DNA ligase to join double-stranded DNA molecules with complementary cohesive termini that base pair with one another and bring together 3′-OH and 5′-P termini. *Compare with* blunt-end ligation.

cohesive ends *See* restriction endonuclease, sticky ends.

cohort a group of individuals of similar age within a population.

coilin a protein that is concentrated in Cajal bodies (*q.v.*) and serves as a cytological marker for this organelle. Coilin shuttles between the nucleus and cytoplasm and may function as part of a transport system among the cytoplasm, Cajal bodies, and nucleoli.

coincidence, coefficient of an experimental value equal to the observed number of double crossovers divided by the expected number.

coincidental evolution *See* concerted evolution.

cointegrate structure the circular molecule formed by fusing two replicons, one possessing a transposon, the other lacking it. The cointegrate structure has two copies of the transposon located at both replicon junctions, oriented as direct repeats. The formation of a cointegrate structure is thought to be an obligatory intermediate in the transposition process. The donor molecule (containing the transposon) is nicked in opposite strands at the ends of the transposon by a site-specific enzyme. The recipient molecule is nicked at staggered sites. Donor and recipient strands are then ligated at the nicks. Each end of the transposon is connected to one of the single strands protruding from the target site, thereby generating two replication forks. When replication is completed, a cointegrate structure is formed, which contains two copies of the transposon oriented as direct repeats. An enzyme required in the formation of the cointegrate structure is called a *transposase*. The cointegrate can be separated into donor and recipient units each of which contains a copy of the transposon. This process is called *resolution* of the cointegrate, and it is accomplished by recombination between the transposon copies. The enzyme involved in resolution is called a *resolvase*.

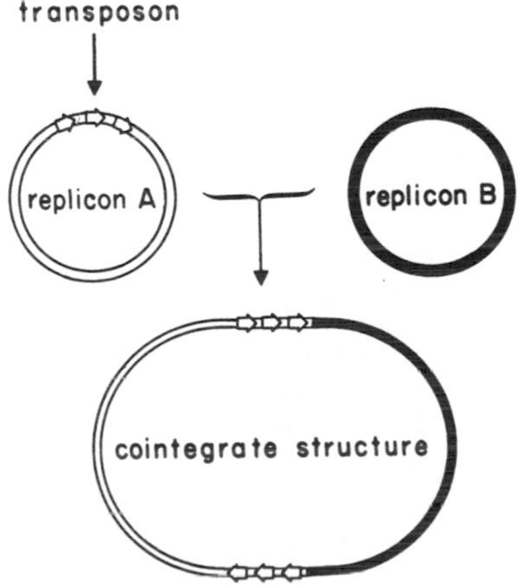

coisogenic designating inbred strains of organisms that differ from one another only by a single gene as the result of mutation. *Contrast with* **congenic strains.**

coitus copulation; sexual intercourse in vertebrates.

Colcemid trademark for a colchicine derivative extensively used as a mitotic poison.

colchicine an alkaloid that inhibits the formation of the spindle and delays the division of centromeres. Colchicine is used to produce polyploid varieties of horticulturally important species. It is also used in medicine in the treatment of gout. Also used to stop mitosis at metaphase (when chromosomes are maximally condensed) for preparation of karyotypes. *See* Appendix C, 1937, Blakeslee and Avery; *Colchicum*, haploid sporocytes, paclitaxel.

```
                 H
                 |
  CH3O           C   CH2 — CH2
      \        //  \ |       |
       C           C        CH — R
       |           ||        |
       C           C         C
      / \\        / \       / \\
  CH3O    C      C          CH
          |      ||          |
    CH3 — O      HC         C = O
                 |           |
                 HC ====  C — O — CH3
```

R = $NHCOCH_3$ for colchicine.

R = $NHCH_3$ for Colcemid.

Colchicum the genus of crocuses, including: C. *autumnale*, the autumn crocus, source of colchicine; C. *aureus*, the Dutch crocus; C. *sativus*, the Saffron crocus.

cold-sensitive mutant a gene that is defective at low temperature but functional at normal temperature.

coleoptile the first leaf formed during the germination of monocotyledons.

***col* factors** bacterial plasmids that allow the cell to produce colicins (*q.v.*).

colicin any of a group of proteins produced by certain strains of *E. coli* and related species that have bactericidal effects. Colicinogenic bacteria are immune to the lethal effects of their own colicins.

coliform designating a Gram-negative, lactose-fermenting rod related to *E. coli.*

colinearity correspondence between the location of mutant sites within a bacterial cistron and the location of amino acid substitutions in its translational product. This correspondence is not complete in most eukaryotic genes because of the presence of nontranslated introns (*q.v.*). *See* Appendix C, 1964, Sarabhai *et al.*, Yanofsky *et al.*

coliphage a bacteriophage (*q.v.*) that parasitizes *E. coli. See* lambda phage.

collagen the most abundant of all proteins in mammals; the major fibrous element of skin, bone, tendon, cartilage, and teeth, representing one-quarter of the body's protein. The protein consists of a triple helix 3,000 Ångstroms long and 15 Ångstroms across. Five types of collagen are known, differing in amino acid sequence of the three polypeptide chains. In some molecules of collagen, the three chains are identical; in others, two of the chains are identical and the third contains a different amino acid sequence. The individual polypeptide chains are translated as longer precursors to which hydroxyl groups and sugars are attached. A triple helix is formed and secreted into the space between cells. Specific enzymes trim the ends of each helix. Type 5 collagen is a minor collagen component present in adult skin. It is encoded by the *COL5A1* gene located on human chromosome 9 at q34. The gene contains at least 750 kb of DNA and is made up of 66 exons. Mutations in *COL5A1* cause one form of Ehlers–Danlos syndrome (*q.v.*). Mutations in collagen genes also cause hereditary blistering diseases. *See* cartilage, dominant negative mutation, epidermolysis bullosa. *Compare with* keratin.

collagenase an enzyme that digests collagen.

collenchyma plant tissue composed of cells that fit closely together and have thickened walls; in metazoans, undifferentiated mesenchyme cells lying in a gelatinous matrix.

collision, inelastic the event occurring when a moving electron passes close to an atom. There is repulsion between the moving electron and an atomic electron sufficient to dislodge the latter from its nucleus.

colony in bacteriology, a contiguous group of single cells derived from a single ancestor and growing on a solid surface.

colony bank *See* gene library.

colony hybridization an *in situ* hybridization (*q.v.*) technique used to identify bacteria carrying chimeric vectors whose inserted DNA is homologous with the sequence in question. Colony hybridization is accomplished by transferring bacteria from a petri plate to a nitrocellulose filter. The colonies on the filter are then lysed, and the liberated DNA is fixed to the filter by raising the temperature to 80°C. After hybridization with a labeled probe, the position of the colonies containing the sequence under study is determined by autoradiography (*q.v.*). *See* Appendix C, 1975, Grunstein and Hogness.

color blindness defective color vision in humans due to an absence or reduced amount of one of the three visual pigments. The pigments *chlorolabe, erythrolabe,* and *cyanolabe* absorb green, red, and blue light, respectively. The pigments are made of three different opsins combined with vitamin A aldehyde. The green-blind individual suffers from *deuteranopia* and cannot make chlorolabe. The red-blind individual suffers from *protanopia* and cannot make erythrolabe. The blue-blind individual suffers *tritanopia* and cannot make cyanolabe. *Deuteranomaly, protanomaly* and *tritanomaly* are conditions caused by reduced amounts of the visual pigments in question, rather than their complete absence. The *protan* and *deutan* genes for red and green blindness, respectively, occupy different loci on the X chromosome. Blue blindness is quite rare and due to an autosomal gene. The famous physicist John Dalton was color blind, and he gave the earliest account of this condition. For this reason, protanopia is sometimes called *daltonism. See* cone pigment genes (CPGs), dalton, opsin, retina.

Colton blood group antigens produced by a gene at human chromosome 7p14. The gene product belongs to the aquaporin (*q.v.*) family of water channel proteins. For this reason, the gene is symbolized *AQP1*. The gene has been cloned and is 13 kb long and contains 4 exons. *See* Appendix C, Preston and Agre.

Columba livia the domesticated pigeon. In nature the species is called the rock dove. Its haploid chromosome number is 40, and its C value is 1.46 pg of DNA. Charles Darwin bred pigeons, and in his *Origin of Species* described the numerous variations that had been selected by pigeon fanciers. There are currently at least 200 breeds, including those selected for homing and racing. *See* Appendix A, Aves, Columbiformes.

column chromatography the separation of organic compounds by percolating a liquid containing the compounds through a porous material in a cylinder. The porous material may be an ion exchange resin. *See* chromatography.

combinatorial association within a pool of immunoglobulin molecules, the association of molecules from any class of heavy chain with molecules from any type of light chain. Within a given immunoglobulin molecule, however, there is only one class of heavy and one type of light chain. *See* immunoglobulin.

combinatorial translocation for an immunoglobulin chain (heavy or light), the association of any variable region gene with any constant region gene within that same multigene family (*q.v.*). The two genes are brought together by interstitial translocation that may involve deletion of intervening genetic material. *See* immunoglobulin.

combining ability 1. *general:* average performance of a strain in a series of crosses. 2. *specific:* deviation of a strain's performance in a given cross from that predicted on the basis of its general combining ability.

comb shape a character influenced by two nonallelic gene pairs, *Rr* and *Pp*, in the chicken. One of the early examples of gene interaction (*q.v.*). *See* *Gallus domesticus*.

commaless genetic code successive codons that are contiguous and not separated by noncoding bases or groups of bases. Bacterial genes do not have introns (*q.v.*), and therefore the sequences of amino acids in polypeptides and of codons in the gene are colinear. Most eukaryotic genes contain coding regions called exons (specifying amino acids) interrupted by noncoding regions called introns, and in these situations the code is said to contain commas.

commensalism the association of organisms of different species without either receiving benefits essential or highly significant to survival. The species may live in the same shell or burrow, or one may be attached to or live within the other. *See* parasitism, symbiosis.

community interacting populations of individuals belonging to different species and occupying a given region or a distinctive range of habitat conditions.

community genome sequencing the simultaneous sequencing of the genomes present in a specified community of microorganisms. A sample is taken of the species living in specific habitat, and the DNA isolated from that sample is subjected to shotgun sequencing (*q.v.*). Advanced computer programs then allow the sequenced fragments to be assembled into groupings that represent the genomes of the most common species in the sample. About 99% of all known bacteria have never been cultured in laboratories, and genomes have been constructed for only a small subset of the cultured species. Community genome sequencing allows the construction of genomes for species that have never been cultured. Such a study was made of samples from a biofilm in a California mine. The film forms a scum on the surface of a pool that has formed in a shaft that extends 1,400 feet into Iron Mountain. The water is hot (106°F), very acidic, and rich in toxic metals. There is little light or oxygen. Eighty mbp of DNA was recovered from a sample of this biofilm and assembled into five genomes. The two most complete genomes belonged to extremophiles (*q.v.*). These species were placed in the *Leptospirillum* and *Ferroplasma* groups of the Archaea. *See* Appendix C, 2004, Tyson *et al.*, Venter *et al.*; Appendix F.

compartmentalization a phenomenon discovered in *Drosophila* by Garcia-Bellido during investigations of the distribution of genetically marked clones of

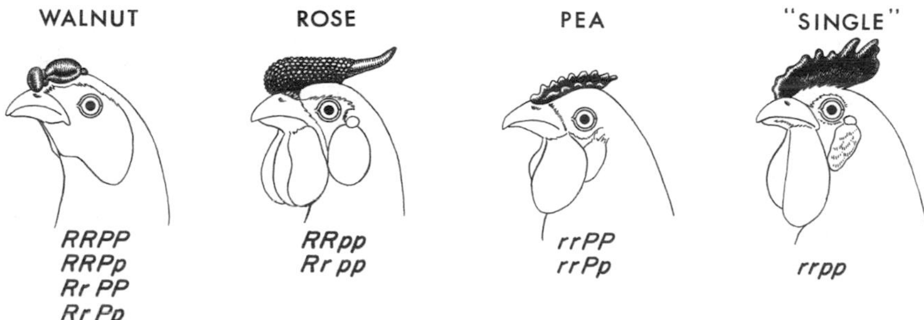

Comb shape

cells during the development of imaginal discs (*q.v.*). When such clones are analyzed it is found that they do not randomly overlap any area of the disc, but are confined to compartments and never cross the borders between compartments. A compartment contains all the descendants of a small number of founder cells called a *polyclone*. As development proceeds, large compartments are progressively split into smaller ones. The founder cells are related to each other by position, not ancestry; the progeny of the founder cells form the compartment under consideration, and no other cells contribute to it. The developmental prepattern (*q.v.*) of each compartment is controlled by a selector gene. When selector genes mutate, the cells in a compartment may develop a pattern of cell types appropriate for another compartment. The homeotic mutants (*q.v.*) are examples of mutated selector genes. *See* Appendix C, 1973, Garcia-Bellido *et al.*; 1975, Morata and Lawrence; *engrailed.*

compatibility test any serological assay designed to detect whether blood or tissue from a prospective donor(s) can be transfused or transplanted without immunological rejection. *See* cross matching, major histocompatibility complex.

compensator genes sex-linked genes in *Drosophila* that when present in the female (in double dose) reduce the activity of her two doses of a given "primary sex-linked gene" so as to make the total phenotypic effect equivalent to that seen in the male (which has one dose of both the primary gene and the compensator gene).

competence 1. the state of a part of an embryo that enables it to react to a given morphogenetic stimulus by determination and subsequent differentiation, in a given direction. **2.** in bacterial or eukaryotic cells, the ability of cells to bind and internalize exogenous DNA molecules, thereby allowing transformation (*q.v.*). Many bacteria are naturally competent at some stage of their life cycle, while others are not. Bacteria in the latter category and eukaryotic cells can be induced to become competent through biochemical or physical treatments, such as exposure to high concentration of calcium (Ca^{2+}) ions or electroporation (*q.v.*).

competition 1. the mutually exclusive use of the same limited resources (for example, food or a place to live, to hide, or to breed) by two or more organisms; **2.** the mutually exclusive binding of two different molecules to the same site on a third molecule. For example, folic acid and aminopterin compete for combination sites on various folic acid-dependent enzymes.

competitive exclusion principle the assumption that two species with identical ecological requirements cannot coexist in the same niche in the same location indefinitely. One species will eventually supplant the other, unless they evolve adaptations that allow them to partition the niche and thereby reduce competition.

complement a group of at least nine proteins (C1, C2, . . . C9) normally found in vertebrate blood serum that can be activated immunologically (by antibodies of immunoglobulin classes IgG or IgM) or nonimmunologically (by bacterial lipopolysaccharides and other substances) through an alternate (properdin) pathway. Activation of the system involves sequential conversion of proenzymes to enzymes in a manner analogous to the formation of fibrin through the blood-clotting sequence. Some activated complement components enhance phagocytosis (opsonic activity), some make antigen-antibody-complement complexes "sticky" and cause them to become affixed to endothelial tissues or blood cells (serological adhesion or immune adherence), some cause release of vasoactive amines from blood basophils or tissue mast cells (anaphylotoxins), and some cause dissolution of bacterial cells (bacteriolysis). *See* HLA complex, immune response.

complementarity-determining region the segment of the variable region of an immunoglobulin or T cell receptor molecule that contains the amino acid residues that determine the specificity of binding to the antigen. *See* paratope.

complementary base sequence a sequence of polynucleotides related by the base-pairing rules. For example, in DNA a sequence A-G-T in one strand is complementary to T-C-A in the other strand. A given sequence defines the complementary sequence. *See* deoxyribonucleic acid.

complementary DNA *See* cDNA.

complementary factors complementary genes.

complementary genes nonallelic genes that complement one another. In the case of dominant complementarity, the dominant alleles of two or more genes are required for the expression of some trait. In the case of recessive complementarity, the dominant allele of either gene suppresses the expression of some trait (i.e., only the homozygous double recessive shows the trait).

complementary interaction the production by two interacting genes of effects distinct from those produced by either one separately.

complementary RNA *See* cRNA.

complementation appearance of wild-type phenotype in an organism or cell containing two different mutations combined in a hybrid diploid or a heterokaryon. *See* allelic complementation, complementation test.

complementation group mutants lying within the same cistron; more properly called a *noncomplementation group*.

complementation map a diagrammatic representation of the complementation pattern of a series of mutants occupying a short chromosomal segment. Mutually complementing mutants are drawn as nonoverlapping lines, and noncomplementing mutants are represented by overlapping, continuous lines. Complementation maps are generally linear, and the positions of mutants on the complementation and genetic maps usually agree. A complementation map is thought to show sites where lesions have been introduced into the polypeptides coded for by the DNA segment under study.

complementation test the introduction of two mutant chromosomes into the same cell to see if the mutations in question occurred in the same gene. If the mutations are nonallelic, the genotype of the hybrid may be symbolized (*a*+/+*b*). The wild phenotype will be expressed, since each chromosome "makes up for" or "complements" the defect in the other. *See also* allelic complementation, *cis-trans* test.

complete dominance *See* dominance.

complete linkage a condition in which two genes on the same chromosome fail to be recombined and therefore are always transmitted together in the same gamete.

complete medium in microbiology a minimal medium supplemented with nutrients (such as yeast extract, casein hydrolysate, etc.) upon which nutritional mutants can grow and reproduce.

complete metamorphosis *See* Holometabola.

complete penetrance the situation in which a dominant gene always produces a phenotypic effect or a recessive gene in the homozygous state always produces a detectable effect.

complete sex linkage *See* sex linkage.

complexity in molecular biology, the total length of different sequences of DNA present in a given preparation as determined from reassociation kinetics; usually expressed in base pairs, but the value may also be given in daltons or any other mass unit.

complex locus a closely linked cluster of functionally related genes: e.g., the human hemoglobin gene complex or the *bithorax* locus in *Drosophila*. *See* pseudoalleles.

composite a plant of the immense family Compositae, regarded as comprising the most highly developed flowering plants. The family has over 25,000 species including asters, chrysanthemums, dandelions, daisies, lettuce, sunflowers, and thistles. *See* Appendix A, Angiospermae, Dicotyledonae, Asterales.

composite transposon a DNA segment flanked on each end by insertion sequences (*q.v.*), either or both of which allow the entire element to transpose.

compound eye the multifaceted kind of eye of insects. In *Drosophila* each compound eye contains nearly 800 ommatidia. *See Bar, eyeless.*

compound heterozygote an individual that carries two different recessive mutations at the same locus, one on each homolog. An example would be an individual with a different mutation in each beta chain gene. *See* hemoglobin genes.

compound X *See* attached X chromosomes.

Compton effect an attenuation process observed for x- or gamma-radiation. An incident photon interacts with an orbital electron of an atom to eject an electron and a scattered photon of less energy than the incident photon. A. H. Compton discovered the phenomenon in 1923.

concanavalin A a lectin (*q.v.*) derived from *Canavalia ensiformis*, the jack bean; abbreviated conA. The compound stimulates T lymphocytes to enter mitosis. *Compare with* pokeweed mitogen.

concatemer the structure formed by concatenation of unit-sized components.

concatenation linking of multiple subunits into a tandem series or chain, as occurs during replication of genomic subunits of T7 phage.

conception *See* syngamy.

concerted evolution the production and maintenance of homogeneity within families of repetitive DNAs. An example of concerted evolution would be the maintenance of the pattern of nucleotide sequences in each of the hundreds of tandemly arranged, rRNA genes of *Xenopus laevis*. These repetitive DNAs differ in sequence from the rRNA genes of *X. borealis*, which are also homogeneous within this species. *See* Appendix C, 1972, Brown, Wensink and Jordan; *Xenopus.*

concordance the identity of traits displayed by members of matched pairs of individuals. Twins, for example, are said to be concordant for a given trait if both exhibit the trait. Monozygotic twins are expected to show 100% concordance for genetically determined traits. If the trait is influenced by environmental factors, the degree of concordance will be lowered. *See* twin studies.

consilience the bringing together of knowledge from diverse disciplines to elucidate a complex problem. For example, determining the history of the domestication of maize requires the consilience of archaeology, biochemistry, botany, computer technology, evolutionary theory, and genetics.

conditional gene expression *See* gene expression.

conditional mutation a mutation that exhibits wild phenotype under certain (permissive) environmental conditions, but exhibits a mutant phenotype under other (restrictive) conditions. Some bacterial mutants are conditional lethals that cannot grow above 45°C, but grow well at 37°C.

conditional probability the probability of an event that depends upon whether or not some other event has occurred previously or simultaneously. For example, the probability of a second crossover between two linked genes is usually greater as its distance from the first crossover increases. *See* positive interference.

conditioned dominance referring to an allele that may or may not express itself depending upon environmental factors or the residual genotype. Thus, under different conditions the gene may behave as a dominant or as a recessive in heterozygotes.

cone 1. one of the elongate, unicellular photoreceptors in the vertebrate retina (*q.v.*) involved with vision in bright light and color recognition. 2. ovule- or pollen-bearing scales, characteristic of conifers.

cone pigment genes (CPGs) the genes that encode the opsins (*q.v.*) synthesized by the cone cells of the retina (*q.v.*). The table compares the three CPGs of humans. The green and red CPGs differ by only 15 codons. The genes lie in tandem, with the red CPG farther away from the telomere of Xq. The green CPG may be present in two or three copies, the result of unequal crossing over (*q.v.*). This has also produced hybrid genes that contain coding sequences from both red and green CPGs. An upstream sequence, 4 kilobases from the red gene and 43 kilobases from the green gene, is essential for the activity of both. Deletions of this 580 base pair element result in a rare form of X-linked color blindness that is characterized by the absence of both red and green cone sensitivities. The blue opsin shows about 40% amino acid sequence identity with either green or red opsins, and the identity value is similar when the blue opsin and the rod cell opsin are compared. All catarrhine primates (*q.v.*) have the three cone pigment genes mentioned. In contrast, platyrrhine monkeys (*q.v.*) have only one X-linked and one autosomal color photopigment gene. *See* Appendix C, 1986, Nathans *et al.*; color blindness, rhodopsin (RHO).

confidence limits the range of values about the true value of a population parameter (such as the mean) within which sample values (e.g., means) may fall and from which inferences may be made, with any desired degree of confidence, such that it is unlikely that a disproportionate number of rejections of a true hypothesis should occur. Conventionally, the 95% confidence limits (also known as *fiducial limits*) around the mean are $\pm 1.96\ s\sqrt{n}$, where *s* is the standard deviation (*q.v.*) and $\sqrt{n}$ is the standard error (*q.v.*). *See* statistical errors.

confocal microscopy a form of light microscopy that allows successively deeper layers of a fixed or living specimen to be viewed with astonishing clarity. In conventional light microscopy, the specimen is illuminated throughout its depth, while the objective lens is focused only on a narrow plane within the specimen. Therefore, an out-of-focus blur generated by regions above and below the focal plane of interest reduces the contrast and resolution of the image. Confocal microscopes are designed so that mainly signals from the elements in the focal plane

Cone Pigment Genes

CPGs	Locus	Total exons	Coding length (bp)	Total intron length (bp)
blue	7q 31.3–32	5	1,044	2,200
green	Xq 28	6	1,092	12,036
red	Xq 28	6	1,092	14,000

are detected, and signals originating from above and below the focal plane are filtered out. The name *confocal* indicates that the illumination, specimen, and detector all have the same focus. The specimen volume assayed can be as small as 0.25 nanometers in diameter and 0.5 nanometers in depth. The two-dimensional image is built up by collecting and integrating signals as the light beam moves horizontally and vertically across the specimen, and the image elements are converted into a video signal for display on a computer screen. Subsequently, stacks of optical sections can be reconstructed to produce a view of the specimen in three dimensions.

confusing coloration a form of protective coloration that tends to confuse the predator by having a different appearance according to whether its possessor is at rest or in motion.

congenic strains strains that differ from one another only with respect to a small chromosomal segment. Several congenic mouse strains differ only in the major or minor histocompatibility loci they contain. *Contrast with* coisogenic.

congenital existing at birth. Congenital defects may or may not be of genetic origin.

congression *See* chromosome congression.

congruence in cladistics, congruent characters are shared features whose distribution among organisms fully corresponds to that in the same cladistic grouping. The most likely cladogram is the one that provides the maximum congruence between all of the characters involved.

conidium an asexual haploid spore borne on an aerial hypha. In *Neurospora* two types of conidia are found: oval *macroconidia*, which are multinucleate, and *microconidia*, which are smaller, spherical, and uninucleate. When incubated upon suitable medium, a conidium will germinate and form a new mycelium.

conifers plants characterized by needle-shaped leaves and which bear their microsporangia and megasporangia in cones. The class contains 50 genera and 550 species, and among them are the largest (the Pacific coast redwood, *Sequoia gigantia*) and the longest-lived (the bristlecone pine, *Pinus aristata*) species. *See* Appendix A, Trachaeophyta, Pteropsida, Gymnospermae, Coniferophyta.

conjugation a temporary union of two single-celled organisms or hyphae with at least one of them receiving genetic material from the other. **1.** In bacteria, the exchange is unidirectional with the "male" cell extruding all or a portion of one of its chromosomes into the recipient "female" (*see* F factor, pilus). **2.** In *Paramecium*, as shown in the illustration on page 98, entire nuclei are exchanged. *See* nuclear dimorphism. **3.** In fungi, conjugation also occurs between hyphae of opposite mating type to produce heterokaryons (*q.v.*).

conjugation tube *See* pilus.

conjugon a genetic element essential for bacterial conjugation. *See* fertility factor.

conodonts **1.** the earliest known fossil elements that contain dentine, ranging in age from 510 to 220 million years old; thought to have functioned as teeth. **2.** the earliest known vertebrates, intermediate between the more primitive hagfish/lamprey lineages and the more complex armored jawless fishes (ostracoderms). Mineralization apparently began in the mouths of conodonts, rather than in the skin of ostracoderms.

consanguinity genetic relationship. Consanguineous individuals have at least one common ancestor in the preceding few generations. *See* isonymous marriage.

consecutive sexuality the phenomenon in which most individuals of a species experience a functional male phase when young, and later change through a transitional stage to a functional female phase. A situation common in some molluscs.

consensus sequence synonymous with canonical sequence (*q.v.*).

conservative recombination breakage and reunion of preexisting strands of DNA in the absence of DNA synthesis.

conservative replication an obsolete model of DNA replication in which both old complementary polynucleotides are retained in one sibling cell, while the other gets the two newly synthesized strands. *Compare with* semiconservative replication.

conservative substitution replacement of an amino acid in a polypeptide by one with similar characteristics; such substitutions are not likely to change the shape of the polypeptide chain, e.g., substituting one hydrophobic amino acid for another.

conserved sequence a sequence of nucleotides in genetic material or of amino acids in a polypeptide chain that either has not changed or that has changed only slightly during an evolutionary period of time. Conserved sequences are thought to generally regulate vital functions and therefore have been selectively preserved during evolution.

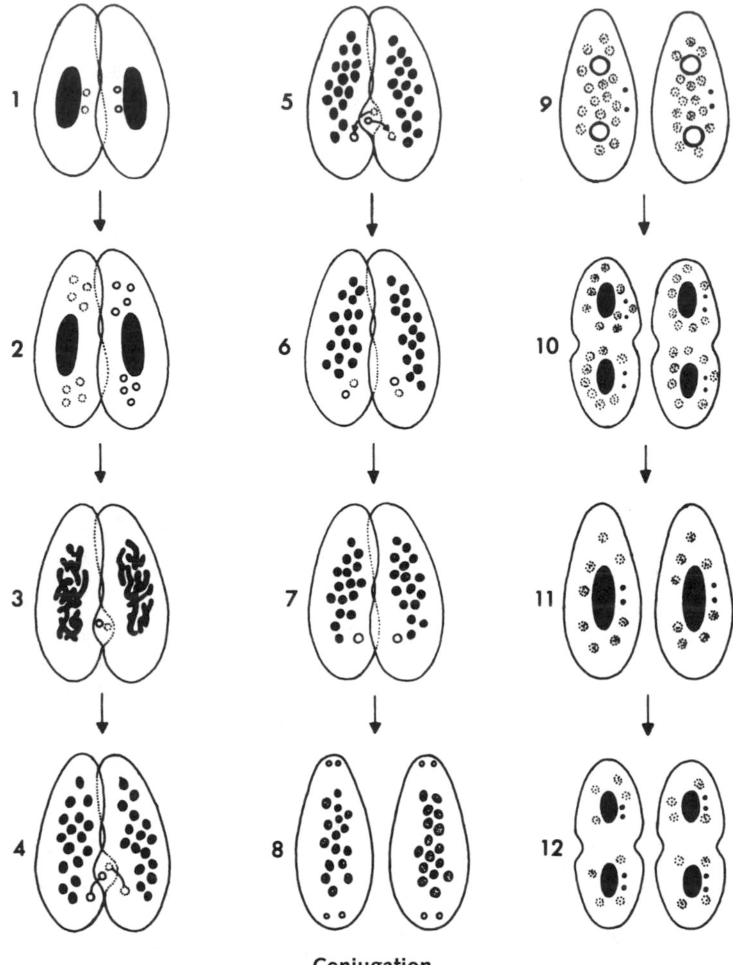

Conjugation

Nuclear changes that accompany conjugation in *Paramecium aurelia*. (1) Two parental animals, each with one macronucleus and two diploid micronuclei. (2) Formation of eight haploid nuclei from the micronuclei of each conjugant. (3) Seven nuclei in each conjugant disappear, the remaining nucleus resides in the paroral cone; the macronucleus breaks up into fragments. (4–7) The nuclei in the paroral cones divide mitotically, forming "male" and "female" gamete nuclei. The female nuclei pass into the interior of the parental animals, while the male nuclei are transferred to the partners. Male and female haploid nuclei fuse. (8) Each fusion nucleus divides twice mitotically. (9) Two of the four nuclei so formed differentiate into macronuclear anlagen (white circles) while the other two become micronuclei. (10–12) Each micronucleus divides, and transverse fission of the exconjugants produces four animals (two of which are shown in 11). Fragments of the old macronucleus are gradually lost. In (12) transverse fission begins in the animals seen previously in (11).

conspecific belonging to the same species.

constant region *See* immunoglobulin.

constitutive enzyme an enzyme that is always produced irrespective of environmental conditions. *See* Appendix C, 1937, Karström.

constitutive gene a gene whose activity depends only on the efficiency of its promoter in binding RNA polymerase.

constitutive gene expression *See* gene expression.

constitutive heterochromatin *See* heterochromatin.

constitutive mutation a mutation that results in an increased constitutive synthesis by a bacterium of several functionally related, inducible enzymes. Such a mutation either modifies an operator gene so that the repressor cannot combine with it or modifies the regulator gene so that the repressor is not formed. *See* regulator gene.

constriction an unspiralized region of a metaphase chromosome. Kinetochores and nucleolus organizers are located in such regions.

contact inhibition the cessation of cell movement on contact with another cell. It is often observed when freely growing cells, tissue cultured on a petri plate, come into physical contact with each other. Cancer cells lose this property and tend to pile up in tissue culture to form multilayers called *foci*.

containment in microbiology, measures taken to diminish or prevent the infection of laboratory workers by the products of recombinant DNA technology and the escape of such products from the laboratory. *Biological containment* is accomplished by using genetically altered bacteria, phages, and plasmids that are unable to carry out certain essential functions (e.g., growth, DNA replication, DNA transfer, infection, and propagation) except under specific laboratory conditions. *Physical containment* is accomplished by design and use of special facilities and laboratory procedures such as limited access, safety hoods, aerosol control, protective clothing, pipeting aids, etc.

contig an abbreviation for *contiguous*. When a collection of cloned DNA fragments can be arranged so that they overlap to provide gap-free coverage of a chromosome, one has assembled the fragments (contigs) into a contig map. *See* genomic library, reads, scaffold, shotgun sequencing.

contiguous gene syndrome a situation where a patient suffers from two or more genetic diseases simultaneously because of the simultaneous deletion of neighboring genes. A classic example is an unfortunate boy who had a deletion in Xp 21. This removed a contiguous set of genes and resulted in a complex array of syndromes, including Duchenne muscular dystrophy (DMD) and retinitus pigmentosum (*both of which see*). DNA from this patient provided investigators with a means to clone the DMD gene and subsequently to isolate dystrophin. *See* Appendix C, 1987, Hoffman, Brown, and Kunkel.

continental drift the concept that the world's continents once formed a part of a single mass and have since drifted into their present positions. The modern concept of plate tectonics (*q.v.*) has refined the theory of continental drift by placing the continents on larger sections of the earth's crust (tectonic plates) that are in motion. Continental drift began in the Proterozoic era; the continents collided to form a giant land mass named Pangaea in the late Permian period, and redispersed in the Mesozoic era. *See* Appendix C, 1912, Wegener; 1927, du Toit; geological time divisions, sea floor spreading.

continental island an island assumed to have been once connected to a neighboring continent. *Contrast with* oceanic island.

continuous distribution a collection of data that yield a continuous spectrum of values. For example, measurements such as height of plant or weight of fruit, carried out one or more decimal places. *See* discontinuous distribution.

continuous fibers the microtubules that connect the two poles of the mitotic apparatus, as distinct from traction fibers and astral fibers. *See* mitotic apparatus.

continuous variation the phenotypic variation exhibited by quantitative traits that vary by imperceptible degrees from one extreme to another. In human populations, phenotypes like body weight, height, and intelligence show continuous variation. *See* discontinuous variation, quantitative inberitance.

contractile ring a transitory organelle that, during late anaphase and telophase, assumes the form of a continuous equatorial annulus beneath the plasma membrane of the cleavage furrow. The ring is composed of an array of actin microfilaments aligned circumferentially along the equator of the cell. An interaction between cytoplasmic myosin and these actin molecules causes them to slide past one another, closing down the contractile ring, and producing the cleavage furrow.

contractile vacuole an organelle found in fresh water protozoans, like *Paramecium*. Contractile vacuoles are pumping structures that fill and then contract to expel excess water from the cell.

control a standard of comparison; a test or experiment established as a check of other experiments, performed by maintaining identical conditions except for the one varied factor, whose causal significance can thus be inferred.

controlled pollination a common practice in plant hybridization of bagging the pistillate flowers to protect them from undesired pollen. When the pistillate flowers are in a receptive condition, they are dusted with pollen of a specified type.

controlling elements a class of genetic elements that renders target genes unstably hypermutable, as in the *Dissociation-Activator* system of corn (*q.v.*). They include receptors and regulators. The receptor element is a mobile genetic element that, when inserted into the target gene, causes it to become inactivated. The regulator gene maintains the mutational instability of the target gene, presumably by its capacity to release the receptor element from the target gene and thus return that locus to its normal function. *See* transposable elements.

controlling gene a genomic sequence that can switch on or off the transcription of one or more separate *structural genes* (*q.v.*). Also called a *regulator gene* (*q.v.*).

convergence the evolution of unrelated species occupying similar adaptive zones, resulting in structures bearing a superficial resemblance (e.g., the wings of birds and insects).

convergent evolution *See* convergence.

conversion *See* gene conversion.

conversion factors for RNA and protein molecules the average relative molecular mass (Mr) for an amino acid is 110, while the Mr for a nucleotide is 330. With these values one can estimate, for example, the approximate Mr of a protein containing a given number of amino acids or the number of nucleotides in a specific RNA with a known Mr value.

Cooley anemia the most severe form of beta thalassemia (sometimes called *thalassemia major*). Few or no functional beta globin chains are made, and this results in a life-threatening anemia. The patient requires lifelong blood transfusions, and these lead to an iron overload that requires chelation therapy. The eponym refers to Thomas Benton Cooley, an American pediatrician who first described the condition in 1925. *See* Desferal, thalassemia.

coordinated enzymes enzymes whose rates of production vary together. For example, the addition of lactose to the medium causes the coordinated induction of beta-galactosidase *and* beta-galactoside permease in *E. coli.* Such enzymes are produced by cistrons of the same operon. *See* regulator genes.

Cope law of the unspecialized the generalization put forward by the nineteenth-century paleontologist Edward Drinker Cope that the evolutionary novelties associated with new major taxa are more likely to originate from a generalized, rather than from a specialized, member of an ancestral taxon.

Cope size rule the general tendency for animals to increase in body size during the course of phyletic evolution (*q.v.*).

copia elements retrotransposons (*q.v.*) of *Drosophila* existing as a family of closely related base sequences that code for abundant mRNAs. There are usually between 20 and 60 copia elements per genome. The actual number depends on the *Drosophila* strain employed. The copia elements are widely dispersed among the chromosomes, and the sites occupied vary between strains. Each copia element ranges in size from 5 to 9 kilobase pairs, and it carries direct terminal repeats about 280 base pairs long. Several *Drosophila* mutations have been found that result from the insertion of copia-like elements. *See* Appendix C, 1985, Mount and Rubin.

copolymer a polymeric molecule containing more than one kind of monomeric unit. For example, copolymers of uridylic and cytidylic acids (poly UC) are used as synthetic messengers.

copper a biological trace element. Atomic number 29; atomic weight 63.45; valance 1^+, 2^+; most abundant isotope ^{63}Cu; radioisotope ^{64}Cu, half-life 12.8 hours, radiations emitted—gamma rays, electrons, and positrons. *See* Wilson disease.

copy-choice hypothesis an explanation of genetic recombination based on the hypothesis that the new strand of DNA alternates between the paternal and maternal strands of DNA during its replication. Therefore crossing over involves a switch by the forming-strand between templates and does not require physical breakage and reunion of exchanged strands.

copy DNA synonymous with complementary DNA. *See* cDNA.

copy error a mutation resulting from a mistake during DNA replication.

core **1.** the region of a nuclear reactor containing the fissionable material. **2.** synaptonemal complex.

core DNA the segment of DNA in a nucleosome (*q.v.*) that wraps around a histone octamer.

core granule RNP granules in the ommatidia of *Drosophila.* The xanthomatins and drosopterins are normally bound to these granules.

core particle a structural unit of eukaryotic chromosomes revealed by digestion with micrococcal nuclease to consist of a histone octamer and a 146 base pair segment of DNA. *See* nucleosome.

corepressor in repressible genetic systems, the small effector molecule (usually an end product of a metabolic pathway) that inhibits the transcription of genes in an operon by binding to a regulator protein (aporepressor). Also called a *repressing metabolite.*

corm a swollen vertical underground stem base containing food material and bearing buds. It can function as an organ of vegetative reproduction. The crocus and gladiolus have corms.

corn *Zea mays*, the most valuable crop plant grown in the United States, it ranks along with wheat, rice, and potatoes, as one of the four most important crops in the world. Corn is generally classified in five commercial varieties on the basis of kernel morphology: (1) Dent corn (var. *indentata*), the most common variety of field corn; the kernel is indented, from drying and shrinkage of the starch in the summit of the grain. (2) Flint corn (var. *indurata*), the kernel is completely enclosed by a horny layer and the grain is therefore smooth and hard; flint corn is the fastest to mature. (3) Sweet corn (var. *saccharata*), which is grown for human consumption, is picked when it is filled with a milky fluid, before the grain hardens. (4) Popcorn (var. *everta*), the ear is covered with small kernels that are enclosed by a tough coat; when heated the contained moisture is turned to steam and the kernel explodes. (5) Flour corn (var. *amylacea*), the ear contains soft, starchy kernels; it requires a long growing season and is therefore grown primarily in the tropics. *See* double cross, hybrid corn, pod corn, quantitative inheritance, teosinte, *Zea mays*.

coronavirus a family of viruses that have a positive-strand RNA genome and are characterized by a viral envelope from which petal-shaped spikes protrude. The virus causing severe acute respiratory syndrome (SARS) in humans belongs to this family. Its genome contains 27,727 nucleotides. *See* Baltimore classification of viruses, enveloped viruses.

corpus allatum an endocrine organ in insects that synthesizes the allatum hormone (*q.v.*). In the larvae of cyclorrhaphous diptera, the corpus allatum forms part of the ring gland (*q.v.*).

corpus cardiacum an endocrine organ in insects consisting of a central bundle of axons enveloped by cortical cells. Axons from the corpus cardiacum enter the corpus allatum. Most axons associated with the corpus cardiacum have their cell bodies in the pars intercerebralis. The cortical cells and many of the axons contain numerous neurosecretory spheres (*q.v.*).

corpus luteum a mass of yellowish tissue that fills the cavity left after the rupture of the mature ovum from the mammalian ovary.

correction in a hybrid DNA sequence, the replacement (i.e., by excision and repair) of illegitimate nucleotide base pairs by bases that pair properly.

correlated response the change in one character occurring as an incidental consequence of the selection for a seemingly independent character. For example, reduced fertility may accompany selection for increased bristle number in *Drosophila*. *See* pleiotropy.

correlation the degree to which statistical variables vary together. It is measured by the *correlation coefficient (r)*, which has a value from zero (no correlation) to −1 or +1 (perfect negative or positive correlation, respectively).

corresponding genes *See* gene-for-gene hypothesis.

corridor a migration route allowing easy dispersal for certain species.

corticosterone one of a family of adrenal cortical hormones influencing glucose metabolism.

corticotropin *See* adrenocorticotropic hormone.

Corynebacterium diphtheriae the bacterium responsible for diphtheria. This was the first disease shown to be caused by a toxin secreted by a bacterium and the first to be successfully treated by an antitoxin (*q.v.*). *See* Appendix A, Bacteria, Actinobacteria; Appendix C, 1888, Roux and Yersin; 1890, vonBehring; diphtheria toxin.

COS cells a monkey cell line that has been transformed by an SV40 viral genome containing a defective origin of viral replication. When introduced into COS cells, recombinant RNAs containing the SV40 origin and a foreign gene should replicate many copies.

cosmic rays high-energy particulate and electromagnetic radiations originating outside the earth's atmosphere.

cosmid plasmid vectors designed for cloning large fragments of eukaryotic DNA (i.e., in the size range of 40–45 kilobases). The term signifies that the vector is a plas*mid* into which phage lambda *cos* sites (*q.v.*) have been inserted. As a result, the plasmid DNA can be packaged in a phage coat *in vitro*. *See* Appendix C, 1977, Collins and Holm; genomic library, physical map.

cos sites *co*hesive end sites, nucleotide sequences that are recognized for packaging a phage DNA molecule into its protein capsule.

cot the point (symbolized by $C_0t_{1/2}$) in a reannealing experiment where half of the DNA is present as double-stranded fragments; also called the *half* *reaction time.* If the DNA fragments contain only unique DNA sequences and are similar in length, then $C_0t_{1/2}$ varies directly with DNA complexity (*q.v.*). *See* reassociation kinetics.

cotransduction the simultaneous transduction of two or more genes because the transduced element contains more than one locus.

cotransformation 1. the simultaneous transformation of two or more bacterial genes; the genes cotransformed are inferred to be closely linked because transforming DNA fragments are usually small. Also called *double transformation.* 2. in molecular biology, introduction of two physically unlinked sets of genes, one of which codes for a selectable marker, into a cell. This technique is useful in animal cells in which the isolation of cells transformed with a gene that does not code for a selectable marker has been problematic.

cotranslational sorting *See* protein sorting.

cotton *See Gossypium.*

Coturnix coturnix japonica the Japanese quail, a small bird domesticated since the 12th century, and currently used as a laboratory animal. The European quail is *Coturnix coturnix coturnix. See* Appendix A, Aves, Neognathae, Galliformes.

cot value *See* cot.

cotyledon the leaf-forming part of the embryo in a seed. Cotyledons may function as storage organs from which the seedling draws food, or they may absorb and pass on to the seedling nutrients stored in the endosperm. Once the cotyledon is exposed to light, it develops chlorophyll and functions photosynthetically as the first leaf.

counteracting chromatographic electrophoresis a group of methods for purifying specific molecules from a mixture by the application of two counteracting forces; specifically, the chromatographic flow of a solute down a separation column vs. solute electrophoresis in the opposite direction.

countercurrent distribution apparatus an automated apparatus used for separating mixtures. The method takes advantage of differences in the solubilities of the components of the mixture in two immiscible solvents. An example of the usefulness of the technique is in the separation of different transfer RNA molecules.

counterselection a technique used in bacterial conjugation experiments to allow recovery of recombinant F^- cells, while at the same time selecting against (preventing growth of) Hfr donor cells. For example, suppose the Hfr donor strain is susceptible to an antibiotic (such as streptomycin) and can synthesize histidine; the streptomycin locus must be so far from the origin of chromosome transfer that the mating pairs, which inevitably break apart, have separated before the *str* locus has been transferred. Suppose further that the recipient F^- cell cannot make histidine (*his*$^-$) but is resistant to the antibiotic (*Str*^{r}). Only His$^+$*Str*^{r} recombinants can survive on a medium lacking histidine and containing streptomycin. The desired gene (*His*$^+$ in this case) is called a *selected marker*; the gene that prevents growth of the male (*str*^{s} in this case) is called the *counterselective marker.*

coupled reactions chemical reactions having a common intermediate and therefore a means by which energy can be transferred from one to the other. In the following pair of enzyme catalyzed reactions, glucose-1-phosphate is the common intermediate that is formed in the first reaction and used up in the second:

1. ATP + glucose → ADP + glucose-1-phosphate
2. glucose-1-phosphate + fructose →
sucrose + phosphate

A molecule of sucrose is synthesized from glucose and fructose at the expense of the energy stored in ATP and transferred by glucose-1-phosphate.

coupled transcription-translation a characteristic of prokaryotes wherein translation begins on mRNA molecules before they have been completely transcribed.

coupling, repulsion configurations when both nonallelic mutants are present on one homolog and the other homologous chromosome carries the plus alleles (*a b*/++), the genes are said to be in the *coupling* configuration. The *repulsion* configuration refers to a situation in which each homolog contains a mutant and a wild-type gene (*a*+/+*b*). *See cis-trans* configurations.

courtship ritual a characteristic genetically determined behavioral pattern involving the production and reception of an elaborate sequence of visual, auditory, and chemical stimuli by the male and female prior to mating. Such rituals are interpreted as ensuring that mating will occur only between the most fit individuals of the opposite sex and the same species. *See* allesthetic trait, gustatory receptor (GR) genes, mate choice, sexual selection.

cousin the son or daughter of one's uncle or aunt. The children of siblings are *first* cousins. Children of *first* cousins are second cousins. Children of second cousins are *third* cousins, etc. The child of a first cousin is a *cousin once removed* of his father's or mother's first cousin. *See* consanguinity, inbreeding.

covalent bond a valence bond formed by a shared electron between the atoms in a covalent compound. *See* disulfide linkage, glycosidic bonds, high-energy bond, peptide bond, phosphodiester.

covariance a statistic employed in the computation of the correlation coefficient between two variables; the covariance is the sum of $(x - \bar{x})$ $(y - \bar{y})$ over all pairs of values for the variables x and y, where $\bar{x}$ is the mean of the x values and $\bar{y}$ is the mean of all y values.

cpDNA chloroplast DNA. Also abbreviated *ct-DNA*. *See* chloroplast.

CPEB protein *c*ytoplasmic *p*olyadenylation *e*lement-*b*inding protein. A protein first identified in *Xenopus* oocytes, where it activates dormant mRNAs by elongating their poly(A) tails. Homologs of this protein exist in humans, mice, flies, and marine mollusks. In *Aplysia*, a neuronal isoform of CPEB protein is expressed in central nervous system synapses and functions to regulate local protein synthesis at activated synapses and to strengthen these synapses during long-term memory formation. This protein is remarkable in that it has prion-like properties in its biologically functional state and it is thought that this prion-like transformation is required for maintaining synaptic changes associated with long-term memory storage. *See* Appendix C, 2003, Si *et al.*; memory, prions.

CpG island *See* DNA methylation.

Craniata the subphylum of the Chordata containing animal species with a true skull. *See* Appendix A.

Crassostrea virginica *See* Pelecypoda.

creationism the belief that the universe and living organisms originate from specific acts of divine creation, as in the biblical account, rather than by natural processes such as evolution; another term for creation science. *See* fundamentalism.

CREBs The abbreviation for *c*yclic AMP *r*esponse *e*lement *b*inding proteins. These proteins are required for the consolidation of short-term memory into long-term memory. Genes encoding CREBs have been cloned in *Drosophila*. One of these genes, *dCREB2*, encodes a protein, dCREBa, that activates transcription of genes that enhance the ability of the fly to consolidate short-term memory into long-term memory. The isoform dCREBb inhibits this process. *See* Appendix C, 1982, Kandel and Schwartz; 1994, Tully *et al.*, cyclic AMP.

Cretaceous the most recent of the Mesozoic periods, during which the dinosaurs continued to diversify. The first angiosperms appeared during the late Cretaceous, and pollination interactions with insects developed. The first marsupials and placental mammals arose. At the end of the Cretaceous, there occurred the second most severe of all mass extinctions. About half of all animal families were wiped out. All of the dinosaurs became extinct. The continents formed from Pangea were now widely separated. *See* continental drift, geologic time divisions, impact theory.

cretinism a stunting of bodily growth and mental development in humans due to a deficiency of thyroid hormones. Hereditary cretinism, which is often accompanied by goiter (*q.v.*) and deafness, consists of a group of metabolic disorders that results in a failure in the formation of sufficient thyroxine and triiodothyronine. The defects include the inability of the thyroid gland to accumulate sufficient iodine, to convert it into organically bound iodine, and to couple iodotyrosines to form iodothyronines. All hereditary defects in thyroid hormonogenesis are inherited as autosomal recessives. *See* thyroid hormones.

Creutzfeldt-Jakob disease a fatal degenerative nervous disorder caused by infectious proteins called prions (*q.v.*). The prion protein PrP is a normal protein of the nervous system, and it is encoded by a gene located in chromosome 20 p12-pter. The disease is caused by a mutant form of the protein, which is infectious, and contains no detectable nucleic acid. The syndrome gets its name from the German physicians Hans G. Creutzfeldt and Alfons M. Jakob, who gave the first description of patients with the condition in 1920 and 1921, respectively.

Cricetulus griseus the Chinese hamster. The rodent is a favorite for cytogenetic studies because of its small chromosome number (N = 11). A total of about 40 genetic loci have been assigned to specific chromosomes. *See* CHO cell line.

cri du chat syndrome *See* cat cry syndrome.

Crigler-Najjar syndrome a defect in the metabolism of bilirubin (*q.v.*) caused by a recessive mutation in a gene located between q21 and 23 on human chromosome 1. The condition was first described in 1952 by John Crigler and Victor Najjar. Patients lack hepatic bilirubin UDP-glucuronyl transferase. This enzyme functions to conjugate bilirubin with glucuronic acid prior to biliary excretion. In the absence of the enzyme, excess bilirubin builds up in all tissues causing jaundice, brain damage, and death.

crisis period the time interval of a primary cell culture, following a number of cell divisions, during which most secondary progeny die even though culture conditions are adequate to initiate a new primary culture of low cell density from a fresh isolate. *See* Hayflick limit, tissue culture.

criss-cross inheritance referring to the passage of sex-linked traits from mother to son and from father to daughter.

cristae elaborate invaginations of the inner mitochondrial membrane.

CRM cross-reacting material (*q.v.*).

cRNA synthetic transcripts of a specific DNA molecule or fragment made by an *in vitro* transcription system. This cRNA can be labeled with radioactive uracil and then used as a probe (*q.v.*).

Cro-Magnon man *Homo sapiens sapiens* living in the upper Pleistocene. Cro-Magnon replaced the Neanderthal (*q.v.*) throughout its range.

cro repressor the protein encoded by the *cro* regulator gene of lambda (λ) bacteriophage (*q.v.*). The *cro* gene lies alongside C_I gene, with its promoter to the immediate left. During transcription the host transcriptase moves to the right. Movement of this enzyme can be blocked by the lambda repressor (*q.v.*), which binds to an operator that overlaps the *cro* promoter. The *cro* repressor contains 66 amino acids. The monomers associate in pairs to form the active repressor, which binds to DNA via a helix-term-helix motif (*q.v.*). The *cro* repressor is required for the virus to enter the lytic cycle (*q.v.*). Since the *cro* repressor represses the lambda repressor, it is sometimes called an *antirepressor*. *See* Appendix C, 1981, Anderson *et al.*; regulator gene.

cross in higher organisms, a mating between genetically different individuals of opposite sex. In microorganisms, genetic crosses are often achieved by allowing individuals of different mating types to conjugate. In viruses, genetic crossing requires infecting the host cells with viral particles of different genotypes. The usual purpose of an experimental cross is to generate offspring with new combinations of parental genes. *See* backcross, conjugation, dihybrid, E_1, F_1, I_1, monohybrid cross, P_1, parasexuality, test cross.

cross-agglutination test one of a series of tests commonly employed in blood typing in which erythrocytes from a donor of unknown type are mixed with sera of known types.

crossbreeding outbreeding (*q.v.*).

cross-fertilization union of gametes that are produced by different individuals. *Compare with* self-fertilization.

cross hybridization (molecular) hybridization of a probe (*q.v.*) to a nucleotide sequence that is less than 100% complementary.

cross-induction the induction of vegetative phage replication in lysogenic bacteria in response to compounds transferred from UV-irradiated F^+ to nonirradiated F^- bacteria during conjugation.

crossing over the exchange of genetic material between homologous chromosomes. Meiotic crossing over occurs during pachynema and involves the nonsister strands in each meiotic tetrad. Each exchange results in a microscopically visible chiasma (*q.v.*). In order for proper segregation of homologs at the first meiotic division, each tetrad must have at least one chiasma. For this reason meiotic recombination is enhanced in very short chromosomes. Crossing over can also occur in somatic cells during mitosis. In suitable heterozygotes this may result in twin spots (*q.v.*). Exchange between sister chromatids can also occur and is a sensitive indicator of DNA damage caused by ionizing radiations and chemical mutagens. Sister chromatid exchanges normally do not result in genetic recombination. *See* Appendix C, 1912, Morgan; 1913, Tanaka; 1931, Stern; 1931, Creighton and McClintock; 1961, Meselson and Weigel; 1964, Holliday; 1965, Clark; 1971, Howell and Stern; 1989, Kaback, Steensma, and De Jonge; 1992, Story, Weber, and Steitz; genetic recombination, Holliday model, human pseudoautosomal region, meiosis, Rec A protein, site-specific recombination.

crossing over within an inversion *See* inversion.

cross-linking formation of covalent bonds between a base in one strand of DNA and an opposite base in the complementary strand by mitotic poisons such as the antibiotic mitomycin C or the nitrite ion.

cross-matching *See* cross-agglutination test.

crossopterygian a lobe-finned bony fish, one group of which was ancestral to the amphibians. *See* living fossil.

crossover fixation the spreading of a mutation in one member of a tandem gene cluster through the entire cluster as a consequence of unequal crossing over.

crossover region the segment of a chromosome lying between any two specified marker genes.

crossover suppressor a gene, or an inversion (*q.v.*), that prevents crossing over in a pair of chromosomes. The *Gowen crossover suppressor* gene (*q.v.*) of *Drosophila* prevents the formation of synaptonemal complexes (*q.v.*).

crossover unit a 1% crossover value between a pair of linked genes.

cross-pollination the pollination of a flower with pollen from a flower of a different genotype.

cross-reacting material any nonfunctional protein reactive with antibodies directed against its functional counterpart. For example, some patients with classical hemophilia (*q.v.*) produce a CRM that reacts with anti-AHF serum, but this protein has lost its ability to take part in the blood-clotting process.

cross-reaction, serological union of an antibody with an antigen other than the one used to stimulate formation of that antibody; such cross-reactions usually involve antigens that are stereochemically similar or those that share antigenic determinants.

cross reactivation *See* multiplicity reactivation.

crown gall disease an infection caused by the soil-borne bacterium *Agrobacterium tumefaciens* (*q.v.*) that is characterized by tumor-like swellings (galls) that often occur on the infected plant at the stalk just above soil level (the crown). The neoplastic outgrowths are composed of transformed cells which synthesize metabolites specifically used by the bacterium. Plants belonging to more than 90 different families are susceptible to crown gall disease. *See* Appendix C, 1907, Smith and Townsend; promiscuous DNA, selfish DNA, Ti plasmid.

crozier the hook formed by an ascogenous hypha of *Neurospora* or related fungi previous to ascus development. The hook is formed when the tip cell of an ascogenous hypha grows back upon itself. Within the arched portion of the hypha, cell walls are subsequently laid down in such a way that three cells are formed. The terminal cell of the branchlet is uninucleate, the penultimate one is binucleate, and the antipenultimate one is uninucleate. Fusion of the haploid nuclei of different mating types in the penultimate cell occurs, and it enlarges to form the ascus in which meiosis occurs.

CRP cyclic AMP receptor protein. *See* catabolite activating protein (CAP).

cruciform structure a cross-shaped configuration of DNA produced by complementary inverted repeats pairing with one another on the same strand instead of with its normal partner on the other strand. *See* palindrome.

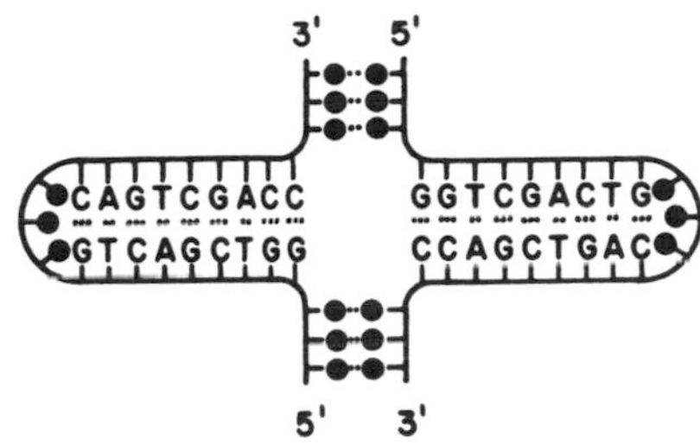

cryostat a device designed to provide low-temperature environments in which operations (like sectioning frozen tissues) may be carried out under controlled conditions.

crypsis those mechanisms which enable a species to remain hidden from its predators. Cryptic coloration is a form of camouflage that makes the species inconspicuous against its natural backgrounds. Behavioral crypsis includes stealthy movements and unflinching inactivity when a predator is nearby. *See* adaptive melanism.

cryptic coloration *See* crypsis.

cryptic gene a gene that has been silenced by a single nucleotide substitution, that is present at a high frequency in a population, and that can be reactivated by a single mutational event.

cryptic prophage a prophage that has lost certain functions essential for lytic growth and the production of infectious particles. However, these defective viruses still retain some functional genes, and therefore they can rescue mutations in related bacteriophages by recombining with them to generate viable hybrids. *See Escherichia coli.*

cryptic satellite a satellite DNA sequence that cannot be separated from the main-band DNA by density gradient ultracentrifugation. Cryptic satellite DNA can be isolated from the main-band DNA by its unique features (e.g., by the more rapid reannealing of the highly repeated segments that constitute the satellite).

cryptic species phenotypically similar species that never form hybrids in nature. *See* sibling species.

cryptogam a spore-bearing rather than a seed-bearing plant. In older taxonomy a member of the Cryptogamia, including the ferns, mosses, algae, and fungi. *See* phanerogam.

cryptomonads a group of single-celled algae characterized by a "nucleomorph," sandwiched between the membranes that surround the chloroplast. Cryptomonads are thought to have arisen hundreds of millions of years ago by the fusion of a red algal symbiont and a biflagellated protozoan. The protozoan was the source of the conventional nucleus, whereas the nucleomorph is all that remains of the symbiont's nucleus. It contains three minute linear chromosomes with telomeres and densely packed genes and is surrounded by a double envelope with characteristic pores. Nucleomorphs represent nuclei that have undergone the greatest genomic reduction in the history of eukaryotes, *See* Appendix A, Protoctista, Cryptomonada; Appendix C, 1999, Beaton and Cavalier-Smith; C value paradox, serial symbiosis theory, skeletal DNA hypothesis.

Cryptosporidium a genus of protozoan parasites that cause gastrointestinal diseases of medical and veterinary importance. These protoctists are placed in the same phylum as the malaria parasites. However, they lack apicoplasts and have no second host, unlike *Plasmodium (q.v.)*. Cryptosporidia have complex life cycles with motile and non-motile forms in both asexual and sexual stages. They spend most of their lives within the epithelial cells of the gut or in its lumen. The infective phase of *Cryptosporidium* is a cyst that contains several haploid sporozoites enclosed in a thick capsule. The cysts are about 3 μm in diameter, are easily spread via water, are resistant to most chemical disinfectants, and can be removed from drinking water only by filtration. C. *parvum*, the cause of cryptosporidosis in humans, has a genome size of 9 million base pairs distributed among 8 chromosomes. *See* Appendix A, Protoctista, Apicomplexa.

Cryptozoic a synonym for Precambrian (*q.v.*).

crystallins a family of structural proteins in the lens of the vertebrate eye. However, some crystallins play an enzymatic role in other tissues. For example, in reptiles and birds a form of crystallin is found in heart muscle, where it functions as a lactic dehydrogenase.

c-src a cellular gene, present in various vertebrates, that hybridizes with *src*, the oncogene of the Rous sarcoma virus (*q.v.*). The *c-src* genes code for pp60c-Src proteins that resemble pp60v-*src* proteins in their enzymatic properties.

CTCF protein a highly conserved and ubiquitous DNA binding protein of vertebrates. CTCF is an 82 kDa protein with 11 zinc fingers, and it binds to DNA segments that contain the sequence CCCTC. The CTCF protein functions to silence transcription by preventing enhancers from interacting with promoters of genes on the other side of domain boundaries. *See* Appendix C, 2000, Bell and Felsenfield; *H19*, insulator DNAs.

ctDNA chloroplast DNA. Also abbreviated *cpDNA*. *See* chloroplast.

C-terminus that end of the peptide chain that carries the free alpha carboxyl group of the last amino acid. By convention, the structural formula of a peptide chain is written with the C-terminus to the right. *See* translation.

"C"-type particles a group of RNA viruses with similar morphologies under the electron microscope, having a centrally placed, spherical RNA-containing nucleoid. These viruses are associated with many sarcomas and leukemias. The "C" refers to "cancer."

Cucumis a genus of nearly 40 species including several of considerable economic importance, such as the cucumber *(C. sativus)* and the muskmelon *(C. melo)*. Considerable genetic information is available for both these species.

Cucurbita a genus of about 27 species, including 5 that are extensively cultivated: C. *pepo*, summer squash; C. *mixta*, cushaws: C. *moschata*, winter squash; C. *maxima*, Hubbard squash; and C. *ficifolia*, Malabar gourds. Most genetic information is available for C. *pepo* and C. *maxima*.

Culex pipiens the most widely distributed species of mosquito in the world. The genetics of insecticide resistance has been intensively studied in this species. Giant polytene chromosomes occur in the salivary gland and Malpighian tubule cells of larvae.

cull to pick out and discard inferior animals or plants from a breeding stock.

cultigen a plant that is known only under cultivation and whose place and method of origin is unknown.

cultivar a variety of plant produced through selective breeding by humans and maintained by cultivation. *See* strain.

curie the quantity of a radioactive nuclide disintegrating at the rate of 3.700×10^{10} atoms per second. Abbreviated Ci. 1 Ci = 3.7×10^{10} Bq.

cut a double-strand incision in a duplex DNA molecule. *Compare with* nick.

cut-and-patch repair repair of damaged DNA molecules by the enzymatic excision of the defective single-stranded segments and the subsequent synthesis of new segments. Using the complementary strand as a template, the correct bases are inserted and are interlinked by a DNA polymerase. A DNA ligase joins the two ends of the "patch" to the broken strand to complete the repair. *See* AP endonuclease, repair synthesis, thymine dimer, xeroderma pigmentosum.

cuticle the chitinous, acellular outer covering of insects.

Cu Zn SOD *See* superoxide dismutase.

C value the amount of DNA that comprises the haploid genome for a given species. Diploid cells that result from fertilization have the 2C value until they enter the S phase of their cell cycle (*q.v.*). Following S, they will have the 4C amount until mitosis produces two sibling nuclei, each with 2C. In species where females are XXAA and males XYAA (A = one set of autosomes), the diploid nuclei of females usually contain more DNA than male nuclei because the X chromosome has more DNA than the Y. In *Drosophila melanogaster*, for example, measurements reported in 1980 by P. K. Mulligan and E. M. Rasch show that male nuclei have about 90% the amount of DNA contained in female nuclei. The genome sizes published for most organisms do not differentiate separate values for the two sexes. The table illustrates the large range in the C values found among multicellular organisms. *See* Appendix C, 1948, Boivin, Vendrely, and Vendrely; 1950, Swift; Appendix F; cell cycle, chromosome set, C value paradox, genome size.

C value paradox the paradox that there is often no correlation between the C values of species and their evolutionary complexity. For example, the C values for mammals fall into a narrow range (between 2 and 3 pg). By contrast, the C values for amphibia vary from 1 to 100 pg. However, the minimum C values reported for species from each class of eukaryotes does increase with evolutionary complexity. In species with C values above the expected range, there is a greater amount of noncoding DNA. Much of this DNA is repetitive and may result from the replication of transposable elements (*q.v.*). *See* Appendix C, 1971, Thomas; chromatin diminution, repetitious DNA, selfish DNA, skeletal DNA hypothesis.

CVS chorionic villi sampling (*q.v.*).

cyanelles organelles that allow glaucocystophytes (*q.v.*) to perform photosynthesis. Cyanelles occupy an intermediate level of symbiotic integration between free-living cyanobacteria (*q.v.*) and chloroplasts (*q.v.*). Both cyanobacteria and cyanelles contain chlorophyll a. The genomes of cyanelles are about one-tenth the size of free-living cyanobacteria, but they are similar in size to the genomes of the chloroplasts of plants. The DNA genome in each cyanelle is present in about 60 copies. Unlike the situation in plants, where the large subunit of RuBisCO is encoded by chloroplast genes and the small subunit by nuclear genes, both subunits are encoded by the cyanelle genomes. *See* ribulose-1, 5-bisphosphate carboxylase-oxygenase (RuBisCO), serial symbiosis theory.

Cyanidioschyzon merolae a red alga about 2 mμ in diameter that inhabits sulfate-rich hot springs (pH 1.5, 45° C). The whole-genome shotgun (WGS) assembly (*q.v.*) method has been used to determine its nuclear genome. This contains 16,520,305 bp of DNA distributed among 20 chromosomes. The genome is unique in that only 26 of its 5,331 genes contain introns. *C. merolae* has the smallest genome of all photosynthetic eukaryotes so far studied. This protoctist also has the smallest set of rRNA genes known for any eukaryote. Each cell contains one mitochondrion and one chloroplast. Both organelles have had their DNAs sequenced, and the mitochondrion contains 32,211 bp and the chloroplast 149,987 bp of DNA. *See* Appendix A, Protoctista, Rhodophyta; Appendix C, Matsuzaki *et al.*; Appendix F; division rings, dynamin.

Cyanobacteria a phylum in the kingdom Eubacteria (*see* Appendix A). The cyanobacteria produce oxygen gas, an ability that distinguishes them from

Species	Common names	C values (gbp of DNA)
Takifugu rubripes	pufferfish	0.4
Homo sapiens	humans	3.2
Necturus punctatus	salamanders	119
Fritillaria uva-vulpis	lilies	125
Protopterus aethiopicus	lungfish	127

Note that the smallest and the largest C values so far recorded belong to fish species. There are Web sites for animal C values (http://www.genomesize.com) and plant C values (http://www.kew.org.cval.org/database1.html).

other photosynthetic bacteria. In the older literature, these bacteria were misclassified as blue-green algae and placed in the phylum Cyanophyta. The ancestors of present-day cyanobacteria were the dominant life form in the Proterozoic era, and the oxygen they generated from photosynthesis caused a transformation some 2 billion years ago of the earth's atmosphere from a reducing to an oxidizing one. The serial symbiosis theory (*q.v.*) derives chloroplasts from cyanobacteria. *See* chlorophyll, cyanelle, photosynthesis, *Prochloron*, stromatolites, *Synechocystis.*

cyanocobalamin cobalamin.

cyanogen bromide BrCN, a reagent used for splitting polypeptides at methionine residues; commonly used in studies of protein structure and the determination of amino acid sequences.

cyanolabe *See* color blindness.

cyanophage a virus that has a cyanobacterium as its host.

Cyanophyta *See* Cyanobacteria.

cyclically permuted sequences DNA sequences of the same length containing genes in the same linear order, but starting and ending at different positions, as in a circle. For example, the genes ABCDEFG can be circularly permuted to give BCDEFGA, CDEFGAB, DEFGABC, and so forth. In T4 DNA, each phage contains a different cyclically permuted sequence that is also terminally redundant. Cyclic permutation is a property of a population of phage DNA molecules, whereas terminal redundancy is a property of an individual phage DNA molecule. *See* headful mechanism, terminal redundancy.

cyclical selection selection in one direction followed by selection in the opposite direction resulting from cyclical environmental fluctuations, such as seasonal temperature changes. If the generation time is short relative to the environmental cycle, different genotypes will be selected at different times, and the population will remain genetically inhomogeneous.

cyclic AMP adenosine monophosphate with the phosphate group bonded internally to form a cyclic molecule; generated from ATP by the enzyme adenylcyclase; abbreviated cAMP. Likewise, guanosine monophosphate (GMP) can become a cyclic molecule by a phosphodiester bond between 3′ and 5′ atoms. Cyclic AMP has been shown to function as an acrasin in slime molds and to be active in the regulation of gene expression in both prokaryotes and eukaryotes. In *E. coli*, cyclic AMP is required for the transcription of certain operons. *See* Appendix C, 1957, Sutherland and Rall; adenylcyclase, catabolite repression, cellular signal transductions, CREBs, G proteins, protein kinase, second messenger.

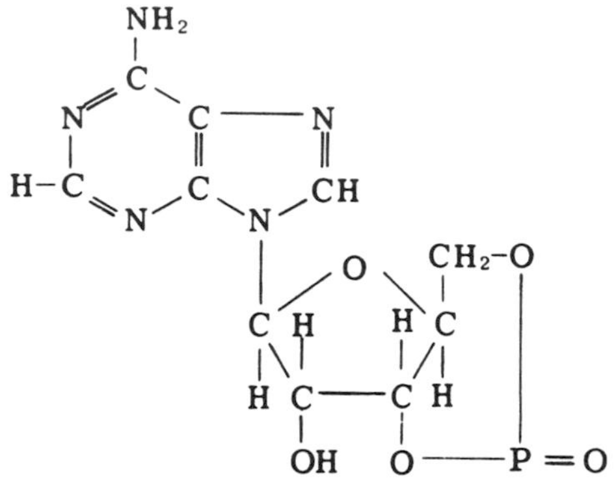

cyclins a family of proteins whose concentrations rise and fall during the cell cycle (*q.v.*). Cyclins form complexes with specific protein kinases, thereby activating them and regulating the passage of the cell through the cell division cycle. The protein kinases are called cyclin-dependent kinases (cdks) or cell-division cycle (cdc) kinases. There are two main classes of cyclins: G_1 cyclins, which bind cdks during G_1 and are necessary for entry into the S phase, and mitotic cyclins, which bind cdks during G_2 and signal entry into mitosis. Mitotic cyclins are destroyed at the subsequent anaphase. Near their *N*-terminal ends, all cyclin proteins contain a *destruction box.* This refers to a sequence of amino acids that determines whether or not the cyclin will be degraded at anaphase. Cyclins are posttranslationally modified by the covalent attachment of multiple copies of ubiquitin (*q.v.*) to a lysine residue to the right of the destruction box. Polyubiquitin-containing proteins are degraded by large protein complexes called *proteasomes.* The attachment of ubiquitin to mitotic cyclins requires the enzyme *ubiquitin ligase* and a *recognition protein* that attaches to the destruction box. G_1 cyclins combine with different kinases than do mitotic cyclins. The result is a start kinase, which induces chromosome replication. *See* Appendix C, 1983, Hunt *et al.*; checkpoint, cyclin-dependent kinase 2 (Cdk2), maturation promoting factor (MPF), protein kinase.

cycloheximide an antibiotic synthesized by *Streptomyces griseus*. The drug inhibits translation on 80S ribosomes. Therefore, it suppresses cytosolic protein synthesis without affecting the synthesis of proteins in mitochondria or chloroplasts. Protein synthesis in these organelles can be specifically inhibited by chloramphenicol, erythromycin, or tetracycline. *See* ribosome, ribosomes of organelles.

```
        H  CH3
        |  |
        C=C
       /    \
H3C—C        C=O          H  O
     \\     /             |  ||
      C—C                 C—C
      |  |\             //    \
      H  H CH—CH2—C            NH
           |        \         /
           OH        C—C
                    /|  ||
                   H H  O
```

cyclorrhaphous diptera flies belonging to the suborder Cyclorrhapha, which contains the most highly developed flies. It includes the hover flies, the drosophilids, house flies, blow flies, etc.

cyclosis cytoplasmic streaming.

cyclotron *See* accelerator.

cys cysteine (*q.v.*).

cysteine a sulfur-bearing amino acid found in biological proteins. It is important because of its ability to form a disulfide cross-link with another cysteine, either in the same or between different polypeptide chains. *See* amino acid, cystine, insulin.

csyteine proteases proteolytic enzymes in which a cysteine residue resides in the catalytic domain and is required for enzymatic activity. These enzymes form four large superfamilies consisting of at least 30 families, each of which has evolutionarily conserved sequence domains. Examples of cysteine proteases include papain, caspases, cathepsins, and various deubiquitinating enzymes (all of which *See*).

cystic fibrosis (CF) the most common hereditary disease of Caucasians. In the United States, the frequency of homozygotes is 1/2,000, while heterozygotes make up about 5% of the population. CF is a generalized multiorgan system disease arising from viscous mucous secretions that clog the lungs and digestive tract. The disease is inherited as an autosomal recessive and is caused by mutations in a gene residing on the long arm of chromosome 7 in region 31–32. The *CF* gene is approximately 250 kilobases long, and its 27 exons encode a protein containing 1,480 amino acids. This has been named the cystic fibrosis transmembrane-conductance regulator (CFTR). The *CF* gene is expressed predominantly in mucus-secreting epithelial cells, such as those of the submucosal glands of the bronchi, the salivary glands, the sweat glands, pancreas, testes, and intestines. The CFTR functions as a channel for chloride ions. Proper chloride transport is necessary for diluting and flushing mucus downstream from mucus-secreting glands. Frameshift, missense, nonsense, and RNA splicing mutations have been isolated from victims of the disease. The most common mutation is ΔF508. The abbreviation indicates that there is a deletion (Δ) of phenylalanine (F) at position 508. This mutation is present in 60–70% of the CF chromosomes from North American Caucasians. A study of ΔF508 chromosomes in European families indicates that the mutation arose during paleolithic times in a population resembling the present-day Basques (*q.v.*). ΔF508 results in a temperature-sensitive defect in protein processing. At 27°C the chloride channels are normal, but at 37°C transport of CFTR from the endoplasmic reticulum to the cell membrane never occurs. Therefore, Cl^- channels cannot form, and a very severe form of CF results. The diagram of the CFTR molecule shows that the ΔF508 mutation resides in the first of two nucleotide-binding domains (NBDs). The regulatory domain (RD) is a region

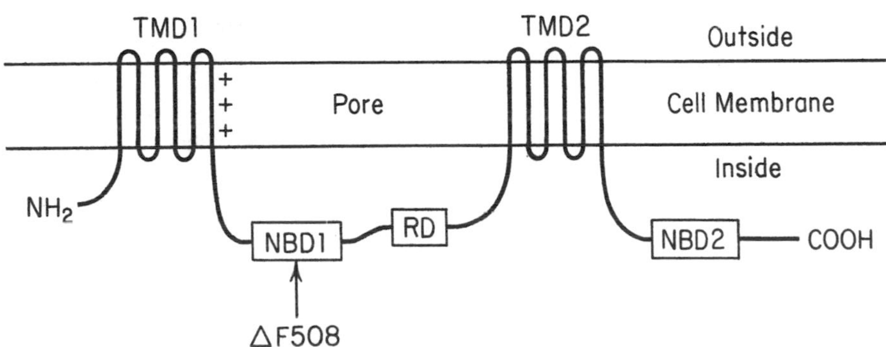

Cystic fibrosis transmembrane-conductance regulator (CFTR)

that controls the response of CFTR to protein kinases (*q.v.*). There are two transmembrane domains (TMDs) where the protein folds back and forth, spanning the lipid bilayer of the cell membrane six times. Positively charged arginine and lysine molecules (indicated by pluses in the diagram) are essential for the passage of anions through the pore. Missense mutations that replace these with neutral amino acids also cause CF. CF heterozygotes appear to be resistant to cholera, which may explain why the mutants like ΔF508 have been retained in human populations. *See* Appendix C, 1989, Tsui *et al.*, 1993; Tabcharani *et al.*; 1994, Morral *et al.*, Gabriel *et al.*; ABC transporter, calnexin, cellular signal transduction, cholera, gene. http://www.cff.org

cystine a derived amino acid formed by the oxidation of two cysteine thiol side chains, which join to form a disulfide covalent bond. Such bonds play an important role in stabilizing the folded configurations of proteins. *See* cysteine, insulin, posttranslational processing.

cystoblast *See* cystocyte divisions.

cystocyte divisions the series of mitotic divisions which generate the nurse cell/oocyte clones that characterize insects with polytrophic meroistic ovaries (like *Dorsophila*). In *D. melanogaster* two or three stem-line oogonia reside in each germarium (*q.v.*). Each stem cell (S) divides into two daughter cells. One behaves like its parent, and the other differentiates into a cystoblast (C_b). This cell, by a series of four mitoses (M_1–M_4), each followed by incomplete cytokinesis, produces a branching chain of 16 interconnected cells. In the diagram here, cystocytes (represented by open circles) belong to the first, second, third, or fourth generation. The area in each circle is proportional to the volume of the cell. The number of lines connecting any two cells shows the division at which the ring canal (*q.v.*) joining them was formed. Cells 1^4 and 2^4 enter the oocyte developmental pathway and form synaptonemal complexes (*q.v.*). These cells are therefore called pro-oocytes (*q.v.*). *See* insect ovary types, polyfusome, stem cell.

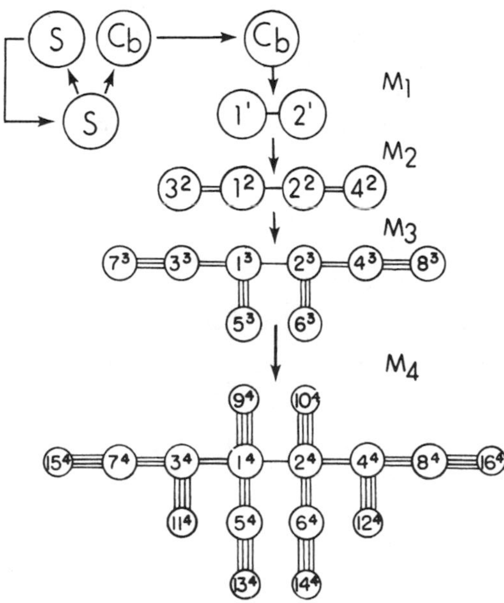

cytidine *See* nucleoside.

cytidylic acid *See* nucleotide.

cytochalasin B a mold antimetabolite that prevents cells from undergoing cytokinesis. *See* actin, contractile ring.

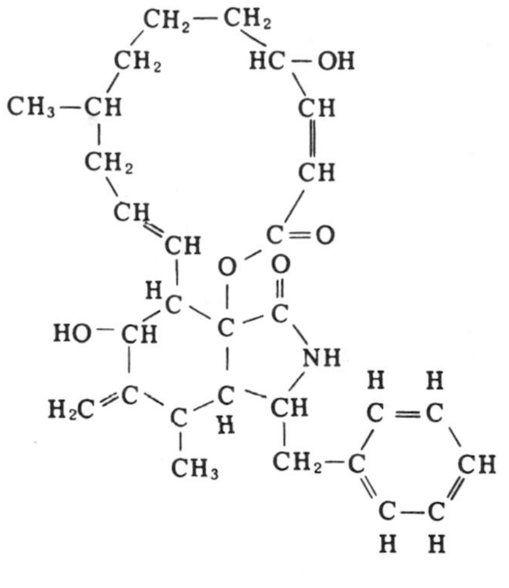

cytochromes a family of heme-containing proteins that function as electron donors and acceptors during the chains of reactions that occur during respiration and photosynthesis. Electron transport depends upon the continued oxidation and reduction of the iron atom contained in the center of the porphyrin prosthetic group (*see* heme). The first cytochrome is thought to have arisen about two billion years ago, and the genes that encode cytochromes have been modified slowly by base substitutions since then. The cytochromes were the first group of proteins for which amino acid sequence data allowed the construction of an evolutionary tree. *See* Appendix C, 1963, Margoliash; 1967, Fitch and Margoliash.

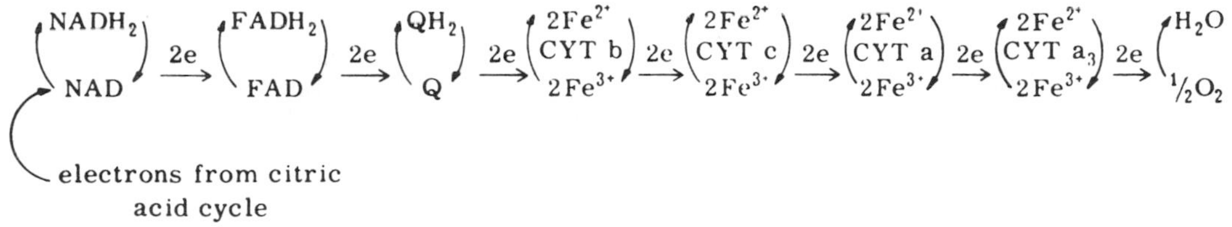

$NADH_2$ / NAD $\xrightarrow{2e}$ $FADH_2$ / FAD $\xrightarrow{2e}$ QH_2 / Q $\xrightarrow{2e}$ $2Fe^{2+}$ CYT b $2Fe^{3+}$ $\xrightarrow{2e}$ $2Fe^{2+}$ CYT c $2Fe^{3+}$ $\xrightarrow{2e}$ $2Fe^{2+}$ CYT a $2Fe^{3+}$ $\xrightarrow{2e}$ $2Fe^{2+}$ CYT a_3 $2Fe^{3+}$ $\xrightarrow{2e}$ H_2O / $\frac{1}{2}O_2$

electrons from citric acid cycle

Cytochrome system

cytochrome system a chain of coupled oxidation/reduction reactions that transports the electrons produced during the oxidations occurring in the citric acid cycle (*q.v.*) to the final hydrogen and electron acceptor, oxygen, to form water. The molecules involved in this chain are NAD (*q.v.*), FAD (*q.v.*), coenzyme Q (*q.v.*), and cytochromes b, c, a, and a_3. The sequence of reactions is diagrammed above. *See* ATP synthase, electron transport chain, Leber hereditary optic neuropathy (LHON), Leigh syndrome, mitochondrial proton transport.

cytogamy synonymous with autogamy (*q.v.*).

cytogenetic map a map showing the locations of genes on a chromosome.

cytogenetics the science that combines the methods and findings of cytology and genetics. *See* symbols used in human cytogenetics.

cytohet a eukaryotic cell containing two genetically different types of a specific organelle; the term is an abbreviation for *cyto*plasmically *het*erozygous. For example, in the single-celled alga *Chlamydomonas*, the frequency of rare cytohets (containing chloroplasts from both parents) can be greatly increased by treatment of one parent (mating type +) with ultraviolet light. *See* mitotic segregation.

cytokines a group of small proteins (5–20 kilodaltons) involved primarily in communication between cells of the immune system. Unlike hormones of the endocrine system, which can exert their effects over long distances, cytokines usually act locally on nearby cells. The term includes interleukins, interferons, lymphokines, and tumor necrosis factors (*all of which see*). *Compare with* autocrine.

cytokinesis cytoplasmic division as opposed to karyokinesis (*q.v.*) *See* cleavage, contractile ring.

cytokinins a family of *N*-substituted derivatives of adenine (*q.v.*) synthesized mainly in the roots of higher plants. Cytokinins (also called kinins and phytokinins) promote cell division and the synthesis of RNA and protein. The first molecule with these properties was called kinetin (*see* Appendix C, 1956, Miller). The first cytokinin obtained from a plant was zeatin. It was isolated from maize kernels in 1964.

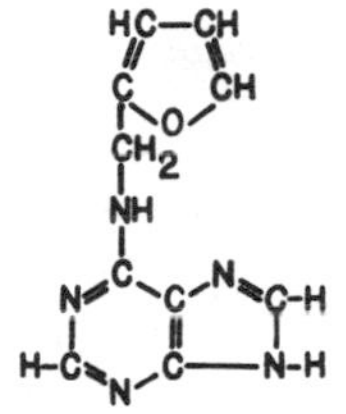

Kinetin 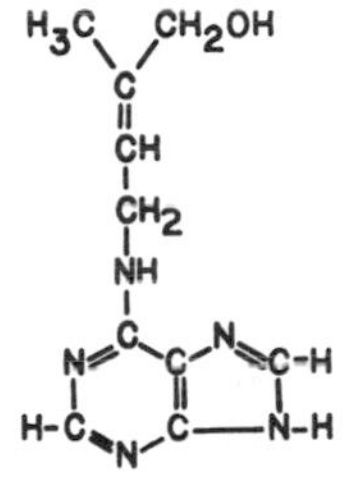 **Zeatin**

cytological hybridization synonymous with *in situ* hybridization (*q.v.*).

cytological map a diagrammatic representation of the physical location of genes at specific sites, generally on dipteran giant polytene chromosomes or on human mitotic chromosomes.

cytology the branch of biology dealing with the structure, function, and life history of the cell. *See* Appendix C, 1838, Schleiden and Schwann; 1855, Virchow; 1896, Wilson.

cytolysis the dissolution of cells.

cytophotometry quantitative studies of the localization within cells of various organic compounds using microspectrophotometry. Cytophotometric techniques are employed, for example, to determine changes in the DNA contents of cells throughout their life cycle. *See* Appendix C, 1936, Caspersson; microspectrophotometer.

cytoplasm the protoplasm exclusive of that within the nucleus (which is called nucleoplasm).

cytoplasmic asymmetry uneven distribution of cytoplasmic components in a cell. *See* cytoplasmic determinants, cytoplasmic localization.

cytoplasmic determinants molecules that are localized in specific cytoplasmic regions of the unfertilized egg or zygote and affect cell fate decisions by segregating into different embryonic cells and controlling distinct gene activities in these cells. In the egg, such determinants are usually maternal mRNAs and proteins. Cytoplasmic determinants are also found in some post-embryonic cells, where they produce cytoplasmic asymmetry (*q.v.*). In dividing cells, this leads to asymmetric cell division in which each of the daughter cells differentiates into a different cell type. *Also called* localized cytoplasmic determinants or morphogenetic determinants. *See* *bicoid*, cytoplasmic localization, maternal effect gene, maternal polarity mutants, pole plasm.

cytoplasmic inheritance non-Mendelian heredity involving replication and transmission of extrachromosomal genetic information found in organelles such as mitochondria and chloroplasts or in intracellular parasites such as viruses; also called *extranuclear inheritance*. *See* Appendix C, 1909, Correns and Bauer; mtDNA lineages.

cytoplasmic localization the process whereby maternally or zygotically synthesized molecules become situated in specific spatial locations in the egg or zygote. This has been most widely examined in the *Drosophila* egg (e.g., in formation of the pole plasm (*q.v.*) or in positioning of cytoplasmic determinants (*q.v.*) that are later required for embryonic body pattern formation) and thought to be a stepwise process involving synthesis of the maternal product, its transport to the desired location, anchoring, and maintenance of localization. These steps are dependent upon sequential gene expression, cytoskeletal elements, and cell organelles. *See* Balbiani body, *bicoid*, maternal effect gene, maternal polarity mutants, mitochondrial cloud, sponge body.

cytoplasmic male sterility (CMS) pollen abortion due to cytoplasmic factors that are maternally transmitted, but that act only in the absence of pollen-restoring genes. Such sterility can also be transmitted by grafting. In maize, pollen death is due to "abortion proteins" secreted by mitochondria, and the genes required to restore pollen fertility lower the abundance of abortion proteins by reducing rates of transcription of their mRNAs. Hybrid corn seed is produced commercially by a breeding system involving CMS. Unfortunately, the abortion proteins also enhance susceptibility of the plants to fungal toxins. *See* Appendix C, 1987, Dewey, Timothy, and Levings, *Bipolaris maydis*, hybrid corn.

cytoplasmic matrix *See* microtrabecular lattice.

cytoplast the structural and functional unit of an eukaryotic cell formed by a lattice of cytoskeletal proteins to which are linked the nucleus and the cytoplasmic organelles.

cytosine *See* bases of nucleic acids, 5-hydroxymethylcytosine.

cytosine deoxyriboside *See* nucleoside.

cytoskeleton an internal skeleton that gives the eukaryotic cell its ability to move, to assume a characteristic shape, to divide, to undergo pinocytosis, to arrange its organelles, and to transport them from one location to another. The cytoskeleton contains microtubules, microfilaments, and intermediate filaments.

cytosol the fluid portion of the cytoplasm exclusive of organelles; synonymous with hyaloplasm. *See* cell fractionation.

cytostatic referring to any agent that suppresses cell multiplication and growth.

cytotaxis the ordering and arranging of new cell structure under the influence of preexisting cell structure. The information controlling the three-dimensional architecture of the eukaryotic cell is thought to reside in the structure of the cytoplasmic ground substance. Evidence for this comes from microsurgical experiments on *Paramecium*. Cortical segments reimplanted with inverted polarity result in a changed pattern that is inherited through hundreds of generations. *See* microtrabecular lattice.

cytotoxic T lymphocyte a lymphocyte that binds to a foreign cell and kills it. Such lymphocytes recognize target cells on the basis of the antigenic properties of their class I histocompatibility molecules. *See* helper T lymphocyte, T lymphocyte.

D

d 1. dextrorotatory. 2. the dalton unit.

2,4D 2,4 dichlorophenoxyacetic acid (*q.v.*).

daf-2 a gene in *Caenorhabditis* that regulates its life span. *See* insulin-like growth factors 1 and 2 (IGF-1 and IGF-2).

dalton a unit equal to the mass of the hydrogen atom (1.67×10^{-24} g) and equal to 1.0000 on the atomic mass scale. The unit is named after John Dalton (1766–1844), who developed the atomic theory of matter. Abbreviated Da.

daltonism *See* color blindness.

dam the female parent in animal breeding. *Compare with* sire.

Danaus plexippus the Monarch butterfly. *See* automimicry.

Danio rerio the fish that has become a model organism for the genetic study of vertebrate development. The fish has a 3-month life cycle and produces large, transparent embryos. Large-scale mutagenesis experiments have generated a wealth of mutations that produce a dazzling array of abnormal phenotypes. The genome contains about 1,700 mbp of DNA distributed among 25 chromosomes. *See* Appendix A, Chordata, Osteichthyes, Neopterygii, Cyprinidontiformes; Appendix C, 1993, Mullins and Nüsslein-Volhard; Appendix E.

DAPI 4′,6-diamidino-2-phenylindole, a fluorescent dye that binds to DNA. DAPI-staining of chromosomes within nuclei can be followed with the collection of three dimensional data sets obtained by recording serial images at 0.25 μm intervals. From these, linearized maps of all the chromosomes can be constructed. The structure of the DAPI molecule is shown below.

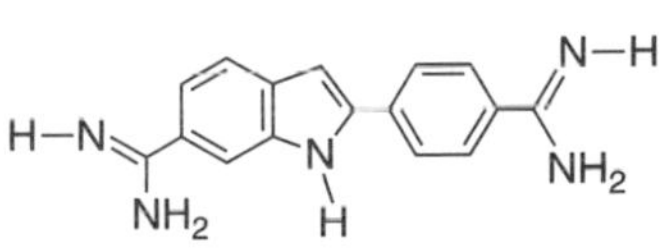

dark-field microscope a microscope designed so that the entering center light rays are blacked out and the peripheral rays are directed against the object from the side. As a result, the object being viewed appears bright upon a dark background.

dark reactivation repair of mutagen-induced genetic damage by enzymes that do not require light photons for their action. *See* photoreactivating enzyme.

Darwinian evolution *See* Darwinism.

Darwinian fitness synonymous with adaptive value (*q.v.*).

Darwinian selection synonymous with natural selection (*q.v.*).

Darwinism the theory that the mechanism of biological evolution involves natural selection of adaptive variations. *See* gradualism, *Origin of Species*.

Darwin on the Web the most extensive collection of Darwin's writings (http://pages.britishlibrary.net/charles.darwin/).

Darwin's finches a group of finches observed and collected by Charles Darwin during his visit to the Galapagos Islands in 1835. Birds of all 14 species are seed eaters, but they are subdivided into one genus of ground finches (*Geospiza*) and two genera of tree finches (*Camarhynchus* and *Cactospiza*). The species differ in beak morphology, coloration of plumage, size, and habitat preferences. Darwin was the first to suggest that the modern populations of these birds are the end product of an adaptive radiation from a single ancestral species. The evolutionary divergences resulted from adaptations that allowed different populations to utilize different food sources on different islands and to avoid competition. This adaptive radiation occurred in less than 3 million years. Recent DNA analyses suggest that the ancestor to Darwin's finches was phenotypically similar to a warbler finch, *Certhidea olivacea*, that currently inhabits many of the islands. *See* Appendix C, 1835, Darwin; 1947, Lack; 1999, Petren, Grant, and Grant.

Dasypus a genus of armadillos that contains six species, all of which are always polyembryonic, producing four genetically identical offspring per litter. The nine-banded armadillo, *Dasypus novemcinctus*, is the most studied species. *See* cloning.

Datura stramonium the Jimson weed, a species belonging to the nightshade family of plants. It is found all over North America as a roadside weed. The plant is dangerous to eat, since it synthesizes a variety of toxic and hallucinogenic alkaloids (*q.v.*). *D. stramonium* has 12 pairs of chromosomes. A set of trisomics was developed, each with a different chromosome in triplicate. Each primary trisomic differed from normal and from each other in characteristic ways. This suggested that each chromosome contained genes with morphogenetic effects and that the abnormal phenotype that characterized each trisomic was the result of increases in the relative doses of these genes. *See* Appendix A, Plantae, Angiospermae, Dicotyledonae, Solanales; Appendix C, 1920, Blakeslee, Belling, and Farnham; aneuploidy, haploid sporophytes, haploidy, polyploidy.

dauermodification an environmentally induced phenotypic change in a cell that survives in the generative or vegetative descendants of the cell in the absence of the original stimulus. However, with time the trait weakens and eventually disappears.

daughter cells (nuclei) the two cells (nuclei) resulting from division of a single cell (nucleus). Preferably called *sibling* or *offspring cells* (nuclei).

day-neutral referring to a plant in which flowering is not controlled by photoperiod. *See* phytochrome.

DBM paper diazobenzyloxymethyl paper that binds all single-stranded DNA, RNA, and proteins by means of covalent linkages to the diazonium group; used in situations where nitrocellulose blotting is not technically feasible. *See* Appendix C, 1977, Alwine *et al.*

DEAE-cellulose diethylaminoethyl-cellulose, a substituted cellulose derivative used in bead form for chromatography of acidic or slightly basic proteins at pH values above their isoelectric point.

deamination the oxidative removal of NH_2 groups from amino acids to form ammonia.

decarboxylation the removal or loss of a carboxyl group from an organic compound and the formation of CO_2.

decay of variability the reduction of heterozygosity because of the loss and fixation of alleles at various loci accompanying genetic drift.

deciduous 1. designating trees whose leaves fall off at the end of the growing season, as opposed to evergreen. 2. designating teeth that are replaced by permanent teeth.

decoy protein *See* sporozoite.

dedifferentiation the loss of differentiation, as in the vertebrate limb stump during formation of a blastema. In the regenerating mammalian liver, cells undergo partial dedifferentiation, allowing them to reenter, the cell cycle while maintaining all critical differentiation functions. *See* differentiation, regeneration.

defective virus a virus that is unable to reproduce in its host without the presence of another "helper" virus (*q.v.*).

deficiency in cytogenetics, the loss of a microscopically visible segment of a chromosome. In a structural heterozygote (containing one normal and one deleted chromosome), the nondeleted chromosome forms an unpaired loop opposite the deleted segment when the chromosomes pair during meiosis. *See* Appendix C, 1917, Bridges; cat cry sydrome.

deficiency loop in polytene chromosomes, deficiency loops allow one to determine the size of the segment missing. The illustration on page 115 shows a portion of the X chromosome from the nucleus of a salivary gland cell of a *Drosophila* larva structurally heterozygous for a deficiency. Note that bands C2–C11 are missing from the lower chromosome.

defined medium a medium for growing cells, tissues, or multicellular organisms in which all the chemical components and their concentrations are known.

definitive host the host in which a parasite attains sexual maturity.

deformylase an enzyme in prokaryotes that removes the formyl group from the *N*-terminal amino acid; fMet is never retained as the *N*-terminal amino acid in functional polypeptides. *See* start codon.

degenerate code one in which each different word is coded by a variety of symbols or groups of letters. The genetic code is said to be degenerate because more than one nucleotide triplet codes for the same amino acid. For example, the mRNA triplets GGU, GGC, GGA, and GGG all encode glycine. When two codons share the same first two nucleotides they will encode the same amino acids if the third nucleotide is either U or C and often if it is A or G. *See* amino acids, codon bias, genetic code, wobble hypothesis.

degrees of freedom the number of items of data that are free to vary independently. In a set of quantitative data, for a specified value of the mean, only $(n - 1)$ items are free to vary, since the value of the

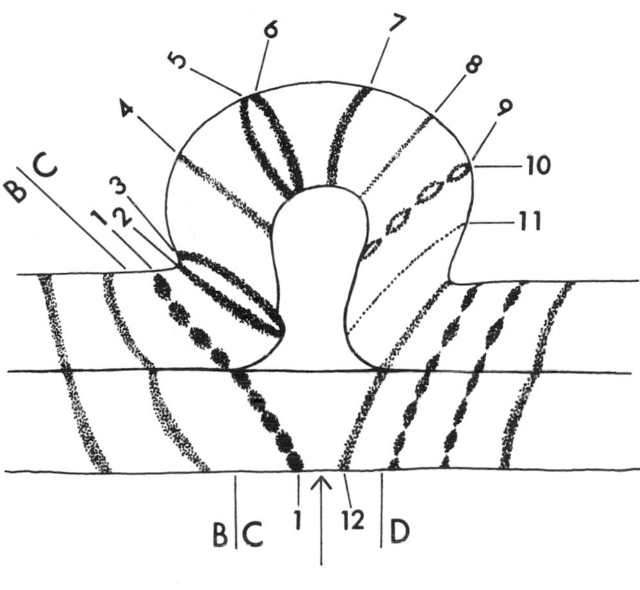

Deficiency loop

*n*th item is then determined by the values assumed by the others and by the mean. In a chi-square test (*q.v.*) the number of degrees of freedom is one less than the number of phenotypic classes observed.

dehiscent designating fruit that opens when ripe to release seeds.

Deinococcus radiodurans a Gram-positive red-pigmented, nonmotile, aerobic bacterium that is extremely resistant to a number of agents that damage DNA (ionizing radiation, ultraviolet radiation, and hydrogen peroxide). *D. radiodurans* can tolerate 3 million rads of ionizing radiation (the human lethal dose is about 500 rads). The *D. radiodurans* genome is composed of four circular molecules: chromosome 1 (2,649 kb), chromosome 2 (412 kb), a megaplasmid (177 kb), and a plasmid (46 kb). The genome contains 3,187 ORFs, with an average size of 937 kb, and these occupy 91% of the genome. The species possesses a highly efficient DNA repair system that involves about 40 genes, many of which are present in multiple copies. *See* Appendix A, Bacteria, Deinococci; Appendix C, 1999, White *et al.*; Appendix E; haploidy.

delayed dominance *See* dominance.

delayed hypersensitivity a cell-mediated immune response manifested by an inflammatory skin response 24–48 hours after exposure to antigen. *Compare with* immediate hypersensitivity.

delayed Mendelian segregation *See Lymnaea peregra.*

deletion the loss of a segment of the genetic material from a chromosome. The size of the deletion can vary from a single nucleotide to sections containing a number of genes. If the lost part is at the end of a chromosome, it is called a terminal deletion. Otherwise, it is called an intercalary deletion. *See* indels.

deletion mapping **1.** the use of overlapping deletions to localize the position of an unknown gene on a chromosome or linkage map. **2.** the establishment of gene order among several phage loci by a series of matings between point mutation and deletion mutants whose overlapping pattern is known. Recombinants cannot be produced by crossing a strain bearing a point mutant with another strain carrying a deletion in the region where the point mutant resides. *See* Appendix C, 1938, Slizynska; 1968, Davis and Davidson.

deletion method a method of isolating specific messenger RNA molecules by hybridization with DNA molecules containing genetic deletions.

deletion-substitution particles a specialized transducing phage in which deleted phage genes are substituted by bacterial genes.

Delta the capital Greek letter (Δ) used in molecular biology to indicate a deletion of one or more amino acids in a polypeptide chain. *See* cystic fibrosis (CF).

delta chain a component of hemoglobin A_2. *See* hemoglobin.

delta ray the track or path of an electron ejected from an atomic nucleus when an ionizing particle passes through a detection medium, especially through a photographic emulsion.

delta T50H the difference between the temperature at which DNA homoduplexes and DNA heteroduplexes undergo 50% dissociation. The statistic is often used to measure the genetic relationship between the nucleotide sequences of two or more species. A ΔT50H value can be converted into an absolute time interval if the fossil record can provide an independent dating estimate. In primates, a ΔT50H value of 1 equals about 11 million years. If repetitive sequences have been removed from the DNAs, then a ΔT50H value of 1 represents about a 1% difference in single-copy genes between the samples. *See* DNA clock hypothesis, reassociation kinetics.

deme a geographically localized population within a species.

denaturation the loss of the native configuration of a macromolecule resulting from heat treatment, extreme pH changes, chemical treatment, etc. Denaturation is usually accompanied by loss of biological activity. Denaturation of proteins often results in an unfolding of the polypeptide chains and renders the molecule less soluble. Denaturation of DNA leads to changes in many of its physical properties, including viscosity, light scattering, and optical density. This "melting" occurs over a narrow range of temperatures and represents the dissociation of the double helix into its complementary strands. The midpoint of this transition is called the melting temperature. *See* T_m.

denaturation map a map, obtained through electron microscopy using the Kleinschmidt spreading technique (*q.v.*), of a DNA molecule that shows the positions of denaturation loops. These are induced by heating the molecules to a temperature where segments held together by A=T bonds detach while those regions held together by G≡C base pairs remain double-stranded. Formaldehyde reacts irreversibly with bases that are not hydrogen bonded to prevent reannealing. Thus, after the addition of formaldehyde the DNA molecule retains its denaturation loops when cooled. Denaturation maps provide a unique way to distinguish different DNA molecules.

denatured DNA *See* denaturation.

denatured protein *See* denaturation.

dendrite one of the many short, branching cytoplasmic projections of a neuron. Dendrites synapse with and receive impulses from the axons of other neurons. These impulses are then conducted toward the perikaryon.

Denhardt solution a solution consisting of Ficoll, polyvinylpyrrolidone, and bovine serum albumin, each at a concentration of 0.02% (w/v). Preincubation of nucleic acid–containing filters in this solution prevents nonspecific binding of single-stranded DNA probes. The solution is named after David T. Denhardt who formulated it in 1966.

de novo **1.** arising from an unknown source. **2.** denoting synthesis of a specified molecule from very simple precursors, as opposed to the formation of the molecule by the addition or subtraction of a side chain to an already complex molecule.

***de novo* mutation** a mutation that occurs in one member of a family as a result of a mutation in a germ cell of a parent or in a fertilized egg. For example, there is no history of achondroplasia (*q.v.*) in 90% of the families that have one child with the condition. In these situations a *de novo mutation* occurred, and the parents have a very low chance of having a second child with achondroplasia.

***de novo* pathway** a process for synthesizing ribonucleoside monophosphates from phosphoribosylpyrophosphate, amino acids, CO_2, and NH_3, rather than from free bases, as in the salvage pathway.

densitometer an instrument used for measuring the light transmitted through an area of interest. Densitometers are used for scanning chromatograms and electropherograms and for measuring the blackening of photographic films.

density-dependent factor an ecological factor (e.g., food) that becomes increasingly important in limiting population growth as the population size increases.

density-dependent selection selection in which the values for relative fitness depend upon the density of the population.

density gradient equilibrium centrifugation *See* centrifugation separation.

density gradient zonal centrifugation *See* centrifugation separation.

density-independent factor an ecological factor (e.g., temperature) that is uncorrelated with variations in size of a population.

dent corn *See* corn.

deoxyadenylic, deoxycytidylic, deoxyguanylic acids *See* nucleotide.

deoxyribonuclease any enzyme that digests DNA to oligonucleotides or nucleotides by cleaving the phosphodiester bonds. *See* deoxyribonucleic acid, DNAase footprinting, DNAase protection, nucleotide, oligonucleotide.

deoxyribonucleic acid DNA, the molecular basis of heredity. DNA consists of a polysugar-phosphate backbone from which the purines and pyrimidines project. The backbone is formed by bonds between the phosphate molecule and carbon 3 and carbon 5 of adjacent deoxyribose molecules. The nitrogenous base extends from carbon 1 of each sugar. According to the Watson-Crick model, DNA forms a double helix that is held together by hydrogen bonds between specific pairs of bases (thymine to adenine and cytosine to guanine). Each strand in the double helix is complementary to its partner strand in terms of its base sequence. The diagram on page 118 shows that the two strands are aligned in opposite directions. Thymine and guanine are connected in the 3′ → 5′ direction and the O atom of deoxyribose points down. Adenine and cytosine are linked in a 5′ → 3′ direction, and the O atom of the pentose points up. The antiparallel strands form a right-handed helix that undergoes one complete revolution with each 10 nucleotide pairs. DNA molecules are the largest biologically active molecules known, having molecular weights greater than 1×10^8 daltons. In the adjacent diagram, only five nucleotide pairs of the ladderlike DNA molecule are shown. "Uprights" of the ladder consist of alternating phosphate (P) and deoxyribose sugar (S) groups. The "cross rungs" consist of purine-pyrimidine base pairs that are held together by hydrogen bonds (represented here by dashed lines). A, T, G, and C represent adenine, thymine, guanine, and cytosine, respectively. Note that the AT pairs are held together less strongly than the GC pairs. In reality, the ladder is twisted into a right-handed double helix, and each nucleotide pair is rotated 36° with respect to its neighbor. A DNA molecule of molecular weight 2.5 $\times 10^7$ daltons would be made up of approximately 40,000 nucleotide pairs. The type of DNA described here is the B form that occurs under hydrated conditions and is thought to be the principal biological conformation. The A form occurs under less hydrated conditions. Like the B form, it, too, is a right-handed double helix; however, it is more compact, with 11 base pairs per turn of the helix. The bases of the A form are tilted 20° away from perpendicular and displaced laterally in relation to the diad axis. The Z form of DNA is a left-handed double helix. It has 12 base pairs per turn of the helix, and presents a zigzag conformation (hence the symbolic designation). Unlike B DNA, Z DNA is antigenic. *See* Appendix C, 1871, Meischer; 1929, Levine and London; 1950, Chargaff; 1951, Wilkins and Gosling; 1952, Franklin and Gosling; Crick; Brown and Todd; 1953, Watson and Crick; 1961, Josse, Kaiser, and Kornberg; 1973, Rossenberg *et al.*; 1976, Finch and Klug; Appendix E; antiparallel, C value, DNA grooves, genome size, hydrogen bond, nucleic acid, nucleosome, photograph 51, promiscuous DNA, solenoid structure, strand terminologies, zygotene (zy) DNA.

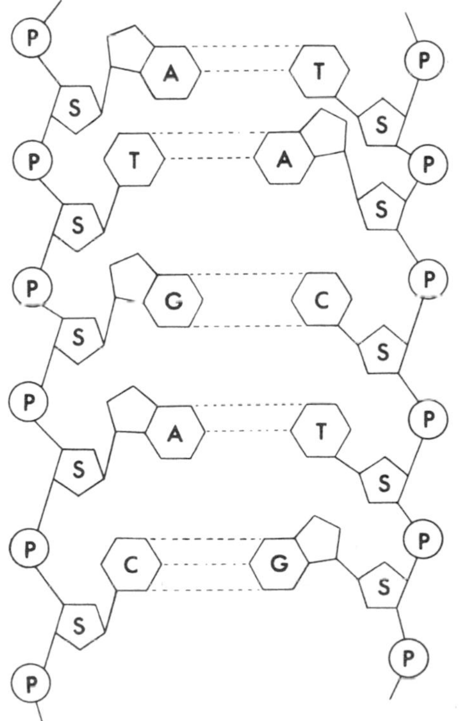

deoxyribonucleoside a molecule containing a purine or pyrimidine attached to deoxyribose.

deoxyribonucleotide a compound consisting of a purine or pyrimidine base bonded to deoxyribose, which in turn is bound to a phosphate group.

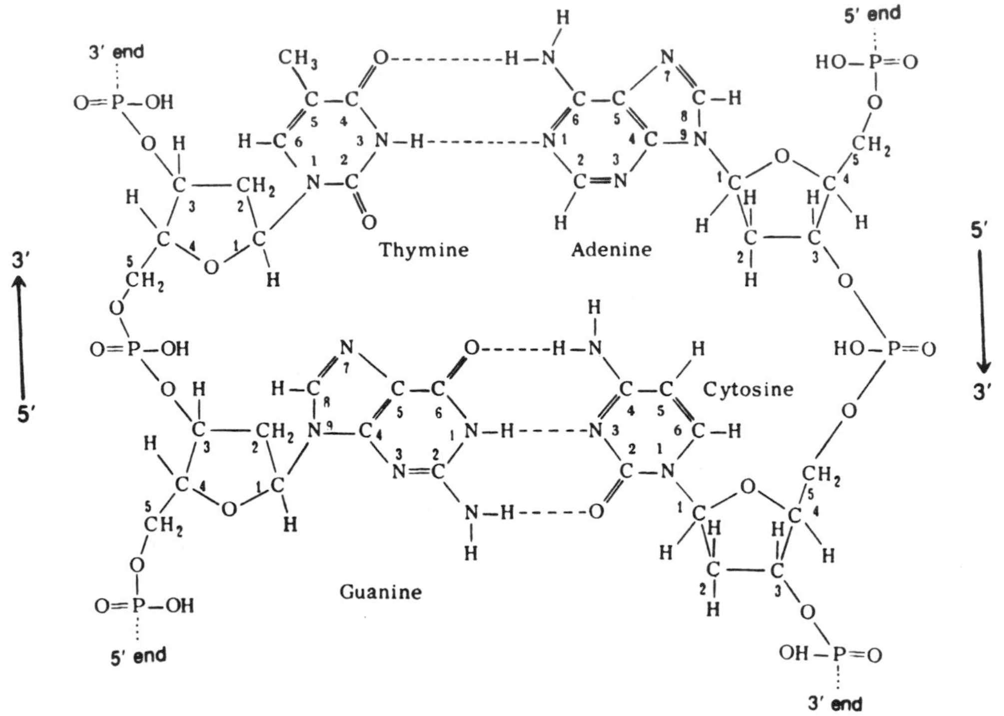

Deoxyribonucleic acid

deoxyribose the sugar characterizing DNA.

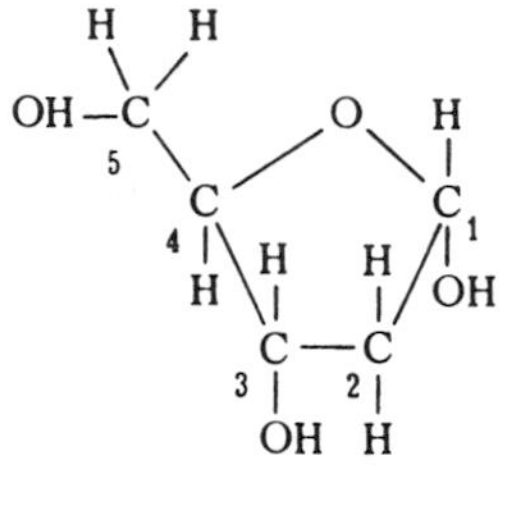

dependent differentiation differentiation of an embryonic tissue caused by a stimulus coming from other tissue and dependent on that stimulus.

depolymerization the breakdown of an organic compound into two or more molecules of less complex structure.

derepression an increased synthesis of gene product accomplished by preventing the interaction of a repressor with the operator portion of the operon in question. In the case of inducible enzyme systems, the inducer derepresses the operon. A mutation of the regulatory gene that blocks synthesis of the repressor or a mutation of the operator gene that renders it insensitive to a normal repressor will also result in derepression. *See* regulator gene.

derived the more recent stages or conditions in an evolutionary lineage; the opposite of primitive.

dermatoglyphics the study of the patterns of the ridged skin of the palms, fingers, soles, and toes.

Desferal the trade name for desferrioxamine, an iron chelator. Children with hereditary diseases that cause red blood cells to die at an accelerated pace receive frequent blood transfusions. Eventually their systems become overloaded with iron, and this can damage both heart and liver. Such children are often fitted with an intravenous Desferal pump. This infuses them with the chelator, which leaches the excess iron out of their bodies. *See* Cooley anemia, thalassemia.

desmids green algae that exist as pairs of cells with their cytoplasms joined at an isthmus that contains a single shared nucleus. *See* Appendix A, Protoctista, Gamophyta; *Micrasterias thomasiana*.

desmin a 51,000-dalton cytoskeletal protein. Desmin molecules fall into the intermediate filament class and are found in glial and muscle cells.

desmosome an intercellular attachment device. It is a discontinuous button-like structure consisting of two dense plaques on the opposing cell surfaces, separated by an intercellular space about 25 nanometers wide. On each symmetrical half-desmosome a thin layer of dense material coats the inner leaf of the cell membrane, and bundles of fine cytoplasmic filaments converge upon and terminate in this dense substance.

desoxyribonucleic acid an obsolete spelling of *deoxyribonucleic acid* found in older literature.

destruction box *See* cyclins.

desynapsis the failure of homologous chromosomes that have synapsed normally during pachynema to remain paired during diplonema. Desynapsis is usually the result of a failure of chiasma formation. *Contrast with* asynapsis.

detached X an X chromosome formed by the detaching of the arms of an attached X chromosome (*q.v.*), generally through crossing over with the Y chromosome.

determinant 1. in immunology, the portion of the antigen that is responsible for the specificity of the response and that is recognized by the binding sites of immunoglobulins and antigen-recognizing lymphocytes. 2. a factor that signals a cell to follow a particular developmental pathway. *See* cytoplasmic determinant.

determinant cleavages a successive series of cleavages that follow a specific three-dimensional pattern such that with each division, cells are produced, each of which can be shown to serve as the progenitor of a specific type of tissue. In developing mollusc eggs, for example, cell 4d, which is formed at the sixth cleavage, is always the progenitor of all primary mesodermal structures.

determinate inflorescence an inflorescence in which the first flowers to open are at the tip or inner part of the cluster, and the later ones are progressively lower or farther out.

determination the establishment of a single kind of histogenesis for a part of an embryo, which it will perform irrespective of its subsequent situations. *Compare with* differentiation.

deubiquitinating enzymes a large and heterogeneous group of cysteine proteases (*q.v.*) that specifically cleave off polyubiquitin chains from ubiquitin-conjugated proteins or generate ubiquitin monomers from polyubiquitin chains. These enzymes are thought to have a broad range of substrate specificities and may play a regulatory role in protein ubiquitination-related processes. *See* *otu*, ubiquitin, ubiquitin-proteasome pathway (UPP).

deutan *See* color blindness.

deuteranomaly *See* color blindness.

deuteranopia *See* color blindness.

deuterium *See* hydrogen.

deuteron the nucleus of a deuterium atom, containing one proton and one neutron.

Deuterostomia one of the two subdivisions of the Bilateria. It contains the echinoderms, the chordates, and a few smaller phyla. The deuterostome egg undergoes radial cleavage (*q.v.*), and the cells produced in early cleavage divisions retain the ability to develop into the complete embryo. The blastopore (*q.v.*) becomes the anus, and the coelom arises as pouches from the primitive gut. *Compare with* Protostomia. *See* Appendix A.

deuterotoky parthenogenesis in which both males and females are produced.

developer a chemical that serves as a source of reducing agents that will distinguish between exposed and unexposed silver halide and convert the exposed halide to metallic silver, thus producing an image on a photographic film.

development an orderly sequence of progressive changes resulting in an increased complexity of a biological system. *See* determination, differentiation, morphogenesis.

developmental control genes genes which control the developmental decisions of other genes. Such genes have been extensively studied in *Drosophila, Caenorhabditis, Danio, Mus,* and *Arabidopsis*. *See* Appendix C, 1978, Lewis; 1980, Nüsslein-Volhard and Weischaus; 1981, Chalfie and Sulston; 1983, Bender *et al.*, Scott *et al.*; 1984, McGinnis *et al.*; 1986, Tomlinson and Ready, Noll *et al.*; 1987, Nüsslein-Volhard *et al.*; 1988, Macdonald and Struhl, Herr *et al.*; 1989 Driever and Nüsslein-Volhard, Zink and Paro; 1990, Malicki *et al.*; 1993, Mullins and Nüsslein-Volhard; 1994, Bollag *et al.*; 1995, Halder *et al.*; 1996, Dubnau and Struhl, Krizek and Meyerowitz; cell lineage mutants, compartmentalization, floral identity mutations, gene networking, *Hox* genes, metamerism, selector genes, T box genes, zygotic segmentation mutations.

developmental genetics the study of mutations that produce developmental abnormalities in order to gain understanding of how normal genes control growth, form, behavior, etc.

developmental homeostasis canalization (*q.v.*).

developmental homology anatomical similarity due to derivation from a common embryological source; e.g., the halteres of flies are developmentally homologous to the hind wings of moths.

deviation the departure of a quantity (derived from one or more observations) from its expected value (usually the mean of a series of quantities).

Devonian the Paleozoic period during which cartilagenous and bony fishes evolved. On land, lycopods, sphenophytes, and ferns were the abundant plants and amphibians and wingless insects the most common animals. A mass extinction occurred late in the period. *See* geologic time divisions.

dex dextrorotatory. *See* optical isomers.

dextran a polysaccharide (composed of repeating D-glucose subunits) synthesized by certain lactic acid bacteria.

dextrose glucose (*q.v.*).

df, d.f., D/F degrees of freedom (*q.v.*).

DHFR dihydrofolate reductase (*q.v.*). *See* amplicon.

diabetes insipidus (DI) excessive excretion of normal urine; brought about because of inadequate output of vasopressin (*q.v.*) or its receptor. In humans, autosomal dominant DI is caused by mutations in a gene that encodes the vasopressin precursor protein. DI inherited as an X-linked recessive is due to mutations in a gene that encodes a vasopressin receptor. This belongs to the family of G protein–coupled receptors. *See* aquaporins, G proteins.

diabetes mellitus a disease in humans marked by glucose intolerance. It exists in two forms: type 1, insulin-dependent diabetes mellitus (IDDM), and type 2, non-insulin-dependent diabetes mellitus (NIDDM). Since type 1 diabetes usually occurs before age 20, it is often called "juvenile-onset diabetes." It is usually caused by the autoimmune destruction of the beta cells of the pancreas, which secrete insulin (*q.v.*). Type 1 diabetes can also result from mutations in the coding region of the insulin gene and from variations in the number of tandem repeats of a segment containing 14 to 15 nucleotides that resides upstream of the coding region. This region may regulate the rate of transcription of insulin mRNA. Since type 2 diabetes usually begins between the ages of 40 and 60, it is often called "maturity-onset diabetes." Genes on at least 10 different chromosomes have been identified that increase susceptibility to NIDDM. This disease is far more common than IDDM, and its prevalence is rising in affluent societies throughout the world where people get little exercise, overeat, and tend to become obese. J. V. Neel's "thrifty gene hypothesis" (*q.v.*) provides an evolutionary explanation for the presence in human populations of genes that predispose their bearers to type 2 diabetes. *See* Appendix C, 1962, Neel; *obese*.

diakinesis *See* meiosis.

diallelic referring to a polyploid in which two different alleles exist at a given locus. In a tetraploid, $A_1A_1A_2A_2$ and $A_1A_2A_2A_2$ would be examples.

dialysis the separation of molecules of differing size from a mixture by their differential diffusibility through a porous membrane. In the procedure knowns as *equilibrium dialysis*, soluble molecules of the same size are allowed to reach equivalent concentrations on either side of a semipermeable membrane. At equilibrium, if more molecules are detected on one side of the membrane, it indicates that they have become bound to some other larger molecules (e.g., repressor proteins, transport proteins, antibodies, etc.) present only on that side of the membrane, and thus are too large to pass through the pores of the membrane. This procedure is also used in immunology as a method of determining association constants for hapten-antibody reactions.

2,6-diaminopurine a mutagenically active purine analog. *See* bases of nucleic acids.

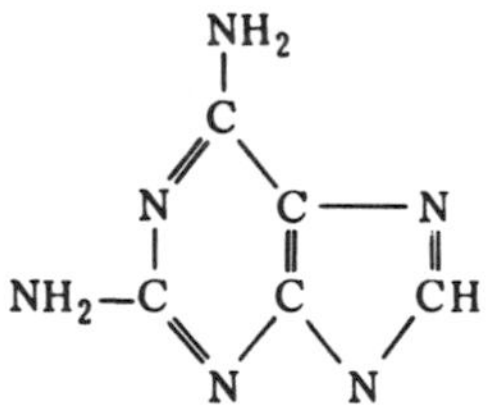

```
            NH2
             |
             C
           // \
          N    C —— N
          |    ||   ||
   NH2 —C     C    CH
          \\  / \  /
           N     N
```

diapause a period of inactivity and suspension of growth in insects accompanied by a greatly decreased metabolism. In a given species, diapause usually takes place in a specific stage in the life cycle, and it often provides a means of surviving the winter.

diaspora the dispersion of an originally homogeneous group of people from their homeland into foreign territories. Also, the people who have dispersed from their homelands (e.g., those Jews who live in communities outside the biblical land of Israel).

diasteromer epimer (*q.v.*).

diauxy the adaptation of a microorganism to culture media containing two different sugars. The organism possesses constitutive enzymes for one of the sugars, which it utilizes immediately. Induced enzyme synthesis is required before the second sugar can be metabolized.

dicentric designating a chromosome or chromatid having two centromeres.

Dicer a nuclease (*q.v.*) that processes endogenous or exogenous double-stranded RNA (dsRNA) (*q.v.*) precursors to 22 nucleotides-long RNAs, such as small interfering RNAs (*q.v.*) or small temporal RNAs (*q.v.*). The Dicer protein is evolutionarily conserved and is found in fungi, plants, worms, flies, and humans. The enzyme structure includes a helicase domain, domains related to the bacterial dsRNA-specific endonuclease, RNase III, and RNA-binding domains. Inactivation of Dicer in vertebrates results in the cessation of microRNA (*q.v.*) production, leading to early developmental arrest or lethality. In *Caenorhabditis elegans* inactivation of the Dicer gene causes developmental timing defects. *See* RNA interference (RNAi).

dichlorodiphenyltrichloroethane (DDT) an insecticide to which many insect species have developed resistant races.

```
                     Cl
        H  H     Cl | Cl     H  H
        |  |       \|/       |  |
        C=C         C        C=C
       /   \        |       /   \
  Cl--C     C---C---C      C     C--Cl
       \\  //       |      \\   //
        C--C        H       C--C
        |  |                |  |
        H  H                H  H
```

2,4-dichlorophenoxyacetic acid (2,4-d) a phytohormone used as a weed killer.

```
          H  H
          |  |
          C--C
        //    \\
  Cl--C         C--O--CH2--COOH
        \      /
          C--C
         /    \
        H      Cl
```

2,6-dichlorophenoxyacetic acid an antiauxin (*q.v.*).

dichogamous referring to flowers or hermaphroditic animals characterized by male and female sex organs that become mature at different times.

dichroism *See* circular dichroism.

Dicotyledoneae one of the two classes of flowering plants (*see* Appendix A, Kingdom 5, Plantae). The seeds of all dicots produce two primary leaves. *See* cotyledon, Monocotyledoneae.

dictyosome 1. a synonym for Golgi apparatus (*q.v.*). 2. one of the flattened vesicles that make up the Golgi apparatus. Most eukaryotes contain a Golgi of stacked dictyosomes, but fungal cells generally contain dispersed dictyosomes.

Dictyostelium discoideum a protoctist that has the ability to alternate between unicellular and multicellular life-styles. Individual *Dictyostelium* amoebas live in forest soil and eat bacteria and yeasts. However, when challenged by adverse conditions, such as starvation, groups of up to 100,000 cells signal each other by secreting acrasin (*q.v.*). This chemical attractant causes the amoebas to aggregate, forming a motile slug that is surrounded by a slimy extracellular matrix. At the apex of the mound, a fruiting body that produces spores differentiates. Dictyostelids are placed in the phylum Acrasiomycota (*q.v.*) and go by common names such as slime molds, social amoebas, or amoebozoans. They represent one of the earliest branches from the last common ancestor of all eukaryotes. Slime molds diverged after the split between the plants and opisthokonts (*q.v.*), but before the split of the fungi and animals. Therefore the slime molds, fungi, and metazoa are sister groups. *D. discoidium* has a genome size of 34 mb of DNA distributed among six chromosomes. The number of genes is about 12,500, and many of these have orthologs among the genes of opisthokonts. For example, there are 64 genes that are orthologs of human disease genes, such as Tay-Sachs, G6PD deficiency, and cystic fibrosis. The *Dictyostelium* genome contains genes that encode cell adhesion and signaling molecules (normally exclusive to animals) and genes that encode proteins controlling cellulose deposition and metabolism (normally exclusive to plants). *See* Appendix C, 2005, Eichinger *et al.*

dictyotene stage a prolonged diplotene stage of meiosis seen in oocytes during vitellogenesis. The chromosomes that have already undergone crossing over may remain in this stage for months or even years in long-lived species.

2′,3′-dideoxynucleoside triphosphates analogs of normal 2′-deoxyribonucleoside triphosphates used in a modified "minus" technique for base sequencing of DNA molecules. Because these analogs have no oxygen at the 3′ position in the sugar, they act as specific chain-terminators (*q.v.*) for primed synthesis techniques (*see* DNA sequencing techniques). Nucleotides in which arabinose is substituted for deoxyribose also exhibit this chain-terminating effect.

differential affinity the failure of two partially homologous chromosomes to pair during meiosis when a third chromosome is present that is more completely homologous to one of the two. In its absence, however, pairing of the partially homologous chromosomes can occur. *See* autosyndesis, homoeologous chromosomes.

differential gene expression the principle that all the cells of a multicellular organism have the same genetic content, but differ from one another in the sets of genes that they express.

differential segment *See* pairing segment.

differential splicing *See* alternative splicing.

differentiation the complex of changes involved in the progressive diversification of the structure and functioning of the cells of an organism. For a given line of cells, differentiation results in a continual restriction of the types of transcription that each cell can undertake. *See* development, morphogenesis. *Compare with* dedifferentiation, determination.

differentiation antigen a cell-surface antigen that is expressed only during a specific period of embryological differentiation.

diffuse centromere (kinetochore) *See* centromere.

diffusion the tendency for molecules because of their random heat motion to move in the direction of a lesser concentration, and so make the concentration uniform throughout the system.

digenetic descriptive of organisms of the subclass Digenea of the class Trematoda within the flatworm phylum Platyhelminthes. The term means "two beginnings," referring to a life cycle with alternation of generations, one parasitic and the other free-living. Digenea is the largest group of trematodes and the most important medically and economically. All members are endoparasitic with two or more hosts in the life cycle, the first host usually being a mollusc. The digenetic flukes include blood flukes and schistosomes that are generally considered to be the most serious helminthic human parasite. *See* Appendix A; schistosomiasis.

dihaploid a diploid cell, tissue, or organism having arisen from a haploid cell by chromosome doubling.

dihybrid a genotype characterized by heterozygosity at two loci. Mendel found that crosses between pure lines of peas that differed with respect to two unrelated traits produced genetically uniform F_1 dihybrid offspring. Intercrossing F_1 dihybrids produced parental and recombinant types in the F_2 population.

dihydrofolate reductase (DHFR) an enzyme essential for *de novo* thymidylate synthesis. It regenerates an intermediate (tetrahydrofolate) in thymidylate synthesis and is also essential for other biosynthetic events that depend on tetrahydrofolate, such as the synthesis of purines, histidine, and methionine. *See* amplicon, folic acid.

dihydrouridine *See* rare bases.

2,5-dihydroxyphenylacetic acid homogentisic acid (*q.v.*).

dimer a chemical entity consisting of an association of two monomeric subunits; e.g., the association of two polypeptide chains in a functional enzyme. If the two subunits are identical, they form a homodimer; if nonidentical, they form a heterodimer. Hexosaminidase (*q.v.*) is an example of a heterodimeric enzyme.

dimethylguanosine *See* rare bases.

dimethyl sulfate protection a method for identifying specific points of contact between a protein (such as RNA polymerase) and DNA based on the principle that, within an endonuclease-protected region (*see* DNAase protection), the adenines and guanines in the site of contact are not available to be methylated by exposure to dimethyl sulfate.

dimorphism the phenomenon of morphological differences that split a species into two groups, as in the sexual dimorphic traits distinguishing males from females.

dinitrophenol (DNP) a metabolic poison that prevents the uptake of inorganic phosphate and the production of energy-rich phosphorus compounds like ATP. DNP is a commonly used hapten in immunological experiments.

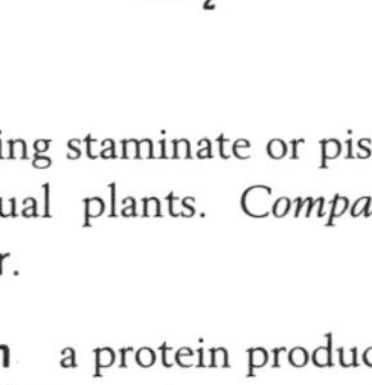

```
        OH
        |
        C     NO2
      //  \  /
    HC      C
    |       ||
    HC      CH
      \\  /
        C
        |
       NO2
```

dioecious having staminate or pistillate flowers on separate unisexual plants. *Compare with* monoecious. *See* flower.

diphtheria toxin a protein produced by certain lysogenic strains of *Corynebacterium diphtheriae* that is responsible for the symptoms of diphtheria. The structural gene for the toxin is carried by certain bacteriophages (e.g., corynephages beta, omega, and gamma). The host bacterium regulates the expression of the gene. No syntheses of the toxin occur until the intracellular level of iron falls below a certain threshold. *See* Appendix C, 1888, Roux and Yersin; 1971, Freeman; prophage-mediated conversion.

diploblastic having a body made of two cellular layers only (ectoderm and endoderm), as the coelenterates.

diplochromosome a chromosome arising from an abnormal duplication in which the centromere fails to divide and the daughter chromosomes fail to move apart. The resulting chromosome contains four chromatids.

Diplococcus pneumoniae the former designation given for *Streptococcus pneumoniae*, the cause of bacterial pneumonia. *See Streptococcus.*

diplo-haplont an organism (such as an embryophyte) in which the products of meiosis form haploid gametophytes that produce gametes. Fertilization generates a diploid sporophyte in which meiosis takes place. Thus, diploid and haploid generations alternate. *Contrast with* diplont, haplont.

diploid or **diploidy** referring to the situation or state in the life cycle where a cell or organism has two sets of chromosomes: one from the mother and one from the father. Diploidy results from the fusion of the haploid egg nucleus and a haploid sperm nucleus. *See* autosome, C value, merozygote, N value, polyploidy, sex chromosome, syngamy.

diplonema *See* meiosis.

diplont an organism (such as any multicellular animal) characterized by a life cycle in which the products of meiosis function as gametes. There is no haploid multicellular stage as in a diplo-haplont and haplont (*q.v.*).

diplophase the diploid phase of the life cycle between the formation of the zygote and the meiosis.

diplospory a type of apomixis in plants in which a diploid gametophyte is formed after mitotic divisions of the spore-forming cells.

diplotene *See* meiosis.

Dipodomys ordii a species of jumping rodent found in arid and desert regions of North America. This kangaroo rat is famous for the large amount of repetitious DNA (*q.v.*) in its genome.

dipole a molecule carrying charges of opposite sign at opposite poles.

Diptera an insect order containing midges, mosquitoes, and flies. *See* Appendix A, Animalia, Arthropoda.

directional selection selection resulting in a shift in the population mean in the direction desired by the breeder or in the direction of greater adaptation by nature. For example, the breeder might select for a number of generations seeds from only the longest ear of corn in the population. *See* disruptive selection.

direct repeats identical or closely related DNA sequences present in two or more copies in the same orientation in the same molecule, although not necessarily adjacent.

DIS *Drosophila Information Service* (*q.v.*).

discoidal cleavage cleavage occurring at the surface of an enormous yolk mass.

discontinuous distribution a collection of data recorded as whole numbers, and thus not yielding a continuous spectrum of values; e.g., the number of leaves per plant in a population of plants. *See* continuous distribution.

discontinuous replication *See* replication of DNA.

discontinuous variations variations that fall into two or more non-overlapping classes.

discordant twins are said to be discordant with respect to a trait if one shows the trait and the other does not.

disequilibrium *See* gametic disequilibrium, linkage disequilibrium.

disjunction the moving apart of chromosomes during anaphase of mitotic or meiotic divisions.

disomy the presence in a cell of a pair of chromosomes of a specified kind. The normal condition for a diploid cell is *heterodisomy*, where one member of each autosomal pair is of maternal and the other of paternal origin. If both chromosomes are inherited from the same parent, the term *uniparental disomy* is used. There are instances reported where a child suffering from cystic fibrosis (*q.v.*) has two copies of chromosome 7, both containing the *CF* gene from a heterozygous mother. Here it is assumed that a disomic egg produced by nondisjunction (*q.v.*) was fertilized by a normal sperm to produce a trisomic, but the paternal chromosome 7 was lost early in development and only the disomic, diploid cell line survived.

dispersal mechanism any means by which a species is aided in extending its range. For example, sticky seeds can cling to animals and be transported by them to new regions.

dispersive replication an obsolete model of DNA replication in which parental and newly synthesized daughter molecules are interspersed in an essentially random fashion.

disruptive selection the selection of divergent phenotypic extremes in a population until, after several generations of selection, two discontinuous strains are obtained. For example, the breeder might select for a number of generations seeds from the longest and the shortest ears of corn in a population. *See* directional selection.

disseminule a plant part that gives rise to a new plant.

***Dissociation-Activator* system** *See Activator-Dissociation* system.

distal situated away from the place of attachment. In the case of a chromosome, the part farthest from the centromere.

distributive pairing the pairing of chromosomes at metaphase I of meiosis that leads to their proper distribution to daughter cells. Synaptonemal complexes play no role in this type of chromosomal association.

distylic species a plant species composed of two types of individuals each characterized by a different flower morphology.

disulfide linkage the sulfur-to-sulfur bonding of adjacent cysteine residues in or between protein molecules.

diurnal **1.** pertaining to the daytime. **2.** recurring in the period of a day; daily.

divergence in molecular biology, the percent difference between nucleotide sequences of two related DNA segments or between amino acid sequences of the two related polypeptide chains.

divergence node the branching point in an evolutionary tree. The place where two lineages diverge from a common ancestor. *See* cladogram, node.

divergent transcription the transcriptional orientation of different DNA segments in opposite directions from a central region.

diversity in ecology, the number of species or other taxa in a particular ecological unit.

Division *See* Appendix A: Classification.

dizygotic twins *See* twins.

D loop **1.** a displacement loop formed early in the replication of duplex DNA (either circular or linear) consisting of a single, unreplicated, parental strand on one side, and a double-stranded branch (composed of one parental strand paired with the leading strand, *q.v.*) on the other side. Because the leading strand displaces the unreplicated parental strand, the replication "bubble" or "eye" is called a displacement or D loop. **2.** a region of vertebrate mtDNA that is noncoding but contains promoters and an origin for the replication of mtDNA. Shortly after replication is initiated, a temporary arrest in DNA elongation creates this displacement loop. The D loop is a bubble in which one strand of the control region has been copied and the other displaced. This D loop has been used as a target region for sequence comparisons when erecting phylogenetic trees. *See* Neandertal.

DNA deoxyribonucleic acid (*q.v.*). *Also see* insulator DNA, promiscuous DNA.

DNA adduct *See* adduct.

DNA-agar technique a technique for testing the degree of homology between nucleic acid molecules from different sources by allowing fragments of radioactive nucleic acid from one source to react with nonlabeled nucleic acids from another source trapped in an agar gel. This procedure binds to the gel radioactive polynucleotide fragments that are complementary to those trapped in the agar. *See* Appendix C, 1963, McCarty and Bolton; hybrid duplex molecule.

DNA amplification *See* amplicon, polymerase chain reaction.

DNAase symbol for deoxyribonuclease (*q.v.*).

DNAase footprinting a technique for determining the sequence of a DNA segment to which a DNA-binding protein binds. In this technique, a double-stranded DNA fragment is radioactively labelled at the 5′ end, partially digested with DNAase (*q.v.*) in the presence and absence of the binding protein, and the resulting fragments compared by electrophoresis (*q.v.*) and autoradiography (*q.v.*) on a gel that also runs in parallel the reaction products of a sequencing reaction performed on the unprotected sample of DNA. This produces an autoradiograph with ladders of oligonucleotides of varying lengths, increasing in single-nucleotide increments. The DNA region covered by the binding protein is protected from DNAase degradation and appears as a gap, or a *footprint*, that is missing from the sample lacking the protective protein. The footprint-containing ladder aligned with DNA sequencing ladders then identifies the exact sequence of bases in the footprint. *See* dimethyl sulfate protection, DNAase protection, DNA sequencing techniques.

DNAase protection the shielding of DNA sequences bound by a protein from degradation by an endonuclease. DNAase protection is used to characterize a DNA segment that binds a specific protein. In this approach, the protein in question is bound to the DNA, DNAase added to degrade the surrounding unprotected sequences, and the remaining bound DNA isolated and analyzed. *See* dimethyl sulfate protection, DNAase footprinting.

DNA-binding motifs sites on proteins which facilitate their binding to DNA. *See* DNA Grooves, DNA methylation, helix-turn-helix motif, homeobox, leucine zipper, POU genes, T box genes, zinc finger proteins.

DNA chip *See* DNA microarray technology.

DNA clock hypothesis the postulation that, when averaged across the entire genome of a species, the rate of nucleotide substitutions in DNA remains constant. Hence the degree of divergence in nucleotide sequences between two species can be used to estimate their divergence node (*q.v.*). *See* Appendix C, 1983, Kimura and Ohta.

DNA clone a DNA segment that has been inserted via a viral or plasmid vector into a host cell with the following consequences: the segment has replicated along with the vector to form many copies per cell, the cells have mutiplied into a clone, and the insert has been magnified accordingly.

DNA complexity a measure of the amount of nonrepetitive DNA characteristic of a given DNA sample. In an experiment involving reassociation kinetics (*q.v.*), DNA complexity represents the combined length in nucleotide pairs of all unique DNA fragments. The DNA of evolutionarily advanced species is more complex than that of primitive species.

DNA damage checkpoint a system that checks for regions where DNA has single-stranded or mismatched regions or stalled replication forks. Further progress through mitosis is then halted until the damage is corrected. If it cannot be rectified, the cell is diverted to apoptosis (*q.v.*). *See* Appendix C, 1989, Hartwell and Weinert; Adriamycin, ATM kinase, *RAD*.

DNA-dependent RNA polymerase RNA polymerase (*q.v.*). *Contrast with* RNA-dependent DNA polymerase.

DNA-driven hybridization reaction a reaction involving the reassociation kinetics of complementary DNA strands when DNA is in great excess of a radioactive RNA tracer; employed in cot analysis to determine the repetition frequencies of the corresponding genome sequences. *See* reassociation kinetics.

DNA duplex a DNA double helix. *See* deoxyribonucleic acid.

DNA fiber autoradiography light microscopic autoradiography of tritiated thymidine-labeled DNA molecules attached to millipore filters. The technique was devised by Cairns (1963) for studying DNA replication in *E. coli* and later adapted by Huberman and Riggs (1968) for visualizing the multiple replicons of mammalian chromosomes.

DNA fingerprint technique a technique (more properly termed DNA typing) that relies on the presence of simple tandem-repetitive sequences that are scattered throughout the human genome. Although these regions show considerable differences in lengths, they share a common 10 15 base pair core sequence. DNAs from different individual humans are enzymatically cleaved and separated by size on a gel. A hybridization probe containing the core sequence is then used to label those DNA fragments that contain complementary sequences. The pattern displayed on each gel is specific for a given individual. The technique has been used to establish family relationships in cases of disputed parentage. In violent crimes, blood, hair, semen, and other tissues from the assailant are often left at the scene. The DNA fingerprinting technique provides the forensic scientist with a means of identifying the assailant from a group of suspects. *See* Appendix C, 1985, Jeffries, Wilson, and Thien; alphoid sequences, DNA forensics, fingerprinting technique, oligonucleotide fingerprinting, restriction fragment length polymorphisms, VNTR locus.

DNA forensics the use of DNA technology during the evidence-gathering phases of criminal investigations, as well as any use of DNA evidence in the legal system. The first case in which a person was convicted of a crime on the basis of DNA evidence occurred in 1987. In 1989, the first conviction was overturned on the basis of DNA evidence. DNA profiling has also been successful in identifying victims of catastrophes, in establishing paternity, and in determining the bacteria or viruses responsible for outbreaks of infectious diseases. *See* CODIS, DNA fingerprint technique, Romonov, STR analysis.

DNA glycosylases a family of enzymes, each of which recognizes a single type of altered base in DNA and catalyzes its hydrolytic removal from the sugar-phosphodiester backbone. *See* AP endonucleases.

DNA grooves two grooves that run the length of the DNA double helix. The major groove is 12 Angstroms wide, while the minor groove is 6 Angstroms wide. The major groove is slightly deeper than the minor groove (8.5 versus 7.5 Angstroms). The grooves have different widths because of the asymmetric attachment of the base pairs to the sugar-phosphate backbone. As a result the edges of the base pairs in the major groove are wider than those in the minor groove. Each groove is lined by potential hydrogen-bond donor and acceptor atoms, and these interact with DNA-binding proteins that recognize specific DNA sequences. For example, endonucleases bind electrostatically to the minor groove of the double-helical DNA. The figure below shows the binding of a helix-turn-helix motif (*q.v.*) to a DNA segment. The minor groove, the major groove, and the recognition helix are labeled G^m, G^M, and *RH*, respectively. *See Antennapedia*, deoxyribonuclease.

DNA gyrase *See* gyrase.

DNA helicase *See* helicase.

DNA hybridization a technique for selectively binding specific segments of single-stranded DNA or RNA by base pairing to complementary sequences on ssDNA molecules that are trapped on a nitrocellulose filter (*q.v.*). **1.** DNA-DNA hybridization is commonly used to determine the degree of sequence identity between DNAs of different species. **2.** DNA-RNA hybridization is the method used to select those molecules that are complementary to a specific DNA from a heterogeneous population of RNAs. *See* Appendix C, 1960, Doty *et al.*; 1963, McCarty and Bolton; 1972, Kohne *et al.*; *in situ* hybridization, reassociation kinetics.

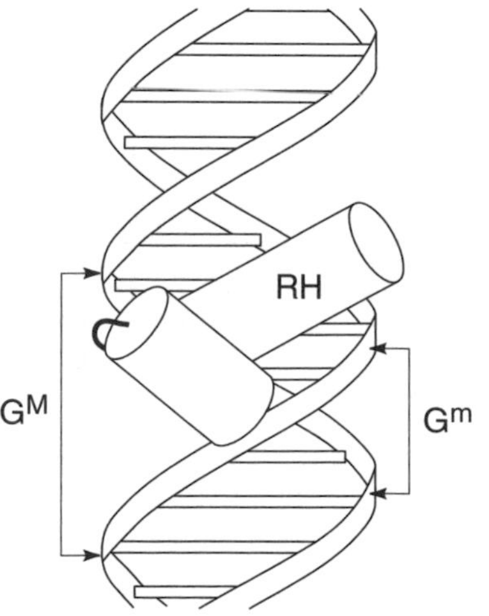

DNA library *See* genomic library.

DNA ligase genes in humans three genes have been identified that encode DNA ligases. *LIG1* (19q13.2–13.3) joins Okazaki fragments during DNA replication (*q.v.*). *LIG3* (17q11.2–q12) seals chromosome breaks produced during meiotic recombination. *LIG4* (13q22–q34) functions during V(D)J recombination (*q.v.*).

DNA ligases enzymes that catalyze the formation of a phosphodiester bond between adjacent 3′-*OH* and 5′-*P* termini in DNA. DNA ligases function in DNA repair to seal single-stranded nicks between adjacent nucleotides in a duplex DNA chain. *See* Appendix C, 1966, Weiss and Richardson; blunt-end ligation, cohesive-end ligation, cut-and-patch repair, mismatch repair, replication of DNA.

DNA looping a phenomenon that involves proteins that bind to specific sites on a DNA molecule while also binding to each other. The DNA loops that form as a result stimulate or inhibit the transcription of associated genes. Enhancer (*q.v.*) sequences may represent DNA segments involved in DNA looping.

DNA methylase *See* methyl transferase.

DNA methylation the addition of methyl groups to specific sites on a DNA molecule. Between 2 and 7% of the cytosines in the DNA of animal cells are methylated, and the methylated cytosines are found in CG doublets (often called CpG islands). The Cs on both strands of a short palindromic sequence are often methylated, giving a structure

5′ *CpG *CpG 3′
3′ GpC* GpC* 5′

where asterisks represent methylated sites. Upstream elements that control the expression of genes contain repeated CG doublets that may be methylated or unmethylated. The absence of methyl groups is associated with the ability to be transcribed, while methylation results in gene inactivity. Methylation occurs immediately after replication. Methylation of cytosine prevents transcription, and it has been suggested that methylation is a mechanism that evolved to suppress transcription by transposons and forms of selfish DNA (*q.v.*). Proteins which recognize CpG islands have been isolated from many animal and plant species. These proteins have a methyl-CpG-binding domain (MBD) that is about 70 amino acids long. Such MBD proteins function as transcription repressors. Some of the genes encoding MBD proteins have been localized in mice and humans. One

MBD gene occurs on the X chromosome at the same site in both species. In humans progressive neurologic developmental disorders occur in individuals hemizygous or homozygous for mutations in this gene. *See* Appendix C, 1997, Yoder, Walsh, and Bestor; 2000, Bell and Felsenfeld; *H19*, 5-methylcytosine (5-mCyt), methyl transferase, parental imprinting, telomeric silencing.

DNA microarray technology a technique which allows the analysis of gene expression or gene structure in hundreds to thousands of genes simultaneously by measuring the extent of nucleic acid hybridization (*q.v.*) in DNA microarrays. DNA microarrays are small glass microscope slides, silicone chips, or specialized membranes containing hundreds or thousands of closely spaced spots, to each of which are bound short, single-stranded gene sequences. The DNA for the arrays is derived either from genomic DNA or cDNA (*q.v.*) and applied with a robotic instrument. Expression of genes represented on a DNA microarray can be assayed by hybridization with fluorescently- or radioactively-labeled cDNA (*q.v.*) or mRNA (*q.v.*) probes (*q.v.*) and quantitative analysis of the extent of nucleic acid hybridization (i.e., fluorescence or radioactivity) on each spot on the microarray. This approach can be used to identify transcribed regions in a genome (*q.v.*). By simultaneous hybridization with different-colored fluorescent probes derived from different sources, one can compare gene expression in different cell types, examine temporal and spatial expression patterns, or identify genetic variations associated with disease. In an alternative approach, the DNA for the microarray is synthesized directly on the microarray support, using as templates (*q.v.*) single-stranded oligonucleotides (*q.v.*) which have been annealed to the support and which are derived from individual genes. Hybridization of oligonucleotide arrays (often called *DNA chips*) with genomic DNA probes can detect mutations or polymorphisms in gene sequences. A helpful Web site for microarray users is www.biochipnet.de. One of the earliest studies utilizing this technology involved *Arabidopsis* (*q.v.*). The differential expression of 45 genes was measured with a microarray of 45 cDNAs. *See* Appendix C, 1995, Schena *et al.*; 1999, Evans and Wheeler.

DNA modification *See* modification.

***dna* mutations** mutations of *E. coli* that influence DNA replication. The *dna A*, *dna B*, and *dna C* mutations are defective in proteins that interact with replication origins. The *dna E*, *dna X*, and *dna Z* genes encode subunits of DNA polymerase III, and *dna* G encodes primase (*q.v.*). Mutation in *dna*Q damages the proofreading (*q.v.*) ability of the epsilon subunit of DNA polymerase III and greatly increases the mutation rate. *See* DNA polymerase, mutator gene, repair synthesis, replisome.

DNA polymerase an enzyme that catalyzes the formation of DNA from deoxyribonucleoside triphosphates, using single-stranded DNA as a template. Three different DNA polymerases (pol I, pol II, and pol III) have been isolated from *E. coli*. Pol III is the major enzyme responsible for cellular DNA replication in this bacterium. The other two enzymes function primarily in DNA repair. Eukaryotes contain a variety of polymerases that participate in chromosomal replication, repair, and crossing over and also in mitochondrial replication. In mammals, DNA replicase alpha functions in the priming and synthesis of the lagging strand, while replicase delta catalyzes the synthesis of the leading strand. All DNA polymerases extend the DNA chain by adding nucleotides, one at a time, to the 3′ OH end of the growing strand. Each base added must be complementary to the next nucleotide presented by the template strand. To initiate replication, DNA polymerases require a priming RNA molecule. This binds to the template DNA molecule and provides the 3′ OH start point for the enzyme. DNA polymerase III of *E. coli* is the major polymerase involved in replication. It is made up of 18 subunits. The catalytic core contains three proteins (alpha, epsilon, and theta). The epsilon subunit has a proofreading function and is the product of the *dna*Q gene. The mutation frequency increases 10^3- to 10^5-fold in cells carrying mutations in the *dna*Q gene. *See* Appendix C, 1956, Kornberg, Lehrman, and Simms; cut-and-patch repair, *dna* mutations, Klenow fragment, polymerase chain reaction, replication of DNA, replicon, replisome.

DNA probe *See* probe.

DNA puff *See* chromosomal puff.

DNA relaxing enzyme *See* topoisomerase.

DNA repair any mechanism that restores the correct nucleotide sequence of a DNA molecule that has incurred one or more mutations, or that has had its nucleotides modified in some way (e.g., methylation). *See* ATM kinase, cut-and-patch repair, error-prone repair, mismatch repair, photoreactivating enzyme, proofreading, recombination repair, SOS response, thymine dimer, xeroderma pigmentosum.

DNA replication *See* replication of DNA.

DNA restriction enzyme any of the specific endonucleases (*q.v.*) present in many strains of *E. coli* that recognize and degrade DNA from foreign sources. These nucleases are formed under the directions of genes called *restriction alleles*. Other genes called *modification alleles* determine the methylation pattern of the DNA within a cell. It is this pattern that determines whether or not the DNA is attacked by a restriction enzyme. *See* modification methylases, restriction endonuclease.

DNA-RNA hybrid a double helix consisting of one chain of DNA hydrogen bonded to a complementary chain of RNA. Some RNA molecules produced by an immunoglobulin gene remain attached to the gene and mark it for retention after a DNA-cutting enzyme removes all the other genes that code for the constant region (Y-stem) of immunoglobulins during "heavy chain class switching" (*q.v.*). *See* Appendix C, 1961, Hall and Spiegelman.

DNase also symbolized *DNAase*. *See* deoxyribonuclease.

DNase protection *See* DNAase protection.

DNA sequencing techniques 1. the method developed by F. Sanger and A. R. Coulson (1975) is known as the "plus and minus" method or the "primed synthesis" method. DNA is synthesized *in vitro* in such a way that it is radioactively labeled and the reaction terminates specifically at the position corresponding to a given base. After denaturation, fragments of different lengths are separated by electrophoresis and identified by autoradiography. In the "plus" protocol, only one kind of deoxyribonucleoside triphosphate (dNTP) is available for elongation of the ^{32}P-labeled primer. In the "minus" protocol, one of the four dNTPs is missing; alternatively, specific terminator base analogs (2′,3′-dideoxyribonucleoside triphosphates, *q.v.*) can be used instead of the "minus" technique. 2. in the 1977 procedure of A. M. Maxam and W. Gilbert (the "chemical" method), single-stranded DNA (derived from double-stranded DNA and labeled at the 5′ end with ^{32}P) is subjected to several chemical (dimethyl sulfate-hydrazine) cleavage protocols that selectively make breaks on one side of a particular base; fragments are separated according to size by electrophoresis on acrylamide gels and identified by autoradiography.

DNA sequencers commercial robotic machines that take the drudgery out of sequencing. The *ABI PRISM 3700* is an example. It is the invention of Michael Hunkapillar, president of Applied Biosystems, Inc., and it can produce as much as 1 million bases of DNA sequence per day. Using 300 of these synthesizers, Celera Genomics (*q.v.*) sequences more than 1 billion nucleotides per month. *See* Appendix C, 1986, Hood *et al.*

DNA topoisomerase *See* topoisomerase.

DNA typing *See* DNA fingerprint technique.

DNA unwinding protein a protein that binds to single-stranded DNA and facilitates the unwinding of the DNA duplex during replication and recombination. *See* gene 32 protein.

DNA vaccines *See* vaccine.

DNA vector a replicon, such as a small plasmid or a bacteriophage, that can be used in molecular cloning experiments to transfer foreign nucleic acids into a host organism in which they are capable of continued propagation. *See* bacterial artificial chromosomes (BACs), cosmid, lambda cloning vector, P1 artificial chromosomes (PACs), pBR322, P elements, plasmid cloning vector, shuttle vector, Ti plasmid, yeast artificial chromosomes (YACs).

DNP 1. 2 : 4 dinitrophenol. 2. DNA-protein complex.

docking protein *See* receptor mediated translocation.

dog breeds any of about 400 described breeds of the species *Canis familiaris* (*q.v.*). According to the "breed barrier" rule followed since the mid-19th century, no dog may become a registered member of a breed unless both dam and sire are registered members. As a result, purebred dogs belong to a closed gene pool. Some popular breeds include TERRIERS: Welsh, Bedlington, Dandie Dinmont, West Highland White, Skye, Cairn, Scottish, Sealyham, Fox (Smooth), Fox (Wire), Schnauzer, Airedale, Irish, Kerry Blue, Bull, Manchester. POINTERS: German Shorthaired Pointer, Irish Setter, English Setter, Gordon Setter, Weimaraner, Pointer, Brittany Spaniel. COURSING HOUNDS: Irish Wolfhound, Scottish Deerhound, Greyhound, Whippet, Borzoi, Saluki, Afghan. TRAILING HOUNDS: Basenji, Bloodhound, Dachshund, Bassett, Beagle, Black and Tan Coonhound. MISCELLANEOUS HOUNDS: Otterhound, Norwegian Elkhound. FLUSHING SPANIELS: English Springer, English Cocker, American Cocker, Welsh Springer. RETRIEVERS: Golden Retriever, Labrador Retriever, Chesapeake Bay Retriever, Irish Water Spaniel, Curly-coated Retriever. SHEEP DOGS: Briard, Kuvasz, Shetland Sheepdog, Collie, Belgian Sheepdog. SLED DOGS: Siberian Huskie, Eskimo, Samoyed, Alaskan Malamute. GUARD DOGS: Bouvier de Flandres, Mastiff, Rottweiller, Boxer, Great Dane, Bull Mastiff, Schnauzer, German Shepherd, Dobermann Pinscher. MISCELLANEOUS WORKING DOGS: St. Bernard, Welsh Corgi (Cardigan), Welsh Corgi (Pembroke), Newfoundland, Great Pyrenees. TOYS: Maltese, Pug, Japanese Spaniel, English Toy Spaniel (King Charles), Pekingese, Pomeranian, Yorkshire Terrier, Griffon, Chihuahua, Papillon, Poodle (Toy), Mexican hairless. NONSPORTING BREEDS: Lhasa Apso, Poodle (Standard), Poodle (Miniature), Dalmatian, Chow Chow, Keeshond, Schipperke, English Bulldog, French Bulldog, Boston Terrier.

Dollo law the proposition that evolution along any specific lineage is essentially irreversible. For example, no modern mammal can de-evolve back to a form identical in all respects to the mammal-like, reptilian ancestor from which it was derived. This biological principle was formulated about 1890 by Louis Dollo, a Belgian paleontologist. *See* rachet.

Dolly a sheep (*q.v.*) born in Scotland in 1996 and the first mammal to be experimentally cloned. This was done by fusing the nucleus of an adult somatic cell from one sheep with an enucleated egg from another, followed by implantation into a surrogate mother. Dolly's chromosomes were therefore genetically identical to those of the somatic cell that provided the nucleus. When Dolly was two and a half years old, the lengths of her telomeres were determined. The lengths corresponded to telomeres the age of the nuclear donor, not to telomeres of her chronological age. It was later found that the cloned adult sheep contained mtDNA derived solely from the recipient egg. So Dolly was actually a genetic chimera (*q.v.*). Her cells contained nuclear DNA of somatic origin, while her mitochondria were derived from ooplasm. Dolly was euthanized in February of 2003 after developing progressive lung disease. Dolly's skin was used in a taxidermic mount currently on display at the Royal Museum of Edinburgh. *See* Appendix C, 1997, Wilmut *et al.*; cloning, mitochondrial DNA (mtDNA), nuclear reprogramming, nuclear transfer, telomere.

domain 1. a homology unit; i.e., any of the three or four homologous regions of an immunoglobulin heavy chain that apparently evolved by duplication and diverged by mutation. 2. any discrete, continuous part of a polypeptide sequence that can be equated with a particular function. 3. a relatively short sequence of about 100 amino acids that adopts a defined three-dimensional structure within a protein. Also known as a *module*. 4. any region of a chromosome within which supercoiling is independent of other domains. 5. an extensive region of DNA including an expressed gene that exhibits pronounced sensitivity to degradation by endonucleases. *See* immunoglobulin domain superfamily, *otu* domain, SH domain.

domesticated species *See* Appendix B.

dominance referring to alleles that fully manifest their phenotype when present in the heterozygous, heterokaryotic, or heterogenotic state. The alleles whose phenotypic expressions are masked by dominant alleles are termed *recessive alleles*. Sometimes the dominant allele expresses itself late in development (e.g., Huntington disease, *q.v.*), in which case the allele is said to show *delayed dominance*. *See* codominant, incomplete dominance, semidominance.

dominance variance genetic variance for a polygenic trait in a given population attributed to the dominance effects of contributory genes.

dominant complementarily *See* complementary genes.

dominant gene *See* recessive gene.

dominant negative mutation a mutation which produces a product that binds to the product of the normal allele. The heteropolymer that results damages the cell. A dominant negative mutation therefore has a more severe effect than the deletion of the same gene. Several hereditary human diseases are caused by dominant negative mutations in genes that encode collagens and keratins (both of which *see*).

donkey *Equus asinus*, a close relative of the horse. The female is referred to as a jennet, the male as a jack. *See Equus*, horse-donkey hybrids.

donor splicing site *See* left splicing junction.

DOPA the abbreviation for dihydroxyphenylalanine, a compound derived from the amino acid tyrosine, by the addition of a second hydroxyl group. *See* albino, amino acids (page 21), melanism, tyrosinase.

```
          OH
          |
    OH    C
      \  // \
       C     CH
       |     ||
      HC     CH    NH2
        \\  /      |
          C—CH2—CH—COOH
```

Dopamine the compound derived from DOPA (*q.v.*) by removal of the carboxyl group. *See* Parkinsonism.

dorsoventral genes genes that specify the dorsal or ventral patterning program of the embryonic cells in which they are expressed. In *Drosophila,* the gene *decapentaplegic* specifies dorsal development, whereas its mouse homolog, *BMP4,* specifies ventral development. In *Drosophila,* the gene *short gastrulation* specifies ventral development, whereas the homologous gene in the mouse, *chordin,* specifies the dorsal development pattern. *See* Saint Hilaire hypothesis.

dosage compensation a mechanism that regulates the expression of sex-linked genes that differ in dose between females and males in species with an XX-XY method of sex determination. In *Drosophilia melanogaster,* dosage compensation is accomplished by raising the rate of transcription of genes on the single X chromosome of males to double that of genes on either X chromosome in females. In mammals, the compensation is made by inactivating at random one of the two X chromosomes in all somatic cells of the female. The inactivated X forms the Barr body or sex chromatin. In cases where multiple X chromosomes are present all but one are inactivated. *See* Appendix C, 1948, Muller; 1961, Lyon, Russell; 1962, Beutler, *et al.*; Fabry disease, glucose-6-phosphate dehydrogenase deficiency, Lesch-Nyhan syndrome, Lyon hypothesis, Lyonization, mosaic, MSL proteins, ocular albinism, Ohno hypothesis.

dose 1. gene dose—the number of times a given gene is present in the nucleus of a cell. 2. radiation dose—the radiation delivered to a specific tissue area or to the whole body. Units for dose specifications are the gray, roentgen, red, rep, and sievert.

dose-action curve dose-response curve (*q.v.*).

dose fractionation the administration of radiation in small doses at regular intervals.

dose-response curve the curve showing the relation between some biological response and the administered dose of radiation. *See* extrapolation number.

dosimeter an instrument used to detect and measure an accumulated dosage of radiation.

dot blot *See* dot hybridization.

dot hybridization a semiquantitative technique for evaluating the relative abundance of nucleic acid sequences in a mixture or the extent of similarity between homologous sequences. In this technique, multiple samples of cloned DNAs, identical in amount, are spotted on a single nitrocellulose filter in dots of uniform diameter. The filter is then hybridized with a radioactive probe (e.g., an RNA or DNA mixture) containing the corresponding sequences in unknown amounts. The extent of hybridization is estimated semiquantitatively by visual comparison to radioactive standards similarly spotted.

dot-matrix analysis a graphical method of comparing the nucleotide sequences or amino acid sequences along sections of two polymeric molecules that may or may not be homologous. For example, a comparison could be made of the exons from a gene known for two different animal species (A and B). A dot plot diagram is generated with the A gene on the vertical axis and the B gene on the horizontal axis. Dots are placed within this rectangular array at every place where sequences from the two species match. The technique allows all pairs to be compared simultaneously, and regions of sequence similarity are seen as a series of dots. If there is no homology, the dots form a random pattern. If the dots form a diagonal line, the exons of the two genes have similar sequences and are arrayed in the same order.

Dotted a gene, symbolized by *Dt,* residing on chromosome 9 of maize, that influences the rate at which *a* mutates to *A*. *A* is on chromosome 3 and the gene controls the ability of the cells of the aleurone (*q.v.*) layer of the kernel (*q.v.*) to produce colored pigments. Clones of cells with restored pigment production generate spots on the kernel, as shown in the illustration. *Dt* was the first genetic element to be called a mutator gene (*q.v.*). However, it is now clear that it is a transposable element (*q.v.*) belonging to a different transposon family than *Ac* or *Ds*. *See* Appendix C, 1938, Rhoades; *Activator-Dissociation* system, genetic instability.

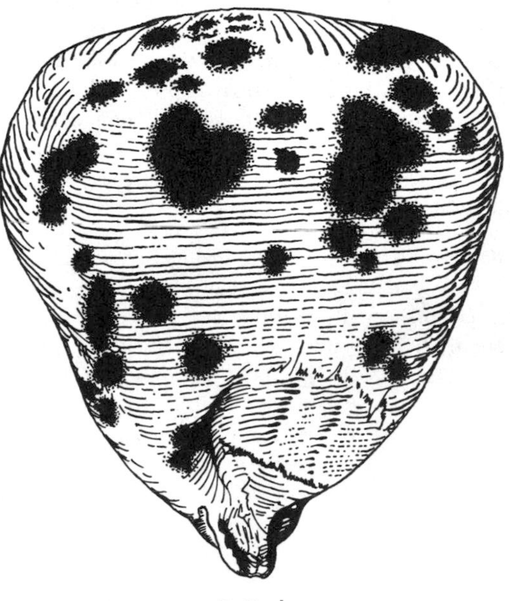

Dotted

double cross the technique used for producing hybrid seed for field corn. Four different inbred lines (A, B, C, and D) are used. A × B → AB hybrid and C × D → CD hybrid. The single-cross hybrids (AB and CD) are then crossed and the ABCD seed is used for the commercial crop.

double crossover *See* double exchange.

double diffusion technique synonymous with Ouchterlony technique (*q.v.*).

double exchange breakage and interchange occurring twice within a tetrad involving two, three, or four of the chromatids.

double fertilization a type of fertilization that distinguishes flowering plants from other seed plants. As shown in the illustration, the pollen grain upon germination forms a male gametophyte that injects two sickle-shaped haploid sperm nuclei into the female gametophyte. The egg nucleus and the polar nuclei are haploid and genetically identical. The union of one sperm nucleus with the egg nucleus produces a diploid nucleus from which the embryo develops. The two polar nuclei fuse with the other sperm nucleus to form a triploid nucleus. The endosperm (*q.v.*) develops by the mitotic activity of this 3N nucleus. *See* Appendix A, Plantae, Pteropsida, Angiospermae; Appendix C, 1898, Navashin; kernel, pollen grain, synergid.

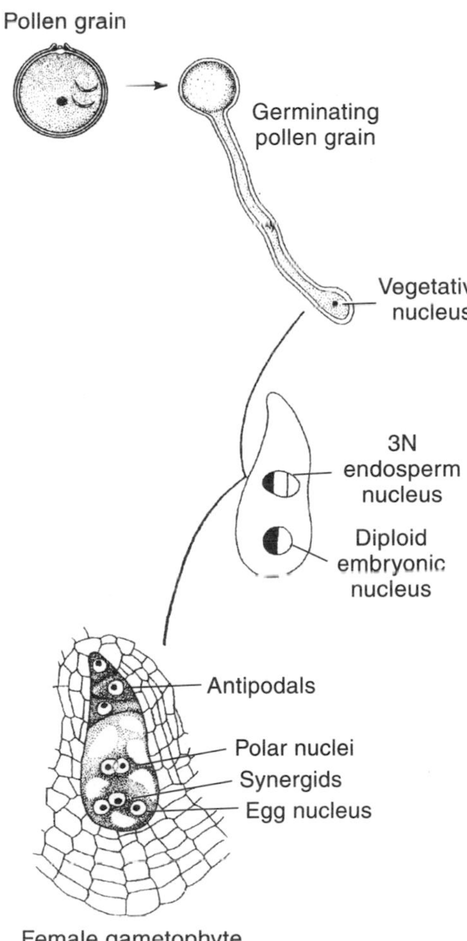

Double fertilization

double haploids plants that are completely homozygous at all gene loci, generated when haploid germ cells, grown in tissue culture, double their chromosome sets. *See* anther culture.

double helix the Watson-Crick model of DNA structure, involving plectonemic coiling (*q.v.*) of two hydrogen-bonded polynucleotide, antiparallel (*q.v.*) strands wound into a right-handed spiral configuration. *See* deoxyribonucleic acid.

double infection infection of a bacterium with two genetically different phages.

double-sieve mechanism a model that explains the rarity of misacylation of amino acids by proposing that an amino acid larger than the correct one is rarely activated because (1) it is too large to fit into the active site of the tRNA synthetase (first sieving), and (2) the hydrolytic site of the same synthetase is too small for the correct amino acid (second sieving). Thus, an amino acid smaller than the correct one can be removed by hydrolysis.

double-stranded RNA (dsRNA) an RNA duplex in which a messenger RNA (*q.v.*) is bound to an antisense RNA (*q.v.*) containing a complementary sequence of bases. Endogenous or exogenous dsRNAs provide a powerful means of silencing gene expression. *See* Morpholinos, RNA interference (RNAi).

double transformation *See* cotransformation.

double X in *Drosophila melanogaster*, an acrocentric, double-length X chromosome arising as a radiation-induced aberration. Such double X chromosomes are superior to the ordinary metacentric, attached X chromosomes (*q.v.*) for most stockkeeping operations, since they do not break up by crossing over with the Y. *See also* detached X.

doubling dose that dose of ionizing radiation that doubles the spontaneous mutation rate of the species under study.

doubling time the average time taken for the cell number in a population to double. The doubling time will equal the generation time (*q.v.*) only if (1) every cell in the population is capable of forming two daughter cells, (2) every cell has the same average generation time, and (3) there is no lysis of cells. The doubling time is generally longer than the generation time.

Dowex trademark of a family of ion-exchange resins.

down promoter mutations promoter alterations that decrease the frequency with which transcription is initiated relative to wild type; promoters with this property are called low-level or weak promoters.

downstream *See* strand terminologies, transcription unit. *Compare with* upstream.

downstream genes genes whose expression is subordinate to developmental control genes (*q.v.*). Downstream genes are switched on or off at various times and in different tissues by transcription factors (*q.v.*) encoded by upstream regulatory genes. A master controlling gene like *eyeless* (*q.v.*) will activate some downstream genes that encode their own transcription factors. The result will be a cascade of proteins that each regulate their own subsets of downstream genes. *See* gene networking, *Hox* genes, selector genes.

Down syndrome a type of mental retardation due to trisomy of autosome 21. The syndrome is named after the British physician, John Langdon Down, who identified it in 1866. The underlying trisomy was discovered 107 years later. Since the eyelid openings of the patient are oblique and the inner corner of the eyelid may be covered by an epicanthic fold, the condition is sometimes called *mongolism*. The frequency of such trisomic births increases with advancing maternal age as shown in the table.

Maternal age	Trisomics per 1,000 live births
16–24	0.58
25–29	0.91
30–34	1.30
35–39	4
40–44	12.5
>45	40

Although most patients with Down syndrome have a complete third copy of chromosome 21, the phenotype is due primarily to a segment, the Down syndrome chromosome region (DCR), between bands 21q 22.2 and 22.3. Within it lie five genes that seem critical to the syndrome. Chromosome 21 and chromosome 22 are similar in size, and both are acrocentrics. Chromosome 21 contains 225 genes, aligned along a 33.8 Mb DNA molecule, whereas chromosome 22 contains 545 genes on a 33.4 Mb DNA molecule. The relatively low density of genes on chromosome 21 is consistent with the observation that trisomy 21 is viable, while trisomy 22 is not. *See* Appendix C, 1959, Lejeune, Gautier, and Turpin; 1968, Henderson and Edwards; 1999, Dunham *et al.*; 2000, Hattori *et al.*; Alzheimer disease; *AML1* gene; translocation Down syndrome. http://www.nads.org.

Down syndrome cell adhesion molecule (DSCAM) a protein encoded by a gene in band 22 of the long arm of human chromosome 21. The gene contains multiple exons which allow multiple mRNAs to be transcribed by alternative splicing (*q.v.*). The transcripts are differentially expressed in different substructures of the adult brain. The DSCAM is a member of the immunoglobulin domain superfamily (*q.v.*). These isoforms may be involved in the patterning of neural networks by selective adhesions between axons. *See* innate immunity.

DPN diphosphopyridine nucleotide. NAD is the preferred nomenclature.

draft sequence in genome sequencing, a preliminary DNA sequence that has enough accuracy and continuity to allow an initial genomic analysis and annotation, but that is incomplete. It is separated by small gaps of unknown sequence, and the order and orientation of all the sequenced fragments are not always fully determined. A draft sequence of the human genome was published in 2001. *Compare with* finished sequence. *See* Appendix C, 2001, The International Human Genome Sequencing Consortium, Venter *et al.*; genomic annotation.

drift genetic drift (*q.v.*).

dRNA DNA-like RNA. RNA molecules that are not included in the rRNA and tRNA classes. Much of the dRNA is of high molecular weight, short half-life, and never leaves the nucleus. *See* hnRNA.

Drosophila a genus of flies containing about 900 described species. The most extensively studied of all genera from the standpoint of genetics and cytology. The genus is subdivided into eight subgenera: (1) *Hirtodrosophila*, (2) *Pholadoris*, (3) *Dorsilopha*, (4) *Phloridosa*, (5) *Siphlodora*, (6) *Sordophila*, (7) *Sophophora*, and (8) *Drosophila*. *D. melanogaster*, the multicellular organism for which the most genetic

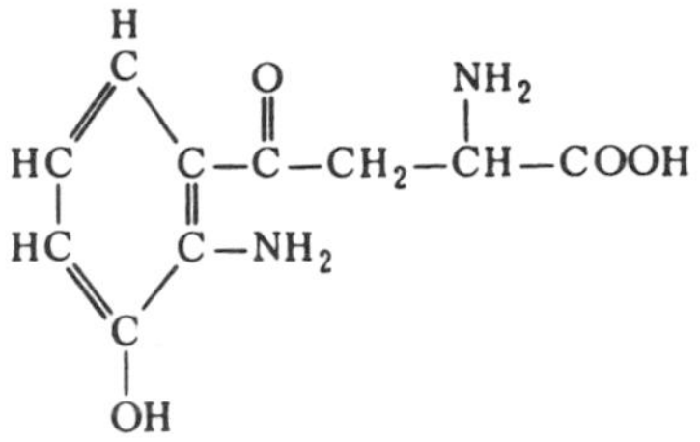

hydroxykynurenine

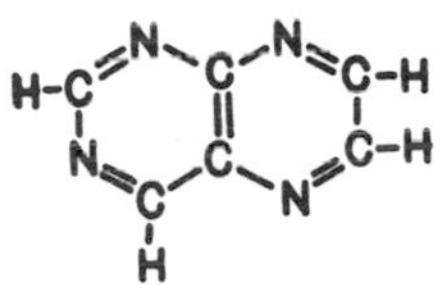

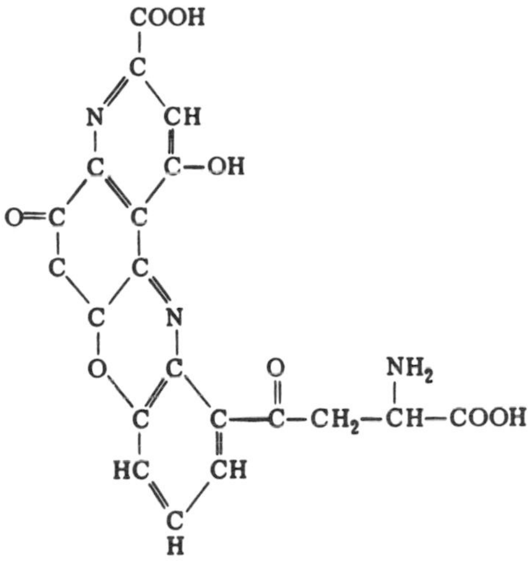

xanthommatin

pteridine

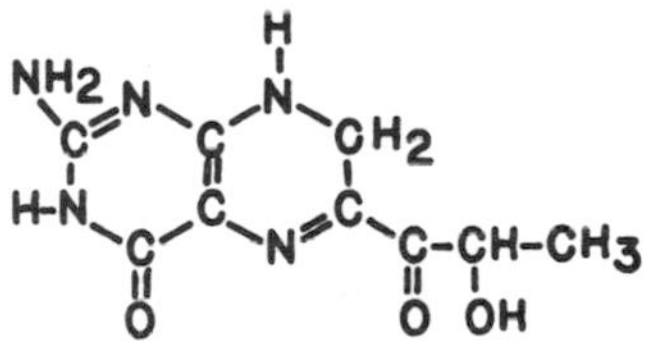

sepiapterin

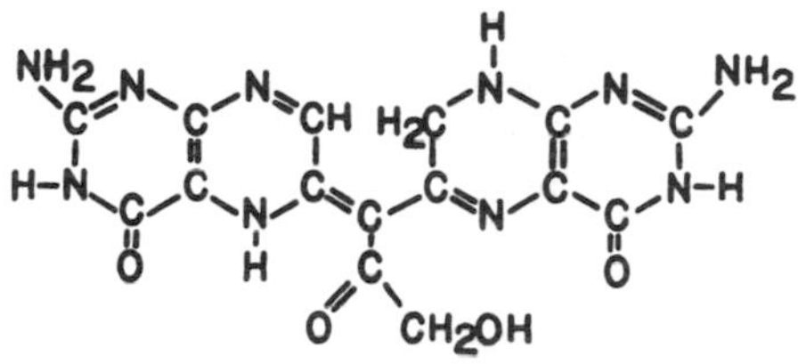

drosopterin

***Drosophila* eye pigments**

information is available, belongs in the subgenus *Sophophora*. *See* Appendix C, 1926, Chetverikov; 1936, Sturtevant and Dobzhansky; 1944, Dobzhansky; 1952, Patterson and Stone; 1985, Carson; *Drosophila virilis*, Hawaiian Drosophilidae.

***Drosophila* databases** *See* Appendix E.

***Drosophila* eye pigments** the ommatidia of the dull red compound eyes of *Drosophila* contain two classes of pigments, one brown (the ommochromes) and one bright red (the drosopterins). Studies of the precursor compounds isolated from eye color mutants played an important role in the development of the one gene–one enzyme concept (*see* Appendix C, 1935, Beadle and Ephrussi). An example of an ommochrome is xanthommatin. Hydroxykynurenine, a compound biosynthesized from tryptophan, serves as a precursor of xanthommatin. Flies lacking the plus allele of the *cinnabar* gene are unable to synthesize hydroxykynurenine, and therefore this is sometimes called the cn^+ substance. Drosopterins are pteridine derivatives. Sepiapterin is a precursor of drosopterin that accumulates in *sepia* mutants. It also gives the *Drosophilia* testis its yellow color. Structural formulas are given above. *See* formylkynurenine.

***Drosophila* immune peptides** antimicrobial peptides produced by *Drosophila* in response to infections. Drosomycin, a potent antifungal agent, is synthesized under the control of *Toll* (*q.v.*), and diptericin, an antibacterial agent, is synthesized under the control of the *immune deficiency* gene (*q.v.*). *See* Appendix C, 1996, Lemaitre *et al.*

Drosophila Information Service a yearly bulletin that lists all publications concerning *Drosophila* that year, the stock lists of major laboratories, the addresses of all *Drosophila* workers, descriptions of new mutants and genetic techniques, research notes, and new teaching exercises. *See* Appendix D.

Drosophila melanogaster commonly called the "fruit fly," this species is a model organism for the study of specific genes in multicellular development and behavior. Its haploid genome contains about 176 million nucleotide pairs. Of these, about 110 million base pairs are unique sequences, present in the euchromatin (*q.v.*). The diagram on page 134

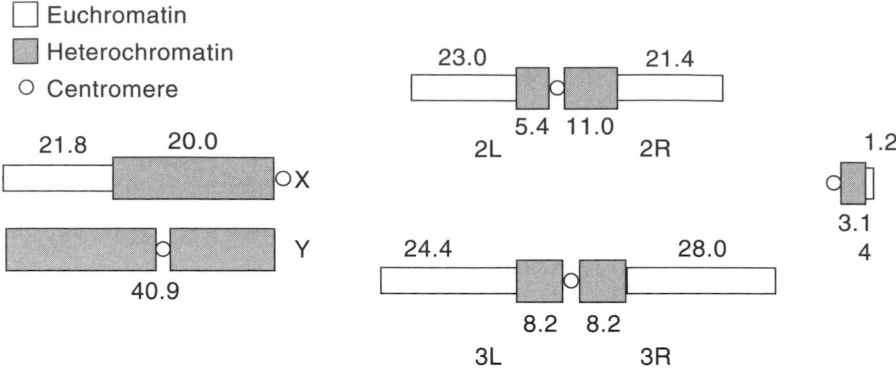

Drosophila melanogaster

shows the relative lengths of the sex chromosomes (X and Y), the major autosomes (2 and 3), and the microchromosome (4) as they appear at the metaphase stage of mitosis. The numbers give the amounts of DNA in megabases for the adjacent segments. About 13,000 genes are located in the euchromatin, and about 20% of these have been defined chemically. The average gene contains four exons, and the average transcript is made up of 3,060 nucleotides. Many *Drosophila* genes show base sequence similarities to human genes. For example, comparative studies of 290 human genes that increase susceptibility to cancer showed that 60% have *Drosophila* orthologs (*q.v.*). *See* Appendix A, Arthropoda, Insecta, Diptera; Appendix C, 1910, 1911, 1912, 1919, Morgan; 1913, 1925, 1926, Sturtevant; 1917, 1919, 1921, 1923, 1925, 1935, Bridges; 1916, 1918, 1927, Muller; 1933, Painter; 1935, Beadle and Ephrussi; 1966, Ritossa *et al.*; 1972, Pardue *et al.*; 1973, Garcia-Bellido *et al.*; 1974, Tissiers *et al.*; 1975, McKenzie *et al.*; 1978, Lewis; 1980, Nüsslein-Volhard and Wieschaus; 1982, Bingham *et al.*, Spradling and Rubin; 1983, Scott *et al.*, Bender *et al.*; 1984, Bargiello and Young; 1987, Nüsslein-Volhard *et al.*; 1988, MacDonald and Struhl; 1990, Milicki *et al.*; 1993, Maroni; 1994, Tully *et al.*, Orr and Sohal; 1995, Halder *et al.*, Zhao, Hart, and Laemmli, Kerrebrock *et al.*; 1996, Dubnau and Struhl, Rivera-Pomar *et al.*; 1998, Lim, Serounde, and Benzer; 2000, Adams *et al.*, Rubin *et al.*; Appendix E, Individual Databases; centromere, *Drosophila* targeted gene expression technique, heterochromatin, shotgun sequencing.

***Drosophila* salivary gland chromosomes** the most extensively studied polytene chromosomes. During larval development, the cells of the salivary gland undergo 9 or 10 cycles of endomitotic DNA replications to produce chromosomes that contain 1,000–2,000 times the haploid amount of DNA. The cytological map of the chromosomes of *D. melanogaster* contains slightly over 5,000 bands. It is divided into 102 divisions, distributed as illustrated.

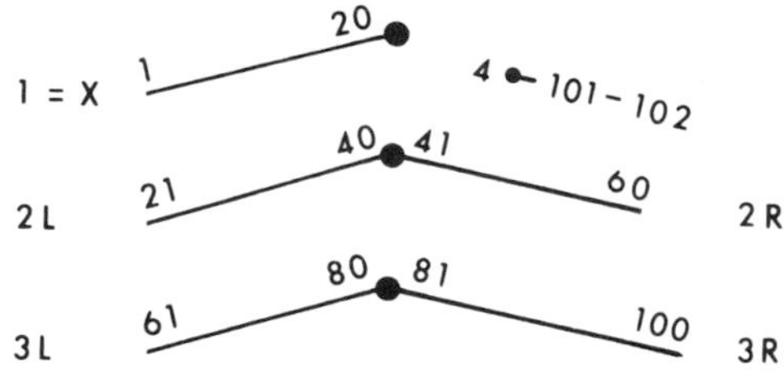

The solid circles represent the centromeres. Each division is subdivided into subdivisions lettered A–F, and the subdivisions contain varying numbers of bands. Genes have been localized within these bands by studying overlapping deficiencies and, more recently, by *in situ* hybridization with labeled probes. Since the number of genes in the euchromatin of *Drosophila* is known to be 13,000, the average band in a giant chromosome must contain two or three genes. The insertion of a transposable element (*q.v.*) can generate new bands and interbands in the salivary chromosomes. The cells of the larval salivary gland are in interphase, and within each nucleus the chromosomes show a typical orientation. The telomeres tend to be on the surface of the nuclear envelope, opposite the portion of the envelope nearest to the nucleolus, where all the centromeres are located. The arms of each autosome remain close together, whereas the relative positions of the arms vary. Different chromosomes are never entangled. The dominant chromosome folding motif is a right-handed coil. *See* Appendix C, 1933, Painter; 1935, Bridges; 1968, Semeshin *et al.*; 1988, Sorsa; biotinylated DNA, chromosomal puff, deficiency loop, *Drosophila virilis*, heterochromatin, insulator DNAs, Rabl orientation, salivary gland chromosomes.

Drosophila pseudoobscura the second *Drosophila* species to have its genome sequenced. When the sequences of *D. pseudoobscura* and *D. melanogaster* were compared, it was observed that the vast majority of genes remained on the same chromosome arms. However, within each arm the gene order was extensively shuffled. *See* Appendix C, 1944, Dobzhansky; 2005, Richards *et al.*; inversion, syntenic genes.

***Drosophila* targeted gene expression technique** an experimental system that allows the selective activation of a chosen regulatory gene in a variety of tissues and organs during *Drosophila* development. For example, the technique allowed the activation of *eyeless* (*q.v.*), a gene that specifies the differentiation of eye tissues in ectopic imaginal discs (*q.v.*). The result was extra eyes developing on antennae, legs, and wings. *See* Frontispiece illustration.

Drosophila virilis a species with polytene chromosomes that are larger than those of *Drosophila melanogaster* and exceptionally favorable for cytological analysis. The *virilis* genome is one of the largest in the genus (313 mb). *D. virilis* and *D. melanogaster* are very distantly related, having diverged about 60 million years ago. The *virilis* karyotype is considered to be ancestral for the genus. Its chromosomes X, 2, 3, 4, 5, 6 correspond to the chromosome arms X, 3R, 3L, 2L, 2R, and 4 of *Drosophila melanogaster*. *See* *Drosophila*, *Drosophila* salivary gland chromosomes.

drosopterins *See* *Drosophilia* eye pigments.

drug resistance genes genes which confer upon certain organisms or cell types resistance to the toxic effects of specific chemicals. Examples of such genes are those conferring antibiotic resistance upon bacteria, insecticide resistance upon insects, or resistance of cancer cells to anti-cancer drugs. *See* *amp*^{R}, chloroquine, dichlorodiphenyltrichloroethane (DDT), meth othrexate, penicillin, *pfcrt* gene.

drug-resistant plasmid *See* R plasmid.

drumstick a small protrusion from the nucleus of the human polymorphonuclear leukocyte, found in 3 to 5% of these cells in females, but not in males. *See* Barr body.

drupe a simple, fleshy fruit, such as an olive, derived from a single carpel and usually single-seeded.

Dryopithecus a genus of fossil primates from which the great apes and humans are thought to have diverged about 25 million years ago.

Dscam the gene in *Drosophila* that encodes a homolog of the Down syndrome cell adhesion molecule (DSCAM) (*q.v.*). This protein plays a role in directing axons to the embryonic cells that will form Bolwig organs (*q.v.*). *See* innate immunity.

dsDNA double-standed DNA.

dsRNA double-stranded RNA (*q.v.*).

D_1 trisomy syndrome *See* Patau syndrome.

dual recognition an immunological model proposing that a T cell has two receptors, both of which must simultaneously bind specific molecules in order to activate the cell; one receptor binds to the antigen, the other binds to a self-molecule of the major histocompatibility system (*q.v.*); a form of *associative recognition* (*q.v.*).

Duchenne muscular dystrophy (DMD) a disease named after G. A. B. Duchenne, who in 1868 published the first histological account of the pathological changes occurring in the muscles of patients. DMD, the most common type in humans, is X-linked, and it affects about 1 in 3,500 boys. The normal gene, *DMD*, is composed of about 2,300 kilobases. It is the largest known gene and contains 79 exons. Over 99% of the gene is made up of introns. It takes RNA polymerase II 16 hours to traverse this giant gene. The processed mRNA is about 14 kilobases, and it specifies a protein named dystrophin (*q.v.*). Victims of Duchenne muscular dystrophy have null mutations in their dystrophin genes. Less severe mutations allow reduced amounts of dystrophin to be produced, resulting in a milder disease called Becker muscular dystrophy (*q.v.*). There is a mutational hot spot between exons 44 and 45. Orthologs of DMD occur in both dogs and cats. *See* Appendix C, Hoffman, Brown, and Kunkel; *Canis familiaris*, calveolins, continuous gene syndrome, hot spot, null allele, RNA polymerase. http://www.mdausa.org.80

Duffy blood group gene the first human genetic locus to be localized on a specific autosome. The gene (symbolized *FY*) is at 1q21–22, and its nucleotide sequence has been determined. The gene encodes a protein that contains 338 amino acids organized into several transmembrane domains. The protein functions as a receptor for various cytokines (*q.v.*) and for *Plasmodium vivax* merozoites. Individuals homozygous for mutations that repress the transcription of Duffy gene in erythrocytes resist the invasion of these malaria parasites. Practically all West Africans are Duffy negative. The blood group was named in 1950 after the patient whose blood contained antibodies against the FY gene product. *See* Appendix C, 1968, Donahue *et al.*; 1976, Miller *et* *al.*; Chaudhuri *et al.*; G proteins, malaria, *Mendelian Inheritance in Man* (MIM), *Plasmodium* life cycle.

Duffy blood group receptor a soluble protein synthesized by *Plasmodium vivax* that binds to the Duffy receptor on human erythrocytes. The receptor protein contains 1,115 amino acids, and it is encoded by a single-copy gene that has been cloned. *See* Appendix C, 1990, Fang *et al.*

duplex *See* autotetraploid.

duplex DNA DNA molecules as described in the Watson-Crick model; that is, with the two polynucleotide chains of opposite 3′-5′ polarity intertwined and annealed.

duplication *See* chromosome aberration, gene duplication, repeat-induced point mutation (RIP).

durum wheat *Triticum durum* (N = 14), an ancient wheat grown since Egyptian times. Used today in the manufacture of macaroni. *See* wheat.

dwarf a short person with abnormal proportions, usually suffering from achondroplasia (*q.v.*). *Compare with* ateliosis, Laron dwarfism, midget, pituitary dwarfism.

dyad **1.** a pair of sister chromatids (*q.v.*). **2.** the products resulting from disjunction of tetrads at the first meiotic division; dyads are contained in the nuclei of secondary gametocytes. *See* meiosis.

dynamins a family of proteins that function in the pinching off of vesicles. Specific dynamins are also recruited for the division rings of mitochondria and chroroplasts.

dynein a complex protein, generally composed of two subunits, found in the armlike extensions of the A microtubules of the nine peripheral doublets of axonemes (*q.v.*). Dynein functions to couple a cycle of ATP binding and hydrolysis with a mechanical binding and release cycle to generate a sliding in relation to one another of the microtubule doublets in axonemes. Cytoplasmic dyneins also exist and are thought to function in the translocation of organelles. *See* Appendix C, 1963, Gibbons.

dysgenesis any morphological or biochemical abnormality that is evident at birth or during neonatal development.

dysplasia the abnormal growth or development of a tissue or organ. Subsequent microscopic study often shows that the component cells have abnormalities, but are not cancerous.

dysploidy the situation where the species in a genus have different diploid chromosome numbers, but the numbers do not represent a polyploid series.

dystrophin a protein encoded by the Duchenne muscular dystrophy (DMD) gene. This molecule contains 3,685 amino acids, and it only represents 0.002% by weight of muscle. Dystrophin is located at the cytoplasmic faces of the plasma membranes of striated, smooth, and cardiac muscle cells. It binds cytoskeletal actin at its N-terminal end and a transmembrane glycoprotein at its C-terminal end. A number of shorter isoforms of dystrophin are localized in brain liver and other tissues. These proteins are translated from mRNAs transcribed from promoters that reside in the DMD gene at positions to the right of the one specific for muscle transcripts. Other isoforms are the result of alternative splicing (*q.v.*). *See* gene, isoforms, muscular dystrophy, utrophin.

E

E₁, E₂, E₃ the first, second, and third generation of organisms following some experimental manipulation, such as irradiation with x-rays.

early genes genes expressed during early development. In T4 bacteriophage, those genetic elements that function in the period of phage infection before the start of DNA replication. *Compare with* late genes.

ecdysis 1. molting, the periodic shedding of the cuticle of arthropods. Ecdysis is of great antiquity, since fossil trilobites (*q.v.*) in exuvial condition have been collected from Cambrian strata. 2. in dinoflagellates, the shedding of the thecal wall layer.

ecdysones steriod hormones produced by the prothoracic gland (*q.v.*) of insects, required for molting and puparium formation; also called *molting hormones*. The structural formulas for alpha and beta ecdysone are illustrated. Beta ecdysone is sometimes called 20-hydroxy-ecdysone (20 HE). The precursor of alpha ecdysone is dietary cholesterol (*q.v.*). The alpha ecdysone synthesized by the prothoracic gland is converted in peripheral tissues, such as the fat body, to beta ecdysone, which is the active molting hormone. The conversion involves the addition of a hydroxyl group at the position indicated by the arrow. The relative concentrations of ecdysone and allatum hormone (*q.v.*) determine whether juvenile or adult development will follow a given molt. Temperature-sensitive ecdysone mutants are known for *Drosophila melanogaster*. *See* Appendix C, 1965, Karlson *et al.*; 1980, Gronemeyer and Pongs; ring gland.

Ecdysozoa a subdivision of the Articulata that contains animals that grow by periodically molting. It represents the largest clade in the animal kingdom, since it contains the Annelid and Arthropod phyla. The small phyla Onycophora and Tardigrada also

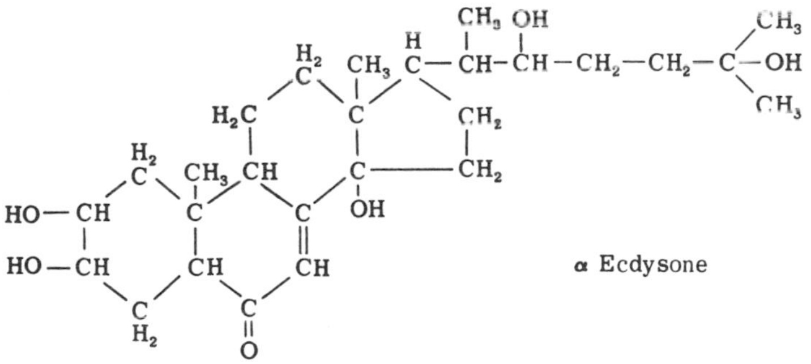

Ecdysones

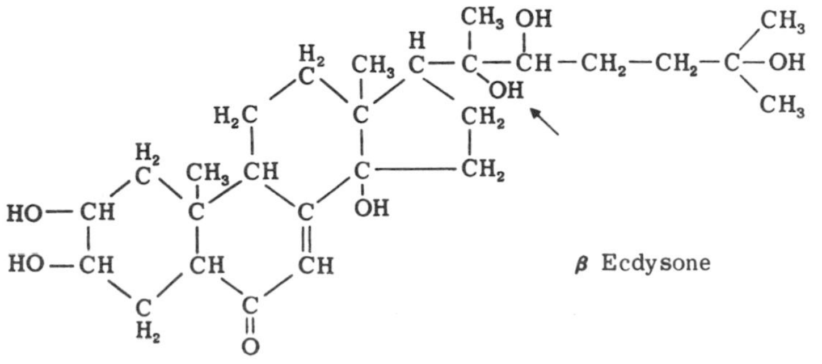

belong to this subdivision. *See* Appendix A, Animalia, Eumetazoa, Bilateria, Protostomia, Coelomata, Articulata.

echinoderm an animal with an external skeleton of calcareous plates and internal water-vascular system (starfish, sea urchins, sea lilies, sea cucumbers, etc.). Echinoderm embryos were used to isolate the first histone genes (*q.v.*) and the first cyclins (*q.v.*). *See* Appendix A, Echinodermata.

eclipsed antigens antigenic determinants of parasite origin that resemble antigenic determinants of their hosts to such a degree that they do not elicit the formation of antibodies by the host. The formation of eclipsed antigens by parasites is termed *molecular mimicry*.

eclipse period in virology, the time interval between infection and the first intracellular reappearance of infective phage particles. *See* one-step growth experiment. *Compare with* latent period.

eclosion the emergence of the adult insect from its pupal case.

ecodeme a deme (*q.v.*) associated with a specific habitat (a cypress swamp, for example).

ecogeographical divergence the evolution from a single ancestral species of two or more different species, each in a different geographical area and each adapted to the local peculiarities of its habitat.

ecogeographic rules any of several generalizations concerning geographic variation within a species that correlate adaptations with climate or other environmental conditions: e.g., Allen's rule, Bergman's rule.

E. coli *Escherichia coli* (*q.v.*).

ecological genetics the analysis of the genetics of natural populations and of the adaptations of these populations to environmental variables.

ecological isolation a premating (prezygotic) isolating mechanism in which members of different species seldom, if ever, meet because each species prefers to live (is adapted to) different habitats.

ecological niche the position occupied by a plant or animal in its community with reference both to its utilization of its environment and its required associations with other organisms.

ecology the study of the relationships between organisms and their environment.

ecophenotype a nongenetic phenotypic modification in response to environmental conditions. *See* phenotypic plasticity.

***Eco*RI** *See* restriction endonuclease.

ecosystem an assemblage of interacting populations of species grouped into communities in a local environment. Ecosystems vary greatly in size (e.g., a small pool *vs.* a giant reef). *See* biome.

ecotype race (within a species) genetically adapted to a certain environment. *See* Appendix C, 1948, Clausen *et al.*

ectoderm a germ layer forming the external covering of the embryo and neural tube (from which develop the brain, spinal cord, and nerves). Ectodermal derivatives include all nervous tissues, the epidermis (including cutaneous glands, hair, nails, horns, the lens of the eye, etc.), the epithelia of all sense organs, the nasal cavity sinuses, the anal canal, and the mouth (including the oral glands, and tooth enamel), and the hypophysis. *See* Appendix C, 1845, Remak.

ectopic out of place; referring to a biological structure that is not in its proper location. In developmental genetics the adjective is sometimes used to describe the expression of a regulatory gene in the wrong place. An example is the cluster of ommatidia on the antenna of the fly head in the Frontispiece illustration.

ectopic pairing nonspecific pairing of intercalary and proximal heterochromatic segments of *Drosophila* salivary chromosomes.

ectoplasm the superficial cytoplasm of ciliates.

ectotherm vertebrates such as fishes, amphibians, and reptiles that have little or no endogenous mechanisms for controlling their body temperature. Their temperature is determined by environmental conditions. *Contrast with* endotherm.

editing *See* proofreading, RNA editing.

EDTA an abbreviation for ethylene diaminetetracetic acid. A molecule capable of reacting with metallic ions and forming a stable, inert, water-soluble complex. EDTA is used to remove metals that occur in minute amounts even in distilled water.

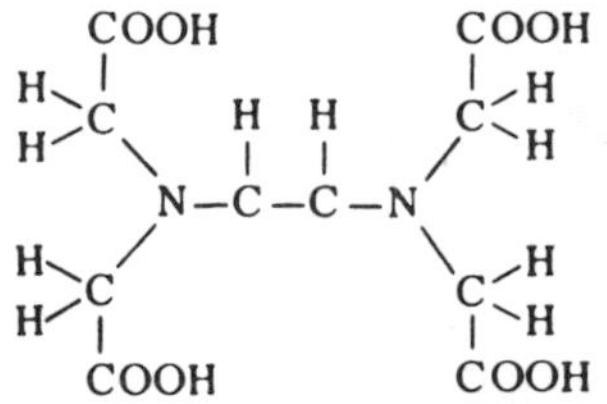

```
     COOH               COOH
      |                  |
  H\  |                  |  /H
    >C     H   H        C<
  H/  \    |   |       /  \H
       N — C — C — N
  H\  /               \  /H
    >C                  C<
  H/  |                  |  \H
     COOH               COOH
```

Edwards syndrome a well-defined set of congenital defects in humans caused by the presence of an extra chromosome 18. Also called trisomy 18 syndrome or E_1 trisomy syndrome. The mean survival is two months, and the incidence is 1/3,500 live births. The condition was first described in 1960 by J. H. Edwards and four colleagues. *See* human mitotic chromosomes.

EEG electroencephalogram (*q.v.*).

Ef-1α, Ef-Tu *See* translation elongation factors.

effective fertility *See* fertility.

effective lethal phase the stage in development at which a given lethal gene generally causes the death of the organism carrying it.

effective population size the average number of individuals in a population that contribute genes to succeeding generations. If the population size shows a cyclical variation as a function of season of the year, predation, parasitism, and other factors, the effective population size is closer to the number of individuals observed during the period of maximal contraction.

effector 1. a molecule that affects (positively or negatively) the function of a regulatory protein. 2. an organ or cell that reacts to a nervous stimulus by doing chemical or mechanical work; for example, muscles, glands, electric organs, etc.

effector cell in immunology, a cell (usually a T lymphocyte) that carries out cell-mediated cytotoxicity.

effector molecules small molecules that combine with repressor molecules and activate or inactivate them with respect to their ability to combine with an operator gene. *See* inducible system, repressible system, regulator gene.

efferent leading away from an organ, cell, or point of reference. In immunology, the *efferent* branch of the immune response includes the events occurring after the activation of the immune system (e.g., antibodies combining with antigens or cytokines stimulating specific cells). *Compare with* afferent.

efflux pump any cellular mechanism that expels drugs (such as antibiotics) or other environmental toxins from the cell.

EGF epidermal growth factor (*q.v.*).

egg a female gamete; an ovum. *See* Appendix C, 1651, Harvey; 1657, de Graff; 1827, von Baer.

egg chamber the insect ovarian follicle. In *Drosophila,* it consists of a cluster of 16 interconnected cystocytes surrounded by a monolayer of follicle cells. The oocyte is the most posterior cystocyte. The remaining 15 cystocytes function as nurse cells (*q.v.*).

Ehlers-Danlos syndrome a family of hereditary diseases characterized by overelasticity and brittleness of the skin and by excessive extensibility of the joints. The underlying defects involve blocks in the synthesis of collagen (*q.v.*). In type VI, for example, a hydroxylysine-deficient collagen is produced that is unable to form intermolecular cross-links. Both X-linked and autosomal genes are involved. The syndrome gets its name from Edward Ehlers, a Danish dermatologist, and Henri Danlos, a French physician, who published descriptions of the condition in 1901 and 1908, respectively.

einkorn wheat *Triticum monococcum* (N = 7), "one-grained" wheat, so called because it has a single seed per spikelet. A wheat in cultivation since stone age times. *See* wheat.

ejaculate 1. the process of semen release in higher vertebrates; also called *ejaculation.* 2. the semen released in a given copulatory interaction or in a given artificially induced response.

elaioplast an oil-rich plastid.

elastin a rubber-like, 70 kilodalton glycoprotein that is the main component of the elastic fibers found in tendons, ligaments, and the walls of bronchi and arteries.

electroblotting *See* blotting.

electrode either terminal of an electrical apparatus.

electroencephalogram the record of the rhythmical changes in the electrical potential of the brain.

electrofusion *See* Zimmermann cell fusion.

electrolyte a substance that when dissolved in water conducts an electric current.

electron a negatively charged particle that is a constituent of every neutral atom. Its mass is 0.000549 atomic mass units. *Compare with* positron.

electron carrier an enzyme, such as a flavoprotein or cytochrome, that can gain and lose electrons reversibly.

electron-dense said of a dense area seen on an electron micrograph, since the region has prevented electrons from passing through it. High electron density may mean that the area contains a high concentration of macromolecules or that it has bound the heavy metals (Os, Mn, Pb, U) used as fixatives and/or stains.

electron microscope a magnifying system that uses beams of electrons focused in a vacuum by a series of magnetic lenses and that has a resolving power hundreds of times that of the best optical microscopes. Two main types of electron microscopes are available. In the *transmission electron microscope* (TEM), the image is formed by electrons that pass through the specimen. In the *scanning electron microscope* (SEM), image formation is based on electrons that are reflected back from the specimen. Thus, the TEM resembles a standard light microscope, which is generally used to look at tissue slices, and the SEM resembles a stereoscopic dissecting microscope, which is used to examine the surface properties of biological materials. *See* Appendix C, 1932, Knoll and Ruska; 1963, Porter and Bonneville.

electron microscope techniques *See* freeze-etching, freeze-fracture, Kleinschmidt spreading technique, negative staining.

electron pair bond convalent bond.

electron transport chain a chain of molecules localized in mitochondria and acting as hydrogen and electron acceptors. The chain functions to funnel electrons from a given substrate to O_2. The energy released is used to phosphorylate ADP. *See* cytochrome system, mitochondrial proton transport.

electron volt a unit of energy equivalent to the amount of energy gained by an electron passing through a potential difference of one volt. Larger multiple units of the electron volt (eV) are frequently used: keV = kilo (thousand) eV; MeV = mega (million) eV; and GeV = giga (billion) eV.

electropherogram a supporting medium containing a collection of molecules that have been separated by electrophoresis. The medium is generally in the form of a sheet, much longer and wider than it is thick, and it is often a gel such as agarose.

electrophoresis the movement of the charged molecules in solution in an electrical field. The solution is generally held in a porous support medium such as filter paper, cellulose acetate (rayon), or a gel made of starch, agar, or polyacrylamide. Electrophoresis is generally used to separate molecules from a mixture, based upon differences in net electrical charge and also by size or geometry of the molecules, dependent upon the characteristics of the gel matrix. The SDS-PAGE technique is a method of separating proteins by exposing them to the anionic detergent sodium dodecyl sulfate (SDS) and polyacrylamide gel electrophoresis (PAGE). When SDS binds to proteins, it breaks all noncovalent interactions so that the molecules assume a random coil configuration, provided no disulfide bonds exist (the latter can be broken by treatment with mercaptoethanol). The distance moved per unit time by a random coil follows a mathematical formula involving the molecular weight of the molecule, from which the molecular weight can be calculated. *See* Appendix C, 1933, Tiselius; pulsed field gradient gel electrophoresis, zonal electrophoresis.

electroporation The application of electric pulses to animal cells or plant protoplasts to increase the permeability of their membranes. The technique is used to facilitate DNA uptake during transformation experiments.

electrostatic bond the attraction between a positively charged atom (cation) and a negatively charged atom (anion), as in a crystal of common table salt (NaCl); also known as an *ionic bond*. Electrostatic bonds are called *salt linkages* in the older literature. In proteins, the positively charged side groups of lysine and arginine form electrostatic bonds with the negatively charged side groups of aspartic and glutamic acids. This stabilizes the tertiary and quaternary structures of these molecules. *See* amino acids, protein structure.

element a pure substance consisting of atoms of the same atomic number and that cannot be decomposed by ordinary chemical means. *See* chemical elements, periodic table.

ELISA enzyme-linked immunosorbent assay (*q.v.*).

Ellis-vanCreveld (EVC) syndrome a hereditary disease characterized by shortening of the forearms and lower legs, extra fingers, and congenital heart malformations. The condition was first described by Richard Ellis and Simon vanCreveld in 1940. Fifty cases of EVC syndrome that occurred in the Amish (*q.v.*) community of Lancaster County have been traced back to a single couple who emigrated to Pennsylvania in 1744. The disease has been shown to be due to a mutation in an intron of the EVC gene that causes aberrant splicing. This gene has been mapped to 4p16, and it encodes a protein that contains 992 amino acids. The function of the EVC protein has not been determined. *See* posttranscriptional processing.

Elodea canadensis a common pond weed that behaves as a facultative apomict (*q.v.*). *See* Appendix C, 1923, Santos.

elongation factors proteins that complex with ribosomes to promote elongation of polypeptide chains; they dissociate from the ribosome when translation is terminated. Elongation factor G (EF-G), also called *translocase,* associated with the movement of the peptidyl tRNA from the "A" site to the "P" site of the ribosome. Elongation factor T (EF-T) is responsible for alignment of the AA ~ tRNA complex in the "A" site of the ribosome. *See* translation.

emasculation 1. the removal of the anthers from a flower. 2. castration.

EMB agar a complex bacterial growth medium containing two pH-sensitive dyes, eosin and methylene blue, and a sugar such as lactose. Lac^+ cells preferentially metabolize lactose oxidatively, resulting in excretion of hydrogen ions; under low pH, the dyes become deep purple. Lac^- cells that use amino acids as a source of energy release NH_3, thus raising pH; this decolorizes the dyes and produces a white colony. These color differences occur for all metabolizable sugars and also for many other carbon sources.

EMBL Data Library European Molecular Biology Laboratory Data Library. *See* Appendix E.

embryo a rudimentary animal or plant in the earliest stages of development, produced by zygotic cleavages and dependent upon nutrients stored within the membranes that enclose it (e.g., those covering an egg or a seed). In humans, embryonic development begins with the first zygotic division and lasts until approximately the eighth week of gestation, when an embryo becomes a fetus (*q.v.*). Early development in viviparous (*q.v.*) animals is sometimes divided into two distinct stages, pre-embryonic and embryonic, which are separated by the commencement of organ differentiation or by implantation (*q.v.*). In humans, the cell mass resulting from zygotic cell divisions up to about the fourteenth day of gestation is called a *pre-embryo,* although the use of this term is controversial. The moral status of a human embryo is a major issue area, particularly in embryonic stem cell research and in *in vitro* fertilization (*q.v.*), where surplus embryos may have to be destroyed. *See* gestation period, stem cells.

embryonic induction *See* induction.

embryonic stem cells *See* stem cells.

embryo polarity mutants *See* maternal polarity mutants.

embryo sac the female gametophyte of angiosperms. It contains several haploid nuclei formed by the division of the haploid megaspore nucleus. *See* synergid.

embryo transfer (transplantation) artificial introduction of an early embryo into the oviduct or uterus of the biological mother or of a surrogate mother. For example, a woman with blocked Fallopian tubes may choose to have an egg fertilized *in vitro,* and the resulting embryo may be introduced into her uterus for implantation. In the case of a superior dairy or beef cow, many eggs can be simultaneously artificially induced (superovulation), flushed from the oviducts, fertilized *in vitro,* and each embryo transferred to a different surrogate mother for subsequent development.

emergent properties in a hierarchical system, properties that are manifest at higher levels because they are formed by measures or processes involving aggregations of lower-level units. For example, in an ecological hierarchy, populations have properties not expressed by an individual; a community has properties not expressed in a population, etc.

emmer wheat *Triticum dicoccum* (N = 14) (also called *starch wheat* or *two-grained spelt*), a wheat cultivated since Neolithic times. *See* wheat.

EMS ethyl methane sulfonate (*q.v.*).

enantiomers, enantiomorphs compounds showing mirror-image isomerism.

Encephalitozoon cuniculi an intracellular parasite that infects a wide range of eukaryotes from protozoans to humans. Its genome is the smallest sequenced to date (2.71 Mbp). Its 11 chromosomes range in size from 217 to 315 kbp. The genome has been compacted by the shortening of intergenic spacers and of the genes themselves. For example, 85% of the proteins translated by *Encephalitozoon* are shorter than the homologous proteins of *Saccharomyces,* with a relative mean size difference of 15%. *E. cuniculi* lacks mitochondria but contains genes that have mitochondria-related functions. This suggests that mitochondria were lost during the evolution of its parasitic way of life. *See* Appendix A, Protoctista, Microspora; Appendix F.

endemic in epidemiology, referring to a disease that occurs continually in a given geographical area. In ecology, referring to a species that is native to a region or place; not introduced. In its strictest sense, endemic means entirely restricted to the area in question. For example, more than 90% of the insect species on the high Hawaiian islands are endemic. *Contrast with* indigenous.

endergonic reaction a reaction requiring energy from the outside before reactants will form products.

end labeling the attachment of a radioactive chemical (usually ^{32}P) to the 5′ or 3′ end of a DNA strand.

endocrine system a system of ductless glands that controls metabolism through the synthesis and release into the bloodstream of hormones (*q.v.*). *See* autocrine.

endocytosis the uptake by a cell of particles, fluids, or specific macromolecules by phagocytosis, pinocytosis, or receptor-mediated endocytosis, respectively. The functions in which endocytosis plays a role include antigen presentation, nutrient acquisition, clearance of apoptotic cells, pathogen entry, receptor regulation, and synaptic transmission.

endocytotic vesicles *See* receptor-mediated endocytosis.

endoderm the primary germ layer lining the primitive gut (archenteron) in an early embryo, beginning in the gastrula stage. The endoderm forms the epithelial lining of the intestine and all of the outgrowths of the intestine (gill pouches and gills, the larynx, wind pipe and lungs, the tonsils, thyroid, and thymus glands, the liver, the gall bladder and bile duct, the pancreas, and the urinary bladder and adjacent parts of the urogenital system). In the older literature, this layer is called *entoderm*. *See* Appendix C, 1845, Remak.

endogamy the selection of a mate from within a small kinship group. Inbreeding. *Contrast with* exogamy.

endogenote that portion of the original chromosome of a merozygote that is homologous to the exogenote (*q.v.*).

endogenous originating within the organism.

endogenous virus an inactive virus that is integrated into the chromosome of its host cell, and can therefore exhibit vertical transmission (*q.v.*).

endokaryotic hypothesis the theory that the nucleus of eukaryotes arose by capture of a guest prokaryote by an engulfing host prokaryote. Thus the inner nuclear membrane would originally be the guest's plasma membrane and the outer nuclear membrane would be the invaginated host's cell membrane. Phylogenetic studies of translation elongation factors (*q.v.*) suggest that the eukaryotic nucleus arose from a guest bacterium similar to an eocyte, one of the sulfur-dependent thermophiles (*q.v.*). *See* Appendix A, Archaea Crenarchaeota; Appendix C, 1992, Rivera and Lake; endosymbiont theory.

endometrium the glandular lining of the uterus of mammals that undergoes cyclical growth and regression during sexual maturity.

endomitosis somatic polyploidization taking place within an intact nuclear envelope. No stages comparable to the normal mitotic cycle are observed, but the DNA content increases in multiples of the haploid value. If the reduplicated chromosomes fail to separate and remain in register, polytene chromosomes (*q.v.*) are formed.

endomixis a process of self-fertilization in which the sperm and egg nuclei from one individual unite.

endonuclease an enzyme that breaks the internal phosphodiester bonds in a DNA molecule. Endonucleases of somatic tissue hydrolyze DNA by introducing double-strand breaks. Endonucleases isolated from cells in meiotic prophase produce single-strand breaks in the DNA with 5′ hydroxy termini. Single-strand breaks are essential first steps in replication and recombination. *See* Appendix C, 1971, Howell and Stern; 1972, Hedgepeth, Goodman, and Boyer; restriction endonuclease.

endoplasmic reticulum a membranous system organized into a net-like labyrinth of branching tubules and flattened sacs in the cytoplasm of most eukaryotic cells. In places the endoplasmic reticulum (ER) is continuous with the plasma membrane or the outer membrane of the nuclear envelope. If the outer surfaces of the ER membranes are coated with ribosomes, the ER is said to be *rough-surfaced;* otherwise it is said to be *smooth-surfaced*. The membranes of the ER constitute over half of the total membranes in the average animal cell, and the highly convoluted space enclosed by the ER (the ER cisternal space) occupies 10% or more of the total cell volume. A protein synthesized on ribosomes attached to the ER has one of two fates: (1) Water-soluble proteins are fully translocated across the ER membrane and released into its lumen. These proteins may end up in the secretion granules or in lysosomes or peroxisomes (*q.v.*). (2) Transmembrane proteins remain embedded in the ER membrane. As the ER expands, it becomes the source of the membranes of many of the cell's organelles, including the Golgi apparatus (*q.v.*), the mitochondrion (*q.v.*), the lysosome (*q.v.*), the peroxisome (*q.v.*), and the plasma membrane (*q.v.*). The ER was the first cellular component to be revealed by the electron microscope that had not already been identified by light microscopists. *See* Appendix C, 1953, Porter; 1960, Siekevitz and Palade; 1965, Sabatini *et al.*; 1991, Simon and Blobel; chloroplast er, leader sequence peptide, microsome fraction, protein sorting, receptor-mediated translocation, signal peptide, smooth endoplasmic reticulum (SER), sorting signals, translation, translocon.

endopolyploidy the occurrence in a diploid individual of cells containing 4C, 8C, 16C, 32C, etc., amounts of DNA in their nuclei. The nurse-cell nuclei in *Drosophila* egg chambers are good examples of endopolyploidy. The highest level of endopolyploidy known occurs in the larva of the silkworm moth. The silk-producing cells have nuclei that are at least 1 million–ploid. *See* Appendix C, 1979, Perdix-Gillot; *Bombyx mori*, silk.

Endopterygota Holometabola (*q.v.*).

endoreduplication cycles cycles of DNA replication in the absence of cell division, resulting in polyploidy (*q.v.*). Almost all plants and animals produce specific subpopulations of polyploid cells by this process. For example, the ploidy of megakaryocytes (the progenitor cells of blood platelets) ranges from 16N to 64N, that of cardiomyocytes (heart muscle cells) from 4N to 8N, and that of hepatocytes (liver cells) from 2N to 8N. *Compare with* polyteny.

endorphins one of a group of mammalian peptide hormones liberated by proteolytic cleavage from a 29-kilodalton prohormone called pro-opiocortin. Beta endorphins are found in the intermediate lobe of the pituitary gland and are potent analgesics (pain suppressors). The name "endorphin" is a contraction of "endogenous morphine" because it is produced within the body (endogenously), attaches to the same neuron receptors that bind morphine, and produces similar physiological effects.

endoskeleton an internal skeleton, as in the case of the bony skeleton of vertebrates.

endosperm a tissue found only in flowering plants that acquires resources from the maternal sporophyte and subsequently is digested by the developing embryo in its own seed. Endosperm develops from a triploid cell formed by a double fertilization (*q.v.*). The nucleus of each endosperm cell contains one copy of the paternal genetic contribution to the associated embryo and two copies of the maternal contribution. *See* kernel, parental imprinting.

endosymbiont theory the proposal that self-replicating eukaryotic organelles originated by fusion of formerly free-living prokaryotes with primitive nucleated cells. *See* Appendix C, 1972, Pigott and Carr; 1981, Margulis; serial symbiosis theory.

endotherm vertebrates that control their body temperature by endogenous mechanisms (e.g., birds and mammals). *Contrast with* ectotherm.

endotoxins complex lipopolysaccharide molecules that form an integral part of the cell wall of many Gram-negative bacteria and are released only when the integrity of the cell is disturbed. Endotoxins are capable of nonspecifically stimulating certain immune responses. *Compare with* exotoxin.

end product a compound that is the final product in a chain of metabolic reactions.

end product inhibition the situation in which the activity of an initial enzyme in a chain of enzymatic reactions is inhibited by the final product; feedback inhibition.

end product repression the situation in which the final product in a chain of metabolic reactions functions as a corepressor and shuts down the operon turning out the enzymes that control the reaction chain.

energy-rich bond a chemical bond that releases a large amount of energy upon its hydrolysis. The ATP molecule has an energy-rich phosphate bond, for example.

***engrailed* (*en*)** a segment polarity gene in *Drosophila* located at 2–62.0. In an organ that has undergone developmental compartmentalization (*q.v.*), such as an imaginal disc (*q.v.*) of a wing, en^+ functions as a selector gene (*q.v.*) that directs cells in a posterior compartment to form the posterior structures, rather than anterior structures, and to avoid mixing with anterior cells. Cells in the anterior compartment do not require en^+. The en^+ gene encodes a homeobox (*q.v.*) protein that contains 552 amino acids and acts as a transcription regulator. The gene is expressed in many different tissues of the embryo and larva. Homologs of *en* occur in *Caenorhabditis*, the mouse, and humans. *See* Appendix C, 1975, Morata and Lawrence.

enhancers sequences of nucleotides that potentiate the transcriptional activity of physically linked genes. The first enhancer to be discovered was a 72-base pair, tandem repeat located near the replication origin of simian virus 40. Enhancers were subsequently found in the genomes of eukaryotic cells and in RNA viruses. Some enhancers are constitutively expressed in most cells, while others are tissue specific. Enhancers act by increasing the number of RNA polymerase II molecules transcribing the linked gene. An enhancer may be distant from the gene it enhances. The enhancer effect is mediated through sequence-specific DNA-binding proteins. These observations have led to the suggestions that, once the DNA-binding protein attaches to the enhancer element, it causes the intervening nucleotides to loop out to bring the enhancer into physical contact with the promoter of the gene it enhances. This loop structure then facilitates the attachment of polymerase molecules to the transcribing gene. Enhancers that specify the type of cell in which a given gene will be expressed have been located within the first introns of genes such as those that encode the heavy-chain immunoglobin proteins and certain collagens. *See* Appendix C, 1981, Banjerji *et al.*; 1983, Gillies *et al.*; CTCF protein, DNA looping, *H19*, immunoglobin genes, insulator DNAs, intron.

enhancer trap a technique used in *Drosophila* to demonstrate the occurrence of enhancers that switch on genes in specific groups of cells during certain developmental periods. A reporter gene (*q.v.*) is used, which has a promoter that requires the assistance of an enhancer to be activated. The reporter gene, together with its "weak" promotor, is spliced into a transposable element (*q.v.*). In *Drosophila*, P elements (*q.v.*) are used, and these insert themselves at various chromosomal sites. When insertion occurs near an enhancer that normally activates genes in specific areas of a developing embryo, the reporter gene intercepts the signal and reports the position of the cells by its activity. In cases where the reporter is *lacZ*, blue pigment appears in specific cells; for example, in localized regions of the developing central nervous system. Once insertions that generate interesting staining patterns are identified, stable strains of flies carrying the insertions can be produced. Such lines are called *transposants*. Cytological mapping has shown that insertions occur throughout the genome. Since the enhancers are generally positioned within a few hundred base pairs of the start site of their target genes, these can be subsequently cloned and sequenced. *See* Appendix C, 1987, O'Kane and Gehring.

enkephalins pentapeptides with opiate-like activity (*compare with* **endorphins**), first isolated in 1975 from pig brain. Met-enkephalin has the amino acid sequence Tyr-Gly-Gly-Phe-Met; Leu-enkephalin has the sequence Tyr-Gly-Gly-Phe-Leu. *See* polyprotein.

enol forms of nucleotides *See* tautomeric shift.

enrichment methods for auxotrophic mutants *See* filtration enrichment, penicillin enrichment technique.

Ensembl genome database a joint project between the European Bioinformatics Institute, the European Molecular Biology Laboratory, and the Wellcome Trust-Sanger Institute. Ensemble continues to update the annotations of the genomes of a large sample of eukaryotes including *Homo sapiens*, chimpanzee, mouse, rat, dog, chicken, *Xenopus*, pufferfish, fruitfly, mosquito, honeybee, *Caenorhabditis*, and *Ciona*. http://www.ensembl.org/.

enterovirus a member of a group of RNA-containing viruses (including poliomyelitis virus) inhabiting the human intestine.

entoderm endoderm (*q.v.*).

entomophilous designating flowers adapted for pollination by insects.

entrainment dragging along after itself. In reference to circadian rhythm (*q.v.*), the resetting of an endogenous clock to exactly 24 hours by environmental cues, such as 12 hours of light (or warmth), followed by 12 hours of darkness (or cold).

Entrez a powerful search engine allowing access to the databases maintained by the National Center for Biotechnology Information. *See* Appendix E, Master Web Sites.

enucleate to remove the nucleus from a cell.

enucleated lacking a nucleus.

enveloped viruses viruses that have envelopes surrounding their nucleocapsids. Each viral envelope is derived from the host cell, and as a result the viruses will contain lipids, as well as nucleic acids and proteins. Spikes often project from the envelopes, and these contain virus-encoded proteins. The corona viruses, herpes viruses, HIVs, influenza viruses, Sendai viruses, and smallpox viruses are examples of viruses with envelopes.

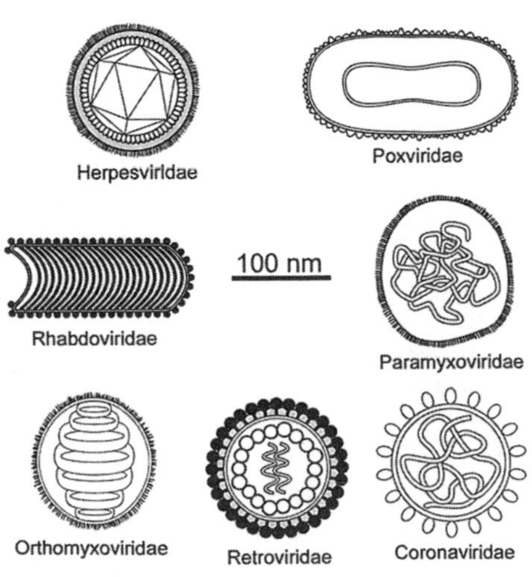

Enveloped viruses

environment the complex of physical and biotic factors within which an organism exists.

environmental sex determination any mechanism of sex determination (*q.v.*) in which the sex of an organism is permanently established not by its genetic constitution, but by environmental conditions (such as temperature, local hormonal concentrations, or population density) in which it develops. For example, sex is determined in turtles and crocodiles by the prevailing temperature during the incubation of the fertilized eggs. *Compare with* genotypic sex determination.

environmental variance that portion of the phenotypic variance caused by differences in the environments to which the individuals in a population have been exposed.

enzyme fusion *See* histidine operon.

enzymes protein catalysts. Enzymes differ from inorganic catalysts in their extreme specificity. They catalyze reactions involving only one or a few closely related compounds, and they are able to distinguish between stereoisomers. All chemical reactions have potential energy barriers. In order to pass this barrier, the reactant must be activated to reach a transition state from which the products of the reaction can be released. During an enzyme-catalyzed reaction, an enzyme-substrate complex is formed. The E-S complex has a lowered activation energy, and this allows the reaction to occur at body temperature. A systematic classification of enzymes has been established by the Commission on Enzymes of the International Union of Biochemistry. All known enzymes have been grouped into six classes: (1) oxidoreductases, (2) transferases, (3) hydrolases, (4) lyases, (5) isomerases, and (6) ligases. Each specific enzyme is given a classification code by the Enzyme Commission, which consists of EC and four figures. For example, EC 3.1.21.1 is the code designation for deoxyribonuclease 1. The first three numbers define various properties of the subfamily of enzymes to which DNase 1 belongs, while the last number uniquely specifies this enzyme. *See* Appendix C, 1876, Kühne; 1926, Sumner; Appendix E, Individual Databases; abzymes, DNA glycosylases, DNA ligases, DNA methylase, DNA polymerase, DNA restriction enzyme, DNase, endonuclease, exonuclease, first-order kinetics, kinase, papain, pepsin, peptidyl transferase, phosphodiesterase I, ribozyme, RNA ligase, RNA polymerase, RNA replicase, RNase, substrate, topoisomerase, trypsin, triptophan synthetase, tyrosinase, zymogen.

enzyme-linked immunosorbent assay (ELISA) an immunochemical technique that avoids the hazards of radiochemicals and the expense of fluorescence detection systems. Instead, the assay uses enzymes as indicators. This ELISA technique can be illustrated by its use in detecting the AIDS virus in serum samples. First, antibodies are prepared in rabbits against specific proteins present in the coat of the virus. Next, a reporter antibody is prepared in goats against rabbit immunoglobulins. Then, the enzyme horseradish peroxidase (*q.v.*) is covalently attached to the goat-antirabbit Ig-antibody. Serum from a donor is incubated in a plastic dish. The proteins in the serum sample bind to the surface of the dish. Now the AIDS-specific rabbit antibody is added. If virus coat proteins are present, they have been bound to the dish and the antibody now binds to them. The dish is washed to remove any unbound rabbit antibodies. Next, the goat antibody-peroxidase conjugate is added, and it binds to the rabbit antibody-viral protein complex. A reagent is then added, which becomes colored when dehydrogenated by the peroxidase. The amount of color developed in a specified time is a measure of the amount of virus in the serum sample. *See* HIV.

eobiogenesis the first instance of the generation of living matter from inorganic material.

Eocene the second epoch in the Tertiary period. The angiosperms and gymnosperms were the dominant plants, and representatives of all mammalian orders were present. Huge flightless birds were major predators. Whales evolved in the oceans. *See* geologic time divisions, *Hyracotherium*.

eocytes *See* sulfur-dependent thermophiles.

Eohippus *See Hyracotherium*.

eon the most inclusive of the divisions of geologic time. *See* geologic time divisions.

Ephestia kühniella the Mediterranean meal moth. Numerous mutants affecting the color pattern of the forewing and eye pigmentation have been studied in this species. It is parasitized by *Habrobracon juglandis* (*q.v.*). *See* Appendix C, 1935, Kuhn and Butenandt.

epicanthus a fold of skin extending over the inner corner of the eye characteristic of members of the Mongoloid race.

epicotyl the portion of the seedling above the cotyledons that develops into the shoot and its derivatives.

epidemiology the study of disease epidemics, with an effort to tracing down the cause.

epidermal growth factor (EGF) a mitogen that stimulates cell proliferation in a wide variety of eukaryotic cells. It is a protein containing 53 amino acids. The receptor for EGF is a protein kinase (*q.v.*), which is activated when the mitogen attaches. Once bound to the cell surface, the complex is internalized. EGF is generated by enzymatic cleavage between repeated EGF domains carried in an EGF precursor protein, which is embedded in cell membranes by a membrane-spanning domain. The EGF precursor molecule of humans contains 1,207 amino acids. EGF repeats facilitate interactions between proteins and calcium binding, and they are found in a wide variety of proteins. These include transmembrane proteins like the low-density lipoprotein receptor (*q.v.*), extracellular matrix proteins like fibrillin (*q.v.*), and soluble, secreted proteins like the plasma thromboplastin component (*q.v.*). *See* Appendix C, 1962, Cohen; familial hypercholesterolemia, neuregulins (NRGs), *Notch*.

epidermolysis bullosa a group of diseases usually showing autosomal dominant inheritance. They are characterized by blistering owing to fragility of the skin and mucous membranes. The blistering is due to separation of epidermal tissues from the underlying dermis. In *EB simplex*, tissue separation occurs within the basal layer of the epidermis, and the defect arises from mutations in a gene located at 12q11–13. This encodes keratins, which are synthesized by basal keratinocytes. In *junctional EB*, the separation occurs within the basement membrane zone of the lamina lucida, and the defect arises from mutations in a gene located at 1q25–31. This encodes a laminin protein component of the basement membrane. In *dystrophic EB*, the separation is in the papillary dermis at the level of the fibrils that connect the lamina lucida to the dermis. The defect arises from mutations in a gene located at 3p21, which encodes a type of collagen that is a major component of the anchoring fibrils.

epigamic serving to attract or stimulate members of the opposite sex during courtship.

epigamic selection *See* sexual selection.

epigenesis the concept that an organism develops by the new appearance of structures and functions, as opposed to the hypothesis that an organism develops by the unfolding and growth of entities already present in the egg at the beginning of development (preformation).

epigenetics the study of the mechanisms by which genes bring about their phenotypic effects. Clones of cells may inherit phenotypic changes that are not due to changed nucleotide sequences in the genome. DNA-binding proteins encoded by regulatory genes sometimes produce epigenetic changes that can be preserved during the mitotic division of the cells of different somatic tissues in a multicellular organism. However, meiosis seems to reset the genome to some baseline epigenetic state, so that the developmental program of the species unfolds anew with each new generation.

epigenotype the series of interrelated developmental pathways through which the adult form is realized.

epimers organic compounds that are partial isomers in that they differ from each other only in the three-dimensional positioning of atoms about a single asymmetric carbon atom.

epinephrine a hormone from the adrenal medulla that elevates blood glucose by mobilizing glycogen reserves.

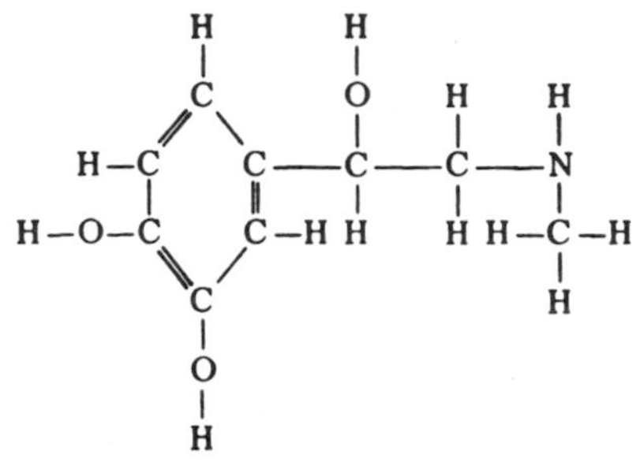

episome a class of genetic elements of which phage lambda and sex factor F are examples in *E. coli*. Episomes may behave (1) as autonomous units replicating in the host independent of the bacterial chromosome, or (2) as integrated units attached to the bacterial chromosome and replicating with it. *Compare with* plasmid.

epistasis the nonreciprocal interaction of nonallelic genes. The situation in which one gene masks the expression of another. The recessive gene *apterous* (*ap*) in *Drosophila* produces wingless homozygotes. In such individuals, any other recessive gene affecting wing morphology will have its action masked. The *apterous* gene is said to be epistatic to a gene like *curled wing*, which is hypostatic to *ap*. *See* Appendix C, 1900, Bateson; Bombay blood group.

epistatic gene *See* epistasis.

epithelium a tissue that forms the surface of an organ or organism. For example, the outer skin is an epithelial tissue; the cells lining the gut and respiratory cavities are also epithelial cells.

epitope the antigenic determinant on an antigen to which the paratope on an antibody binds.

epizoite a nonparasitic sedentary protoctist or animal living attached to another animal.

epoch in geological time, a major subdivision of a geological period. *See* geologic time divisions.

eponym a word, phrase, or abbreviation derived from the name of a person or place. Eponyms often honor the discoverers of laws or other phenomena and the inventors of methods or techniques. Examples are Hardy-Weinberg law, Lyon hypothesis, Haldane rule, Batesian mimicry, Rabl orientation, Ouchterlony technique, and Miller spread. Genetic diseases are often named after the physicians who first described them (e.g., Down syndrome and Tay-Sachs disease), or, more rarely, after the patients in which the conditions were first recognized (e.g., Christmas disease and hemoglobin Lepore). Names of places and institutions are sometimes used as medical eponyms (e.g., Bombay blood group and hemoglobin Barts). The names of scientists who gave the first descriptions of various organs, organelles, and submicroscopic life forms are also honored with eponyms (e.g., Malpighian tubule, Golgi apparatus, Cajal body, and Epstein-Barr virus). In most eponyms the possessive form ('s) is avoided. There are about 120 eponyms distributed among the definition headings of this dictionary.

epoxide a family of chromosome-breaking, alkylating agents. Di(2,-3 epoxy)propyl ether is an example.

$$O(CH_2-\overset{O}{\overbrace{CHCH_2}})(CH_2-\underset{O}{\underbrace{CHCH_2}})$$

epoxide group

Epstein-Barr virus (EBV) a DNA virus of the herpes group discovered in 1964 by M. A. Epstein and Y. M. Barr in cultures of Burkitt lymphoma cells. EBV is the cause of infectious mononucleosis, and it has an integration site on human chromosome 14. *See* Burkitt lymphoma, virus.

equational division a division of each chromosome into equal longitudinal halves that are incorporated into two daughter nuclei. It is the type of division seen in mitosis.

equatorial plate *See* mitosis.

Equidae the horse family and its extinct ancestors. Living species include the domesticated horse, *Equus caballus;* the donkey, *E. asinus;* and three species of zebras. The earliest horse genus in the fossil record is *Hyracotherium* (*q.v.*). *See* Appendix A, Mammalia, Perissodactyla; horse; horse-donkey hybrids.

equilibrium centrifugation *See* centrifugation separation.

equilibrium dialysis *See* dialysis.

equilibrium population a population in which the allelic frequencies of its gene pool do not change through successive generations. An equilibrium can be established by counteracting evolutionary forces (e.g., a balance between selection and mutation pressures) or by the absence of evolutionary forces. *See* Hardy-Weinberg law.

equine referring to members of the horse family, especially the domestic horse *Equus caballus.*

Equus the genus that contains two domesticated species: *E. caballus*, the horse (2N – 64), and *E. asinus*, the donkey (2N = 62). *E. przewalskii*, the Asiatic wild horse or taki (2N = 66), can produce fertile hybrids with domesticated horses. The species name comes from Nikolai Przewalski, the Russian naturalist who described takis in 1883. Takis are descendants of the horses like those shown in the ancient cave paintings of Lascaux, France. About 50 genes have been mapped for *E. caballus*. *See* Appendix E; horse (for a listing of breeds), horse-donkey hybrids.

ER endoplasmic reticulum (*q.v.*).

era in geological time, a major division of a geological eon. *See* geologic time divisions.

ergastoplasm rough-surfaced endoplasmic reticulum.

ergosome polysome (*q.v.*).

error-prone repair a process first demonstrated in *E. coli* exposed to UV light that allows DNA replication to occur across thymine dimers (*q.v.*), but at the cost of fidelity of replication. The resulting strand often has incorrect nucleotides inserted opposite the UV photoproducts in the template. *See* Appendix C, 1967, Witkin; SOS response.

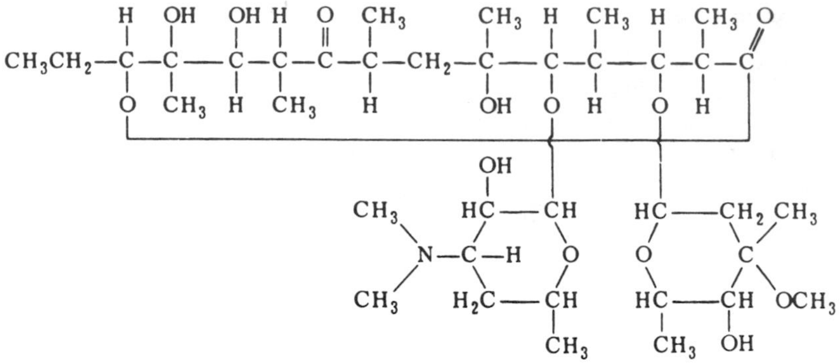

Erythromycin

errors *See* statistical errors.

erythroblastosis the discharge of nucleated red blood corpuscles from the blood-forming centers into the peripheral blood.

erythroblastosis fetalis a hemolytic disease of infants due to Rh incompatibilities between the fetus and its mother. *See* Appendix C, 1939, Levine and Stetson; Rh factor, RhoGAM.

erythrocyte the hemoglobin-containing cell found in the blood of vertebrates.

erythrolabe *See* color blindness.

erythromycin an antibiotic (illustrated above) produced by *Streptomyces erythieus*. It inhibits protein synthesis by binding to the 50S subunits of 70S ribosomes. *See* cyclohexamide, ribosome, ribosomes of organelles.

erythropoiesis the production of erythrocytes (*q.v.*).

erythropoietin a glycoprotein cytokine (*q.v.*) produced by the kidney; it regulates the production of red blood cells. *See* Janus kinase 2.

escaper *See* breakthrough.

Escherichia coli the "colon bacillus," the organism about which the most molecular genetics is known. The genus is named after Theodore Escherich, the German bacteriologist who first described *E. coli* in 1885. The distances between genes presented on its circular linkage map are measured in minutes, based on interrupted mating experiments (*q.v.*). *E. coli* is of preeminent importance in recombinant DNA research, since it serves as a host for a wide variety of viral, plasmid, and cosmid cloning vectors. Given glucose and inorganic salts, *E. coli* can synthesize all the compounds it requires for life. Its chromosome is a circular DNA molecule that contains 4,639,221 base pairs. Some 4,288 ORFs have been identified. The functions of 40% of these are unknown. The average distance between genes is 118 bp. The average size of a protein encoded by an ORF is 317 amino acids. The arrangement of genes that control flagellar synthesis is nearly identical to that of *Salmonella typhimurium*. The largest functional group of genes consists of 281 that encode transport and binding proteins. Of these, 54 are ABC transporters (*q.v.*). The *E. coli* chromosome is a host for many prophages and cryptic prophages (*q.v.*). *See* Appendix A, Bacteria, Proteobacteria; Appendix C, 1946, Lederberg and Tatum; 1953, Hayes; 1956, 1958, Jacob and Wollman; 1961, Jacob and Monod, Nirenberg and Matthaei; 1963, Cairns, Jacob and Brenner; 1969, Beckwith *et al.*; 1972, Jackson *et al.*; 1973, Cohen *et al.*; 1997, Blattner *et al.*, Lawrence and Oschman; Appendix E, Individual Databases; insertion sequence, lambda (λ) bacteriophage, operon, *Salmonella*, sympatric speciation, virulence plasmids.

***Escherichia coli* databases** *See* Appendix E.

essential amino acids amino acids required in the diet of a species because these molecules cannot be synthesized from other food materials; in contrast to nonessential amino acids that can be synthesized by normal members of the species. *See* amino acid.

EST expressed sequence tag. *See* sequence tagged site (STS).

established cell line a cell line (*q.v.*) that demonstrates the potential to be subcultured indefinitely *in vitro*. HeLa cells (*q.v.*) represent an established cell line.

estivate aestivate (*q.v.*).

estradiol a steroidal estrogen.

estrogen an ovarian hormone that prepares the mammalian uterus for implantation of an embryo; also responsible for development of secondary sexual characteristics in females.

Estradiol

estrous cycle a seasonal cycle of reproductive activity dependent upon endocrine factors. If the organism has one estrous period per year, it is called monestrous, if more than one, polyestrous.

estrus **1.** the period of reproductive activity. **2.** the estrous cycle.

ethidium bromide a compound used to separate covalent DNA circles from linear duplexes by density gradient centrifugation. Because more ethidium bromide is bound to a linear molecule than to a covalent circle, the linear molecules have a higher density at saturating concentrations of the chemical and can be separated by differential centrifugation. It is also used to locate DNA fragments in electrophoretic gels because of its fluorescence under ultraviolet light.

Ethiopian designating or pertaining to one of the six biogeographic realms (*q.v.*) of the world; it includes Africa, Iraq south of the Tropic of Cancer, Madagascar, and the adjacent islands.

ethological behavioral.

ethological isolation the failure of related species or semispecies to produce hybrid offspring because of differences in their mating behaviors. *See* allesthetic trait, courtship ritual, mate choice, species recognition.

ethology the scientific study of animal behavior, particularly under natural conditions.

ethylene a gas ($H_2C = CH_2$) generated by plant cells, which functions as a growth regulator. It inhibits the elongation of stems, while promoting their growth in diameter. It also accelerates the ripening of fruit and, in conjuction with auxin (*q.v.*), can stimulate the growth of lateral roots. In *Arabidopsis thaliana*, mutations of the *ETRI* gene confer resistance to ethylene. The protein encoded by *ETRI* contains ethylene-binding sites and so plays a role in transduction of the ethylene signal.

ethylene dinitrilotetra-acetic acid *See* EDTA.

ethyl methane sulfonate (EMS) one of the most commonly used mutagenic alkylating agents. The most common reaction of EMS is with guanine (*q.v.*), where it causes an ethyl group to be added to the number 7 nitrogen. Alkylation of guanine allows it to pair with thymine. Then, during replication, the complementary strand receives thymine rather than cytosine. Thus, EMS causes base substitutions of the transition type.

$$CH_3CH_2-O-\overset{\overset{\displaystyle O}{\|}}{\underset{\underset{\displaystyle O}{\|}}{S}}-CH_3$$

etiolation a plant syndrome caused by suboptimal light, consisting of small, yellow leaves and abnormally long internodes.

etiology the study of causes, especially of disease.

E_1 trisomy syndrome *See* Edwards syndrome.

Euarchontoglires a clade which groups together five mammalian orders because of similarities in the sequences of the nucleotides in a sample of their genes. It contains the Rodentia, Lagomorpha, Scandentia, Dermoptera, and Primates. *See* Appendix A.

Eubacteria a subkingdom of the Prokaryotae (*see* Appendix A) composed of bacteria that, unlike the archaebacteria, contain neuraminic acid (*q.v.*) in their cell walls. They also differ from archaebacteria in the composition of their tRNAs, rRNAs, and their RNA polymerases. *See* TATA box-binding protein.

eucaryote *See* eukaryote.

Eucaryotes *See* Eukaryotes.

euchromatic containing euchromatin.

euchromatin the chromatin that shows the staining behavior characteristic of the majority of the chromosomal complement. It is uncoiled during interphase and condenses during mitosis, reaching a maximum density at metaphase. In polytene chromosomes, the banded segments contain euchromatin. *See* heterochromatin.

eugenics the improvement of humanity by altering its genetic composition by encouraging breeding of those presumed to have desirable genes (positive eugenics), and discouraging breeding of those presumed to have undesirable genes (negative eugenics). The term was coined by Francis Galton.

Euglena gracilis a flagellated protoctist belonging to the Euglenida (*See* Appendix A). It was in this species that circular molecules of DNA from chloroplasts were first observed. Later when the complete nucleotide sequence of chDNA was determined, genes were identified with introns and twintrons (introns within introns). *See* Appendix C, 1961, Manning *et al.*; 1972, Pigott and Carr; 1993, Hallick *et al.*

eukaryon the highly organized nucleus of an eukaryote.

eukaryote a member of the superkingdom Eukaryotes (*q.v.*).

Eukaryotes the superkingdom containing all organisms that are, or consist of, cells with true nuclei bounded by nuclear envelopes and that undergo meiosis. Cell division occurs by mitosis. Oxidative enzymes are packaged within mitochondria. The superkingdom contains four kingdoms: the Protoctista, the Fungi, the Animalia, and the Plantae. *See* Appendix A, Eukaryotes; Appendix C, 1937, Chatton; TATA box-binding protein. *Contrast with* Prokaryotes.

eumelanin one of the pigment molecules found in the coat and pigmented retinal epithelium of mammals. It is derived from the metabolism of tyrosine and is normally black in color. Its precise coloration is affected by a large number of mutant genes. *See* agouti, *MC1R* gene, melanin.

Eumetazoa the subdivision of the animal kingdom containing organisms possessing organ systems, a mouth, and digestive cavity. *See* Appendix A.

euphenics the amelioration of genotypic maladjustments brought about by efficacious treatment of the genetically defective individuals at some time in their life cycles.

euploid a polyploid cell or organism whose chromosome number is an exact multiple of the basic number of the species from which it originated. *See* polyploidy.

eupyrene sperm *See* sperm polymorphism.

eusocial a social system in which certain individuals incur obligate sterility, but enhance their fitness by aiding their collateral kin to rear their offspring. For example, sterile female worker bees may rear the offspring of their fertile sister queens. Eusocial animals exhibit (1) cooperative care of the young; (2) a reproductive division of labor (i.e., certain castes are sterile or less fecund); and (3) an overlap of at least two generations (i.e., the offspring assist the parents during some period of their lives). Many eusocial species, such as bees and ants, are haplodiploid (*q.v.*). However, in eusocial naked mole rats and snapping shrimp (*Synalpheus regalis*) both males and females develop from fertilized (diploid) eggs. *See* inclusive fitness.

euthenics the control of the physical, biological, and social environments for the improvement of humanity.

eV electron volt.

evagination an outpocketing.

eversporting referring to a strain characterized by individuals that, instead of breeding true, produce variations of a specific sort in succeeding generations. Such strains generally contain mutable genes.

evocation the morphogenetic effect produced by an evocator (*q.v.*).

evocator the morphogenically active chemical emitted by an organizer.

evolution those physical and biological changes that take place in the environment and the organisms occupying it during geological time. The organisms undergo hereditary transformations in their form, functioning, and behavior making their descendants different from their ancestors in adaptive ways. Therefore lineages accumulate successive genetic adaptations to the way of life that characterizes the species. Potentially reversible changes in the gene pool of a population constitute microevolution. Irreversible changes result in either anagenesis (*q.v.*) or cladogenesis (*q.v.*). The production of novel adaptive forms worthy of recognition as new taxa (e.g., the appearance of feathers on a reptilian ancestor) is considered to define a new taxon—Aves (birds). Such changes are classified as macroevolutionary. *See* adaptive radiation, *Archaeopteryx*, cladogram, concerted evolution, Darwin's finches, Dollo law, founder effect, genetic drift, gradualism, intron origins, *in vivo* evolution, Linnean Society of London, mtDNA lineages, natural selection, *Origin of Species*, orthogenesis, parasite theory of sex, punctuated equilimbrium, sexual selection, speciation.

evolutionarily derived character *See* phylogenetic classification.

evolutionarily primitive character *See* phylogenetic classification.

evolutionary clock *See* DNA clock hypothesis, protein clock hypothesis.

evolutionary rate rapid evolution is called *tachytelic,* slow evolution is called *bradytelic,* and evolution at an average rate is called *horotelic.*

evolutionary studies: milestones *See* Appendix C, 1858, Darwin and Wallace; 1859, Darwin; 1868, Huxley; 1872, Gulick; 1876, Wallace; 1891, Tutt; 1911, Robertson; 1917, Winge; 1930, Fisher; 1931, Wright; 1932, Haldane; 1936, Sturtevant and Dobzhansky; 1937, Dobzhansky, Chatton; 1945, White; 1947, Mourant; 1948, Clausen, Keck, and Hiesey; 1952, Patterson and Stone, Bradshaw; 1955, Flor; 1956, Brown and Wilson; 1958, Kettlewell; 1961, von Eherenstein and Lipman; 1962, Zukerkandl and Pauling; 1963, Margoliash, Mayr; 1964, Hamilton; 1967, Spiegelman, Mills, and Peterson, Fitch and Margoliash; 1968, Kimura, Wright; 1972, Kohne, Chisson, and Hoger; 1974, Stebbins; 1977, Woese and Fox, 1978, Schwartz and Dayhoff; 1981, Margulis; 1983, Kimura and Ohta; 1985, Carson; 1987, Cann, Stoneking and Wilson; 1988, Kazazian *et al.*; 1990, Maliki, Schughart and McGinnis; 1991, Baldwin *et al.*, Sogin, Ijdo *et al.*; 1992, Rivera and Lake, Haig; 1993, Baldauf and Palmer; 1994, Morral *et al.*; 1995, Nilsson and Pelger, Wilson and Szostak, Horai *et al.*; 1996, Burgher *et al.*; 1997, Krings *et al.*; 1999, Petren, Grant, and Grant; 2000, Lemieux, Otis and Thurmel, Singh and Kulathinal; 2001, Masden *et al.*, Nachman, Hoehstra and D'Agostino; 2004, Rice et al.

ewe a female sheep. *See* ram.

exaptation a character that provides a selective advantage under current conditions but had a different original function. For example, the metabolic enzymes aldehyde dehydrogenase, glutathione transferase, and transketolase have been appropriated (have undergone exaptation) as lens crystalins. Parts of the old reptilian jaw became the ear bones of mammals. The term *predaption* (*q.v.*) is also used for molecules or more complex structures that were present earlier and then switched function in response to changed selective pressures. *Exaptation* is the preferred term, since it does not suggest foresight or preplanning for the ultimate use of the structure. *See* aptation.

exchange pairing the type of pairing of homologous chromosomes that allows genetic crossing over to take place. Synaptonemal complexes (*q.v.*) play a critical role in exchange pairing.

excision the enzymatic removal of a polynucleotide segment from a nucleic acid molecule. *See* imperfect excision.

excisionase an enzyme required (in cooperation with an integrase) for deintegration of prophage from the chromosome of its bacterial host.

excision repair *See* cut-and-patch repair, repair synthesis.

excitation the reception of a quantum of energy by an atomic electron: an altered arrangement of planetary electrons in orbit about an atom resulting from absorption of electromagnetic energy.

exclusion principle the principle according to which two species cannot coexist in the same locality if they have identical ecological requirements.

exclusion reaction the healing reaction of a phage-infected bacterium that strengthens its envelope and prevents entry of additional phages.

exconjugant **1.** ciliates (e.g., paramecia) that were partners in conjugation and therefore have exchanged genetic material. **2.** a female (F^-) recipient bacterial cell that has separated from a male (Hfr) donor partner after conjugation and therefore contains some of the donor's DNA.

exergonic reaction a reaction proceeding spontaneously and releasing energy to its surroundings.

exocrine referring to endocrine glands that release their secretory products into ducts that open on an epithelial surface. Examples would be the sweat glands and glands secreting mucus. *Compare with* endocrine system.

exocytosis the discharge from a cell of materials by reverse endocytosis (*q.v.*).

exogamy the tendency of an individual to mate selectively with nonrelatives. *Contrast with* endogamy.

exogenic heredity transmission from generation to generation of information in the form of knowledge and various products of the human mind (i.e., books, laws, inventions, etc.).

exogenote the new chromosomal fragment donated to a merozygote (*q.v.*).

exogenous DNA DNA that originates outside an organism (e.g., from another cell or virus).

exogenous virus a virus that replicates vegetatively (productively in lytic cycle) and is not vertically transmitted in a gametic genome.

exon a portion of a split gene (*q.v.*) that is included in the transcript of a gene and survives processing of the RNA in the cell nucleus to become part of a spliced messenger of a structural RNA in the cell cytoplasm. Exons generally occupy three distinct regions of genes that encode proteins. The first, which is not translated into protein, signals the beginning of RNA transcription and contains sequences that direct the mRNA to the ribosomes for protein synthesis. The exons in the second region contain the information that is translated into the amino acid sequence of the protein. Exons in the third region are transcribed into the part of the mRNA that contains the signals for the termination of translation and for the addition of a polyadenylate tail. *See* Appendix C, 1978, Gilbert; alternative splicing, intron, leader sequence, polyadenylation, posttranscriptional processing, terminators.

exon shuffling the creation of new genes by bringing together, as exons of a single gene, several coding sequences that had previously specified different proteins or different domains of the same protein, through intron-mediated recombination.

exonuclease an enzyme that digests DNA, beginning at the ends of the strands.

exonuclease III an enzyme from *E. coli* that attacks the DNA duplex at the 3′ end on each strand; used together with S1 nuclease (*q.v.*) to create deletions in cloned DNA molecules. *Compare with* Bal 31 exonuclease.

exonuclease IV an enzyme that specifically degrades single-stranded DNA. It initiates hydrolysis at both the 3′ and 5′ ends to yield small oligonucleotides. This enzyme is active in the presence of EDTA.

exopterygota hemimetabola (*q.v.*).

exoskeleton a skeleton covering the outside of the body, characteristic of arthropods.

exotoxin a poison excreted into the surrounding medium by an organism (e.g., certain Gram-positive bacteria such as those causing diphtheria, tetanus, and botulism). Exotoxins are generally more potent and specific in their action than endotoxins (*q.v.*).

experimental error 1. the chance deviation of observed results from those expected according to a given hypothesis; also called random sampling error. 2. uncontrolled variation in an experiment. *See* analysis of variance.

explant an excised fragment of a tissue or an organ used to initiate an *in vitro* culture.

exponential growth phase that portion of the growth of a population characterized by an exponential increase in cell number with time. *See* stationary phase.

exponential survival curve a survival curve without a shoulder, or threshold region, and that plots as a straight line on semilog coordinates.

expressed sequence tags (ESTs) partial cDNA (*q.v.*) sequences, generally 200–400 base pairs in length, which are used as "tags" to isolate known or new genes from genomic DNA. ESTs from hundreds of organisms, including model organisms such as the fruitfly, zebrafish, and mouse, are screened, annotated, and stored in a publicly accessible computer database, called *dbEST*. A database search using tools such as BLAST (*q.v.*) can be undertaken for sequence similarity between the EST and a putative disease gene or between the EST's amino acid sequence and partial amino acid sequence from a protein of interest. If sequence similarity is found, the EST can be used as a probe (*q.v.*) to screen a library (*q.v.*) for the gene in question. ESTs are a powerful tool for quickly isolating a gene that codes for a known protein, for identifying new genes, for identifying transcription units in genomic sequences and on physical maps (*q.v.*), and for gene comparisons between organisms. *See* Appendix E, Individual Databases; sequence tagged sites (STS).

expression vector cloning vehicles designed to promote the expression of gene inserts. Typically, a restriction fragment carrying the regulatory sequences of a gene is ligated *in vitro* to a plasmid containing a restriction fragment possessing the gene but lacking its regulatory sequences. The plasmid with this new combination of DNA sequences is then cloned under circumstances that promote the expression of the gene under the control of the regulatory sequences.

expressivity the range of phenotypes expressed by a given genotype under any given set of environmental conditions or over a range of environmental conditions. For example, *Drosophila* homozygous for the recessive gene "eyeless" may have phenotypes varying from no eyes to completely normal eyes, but the usual condition is an eye noticeably smaller than normal.

expressor protein the product of a regulatory gene, necessary for the expression of one or more other genes under its positive transcriptional control.

extant living at the present time, as opposed to extinct.

exteins *ex*ternal pro*tein* sequences that flank an intein (*q.v.*) and are ligated during protein splicing (*q.v.*) to form a mature protein.

extended phenotype *See* phenotype.

extinction termination of an evolutionary lineage without descendants. *See* mass extinction, pseudoextinction, taxonomic extinction.

extrachromosomal inheritance *See* extranuclear inheritance.

extragenic reversion a mutational change in a second gene that eliminates or suppresses the mutant phenotype of the first gene. *See* suppressor mutation.

extranuclear inheritance non-Mendelian heredity attributed to DNA in organelles such as mitochondria or chloroplasts; also called extrachromosomal inheritance, cytoplasmic inheritance, maternal inheritance.

extrapolation number in target theory (*q.v.*) the intercept of the extrapolated multitarget survival curve with the vertical logarithmic axis specifying the survival fraction. The extrapolation number gives the number of targets that each must be hit at least once to have a lethal effect on the biological system under study.

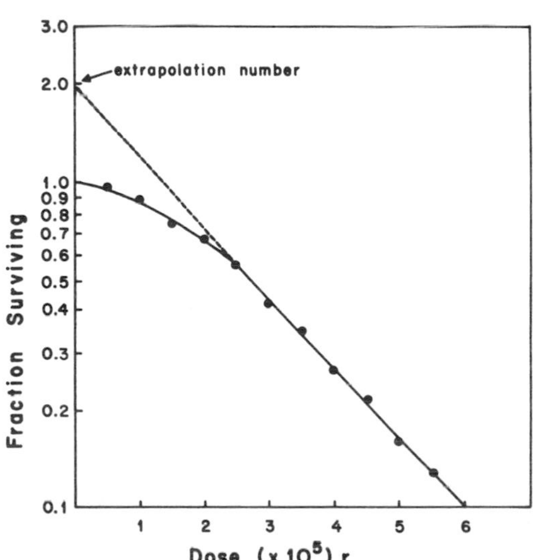

extremophiles archaean microbes that not only can survive in environments that are too hot, cold, salty, acidic, or alkaline for most bacteria (and all eukaryotes) but also in some cases require one or more of these extremes to survive. Heat-loving microbes are referred to as *thermophilic* (*q.v.*). Those that prefer both heat and acid conditions are called *thermoacidophiles* (*q.v.*). Psychrophils are cold lovers, barophils survive at high pressures, and alkaliphils survive best at a pH of about 8 (above pH 8, RNA molecules begin to break down). Salt-loving microbes are termed *halophiles* (*q.v.*). Some of the enzymes produced by extremophiles can function well outside the range of environments in which the enzymes of most other organisms would be inactivated. These catalysts are termed *extremozymes*. Extremophiles use a variety of processes to maintain a less harsh milieu within their cells. For example, acidophils tend to prevent acid from entering the cell and may also produce some molecules that reside on the cell wall and underlying cell membranes that provide protection from low pH in the external environment. *Taq* DNA polymerase (*q.v.*) is an extremozyme that allowed automation of the polymerase chain reaction (*q.v.*). More recently, *Taq* polymerase has been replaced by *Pfu* DNA polymerase (*q.v.*) from the hyperthermophile *Pyrococcus furiosus* because it works best at 100°C and has superior proofreading activity that produces high-fidelity amplification.

extremozymes *See* extremophiles.

exuvial an adjective describing an animal in the process of molting. The word is derived from *exuviae*, the cast-off exoskeletons of arthropods. *See* ecdysis.

ex vivo referring to experiments in which genetically defective cells from an individual are removed and cultured *in vitro*. After a period of multiplication, wild-type genes are added to the cells to correct the defect. These modified cells are then returned to the donor. *See* gene therapy.

eye evolution modeling Nilsson-Pelger model of eye evolution (*q.v.*).

eyeless (ey) a *Drosophila* gene located at 4–2.0 that controls eye development. Mutations result in the complete loss or the reduction in size of the compound eye. Also severe defects occur in the brain structures essential for vision, olfaction, and the coordination of locomotion. The most studied mutation, ey^2, is caused by the insertion of a transposable element (*q.v.*) in the first intron of the gene. The ey^2 phenotype is the result of cell death in the eye imaginal discs (*q.v.*) during the third larval instar. The *ey* gene is expressed in the eye primordium in the embryo during the development of the eye imaginal disc and in the pupa when rhodopsin (*q.v.*) is being synthesized. Also, *eyeless* is expressed in the cells of Bolwig organs (*q.v.*) which serve as larval eyes. The *ey* transcription unit is about 16 kb long. Two transcripts are produced by alternative splicing (*q.v.*). The Ey proteins are 838 and 857 amino acids long, and they function as transcription factors (*q.v.*). The proteins contain an N-terminal paired (*q.v.*) domain and a homeobox (*q.v.*). The wild-type allele of *eyeless* is a master control gene for eye morphogenesis in this species. This has been shown by experimentally activating the gene in imaginal discs that normally give rise to organs other than eyes. After eclosion, flies appear with organized clusters of ommatidia on their antennae, legs, halteres, and wings. A *Drosophila* head that bears a "minieye" on one antenna is shown in the frontispiece. Since eye development in *Drosophila* is controlled by more than 2,500 genes, most of these must be under the direct or indirect control of ey^{+}. The *Small eye* (*Sey*) (*q.v.*) genes of mice and rats and the *Aniridia* (*q.v.*) gene of humans are homologs of the *eyeless* gene. *Ey* homologs have also been found in the zebrafish, quail, chicken, sea urchin, and squid. The squid homologous gene when activated in developing *Drosophila* also initiates the formation of ectopic eyes. These findings suggest that all metazoans share the same master control genes for eye morphogenesis. *See* Appendix C, 1995, Halder *et al.*; developmental control genes, downstream genes, *Drosophila* targeted gene expression technique.

F

F 1. Wright's inbreeding coefficient 2. fluorine. 3. Fahrenheit. 4. fertility factor.

F_1 first filial generation; the offspring resulting from the first experimental crossing of plants or animals. The parental generation with which the genetic experiment starts is referred to as P_1 (*q.v.*). *See* Appendix C, 1900, Bateson.

F_2 the progeny produced by intercrossing or self-fertilization of F_1 individuals.

Fab that fragment of a papain-digested immunoglobulin molecule that bears a single antigen-binding site and contains one intact light chain and a portion of one heavy chain. *See* immunoglobulin.

$F(ab)_2$ fragment that fragment of a pepsin-digested immunoglobulin molecule that contains portions of two heavy chains and two intact light chains and bears two antigen-binding sites. *See* immunoglobulin.

Fabry disease a hereditary disease of glycosphingolipid metabolism in humans. The disease results from a mutation in the X-linked gene that codes for lysosomal *alpha galactosidase A* (*q.v.*). The enzyme defect causes the accumulation of glycosphingolipids in lysosomes of blood vessels and leads to vascular malfunctions. Fabry disease can be detected *in utero* by demonstration of a lack of alpha galactosidase A activity in amniocytes of XY fetuses. Heterozygous females can be documented by cloning cultured skin fibroblasts followed by the demonstration of two cell populations, one normal and one lacking enzyme activity (*see* dosage compensation). Hemizygotes survive to sexual maturity, so the mutation is transmitted by both heterozygotes and hemizygotes. Prevalence 1/40,000 males. The disease was described in 1898 by the German dermatologist Johannes Fabry. An English surgeon, William Anderson, discovered it independently in the same year. The condition is most commonly referred to as Fabry disease but sometimes as Fabry-Anderson or Anderson-Fabry disease or syndrome.

facet ommatidium (*q.v.*).

factor 8 *See* antihemophilic factor.

factor 9 *See* plasma thromboplastin component.

factorial a continuing product of factors, calling for the multiplication together of all the integers from one to the given number. Factorial four means $1 \times 2 \times 3 \times 4 = 24$. The usual factorial symbol is an exclamation point (!) following the largest number in the series. Thus factorial four is written 4!

facultative having the capacity to live under more than one specific set of environmental conditions. For example, a facultative parasite need not live as a parasite (if another food source is available). A facultative apomict may reproduce sexually or asexually depending upon environmental conditions. *Compare with* obligate.

facultative heterochromatin *See* heterochromatin.

FAD flavin adenine dinucleotide, a coenzyme (*q.v.*).

Fagus sylvatica the European beech tree.

Fahrenholz rule the hypothesis that in groups of permanent parasites the classification of the parasites corresponds directly with the natural relationships of their hosts. For example, closely related species of mammals are generally parasitized by closely related species of lice. The rule is based on the assumption that the intimate associations of parasites with their hosts necessitate that they evolve and speciate in harmony with their hosts. As a result of this coevolution, speciation and patterns of divergence in host taxa are paralleled by their parasites. An underlying assumption of the Fahrenholz rule is that there is no dispersal of parasites between unrelated hosts. The eponym honors the German entomologist, Heinrich Fahrenholz, who formulated the concept. *See* resource tracking.

falciparum malaria *See* malaria.

fallout the radioisotopes generated by nuclear weapons that are carried aloft and eventually fall to the earth's surface and contaminate it.

false negative, false positive *See* statistical errors.

familial Down syndrome *See* translocation Down syndrome.

familial amyotrophic lateral sclerosis (FALS) the hereditary form of amyotrophic lateral sclerosis (ALS) (*q.v.*). Five to ten percent of the cases of ALS are hereditary and about 20% of these are due to *SOD* mutations. *See* superoxide dismutase.

familial hypercholesterolemia (FH) a human hereditary disease characterized by an elevation in the plasma concentration of low-density lipoproteins (LDLs). FH is inherited as an autosomal dominant, and the gene responsible resides on the short arm of chromosome 19 at region 13.2–3. The prevalence of heterozygotes is about 1/500 among American, European, and Japanese populations, and this makes FH among the most common hereditary diseases. Homozygotes are rare (1 per million in the U.S.A.). The gene is 45 kilobase pairs long and contains 18 exons that encode the low-density lipoprotein receptor (LDLR). This glycoprotein is made up of 839 amino acids, which are grouped into five domains. These are (1) the ligand-binding domain, (2) the domain showing homology to the precursor of the epidermal growth factor (*q.v.*), (3) the domain containing bound chains of carbohydrates, (4) the transmembrane domain, and (5) the cytoplasmic domain. The first domain is negatively charged, and it binds to the positively charged LDL particle. The second is essential for the normal recycling of the receptors. The third stabilizes the receptor. The fourth contains hydrophobic amino acids, and it spans the plasma membrane and anchors the LDLR to the cell. The fifth is necessary for clustering the receptors into coated pits. There is a direct relationship between the structural domains and exon sequences. For example, domain 2 is encoded by exons 7–14 and domain 5 by exons 17 and 18. Mutations of the LDLR gene result from insertions, deletions, and missense or nonsense mutations. Most mutations prevent the synthesis of LDLRs. The second most common mutant class generates LDLRs that cannot exit the endoplasmic reticulum. In a third mutant class LDLRs are transported to the cell surface, but cannot bind LDL particles. Finally, there is a rare group of mutant LDLRs that can bind LDL particles, but cannot cluster into coated pits. In most heterozygotes, receptor activity is one-half normal and in homozygotes LDLRs are absent. Deficient receptor-mediated endocytosis (*q.v.*) causes LDLs to accumulate in the plasma. Cholesterol is deposited on arterial walls and atherosclerosis results. The disease is far more serious in homozygotes than in heterozygotes. *See* Appendix C, 1975, Goldstein and Brown; plasma lipoproteins, WHHL rabbit.

family **1.** in taxonomy, a cluster of related genera. **2.** a set of parents (mother and father, sire and dam, etc.) together with their children (progeny, offspring) constitutes a *nuclear family*; an *extended family* could include half-sibs, aunts, uncles, grandparents, and/or other relatives. **3.** gene family. *See* multigene family.

family selection artificial selection of an individual(s) to participate in mating(s) based on the merits of other members of the same family exclusive of parents and ancestors (e.g., full sibs or half sibs).

Fanconi anemia the first inherited disease in which hereditary chromosome fragility was established. The disease is characterized by a pronounced reduction in the number of erythrocytes, all types of white blood cells, and platelets in the circulating blood. Chromosome aberrations are common and usually involve nonhomologous chromosomes. In the Bloom syndrome (*q.v.*), most chromosome interchanges are between homologous chromosomes. The disease may result from a defect in the transport of enzymes functioning in DNA repair from cytoplasm to nucleus.

fast component **1.** in reassociation kinetics, the first components to renature, containing highly repetitive DNA. **2.** in electrophoresis, the molecules that move farthest in a given time from the origin.

fast green an acidic dye used in cytochemistry. The structure is shown on page 157.

fat a glycerol ester of fatty acids. Glycerol tripalmitate may be taken as an example:

$$\begin{array}{l} H_2{-}C{-}O{-}CO{-}(CH_2)_{14}CH_3 \\ \quad\;\; | \\ \;H{-}C{-}O{-}CO{-}(CH_2)_{14}CH_3 \\ \quad\;\; | \\ H_2{-}C{-}O{-}CO{-}(CH_2)_{14}CH_3 \end{array}$$

fat body the adipose tissue found in larval and adult insects. These organs behave in many ways like the livers of vertebrates. The fat body is involved in the metabolism of glycogen and fats. It synthesizes and secretes into the hemolymph proteins such as vitellogenin (*q.v.*) which are taken up by the oocyte. The fat body responds to different infections by synthesizing and secreting peptides with antifungal properties like drosomycin or antimicrobial properties like diptericin. *See* innate immunity.

fate map a map of an embryo in an early stage of development that indicates the various regions whose prospective significance has been established by marking methods.

fatty acid an acid present in lipids, varying in carbon content from C_2 to C_{34}. Palmitic acid may be taken as an example: $CH_3(CH_2)_{14}COOH$.

fauna the animal life in a given region or period of time.

favism a hemolytic response to the consumption of beans produced by *Vicia faba* (*q.v.*). Breast-fed babies whose mothers had ingested fava beans have

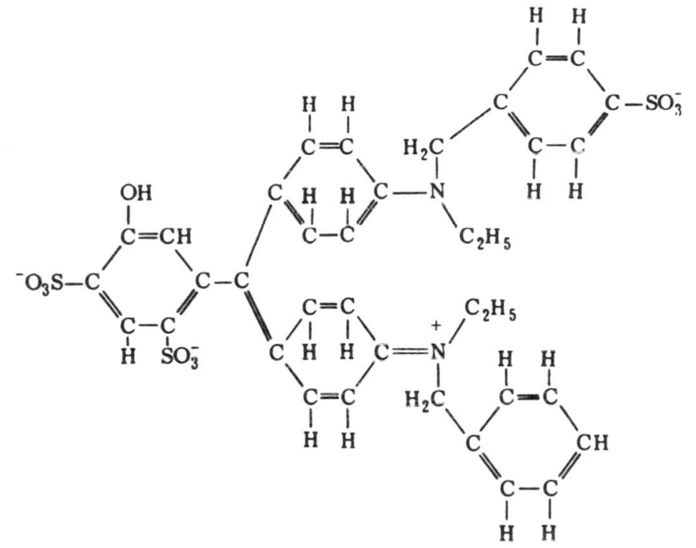

Fast green

also developed favism. Various compounds in fava beans are enzymatically hydrolyzed to quinones, which generate oxygen radicals. Red blood cells deficient in glucose-6-phosphate dehydrogenase (*q.v.*) have a marked sensitivity to oxidating agents and lyse when oxygen radicals are abundant. The toxic properties of fava beans have been known for centuries in Mediterranean cultures where glucose-6-phosphate dehydrogenase deficiency (*q.v.*) is a common hereditary disease. Favism generally occurs in males hemizygous for the A^- or M alleles of the *Gd* gene.

FBNI *See* fibrillin.

F^+ cell a bacterial cell possessing a fertility (F) factor extrachromosomally in a plasmid. An F^+ cell can donate the F factor to an F^- cell during conjugation. If the F factor integrates into the bacterial chromosome, the cell becomes an Hfr (*q.v.*), capable of transferring chromosomal genes. *See* F factor.

F^- cell a bacterial cell devoid of an F factor and that therefore acts only as a recipient ("female") in bacterial conjugation.

Fc fragment that *c*rystallizable *f*ragment (hence the name) of a papain-digested immunoglobulin molecule that contains only portions of two heavy chains and no antibody binding sites. This fragment does, however, bind complement and is responsible for the binding of immunoglobulin to various types of cells in a non-antigen-specific manner. *See* immunoglobulin.

F_c receptor a cell-surface component of many cells of the immune system responsible for binding the F_c portion of immunoglobulin molecules.

F-duction *See* sexduction.

fecundity potential fertility or the capability of repeated fertilization. Specifically, the term refers to the quantity of gametes, generally eggs, produced per individual over a defined period of time. *See* fertility.

feedback the influence of the result of a process upon the functioning of the process.

feedback inhibition end product inhibition (*q.v.*).

feeder cells irradiated cells, capable of metabolizing but not of dividing, that are added to culture media to help support the growth of unirradiated cells.

feline leukemia virus an oncogenic RNA virus.

Felis catus the house cat, a domesticated version of the African wild cat *Felis silvestris*. Its haploid chromosome number is 19, and about 600 genes have been mapped. Of all the non-primate species for which DNA sequences are available, the cat's genome is the closest to the human. Cats and humans have about 200 disease-associated genes that are orthologs. The complete 17,009 bp sequence is known for the mtDNA of cats. *See* Appendix A, Animalia, Eutheria, Carnivora; Appendix E; cat for a listing of breeds, tortoiseshell cat.

female carrier in human pedigrees, a woman who is heterozygous for a recessive, X-chromosomal gene.

female gonadal dysgenesis Turner syndrome (*q.v.*).

female pronucleus the haploid nucleus of a female gamete, which functions in syngamy.

female-sterile mutation one of a class of mutations that cause female sterility generally because of a developmental block during oogenesis. Recessive female-sterile mutations are common in *Drosophila melanogaster* and *Bombyx mori*. Dominant female-steriles are much rarer.

female symbol ♀ the zodiac sign for Venus, the goddess of love and beauty in Roman mythology. The sign represents a looking glass.

F-episome *See* fertility factor.

F′-episome an F-episome carrying a genetically recognizable fragment of bacterial chromosome.

feral pertaining to formerly domesticated animals now living in a wild state.

fermentation an energy-yielding enzymatic breakdown of sugar molecules that takes place in bacteria and yeasts under anaerobic conditions. *See* Appendix C, 1861, Pasteur.

ferritin an iron storage protein found in the liver and spleen, containing up to 20% of its weight in the form of iron. It consists of a protein component (apoferritin) and colloidal micelles of ferric hydroxide-ferric phosphate. Ferritin is often conjugated to proteins such as immunoglobulins, thus enabling their locations within tissues to be determined in electron micrographs due to the great electron scattering property of the iron atoms. *See* Appendix C, 1959, Singer.

ferritin-labeled antibodies *See* ferritin.

fertile crescent a crescent-shaped region in the Middle East that in ancient times stretched along the eastern shore of the Mediterranean Sea through the valleys of the Tigris and Euphrates Rivers to the Persian Gulf. This region was a very early center of agriculture and was the place where most of the major domesticated animals and several of the world's crops were first farmed (9,000 to 4,500 B.C.).

fertility the productivity of an individual or population in terms of generating viable offspring. The term is often used to refer to the number of offspring generated by a female during her reproductive period. In human genetics, the term *effective fertility* is used to refer to the mean number of offspring produced by individuals suffering from a hereditary disease as compared to the mean number of offspring produced by individuals free from the disease, but otherwise very similar. Effective fertility thus gives an indication of the selective disadvantage of the disease. *Compare with* fecundity.

fertility factor F factor (*q.v.*).

fertility restorer a dominant nuclear gene in corn that nullifies the effect of a cytoplasmic male-sterility factor.

fertilization the union of two gametes to produce a zygote. *See* Appendix C, 1769, Spallanzani; 1875, Hertwig; double fertilization, syngamy.

fertilization cone a conical projection protruded from the surface of certain eggs at the point of contact with the fertilizing sperm.

fertilization membrane a membrane that grows outward from the point of contact of the egg and sperm and rapidly covers the surface of the egg.

fertilizin a substance secreted by the ovum of some species, that attracts sperm of the same species.

fetus (*alternative spelling,* **foetus**) a post-embryonic, unborn or unhatched vertebrate that has developed to a stage where its major body parts resemble those of the mature animal. The term is usually applied to viviparous (*q.v.*) animals. In humans, a fetus is identified from approximately the eighth week of gestation until birth. Determination of the legal and moral status of a human fetus is a major area of debate in ethics, law, and public policy. *See* embryo, gestation period.

Feulgen-positive stained by the Feulgen procedure and therefore containing DNA.

Feulgen procedure a cytochemical test that utilizes the Schiff reagent (*q.v.*) as a stain and is specific for DNA. *See* Appendix C, 1924, Feulgen and Rossenbeck; 1950, Swift; 1951, Chiba; 1959, Chèvremont *et al.*

F factor (fertility factor) a supernumerary sex chromosome, symbolized by F, that determines the sex of *E. coli*. In the presence of the F episome, the bacterium functions as a male. F is a circular DNA molecule made up of about 94,000 base pairs, about 2.5% the amount in the *E. coli* chromosome. About one-third of the genes in the F chromosome are involved in the transfer of male genetic material to the female, including the production of the F-pilus, a hollow tube through which DNA is transferred during conjugation. *See* circular linkage map, F-prime factor, Hfr strain, MS2, pilus.

F′ factor *See* F-prime factor.

fibrillin a 2,871 amino acid glycoprotein associated with microfibrils about 10 nanometers in diameter. The protein is found in skin, tendon, bone, muscle, lung, kidney, blood vessels, and the suspensory ligament of the lens. Fibrillin contains 49 EGF domains. The protein is encoded by a gene symbolized by *FBN1*, and mutations in it are responsible for an inherited disease called Marfan syndrome (*q.v.*). *See* epidermal growth factor.

fibrin *See* blood clotting.

fibrinogen *See* blood clotting.

fibroblasts spindle-shaped cells responsible for the formation of extracellular fibers such as collagen (*q.v.*) in connective tissues.

fibroin the major protein component of silk (*q.v.*).

fibronectin a dimer made up of two similar protein subunits. Each has an Mr of 250,000, and the two are joined at one end by disulfide bonds. The proteins are modular in the sense that they are divided into a series of domains, each with specific binding properties. For example, there are different domains that bind specifically to actin, to collagen, and to certain receptor proteins embedded in the plasma membranes of cells. Fibronectin mediates the attachment of cells to collagenous substrates, participates in the organization of stress fibers, and facilitates cell-to-cell adhesions. The fibronectin gene contains a series of exons, and there is one-to-one correspondence of exons to the protein-binding domains. Fibronectins exist in a variety of isoforms, many of which result from alternative splicing (*q.v.*). *See* peripheral protein.

fiducial limits *See* confidence limits.

field *See* prepattern.

filaform thread-shaped.

filamentous phage a bacterial virus (e.g., M13, fd) that specifically infects male (donor) cells and carries a single strand of DNA within a filamentous protein coat. A filamentous phage forms a double-stranded replicative form (*q.v.*) during its life cycle.

filial generations any generation following the parental generation. Symbolized F_1, F_2, etc.

filopodia very thin, fingerlike extensions of the plasma membrane; used by cells that move by amoeboid locomotion.

filter hybridization exposing DNA, denatured to single strands and immobilized on a nitrocellulose filter, to a solution of radioactively labeled RNA or DNA; only hybrid double-stranded molecules remain on the filter after washing. *Compare with* liquid hybridization. *See* Appendix C, 1975, Benton and Davis.

filter route a migration path along which only a few species can easily disperse.

filtration enrichment a method for the isolation of nutritional mutants in fungal genetics. Mutagenized spores are placed upon a minimal medium. Normal spores germinate and send out an extensive mycelial network. These colonies are then filtered off, and the remaining germinated spores that show poor mycelial development are grown upon a supplemented medium, where each produces enough mycelia to allow further propagation and study.

fimbria (*plural* **fimbriae**) a thin filament that extends from the surface of a microorganism and functions to facilitate the adhesion of the cell to other cells or to the substratum. Fimbriae occur in large numbers on a given cell, and they are not to be confused with conjugative pili. Fimbriae occur on Gram-negative bacteria and certain fungi. Each fimbria consists of linear repeating molecules of a protein called fimbrillin. *See* P blood group, pilus.

fimbrillins adhesive proteins of fimbriae (*q.v.*). Bacteria that are able to synthesize fimbrillin H avoid destruction by phagolysosomes (*q.v.*). This is because endocytosis by caveolae (*q.v.*) is triggered when FimH binds to a receptor in the caveolae in the host cell.

finalism a philosophy that views evolution as being directed (by some rational force) toward an ultimate goal. *See* teleology (of which this is a special case); *see also* orthogenesis.

fine-structure genetic mapping the high-resolution analysis of intragenic recombination down to the nucleotide level.

fingerprinting technique in biochemistry, a method employed to determine differences in amino acid sequences between related proteins. The protein under study is enzymatically cleaved into a group of polypeptide fragments. These are separated in two dimensions: first by paper electrophoresis that separates peptides on the basis of net charge and second by partition chromatography that separates peptides on the basis of their degree of polarity (affinity for the hydrated cellulose support, which is highly polar). The result will be a two-dimensional array of spots, the "fingerprint." This is compared to the standard fingerprint. The difference in the position of one spot in the case of the Hb^S and Hb^A fingerprints led to the discovery that the normal and mutant hemoglobins differed in a single amino acid substitution. *See* Appendix C, 1957, Ingram; nucleic acid fingerprinting.

finished sequence in genome sequencing, a high-quality DNA sequence that is contiguous (or nearly contiguous) and has a high rate of sequence accuracy. With an error rate of 0.01%, i.e., just one error per 10,000 bases, The Human Genome Project (*q.v.*) produced data with 99.99% accuracy. The finished sequence of the human genome was ready in 2003. *Compare with* draft sequence. *See* Appendix C, 2003, The International Human Genome Sequencing Consortium.

fireflies beetles belonging to the family Lampyridae. They are unique in being able to flash their abdominal luminescent organs on and off. The rhythm of the flashes is species specific. Males fly about in the evening flashing their signals. Receptive females on the ground respond with a delayed signal which guides males to them. *See* luciferase, mimicry.

first-arriver principle a theory proposing that the first individuals to colonize a new environment or to become adapted to a specific niche acquire thereby a selective advantage over later arrivals, merely because they got there first; also known as "king-of-the-mountain" principle.

first cousin *See* cousin.

first-degree relative when referring to a specific individual in a pedigree (*q.v.*), any relative who is only one meiosis away from that individual (a parent, a sibling, or an offspring). Any relative with whom one half of one's genes are shared. *Contrast with* second-degree relative.

first-division segregation ascus pattern in ascomycetes, a 4–4 linear order of spore phenotypes within an ascus. This pattern indicates that a pair of alleles (e.g., those controlling spore pigmentation) separated at the first meiotic division, because no crossovers occurred between the locus and the centromere. *See* ordered tetrad.

first-order kinetics the progression of an enzymatic reaction in which the rate at which the product is formed is proportional to the prevailing substrate concentration, with the result that the rate slows gradually, and the reaction never goes to completion. *See* zero-order kinetics.

FISH fluorescence *in situ* hybridization (*q.v.*).

fission 1. binary fission (*q.v.*). 2. nuclear fission (*q.v.*).

fitness the relative ability of an organism to survive and transmit its genes to the next generation.

fixation the first step in making permanent preparations of tissues for microscopic study. The procedure aims at killing cells and preventing subsequent decay with the least distortion of structure. *See* fixative, genetic fixation, nitrogen fixation.

fixative a solution used for the preparation of tissues for cytological or histological study. It precipitates the proteinaceous enzymes of tissues and so prevents autolysis, destroys bacteria that might produce decay of the tissue, and causes many of the cellular constituents to become insoluble.

fixing in photography, the removal of the unchanged halide after the image is developed. An aqueous solution of sodium thiosulfate (hypo) is used.

flagellate a protoctist belonging to the Zoomastigina or Euglenophyta. *See* Appendix A.

flagellin a member of a family of proteins that are a major component of the flagellae of prokaryotes. The flagellins found in the archaea are distinct in composition and assembly from those of bacteria.

flagellum (*plural* flagella) 1. in prokaryotes, a whip-like motility appendage present on the surface of some species. Bacterial flagellae range in length from 2 to 20 nanometers. Bacteria having a single flagellum are called *monotrichous;* those with a tuft of flagella at one pole are called *lophotrichous;* and those with flagella covering the entire surface are called *peritrichous*. Antigens associated with flagella are called *H antigens*. *Compare with* pilus. 2. in eukaryotes, flagellum refers to a threadlike protoplasmic extension containing a microtubular axoneme (*q.v.*) used to propel flagellates and sperm. Flagella have the same basic structure as cilia (*q.v.*), but are longer in proportion to the cell bearing them and are present in much smaller numbers (most sperm are monotrichous). In recent literature, *flagellum* is restricted to prokaryotic mobility appendages, and the term *flagellum* in eukaryotes has been replaced by *undulipodium* (*q.v.*).

flanking DNA nucleotide sequences on either side of the region under consideration. For example, the hallmarks of a transposon (*q.v.*) are (1) it is flanked by inverted repeats at each end, and (2) the inverted repeats are flanked by direct repeats.

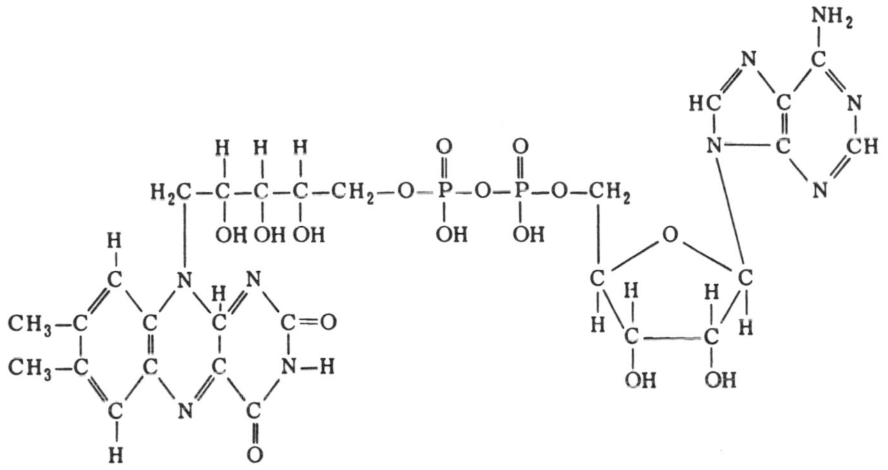

Flavin adenine dinucleotide

flavin adenine dinucleotide (FAD) a coenzyme composed of riboflavin phosphate and adenylic acid. FAD forms the prosthetic group of enzymes such as d-amino acid oxidase and xanthine oxidase.

flavin mononucleotide riboflavin phosphate, a coenzyme for a number of enzymes including l-amino acid oxidase and cytochrome *c* reductase.

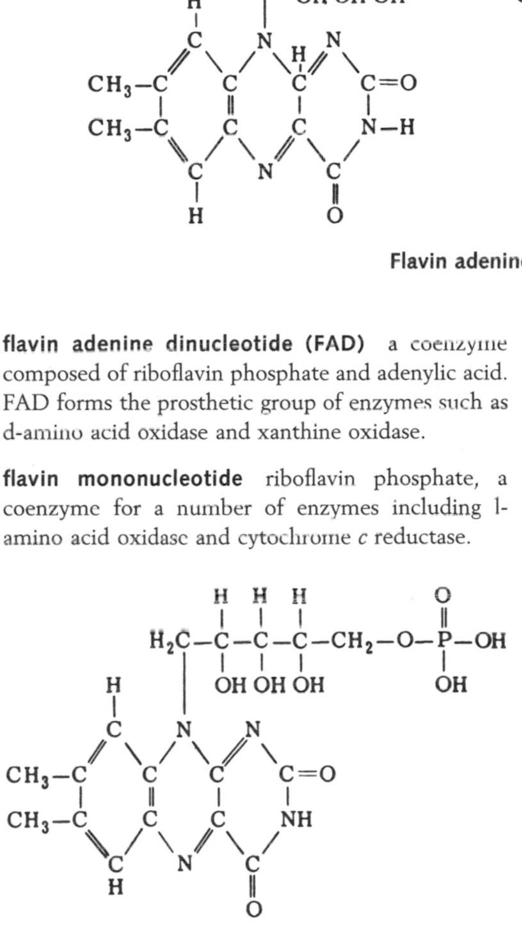

flavoprotein a protein requiring FMN or FAD to function.

flint corn *See* corn.

flora the plant life in a given region or period of time.

floral organ identity mutations homeotic mutations (*q.v.*) in which one floral organ has been replaced by another. For example, in *Arabidopsis thaliana* (*q.v.*) the *apetala 2* mutations have sepals converted into carpels and petals into stamens. In *apetala 3* mutants, petals are converted into sepals and stamens into carpels. In the *agamous* mutant, stamens are converted to petals and carpels to sepals. The ingenious "ABC model" illustrated on page 162 explains the bizarre phenotypes of the mutant flowers. In the normal flower (matrix I) a class A gene (*APETALA 2*) produces an A morphogen, a class B gene (*APETALA 3*) produces a B morphogen, and a Class C gene (*AGAMOUS*) produces a C morphogen. Whorl 1 (W_1) meristems produce sepals when A morphogen is present. Whorl 2 (W_2) meristems produce petals when A and B morphogens are present. W_3 meristems produce stamens when B and C are present, and W_4 meristems produce carpels when only C is present. The *apetala 2* mutant lacks morphogen A (matrix II), the *apetala 3* mutant lacks morphogen B (matrix III), and the *agamous* mutant lacks morphogen C (matrix IV). Furthermore, C genes must inhibit A genes in W_3 and W_4 (matrix IV), and A genes must inhibit C genes in W_1 and W_2 (matrix II). Homeotic mutations allow the genes, normally inactivated, to express themselves. The positions in the matrices where this happens are starred. The proteins encoded by homeotic floral identity genes contain a conserved sequence of 58 amino acids. This presumably binds to DNA in much the same way as the homeobox of *Hox* genes (*q.v.*). *See* Appendix C, 1996, Krizek and Mayerowitz; *Antennapedia, Antirrhinum,* cadastral genes, floral organ primordia, meristems.

floral organ primordia meristematic cells that give rise to the organs of flowers. The primordia are arranged into four concentric whorls as shown on page 162. The leaf-like sepals form whorl number 1 at the periphery. The showy petals form whorl 2, inside the first. The stamens, the male reproductive organs,

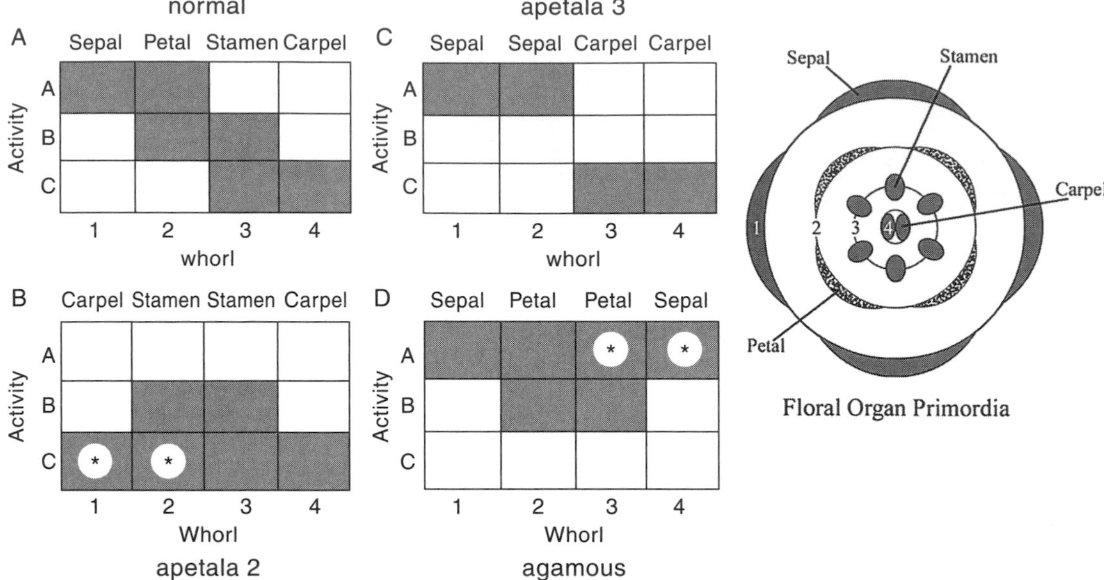

form whorl 3 (inside whorl 2), and the carpels, the female reproductive organs, form central whorl 4. *See* Appendix A, Plantae, Angiospermae; *Arabidopsis thaliana,* floral identity mutations, meristems.

floret a small flower from an inflorescence, as in a grass panicle.

flour corn *See* corn.

flow cytometry a technology that utilizes an instrument in which particles in suspension and stained with a fluorescent dye are passed in single file through a narrow laser beam. The fluorescent signals emitted when the laser excites the dye are electronically amplified and transmitted to a computer. This is programmed to instruct the flow cytometer to sort the particles having specified properties into collecting vessels. Human mitotic chromosomes can be sorted to about 90% purity by this technique.

flower the specialized reproductive shoot of angiosperms (*q.v.*). Perfect flowers bear both pistils and stamens. Imperfect flowers bear either pistils or stamens. A plant bearing solely functional staminate or pistillate flowers is said to be *dioecious.* A plant bearing (1) imperfect flowers of both sexual types (corn is an example), or (2) perfect flowers (the garden pea), or (3) staminate, pistillate, and perfect flowers (the red maple, for example) is said to be *monoecious. See* androgynous.

FLP/FRT recombination a system of site-specific recombination (*q.v.*) found in the yeast and based on the yeast two-micron plasmid. This plasmid is an autonomously replicating, circular plasmid of 6,318 base pairs, which exists in many copies in most strains of *S. cerevisiae* (*q.v.*). It encodes a site-specific recombinase (*q.v.*) called the FLP (pronounced 'flip') protein. FLP acts on the *F*LP *r*ecombination *t*arget (FRT) located within two 599-base pair inverted repeats in the plasmid DNA and catalyzes recombination between the inverted repeats. This recombination event inverts, or 'flips', part of the plasmid with respect to the remainder, and this is essential for plasmid amplification. FLP can also induce recombination between direct FRT repeats or between FRTs on different DNA molecules. In the former case, the recombination event results in excision of intervening DNA and an FRT. This fact has been exploited to experimentally induce site-specific recombination in *Drosophila* and other organisms. In *Drosophila* this is done by placing FLP under the control of a heat shock protein promoter, and the gene under study between two flanking FRT direct repeats. These DNAs are then integrated into the *Drosophila* genome by germ line transformation (*q.v.*). Under heat-shock conditions, FLP catalyzes recombination between the FRTs, resulting in excision of the gene of interest. The descendants of cells without such a gene show a null phenotype. This system can thus produce somatic or germ line mosaics for specific genes.

fluctuation test a statistical analysis first used by Luria and Delbrück to prove that selected variants (*q.v.*), such as bacteriophage-resistant bacteria, are spontaneous mutants that arose prior to the exposure to the selective agent. They reasoned that if bacterial mutation is an event that is rare, discontinuous, and random, then there should be a marked fluctuation in the number of resistant variants present, at a given time, in a large number of independent cultures, each of which was grown from a small inoculum. The number of variants per sample would fluctuate because some cultures would contain large numbers of variants arising from the division of an early mutant, while in other cultures mutants occurring later during the growth of the culture would produce clones of smaller size. Conversely, in separate samples taken from a single culture inoculated under identical conditions the variability in the number of mutants should be much less. However, if the agent used for selection induced the mutations, then the distribution of the number of mutants in any population of samples should be independent of the previous history of the culture. Since it was found that the variance was much larger when samples came from independent cultures than when they were taken from the same culture, it was concluded that spontaneous mutation was the source of the variants. *See* Appendix C, 1943, Luria and Delbrück.

fluid mosaic concept a model in which the cell membrane is considered to be a two-dimensional viscous solution consisting of a bilayer of highly oriented lipids. The layer is discontinuous, being interrupted by protein molecules that penetrate one or both layers. *See* Appendix C, 1972, Singer and Nicholson; lipid bilayer model.

fluke a common name for flatworms belonging to the class Trematoda. Flukes of medical importance are members of the order Digenea. These parasites have molluscs as intermediate hosts. *See* schistosomiasis.

fluourescein an orange-red compound that yields a bright-green fluorescence when exposed to ultraviolet light. When conjugated to a specific antibody, this dye provides a means of localizing the antigen when the stained cell is viewed with a fluorescence microscope.

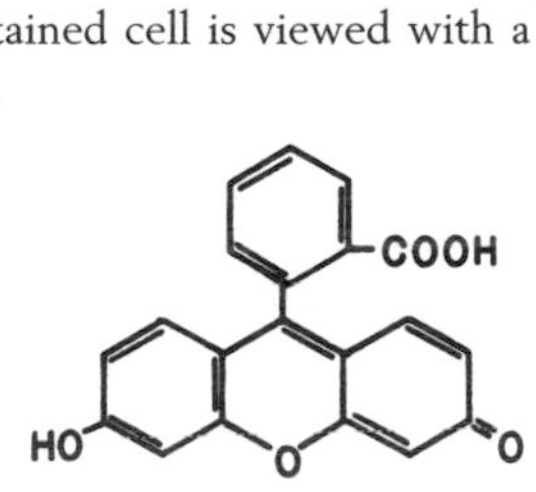

fluorescence *See* luminescence.

fluorescence *in situ* hybridization (FISH) this technique uses synthetic polynucleotide strands that bear sequences known to be complementary to specific target sequences at specific chromosomal sites. The polynucleotides are bound via a series of linking molecules to a fluorescent dye that can be detected by a fluorescence microscope. This probe is then *in situ* hybridized to the cells to be tested. The fluorescence signals observed under the microscope permit the number, size, and location of the target sequences to be determined with speed and precision. *See* chromosome painting, *in situ* hybridization.

fluorescence microscopy the usual methods of microscopical examination are based on observing the specimen in the light transmitted or reflected by it. Fluorescence microscopical preparations are self-luminous. In most biological preparations the tissue sections are stained with a *fluorochrome*, a dye that emits light of longer wavelength when exposed to blue or ultraviolet light. The fluorescing parts of the stained object then appear bright against a dark background. The staining techniques are extremely sensitive and can often be used in living materials. *See* Appendix C, 1970, Caspersson, Zech, and Johansson.

fluorescent antibody technique a method for localizing a specific protein or other antigen in a cell by staining a section of the tissue with an antibody specific for that antigen. The antibody is tagged directly or indirectly by a fluorochrome for detection under a fluorescence microscope. *See* immunofluorescence.

fluorescent screen a sheet of material coated with a substance such as calcium tungstate or zinc sulfide that will emit visible light when irradiated with ionizing radiation. Such screens are used in TV sets and as the viewing screens of electron microscopes.

fluorine a biological trace element. Atomic number 9; atomic weight 18.9984; valence 1^-; most abundant isotope ^{19}F.

fluorochrome a fluorescent dye that can be conjugated to a compound that binds to a specific cell component. An example would be a fluorescein-labeled antibody of rodaminylphalloidin (*q.v.*).

flush end synonymous with blunt end. *See* restriction endonuclease.

flying spot cytometer an instrument used in cytometric DNA measurements. The heterogenous distribution of Feulgen stain within a nucleus leads to distributional errors when a single measurement is made of light transmission to estimate the amount of dye bound in the nucleus. Flying spot cytometers scan a defined microscopic area while making thousands of measurements with a minute measuring spot. The sum of these point absorbance measurements as determined by a built-in computer is proportional to the true absorbance of the specimen. *See* Feulgen procedure, microspectrophotometer.

F-mediated transduction sexduction.

f-met *N*-formylmethionine (*q.v.*)

f-met-tRNA the complex between *N*-formylmethionine and its transfer RNA.

FMN flavine mononucleotide, a coenzyme.

FMR-1 gene *See* fragile X-associated mental retardation.

foci **1.** regions of growth of tumor cells appearing as raised clusters above a confluent monolayer of cells in tissue culture. **2.** opaque pocks appearing on the chorioallantoic membrane of a developing chick embryo that has been infected with certain viruses such as herpes viruses.

focus map a fate map (*q.v.*) for regions of the *Drosophila* blastoderm determined to become adult structures, inferred from the frequencies of specific kinds of mosaics.

foldback DNA single-stranded regions of DNA that have renatured by intrastrand reassociation between inverted repeats; hairpin DNA.

folic acid the anti-pernicious-anemia vitamin. It is a compound made up of three components: a pteridine (*q.v.*), *p*-aminobenzoic acid, and glutamic acid.

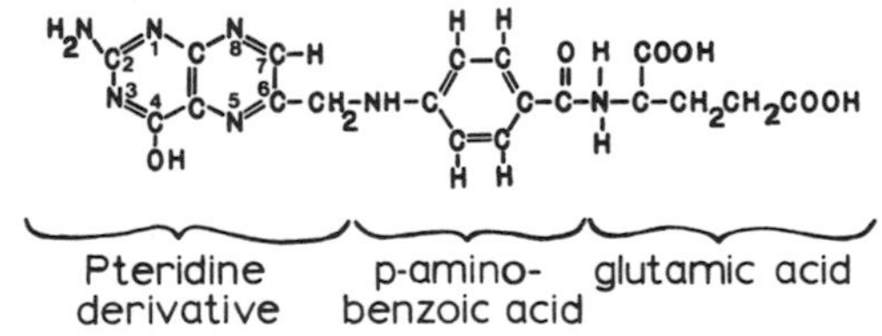

Pteridine derivative | p-amino-benzoic acid | glutamic acid

The active form of folic acid is tetrahydrofolate. This compound contains hydrogen atoms attached to nitrogens 5 and 8 and carbons 6 and 7. The enzyme dihydrofolate reductase catalyzes certain of these addition reactions. Tetrahydrofolate is an essential coenzyme in the biosynthesis of thymidylic acid. Thus, folic acid analogs like aminopterin and methotrexate block nucleic acid synthesis. Exposure to ultraviolet radiation causes the breakdown of folate. In light-skinned people, half the folate in the bloodstream can be lost during an hour's exposure to sunlight. Women should take folic acid before and during pregnancy because too low folate levels can cause neural tube defects in fetuses.

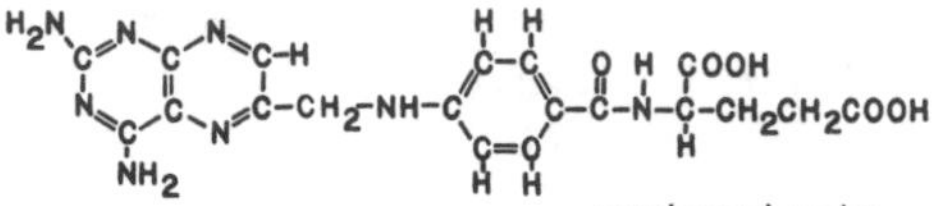

aminopterin

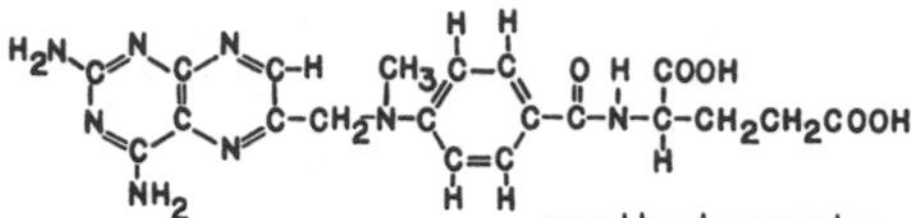

methotrexate

follicle-stimulating hormone (FSH) a glycoprotein hormone that stimulates the growth of ovarian follicles and estrogen secretion. It is produced by the adenohypophysis of vertebrates.

Følling disease *See* phenylketonuria.

footprinting a technique for identifying a segment of a DNA molecule that is bound to some protein of interest, on the principle that the phosphodiester bonds in the region covered by the protein are protected against attack by endonucleases. A control sample of pure DNA and one of protein-bound DNA are subjected to endonuclease attack. The resulting fragments are electrophoresed on a gel that separates them according to their lengths. For every bond position that is susceptible, a band is found on the control gel. The gel prepared from the protein-bound DNA will lack certain bands, and the missing bands identify the length of the site covered by the protein.

Forbes disease a hereditary glycogen storage disease in humans arising from a deficiency of the enzyme amylo-1,6-glucosidase. Inherited as an autosomal recessive. Prevalence 1/100,000.

forensic DNA analysis *See* DNA forensics.

formaldehyde CH_2O, a colorless gas readily soluble in water and having mutagenic properties. *See* Appendix C, 1946, Rapoport; formalin.

formalin an aqueous solution of formaldehyde (*q.v.*) commonly used as a fixative, which functions through cross linking protein molecules.

formamide a small organic molecule that combines with the free NH_2 groups of adenine and prevents the formation of A-T base pairs, thereby causing denaturation of double-stranded DNA. *See* stringency.

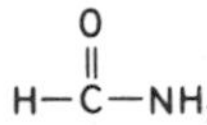

$$H-\overset{O}{\overset{\|}{C}}-NH_2$$

formylkynurenine the vermilion-plus substance (*q.v.*) in *Drosophila melanogaster*. *See Drosophila* eye pigments.

```
       H
       |
       C     O          NH2
     //  \   ||          |
   HC     C--C--CH2--CH--COOH
   |      ||
   HC     C--N--C=O
     \\  /   |  |
       C     H  H
```

formylmethionine *See* N-formylmethionine.

forward genetics *See* reverse genetics.

forward mutation a change in a gene from wild-type (normal) allele to a mutant (abnormal) allele.

fossil any remains or traces of former life, including shells, bones, footprints, etc. exposed in rocks. *Compare with* living fossil.

founder cells *See* compartmentalization.

founder effect the principle that when a small sample of a larger population establishes itself as a newly isolated entity, its gene pool carries only a fraction of the genetic diversity represented in the parental population. The evolutionary fates of the parental and derived populations are thus likely to be set along different pathways because the different evolutionary pressures in the different areas occupied by the two populations will be operating on different gene pools. *See* Appendix C, 1947, Mourant; 1980, Templeton; cystic fibrosis (CF), Huntington disease (HD), peripatric speciation, porphyrias, Rh factor.

fox *See Vulpes vulpes.*

FOXP2 a gene required for the proper development of speech in children. *See* speech-language disorder 1 (SPCH1).

fowl *See* poultry breeds.

fowl achondroplasia hereditary chondrodystrophy affecting certain breeds of chickens (Scots, Dumpies, Japanese Bantams). The condition is inherited as an autosomal dominant. Homozygotes die as embryos. *See* achondroplasia.

fowl leukosis *See* leukemia.

F-pilus *See* F factor.

F-prime (F′) factor a bacterial episomal fertility (F) factor containing an additional portion of the bacterial genome. F-prime factors have been most extensively studied in *E. coli.*

fractionated dose the treatment of an organism by a series of short exposures to mutagenic radiations.

fraction collector an automated instrument that collects consecutive samples of fluids percolating through a column packed with porous material.

Frageria a genus of perennial, stoloniferous herbs belonging to the order Rosales. Various species are cultivated for their delicious fruits. *Frageria vesca*, the wood strawberry, is diploid, while *F. ananassa*, the garden strawberry, and *F. chiloensis*, the Chilean strawberry, are both octoploids. The haploid chromosome number is 7, and the haploid genome is 164 Mbp of DNA. *See* modular organism, runner.

fragile chromosome site a nonstaining gap of variable width that usually involves both chromatids and is always at exactly the same point on a specific chromosome derived from an individual or kindred. Such fragile sites are inherited in a Mendelian co-dominant fashion and exhibit fragility as shown by the production of acentric fragments and chromosome deletions. In cultured human cells, fragile sites are expressed when the cells are deprived of folate or thymidine or if methotrexate is added to the medium. *See* folic acid, fragile X–associated mental retardation. http://www.fraxa.org.

fragile X–associated mental retardation a moderate degree of mental retardation (IQs around 50) found in males carrying an X chromosome that has a fragile site at the interface of bands q27 and q28. The frequency of such hemizygotes is about 1.8 per 1,000. X-linked mental retardation accounts for about 25% of all mentally retarded males. The fragile X site contains a gene that is expressed in human brain cells. The gene designated *FMR-1* (fragile X mental retardation 1) generates a 4.8-kilobase mRNA, which encodes a protein containing 657 amino acids. Upstream of the coding region of the gene is a segment in which the CGG triplet is repeated about 30 times. In cases where the CGG repeat is expanded 50–200 times, the cytologically detectable fragile X phenotype appears. Males with this X^F show normal intelligence, but are *transmitters*. F_1 daughters who receive this X^F also have normal intelligence; however, the expansion of the CGG repeat is activated in the X^F once it is transmitted to the next generation. Since the expansion occurs during early development, the F_2 children are genetic mosaics. Their germ cells contain maternal X^Fs with 50–200 repeats, but cells in other tissues may have thousands of repeats. The FMR-1 gene is evidently inactivated under such circumstances, and mental development is impaired. Some 30% of the females and 50% of the males in this F_2 generation are retarded. *See* Appendix C, 1969, Lubs; 1991, Verkerk *et al.*; CpG island, DNA methylation, fragile chromosome site, parental imprinting, trinucleotide repeats.

fragile X syndrome *See* fragile X–associated mental retardation.

frameshift mutation *See* reading frame shift.

framework region the highly conserved, relatively invariant portion of the variable (V) region of an immunoglobulin chain, as distinguished from the hypervariable segments of the V region.

fraternal twins *See* twins.

free energy that component of the total energy of a system that can do work. *See* thermodynamics, second law of.

freemartin a mammalian intersex arising due to the masculinization of a female twin by hormones from its male sibling when the fetal circulations are continuous.

free radical an unstable and highly reactive molecule, bearing an atom with an unpaired electron, that nonspecifically attacks a variety of organic structures, including DNA. The interaction of ionizing radiation with water can generate hydroxyl and hydroperoxyl groups (free radicals that are potent oxidizing agents). *See* superoxide anion.

free radical theory of aging aging due to the production of reactive free radicals of oxygen that inflict molecular damage to cell organelles, especially DNA. These lesions accumulate with time and cause a progressive reduction in viability. *See* antioxidant enzymes.

freeze-drying a method of dehydrating a cell or solution by rapidly freezing its moisture content to ice. The solid material is then dried in the frozen state under vacuum, so that ice sublimes directly to water vapor with a minimization of shrinkage. *See* lyophilize.

freeze-etching a technique for preparing biological material for electron microscopy. Live or fixed specimens are frozen in a liquid gas, such as freon or nitrogen, and then placed in a Balzer freeze-fracture apparatus. This is an instrument that allows frozen tissues to be sectioned in a vacuum. The exposed surface is allowed to sublime slightly (to etch), so that surface irregularities that reflect the type and distribution of cell constituents are accentuated. The surface is then replicated, and the replica is stripped away and viewed under the electron microscope. Preparations made in this way provide useful information concerning three-dimensional organization of protein particles embedded in the lipoidal membranes of cells.

freeze fracture a method for preparing samples for electron microscopy; frozen samples are fractured with a knife and the complementary surfaces are cast in metal. *See* Appendix C, 1961, Moor *et al.*

frequency the most studied gene, which controls the biological clock of *Neurospora*. Mutations of *frq* either shorten or lengthen the period between conidiations. The *frq* gene has been cloned, and it gives rise to at least two processed transcripts. The longer transcript arises from an ORF of 2,364 base pairs. Although the predicted protein product of *frq* has an amino acid sequence that shows no extended similarities to any previously described protein, there is a segment about 50 amino acids long with sequence similarities to the *per* gene of *Drosophila*. *See period.*

frequency-dependent fitness a phenomenon in which the adaptive value of a genotype varies with changes in allelic frequencies. For example, in Batesian mimicry, mimics have greater fitness when they are rare relative to their models. *See* mimicry.

frequency-dependent selection selection involving frequency-dependent fitness (*q.v.*). *See* Appendix C, 1937; L'Héritier and Tiessier; 1951, Petit; minority advantage.

Freund adjuvant a widely used adjuvant containing killed, dried mycobacteria suspended in the oil phase of a water-in-oil emulsion. The bacteriologist Jules T. Freund developed this infusion and found that it prolonged antibody synthesis when injected together with the antigen.

Friend leukemia virus (FLV) a virus inducing leukemia in mice and rats. It was discovered in 1956 by Charlotte Friend, who later showed that it was a retrovirus (*q.v.*). It proved to be superior to work with than the two other mammalian cancer viruses previously discovered (mouse mammary tumor virus and Gross mouse leukemia virus).

Fritillaria a genus of lilies. Species of this genus are widely used in cytogenetic investigations because of their large chromosomes. In fact the largest C value (*q.v.*) so far recorded for a plant is 12.5×10^{10} bp of DNA for *Fritillaria uva-vulpis*.

frizzle a feather mutation in domestic fowl. *FF* individuals are "extreme frizzle," with bristle feathers that wear off easily; whereas *Ff* individuals are "mild frizzle," with more normal curly feathers. Frizzle feather keratin shows a poorly ordered crystalline structure, and its amino acid composition is abnormal.

fructification 1. a reproductive organ or fruiting body. 2. the generation of fruit or spore-producing structures by plants.

fructose a six-carbon hexose sometimes called levulose. It is a component of sucrose.

```
HO—CH2   O        OH
   |                |
   C                C
   |\ H      OH   / |
   H \ |      |  /  CH2OH
      C ———— C
      |      |
      OH     H
```

fructose intolerance a disorder of carbohydrate metabolism inherited as an autosomal recessive. Patients lack fructose-1,6-diphosphatase. Symptoms disappear if dietary fructose is restricted.

fruit the ripened ovary of the flower that encloses the seeds.

fruit fly *See Drosophila.*

FSH follicle-stimulating hormone (*q.v.*).

F^- strain *Escherichia coli* behaving as recipients during unidirectional genetic transfer.

F^+ strain *Escherichia coli* behaving as donors during unidirectional genetic transfer. *See* F factor.

***F* test** *See* analysis of variance.

Fugu rubripes *Takifugu rubripes* (*q.v.*).

functional cloning in human genetics, the identification of the gene responsible for a disease from a knowledge of the underlying molecular defect. If the protein encoded by the gene is known, it is often possible to isolate the appropriate mRNAs and to use them, or cDNAs derived from them, as probes for the gene. This was the first method used successfully to clone genes responsible for certain hereditary diseases, such as sickle cell anemia, Tay-Sachs disease, and phenylketonuria. *Contrast with* positional cloning. *See* cDNA, hereditary disease, mRNA, probe.

fundamentalism a conservative religious ideology that holds the origin and diversity of life is by divine creation, based upon a literal interpretation of the biblical account of Genesis. *See* creationism.

fundamental theorem of natural selection a theorem developed by R. A. Fisher according to which the increase in fitness of a population at any given time is directly proportional to the genetic variance in fitness of its members.

Fungi the kingdom that contains yeasts, molds, smuts, rusts, mushrooms, and other saprophytes. These organisms are placed at the bottom of most phylogenies that show the evolution of the eukaryotic kingdoms. Their primitive characters include mitochondria with plate-shaped cristae, a Golgi made up of individual dictyosomes, and mitosis with an endonuclear spindle. Fungi cannot perform endocytosis, and they lack undullipodia and centrioles. They reproduce by forming spores. *See* Appendix A, Kingdom 3; opisthokonta.

funiculus the plant stalk bearing on ovule.

fused gene *See* fusion gene.

fused protein a hybrid protein molecule produced when a gene of interest is inserted by recombinant DNA techniques into a recipient plasmid and displaces the stop codon for a plasmid gene. The fused protein begins at the amino end with a portion of the plasmid protein sequence and ends with the protein of interest. *Compare with* polyprotein. *See* Appendix C, 1970, Yourno *et al.*

fushi tarazu (ftz) one of the pair rule selector genes of *Drosophila*. The name means *segment deficient* in Japanese. The *ftz* gene is located at 3–47.5, within the *Antennapedia* complex, and it is first expressed within 7 vertical stripes of cells in the early embryo. It has later functions during metamorphosis when it specifies the identities of individual neurons in the developing central nervous system. FTZ, the protein encoded by *ftz*, contains a PEST sequence (*q.v.*) and a homeobox (*q.v.*). FTZ functions as an activator of the transcription of segment polarity genes such as *engrailed*. But FTZ can also function as a suppressor of transcription for genes such as *wingless*. *See* zygotic segmentation mutants.

fusidic acid an antibiotic that prevents translation by interfering with elongation factor G.

fusion gene 1. a hybrid gene, composed of parts of two other genes, arising from deletion of a chromosomal segment between two linked genes or by unequal crossing over. Hemoglobin Lepore (*q.v.*) is an example of such a fused gene. *See* cone pigment genes (CPGs), Philadelphia chromosome. 2. a laboratory construct consisting of regulatory elements from one gene ligated to the structural elements of another. Transgenic animals (*q.v.*) often carry fused genes.

fusome a cytoplasmic organelle that is required for the proper formation of germ line syncytia during gametogenesis in both male and female insects. The fusome arises from endoplasmic reticulum that traverses the ring canals (*q.v.*) formed during successive cycles of incomplete mitotic germ cell divisions. After each division, a plug of fusomal material accumulates in each newly-formed ring canal. This material then fuses with the fusome(s) formed from the previous division(s), and ultimately a mature, branched structure called a polyfusome is produced. The polyfusome entry illustrates how this structure directs the pattern of cystocyte interconnections by anchoring one pole of each mitotic spindle, thus orienting the plane of cell division. The first *Drosphila* mutant shown to have fragmented fusomes was called *otu* because it formed *o*varian *tu*mors made up of hundreds of cells, most of which were not connected by ring canals and which never differentiated into either oocytes or nurse cells. Mitotic effectors such as cyclin A (*q.v.*) have been shown to bind transiently with fusomes during G2 and prophase, and this suggests that the fusomal system plays a role in the timing, synchronization, and eventual cessation of cystocyte divisions (*q.v.*). Among the other components identified in fusomes are alpha and beta spectrins, ankyrin, an adducin-like protein, dynein, and a protein encoded by the *bag of marbles* gene. Mutations in this gene also produce ovarian tumors. *See* adducin, *bag of marbles (bam), hu-li tai shao (hts).*

G

g gravity; employed in describing centrifugal forces. Thus, 2,000 × g refers to a sedimenting force 2,000 times that of gravity.

G guanine or guanosine.

G_0, G_1, G_2 *See* cell cycle.

ga/gigaannum one billion years. The age of the earth is 4.6 ga. *See* Appendix C, 1953, Patterson.

gain of function mutation a genetic lesion that causes a gene to be overexpressed or expressed at the wrong time. Such mutations often affect upstream elements that control the time in the life cycle when a gene is turned on or the specific tissue in which it is expressed. Gain of function mutations are often dominant. *Contrast with* loss of function mutations.

GAL4 a transcriptional activator protein encoded by the *gal4* gene of yeast and required for the expression of genes encoding galactose-metabolizing enzymes. The GAL4 protein consists of two separable but essential domains (*q.v.*): an N-terminal (*q.v.*) domain which binds to specific DNA sequences upstream (*q.v.*) from the various target genes, and a C-terminal (*q.v.*) domain which is required to activate transcription. These properties of GAL4 have been exploited to develop the *yeast two-hybrid system* (*q.v.*) for detecting protein-protein interactions and the *Drosophila targeted gene expression technique* (*q.v.*) for studying the functioning of master control genes in various targeted tissues. *See* Appendix C, 1989, Field and Song; 1995, Halder *et al.*

galactose a six-carbon sugar that forms a component of the disaccharide lactose and of various cerebrosides and mucoproteins. *See* beta galactosidase.

```
          CH2OH
           |
           C——O
    OH   / |     \    OH
    |   /  |      \   |
    C      H        C
    | \   OH  H    /  |
    H  \  |   |   /   H
        C——C
        |     |
        H    OH
```

galactosemia a hereditary disease in humans inherited as an autosomal recessive due to a gene on the short arm of chromosome 9. Homozygotes suffer from a congenital deficiency of the enzyme galactosyl-1-phosphate uridyl-transferase, and galactose-1-phosphate accumulates in their tissues. They exhibit enlargement of the liver and spleen, cataracts, and mental retardation. Symptoms regress if galactose is removed from the diet. Prevalence 1/62,000. *See* Appendix C, 1971, Meril *et al.*

galactosidase *See* alpha galactosidase, beta galactosidase.

Galapagos finches *See* Darwin's finches.

Galapagos Islands a cluster of 14 islands that straddle the equator 650 miles west of Ecuador. Many of its species are found nowhere else in the world, such as marine iguanas, flightless cormorants, giant tortoises, and a special group of finches. The five weeks Darwin spent exploring these islands in 1835 were the crucial weeks of his scientific life. *See* Appendix C, 1837, Darwin; Darwin's finches; hot spot archipelago.

Galapagos rift *See* rift.

gall an abnormal growth of plant tissues.

gallinaceous resembling domestic fowl.

Gallus gallus domesticus the domesticated chicken, the bird for which the most genetic information is available. Its ancestor is the Asian red jungle fowl, from which it was domesticated around 8,000 BC. Its genome size is about 1 gbp and its number of protein coding genes is ~23,000. Birds, snakes, and lizards have two classes of chromosomes: macrochromosomes and microchromosomes. In chickens there are 10 macrochromosomes and 29 microchromosomes. The fifth chromosome in length is the metacentric Z, which occurs in duplicate in males. The female has one Z and a smaller w chromosome. The macrochromosomes including the Z replicate synchronously. All the microchromosomes replicate late and so does the w of females. Roughly 60% of all the protein coding genes in the chicken have human orthologs. The haploid chromosome number is 39. The female is the heterogametic sex (ZW), whereas the male is the homogametic sex (ZZ). There are over 200 genes that have been mapped. Estimated genome size is 1.125×10^9 base pairs. *See* Appendix A, Chordata, Aves, Galliformes; Appendix E; comb shape, plumage pigmentation genes, poultry breeds.

Galton apparatus an apparatus invented by Francis Galton (illustrated below) consisting of a glass-faced case containing an upper reservoir where balls are stored. Below the reservoir are arranged row after row of equally spaced pegs that stand out from the wall, and below these is a series of vertical slots. The balls are allowed to fall one at a time through a central opening at the bottom of the reservoir. Since each ball after striking a peg has an equal probability of bouncing to the left or right, most will follow a zigzag course through the pegs and will eventually land in a central slot. The final distribution of balls in the slots will be a bell-shaped one. The apparatus demonstrates how the compounding of random events will generate a family of bell-shaped curves. *See* Appendix C, 1889, Galton.

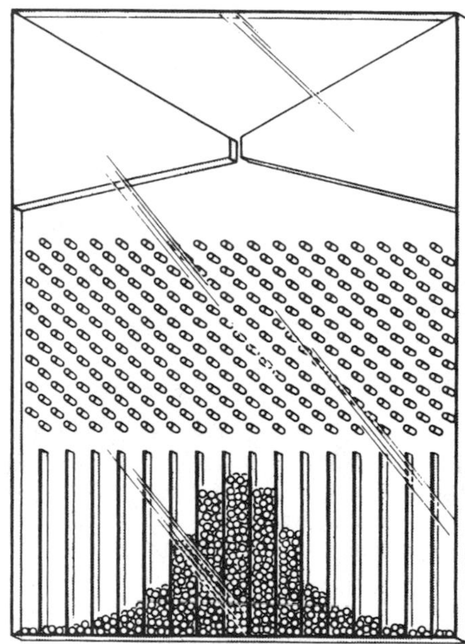

Galton apparatus

gametangium an organ in which gametes are formed. *See* antheridium, oogonium.

gamete a haploid germ cell. *See* Appendix C, 1883, van Beneden.

game theory a mathematical theory dealing with the determination of optimum strategies where the policies adopted depend on the most likely behaviors of two or more competitors. Game theory is employed in mathematical models of species competition.

gametic disequilibrium the nonrandom distribution into the gametes in a randomly mating population of the alleles of genes occupying different loci. The nonrandom distribution may result from linkage of the loci in question or because the loci interact with respect to their effects on fitness. *See* linkage disequilibrium.

gametic meiosis *See* meiosis.

gametic mutation any mutation in a cell destined to become a gamete, and therefore potentially hereditary. *Compare with* somatic mutation.

gametic number the haploid number of chromosomes (symbolized by N) characterizing a species.

gametoclonal variation the appearance of new traits in haploid plants that grow in tissue culture from anthers or other reproductive material rather than from diploid body tissue as in somatoclonal variation (*q.v.*).

gametocyte a cell that will form gametes through division; a spermatocyte or oocyte.

gametogamy the fusion of gamete cells or nuclei.

gametogenesis the formation of gametes.

gametophore a branch bearing a gametangium or gametangia.

gametophyte the haploid phase (of the life cycle of plants undergoing an alternation of generations) during which gametes are produced by mitosis. *See* sporophyte.

gamma chain one of the two polypeptides found in fetal hemoglobin (*q.v.*).

gamma field a field where growing plants may be exposed to chronic irradiation from a centrally placed multicurie ^{60}Co gamma-ray source.

gamma globulin an antibody-containing protein fraction of the blood. *See* Appendix C, 1939, Tiselius and Kabat.

gamma ray an electromagnetic radiation of short wavelength emitted from an atomic nucleus undergoing radioactive decay.

gamogony a series of cell or nuclear divisions that eventually lead to the formation of gametes.

gamone a compound produced by a gamete to facilitate fertilization. Chemotactic sperm attractants produced by eggs are examples.

gamont the haploid adult form of those protoctists that have both haploid and diploid phases in their life cycles. Gamonts function in sexual reproduction; they undergo gametogony to produce diploid agamonts. Meiosis takes place in agamonts, and the haploid agametes that result disperse, undergo mitotic divisions, and differentiate into gamonts, completing the cycle.

gamontogamy the aggregation of gamonts during sexual reproduction and the fusion of gamont nuclei to produce agamonts.

ganglion a small nervous-tissue mass containing numerous cell bodies.

ganglioside a family of complex lipids containing sphingosine, fatty acids, carbohydrates, and neuraminic acid. The Gm2 ganglioside that accumulates in the brain of patients with Tay-Sachs disease (*q.v.*) is shown below. *See* Appendix C, 1935, Klenk.

gap the position where one or more nucleotides are missing in a double-stranded polynucleotide containing one broken chain.

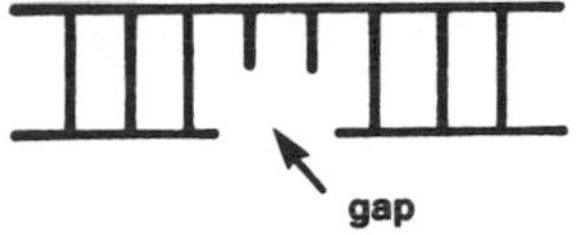

gap genes a class of *Drosophila* genes that control embryogenesis. Loss of function mutations result in the loss of contiguous body segments, and therefore gaps appear in the normal pattern of segmented structures in the embryo. *See* zygotic segmentation mutants.

gargoylism a term covering two genetically distinct hereditary diseases of connective tissue in humans, Hunter syndrome (*q.v.*) and Hurler syndrome (*q.v.*).

Garrod disease *See* alkaptonuria.

gas chromatography a chromatographic technique in which an inert gas is used to sweep through a column the vapors of the materials to be separated.

```
                         Stearic Acid
                              ↘
                    CH3—(CH2)16—CO
                                |
                                NH
                                |
CH3—(CH2)12—CH=CH—CHOH—CH—CH2O
               ↗                   |            Glucose
         Sphingosine               HC
                                   |
                                H—C—OH                 Galactose
                                   |
                               HO—C—H   O
                                   |
                                H—C———————O——CH
                                   |           |
                                H—C        H—C—OH
                                   |
                                  CH2OH
            COOH  O————————————————————————C—H   O
              |                                |
              C                       O————————C—H
              |                                |
             CH2                            H—C
              |                                |
   O     HO—C—H                              CH2OH
              |
 CH3CO—HN—C—H                    ←—Cleaved by hexosaminidase A
              |
             C—H                    HC
              |                      |
          H—C—OH                 H—C—NH—COCH3
              |                      |
          H—C—OH                 H—C—H          O
              |                      |
             CH2OH              HO—C—H
                                     |
  N-acetylneuraminic acid         H—C
                                     |
                                    CH2OH      N-acetylgalactosamine
```

Gm2 ganglioside

gas-flow radiation counter a counter in which an appropriate atmosphere is maintained in the sensitive volume by allowing a suitable gas to flow slowly through it.

gastrin a hormone secreted by the stomach that causes secretion of digestive enzymes by other stomach cells.

Gastropoda the class of molluscs containing the snails. *See* Appendix A.

gastrula the stage of embryonic development when the gastrulation movements occur.

gastrulation the complex cell movements that carry those cells whose descendants will form the future internal organs from their largely superficial position in the blastula to approximately their definitive positions inside the animal embryo. Prior to gastrulation, the amphibian embryo relies on RNA molecules pre-loaded into the ooplasm during oogenesis. During gastrulation, however, newly synthesized nuclear gene products are required.

Gaucher disease the most common hereditary disorder of glycolipid metabolism, due to mutations in a gene on chromosome 1 at position q21. The gene contains 11 exons, that encode a 7,500 base pair transcript that specifies the enzyme glucocerebrosidase. This cleaves glucose from cerebrosides, and glucocerebrosides accumulate in lysosomes when the enzyme is defective. Although the enzyme deficiency exists in all the cells of persons with the disease, the cell primarily responsible for the syndrome is the macrophage (*q.v.*). Macrophages swollen by the accumulation of glucocerebrosides are called Gaucher cells. An effective treatment involves infusing the patient with a genetically engineered glucocerebrosidase, which is specifically targeted to Gaucher cells. Unfortunately, the treatment is tremendously expensive, amounting to $200,000 per year or more. The most common mutations of the cerebrosidase gene are missense mutations, and the positions of the amino acid substitutions often determine whether the condition will be mild or severe. Five common mutations collectively account for over 95% of the cases of Gaucher disease in the Ashkenazi Jewish population. These alleles occur in a frequency far higher than could be sustained by mutation. Heterozygotes can be identified because their peripheral leukocytes show lower enzyme levels. Homozygotes can be detected *in utero* by assays run on cells obtained by amniocentesis. Adjacent to the Gaucher disease gene is a pseudogene that also contains 11 exons. In the regions present in both sequences, 96% of the nucleotides are identical. The pseudogene is smaller because of large deletions in four of the introns and small deletions in two of the exons. Some alleles responsible for Gaucher disease symptoms appear to have arisen by rearrangements between the structural gene and the pseudogene. The first description of the disease was published in 1882 by Dr. Philippe Gaucher, hence the eponym. *See* Appendix C, 1989, Horowitz *et al.*; Ashkenazi, cerebrosidase, lysosomal storage diseases, Tay-Sachs disease. http://www.gaucherdisease.org.

Gaussian curve *See* normal distribution.

G banding *See* chromosome banding techniques.

gb, gbp *See* gigabase, gigabase pairs.

GDB Genome Data Base (human) *See* Appendix E.

Geiger-Mueller (G-M) counter a sensitive gas-filled radiation-measuring device.

gel diffusion technique *See* immunoelectrophoresis, Ouchterlony technique, Oudin technique.

gemma (*plural*, gemmae) a multicellular, asexual reproductive structure, such as a bud or a plant fragment.

gemmules pangenes. *See* pangenesis.

gene the definition of a gene changes as more of its properties are revealed. In the classical literature it is defined as a hereditary unit that occupies a spe-

Characteristics of Genes Responsible for Five Human Diseases

Genetic Disease	Protein Product	Gene Size (Kbp)	mRNA Size (Kb)	No. of Introns
sickle-cell anemia	beta chain of hemoglobin	1.6	0.6	2
hemophilia B	factor 9	34	1.4	7
phenylketonuria	phenylalanine hydroxylase	90	2.4	12
cystic fibrosis	CF transmembrane-conductance regulator	250	6.5	26
Duchenne muscular dystrophy	dystrophin, plus shorter isoforms	2300	14	78

cific position (locus) within the genome or chromosome; a unit that has one or more specific effects upon the phenotype of the organism; a unit that can mutate to various allelic forms; and a unit that recombines with other such units. Two classes are now recognized: (1) genes that are transcribed into mRNAs, which enter ribosomes and are translated into polypeptide chains, and (2) genes whose transcripts (tRNAs, rRNAs, snRNAs, etc.) are used directly. Class 1 genes are called *structural genes* or *cistrons* in the earlier literature. As shown in the table on page 172, structural genes vary greatly in size. Some genes that are transcribed into mRNAs can undergo alternative splicing and generate a series of structurally related proteins. There are also shorter DNA segments that are not transcribed but serve as recognition sites for enzymes and other proteins that function during transcription or replication. Some of these elements (i.e., operators) fulfill the classical definition of genes, but now they are generally called *regulatory sequences*. These should not be confused with regulatory genes, which encode (1) proteins that bind to regulatory sequences in other parts of the genome or (2) RNAs that inactivate entire chromosomes. The *i* gene of the *lac* operon is an example of a regulatory gene of the first type. The *Xist* gene is an example of the second type. It encodes an RNA that inactivates an entire X chromosome and is responsible for dosage compensation in female mammals. *See* Appendix C, 1909, Johannsen; 1933, Morgan; 1955, Benzer; 1961, Jacob and Monod; 1975, King and Wilson; dosage compensation, isoform, *lac* operon, replicon, selector gene, transcription unit, *Xist*.

gene activation *See* genetic induction.

genealogy a record of the descent of a family, group, or person from an ancestor or ancestors; lineage; pedigree.

gene amplification any process by which specific DNA sequences are replicated to a disproportionately greater degree than their representation in the parent molecules. During development, some genes become amplified in specific tissues; e.g., rRNA genes amplify and become active during oogenesis, especially in some amphibian oocytes (*see* rDNA amplification, *Xenopus*). Genes encoding *Drosophila* chorion proteins are also amplified in ovarian follicle cells. Gene amplification can be induced by treating cultured cells with drugs like methotrexate (*q.v.*). Gene amplification is a common and critically important defect in cancer cells. For example, a specific set of autosomal genes is overexpressed in oral squamous cell carcinomas. *See* Appendix C, 1968, Gall, Brown, and Dawid; 1978, Schimke *et al.*; 2002, Huang *et al.*; amplicon, genomic equivalence, *Podospora anserina, Rhynchosciara, TAOS 1.*

gene bank *See* genomic library.

GeneCards a database of human genes, their products, and their involvement in diseases assembled by M. Rebhan and three colleagues at the Weizmann Institute of Science, Rehovot, Israel. http://bioinformatics.weizmann.ac.il/cards.

gene cloning creation of a line of genetically identical organisms, containing recombinant DNA molecules, which can be propagated and grown in bulk, thus amplifying the recombinant molecules.

gene cloning vehicle *See* lambda cloning vector, plasmid cloning vector.

gene cluster *See* multigene family.

gene conversion a situation in which the products of meiosis from an *AA′* individual are 3*A* and 1*A′* or 1*A* and 3*A′*, not 2*A* and 2*A′* as is usually the case. Thus, one gets the impression that one *A* gene has been converted to an *A′* gene (or vice versa). Gene conversion is thought to involve a rare error in DNA repair that occurs while recombination is going on during meiotic prophase. A double-strand break in one bivalent is enlarged to eliminate one allele of the sister strand. When the gap is repaired, a non-sister strand carrying the alternate allele is used as a template, with the result that the tetrad comes to contain three copies of one allele and one of the other. Therefore gene conversion leads to the unequal recovery of alleles from DNA molecules that each carry a pair of alleles, one normal and one defective. The human Y chromosome contains several palindromes (*q.v.*) in which are imbedded structural genes that function in spermatogenesis. Within each palindrome, multiple copies of structural genes function as templates for repair of mutated genes. In this way gene conversion prevents the Y from accumulating sterility mutations. *See* Appendix C, 1935, Lindgren; 2003, Skaletsky *et al.*

gene dosage the number of times a given gene is present in the nucleus of a cell.

gene duplication the production of a tandem repeat of a DNA sequence by unequal crossing over (*q.v.*) or by an accident of replication. Duplicated genes created in these ways may subsequently evolve new functions. *See* hemoglobin genes, repeat-induced point mutation (RIP).

gene expression the display of genetic activity by the synthesis of gene products that affect the phenotype. Some genes are active throughout the life of the cell or organism. Such genes, which are continually transcribed, show *constitutive expression*. Other genes are transcribed only under certain environmental conditions or at specific times during development. These genes show *conditional expression*. Most genes are expressed in direct proportion to their ploidy levels in the nucleus. However, there is a small number of genes whose transcription rates are increased disproportionally, or even decreased, as polyploidy levels rise. *See* Appendix C, 1999, Galitski *et al.*; constitutive mutation, derepression, DNA methylation, inducible system, operon, parental imprinting, repressible system, selector gene, selfish operon.

gene family *See* multigene family.

gene flow the exchange of genes between different populations of the same species produced by migrants, and commonly resulting in simultaneous changes in gene frequencies at many loci in the recipient gene pool.

gene-for-gene hypothesis the proposal that during their evolution a host and its parasite develop complementary genetic systems, with each gene that provides the host with resistance matched by a gene in the parasite that confers susceptibility. The interacting genes from the two species are called *corresponding genes*, since for each gene that conditions resistance in the host there is a corresponding gene that conditions avirulence in the parasite, and the products of the two genes interact. The product of the resistance gene serves as a receptor for a ligand produced by the parasite, directly or indirectly through expression of an avirulence gene. The binding of receptor and ligand is the recognition event that elicits through cellular signal transduction (*q.v.*), a cascade of defense responses that constitute the resistant phenotype. *See* Appendix C, 1955, Flor; coevolution, *Linum usitatissimum, Melampsora lini.*

gene frequency the percentage of all alleles at a given locus in a population represented by a specific allele. Also referred to as *allelic frequency* (*q.v.*).

gene fusion the union by recombinant DNA techniques of two or more genes that code for different products, so that they are subject to control by the same regulatory systems.

gene gun *See* particle-mediated gene transfer.

gene insertion any technique that inserts into a cell a specific gene or genes from an outside source, including cell fusion, gene splicing, transduction, and transformation.

gene interaction interaction between different genes residing within the same genome in the production of a particular phenotype. Such interactions often occur when the products of the nonallelic genes under study function at steps in a sequence of reactions that result in compounds which generate the phenotype in question. These interactions can produce variations from the classical genetic ratios. An example would be the inheritance of aleurone color in *Zea mays*. In order for the corn kernel to possess colored aleurone, at least one *A* and one C gene must be present. Given *A* and C in the heterozygous or homozygous condition and, in addition, *R* in the heterozygous or homozygous condition, then a red pigment is produced. Purple pigment is synthesized if *P* is present in addition to *A*, C, and *R*. All four genes reside on different chromosomes. Thus, if a plant of genotype *AaCCRRPp* is self-pollinated, the offspring will contain the following aleurone classes: purple, red, and white in a 9 : 3 : 4 ratio. Here the 9 : 3 : 3 : 1 ratio has been converted into a 9 : 3 : 4 ratio, because the *P* gene cannot be expressed in the absence of *A*.

gene knockout *See* knockout.

gene library *See* genomic library.

gene machine an automated DNA synthesizer for producing short DNA probes (generally 15–30 base pairs long) or primer DNA (*q.v.*) for use in a polymerase chain reaction (*q.v.*).

gene manipulation the formation of new combinations of genes *in vitro* by joining DNA fragments of interest to vectors so as to allow their incorporation into a host organism where they can be propagated. *See* DNA vector, genetic engineering.

gene mapping assignment of a locus to a specific chromosome and/or determining the sequence of genes and their relative distances from one another on a specific chromosome.

gene networking the concept that there exist functional networks of genes which program early development, and that genes which encode proteins with multiple conserved domains serve to cross-link such networks. Thus, a set of genes containing domain A and a set containing domain B are linked by genes containing both domains. The segmentation gene *paired* (*prd*) of *Drososphila* illustrates the theory. It contains a homeobox (*q.v.*) and a histidine-proline repeat domain. This *prd*-specific repeat occurs in at least 12 other genes, while the homeobox defines a second gene set. Presumably, the *prd* product can interact with products of genes containing only the homeobox sequence or the histidine-proline repeat, or both. The conserved domains are thought to serve as the sites to which the proteins bind to specific chromosomal regions to regulate neighboring genes. *See* Appendix C, 1986, Noll *et al.*; *eyeless.*

Genentech, Inc. the first company to specialize in recombinant DNA technology. It is located in South San Francisco, California, and its name is a contraction of *Gen*etic *En*gineering *Tech*nology. *See* Appendix C, 1976, Boyer and Swanson; 1982, Eli Lilly.

gene pair in a diploid cell, the two representative genes (either identical or nonidentical alleles) at a given locus on homologous chromosomes.

gene pool the total genetic information possessed by the reproductive members of a population of sexually reproducing organisms.

gene probe *See* probe.

gene product for most genes, the polypeptide chain translated from an mRNA molecule, which in turn is transcribed from a gene; if the RNA transcript is not translated (e.g., rRNA, tRNA), the RNA molecule represents the gene product.

gene 32 protein the first DNA unwinding protein (*q.v.*), to be isolated. It is the product of gene 32 of phage T4 and is essential for its replication. The protein has a molecular weight of 35,000 daltons and binds to a stretch of DNA about 10 nucleotides long. *See* Appendix C, 1970, Alberts and Frey.

generalized in evolution theory, an unspecialized condition or trait, usually considered to have a greater potential for evolving into a variety of alternative conditions than that possessed by a highly specialized one. Primitive traits tend to be generalized; derived or advanced traits tend to be more specialized.

generalized transduction *See* transduction.

generation time (Tg) the time required for a cell to complete one growth cycle. *See* doubling time.

gene redundancy the presence in a chromosome of many copies of a gene. For example, the nucleolus organizer of *Drosophila melanogaster* contains hundreds of duplicate copies of the cistrons that code for the 18S and 28S rRNA molecules.

gene silencing a phenomenon in which genes near certain chromosomal regions, such as centromeres or telomeres, are rendered transcriptionally inactive. *See* antisense RNA, DNA methylation, heterochromatin, position effect, RNA interference, somatoclonal variation, telomeric silencing.

gene splicing *See* recombinant DNA technology.

gene substitution the replacement of one gene by its allele, all the other genes (or all other relevant genes) remaining unchanged.

gene superfamily a collection of genes that are all products of gene duplication and have diverged from one another to a considerable degree. The repeated copies of an ancestral gene can follow three evolutionary pathways: (1) they can be inactivated by mutation, (2) they can acquire new functions, or (3) they can retain their original function. The globin superfamily of genes provides examples of genes that (1) mutated to pseudogenes (*q.v.*), (2) acquired new functions (the gene for myoglobin [*q.v.*] versus the alpha chain gene of hemoglobin), and (3) retained their original functions (the Gγ and Aγ genes). *See* hemoglobin genes.

gene targeting a technique for inserting into laboratory mice genetic loci modified in desired ways. Standard recombinant DNA techniques are used to introduce desired chemical changes into cloned DNA sequences of a chosen locus. The mutated sequence is then transferred into an embryo-derived, stem-cell genome, where it is allowed to undergo homologous recombination (*q.v.*). Microinjection of mutant stem cells into mouse blastocysts is then performed to generate chimeras. The stem cells come from a black mouse line, and the recipient embryos are from a white strain. Therefore, chimeras can be identified by their variegated coat colors. Large numbers of these chimeras are mated together, and in the F_1 rare black progeny are observed. Some of these will be homozygous for the targeted gene. If the mutation represents a null allele (*q.v.*), the functions of the normal allele can be inferred from the abnormal phenotypes shown by the homozygotes. *See* Appendix C, 1988, Mansour *et al.*; knockout.

gene therapy addition of a functional gene or group of genes to a cell by gene insertion (*q.v.*) to correct a hereditary disease. *See* Appendix C, 1983, Mann, Mulligan, and Baltimore; 1990, Anderson; 1996, Penny *et al.*; *ex vivo.*

genetic anticipation the occurrence of a hereditary disease with a progressively earlier age of onset in successive generations. In those diseases caused by expansion of trinucleotide repeats (*q.v.*) anticipation results from an intergenerational increase in repeat lengths. However, reversion of the triplet repeat lengths to the normal size range can also occur, although this happens less often.

genetic assimilation the process by which a phenotypic character initially produced only in response to some environmental influence becomes, through a process of selection, taken over by the genotype, so that it is formed even in the absence of the environmental influence that at first had been necessary.

genetic background all genes of the organism other than the one(s) under consideration; also known as the *residual genotype*.

genetic block a block in a biochemical reaction generally due to a mutation that prevents the synthesis of an essential enzyme or results in the formation of a defective enzyme. If the defective enzyme has limited activity, the block may be a partial one, and the mutant is referred to as "leaky."

genetic bottleneck *See* bottleneck effect.

genetic burden *See* genetic load.

genetic coadaptation *See* coadaptation.

genetic code the consecutive nucleotide triplets (codons) of DNA and RNA that specify the sequence of amino acids for protein synthesis. The code shown below is used by most organisms, but there are exceptions (*see* universal code theory). The mRNA nucleotide sequences are written 5′ to 3′ left to right, respectively, because that is the direction in which translation occurs. Thus, an mRNA segment specifying proline-tryptophan-methionine would be (5′) CCU-UGG-AUG (3′), whereas its complementary, antiparallel DNA template strand would be (3′) GGA-ACC-TAC (5′). The code is degenerate in that all amino acids, except methionine and tryptophan, are specified by more than one codon. Most of the degeneracy involves the third nucleotide at the 3′ end of the codon (*see* wobble hypothesis). The code is read from a fixed starting point, in one direction, in groups of three consecutive nucleotides. The start codon is AUG, and in bacteria it specifies the insertion of *N*-formylmethionine (*q.v.*). When AUG occupies an internal position in the mRNA, it specifies methionine. *See* Appendix C, 1961, von Ehrenstein and Lipmann, Crick *et al.*, Nirenberg and Matthaei; 1966, Terzaghi *et al.*; 1967, Khorana; 1968, Holley *et al.*; 1979, Barrell *et al.*; 1985, Horowitz and Gorowski, Yamao; codon bias, strand terminologies, transcription unit.

genetic code dictionary *See* amino acid, universal code theory.

genetic colonization introduction of genetic material from a parasite into a host, thereby inducing the host to synthesize products that only the parasite can use. *See Agrobacterium tumefaciens*, opine.

genetic counseling the analysis of risk of producing genetically defective offspring within a family, and the presentation to family members of available options to avoid or ameliorate possible risks. *See* informed consent.

Genetic Code

FIRST BASE	SECOND BASE: U	C	A	G	THIRD BASE
U	phe	ser	tyr	cys	U
	phe	ser	tyr	cys	C
	leu	ser	Ter	Ter	A
	leu	ser	Ter	trp	G
C	leu	pro	his	arg	U
	leu	pro	his	arg	C
	leu	pro	gln	arg	A
	leu	pro	gln	arg	G
A	ile	thr	asn	ser	U
	ile	thr	asn	ser	C
	ile	thr	lys	arg	A
	S, I	thr	lys	arg	G
G	val	ala	asp	gly	U
	val	ala	asp	gly	C
	val	ala	glu	gly	A
	val	ala	glu	gly	G

See amino acids entry for three-letter symbols. S = Start codon for met for eukaryotes (*N*-formylmethionine for prokaryotes). I = Internal codon for met. Ter = Termination codon.

genetic death death of an individual without reproducing. *See* reproductive death.

genetic detasseling a breeding technique used in the commercial production of corn seed. The breeding scheme produces pollen abortion with the result that the plants are no longer hermaphroditic and can only be cross-fertilized.

genetic differentiation the accumulation of differences in allelic frequencies between isolated or semi-isolated populations due to various evolutionary forces such as selection, genetic drift, gene flow, assortative mating, etc.

genetic dissection analysis of the genetic basis of a biological phenomenon through the study of mutations that affect that phenomenon. For example, spermatogenesis can be "genetically dissected" by inducing and then characterizing mutations that sterilize male *Drosophila*.

genetic distance **1.** a measure of the numbers of allelic substitutions per locus that have occurred during the separate evolution of two populations or species. **2.** the distance between linked genes in terms of recombination units or map units.

genetic divergence *See* genetic differentiation.

genetic drift the random fluctuations of gene frequencies due to sampling errors. While drift occurs in all populations, its effects are most evident in very small populations. *See* Sewall Wright effect.

genetic engineering an all-inclusive term to cover all laboratory or industrial techniques used to alter the genetic machinery of organisms so that they can subsequently synthesize increased yields of compounds already in their repertoire, or form entirely new compounds, adapt to drastically changed environments, etc. Often, the techniques involve manipulating genes in ways that bypass normal sexual or asexual transmission. The vector of choice in plant genetic engineering is the Ti plasmid of *Agrobacterium tumefaciens* (*q.v.*). Genes of commercial importance are inserted into Ti DNA under laboratory conditions, and they become integrated into the genomes of host plants when the Ti DNA is transfected. *See* biotechnology, recombinant DNA technology.

genetic equilibrium the situation reached in a population containing, as an example, the allelic genes *A* and *a*, where the frequencies of both alleles are maintained at the same values generation after generation. *See* Hardy-Weinberg law.

genetic fine structure *See* fine-structure genetic mapping.

genetic fingerprinting *See* DNA fingerprint technique.

genetic fitness the contribution to the next generation of a specified genotype in a population in relation to the contributions of all other genotypes of that same population.

genetic fixation the status of a locus in which all members of a population are homozygous or hemizygous for a given allele; the frequency of the fixed allele is 1.0; all other alleles at that locus have been lost, and therefore their frequencies are zero. *See* monomorphic population.

genetic hitchhiking *See* hitchhiking.

genetic homeostasis the tendency of a population to equilibrate its genetic composition and to resist sudden changes.

genetic identity a measure of the proportion of genes that are identical in two populations.

genetic induction the process of gene activation by an inducer molecule, resulting in transcription of one or more structural genes. *See* inducible system.

genetic information the information contained in a sequence of nucleotide bases in a nucleic acid molecule. *See* exon, intron.

genetic instability a term that generally refers to chromosomal or other wide-scale genetic alterations that vary from cell to cell. They often arise from an overall deficit in systems that control the replication or repair of DNA or the checkpoints (*q.v.*) of the cell cycle. Widespread instabilities may also be caused by transposable elements (*q.v.*) or breakage-fusion-bridge cycles (*q.v.*) due to chromosome aberrations. *See* DNA repair, *Dotted*, helicase, mitotic arrest-deficient (mad) mutations, mutator genes, *RAD9*. *Contrast with* genomic instability.

genetic load **1.** the average number of lethal equivalents per individual in a population. **2.** the relative difference between the actual mean fitness of a population and the mean fitness that would exist if the fittest genotype presently in the population were to become ubiquitous. The genetic load of a given species may contain several components. The *mutational load* is due to recurrent mutations occurring in beneficial loci. Most new mutations are recessive, hypomorphic, and slow to be eliminated. The *segregational load* is caused by genes segregating from favored heterozygotes that generate less fertile or less viable homozygotes. There may also be an *input load* due to migrant individuals with an average fitness less than that of the original population. *See* heterozygote advantage, migrant selection, substitutional load.

genetic map the linear arrangement of mutable sites on a chromosome as deduced from genetic recombination experiments. *See* Appendix C, 1913, Sturtevant. *Compare with* physical map.

genetic marker a gene, whose phenotypic expression is usually easily discerned, used to identify an individual or a cell that carries it, or as a probe to mark a nucleus, chromosome, or locus.

genetic polymorphism the long-term occurrence in a population of two or more genotypes in frequencies that cannot be accounted for by recurrent mutation. Such polymorphism may be due to mutations that are (a) advantageous at certain times and under certain conditions and (b) disadvantageous under other circumstances, and which exist in habitats where situations (a) and (b) are encountered frequently. Genetic polymorphism may also result if genotypes heterozygous at numerous loci are generally superior to any homozygous genotype. *See* balanced polymorphism, transient polymorphism.

genetic predisposition the increased susceptibility to a specific pathological condition due to the presence of one or more mutated genes or a combination of alleles. In some cases all that is known is that there is a family history which indicates a genetic susceptibility to the condition.

genetic recombination in eukaryotic organisms the occurrence of progeny with combinations of genes other than those that occurred in the parents, due to independent assortment (*q.v.*) or crossing over (*q.v.*). In bacteria recombination of genes may occur as a result of conjugation, sexduction, transduction, or transformation (*all of which see*). In bacterial viruses an infection of a host by two or more genetically distinct bacteriophages may result in production of recombinant phage. *See* Appendix C, 1961, Meselson and Weigle; beads on a string, Holliday model, Visconti-Delbrück hypothesis.

genetics the scientific discipline dealing with **1.** the study of inheritance and variation of biological traits, and **2.** the study of genes, including their structure, function, variation, and transmission. *See* Appendix C, 1856, 1865, 1866, Mendel; 1900, Bateson; biochemical genetics, gene, heredity, Mendelian genetics, molecular genetics, population genetics, trait, transmission genetics, variation.

genetic sex determination genotypic sex determination (*q.v.*).

genetic surgery replacement of one or more genes of an organism with the aid of plasmid vectors, or the introduction of foreign genetic material into cells by microsyringes or micromanipulators.

genetic variance the phenotypic variance of a trait in a population attributed to genetic heterogeneity.

genic balance a mechanism of sex determination, originally discovered in *Drosophila*, that depends upon the ratio of X chromosomes to sets of autosomes (A). Males develop when the X/A ratio is 0.5 or less; females develop when the X/A ratio is 1.0 or greater; an intersex develops when the ratio is between 0.5 and 1.0. *See* Appendix C, 1925, Bridges; metafemales, metamales, sex determination.

genital disc the imaginal disc from which the reproductive duct system and the external genitalia are derived in *Drosophila*.

genome in prokaryotes and eukaryotes, the total DNA in a single chromosome and in a haploid chromosome set (*q.v.*), respectively, or all of the genes carried by this chromosome or chromosome set; in viruses, a single complement of DNA or RNA, or all of the genes encoded therein. *See* C value, eukaryote, haploid or haploidy, genome size, Human Genome Project, metabolic control levels, prokaryote, virus.

genome annotation conversion of raw sequence data into useful information that concerns the positions of structural genes on each chromosome, the methods by which they are switched on and off, and the functions of their products. Genes whose end products are RNA molecules must also be annotated. The sequence organization of specialized chromosomal regions, like centromeres, replicons, and telomeres, must be worked out. Finally, there is the puzzle of annotating chromosomal sequences that contain repetitive sequences that function somehow to facilitate the shortening of chromosomes during mitotic prophase, their pairing during synaptonemal complex formation, and the condensation of an X chromosome in the somatic cells of mammalian females. *See* centromere, gene, heterochromatin, insulator DNAs, meiosis, mitosis, repetitious DNA, replicon, selfish DNA, shotgun sequencing, telomere, *XIST*.

Genome Sequence Database accession number *See* Appendix E, *Haemophilus influenzae.*

genome size the amount of DNA in the haploid genome. It is often measured in picograms, kilobases, megabases, or gigabases when organisms the size range of eukaryotes are being considered. Prokaryotic genomes are smaller and are sometimes measured in daltons (*q.v.*). When one refers to viral genomes or those of mitochondria or chloroplasts, the size is generally given in kilobase pairs. One kbp of DNA equals 1.02×10^{-6} picograms or 618,000 daltons. One pg of double-stranded DNA equals

0.98×10^6 kbp or 6.02×10^{11} daltons. *See* Appendix F; C value, gigabase, kilobase, megabase, picogram.

genomic blotting *See* Southern blotting.

genomic equivalence the theory that all cells of an organism contain an equivalent complement of genetic information. Genomic equivalence has been confirmed for most cells, but exceptions occur in some animal cells where loss, gain, or rearrangement of nuclear DNA has been observed. Examples of such exceptions include chromatin diminution (*q.v.*) in somatic cells of some nematodes, selective amplification of rRNA genes in *Xenopus* oocytes, and DNA excision and rearrangement during mammalian lymphocyte maturation that result in the generation of antibody diversity. *See* gene amplification, immunoglobulin genes.

genomic exclusion an abnormal form of conjugation occurring in *Tetrahymena pyriformis* between cells with defective micronuclei and normal cells. The progeny are heterokaryons; each has an old macronucleus but a new diploid micronucleus derived from one meiotic product of the normal mate.

genomic formula a mathematical representation of the number of genomes (sets of genetic instructions) in a cell or organism. Examples: N (haploid gamete or monoploid somatic cell), 2N (diploid), 3N (triploid), 4N (tetraploid), 2N – 1 (monosomic), 2N + 1 (trisomic), 2N – 2 (nullisomic), etc.

genomic imprinting *See* parental imprinting.

genomic instability the term is generally used to refer to a localized instability due to a chromosomal site that contains a small number of nucleotides repeated in tandem. As a result of slippage during replication the number of repeated sites may be increased or decreased. If this repeated segment is in a structural gene, it may be converted to a poorly or nonfunctioning allele. The term *microsatellite instability* is used synonymously. *See* Huntington disease (HD). *Contrast with* genetic instability.

genomic library a collection of cloned DNA fragments representing all the nucleotide sequences in the genome (*q.v.*) of an organism. A genomic library is usually constructed by cutting genomic DNA into random fragments, ligating the resulting fragments into a suitable cloning vector, and transforming a host cell. The library can then be screened with a molecular probe (*q.v.*) to identify a clone of interest, or used for sequence analysis. *Compare with* cDNA library. *See* Appendix C, 1978, Maniatis *et al.*; bacterial artificial chromosomes (BACs), cosmid, DNA vector, lambda phage vector, plasmid cloning vector, P1 artificial chromosomes (PACs), restriction endonuclease, transformation, yeast artificial chromosomes (YACs).

genomic RNA the genetic material of all viruses that do not use DNA as genetic material. All cells and the vast majority of viruses use DNA as genetic material, but some bacteriophages and a few plant and animal viruses use RNA as genetic material. Examples include tobacco mosaic virus (TMV) and all retroviruses such as the human immunodeficiency virus (HIV) responsible for AIDS (acquired immunodeficiency syndrome).

genomics the scientific study of the structure and functioning of the genomes of species for which extensive nucleotide sequences are available. The genome includes both nuclear DNA and that of mitochondria and chloroplasts. In the case of structural genomics, high-resolution genetic, physical, and transcript maps are constructed for each species. Functional genomics expands the scope of research to the simultaneous study of large numbers of structural genes that respond to a suitable stimulus. Evolutionary genomics contrasts the genomes of different species to follow evolutionary changes in genome organization. A common exercise in genomics is the *in silico* (*q.v.*) investigation of orthologs (*q.v.*). A recent study of comparative genomics showed that *Drosophila* has orthologs to 177 of 289 human disease genes. *See* Appendix C, 1997, Lawrence and Oschman; 1999, Galitski *et al.*; 2000, Rubin *et al.*; DNA chip, genomic RNA, genome annotation.

genopathy a disease resulting from a genetic defect.

genophore the chromosome equivalent in viruses, prokaryotes, and certain organelles (e.g., the discrete, ringlike structure occurring in some algal chloroplasts). Genophores contain nucleic acids but lack associated histones.

genotype the genetic constitution of a cell or an organism, as distinguished from its physical and behavioral characteristics, i.e., its phenotype (*q.v.*). *See* Appendix C, 1909, Johannsen.

genotype-environment interaction an inference drawn from the observation that the phenotypic expression of a given genotype varies when measured under different environmental conditions.

genotype frequency the proportion of individuals in a population that possess a given genotype.

genotypic sex determination any mechanism of sex determination (*q.v.*) in which a genetic factor (such as the nature of the sex chromosomes in the fertilized egg or the X:A ratio (*q.v.*) of the embryo) is the primary sex-determining signal. Also called *genetic sex determination. Compare with* environmental sex determination.

genotypic variance the magnitude of the phenotypic variance for a given trait in a population attributable to differences in genotype among individuals. *See* heritability.

genus (*plural* genera) a taxon that includes one or more species presumably related by descent from a common ancestor. *See* hierarchy.

geochronology a science that deals with the measurement of time in relation to the earth's evolution. *See* Appendix C, 1953, Patterson; 1954, Barghoorn and Tyler; 1980, Lowe.

geographical isolate a population separated from the main body of the species by some geographical barrier.

geographic speciation the splitting of a parent species into two or more daughter species following geographic isolation of two or more parental populations; allopatric speciation.

geologic time divisions *See* page 181.

geometric mean the square root of the product of two numbers; more generally, the *n*th root of the product of a set of *n* positive numbers.

geotropism the response of plant parts to the stimulus of gravity.

germarium the anterior, sausage-shaped portion of the insect ovariole. It is in the germarium that cystocyte divisions occur and the clusters of cystocytes become enveloped by follicle cells. *See* insect ovary types, vitellarium.

germ cell a sex cell or gamete; egg (ovum) or spermatozoan; a reproductive cell that fuses with one from the opposite sex in fertilization to form a single-celled zygote.

germinal cells cells that produce gametes by meiosis: e.g., oocytes in females and spermatocytes in males.

germinal choice the concept advocated by H. J. Muller of progressive human evolution by the voluntary choice of germ cells. Germ cells donated by individuals possessing recognized superior qualities would be frozen and stored in germ banks. In subsequent generations these would be available for couples who wished to utilize these rather than their own germ cells to generate a family. Such couples are referred to as "preadoptive" parents.

germinal mutations genetic alterations occurring in cells destined to develop into germ cells.

germinal selection 1. selection by people of the germ cells to be used in producing a subsequent generation of a domesticated species. Such selection has been suggested for human beings. 2. selection during gametogenesis against induced mutations that retard the proliferation of the mutated cells. Such selection introduces errors in estimating the frequency of mutations induced in gonial cells.

germinal vesicle the diploid nucleus of a primary oocyte during vitellogenesis. The nucleus is generally arrested in a postsynaptic stage of meiotic prophase. *See* Appendix C, 1825, Purkinje.

germination inhibitor any of the specific organic molecules present in seeds that block processes essential to germination and therefore are often the cause of dormancy.

germ layers three primordial cell layers from which all tissues and organs arise. *See* Appendix C, 1845, Remak; ectoderm, endoderm, mesoderm.

germ line pertaining to the cells from which gametes are derived. When referring to species, the cells of the germ line, unlike somatic cells, bridge the gaps between generations. *See* Appendix C, 1883, Weismann.

germinal granules polar granules (*q.v.*).

germ line sex determination the genetic and developmental process that specifies sexual identity and sex-specific development of the germ line cells of an organism. *Compare with* somatic sex determination. *See* sex determination.

germ line transformation *See* transformation.

germ plasm 1. a specialized cytoplasmic region of the egg or zygote (*q.v.*) in many vertebrate and invertebrate species where germ cell determinants are localized. Removal or destruction of the germ plasm results in the absence of germ cells in the embryo. The germ plasm was first identified by Theodore Boveri in the nematode *Ascaris* in the late 1800s. 2. The hereditary material transmitted to the next generation through the germ cells or used for plant propagation through seeds or other materials. *See* Appendix C, 1883, Weismann; pole plasm.

gerontology the study of aging.

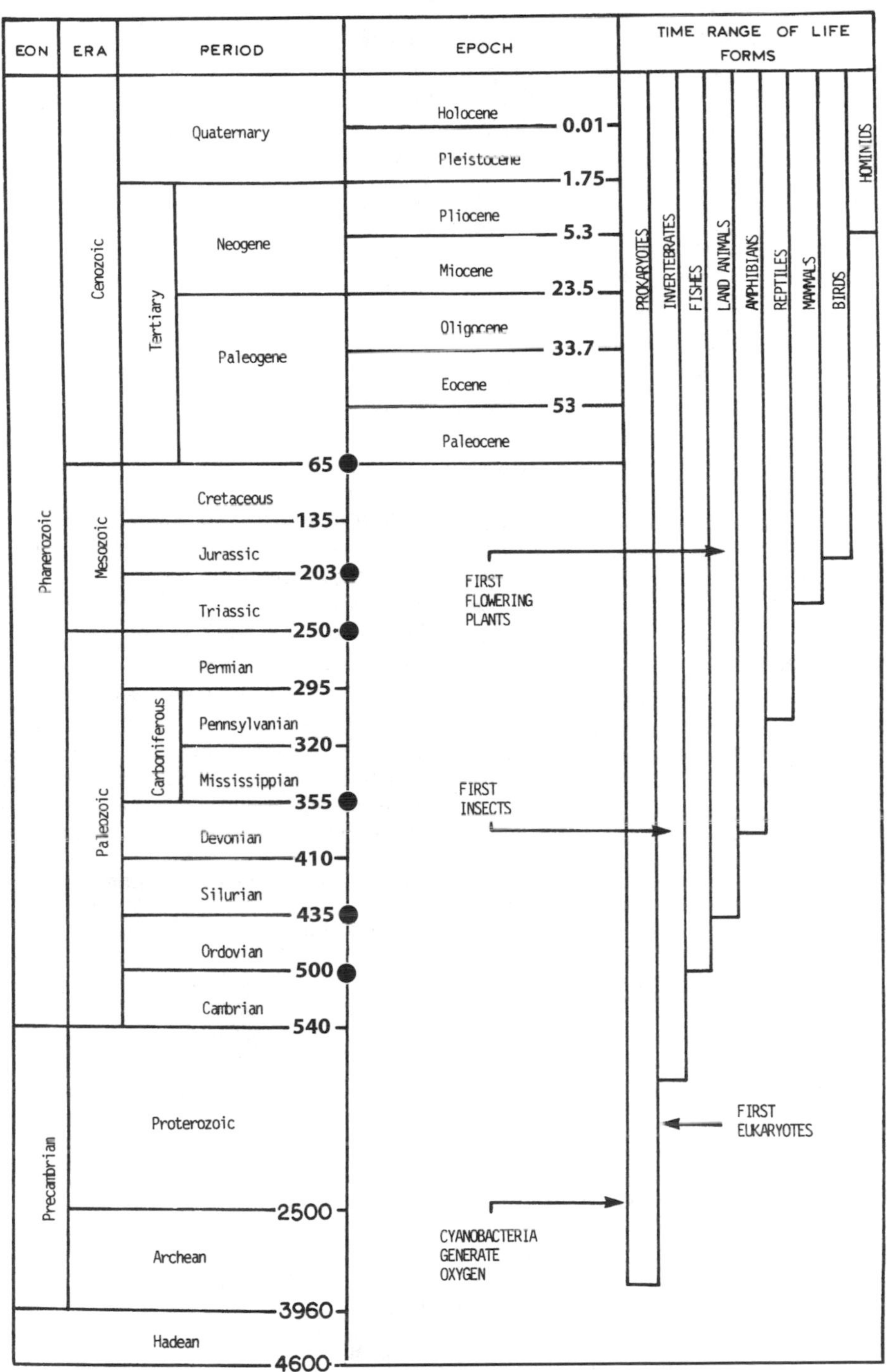

Geologic time divisions

Each number gives the date in millions of years before present. In the figure, the relative lengths of each time division are not proportional to the absolute time they represent. Solid circles mark the occurrence of mass extinctions (*q.v.*).

gestation period in a viviparous animal, the time from conception to birth. The average gestation periods in days for representative mammals are mouse 19, rat 21, rabbit 31, dog 61, cat 63, chimpanzee 238, woman 267, cow 284, mare 340, and elephant 624.

GFP green fluorescent protein (*q.v.*).

GH, GHIF, GHRH growth hormone, growth hormone inhibiting factor, growth hormone releasing hormone. *See* human growth hormone.

Giardia lamblia a species of protoctists (sometimes called *G. intestinalis*) that is the most common flagellated protozoan found in the human digestive tract and is often a cause of diarrhea in hikers. *Giardia* was first described about 1681 by van Leeuwenhoek. The parasite is tear-drop shaped, about 12 μm long, and is binucleate. The stationary phase trophozoites arrest in the G2 phase with a ploidy of 8N (2 nuclei, each with 4N ploidy). The haploid number of chromosomes is 5 and the C value is 10–12 Mb. Each cell has 4 pairs of flagella, but lacks conventional mitochondria. However, the cells each contain several dozen mitosomes (*q.v.*). These provide clusters of iron and sulfur atoms that the *Giardia* need to synthesize ATP (*q.v.*). *Giardia* trophozoites form a Golgi apparatus as they differentiate into cysts, but no Golgi can be identified within noncysting cells. Comparative analyses of the 16 S-like RNAs of *Giardia* and several other eukaryotes show that *Giardia* represents the earliest diverging lineage in the eukaryotic line of descent yet encountered. *See* Appendix A, Protoctista, Archaeprotista; Appendix C, 1677, van Leeuwenhoek; Archaeprotista, ribosomes.

gibberellin one of a family of phytohormones of widespread distribution in plants. Many single-gene, dwarf mutants in *Pisum*, *Vicia*, and *Phaseolus* are cured by the application of gibberellins. Gibberellins also promote seed germination, the breaking of dormancy, and flowering. In maize, hybrids contain higher concentrations of gibberellins than their homozygous parents. This suggests that heterosis (*q.v.*) has a phytohormonal basis.

gibbon *See Hylobates.*

gigabase a unit of length for DNA molecules, consisting of one billion nucleotides; abbreviated *gb*, or *gbp* for gigabase pairs. The genomes for *Homo sapiens* and *Pisum sativum* are 3.2 and 4.1 gigabases, respectively. *See* base pair, megabase.

gland an organ that synthesizes specific chemical compounds (secretions) that are passed to the outside of the gland. *See* endocrine system, exocrine.

glaucocystophytes a group of protoctists that are able to live photoautotrophically with the aid of cyanelles (*q.v.*). Glaucocystophytes are all freshwater organisms that are rarely seen in nature. The most frequently encountered species belong to the genera *Glaucocytis* (see illustration on page 183) and *Cyanospora*. *C. paradoxa* has provided most of the data on cyanelle DNA. These species are of interest because of the support they give to the theory of the evolution of eukaryotic cells by symbiogenesis. *See* Appendix A, Protoctista, Glaucocystophyta; serial symbiosis theory.

glaucophytes synonym for *glaucocystophytes* (*q.v.*).

gln glutamine. *See* amino acid.

globins a widespread group of respiratory proteins including the tetrameric hemoglobins of protochordates and various invertebrates, the monomeric myoglobins, and the leghemoglobins. In the case of myoglobins and the vertebrate hemoglobins, phylogenetic trees have been constructed showing that the

Gibberellin A_1

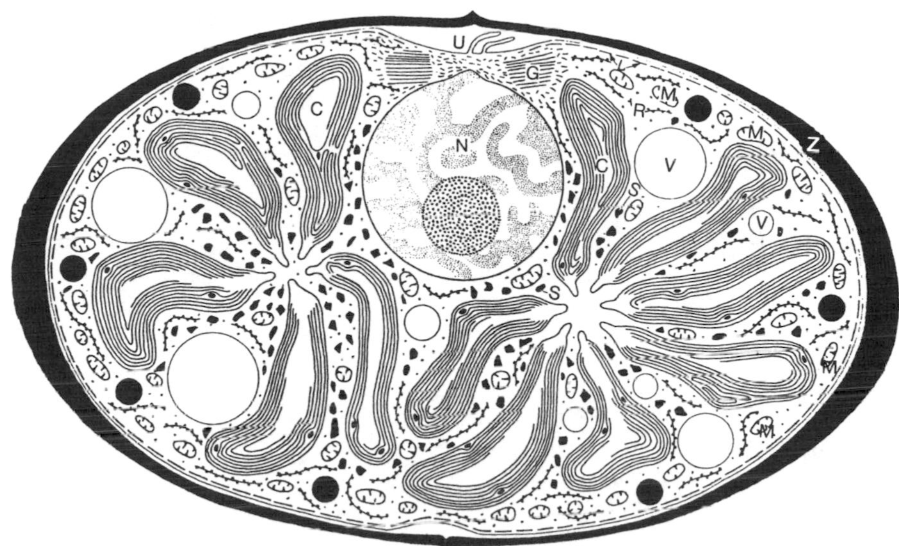

Glaucocystophyte

Glaucocystis nostochinearum. C, cyanelle; G, Golgi; M, mitochondrion with tubular cristae; N, nucleus; R, endoplasmic reticulum; S, starch granule; U, paired, short undulipodia; V, vacuole; Z, cell wall.

globin sequences have diverged during evolution at a constant rate. From these data, the timing of the gene duplications that gave rise to the family of hemoglobins has been deduced. *See* hemoglobin genes, myoglobin gene.

globulin any of certain proteins that are insoluble in distilled water, but soluble in dilute aqueous salt solution. *See* albumin, immunoglobulin.

Glossina a genus of viviparous dipterans that serve as vectors of trypanosomes. *G. morsitans* is the African tsetse fly. *See Trypanosoma.*

glu glutamate; glutamic acid. *See* amino acid.

glucagon a polypeptide hormone from the alpha cells of the pancreas that promotes the breakdown of liver glycogen and the consequent elevation of blood glucose.

glucocerebroside any of certain compounds related to sphingomyelins. They differ in that the phosphorylcholine seen in sphingomyelin (*q.v.*) is replaced by glucose. *See* cerebroside, Gaucher disease.

glucocorticoids steroid hormones produced by the adrenal cortex having effects on intermediary metabolism, such as stimulating glycogen deposition by the liver. Some glucocorticoids, such as cortisone, produce anti-inflammatory effects.

gluconeogenesis the formulation of glucose or other carbohydrates such as glycogen (glyconeogenesis) from noncarbohydrate precursors such as glycogenic amino acids, lactate, and Krebs TCA cycle intermediates. Gluconeogenesis occurs in mammalian liver under conditions such as starvation or low carbohydrate intake.

glucose a six-carbon sugar widely distributed in plants, animals, and microorganisms. In aqueous solution, less than 1% of the glucose molecules are in an open chain form, the majority being in a cyclic form. The C1 aldehyde in the linear glucose molecule reacts with the C5 hydroxyl group, resulting in the formation of the glucose ring. The structure on page 184 represents the chair conformation of β-D-glucose, the predominant form of glucose. The heavy lines in the ring are closest to the reader. This structure shows that the glucose ring is not planar; rather, C1 and C4 atoms lie either above or below the average plane defined by the C2, C3, C5, and O5 atoms of the ring. The labels α and β specify that the hydroxyl group attached to C1 is below or above the plane of the ring, respectively. The α and β forms of glucose interconvert rapidly by way of the open-chain. The D designation indicates that glucose is dextrorotary, i.e., it rotates polarized light to the right. For this reason, glucose is also known as *dextrose*. Glucose is a basic monosaccharide unit in some disaccharides (e.g., maltose, lactose, and sucrose). It is also the repeating monomer from which some polysaccharides (e.g., starch, cellulose, and glycogen) are made.

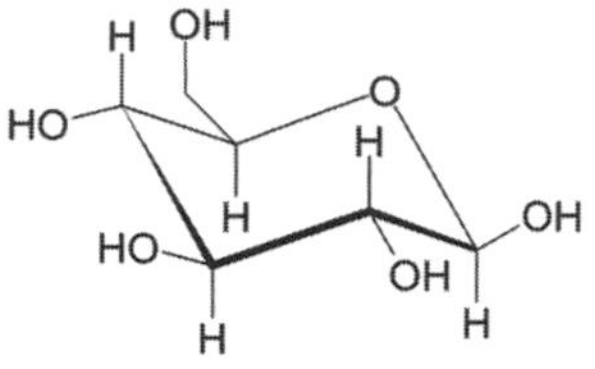

Glucose

glucose-6-phosphate dehydrogenase (G6PD) an enzyme catalyzing the conversion of glucose-6-phosphate to 6-phosphogluconate. G6PD from human erythrocytes is a dimer consisting of identical subunits, each containing 515 amino acids. About 400 genetic variants of this enzyme are known. The four most common forms of the enzyme are

ENZYME	ACTIVITY	SOURCE
B	100%	widespread
A	90%	African blacks
A^-	8–20%	African blacks
M	0–7%	Mediterranean whites

glucose-6-phosphate dehydrogenase deficiency (G6PD) the most common disease-producing enzyme deficiency of humans. It affects about 400 million humans. The gene (Gd^+) encoding G6PD resides at Xq28. The red blood cells of males with the A^- or M forms of the enzyme have a reduced life span, and exposure to antimalarial drugs such as primaquine results in life-threatening hemolysis. As a result of random X chromosome inactivation, females heterozygous for the mutant G6PD gene have some erythrocytes with the normal enzyme and some with the defective enzyme. Mutant G6PD alleles are kept in some human populations because of the heterozygote advantage (*q.v.*) occurring during infections by *Plasmodium falciparum*. Gd^+/Gd females show lower levels of parasitemia than Gd^+/Gd^+ females. *See* Appendix C, 1962, Beutler *et al.*, dosage compensation, favism, malaria.

glucose-sensitive operons bacterial operons whose activity is inhibited by the presence of glucose. This lowers the level of cyclic AMP, thereby blocking a required positive control signal.

glucosylceramide lipidosis Gaucher disease (*q.v.*).

glume a chaffy bract, pairs of which enclose the base of grass spikelets.

glutamic acid *See* amino acid.

glutamine *See* amino acid.

glutathione a tripeptide containing glutamic acid, cysteine, and glycine and capable of being alternately oxidized and reduced. Glutathione plays an important role in cellular oxidations.

gly glycine. *See* amino acid.

glycerol a trihydric alcohol, that combines with fatty acids to form fats (*q.v.*).

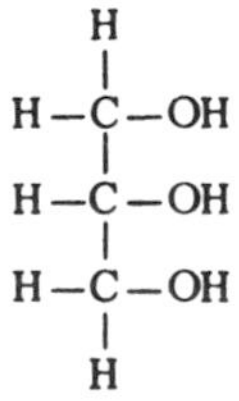

Glycine max the soybean, a legume native to China. Its haploid chromosome number is 20, and its genome size is 1.1 gigabases. The bean is the source of oil used in industry and protein used in human and livestock consumption. The soft, cheese-like bean curd is known as *tofu* or *tubu*. The soybean is an example of a nodulating legume (*q.v.*). *See* Appendix A, Plantae, Tracheophyta, Angiospermae, Dicotyledonae, Leguminales; Appendix E, *Rhizobium*.

glycogen a soluble polysaccharide (see formula on page 165) built up of numerous glucose molecules. Carbohydrate is stored as glycogen by vertebrates, especially in liver and muscles.

glycogenesis glycogen synthesis from carbohydrates. *See* glyconeogenesis.

glycogenolysis the liberation of glucose from glycogen.

glycogenosis glycogen storage disease (*q.v.*).

glycogen storage disease any of a group of congenital and familial disorders characterized by the deposition of either abnormally large or abnormally small quantities of glycogen in the tissues. Anderson, Forbes, von Gierke, Hers, McArdles, and Pompe diseases are examples.

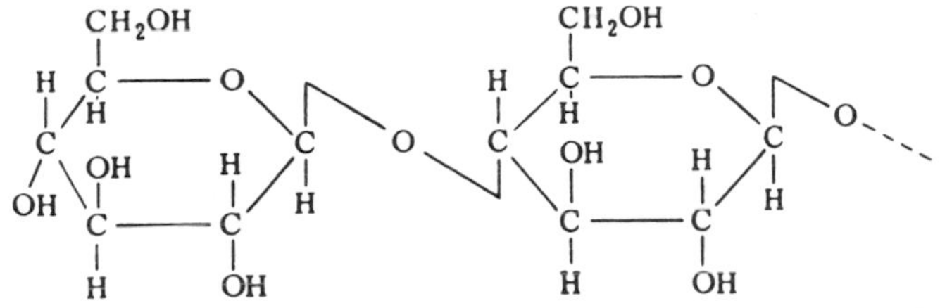

Glycogen

glycolipid a lipid containing carbohydrate.

glycolysis the sequential series of anaerobic reactions diagrammed on page 186 found in a wide variety of tissues that constitutes the principal route of carbohydrate breakdown and oxidation. The process starts with glycogen, glucose, or fructose and ends with pyruvic or lactic acids. The conversion of a molecule of glucose to two molecules of pyruvic acid generates two ATP molecules. Under aerobic conditions, the yield is eight ATP molecules. The pyruvic acid formed is broken down by way of the citric acid cycle (*q.v.*). *See* pentosephosphate pathway.

glycolytic participating in glycolysis.

glyconeogenesis *See* gluconeogenesis.

glycophorin A a protein containing 131 amino acids that is encoded by the *GYPA* gene. It spans the plasma membrane of the human red blood cell once and presents its amino terminal end to the extracellular surface. There are two polymorphic versions of glycophorin A in humans. The only differences are at positions 1 and 5 in the extracellular portion of the molecule. The M antigen contains serine and glycine at positions 1 and 5 in the chain, while the N antigen has leucine and glutamic acid at these positions. *See* MN blood groups.

glycoprotein an organic molecule consisting of a protein covalently linked to one or more carbohydrate molecules. The carbohydrate components of glycoproteins are usually small sugars, which attach to the protein post-translationally at either an asparagine (N-glycans), or at a serine or threonine (O-glycans). The attachment of the carbohydrate(s) is thought to affect protein folding, its stability and physical properties, and its recognition by other molecules. *See* amino acids, carbohydrate, glycosylation, post-translational processing, mucoprotein.

glycoside a compound yielding a sugar upon enzymatic hydrolysis.

glycosidic bonds the bonds coupling the monosaccharide subunits of a polysaccharide.

glycosome a microbody (*q.v.*), found in parasitic protozoa like trypanosomes, which contains most of the enzymes functioning in the conversion of glucose to 3-phosphoglyceric acid. *See* glycolysis.

glycosylation addition of one or more sugars to other molecules such as lipids and proteins; these molecules so modified are called *glycolipids* and *glycoproteins*, respectively. Glycoproteins appear to be universal in eukaryotic cells but are rare in or absent from bacteria. Glycosylation of proteins occurs within the lumen of the endoplasmic reticulum (ER) by glycosyltransferase enzymes. These glycoproteins are transported by clathrin-coated vesicles to the Golgi apparatus, where some sugars are removed and new ones added. This last phase of modification is termed *terminal glycosylation* to distinguish it from *core glycosylation* that occurs in the ER. In most glycoproteins, sugars are usually added to the hydroxy-group oxygen (*O-linked*) of serine or threonine; *N-linked* sugars are attached to amide nitrogens of asparagine. Most proteins in the cytosol or nucleus are not glycosylated. *See* oligosaccharide.

glyoxylate cycle an alternative to the citric acid cycle. This series of metabolic reactions is catalyzed by enzymes localized to glyoxysomes (*q.v.*). The cycle plays an important role in the photorespiration of plants and in the utilization of fat reserves by seedlings. *See* peroxisomes.

glyoxysome a membrane-bound organelle, found in germinating seeds and other plant tissues; it contains enzymes of the glyoxylate cycle (*q.v.*). *See* microbody.

Glyptotendipes barbipes a midge possessing exceptionally large polytene chromosomes in the salivary glands of larvae and therefore a favorite for cytological studies.

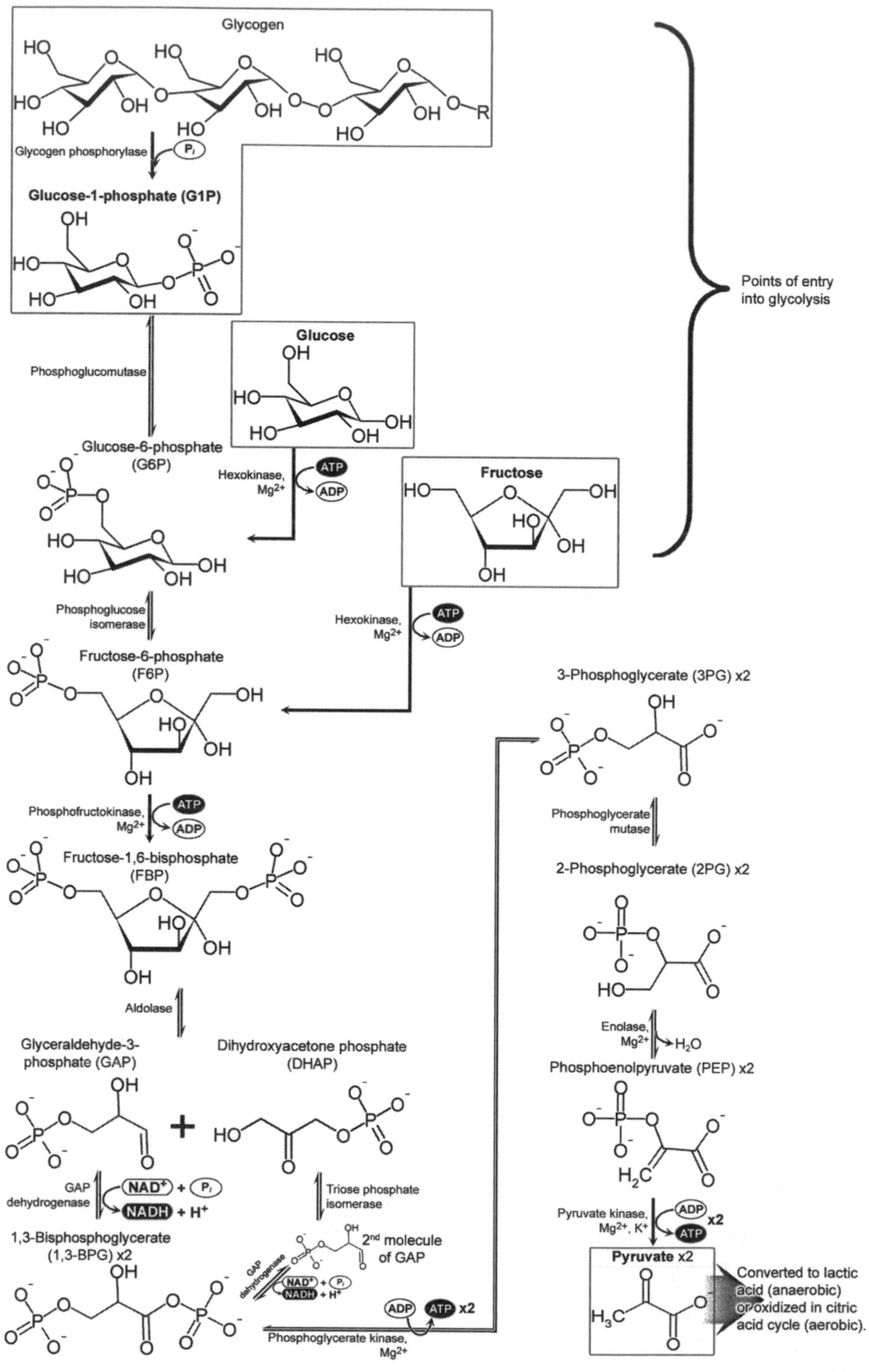
Glycogen
Glycogen phosphorylase
Glucose-1-phosphate (G1P)
Points of entry into glycolysis
Phosphoglucomutase
Glucose
Glucose-6-phosphate (G6P)
Hexokinase, Mg2+
ATP
ADP
Fructose
Phosphoglucose isomerase
Hexokinase, Mg2+
Fructose-6-phosphate (F6P)
3-Phosphoglycerate (3PG) x2
Phosphofructokinase, Mg2+
Phosphoglycerate mutase
Fructose-1,6-bisphosphate (FBP)
2-Phosphoglycerate (2PG) x2
Aldolase
Enolase, Mg2+
Glyceraldehyde-3-phosphate (GAP)
Dihydroxyacetone phosphate (DHAP)
Phosphoenolpyruvate (PEP) x2
GAP dehydrogenase
NAD+ + Pi
NADH + H+
Triose phosphate isomerase
Pyruvate kinase, Mg2+, K+
x2
1,3-Bisphosphoglycerate (1,3-BPG) x2
2nd molecule of GAP
Pyruvate x2
Converted to lactic acid (anaerobic) or oxidized in citric acid cycle (aerobic).
ATP x2
Phosphoglycerate kinase, Mg2+

Glycolysis

GMO an acronym for genetically modified organisms (i.e., transgenic bacteria, plants, or animals). *See* Bt designer plants, promoter 35S, Roundup, transformation.

gnotobiosis the rearing of laboratory animals in a germ-free state or containing only microorganisms known to the investigator.

gnotobiota the known microfauna and microflora of a laboratory animal in gnotobiosis (*q.v.*).

goiter a chronic enlargement of the thyroid gland that is due to hyperplasia, not neoplasia.

Golgi apparatus (or **body** or **complex** or **material)** a cell organelle identified in electron micrographs as a complex made up of closely packed broad cisternae and small vesicles. The Golgi apparatus is distinguished from the endoplasmic reticulum by the arrangement of the membranous vesicles and by the lack of ribosomes. The Golgi apparatus functions to collect and sequester substances synthesized by the endoplasmic reticulum. The Golgi apparatus in a typical animal cell appears as a stack of six to eight flattened membranous sacs. Plant Golgi often have 20 or more of these cisternae. The Golgi apparatus is differentiated into a cis face, the receiving surface, which is closest to the ER. Protein-rich vesicles bud off the ER and fuse with the cis face of the Golgi. While in the medial region of the Golgi, the carbohydrate units of glycoproteins are chemically modified in various ways that target them to their final destinations. Eventually the finished proteins are exported from the cisterna farthest from the ER. They are enclosed in vesicles that bud off the trans face and function as secretory vesicles (*q.v.*) or are retained as lysosomes (*q.v.*). *See* Appendix C, 1898, Golgi; 1954, Dalton and Felix; dictyosome, protein sorting signal peptide.

gonad a gamete-producing organ of an animal; in the male, the testis; in the female, the ovary.

gonadotropic hormones pituitary hormones (such as LH and FSH, *q.v.*) that stimulate the gonads; also called *gonadotropins*.

gonadotropin-releasing hormone (GnRH) a neurohormone from the hypothalamus that stimulates release of luteinizing hormone (LH) and follicle-stimulating hormone (FSH) from the pituitary gland.

gonochorism a sexual system in which each individual is either a male or a female. *Compare with* hermaphrodite, monoecy.

gonophore 1. in sessile coelenterates, the bud producing the reproductive medusae. 2. in higher animals, any accessory sexual organ, such as an oviduct, or a sperm duct. 3. in plants, a stalk that bears stamens and pistils.

gonomery the separate grouping of maternal and paternal chromosomes during the first few mitoses following fertilization as occurs in some insect embryos.

gonotocont an auxocyte (*q.v.*).

goosecoid a gene that maps to 14q32.1 in humans. It encodes a homeodomain transcription factor with a DNA-binding specificity identical to the morphogen encoded by the *Drosophila* gene *bicoid* (*q.v.*). Microinjection of *gsc* mRNAs into the ventral sides of *Xenopus* embryos leads to the formation of an additional complete body axis. Therefore the product of the *gsc* gene can mimic the natural organizer (*q.v.*). Genes homologous to gsc have been isolated from the mouse, chick, and zebrafish. *See* homeobox, Spemann-Mangold organizer.

Gorilla gorilla the gorilla, a primate with a haploid chromosome number of 24. About 40 biochemical markers have been found to be distributed among 22 linkage groups. *See* Hominoidea.

Gossypium a genus of plants composed of about 50 diploid and tetraploid species, four of which have been domesticated and produce most commercial cotton worldwide. These species are *G. herbaceum* (African-West Asian cotton), *G. arboreum* (Pakistani-Indian cotton), *G. hirsutum* (Mexican cotton) and *G. barbadense* (South American cotton). Of these, *G. hirsutum* and *G. barbadense* are the predominant species for commercial cotton production. The comparative genetics of these species has been intensively studied. *See* Appendix A, Plantae, Tracheophyta, Angiospermae, Dicotyledonae, Malvales; Appendix E.

gout a hereditary disorder of purine metabolism characterized by increased amounts of uric acid in the blood and recurrent attacks of acute arthritis.

Gowen crossover suppressor a recessive gene on the third chromosome of *Drosophila melanogaster* symbolized by *c(3)G*. It was discovered in 1917 by Marie and John Gowen, who showed that crossing over was suppressed in homozygous females. Later it was observed that mutant oocytes lacked synaptonemal complexes (*q.v.*), and the wild-type allele was found to encode a protein component of that organelle. *C(3)G* and the *Zip 1* gene of *Saccharomyces cerevisiae* are orthologs. *See* centromeric coupling.

GPCRs G protein-coupled receptors (*q.v.*).

G6PD glucose-6-phosphate dehydrogenase (*q.v.*).

G6PD deficiency *See* glucose-6-phosphate dehydrogenase deficiency.

G_0, G_1, G_2 phases *See* cell cycle.

G proteins guanine nucleotide-binding regulatory proteins. They are activated by the binding of a signaling ligand, such as a hormone, to a transmembrane receptor protein. This interaction causes the receptor to change its shape so that it can now react with a G protein. G proteins are heterotrimeric molecules made up of alpha, beta, and gamma chains. G proteins are active when GTP is bound to them and inactive when GDP is present instead. Activated G proteins dissociate from their receptors and activate effector proteins that control the level of second messengers (*q.v.*). If, for example, adenylcyclases were the effector proteins, cAMP would be generated. *See* Appendix C, 1970, Rodbell and Birmbaumer; 1977, Ross and Gilman; cellular signal transduction, cholera, diabetes insipidus, G protein-coupled receptors (GPCRs).

G protein-coupled receptors (GPCRs) integral proteins belonging to a large superfamily of cell surface receptors, each of which contains 7 membrane-spanning, alpha helix (*q.v.*) domains and a bound G protein. GPCRs bind ligands, such as hormones, odorants, growth factors, and neurotransmitters on the cell surface, and transmit signals to the interior of the cell to regulate cell functions, such as hormonal regulation, olfactory perception, cell growth, and transmission of nerve impulses. *See* cellular signal transduction, G proteins, growth factor, hormone, integral protein, ligand, odorant, odorant receptor, seven transmembrane domain (7 TM) receptor, taste receptor gene.

G quartet *See* guanine quartet model.

Graafian follicle a fluid-filled spherical vesicle in the mammalian ovary, containing an oocyte attached to its wall. *See* Appendix C, 1657, de Graaf.

grade a stage of evolutionary advance. A level reached by one or more species in the development of a structure, physiological process, or behavioral character. Different species may reach the same grade because they share genes that respond in the same way to an environmental change. When two species that do not have a common ancestry reach the same grade, the term *convergence* is used to describe this evolutionary parallelism.

gradient a gradual change in some quantitative property of a system over a specific distance (e.g., a clinal gradient; density gradient).

gradualism a model explaining the mechanism of evolution that represents an updating of the original ideas set forth by Charles Darwin. According to this model, those individuals with hereditary traits that best adapt them to their habitat are most likely to survive and to transmit these adaptive genes to their offspring. As a result, with the passage of time the frequencies of beneficial genes rise in the population, and when the composition of the gene pool of the evolving population becomes sufficiently different from that of the original population, a new species will have arisen. Since a beneficial mutation must spread through an entire population to produce detectable evolutionary changes, speciation will be a gradual and continuous process. *Contrast with* punctuated equilibrium.

Graffi leukemia virus a virus that induces myeloid leukemia in mice and rats.

graft **1.** a relatively small piece of plant or animal tissue implanted into an intact organism. **2.** the act of transferring a part of an organism from its normal position to another position in the same or another organism. In the case of relatively giant cells, cytoplasmic regions of characteristic morphology can be transplanted. Grafts between species of *Acetabularia* (*q.v.*) are examples. *See* allograft, autograft, heterograft, homograft, scion, stock, transplantation, xenograft.

graft hybrid a plant made up of two genetically distinct tissues due to fusion of host and donor tissues after grafting.

graft rejection a cell-mediated immune response to transplanted tissue that causes destruction of a graft. Rejection is evoked by the histocompatibility antigens (*q.v.*) of the foreign cells.

graft-versus-host reaction a syndrome arising when an allograft, containing immunocompetent cells, mounts an immune response against a host that is unable to reject it because the host is immunologically immature or immunologially compromised or suppressed (e.g., by radiation or drugs); synonymous with allogeneic disease, runt disease (*q.v.*).

gram atomic weight the quantity of an element that has a mass in grams numerically equal to its atomic weight.

gram equivalent weight the mass of an acid or a base that will release or neutralize one gram molecule (mole) of hydrogen ion. A 1-mole solution of H_2SO_4 contains 2 gram equivalents. A thousandth of a gram equivalent weight is a millequivalent. *See* normal solution.

gramicidin S a cyclic antibiotic synthesized by *Bacillus brevis*. The molecule has the structure

$$\left[\begin{matrix} d\text{-phe–pro–val–orn–leu} \\ \text{leu–orn–val–pro–}d\text{–phe} \end{matrix}\right]$$

and it contains amino acids not usually found in proteins—namely, ornithine (orn) and *d*-phenylalanine (rather than the usual *l*-isomer). The synthesis of gramicidin S constitutes one of the best understood examples of a polypeptide that is not synthesized on a ribosome. Two enzymes are required (E_1 and E_2), which are bound together, forming a unit called gramicidin synthetase. One molecule each of proline, valine, ornithine, and leucine bind in that sequence to sulfhydryl groups of E_1. E_2 functions to isomerize *l*- to *d*-phenylalanine and to transfer it to the proline attached to E_1. The two identical polypeptides are then joined head to tail to form a decapeptide. This rare type of polypeptide synthesis is very uneconomical and cannot generate molecules greater than 20 amino acids long.

gram molecular weight the quantity of a compound that has a mass in grams numerically equal to its molecular weight. Gram molecular weight is often shortened to *gram-mole* or *mole*.

Gram staining procedure a staining technique that allows bacteria to be divided into two groups, Gram-positive (which stain deep purple) and Gram-negative (which stain light pink). The staining differences lie in the permeability properties of the cell walls of the two groups of bacteria. *Agrobacterium, Escherichia, Haemophilus, Salmonella, Serratia, Shigella*, and *Vibrio* are all Gram-negative; *Bacillus, Mycobacterium, Staphylococcus*, and *Streptococcus* are examples of Gram-positive bacterial genera. *See* Appendix C, 1884, Gram.

grana (*singular* granum) long columns of dense discs found in chloroplasts (*q.v.*). Each disc contains a double layer of quantasomes (*q.v.*).

grandchildless genes genes that cause female mutants to produce progeny that are sterile and often show other developmental abnormalities. *See* maternal effect gene, pole plasm.

granulocytes white blood cells possessing distinct cytoplasmic granules and a multilobed nucleus; includes basophils, eosinophils, and neutrophils; also called *polymorphonuclear leukocytes*.

grass any species of monocotyledon belonging to the family Gramineae. Such species are characterized by leaves with narrow, spear-shaped blades, and flowers borne in spikelets of bracts.

gratuitous inducer a compound not found in nature that acts as an inducer although it cannot be metabolized. *See* IPTG, ONPG.

gravid an animal that is swollen from accumulated eggs or embryos.

Gray (Gy) a unit that defines that energy absorbed from a dose of ionizing radiation equal to one joule per kilogram. 1 Gy = 100 rad.

green fluorescent protein (GFP) a protein produced by the jellyfish *Aequorea victoria*. It is made up of 238 amino acids and produces a green emission when it is excited by blue light. This green fluorescence is stable and shows very little photobleaching. GFP provides an excellent means for cytologically localizing the product from any foreign gene that can be spliced to the GFP open reading frame. The fused protein is often fully functional and can be localized to its normal site in the cell by its green fluorescence. *See* Appendix C, 1994, Chalfie *et al.*. *Compare with* luciferase.

grid 1. a network of uniformly spaced horizontal and vertical lines. 2. a specimen screen used in electron microscopy.

griseofulvin an antibiotic synthesized by certain *Penicillium* species that is used as a fungicide.

gRNA guide RNA. *See* RNA editing.

groove *See* DNA grooves.

Gross mouse leukemia virus (GMLV) a filterable RNA virus discovered by Ludwig Gross in 1951. It caused leukemia when injected into newborn mice.

group selection natural selection acting upon a group of two or more individuals by which traits are selected that benefit the group rather than the individual. *See* Hamilton's genetical theory of social behavior.

group-transfer reactions chemical reactions involving the exchange of functional groups between molecules (excluding oxidations or reductions and excluding water as a participant). The enzymes that catalyze group-transfer reactions are called transferases or synthetases. For example, activation of an amino acid involves transfer of an adenosine monophosphate group from ATP to the COO^- group of the amino acid.

growth curve in microbiology, a curve showing the change in the number of cells in a growing culture as a function of time.

growth factor a specific substance that must be present in the growth medium to permit cell multiplication.

growth hormone *See* human growth hormone.

growth hormone deficiencies *See* hereditary growth hormone deficiencies.

GSH reduced glutathione.

GT-AG rule intron junctions start with the dinucleotide GT and end with the dinucleotide AG, corresponding to the left and right ("donor and acceptor") splicing sites, respectively.

GTP guanosine triphosphate (*q.v.*).

guanine *See* bases of nucleic acids.

guanine deoxyriboside *See* nucleoside.

guanine-7-methyl transferase *See* methylated cap.

guanine quartet model a three-dimensional arrangement of guanine molecules which explains the interactions occurring within DNA strands that contain repeating units rich in guanine, such as telomeric repeats. One, two, or four DNA strands can fold into a compact unit containing a planar array of four guanines, as shown above. Each guanine serves both as a proton donor and acceptor with its neighbors, and so the quartet is held together by eight hydrogen bonds. The quartets can stack one above another. This stacking is facilitated by axially located monovalent cations. For example, Na^+ fits exactly in the central cavity, and the slightly larger potassium ion can fit by binding in the cavity between adjacent quartets. *See* Appendix C, 1989, Williamson, Raghuraman, and Czech; hydrogen bond, telomere.

guanine tetraplex *See* guanine quartet model.

guanosine *See* nucleoside.

guanosine triphosphate an energy-rich molecule (analogous to ATP) that is required for the synthesis of all peptide bonds during translation.

guanylic acid *See* nucleotide.

guanylyl transferase *See* methylated cap.

guide RNA *See* RNA editing.

guinea pig *See* *Cavia porcellus.*

guppy *See* *Lebistes reticularis.*

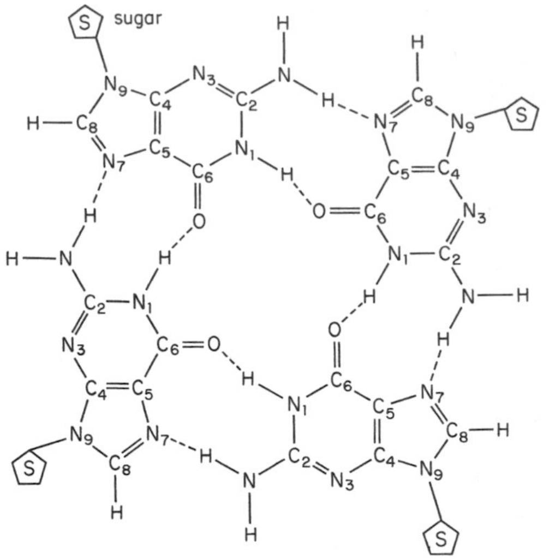

Guanine quartet model

gustatory receptor (GR) genes genes which encode in chemosensory neuron proteins that function as taste receptors. For example, in *Drosophila melanogaster* there are male-specific GR genes that are expressed on appendages including the labial palps of the proboscis, the taste bristles of the fore tarsi, and the maxillary palps. They perceive pheromones (*q.v.*) that are utilized during mating displays. *See* courtship rituals.

Guthrie test a bacterial assay for phenylalanine developed by R. Guthrie. He not only developed this methodology of newborn screening but also promoted the passage of state laws that mandated screening. Newborn screening and dietary treatment that begins within the first weeks of life have virtually eliminated mental retardation from phenylketonuria (*q.v.*) in the United States. *See* Appendix A, 1961, Guthrie.

Gy abbreviation for gray (*q.v.*).

gymnosperm a primitive plant having naked seeds (conifers, cycads, ginkgos, etc.).

gynander synonymous with gynandromorph (*q.v.*).

gynandromorph an individual made up of a mosaic of tissues of male and female genotypes. The fruit fly illustrated on page 191 is a bilateral gynandromorph, with the right side female and the left side male. The zygote was ++/*w m*. Loss of the X chromosome containing the dominant (+) genes occurred at the first nuclear division. The cell with the single X chromosome containing the recessive marker genes gave rise to the male tissues. There-

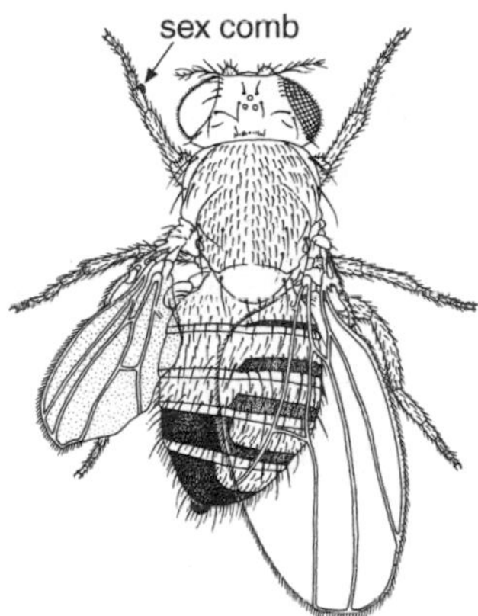

fore, the left eye is white and the left wing is miniature. Note the male abdominal pigmentation and the sex comb.

gynodioecy a sexual dimorphism in plants having both bisexual and separate female individuals.

gynoecium a collective term for all the carpels of a flower.

gynogenesis 1. reproduction by parthenogenesis requiring stimulation by a spermatozoan for the activation of the egg; synonymous with pseudogamy (*q.v.*). 2. production of a diploid embryo having two sets of maternal chromosomes by nuclear transfer (*q.v.*). *Compare with* androgenesis.

gynogenote a cell or embryo produced by gynogenesis. *Compare with* androgenote.

GYPA *See* glycophorin A.

gyrase the colloquial name for a type II topoisomerase (*q.v.*) of *E. coli* that converts relaxed closed-circular, duplex DNA to a negatively superhelical form both *in vitro* and *in vivo*. This enzyme prepares the DNA for proteins that require unwinding of the duplex or single-stranded regions in order to participate in such processes as replication, transcription, repair, and recombination. Several drugs are known to inhibit gyrase, including adriamycin, naladixic acid, and novobiocin. *See* Appendix C, 1976, Gellert *et al.*; replisome.

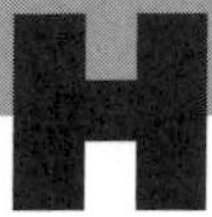

H 1. symbol for broad heritability (also symbolized H^2); h^2 for narrow heritability. 2. symbol for hydrogen.

H1, H2A, H2B, H3, H4 *See* histones.

H19 a gene that encodes one of the most abundant RNAs transcribed by the mouse embryo. The product of *H19* is not a protein but an RNA particle with a sedimentation coefficient of 28S. *H19* lies on chromosome 7, about 10 kb downstream from *Igf2*, the gene that encodes a protein called the insulin-like growth factor 2. The genes are oppositely imprinted, with *H19* and *Igf2* being expressed in the maternal and paternal homologs, respectively. Imprinting is controlled by the imprinting control region (ICR). The mechanism for parental imprinting of *H19* and *Igf2* is illustrated here. Each embryonic cell has two homologs of chromosome 7, one maternally derived (M) and one paternally derived (P). ICR, lying between the two genes, contains binding sites (s1-s4) for the CTCF protein (*q.v.*). Methylation (CH_3) of these sites prevents attachment of CTCF to the ICR. The enhancer (E) is free to loop over to the promoter of *Igf2* and facilitate its transcription (II). When ICR is unmethylated (I), CTCF binds to it and insulates (i) *Igf2* from E. So *Igf2* is silenced, and *H19* is switched on by its interaction with E. *See* Appendix C, 2000, Bell and Felsenfeld; enhancers, insulator DNAs, insulin-like growth factors 1 and 2 (IGF-1 and IGF-2), parental imprinting.

^{3}H *See* tritium.

habitat the natural abode of an organism.

Habrobracon juglandis *See Microbracon hebetor.*

Hadean the geologic eon beginning with the origin of the earth about 4.6 billion years ago and ending with the formation of the earliest rocks about 4 billion years ago. *See* geologic time divisions.

Haemanthus katherinae the African blood lily. A favorite species used for the time-lapse photographic study of mitotic endosperm cells.

haemoglobin *See* hemoglobin.

Haemophilus influenzae a Gram-negative bacterium parasitic on the mucous membranes of the human respiratory tract. It is the most common cause of middle ear infections in children. This bacterium is the first free-living organism to have its entire genome sequenced. Therefore, L 42023, which represents its *Genome Sequence Database accession number* (*q.v.*), is of historic significance. The *Haemophilus*

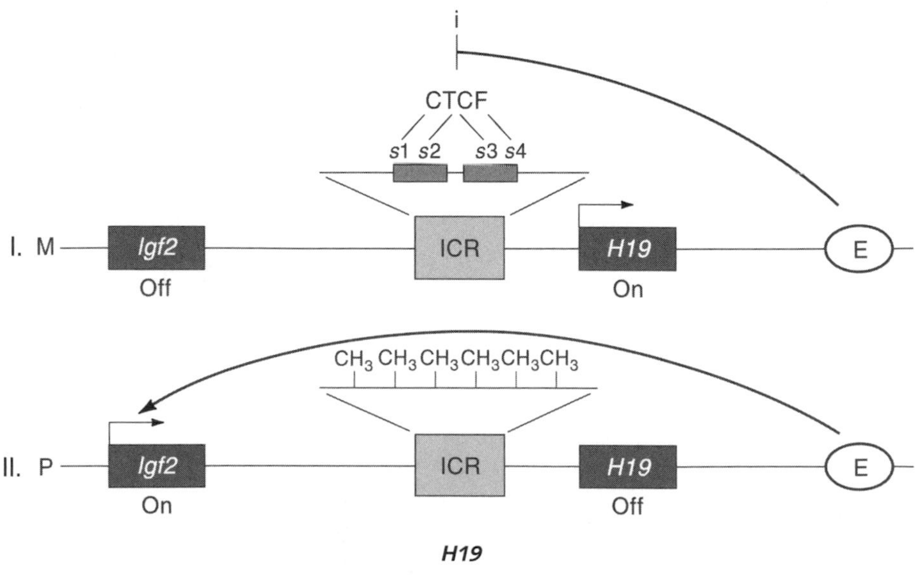

H19

genome consists of a circular chromosome containing 1,830,173 base pairs. The number of genes it contains is estimated to be 1,743, and 60% of these have sequences similar to genes previously described in other bacteria. The remainder have unknown functions. *See* Appendix A, Bacteria, Proteobacteria; Appendix C, 1995, Fleischmann, Venter *et al.*; Appendix E; shotgun sequencing, TIGR.

hairpin loops any double-helical regions of DNA or RNA formed by base pairing between adjacent inverted complementary sequences on the same strand. *See* attenuator, palindrome, terminators.

hairpin ribozyme the catalytic center of the 359-base, negative strand of the satellite RNA (*q.v.*) of the tobacco ringspot virus; it consists of a catalytic segment 50 bases long and a 14-base substrate. The catalytic RNA forms a closed loop during the cleavage reaction, hence the term *hairpin ribozyme* (HR). The HR has been engineered to bind to and cleave specific foreign RNAs. One of these is the transcript from a gene of the HIV-I virus that is essential for its replication. Suitably engineered ribozymes may someday play an important role in AIDS therapy. *See* AIDS, HIV, plus (+) and minus (–) viral strands, ribozyme.

Haldane rule the generalization that when one sex is absent, rare, or sterile, in the offspring of two different animal races or species, that sex is the heterogametic sex. The Haldane rule is known to apply for various species of mammals, birds, and insects. In *Drosophila* and *Mus*, the X and Y chromosomes interact during spermatogenesis, with the Y repressing the transcription of certain X-linked loci. Presumably, when the X and Y chromosomes are from different species, such regulation does not take place and sterility results. Thus, the Haldane rule may be explained by the nonharmonious interaction of X- and Y-linked fertility genes in the hybrid. *See* Appendix C, 1922, Haldane.

half-chromatid conversion *See* chromatic conversion.

half-life 1. **biological** the time required for the body to eliminate one-half of the dose of a given substance. This time is approximately the same for both stable and radioactive isotopes of any element. 2. **radioactive** the time required for half the mass of a radioactive substance to decay into another substance. Each radionuclide has a unique half-life.

half-sib mating mating between half brother and half sister. Such individuals have one parent in common.

half-tetrad analysis recombinational analysis where two of the four chromatids of a given tetrad can be recovered, as in the case of attached X chromosomes in *Drosophila*.

half-value layer the thickness of a specified material that reduces the flux of radiation by one-half.

halide a fluoride, chloride, bromide, or iodide.

***Halobacterium* species NRC-1** an archaeon species that can be easily cultured. It grows readily at 40°C–50°C on a well aerated tryptone-yeast extract-salt medium. Its genome has been recently sequenced and shown to consist of a large chromosome (2,014,239 bp) and two minichromosomes (191,346 bp and 365,425 bp). The chromosomes are a home for a total of 91 insertion sequences (*q.v.*) belonging to 12 different families. The genome contains 2,682 coding genes, and 972 of these are URFs. Among the genes with known functions are many that encode proteins which control the active transport of anions and cations. *See* Appendix A, Archaea, Euryarchaeota; Appendix C, 2000, Ng *et al.*; halophiles.

halogen fluorine (F), chlorine (Cl), bromine (Br), or iodine (I).

halophiles bacteria that require high salt concentrations in order to survive. They are found in salt lakes and evaporating brine.

halteres paired club-shaped appendages that extend from the metathorax of Dipterans. They serve as gyroscopic sense organs adapted to perceive deviations from the plane of their vibration. Halteres are evolutionarily equivalent to the hind pair of wings in other insects. In *Drosophila*, certain homeotic mutations (*q.v.*) convert halteres into wings and vice versa. *See* bithorax.

Hamilton genetical theory of social behavior a theory put forth by W. D. Hamilton to explain how altruism can evolve when it increases the fitness of relatives. The theory proposes that a social act is favored by natural selection if it increases the *inclusive fitness* of the performer. Inclusive fitness consists of the individual's own fitness as well as his effects on the fitness of any genetically related neighbors. The idea is that genetic alleles change in frequency in a population owing to effects on the reproduction of relatives of the individual in which the character is expressed, rather than on the personal reproduction of the individual. According to Hamilton, altruism is favored when k is greater than 1 divided by r, where r is the relatedness between individuals and k is the ratio of gain to loss of the behavior being studied. The Hamilton theory is often referred to as *kin selection*. For example, a mutation that affected the behavior of a sterile worker bee so that she fed her fertile queen but starved herself would increase the inclusive fitness of that worker because, while her own fitness decreased, her actions increased the fitness of a close relative. *See* Appendix C, 1954, Hamilton.

hammerhead ribozyme a folded RNA structure found in the genome of the satellite tobacco ring spot virus and other RNA viruses of plants. The virus replicates by a rolling circle mechanism to produce a concatomeric molecule. This is self-cleaved by the hammerhead to generate single genomic molecules. *See* hairpin ribozyme, rolling circle, viroid.

hamster common laboratory rodent. *See Cricetulus griseus, Mesocricetus auratus.*

hanging drop technique a method for microscopic examination of organisms suspended in a drop on a special concave microscope slide. The technique was invented by Robert Koch in 1878.

Hansen disease the preferred term for leprosy (*q.v.*). The eponym honors the scientist who discovered the leprosy bacterium.

Hansenula wingei a yeast that has provided information concerning the genetic control of mating type (*q.v.*).

H antigens **1.** histocompatibility antigens governed by histocompatibility genes (*q.v.*). **2.** flagellar protein antigens of motile Gram-negative enterobacteria. *Compare with* O antigens.

haplo- the prefix haplo-, when followed by a symbol designating a particular chromosome, indicates an individual whose somatic cells lack one member of the designated chromosome pair. Thus, in *Drosophila*, haplo-IV means a fly that is monosomic for chromosome 4.

haplodiploidy a genetic system found in some animals (such as the honey bee) in which males develop from unfertilized eggs and are haploid, whereas the females develop from fertilized eggs and are diploid.

haplodiplomeiosis *See* meiosis.

haploid cell culture *See* anther culture.

haploidization a phenomenon taking place during the parasexual cycle in certain fungi during which a diploid cell is transformed into a haploid cell by the progressive loss of one chromosome after another by nondisjunction.

haploid or haploidy referring to the situation or stage in the life cycle where a cell or organism has a single set of chromosomes. *N* refers to the normal haploid chromosome number for a species, while C is the haploid amount of DNA. Prokaryotic cells are characterized by a single large chromosome and are therefore haploid. However, *Deinococcus radiourans* (*q.v.*) is an exception. It has multiple chromosomes each of which is present in quadruplicate in each cell during the stationary phase (*q.v.*). *See* autosome, chromosome set, diploid or diploidy, merozygote, sex chromosome.

haploid parthenogenesis the situation in which a haploid egg develops without fertilization, as in the honey bee.

haploid sporophytes flowering plants have a reproductive cycle in which a diploid sporophytic phase alternates with a haploid gametophytic phase. The plant represents the sporophyte, and the gametophytes are microscopic. As illustrated in the double fertilization entry on page 131, the male gametophyte is the pollen grain and the female gametophyte is the embryo sac. However, parthenogenetic development of unfertilized eggs can occur, but it is very rare. Haploid sporocytes were first reported in *Datura stramonium* (*q.v.*). The frequency of haploid plants can be greatly increased using anther culture (*q.v.*). Homozygous diploid plants can then be generated by treatment with colchicine (*q.v.*). *See* Appendix C, 1922, Blakeslee *et al.*; alternation of generations.

haploinsufficiency the situation in which a single copy of a normal gene is not enough to ensure the normal phenotype. Therefore the heterozygote or an individual carrying a deletion of the gene in one homolog is detectably abnormal.

haplont an organism in which only the zygote is diploid (as in the algae, protozoa, and fungi). It immediately undergoes meiosis to give rise to the haplophase. *See* diplo-haplont, diplont.

Haplopappus gracilis a species of flowering plant showing the lowest number of chromosomes (N = 2), and therefore studied by cytologists. *See* Appendix A, Dicotyledoneae, Asterales.

haplophase the haploid phase of the life cycle of an organism, lasting from meiosis to fertilization.

haplosis the establishment of the gametic chromosome number by meiosis.

haplotype the symbolic representation of a specific combination of linked alleles in a cluster of related genes. The term is a contraction of *hap*loid genotype and is often used to describe the combination of alleles of the major histocompatibility complex (*q.v.*) on one chromosome of a specific individual. *Compare with* phenogroup.

hapten an incomplete antigen; a substance that cannot induce antibody formation by itself, but can be made to do so by coupling it to a larger carrier molecule (e.g., a protein). *Complex haptens* can react with specific antibodies and yield a precipitate; *simple haptens* behave as monovalent substances that cannot form serological precipitates.

haptoglobin a plasma glycoprotein that forms a stable complex with hemoglobin to aid the recycling of heme iron. In man this protein is encoded by a gene on chromosome 16.

Hardy-Weinberg law the concept that both gene frequencies and genotype frequencies will remain constant from generation to generation in an infinitely large, interbreeding population in which mating is at random and there is no selection, migration, or mutation. In a situation where a single pair of alleles (*A* and *a*) is considered, the frequencies of germ cells carrying *A* and *a* are defined as p and q, respectively. At equilibrium the frequencies of the genotypic classes are p^2 (*AA*), 2 pq (*Aa*), and q^2 (*aa*). *See* illustration below and Appendix C, 1908, Hardy, Weinberg.

harlequin chromosomes *See* 5-bromodeoxyuridine.

Harvey rat sarcoma virus a virus carrying the oncogene *v-ras* that was discovered in 1964 by J. J. Harvey and is homologous to its cellular proto-oncogene *c-ras*. Human DNA sequences homologous to

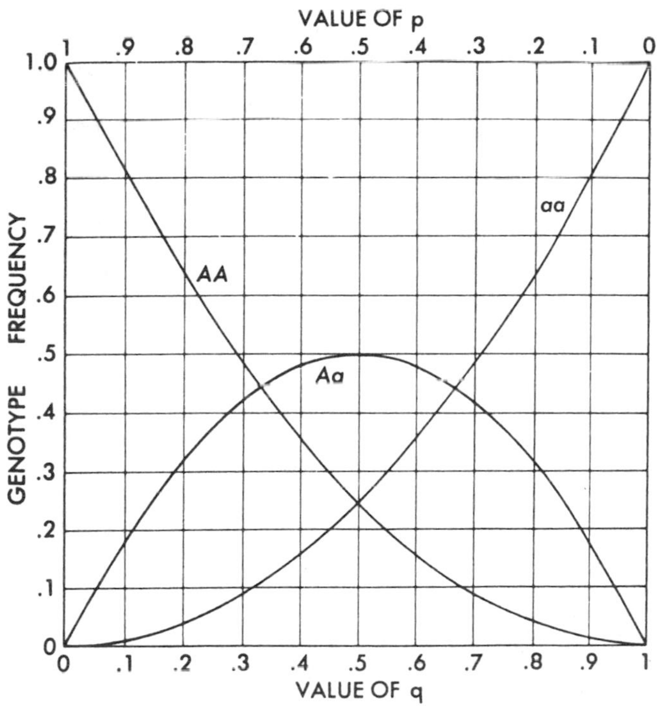

Hardy-Weinberg law

The relationships between the frequencies of genes *A* or *a* and the frequencies of genotypes *AA*, *Aa*, and *aa* as predicted by the Hardy-Weinberg law.

the *c-ras* gene have been identified, and the homolog has been mapped to human locus 11p15.5. *See* retrovirus, T24 oncogene.

HAT medium a tissue culture medium containing *h*ypoxanthine, *a*minopterin, and *t*hymidine. Mutant cells deficient in or lacking the enzymes thymidine kinase (TK^-) and hypoxanthine-guanine-phosphoribosyl transferase ($HGPRT^-$) cannot grow in HAT medium because aminopterin blocks endogenous (*de novo*) synthesis of both purines and pyrimidines. Normal TK^+ $HGPRT^+$ cells can survive by utilizing the exogenous hypoxanthine and thymidine via the *salvage pathway* (*q.v.*) of nucleotide synthesis. HAT medium has been used to screen for hybridomas (*q.v.*) by mixing TK^+ $HGPRT^-$ myeloma cells with antigen-stimulated TK^- $HGPRT^+$ spleen cells. The hybrid TK^+ $HGPRT^+$ clones that survive in HAT medium are then assayed for monoclonal antibodies specific to the immunizing antigen. *See* Appendix C, 1964, Littlefield; 1967, Weiss and Green.

Hawaiian Drosophilidae about a thousand species of *Drosophila* living on the Hawaiian islands that have undergone a remarkable adaptive radiation (*q.v.*). Over one hundred of these species are much larger than familiar fruit flies and are called *picture-winged* because of the striking pattern of spots they have on their wings. Over 90% of them are single island endemics, and all are derived from one or two ancestral lines. All have undergone parapatric speciation (*q.v.*). *See* Appendix A, 1985, Carson; courtship ritual, hot spot archipelago, mate choice, sexual selection.

Hayflick limit an experimental limit to the number of times a normal animal cell seems capable of dividing; mouse and human cells divide 30 to 50 times before they enter the "crisis period" (*q.v.*). *See* Appendix C, 1961, Hayflick and Moorehead; marginotomy, passage number, telomerase.

Hb hemoglobin. Hb^A symbolizes normal hemoglobin; Hb^F, fetal hemoglobin: Hb^S, sickle hemoglobin, etc.

HbO_2 oxyhemoglobin (*q.v.*).

H chain *See* immunoglobulin.

H-2 complex the major histocompatibility complex of the mouse lying in a segment of chromosome 17, which carries a number of polymorphic loci associated with various aspects of the immune system. It consists of four regions (K, I, S, D) that contain genes coding for classical transplantation antigens, Ia antigens and complement components, as well as immune response genes. The I region is further subdivided (*see* I region).

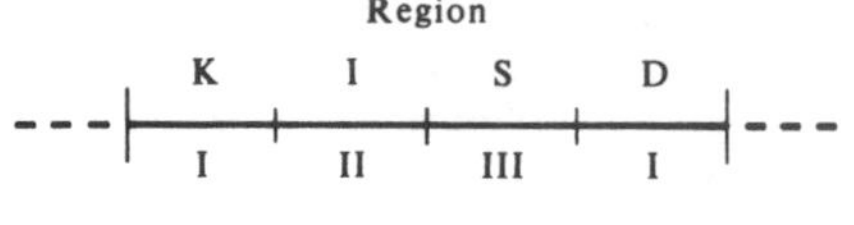

H-2 complex

hCS somatomammotropin. *See* human growth hormone.

headful mechanism a mechanism of packaging DNA in a phage head (e.g., T4) in which concatemeric DNA is cut, not at a specific position, but rather when the head is filled. This mechanism accounts for the observations of terminal redundancy and cyclic permutation in T4.

heat in reproductive biology, that period of the sexual cycle when female mammals will permit coitus; estrus.

heat-shock proteins proteins synthesized in *Drosophila* cells within 15 minutes after a heat shock. These proteins are named in accordance with their molecular weights (in kilodaltons). Hsp 70 is the major protein involved in tolerance to high temperatures in *Drosophila*. Similar proteins have been isolated from bacteria, fungi, protozoa, birds, and mammals, including humans. It now appears that heat-shock proteins combine with target proteins and prevent their aggregation and denaturation at high temperatures, and they promote the normal refolding of the target proteins when cells are returned to normal temperatures. Comparative studies of the base sequences of genes encoding heat-shock proteins from diverse organisms show that these genes have been highly conserved during evolution. *See* Appendix C, 1974, Tissiers *et al.*; 1975, McKenzie *et al.*; chaperones.

heat-shock puffs a unique set of chromosomal puffs induced in *Drosophila* larvae exposed to elevated temperatures (for example, 40 minutes at 37°C). In *D. melanogaster*, there are nine heat-inducible puffs. Heat shock results in the transcription of a specific set of mRNAs by the genes that form the puffs in polytene chromosomes. In culture, heat-shocked *Drosophila* cells also transcribe these same mRNAs into specific heat-shock proteins. *See* Appendix C, 1962, Ritossa.

heat-shock response the transcriptional activity that is induced at a small number of chromosomal loci following exposure of cells to a brief period of elevated temperature. At the same time, other loci that were active prior to the heat shock are switched off. This phenomenon appears to be universal, since it has been observed in *Drosophila, Tetrahymena,* sea urchin embryos, soybeans, chick fibroblasts, etc. *See* ubiquitin.

heavy chain in a heteromultimeric protein, the polypeptide chain with the higher molecular weight (e.g., in an immunoglobulin molecule, the heavy chains are about twice the length and molecular weight of the light chains); abbreviated *H chain;* the smaller molecules are called *light (L) chains.* The heavy chain determines the class to which the immunoglobulin belongs.

heavy chain class switching the switching of a B lymphocyte from synthesis of one class of antibodies to another. For example, a B lymphocyte first synthesizes IgM, but it may later switch to the synthesis and secretion of IgG, and both antibodies will have the same antigen specificity. Thus, only the constant regions of the heavy chains differ between these two classes of antibodies. Class switching of this type involves both somatic recombination prior to transcription and processing of transcripts by eliminating some segments and splicing the fragments. *See* somatic recombination, V(D)J recombination.

heavy isotope an atom, such as ^{15}N, that contains more neutrons than the more frequently occurring isotope, and thus is heavier.

heavy metals elements such as cobalt, nickel, manganese, copper, zinc, arsenic, antimony, mercury, lead, and bismuth, which are toxic pollutants found in the tailings left behind in gold, silver, and iron mines. The development of resistance to such heavy metals in plants growing in mine entrances is an example of recent evolution by natural selection. *See* Appendix C, 1952, Bradshaw.

heavy-metal stain one of the elements of high atomic weight, often used as stains in electron microscopy (U, Pb, Os, Mn).

heavy water nontechnical term for deuterium oxide.

hedgehog (hh) a segment polarity gene of *Drosophila* located at 3–81.2. This selector gene (*q.v.*) is responsible for the proper development of both wings and legs. Homologs of *hh* have been found in fish, birds, and mammals. *See Sonic hedgehog (Shh).*

HeLa cells an established cell line (*q.v.*) consisting of an aneuploid strain of human epithelial-like cells, maintained in tissue culture since 1951; originating from a specimen of tissue from a carcinoma of the cervix in a patient named *He*nrietta *La*cks. H. Lacks eventually died from her cancer. However, her cancer cells are subcultured in laboratories all over the world. It is estimated that the combined weight of these cells is now 400 times Henrietta's adult body weight. HeLa cells have been used to investigate the cell cycle (*q.v.*), to develop the polio vaccine, and to study the behavior of human cells in the environment of outer space. Because of sloppy subculturing, HeLa cells contaminate other cell lines in many laboratories throughout the world. *See* Appendix C, 1951, Gey; 1968, Gartler; 1995, Feng *et al.*

Helianthus annuus the sunflower.

helicase a DNA-binding protein that functions to unwind the double helix. Such unwinding is necessary at replication forks so that DNA polymerases can advance along single strands. Unwinding is also necessary for cut and patch repair (*q.v.*). Loss of function mutations in genes that encode helicases can result in cancer or premature aging. For example, mutations of genes that encode enzymes belonging to the RecQ family of human DNA helicases cause Bloom syndrome and Werner syndrome (both of which *see*). *See* nucleolus, replisome, xeroderma pigmentosum.

Helicobacter pylori a bacterium that is thought to infect the gastric mucosa of over half the adults over the age of 60 in industrial countries. Attachment of the parasite to the stomach wall is facilitated by receptors on the surface of the mucous cells to which the bacterium binds. In humans most stomach ulcers are caused by *H. pylori.* It is but one of a dozen species of *Helicobacter* that inhabit the gastrointestinal tract of birds and mammals. Humans with AB antigens (*q.v.*) are less susceptible to gastric ulcers than type O individuals. *See* Appendix A, Bacteria, Proteobacteria; Appendix E; Lewis blood group.

heliotropism a synonym for phototropism (*q.v.*).

helitron a DNA segment that transposes as a rolling-circle transposon. Unlike other transposons (*q.v.*), helitrons lack terminal repeats and do not duplicate the insertion sites of their hosts. They carry 5′ TC and 3′ CTTR termini and are always inserted between nucleotides A and T of the host DNA. Helitrons were uncovered by an *in silico* analysis of genomic sequences from *Arabidopsis,* rice, and *Caenorhabditis,* and a helitron was later isolated as an insertion in a mutated *Sh2* gene of *Zea mays. See* Appendix C, 2001, Kapitonov and Jura; rolling circle.

helix a curve on the surface of a cylinder or cone that cuts all the elements of the solid at a constant angle. Applied especially to the *circular helix* on a right circular cylinder. It resembles the thread of a bolt.

Helix a genus of gastropod containing common garden snails. Species from this genus are favorite experimental animals for the experimental analysis of molluscan development.

helix-destabilizing proteins any protein that binds to single-stranded regions of duplex DNA created by "breathing" (*q.v.*), and thereby causes unwinding of the helix; e.g., helicase (*q.v.*); also called *relaxation proteins* or *melting proteins*.

helix-turn-helix motif a term describing the three-dimensional structure of a segment that characterizes certain DNA-binding proteins. Regulatory proteins like the lambda and *cro* repressors (*q.v.*) are examples. The protein bends so that two successive alpha helices are held at right angles by a turn that contains four amino acids. One of the two helices lies across the major groove of the DNA, while the other sits within the major groove, where it can make contacts with these nucleotides of the DNA that are in a specific sequence. This "recognition helix" modulates the expression of the gene to which it binds. *See Antennapedia*, DNA grooves, motifs, *myc*.

Helminthosporium maydis *See Bipolaris maydis* (the preferred nomenclature).

helper T lymphocytes a T lymphocyte that amplifies the activity of B lymphocytes, other T lymphocytes, and macrophages. Once a helper T lymphocyte recognizes an antigen, it divides and its progeny start to synthesize and secrete a variety of lymphokines (*q.v.*). These cause B lymphocytes to divide and differentiate into plasma cells, inactive precursor T cells to develop into cytotoxic T lymphocytes, and macrophages to be recruited and activated. *See* AIDS, cytotoxic T lymphocyte, T lymphocyte.

helper virus a virus that, by its infection of a cell already infected by a defective virus, is able to supply something the defective virus lacks, thus enabling the latter to multiply. *See* satellite RNA.

HEMA the gene, located at Xq28, which encodes factor 8, the antihemophilic factor (*q.v.*). *HEMA* contains 189 kilobases of DNA and is one of the largest human genes. It has 26 exons and 25 introns. One of these (number 22) is gigantic (32 kb), and it contains a CpG island (*q.v.*). This serves as a bidirectional promoter for two genes (*A* and *B*). Gene *A* contains no introns and is transcribed in the opposite direction from factor 8. Gene *B* is transcribed in the same direction as factor 8. Its first exon is in intron 22, and this exon is spliced to exons 23–26 to form the complete transcript. The *A* and *B* transcripts are found in many different tissues, but the functions of the proteins they encode are not known. Two additional copies of gene *A* also lie about 500 kilobases upstream of *HEMA*. Intrachromosomal recombination between the *A* gene within intron 22 and either of these duplicates can yield an inversion that disrupts *HEMA*. Such inversions are thought to cause 25% of all cases of hemophilia A. The gene is expressed in the liver. The first human mutations caused by transposable elements were discovered in *HEMA*. *See* Appendix C, 1988, Kazazian *et al.*; hemophilia.

hemagglutinins **1.** antibodies involved in specific aggregation of red blood cells. **2.** glycoproteins, formed on the surfaces of cells infected with certain viruses (e.g., influenza virus) or on the surfaces of enveloped viruses released from such cells, that can aggregate red blood cells of certain species. **3.** lectins (*q.v.*).

hematopoiesis the formation of red blood cells.

hematopoietic pertaining to hematopoiesis.

HEMB the gene, located at Xq27, which encodes factor 9, the plasma thromboplastin component (*q.v.*). *HEMB* is much smaller than *HEMA*. *HEMB* contains 34 kilobases of DNA and is made up of 8 exons. Most individuals suffering from hemophilia B have deletions in the factor 9 gene. The liver is the site of transcription and translation of the *HEMB* mRNA. *See* gene, hemophilia.

heme an iron-containing molecule that forms the oxygen-binding portion of hemoglobin (*q.v.*). Heme consists of a tetrapyrrole ring, named protoporphorin IX, and a ferrous ion. An oxygen molecule can reversibly bind to the iron atom which in turn is covalently attached to the nitrogens of the four pyrrole molecules. Heme-containing proteins also occur in bacteria and archaeans, where they generate signals in response to their binding oxygen. Presumably the signals are of adaptive significance to these microorganisms. *See* bilirubin.

Hemimetabola a superorder of primitive insects having incomplete metamorphosis (mayflies, dragon flies, stoneflies, roaches, etc.). Synonymous with Exopterygota. *See* Appendix A.

hemimetabolous referring to those insects in which metamorphosis is simple and gradual. Each insect gradually acquires wings during a period of growth interrupted by several molts. Immature forms are called nymphs if terrestrial, naiads if aquatic. *Contrast with* holometabolous.

hemipteran a true bug; a member of the order Hemiptera. *See* Appendix A.

hemizygous gene a gene present in single dose. It may be a gene in a haploid organism, or a sex-linked gene in the heterogametic sex, or a gene in the appropriate chromosomal segment of a deficiency heterozygote.

hemochromatosis impaired functioning of a tissue or organ due to the storage of excessive amounts of iron in its cells. Hereditary hemochromatosis is the result of homozygosity of a recessive gene (*HLA-H*) on the short arm of human chromosome 6. The gene appears to be one of the most common disease-producing genes, at least in populations of European origin. People with one copy of HLA-H absorb increased amounts of iron from their diets but do not show symptoms of hemochromatosis. When both copies of the flawed gene are present the disease occurs, but usually after the reproductive period is over. Premenopausal females rarely show symptoms, possibly because of iron loss due to bleeding during menstruation. It is possible in the past that heterozygotes had a selective advantage in regions where their diets contained suboptimal amounts of iron.

hemocoel a body cavity of arthropods and molluscs that is an expanded part of the blood system. The hemocoel never communicates with the exterior, and never contains germ cells.

hemocyte an amoeboid blood cell of an insect. It is analogous to a mammalian leukocyte.

hemoglobin the oxygen-carrying pigment (molecular weight 64,500 daltons) of red blood cells. Each hemoglobin molecule is a spheroid with dimensions 50 × 55 × 65Å. In red blood cells where hemoglobin is very concentrated, the molecules are only about 10Å apart. However, they can still rotate and flow past one another. Hemoglobin is a conjugated protein composed of four separate chains of amino acids and four iron-containing ring compounds (heme groups). The protein chains of adult hemoglobin (HbA) occur as two pairs, a pair of alpha chains and a pair of beta chains. Each alpha chain contains 141 amino acids, and each beta chain contains 146 amino acids. Normal human adults also have a minor hemoglobin component (2%) called A_2. This hemoglobin has two alpha chains and two delta peptide chains. The delta chains have the same number of amino acids as the beta chains, and 95% of their amino acids are in sequences identical to those of the beta chains. The hemoglobin of the fetus (HbF) is made up of two alpha chains and two gamma chains. Each gamma chain also contains 146 amino acids. There are two types, G gamma and A gamma, which differ only in the presence of glycine or alanine, respectively, at position 136. The earliest embryonic hemoglobin tetramer consists of two zeta (alphalike) and two epsilon (betalike) chains. Beginning about the eighth week of gestation, the zeta and epsilon chains are replaced by alpha and gamma chains, and just before birth the gamma chains begin to be replaced by beta chains. *See* Appendix C, 1960, Perutz *et al.*; 1961, Dintzis, Ingram; 1962, Zuckerkandl and Pauling; leghemoglobin.

hemoglobin Bart's hemoglobin made up of homotetramers of gamma globin chains. In alpha thalassemia, the underproduction of alpha globin chains stimulates the expression of gamma globin genes. As a result, gamma globin tetramers become predominant. In fetuses homozygous for deletions in alpha hemoglobin genes, the level of Bart's hemoglobin is 80%–100%, and hydrops fetalis (*q.v.*) results. The hemoglobin was identified at St. Bartholomew's Hospital in London. This venerable hospital, founded in 1123, is nicknamed "Bart's." *See* hemoglobin fusion genes.

hemoglobin C a hemoglobin with an abnormal beta chain. Lysine is substituted for glutamic acid at position 6. Hemoglobin C is thought to protect against malaria by interfering with the production of adhesion knobs on the surface of infected red blood cells. AC and CC erythrocytes show reduced adhesion to endothelial monolayers. Therefore parasitized cells are more likely to be swept into the spleen and destroyed. *See* hemoglobin S.

hemoglobin Constant Spring a human hemoglobin possessing abnormal alpha chains that contain 172 amino acids, rather than the normal number of 141. Hb Constant Spring seems to be the result of a nonsense mutation that converted a stop codon into

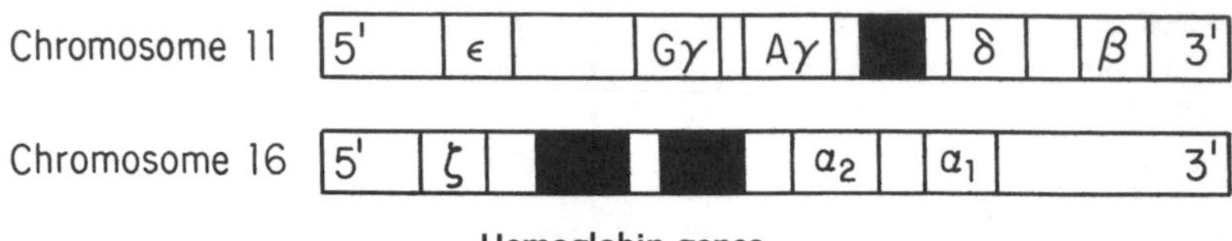

Hemoglobin genes

a codon specifying an amino acid (glutamine, in this case). The 30 additional amino acids must be coded by adjacent base sequences that normally are not transcribed, or, if transcribed, not translated. The designation *Constant Spring* comes from the region of the island of Jamaica where the mutant hemoglobin was discovered.

hemoglobin fusion genes abnormal hemoglobin genes arising as a result of unequal crossing over between genes sharing homologous nucleotide sequences. Examples are the fused gamma-beta gene that codes for the non-alpha chain of Hb Kenya and the variety of fused delta-beta genes that code for the non-alpha chains of the Hb Lepore (*q.v.*).

hemoglobin genes the genes coding for human hemoglobins are located on chromosomes 11 and 16. As illustrated above, chromosome 11 contains the epsilon chain gene, two gamma chain genes (symbolized by Gγ and Aγ), a delta chain gene, and a beta chain gene. Chromosome 16 contains a zeta gene and two alpha chain genes (α_1 and α_2). There are also DNA segments that are structurally similar to the expressed hemoglobin genes. These *pseudogenes* were probably functional in the past but are now incapable of generating a product because of mutations they contain. These pseudogenes are represented by black boxes in the diagram above. The globin genes on chromosome 16 (the alpha-like cluster) and on chromosome 11 (the beta-like cluster) are thought to have arisen by the duplication of a single ancestral gene some 450 million years ago. The table below shows the segmental organization of the transcription units (TUs) from the globin genes. All contain 5′ and 3′ ends (which are not translated), three exons (E_1, E_2, E_3), and two introns (I_1 and I_2) that are spliced out of the mature mRNA. The numbers represent base pairs. The major difference between the two gene clusters is that the second introns of the beta-like genes contain about six times as many base pairs as the second introns of alpha-like genes. Transcription begins 40 to 50 base pairs upstream of E_1, and ends about 100 base pairs downstream from E_2. Subsequently the 5′ end receives a methylated cap (*q.v.*), and the 3′ end loses a segment of about 25 nucleotides but gains a tail of 50 or more adenylic acids. *See* Appendix C, 1961, Ingram; 1976, Efstratiadis *et al.*; 1977, Tilghman *et al.*; 1979, Fritsch *et al.*; 1985, Saiki *et al.*; 1986, Constantini *et al.*; compound heterozygote, gene, hereditary persistence of hemoglobin F, myoglobin gene, polyadenylation, posttranscriptional processing, sickle-cell anemia, thalassemias, transcription unit.

hemoglobin H *See* hemoglobin homotetramers, thalassemias.

hemoglobin homotetramers abnormal hemoglobins made up of four identical polypeptides. Examples are Hb Bart's (only gamma chains) and Hb H (only beta chains).

hemoglobin Kenya *See* hemoglobin fusion genes.

hemoglobin Lepore an abnormal hemoglobin named after the Italian family in which it was first found. The protein contains a pair of normal alpha chains and a pair of abnormal chains, each chain made up of 146 amino acids. Each abnormal chain appears to be a hybrid molecule consisting of an *N*-terminal end containing amino acids in a sequence characteristic of the left end of the delta chain and a C-terminal end containing amino acids in a sequence characteristic of the right end of a beta chain. Hemoglobin Lepore is presumed to have arisen by un-

Globin TU	5′ end	E_1	I_1	E_2	I_2	E_3	3′ end
alpha	37	93	113	204	141	129	112
beta	50	93	130	222	850	126	132
gamma	53	93	122	222	886	126	87

equal crossing over between a mispaired delta and beta cistron.

hemoglobin S a hemoglobin with a glutamic acid replaced by a valine at position 6 of the beta chains. The substitution replaces a charged side chain with a hydrophobic residue, and this results in a hydrophobic patch on the surface of the molecule. This patch can fit into a pocket in an adjacent hemoglobin S, and therefore adjacent molecules stack up to form fibers that deform the cell. The stacking only occurs when hemoglobin is in the deoxygenated state, and it occurs to only a minor degree in the cells of heterozygotes. Misshapen red blood cells in venous blood are a diagnostic feature of sickle-cell anemia (*q.v.*). *See* Appendix C, 1949, Pauling; 1957, Ingram; 1973, Finch *et al.*; gene, sickle-cell trait.

hemoglobinuria the excretion of hemoglobin in solution in the urine.

hemolysis the rupturing of blood cells.

hemolytic anemia *See* anemia.

hemophilia a hereditary disease characterized by a defect in the blood-clotting mechanism. Hemophilia was the first human trait recognized to show sex-linked inheritance. Hemophilia A (classical hemophilia) is due to a deficiency of functional *antihemophilic factor* (factor 8). Hemophilia B (Christmas disease) results from a deficiency of *plasma thromboplastin component* (factor 9). The clotting factors are proteins encoded by genes (*HEMA* and *HEMB*) that reside at a considerable distance from one another on the X chromosome. The hemophilic population consists of hemizygotes for *HEMA* and for *HEMB* in a ratio of 4 : 1. The hemoglobin genes have also mutated in various domesticated animals, since hemophilia A occurs in cats, dogs, horses, and sheep, and hemophilia B is known in cats and dogs. *See* Appendix C, 1820, Nasse; 1988, Kazazian *et al.*; anti-hemophilic factor, blood clotting, *HEMA*, *HEMB*, plasma thromboplastin component, royal hemophilia, von Willebrand disease. http://www.hemophilia.org.

hemopoiesis *See* hematopoiesis.

hemopoietic *See* hematopoietic.

hemopoietic histocompatibility (Hh) used in reference to transplantation of bone marrow in inbred strains of mice and their hybrids. As measured by the growth of the transplanted marrow cells in the spleen, a unique form of transplantation genetics seems to apply whereby hybrid recipients resist parental bone marrow grafts, but parental recipients accept hybrid bone marrow.

hemozoin a non-toxic form of hemoglobin deposited in the food vacuole of the *Plasmodium* trophozoite (*q.v.*). *See* malaria, vacuoles.

herb a plant with no persistent parts above ground, as distinguished from a shrub or tree.

herbage herbs collectively, especially the aerial portion.

herbarium a collection of dried and pressed plant specimens.

herbicide a chemical used to kill herbaceous plants. *See* Roundup.

herbivore a plant eater.

hereditary disease a pathological condition caused by a mutant gene. *See* Appendix C, 1966, McKusick, human genetic diseases.

hereditary growth hormone deficiencies a deficiency in human growth hormone (*q.v.*) often due to deletions or point mutations at the *GH1* locus. Deletions result in severe dwarfism by 6 months of age. hGH replacement is initially beneficial, but anti-hGH antibodies subsequently develop and may arrest the response to exogenous hGH. The lack of immunological tolerance to hGH occurs because the mutant fetus never produced hGH during the period when its immune system was being programmed not to respond to self-proteins. In cases where a biologically inactive, mutant hGH is produced, treatment with synthetic hGH is usually effective. *See* human growth hormone gene, human growth hormone receptor.

hereditary hemochromatosis *See* hemochromatosis.

hereditary nonpolyposis colorectal cancer (HNPCC) a human cancer syndrome due to mutations in the *MSH2* gene. This gene has been mapped to 2p22-p21, and it encodes an enzyme that functions in mismatch repair (*q.v.*).

hereditary persistence of hemoglobin F (HPHF) the continued synthesis of fetal hemoglobin in adults. One form of HPHF results from large deletions that remove the δ and β globin genes from chromosome 11. Under these circumstances, the Gγ and $A\gamma$ genes are switched on and γ globin chains are synthesized. In the nondeletion form of HPHF, missense mutations have been found 158–202 base pairs upstream of the $G\gamma$ chain ORF. These base substitutions presumably prevent a control element from switching off production of gamma chains. *See* Appendix C, 1984, Collins *et al.*; hemoglobin, hemoglobin genes.

hereditary spherocytosis (HS) a highly variable genetic disorder affecting the red blood cell membrane and characterized by anemia (*q.v.*), jaundice (*q.v.*), enlargement of the spleen, and hemochromatosis (*q.v.*). HS results from a defect or deficiency in one of the cytoskeletal proteins associated with the erythrocyte membrane. 35%–65% of all HS cases are associated with mutations in the *ANK1* gene, and this form of HS is also called *ankyrin-1 deficiency*, *ankyrin-R deficiency*, or *erythrocyte ankyrin deficiency*. Erythrocytes in HS patients are smaller, rounder, and more fragile than normal, get trapped in narrow blood vessels, particularly in the spleen, and hemolyze, resulting in HS symptoms. This disorder is most common in people of northern European ancestry. *See* ankyrin.

heredity a familial phenomenon wherein biological traits appear to be transmitted from one generation to another. The science of genetics has shown that heredity results from the transmission of genes from parents to offspring. The genes interact with one another and with their environment to produce distinctive characteristics or phenotypes. Offspring therefore tend to resemble their parents or other close relatives rather than unrelated individuals who do not share as many of the same kinds of genes.

heritability an attribute of a quantitative trait in a population that expresses how much of the total phenotypic variation is due to genetic variation. In the broad sense, heritability is the degree to which a trait is genetically determined, and it is expressed as the ratio of the total genetic variance to the phenotypic variance (V_G/V_P). In the narrow sense, heritability is the degree to which a trait is transmitted from parents to offspring (i.e., breeding value), and it is expressed as the ratio of the additive genetic variance to the total phenotypic variance (V_A/V_P). The concept of additive genetic variance makes no assumption concerning the mode of gene action involved. Heritability estimates are most commonly made by (1) regression-correlation analyses of close relatives (e.g., parent-offspring, full sibs, half sibs), (2) experiments involving response to selection, and (3) analysis of variance components. Traits with high heritabilities respond readily to selection.

hermaphrodite an individual having both male and female reproductive organs. A *simultaneous hermaphrodite* has both types of sex organs throughout life. A *sequential hermaphrodite* may have the ovary first (protogyny), to be replaced by a testis later, or may develop the testis first (protandry), to be replaced later by an ovary. *See* consecutive sexuality, sperm sharing.

herpes virus one of a group of animal viruses having a duplex DNA molecule within an icosahedral capsid. They range in size from 180 to 250 nanometers and produce internuclear inclusions in host cells. If the virus infects humans, the acronym is given an initial "H." HHV1 (herpes simplex virus) causes cold sores. HHV2 (genital herpes virus) causes lesions and is sexually transmitted. HHV3 (varicella zoster virus) causes chicken pox and shingles. *See* enveloped viruses, Epstein-Barr virus (EBV), human cytomegalovirus (HCMV), human papillomavirus, virus.

Hers disease a hereditary glycogen storage disease in humans arising from a deficiency of the enzyme hepatic phosphorylase. It is inherited as an autosomal recessive trait with an incidence of 1/200,000.

het a partially heterozygous phage.

heterauxesis the relation of the growth rate of a part of a developing organism to the growth rate of the whole or of another part. If the organ in question grows more rapidly than the organism as a whole, it shows tachyauxesis; if less rapidly, bradyauxesis; if at the same rate, isauxesis. *See* allometry.

heteroalleles alternative forms of a gene that differ at nonidentical muton sites. Intragenic recombination between mutant heteroalleles can yield a functional cistron. *Compare with* homoalleles. *See* Appendix C, 1955, Pritchard; 1962, Henning and Yanofsky.

heterobrachial inversion pericentric inversion.

hetrocapsidic virus *See* segmented genome.

heterocaryon heterokaryon.

heterochromatin chromosomal material that, unlike euchromatin (*q.v.*) shows maximal condensation in nuclei during interphase. Chromosomal regions behaving in this way are said to show *positive heteropycnosis*. When entire chromosomes (like some Y chromosomes) behave this way, they are called *heterochromosomes*. In polytene chromosomes, heterochromatic regions adjacent to the centromeres of all chromosomes tend to adhere, forming a *chromocenter*. Heterochromatin is composed of repetitious DNA, is late to replicate, and is transcriptionally inactive. Such heterochromatic segments are often called *constitutive* to distinguish them from chromosomal segments or whole chromosomes that become dense and compact at different developmental stages. In such cases, one homologous chromosome may behave differently from the other. An example is the condensed, inactivated X chromosome characteristic of the diploid somatic cells of mammalian females. Such chromosomes are sometimes said to contain *facultative heterochromatin,* although there is no evidence that they contain a type of DNA different from euchromatin. The relative amounts of heterochromatin in each chromosome are shown in the *Drosophila melanogaster* entry. The pericentric heterochromatin is underreplicated in the polytene chromosomes of *Drosophila,* and this underreplication is under genetic control. *See* Appendix C, 1928, Heitz; 1936, Schultz; 1959, Lima-de-Faria; 1970, Pardue and Gall, 1998, Belyaeva *et al.*; centromere, euchromatin, position effects.

heterochromatin protein 1 (HP1) a nonhistone protein that binds to heterochromatin and to the telomeres of *Drosophila* polytene chromosomes. It is a highly conserved protein with homologs in yeast, plants, and mammals. HP1 is required for the proper segregation of chromosomes during mitosis in embryos. It can enter nuclei and bind to specific chromosomal sites where it causes localized condensation of DNA and represses the transcriptional activity of neighboring genes. The gene *Suppressor of variegation 205* encodes HP1. It is located on the left arm of chromosome 2 at 31.1. Loss of function mutant alleles are recessive lethals and cause end-to-end fusions of telomeres. Reduced function alleles cause variegation. *See* Appendix C,1986, James and Elgin.

heterochromatization the situation where euchromatic chromosome regions acquire the morphology of heterochromatin when juxtaposed to heterochromatin. *See* Appendix C, 1939, Prokofyeva-Belgovskaya; position effects

heterochromosome *See* B chromosomes, heterochromatin.

heterochronic genes genes whose products are required for the timing of specific developmental events. Mutations in these genes disrupt the temporal pattern of development and can also lead to phylogenetic variation. In *Caenorhabditis elegans* mutations in single heterochronic genes affect the timing of cell division and differentiation in different postembryonic cell lineages (without affecting other lineages), resulting in the generation of cell patterns normally associated with earlier or later developmental stages. Two of the heterochronic genes in *C. elegans, lin-4* and *let-7,* encode small temporal RNAs (*q.v.*), which control developmental timing by inhibiting the translation of target messenger RNAs. Other heterochronic genes affect the timing of neuroblast proliferation (*Drosophila*), scale cell maturation (*Papilio*), sporulation (*Dictyostelium*), post-embryonic shoot development (maize), and flower formation (*Arabidopsis*). In many eukaryotes the timing of many developmental events is controlled by hormones, and numerous genes affecting hormone function have been described. However, little is known about what controls the timing of hormone secretion. *See* Appendix C, 1984, Ambros and Horvitz; heterochrony.

heterochrony a change in the timing of a developmental event (e.g., time of appearance of a morphological trait or of expression of a gene) relative to other developmental event(s). Heterochrony has been well documented at the genetic and molecular level in *Caenorhabditis elegans,* where mutations in known heterochronic genes (*q.v.*) result in precocious or retarded development in specific postembryonic cell lineages. In addition to its role in regulating the temporal pattern of development, heterochrony is thought to be a major force for phylogenetic variation (e.g., variation between an ancestral and descendant species). A commonly cited example of heterochrony is neoteny (*q.v.*) in *Ambystoma mexicanum* (*q.v.*). The sexually mature adult of this aquatic salamander is a chimera (*q.v.*) consisting of adult and larval tissues (including gills), compared to the adult of the land-dwelling species. This is due to a genetic defect in thyroid hormone production, which retards the maturation of juvenile somatic tissues relative to gonadal tissues. The term *heterochrony* was originally coined by the German physician and biologist, Ernst Haeckel (1834–1919), but has evolved over the years to its present form.

heterocyclic referring to any organic compound forming a ring made up of carbon atoms and at least one atom other than carbon. Examples of heterocyclic amino acids are proline, hydroxyproline, tryptophan, and histidine (*see* amino acid). Purines and pyrimidines are heterocyclic compounds (*see* bases of nucleic acids). The porphyrin portion of heme is made up of heterocyclic rings (*see* heme).

heterodimer a protein made up of paired polypeptides that differ in their amino-acid sequences.

heterodisomy *See* disomy.

heteroduplex 1. a DNA generated during genetic recombination by base pairing between complementary single strands from different parental duplex molecules. **2.** a double-stranded nucleic acid in which each chain has a different origin and therefore is not perfectly complementary; e.g., the hybrid molecule generated by *in vitro* annealing of eukaryotic mRNA with its DNA. *See* Appendix C, 1969, Westmoreland *et al. See* R-loop mapping.

heteroecious referring to a parasite requiring two or more hosts to complete its life cycle, as with some rust fungi and insects.

heterofertilization double fertilization of angiosperms in which the endosperm and egg are derived from genetically different sperm nuclei.

heterogametic sex the sex that produces gametes containing unlike sex chromosomes (e.g., male mammals produce X- and Y-bearing sperm, usually in equal proportions). Crossing over is often suppressed in the heterogametic sex. *See* Appendix C, 1912, Morgan; 1913, Tanaka. *Compare with* homogametic sex.

heterogamy the alternation of bisexual reproduction with parthenogenetic reproduction.

heterogeneous nuclear RNA (hnRNA) the pool of extrachromosomal RNA molecules found in the nucleus, consisting of a heterogeneous mixture of primary transcripts, partly processed transcripts, discarded intron RNA, and small nuclear RNA. The term is often used to refer to the primary transcripts or to their modified products alone. *See* Appendix C, 1961, Georgiev and Mantieva; posttranscriptional processing, spliceosome, Usn RNAs.

heterogenetic antigens the same or similar (cross-reacting) antigens shared by several species (e.g., Forssman antigen, *q.v.*). Antibodies produced against one of these antigens will also react with the other antigens of the system even though these are derived from a different species; such antibodies are also called *heterophile antibodies*. *See* heterophile antigen.

heterogenote *See* heterogenotic merozygote.

heterogenotic merozygote a partially heterozygous bacterium carrying an exogenote containing alleles differing from those on the endogenote.

heterogony cyclical parthenogenesis, when one or more parthenogenetic generations alternate with an amphimictic one, usually in an annual cycle. Aphids, gall wasps, and rotifers are examples of animals undergoing heterogony.

heterograft a graft between members of different species; also known as a xenograft. *Compare with* allograft, isograft.

heterokaryon a somatic cell that contains nuclei derived from genetically different sources. The nuclei do not fuse, but divide individually and simultaneously to form new cells, as commonly occurs in fungal hyphae. Ciliate protozoans, such as *Tetrahymena*, may be functional heterokaryons, if they carry a drug-resistant gene in the transcriptionally inactive, diploid, germinal micronucleus and a drug-sensitive allele in the transcriptionally active, highly polyploid, somatic macronucleus. *See* interspecific heterokaryon.

heterokaryon test a test for organelle mutations based upon the appearance of unexpected phenotypes in uninucleate cells derived from specially marked heterokaryons. For example, suppose that heterokaryons form between colonies thought to carry a mitochondrial mutant (A) and colonies carrying a known nuclear mutation (B). If uninucleate progeny cells or spores exhibiting both the A and the B phenotypes can be derived from the heterokaryons, then the A mutation is probably in an extranuclear gene, because recombination of nuclear genes does not occur in heterokaryons.

heterokaryosis the condition in which fungus hyphae contain haploid nuclei of different genotypes, as a result of nonsexual fusion of the different types of hyphae.

heterokaryotypic referring to an individual carrying a chromosomal aberration in the heterozygous condition.

heterologous **1.** in immunology, referring to an antibody and an antigen that do not bind to one another; either may be said to be heterologous to the other. **2.** in transplantation studies, referring to a graft originating from a donor belonging to a species different from the host's. **3.** in nucleic acid studies, referring to a DNA of a different source from the rest. Thus, a rabbit hemoglobin gene used as a probe to detect a hemoglobin gene from a mouse gene library represents heterologous DNA. **4.** In transposon tagging (*q.v.*), the use of a transposon from one species to tag a gene from another—for example, using the maize *Activator* transposon to tag and clone a gene conferring disease resistance in tobacco.

heterologous chimera a chimera (*q.v.*) formed by cells or tissues from two different species.

heteromeric *See* heteropolymeric protein.

Heterometabola Hemimetabola (*q.v.*).

heteromixis in fungi, referring to the mating system where sexual reproduction involves the fusion of genetically different nuclei each from a different thallus. *Contrast with* homomixis. *See* Appendix C, 1904, Blakeslee.

heteromorphic bivalent a bivalent made up of chromosomes that are structurally different and consequently are only partly homologous (the XY bivalent is an example). *Contrast with* homomorphic bivalent.

heteromorphic chromosomes homologous chromosomes that differ morphologically.

heteromorphosis (*also* homeosis) the formation, whether in embryonic development or in regeneration, of an organ or appendage inappropriate to its site (for example, an antenna instead of a leg).

heteromultimeric protein *See* heteropolymeric protein.

heterophile antigen a substance that stimulates production in a vertebrate of antibodies capable of reacting with tissue components from other vertebrates or even from plants.

heteroplasmy the phenomenon where two distinct populations of mitochondria exist in the same invidiual. In those cases where a mutant in mtDNA is responsible for a disease, the larger the proportion of mutant mitochondria, the more likely the person will show symptoms of the disease. *See* Leber hereditary optic neuropathy (LHON), Leigh syndrome.

heteroplastic transplantation a transplantation between individuals of different species within the same genus.

heteroplastidy having two kinds of plastids, specifically, chloroplasts and starch-storing leukoplasts.

heteroploid referring in a given species to the chromosome number differing from the characteristic diploid number (or haploid number, if the species has a predominating haplophase).

heteropolymeric protein a protein made up of more than one kind of polypeptide (e.g., hemoglobin).

heteropycnosis referring to the appearance of chromosomes or chromosomal regions that have a coiling cycle out of phase with the rest of the genome. Positively heteropycnotic segments are more tightly coiled and negatively heteropycnotic segments are more loosely coiled than the rest of the chromosomal complement. *See* allocycly, isopycnotic.

heterosis the greater vigor in terms of growth, survival, and fertility of hybrids, usually from crosses between highly inbred lines. Heterosis is always associated with increased heterozygosity. *See* gibberellin.

heterosomal aberration *See* chromosomal aberration.

heterospory in plants, the existence (within the same species or within an individual organism) of two kinds of meiocytes (megasporocytes and microsporocytes) that produce two kinds of meiospores. *Compare with* homospory.

heterostyly a polymorphism of flowers that ensures cross-fertilization by producing flowers having stamens and styles of unequal lengths.

heterothallic fungus a fungal species producing a sexual spore that results from the fusion of genetically different nuclei which arose in different thalli. *Compare with* homothallic fungus.

heterotopic transplantation transplantation of tissue from one site to another on the same organism.

heterotrophs organisms that require complex organic molecules such as glucose, amino acids, etc., from which to obtain energy and to build macromolecules. *Contrast with* autotrophs.

heterozygosis heterozygosity.

heterozygosity the condition of having one or more pairs of dissimilar alleles.

heterozygote a diploid or polyploid individual that has inherited different alleles at one or more loci and therefore does not breed true. In diploids the heterozygote usually has a normal dominant allele and a mutant recessive allele and is sometimes called a *carrier*. *Contrast with* homozygote. *See* Appendix C, 1900, Bateson; compound heterozygote.

heterozygote advantage the situation where the heterozygote has a greater fitness than either homozygote. *See* cystic fibrosis, glucose-6-phosphate dehydrogenase deficiency, hemochromatosis, *Indy*, overdominance, sickle-cell trait.

HEXA *See* Tay-Sachs disease.

hexaploid a polyploid possessing six sets (6N) of chromosomes or genomes; e.g., bread wheat is thought to have originated by hybridizations involving three different species, each of which contributed two genomes to the allohexaploid. *See* wheat.

HEXB *See* Sandhoff disease.

hexosaminidase an enzyme functioning in the catabolism of gangliosides (*q.v.*). Hexosaminidase A is composed of alpha and beta subunits coded for by genes on human autosomes 15 and 5, respectively. Mutations at these loci result in Tay-Sachs and Sandhoff diseases (*q.v.*).

hexose monophosphate shunt *See* pentose phosphate pathway.

Hfr strain a strain of *Escherichia coli* that shows high frequencies of recombination (hence the abbreviation). In cells from such a strain, the F-factor is integrated into the bacterial chromosome. *See* Appendix C, 1953, Hayes; circular linkage map.

hGH human growth hormone (*q.v.*).

hGHR human growth hormone receptor (*q.v.*).

HGPRT hypoxanthine-guanine-phosphoribosyltransferase (*q.v.*); an enzyme involved in the *salvage pathway* (*q.v.*) of nucleotide synthesis. *See* HAT medium.

hibernate to be dormant during winter. Many mammals, reptiles, amphibians, and certain invertebrates hibernate. *See also* aestivate.

hierarchy an organization pattern involving groups within groups, as exemplified by the taxonomic hierarchy of organisms. *See* classification.

high-energy bond a covalent chemical bond (e.g., the terminal phosphodiester bond of adenosine triphosphate) that liberates at least 5 kcal/mol free energy upon hydrolysis.

high-energy phosphate compound a phosphorylated molecule that upon hydrolysis yields a large amount of free energy. *See* ATP, phosphate bond energy.

high frequency of recombination cell *See* Hfr strain.

highly repetitive DNA the fast component in reassociation kinetics, usually equated with satellite DNA. *See* repetitive DNA.

high-resolution chromosome studies analysis of the banding pattern of human chromosomes at prometaphase, when they are longer than at metaphase. Such analyses reveal 700–1200 bands and allow a more detailed localization of adjacent genes than is possible for metaphase chromosomes, where 300–600 bands are typically observed. *See* human chromosome band designations.

***Himalayan* mutant** an allele at the albino locus (with known examples in the mouse, rat, rabbit, guinea pig, hamster, and cat) associated with a very lightly pigmented body and somewhat darker extremities. The form of tyrosinase encoded by this gene is temperature sensitive and normally functions well only in the extremities where the body temperature is lower. Such animals raised in cold environments show darker pigmentation. *See* albinism, temperature-sensitive mutation, tyrosinase.

***Hin*dII** *See* restriction endonuclease.

hinge region *See* immunoglobulin.

hinny *See* horse-donkey hybrids.

his histidine. *See* amino acid.

histidine *See* amino acid.

histidinemia a hereditary disease in humans arising from a deficiency of the enzyme histidase.

histidine operon a polycistronic operon of *Salmonella typhimurium* containing nine genes involved in the synthesis of histidine. Fused genes from this operon encoded a fused enzyme. *See* Appendix C, 1970, Yourno, Kohno, and Roth.

histochemistry the study by specific staining methods of the distribution of particular molecules within sections of tissues. *See* Appendix C, 1825, Raspail.

histocompatibility antigen genetically encoded cell-surface alloantigens that can cause the rejection of grafted tissues, cells, and tumors bearing them. *See* Appendix C, 1937, Gorer.

histocompatibility gene a gene belonging to the major histocompatibility (MHC) system or to any of numerous minor histocompatibility systems responsible for the production of histocompatibility antigens (*q.v.*) *See* Appendix C, 1948, Snell.

histocompatibility molecules genetically encoded cell-surface alloantigens that can cause the rejection of grafted tissues, cells, and tumors bearing them. These cell-membrane glycoproteins are grouped into two classes. Class I molecules are found on the surfaces of all mammalian cells (except trophoblasts and spermatozoa). T lymphocytes (*q.v.*) of the $CD8^+$ subgroup recognize antigenic determinants of foreign class I histocompatibility molecules. Class II histocompatibility molecules are abundant on the surfaces of B lymphocytes (*q.v.*). T lymphocytes of the $CD4^+$ subgroup recognize antigenic determinants of foreign class II histocompatibility molecules. These T cells subsequently divide and secrete lymphokines (*q.v.*), which are important for B cell growth and differentiation. Class I histocompatibility molecules are heterodimers made up of heavy (alpha) and light (beta) polypeptide chains. The class I chains are encoded by genes residing in the right portion of the HLA complex (*q.v.*). The alpha chain contains regions showing sequence diversity, whereas the beta chain has an invariant amino acid composition. The class II histocompatibility molecules are also dimers composed of heavy alpha and light beta chains. These are encoded by genes in the left portion of the HLA complex. Most of the sequence diversity of class II histocompatibility molecules is localized within a segment of the beta chain. Class II histocompatibility dimers are associated with a third polypeptide chain that exhibits no polymorphism. *See* Appendix C, 1937, Gorer; 1987, Wiley *et al.*; major histocompatibility complex.

histogenesis the development of histologically detectable differentiation.

histogenetic antigens or responses defined by means of cell-mediated immunity.

histogram a bar graph.

histoincompatibility intolerance to transplanted tissue.

histology the study of tissues.

histolysis tissue destruction.

histone genes in both the sea urchin and *Drosophila*, these genes are repetitive and clustered. In *Strongylocentrotus purpuratus*, the genes for H4, H2B, H3, H2A, and H1 lie in a linear sequence. They are transcribed from the same DNA strand to form a polycistronic message, and are separated by DNA spacers of similar lengths. In *Drosophila* there are about 110 copies of the histone genes, and these are localized in a four-band region in the left arm of chromosome 2. The gene sequence is *H3, H4, H2A, H2B, H1*. Two genes are transcribed from one DNA strand, three from the other. The gene order in *Xenopus laevis* is the same as in *Drosophila*. In humans, histone genes show less clustering, and they are located on a variety of chromosomes (1, 6, 12, and 22). *Notophthalmus viridescens* (*q.v.*) has 600 to 800 copies of the histone gene repeat. Transcribing histone genes have been detected in loops at specific loci on lampbrush chromosomes 2 and 6. Most histone genes lack introns (*q.v.*). *See* Appendix C, 1972, Pardue *et al.*; 1981, Gall *et al.*; *Strongylocentrotus purpuratus*.

histones small DNA-binding proteins. They are rich in basic amino acids and are classified according to the relative amounts of lysine and arginine they contain (see table below). Histones are conserved during evolution. For example, H4 of calf thymus and pea differ at only two sites. The nucleosome (*q.v.*) contains two molecules each of H2A, H2B, H3, and H4. A single H1 molecule is bound to the DNA segment that lies between nucleosomal cores. The histone H1 molecules are responsible for pulling nucleosomes together to form a 30 nm fiber. It is this fiber that is seen when nuclei are lysed and their contents are viewed under the electron microscope. During the maturation of sperm heads, the histones break down and are replaced with protamines (*q.v.*). *See* Appendix C, 1884, Kossel; 1974, Kornberg; 1977, Leffak *et al.*; CENP-A, chromatin fibers, chromatosome, histone genes, nucleoprotein, nucleosome, *Polycomb (Pc)*, solenoid structure, ubiquitin.

hitchhiking the spread of a neutral allele through a population because it is closely linked to a beneficial allele and therefore is carried along as the gene that is selected for increases in frequency.

HIV the *h*uman *i*mmunodeficiency *v*irus, a human RNA retrovirus known to cause AIDS. HIV is now referred to as HIV-1, to distinguish it from a related but distinct retrovirus, HIV-2, which has been isolated from West African patients with a clinical syndrome indistinguishable from HIV-induced AIDS. Each viron contains two molecules of single-stranded, positive sense RNA (held together by hydrogen bonds to form a dimer). Each linear molecule is made up of 9.3×10^3 nucleotides. It contains *gag*, *pol*, and *env*, the same genes first observed in the Rous sarcoma virus (*q.v.*). There are seven other

Histones

HISTONE SYNONYMS	MOLECULAR WEIGHT (Da)	TOTAL AMINO ACIDS	% LYSINE	% ARGININE	RELATIVE AMOUNT PER 200 bp DNA
H1 = I = F1	21,000	207	27	2	1
H2A = IIbl = F2A2	14,500	129	11	9	2
H2B = IIb2 = F2B	13,700	125	16	6	2
H3 = III = F3	15,300	135	10	15	2
H4 = IV = F2A1	11,300	102	10	14	2

genes as well, and some encode multiple proteins. *See* Appendix C, 1983, Montaigner and Gallo; AIDS, enzyme-linked immunosorbant assay, hairpin ribozyme, retroviruses, RNA vector, simian immunodeficiency viruses (SIVs).

HLA human leukocyte antigens concerned with the acceptance or rejection of tissue or organ grafts and transplants. These antigens are on the surface of most somatic cells except red blood cells, but are most easily studied on white blood cells (hence the name). *See* Appendix C, 1954, Dausset.

HLA complex the major histocompatibility gene complex of humans. The complex occupies a DNA segment about 3,500 kilobase pairs long on the short arm of chromosome 6. The portion of the HLA complex closest to the telomere contains the genes that encode the class I histocompatibility molecules (HLA-B, -C, and -A). The portion closest to the centromere contains the genes encoding the class II histocompatibility molecules (DP, DQ, and DR). Genes encoding components of the complement (*q.v.*) system lie in the midregion of the complex. *See* major histocompatibility complex (MHC).

***H* locus** in humans, a genetic locus that encodes a fucosyl transferase enzyme that is required during an early step in the biosynthesis of the antigens of the ABO blood group system. *See* A, B antigens, Bombay blood group.

hnRNA heterogeneous nuclear RNA (*q.v.*).

hog *See* swine.

Hogness box a segment 19–27 base pairs upstream from the startpoint of eukaryotic structural genes to which RNA polymerase II binds. The segment is 7 base pairs long, and the nucleotides most commonly found are TATAAAA; named in honor of David Hogness. *See* canonical sequence, Pribnow box, promoter, TATA box-binding protein.

holandric appearing only in males. Said of a character determined by a gene on the Y chromosome. *See* hologynic.

holism a philosophy maintaining that the entirety is greater than the sum of its parts. In biology, a holist is one who believes that an organism cannot be explained by studying its component parts in isolation. *See* reductionism.

Holliday intermediate *See* Holliday model.

Holliday model a model that describes a series of breakage and reunion events occurring during crossing over between two homologous chromosomes. The diagram on page 209 illustrates this model. In (a) the duplex DNA molecule from two nonsister strands of a tetrad have aligned themselves in register so that the subsequent exchange does not delete or duplicate any genetic information. In (b) strands of the same polarity are nicked at equivalent positions. In (c) each broken chain has detached from its partner and paired with the unbroken chain in the opposite duplex. In (d) ligases have joined the broken strands to form an internal branch point. The branch is free to swivel to the right or left, and so it can change its position. This movement is called *branch migration*, and in (e) the branch is shown as having moved to the right. In (f) the molecule is drawn with its arms pulled apart, and in (g) the a b segment has been rotated 180° relative to the A B segment. The result is an x-shaped configuration, and in it one can readily see the four sectors of single-stranded DNA in the branch region. To separate the structure into two duplexes, cuts must be made across the single strands at the branch region. The cuts can occur horizontally (h) or vertically (i). In the horizontal case, the separated duplexes will appear as shown in (j), and when the nicks are ligated, the duplexes will each contain a patch in one strand, as shown in (k). In the vertical case, the separated duplexes will appear as shown in (l), and when the nicks are ligased, the duplexes will contain splices in every strand, as shown in (m). It is only in this case that single crossover chromatids (Ab and aB) will be detected. Structures similar to the molecule drawn in (g) have been observed under the electron microscope and have been named *Holliday intermediates*. *See* Appendix C, 1964, Holliday; chi structure, homologous recombination.

holoblastic cleavage cleavage producing cells of approximately equivalent size.

Holocene the epoch of the Quaternary period from the end of the Pleistocene to the present time. Neolithic to modern civilization. *See* geologic time divisions.

holocentric referring to chromosomes with diffuse centromeres. *See* centromere.

holoenzyme the functional complex formed by an apoenzyme (*q.v.*) and its appropriate coenzyme (*q.v.*).

hologynic appearing only in females. Said of a trait passed from a P_1 female to all daughters occurring, for example, in the case of a gene linked to the W chromosome (*q.v.*). *See* holandric, matrocliny.

Holometabola the superorder of insects containing species that pass through a complete metamorphosis.

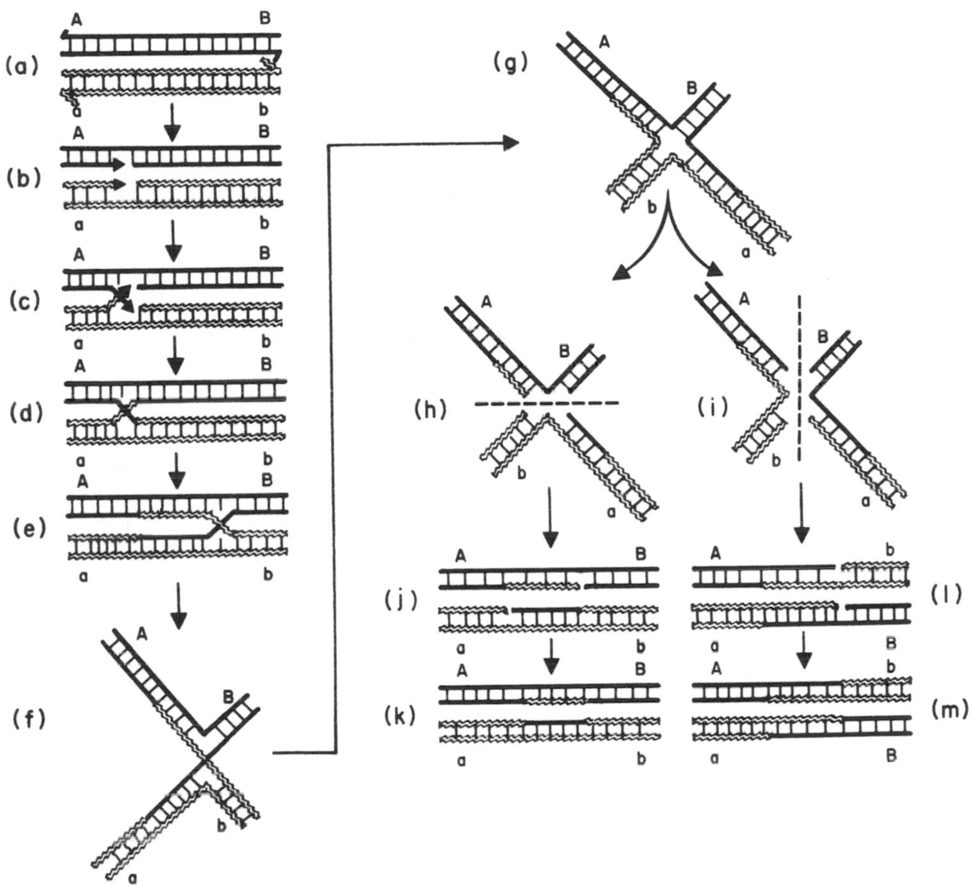

Holliday model

holometabolous referring to those insects in which larval and pupal stages are interposed between the embryo and the adult. *Contrast with* hemimetabolous.

holophyletic an evolutionary lineage consisting of a species and all of its descendants.

holophytic nutrition nutrition requiring only inorganic chemicals, as that of photosynthetic plants.

holorepressor *See* aporepressor.

holotype the single specimen selected for the description of a species.

holozoic nutrition nutrition requiring complex organic foodstuffs, as that of organisms other than photosynthetic plants and protoctists.

homeobox a sequence of about 180 base pairs near the 3′ end of certain homeotic genes. The 60 amino acid segment encoded by the homeobox is a DNA-binding domain with a helix-turn-helix motif (*q.v.*). Homeobox proteins can bind to and regulate the transcription of genes that contain homeobox responsive elements (HREs). Homeobox proteins can also regulate translation by binding to mRNAs that also contain HREs. *See* Appendix C, 1984, McGinnis *et al.*; 1984, Shepard *et al.*; 1989, Qian *et al.*; 1990, Malicki *et al.*; 1996, Dubnau and Struhl, Rivera-Pomar *et al.*; *Antennapedia, apetala-2, bicoid, bithorax, engrailed,* floral identity mutations, metamerism, *proboscipedia, spineless-aristapedia*, homeotic mutations, *Hox* genes, *RAG-1* and *RAG-2*.

homeodomain synonym for homeobox.

homeologous chromosomes *See* homoeologous chromosomes.

homeoplastic graft (*also* homoeoplastic) a graft of tissue from one individual to another of the same species.

homeosis (*also* homoeosis) heteromorphosis (*q.v.*).

homeostasis (*also* homoeostasis) a fluctuation-free state. *See* developmental homeostasis.

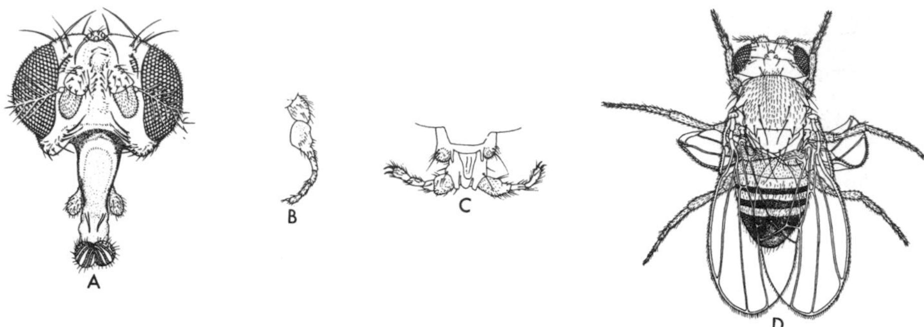

Homeotic mutations

homeotic mutations those in which one developmental pattern is replaced by a different, but homologous one. The homeotic mutations of *Drosophila* cause an organ to differentiate abnormally and to form a homologous organ that is characteristic of an adjacent segment. Three such mutations are illustrated above. (A) A frontal view of the normal head. (B) The leg-like antenna of an *ssa* mutant, (C) A *pb* mutant with its proboscis transformed into legs. (D) A *bx* male in which halteres are changed into wing-like appendages. *Bithorax* was the first homeotic mutation to be discovered. Homeotic genes were subsequently shown to control segment identity and were found to contain conserved segments called *homeoboxes* (*q.v.*). Homeotic mutations in which one floral organ is replaced by another occur in plants. *See* Appendix C, 1915, Bridges; 1978, Lewis; 1983, Hafen, Levine, and Gehring; 1984, McGinnis *et al.*; 1996, Krizek and Meyerowitz; *Arabidopsis thaliana*, *bithorax*, floral identity mutations, *Hox* genes, metamerism, *proboscipedia*, *spineless-aristapedia*.

homing endonuclease a site-specific DNA endonuclease, usually encoded by an open reading frame (*q.v.*) within an intron (*q.v.*) or an intein (*q.v.*) sequence, which mediates the horizontal transfer of the DNA sequence that encodes it to a new genomic location. It does this by introducing double-stranded breaks at or near the insertion site (*home*) of an intron or intein DNA in alleles that lack these intervening genetic elements. Homing endonucleases can be distinguished from other site-specific endonucleases based on their structural and biochemical properties.

hominid a member of the family Hominidae including humans and related fossil species. Two genera are recognized: *Australopithecus* and *Homo*.

hominoid a member of the superfamily of primates, the Hominoidea. Its living species include humans, two species of chimpanzee, the gorilla, the orangutan, and nine species of gibbon. *See* Appendix C, 1967, Sarich and Wilson; Cercopithecoidea.

Homo a genus in which humans are placed. It contains two fossil species: *H. habilis*, which lived 2.3 to 1.5 million years ago, and *H. erectus*, which lived 1.5 to 0.3 million years ago. A third fossil species, *Homo neandertalensis*, became extinct about 30,000 years ago. It overlapped *Homo sapiens* (*q.v.*), which dates from about 300,000 years ago to the present. *See* Neandertal.

homoalleles alternative forms of a gene that differ at the same muton site. Intragenic recombination between homoalleles is not possible. *Compare with* heteroalleles.

homoallelic referring to allelic mutant genes that have mutations at the same site (*q.v.*). A functional cistron cannot be generated by intragenic recombination between homoalleles. *Contrast with* heteroallelic.

homobrachial inversion paracentric inversion (*q.v.*)

homocaryon homokaryon (*q.v.*).

homocystinuria a hereditary disease in man arising from a deficiency of the enzyme serine dehydratase.

homodimer a protein made up of paired identical polypeptides.

homoeologous chromosomes chromosomes that are only partially homologous. Such chromosomes are derived from ancestral chromosomes that are believed to have been homologous. Evolutionary divergence has reduced the synaptic attraction of homoeologs. *See* Appendix C, 1958, Okamoto; differential affinity, isoanisosyndetic alloploid.

homoeosis homeosis. *See also* homeoplastic graft, homeostasis, homeotic mutations.

homoeotic mutations homeotic mutations (*q.v.*).

homogametic sex the sex that produces gametes all of which carry only one kind of sex chromosome; e.g., the eggs of female mammals carry only an X chromosome. *Compare with* heterogametic sex.

homogamy the situation in which the male and female parts of a flower mature simultaneously.

homogenote a partial diploid (merozygote) bacterium in which the donor (exogenote) chromosomal segment carries the same alleles as the chromosome of the recipient (endogenote) cell.

homogentisic acid a compound derived from the metabolic breakdown of the amino acid tyrosine. *See* alkaptonuria.

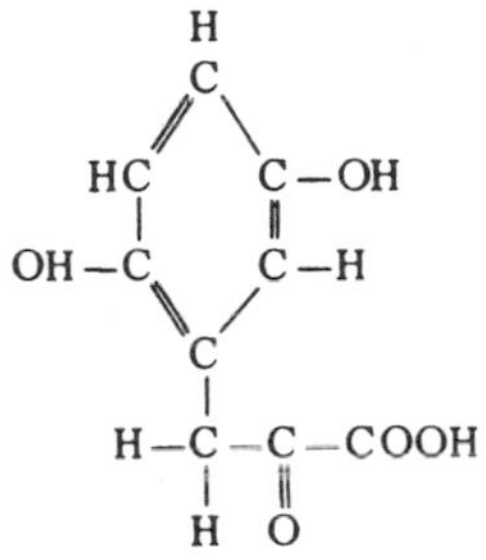

```
          H
          C
         // \
      HC      C—OH
      |       ||
  OH—C        C—H
       \\    /
          C
          |
     H—C—C—COOH
       |   ||
       H   O
```

homograft homeoplastic graft (*q.v.*).

homoimmunity the resistance of a lysogenic bacterium (harboring a prophage, *q.v.*) to superinfection by phage of the same kind as that carried in the prophage state. The excess repressor molecules generated by the prophage bind to operators on the infecting DNA molecules and prevent their transcription.

homokaryon a dikaryotic mycelium in which both nuclei are of only one genotype.

homokaryotypic referring to an individual carrying a chromosomal aberration in the homozygous condition.

homolog **1.** in classification, a character that defines a clade. **2.** in evolution, homologs are characteristics that are similar in different species because they have been inherited from a common ancestor. **3.** in cytology, *See* homologous chromosomes.

homologous referring to structures or processes in different organisms that show a fundamental similarity because of their having descended from a common ancestor. Homologous structures have the same evolutionary origin although their functions may differ widely: e.g., the flipper of a seal and the wing of a bat. *See* analogous.

homologous chromosomes chromosomes that pair during meiosis. Each homolog is a duplicate of one of the chromosomes contributed at syngamy by the mother or father. Homologous chromosomes contain the same linear sequence of genes and as a consequence each gene is present in duplicate. *See* synaptonemal complex.

homologous recombination the exchange or replacement of genetic material as a result of crossing over (*q.v.*) or gene conversion (*q.v.*), respectively. Homologous recombination occurs between two long stretches of DNA with similar sequences, which may be present in two homologous chromosomes, in non-homologous chromosomes, or within a chromosome. Meiotic and mitotic recombination (*q.v.*) are examples of homologous recombination. *Compare with* site-specific recombination. *See* Holliday model.

homologue alternative spelling of homolog (*q.v.*).

homology the state of being homologous. In molecular biology, the term is often misused when comparing sequences of nucleotides or amino acids from nucleic acids or proteins obtained from distantly related species. In such instances it is preferable to refer to sequence identities or similarities rather than "homologies."

homomeric protein referring to a protein made up of two or more identical polypeptide chains. An example would be beta galactosidase (*q.v.*), which is an aggregate of four identical polypeptides.

homomixis referring in fungi to the mating system in which sexual reproduction involves the fusion of genetically similar nuclei derived from one thallus.

homomorphic bivalent a bivalent made up of homologues of similar morphology. *See* heteromorphic bivalent.

homomultimer *See* homopolymer.

homoplasy parallel or convergent evolution; structural similarity in organisms not due directly to inheritance from a common ancestor or development from a common anlage.

homopolar bond covalent bond (*q.v.*).

homopolymer a polymer composed of identical monomeric units (poly U, for example).

homopolymer tails a segment containing several of the same kind of deoxyribonucleotides arranged in tandem at the 3′ end of a DNA strand. *See* Appendix C, 1972, Lobban and Kaiser; terminal transferase.

Homo sapiens the species to which human beings belong. The scientific name was coined in 1758 by Linnaeus. The estimated genome size is 3.2×10^9 base pairs. *See* Appendix A, Chordata, Mammalia, Primates; Appendix C, 1735, Linné; 1966, McKusick; 1991, Ijdo *et al.*; Appendix E, Appendix F; hominid, Hominoidea, human chromosome band designations, human gene maps, human genetic databases, Human Genome Project, human mitotic chromosomes, shotgun sequencing, symbols used in human cytogenetics.

homosequential species species with identical karyotypes, as found in many species of *Drosophila* endemic to the Hawaiian Islands.

homosomal aberration *See* chromosomal aberration.

homospory in plants of both mating types or sexes, the production of meiospores of equivalent size. *Compare with* isogamy.

homothallic fungus a fungal species producing a sexual spore that results from the fusion of genetically different nuclei derived from the same thallus. *Compare with* heterothallic fungus.

homozygosity the condition of having identical alleles at one or more loci in homologous chromosome segments.

homozygote an individual or cell characterized by homozygosity. *See* Appendix C, 1900, Bateson.

homozygous having identical rather than different alleles in the corresponding loci of homologous chromosomes and therefore breeding true. *See* heterozygosity.

homunculus a miniature individual imagined by early biologists to be present in a sperm.

Hoogsteen base pairs *See* hydrogen bond.

hopeful monster *See* saltation.

***Hordeum vulgare* L.** barley, a cereal grain domesticated about 10,000 years ago in the fertile crescent (*q.v.*). It is the fourth most important cereal crop (coming after wheat, maize, and rice). Its haploid chromosome number is 7, and its genome size is 4.8 gigabase pairs. Malted barley is used in the brewing industry, and food-grade malts are components in the formulation of baked goods and breakfast cereals. *See* Appendix A, Plantae, Tracheophyta, Angiospermae, Monocotolyledoneae, Graminales.

horizontal classification a system of evolutionary classification that tends to unite transitional forms with their ancestors; the opposite of *vertical classification* (*q.v.*).

horizontal mobile elements (HMEs) genetic elements capable of horizontal transmission (*q.v.*). Such HMEs enter host cells during conjugation or infection and subsequently insert their DNAs into the chromosomes of the hosts. In prokaryotes, examples of HMEs are F factors, R (resistance) plasmids, and prophages (*all of which see*). The transposable elements (*q.v.*) first identified in corn are examples of eukaryotic HMEs. The Ti plasmid (*q.v.*) can accomplish transkingdom DNA exchanges. *See* Appendix C, 1977, Lawrence and Oschman; episome, selfish DNA, sympatric speciation.

horizontal transmission the transfer of genetic information from one cell or individual organism to contemporary cells or organisms by an infection-like process, in contrast to vertical transmission (*q.v.*). *See* Appendix C, 1999, Nelson *et al.*; *Bacillus*, mariner elements, P elements, *Wolbachia.*

hormone an organic compound produced in one part of an organism and transported to other parts, where it exerts a profound effect. Mammalian hormones include ACTH, endorphins, epinephrine, FSH, glucagon, GH, LH, insulin, intermedin, oxytocin, progesterone, prolactin, secretin, somatostatin, testosterone, thyroxin, and vasopressin. *See also* allatum hormones, ecdysterone, endocrine system, plant growth regulators.

hormone receptors molecules located on the surface or within target cells to which hormones bind. When such receptors are absent or defective, the hormones are without effect. Examples of genetic diseases caused by mutations of genes that encode hormone receptors are *androgen-insensitivity syndrome* (*q.v.*) and *vitamin D-resistant rickets* (*q.v.*). *See* androgen receptor, steroid receptor, vitamin D receptor.

horotelic evolution *See* evolutionary rate.

horse any of a number of domesticated breeds of the species *Equus caballus*. Popular breeds include: DRAFT HORSES: Belgian, Clydesdale, Percheron, Shire, Suffolk. COACH HORSES: Cleveland Bay, French Coach, German Coach, Hackney. LIGHT HARNESS HORSES: American Trotter. SADDLE HORSES American Saddle Horse, American Quarter Horse, Appaloosa, Arabian, Morgan, Palomino, Tennessee

Walking Horse, Thoroughbred, Lippizzaner. PONIES: Hackney Pony, Shetland Pony, Welsh Pony. *See* Equidae.

horse bean broad bean (*q.v.*).

horse-donkey hybrids the horse female X donkey male cross produces a mule; the reciprocal cross produces a hinny. These hybrids have 63 chromosomes and their mitochondria are of maternal origin. Both mules and hinnies are sterile, but mules have been successfully cloned. The procedure involved implanting a nucleus from a mule fetus into the enucleated egg from a horse. Therefore cloned mules also have horse mitochondria. *See* Appendix C, 1974, Hutchison; *Equus*, donkey.

horseradish peroxidase the most widely studied of the peroxidase enzymes that utilize hydrogen peroxide as an oxidant in the dehydrogenation of various substrates. The reaction catalyzed is $AH_2 + H_2O_2 \rightarrow A + 2H_2O$, where A is the substrate. Horseradish peroxidase reacts with a variety of organic reagents to produce a colored product. So when this enzyme is covalently linked to a reporter antibody, the position or amount of the complementary antigen can be determined by the location or amount of the dye generated. *See* enzyme linked immunosorbant assay.

host 1. an organism infected by a parasite. 2. the recipient of a graft from a donor individual.

host-cell reactivation cut-and-patch repair (*q.v.*) of UV-induced lesions in the DNAs of bacteriophages once they infect a host cell. Host-cell reactivation does not occur in the case of ssDNA or RNA viruses.

host-controlled restriction and modification *See* DNA restriction enzyme, restriction and modification model.

host range the spectrum of strains of a species of bacterium that can be infected by a given strain of phage. The first mutations to be identified in phage involved host range. The term is used more generally to refer to the group of species that can be attacked by a given parasite. Some bacteria can parasitize certain mammals, but only at specific developmental stages. For example, the K99 strain of *E. coli* can infect calves, lambs, and piglets, but not adult cattle, sheep, and swine. The resistance shown by the adult farm animals is due to the replacement of cell surface receptors by molecules to which the bacterium does not bind. *See* Appendix C, 1945, Luria; Duffy blood group gene.

host-range mutation a mutation of a phage that enables it to infect and lyse a previously resistant bacterium.

hot spot 1. a site at which the frequencies of spontaneous mutation or recombination are greatly increased with respect to other sites in the same cistron. Examples are in the *rII* gene of phage T4 and in the *lacZ* and *trpE* genes of *E. coli*. 2. a chromosomal site at which the frequencies of mutations are differentially increased in response to treatment with a specific mutagen. *See* Appendix C, 1961, Benzer; 5-methyl cytosine, muscular dystrophy.

hot spot archipelago a chain of islands formed when a tectonic plate moves over a stationary plume of lava arising from the earth's deep mantle. For example, the Hawaiian archipelago was formed as the Pacific tectonic plate moved northwestward (at a rate estimated to be about 9 cm per year) over just such a hot spot. This erupted periodically, perforating the plate with volcanoes that grew until some of them rose above sea level to form an island chain. Kauai was the first of the large islands to be formed, 5 million years ago. Hawaii, the youngest island, was formed about 500,000 years ago, and its three active volcanoes sit above the hot spot. An undersea volcano, Loihi, is also erupting 30 km off shore on the southeast flank of Mauna Loa.

The hot spot hypothesis, put forth in 1963 by J. Tuzo Wilson, is now the main unifying theory that explains the origin of many groups of oceanic islands worldwide. The Galapagos islands (*q.v.*) are also a hot spot archipelago. They began as a group of submarine volcanoes that grew progressively from the ocean floor until about 4.5 million years ago, when they emerged above sea level. The northwest islands of Fernandina and Isabela, which currently lie above the hot spot, have active volcanoes. *See* plate tectonics.

housekeeping genes constitutive loci that are theoretically expressed in all cells in order to provide the maintenance activities required by all cells: e.g., genes coding for enzymes of glycolysis and the citric acid cycle.

***Hox* genes** genes that contain *ho*meobo*x*es (*q.v.*). The *Hox* genes of most animals have three properties: (1) the genes are arranged along a chromosome in tandem arrays, (2) they are expressed sequentially along the chromosome, and (3) the sequence of activation of the genes corresponds to the relative order in which the genes are expressed in tissues along the main anterior-posterior axis of the organism. The situation is illustrated for some animal species that show bilateral symmetry (*see* Appendix A, Eumeta-

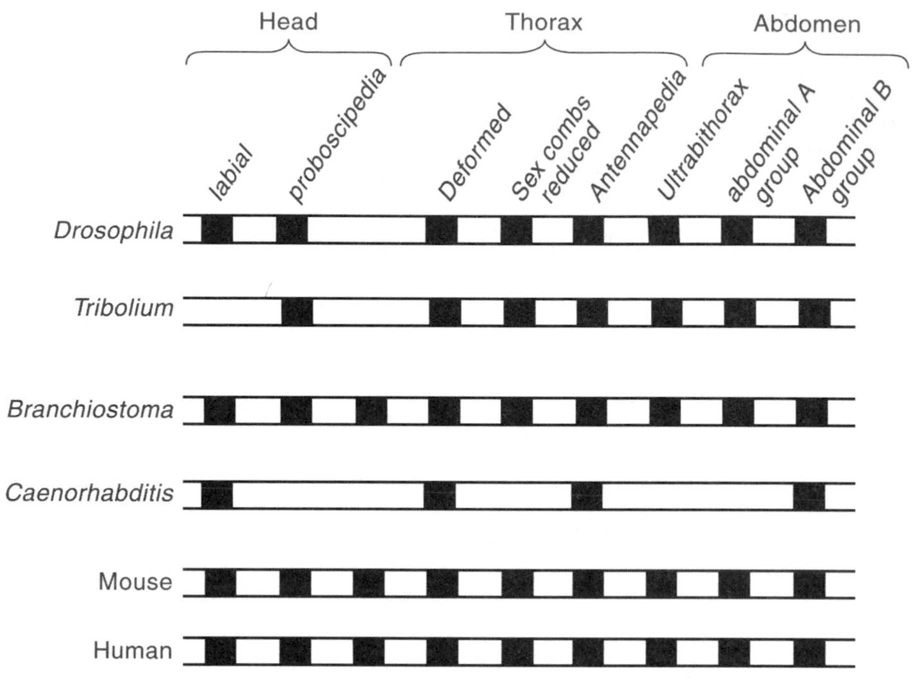

***Hox* genes**

zoa, Bilateria). In the case of *Drosophila melanogaster*, where *Hox* genes were discovered, there are eight such genes, and they all reside on chromosome 3. The genes *lab, pb, Dfd, Scr,* and *Antp* belong to the *Antennapedia complex*, while farther down the chromosome are the genes of the *bithorax complex* (*Ubx, abd-A,* and *Abd-B*). The location of tissues along the anterior-posterior axis of the embryo where the genes are active is shown at the top of the figure. Several *Hox* genes also occur in a similar sequence on specific chromosomes of *Tribolium, Branchiostoma,* and *Caenorhabditis* (*all of which see*). In mammals like the laboratory mouse and humans there are 38 or 39 *Hox* genes grouped in clusters on 4 different chromosomes. The locations of nine genes of the Hox-2 complex on mouse chromosome 11 are shown below the homologous genes from the fly. Moving from left to right in the mouse Hox-2 complex, each gene is expressed progressively later in development. Human chromosome 7 contains the Hox-1 cluster of nine genes, which are homologs of the aligned genes of the Hox-2 cluster of the mouse. *See* Appendix C, 1978, Lewis; 1983, Hafen, Levine, and Gehring; 1984, McGinnis *et al.*; *Antennapedia, bicoid, bithorax,* floral identity mutations, homeotic mutations, *polycomb*, segment identity genes, selector genes.

HP1 heterochromatin-associated protein 1 (*q.v.*).

HPHF hereditary persistence of hemoglobin F.

HPRL human prolactin. *See* human growth hormone.

HPRT hypoxanthine-guanine-phosphoribosyl transferase (*q.v.*).

H substance a precursor polysaccharide for production of A and B antigens of the ABO blood group system. It is usually unmodified on group O cells, but is modified by the addition of different sugars to produce the A or B antigens. The molecule is specified by a gene on human chromosome 19. *See* Bombay blood group.

hts *See hu-li tai shao (hts).*

hu-li tai shao (hts) a second chromosomal gene in *Drosophila melanogaster* which encodes isoforms of adducin-related protein. Some are expressed during pre-adult development (e.g., in the embryonic central nervous system and epidermis and in the digestive system of embryos and larvae). Two isoforms are expressed in the germ line. One localizes in the spectrosome (*q.v.*) and the fusome (*q.v.*) and is homologous to mammalian adducin (*q.v.*), while the other coats ring canals (*q.v.*). Female sterile mutations of *hts* eliminate the formation of spectrosomes and fusomes, and produce egg chambers with reduced numbers of nurse cells, which often lack oocytes (hence the name *hu-li tai shao*, which means *too little nursing* in Chinese). *See* cystocyte divisions.

human adenovirus 2 (HAdV-2) an adenovirus (*q.v.*) that causes respiratory infections in children. The genome is a double-stranded DNA molecule that contains 35,937 bp and includes 11 transcription units. Split genes (*q.v.*) and alternative splicing (*q.v.*) were found in HAdV-2.

human chromosome band designations quinacrine and Giemsa-stained human metaphase chromosomes show characteristic banding patterns, and standard methods have been adopted to designate the specific patterns displayed by each chromosome. The X chromosome shown here illustrates the terminology. In the diagram, the dark bands represent those regions that fluoresce with quinacrine or are darkened by Giemsa. The short (p) arm and the longer (q) arm are each divided into two regions. In the case of longer autosomes, the q arm may be divided into three or four regions and the p arm into three regions. Within the major regions, the dark and light bands are numbered consecutively. To give an example of the methods used for assigning loci, the *G6PD* gene is placed at q28, meaning it is in band 8 of region 2 of the q arm. The color-blindness genes are both assigned to q27–qter. This means they reside somewhere between the beginning of q27 and the terminus of the long arm. *See* Appendix C, 1970, Casperson, Zech, and Johansson; 1971, O'Riordan *et al.*; high resolution chromosome studies, human mitotic chromosomes.

human cytogenetics *See* human mitotic chromosomes, symbols used in human cytogenetics.

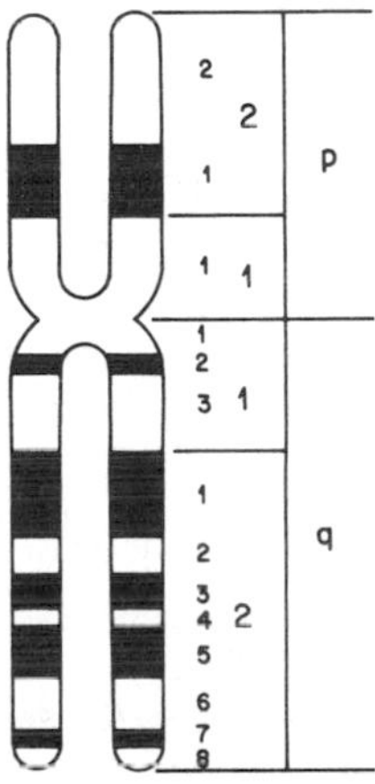

human cytomegalovirus (HCMV) a giant virus (240 nm in diameter) belonging to the Herpesviridae. Most of the congenital viral infections in humans are caused by this virus. The genome is a 230 kbp linear dsDNA in which reside at least 208 ORFs. HCMV is covered by a phospholipid envelope. Between this and the capsid is a compartment (the tegument) which contains 20 different proteins encoded by the viral genome. Four different mRNAs are packaged inside the virion. *See* Appendix F, virus.

human gene maps the total human genome contains 3.2 Gbp of DNA, and 2.95 Gbp of this represents euchromatin (*q.v.*). There are about 31,000 genes that transcribe mRNAs and at least 750 genes that encode other RNAs. For example, there are about 500 different tRNA genes. The genes are distributed over 22 pairs of chromosomes that range in size from autosome 1 (263 Mbp) to autosome 21 (50 Mbp). The X and Y chromosomes are 164 and 59 Mbp, respectively. The most gene-rich chromosome is autosome 19 (with 23 genes/Mbp), while the least gene-rich chromosomes are autosome 13 and the Y (each with 5 genes/Mbp). Genes (or at least their coding regions) make up only 1–2% of the genome. Just over 40% of the genes that encode proteins have orthologs in *Drosophila* and *Caenorhabditis*. Hundreds of human genes have resulted by horizontal transfer from bacteria at some point during the evolution of vertebrates. More than half of the genome is composed of repetitious DNA (*q.v.*). About 45% of the genome is derived from transposable elements (*q.v.*). Although the human genome has only about twice as many protein-coding genes as *Drosophila* or *Caenorhabditis*, human genes are more complex and often undergo alternative splicing (*q.v.*) to produce an array of different transcripts. More than 1,000 genes that cause specific diseases when they mutate have been mapped and are listed in OMIM (*q.v.*). The human mitochondrial chromosome is sometimes referred to as chromosome 25 or M. *See* Appendix C, 2001, Collins and Venter *et al.*; 2003, International Human Genome Sequencing Consortium; Down syndrome, gene, genome annotation, horizontal transmission, human mitochondrial DNA, repetitious DNA.

human genetic databases *See* Appendix E.

human genetic diseases maladies caused by defects in the human genome. A *single-gene disorder* is caused by a single mutant gene (see the table on page 172). *Chromosomal disorders* arise from disruptions in development produced by an excess or deficiency of whole chromosomes or chromosome segments (e.g., Down syndrome). Chromosomal disorders usually do not run in families because their effects are so detrimental or lethal. They affect about seven individuals per thousand births and account for about half of all spontaneous first-trimester abortions. *Polygenic disorders* are determined by combined action of a number of genetic loci, each with a small effect. Single-gene disorders are cataloged in two online databases. *See* GeneCards, OMIM.

Human Genome Project a successful effort by an international coalition of scientists to complete the sequence and assembly of the human genome. Evaluating the evolution and functioning of these genes is currently under way. Once groups led independently by Francis Collins and J. Craig Venter finished a draft sequence of the genome, a celebration was held on June 26, 2000, in the East Room of the White House. In his opening remarks President Clinton praised the draft saying "without a doubt, this is the most wondrous map ever produced by mankind." By mid-February of 2001, annotated maps were published in *Nature* and *Science*, and The International Human Genome Sequencing Consortium published the completed sequence in 2003. *See* Appendix C, 2001, Collins and Venter *et al.*; 2003, International Human Genome Sequencing Consortium; Celera Genomics, DNA sequencers, shotgun sequencing.

human growth hormone (hGH) a protein (also called somatotropin) secreted by cells located in the anterior lobe of the pituitary gland. The hGH molecule folds upon itself and is held together by two disulfide bridges. There are two other hormones that have structures similar to hGH. The first is chorionic somatomammotropin (hCS). Both hGH and hCS contain 191 amino acids, and hCS has two disulfide bridges in the same positions as hGH. There is an 85% identity between the amino acid sequences of hGH and hCS. The second hormone is prolactin (hPRL). It contains 199 amino acids, and the sequence identities are lower (35% with hGH and 13% with hCS). The hypothalamus (*q.v.*) produces two hormones that modulate the secretion of hGH. Secretion is promoted by a 44 amino acid polypeptide named *growth hormone releasing hormone* (GHRH), and secretion is inhibited by a 14 amino acid polypeptide named *growth hormone inhibiting factor* (GHIF). GHRH and GHIF are also called *somatocrinin* and *somatostatin*, respectively. Limitations upon the supplies of hGH extracted from cadavers restricted its clinical use in the treatment of pituitary dwarfism (*q.v.*). These problems were overcome with the advent of recombinant DNA–derived hGH. *See* Appendix C, 1979, Goeddel *et al.*, human growth hormone receptor.

human growth hormone gene a gene located at 17q21. All primates contain five tandemly linked genes of similar structure. In humans, the sequence is 5′-*GH1-CSHP1-CHS1-GH2-CSH2*-3′. *GH1* codes for the hGH secreted by the pituitary gland. The other four genes are expressed in the placenta. *GH2* is a pseudogene (*q.v.*), which codes for an inactive form of hGH that differs from normal hGH by 13 amino acids. *GSH1* and *CSH2* encode hCS, and *CSHP1* encodes a variant form of hCS. Each of the five genes contains five exons, which are interrupted at identical positions by small introns. All have similar promotors and poly-A addition sites. The five genes presumably evolved from duplications of a common ancestral gene, followed by codon mutations. *See* hereditary growth hormone deficiency, pituitary dwarfism.

human growth hormone receptor (hGHR) a receptor protein that is activated on binding with growth hormone (GH) to stimulate the growth and metabolism of muscle, bone, and cartilage cells. The extracellular domains of two receptor proteins bind one molecule of GH. The hGHR protein is encoded by a gene in region 12–13.1 on the short arm of chromosome 5. Mutations in these genes cause *Laron dwarfism*, a hereditary form of ateliosis inherited as an autosomal recessive.

human immunodeficiency virus *See* AIDS, HIV.

human mitochondrial DNA the mt chromosome is a circular DNA molecule that contains 16,560 bp. The genome is extremely compact. Coding sequences make up 93% of the total genome, and all genes lack introns. There are 37 genes; 28 are encoded by the heavy strand and 9 by the light. There are 2 rRNA genes, 22 tRNA genes, and 13 genes that encode proteins, which function as subunits in enzymes that synthesize ATP. The original DNA sequence published in 1981 is referred to as the *Cambridge reference sequence*. It contains errors that were corrected in 1999. After single base pair corrections were made at 10 sites, the revised Cambridge reference sequence corresponds to European haplogroup H. *See* Appendix C, 1979, Barrell, Banker, and Drouin; 1981, Andrews *et al.*; ATP synthase, mitochondrial DNA (mtDNA), mitochondrial syndromes, mitochondrion, mtDNA lineages.

human mitotic chromosomes the number of human chromosomes observed at mitosis was incorrectly reported as 24 pairs by W. Flemming in 1898. It was not until 1956 that the correct number was determined as 23 pairs by J. H. Tjio and A. Levan. The mitotic chromosomes are generally grouped into seven classes (A-G) according to the following cytological criteria: Group A (chromosomes 1–3) large chromosomes with approximately median centromeres. Group B (chromosomes 4–5)—large chromosomes with submedian centromeres. Group C (chromosomes 6–12 and the X chromosome)—medium-sized chromosomes with submedian centromeres. Group D (chromosomes 13–15)—medium-sized acrocentric chromosomes. Chromosome 13 has a prominent satellite on the short arm. Chromosome 14 has a small satellite on the short arm. Group E (chromosomes 16–18)—rather short chromosomes with approximately median (in chromosome 16) or submedian centromeres. Group F (chromosomes 19 and 20)—short chromosomes with approximately median centromeres. Group G (chromosomes 21, 22, and the Y chromosome)—very short acrocentric chromosomes. *See* Appendix C, 1956, Tjio and Levan; 1971, O'Riordan; 1981, Harper and Saunders; 1991, Ijdo *et al.*; high-resolution chromosome studies, human chromosome band designations, symbols used in human cytogenetics.

human papilloma virus (HPV) a nonenveloped virus with a genome of 8kb of ds DNA contained in an icosahedral capsid 55 nm in diameter. The genome is divisible into an early (E) region, a late (L) region, and a C region. The E region contains 4 kb of DNA, the L region 3kb, and the C region about 1kb. The L genes encode the two proteins of the capsid, and the C segment is a noncoding control region. The genes of the E region function during viral replication. There are over 200 different strains of HPVs. Most cause localized benign warts, but a few induce basal cells in the mucosa of the cervix to undergo uncontrolled proliferation. The potential to induce cervical cancer depends upon the genome of the specific viral strain. HPV carcinogenesis requires the integration of viral genomes into human chromosomes. It is the subsequent expression of E genes in the host cells that causes their transformation. E proteins modify the progression through the cell cycle, telomeres malfunction, and chromosomes segregate improperly, leading to aneuploidy.

human pseudoautosomal region segments containing 2.5 Mbp of DNA in the distal parts of the short arms of the human X and Y chromosomes that share homologous genes. These segments pair during meiosis, and obligatory crossing over takes place, so that genes in this region segregate like autosomal loci, rather than showing X or Y linkage. The gene *MIC2* (*q.v.*) resides in this region. In the mouse, the steroid sulfatase gene (*Sts*) is pseudo-autosomal. The obligatory crossing over that occurs in the pseudoautosomal regions ensures that the X and Y chromosomes will segregate properly at the first meiotic division. *See XG.*

human X chromosome the larger of the two sex chromosomes in humans. Females are XX; males XY. The X is composed of 155 mbp of DNA. It contains 1,098 genes, but only the 54 residing in the human pseudoautosomal region (*q.v.*) have alleles on the Y. In the somatic cells of females, one of the two X chromosomes is inactivated. In the inactivated X, 75% of the genes are transcriptionally silenced, but 15% escape inactivation. The other 10% vary in their activity from one woman to another. *See* Appendix C, 2005, Ross *et al.*; dosase compensation, sex determination, X chromosome inactivation.

human Y chromosome the homolog of the X chromosome in males. The majority of the Y chromosome never undergoes crossing over with the X. Crossing over only occurs within a pairing segment at the distal end of the short arm (Yp), called the *human pseudoautosomal region* (*q.v.*). There is a similar pseudoautosomal pairing segment on the X chromosome. The region of the Y that never undergoes crossing over is called the male specific region (MSY), and it comprises 95% of the chromosome's length. Heterochromatin (*q.v.*) occupies the distal 2/3 of the long arm (Yq). Euchromatin (*q.v.*) comprises 23 mbp of DNA (8 mbp in Yp and 15 mbp in Yq). Genes in the euchromatin encode about 30 different proteins or protein families. About half of these genes are expressed predominantly or exclusively in the testis and half are expressed ubiquitously. *See* Appendix C, 2003, Skaletsky *et al.*; gene conversion, palindrome, *SRY*, *TSPY* genes, XXY trisomic.

humoral immunity immunoglobulin (antibody)-mediated immune responses.

humulin the trade name for a human insulin made in *E. coli* utilizing recombinant DNA techniques and marketed by Eli Lilly and Company. *See* Appendix C, 1982, Eli Lilly.

hunchback (hb) a *Drosophila* gene located at 3-48.3. It produces both maternal and zygotic transcripts, and these are translated into zinc finger proteins (*q.v.*) that regulate the transcription of other genes, especially those involved in the development of the head and thoracic segments of the embryo. The expression of *hb* is stimulated by a transcription factor encoded by *bicoid* (*q.v.*). Within the embryo this factor exists in a gradient where the concentration is highest anteriorly and then trails off. The posterior expression of *hb* is controlled by a gene called *nanos* (*q.v.*). This encodes a translation repressor that is localized in the posterior pole of the embryo. *See* Appendix C, 1989, Driever and Nusslein-Volhard; zygotic segmentation mutants.

Huntington disease (HD) a human neurological disease characterized by irregular, spasmotic, involuntary movements of the limbs and facial muscles, mental deterioration, and death, usually within 20 years of the onset of symptoms. This neuropathy is named after George Huntington, the American physician who published the first detailed account of the disorder in 1872. HD is an autosomal dominant disorder with complete penetrance. Symptoms do not usually appear until the victim is between 35 and 40 years of age. The *HD* gene is on chromosome 4 at p16.3. It generates a 10-kilobase transcript that encodes *huntingtin,* a protein of unknown function. The mutation occurs in a polymorphic CAG repeat at the 5′ end of the gene. The CAG repeat is unstable and is expanded in HD patients. The age of onset is inversely correlated with repeat length. The CAG repeats are translated into polyglutamine tracts, and the polyglutimine repeat sequences in mutant huntington molecules make them aggregate. Caspases (*q.v.*) cleave these mutant proteins and generate fragments that also contain polyglutamine tracts. These activate other caspases and so create a caspase cascade, ending in the death of brain cells. Genes homologous to human *HD* have been identified in the mouse and the puffer fish. The *Takifugu HD* gene spans only 23 kb of genomic DNA, compared to the 170 kb human gene, and yet all 67 exons are conserved. These exons are similar in size in puffer fish and humans, but the introns are much smaller in *Fugu.* In this fish the introns range in size from 47 to 1,476 bp, compared to human intron sizes of 131 to 12,286 bp. *See* Appendix C, 1979, Wexler *et al.*; 1993, MacDonald; 1995, Baxendale *et al.*; genetic instability, parental imprinting, *Takifugu rubripes,* trinucleotide repeats.

Hutchinson-Gilford syndrome a hereditary form of progeria (*q.v.*) described by the physicians Jonathan Hutchinson and Hastings Gilford in 1886 and 1897, respectively.

HVL half-value layer (*q.v.*).

Hyalophora cecropia the giant cecropia moth; because of its large size a favorite experimental insect. *See* Appendix C, 1966, Roller *et al.*

hyaloplasm cytosol.

hyaluronic acid a mucopolysaccharide that is abundant in the jelly coats of eggs and in the ground substance of connective tissue. As shown below, hyaluronic acid is a polymer composed of glucosamine and glucuronic acid subunits.

hyaluronidase an enzyme that digests hyaluronic acid.

H-Y antigen an antigen detected by cell-mediated and humoral responses of homogametic individuals against heterogametic individuals of the same species, which are otherwise genetically identical. Antigenic responses of this sort have been demonstrated in mammals, birds, and amphibians. In mammals, the antigen is called H-Y because it acts as a *H*istocompatibility factor determined by the *Y* chromosome. The location of the gene encoding the H-Y antigen is not known. However, the gene that induces synthesis of the H-Y antigen in humans is located on the Y chromosome. A homologous locus, which suppresses H-Y production, lies on the distal end of the short arm of the X. The H-Y locus is one of the areas that escapes X-chromosome inactivation (*q.v.*).

hybrid **1.** a heterozygote (e.g., a monohybrid is heterozygous at a single locus: a dihybrid is heterozygous at two loci; etc.). **2.** an offspring from geneti-

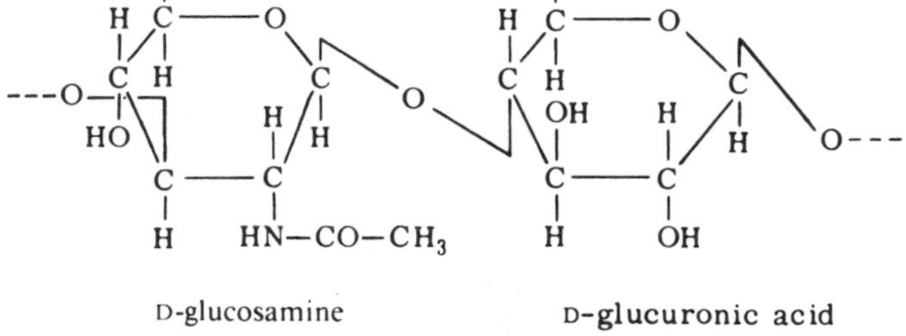

D-glucosamine D-glucuronic acid

Hyaluronic acid

cally dissimilar parents, even different species. *See* *Raphanobrassica*, mullato, mule.

hybrid arrested translation a method for identifying the cDNA corresponding to an mRNA that depends upon the ability of cDNA to hybridize with its mRNA and thereby to inhibit its translation in an *in vitro* system; the disappearance of the translation product from the system indicates the presence of the cDNA.

hybrid breakdown the reduction in fitness of F_2 and/or backcross populations from fertile hybrids produced by intercrossing genetically disparate populations or species; a postzygotic reproductive isolating mechanism.

hybrid corn commercial corn grown from seed produced by the "double cross" (*q.v.*) procedure. Such corn is characterized by its vigor and uniformity. *See* Appendix C, 1909, Shull.

hybrid DNA model a model used to explain both crossing over and gene conversion by postulating that a short segment of heteroduplex (hybrid) DNA is produced from both parental DNAs in the neighborhood of a chiasma. *See* Holliday model.

hybrid duplex molecule an experimentally reconstituted molecule containing a segment of single-stranded DNA hydrogen bonded to a second RNA or DNA molecule of complementary base sequence.

hybrid dysgenesis a syndrome of correlated genetic abnormalities that occurs spontaneously when hybrids are formed between certain strains of *Drosophila*. The hybrids show germ line defects including chromosomal aberrations, high frequencies of lethal and visible mutations, and in most extreme cases, sterility. The cause of PM hybrid dysgenesis is a transposable element named P. Long-established laboratory strains lack P elements. Those strains susceptible to P elements are called M strains. P elements are often present in strains established from newly captured wild flies. The cross P male × M female generates dysgenic F_1 individuals; the reciprocal cross does not. The P elements do not produce dysgenesis within P strains. P elements are dispersed over all the chromosome arms, and in P strains their transposition is repressed. When chromosomes carrying P elements are placed in M cytoplasm by suitable crosses, the P elements become derepressed and transpose at high rates, disrupting genetic loci and causing the dysgenic syndrome. *See* Appendix C, 1982, Bingham *et al.*; dysgenesis, P elements.

hybrid inviability a postmating (postzygotic) reproductive isolating mechanism in which hybrids between disparate populations fail to survive to reproductive age.

hybridization **1.** the mating of individuals belonging to genetically disparate populations or to different species. **2.** in Mendelian terms, the mating of any two unlike genotypes or phenotypes. **3.** the pairing of complementary RNA and DNA strands to produce an RNA-DNA hybrid, or the pairing of complementary DNA single strands to produce a DNA-DNA hybrid. *See* introgressive hybridization.

hybridization competition a technique for distinguishing different mRNA molecules, using a variation of the basic filter hybridization (*q.v.*) technique. A specific DNA sequence trapped on a nitrocellulose filter is exposed to a tritiated RNA known to be complementary to that DNA. An unlabeled RNA of unknown specificity is added and, if complementary to the DNA, it will compete with the labeled molecules for hybridization to the DNA. Any diminution of labeling in the hybrid after equilibrium is reached is attributed to displacement by the unlabeled molecules.

hybridogenesis a form of clonal reproduction in species hybrids whose gametes carry only the nuclear genome derived from one of the parental species. For example, a hybrid species of frog, *Rana esculenta*, is derived from crosses between *R. lessonae* and *R. ridibunda*, but only *ridibunda* chromosomes (and *lessonae* mitochondria) are found in gametes of *esculenta*.

hybridoma a cell resulting from the fusion of an antibody-producing plasma cell (a B lymphocyte) and a myeloma (a bone-marrow cancer) cell. Such a hybrid cell produces a clone that can be maintained in tissue culture or as an animal tumor, and the clone may secrete only a single kind of antibody. Such monoclonal antibodies are used as probes in western blot experiments or in the histochemical localization of antigens of interest. *See* Appendix C, 1975, Köhler and Milstein; 1980, Olsson and Kaplan; immunofluorescence, transfectoma.

hybrid resistance the phenomenon whereby tumors may grow more readily in homozygous recipients than in heterozygous recipients even though the tumor may be genetically histocompatible with both types of recipients.

hybrid sterility the failure of hybrids between different species to produce viable offspring. *See* Haldane rule for hybrids.

hybrid swarm a continuous series of morphologically distinct hybrids resulting from hybridization of two species followed by crossing and backcrossing of subsequent generations.

hybrid vigor heterosis (*q.v.*).

hybrid zone a geographical zone where hybrids between two geographical races are observed. *See* Appendix C, 1973, Hunt and Selander.

hydrocarbon an organic compound composed only of carbon and hydrogen atoms.

hydrogen the most abundant of the biologically important elements. Atomic number 1; atomic weight 1.00797; valence 1^+, most abundant isotope ^{1}H, heavy isotope ^{2}H (deuterium); radioisotope ^{3}H (tritium, *q.v.*).

hydrogen bond a weak (relative to a covalent bond) electrostatic attraction between an electronegative atom (such as O or N) in one chemical group and a hydrogen atom covalently bound to an electronegative atom in another group. Hydrogen bonds between purines and pyrimidines are important because they are the basis for molecular recognition during replication and transcription. According to the DNA model proposed by Watson and Crick, adenosine (A) pairs with two hydrogen bonds to thymine (T) and guanine (G) pairs with three hydrogen bonds to cytosine (C). In the diagram below, the hydrogen bonds are dashed lines, open circles represent oxygens, and closed circles represent nitrogens. The amino groups of adenine and cytosine serve as hydrogen donors, and nitrogen atoms (N-1 in adenine and N-3 in cytosine) serve as hydrogen acceptors. The oxygen of the C-2 of cytosine is also a hydrogen acceptor. In the case of thymine, hydrogen is donated by N-3 and accepted by the oxygen of C-4. Guanine has two hydrogen donors (N1 and the amino group of C-2) and one hydrogen acceptor (the oxygen of C-6). In RNA the hydrogen-bonding abilities of uracil mimic those of thymine. K. Hoogsteen proved in 1963 that the bases can adopt a different conformation, as shown in the upper part of the diagram. Hoogsteen base pairs do not occur in DNA, except in the regions of the telomeres (*q.v.*). *See* Appendix C, 1973, Rosenberg *et al.*; alpha helix, bases of nucleic acids, beta pleated sheet, deoxyribonucleic acid, guanine quartet model.

hydrogen ion concentration expressed as the logarithm of the reciprocal of the concentration of hydrogen ions in grams per liter of solution; abbreviated as pH, the scale runs from 0 to 14, with values above 7 basic; those below, acidic.

hydrogenosomes organelles that resemble mitochondria, but generally lack genomic DNA. They generate hydrogen and ATP and are found in various unrelated protoctists, such as anaerobic ciliates and flagellates. However, one ciliate (*Nyctotherus ovalis*), living in the hindgut of cockroaches, has a rudimentary genome which encodes components of the mitochondrial electron transport chain. This finding supports the idea that most hydrogenosomes have evolved from mitochondria. *See* adenosine phosphate.

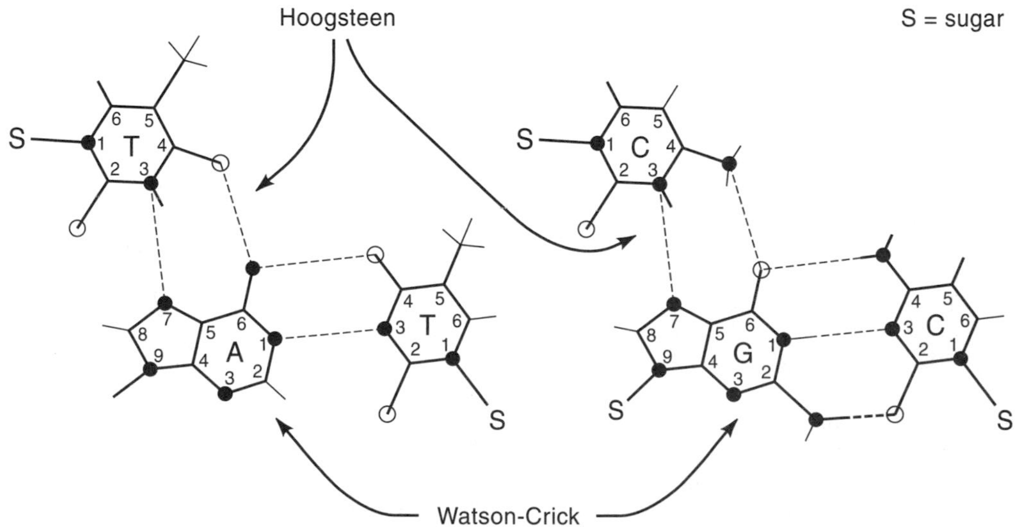

Hydrogen bond

hydrogen peroxide H_2O_2. *See* catalase, horseradish peroxidase, peroxisomes.

hydrolases enzymes that catalyze the transfer of water between donor and receptor molecules. Proteolytic enzymes are a special class of hydrolases.

hydrolysis the splitting of a molecule into two or more smaller molecules with the addition of the elements of water.

hydroperoxyl radical HO_2, an oxidizing agent formed during the interaction of ionizing radiation with oxygenated water. *See* free radical.

hydrophilic water attracting; referring to molecules or functional groups in molecules that readily associate with water. The carboxyl, hydroxyl, and amino groups are hydrophilic.

hydrophobic water repelling; referring to molecules or functional groups in molecules (such as alkyl groups) that are poorly soluble in water. Populations of hydrophobic groups form the surface of water-repellent membranes.

hydrophobic bonding the tendency of nonpolar groups to associate with each other in aqueous solution, thereby excluding water molecules.

hydrops fetalis a fetal syndrome characterized by the accumulation of fluids in the tissues, jaundice, and erythroblastosis due to hypoxia. It occurs in individuals homozygous for deletions of alpha hemoglobin genes (α^0/α^0). The mothers are generally of Southeast Asian or Mediterranean origin. Affected infants die *in utero* between 20 and 40 weeks of gestation or soon after birth. When obstetric assistance is not available, 20–50% of mothers with a hydropic fetus suffer lethal complications. *See* thalassemias.

hydroquinone *See* quinone.

hydrothermal vent a fissure in the ocean floor out of which hot water flows. *See* hyperthermophile, undersea vent communities.

hydroxykynurenine *See Drosophila* eye pigments.

hydroxylamine NH_2OH, a mutagen that converts the NH_2 of cytosine to NHOH, which pairs only with adenine.

hydroxylapatite a hydrated form of calcium phosphate that is naturally found in bone and tooth enamel. A synthetic form of this compound is used by molecular biologists to separate single-stranded nucleic acids from double-stranded ones. The latter bind to a column of hydroxylapatite, while single-stranded nucleic acids pass through it. Also called *hydroxyapatite*.

5-hydroxymethyl cytosine a pyrimidine found instead of cytosine in the DNA of T-even coliphages. 5-Hydroxymethyl cytosine pairs with guanine. It is postulated that the phage-specific DNase, which breaks down the host DNA, attacks DNA molecules containing cytosine. *See* T phages.

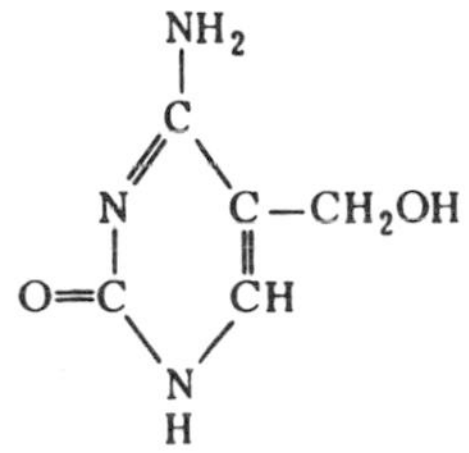

```
          NH2
          |
          C
        // \
       N    C—CH2OH
       |    ||
    O=C     CH
        \  /
         N
         H
```

hydroxyurea a compound that inhibits semiconservative DNA replication, but not repair synthesis.

Hylobates a primate genus that contains nine species of gibbons. *H. concolor* is the best known from the genetic standpoint. Its haploid chromosome number is 26, and 20 genes have been assigned to 10 syntenic groups.

hymenopteran an insect belonging to the order Hymenoptera (which includes bees, ants, wasps, etc.). *See* Appendix A.

hyparchic genes *See* autarchic genes.

hyperammonemia a hereditary disease in man arising from a deficiency of the enzyme ornithine carbamoyl transferase.

hypercholesterolemia *See* familial hypercholesterolemia.

hyperchromic shift an increase in the absorbtion of ultraviolet light by a solution of DNA as these molecules are subjected to heat, alkaline conditions, etc. The shift is caused by the disruption of the hydrogen bonds of each DNA duplex to yield single-stranded structures.

hyperdontia the hereditary presence of one or more additional teeth.

hyperglycemia an increased glucose content in the blood.

hyperlipemia an increased concentration of neutral fat in the blood serum.

hypermorph a mutant gene whose effect is similar to, but greater than, that of the standard or wild-type gene.

hyperplasia an increase in amount of tissue produced by an increase in the number of cells. Hyperplasia often accompanies the regeneration of a damaged organ. *See* hypertrophy.

hyperploid referring to cells or individuals containing one or more chromosomes or chromosome segments in addition to the characteristic euploid number.

hyperprolinemia a hereditary disease in man arising from a deficiency of the enzyme proline oxidase.

hypersensitivity the characteristic of responding with clinical symptoms to allergens in amounts that are innocuous to most individuals. *See* allergy.

hypertension an increased blood pressure.

hyperthermophile a prokaryote that flourishes at very high temperatures. Some live under high pressures at great ocean depths and in the absence of sunshine. They grow in tectonically active rift zones around volcanic vents. Some live at temperatures as high as 113°C! The group contains archaeons like *Archaeoglobus fulgidus* and *Methanococcus jannaschii* and bacteria like *Thermotoga maritima* (*see* entry for each species). The species that have been placed closest to the trunk of the "universal tree of life" (*q.v.*) are all hyperthermophiles, and this suggests that the common ancestor of all prokaryotes may also have been a hyperthermophile. *See* extremophiles, plate tectonics, undersea vent communities.

hypertrophy an increase in the size of a tissue or organ because of the increased volume of the component cells. *See* hyperplasia.

hypervariable (hv) sites amino acid positions within the variable region of an immunoglobulin light chain or heavy chain, exhibiting great variation among antibodies of different specificity; these noncontiguous sites are brought together in the active site where antigen is bound (a paratope) by complex folding of the polypeptide chain. *See* immunoglobulin.

hyphae branched or unbranched filaments that together form the mycelium (*q.v.*) of a fungus. A single filament is a hypha.

hypo *See* fixing.

hypochromic anemia *See* anemia.

hypochromic shift reduction in the absorption of ultraviolet light as complementary single strands of DNA unite to form duplexes. *See* hyperchromic shift.

hypodontia the congenital absence of teeth.

hypoglycemia a decrease in sugar content of the blood serum.

hypomorph any allele that permits a subnormal expression of the normal phenotype. For example, a mutated allele may encode an enzyme that is unstable. However, enough functional enzyme may be generated so that the reaction proceeds, but slowly. Since the genetic block is incomplete, a hypomorphic allele is sometimes called "leaky."

hypophosphatasia a hereditary disease in humans arising from a deficiency of the enzyme alkaline phosphatase.

hypophosphatemia a decreased concentration of inorganic phosphate in the blood serum.

hypophysis the pituitary gland.

hypoplasia an arrested development of an organ or part. The opposite of hyperplasia (*q.v.*).

hypoploid referring to cells or individuals containing one or more fewer chromosomes or chromosome segments than the characteristic euploid number.

hypostatic gene *See* epistasis.

hypothalamus the floor and sides of the vertebrate brain just behind the attachment of the cerebral hemispheres. The hypothalamus controls the secretion of a variety of releasing hormones. These are transported down a closed portal system to the pituitary gland. Here releasing hormones bind to receptors on cells in the anterior lobe. These cells then secrete hormones into the circulatory system that eventually bind to receptors in specific tissues. Hypothalamic releasing hormones include: prolactin-releasing factor, somatostatin, somatocrinin, thyrotropin-releasing hormone, and gonadotropin-releasing hormone. *See* human growth hormone (hGH).

hypothyroidism a diminished production of thyroid hormone.

hypoxanthine 6-hydroxypurine. *See* purine.

hypoxanthine-guanine-phosphoribosyl transferase the enzyme that catalyzes the transfer of the phosphoribosyl moiety of 5-phosphoribosyl-1-pyrophosphate to the 9 position of hypoxanthine and guanine to form inosine monophosphate and guanosine monophosphate. Abbreviated HPRT or HGPRT. The Lesch-Nyhan syndrome (*q.v.*) is caused by deficiency of HPRT. *See* Appendix C, 1987, Kuehn *et al.*; HAT medium.

Hyracotherium the genus that contains the earliest ancestors of the horse. Adults of the smallest species in this genus were only about 10 inches high at the shoulders. These fossils were first described in Eocene deposits in Europe. In the North American Eocene rocks, where fossils of horses were more abundant, the genus was named *Eohippus* (dawn horse). When it later became clear that the two genera represented the same animals, *Hyracotherium* was chosen as the correct scientific name, since it had been coined earlier (1840, rather than 1876). *See* Linnean system of binomial nomenclature.

I

i the regulator gene of the lactose operon in *E. coli*. *See* regulator genes.

I iodine.

I_1, I_2, I_3, etc. the first, second, third, etc., generations obtained by inbreeding.

I^A, I^B, i the allelic genes responsible for the ABO blood group system. *See* A, B antigens.

IAA indole acetic acid (*q.v.*).

Ia antigens alloantigens encoded by the Ia region of the mouse major histocompatibility complex (H-2). They are defined by serological methods and are found predominantly (but not exclusively) on B lymphocytes and macrophages.

ICM inner cell mass (*q.v.*).

icosahedron a regular geometric polyhedron composed of 20 equilateral triangular faces with 12 corners. The capsids of many spherical eukaryotic viruses and bacteriophages are icosahedral. *See* adenovirus, enveloped viruses, herpesvirus, polio virus, Q beta (Qβ) phage, Shope papillomavirus, virus.

ICSH interstitial cell-stimulating hormone. Identical to LH (*q.v.*).

icterus jaundice (*q.v.*).

identical twins *See* twins.

idiocy the most severe degree of mental retardation. An idiot reaches an intelligence level below that of a two-year-old child.

idiogram a diagrammatic representation of the karyotype (*q.v.*) of an organism.

idiotypes antigenic determinants characteristic of a particular variable domain of a specific immunoglobulin or T cell receptor molecule. The idiotype is a unique attribute of a particular antibody from a specific individual. *Contrast with* allotypes, isotypes.

idling reaction production of ppGpp and pppGpp by ribosomes when an uncharged tRNA is present in the A site. *See* translation.

IF initiation factor (*q.v.*).

IFNs interferons (*q.v.*).

Ig immunoglobulin (*q.v.*).

IgA human immunoglobulin A, found as a 160-kilodalton monomer or as a 320-kilodalton dimer in mucus and secretory fluids and on the surface of cell membranes.

IgD human immunoglobulin D, found as a 185-kilodalton monomer on the surface of lymphocytes.

IgE human immunoglobulin E, found as a 200-kilodalton monomer and involved in allergic reactions. It forms a complex with antigen and then binds to the surface of mast cells, triggering the release of histamine.

Igf 2 insulin growth factor 2. *See* H19.

IgG human immunoglobulin G, found as a 150-kilodalton monomer, which is the predominant molecule involved in secondary immune responses. It fixes complement and is the only immunoglobulin that crosses the placenta. *See* Appendix C, 1969, Edelman *et al.*; immune response.

IgM human immunoglobulin M, found as a 900-kilodalton pentamer that is the predominant molecule involved in the primary immune response. It fixes serum complement and agglutinates effectively.

ile isoleucine. *See* amino acid.

imaginal discs inverted thickenings of epidermis containing mesodermal cells found in a holometabolous insect. During the pupal stage, the imaginal discs give rise to the adult organs, and most larval structures are destroyed. *See* Appendix C, 1973, Garcia-Bellido *et al.*; 1975, Morata and Lawrence; compartmentalization, *in vivo* culturing of imaginal discs.

imino forms of nucleotides *See* tautomeric shifts.

immediate hypersensitivity a type of hypersensitivity reaction that is mediated by antibodies and that occurs within minutes after exposure to the allergen or antigen in a previously sensitized individual. *Compare with* delayed hypersensitivity.

immortalizing genes genes carried by oncogenic viruses that confer upon cultured mammalian cells the ability to divide and grow indefinitely, thereby overcoming the Hayflick limit (*q.v.*).

immune competent cell a cell capable of producing antibody in response to an antigenic stimulus.

immune decoy protein *See* sporozoite.

immune globulins *See* antibody.

immune response the physiological response(s) stemming from activation of the immune system by antigens, including beneficial immunity to pathogenic microorganisms, as well as detrimental autoimmunity to self-antigens, allergies, and graft rejection. The cells mainly involved in an immune response are T and B lymphocytes and macrophages. T cells produce lymphokines (*q.v.*) that influence the activities of other host cells, whereas B cells mature to produce immunoglobulins (*q.v.*) or antibodies that react with antigens. Macrophages "process" the antigen into immunogenic units that stimulate B lymphocytes to differentiate into antibody-secreting plasma cells, and stimulate T cells to release lymphokines. Complement (*q.v.*) is a group of normal serum proteins that can aid immunity by becoming activated as a consequence of antigen-antibody interactions. The first contact with an antigen "sensitizes" the animal and results in a *primary immune response*. Subsequent contact of the sensitized animal with that same antigen results in a more rapid and elevated reaction, called the *secondary immune response* (also known as the "booster response" or the "anamnestic reaction"), which is most easily demonstrated by monitoring the level of circulating antibodies in the serum. The immune response can be transferred from a sensitized to an unsensitized animal via serum or cells. It is highly specific for the inciting antigen, and is normally directed only against foreign substances. *See* adenosine deaminase deficiency.

immune response (*Ir*) gene any gene that determines the ability of lymphocytes to mount an immune response to specific antigens. In the major histocompatibility complex of the mouse (the H-2 complex), the I region contains *Ir* genes and also codes for Ia (immune-associated) antigens found on B cells and on some T cells and macrophages. In humans, the HLA D (DR) region is the homolog of the mouse H-2 I region. *See* Appendix C, 1948, Snell; 1963, Levine *et al.*; 1972, Benacerraf and McDevitt.

immune system the organs (e.g., thymus, lymph nodes, spleen), tissues (e.g., hematopoietic tissue of bone marrow, mucosal and cutaneous lymphoid tissues), cells (e.g., thymocytes, blood and tissue lymphocytes, macrophages), and molecules (e.g., complement, immunoglobulins, lymphokines) responsible for immunity (protection against foreign substances).

immunity 1. the state of being refractive to a specific disease, mediated by the immune system (T and B lymphocytes and their products—lymphokines and immunoglobulins, respectively). *Active immunity* develops when an individual makes an immune response to an antigen; *passive immunity* is acquired by receiving antibodies or immune cells from another individual. 2. the ability of a prophage to inhibit another phage of the same type from infecting a lysogenized cell (phage immunity). 3. the ability of a plasmid to inhibit the establishment of another plasmid of the same type in that cell. 4. the ability of some transposons to prevent others of the same type from transposing to the same DNA molecule (transposon immunity). 5. phage-resistant bacteria are usually "immune" to specific phages because they lack the cell-surface receptors that define the host range of that phage. *See* innate immunity.

immunity substance a cytoplasmic factor produced in lysogenic bacteria that prevents them from being infected by bacteriophages of the same type as their prophages and also prevents the vegetative replication of said prophages.

immunization administration of an antigen for the purpose of stimulating an immune response to it. Also known as *inoculation* or *vaccination*.

immunochemical assay any technique that uses antigen-antibody reactions to detect the location of or to determine the relative amounts of specific antibodies or antigenic substances. *See* enzyme-linked immunosorbent assay, immunofluorescence assay.

immunocompetent (immune competent) cell a cell capable of carrying out its immune function when given the proper stimulus.

immunodominance within a complex immunogenic molecule, the ability of a specific component (1) to elicit the highest titer of antibodies during an immune response, or (2) to bind more antibodies from a given polyvalent antiserum than any other component of that same molecule. For example, in a glycoprotein antigen, a specific monosaccharide may be the most highly antigenic component of the entire molecule and therefore exhibits immunodominance over other components of the same molecule.

immunoelectrophoresis a technique that first separates a collection of different proteins by electrophoresis through a gel and then reacts them with a specific antiserum to generate a pattern of precipitin arcs. The proteins can thus be identified by their electrophoretic mobilities and their antigenic properties. *See* Appendix C, 1955, Grabar and Williams.

immunofluorescence assay a visual examination of the presence and the distribution of particular antigens on or in cells and tissues using antibodies that have been coupled with fluorescent molecules such as rhodamine and fluorescein. In the *direct method,* the fluorescent probe combines directly to the antigen of interest. In the *indirect method,* two antibodies are used in sequence. The first is the one specifically against the antigen under study. Subsequently, the tissue is incubated with a second antibody, prepared against the first antibody. The second antibody has been conjugated previously with a fluorescent dye, which renders the complex visible. The indirect method is often preferred because, if one wants to localize more than one antigen, only one fluorescently labeled antigen need be used, provided the first antibody in each case is from the same species of animal. The second fluorescent antibody is generally commercially available. *See* Appendix C, 1941, Coons *et al.*

immunogen a substance that causes an immune response. Foreign proteins and glycoproteins generally make the most potent immunogens. *See* antigen.

immunogene any genic locus affecting an immunological characteristic; examples: immune response genes, immunoglobulin genes, genes of the major histocompatibility complex (*all of which see*).

immunogenetics studies using a combination of immunologic and genetic techniques, as in the investigation of genetic characters detectable only by immune reactions. *See* Appendix C, 1948, Snell; 1963, Levine *et al.*; 1972, Benacerraf and McDevitt.

immunogenic capable of stimulating an immune response.

immunoglobulin an antibody secreted by mature lymphoid cells called plasma cells. Immunoglobulins are Y-shaped, tetrameric molecules consisting of two relatively long polypeptide chains called heavy (H) chains and two shorter polypeptide chains called light (L) chains (see illustration on page 227). Each arm of the Y-shaped structure has specific antigen-binding properties and is referred to as an antigen-binding fragment (Fab). The tail of the Y structure is a crystallizable fragment (Fc). Five H chain classes of immunoglobulin are based upon their antigenic structures. Immunoglobulin class G (IgG) is the most common in serum and is associated with immunological "memory"; class IgM is the earliest to appear upon initial exposure to an antigen. Class IgA can be secreted across epithelial tissues and seems to be associated with resistance to infectious diseases of the respiratory and digestive tracts. The antibodies associated with immunological allergies belong to class IgE. Not much is known about the functions of IgD. Antibodies of classes IgG, IgD, and IgE have molecular weights ranging from 150,000 to 200,000 daltons (7S); serum IgA is a 7S monomer, but secretory IgA is a dimer (11.4S); IgM is a pentamer (19S; 900,000 daltons) of five 7S-like monomers.

In the case of IgG, each heavy chain consists of four "domains" of roughly equal size. The variable (V_H) domain at the amino (*N*-terminus) end contains different amino acid sequences from one immunoglobulin to another, even within the same H chain class. The other three domains have many regions of homology that suggest a common origin by gene duplication and diversification by mutation. These "constant" domains (C_H1, C_H2, C_H3) are essentially invariate within a given H chain class. An L chain is about half as long as an H chain. Its amino end has a variable region (V_L); its carboxyl end has a constant region (C_L). An Fab fragment consists of an L chain and an Fd segment of an H chain ($V_H + C_H1$). Within a tetrameric immunoglobulin molecule, the two L chains are identical and the two H chains are identical. The Fc fragment consists of carboxy-terminal halves of two H chains ($C_H2 + C_H3$). The region between C_H1 and C_H2 is linear rather than globular, and is called the *hinge region.* Crystallographic studies of human IgG show that the oligosaccharide chains (OC) that are attached to the C_H2 regions provide surfaces that bind these regions to each other and to the Fab units. Each mature antibody-synthesizing plasma cell produces a single species of immunoglobulin, all of which con-

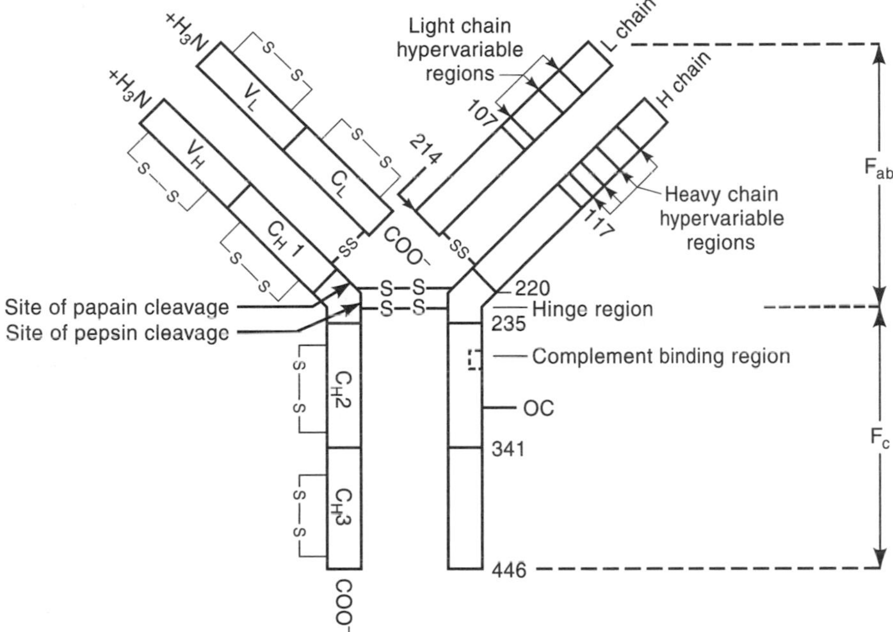

Immunoglobulin

Diagram of a typical IgG molecule. Within each immunoglobulin molecule, the two L chains are identical and the two H chains are identical. Numbers represent approximate amino acid residues from the *N* terminus of the respective chain.

tain identical L and H chains. *See* Appendix C, 1939, Tiselius and Kabat; 1959, Edelman; 1962, Porter; 1965, Hilschmann and Craig; 1969, Edelman *et al.*; 1976, Hozumi and Tonegawa; 1977, Silverton, Navia, and Davies; 1987, Tonegawa; abzymes, B lymphocyte, glycosylation, hybridoma, IgA, IgM, immune response, monoclonal antibodies, oligosaccharide, V(D)J recombination.

immunoglobulin chains the components of the heteropolymeric immunoglobulin molecules. There are five groups of heavy chains, each characteristic of a specific class of immunoglobulin: gamma (IgG), mu (IgM), epsilon (IgE), alpha (IgA), and delta (IgD). The genes encoding all the heavy immunoglobulin chains are located on human chromosome 14. The constant region of each heavy chain makes up about three-fourths of the molecule, and the gene segments encoding the constant regions are arranged in the sequence mu, delta, gamma, epsilon, and alpha in both humans and mice. There are two groups of light chains: kappa chains, encoded by gene segments on human chromosome 2, and lambda chains, encoded by gene segments on chromosome 22. *See* immunoglobulin genes.

immunoglobulin domain superfamily a group of glycoproteins that are embedded in the surface of the membranes of certain cells and which have one or more immunoglobulin domains. Each domain is a chain of about 100 amino acids that folds back and forth upon itself to form a sandwich of two pleated sheets linked by a disulfide bond. Included in the superfamily are the immunoglobulins (*q.v.*) with up to 12 domains per molecule, the T cell receptors (*q.v.*), and the MHC receptors (*q.v.*), each with two domains per molecule, and the CD4 and CD8 receptors (*q.v.*) with four domains and a single domain, respectively. The genes that encode these proteins are assumed to have evolved from a common ancestral gene over a period of hundreds of millions of years.

immunoglobulin genes genes encoding the light and heavy chains of the immunoglobulins. These genes are remarkable in that they are made up of segments that are shuffled as the B lymphocytes mature. The light chains contain segments that can be symbolized L-V, J, and C. The V, or variable, segment codes for the first 95 amino acids of the chain, whereas the C, or constant, segment codes for amino acids 108 to 214. The joining segment, J, codes for amino acids 96 to 107. L codes for a leader sequence 17–20 amino acids long; it functions in the transport of the molecule through the plasmalemma and is cleaved off the molecule in the process. There are about 300 L-V segments per light chain gene, and each of the V segments has a different base sequence. In the kappa gene, there are six J segments, each with a different base sequence, and one C segment. During differentiation of a given B lymphocyte stem cell, an immunoglobulin gene is assembled containing one L-V, one J, and one C segment, and this gene is transcribed by the lymphocyte and all of its progeny. The lambda gene also contains about 300 L-V segments, but each of the six J segments has its own adjacent C segment. The heavy chain gene is over 100,000 nucleotides long and contains a series of segments that can be symbolized L-V, D, J, C_{μ}, C_{δ}, $C_{\gamma3}$,$C_{\gamma1}$,$C_{\gamma2b}$, $C_{\gamma2a}$, C_{ε}, and C_{α}. There are about 300 L-V segments, 10–50 D segments, 4 J segments, and one each of the C segments. Each D segment codes for about 10 amino acids. During differentiation the segments are shuffled so that the variable region of a heavy chain is encoded by a segment that contains one L-V, one D, and one J segment. The gene also contains mu, delta, gamma, epsilon, and alpha subsegments, and which one of these is transcribed determines the class to which the antibody will belong. *See* Appendix C, 1965, Dreyer and Bennett; 1976, Hozumi and Tonegawa; 1981, Sakano *et al.*.; 1987, Tonegawa; allelic exclusion, genomic equivalence, heavy chain class switching, immunoglobulin chains, transfectoma, V(D)J recombination.

immunological memory the capacity of the immune system to respond more rapidly and more vigorously to the second contact with a specific antigen than observed in the primary response to the first contact; the booster or anamnestic response.

immunological suppression a genetic or induced condition in which the ability of an individual's immune system to respond to most or all antigens is impaired. *See* specific immune suppression.

immunological surveillance theory the theory that the cell-mediated immune system evolved primarily to continuously monitor the body for spontaneously arising cancerous cells or those containing foreign pathogens and to destroy them.

immunological tolerance a state of nonreactivity toward a substance that would normally be expected to elicit an immune response. Tolerance to specific foreign antigens can be induced by the exposure of a bird or mammal to the foreign antigens during embryonic or neonatal (depending upon species) life. In adults, tolerance (usually of shorter duration) can be induced by using particular routes of administration for the antigens or administration of agents that are particularly effective against cells proliferating in response to antigen. Mechanisms may include actual deletion of potentially reactive lymphocytes or their "inactivation" by immunological suppression. *See* Appendix C, 1945, Owen; 1953, Billingham *et al.*

immunology the science dealing with immunity, serology, immunochemistry, immunogenetics, hypersensitivity, and immunopathology. *See* Appendix C, 1778, Jenner; 1900, Ehrlich; 1930, Landsteiner; cellular immunity.

immunoselection a method for isolating cell-line variants lacking certain antigens, such as those of the major immunogene complex. By treating cells with a specific antiserum and complement, all cells die, except a few spontaneously arising variants. These do not express the corresponding antigen, and therefore they live and can be isolated. Many of these variants appear to be due to deletion mutations rather than to epigenetic changes or mitotic crossing-over. *Compare with* antigenic conversion.

impact theory a proposal, published in 1984 by Walter Alvarez and five colleagues, that the mass extinction of various groups of organisms that occured at the end of the Cretaceous (*q.v.*) resulted from the collision of the earth with an asteroid or comet. Rocks at the Cretaceous-Tertiary boundary have high iridium concentrations, and this iridium is postulated to have arisen from the pulverized asteroid.

impaternate offspring an offspring from parthenogenetic reproduction in which no male parent took part. *See* parthenogenesis.

imperfect excision the release of a genetic element (e.g., an insertion sequence or prophage) from a DNA molecule in a way that either includes more than or less than the element itself.

imperfect flower *See* flower.

implant material artificially placed in an organism, such as a tissue graft, an electronic sensor, etc.

implantation 1. attachment of a mammalian embryo to the uterine wall. 2. the addition of tissue grafts to an organism without the removal of anything from it.

imprinting 1. the imposition of a stable behavior pattern in a young animal by exposure, during a particular period in its development, to one of a restricted set of stimuli. 2. *See* parental imprinting.

inactivation center a region of the mouse X chromosome that governs the degree to which translocated autosomal genes are inactivated when the associated X-linked genes are inactivated as the result of random X-inactivation. *See* Cattanach's translocation, Lyon hypothesis.

inactive X hypothesis Lyon hypothesis (*q.v.*).

Inarticulata a division of invertebrates containing the unsegmented, coelomate protostomes, such as sipunculids and molluscs. *See* Appendix A.

inborn error a genetically determined biochemical disorder resulting in a metabolic defect that produces a metabolic block having pathological consequences. *See* Appendix C, 1909, Garrod.

inbred strain a group of organisms so highly inbred as to be genetically identical, except for sexual differences. *See* isogenic, mouse inbred lines.

inbreeding the crossing of closely related plants or animals.

inbreeding coefficient *See* Wright's inbreeding coefficient.

inbreeding depression decreased vigor in terms of growth, survival, or fertility following one or more generations of inbreeding.

incapsidation the construction of a capsid around the genetic material of a virus.

inclusive fitness *See* Hamilton's genetical theory of social behavior.

incompatibility in immunology, genetic or antigenic differences between donor and recipient tissues that evoke an immunological rejection response.

incomplete dominance failure of a dominant phenotype to be fully expressed in an organism carrying a dominant and a recessive allele. The result is usually a phenotype that is intermediate between the homozygous dominant and the recessive forms. the term is synonymous with *partial dominance* and *semidominance*. *See* dominance.

incompletely linked genes genes on the same chromosome that can be recombined by crossing over.

incomplete metamorphosis *See* Hemimetabola.

incomplete sex linkage the rare phenomenon of a gene having loci on the homologous segments of both X and Y chromosomes. *See* XY homology.

incross mating between individuals from the same inbred line or variety, often of the same genotype.

incubation period the period over which eggs, cells, and so forth are incubated; the period between exposure to an infection and the appearance of the first symptoms.

indels an acronym for *in*sertions and *del*etions. The term often appears in studies of the mechanisms that cause genomic divergence between related species. *See* sequence similarity estimates.

independent assortment the random distribution to the gametes of genes located on different chromosomes. Thus, an individual of genotype *Aa Bb* will produce equal numbers of four types of gametes: *AB*, *Ab*, *aB*, and *ab*. *See* Mendel's laws.

independent probabilities in a group of events, the occurrence of any one event having no influence on the probability of any other event. For example, the orientation of one pair of homologous chromosomes on the first meiotic metaphase plate does not influence the orientation of any other pair of homologs. *See* independent assortment.

indeterminant inflorescence an inflorescence, such as a raceme (*q.v.*), in which the first flowers to open are at the base and are followed upward by progressively younger ones.

index case a synonym for propositus (*q.v.*).

index fossil a fossil that appears only in rocks of a relatively limited geological age span.

indigenous referring to a species that colonized a specific area, such as an island, without human intervention. However, the species lives naturally in other places as well. *Contrast with* endemic.

indirect immunofluorescence microscopy *See* immunofluorescence.

indoleacetic acid auxin, a phytohormone. *See* auxins.

```
        H
        C
      //  \
   HC       C ——— CH
   |        ||     ||
   HC       C      CH
     \\   /   \   /
        C       N
        H       H
```

indolephenoloxidase an earlier name for the enzyme now called superoxide dismutase (*q.v.*).

Indrichotherium the largest mammal ever to walk the earth. It belonged to the rhinoceros family and lived in Asia during the Oligocene (*q.v.*).

induced mutation a genetic alteration produced by exposure to a mutagen. *Compare with* spontaneous mutation.

inducer any of the small organic molecules that cause the cell to produce larger amounts of the enzymes involved in their metabolism. Inducers are a class of effector molecules (*q.v.*). *See* gratuitous inducer, regulator genes.

inducible enzyme an enzyme synthesized only in response to an inducer. *See* adaptive enzyme, regulator gene.

inducible system a regulatory system in which the product of a regulator gene (the repressor) is active and blocks transcription of the operon. The effector (called an inducer) inactivates the repressor and therefore allows mRNA synthesis to occur. Thus, transcription occurs only in the presence of effector molecules. *See* regulator gene. *Compare with* repressible system.

induction 1. the determination of the developmental fate of one cell mass by another. The morphogenic effect is brought about by an evocator acting upon competent tissue. 2. the stimulation of a lysogenized bacterium to produce infective phage. 3. the stimulation of synthesis of a given enzyme in response to a specific inducer. *See* Appendix C, 1924, Spemann and Mangold.

inductor any substance that carries out an induction similar to that performed by an organizer (*q.v.*).

industrial melanism the phenomenon where melanic morphs appear among the animals that live in industrial areas. As soot from factory smoke darkens the landscape, the frequency of melanic offspring increases until the original lighter forms become a minority. The pigments involved are melanins (*q.v.*), and the species undergoing melanism rely on crypsis (*q.v.*) to avoid being eaten. The most famous example of industrial melanism involves the moth *Biston betularia* (*q.v.*). *See* Appendix C, 1891, Tutt; 1958, Kettlewell; Bibliography, 2003, Hooper.

Indy a gene in *Drosophila* that has a profound effect upon life span. The gene symbol is an acronym for *I*'m *n*ot *d*ead *y*et. The gene encodes a protein, localized in the plasma membranes of cells of the fat body (*q.v.*), that transports molecules generated during the citric acid cycle (*q.v.*). Flies with two normal alleles have a mean life span of 37 days. Flies heterozygous for *Indy* mutants have a mean life span of 70 days. *Indy* homozygotes show only a 15% increase in life span. *See* heterozygote advantage.

inelastic collision *See* collision.

infectious nucleic acid purified viral nucleic acid capable of infecting a host cell and causing the subsequent production of viral progeny.

infectious transfer the rapid spread of extrachromosomal episomes (plus any integrated chromosomal genes) from donor to recipient cells in a bacterial population.

inflorescence 1. a flower cluster. 2. the arrangement and mode of development of the flowers on a floral axis. *See* determinant inflorescence, indeterminant inflorescence.

influenza viruses viruses that belong to the Orthomyxoviridae and cause epidemics of influenza in humans, pigs, horses, and birds. The last great epidemic occurred during the first world war. Between 1918 and 1919 there were 50 million deaths worldwide from influenza. The genome consists of eight molecules of linear negative-sense ssRNA, which form a helical complex with a protein called NP. Several other proteins form spikes and mushroom-shaped projections that radiate from the outer surface of the viral envelope. These viruses undergo frequent changes in their antigenic properties as a result of small mutational changes. *See* enveloped viruses, virus, zoonotic viruses.

informed consent the permission given by an individual that allows a previously discussed procedure to be performed in the future. Amniocentesis (*q.v.*) would be an example of such a procedure. It is known that the individual has been made aware of the risks and benefits of the procedure and the implications of the findings. *See* genetic counseling.

in-frame mutation a mutation, generally of the missense type, that does not cause a reading frame shift (*q.v.*).

inheritance of acquired characteristics *See* acquired characteristics, Lamarckism.

initiation codon *See* start codon.

initiation factors proteins required for the initiation of protein synthesis. One (protein IF3) is required for the binding of the 30S particle to mRNA. A second (protein IFI) binds to f-met-tRNA and helps it attach to the 30S mRNA initiation complex. A third protein (IF2) is required, although its precise function is unclear. Initiation factors are symbolized IF in prokaryotes and elF in eukaryotes, followed by a number. *See* *N*-formylmethionine, scanning hypothesis, translation.

initiator a molecule that initiates replication once it binds to a replicator. *See* replicon.

initiator tRNA the special tRNA molecule which provides the amino acid that starts the protein chain. In the case of prokaryotes, the initiator tRNA carries *N*-formylmethionine, while eukaryotic initiators carry methionine. *See* transfer RNA.

innate immunity an immunse response activated by receptors that recognize foreign molecules, such as lipopolysaccharides (*q.v.*), attached to the surfaces of common microorganisms. *Drosophila* combats microbial infections by having receptors of this sort. Activation of the receptors induces cells of the fat body (*q.v.*) to synthesize antimicrobial peptides. DNA chip (*q.v.*) technologies have been used to identify over 400 genes in *Drosophila* that play roles in innate immunity. *Drosophila* immune-competent cells can express more than 18,000 isoforms containing immunoglobulin receptor domains by alternative splicing (*q.v.*) of the *Dscam* gene. *Contrast with* adaptive immunity.

inner cell mass (ICM) in mammals, a clump of embryonic cells that attaches itself to the inside of the trophoblast (*q.v.*) during blastocyst (*q.v.*) formation and from which the fetus (*q.v.*) develops. The ICM is one of the sources of embryonic stem cells (*q.v.*).

innervation the nerve supply to a particular organ.

inoculum a suspension of cells introduced into a nutrient medium to start a new culture.

input load *See* genetic load.

inosine hypoxanthine riboside. *See* rare bases.

inquiline an animal that lives in the abode of another species.

insect ovary types three types of ovaries are found among insects. The *panoistic* ovary appears to be the ancestral type. Here, all oogonia (except stemline oogonia) are eventually transformed to oocytes. In *meroistic* ovaries, both oocytes and nurse cells (*q.v.*) are generated. These may be organized within the ovariole in two ways. In the *polytrophic meroistic* ovary, the nurse cells and oocytes alternate along the length of the ovariole. In the *telotrophic meroistic* ovary, the nurse cells are restricted to the germarium and are connected to oocytes in early stages of their development by cytoplasmic processes called nutritive chords. Panoistic ovaries are found in insects belonging to the more primitive orders (Archeognatha, Zygentoma, Ephemeroptera, Odonata, Plecoptera, Phasmida, Orthoptera, and Dictyoptera). Polytrophic meroistic ovaries occur in the Psocoptera, Phthiraptera, Hymenoptera, Trichoptera, Lepidoptera, and Diptera. Telotrophic ovaries occur in the Hemiptera, Coleoptera, Raphidioptera, and Megaloptera.

insertion the addition of one or more base pairs into a DNA molecule; a type of mutation commonly induced by acridine dyes or by mobile insertion sequences (*q.v.*). *See* indels.

insertional inactivation abolition of the functional properties of a gene product by insertion of a foreign DNA sequence into that gene's coding sequence; used in genetic engineering as a means of detecting when a foreign DNA sequence has become integrated into a plasmid or other recipient molecule of interest.

insertional mutagenesis alteration of a gene as a consequence of inserting unusual nucleotide sequences from such sources as transposons, viruses, transfection, or injection of DNA into fertilized eggs. Such mutations may partially or totally inactivate the gene product or may lead to altered levels of protein synthesis. *See* insertional inactivation, insertion sequences, transgenic animals.

insertional translocation *See* translocation.

insertion sequences transposable elements (*q.v.*) first detected as the cause of spontaneous mutations in *E. coli*. The majority of IS elements studied so far range in size from 0.7 to 1.8 kilobase pairs. IS termini carry inverted repeats of about 10 to 40 base pairs, which are believed to serve as recognition sequences for a transposase (*q.v.*). The IS also contains a gene that encodes the transposase. The genome of the *E. coli* strain sequenced in 1997 contained 10 different insertion sequences, and most of these were present at multiple sites along the chromosome. *See* Appendix C, 1969, Shapiro; 1997, Blattner *et al.*

insertion vector *See* lambda cloning vehicle.

in silico a term that refers to inferred relationships and hypotheses that are generated from the analysis of information retrieved from computer-based data banks that contain amino acid or nucleotide sequences. The information analyzed resides in silicon chips, hence the use of *silico* in the term.

in situ "in place"; in the natural or original position.

***in situ* hybridization** a technique utilized to localize, within intact chromosomes, eukaryotic cells, or bacterial cells, nucleic acid segments complementary to specific labeled probes. To localize specific DNA sequences, specimens are treated so as to denature DNAs and to remove adhering RNAs and proteins. The DNA segments of interest are then detected via hybridization with labeled nucleic acid probes. The distribution of specific RNAs within intact cells or chromosomes can be localized by hybridization of squashed or sectioned specimens with an appropriate RNA or DNA probe. *See* Appendix C, 1969, Gall and Pardue; 1975, Grunstein and Hogness; 1981, Harper and Saunders; 1983, Hafen, Levine, and Gehring; chromosome painting, fluorescence *in situ* hybridization (FISH).

instar the period between insect molts.

instinct an unlearned pattern of behavior.

instructive theory an early immunological theory in which it was believed that the specificity of antibody for antigen was conferred upon it by its initial contact with the antigen. This theory has been discarded in favor of the clonal selection theory (*q.v.*), in which specificity exists prior to contact with antigen.

insulator DNAs segments of DNA that serve to isolate neighboring genes within a specific domain by blocking interactions between enhancers (*q.v.*) on one side of a domain from the inappropriate target promotors of neighboring genes belonging within an adjacent domain. Specific proteins that bind to insulator DNA segments are responsible for defining certain chromosomal regions, such as the interbands and puff boundaries in polytene chromosomes. *See* Appendix C, 1995, Zhao, Hart, and Laemmli; 2000, Bell and Felsenfeld; *H19*, matrix attachment regions, parental imprinting.

insulin a polypeptide hormone produced by the beta cells in the islets of Langerhans (*q.v.*). Insulin causes a fall in the sugar concentration of the blood, and its deficiency produces the symptoms of diabetes mellitus. Beef insulin was the first protein to have its amino acid sequence determined. This molecule (illustrated below) is made up of an A polypeptide (21 amino acids long) and a B peptide (containing 30 amino acids) joined by two disulfide bridges. Sanger's analysis showed that proteins had chemical structures in the form of specific sequences of amino acids. In humans, insulin is encoded by a gene on the short arm of chromosome 11 at band 15.5. A genetically engineered form of human insulin goes under the trade name *humulin* (*q.v.*). *See* Appendix C, 1921, Banting and Best; 1955, Sanger *et al.*; 1964, Hodgkin; 1977, Gilbert; 1982, Eli Lilly; diabetes mellitus, proinsulin.

insulin-like growth factors 1 and 2 (IGF-1 and IGF-2) single-chain protein growth factors that closely resemble insulin and each other in their amino acid sequences. Both IGF-1 and IGF-2 and their receptors are present as early as the eight-cell stage in the mouse, and growth is retarded if either *igf-1* or *igf-2* gene is inactivated. IGF-2 is essential for early embryonic growth in the mouse, but IGF-1 seems more important in the later development. The IGF-2 gene shows parental imprinting (*q.v.*). In *Caenorhabditis* the gene for the IGF-1 receptor is encoded by *daf-2*, and mutations of this gene cause a two- to threefold increase in the worm's normal (10-day) life span. *See H9*.

integrase an enzyme that catalyzes a site-specific recombination (*q.v.*) by which a prophage becomes integrated into or excised (deintegrated) from a bacterial chromosome; an excisionase enzyme is also required for the excision process. *See* lambda (λ) bacteriophage. The term also refers to a family of site-specific recombinases. *See* site-specific recombinase.

integral protein an amphipathic (*q.v.*) protein that is firmly embedded in the plasma membrane. Integral proteins interact with both the hydrophobic (*q.v.*) and the hydrophilic (*q.v.*) components of the phospholipid bilayer and are difficult to isolate. *Compare with* peripheral protein. *See* aquaporins, calnexin, lipid bilayer model.

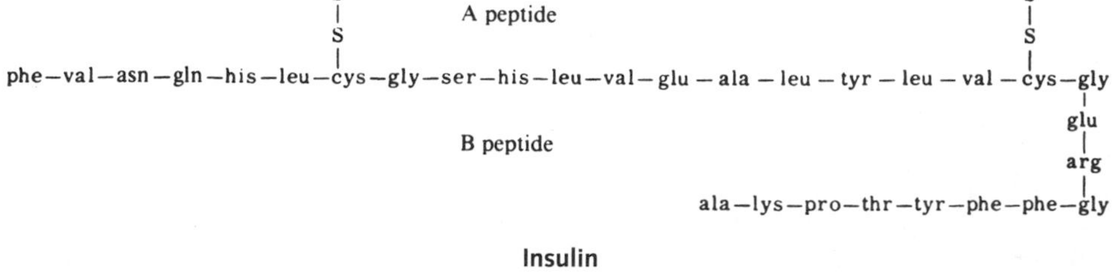

```
                              S ─────────────────── S
                              |                     |
gly─ile─val─glu─gln─cys─cys─ala─ser─val─cys─ser─leu─tyr─gln─leu─glu─asn─tyr─cys─asn
                        |                                                   |
                        S                                                   S
                        |          A peptide                                |
                        S                                                   S
                        |                                                   |
phe─val─asn─gln─his─leu─cys─gly─ser─his─leu─val─glu─ala─leu─tyr─leu─val─cys─gly
                                                                            |
                                                                           glu
                                   B peptide                                |
                                                                           arg
                                                                            |
                                          ala─lys─pro─thr─tyr─phe─phe─gly
```

Insulin

integration efficiency the frequency with which a foreign DNA segment is incorporated into the genotype of a recipient bacterium, particularly with reference to transformation.

integrins a large family of cell-surface receptor proteins that bind to the components of the extracellular matrix and function as cellular "glue," facilitate cellular migrations (in embryological development or in cells of the immune system of adults), and activate signal transduction pathways within cells. Integrins are structurally related to one another, and one or more of them appear on virtually every cell type in the animal kingdom. Integrins consist of two protein chains (α, β). There are at least 15 variants of the α chain and 8 of the β chains, which combine into at least 20 functional dimers (e.g., α2β1, αIIβ3). The family name is apropos because of the importance of these molecules to the structural integrity of cells and tissues and because of their integrative functions among the diverse signals that impinge on cells. The extracellular matrix consists primarily of gel-like chains of sugars and interconnected fibrous proteins (including laminin, fibronectin, and collagen). Integrins are connected intracellularly via one or more intermediary molecules (e.g., talin, vinculin, paxillin, tensin) in a focal adhesion complex to actin molecules of the cytoskeleton. Most integrins interact with the extracellular matrix, but some participate in adhesion between cells. The molecules responsible for most cell-to-cell adhesions belong to such groups as the cadherin, selectin, and immunoglobulin families. Some microbes enter cells at least in part by attaching to integrins.

integron any transposon responsible for the horizontal transfer of genes between different species of bacteria. Integrons are DNA segments that (1) encode a site-specific recombinase, (2) include integrase-specific recombination sites, and (3) include a promoter that expresses one or more genes that confer antibiotic resistance or other adaptive traits on the host. An integron is often incorporated within a plasmid (*q.v.*). *See* R (resistance) plasmid.

intein an *in*ternal pro*tein* sequence that is translated in-frame in a precursor polypeptide and is excised during protein splicing (*q.v.*).

intelligence quotient (IQ) an individual can be assigned to a "mental age" group on the basis of performance on standardized intelligence tests. This mental age divided by the individual's chronological age and multiplied by 100 is the IQ.

intelligence quotient classification according to the Binet-Simon classification, intelligence quotients can be grouped as follows: genius, 140 and over; very superior, 120–139; superior, 110–119; average, 90–109; dull, 80–89; borderline, 70–79; mild retardation (moron), 50–69; moderate retardation (imbecile), 25–49; and severe retardation (idiot), 0–24.

interallelic complementation referring to the change in the properties of a multimeric protein as a consequence of the interaction of subunits coded by two different mutant alleles (in contrast to the protein consisting of subunits derived from a single mutant allele). The mixed protein (heteromultimer) may exhibit more activity (positive complementation) or less activity (negative complementation, *q.v.*) than the homomultimer. Also known as *intragenic complementation, allelic complementation.*

interbands the regions between bands in a polytene chromosome (*q.v.*). The DNA concentration in interbands is only a fraction of that in bands.

intercalary deletion *See* deletion.

intercalating agent a substance (e.g., acridine dyes) that inserts between base pairs in a DNA molecule, often disrupting the alignment and pairing of bases in the complementary strands. By causing addition or deletion of one or more base pairs during replication, a reading frame shift (*q.v.*) is often induced. *See* proflavin.

intercellular between cells.

interchange an exchange of segments between nonhomologous chromosomes resulting in translocations.

interchromosomal translocation *See* translocation.

intercistronic region the segment between the termination codon of one gene and the initiation codon of the next gene in a polycistronic transcription unit. *See* spacer DNA.

intercross mating of heterozygotes (*a*/+ × *a*/+).

interference *See* positive interference.

interference filter a filter used to produce a monochromatic light source.

interference microscope like the phase microscope, the interference microscope is used for observing transparent structures. However, with the interference microscope *quantitative* measurements of the relative retardation of light by various objects can be made. Such measurements can be used to determine the dry mass per unit area of specimen or the section thickness.

interferons (IFNs) a family of small glycoproteins produced by mammalian cells, often in response to viral infections. Type 1 IFNs are monomeric proteins that are produced by a wide variety of virus-infected cells. These IFNs induce synthesis of enzymes that inhibit viral proliferation. Type 2 IFNs consist of dimers of identical proteins unrelated to Type 1 IFNs. Type 2 IFNs are synthesized by T lymphocytes and natural killer cells and function to destroy certain cancer cells and cells infected by parasites. *See* Appendix C, 1977, Gilbert; leader sequence peptide, lymphocytes.

intergenic suppression *See* suppression.

interkinesis the abbreviated interphase between the first and second meiotic division. No DNA replication occurs during interkinesis, unlike a premitotic interphase.

interleukins a group of at least 15 soluble proteins, secreted by leukocytes, that function to promote the growth and differentiation of cells of the immune system. The different interleukins are designated IL1, IL2, etc., in order of their discovery. Most interleukins are products of single genes. Some ILs consist of two amino acid chains, but these result from the posttranslational cleavage of a single precursor protein. There is one exception, IL12, which has two chains (p35 and p40), each encoded by a different gene.

intermediary metabolism the chemical reactions in a cell that transform ingested nutrient molecules into the molecules needed for the growth of the cell.

intermediate filaments cytoplasmic filaments with diameters between 8 and 12 nanometers. They comprise a heterogeneous class of cytoskeletal proteins. In general, a given class of intermediate filaments is characteristic of a specific cell type. For example, keratin filaments are characteristic of epithelial cells, neurofilaments of neurons, vimentin filaments of fibroblasts, and desmin filaments of glial cells.

intermediate host a host essential to the completion of the life cycle of a parasite, but in which it does not become sexually mature.

intermedin a polypeptide hormone from the intermediate lobe of the pituitary gland that causes dispersion of melanin in melanophores. Also called melanocyte-stimulating hormone or MSH.

internal radiation the exposure to ionizing radiation from radioelements deposited in the body tissues.

interphase the period between succeeding mitoses. *See* cell cycle.

interrupted genes *See* split genes.

interrupted mating experiment a genetic experiment in which the manner of gene transfer between conjugating bacteria is studied by withdrawing samples at various times and subjecting them to a strong shearing force in an electric blender. *See* Appendix C, 1955, Jacob and Wollman; Waring blender.

intersex a class of individuals of a bisexual species that have sexual characteristics intermediate between the male and female. *See* Appendix C, 1915, Goldschmidt.

interspecific heterokaryons cells containing nuclei from two different species produced by cell fusion (*q.v.*). *See* Appendix C, 1965, Harris and Watkins.

interspersed elements *See* repetitious DNA.

interstitial cells cells that lie between the testis tubules of vertebrates and secrete testosterone.

intervening sequence *See* intron.

intra-allelic complementation *See* allelic complementation.

intrachromosomal aberration *See* translocation.

intrachromosomal recombination sister chromatid exchange (*q.v.*).

intrachromosomal translocation *See* translocation.

intragenic complementation *See* interallelic complementation.

intragenic recombination recombination between mutons of a cistron. Such recombination is characterized by negative interference and by nonreciprocality (recovery of either wild-type or double-mutant recombinants, but not both from the same tetrad).

intragenic suppression *See* suppression.

intrasexual selection *See* sexual selection.

introgression *See* introgressive hybridization.

introgressive hybridization the incorporation of genes of one species into the gene pool of another. If the ranges of two species overlap and fertile hybrids are produced, they tend to backcross with the more abundant species. This process results in a population of individuals, most of which resemble the more abundant parents but that also possess some characters of the other parent species. Local habitat modification can lead to mixing of previously distinct gene pools. Introduced species (or subspecies) can generate extinction of the older species by hybridization and introgression. The use of molecular markers has greatly increased the ability to detect and quantify interspecific gene exchanges. For example, phylogenetic trees based on chDNAs have shown many examples of both recent and ancient exchanges of chloroplasts between sympatric species. *See* chloroplast DNA (chDNA), wolf.

intromittent organ any male copulatory organ that implants sperm within the female.

intron in split genes (*q.v.*), a segment that is transcribed into nuclear RNA, but is subsequently removed from within the transcript and rapidly degraded. Most genes in the nuclei of eukaryotes contain introns. The number of introns per gene varies greatly, from one in the case of rRNA genes to more than 30 in the case of the yolk protein genes of *Xenopus*. Introns range in size from less than 100 to more than 10,000 nucleotides. There is little sequence homology among introns, but there are a few nucleotides at each end that are nearly the same in all introns. These boundary sequences participate in excision and splicing reactions. The first introns of some genes have been shown to contain tissue-specific enhancers. The splicing reactions involving the introns of nuclear mRNAs take place within a spliceosome (*q.v.*). However, the introns of mitochondrial and chloroplast DNAs are self-splicing. *See* Appendix C, 1977, Roberts and Sharp; 1978, Gilbert; 1983, Gillies *et al.*; alternative splicing, *Caenorhabditis elegans*, enhancers, *Euglena gracilis*, exon, GT-AG rule, posttranscriptional processing, R-loop mapping, splice junctions, transcription unit.

intron dynamics the lengthening or shortening of the non-coding regions of specific genes during their evolution. A pairwise alignment of about 6,000 orthologous genes demonstrates that the equivalent introns in genes of *Drosophila* are only half the length of *Anopheles* introns, whereas the exon lengths and intron frequencies are similar in both insects. Therefore a change in intron lengths has occurred during the period since the divergence of fruit flies and mosquitoes from a common dipteran ancestor (about 250 million years ago). This difference in intron lengths explains why the *A. gambiae* genome is larger than the *D. melanogaster* genome (278 *vs* 180 mbp), although both species have 13,000–14,000 genes. *See* Appendix C, 2002, Holt *et al.*

intron intrusion the disruption of a preexisting gene by the insertion of an intron into a functional gene. Intron intrusion and the exon shuffling (*q.v.*) along with junctional sliding (*q.v.*), have been proposed as mechanisms for evolutionary diversification of genes.

intron-mediated recombination *See* exon shuffling.

intron origins two conflicting hypotheses have been proposed to explain the origin of introns. The *introns early hypothesis* assumes that the DNA molecules in which genes originated initially contained random sequences of nucleotides. The random distribution of stop codons permitted only short reading frames to accumulate. Next, a mechanism arose that allowed splicing out regions containing stop codons from the primary message, and so proteins of greater length and with more useful biochemical functions could be translated and selected. The original short reading frames became the exons of present-day genes, while the introns represent segments containing splice junctions originally designed to remove deleterious stop signals. The *introns late hypothesis* assumes that genes arose from short reading frames that grew larger by duplications and fusions. Introns arose secondarily as a result of insertions of foreign DNA into these genes. Thus, present-day introns are the descendants of ancient transposons (*q.v.*).

intussusception 1. the growth of an organism by the conversion of nutrients into protoplasm. 2. the deposition of material between the microfibrils of a plant cell wall. 3. the increase in surface area of the plasmalemma by intercalation of new molecules between the existing molecules of the extending membrane.

in utero within the uterus.

inv *See* symbols used in human cytogenetics.

in vacuo in a vacuum.

invagination an inpocketing or folding in of a sheet of cells or a membrane.

inversion chromosome segments that have been turned through 180° with the result that the gene sequence for the segment is reversed with respect to that of the rest of the chromosome. Inversions may include or exclude the centromere. An inversion

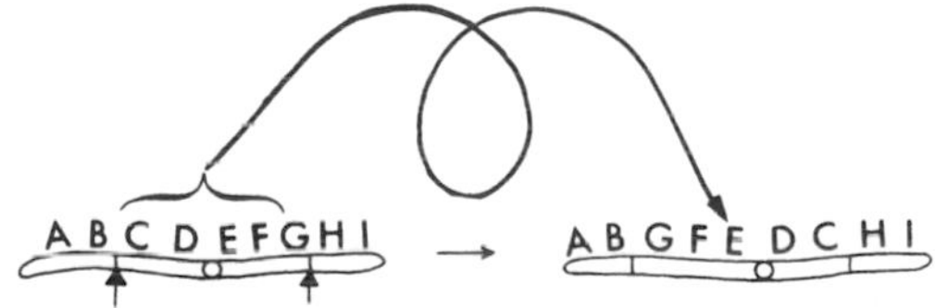

that includes the centromere is called *pericentric* or *heterobrachial*, whereas an inversion that excludes the centromere is called *paracentric* or *homobrachial*. Paracentric inversions are found more often in nature than pericentric inversions. A paracentric inversion heterozygote forms a reverse loop pairing configuration during pachynema.

inversion heterozygote an organism in which one of the homologs has an inverted segment while the other has the normal gene sequence. The results of single and double exchanges within an inversion heterozygote are shown on page 237. Note that no monocentric, single-crossover chromatids are produced. For this reason, inversions give the impression of being *crossover suppressors*, and it was their action on crossing over that led to their discovery. *See* Appendix C, 1926, Sturtevant; 1933, McClintock; 1936, Sturtevant and Dobzhansky.

invertebrate an animal without a dorsal column of vertebrae; nonchordate metazoans.

inverted repeats (IR) two copies of the same DNA sequence orientated in opposite directions on the same molecule. IR sequences are found at opposite ends of a transposon (*q.v.*). *See* palindrome.

inverted terminal repeats short, related, or identical sequences oriented in opposite directions at the ends of some transposons (*q.v.*).

in vitro designating biological processes made to occur experimentally in isolation from the whole organism; literally "in glass," i.e., in the test tube. Examples: tissue cultures, enzyme-substrate reactions. *Contrast with in vivo, ex vivo.*

***in vitro* complementation** *See* allelic complementation.

***in vitro* evolution** experiments designed to study the evolution of self-duplicating nucleic acid molecules outside of living cells. A classic example is a study that involved the synthesis of RNA molecules using Qβ replicase and the RNA genome of Qβ phage. Serial transfer experiments were performed in which the intervals of synthesis were adjusted to select the earliest molecules completed. As the experiment progressed, the rate of RNA synthesis increased, while the product became smaller. By the 74th transfer, an RNA molecule had evolved that was only 17% of its original size and constituted the smallest known self-duplicating molecule. While it had a very high affinity for the replicase, it was unable to direct the syntheses of viral particles. Ribozymes (*q.v.*) have also been shown to undergo *in vitro* evolution. *See* Appendix C, 1967, Spiegelmann, Mills, and Peterson; 1973, Mills, Kramer, and Spiegelmann; 1995, Wilson and Szostak.

***in vitro* fertilization** experimental fertilization of an egg outside the female body. In humans, this is usually done because the woman's Fallopian tubes are blocked. The resulting embryo can then be inserted into the uterus for implantation. *See* embryo transfer.

***in vitro* marker** a mutation induced in a tissue culture that allows subsequent phenotypic detection. Human *in vitro* markers include genes conferring resistance to various viruses, aminopterin, and purine analogs.

***in vitro* mutagenesis** experiments in which segments of genomic DNA are treated with reagents that produce localized chemical changes in the molecule. The subsequent ability of the mutated molecules to function during replication, transcription, etc., is assayed either by using cell-free systems or *in vivo*, after splicing the fragment into an appropriate plasmid.

***in vitro* packaging** the production of infectious particles from naked DNA by incapsidation of the DNA in question after supplying lambda phage packaging proteins and preheads.

***in vitro* protein synthesis** the incorporation in a cell-free system of amino acids into polypeptide chains. *See* Appendix C, 1976, Pelham and Jackson.

in vivo within the living organism. *Contrast with in vitro, ex vivo.*

***in vivo* culturing of imaginal discs** the technique developed by Hadorn in which an imaginal disc is removed from a mature *Drosophila* larva, cut in half, and the half organ implanted into a young larva. Here regenerative growth occurs, and once the host larva has reached maturity the implant is removed once again, bisected, and one of the halves transplanted to a new host. By multiple repetition of this procedure, the cells are subjected to an abnormally long period of division and growth in a larval environment. If the regenerated disc is finally allowed to undergo metamorphosis, it shows an abnormally high probability of producing structures characteristic of different discs. A regenerated genital disc may produce antennae, for example. Hadorn terms such differentiation *allotypic*. Since the allotypic organs appear in the offspring of cells that were previously determined to form genital structures, a change in determination must be postulated. This event is called *transdetermination*. *See* Appendix C, 1963, Hadorn.

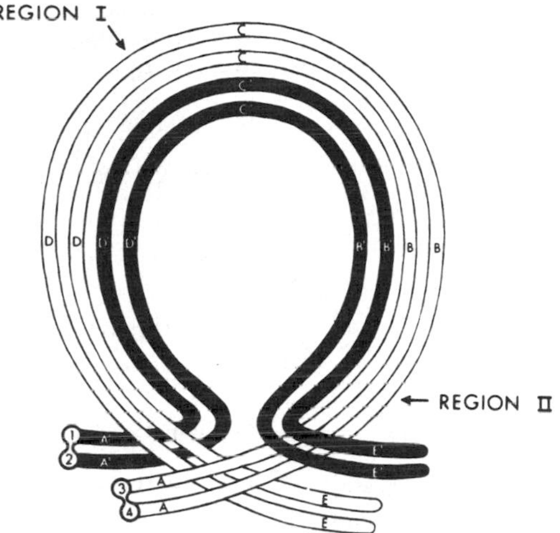

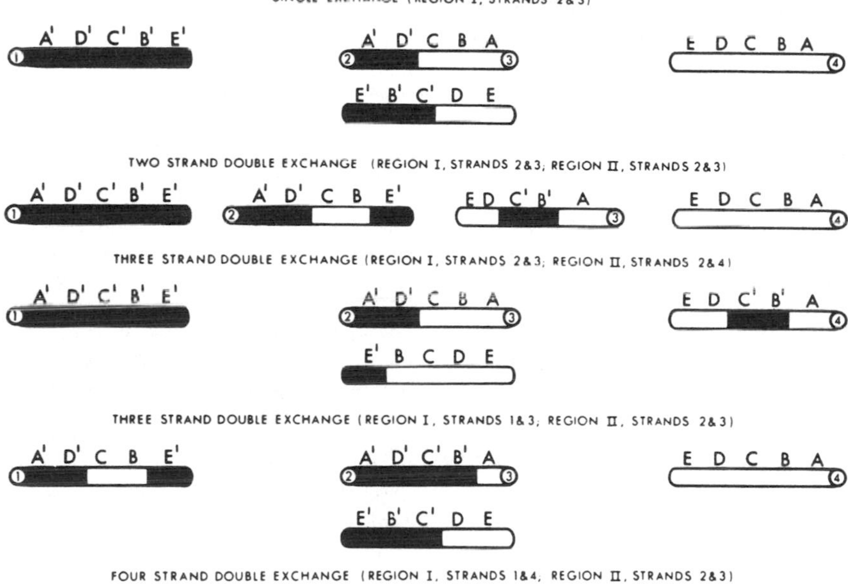

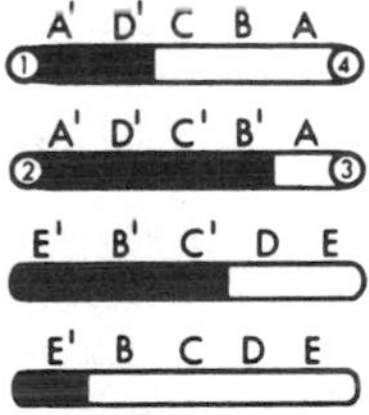

Inversion heterozygote

***in vivo* marker** a naturally occurring mutant mammalian gene that allows phenotypic detection of the tissue-cultured cells bearing it. Examples are the genes causing galactosemia and glucose-6-phosphate dehydrogenase deficiency in man, and the genes producing certain cell surface antigens in the mouse.

iodine a biological trace element. Atomic number 53; atomic weight 129.9044; valence 1 ; most abundant isotope ^{127}I, radioisotopes ^{125}I (half-life 60 days) and ^{131}I (half-life 8 days), radiations: beta particles and gamma rays. Radioisotopes of iodine are commonly used in radioimmunoassay (*q.v.*).

iojap a mutant nuclear gene in maize that induces changes in chloroplast characters. The mutant plastids behave autonomously thereafter.

ion exchange column a column packed with an ion exchange resin. *See* chromatography, column chromatography.

ion exchange resin a polymeric resin that has a higher affinity for some charged groups than it has for others. For example, resins with fixed cation groups will bind anions and thus can be used in column separation procedures. *See* molecular sieve.

ionic bond electrostatic bond (*q.v.*).

ionization any process by which a neutral atom or molecule acquires a positive or negative charge.

ionization chamber any instrument designed to measure the quantity of ionizing radiation in terms of the charge of electricity associated with ions produced within a defined volume.

ionization track the trail of ion pairs produced by an ionizing radiation during its passage through matter.

ionizing energy the average energy lost by an ionizing radiation in producing an ion pair in a given gas. The average ionizing energy for air is about 33 eV.

ionizing event the occurrence of any process in which an ion or group of ions is produced.

ionizing radiation electromagnetic or corpuscular radiation that produces ion pairs as it dissipates its energy in matter.

ionophores a class of antibiotics of bacterial origin that facilitate the movement of monovalent and divalent cations across biological membranes. Some of the major ionophores and the ions they transport are valinomycin (K^+, Rb^+), A 23187 (Ca^{++}, $2H^+$), nigericin (K^+, H^+), and gramicidin (H^+, Na^+, K^+, Rb^+).

ion pair the electron and positive atomic or molecular residue resulting from the interaction of ionizing radiation with the orbital electrons of atoms.

IPTG isopropylthiogalactoside; a gratuitous inducer for the *E. coli lac* operon (*q.v.*). *See* ONPG.

IQ intelligence quotient (*q.v.*).

I region one of the central regions of the major histocompatibility complex (H-2) of the mouse. It contains genes coding for Ia antigens and controlling various immune responses. It has five subregions (A, B, J, E, C) and may be the equivalent of the D/DR region of the human major histocompatibility complex. *See* HLA complex.

IR inverted repeat (*q.v.*).

Ir gene *See* immune response gene.

iron a biological trace element. Atomic number 26; atomic weight 55.847; valence $2,3^+$; most abundant isotope ^{56}Fe; principal radioisotope ^{59}Fe, half-life 46 days, radiation emitted—beta particle.

isauxesis *See* allometry, heterauxesis.

IS element *See* insertion sequences.

islets of Langerhans clusters of hormone-secreting cells located in the pancreas of vertebrates. Two types of cells are found: alpha cells, which secrete glucagon (*q.v.*), and beta cells, which secrete insulin (*q.v.*).

isoacceptor transfer RNA one of a group of different tRNAs that accept the same amino acid but possess different anticodons. Higher organisms contain two to four isoacceptor tRNAs for certain amino acids. *See* amino acid.

isoagglutinin an antibody directed against antigenic sites on the red blood corpuscles of the same species and that causes agglutination.

isoagglutinogen an antigenic factor on the surface of cells that is capable of inducing the formation of homologous antibodies (isoagglutinins) in some members of the same species.

isoallele an allele whose effect can only be distinguished from that of the normal allele by special tests. For example, two + alleles $+^1$ and $+^2$ may be indistinguishable (i.e., $+^1/+^1$, $+^2/+^2$, and $+^2/+^1$ individuals are phenotypically wild type). However, when compounded with a mutant allele *a*, $+^1$ and $+^2$ prove to be distinguishable (i.e., $a/+^1$ and $a/+^2$ individuals are observably different).

isoanisosyndetic alloploid an allopolyploid in which some chromosomes derived from both species are homoeologous and undergo a limited synapsis. *See* isosyndetic alloploid.

isoantibody an antibody formed in response to immunization with tissue constituents derived from an individual of the same species as the recipient.

isocapsidic viruses *See* segmented genome.

isochore a segment of DNA that has a uniform base composition that is different from adjacent segments. The DNA of vertebrates and plants are mosaics of such isochores. In humans, isochores are about 300 kb in length and consist of five classes. The AT-rich isochores are called *L1* and *L2*, and the GC-rich isochores are *H1*, *H2*, and *H3*. Although H3 makes up only 3% of the total DNA, it contains over 25% of the ORFs. *See* major histocompatibility complex (MHC).

isochromatid break an aberration involving breaks in both sister chromatids at the same locus, followed by lateral fusion to produce a dicentric chromatid and an acentric fragment.

isochromosome a metacentric chromosome produced during mitosis or meiosis when the centromere splits transversely instead of longitudinally. The arms of such a chromosome are equal in length and genetically identical. However, the loci are positioned in reverse sequence in the two arms.

isocoding mutation a point mutation that alters the nucleotide sequence of a codon but, because of the degeneracy of the genetic code, does not change the amino acid that the codon specifies.

isoelectric point the pH at which the net positive and negative charge on a protein is zero.

isoenzymes isozymes (*q.v.*).

isofemale line a genetic lineage that began with a single inseminated female.

isoforms families of functionally related proteins that differ slightly in their amino acid sequences. Such proteins may be encoded by different alleles of the same structural gene. The D and d isoforms encoded by the *MC1R* gene (*q.v.*) are examples. Other proteins are encoded by genes that are now located at different chromosomal positions but are believed to be derived from a single ancestral gene. Isoforms may also be generated by mRNAs transcribed from different promoters located in the same gene or by alternative splicing (*q.v.*). *See* actin, actin genes, fibronectin, multigene family, muscular dystrophy, myosin, myosin genes, tropomyosin, tubulin.

isogamy that mode of sexual reproduction involving sex cells of similar size and morphology but opposite mating types. *See* anisogamy.

isogeneic referring to a graft involving genetically identical donor and host; an isograft.

isogenic genetically identical (except possibly for sex); coming from the same individual or from a member of the same inbred strain.

isograft a tissue graft between two individuals of identical genotype.

isohemagglutinin isoagglutinin (*q.v.*).

isoimmunization antibody formation in reaction to antigens of the same species.

isolonic point isoelectric point (*q.v.*).

isolabeling labeling of both, or parts of both, daughter chromatids at the second metaphase after one replication in tritiated thymidine, as a result of sister chromatid exchange. In the absence of sister chromatid exchange, both daughter chromatids are labeled at metaphase I, but only one is labeled at metaphase II.

isolate a segment of a population within which assortative mating occurs.

isolating mechanism a cytological, anatomical, physiological, behavioral, or ecological difference, or a geographical barrier that prevents successful mating between two or more related groups of organisms. *See* postzygotic isolation mechanism, prezygotic isolation mechanism, Wallace effect.

isolecithal egg one in which the yolk spheres are evenly distributed throughout the ooplasm. *See* centrolecithal egg, telolecithal egg.

isoleucine *See* amino acid.

isologous synonymous with isogeneic (*q.v.*).

isologous cell line cell lines derived from identical twins or from highly inbred animals.

isomerases a heterogeneous group of enzymes that catalyze the transfer of groups within molecules to yield isomeric forms. An example would be *racemase*, which interconverts D-lactic acid and L-lactic acid.

isomers compounds with the same molecular formula but with different three-dimensional molecular shapes or orientations in space.

isometry isauxesis.

isomorphous replacement a technique that allows specific atoms in a complex molecule to be replaced with atoms of higher atomic number. Differences can then be seen in the intensities of specific spots on diffraction patterns when crystals of these molecules are x-rayed. *See* x-ray crystallography.

isonymous marriage marriage between persons with the same surname. Isonymous marriages are used as indications of consanguinity in population genetics.

isophene a line on a map which connects points of equal expression of a character that varies clinally.

isoprenoid lipid a family of lipid molecules made up of linear arrays of multiple isoprene units. Isoprene has the formula

$$\begin{array}{c} \quad CH_3 \\ \quad | \\ CH_2{=}CH{=}CH_2. \end{array}$$

The fat-soluble vitamins (A, D, K, and E) contain multiple isoprene units.

isopropylthiogalactoside a gratuitous inducer of the *lac* operon (*q.v.*); abbreviated IPTG.

isopycnic having the same density; used to refer to cell constituents having similar buoyant densities. *See* centrifugation separation.

isopycnotic referring to chromosomal regions or entire chromosomes that are not heteropycnotic, that is, are the same in appearance as the majority of the chromosomes. *See* heteropycnosis.

isoschizomers two or more restriction endonucleases (*q.v.*) isolated from different sources that cleave DNA within the same target sequences.

isosyndetic alloploid an allopolyploid where synapsis is restricted to the homologs derived from one species. *See* isoanisosyndetic alloploid.

isotonic solution a solution having the same osmotic pressure as another solution with which it is compared (usually blood or protoplasm).

isotope one of the several forms of a chemical element. Different isotopes have the same number of protons and electrons, but differ in the number of neutrons contained in the atomic nucleus. Hence they have identical chemical properties, but differ in atomic weights. *See* Appendix C, 1942, Schoenheimer; radioactive isotope.

isotopic dilution analysis a method of chemical analysis for a component of a mixture. The method is based on the addition to the mixture of a known amount of labeled component of known specific activity, followed by isolation of a quantity of the component and measurement of the specific activity of that sample.

isotopically enriched material material in which the relative amount of one or more isotopes of a constituent has been increased.

isotropic *See* anisotropy.

isotype exclusion synthesis of only kappa or lambda light chains by a given plasma cell as a consequence of allelic exclusion (*q.v.*). *See* immunoglobulin.

isotypes antigenic determinants shared by all individuals of a given species, but absent in individuals of other species. *Compare with* allotypes, idiotypes.

isozymes multiple forms of a single enzyme. While isozymes of a given enzyme catalyze the same reaction, they differ in properties such as the pH or substrate concentration at which they function best. Isozymes are complex proteins made up of paired polypeptide subunits. The lactic dehydrogenases, for example, are tetramers made up of two polypeptide units, A and B. Five isozymes exist and can be symbolized as follows: AAAA, AAAB, AABB, ABBB, and BBBB. Isozymes often have different isoelectric points and therefore can be separated by electrophoresis. The different monomers of which isozymes like lactic dehydrogenase are composed are specified by different gene loci. The term *allozyme* is used to refer to variant proteins produced by allelic forms of the same locus. *See* allozymes.

iteroparity repeated periods of reproduction during the life of an individual. *Compare with* semelparity.

IVS intervening sequence. *See* intron.

J

Janus kinase 2 a protein kinase (*q.v.*) that functions in cellular signal transduction (*q.v.*). Janus kinases phosphorlyate specific tyrosine residues in substrate proteins. The gene (*JAK 2*) which encodes the enzyme is located at 9p24. One function of the JAK 2 protein is to control the responses of erythroblasts to erythropoietin (*q.v.*). Base substitutions at certain positions in the *JAK 2* gene cause polycythemia vera (*q.v.*).

Japanese quail *See Coturnix coturnix japonica.*

jarovization synonym for vernalization (*q.v.*).

jaundice yellowing of the skin, whites of the eyes, and certain body fluids due to abnormally high levels of bilirubin (*q.v.*) in the blood. *See* Crigler-Najjar syndrome, hereditary spherocytosis (HS).

Java man an extinct subspecies of primitive man known from fossils obtained in central Java. Now classified as *Homo erectus erectus*, but formerly referred to as *Pithecanthropus erectus*.

J chain a small protein of about 15,000 daltons that holds the monomeric units of a multimeric immunoglobulin together, as occurs in the classes IgM and IgA.

Jews those people who belong to the ancient Near Eastern Hebrew tribe, the Israelites; also those who trace their descent to Israelites by genealogy or religious conversion to Judaism. Recently an analysis was made of Y-chromosome markers from males who belonged to various Jewish and Moslem populations residing in Europe and North Africa and on the Arabian Peninsula. The results show that the geographically dispersed Jewish communities closely resemble not only one another but also Palestinians, Syrians, and Lebanese. This analysis suggests that all populations are descended from a common ancestral tribe that lived in the Middle East about 4,000 years ago. *See* Appendix C, 2000, Hammer *et al.*; Ashkenazi.

J genes a tandem series of four or five homologous nucleotide sequences coding for part of the hypervariable regions of light and heavy chains of mouse or human immunoglobulins; so named because they help *join* one of the genes for the variable region upstream to one of the genes for the constant region downstream and therefore are an important part of the mechanism generating antibody diversity.

JH juvenile hormone. *See* allatum hormones.

Jordan rule an evolutionary principle put forth by the German entomologist Karl Jordan in 1905. It states that closely related species or subspecies are generally adjacent, but separated by a natural barrier (such as a river) that neither can cross easily.

jumping genes mobile or "nomadic" genetic entities such as insertion elements and transposons.

junctional complex a term used in electron microscopy to refer to any specialized region of intercellular adhesion, such as a desmosome (*q.v.*).

junctional sliding a term descriptive of the fact that the location of intron-exon junctions is not constant within members of a gene family, such as the serine proteases. Some variation in length of such gene products can be attributed to extension or contraction of exons at the intron junctions.

junk DNA a term sometimes used to refer to the majority of the DNA in most eukaryotic genomes which does not seem to have a coding or regulatory function. Like "junk," it is of questionable value, but is not thrown out. *See* Alu family, selfish DNA, skeletal DNA hypothesis.

Jurassic the middle period in the Mesozoic era, during which the dinosaurs became the dominant land vertebrates. Flying reptiles called pterosaurs evolved, and the first birds appeared. Archaic mammals persisted. Ammonites underwent great diversification, and teleost fishes made an appearance. The fragments formed from Pangea began to separate. *See* Archaeopteryx, continental drift, geologic time divisions.

juvenile hormone *See* allatum hormones.

K

K **1.** degrees Kelvin. *See* temperature. **2.** Cretaceous. **3.** potassium. **4.** the gene in *Paramecium aurelia* required for the maintenance of kappa. **5.** carrying capacity (*q.v.*).

kairomone a *trans*-specific chemical messenger the adaptive benefit of which falls on the recipient rather than the emitter. Kairomones are commonly nonadaptive to the transmitter. For example, a secretion that attracts a male to the female of the same species may also attract a predator. *See* allomone.

Kalanchoe a genus of succulent plants studied in terms of the genetic control of photoperiodic flowering response. *See* phytochrome.

kanamycin an antibiotic that binds to the 70S ribosomes of bacteria and causes misreading of the mRNA.

K and r selection theory *See* r and K selection theory.

kangaroo rat *See Dipodomys ordii.*

K antigens *See* O antigens.

kappa symbiont *See* killer paramecia.

karyogamy the fusion of nuclei, usually of the two gametes in fertilization; syngamy.

karyokinesis nuclear division as opposed to cytokinesis (*q.v.*).

karyolymph nucleoplasm (*q.v.*).

karyon nucleus (*q.v.*).

karyoplasm nucleoplasm (*q.v.*).

karyosome a Feulgen-positive body seen in the nucleus of the *Drosophila* oocyte during stages 3–13. During stages 3–5, it contains synaptonemal complexes.

karyosphere the condensed Feulgen-positive mass seen in the anterior, dorsal portion of the mature primary oocyte of *Drosophila melanogaster*. This mass of DNA is not surrounded by a nuclear envelope. The tetrads subsequently emerge from the karyosphere and enter metaphase of the first meiotic division. The karyosphere stage of oogenesis is the most radiation-sensitive one.

karyotheca nuclear envelope (*q.v.*).

karyotype the chromosomal complement of a cell, individual, or species. It describes the light microscopic morphology of the component chromosomes, so that their relative lengths, centromere positions, and secondary constrictions can be identified. Attention is called to heteromorphic sex chromosomes. The karyotype is often illustrated with a figure showing the chromosomes placed in order from largest to smallest. This illustration, called an *idiogram*, may be constructed by aligning photomicrographs of individual chromosomes, or it may be an inked drawing summarizing the data from a series of analyses of chromosome spreads. *See* human mitotic chromosomes.

kb *See* kilobase.

KB cells a strain of cultured cells derived in 1954 by H. Eagle from a human epidermoid carcinoma of the nasopharynx.

kbp kilobase pairs.

K cells killer cells that mediate *a*ntibody-*d*ependent *c*ellular *c*ytotoxicity (ADCC). These cells and natural killer (NK) cells have many similar properties, and may belong to the same cell lineage (lymphocyte or monocyte). Neither K nor NK cells have surface markers characteristic of either T cells (sheep red blood cell receptors) or B cells (endogenous surface immunoglobulins). Both K and NK cells possess Fc receptors for class Ig immunoglobulins and thus may acquire membrane-bound antibodies that react with target cells bearing the corresponding antigens. K cells cannot exhibit cytotoxicity without their bound antibodies; NK cells are not so restricted. Prior contact with the antigen is required by the host in order to arm its K cells with antibodies effective in ADCC.

kDa kilodalton. *See* dalton.

kDNA kinetoplast DNA. *See* kinetoplast.

***kelch* (*kel*)** a *Drosophila* gene on chromosome 2L at 36D3. It encodes an actin-binding protein that is characterized by the presence of six 50-amino acid *kelch* repeats. The phosphorylation of the kelch protein is required for the proper morphogenesis of ovarian ring canals (*q.v.*), and a protein kinase encoded by *Src* gene catalyzes this reaction. *See* actin, *Src*.

Kell-Cellano antibodies antibodies against the red-cell antigens specified by the *K* gene, named for the first patient known to produce them. *See* blood groups.

kelp the largest of the seaweeds. The giant kelps of the genus *Macrocystis* reach lengths of 100 meters and form great forests in shallow oceans. *See* agar, agarose, Phaeophyta.

keratins a family of insoluble, cystine-rich intracellular proteins that are a major component of epidermal coverings such as hair, fur, wool, feathers, claws, hoofs, horns, scales, and beaks. There are many types of keratins, encoded by a large family of genes. Epidermal cells produce a sequence of different keratins as they mature. In humans, there are more than 20 different keratins synthesized by epithelial cells. Mutations in keratin genes cause hereditary blistering diseases in humans and feather defects in chickens. The fibroin of insect silk also belongs to the keratin family. *See* dominant negative mutation, epidermolysis bullosa, frizzle, intermediate filaments, silk.

kernel the seed of a cereal plant such as corn or barley. The sectioned kernel shown below could be any one of the thousand or so found on a corn ear. Each kernel consists of a relatively small diploid embryo, a triploid endosperm, and a tough diploid layer of maternal origin, the pericarp. The surface cells of the endosperm contain aleurone grains and oil. The remaining cells contain starch. The scutellum serves to digest and absorb the endosperm during the growth of the embryo and seedling. *See* anthocyanines, vacuoles.

keto forms of nucleotides *See* tautomeric shift.

keV *See* electron volt.

Kidd blood group a blood group defined by a human red cell antigen encoded by the *JK* gene at 18q 11–12. It is about 30 kb long and encodes an integral membrane glycoprotein that functions in the transport of urea. The antibody was discovered in 1951 and given the family name of the female patient who produced it.

killer paramecia paramecia that secrete into the medium particles that kill other paramecia. The killer trait is due to kappa particles, which reside in the cytoplasm of those strains of *Paramecium aurelia* syngen 2 that carry the dominant K gene. Later it was found that kappa particles were symbiotic bac-

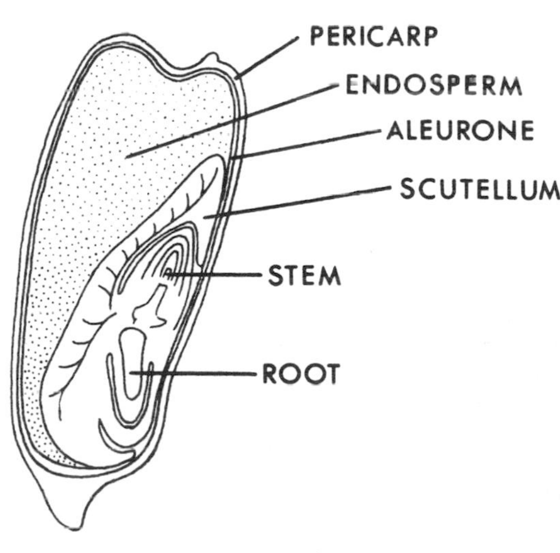

Kernel

teria and that the particles with killing activity were defective DNA phages. The lysogenic, symbiotic bacterium has been named *Caenobacter taenospiralis*, and it is but one of many kinds of bacterial endosymbionts that occur in over 50% of the *P. aurelia* collected in nature. *See* Appendix C, 1938. Sonneborn; *Paramecium aurelia.*

killer particle in yeast, a double-stranded RNA plasmid containing 10 genes for replication and several others for synthesis of a killer substance similar to bacterial colicin (*q.v.*). It is the only known plasmid that does not contain DNA.

killifish a common name for *Oryzia latipes* (*q.v.*).

kilobase a unit of length for nucleic acids consisting of 1,000 nucleotides; abbreviated kb, or kbp for kilobase pairs (DNA).

kilovolt a unit of electrical potential equal to 1,000 volts, symbolized by kV.

kinase an enzyme that catalyzes the transfer of a phosphate group from ATP to a second substrate. *See* protein kinases.

kindred a group of human beings each of which is related, genetically or by marriage, to every other member of the group.

kinesins a superfamily of proteins found in cells that contain microtubules. One function of the kinesin motor protein is to move vesicles and particles laterally along these tubules toward their distal ends. Kinetochores also contain specific kinesins that drive each kinetochore along the spindle microtubules toward the poles.

kinetic complexity *See* DNA complexity.

kinetin *See* cytokinins.

kinetochore *See* centromere, *MAD* mutations.

kinetoplast a highly specialized mitochondrion associated with the kinetosome of trypanosomes. Kinetoplast DNA is the only DNA known in nature that is in the form of a network consisting of interlocked circles. There are about 50 maxicircles and 5,000 minicircles per kinetoplast network. The maxicircles contain the genes essential for mitochondrial biogenesis. Unlike the maxicircles, the minicircles are not transcribed, and their function is unknown.

kinetosome a self-duplicating organelle homologous to the centriole. Kinetosomes reside at the base of undulipodia (*q.v.*) and are responsible for their formation. *See* Appendix C, 1976, Dippell; basal body (granule), centriole.

kinety a row of interconnected kinetosomes on the surface of a ciliate.

kingdom systems *See* classification.

king-of-the-mountain principle *See* first-arriver principle.

kinin *See* cytokinins.

kin selection a term invented by John Maynard Smith. *See* Hamilton genetical theory of social behavior.

Kjeldahl method a technique often used for the quantitative estimation of the nitrogen content of biological material.

Kleinschmidt spreading technique a procedure developed by A. K. Kleinschmidt that allows DNA molecules to be viewed under the electron microscope. The DNA is mounted in a positively charged protein film formed on the surface of an aqueous solution. The film of protein serves to hold the DNA in a relaxed but extended configuration and allows the sample to be transferred to a hydrophobic electron microscope grid when it is touched to the surface film. *See* denaturation map.

Klenow fragment the larger of two fragments obtained by the enzymatic cleavage of the DNA polymerase I of *E. coli*. The drawing below shows that the Klenow fragment lacks 5′ → 3′ exonuclease activity. Because of this, the Klenow fragment has various uses in genetic engineering techniques and in DNA sequencing methodologies where the 5′ → 3′

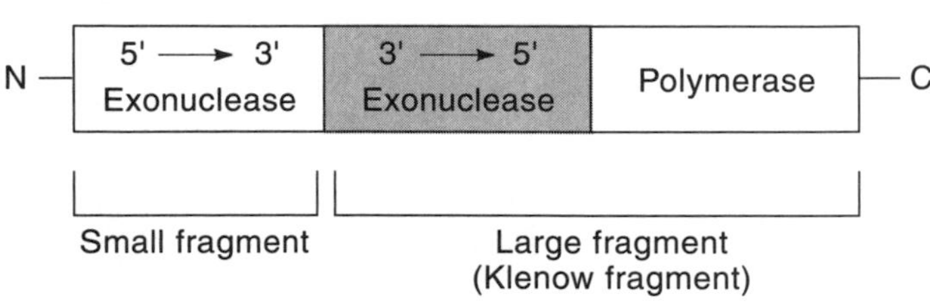

Klenow fragment

exonuclease activity of intact DNA polymerase I is disadvantageous. *See* Appendix C, 1971, Klenow; DNA polymerase.

Klinefelter syndrome (KS) a genetic disease that produces sterile males with small testes lacking sperm. Dr. Harry Klinefelter accurately described the condition in 1943, but the underlying chromosomal abnormality was not discovered until 1959. The most common karyotype is XXY AA. Less common variations such as XXYY and XXXY also occur. KS demonstrated that maleness in humans depends on the presence of the Y, not the number of X chromosomes. KS appears once in every 500 to 1,000 live-born males. Aside from their sterility, most KS males lead normal healthy lives, although some are mildly retarded. *See* Appendix C, 1959, Jacobs and Strong.

Km the Michaelis constant (*q.v.*).

knife breaker a mechanical apparatus that provides a method for breaking strips of plate glass first into squares and then into triangles. These are used as knives against which plastic-embedded tissues are cut into ultrathin sections for observation under the electron microscope. *See* ultramicrotome.

knob in cytogenetics, a heavily staining enlarged chromomere that may serve as a landmark, allowing certain chromosomes to be identified readily in the nucleus. In maize, knobbed chromatids preferentially enter the outer cells of a linear set of four megaspores during megasporogenesis and are therefore more likely to be included in the egg nucleus (*see* meiotic drive); genetic markers close to a knob tend to appear more frequently in gametes than those far from a knob.

knockout an informal term coined for the generation of a mutant organism (generally a mouse) containing a null allele of a gene under study. Usually the animal is genetically engineered with specified wild-type alleles replaced with mutated ones. The URL for the Mouse knockout database is http://research.bmn.com/mkmd. *See* gene targeting, homologous recombination.

Knudson model the "two hit" model of carcinogenesis invented by Alfred K. Knudson to explain clinical-epidemiological observations on retinoblastoma (*q.v.*). This revolutionary concept proposed that certain cancers were caused, not by the presence of an oncogene, but by the absence of an anti-oncogene. In the case of retinoblastoma, children with both eyes affected had a germ-line mutation that predisposed them to the disease. However, a second "hit" or mutation was needed to produce the cancer. Patients without the germ-line mutation required two hits. Extra time was required to acquire the first of the two mutations, and so children with hereditary retinoblastoma developed the disease in both eyes or at multiple sites in one eye, whereas children lacking the RB gene developed single tumors and at a later age. *See* Appendix C, 1971, Knudson.

Kornberg enzyme the DNA polymerase isolated from *E. coli* in 1959 by a group led by A. Kornberg; now called *DNA polymerase I*; it functions mainly in repair synthesis (*q.v.*).

Krebs cycle a synonym for the citric acid cycle (*q.v.*). It was named to honor Hans Adolph Krebs, the biochemist who discovered it. *See* Appendix C, 1937, Krebs.

K strategy a type of life cycle relying on finely tuned adaptation to local conditions rather than on high reproductive rate. *See* r and K selection theory.

Kupffer cells phagocytotic macrophages residing in the liver and first described by the German histologist K. W. von Kupffer in 1876.

kurtosis the property of a statistical distribution that produces a steeper or shallower curve than a normal distribution (*q.v.*) with the same parameters.

kuru a chronic, progressive, degenerative disorder of the central nervous system found in the Fore natives living in a restricted area of New Guinea. The disease was at one time thought to be genetically determined, but it is now believed to be caused by a prion (*q.v.*).

kV kilovolt (*q.v.*).

kwashiorkor a severe nutritional disorder due to a deficiency of certain amino acids (especially lysine). Kwashiorkor occurs in humans that subsist on a diet of cereal proteins deficient in lysine. *See* *opaque-2*.

L

L 1. line. 2. levorotatory. 3. liter.

label the attachment of any substance to a cell or molecule of interest that allows these targets to be readily identified, quantitated, and/or isolated from all other objects in either an *in vitro* or an *in vivo* system. Commonly used labels are dyes, fluorescent compounds, enzymes, antibodies, and radioactive elements of compounds. Labels are sometimes referred to as *tags*.

label, electron dense *See* ferritin.

label, heavy a heavy isotopic element introduced into a molecule to facilitate its separation from otherwise identical molecules containing the more common isotope. *See* Appendix C, 1958, Meselson and Stahl.

lac operon in *E. coli*, a DNA segment about 6,000 base pairs long that contains an operator sequence and the structural genes *lac Z, lac Y*, and *lac A*. The structural genes code for beta galactosidase, beta galactoside permease, and beta galactoside transacetylase, respectively. The three structural genes are transcribed into a single mRNA from a promoter lying to the left of the operator. Whether or not this mRNA is transcribed depends upon whether or not a repressor protein is bound to the operator, a regulatory sequence of 24 base pairs. The repressor protein is encoded by *lac I*, a gene lying to the left of the *lac* promoter. Beta galactosidase (*q.v.*) catalyzes the hydrolysis of lactose (*q.v.*) into glucose and galactose. After glucose and galactose are produced, a side reaction occurs, forming allolactose. This is the inducer that switches on the *lac* operon. It does so by binding to the repressor and inactivating it. *See* Appendix C, 1961, Jacob and Monod; 1969, Beckwith *et al*; IPTG, *lac* repressor, ONPG, polycistronic mRNA, regulator gene, reporter gene.

lac repressor the protein that regulates the *lac* operon in *E. coli*. The protein is the product of the *lac I* gene and functions as a molecular switch in response to inducer molecules. A single bacterium contains only 10 to 20 *lac* repressor molecules. Each is a homotetramer of Mr 154,520. The monomeric subunit has 360 amino acids, and its structure is diagrammed in A, below. It is composed of four functional domains: the head piece (HP), core domains 1 and 2 (CD1 and CD2), and the tail piece (TP). The HP is at the N terminus (NT), and it contains four alpha helices (represented by circles 1–4), which function in DNA binding. Together, the core domains contain 12 beta sheets (represented by squares A–L) sandwiched between nine alpha helices (circles 5–13). The tail piece contains an alpha

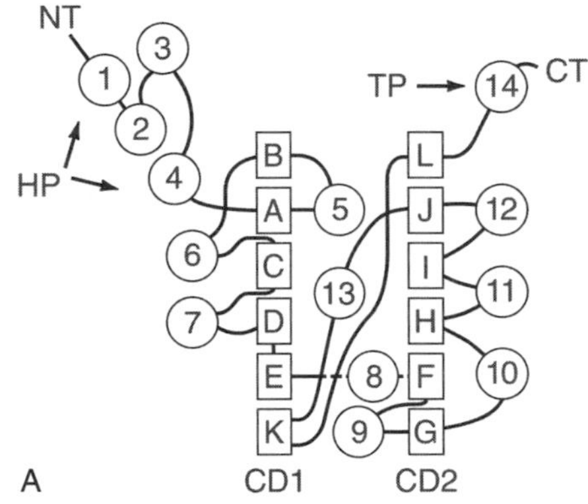

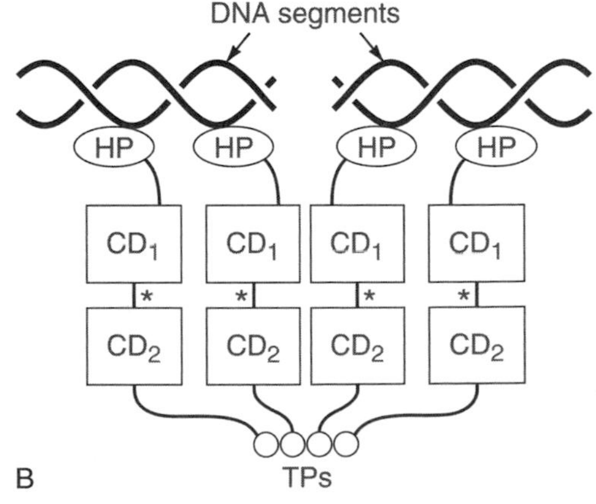

(Reprinted with permission from M. Lewis et al., 1996, Crystal structure of the lactose operon repressor and its complexes with DNA and inducer. *Science* 271 [5253]: 1247–1254. © 1996 American Association for the Advancement of Science.)

helix (circle 14) near the C terminus (CT) of the protein. The average alpha helix contains 11 amino acids; the average beta sheet, 4 or 5. As shown in diagram B on page 246, the repressor is a *V*-shaped molecule. Paired dimers make up the arms of the *V*, but all four chains are bound together at their C termini. The N termini of each dimer bind to different DNA segments. Inducer molecules can bind to the core in the starred regions. The shape change that follows moves the head piece out of the contact with its binding site on the DNA, and the RNA polymerase can now transcribe the structural genes of the operon. *See* Appendix C, 1966, Gilbert and Müller-Hill; 1996, Lewis *et al.*

lactamase *See* penicillin.

lactic dehydrogenase *See* isozymes.

lactogenic hormone a protein hormone secreted by the anterior lobe of the pituitary that stimulates milk production in mammals and broodiness in birds. *See* human growth hormone.

lactose 4-(β-D-galactoside)-D-glucose. A disaccharide made up of two hexoses joined by a beta galactoside linkage. It is split into galatose and glucose by the enzyme beta galactosidase. Lactose differs from allolactose in that in lactose the galactose and glucose moieties are joined by a 1–4 linkage, whereas in allolactose the linkage is 1–6. As its name implies, lactose is abundant in the milk of mammals. *See lac operon.*

```
   6 CH2OH                          6 CH2OH
      | 5                              |
      C ------ O               H   C ------ O     H
 HO  /|         \   ____       |  /| 5       \    |
    / H          1 C    |      C 4 H           1 C
 C 4 OH     H   /  |    O      \  OH      H   /  |
  | \ | 3  2 | /   H    |____   \ | 3    2 | /   OH
 H   C ------ C               C ------ C
     |        |               |        |
     H        OH              H        OH
```

lagging delayed movement from the equator to the poles at anaphase of a chromosome so that it becomes excluded from the daughter nuclei.

lagging strand the discontinuously synthesized strand of DNA containing ligated Okazaki fragments (*q.v.*). *See* leading strand, replication of DNA.

lag growth phase a period of time in the growth of a population during which little or no increase in the number of organisms occurs. The lag period precedes the exponential growth phase (*q.v.*).

lag load a measure of the distance of a species from its local adaptive peak. The greater the lag load of a species, the more selective pressure is applied to the species, and hence the more rapid the rate of evolution it is likely to be experiencing. Also called *evolutionary lag*. *See* Red Queen hypothesis.

Lamarckism a historically important, but no longer credited, theory that species can change gradually into new species by the willful striving of organisms to meet their own needs, together with the cumulative effects of use and disuse of body parts. All such acquired characteristics were thought to become part of the individual's heredity and as such could be transmitted to their offspring; otherwise known as the inheritance of acquired characteristics. *See* Appendix C, 1809, Lamarck.

lambda (λ) bacteriophage a double-stranded DNA virus that infects *E. coli.* The head of the virus contains a linear DNA molecule 48,514 bp long. However, upon entering the bacterium, the two ends of the DNA molecule become covalently joined to form a circle. Once inside the host cell, the virus can enter either a lytic developmental cycle or a lysogenic cycle. Specific repressors control the switch for either cycle. If the lysogenic cycle is chosen, the virus is eventually integrated into the *E. coli* chromosome at a specific site. *See* Appendix C, 1950, Lederberg; 1961, Meselson and Weigle; 1965, Rothman; 1967, Taylor, Hradecna, and Szybalski; 1968, Davis and Davidson; 1969, Westmoreland *et al.*, 1974, Murray and Murray; Appendix F; *cro* repressor, lambda cloning vector, lambda repressor, lysogenic cycle, lytic cycle, prophage, site-specific recombination, virus.

lambda cloning vector a lambda phage that is genetically engineered to serve as a receptor for foreign DNA fragments in recombinant DNA experiments. Vectors that have a single target site at which foreign DNA is inserted are called *insertion vectors*, those having a pair of sites that span a DNA segment that can be exchanged with a foreign DNA fragment are called *replacement* or *substitution vectors*. *See* Appendix C, 1977, Tilghman *et al.*

lambda d *gal* (λd*gal*) a lambda (λ) phage carrying a gene for galactose fermentation (*gal*) and also defective (d) for some phage function (usually lacking genes for making tails).

lambda phage genome the chromosome of this temperate bacteriophage can be represented by a circle with the genes that regulate the lytic versus the lysogenic pathways (C_I and *cro*) at 12 o'clock. Genes that are required for the lytic cycle are located in clockwise order to the right. These include genes that function in DNA replication, those that encode the enzymes that cause lysis of the host, and genes that specify proteins used in constructing the virus's head and tail. Genes required for the lysogenic cycle are arranged in counterclockwise order to the left. These include genes that encode recombination proteins and integrases (*q.v.*). *See* *cro* repressor, lambda repressor, lysogenic cycle.

lambda repressor the protein encoded by the C_I regulator gene of lambda (λ) bacteriophage (*q.v.*). This DNA-binding protein represses the transcription of genes that facilitate the lytic state and therefore keeps the virus in the lysogenic condition. The promoter for C_I is to the immediate right of the gene, and during translation the host transcriptase moves to the left. Movement of this enzyme can be blocked by the *cro* repressor (*q.v.*), which binds to an operator that overlaps the C_I promoter. The lambda repressor forms a dimer, which binds to DNA via a helix-turn-helix motif (*q.v.*). *See* Appendix C, 1966, Ptashne; 1987, Anderson, Ptashne, and Harrison; regulator gene.

lamellipodia extensive, lamellar cellular projections involved in attachment of eukaryotic cells such as fibroblasts to solid surfaces. Lamellipodia mark the forward edges of moving cells such as macrophages and are also called *ruffled edges*.

lamins proteins belonging to the intermediate filament (*q.v.*) class. Lamins are major structural proteins of the nuclear lamina, a structure that lines the nucleocytoplasmic surface of the inner nuclear membrane in eukaryotes. The nuclear lamina is a meshwork of 10-nm filaments. These support the nuclear envelope and nuclear pore complexes (*q.v.*). Lamins also interact with interphase DNA, especially telomeric and centromeric sequences. In humans the lamin A gene (*LMNA*) maps to chromosome 1q21.2-q21.3. LMNA has 12 exons and measures 57.6 kb. Progeria (*q.v.*) is caused by mutations at the *LMNA* locus.

laminin a fibrous glycoprotein forming a major component of basement membranes and serving as an adhesive surface for epithelial cells. Mutations in laminin genes cause hereditary blistering diseases in humans. *See* epidermolysis bullosa, integrins.

lampbrush chromosome a chromosome characteristic of primary oocytes of vertebrates. The LBCs found in the oocytes of salamanders are the largest known chromosomes and are generally studied in diplonema. They have a rather fuzzy appearance when viewed at low magnification with a light microscope, and this led early cytologists to suggest they resembled the brushes used to clean the chimneys of kerosene lamps. Hundreds of paired loops extend laterally from the main axis of each LBC, and large numbers of RNA polymerase II molecules move in single file along each of the loops while transcribing pre mRNAs. These transcripts extend laterally from the DNA axis of the loop. The loops also stain strongly with antibodies against splicing snRNAs, so that as the mRNAs are transcribed they are also processed in a variety of ways. If sperm chromosomes are injected into *Xenopus* germinal

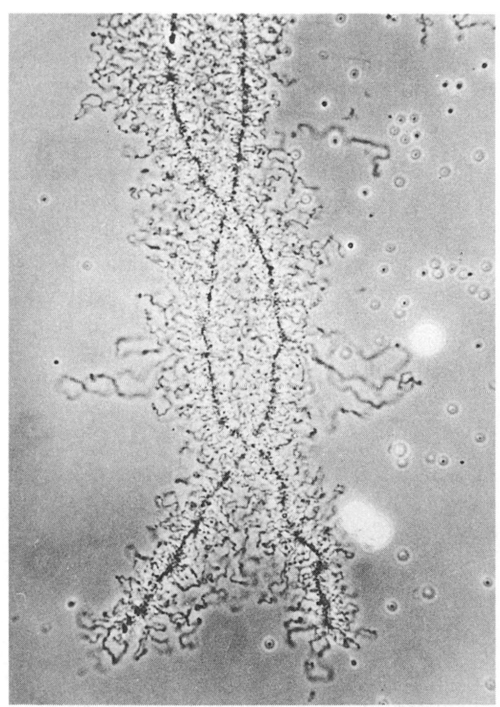

Lampbrush chromosome from a *Notophthalamus viridescens* oocyte

vesicles, they will transform within hours to transcriptionally active LBCs. However, sperm LBCs contain single unreplicated chromatids, and, as expected, their loops are single, never paired. Lampbrush chromosomes are also seen in the Y chromosomes (*q.v.*) of primary spermatocytes of males from a large number of *Drosophila* species. *See* Appendix C, 1882, Flemming; 1958, Callan and MacGregor; 1963, Gall; 1968, Davidson *et al.*, Hess and Meyer; 1977, Old *et al.*; 1993, Pisano *et al.*; 1998, Gall and Murphy; *Ambystoma mexicanum*, posttranscriptional processing, premessenger RNA, *Triturus*, *Xenopus*.

Laplacian curve *See* normal distribution.

large angle x-ray diffraction a technique for the analysis of the small distances between individual atoms. *See* small angle x-ray diffraction, x-ray crystallography.

lariat an RNA intermediate formed during posttranscriptional processing (*q.v.*). During excision, the intron is first cleaved at the donor junction. The 5′ terminus at the left end of the cleaved intron becomes linked to an adenosine in a region about 30 bases upstream of the acceptor junction. The sequence to which the intron becomes attached is called the *branch site*. Next, the intron is cut at the acceptor junction and released as a noose-shaped molecule. The left and right introns are then ligated, and the lariat-shaped intron opens up and is degraded.

Laron dwarfism human growth hormone insensitivity syndrome due to defects in the synthesis of hGHR. Many mutations have been reported in the hGHR gene including R43X, a recurrent mutation occurring in a CpG dinucleotide hot spot. This type of familial dwarfism was first reported by Zvi Laron. *See* DNA methylation, human growth hormone receptor (hGHR).

larva the preadult form in which some animals hatch from the egg. A larva is capable of feeding, though usually on a diet different from the adult, and is usually incapable of sexual reproduction.

larviparous depositing larvae rather than eggs. Fertilized eggs develop internally up to the larval stage. The female then lays these larvae. Some blowflies, for example, are larviparous.

laser an electronic device that generates and amplifies light waves coherent in frequency and phase in a narrow and extremely intense beam of light. The word is an acronym for *l*ight *a*mplification by *s*timulated *e*mission of *r*adiation. Microlaser beams are sometimes used to perform microcautery in experimental studies of cell division or morphogenesis.

laser microprobe a technique that uses a laser beam focused by a microscope to vaporize a minute tissue area. The vapor is then analyzed spectrographically.

late genes genes expressed late in the life cycle. In T4 bacteriophage, those genes responsible for making capsid proteins, lysozyme, and other proteins after the initiation of phage DNA replication. *Compare with* early genes.

latent image the pattern of changes occurring in a surface of silver halide crystals when they absorb photons (from light) or ionizing particles. The pattern may be developed into a photographic image by chemical treatment.

latent period **1.** the period during an infection when the causative agent cannot be detected by conventional techniques (e.g., the time from entry of phage DNA into a host cell until the release of infective phage progeny); prepatent period. *Compare with* eclipse period, patent period. **2.** the time between infection and development of disease symptoms; period of incubation. *See* Appendix C, 1939, Ellis and Delbrück, one step growth experiment.

lateral element *See* synaptonemal complex.

lateral gene transfer *See* horizontal transmission.

late-replicating DNA *See* zygotene DNA.

late-replicating X chromosome in the mammalian somatic cell nucleus, all X chromosomes but one coil up into a condensed mass (the Barr body or sex chromatin body) and do not function in transcription. Such X chromosomes complete their replication later than the functional X and the autosomes. *See* Barr body, Lyon hypothesis, sex chromatin.

Latimeria a genus of coelacanths containing the only living species belonging to the order Actinista (*see* Appendix A, Osteichthyes, Crossopterygii). *L. menaodensis* has a cluster of four *Hox* genes that are more similar to mammals than to ray-finned fishes. It also has an additional gene *HoxC1*, which was lost during the evolution of mammals from lobed-finned fish. *See* crossopterygians, *Hox* genes, living fossil.

Latin square (*also* Roman square) a set of symbols arranged in a checkerboard in such a fashion that no symbol appears twice in any row or column. The Latin square, long a mathematical curiosity, was discovered to be useful for subdividing plots of land for agricultural experiments so that treatments could be tested even though the field had soil conditions that might vary in an unknown fashion in different areas. The technique required that the field be subdivided by a grid into subplots and the differing treatments be performed at consecutive intervals to plants from different subplots. Thus, if the subplots are A, B, C, and D, the experiments could be run as illustrated.

treatment	*day 1*	*day 2*	*day 3*	*day 4*
1	A	B	C	D
2	B	C	D	A
3	C	D	A	B
4	D	A	B	C

latitudinal parallel to the equator.

lattice a structure composed of elements arranged in a geometrical pattern with spaces between them.

lawn a continuous layer of bacteria on the surface of an agar plate.

law of parsimony *See* Occam's razor.

lazy maize a maize mutant characterized by a stalk that grows flat on the ground like a vine.

L cell *See* mouse L cells.

L chain *See* immunoglobulin.

LCR ligase chain reaction (*q.v.*).

LD50 the radiation dose required to kill half of a population of organisms within a specified time. Synonymous with median lethal dose.

LDH lactate dehydrogenase. *See* isozyme.

leader protein *See* leader sequence peptide.

leader sequence the segment of an mRNA molecule from its 5′ end to the start codon of the structural gene. The leader contains ribosome-binding sites and facilitates the initiation of genetic translation. In eukaryotes, the leader sequence is not usually translated. In prokaryotes, the leader may contain an attenuator (*q.v.*) segment that is translated. The resulting peptide functions to terminate transcription before the RNA polymerase reaches the first structural gene of the operon. *See* exon, trailer sequence.

leader sequence peptide a sequence of 16 to 20 amino acids at the *N*-terminus of some eukaryotic proteins that determines their ultimate destination. Proteins that are made and function in the cytosol lack leader sequences. Proteins destined for specific organelles require signal sequences appropriate for each organelle. The leader sequence for a protein destined to enter the endoplasmic reticulum always contains hydrophobic amino acids that become embedded in the lipid bilayer membrane, and it functions to guide the nascent protein to a receptor protein that marks the position of a pore in the membrane. Once the protein passes into the cisternal lumen through the pore, the leader segment is cleaved from the protein. For example, the leader sequence peptide of an interferon protein allows the cell to secrete the interferon, but is removed from the mature molecule during the secretion process. All of the leader sequences of mitochondrial proteins investigated thus far are strikingly basic, but otherwise have no similarities. The leader sequence peptide is also known as the signal peptide (*q.v.*). *See* antihemophilic factor, receptor-mediated translocation.

leading strand the DNA strand synthesized with few or no interruptions; as opposed to the *lagging strand* which is produced by ligation of Okazaki fragments (*q.v.*). The leading strand is synthesized 5′ to 3′ toward the replication fork, whereas the lagging strand is synthesized 5′ to 3′ away from the replication fork. *See* replication of DNA.

leaky gene *See* hypomorph.

leaky protein a mutant protein that has a subnormal degree of biological activity.

least squares, method of a method of estimation based on the minimization of sums of squares. *See* line of best fit.

Leber hereditary optic neuropathy (LHON) a maternally inherited disease of young adults that causes blindness due to the death of the optic nerve. Most cases of LHON are due to missense mutations in mtDNA. The sites of the base substitutions are in genes that encode cytochromes (*q.v.*). The condition is named after Theodore Leber, the German ophthalmologist who provided the first clinical description in 1871. *See* Leigh syndrome, Appendix C, 1988, Wallace *et al.*; Leigh syndrome.

Lebistes reticularis the guppy, a well-known tropical aquarium fish. The genetic control of sexuality has been extensively studied in this species. *See* Appendix A, Chordata, Osteichythes, Neopterygii, Cyprinidontiformes.

lectins proteins capable of agglutinating certain cells, especially erythrocytes, by binding to specific carbohydrate receptors on the surfaces of these cells. This agglutination behavior resembles antibody-antigen reaction but is not a true immunological reaction. Those lectins extracted from plant seeds have been called phytohemagglutinins. However, lectins have also been isolated from sources as varied as bacteria, snails, horseshoe crabs, and eels. *See* concanavalin A, phytohemagglutinin, pokeweed mitogen.

left splicing junction the boundary between the left (5′) end of an intron and the right (3′) end of an adjacent exon in mRNA; also termed the donor splicing site.

leghemoglobin an oxygen-binding protein found in the root nodules of leguminous plants. Leghemoglobin is structurally and functionally related to the myoglobins and hemoglobins of vertebrates. The heme portion of leghemoglobin is synthesized by the *Rhizobium* bacteroids, while the protein is encoded by the plant genome. Within the root nodule, bacteroids are bathed in a solution of leghemoglobin that serves to supply them with oxygen.

Leigh syndrome a hereditary, early-onset degeneration of the central nervous system first described in 1951 by the English physician Denis Leigh. The disease is caused by mutations which encode proteins that are essential components for the enzymes of the cytochrome system (*q.v.*) or for ATP synthase (*q.v.*). Since genes responsible for LS have been mapped to the X, to chromosomes 2, 5, 7, 10, 11, and 19, and to mtDNA, the condition may show sex-linked, autosomal, or maternal inheritance. *See* heteroplasmy.

Leishmania a genus of trypanosomatid parasites (order Kinetoplastida; family Trypanosomatidae) that are transmitted by sand flies and are responsible for various diseases in vertebrates. *Leishmania major* causes leishmaniasis (*q.v.*) in humans, and its 32.8 mb genome has been sequenced and predicted to contain 8,272 protein-coding genes on its 36 chromosomes. *See* Appendix A, Protoctista, Zoomastigina; Appendix C, 2005, Ivens *et al.*; *Trypanosoma.*

leishmaniasis any of a variety of diseases caused by pathogenic species of the trypanosomatid genus *Leishmania* (*q.v.*). Leishmaniasis affects approximately 2 million people annually and is prevalent in tropical and sub-tropical regions of Africa, Asia, and South and Central America.

lek a traditional Scandinavian term used for a territory in which male birds performed epigamic displays. The term has been broadened to include mating assemblages of any animal species. The site may be fixed or mobile, and males gather there to determine which are dominant or to compete in attracting females. Such behavior is often called "lekking." Leks provide no resources to the females except mating opportunities. An insect example of lekking is *Drosophila silvestris*, where dominant males patrol stalks of Hawaiian tree ferns and drive other males away. *See* arena-breeding birds, Hawaiian Drosophilidae.

lentivirus *See* retrovirus.

leprosy a chronic infectious disease known and dreaded since biblical times because of the severe deformities it can cause. The preferred name for the condition is Hansen disease. This eponym honors Armauer Hansen, the Norwegian physician who discovered the leprosy bacillus, *Mycobacterium leprae*, in 1879. The leprosy bacterium and the tuberculosis bacterium, *Mycobacterium tuberculosis* (*q.v.*), evolved from a common ancestor. Their genomes are compared below.

Feature	*M. leprae*	*M. tuberculosis*
Genome size (Mbp)	3.2	4.4
Protein-coding genes	1,604	3,942
Pseudogenes	1,116	6

About 27% of the genome of *M. leprae* is made up of pseudogenes (*q.v.*) which have functional counterparts in *M. tuberculosis*. Of the 1,604 active genes of *M. leprae*, 1,440 are also found in *M. tuberculosis*. It follows that *M. leprae* has undergone a far more reductive evolution (*q.v.*) than *M. tuberculosis*.

leptin a circulating hormone synthesized by adipose tissues. Leptin is symbolized *OB protein*, because it is encoded by the gene *obese* (*q.v.*).

leptin receptor (OB-R protein) a protein containing 894 amino acids which appears to be a receptor for leptin, the product of the obesity gene, *ob*. The protein has the properties of a membrane-bound receptor. OB-R mRNAs occur in the hypothalamus, a region controlling energy balance, and other areas of the mouse brain. The OB-R protein is encoded by a gene on mouse chromosome 4, and the homologous human gene has also been identified. The fat cells of massively obese humans, who appear to carry ob^+ alleles, contain elevated amounts of leptin mRNA. Therefore, these individuals seem to lack OB-Rs. *See* Appendix C, 1995, Tartaglia *et al.*; diabetes mellitus.

leptonema *See* meiosis.

leptotene stage *See* meiosis.

Lesch-Nyhan syndrome the most common human hereditary defect in purine metabolism, due to a recessive gene on the long arm of the X chromosome. The 1964 paper by Michael Lesch and William Nyhan established the disease as a disorder of uric acid metabolism. Hemizygotes lack hypoxanthine-guaninephosphoribosyl transferase (*q.v.*). The disease is characterized by excessive production of uric acid, developmental and mental retardation, and death before sexual maturity. As a result of random X-chromosome inactivation (*see* dosage compensation), female heterozygotes are mosaics. About 60% of their cloned fibroblasts show HGPRT activity, whereas 40%, have no detectable activity. The disease is transmitted via heterozygous mothers. Prevalence is 1/10,000 males.

LET linear energy transfer (*q.v.*).

lethal equivalent value the average number of recessive deleterious genes in the heterozygous condition carried by each member of a population of diploid organisms times the mean probability that each gene will cause premature death when homozygous. Thus a genetic burden of eight recessive semilethals each of which, when homozygous, produced only a 50% probability of premature death would be classified as a burden of four "lethal equivalents."

lethal mutation a mutation that results in the premature death of the organism carrying it. Dominant lethals kill heterozygotes, whereas recessive lethals kill only homozygotes. *See* Appendix C, 1905, Cuénot; 1910, Castle and Little; 1912, Morgan; aphasic lethal, monophasic lethal, polyphasic lethal.

leu leucine. *See* amino acid.

leucine *See* amino acid.

leucine zipper a region in DNA-binding proteins spanning approximately 30 amino acids that contains a periodic repeat of leucines every seven residues. The region containing the repeat forms an alpha helix, with the leucines aligned along one face of the helix. Such helices tend to form stable dimers with the helices aligned in parallel. Leucine zippers occur in a number of transcriptional regulators. *See* Appendix C, 1988, Landschulz *et al.*; motifs, *myc*.

leucocyte a variant spelling of leukocyte (*q.v.*).

leukemia a generally fatal disease characterized by an overproduction of white blood cells, or a relative overproduction of immature white cells. Leukemia is a common disease in the cow, dog, cat, mouse, guinea pig, and chicken, as well as humans. Many virus-induced leukemias are known in the mouse and chicken. *See* myeloid leukemia, retroviruses.

leukocyte a white blood cell.

leukopenia a decrease in the number of white blood corpuscles.

leukoplasts colorless plastids of tubers, endosperm, and cotyledons.

leukosis proliferation of leukocyte-forming tissue; leukemia (*q.v.*).

leukoviruses *See* retroviruses.

lev levorotatory. *See* optical isomers.

levulose fructose.

Lewis blood group a blood group determined by an antigen specified by the *Le* gene on human chromosome 19. The Lewis antigen is expressed on epithelial surfaces as well as erythrocytes. The Lewis gene encodes a fucosyltransferase, which adds fucose to the same heterosaccharoside precursor targeted by the glycotransferases encoded by the genes of the ABO blood group system. The bacterium that causes gastric ulcers attaches to the gastric mucosa by binding to receptors containing fucose. Therefore, individuals with the Lewis antigen on their mucosal surfaces are more susceptible to this disease. *See* AB antigens, blood group, *Helicobacter pylori*, Lutheran blood group.

LexA repressor *See* SOS response.

L forms bacteria that have lost their cell walls. *See* protoplast.

LH luteinizing hormone (*q.v.*).

LHON Leber hereditary optic neuropathy (*q.v.*).

library *See* genomic library.

lichen a composite organism that consists of a fungus (usually an ascomycete) and a photosynthetic alga or cyanobacterium, living in symbiosis. Many different fungi have evolved symbiotic relationships with members of the same genera of algae or cyanobacteria. Therefore, most lichens are classified with the group to which the fungus belongs. *See* Appendix A, Fungi, Ascomycota.

life cycle the series of developmental changes undergone by an organism from fertilization to reproduction and death.

life history strategies evolutionary adaptations in a biological lineage involving the timing of reproduction, fecundity, longevity, etc. A grasshopper using its resources to make hundreds of fertilized eggs, each with a low probability of survival, represents a life history strategy different from a bird which uses its resources to produce a few offspring, each with a much greater chance of surviving.

Li-Fraumeni syndrome (LFS) a predisposition to a variety of cancers (breast cancers, soft tissue sarcomas, brain tumors, osteosarcomas, leukemias, lymphomas, and adrenal carcinomas) inherited as an autosomal dominant. Fredrick P. Li and Joseph F. Fraumeni gave the first report of this condition in 1969. The majority of patients with LFS have germline mutations in *TP53* (*q.v.*). *See* Appendix C, 1990, Malkin *et al.*

ligand a molecule that will bind to a complementary site on a given structure. For example, oxygen is a ligand for hemoglobin and a substrate of an enzyme molecule is a specific ligand of that molecule.

ligase chain reaction (LCR) a technique that allows a specific region of a DNA molecule to be screened for mutations. The targeted region is denatured and reannealed to a set of four oligonucleotides. For each strand of the target DNA, two complementary oligonucleotides are designed, which will base pair in tandem with the strand, leaving only a gap where the right end of one molecule abuts the left end of the other. This gap can be sealed by a DNA ligase, and subsequently the complete strand can serve as a substrate against which new pairs of nucleotides can anneal and be ligated. As one cycle follows another, the ligated pairs of oligonucleotides are successively amplified. A thermal cycler is used to alternatively heat and cool the reaction mixture to allow first separation and then binding of complementary molecules. A thermostabile ligase isolated from *Thermus aquaticus* (*q.v.*) serves as the sealing agent. One can now test other DNAs to see if they contain base sequence variations in the targeted region. Any mutation that interfers with the base pairing of the oligonucleotides when they bind to the targeted sequence will prevent the ligase from joining the right and left ends of these two molecules. Therefore no amplification will take place. So the failure of amplification shows that the test DNA contains nucleotide sequences that do not match oligonucleotide sequence sites critical for ligation. The LCR therefore provides a rapid, accurate way for screening the patients suffering from a genetic disease for mutant alleles of a targeted gene. *See* Appendix C, 1990, Barany.

ligases enzymes that form C–C, C–S, C–O, and C–N bonds by condensation reactions coupled to ATP cleavage. *See* DNA ligase.

ligation formation of a phosphodiester bond to join adjacent nucleotides in the same nucleic acid chain (DNA or RNA).

light chain *See* heavy chain.

lightning bugs a synonym for fireflies (*q.v.*).

light repair *See* dark reactivation, photoreactivation.

Lilium the genus containing *L. longiflorum*, the Easter lily and *L. tigrinum*, the tiger lily, favorite species for cytological and biochemical studies of meiosis.

limited chromosome a chromosome that occurs only in nuclei of the cells of the germ line and never in somatic nuclei. *See* chromosome diminution.

Limnaea peregra a freshwater snail upon which were first performed classical studies on the inheritance of the direction of the coiling of the shell. The trait showed delayed Mendelian segregation, since the phenotype of the snail is determined by the genotype of the maternal parent. *See* Appendix A, Mollusca, Gastropoda; Appendix C, 1923, Boycott and Diver, Sturtevant.

lin-12 *See* developmental control genes.

line a homozygous, pure-breeding group of individuals that are phenotypically distinctive from other members of the same species. *See* inbred strain, pure line, true breeding line.

lineage a linear evolutionary sequence from an ancestral species through all intermediate species to a particular descendant species.

linear accelerator *See* accelerator.

linear energy transfer the energy, in electron volts, dissipated per micron of tissue traversed by a particular type of ionizing particle.

linear regression regression line (*q.v.*).

linear tetrad a group of four meiotic products aligned linearly in such a way that sister products remain adjacent to one another. Ascospores show this order because the confines of the fungal ascus prevent nuclei from sliding past one another. *See* ascus.

line of best fit a straight line that constitutes the best moving average for a linear group of observed points. This requires that the sum of the squares of the deviations of the observed points from the moving average be a minimum.

LINEs *See* repetitious DNA.

linkage the greater association in inheritance of two or more nonallelic genes than is to be expected from independent assortment. Genes are linked because they reside on the same chromosome. *See* Appendix A, 1906, Bateson and Punnett; 1913, Sturtevant; 1915, Haldane *et al.*; 1951, Mohr.

linkage disequilibrium the nonrandom distribution into the gametes of a population of the alleles of genes that reside on the same chromosome. The simplest situation would involve a pair of alleles at each of two loci. If there is random association between the alleles, then the frequency of each gamete type in a randomly mating population would be equal to the product of the frequencies of the alleles it contains. The rate of approach to such a random association or equilibrium is reduced by linkage and hence linkage is said to generate a disequilibrium. *See* gametic disequilibrium.

linkage group the group of genes having their loci on the same chromosome. *See* Appendix C, 1919, Morgan.

linkage map a chromosome map showing the relative positions of the known genes on the chromosomes of a given species.

linked genes *See* linkage.

linker DNA **1.** a short, synthetic DNA duplex containing the recognition site for a specific restriction endonuclease. Such a linker may be connected to ends of a DNA fragment prepared by cleavage with some other enzyme. **2.** a segment of DNA to which histone H1 is bound. Such linkers connect the adjacent nucleosomes of a chromosome.

linking number the number of times that the two strands of a closed-circular, double-helical molecule cross each other. The *twisting number* (T) of a relaxed closed-circular DNA is the total number of base pairs in the molecule divided by the number of base pairs per turn of the helix. For relaxed DNA in the normal B form, L is the number of base pairs in the molecule divided by 10. The *writhing number* (W) is the number of times the axis of a DNA molecule crosses itself by supercoiling. The linking number (L) is determined by the formula: $L = W + T$. For a relaxed molecule, $W = 0$, and $L = T$. The linking number of a closed DNA molecule cannot be changed except by breaking and rejoining of strands. The utility of the linking number is that it is related to the actual enzymatic breakage and rejoining events by which changes are made in the topology of DNA. Any changes in the linking number must be by whole integers. Molecules of DNA that are identical except for their linking numbers are called *topological isomers*.

Linnean Society of London a society that takes its name from Carl Linné, the Swedish naturalist and "Father of Taxonomy." The Society was founded in 1788 (ten years after Linné's death) for "the cultivation of the Science of Natural History in all its branches." Subsequently the first president of the Society purchased Linné's botanical and zoological collections, and they are held in the Society's museum. An early publication of the Society contains Brown's discovery of the cell nucleus. In 1858 essays by Darwin and Wallace presenting the theory of evolution by natural selection were first published in the Society's Proceedings. The next year another essay by Wallace was published by the Society. In this account of the zoological geography of the Malay Archipelago, the first description of the Wallace line (*q.v.*) was given. *See* Appendix C, 1735, Linné; 1831, Brown; 1858, Darwin and Wallace; 1859, Wallace.

Linnean system of binomial nomenclature a naming system in which each newly described organism is given a scientific name consisting of two Latin words. For example, in the case of the fruitfly, *Drosophila melanogaster* Meigen, the first name (the genus) is capitalized and the second (the species) is not. The scientific name is italicized, and the author who named and described the species is sometimes given. In instances where the scientific name is followed by an "L" (e.g., *Canis familiaris* L), the species was named by Linné himself. Although Linné was not the first to use binomial names, he was the first to employ them in constructing a taxonomy of plants and animals. When two plants or animals are accidently given the same species name, it remains valid for only the first one named. However, the same name has been allowed for a plant and an animal. A good example is *Cereus*, a name for a well-known genus of sea anemonies and cacti. *See* Appendix C, 1735, Linné; *Hyracotherium*, Linnean Society of London, *Takifugu ruripes*.

Linum usitatissimum the cultivated flax plant, the source of fiber for linen and stationery and flaxseed for linseed oil. Classic studies on coevolution (*q.v.*) have utilized the genes of flax and its parasite, the rust fungus *Melampsora lini*. *See* Appendix A, Plantae, Tracheophyta, Angiospermae, Dicotyledonae, Linales; gene-for-gene hypothesis.

lipase an enzyme that breaks down fats to glycerol and fatty acids.

LIPED *See* lod.

lipid any of a group of biochemicals that are variably soluble in organic solvents like alcohol and barely soluble in water (fats, oils, waxes, phospholipids, sterols, carotenoids, etc.).

lipid bilayer model a model for the structure of cell membranes based upon the hydrophobic properties of interacting phospholiplids. The polar head groups face outward to the solvent, whereas the hydrophobic tails face inward. Proteins (p) are embedded in the bilayer, sometimes being exposed on the outer surface, sometimes on the inner surface, and sometimes penetrating both surfaces. The proteins exposed on the outer surfaces of cells commonly serve as distinctive antigenic markers. Membrane proteins may serve a variety of functions such as communication, energy transduction, and transport of specific molecules across the membrane. *See* fluid mosaic concept.

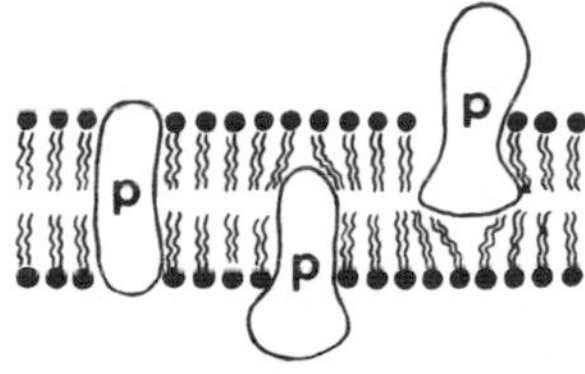

lipopolysaccharide the active component of bacterial endotoxins associated with the cell walls of many Gram-negative species; a B-cell mitogen in some animal species.

liposomes synthetic vesicles surrounded by bilayers of phospholipids. They have been used as models of cell membranes and as therapeutic delivery systems in which the drugs employed are encapsulated within liposomes. In studies of genetic transformation, the synthetic vesicles are used to protect naked DNA molecules and to allow their passage into tissue-cultured animal cells or into protoplasts from plants or bacteria. This technique is called *liposome-mediated gene transfer*.

lipovitellin a lipoprotein of relative molecular mass 150,000 found in amphibian yolk platelets. *See* phosvitin.

liquid-holding recovery a special form of dark reactivation (*q.v.*) in which repair of ultraviolet damage to DNA is enhanced by delaying bacterial growth and DNA replication through postirradiation incubation of cells in a warm, nutrient-free buffer for several hours before plating them on nutrient agar.

liquid hybridization formation of double helical nucleic acid chains (DNA with DNA, DNA with RNA, or RNA with RNA) from complementary single strands in solution. *Compare with* filter hybridization.

liquid scintillation counter an electronic instrument for measuring radioisotopes dissolved in a solvent containing a fluorescent chemical that emits a flash of light (a scintillation) when struck by an ionizing particle or photon of electromagnetic radiation. The flash is captured by a photomultiplier tube, transformed into an electric pulse, amplified, routed through a scaler, and counted.

liter standard unit of capacity in metric system, equal to 1 cubic decimeter.

lithosphere *See* plate tectonics.

lithotroph a prokaryote that uses an inorganic substance as a substrate in its energy metabolism. For example, energy may be obtained by the oxidation of H_2, NH_3, sulphur, sulphide, thiosulphate, or Fe^{++}. Lithotrophic metabolism is independent of light. However, a *photolithotroph* can use its inorganic substrates as electron donors in photosynthesis. *See* chemolithoautotroph.

litter animals of one multiple birth.

littoral pertaining to the shore.

Liturgosa a genus of mantid. The evolution of karyotype has been extensively studied in this genus.

living fossil literally referring an organism that belongs to a group recognized on the basis of fossils and only subsequently found to be extant. The classic example of such a living fossil would be *Latimeria chalumnae*, the sole living representative of an ancient group of bony fishes called coelacanths. These belong to the Crossopterygii and are allies of the lung fishes and near the ancestry of the amphibians. The first fossil Permian *Coelacanthus* was named in 1839, while the first specimen of *Latimeria* was captured in 1938. The term is also used to refer to a living organism that is regarded as morphologically similar to a hypothetical ancestral missing link, even though the group has little or no fossil record. *Peripatus* is such an example, since it and other onycophorans have characteristics of both arthropods and annelids. Finally, *living fossil* is used to refer to the end member of a clade that has survived a long time and undergone little morphological change. This is what Charles Darwin had in mind when he coined the term back in 1859. The externally shelled cephalopod, *Nautilus*, would be an example of a living fossil by this criterion. *See Mesostigma viride.*

load *See* genetic load.

local population a group of conspecific individuals together in an area within much most of them find their mates; synonymous with *deme*, and *Mendelian population*. *See* subpopulations.

locus (*plural* **loci**) the position that a gene occupies in a chromosome or within a segment of genomic DNA.

Locusta migratoria the Old World "plague" locust.

lod the abbreviation for "*l*ogarithm of the *od*ds favoring linkage." The lod score method is employed in the statistical analyses of linkage. In the calculations, a pedigree is analyzed to determine the likelihood or probability (*Pr*) that two genes show a specified recombination value (*r*). Next, the likelihood (*Pi*) is calculated under the assumption that the genes assort independently. The lod score $Z = \log_{10}(Pr/Pi)$. The advantage of using logarithms is that the scores from new pedigrees can be added to earlier *Z* values as they become available. A *Z* score of +3 is considered evidence for linkage. A computer program such as LIPED makes laborious manual computations unnecessary and is generally used in analyzing a collection of complex pedigrees. *See* Appendix C, 1955, Morton; 1974, Ott.

logarithmic phase the growth stage during which organisms are doubling their number each time a specified period of time elapses.

long-day plant a plant in which the flowering period is initiated and accelerated by a daily exposure to light exceeding 12 hours.

long period interspersion a genomic pattern in which long segments of moderately repetitive and nonrepetitive DNA sequences alternate.

long terminal repeats (LTRs) domains of several hundred base pairs at the ends of integrated retroviruses. Each retrovirus has to be copied by a reverse transcriptase, and the DNA strand is then replicated to form a double-stranded DNA. It is this segment that is integrated into the chromosome of the mammalian host cell. LTRs are required for both the replication of the viral DNA and its integration.

long-term memory *See* CREBs, spaced training.

Lophotrochozoa a clade that contains the phyla Platyhelminthes, Annelida, and Mollusca. These animals all have quite similar nucleotide sequences in their 18S rRNA genes (*q.v.*), and they share a cluster of *HOX* genes (*q.v.*) quite different from the clusters found in ecdysozoans (*q.v.*) and deuterostomes (*q.v.*).

loss of function mutation a genetic lesion that prevents the normal gene product from being produced or renders it inactive. An example of a loss of function mutation would be a nonsense mutation that causes polypeptide chain termination during translation. Loss of function mutations are generally recessive. *Contrast with* gain of function mutation.

loss of heterozygosity the loss of the normal allele, as the result of a deletion or inactivating mutation, in a heterozygous cell which already carries a mutant allele in the other homolog. Since there is now no functioning allele in the cell, it and its descendents will show the mutant phenotype.

Lou Gehrig disease *See* amyotrophic lateral sclerosis (ALS).

low-density lipoprotein receptor (LDLR) *See* familial hypercholesterolemia.

low-energy phosphate compound a phosphorylated compound yielding relatively little energy upon hydrolysis.

lozenge a sex-linked gene in *Drosophila* named for its effect on the compound eyes. These have a disturbed surface because of the fusion of adjacent facets. However, various alleles also show defects in the differentiation of hemocytes and abnormalities in the development of antennae, claws, and reproductive systems. Molecular studies show that the gene is made up of two regions. Mutations in the anterior controlling region influence the tissue-specific behavior of the gene, while the posterior region encodes a protein with a DNA-binding domain that functions as a transcription factor (*q.v.*). *See runt* (*run*).

LTH lactogenic hormone (*q.v.*).

luciferase an enzyme, produced in the abdomen of lightning bugs (*Photinus pyralis*), whose gene has been used to determine when a target gene has been inserted into plant genomes. In the presence of oxygen, ATP, and a substrate called *luciferin*, plants genetically engineered to contain the luciferase gene glow in the dark. *Compare with* green fluorescent protein (GFP). *See* fireflies.

Lucilia cuprina the Australian blowfly, a major pest of the sheep industry in Australia. Good polytene chromosome preparations can be obtained from pupal trichogen cells (*q.v.*), and extensive cytogenetic studies have been made of chromosome rearrangements. The concept of pest control through the introduction into the field of chromosomally altered strains whose descendants manifest sterility has been tested in this species.

Lucké virus a virus causing renal cancer in frogs.

Lucy *See* Australopithecine.

Ludwig effect a generalization offered by W. Ludwig in 1950 that a species tends to be more diversified (polymorphic) both morphologically and chromosomally in the center of an old established range than at its margins.

luminescence light emission that cannot be attributed to the temperature of the emitting body. The light energy emitted may result from a chemical reaction going on within the emitter or it may be initiated by the flow of some form of energy into the body from the outside. The slow oxidation of phosphorus at room temperature is an example of the former, while luminescence resulting from electron bombardment of gaseous atoms in a mercury vapor lamp is an example of the latter. *Fluorescence* is defined as a luminescence emission that continues after the source of exciting energy is shut off. This afterglow is temperature independent. In the case of *phosphorescence*, the afterglow duration becomes shorter with increasing temperature.

lungfish a group of lobed finned fishes belonging to the order Dipnoi (*see* Appendix A, Osteichthyes, Crossopterygii). They flourished during the Paleozoic but are now represented by only three genera, Lepidosiren of South America, Neoceratodus of Australia, and Protopterus of Africa. *P. aethiopicus* has a C value of 127 gbp of DNA, a record for an animal. Lungfish have gills and also lungs and can breathe air.

lupus erythematosus a connective tissue disorder characterized by autoantibody production against cellular components that are abundant and highly conserved. Among the antibodies produced by patients with lupus are those that recognize U 1 RNA. *See* autoimmune disease, Usn RNAs.

luteinizing hormone a glycoprotein hormone that stimulates ovulation, growth of the corpus luteum, and secretion of estrogen. LH is secreted by the adenohypophysis of vertebrates. Identical to ICSH.

luteotropin lactogenic hormone (*q.v.*).

Lutheran blood group a blood group determined by a red cell antigen specified by the *Lu* gene on human chromosome 19. Autosomal linkage in humans was first demonstrated between *Lu* and *Le*. The eponym Lutheran arose from a misinterpretation of the name of the patient (Luteran) who provided the antigen. *See* blood group, Lewis blood group.

luxuriance a high degree of vegetative development, often seen in species hybrids; a special feature of heterosis (*q.v.*).

luxury genes genes coding for specialized (rather than "household") functions. Their products are usually synthesized in large amounts only in particular cell types (e.g., hemoglobin in erythrocytes; immunoglobulins in plasma cells).

Luzula a genus of plants containing the wood rushes. Many species in this genus have diffuse centromeres. *See* centromere.

lyases enzymes that catalyze the addition of groups to double bonds or the reverse.

Lycopersicon esculentum the cultivated tomato. The haploid chromosome number is 12. There are extensive linkage maps for each chromosome and also cytological maps for pachytene chromosomes. *See* Appendix A, Angiospermae, Dicotyledonae, Solanales.

lycopods the most primitive trees. They arose in the Devonian and were the dominant plants in the Carboniferous coal swamps. The only lycopods that survived to the present are tiny club mosses like *Lycopodium*.

Lymantria dispar the Gypsy moth. Classical studies on sex determination were performed on this species. *See* Appendix A, Arthropoda, Insecta, Lepidoptera; Appendix C, 1915, Goldschmidt.

lymphatic tissue tissues in which lymphocytes are produced and/or matured, including the thymus, spleen, lymph nodes and vessels, and the bursa of Fabricius (*q.v.*).

lymphoblast the larger, cytoplasm-rich cell that differentiates from antigenically stimulated T lymphocytes.

lymphocyte a spherical cell about 10 micrometers in diameter found in the lymph nodes, spleen, thymus, bone marrow, and blood. Lymphocytes, which are the most numerous cells in the body, are divided into two classes: B cells, which produce antibodies, and T cells, which are responsible for a variety of immunological reactions, including graft rejections. *See* Appendix C, 1962, Miller, Good *et al.*, Warner *et al.*; B lymphocyte, cytotoxic T lymphocyte, helper T lymphocyte, immunoglobulin, lymphokines, T lymphocyte, tumor necrosis factors.

lymphokines a heterogeneous group of glycoproteins (relative molecular mass 10,000–200,000) released from T lymphocytes after contact with a cognate antigen. Lymphokines affect other cells of the host rather than reacting directly with antigens. Various lymphokines serve several major functions: (1) recruitment of uncommited T cells; (2) retention of T cells and macrophages at the site of reaction with antigen; (3) amplification of "recruited" T cells; (4) activation of the retained cells to release lymphokines; and (5) cytotoxic effects against cells bearing foreign antigens (including foreign tissue grafts and cancer cells). Examples of lymphokines produced by T cells following *in vitro* antigenic stimulation are as follows: migration inhibition factor (MIF) prevents migration of macrophages; lymphotoxins (LT) kill target cells; mitogenic factor (MF) stimulates lymphocyte division; interleukin 2 (IL2) is required for T helper activity; interferons (IFN) promote antiviral immunity; chemotactic factors attract neutrophils, eosinophils, and basophils; macrophage activation factor (MAF) stimulates macrophages. *See* histocompatibility molecules.

lymphoma cancer of lymphatic tissue.

Lyon hypothesis the hypothesis that dosage compensation (*q.v.*) in mammals is accomplished by the random inactivation of one of the two X chromosomes in the somatic cells of females. *See* Appendix C, 1961, Lyon and Russell; 1962, Beutler *et al.*; *XIST*.

Lyonization a term used to characterize heterozygous females that behave phenotypically as if they were carrying an X-linked recessive in the hemizygous condition. An example would be a hemophilic mother that has produced nonhemophilic sons. It is assumed, according to the Lyon hypothesis, that occasionally a particular tissue comes to be made up entirely of cells that contain inactivated X chromosomes carrying the normal allele. The phenotype of such a female heterozygote would then resemble that of mutant males. *See* manifesting heterozygote.

lyophilize to render soluble by freeze drying.

lys lysine. *See* amino acid.

lysate a population of phage particles released from host cells via the lytic cycle.

Lysenkoism a school of pseudoscience that flourished in the Soviet Union between 1932 and 1965. Its doctrines were advanced by T. D. Lysenko (1898–1976), who did not accept the gene concept and believed in the inheritance of acquired characteristics. Lysenko became the dominant figure in Soviet agriculture, after winning the support of Stalin. N. V. Vavilov, the great geneticist and leader of Soviet agriculture who Lysenko replaced, was later arrested for "sabotaging Soviet science" and thrown into prison, where he died of starvation in 1943. *See* Appendix C, 1926, Vavilov.

lysine *See* amino acid.

lysis *See* cell lysis.

lysochrome a compound that colors lipids by dissolving in them. *See* Nile blue, Sudan black B.

lysogen a lysogenic bacterium.

lysogenic bacterium one carrying a temperate virus in the prophage state.

lysogenic conversion the change in phenotype of a bacterium (in terms of its morphology or synthetic properties) accompanying lysogeny (*q.v.*). Lysogenic cells exhibit immunity to superinfection by the same phage as that in the prophage state. Toxin production by *Corynebacterium diphtheriae* only occurs in strains that are lysogenic for phage beta.

lysogenic cycle a method of temperate phage reproduction in which the phage genome is integrated into the host chromosome as a prophage and replicates in synchrony with the host chromosome. Under special circumstances (e.g., when growth conditions for the host are poor), the phage may leave the host chromosome (deintegration or excision) and enter the vegetative state or the lytic cycle that produces progeny phage. In lambda (λ) bacteriophage (*q.v.*), the O_R region of the genome acts as a genetic switch between the lytic and lysogenic life cycles. O_R contains operator sites for C_I and *cro* genes, which encode the lambda repressor (*q.v.*) and the *cro* repressor (*q.v.*), respectively. At any given time the developmental pathway followed by the virus depends on the relative concentrations of these repressors. *See* Appendix C, 1950, Lwoff and Gutman; lambda phage genome.

lysogenic immunity a phenomenon in which a prophage prevents another phage of the same type from becoming established in the same cell.

lysogenic repressor a phage protein responsible for maintenance of a prophage and lysogenic immunity.

lysogenic response the response following infection of a nonlysogenic bacterium with a temperate phage. The infecting phage does not multiply but rather behaves as a prophage. *See* lytic response.

lysogenic virus a virus that can become a prophage.

lysogenization the experimental production of a lysogenic strain of bacteria by exposing sensitive bacteria to a temperate phage.

lysogenized bacterium a bacterium harboring an experimentally introduced, temperate phage.

lysogeny the phenomenon in which genetic material of a virus and its bacterial host are integrated.

lysosomal storage diseases hereditary diseases characterized by abnormal lipid storage due to defects in lysosomal enzymes. The accumulation of the trapped intermediates of catabolism results in the cytoplasmic storage of these complex molecules. The storage material is often desposited in myriads of concentrically arranged lamellar structures that accumulate throughout the cytoplasm of certain cell types. For example, in Tay-Sachs disease (*q.v.*) the targets are ganglion cells, while in Gaucher disease (*q.v.*) macrophages are the storage sites. *See* Ashkenazi, lysosomes, Wolman disease.

***lysosomal trafficking regulator (LYST)* gene** a human gene residing at 1q 42.1-2 that contains 55 exons and is thought to undergo a complex pattern of alternative RNA splicing. One 13.5 kb mRNA encodes a protein containing 3,801 amino acids. Mutations which cause truncations of this protein result in severe Chédiak-Higashi syndrome (*q.v.*). Proteins encoded by the *LYST* gene play a role in the sorting of resident proteins within lysosomes and melanosomes (*both of which see*).

lysosome a membrane-enclosed intracellular vesicle that acts as the primary component for intracellular digestion in all eukaryotes. Lysosomes are known to contain at least 50 acid hydrolases, including phosphatases, glycosidases, proteases, sulfatases, lipases, and nucleases. Lysosomes primarily process exogenous proteins that are taken into the cell by endocytosis (*q.v.*) and proteins of the cell surface that are used in receptor-mediated endocytosis (*q.v.*). *See* Appendix C, 1955, deDuve *et al.*; lysosomal storage diseases, protein sorting, sorting signals, ubiquitin-proteasome pathway (UPP).

lysozyme an enzyme digesting mucopolysaccharides. Lysozymes having a bacteriolytic action have been isolated from diverse sources (tears, egg white, etc.). An important lysozyme is the enzyme synthesized under the direction of a phage that digests the cell wall of the host from within and thus allows the escape of the phage progeny. Lysozyme was the first enzyme whose three-dimensional structure was determined by x-ray crystallography. *See* Appendix C, 1966, Terzaghi; 1967, Blake *et al.*

lytic cycle the vegetative life cycle of a virulent phage, by which progeny phage are produced and the host is lysed. Temperate phage (*q.v.*) have the option of becoming a prophage (usually when growth conditions for its host cell are good) or entering the lytic cycle when growth conditions for its host are poor. *See cro* repressor.

lytic response lysis following infection of a bacterium by a virulent phage, as opposed to lysogenic response (*q.v.*).

lytic virus a virus whose intracellular multiplication leads to lysis of the host cell.

m 1. mole (expressed in daltons or kilodaltons, *q.v.*). *See* gram molecular weight. *Compare with* gram equivalent weight. 2. molar. *Compare with* normal solution.

M 1. the stage of the cell cycle (*q.v.*) where mitosis occurs. 2. molal (*q.v.*). *Compare with* molar.

M13 a single-stranded bacteriophage cloning vehicle, with a closed circular DNA genome of approximately 6.5 kilobase pairs. The major advantage of using M13 for cloning is that the phage particles released from infected cells contain single-stranded DNA that is homologous to only one of the two complementary strands of the cloned DNA, and therefore it can be used as a template for DNA sequencing analysis.

Ma megaannum (*q.v.*).

Macaca mulatta the rhesus monkey. The catarrhine primate most used in the laboratory and the original source of the Rh antigen. Its haploid chromosome number is 21, and about 30 genes have been located on 14 linkage groups. *See* Rh factor.

macaroni wheat *Triticum durum* (N = 14).

macha wheat *Triticum macha* (N = 21).

macroconidia *See* conidia.

macroevolution a large evolutionary pattern usually viewed through the perspective of geologic time, such as the evolution of the horse from *Eohippus* to *Equus*. Macroevolution includes changes in taxonomic categories above the species level, and events that result in the origin of a new higher taxon. *See* microevolution.

macromolecule a molecule of relative molecular mass ranging from a few thousand to hundreds of millions (proteins, nucleic acids, polysaccharides, etc.).

macromutation *See* evolution.

macronucleus the larger of the two types of nuclei in the ciliates; the "vegetative" nucleus. Macronuclei contain many copies of each gene and are transcriptionally active. *See* micronucleus, nuclear dimorphism, *Tetrahymena*.

macrophage a large, phagocytic, mononuclear leukocyte found in tissues, but derived from blood monocytes. Macrophages are called histiocytes in connective tissues, Kupffer's cells in the liver, microglial cells in the nervous system, and alveolar macrophages in the lung. To stimulate an immune response, most antigens must be "processed" by macrophages and presented on their surfaces to lymphocytes in association with self-Ia molecules. *See* Gaucher disease, major immunogene complex.

macrophage activation factor a lymphokine (*q.v.*) that activates macrophages.

macroscopic visible to the unaided eye.

macula adherans desmosome (*q.v.*).

***MAD* mutations** *mitotic arrest deficient* mutations. The wild-type alleles of these genes encode proteins that bind to kinetochores until spindle fibers attach to them. If there is a failure of spindle microtubules to attach, the MAD protein remains bound to the kinetochore, and the entry into anaphase is blocked. *See* checkpoints, mitosis.

magic bullets *See* chemotherapy.

magnesium an element universally found in small amounts in tissues. Atomic number 12; atomic weight 24.312; valence 2^+; most abundant isotope ^{24}Mg, radioisotope ^{28}Mg, half-life 21 hrs, radiation emitted—beta particle. *See* ribosome.

mainband DNA the major DNA band obtained by density gradient equilibrium centrifugation of the DNA of an organism.

maize *See Zea mays.*

major gene a gene with pronounced phenotypic effects, in contrast to its modifiers; also called an **oligogene.** *Compare with* polygene.

major histocompatibility complex (MHC) a large cluster of genes on human chromosome 6 and on mouse chromosome 17. The MHC controls many activities of immune cells, including the transplantation rejection process and the killing of virus-infected cells by specific killer T lymphocytes. The

MHC is part of a larger *major immunogene complex* (*q.v.*) with more diverse functions. In different mammals different symbols have been assigned to the MHC; for example: chicken (B), dog (DLA), guinea pig (GPLA), human (HLA), mouse (H-2), and rat (Rt-1). The chromosomal segment (6p21-31) spans 3.6 Mb and contains 128 genes and 96 pseudogenes. The MHC is the segment that contains the densest distribution of genes sequenced in the human genome so far. The MHC is associated with more diseases than any other chromosomal region, including autoimmune conditions like diabetes and rheumatoid arthritis. The gene map is divided into three segments: the genes of class I are in the segment closest to the p telomere, genes of class II are in the segment closest to the centromere, and class III genes lie between blocks I and II. Class II genes are in a low G + C isochore, and class III genes are in a high G + C isochore. *See* Appendix C, 1948, Gorer, Lyman, and Snell; 1953, Snell; 1999, MHC Sequencing Consortium; histocompatibility molecules, HLA complex, isochore, pseudogene, sequence similarity estimates.

major immunogene complex (MIC) a genetic region containing loci coding for lymphocyte surface antigens (e.g., Ia), histocompatibility (H) antigens, immune response (Ir) gene products, and proteins of the complement system. The genes specifying immunoglobulins assort independently of the MIC, but the plasma cells responsible for their production are under the control of the MIC.

malaria a disease caused by species of protozoa belonging to the genus *Plasmodium* (*q.v.*) and transmitted by female mosquitoes belonging to the genus *Anopheles* (*q.v.*). Malaria is the single most critical infectious disease of humankind. There are about 200,000,000 people infected by the parasite, and 2,000,000 die annually. Mortality rates are greatest in Africa, below the Sahara desert, where 90% of the deaths occur in children less than 5 years old. Malaria is the strongest known force in recent history for evolutionary selection within the human genome. There are many lines of evidence for diverse genetic adaptions in different human populations. Since the malaria parasite spends a part of its life cycle in the red blood cell, many mutations conferring malaria resistance involve genes that encode erythrocyte proteins. Among these are hemoglobin S, hemoglobin C, glucose-6-phosphate dehydrogenase deficiency, alpha thalassemia, and the Duffy blood group (all of which *see*). Also *see* Appendix C, 1880, Laveran; 1898, Ross; 1899, Grassi; 1954, Allison; apicoplast, artemisinin, blackwater fever, heterozygote advantage, Minos element, *Plasmodium* life cycle.

male gametophyte *See* pollen grain.

male pronucleus the generative nucleus of a male gamete.

male symbol ♂ the zodiac sign for Mars, the Roman god of war. The sign represents a shield and spear.

maleuric acid $C_5H_6N_2O_4$, a mitotic poison.

malignancy a cancerous growth. *See* cancer.

Malpighian tubule the excretory tubule of insects that opens into the anterior part of the hind gut. The tubule was discovered by Marcello Malpighi (1628–1694), the Italian physician who founded microscopic anatomy.

Malthusian having to do with the theory advanced by the English social economist, T. R. Malthus, in his *An Essay on the Principle of Population*, published in 1798. According to this theory the world's population tends to increase faster than the food supply, and poverty and misery are inevitable unless this trend is checked by war, famine, etc.

Malthusian parameter the rate at which a population with a given age distribution and birth and death rate will increase.

mammary tumor agent *See* mouse mammary tumor virus.

man *See Homo,* human mitotic chromosomes.

Mandibulata a subphylum of arthropods containing those species possessing antennae and a pair of mandibles. *See* Appendix A, Animalia, Arthropoda.

mantid an insect belonging to the orthopteran family Mantidae of the Dictyoptera. *See* Appendix A.

manifesting heterozygote a female heterozygous for a sex-linked recessive mutant gene who expresses the same phenotype as a male hemizygous for the mutation. This rare phenomenon results from the situation where, by chance, most somatic cells critical to the expression of the mutant phenotype contain an inactivated X chromosome carrying the normal allele of the gene. *See* Lyonization.

map *See* genetic map.

map distance the distance between genes expressed as map units or centiMorgans (cM).

mapping function a mathematical formula developed by J. B. S. Haldane that relates map distances to recombination frequencies. The function is represented graphically below. It demonstrates that no matter how far apart two genes are on a chromosome, one never observes a recombination value greater than 50%. It also shows that the relation between recombination frequencies and map distances is linear for genes that recombine with frequencies less than 10%.

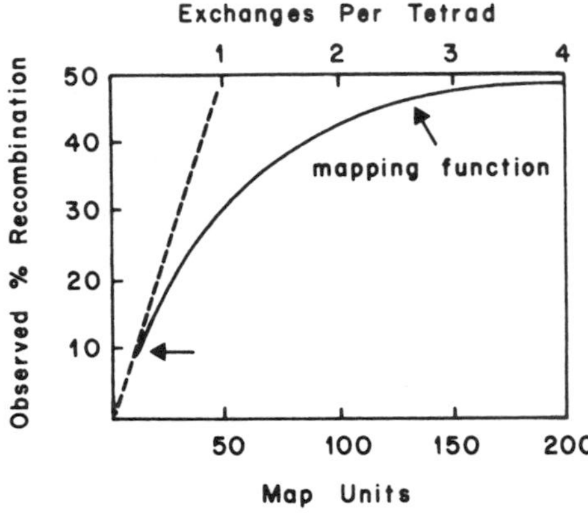

map unit a measure of genetic distance between two linked genes corresponding to a recombination frequency of 1% or 1 centiMorgan (cM). *See* Morgan unit.

mare a female horse.

Marfan syndrome a human hereditary disorder inherited as an autosomal dominant. The disease has a prevalence of 1/20,000 and is caused by mutations in the fibrillin gene (*FBN1*) at band 21.1 on the long arm of chromosome 15. Rupture of the aorta is the major cause of death in Marfan patients, and this is directly linked to defects in fibrillin, which serves as a substrate for elastin in the aorta. *FBN1* is relatively large, occupying 235 kilobases and containing 65 exons. Forty-two of the exons encode cystein-rich, EGF domains that function in calcium binding. The condition was described in 1896 by Antoine B. Marfan and is often called *arachnodactyly* in the earlier literature. *See* Appendix C, 1993, Pereira *et al.*; elastin, epidermal growth factor, fibrillin. http://www.marfan.org

marginotomy a term coined by A. M. Olonikov for the formation of nucleic acid replicas that are shortened relative to their templates. Telomeres (*q.v.*) undergo marginotomy, and the maximum number of mitoses any somatic cell line can undergo is correlated with the lengths of their telomeres and the rate at which they shorten per replication. *See* Appendix C, 1971, Olonikov; Hayflick limit.

mariner elements a family of transposable elements (*q.v.*) found in many species of insects. The mariner element isolated from *Drosophila mauritiana* is short (1,286 base pairs), with 28-bp inverted repeats. Between the repeats is a gene that encodes a transposase of 345 amino acids. Mariner-like transposable elements (MLTEs) occur in *Anopheles* and show promise as vectors for inserting parasite-resistance genes into populations of mosquito species that transmit human diseases. MLTEs appear to have been passed horizontally by insects to other invertebrates and vertebrates. One example of a MLTE in humans is found between bands 11.2 and 12 on the short arm of chromosome 17. This region is also a recombination hot spot. *See* horizontal transmission, hot spot, malaria, Minos element, TC1/mariner element.

marker **1.** a gene with a known location on a chromosome and a clear-cut phenotype, used as a point of reference when mapping a new mutant. **2.** antigenic markers serve to distinguish cell types. **3.** marker DNAs, RNAs, and proteins are fragments of known sizes and/or properties that are used to calibrate an electrophoretic gel.

marker rescue the phenomenon found when bacteria are mixedly infected with two genetically marked phages, of which only one type is irradiated. The progeny phages are predominately of the unirradiated type. However, some recombinants are found that contain genes from the irradiated parent. Such genes have been "rescued" by recombination.

MARs matrix attachment regions (*q.v.*).

marsupial any of a group of primitive mammals, females of which have an abdominal pouch in which the young are carried and nourished (kangaroos, opossums, etc.). *See* Appendix A.

masked mRNA messenger RNA that can be stored in large quantities because it is in some way inactivated and protected against digestion by nucleases (perhaps by being associated with proteins). Sea urchin eggs store mRNA in this masked form.

mass action, law of the principle that the rate of a chemical reaction is proportional to the concentrations of the reacting substances, with each concentration raised to a power equal to the relative number of molecules of each species participating in the reaction.

massed training *See* spaced training.

mass extinctions intervals in the geologic record during which between 25 and 50% of all fossil families became extinct in a geologically short time. During the Phanerozoic Eon, mass extinctions occurred at the close of the Cambrian, Ordovician, Devonian, Permian, Triassic, and Cretaceous periods. A 50% reduction in families is estimated to be equivalent to a reduction of more than 90% in the number of species. *See* geological time divisions, impact theory.

mass number the number of protons and neutrons in the nucleus of an atom.

mass spectrograph an instrument for analyzing a substance in terms of the ratios of mass to charge of its components. It produces a focused mass spectrum of lines on a photographic plate.

mass unit *See* atomic mass unit.

mast cells tissue cells, thought to be the counterparts of blood basophils, that are abundant in lymph nodes, spleen, and bone marrow, in connective tissues, and in the skin. However, blood and lymph do not contain mast cells. The cytoplasm of mast cells contains granules rich in heparin, serotonin, and histamine. Mast cells possess receptors for IgE molecules. When antigen becomes attached to mast-cell-bound antibodies, these vasoactive amines are released and initiate an allergic reaction.

mastigonemes *See* Chromista, Chrysomonads.

mastigote *See* undulipodium.

mate (noun) **1.** an individual who has formed a bond with a member of the opposite sex for breeding. **2.** an individual actually engaging in a sexual union. (verb) **1.** to form an alliance for future breeding. **2.** to engage in a sexual union which may result in the fertilization of eggs.

mate choice the situation where an animal (generally a female) chooses as a mate a specific individual from a small group within the species. Mate choice by females has been observed in insects, fish, reptiles, birds, and mammals. Selection involving mate choice plays a major role in the evolution of many sexually dimorphic characters seen in animals. The genetic changes induced by mate choice occur in quite small subspecific populations, and the factors that control choice behavior generally show quantitative inheritance (*q.v.*). *See* arena-breeding birds, courtship ritual, lek, sexual selection. *Contrast with* random mating.

mate killers *Paramecium aurelia* carrying mu particles. Such ciliates kill or injure sensitives with which they conjugate. The presence of mu in a paramecium protects it from the action of other mate killers.

maternal contamination the situation where blood of maternal genotype contaminates a sample of fetal cells from amniotic fluid, chorionic villi, or umbilical blood. This can confound the interpretation of the genetic analysis.

maternal effect gene a gene that encodes or affects the spatial distribution of a maternal product (mRNA or protein) that is stored in the unfertilized egg and required for normal development in the early embryo. Mothers homozygous for recessive alleles of such genes are viable, but lay eggs that produce phenotypically abnormal embryos. The genotype of the mother thus affects the phenotype of her offsprings. *Compare with* paternal effect gene, zygotic gene. *See* Appendix C, 1987, Nüsslein-Volhard *et al.*; 1988, 1989, Driever and Nüsslein-Volhard; 1988, Macdonald and Struhl; 1995, Lewis *et al.*; *bicoid*, cytoplasmic determinants, cytoplasmic localization, grandchildless genes, maternal polarity mutants.

maternal inheritance phenotypic differences controlled by cytoplasmic genetic factors (e.g., in mitochondria, chloroplasts, or viruses) derived solely from the maternal parent; also known as uniparental heredity, cytoplasmic inheritance, or extranuclear heredity. *Contrast with* paternal inheritance. *See* Appendix C, 1923, Boycott and Diver.

maternal PKU maternal phenylketonuria, a syndrome sometimes shown by infants of mothers who are homozygous for the PKU gene. High concentrations of phenylalanine (PA) damage the developing brain, and over 90% of infants subjected to an *in utero* environment rich in PA are irreversibly retarded, irrespective of their genotypes. Therefore, it is essential that women with PKU who have gone off the low-phenylalanine diet return to it if they become pregnant. *See* 1954, Bickel, Gerrard, and Hickmans; phenylketonuria.

maternal polarity mutants mutations in *Drosophila melanogaster* that are maternally inherited and affect the development of embryos in a polarized manner. These genes are transcribed during oogenesis, and they imprint an anteroposterior pattern on the egg. The genes are grouped phenotypically into *anterior* genes that effect head and thoracic structures, *posterior genes* that define the abdominal region, and *terminal genes* that control the development of the acron and telson. *See* Appendix C, 1987, Nüsslein-Volhard *et al.*; 1988, Driever and Nüsslein-Volhard; *bicoid,* cytoplasmic determinants, cytoplasmic localization, maternal effect gene, zygotic segmentation mutants.

mating type many species of microorganisms can be subdivided into groups (or mating types) on the basis of their mating behavior; only individuals of different mating types will undergo conjugation. Individuals from a given mating type possess on their surfaces proteins that will bind to complementary proteins or polysaccharides found only on the coats of individuals of opposite mating type. *See Paramecium aurelia,* cassettes.

matrilinear inheritance the transmission of cytoplasmic particles only in the female line.

matrix attachment regions (MARs) particular eukaryotic DNA sequences that bind to the nuclear matrix. According to a Domain Model of Chromatin Structure, each chromosome is anchored during interphase to a protein matrix that occupies the periphery of the nucleus. The MARs are located in series at about 50-kb intervals, and each pair defines a DNA loop domain that is physically isolated from the next loop domain. MARs contain predominantly repetitive DNA. *See* insulator DNA.

matroclinous inheritance *See* matrocliny.

matrocliny inheritance in which the offspring resembles the female parent more closely than the male. In *Drosophila* the daughters produced by attached-X females are matroclinous in terms of their sex-linked genes. *See* hologynic.

maturase a protein encoded by an exon-intron combination that helps catalyze the excision of the intron from its own primary transcript. A maturase is probably not an enzyme that catalyzes intron removal and exon splicing, but rather a factor that modifies the specificity of a preexisting splicing enzyme.

maturation divisions a series of nuclear divisions in which the chromosome complement of the nuclei is reduced from diploid to haploid number. *See* meiosis.

maturation promoting factor (MPF) a complex first isolated from eggs of *Xenopus laevis* (*q.v.*). Immature oocytes are normally blocked in the G2 stage of the cell cycle (*q.v.*). However, when the cells are injected with MPF, they proceed through mitosis and stop at the transition between metaphase and anaphase. The MPF causes chromosomal condensation, breakdown of the nuclear envelope, and the inhibition of transcription during mitosis. For these reasons, MPF is often called the *mitosis promoting factor.* This factor is now known to contain two subunits: a protein kinase and a cyclin (*both of which see*). *Also see* Appendix C, 1971, Masui and Smith; cell division genes, checkpoint, p34 (CDC2).

Maxam-Gilbert method *See* DNA sequencing techniques.

maxicells bacterial cells that have degraded chromosomal DNAs because they have been heavily irradiated with ultraviolet light. Replication and transcription in such cells is inhibited because of the inability of the damaged DNA to act as template. However, if the maxicells contain multiple copies of a plasmid, the plasmid molecules that did not receive UV hits continue to replicate and transcribe gene products. Thus, maxicells provide a means of analyzing plasmid-encoded functions, while the products encoded by the host genome are reduced to a minimum. *See* minicells.

maxicircles *See* RNA editing.

maximum permissible dose the greatest amount of ionizing radiation safety standards permit to be received by a person.

maze an experimental device consisting of a network of paths through which a test animal must find its way. The first animal maze was invented by Francis Galton. *See* Appendix C, Galton.

maze-learning ability the speed at which an animal learns to find its way through a maze without entering blind alleys.

mb, mbp *See* megabase.

MBD methyl-CpG-binding domain. *See* DNA methylation.

McArdle disease a hereditary glycogen storage disease in humans with a frequency of 1/500,000. It arises from a deficiency of an enzyme, skeletal muscle glycogen phosphorylase, which is encoded by a gene on chromosome 11.

M chromosome the human mitochondrial chromosome. *See* human gene maps.

MC1R gene a gene containing 954 bp which encodes the melanocyte-1-receptor, a protein that contains 317 amino acids. The dominant allele (*D*) encodes a protein that stimulates the production of black melanin. The recessive allele (*d*) encodes an isoform (*q.v.*) which differs from the D protein by 4 amino acid substitutions. Mice homozygous for *d* are light colored. In *Chaetodipus intermedius* (*q.v.*) the *D* allele is found in those races of mice that live among the dark rocks of ancient lava flows. Mice living in areas of light-colored rocks and sand are homozygous for the *d* allele. *See* Appendix C, 2003, Nachman, Hoekstra, D'Agostino; adaptive melanism, melanin.

MDV-1 a variant molecule derived from the RNA genome of Q beta phage (*q.v.*) by *in vitro* selection experiments. MDV-1 is the shortest replicating molecule known. *See in vitro* evolution.

mean the sum of an array of quantities divided by the number of quantities in the group.

mean free path the average distance a subatomic particle travels between collisions.

mean square variance (*q.v.*).

mechanical isolation reproductive isolation due to the incompatibility of male and female genitalia.

mechanistic philosophy the point of view that holds life to be mechanically determined and explicable by the laws of physics and chemistry. *Contrast with* vitalism.

medaka a common name for *Oryzia latipes* (*q.v.*).

medial complex *See* synaptonemal complex.

median the middle value in a group of numbers arranged in order of size.

median lethal dose the dose of radiation required to kill 50% of the individuals in a large group of organisms within a specified period. Synonymous with LD50.

medium the nutritive substance provided for the growth of a given organism in the laboratory.

megaannum one million years before the present; symbolized Ma.

megabase a unit of length for DNA molecules, consisting of 1 million nucleotides; abbreviated *mb*, or *mbp* for megabase pairs. *See* base pair, gigabase.

megakaryocyte a large cell with a multilobed nucleus present in the bone marrow, but not circulating in the blood. Megakaryocytes bud off platelets (*q.v.*). *See* von Willebrand disease.

megasporangium a spore sac containing megaspores.

megaspore in angiosperms, one of four haploid cells formed from a megasporocyte during meiosis. Three of the four megaspores degenerate. The remaining megaspore divides to produce the female gametophyte or embryo sac (*q.v.*).

megaspore mother cell the diploid megasporocyte in an ovule that forms haploid megaspores by meiotic division.

megasporocyte megaspore mother cell.

megasporogenesis the production of megaspores.

meiocyte an auxocyte (*q.v.*).

meiosis in most sexually reproducing organisms, the doubling of the gametic chromosome number, which accompanies syngamy, is compensated for by a halving of the resulting zygotic chromosome number at some other point during the life cycle. These changes are brought about by a single chromosomal duplication followed by two successive nuclear divisions. The entire process is called meiosis, and it occurs during animal gametogenesis or sporogenesis in plants (see pages 266–267). The prophase stage is much longer in meiosis than in mitosis. It is generally divided into five consecutive stages: *leptonema*, *zygonema*, *pachynema*, *diplonema*, and *diakinesis*. When the stages are used as adjectives, the *nema* suffix is changed to *tene*.

During the leptotene stage the chromosomes appear as thin threads with clearly defined chromomeres. The chromosomes are often all oriented with one or both of their ends in contact with one region of the nuclear membrane, forming the so-called *bouquet configuration*. Although each chromosome appears single, it is actually made up of two chromatids. However, the doubleness of the chromosome does not become obvious until pachynema. The DNA replication that doubles the diploid value occurs before leptonema.

In diploid somatic cells, the 2N chromosomes are present as N pairs, and each chromosome is a replica of one contributed by the male or female parent at fertilization. In the somatic nuclei of most organisms, the homologous chromosomes do not pair. However, during the zygotene stage of meiotic prophase *synapsis* of homologous chromosomes takes place. This pairing begins at a number of points and extends "zipperlike" until complete. Pairing is accompanied by the formation of synaptonemal complexes (*q.v.*). When synapsis is finished, the apparent number of chromosome threads is half what it was

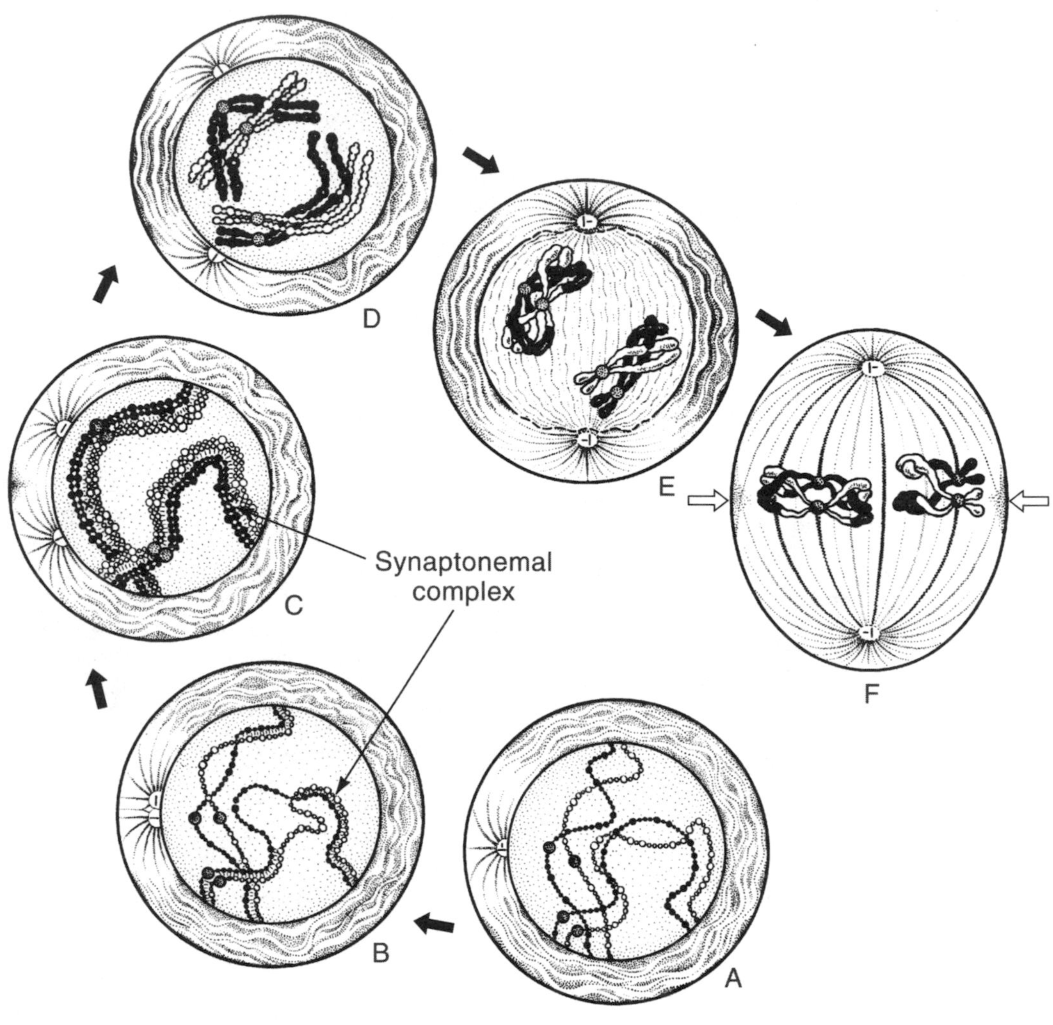

Meiosis

(A) The leptotene stage during spermatogenesis in a hypothetical animal. The organism has two pairs of chromosomes (one metacentric and one submetacentric). The maternal and paternal homologous chromosomes are drawn in light and dark shades to allow them to be distinguished. The centromeres are represented by stippled circles. Note that the chromosomes are oriented with both of their ends attached to the inner surface of the nuclear envelope. The chromosomes are uncoiled and maximally extended. In this state, the chromomeres and chromonemata can be seen readily. A centrosome encompassed by astral rays is present. It contains a mother and daughter centriole, oriented at right angles to each other.
(B) During zygonema, synapsis of homologous chromosomes takes place. This pairing begins at one or more points and extends "zipperlike" until complete. Subsequently, the cell contains two bivalents. The centrosome has divided into two daughter centrosomes, each containing a single centriole.
(C) During pachynema, each chromosome can be seen to be composed of two sister chromatids, except at the region of the centromere. As a result, the bivalents are converted into tetrads. A synaptonemal complex extends the length of the bivalent. Crossing over occurs at the three places marked by the Xs. The centrosomes move apart.
(D) During diplonema, the medial complex of the synaptonemal complex disappears, and in each tetrad one pair of sister chromatids begins to separate from the other pair. However, the chromatids are prevented from separating at places where interchanges have taken place. The points where the chromatids form cross-shaped

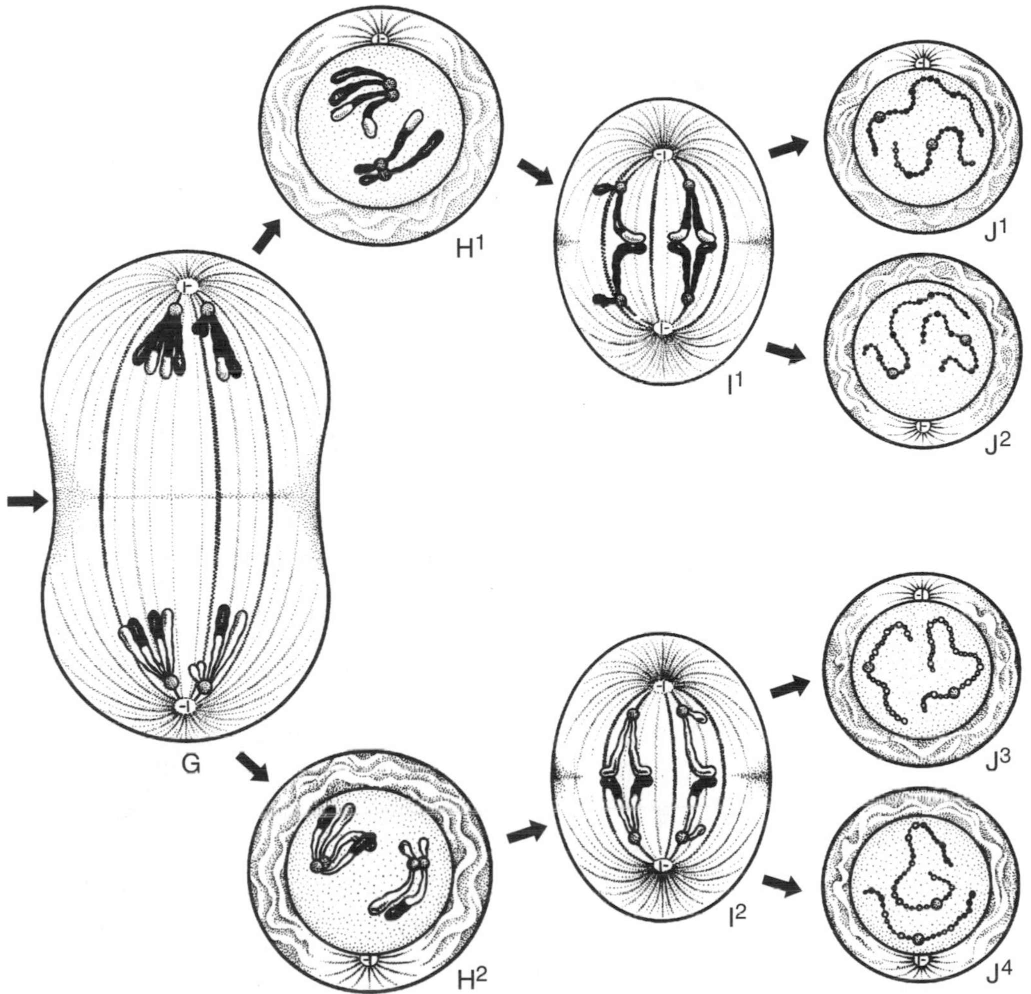

configurations are called chiasmata. As a result of the two-strand single exchange in the submetacentric tetrad, two of the four chromatids have both maternal (light) and paternal (dark) segments in their long arms. The metacentric chromosome has undergone a two-strand double exchange, so two of the four chromatids have light and dark segments in both arms. The reciprocal pattern of the exchanged segments is shown especially clearly in E and F.

(E) At diakinesis, the chromatids are shorter and thicker, and terminalization is occurring. The centrosomes have reached the poles, and the nuclear envelope is beginning to disappear.

(F) At metaphase I, the tetrads are arranged at the equator of the spindle.

(G) At late anaphase I, the homologous chromosomes have separated and have moved to each pole. However, the centromeres have not divided. As a consequence, maternal and paternal chromosomal material has been separated (except in regions distal to points of crossing over).

(H^1, H^2) Secondary spermatocytes containing dyads.

(I^1, I^2) During early anaphse II, centromeres divide and allow separation of sister chromatids.

(J^1, J^2, J^3, J^4) The four spermatids containing monads. Chromosomes once again uncoil and elongate. (J^1) and (J^4) each contain one single and one double crossover chromatid.

before, and the visible bodies in the nucleus are now *bivalents* rather than single chromosomes.

During the pachytene stage, each paired chromosome separates into its two component sister chromatids (except at the region of the centromere). As a result of the longitudinal division of each homologous chromosome into two chromatids, there exist in the nucleus N groups of four chromatids lying parallel to each other called *tetrads*. A type of localized breakage followed by exchange between nonsister chromatids occurs. This process, called crossing over, is accompanied by the synthesis of an amount of DNA constituting less than 1% of the total in the nucleus. The exchange between homologous chromatids results in the production of crossover chromatids containing genetic material of both maternal and paternal origin.

During the diplotene stage one pair of sister chromatids in each of the tetrad begins to separate from the other pair. However, the chromatids are prevented from separating at places where exchanges have taken place. In such regions, the overlapping chromatids form a cross-shaped structure called a *chiasma* (pl. *chiasmata*). The chiasmata slip along laterally toward the ends of the chromatids, with the result that the position of a chiasma no longer coincides with that of the original crossover. This *terminalization* proceeds until during diakinesis all the chiasmata reach the ends of the tetrad, and the homologs can separate during anaphase.

The chromosomes coil tightly during diakinesis and so shorten and thicken to produce a group of compact tetrads lying well spaced out in the nucleus, often near its membrane. Terminalization is completed, and the nucleolus disappears.

The nuclear envelope disappears during division I, and the tetrads are arranged at the equator of the spindle. The chromatids of a tetrad disjoin in such a way that there is separation of maternal from paternal chromosomal material with the exceptions of regions distal to where crossing over has occurred. Division I produces two *secondary gametocytes*, which contain *dyads* surrounded by a nuclear envelope.

Division II commences after a short interphase during which the chromosomes do not uncoil. The nuclear membrane disappears, and the dyads arrange themselves upon the metaphase plate. The chromatids of each dyad are equivalent (again with the exception of regions distal to points of crossing over); the centromere divides and thus allows each chromosome to pass to a separate cell. Thus during meiotic anaphase I centromeres do not have the block to DNA replication removed, as is true for mitotic anaphase, and centromeres remain functionally single. At meiotic anaphase II, however, the block is removed, as in mitosis, and each chromatid is converted into a functional chromosome. In animals, division II produces four *spermatids* (*q.v.*) or *ootids* (*q.v.*) which contain *monads* surrounded by a nuclear membrane. Meiosis therefore provides a mechanism whereby (1) an exchange of genetic material may take place between homologous chromosomes and (2) each gamete receives but one member of each chromosome pair. J. B. Farmer and J. E. S. Moore proposed in 1905 that the name *meiosis* be used for the reduction divisions that precede the formation of gametes. The names *leptotene*, *zygotene*, *pachytene*, and *diplotene* were coined by H. v. Winiwarter in 1900. In the third edition (1925) of his *Cell in Development and Heredity*, E. B. Wilson started the tradition of using these names as adjectives and the *nema* forms as nouns.

The type of meiosis described above immediately precedes gametogenesis. This type (*gametic meiosis*) is characteristic of all animals. Fungi are characterized by *zygotic meiosis*, where meiosis immediately follows zygote formation. *Haplodiplomeiosis* refers to the situation seen in most plants where meiosis intervenes between a prolonged diploid phase and an abbreviated haploid phase. *See* alternation of generations, centromeric coupling, cohesin, interkinesis, oogenesis, recombination nodules, separase, sex, *Sgo*, sister chromatid cohesion, spermatogenesis, telomere-led chromosome movement, zygDNA.

meiosporangium (*plural* meiosporangia) a sporangium in which meiosis occurs. *Compare with* mitosporangium.

meiospore a spore produced by meiosis.

meiotic cycle a first (reductional) division, followed by a second (equational) division. *See* meiosis.

meiotic drive any alteration in meiosis or the subsequent production of gametes that results in preferential transmission of a particular genetic variant. Examples of meiotic drive are known in a wide variety of organisms, including fungi, plants, insects, and mammals. *See* cheating genes, segregation distortion.

mei-W68 a recessive mutation in *D. melanogaster* that abolishes meiotic exchange in the oocytes of homozygotes. *S. cerevisiae* has a homolog called *SPO11*. The *H. sapiens* homolog is located at 20q13.2-q13.3. *See* recombination nodules.

Melampsora lini a rust fungus that parasitizes flax. *See Linum usitatissimum*.

Melandrium album a dioecious plant with a method of sex determination similar to man (that is, maleness is determined by the Y chromosome).

melanin a dark brown to black pigment responsible for the coloration of skin, hair, and the pigmented coat of the retina (*q.v.*). The pigments in bird feathers, reptile skins, insect exoskeletons, and the ink of cephalopods all contain melanins. The macromolecule consists of polymers of indole 5,6-quinone and 5,6-dihydroxyindole-2-carboxylic acid and is formed by the enzymatic oxidation of tyrosine or tryptophan. A segment of the structure is shown below.

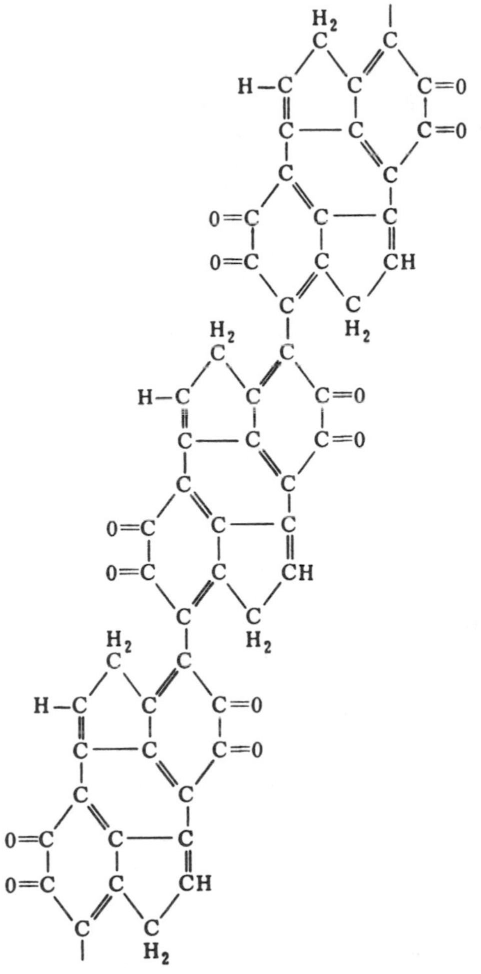

Skin color variation among humans is an evolutionary adaptation using melanin to screen out harmful ultraviolet light. The sunscreen must be dense enough to prevent the photolysis of folic acid (*q.v.*) but transparent enough to allow the synthesis of vitamin D (*q.v.*). For a given human population the trade-off depended on the amount of sunshine in the habitat where it evolved. *See* Appendix C, 2000, Jablonski and Chaplin; agouti, albinism, eumelanin, *MC1R* gene, phaeomelanin, *SLC 24A5*.

melanism the hereditary production of increased melanin resulting in darker coloring. *See* adaptive melanism, albinism, industrial melanism.

melanocytes cells located in the basal cell layer of the skin. Melanocytes form long slender processes which ramify among the adjacent keratinocytes. Eventually about 30 of these are connected to each melanocyte. Melanosomes, the organelles that synthesize melanin, are in the cytoplasm of melanocytes. The larger the melanosomes, the darker the skin pigmentation. Intermedin (*q.v.*) stimulates melanin synthesis in response to UV damage. The melanocytes then transfer small packets of melanin to the keratinocytes, and these move to the surface of the skin forming a sunscreen. The result is tanning. In the mouse the melanocytes that give color to the fur are derived from 34 cells that have their fates determined early during embryogenesis. *See* Appendix C, 1967, Mintz.

melanocyte-1-receptor *See MC1R* gene.

melanocyte-stimulating hormone intermedin (*q.v.*).

melanoma a cancer composed of melanocytes (*q.v.*).

Melanophus femur-rubrum a species of grasshopper widely used for the cytological study of meiosis.

melanosome an intracellular organelle containing aggregations of tyrosinase, found in melanocytes.

melanotic tumor of *Drosophila* *See* pseudotumor.

Meleagris gallopavo the domesticated turkey.

melphalan a mutagenic, alkylating agent. See illustration on page 270.

melting in nucleic acid studies, the denaturation of double-stranded DNA to single strands.

melting-out temperature the temperature at which the bonds joining the constituent strands of a DNA/DNA or DNA/RNA duplex are broken and the molecules dissociate.

melting profile a curve describing the degree of dissociation of the strands in a DNA/DNA or DNA/RNA duplex in a given time as a function of temperature. The stabilities of duplexes are a function of

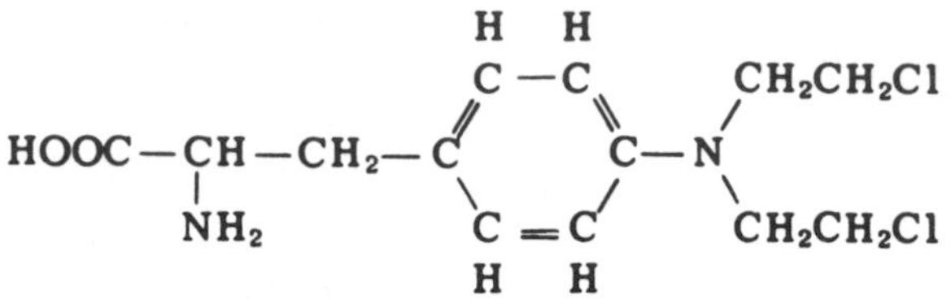

```
                      H  H
                      C—C        CH2CH2Cl
                     //   \\    /
HOOC—CH—CH2—C      C—N
     |            \   /     \
    NH2           C=C        CH2CH2Cl
                      H  H
```

Melphalan

their molecular weights, so that the melting profiles shift to the right as the chain lengths increase.

melting proteins *See* helix-destabilizing proteins.

melting temperature the temperature at which 50% of the double helices have denatured; the midpoint of the temperature range over which DNA is denatured. *See* denaturation, Tm.

membrane 1. *Cytology*: a lipid bilayer that encloses the protoplasm (*q.v.*) of a cell (plasma membrane, *q.v.*) or divides the cell into compartments that are distinctive in morphology and function, such as the membranes of chloroplasts (*q.v.*), mitochondria (*q.v.*), lysosomes (*q.v.*), peroxisomes (*q.v.*), and the cavities of the endoplasmic reticulum (*q.v.*) and Golgi apparatus (*q.v.*). *See* lipid bilayer model. 2. *Anatomy*: any thin, pliable layer of tissue that covers surfaces or separates or connects regions, structures, or organs of a plant or animal, such as the basilar membrane of the inner ear, extra-embryonic membranes, pleuropericardial membranes of the heart, pleuroperitoneal membranes that line the body cavity, and synovial membranes of bony joints. 3. *Biotechnology*: nonlipid membranes of nylon, nitrocellulose, polyvinyladine fluoride (PVDF), or other substances that are used for blotting techniques. *See* northern blotting, Southern blotting, western blotting.

memory the persistent modification of behavior that results from experience. The conversion of short-term memory (STM) to long-term memory (LTM) requires the synthesis of specific gene products. *See* Appendix C, 1982, Kandel and Schwartz; 1994, Tully *et al.*; 2003, Si *et al.*; *Aplysia*, CPEB protein, CREBs, immunological memory, neuregulins, spaced training.

menarche the beginning of the first menstrual cycle during puberty in human females. *Compare with* menopause.

Mendeleev table the periodic table of chemical elements named in honor of Dmitri Mendeleev, the Russian chemist who constructed the earliest form of the periodic table in 1868. The table appears on page 330.

Mendelian character a character that in inheritance follows Mendelian laws (*q.v.*).

Mendelian genetics referring to the inheritance of chromosomal genes following the laws governing the transmission of chromosomes to subsequent generations; also called *Mendelism*.

Mendelian Inheritance in Man *(MIM)* a catalog of human genetic diseases prepared by Victor McKusick. MIM is now in its twelfth edition and is also available electronically. Each gene is given a six digit MIM number, and this is followed by four digits that specify allelic variants. For example, the *BRCA 1* mutation commonly found among Ashkenazi Jews is 113705.0003. In addition to genes that cause hereditary diseases when mutated, MIM also includes polymorphisms that influence disease resistance. The Duffy blood group gene (*q.v.*) (110700.0001), which confers resistance to malaria, is an example. *See* Appendix C, 1966, McKusick; breast cancer susceptibility genes, human genetic diseases, OMIM.

Mendelian laws 1. *The Law of Segregation*. The factors of a pair of characters are segregated. In modern terms, this law refers to the separation into different gametes and thence into different offspring of the two members of each pair of alleles possessed by the diploid parental organism. 2. *The Law of Independent Assortment*. The members of different pairs of factors assort independently. A restatement of the law in modern terms is that the members of different pairs of alleles are assorted independently into gametes during gametogenesis (provided they reside on different chromosomes), and that the subsequent pairing of male and female gametes is at random. *See* Appendix C, 1856, 1865, 1866, Mendel; 1900, de Vries, Correns, Bateson; 1902, Sutton; 1990, Bhattacharyya *et al.*

Mendelian population an interbreeding group of organisms sharing a common gene pool.

mendelize to segregate according to Mendelian laws (*q.v.*).

menopause the cessation of menstrual cycles in human females, usually occurring between age 50 and 60.

meq milliequivalent. *See* gram equivalent weight.

mer *See* mers.

mercaptopurine a synthetic purine analog, one of the first inhibitors of DNA synthesis, shown to suppress the growth of cancer cells.

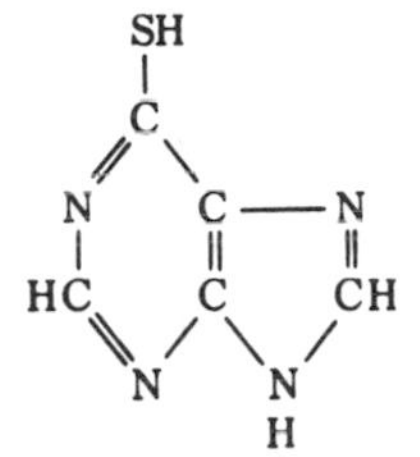

```
       SH
       |
       C
     //  \
    N     C —— N
    |     ||    ||
   HC     C     CH
     \\  / \   /
       N     N
             H
```

Mercenaria mercenaria *See* Pelecypoda.

mericlinal chimera an organism or organ composed of two genetically different tissues one of which partly surrounds the other.

meristems the undifferentiated, mitotically active tissues of plants. The meristems at the tips of the roots and shoots are referred to as *apical meristems*. *See* floral organ primordia.

meristic variation variation in characters that can be counted, like the number of bristles, leaves, scales, etc.

merlin *See* neurofibromatosis.

meroblastic cleavage cleavage producing cells, some of which are larger than others, because of a polarized distribution of yolk.

merogony 1. an asexual replicative cycle occurring during the haploid phase of *Plasmodium* and other apicomplexans. Subsequently the infected hepatocyte or erythrocyte ruptures, releasing hundreds of merozoites. *See* *Plasmodium* life cycle. 2. the development of an experimentally produced egg fragment (containing a diploid nucleus or a haploid nucleus of male or female origin) into a small-sized embryo, called a merogone.

meroistic *See* insect ovary types.

meromixis genetic exchange in bacteria involving a unidirectional transfer of a partial genome. *See* F factor (fertility factor).

merospermy the situation in which the nucleus of the fertilizing sperm does not fuse with the egg nucleus and later degenerates. Subsequent development is by gynogenesis.

merotomy the dissection of cells into several portions, with or without nuclei, as in experimental grafting with *Acetabularia* (*q.v.*).

merozoite *See* *Plasmodium* life cycle.

merozygote a partially diploid bacterial zygote containing an exogenotic chromosomal fragment donated by the F^+ mate. The exogenote may also be introduced during transduction or sexduction. *See* endogenote, exogenote.

mers the unit that defines the number of bases in an oligonucleotide polymer. For example, oligonucleotides that contain 15 or 17 bases are referred to as 15 mers and 17 mers, respectively.

Mertensian mimicry *See* mimicry.

mesenchyme an embryonic type of connective tissue, consisting of amoeboid cells with many processes. Populations of such cells form a loose network. Most mesenchyme is derived from mesoderm. The mesenchyme produces the connective tissue and the circulatory system during the development of vertebrates.

Mesocricetus auratus the golden hamster, a rapidly breeding rodent used in the laboratory. The haploid chromosome number is 22. *Contrast with* *Cricetulus griseus*.

mesoderm the middle layer of embryonic cells between the ectoderm and endoderm in triploblastic animals. Mesoderm forms muscle, connective tissue, blood, lymphoid tissue, the linings of all the body cavities, the serosa of the viscera, the mesenteries, and the epithelia of the blood vessels, lymphatics, kidney, ureter, gonads, genital ducts, and suprarenal cortex. *See* Appendix C, 1845, Remak.

mesokaryotic referring to the dinoflagellate nucleus, in which the chromosomes are in a condensed, discrete condition at all times.

mesophil a microorganism whose optimum growth temperature lies in the range between 20°C and 45°C. Therefore it is easily cultured in the laboratory.

mesosome one of the invaginated segments of the plasma membranes in certain bacteria to which DNA molecules are attached.

Mesostigma viride a unicellular green alga that belongs to the class Prasinophyceae. Phylogenetic trees inferred from the nucleotide sequences of the RNAs of the small ribosomal subunit put the prasinophytes at the base of the Chlorophyta. The entire genome of the chloroplasts of *M. viride* has been sequenced and found to consist of 118,360 base pairs. This chDNA contains 135 ORFs, more genes than reported for any land plant or algal chDNA. *Mesostigma* chloroplasts have a gene organization very similar to that of land plant chloroplasts. This species therefore seems to be a living fossil (*q.v.*) that shows close similarities to the organisms that lived about 800 million years ago and that were the ancestors of both green algae and land plants. *See* Appendix A, Protoctista, Chlorophyta; Appendix C, 2000, Lemieux, Otis, and Turmel; Appendix F; chlorophyte, chloroplast DNA (chDNA).

mesothorax the middle of the three thoracic segments of an insect. It bears a pair of legs and (in winged insects) a pair of wings.

Mesozoic the 180 myr era during which dinosaurs arose, flourished, and became extinct. *See* geologic time divisions.

message transport organizer (METRO) a region of the *Xenopus* mitochondrial cloud (*q.v.*) in which specific RNAs are sorted into distinct spatial locations for translocation to the vegetal pole of the egg.

messenger RNA an RNA molecule that functions during translation (*q.v.*) to specify the sequence of amino acids in a nascent polypeptide. In eukaryotes, mRNA is formed in the nucleus from premessenger RNA molecules (*q.v.*). *See* Appendix C, 1956, Volkin and Astrachan; 1961, Jacob and Monod, Brenner, Jacob, and Meselson; 1964, Marbaix and Burny; 1967, Taylor *et al.*; 1969, Lockard and Lingrel; exon, intron, polyadenylation, polysome, posttranscriptional modification.

met methionine. *See* amino acid.

metabolic block a nonfunctional reaction in a metabolic pathway, as a consequence of a defective (mutant) enzyme whose normal counterpart catalyzes the reaction.

metabolic control levels a hierarchical series of biochemical levels, i.e., the level of the genome, the transcriptome, the proteome, or the metabolome (*all of which See*), at which metabolic controls are studied.

metabolic pathway a series of stepwise biochemical changes in the conversion of some precursor substance to an end product, each step usually catalyzed by a specific enzyme. The citric acid cycle, the ornithine cycle, and the pentosephosphate pathway (*all of which see*) are examples of metabolic pathways. *See* metabolome.

metabolic poison a compound poisoning a metabolic process. *See* dinitrophenol.

metabolism the sum of all the physical and chemical processes by which living cells produce and maintain themselves and by which energy is made available for the use of the organism. *See* chemotrophs, glycolysis, pentosephosphate pathway, photophosphorylation, photosynthesis, phototrophs.

metabolite a chemical compound that is produced or consumed during metabolism. Polymeric biological molecules are excluded from this definition. Metabolites include low molecular weight compounds that are produced or converted by enzymes during metabolism or the precursors or breakdown products of biopolymers. *See* metabolic pathway, metabolome.

metabolome the sum of all small molecular weight metabolites in a biological sample of interest. The metabolome of a given cell will vary greatly depending on its physiological or developmental state, its age, or its response to disease or drugs. *See* metabolic control levels.

metacentric designating a chromosome with a centrally placed centromere.

metachromasy the phenomenon where one dye stains more than one color. A substance that stains metachromatically is called a chromatrope. Chromatropes are high-molecular-weight structures with serially arranged charged groups. Acidic mucopolysaccharides and nucleic acids are prime examples. Azure B (*q.v.*) is an example of a metachromatic dye. The color it produces depends on the way dye molecules stack on the chromatrope. As the amount of stacking increases, the color changes from green, to blue, to red. In an azure B–stained tissue section, chromosomes generally stain green, nucleoli and cytoplasmic ribosomes blue, and mucopolysaccharide-containing deposits red.

metachromatic dye a dye that stains tissues two or more colors. *See* azure B, metachromasy.

metafemale in *Drosophila*, a female phenotype of relatively low viability in which the ratio of X chromosomes to sets of autosomes exceeds 1.0; previously called a superfemale. *See* intersex, metamale.

metagenesis alternation of generations (*q.v.*) among animals. Metagenesis is commonly seen in invertebrates, especially coelenterates. Unlike the situation among plants, both generations are diploid.

metalloenzyme a protein combined with one or more metal atoms that functions as an enzyme.

metalloprotein any protein that requires at least one metal ion for normal functioning. For example, mammalian cytochrome c oxidase is a key enzyme in aerobic metabolism, containing six metal centers (two hemes, two copper centers, Mg, and Zn). The simplest known Mo enzyme (containing molybdenum) is dimethyl sulfoxide (DMSO) reductase of *Rhodobacter spheroides*, which catalyzes the conversion of DMSO to dimethyl sulfide. *See* androgen receptor, ceruloplasmin, metalloenzyme, RING finger, urease, vitamin D receptor, Wilms tumor, Wilson disease, zinc finger proteins.

metallothioneins small proteins that bind heavy metals and thereby protect cells against their toxic effects. Genes coding for metallothioneins are activated by the same metal ions that these proteins bind.

metamale in *Drosophila*, a poorly viable male characterized by cells containing one X and three sets of autosomes; previously called a supermale. *See* intersex, metafemale.

metamerism the serial segmentation of an animal's body along its anterior-posterior axis. A millipede is an example of this kind of organization. The repeated segments are sometimes called *metamers* or *somites*. Metamerism can also occur along the axes of appendages produced by metamers. Metamerism is common in annelids, arthropods, and chordates. *See* homeotic mutations, segment identity genes, selector genes, zygotic segmentation mutants.

metamorphosis the transformation from larval to adult form. In *Drosophila* the destruction of larval tissues and their replacement with adult ones is triggered by 20-hydroxy-ecdysone. *See* ecdysones.

metaphase *See* mitosis.

metaphase arrest referring to the accumulation of metaphase figures in a population of cells poisoned with colchicine, Colcemid, or some other spindle poison.

metaphase plate the grouping of the chromosomes in a plane at the equator of the spindle during the metaphase stage of mitosis (*q.v.*).

metastable state an excited state of an atomic nucleus, which returns to its ground state by the emission of radiation.

metastasis the spread of malignant neoplastic cells from the original site to another part of the body.

metatarsus the basal tarsal segment of the insect leg. In the male of *Drosophila melanogaster*, the metatarsus of each foreleg bears a sex comb.

metathorax the hindmost of the three thoracic segments of an insect. It bears a pair of legs and (in many winged insects) a pair of wings. In flies, it bears the halteres.

Methanococcus jannaschii an anaerobic methanogen that lives at high temperatures and extreme pressures in geothermal marine sediments. This archaeon is a chemolithoautotroph (*q.v.*), and it gains its energy by the reaction $4H_2 + CO_2 \rightarrow CH_4 + 2H_2O$. Its chromosome is a circular DNA molecule made up of 1,644,976 base pairs that contain 1,682 ORFs. It also has two additional genetic elements. The larger is a circle made up of 58,407 bps that contain 44 ORFs, and the smaller circle is made up of 16,550 bps containing 12 ORFs. The genomes of *M. jannaschii* and *H. influenzae* are about the same size, yet they differ in the number of genes of unknown functions that they contain. The bacterium *Haemophilus* has 22% URFs, while the archaeon *Methanoccus* has 62% URFs. Methanococcal genes devoted to transcription, translation, and replication resemble those of eukaryotes more closely than bacteria. *See* Appendix A, Archea, Euryarchaeota; Appendix C, 1996, Bult *et al.*; Appendix E; Archaebacteria, *Archaeoglobus fulgidus*, hyperthermophile.

methanogens prokaryotes that belong to the phylum Euryarcheota of the Archaea (*see* Appendix A). They live in oxygen-free environments and generate methane by the reduction of carbon dioxide. *See* *Methanococcus jannaschii*.

methionine the molecule that is the leading amino acid in a newly formed protein. In many proteins this is subsequently removed by an aminopeptidase and, as a result, the second amino acid becomes *N* terminal. It follows that in eukaryotes the methionine-specifying codon (AUG) is recognized by different specific tRNAs, depending on whether it is at the start of the mRNA or at an internal position. *See* amino acid, initiator tRNA.

method of least squares *See* least squares.

methotrexate a folic acid antagonist that kills cells by inhibiting the enzyme dihydrofolate reductase and thus blocking the synthesis of nucleic acids. Mammalian cells that develop resistance to methotrexate do so by amplifying the genes encoding dihydrofolate reductase. *See* Appendix C, 1978, Schimke *et al.*; folic acid.

methuselah a gene in *Drosophila* that when mutated increases longevity. Normal flies live about 60 days on the average, whereas *methuselah* mutants last 80 days. *See* Appendix C, 1998, Lin, Serounde, and Benzer. *Compare with* progeria.

methyladenine *See* 5-methylcytosine.

methylated cap a modified guanine nucleotide terminating eukaryotic mRNA molecules. The cap is introduced after transcription by linking the 5′ end of a guanine nucleotide to the 5′ terminal base of the mRNA and adding a methyl group to position 7 of this terminal guanine. The addition of the terminal guanine is catalyzed by the enzyme *guanylyl transferase*. Another enzyme, *guanine-7-methyl transferase*, adds a methyl group to the 7 position of the terminal guanine. Unicellular eukaryotes have a cap with this single methyl group (cap 0). The predominant form of the cap in muticellular eukaryotes (cap 1) has another methyl group added to the next base at the 2′-*o* position by the enzyme *2′-o-methyl transferase*. More rarely, a methyl group is also added to the 2′-*o* position of the third base, creating cap 2 type. Capping occurs shortly after the initiation of transcription and precedes all excision and splicing events. The function of the cap is not known, bu it may protect the mRNA from degradation by nucleases or provide a ribosome binding site. *See* posttranscriptional processing.

methylation of nucleic acids the addition of a methyl group ($-CH_3$) to DNA. *See* DNA methylation, restriction and modification model, parental imprinting.

methylcholanthrene a carcinogenic hydrocarbon.

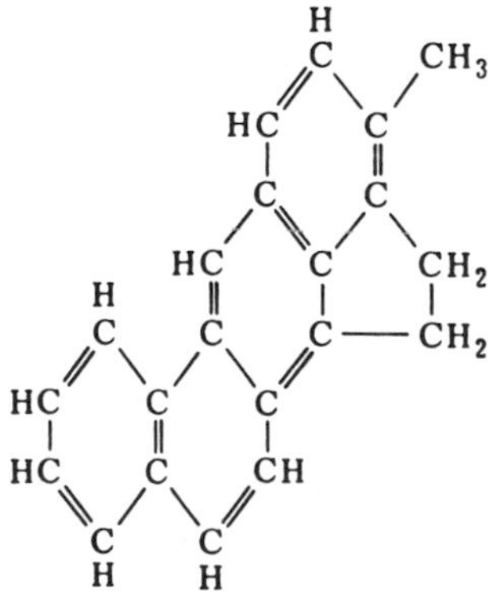

5-methylcytosine (5-mCyt) a modified base found in the DNA of both prokaryotes and eukaryotes. Only 5-mCyt is found in plants and animals, but prokaryotes also contain N^6-methyladenine in their DNA. 5-mCyt is generated by a specific methyl transferase (*q.v.*), which transfers an activated methyl group from a donor molecule (S-adenosylmethionine) to specific cytosines in DNA. In mammalian chromosomes, the bulk of 5-mCyt is in simple sequence repetitive DNAs, but it is also present throughout all sequences. In *E. coli* and its viruses, specific genes contain methylated cytosines. This modified pyrimidine undergoes spontaneous deamination that converts 5-methylcytosine to thymine. Thus, base substitutions occur at elevated rates in these regions and are responsible for mutational hot spots (*q.v.*). *See* Appendix C, 1961, Benzer; 1978, Coulondre *et al.*; DNA methylation.

methyl green a basic dye that can detect DNA. In 2 molar solution of magnesium chloride at pH 5.7 methyl green stains only undegraded DNA. *See* pyronin Y.

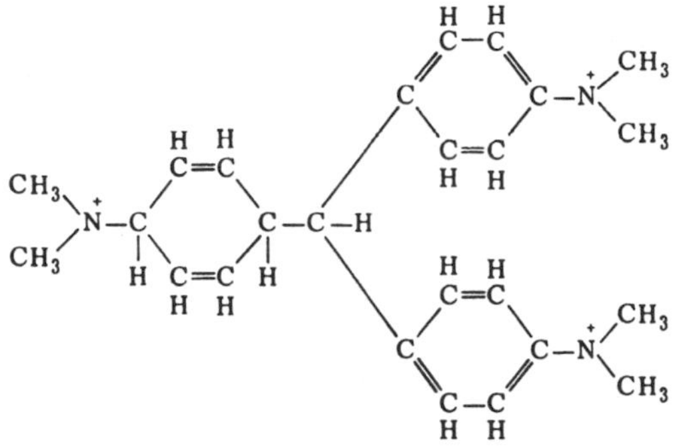

methyl guanosine *See* rare bases.

methyl inosine *See* rare bases.

methyl transferase an enzyme that adds a methyl group to a purine or pyrimidine. Methyl groups are commonly added to guanine or cytosine molecules. As shown on page 275, a fully methylated segment of DNA becomes hemimethylated upon replication. So for a methylated site to be transmitted through a series of mitotic cycles, a mechanism must exist that ensures that methyl transferases recognize hemimethylated sites and convert them to full methylation. *See* Appendix E; Achilles' heel cleavage, bases of nucleic acids, DNA methylation, methylated cap, 5-methylcytosine, parental imprinting, posttranslational processing, restriction, and modification model, telomeric silencing.

metric traits *See* quantitative character, continuous variation.

METRO message transport organizer (*q.v.*).

Mg magnesium.

MHC major histocompatibility complex (*q.v.*).

MIC major immunogene complex (*q.v.*).

met met met
—C—G— —C—G— —C—G— —C—G—
—G—C— —G—C— —G—C— —G—C— new strands
met —C—G—
—G—C—
met

cytosines unmethylated hemi methylated fully methylated

Methyl transferase

MIC2 a gene located distally on the short arm of the X chromosome (at Xp22.23) in the human pseudoautosomal region (*q.v.*). It therefore can undergo crossing over with an allele on the Y (at Yp11.3). *MIC2* encodes a glycoprotein named CD99. This is an integral membrane protein found in hematopoietic cells and skin fibroblasts. *See* XG.

micelle a spherical array of amphipathic molecules in which the nonpolar tails form a hydrocarbon microdroplet enclosed in a shell composed of the polar heads.

Michaelis constant (K_m) the substrate concentration at which the reaction rate of an enzyme is half maximal; the concentration at which half of the enzyme molecules in the solution have their active sites occupied by a substrate molecule.

Michurinism the genetic theories expounded by the Russian horticulturist I. V. Michurin, particularly those dealing with the modification of the genetical constitution of a scion by grafting. Michurin's theories were subsequently incorporated into Lysenkoism (*q.v.*).

micro- 1. a prefix meaning one millionth; used with units of measurements in the metric system. 2. a prefix meaning microscopic or minute.

microarray technology *See* DNA microarray technology.

microbeam irradiation the use of a beam of microscopic diameter to selectively irradiate portions of a cell with ionizing radiation or ultraviolet light.

microbial genetics the genetic study of microorganisms.

microbiology the scientific study of microorganisms.

microbody an organelle, generally less than 1 micrometer in diameter found in the cytoplasm of eukaryotic cells. Microbodies contain a variety of functioanlly related enzymes surrounded by a membranous sac. Examples of microbodies are the glycosome, the glyoxysome and the peroxisome (*all of which see*).

Microbracon hebetor a wasp (called *Habrobracon juglandis* or *Bracon hebetor* in the early genetic literature). It is the arrhenotokous species in which the genetic control of sex determination was elucidated. *See* Appendix A, Arthropoda, Insecta, Hymenoptera.

microcarriers microscopic beads or spheres, composed of dextran or other substances, that are used in tissue culture to attract and hold cells that must be anchored before they can proliferate. *See* anchorage-dependent cells.

microcinematography *See* time-lapse microcinematography.

micrococcal nuclease an endonuclease used to cleave eukaryotic chromatin preferentially between nucleosomes.

microconidia *See* conidium.

microdeletion a chromosomal deletion too short to be identified under the light microscope, but detectable by comparing the base sequences of the normal and deleted DNA segments.

microevolution an evolutionary pattern usually viewed over a short period of time, such as changes in gene frequency within a population over a relatively few generations (industrial melanism, for example). *See* macroevolution.

microfilaments elongated intracellular fibers 5–7 nanometers in diameter, containing polymerized actin, and thought to function in maintenance of cell structure and movement.

Micrographia a book published in 1665 by R. Hooke in which the first description of cells is given.

micromanipulator an instrument allowing surgery on and injection into microscopic specimens. It is also useful in isolating single cells.

micrometer (μm) one-millionth (10^{-6}) meter. The preferred length unit for describing cellular dimensions. The micrometer replaces the micron (μ), an equivalent length unit, found in the earlier literature. *See* nanometer.

micron (μ) *See* micrometer.

micronucleus the smaller reproductive nucleus as distinguished from the larger, vegetative macronucleus (*q.v.*) in the cells of ciliates. Micronuclei are diploid and are transcriptionally inactive. They participate in meiosis and autogamy. *See* nuclear dimorphism, *Tetrahymena.*

microorganism an organism too small to be observed with the unaided eye.

micropyle 1. a canal through the coverings of the nucellus through which the pollen tube passes during fertilization. In a mature seed, the micropyle serves as a minute pore in the seed coat through which water enters when the seed begins to germinate. 2. the pore in the egg membranes of an insect oocyte that allows entry of the sperm.

microRNAs (miRNAs) small, noncoding RNAs, approximately 21–25 nucleotides long, which silence gene expression either by mediating the degradation of or by repressing the translation of protein-coding messenger RNAs (mRNAs) containing complementary or partially complementary sequences. miRNAs are derived from larger precursor RNA transcripts that form hairpin structures, which are processed such that only one strand of these structures gives rise to a mature miRNA molecule. Single-stranded miRNA is bound by a complex and recognizes partially or fully complementary sequences in the target mRNA. Some miRNAs, such as small temporal RNAs (*q.v.*), bind to partially complementary 3′ untranslated regions of target mRNAs to inhibit protein synthesis. Others bind to perfectly complementary sequences in the target mRNA and, like small interfering RNAs (*q.v.*), mediate the cleavage of these mRNAs. Several hundred different miRNAs have been identified in a variety of organisms, including yeast, fungi, plants, nematodes, flies, and mammals. However, the function of most miRNAs is not known. Sequenced miRNAs are stored in a public database. *See* Appendix E, Individual Databases; RNA interference (RNAi).

microsatellites chromosomal sites that contain repeats of a small number of nucleotides arranged one after the other. The tandemly repeated motif usually contains between one and six nucleotides. Microsatellites are also called *short tandem repeats (STRs).* Microsatellites have been found in every species studied so far. They are evenly distributed along the chromosomes, but are rare within coding sequences. During the replication of chromosomal segments containing tandem repeats, the nascent DNA strand may separate from the template strand and then re-anneal so that it is out of phase with the template. When replication resumes, the newly completed strand will be longer or shorter than the template depending on whether the looped-out bases occurred in the template or the nascent strand. Such slippage strand mispairing during the replication of microsatellites is the main mechanism for increasing the length of microsatellites. Some human diseases are caused by expaning the length of trinucleotide repeats (*q.v.*). *See* genetic instability, STR analysis.

microscopy *See* Nomarski differential interference microscope, phase contrast microscope, photographic rotation technique, photomicrography, polarization microscope, SEM, TEM, time-lapse microcinematography, ultraviolet microscope.

microsome fraction a cytoplasmic component, obtained upon centrifugation of homogenized cells, consisting of ribosomes and torn portions of the endoplasmic reticulum. *See* Appendix C, 1943, Claude.

microspectrophotometer an optical system combining a microscope with a spectrophotometer. With this apparatus one can determine the amount of light of specified wavelength passing through a given cytoplasmic region relative to some standard area. From such measurements estimates can be made of the concentrations of dye-binding or ultraviolet-light-absorbing materials present in the cytoplasmic or nuclear area in question. *See* Appendix C, 1950, Swift; flying spot cytometer.

microspore the first cell of the male gametophyte generation of seed plants. Each becomes a pollen grain.

microspore culture *See* Appendix C, 1973, Debergh and Nitsch; anther culture.

microsporidia parasitic species belonging to the phylum Microspora (*see* Appendix A, Protoctista). Microsporidian species of genus *Nosema* are parasites of silkworms, honeybees, and *Drosophila.* A unique characteristic of microsporidians is that the unicellular resting spore contains a coiled hollow tube, the polar filament, which is a specialized product of the Golgi apparatus. During infection, the polar filament turns inside out to form a hollow tube through which the infectious part of the spore is extruded into the new host. While microsporidia lack conventional mitochondria, they have some genes for mitochondrial proteins. On the basis of a reanalysis of the sequence data from their rRNAs, it appears that microsporidia are specialized fungi, and their structural simplicity is now attributed to regressive evolution (*q.v.*). *See* *Encephilitozoon cuniculi*, ribosome.

microsporocyte the pollen mother-cell, which undergoes two meiotic divisions to produce four microspores.

microsporogenesis the production of microspores.

microsporophyll the stamen. *See* flower.

microsurgery surgery done while viewing the object through a microscope and often with the aid of a micromanipulator. *See* Appendix C, 1952, Briggs and King; 1967, Goldstein and Prescott; 1980, Capecchi.

microtome a machine for cutting thin (1 to 10 micrometer thick) slices of paraffin-embedded tissue. These sections are later stained and examined with the light microscope. *See* Appendix C, 1870, His; ultramicrotome.

microtomy the technique of using the microtome in the preparation of sections for study under the microscope.

microtrabecular lattice a network of thin filaments interconnecting the three major cytoskeletal elements (microtubules, microfilaments, and intermediate filaments). This three-dimensional lattice can be visualized in freeze-etched preparations viewed with the electron microscope.

microtubule organizing centers (MTOCs) the structures or loci that give rise to microtubular arrays. Centrioles and kinetosomes function as MTOCs in some organisms. In others, the MTOC merely consists of an amorphous collection of fibrils and granules. MTOCs contain RNA, and RNA replication accompanies their multiplication. *See* Appendix C, 1995, Moritz *et al.*

microtubules long, nonbranching, thin cylinders with an outside diameter about 24 nanometers and a central lumen about 15 nm diameter. The lengths are at least several micrometers. The tubules are composed of strands called protofilaments, and there are usually 13 of these. Each protofilament in turn is composed of a linear array of subunits, and each subunit is a dimer containing an alpha and a beta tubulin molecule. Microtubules play key roles in cell division, secretion, intracellular transport, morphogenesis, and ciliary and flagellar motion. *See* axoneme, centriole, tubulin.

Microtus the genus of meadow mice and voles subjected to extensive cytogenetic study. *M. agrestis*, the field vole, has giant sex chromosomes.

microvillus a fingerlike projection of the plasmalemma of a cell.

mictic referring to an organism or species capable of biparental sexual reproduction.

middle lamella the outermost layer of a plant cell wall that connects it to its neighbor.

midget an abnormally small adult human with normal proportions. *See* ateliosis, human growth hormone. *Contrast with* chrondroplasia, dwarf.

midparent value the mean of two parental values for a quantitative trait in a specific cross. *See* quantitative inheritance.

midpiece the portion of the sperm behind the head containing the nebenkern.

midspindle elongation the elongation during anaphase of the midregion of the mitotic spindle. This elongation serves to draw the chromosomes poleward once the movement of chromosomes along traction fibers has ceased.

migrant selection selection based on the different migratory abilities of individuals of different genotypes. If, for example, individuals carrying gene *M* found new colonies more often than those bearing gene *m*, then gene *M* is said to be favored by migrant selection.

migration in population genetics, the movement of individuals between different populations of a species, resulting in gene flow (*q.v.*).

migration coefficient the proportion of a gene pool represented by migrant genes per generation.

migration inhibition factor a lymphokine that inhibits the movement of macrophages under *in vitro* culture conditions.

Miller spreads chromosomal whole mounts spread for electron microscopy by a method developed by O. L. Miller. In this method, chromosomes from ruptured nuclei are centrifuged through a solution of 10% formalin in 0.1 mole sucrose and onto a membrane-coated grid. The grid is then treated with an agent that reduces the surface tension as the grid dries. After being stained with phosphotungstic acid, the preparation is ready for viewing under the electron microscope.

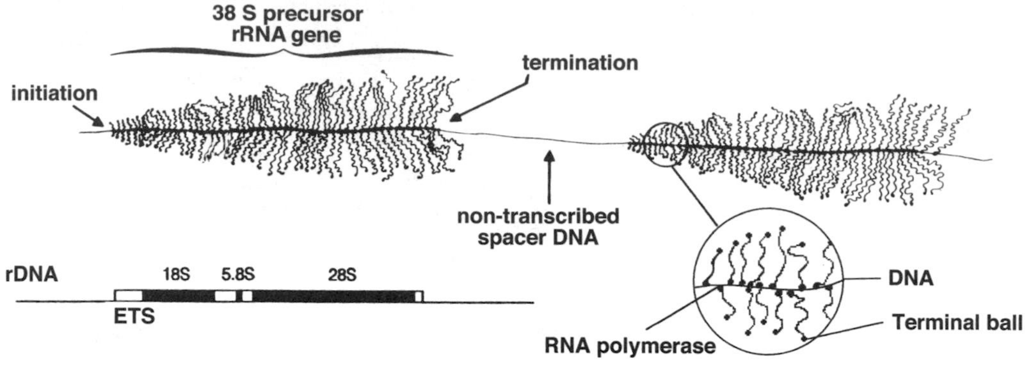

Miller trees

Miller trees transcribing rRNA genes first seen in Miller spreads of the amplified rRNA of extrachromosomal nucleoli from salamander oocytes. Each rRNA transcription unit (rTU) was about 2.5 micrometers long, corresponding to a molecule of 8,000 base pairs. The rRNA molecules attached to the chromatin fiber give rTUs a Christmas tree morphology (hence the term Miller tree), and such rTUs are arranged in tandem, separated by fiber-free spacers. Each nascent transcript ends distally with a terminal ball. This represents a processing complex that forms when the 5′- external transcribed spacer (ETS) region of the rRNA interacts with a group of specific polypeptides. *See* Appendix C, 1969, Miller and Beatty; 1976, Chooi; preribosomal RNA, ribosomal RNA genes, 5.8S rRNA.

milliequivalent *See* gram equivalent weight.

milliliter (ml) one-thousandth of a liter; the volume contained in a cube of side 1 centimeter. One ml of water weighs one gram at 4°C.

millimicron *See* nanometer.

Millipore filter a disc-shaped synthetic filter having holes of specified diameter through its surface. The available pore sizes range between 0.005 and 8 micrometers. The discs are used to filter microorganisms out of nutrient fluids that will not stand sterilization by autoclaving.

mimicry the similarity in appearance of one species of animal to another that affords one or both protection. In *Batesian mimicry*, one of two species is poisonous, distasteful, or otherwise protected from predators, and often conspicuously marked. The mimic is innocuous, gaining protection from predators by its similarity to the model. In *Mullerian mimicry*, both are distasteful to predators, and they gain mutually from having the same warning coloration, since predators learn to avoid both species after tasting one. *Peckhammian mimicry* is an aggressive mimicry in which the predator is the pretender (mimic); e.g., a female firefly of one species mimics the flashing sexual signals of another species, thus attracting a male of that other species which becomes a meal for the pretender. In *Mertensian mimicry*, one species is mildly poisonous (e.g., false coral snakes) and thus serves as a model for a fatally poisonous species (e.g., true coral snakes). Obviously, a predator can learn only if it survives the encounter. *See* automimicry, fireflies, frequency-dependent fitness.

MIM number a catalog number for an inherited human disease that is listed in *M*endelian *I*nheritance in *M*an (*q.v.*).

minicells small anucleate bodies produced by the aberrant division of bacteria. Certain mutant strains of *E. coli* produce large numbers of minicells. Although minicells lack genophores, they may contain one or more copies of the plasmids present in the parent cells. This permits the analysis of the plasmid DNA or of plasmid-encoded gene products in the absence of the host genome and its products. *See* maxicells.

minichromosomes beaded DNA structures of certain viruses, similar to the beaded nucleosomal structures characteristic of eukaryotic chromatin. The circular DNA duplexes of papovaviruses (*q.v.*) are bound throughout the replication cycle to host histones, except H1.

minicircles *See* RNA editing.

minigenes segments on chromosomes that code for the variable regions of immunoglobulin heavy and light chains. There are several hundred of these segments, but during the differentiation of B lymphocytes a single minigene is attached to the gene segment coding for the constant region. *See* Appendix C, 1965, Dreyer and Bennett; immunoglobulin genes.

minimal medium in microbiology a medium providing only those compounds essential for the growth and reproduction of wild-type organisms.

mink long, slender mammals with short legs and partially webbed toes. These carnivores are prized for their luxurious fur. They have adapted well to ranch-rearing, although they have not become tame. Skins from ranch-reared mink are the single most important source of pelts for the fur industry. There are at least three loci controlling coat pigmentation, and mutations of these genes produce color variations, some of which are popular commercially. *See* Aleutian mink, *Mustella*.

minority advantage the phenomenon first observed in multiple choice mating experiments with *Drosophila melanogaster*. Males carrying certain genetic markers mate with relatively greater success when they are rare than when they are common. *See* Appendix C, 1951, Petit; frequency-dependent selection.

Minos element a transposon from *Drosophila hydei* that belongs to the Tc1 family of transposable elements (*q.v.*). Minos elements that carry foreign genes have been experimentally introduced into the germ line of *Anopheles stephensi*. This mosquito is a major carrier of malaria (*q.v.*) in urban areas of the Indian subcontinent. The success in producing transgenic mosquitoes using the Minos vector may eventually lead to the construction of genetically engineered strains of the insect that are resistant to infection by the malaria parasite. *See* Appendix C, 2000, Catteruccia *et al.*; *Anopheles*, mariner elements, *Plasmodium*.

minus (−) viral strands *See* plus (+) and minus (−) viral strands.

Minutes mutations in *Drosophila* that are dominant and lethal when homozygous. Heterozygotes have a prolonged developmental period, and the adults are small, semisterile, with short thin bristles, rough eyes, etched tergites, and abnormal wing venation. Many *Minutes* are small deficiencies. There are at least 60 different *Minute* loci scattered throughout the genome, and some represent genes encoding ribosomal proteins.

Miocene the fourth epoch in the Tertiary period. Representatives of all mammalian families were present. The most advanced family of flowering plants, the Compositae, evolved during this epoch, as did songbirds and rodents. The chimpanzee and hominid lines diverged from a common ancestor late in the Miocene. *See* geologic time divisions, *Pan*.

Mirabilis jalapa the four-o'clock, a variegated dicotyledonous plant extensively studied in terms of plastid inheritance. *See* Appendix A, Plantae, Angiospermae, Dicotyledoneae, Caryophyllales; Appendix C, 1909, Correns and Bauer.

miscarriage a spontaneous abortion (*q.v.*).

mischarged tRNA a tRNA molecule to which an incorrect amino acid is attached.

mismatch *See* mispairing.

mismatch repair a system for correcting mismatches between bases in the parent and daughter strands of DNA that make hydrogen bonding impossible. Recognition of the mismatch requires several proteins, including the one encoded by the *MSH2* gene. After the segment is cut out, the resulting gap is filled by the actions of DNA polymerase I and DNA ligase (*q.v.*). *See* hereditary nonpolyposis colorectal cancer (HNPCC).

mispairing the presence in one chain of a DNA double helix of a nucleotide not complementary to the nucleotide occupying the corresponding position in the other chain.

missense mutant a mutant in which a codon is mutated to one directing the incorporation of a different amino acid. This substitution may result in an inactive or unstable product. *Contrast with* nonsense mutation.

missing link an unknown or postulated intermediate in an evolutionary sequence of fossil forms. Archeopteryx (*q.v.*) provided the missing link between reptiles and birds. *See* living fossil.

Mississippian *See* Carboniferous.

mistranslation the insertion of an incorrect amino acid into a site on a growing polypeptide chain that is the result of environmental factors or mutations that effect either the tRNA, or the enzymes that attach specific amino acids to specific tRNAs, or the ribosome itself.

mitochondrial cloud a mitochondria-surrounded Balbiani body (*q.v.*) found in the pre-vitellogenic *Xenopus* oocyte (*q.v.*) that functions in the assembly and translocation of specific RNAs to the vegetal pole, where the germ plasm (*q.v.*) is formed. The mitochondrial cloud originates near the nuclear membrane during early oogenesis (*q.v.*), increases in size, and subsequently breaks down into smaller "islands" that become localized to the vegetal pole of the egg. *See* cytoplasmic localization, METRO, sponge body.

mitochondrial DNA (mtDNA) the mitochondrial genome consists of a circular DNA duplex, and there are generally 5 to 10 copies per organelle. Human mitochondrial DNA (*q.v.*) is a 16.6 kb circle. The mtDNAs of plants are much larger (the mtDNA of *Arabidopsis* contains 367 kb), and fungi are intermediate (there are 75 kb in the mtDNA circles of *Saccharomyces*). Introns have been found in mitochondrial genes from yeast but not from mammals. The genetic code of mitochondria differs slightly from the "universal" genetic code. Since, of the two gametes, only the egg cell contributes significant numbers of mitochondria to the zygote, mtDNA is maternally inherited. Mitochondrial genomes experience higher mutation rates than those in nuclear genomes and thus have more power to reveal evolutionary differences between closely related species than nuclear genes have. In addition, mtDNAs do not undergo genetic recombination, so their sequences stay the same from one generation to the next, except when altered by mutations. Mitochondrial DNA is usually the only type of DNA to survive in long-dead specimens, due mainly to its abundance (500 to 1,000 copies per cell, rather than only 2 copies as in the nuclear DNA of diploid cells). Mitochondrial DNA sequences are available for about 500 species of eukaryotes (URL http://www.mitomap.org). *See* Appendix C, 1959, Chevremont *et al.*; 1966, Nass; 1968, Thomas and Wilkie; 1974, Dujon *et al.*, Hutchison, *et al.*; 1979, Avise *et al.*; Barrell *et al.*; 1981, Anderson *et al.*; 1999, Andrews *et al.*; ATP synthase, human mitochondrial DNA, mtDNA lineages, Neandertal, petites, *Podospora anserina*, promiscuous DNA, RNA editing, tRNA genes, universal code theory.

mitochondrial Eve *See* mtDNA lineages.

mitochondrial DNA lineages evolutionary trees derived from data on mtDNAs. Human mitochondria are maternally inherited, and therefore mtDNA is contributed by the female parent to the next generation. Furthermore, mitochondrial genes do not undergo recombination. For these reasons, it is much simpler to trace mutations in mtDNA than in genomic DNA. In an analysis of mtDNAs published in1987, the restriction fragment polymorphisms were traced back to an mtDNA molecule carried by a hypothetical woman living in Africa about 200,000 years ago. In 1995, comparative studies of the entire mitochondrial genomes from three humans (one each from Africa, Japan, and Europe) and four apes (a chimpanzee, a bonobo, a gorilla, and an orangutan) resulted in the estimate of 143,000 years as the time for the origin of the ancestral human mtDNA, again in Africa. In the popular press, the woman with this ancestral mtDNA is sometimes called "African Eve" or "mitochondrial Eve." Mitochondrial Eve is our most recent common ancestor *in the purely female line*. Other women were contemporaries of Eve, but their mitochondria have not been retained, usually because their link with today's descendants is blocked because it passes through a male ancestor. *See* Appendix C, 1979, Avise, Lansman, and Shade; 1987, Cann, Stoneking, and Wilson; 1995, Horae *et al.*; mitochondrial DNA (mtDNA); *contrast with* Y chromosomal DNA lineages.

mitochondrial proton transport the pumping of H^+ ions (protons) from the matrix of the mitochondrion through its inner membrane and into the intermembrane space. The energy from NADH is used to transport these protons, and the result is a proton-motive force that drives the synthesis of ATP. The inner mitochondrial membrane is impermeable to protons, but permeable to water, oxygen, and carbon dioxide molecules. However, the protons trapped in the intermembrane space can pass through the proton channel in each ATP synthase (*q.v.*) embedded in the membrane. The reentry of the protons into the matrix supplies the energy to convert ADP to ATP. *See* adenosine phosphate, chemiosmotic theory, cytochrome system, electron transport chain, nicotine-adenine dinucleotide (NAD).

mitochondrial syndromes human disorders caused by mutations that impair the functioning of mitochondrial genomes. Such diseases can affect both sexes, but are passed on only by affected mothers. *See* Leber hereditary optic neuropathy (LHON), Leigh syndrome.

mitochondrial translocases enzymes which form pores in mitochondrial membranes that allow proteins synthesized in the cytosol to enter the organelle. Each mitochondrion imports about 1,000 different proteins. Translocases located in the outer membrane are called TOMs, while those of the inner membrane are called TIMs. The proteins that make up TOMs and TIMs are encoded by nuclear genes and are synthesized in the cytosol. So these precursor proteins must also be imported into the mitochondrion, sorted, and directed to their appropriate locations, where they are assembled into functioning enzymes. The outer mitochondrial membrane contains receptors that recognize those proteins of the cytosol destined to be internalized.

These preproteins contain amino acid sequences to which the receptors bind. The preproteins collected by the receptors are then passed by the TOMs into the fluid-filled space between the membranes. Here appropriate preproteins are passed by TIMs into the mitochondrial matrix for further processing, while some are retained in the inner membrane.

mitochondrion a semiautonomous, self-reproducing organelle that occurs in the cytoplasm of all cells of most, but not all, eukaryotes. For example, microsporidia (*q.v.*) lack mitochondria. Each mitochondrion is surrounded by a double limiting membrane. The inner membrane is highly invaginated, and its projections are called *cristae*. In most eukaryotes, the cristae are platelike, but certain protoctists (*viz.* ciliates, sporozoa, diatoms, chrysophytes) have tubular cristae. Mitochondria are the sites of the reactions of oxydative phosphorylation (*q.v.*), which result in the formation of ATP. Mitochondria contain distinctive ribosomes, tRNAs, aminoacyl-tRNA synthetases, and elongation and termination factors. Mitochondria depend upon genes within the nucleus of the cells in which they reside for many essential mRNAs. The proteins translated from these mRNAs in the cell cytosol are imported into the organelle. The mitochondrion only makes about 1% of its proteins internally. The rest are transported from the cytosol using enzymes called mitochondrial translocases (*q.v.*). Mitochondria are believed to have arisen from aerobic bacteria that established a symbiotic relationship with primitive protoeukaryotes. *See* Appendix C, 1890, Altman; 1898, Benda; 1952, Palade; 1964, Luck and Reich; 1996, Burger *et al.*; 1998, Anderson *et al.*; ATP synthase, chloramphenicol, citric acid cycle, electron transport chain, endosymbiont theory, glycolysis, human mitochondrial DNA, kinetoplast, leader sequence peptide, mitosomes, petites, protein sorting, ribosome, *Rickettsia prowazeki*, serial symbiosis theory, sorting signals.

mitogen a compound that stimulates cells to undergo mitosis. *See* epidermal growth factor.

mitomycin a family of antibiotics produced by *Streptomyces caespitosus*. Mitomycin C (shown above) prevents DNA replication by crosslinking the complementary strands of the DNA double helix.

mitosis mitosis or nuclear division is generally divided into four phases: *prophase, metaphase, anaphase,* and *telophase.*

During mitotic *prophase*, the centriole divides and the two daughter centrioles move apart. The chromosomes become visible within the nucleus because they coil up to produce a series of compact gyres. Each chromosome is longitudinally double except in the region of the centromere, and each replicate strand of a chromosome is called a chromatid. The nucleolus (*q.v.*) and the nuclear envelope break down. During *metaphase*, the chromosomes move about within the spindle and eventually arrange themselves in the equatorial region of the spindle. The two chromatids are now ready to be separated and to move under the action of the traction fibers to the poles of the spindle. During *anaphase*, the block that prevented DNA replication in the centromere region is removed and the centromere becomes functionally double. The chromatids are converted to independent chromosomes that separate and move to opposite poles. During *telophase*, the spindle disappears and reconstruction of nuclear envelopes about the two groups of offspring chromosomes begins. As nuclear envelopes form around each group the chromosomes return to their extended state, and nucleoli reappear.

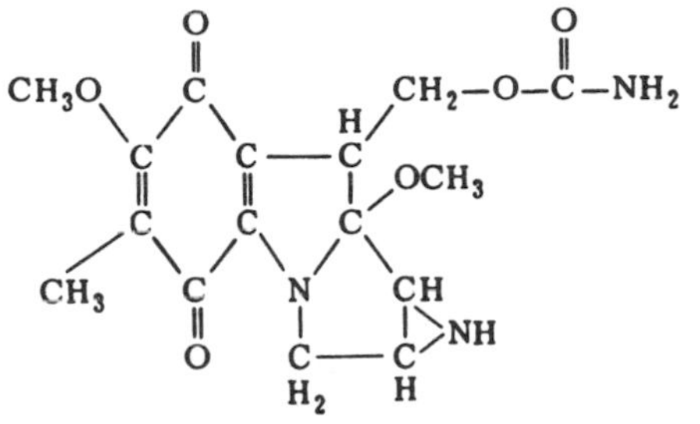

Next, *cytokinesis* occurs, and the cytoplasm is divided into two parts by a cleavage furrow in the case of animal cells or by a cell plate in the case of plant cells. The result of mitosis and cytokinesis is the production of two daughter cells with precisely the same nuclear content and approximately equal amounts of cytoplasm. In contrast to the situation in plants and animals, where the spindle appratus assembles and disassembles at each mitosis, in fungi the spindle with attached chromosomes and centromes persists throughout the cell cycle. *See* Appendix C, 1873, Schneider; 1879, 1882, Flemming; centromere, centrosome, checkpoint, chromosome condensation, cyclins, endomitosis, *MPF*, phragmoplast, sister chromatid cohesion, spindle pole body. *Compare with* meiosis.

Mitomycin C

mitosis promoting factor *See* MPF.

mitosomes organelles found in some protoctists that function in the assembly of iron-sulfur complexes. They are thought to represent mitochondria that have undergone regressive evolution (*q.v.*). *See* *Giardia intestinales*.

mitosporangium a sporangium containing spores produced by mitotic divisions. *Compare with* meiosporangium.

mitospore a spore produced by mitosis and therefore having the same chromosome number as the mother cell.

mitotic apparatus an organelle consisting of three components: (1) the asters, which form about each centrosome, (2) the gelatinous spindle, and (3) the traction fibers, which connect the centromeres of the various chromosomes to either centrosome. *See* Appendix C, 1877, Fol; 1952, Mazia and Dan; mitosis, multipolar spindle.

mitotic center the agent that defines the poles toward which the chromosomes move during anaphase. The centrioles function as mitotic centers in most animal cells. In plants with anastral mitosis, the nature of the mitotic center is unclear.

mitotic chromosome *See* human mitotic chromosomes.

mitotic crossover somatic crossing over (*q.v.*).

mitotic index the fraction of cells undergoing mitosis in a given sample.

mitotic poison any chemical compound that kills dividing cells. Some mitotic poisons are spindle poisons (*q.v.*), and others block topoisomerases (*q.v.*). Poisoned cells are arrested at the spindle or DNA damage checkpoints, and they subsequently undergo apoptosis (*q.v.*).

mitotic recombination a process in which a diploid cell undergoing mitosis gives rise to daughter cells with allele combinations different from that in the parental cell. As in meiotic recombination, mitotic recombination generally involves genetic exchange between chromatids of homologous chromosomes, but occurs less frequently than meiotic recombination. Mitotic recombination can give rise to tissues that are genetic mosaics, including tissues with twin spots (*q.v.*), and can be experimentally induced. *See* Appendix C, 1936, Stern; gene targeting, homologous recombination, somatic crossing over.

mitotic segregation in a eukaryotic cell containing genetically different organelles, their random segregation during mitosis to generate some cells containing only mutant or nonmutant organelles. *See* cytohet, petites.

mixis biparental sexual reproduction.

mixoploidy the presence of more than one chromosome number in a cellular population.

MLD medial lethal dose (*q.v.*).

mM millimolar concentration.

Mn manganese.

MN blood group a human blood group system defined by red cell antigens specified by a gene on the long arm of chromosome 4 between bands 28.2 and 31.1. The MN locus encodes a protein called glycophorin A (*q.v.*), and for this reason the gene symbol has been changed to *GYPA*. Each human red blood cell has about 500,000 of these proteins embedded in its plasma membrane. The M and N antigens are isoforms of glycophorin A that differ in the amino acids present at only two of the 131 sites.

Mo molybdenum.

mobile genetic element *See* transposable genetic elements.

Möbius strip a topological figure, named after the German astronomer, A. F. Möbius, made by putting a 180-degree twist in a long, rectangular strip, then pasting the ends together. The strip has only one surface and one edge. If a Möbius strip is cut longitudinally, it forms a double-sized ring. If a ring with a double twist is cut longitudinally, it forms interlocked rings. A Möbius strip demonstrates the behavior of a twisted ring chromosome when it attempts replication.

modal class the class that contains more individuals than any other in a statistical distribution.

mode modal class.

modification in nucleic acid metabolism, any changes made to DNA or RNA nucleotides after their original incorporation into a polynucleotide chain: e.g., methylation, deamination, formylation, etc. *See* modification methylases.

modification allele *See* DNA restriction enzyme.

modification methylases bacterial enzymes that bind to the DNA of the cell at specific sites determined by specific base sequences. Here they attach methyl groups to certain bases. This methylation pattern is unique to and protects the species from its own restriction endonucleases. Modification methylases are coded for by modification alleles. *See* restriction and modification model.

modified bases postsynthetically altered nucleotides of the four usual bases (A, T, G, and C) of DNA. *See* modification methylases.

modifier referring in the genetic literature to a gene that modifies the phenotypic expression of a nonallelic gene.

modular organisms organisms that consist of populations of units or modules that are attached together, but if separated are capable of independent growth and reproduction. An example would be plants like strawberries where segments of the branching root systems can generate new plants when severed from the parent. In such organisms, a mutation in a meristematic cell may be expressed during the subsequent growth of a module built from the mutant clone, and eventually the mutation may be contained in eggs or pollen. Thus, Weismann's distinction between soma and germ cells does not apply to modular organisms, and evolutionary changes in modular species may originate from both germinal and somatic mutations. *See* Appendix C, 1883, Weismann.

modulating codon specific triplets that code for rare tRNAs. The translation of a mRNA molecule is slowed down when a modulating codon is encountered.

moiety one of two roughly equal parts.

molal descriptive of a solution that contains one mole (*q.v.*) of solute in 1000 grams of solvent, usually water. *Compare with* molar.

molar descriptive of a solution that contains one mole (*q.v.*) of a solute per liter of solvent. *Compare with* molal.

mole the amount of a substance that has a weight in grams numerically equal to the molecular weight of the substance; also called *gram molecular weight* or *gram molecule*. *See* molar.

molecular biology a modern branch of biology concerned with explaining biological phenomena in molecular terms. Molecular biologists often use the biochemical and physical techniques to investigate genetic problems.

molecular clock *See* DNA clock hypothesis, protein clock hypothesis.

molecular cloning *See* gene cloning, recombinant DNA technology.

molecular genetics that subdivision of genetics which studies the structure and functioning of genes at the molecular level.

molecular hybridization base pairing between DNA strands derived from different sources or of a DNA strand with an RNA strand.

molecular mass *See* relative molecular mass (Mr).

molecular mimicry *See* eclipsed antigens.

molecular motors molecules capable of generating torque or movement of other molecules. The smallest known molecular motors are the ATPases responsible for the rotation of bacterial flagellae. The only known enzyme component that must rotate to perform its catalytic function is the F_1-ATPase subunit of the ATP synthase holoenzyme that spans the membranes of mitochondria. F_1-ATPase is less than one-tenth the size of the motors that drive bacterial flagellae. *See* motor proteins.

molecular sieve a crystalline alumino-silicate pellet used to absorb water, carbon dioxide, hydrogen sulfide, and similar gases from gas mixtures and organic solvents. Molecular sieves are also sometimes used as ion exchange media.

molecular weight the sum of the atomic weights of all of the atoms in a given molecule. The term has largely been replaced by *relative molecular mass* (*q.v.*).

molecule that ultimate unit quantity of a compound that can exist by itself and retain all the chemical properties of the compound.

Moloney murine leukemia virus (MoMLV) a virus of mice producing lymphocytic leukemia that can be transmitted from an infected mother to her newborn progeny through her milk. The virus integrates into several locations in the mouse genome, and there is an integration site on human chromosome 5p14. MoMLV is commonly used for gene transfer protocols. First, the viral genes that are required for the production of replication-competent viruses are replaced with those genes that are desired to be transferred. Once the virus is administered, it binds to the host cells and is internalized, and the genes provided by the vector are then integrated randomly into the host chromosomes. Transfers are efficient and the integrations are stable. *See* Appendix C, 1983, Mann, Mulligan, and Baltimore; gene therapy, proto-oncogene, retrovirus, virus.

molting hormone *See* ecdysones.

moltinism a polymorphism in terms of the number of larval molts undergone by a given species. For example, in *Bombyx mori* there are strains that are known to molt three, four, or five times.

molybdenum a biological trace element. Atomic number 24; atomic weight 95.94; most abundant isotopes ^{92}Mo, ^{94}Mo, ^{95}Mo, ^{96}Mo, ^{97}Mo, ^{98}Mo; radio-

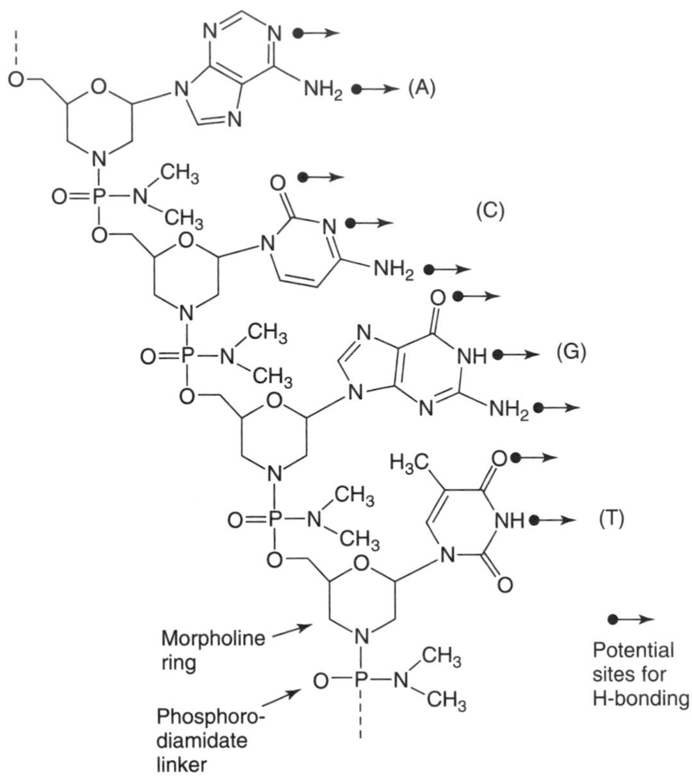

Morpholinos

morphometric cytology the determination of quantitative parameters of cytological structures in tissue sections.

morula an embryo that consists of a cluster of cleaving blastomeres. A stage prior to the blastula.

mosaic an individual composed of two or more cell lines of different genetic or chromosomal constitution, both cell lines being derived from the same zygote; in contrast with a chimera (*q.v.*). *See* Appendix C, 1962, Beutler, Yeh, and Fairbanks; dosage compensation, Lyonization, tortoiseshell cat.

mosaic development in mosaic development, the fates of all the parts of the embryo are already fixed at or before fertilization, so that any localized ablation will later be manifested by the absence of the part in question. *See* regulative development.

mosaic evolution evolutionary change in one or more body parts without simultaneous changes in other parts. For example, in the evolution of birds from dinosaurs, the origin of feathers occurred long before specialized bones (e.g., keel) and powerful flight muscles developed.

motility symbiosis the situation where motility is conferred upon an organism by its symbiont. For example, many protists contain cortical populations of symbiotic spirochaetes whose coordinated movements propel the host through its aqueous medium.

motifs distinctive sequences, on protein or DNA molecules, that have three-dimensional structures that allow binding interactions to occur. *See* DNA-binding motifs, helix-turn-helix motif, leucine zipper, zinc finger proteins.

motor proteins proteins that bind ATP and, after attaching to specific microfilaments or microtubules, are able to move laterally as the ATP is hydrolyzed. *See* actin, dynein, kinesin, myosin.

mouse *See Mus musculus*, oncomouse.

mouse genetic databases *See* Appendix E.

mouse inbred lines laboratory strains of mice propagated by brother-sister matings for many generations and hence highly homozygous and genetically uniform. In some strains, the inbreeding program has been carried out for 40 years. The strains most commonly mentioned in the literature are: albino (A, Ak, BALB, R_{III}), black (C_{57} black, C_{58}), black agouti (CBA, C3H), brown (C_{57} brown), dilute brown (DBA/2), dilute brown piebald (I). *See* Appendix C, 1909, Little; 1942, Snell.

mouse L cells a strain of fibroblastlike cells carried in tissue culture. The cells originated from subcutaneous areolar and connective tissue derived from a male C3H mouse. *See* Appendix C, 1940, Earle.

mouse mammary tumor virus a milk-borne retrovirus that causes mammary cancer in mice of appropriate genotype. The infectious virus is produced by infected epithelial cells of the mammary gland. The virus can also be inherited genetically in the form of an endogenous provirus (*q.v.*) in the germ line. *See* Appendix C, 1936, Bittner; proto-oncogene, retrovirus.

mouse satellite DNA a DNA making up about 10% of the DNA isolated from a wide variety of mouse tissues. It forms a band slightly separated from the main peak when mouse DNA is spun to equilibrium in a CsCl density gradient (*see* centrifugation separation). Mouse satellite DNA consists of about 1 million copies per genome of a sequence some 400 nucleotide pairs in length. *In situ* hybridization experiments show that most of this DNA is located in the pericentric heterochromatin. *See* Appendix C, 1970, Pardue and Gall.

MPD maximum permissible dose (*q.v.*).

MPF maturation promotion factor (*q.v.*).

M phase *See* cell cycle.

Mr relative molecular mass (*q.v.*).

mRNA messenger RNA (*q.v.*).

mRNA coding triplets *See* amino acid, start codon, stop codon.

MS2 one of the smallest autonomous viruses known. It is an RNA bacteriophage of *Escherichia coli* that belongs to the family Leviviridae. Leviviruses are "male-specific" in that they adsorb specifically to the sides of F pili. The MS2 virus was the first to have its genome sequenced. It contains 3,569 nucleotides and encodes four proteins. The molecule is a single-stranded, positive sense RNA that is homologous with the viral mRNA. Therefore, MS2 must synthesize a negative-sense, single-stranded DNA molecule before mRNA can be transcribed. *See* Appendix A, 1973, 1976, Fiers *et al.*; F factor (fertility factor).

MSH melanocyte-stimulating hormone. *See* intermedin.

***MSH2* gene** *See* mismatch repair.

MSL proteins male-specific proteins in *Drosophila* that bind to hundreds of sites along the single X chromosome and increase gene expression to match the two X chromosomes of the female. Null mutations in the genes encoding MSL proteins lead to male lethality; hence, the gene symbol *msl* for *m*ale *s*pecific *l*ethal. *See* dosage compensation.

M strain the *M*aternally contributing strain of *Drosophila* in a P-M hybrid dysgenesis cross. M strains lack P factors. *See* hybrid dysgenesis, P elements, P strain.

MTA mammary tumor agent (*q.v.*).

mtDNA mitochondrial DNA (*q.v.*).

M5 technique a technique used to detect induced sex-linked lethal and viable mutations in *Drosophila melanogaster*. The technique gets its name from the X chromosome used to balance the chromosomes bearing the induced mutations. The M5 or Muller 5 chromosome is the fifth of a series synthesized by H. J. Muller. It contains a complex inversion and the marker genes *B*ar, *a*pricot, and *sc*ute. For this reason, the M5 chromosome is sometimes abbreviated *Basc*. *See* balanced stock.

mtmRNA, mtrRNA, mttRNA symbols for mitochondrial messenger, ribosomal, and transfer RNAs, respectively.

MTOCs *See* microtubule organizing centers.

mu **1.** map unit. **2.** mate killers.

mucopolysaccharide a polysaccharide composed of sugars and sugar derivatives, such as amino sugars and uronic acids. *See* chondroitin sulfuric acid.

mucoprotein a protein containing more than 4% carbohydrate. *See* glycoprotein.

mulatto the hybrid from a Negro-white cross.

mule *See* horse-donkey hybrids.

Mullerian mimicry A form of mimicry (*q.v.*) described in 1878 by the German zoologist Fritz Muller.

Muller rachet the accumulation of deleterious mutations that can lead to the extinction of a population of asexual organisms. H. J. Muller pointed out in 1964 that in asexual species mutations tend to accumulate because of the random loss of individuals with the least mutated genomes. In the absence of back mutation, the process is irreversible. Once mutants replace healthy genes, the good ones never come back and so the process resembles a ratchet wheel, which moves in only one direction. Muller proposed that sexuality arose because it allowed crossing over to occur between homologous chromosomes from parents carrying different mutants. Recombinant offspring with mutant-free genomes could thus arise and halt the action of the rachet. The term *Muller ratchet* was coined by J. Felsenstein in 1974.

multifactorial polygenic.

multiforked chromosome a bacterial chromosome containing more than one replication fork, due to the initiation of a second fork before completion of the first replication cycle.

multigene family a set of genes descended by duplication and variation from some ancestral gene. Such genes may be clustered together on the same chromosome or dispersed on different chromosomes. Examples of multigene families include those that encode the histones, hemoglobins, immunoglobulins, histocompatibility antigens, actins, tubulins, keratins, collagens, heat shock proteins, salivary glue proteins, chorion proteins, cuticle proteins, yolk proteins, and phaseolins. *See* isoforms, reiterated genes.

multimer a protein molecule made up of two or more polypeptide chains, each referred to as a monomer. The terms dimer, trimer, tetramer, pentamer, etc., are used if the number of monomers per multimer is known. *Compare with* monomer, oligomer, polymer.

multiparous bearing or producing more than one offspring at a birth. *See* parity.

multiple allelism *See* allele.

multiple choice mating referring to an experimental design in studies of behavior genetics where a test organism is allowed to choose between two (or more) genetically different mates.

multiple codon recognition *See* wobble hypothesis.

multiple-event curve a curve (relating relative survival to radiation dose) that contains an initial flat portion. This finding indicates that there is little biological effect until a certain dose has accumulated, and suggests that the sensitive target must be hit more than once (or that there must be multiple targets, each of which must be destroyed) to produce a biologically measurable effect. *See* single-event curve, target theory.

multiple factor hypothesis *See* quantitative inheritance.

multiple genes *See* multiple factor hypothesis, polygene, quantitative inheritance.

multiple infection simultaneous invasion of a bacterial cell by more than one phage, often of different genotypes in experiments designed to promote phage recombination; superinfection.

multiple myeloma *See* myeloma.

multiple neurofibromatosis *See* neurofibromatosis.

multiple transmembrane domain proteins protein molecules that contain several segments that lie embedded in the cell membrane. These domains are connected by segments alternately at the cytoplasmic and extracellular surfaces. Rhodopsin (*q.v.*) and the cystic fibrosis transmembrane-conductance regulator are examples of multiple transmembrane domain proteins. *See* cystic fibrosis, opsin.

multiplex PCR a type of polymerase chain reaction (*q.v.*) that is used to sample various regions of a large gene from one end to the other. For example, to analyze the human dystrophin gene, which occupies over 2 million base pairs on the X chromosome, multiplex PCR might involve simultaneous amplification from nine different sets of primers, all within the same reaction test tube. Each set of primers is chosen to produce a different-sized amplification product from a different region of the dystrophin gene. Normal males will display nine characteristic bands after the amplification products are separated by gel electrophoresis. Males with deletions in the dystrophin gene will be missing one or more of these bands. *See* muscular dystrophy.

multiplicity of infection the average number of phages that infect a bacterium in a specific experiment. The fraction of bacteria infected with 0, 1, 2, 3, . . . , *n* phage follows a Poisson distribution.

multiplicity reactivation the production of recombinant virus progeny following the simultaneous infection of each host cell by two or more virus particles, all of which are incapable of multiplying because they carry lethal mutations induced by exposure to a mutagen.

multipolar spindle a spindle with several poles found in cells with multiple centrioles. Such cells are seen infrequently, but they can be produced in large numbers by irradiation. *See* mitotic apparatus.

multitarget survival curve *See* extrapolation number.

multivalent designating an association of more than two chromosomes whose homologous regions are synapsed by pairs (as in autotetraploids and translocation heterozygotes).

multivoltine producing more than one brood in a year, as in certain birds and moths.

Muntiacus the genus containing various species of small deer living in India, Nepal, and southeast Asia. The Indian muntjac, *M. muntjac vaginalis,* is remarkable in having the smallest chromosome number known for any mammal (6 per diploid female). The Chinese muntjac (*M. reevesi*) has a larger diploid value (46). During evolution, the chromosome number of the Indian species appears to have decreased by end-to-end fusions between different chromosomes. The amount of DNA in diploid nuclei from the two species is about the same. *See* Appendix C, 1997, Yang *et al.*; chromosome painting.

***mu* phage** a phage "species" whose genetic material behaves like insertion sequences, being capable of transposition, insertion, inactivation of host genes, and causing rearrangements of host chromosomes.

murine belonging to the family of rodents that contains the mice and rats.

murine mammary tumor virus an oncogenic RNA virus. *See* mammary tumor agent.

Musaceae the family of monocotyledons that contains the major food-producing species of bananas and plantains. From the culinary standpoint, *bananas* refers to fruit eaten fresh and *plantains* to fruit eaten only after cooking. *See* bananas.

Musca domestica the housefly. DDT resistance in this species has been extensively studied by geneticists.

muscular dystrophy a heterogeneous group of hereditary diseases affecting humans and other mammals that cause progressive muscle weakness due to defects in the biochemistry of muscle tissue. *See* Becker muscular dystrophy (BMD), Duchenne muscular dystrophy (DMD).

Mus musculus the laboratory mouse. Its diploid chromosome number is 20, and extensive genetic maps are available for the 19 autosomes and the X chromosome. There are large collections of strains containing neurological mutants, loci associated with oncogenic viruses (especially retroviruses), loci that encode enzymes, and histocompatibility loci. The total number of mapped genes is about 7,000. The mice housed in laboratories throughout the world are all derived from three subspecies. These are all offshoots of an original population that began migrating from northern India about 900,000 years ago. The mouse genome contains 2.5 gbp of DNA and about 30,000 structural genes. Roughly 80% of these have identifiable orthologs in the human genome. *See* Appendix A, Chordata, Mammalia, Rodentia; Appendix C, 1905, Cuénot; 1909, 1914, Little; 1936, Bittner; 1940, Earle; 1942, Snell; 1948, Gorer *et al.*; 1953, Snell; 1967, Mintz; 1972, Benacerraf and McDevitt; 1975, Mintz and Illmensee; 1976, Hozumi and Tonegawa; 1980, Gordon *et al.*; 1987, Kuehn *et al.*; 1988, Mansour, Thomas and Capecchi; 1994, Zhang *et al.*, Arendt and Nübler-Jung; Appendix E, Mouse Databases; *Hox* genes, mouse inbred lines, oncomouse, T complex.

mustard gas sulfur mustard (*q.v.*).

Mustela the genus that includes *M. erminea,* the ermine; *M. lutreola,* the European mink; *M. vison,* the North American mink.

mutable gene in multicellular organisms, a gene that spontaneously mutates at a sufficiently high rate to produce mosaicism.

mutable site a site on a chromosome at which mutations can occur.

mutagen a physical or chemical agent that raises the frequency of mutation above the spontaneous rate.

mutagenesis the production of mutations, generally by the use of agents that interact with nucleic acids. *See* alkylating agent, hot spot, oligonucleotide-directed mutagenesis, radiation genetics.

mutagenic causing mutation.

mutagenize to expose to a mutagenic agent.

mutant an organism bearing a mutant gene that expresses itself in the phenotype of the organism.

mutant hunt the isolation and accumulation of a large number of mutations affecting a given process, in preparation for mutational dissection of the gene(s) governing that process. For example, one might select for mutations that confer phage resistance in *E. coli.*

mutation 1. the process by which a gene undergoes a structural change. 2. a modified gene resulting from mutation. 3. by extension, the individual manifesting the mutation. *See* Appendix C, 1901, de Vries; isocoding mutation, point mutation.

mutational dissection *See* genetic dissection.

mutational hot spot *See* hot spot.

mutational load the genetic disability sustained by a population due to the accumulation of deleterious genes generated by recurrent mutation.

mutation breeding induction of mutations by mutagens to develop new crop varieties that can increase agricultural productivity.

mutation distance the smallest number of mutations required to derive one DNA sequence from another.

mutation event the actual origin of a mutation in time and space, as opposed to the phenotypic manifestation of such an event, which may be generations later.

mutation frequency the proportion of mutants in a population.

mutation pressure the continued production of an allele by mutation.

mutation rate the number of mutation events per gene per unit time (e.g., per cell generation).

mutator gene a mutant gene that increases the spontaneous mutation rate of one or more other genes. Many of the early "mutator genes" have turned out to be transposable elements (*q.v.*). Others are due to mutations in genes that encode helicases (*q.v.*) or proteins that function in proofreading (*q.v.*). *See* *Activator-Dissociation* system, *Dotted*, dna mutations, mismatch repair.

mutein a mutant protein, such as a CRM (*q.v.*).

muton the smallest unit of DNA in which a change can result in a mutation (a single nucleotide). *See* Appendix C, 1955, Benzer.

mutual exclusion a phenomenon observed among ciliary antigens of certain protozoans in which only one genetic locus for a serotype is active at a given time. For example, in *Paramecium primaurelia* and *Tetrahymena pryriformis*, mutual exclusion of serotypes in heterozygotes occurs with allelic genes as well as with nonallelic genes.

mutualism a symbiosis in which both species benefit.

mutually exclusive events a series of alternative events in which only one can occur at a given time.

myc a gene originally described in the avian MC29 myelocytomatosis virus, an oncovirus of the chicken. A homologous gene is located on the long arm of human chromosome 8. The viral gene is often symbolized *v-myc* and the cellular gene *c-myc* (pronounced "see-mick"). The *myc* oncogene encodes a protein which is expressed in proliferating cells in normal embryonic and adult tissues. Its expression is abnormally high in human and rodent tumors. The protein contains a helix-turn-helix motif (*q.v.*) and a leucine zipper (*q.v.*), and it binds to specific target genes when modulating cell proliferation. The gene is activated when it is placed next to certain immunoglobulin genes as a result of a translocation. *See* Burkitt lymphoma, oncogene.

mycelium the vegetative portion of a fungus composed of a network of filaments called hyphae. Tubular hyphae are often divided into compartments by cross walls. However, since there are perforations in the septa, the cytoplasm is continuous. An aerial hypha can constrict to produce a conidium (*q.v.*).

Mycobacterium leprae *See* leprosy bacterium.

Mycobacterium tuberculosis the causative agent of human tuberculosis, a disease with an annual death toll of three million. This human pathogen arose from a soil bacterium and may have subsequently moved to cows and then to humans, following the domestication of cattle. The H37 Rv strain was isolated in 1905, and it is the DNA of bacteria from this strain that was sequenced. The circular chromosome contains 4,411,529 base pairs and 3,924 ORFs. The demonstration that the DNA of *M. tuberculosis* has a high content of GC relative to AT disproved the tetranucleotide hypothesis (*q.v.*). The tubercle bacillus is resistant to many antibiotics, and this natural resistance is mainly due to its hydrophobic cell envelope, which acts as a permeability barrier. Many of its genes are devoted to a synthesis and breakdown of the lipoproteins in this envelope. The genome also contains at least two prophages and over 50 insertion sequences (*q.v.*). *See* Appendix A, Bacteria, Actinobacteria; Appendix C, 1882, 1905, Koch; 1998, Cole *et al.*; Appendix E; Chargaff rule, leprosy bacterium, lysogenic cycle.

Mycoplasma a genus of bacteria that is characterized by the absence of a cell wall. *M. capritolum* is of interest because in this species UGA encodes tryptophan rather than serving as a termination codon. *M. genitalium*, a parasite of the human genital and respiratory tracts, has recently been shown to have a genome of only 580,070 base pairs. Since this is one of the smallest known genomes for any free-living organism, the number of open reading frames reveals the minimal set of genes necessary for independent life. There are only 470 genes (average size, 1040 base pairs), and these comprise 88% of the genome. The related species *M. pneumoniae* has a larger genome (820 kb), and the number of ORFs is 679. All 470 ORFs from the smaller species are found in the larger bacterium, and their protein sequences are 67% identical. *See* Appendix A, Prokaryotae, Aphragmabacteria; Appendix C, 1985, Yamao; 1995, Fraser, Venter *et al.*; Appendix E; bacterial cell wall, pleuropneumonia-like organisms, TIGR, universal code theory.

Mycostatin a trade name for nystatin (*q.v.*).

myelin sheath the insulating covering of an axon formed by the plasma membrane of a Schwann cell.

myeloblasts cells that differentiate by aggregation to form multinucleated, striated muscle cells.

myeloid leukemia *See* Philadelphia (Ph^1) chromosome.

myeloma cancer of plasma cells, presumably due to clonal proliferation of a single plasma cell that escapes the normal control of division. Such cells reproduce and secrete a specific homogeneous protein related to gamma globulins. *See* Bence-Jones proteins, HAT medium, hybridoma.

myeloma protein a partial or complete immunoglobulin molecule secreted by a myeloma (*q.v.*).

myeloproliferative disease any disease caused by the uncontrolled proliferation of blood cells produced in the bone marrow. Leukemias result from proliferating lymphocytes. Lymphomas also contain proliferating lymphocytes, but in this case the sources are lymph nodes. Granulocytes, monocytes, and megakaryocytes (*all of which see*) are the sources of myeloid leukemias. The designation "acute" indicates that the cancer cells remain immature, divide rapidly, and are destined to overwhelm the body; whereas the "chronic" term is used for cells that divide less frequently and mature enough to perform some of their normal functions. *See* Abelson murine leukemia virus, *acute myeloid leukemia 1* gene, Burkitt lymphoma, Philadelphia (Ph^1) chromosome, polycythemia.

Myleran a trade name for busulfan (*q.v.*).

myoglobin the monomeric heme (*q.v.*) protein that stores oxygen in vertebrate muscles. The myoglobin gene is thought to have been derived directly from the ancestral gene that by duplication produced an ORF that evolved into the alpha chain gene of hemoglobin (*q.v.*). The myoglobin and alpha chain genes diverged 600 to 800 million years ago. Human myoglobin contains 152 amino acids. *See* Appendix C, 1958, Kendrew *et al.*; gene superfamily, hemoglobin genes.

myoglobin gene the gene that encodes myoglobin. It is remarkable in that less than 5% of its structure codes for message. All the genes of the alpha and beta hemoglobin families are made up of three coding regions interrupted by two introns. The myoglobin gene contains four exons and three introns, and each of these introns is much longer than any of those found in hemoglobin genes.

myosin the hexameric protein that interacts with actin (*q.v.*) to convert the energy from the hydrolysis of ATP into the force for muscle contraction. Actin functions both as a structural protein and an enzyme. A myosin molecule can catalyze the hydrolysis of 5 to 10 ATP molecules per second. Each myosin consists of a slender stem (about 135 nanometers long) and a globular head region (about 10 nanometers long). The molecule is formed from two identical heavy chains, each possessing about 2,000 amino acids. In the tail region, the heavy chains twist together to form an alpha helix, from which the two globular heads protrude. The C termini are distal to the heads. Two light chain proteins, A_1 (190 amino acids) and A_2 (148 amino acids), attach to the globular heads of each heavy chain. The light chain proteins contain calcium-binding sites. The globular head regions contain the ATPase activity and can bind temporarily to actin to form a complex referred to as actomyosin. In avian and mammalian species, numerous isoforms of both myosin heavy and light chains have been isolated from muscle and nonmuscle tissues.

myosin genes the genes encoding the isoforms of the heavy and light myosin chains. In *Drosophila*, two myosin heavy chain genes have been identified: one encoding a muscle myosin (*Mhc*) and one encoding a cytoplasmic myosin (*Mhc-c*). The transcription unit of *Mhc* is 22 kilobases long and contains 19 different exons. Multiple transcripts are generated by alternative splicing (*q.v.*). Genes for the two light chains are also known. In mammals, the muscle myosin heavy chain isoforms are encoded by a family containing at least 10 genes.

myotonic dystrophy an autosomal dominant disorder due to an unstable *trinucleotide repeat* (*q.v.*). The gene involved encodes a muscle protein kinase, and the trinucleotide repeat is located in the 3′ untranslated region of the gene. In susceptible families, there is an increase in the severity of the disease in successive generations. *See* genetic anticipation.

myria a rarely used prefix meaning 10,000. Used with metric units of measurement.

myriapod an arthropod belonging to the Myriapoda. Millipedes and centipedes were the first animals to colonize the land. *See* Appendix A, Arthropoda, Mandibulata, Myriapoda; metamerism, Silurian.

Mytilus edulis *See* Pelecypoda.

myxomatosis a fatal virus disease affecting rabbits. The virus was introduced into wild populations of rabbits in Australia as a means of controlling them.

Myxomycota the phylum containing the plasmodial slime molds. These protoctists generate multinucleate plasmodia that feed by phagocytosis and subsequently form stalked, funguslike fruiting structures. From the standpoint of genetics, *Physarum polycephalum* is the best-known species.

n neutron (*q.v.*).

N **1.** the haploid chromosome number. **2.** normal solution. **3.** nitrogen.

***N*-acetyl serine** an acetylated serine thought to function in mammalian systems as *N*-formylmethionine does in bacterial translation.

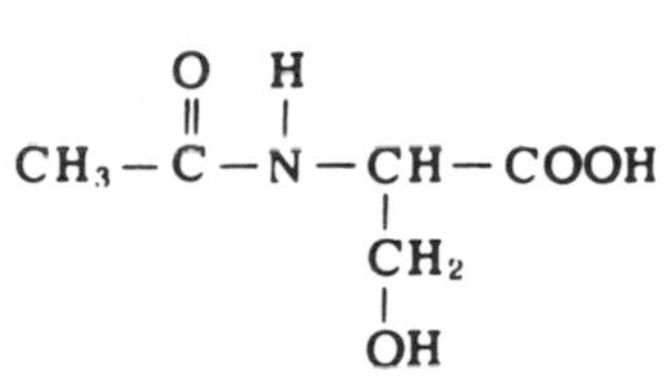

$$CH_3-\overset{\overset{\Large O}{\|}}{C}-\overset{\overset{\Large H}{|}}{N}-\underset{\underset{\underset{\Large OH}{|}}{\underset{\Large CH_2}{|}}}{CH}-COOH$$

NAD nicotinamide-adenine dinucleotide (*q.v.*).

NADP nicotinamide-adenine dinucleotide phosphate (*q.v.*).

Naegleria a genus of soil amoebas capable of transforming into flagellates. Species from this genus are often studied in terms of the morphogenesis of flagella.

nail patella syndrome a hereditary disease in humans. Individuals afflicted with this disorder have misshapen fingernails and small kneecaps, or lack them. The disease is due to a dominant gene residing on chromosome 9.

nalidixic acid an antibiotic that inhibits DNA rep lication in growing bacteria. It specifically inhibits the DNA gyrase of *E. coli.*

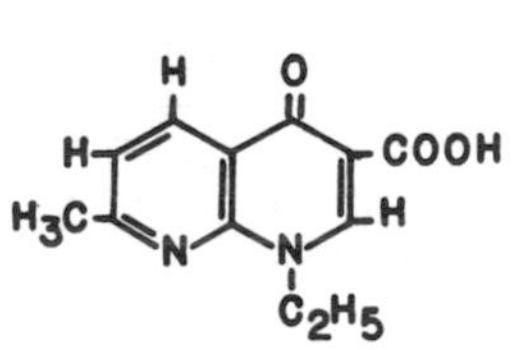

nanometer (nm) one-billionth (10^{-9}) meter. The preferred length unit for describing ultrastructural dimensions (e.g., a ribosome of 15 nm diameter). The nanometer replaces the millimicron (mμ), an equivalent length, found in the earlier literature. Ten nm equals one Angstrom unit.

nanos (nos) a *Drosophila* gene that maps to 3-66.2, and it encodes an RNA-binding protein that blocks translation. This repressor is localized at the posterior pole of embryos where it functions to shut down the translation of mRNAs produced by *hunchback* (*q.v.*).

narrow heritability *See* heritability.

nascent polypeptide chain the forming polypeptide chain that is attached to the 50 S subunit of a ribosome through a molecule of tRNA. The free end of the nascent polypeptide contains the *N*-terminal amino acid. *See* translation.

nascent RNA an RNA molecule in the process of being synthesized (hence incomplete) or a complete, newly synthesized RNA molecule before any alterations have been made (e.g., prior to nuclear processing or RNA editing, *both of which see*).

Nasonia brevicornis another name for *Mormoniella vitripennis* (*q.v.*).

Nasonia vitripennis another name for *Mormoniella vitripennis* (*q.v.*).

native indigenous. A native species is not introduced into an area by humans, either intentionally or accidentally.

natural immunity an outmoded concept that some immunities are inherited in the apparent absence of prior contact with an antigen. The prevailing paradigm is that all immunity ultimately requires contact with a sensitizing antigen and therefore is acquired.

natural killer (NK) cells large leukocytes found in the blood (where they make up about 10% of the total lymphocytes) and in spleen and lymph nodes. They are activated by interferon (*q.v.*), and they attack tumor cells without prior immunization. NK cells are distinct from B lymphocytes and T lymphocytes.

natural selection the differential fecundity (*q.v.*) in nature between members of a species possessing adaptive characters and those without such advantages. *See* Appendix C, 1818, Wells; 1858, Darwin and Wallace; 1859, Darwin; 1934, 1937, L'Héritier and Teissier; 1952, Bradshaw; 1954, Allison; artificial selection, evolution, fundamental theorem of natural selection, heavy metals, selection.

Nautilus *See* living fossil.

n_D refractive index.

Ndj 1 *See* bouquet configuration.

Neandertal a race of humans that roamed through Europe, North Africa, the Near East, Iraq, and Central Asia in the middle and upper Pleistocene (300,000 to 30,000 years ago). The fossils are named after the valley in western Germany where they were first discovered. The ranges of *Homo neandertalensis* and *Homo sapiens* overlapped in Europe during recent millennia, but there seems to have been little interbreeding. Sequence comparisons of the D loop (*q.v.*) region of mtDNA from Neandertal fossilized bones and modern humans show that Neandertals became extinct without leaving a trace of their mtDNAs in modern humans. *See* Appendix C, 1997, Krings *et al.*, *Homo*.

Nearctic one of the six biogeographic realms (*q.v.*) of the earth, comprising North America, Greenland, and extending to the Mexican plateau.

nebenkern a two-stranded helical structure surrounding the proximal region of the tail filament of a spermatozoon. The nebenkern is derived from clumped mitochondria.

negative complementation suppression of the wild-type activity of one subunit of a multimeric protein by a mutant allelic subunit.

negative contrast technique *See* negative staining.

negative eugenics *See* eugenics.

negative feedback the suppression or diminution of an effect by its own influence on the process that gives rise to it.

negative gene control prevention of gene expression by the binding of a specific controlling factor to DNA. For example, in bacterial operons (either inducible or repressible), the binding of a repressor protein to the operator prevents transcription of structural genes in that operon. *See* regulator gene. *Compare with* positive gene control.

negative interference a situation in which the coefficient of coincidence is greater than 1. In such cases, the occurrence of one exchange between homologous chromosomes appears to increase the likelihood of another in its vicinity.

negative regulation *See* negative gene control.

negative sense ssDNA or RNA *See* plus (+) and minus (–) viral strands.

negative staining a staining technique for high-resolution electron microscopy of viruses. A virus suspension is mixed with a phosphotungstic acid solution and poured into an atomizer sprayer. The mixture is then sprayed upon electron microscope grids previously coated with a film of carbon. The phosphotungstic acid enters the contours of the specimen, which is viewed as a light object against a dark background. *See* Appendix C, 1959, Brenner and Horne.

negative supercoiling *See* supercoiling.

neobiogenesis the concept that life has been generated from inorganic material repeatedly in nature.

neo-Darwinism the post-Darwinian concept that species evolve by the natural selection of adaptive phenotypes caused by mutant genes.

Neogene a subdivision of the Tertiary period, incorporating the Pliocene and Miocene epochs. *See* geologic time divisions.

Neolithic pertaining to the later Stone Age, during which agriculture and animal husbandry originated and flourished.

neomorph a mutant gene producing a qualitatively new effect that is not produced by the normal allele.

neomycin an antibiotic produced by *Streptomyces fradiae.*

neontology the study of living (extant) species, as opposed to paleontology (the study of extinct species).

neoplasm a localized population of proliferating cells in an animal that are not governed by the usual limitations of normal growth. The neoplasm is said to be *benign* if it does not undergo metastasis and *malignant* if it undergoes metastasis.

neotenin synonym for allatum hormone (*q.v.*).

neoteny the retention of larval characteristics throughout life with reproduction occurring during the larval period. In *Ambystoma mexicanum*, for example, the gill-breathing, water-dwelling larval salamander matures and reproduces sexually without undergoing metamorphosis to a lung-breathing, land-dwelling, adult form. *See* axolotl.

Neotropical one of the six biogeographic realms (*q.v.*) of the earth, comprising Central and South America (south of the Mexican plateau) and the West Indies.

neuraminic acid a nine-carbon amino sugar widely distributed in living organisms. One of the distinctions between eubacteria and archaebacteria is the presence of neuraminic acid in the cell walls of the former and its absence in the latter. In animals, neuraminic acid is found in mucolipids, mucopolysaccharides, and glycoproteins. Neuraminic acid-containing membrane components play a role in the attachment and penetration of virus particles into animal cells. *See* ganglioside.

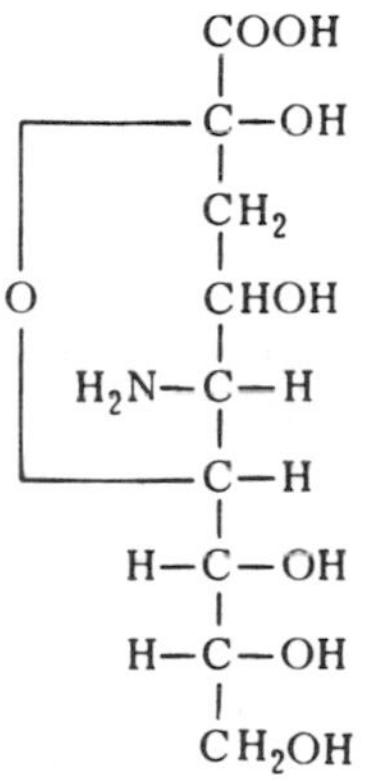

```
            COOH
             |
   ┌──────── C—OH
   |         |
   |         CH2
   |         |
   O         CHOH
   |         |
   |    H2N—C—H
   |         |
   └──────── C—H
             |
          H—C—OH
             |
          H—C—OH
             |
            CH2OH
```

neuregulins (NRGs) a family of structurally related growth and differentiation factors found in the central and peripheral nervous systems, which includes products of the *Nrg1*, *Nrg2*, *Nrg3*, and *Nrg4* genes. NRG1, the most widely studied neuregulin, has 14 different isoforms, produced by alternative splicing of its mRNA, and 7 isoforms of NRG2 have been identified. A variety of proteins identified in various independent studies, including the neu differentiation factor (NDF), heregulin (HRG), glial growth factor 2 (GGF2), and *a*cetylcholine *r*eceptor-*i*nducing *a*ctivity (ARIA), are all isoforms of NRG1, produced by alternatively spliced mRNA. All NRG1 isoforms have in common with each other and with other neuregulins an epidermal growth factor (EGF)-like sequence, which is essential for their function. Some neuregulins have in common a transmembrane domain, followed by a variable intracellular domain, while others differ from one another in their N-terminal domains. These structural features suggest functional similarities as well as distinctions between the neuregulin family members. NRGs interact with a family of receptor tyrosine kinases on target cells to influence a number of cellular processes, including the synthesis of acetylcholine receptors at neuromuscular junctions, the proliferation and survival of oligodendrocytes, and the proliferation and myelination of Schwann cells. *Nrg1* is a candidate gene for schizophrenia (*q.v.*) and is also thought to be linked to age related memory loss. *See* epidermal growth factor (EGF).

neurofibroma a fibrous tumor of peripheral nerves.

neurofibromatosis one of the most common single gene disorders affecting the human nervous system. The disease is characterized by the presence in the skin, or along the course of peripheral nerves, of multiple neurofibromas that gradually increase in number and size. There are two types of neurofibromatosis, abbreviated NF1 and NF2. NF1 (also called von Recklinghausen disease) is one of the most common autosomal dominant disorders of humans, affecting 1/3,000 individuals. The *NF1* gene is located on the long arm of chromosome 17 at 11.2. The gene spans 3×10^5 nucleotides and encodes a protein (neurofibromin) containing 2,818 amino acids. The spontaneous mutation rate of the *NF1* gene is high, and 30–50% of patients carry new *NF1* mutations. Neurofibromatosis 2 is a rarer condition, affecting about 1 in 37,000 individuals. The *NF2* gene is at 22q12, and it encodes a protein (merlin) containing 590 amino acids. Neurofibromin is located in the cytoplasm and apparently functions in signal transduction (*q.v.*), whereas merlin is believed to link the cell membrane to certain cytoskeletal proteins. *See* anti-oncogenes. http://www.nf.org

neurofibromin *See* neurofibromatosis.

neurohormone a hormone synthesized and secreted by specialized nerve cells; e.g., gonadotropin-releasing hormone produced by neurosecretory cells located in the hypothalamus.

neurohypophysis the portion of the hypophysis that develops from the floor of the diencephalon.

neurological mutant a mutant producing malformations of the sense organs or the central nervous system or striking abnormalities in locomotion or behavior. Hundreds of neurological mutants have been collected in *Drosophila*, *Caenorhabditis*, and the mouse. *See* Appendix C, 1969, Hotta and Benzer; 1971, Suzuki *et al.*; 1981, Chalfie and Sulston; 1986, Tomlinson and Ready.

neuron a nerve cell.

neuropathy a collective term for a great variety of behavioral disorders that may have hereditary components.

neurosecretory spheres electron-dense spheres 0.1–0.2 micrometers in diameter, synthesized by and transported in the axoplasm of specialized neurons.

Neurospora crassa the ascomycete fungus upon which many of the classical studies of biochemical genetics were performed. In *Neurospora* each set of meiotic products is arranged in a linear fashion, and therefore the particular meiotic division at which genetic exchange occurs can be determined by dissecting open the ascus and growing the individual ascospores (*see* ordered tetrad). The haploid chromosome number of this species is 7, and seven detailed linkage maps are available. *Neurospora* is estimated to have a genome of 38.6 million nucleotide base pairs. It has around 10,000 genes, but only about 1,400 have counterparts in *Drosophila*, *Caenorhabditis*, or humans. More than half of its genes have no similarity to those in the other fungi that have been sequenced (*Saccharomyces* and *Schizosaccharomyces*). There are about 1.7 introns per gene, with an average intron size of 134 nucleotides. *Neurospora* has a lower proportion of genes in multigene families than any other species for which data are available. This is because it has evolved repeat-induced point mutation (RIP) (*q.v.*), a mechanism for detecting and mutationally inactivating DNA duplications. Dispersed throughout the genome are 424 tRNA genes and 74 5S rRNA genes. There are also 175–200 copies of a tandem repeat that contains the 17S, 5.8S, and the 25S rRNA genes. These are localized in the nucleolus organizer which somehow protects them from RIP. The *Neurospora* mitochondrial DNA contains 60,000 nucleotide pairs. *See* Appendix A, Fungi, Ascomycota; Appendix C, 1927, Dodge; 1941, Beadle and Tatum; 1944, Tatum *et al.*; 1948, Mitchell and Lein; 2003, Galagan *et al.*; Appendix E; Appendix F.

neurula the stage of development of a vertebrate embryo at which the neural axis is fully formed and histogenesis is proceeding rapidly.

neutral equilibrium *See* passive equilibrium.

neutral mutation 1. a genetic alteration whose phenotypic expression results in no change in the organism's adaptive value or fitness for present environmental conditions. 2. a mutation that has no measurable phenotypic effect as far as the study in question is concerned. *See* silent mutation.

neutral mutation–random drift theory of molecular evolution a theory according to which the majority of the nucleotide substitutions in the course of evolution are the result of the random fixation of neutral or nearly neutral mutations, rather than the result of positive Darwinian selection. Many protein polymorphisms are selectively neutral and are maintained in a population by the balance between mutational input and random extinction. Neutral mutations are not functionless; they are simply equally effective to the ancestral alleles in promoting the survival and reproduction of the organisms that carry them. However, such neutral mutations can spread in a population purely by chance because only a relatively small number of gametes are "sampled" from the vast supply produced in each generation and therefore are represented in the individuals of the next generation. *See* Appendix C, 1968, Kimura.

neutron an elementary nuclear particle with a mass approximately the same as that of a hydrogen atom and electrically neutral; its mass is 1.0087 mass units.

neutron contrast matching technique a technique that involves determining the neutron-scattering densities of particles irradiated in solutions containing various concentrations of light and heavy water. This technique was used on nucleosomes (*q.v.*), and it was found that under conditions where neutron scattering from DNA dominated the reaction, the radius of gyration was 50 Ångstroms. When scattering from the histone proteins was dominant, the radius was 30 Ångstroms. The larger radius for DNA proved that it was located on the surface of the nucleosome. *See* Appendix C, 1977, Pardon *et al.*

***N*-formylmethionine** a modified methionine molecule that has a formyl group attached to its terminal amino group. Such an amino acid is "blocked" in the sense that the absence of a free amino group prevents the amino acid from being inserted into a growing polypeptide chain. *N*-formylmethionine is the starting amino acid in the synthesis of all bacterial polypeptides. *See* Appendix C, 1966, Adams and Cappecchi; initiator tRNA, start codon.

$$\begin{array}{l} \quad\;\; O \;\;\; H \\ \quad\;\; \| \;\;\;\; | \\ H-C-N-CH-COOH \\ \qquad\qquad\;\; | \\ \qquad\qquad CH_2-CH_2-CH_2-S-CH_3 \end{array}$$

niacin an early name for nicotinic acid (*q.v.*).

niche from the standpoint of a species, its behavioral, morphological, and physiological adaptations to its habitat. From the standpoint of the environment, the ecological conditions under which the species survives and multiplies. *See* ecological niche, extremophiles.

niche preclusion *See* first-arriver principle.

nick in nucleic acid chemistry, the absence of a phosphodiester bond between adjacent nucleotides in one strand of duplex DNA. *Compare with* cut.

nickase an enzyme that causes single-stranded breaks in duplex DNA, allowing it to unwind.

nick-closing enzyme *See* topoisomerase.

nick translation an *in vitro* procedure used to radioactively label a DNA of interest uniformly to a high specific activity. First, nicks are introduced into the unlabeled DNA by an endonuclease, generating 3′ hydroxyl termini. *E. coli* DNA polymerase I is then used to add radioactive residues to the 3′ hydroxy terminus of the nick, with concomitant removal of the nucleotides from the 5′ side. The result is an identical DNA molecule with the nick displaced further along the duplex. *See* strand-specific hybridization probes.

Nicotiana a genus containing about 60 species, many of which have been intensively studied genetically. Much interest has been generated from the finding that tumors arise spontaneously at high frequency in certain interspecific hybrids, such as those plants produced by the cross *N. langsdorffii* × *N. glauca*. The species of greatest commercial importance is *N. tabacum*, the source of tobacco. *N. tabacum* is an allotetraploid, and *N. sylvestris* and *N. tomentosiformis* are its parental diploids. Analysis of chDNA and mtDNA reveal that tobacco inherited these cytoplasmic organelles from *N. sylvestris*. Tobacco genes that confer resistance to the tobacco mosaic virus (*q.v.*) have been cloned and sequenced utilizing transposon tagging (*q.v.*). *See* Appendix C, 1761, Kölreuter; 1925, Goodspeed and Clausen; 1926, Clausen and Goodspeed; 1986, Shinozaki *et al.*; 1994, Whitham *et al.*

nicotinamide-adenine dinucleotide (NAD) a coenzyme (formerly called DPN or coenzyme 1) functioning as an electron carrier in many enzymatic oxidation-reduction reactions. The oxidized form is symbolized NAD^+, the reduced form NADH (see structural formulas below). *See* citric acid cycle, cytochrome system, mitochondrial proton transport.

nicotinamide-adenine dinucleotide phosphate (NADP) an electron carrier (formerly called TPN or coenzyme 2). The oxidized form is symbolized $NADP^+$, the reduced form NADPH. *See* nicotine-adenine dinucleotide (NAD).

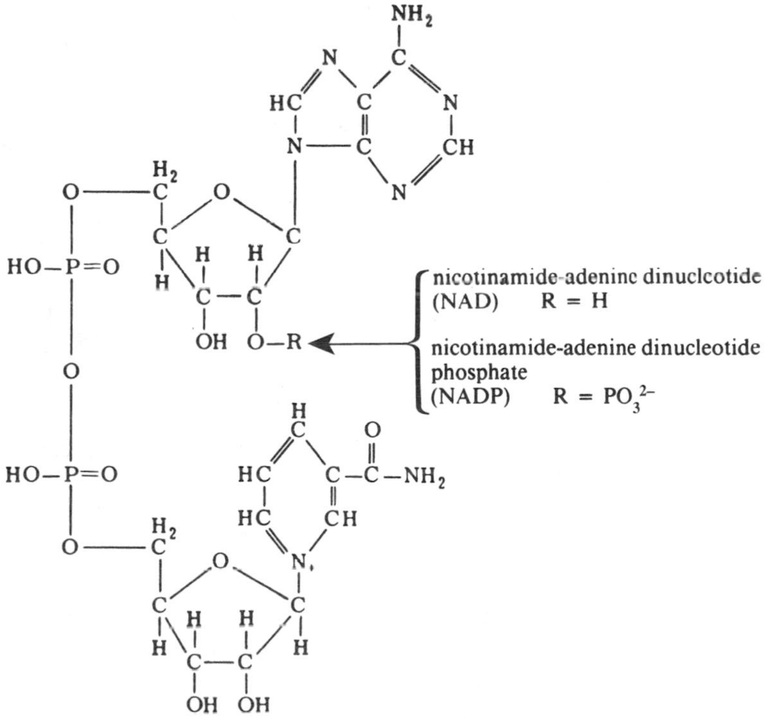

nicotinamide-adenine dinucleotide (NAD) R = H

nicotinamide-adenine dinucleotide phosphate (NADP) $R = PO_3^{2-}$

Nicotinamide-adenine dinucleotide (NAD)/nicotinamide-adenine dinucleotide phosphate (NADP)

nicotine a poisonous, volatile alkaloid present in the leaves of *Nicotiana tabacum* and responsible for many of the effects of tobacco smoking. It functions in the plant as a potent insecticide.

```
      CH        H2C —— CH2
    //  \        |      |
  HC     C —— C        CH2
  |      ||   / \     /
  HC     CH  H    N
    \\  /         |
     N           CH3
```

nicotinic acid one of the B vitamins. Also called niacin in the older literature.

```
      CH
    /    \\
  HC       C–COOH
  ||       |
  HC       CH
    \\   //
      N
```

Niemann-Pick disease a group of human disorders characterized by enlargement of the spleen and liver and by the accumulation of sphingomyelin (*q.v.*) and other lipids throughout the body. Two German pediatricians, Albert Niemann and Ludwig Pick, published accounts of the disease in 1914 and 1927, respectively. The syndrome is due to mutations in a gene at 18q11, q12 that encodes a lysosomal sphingomyelinease. Amniocentesis and testing of fetal cells for sphingomyelinase activity permits monitoring of pregnancies at risk. Heterozygotes can be identified, since their leukocytes contain about 60% the normal activity of sphingomyelinase. *See* sphingomyelin.

***nif* (*nitrogen-fixing*) genes** genes that enable the bacteria containing them to fix atmospheric nitrogen. Such genes are generally carried by the plasmids of nodulating bacteria, and they encode the enzyme nitrogenase. *See* nitrogen fixation, *Rhizobium*.

nigericin *See* ionophore.

Nile blue a mixture of two dyes: Nile blue A, a water-soluble basic dye; and Nile red, a lysochrome formed by spontaneous oxidation of Nile blue A (an example of allochromacy, *q.v.*). Structures are shown on page 299.

Nilsson-Pelger model of eye evolution a computational model (shown below) designed to simulate the evolution of an eye. It starts with at flat sheet of photosensitive cells lying above a flat layer of pigmented cells and below a monolayer of transparent cells. Using a sequence of small modifications in shape, the originally flat patch gradually changes into a cup, which then acquires a lens in its opening. The end result is a focused, light-imaging organ with the geometry typically seen in a fish eye. The mathematical modeling procedure chosen was such that each 1% increment of change produced a maximal

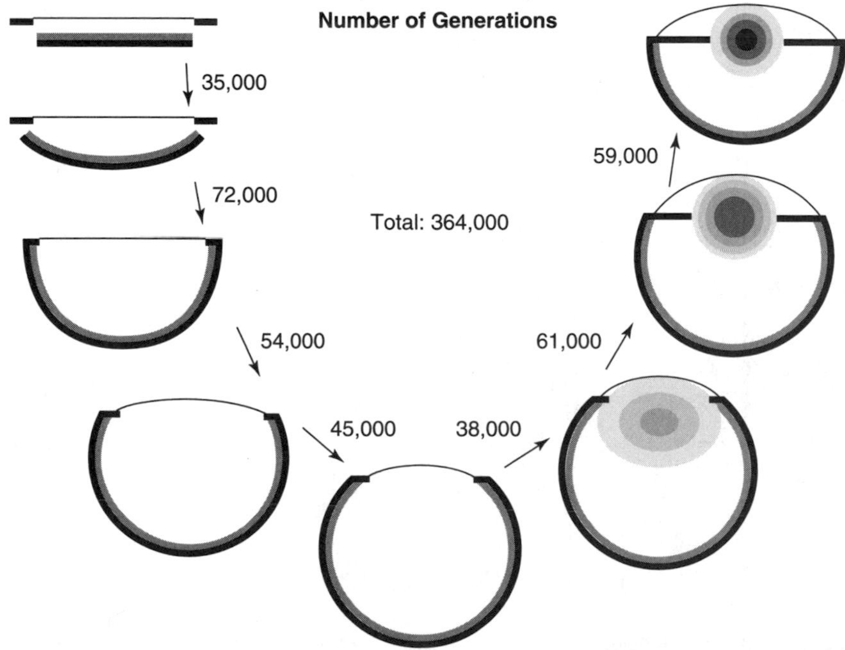

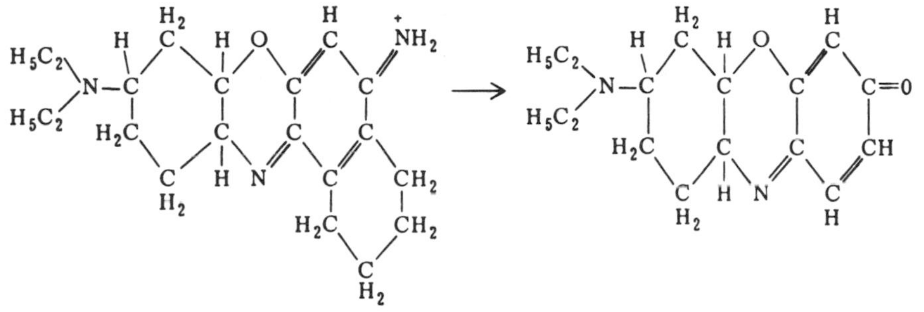

Nile blue A

Nile red

increase in visual acuity. Using conservative assumptions as to heritabilities (*q.v.*) and selection pressures (*q.v.*), the times taken (in generations) to perform each evolutionary stage are shown in the diagram. To complete an eye for a species with one generation per year, it would take a total of only 364,000 years, a relatively short time from a geological prospective. So it is not surprising that eyes have been produced independently at least 40 times during the evolution of Eumetazoa (*q.v.*). *See* Appendix C, 1994, Nilsson and Pelger.

ninhydrin an organic reagent that reacts with and colors amino acids. Ninhydrin solutions are sprayed on chromatographs, and the separated amino acids and polypeptides are then rendered visible as ninhydrin-positive spots.

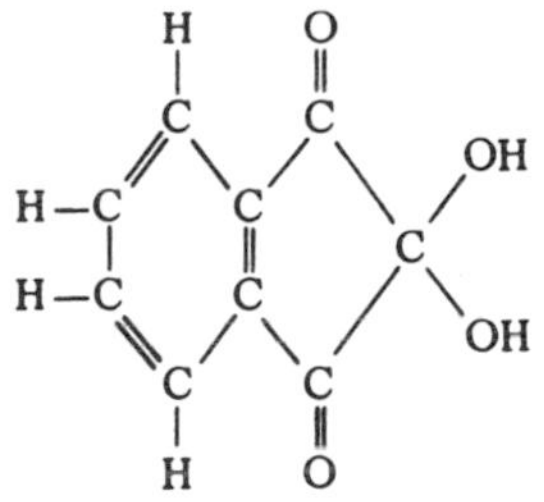

nitrocellulose filter a very thin filter composed of nitrocellulose fibers that selectively bind single-stranded DNA strongly, but not double-stranded DNA or RNA. The ssDNA binds along its sugar-phosphate backbone, leaving its bases free to pair with complementary bases contained in labeled ssDNA or RNA probes. *See* DNA hybridization.

nitrogen the fourth most abundant of the biologically important elements. Atomic number 7; atomic weight 14.0067; valence 3 , 5'; most abundant isotope ^{14}N; heavy isotope ^{15}N. This heavy isotope was used in the famous Meselson-Stahl experiment of 1958. *See* Appendix C.

nitrogen fixation the enzymatic incorporation of nitrogen from the atmosphere into organic compounds. The ability to fix nitrogen is limited to certain bacteria. Sixty five million years ago nodulating legumes (*q.v.*) discovered a way to utilize atmospheric nitrogen directly, through symbiosis with nitrogen-fixing bacteria. *See Azotobacteria, nif* genes, *Rhizobium*.

nitrogen mustard di(2-chloroethyl) methylamine; an alkylating agent (*q.v.*) that is a potent mutagen and chromosome-breaking agent. *See* sulfur mustard.

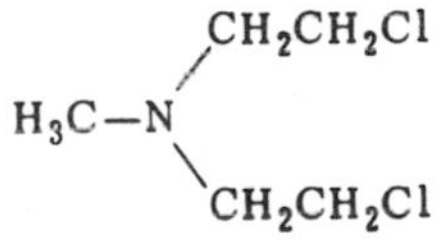

nitrogenous base a purine or pyrimidine; more generally an aromatic, nitrogen-containing molecule that has basic properties (is a proton acceptor).

nitrous acid HNO_2, a mutagen that converts the NH_2 groups of the purines and pyrimidines to OH groups.

NK cells *See* natural killer (NK) cells.

N^6-methyladenine. *See* 5-methylcytosine.

NMR spectroscopy nuclear magnetic resonance spectroscopy (*q.v.*).

node **1.** in vascular plants, a slightly enlarged portion of a stem where leaves and buds arise and where branches originate. **2.** in a circular DNA superhelix, the point of contact in a figure-of-eight; if the left strand in the upper part of 8 is closest to the viewer at the node, it is called a positive node; if the left strand in the upper part of 8 is in back of the other strand at the node, it is called a negative node. **3.** in a cladogram (*q.v.*), a point where one branch splits off from another. Each node represents a common ancestor, and the branches are the lineages derived from it. Also called a *divergence node*. *See* PhyloCode.

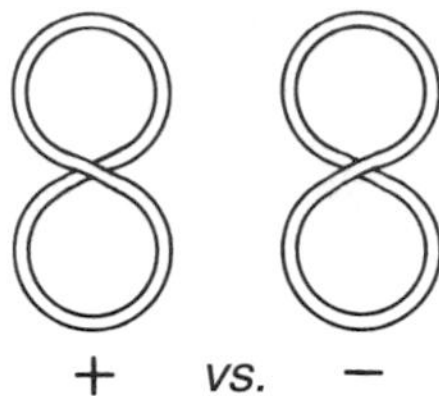

nodulating legumes a large family of dicotyledonous species that form nitrogen-fixing nodules on their roots or stems. Examples are *Glycine max* and *Phaseolus vulgaris* (*both of which see*). *See* Dicotyledoneae, *Rhizobium.*

noise in colloquial usage, variation in an experiment attributed to uncontrolled effects, usually associated with a variance component called experimental error.

Nomarski differential interference microscope an optical system that, like the phase contrast microscope, permits the visualization of transparent structures in a living cell. However, in the Nomarski system the field is quite shallow so that there is freedom from phase disturbances from structures above and below the plane of focus. The observation method is comparable to that with extreme oblique illumination, and the specimen therefore appears in relief.

nomenclature the naming of species according to rules developed by international associations of taxonomists. Several codes have been published, and these continue to be revised. There are five: the International Code of Zoological Nomenclature (ICZN), the International Code of Botanical Nomenclature (ICBN), the International Code of Nomenclature for Cultivated Plants (ICNCP), the International Code for the Nomenclature of Bacteria (ICNB), and the International Code of Virus Classification and Nomenclature (ICVCN). *See* Appendix C, 1735, Linné.

nonautogenous anautogenous (*q.v.*).

nonbasic chromosomal proteins acidic or neutral proteins (therefore not histones) associated with chromosomes: e.g., certain enzymes such as DNA polymerases.

noncoding (of a section of a nucleic acid molecule) not directing the production of a peptide sequence. *See* skeletal DNA hypothesis.

non-Darwinian evolution genetic changes in populations produced by forces other than natural selection; a term usually associated with the neutralist view of evolution. *See* neutral mutation-random drift theory of molecular evolution.

nondisjunction the failure of homologous chromosomes (in meiosis I, primary nondisjunction) or sister chromatids (in meiosis II, secondary nondisjunction; or mitosis) to separate properly and to move to opposite poles. Nondisjunction results in one daughter cell receiving both and the other daughter cell none of the chromosomes in question. *See* Appendix C, 1914, Bridges.

nonessential amino acids *See* essential amino acids.

nonhomologous chromosomes chromosomes that do not synapse during meiosis.

noninducible enzyme constitutive enzyme (*q.v.*).

nonlinear tetrad a group of four meiotic products that are randomly arranged in the ascus. *See* linear tetrad.

non-Mendelian ratio in the progeny of a cross, unusual phenotypic ratios that fail to follow Mendel's laws, suggesting that gene conversion (*q.v.*) or another aberrant mechanism is responsible.

nonparametric statistics *See* statistics.

nonparental ditype *See* tetrad segregation types.

nonpermissive cells *See* permissive cells.

nonpermissive conditions environmental settings in which conditional lethal mutants fail to survive.

nonpolar referring to water-insoluble chemical groups, such as the hydrophobic side chains of amino acids.

nonrandom mating *See* assortative mating, inbreeding, outbreeding.

nonreciprocal recombination *See* unequal crossing over.

nonreciprocal translocation *See* translocation.

nonrecurrent parent the parent of a hybrid that is not again used as a parent in backcrossing.

nonrepetitive DNA segments of DNA exhibiting the reassociation kinetics expected of unique sequences; single sequence DNA.

nonselective medium a growth medium that allows growth of all genotypes present in a recombination or mutation experiment. *Compare with* selective medium.

nonsense codon synonymous with stop codon (*q.v.*).

nonsense mutation a mutation that converts a sense codon to a chain-terminating codon or vice versa. The results following translation are abnormally short or long polypeptides, generally with altered functional properties. *Contrast with* missense mutation.

nonsense suppressor a gene coding for a tRNA that is mutant in its anticodon and therefore able to recognize a nonsense (stop) codon; nonsense suppressors cause extension of polypeptide chain synthesis through stop codons. *See* Appendix C, 1969, Abelson *et al.*; amber suppressor, ochre suppressor, readthrough.

nonspherocytic hemolytic anemia *See* glucose-6-phosphate dehydrogenase deficiency.

nopaline *See* opine.

NOR nucleolar organizer region.

noradrenaline norepinephrine.

norepinephrine a hormone of the adrenal medulla that causes vasoconstriction and raises the blood pressure.

```
         C—H      OH  H   H
       //  \       |   |   |
   H—C      C ——— C — C — N
     |      ||     |   |   |
  HO—C      C—H   H   H   H
      \\   /
        C—OH
```

***n* orientation** one of two possible orientations for inserting a target DNA fragment into a vector; in the *n* orientation, the genetic map of both target and vector have the same orientation; in the *u* orientation, the target and the vector are in different orientations.

normal distribution the most commonly used probability distribution in statistics. The formula for the normal curve is

$$Y = \frac{1}{\sigma\sqrt{2\pi}}\, e^{\frac{-(X-\mu)^2}{2\sigma^2}}$$

where μ = the mean, σ = the standard deviation, e = the base of natural logarithms, $\pi = 3.1416$, and Y = the height of the ordinate for a given value of X. The graph of this formula, the normal curve, also called Laplacian or Gaussian, is bell shaped. The value of m locates the curve along the abscissa and that of σ determines its shape. The larger the standard deviation, the broader the curve. In nature, a vast number of continuous distributions are normally distributed.

normalizing selection the removal of those alleles that produce deviations from the average population phenotype by selection against all deviant individuals. Such selection will reduce the variance in subsequent generations. *Also called* stabilizing selection, centripetal selection.

normal solution one containing a gram-equivalent weight of solute dissolved in sufficient water to make a liter of solution.

norm of reaction the phenotypic variability produced by a given genotype under the range of environmental conditions common to the natural habitat of the species or under the standard culture or experimental conditons. *See* adaptive norm.

northern blotting *See* Southern blotting.

Nosema *See* microsporidia.

Notch a series of overlapping deficiencies of the X chromosomes of *Drosophila melanogaster*. All deficiencies lack the 3C7 band, and females heterozygous for the deficiency show distal notches of the wing. Hemizygous males die as embryos. The wild-type allele of *Notch* is essential for the proper differentiation of ectoderm. Mutations at the *Notch* locus result in a hypertrophy of the embryonic nervous system at the expense of hypodermal structures. N^{+} encodes a transmembrane protein containing 2,703 amino acids. Built into the molecule are 36 EGF repeats, some of which bind calcium while others facilitate the formation of Notch protein dimers. *See* Appendix C, 1938, Slizynska; epidermal growth factor.

Notophthalmus viridescens the common spotted newt of the eastern United States. The amplification of rDNA has been extensively studied using the oocytes of this species. *See* Appendix A, Chordata, Amphibia, Urodela; histone genes, lampbrush chromosome, *Triturus*.

novobiocin an antibiotic produced by *Streptomyces niveus*. (See structure on page 302.)

np nucleotide pair. *See* base pair.

NRG234 *See* sym-plasmid.

nRNA nuclear RNA (*q.v.*).

nt nucleotide. *Compare with* bp.

***N*-terminal end** proteins are conventionally written with the amino (NH_2) end to the left. The assembly of amino acids into a polypeptide starts at the *N*-terminal end. *See* translation.

***N* terminus** *N*-terminal end.

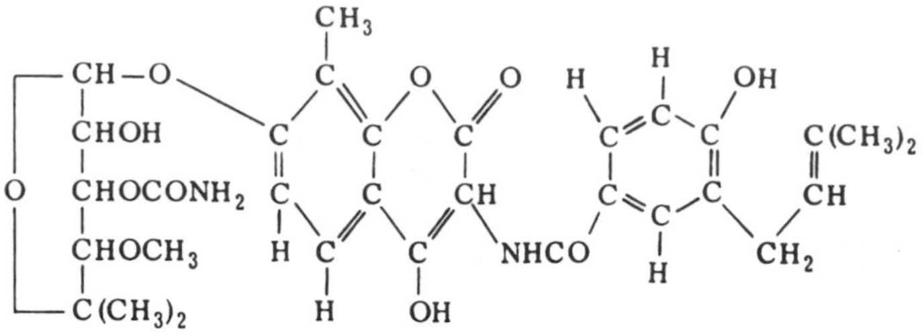

Novobiocin

nu (ν) body particles arranged like beads on a string along interphase chromosomes. These are most clearly seen in electron micrographs of negatively stained Miller spreads (*q.v.*). Nu bodies correspond to the nucleosomes (*q.v.*) of the biochemist.

nuclear dimorphism in ciliated protozoa, the presence of two morphologically and functionally different types of nuclei. The macronucleus is large, highly polyploid, and it contains many nucleoli. Macronuclear DNA functions analogously to the DNA of somatic cells. The micronucleus functions as the germline and is diploid. It is capable of undergoing meiosis during conjugation (*q.v.*). The macronucleus develops from a micronucleus. *See* Appendix A, Protoctista, Ciliophora; Appendix C, 1876, Bütschli.

nuclear duplication mitosis (*q.v.*).

nuclear emulsion a photographic emulsion especially compounded to make visible the individual tracks of ionizing particles.

nuclear envelope an envelope surrounding the nucleus, composed of two membranes enclosing a perinuclear cisterna. The outermost membrane is studded with ribosomes. The perinuclear cisterna is traversed by nuclear pore complexes (*q.v.*). *See* lamins.

nuclear family a pair of parents and their children.

nuclear fission a transformation of atomic nuclei characterized by the splitting of a nucleus into at least two other nuclei and the release of amounts of energy far greater than those generated by conventional chemical reactions.

nuclear fusion the coalescence of two or more atomic nuclei with the release of relatively vast amounts of energy.

nuclear lamina *See* lamins.

nuclear magnetic resonance (NMR) spectroscopy an instrumental technique used to determine the three-dimensional (3-D) structure of biological molecules. NMR spectroscopy and x-ray crystallography (*q.v.*) are the only methods capable of analyzing the structures of proteins and nucleic acids at atomic resolution. NMR spectroscopy exploits the behavior of certain atoms when they are placed in a strong static magnetic field and exposed to short pulses of energy in the radio-wave frequency range. For biological samples, the important atoms are H-1, N-15, and C-13, and the magnets used are 10,000–15,000 times stronger than the earth's magnetic field. To increase the level of N-15 and C-13 in the molecular targets, microorganisms from which the molecules are extracted are grown on media enriched with these isotopes. When placed in a strong magnetic field, the atomic nuclei of these atoms exhibit a property called *nuclear spin*, whereby they behave like tiny compass needles and orient themselves with respect to the magnetic field. When exposed to pulses of radio waves of specific frequencies, the oriented nuclei jump to higher-energy states in which the spin is opposed to the magnetic field. The nuclei are now said to be in *resonance*, and they emit radio frequency radiation when they revert to their lower-energy states. The amount of energy needed to achieve resonance is dependent on the properties of each nucleus and its chemical environment, and plots of the strengths of the resonance signals versus radio-wave frequencies provide information about the nature of atoms and their proximity to one another. NMR data are coupled with computational tools to produce 3-D structures of biomolecules, which are stored in easily accessible databases. The first protein structure determined by NMR spectroscopy was that of a bull seminal proteinase inhibitor. NMR spectroscopy techniques can also be extended to such areas as the study of molecular interactions, molecular motion, and the rate of chemical reactions. *See* Appendix C, 1985, Williamson *et al.*; 1966, Ernst and Anderson; 1991, Ernst; 2002, Wüthrich *et al.*; Appendix E, Individual Databases; *Antennapedia*, proteomics.

nuclear pore complex (NPC) an octagonally symmetrical organelle that allows controlled passage of molecules from nucleus to cytoplasm and vice versa. A typical mammalian nucleus contains between 3,000 and 4,000 NPCs. Each complex is made up of a central core that is formed from eight spokelike structures which encircle a central plug and are sandwiched between two rings. Cytoplasmic fibrils attach to the subunits that make up the outer cytoplasmic ring. A basket-like structure protrudes from the inner nuclear ring. It is composed of struts that connect subunits of this ring to a smaller terminal ring. *See* nucleoporins (Nups).

nuclear processing of RNA *See* posttranscriptional processing, RNA editing.

nuclear reactor the apparatus in which nuclear fission may be sustained in a self-supporting chain reaction. A source of energy and radioisotopes.

nuclear reprogramming modifications of DNA (e.g., by demethylation) and/or chromatin proteins (e.g., by dissociation from DNA) that allow a differentiated nucleus from larval or adult somatic cells to replace an egg nucleus and restore totipotency (*q.v.*) to the transplanted nucleus. *See* Appendix C, 2004, Simonssen and Gurdon; nuclear transfer.

nuclear RNA RNA molecules found in the nucleus either associated with chromosomes or in the nucleoplasm. *See* chromosomal RNA, heterogeneous nuclear RNA.

nuclear targeting signal *See* protein sorting, sorting signals.

nuclear transfer the injection of a diploid somatic nucleus into an enucleated egg. The nature of the ensuing development reveals the developmental potentialities of the implanted nucleus. Various amphibian species were used in early experiments. The number of embryos that survived to tadpoles declined when donor cells were taken from animals at successively more advanced developmental stages. The recent birth of a lamb cloned from the somatic nucleus of an adult attracted worldwide attention. However, Dolly (*q.v.*) was the only successful outcome from 277 nuclear transfer procedures. *See* Appendix C, 1952, Briggs and King; 1962, 1967, Gurdon; 1997, Wilmut *et al.*; cloning, nuclear reprogramming, sheep.

nuclease any enzyme that breaks down nucleic acids.

nucleic acid a nucleotide polymer. In the early literature DNA and RNA were called thymonucleic acid and yeast nucleic acid, respectively. This is because DNA was first isolated from beef thymus glands and RNA from bakers' yeast cultures. *See* deoxyribonucleic acid, ribonucleic acid.

nucleic acid bases *See* bases of nucleic acids.

nucleic acid fingerprinting a method for analyzing digests of DNA or RNA similar to the fingerprinting method for fragmented proteins. *See* Appendix C, 1965, Sanger, Brownlee and Barrell; DNA fingerprint technique, oligonucleotide fingerprinting (OFP).

nuclein the acidic, phosphorus-rich substance isolated from human white blood cells by Miescher. We now know that nuclein was a mixture of nucleic acids and proteins. *See* Appendix C, 1871, Miescher.

nucleocapsid a virus nucleic acid and its surrounding capsid. *See* capsomere.

nucleo-cytoplasmic ratio the ratio of the volume of nucleus to the volume of cytoplasm.

nucleoid 1. a DNA-containing region within a prokaryote, mitochondrion, or chloroplast. **2.** in an RNA tumor virus, the core of genetic RNA surrounded by an icosahedral protein capsid.

nucleolin an acidic phosphoprotein synthesized in the dense fibrillar regions of the nucleolus. Human nucleolin is made up of 707 amino acids. The NCL gene resides at 2q12-qter. It consists of 14 exons with 13 introns and is about 11 kb long. Intron 11 encodes a small nucleolar RNA designated U20. This snoRNA has a region of perfect complementarity with a conserved sequence in the 18S rRNA. It follows that nucleolin is involved in the formation of the small ribosomal subunit. *See* Appendix C, 1989, Srivastava *et al.*

nucleolus an RNA-rich, intranuclear domain found in eukaryotic cells that is associated with the nucleolus organizer (*q.v.*) and is the site of preribosomal RNA (*q.v.*) synthesis and processing (*q.v.*) and of ribosomal particle assembly. The illustration on page 304 shows chromosome 6 of maize (*q.v.*), which contains the nucleolus organizer and its nucleolus as they appear in meiotic prophase. The nucleolus is composed of the primary products of the ribosomal RNA genes (*q.v.*) and a variety of proteins, including RNA polymerases, ribonucleases, molecular chaperones (*q.v.*), helicases, ribosomal proteins, and proteins of unknown function. rRNA genes and their nascent transcripts were first seen as Miller trees (*q.v.*) in nucleoli from salamander oocytes. Under the electron microscope (*q.v.*), the nucleoli of most metazoans contain three major

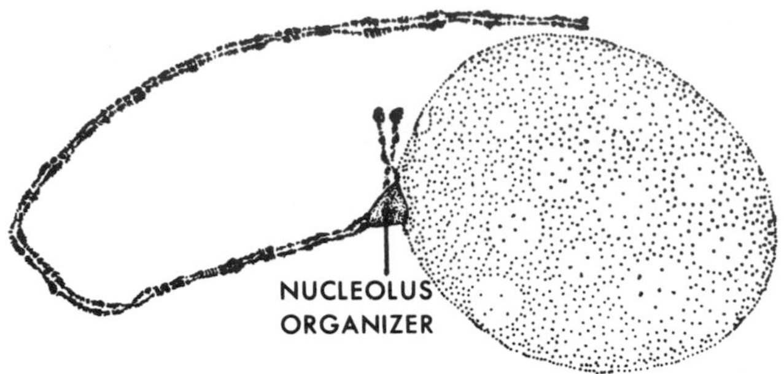

Nucleolus of maize chromosome 6

morphological components: the fibrillar center (FC), the dense fibrillar component (DFC), and the granular component (GC). These represent successive stages in the process of ribosome biogenesis. The FC contains tandem arrays of rRNA genes and is surrounded by the DFC, where newly synthesized pre-rRNA molecules and their associated proteins are found. Later events in posttranscriptional processing (*q.v.*) and assembly of preribosomal particles are associated with the GC that surrounds the DFC. The nucleolus has also been implicated in some non-traditional roles. For example, the yeast Cdc14 (*q.v.*), a protein that promotes the exit from mitosis (*q.v.*), localizes to the nucleolus during the G1 stage of the cell cycle (*q.v.*) and remains there until anaphase (*q.v.*), when it is liberated. Sequestration of this protein by the nucleolus thus prevents the cell from completing mitosis prematurely. *See* Appendix C, 1838, Schleiden; 1934, McClintock; 1965, Ritossa and Spiegelman; 1967, Birnstiel; 1969, Miller and Beatty; 1976, Chooi; 1989, Srivastava *et al.*; Cajal body, Cdc14, helicase, nucleolin, rDNA amplification, ribonuclease, ribosome, RNA polymerase, small nucleolar RNAs.

nucleolus organizer a region of one or more chromosomes that contains the ribosomal RNA genes (*q.v.*) and is associated with the nucleolus (*q.v.*). *Also called* nucleolus organizer region (NOR). *See* Appendix C, 1934, McClintock; 1965, Ritossa and Spiegelman; 1967, Birnstiel; 1969, Miller and Beatty; 1976, Chooi; Cajal body, rDNA amplification, ribosome, RNA polymerase.

nucleolus organizer region (NOR) nucleolus organizer (*q.v.*).

nucleomorph *See* cryptomonads.

nucleon a constituent particle of an atomic nucleus.

nucleoplasm the protoplasmic fluid contained in the nucleus.

nucleoporins (Nups) a family of more than 100 different proteins that are localized in each nuclear pore complex (NPC) (*q.v.*). Some of these proteins are structural components of the organelle, while others facilitate the transport of proteins and RNAs through the NPCs. Some nucleoporins play a role in tethering telomeres (*q.v.*) to the nuclear envelope.

nucleoprotein a compound of nucleic acid and protein. Either one of two main classes of basic proteins are found combined with DNA: one of low molecular weight (protamine) and one of high molecular weight (histone). The basic amino acids of these proteins neutralize the phosphoric acid residues of the DNA. *See* Appendix C, 1866, Miescher.

nucleosidase any enzyme that catalyzes the splitting of nucleosides into bases and pentoses.

nucleoside a purine or pyrimidine base attached to ribose or deoxyribose. The nucleosides commonly found in DNA or RNA are cytidine, cytosine deoxyriboside, thymidine, uridine, adenosine, adenine deoxyriboside, guanosine, and guanine deoxyriboside. Note that thymidine is a deoxyriboside and cytidine, uridine, adenosine, and guanosine are ribosides. *See* rare bases, inosine.

nucleosome a beadlike structure of eukaryotic chromosomes, consisting of a core of eight histone molecules (two each of proteins H2A, H2B, H3, and H4) wrapped by a DNA segment about 150 base pairs in length and separated from adjacent nucleosomes by a "linker" DNA sequence of about 50 base pairs). *See* Appendix C, 1974, Kornberg; 1977, Pardon *et al.*, Leffak *et al.*; chromatosome, histones, solenoid structure.

nucleotide one of the monomeric units from which DNA or RNA polymers are constructed, consisting of a purine or pyrimidine base, a pentose, and a phosphoric acid group. The nucleotides of DNA are deoxyadenylic acid, thymidylic acid, deoxyguanilic acid, and deoxycytidylic acid. The corresponding nucleotides of RNA are adenylic acid, uridylic acid, guanylic acid, and cytidylic acid.

nucleotide pair a hydrogen-bonded pair of purine-pyrimidine nucleotide bases on opposite strands of a double-helical DNA molecule. Normally, adenine pairs with thymine, and guanine pairs with cytosine; also called *complementary base pairs*. *See* Chargaff's rules, deoxyribonucleic acid.

nucleotide pair substitution the replacement of a given nucleotide pair by a different pair, usually through a transition or a transversion (*both of which see*).

nucleotide sequence databases *See* Appendix E.

nucleus the spheroidal, membrane-bounded structure present in all eukaryotic cells which contains DNA, usually in the form of chromatin. Two theories explaining the origin of the nucleus appear below. *See* Appendix C, 1831, Brown; 1991, Sogin; 1992, Rivera and Lake; endokaryotic hypothesis, nuclear envelope, nuclear pore complex (NPC), Sogin's first symbiont.

nuclide a species of atom characterized by the constitution of its nucleus. This is specified by the number of protons and neutrons it contains.

nude mouse a laboratory mouse homozygous for the recessive mutation *nu*, which maps to chromosome 11. Such mice are characterized by the complete absence of hair and thymus glands. Nude mice lack T lymphocytes (*q.v.*), but have natural killer cells (*q.v.*) and B lymphocytes (*q.v.*), and they are unable to reject homografts. The nude mouse serves as a model system for the study of the immunological effects of thymus deprivation. *See* rejection.

null allele an allele that produces no functional product and therefore usually behaves as a genetic recessive. For example, in the human ABO blood group system, the recessive allele (*i*) produces no detectable antigen, either in homozygous condition (blood group O) or in heterozygous condition with allele I^A (blood group A) or with allele I^B (blood group B). *See* silent allele.

null hypothesis method the standard hypothesis used in testing the statistical significance of the difference between the means of samples drawn from two populations. The null hypothesis states that there is no difference between the populations from which the samples are drawn. One then determines the probability that one will find a difference equal to or greater than the one actually observed. If this probability is .05 or less, the null hypothesis is rejected, and the difference is said to be significant.

nulliplex *See* autotetraploidy.

nullosomic lacking both members of a pair of chromosomes.

numerical taxonomy a system of classification using a multitude of characteristics to determine overall phenotypic similarity, each trait being given equal weight and without regard to phylogenetic relationships; also known as *phenetic taxonomy*.

nu particles *See* nucleosomes.

nurse cells cells in the insect ovary that function to nourish the oocyte. In *Drosophila melanogaster* there are 15 nurse cells, and their nuclei undergo endomitosis (*q.v.*). The chromatids generated by the cycles of DNA replication fall apart to produce a tangled mass unsuitable for cytological study. However, in some alleles of the *otu* mutant (*q.v.*), ovarian nurse cells have banded polytene chromosomes suitable for cytological mapping. The nurse-cell chromosomes are active in transcription of a variety of RNA molecules, which enter their cytoplasm and are eventually transported to the oocyte. The nurse cells degenerate after pumping almost all of their cytoplasm to the oocyte. *See* cystocyte divisions, insect ovary types.

nutritional mutant a mutation converting a prototroph into an auxotroph.

nutritive chord *See* insect ovary types.

N value the haploid chromosome number; the number of chromosomes in each germ cell. *See* polyploidy.

nystagmus a jerky twitching of the eye. *See* albinism.

O 1. Ordovician. 2. oxygen.

O antigens polysaccharide antigens of the cell walls of enterobacteria such as *Escherichia* or *Salmonella;* in contrast to the polysaccharide K antigens of bacterial capsules or the protein H antigens of bacterial flagella.

oat *See Avena.*

obese a gene, first identified in the mouse, from a strain where adults were double the normal weight and developed type 2 diabetes. These animals were ob^-/ob^-. The normal allele (ob^+) encodes a 176–amino acid protein. When injected into overweight, ob^- homozygotes, this protein causes a dose-dependent weight loss. For this reason, the protein has been named *leptin* (from the Greek root *leptos,* meaning thin). Leptin injection results in lowering of body weight, percentage of body fat, food intake, and serum concentrations of glucose and insulin. The ob^+ gene is highly conserved among vertebrates, and its human homolog has been identified. *See* Appendix C, 1994, Zhang *et al.*; 1995, Tartaglia *et al.*; diabetes mellitus.

obligate restricted to a specified condition of life. For example, an obligate parasite cannot live in the absence of its host. *See* facultative.

Occam's razor a rule attributed to the medieval philosopher William of Occam. In modern times, the rule states that when there are several possible explanations of a phenomenon, one selects as most probable the explanation that is the simplest and most consistent with the data at hand. Also called the parsimony principle.

oceanic island an island that has risen from the sea. *See* continental island.

ocellus one of the simple eyes located near the compound eyes of an insect; an eyespot in many invertebrates.

ochre codon a triplet of mRNA nucleotides (UAA) usually not recognized by any tRNA molecules; one of three stop codons that normally signal termination of translation. *See* amber codon, opal codon.

ochre mutation one of a group of mutations resulting in abnormally short polypeptide chains. Because of a base substitution, a codon specifying an amino acid is converted to UAA, which signals chain termination. UAA appears to be the codon primarily used for chain termination in *E. coli. See* amber mutation, nonsense mutation.

ochre suppressor any mutant gene coding for a mutant tRNA whose anticodon can respond to the UAA stop codon by the insertion of an amino acid. *See* amber suppressor.

Ockham a variant spelling of *Occam. See* Occam's razor.

octad a fungal ascus containing eight linear ascospores; produced in some ascomycete species when the tetrad of meiospores undergoes a mitotic division following meiosis. *See* ordered tetrad.

octopine *See* opine.

ocular albinism a hereditary eye disease of humans that occurs in two forms, one inherited as an autosomal recessive and the other X-linked. The X-linked condition is the most common form of ocular albinism. In males, the prevalence of the disease is 1/50,000. The normal gene is at Xp22.3, and it encodes a protein that contains 424 amino acids. This is located in the membranes of melanosomes but is not a tyrosinase. Males show a reduced pigmentation of the retina (*q.v.*) and iris but not of the hair and skin. They are extremely sensitive to light and have reduced visual acuity. Patients with ocular albinism or oculocutaneous albinism have a misrouting of the optic tracts, which results in a loss of stereoscopic vision. In heterozygous females, retinas show a mosaic pattern of pigment distribution due to random inactivation of the X chromosomes during the early development of the eye. *See* albinism, dosage compensation.

OD optical density. *See* Beer-Lambert law.

OD_{260} unit one absorbance (OD_{260}) unit is that amount of material per ml of solution that produces an absorbance of 1 in a 1.0-centimeter light path at a wavelength of 260 nanometers. *See* absorbance.

odorant any one of a wide variety of molecules that produces an odor and that binds with an odorant receptor (*q.v.*) to trigger a cascade of signals that eventually allows the brain to recognize the particular odor. Usually, several different odorants combine to produce a single odor. Each type of odorant can bind to several different odorant receptors and each receptor to several different odorants (with varying affinity), thus forming the basis for the wide diversity in odors that can be detected by the brain. The term *odorant* usually precludes pheromones, which elicit endocrine or behavioral responses, and which are detected by two distinct families of odorant receptors specific to cells residing in a distinct region of the olfactory system. *See* pheromone.

odorant receptor a protein molecule that resides on the cell surface of an olfactory receptor neuron (*q.v.*) and which binds an odorant (*q.v.*). Odorant receptors are encoded by distinct families of odorant receptor genes, which encode G protein–coupled receptors (GPCRs) (*q.v.*). The binding of an odorant to an odorant receptor (*q.v.*) causes a structural change in the latter, which leads to the activation of the G protein (*q.v.*) attached to it. The G protein then causes the activation of different intracellular signaling events, which result in the production of an electrical impulse that is transmitted to specific regions of the brain via nerve processes. Here the information from different types of odorant receptors is sorted out, and specific odors are perceived by the organism. *Also called* olfactory receptor. *See* Appendix C, 1991, Buck and Axel; cellular signal transduction, G protein-coupled receptors, G proteins, odorant receptor gene.

odorant receptor gene any one of a family of genes expressed in an olfactory sensory organ and encoding an odorant receptor (*q.v.*). C. *elegans* and several mammalian species have large odorant receptor gene families, which encode approximately 1,000 different genes. This corresponds to approximately 1%–5% of the genes in the euchromatic (*q.v.*) genomes of these organisms. By comparison, *Drosophila* has a family of only 60 such genes (i.e., 0.5% of the total genes). In each organism odorant receptor genes encode a family of related proteins, which have in common the fact that they are G protein–coupled receptors (*q.v.*). Members of a family differ from one another in the extent of sequence similarity. Between organisms, odorant receptor gene families differ vastly in size and sequence composition. Perception of olfaction therefore involves gene families that meet the unique needs of each species. Two additional gene families, one with approximately 35 and another with 150 members, have been identified in some mammals, which are thought to encode pheromone receptors. *See* Appendix C, 1991, Buck and Axel.

Oenothera lamarckiana the evening primrose. During meiosis, plants of this and related species, such as O. *grandiflora*, have their chromosomes arranged in rings rather than pairs. The evolution of this atypical cytogenetic behavior, the result of the accumulation of reciprocal translocations, has been extensively studied. *See* Appendix A, Angiospermae, Dicotyledoneae, Myrtales, Appendix C, 1901, de Vries; 1930, Cleland and Blakeslee; Renner complex.

Ohno hypothesis the proposal advanced by S. Ohno that the unique regulatory features of the X chromosomes dictate the evolutionary conservation of the primordial X-linkage group among mammals. Any translocation between the X chromosome and an autosome would disturb the dosage compensation mechanism, and therefore offspring bearing such a translocation would be eliminated. Therefore, if any gene is found to be sex-linked in a given species, such as *Homo sapiens*, it is likely to be X-linked in all other mammals. *See* Appendix C, 1967, Ohno; dosage compensation.

oil-immersion objective the objective lens system used for highest resolution with the light microscope. The space between the coverslip over the object to be examined and the lens is filled with a drop of oil of the same refractive index as the glass.

Okazaki fragments *See* replication of DNA.

olfactory epithelium in mammals, the tissue located in the nasal cavity that contains olfactory receptor neurons (*q.v.*), which detect and transmit olfactory signals to the brain. In addition to these neurons, the olfactory epithelium also contains supporting cells and stem cells that divide regularly to replace olfactory neurons that die. The corresponding olfactory sensory organs in *Drosophila* are in the antenna and the maxillary palp, where the fly's olfactory receptor neurons are located.

olfactory receptor neurons nerve cells that reside in the olfactory sensory organs and are the primary cells for the detection and transduction of olfactory signals to the brain. In mammalian species, the dendrites of these neurons contain cilia (*q.v.*), on which reside odorant receptors. The binding of an odorant (*q.v.*) to a receptor causes intracellular biochemical changes which result in the generation of an electrical impulse that travels along the axon (*q.v.*) of the olfactory neuron to the olfactory bulb, where the signal is transferred to additional nerve cells for transport to the brain. In rodents and in *Drosophila*, each olfactory receptor neuron is highly specific, in that it expresses only one odorant receptor gene (*q.v.*), but multiple neurons collectively produce a pattern of neuronal activation, which is interpreted by the brain into distinct odors that are perceived by the organism. *Also called* olfactory sensory neuron. *See* anosmia, dendrite, odorant receptor.

Oligocene the third epoch in the Tertiary period. Old World monkeys and apes evolved. Further continental drift (*q.v.*) left South America separated from North America, and Australia separated from Antarctica to which it was fused previously. *See* geologic time divisions, Indrichotherium.

oligo dA (oligo dT) a homopolymer chain of deoxyriboadenylate (or deoxyribothymidylate) subunits of unspecified length, but generally 100–400 residues.

oligogene a gene producing a pronounced phenotypic effect as opposed to a polygene (*q.v.*), which has an individually small effect.

oligomer a molecule made up of a relatively few monomeric subunits.

oligonucleotide a linear sequence of up to 20 nucleotides joined by phosphodiester bonds. *See* allele-specific oligonucleotide testing, polynucleotide.

oligonucleotide-directed mutagenesis a technique that allows a specific mutation to be inserted in a gene at a selected site. An oligonucleotide sequence complementary to the segment of interest, but containing an alteration at a selected site, is chemically synthesized. Next this is hybridized to a complementary wild-type target gene contained in a single-stranded phage such as M13. The hybridized oligonucleotide fragment is then used as a primer by DNA polymerase I, which extends the molecule while taking instructions from the wild-type complementary strand. The result is a double helix containing a mutant and a wild-type strand. The heteroduplex is then used to transform bacterial cells. From these colonies, strains that contain the mutant homoduplexes can be recovered and propagated. This procedure is also called *site-specified mutagenesis*. *See* Appendix C, 1978, Hutchison *et al.*

oligonucleotide fingerprinting (OFP) any technique which produces a "fingerprint" consisting of a distinct oligonucleotide (*q.v.*) pattern representing nucleic acids from a particular source. In the simplest example, the genomes of different strains of an organism may be compared by enzymatic digestion of their genomic DNA (or RNA) to generate oligonucleotide fragments, which, when resolved on a gel by electrophoresis (*q.v.*), produce banding patterns representing fingerprints unique to each strain. A relatively more complex technique uses fingerprints generated by hybridization of oligonucleotides to cDNA (*q.v.*) or genomic libraries to characterize expressed genes at the genome-wide scale, to compare different cDNA libraries, and to select shotgun clones for sequencing. In this method, hundreds of labeled, synthetic oligonucleotides of known sequences, usually 6–10 bp in length, are hybridized to PCR-amplified cDNA or genomic library sequences that have been spotted on parallel DNA microarrays. Each oligonucleotide probe (*q.v.*) is used in a separate hybridization experiment. The extent of hybridization across microarray filters is recorded by a laser scanner and image analysis software. This produces a unique fingerprint of each arrayed DNA fragment, based on the extent of similarity to each oligonucleotide sequence. Using this approach, hundreds of thousands of individual library fragments can be comparatively examined. cDNAs with similar fingerprints are grouped into clusters, and this provides information about the number of expressed genes and their relative expression levels. Individual fingerprints are used for database searches for sequence matches to known genes or to identify new genes. DNA fragments having maximum dissimilarity in their fingerprints (i.e., minimum sequence overlap) are used for genomic sequencing with reduced redundancy. *See* DNA fingerprint technique, DNA microarray technology, genomic library, polymerase chain reaction.

oligopyrene sperm *See* sperm polymorphism.

oligosaccharide a polymer made up of a few (2–10) monosaccharide units. Oligosaccharides are attached to many secreted proteins, such as immunoglobulins and clotting factors. They are also found on the extracellular face of proteins that extend through cell membranes. The lipids of the red cell plasma membrane contain oligosaccharides that specify blood types. Such complex carbohydrates require a different enzyme for each step in their synthesis, and each product serves as the exclusive substrate for the next enzyme in the series. *See* A, B antigens, glycosylation.

oligospermia an abnormally low concentration of sperm in the semen.

OMIA On-line *M*endelian *I*nheritance in *A*nimals, a catalogue of animal species, especially domesticated ones where the molecular basis of genetic diseases has been studied (cat, cattle, chicken, dog, donkey, fox, goat, guinea pig, hamster, horse, llama, mink, mouse, pig, pigeon, rabbit, rat, Rhesus monkey, sheep, turkey, and zebra fish). The database lists such diseases as lysosomal storage diseases, inherited bleeding diseases, dwarfism, retinal defects, sex reversals, and muscular dystrophies. *See* Appendix E, Individual Databases.

OMIM On-line *M*endelian *I*nheritance in *M*an, an electronic catalog of inherited human diseases. The catalog has been available on-line since 1987. It is updated weekly and accessible through the Internet. *See* Appendix E, Individual Databases; human genetic diseases.

ommatidium one of the facets making up the compound eye of insects. The frontispiece illustration shows the right compound eye of a fruit fly. It is composed of a honey comb-like array of facets. An eye contains about 750 ommatidia, and each is made up of 8 photoreceptor cells and 11 accessory cells arranged in a precise three-dimensional pattern. There are 6 outer and 2 inner photoreceptor cells (the outer ones are labeled R1-R6 and the inner ones R7 and R8). Each photoreceptor cell contains a rhabdomere (*q.v.*) in which rhodopsin (*q.v.*) is stored. The rhabdomere functions like the discs in the outer segments of the photoreceptor cells of the vertebrate retina (*see* the illustration on page 385). Overlying the photocells is a quartet of cone cells. Primary pigment cells surround the cone cells, and secondary pigment cells lie between adjacent ommatidia. The ommochrome and drosopterin pigments are stored in the pigment cells. *See Drosophila* eye pigments, *eyeless, sevenless.*

ommochromes *See Drososphila* eye pigments.

omnipotent suppressors nonsense suppressors in yeast that are codon nonspecific, act only upon UAA and UAG mutations, and fall into two complementation groups. They are thought to be mutations of ribosomal components rather than suppressor mutations in tRNAs since these are codon specific.

oncogene a gene that induces uncontrolled cell proliferation. Some oncogenes were originally of cellular origin but now reside in the genomes of retroviruses (*q.v.*). Here they have acquired the ability to transform cells to a neoplastic state. The *v-src* gene of the Rous sarcoma virus (*q.v.*) and the *v-sis* gene of the simian sarcoma virus (*q.v.*) are examples. Oncogenes also have been isolated from tumors that have arisen spontaneously or have been induced by chemical carcinogens. Finally, there are oncogenes that reside in oncogenic viruses with DNA genomes. The polyoma virus (*q.v.*) and simian virus 40 (*q.v.*) are examples. Viral and cellular oncogenes arise from cellular proto-oncogenes (*q.v.*), which play a role in the control of normal cell proliferation. *See* Appendix C, 1981, Parker *et al.*; 1982, Reddy *et al.*; Appendix E; *myc,* oncogenic virus, oncomouse, Ti plasmid, T24 oncogene.

oncogene hypothesis a proposal that carcinogens of many sorts act by inducing the expression of retrovirus genes already resident in the target cell. It is now known that while cells from different species harbor genes homologous to retrovirus oncogenes, the cellular genes were the progenitors of the viral oncogenes. The cellular genes are now called *proto-oncogenes* (*q.v.*) and they evidently function in the normal physiology of cells from evolutionarily diverse species. *See* Appendix C, 1969, Huebner and Todaro.

oncogenic virus a virus that can transform the cells it infects so that they proliferate in an uncontrolled fashion. *See* Appendix C, 1910, Rous; 1981, Parker *et al.*; 1983, Doolittle *et al.*; Abelson mouse leukemia virus, Friend leukemia virus, Gross mouse leukemia virus, Harvey rat sarcoma virus, human papillomavirus, Moloney leukemia virus, mouse mammary tumor virus, polyoma virus, Rauscher leukemia virus, retroviruses, Rous sarcoma virus, Shope papilloma virus, simian sarcoma virus, simian virus 40, transformation.

oncolytic capable of destroying cancer cells.

oncomouse a laboratory mouse carrying activated human cancer genes. Du Pont started selling oncomice late in 1988. They were the first transgenic animals to be patented. These mice carry the *ras* oncogene plus a mouse mammary tumor virus promoter. This ensures that the oncogene is activated in breast tissue, and the mice develop breast cancer a few months after birth. *See* Appendix C, 1988, Leder and Stewart.

oncornavirus an acronym for *oncogenic RNA virus*. *See* retrovirus.

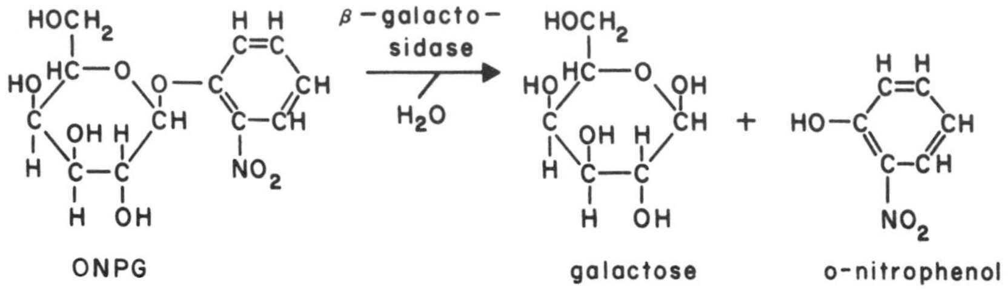

ONPG + H_2O —β-galactosidase→ galactose + o-nitrophenol

ONPG

one gene–one enzyme hypothesis the hypothesis that a large class of genes exists in which each gene controls the synthesis or activity of but a single enzyme. *See* Appendix C, 1941, Beadle and Tatum; 1948, Mitchell and Lein.

one gene–one polypeptide hypothesis the hypothesis that a large class of genes exists in which each gene controls the synthesis of a single polypeptide. The polypeptide may function independently or as a subunit of a more complex protein. This hypothesis replaced the earlier one gene–one enzyme hypothesis once heteropolymeric enzymes were discovered. For example, hexosaminidase (*q.v.*) is encoded by two genes. *See* two genes–one polypeptide chain.

one-step growth experiment the classic procedure that laid the foundation for the quantitative study of the life cycle of lytic bacterial viruses. A suspension of bacteria was mixed with enough viruses to ensure that a virus attached to each host cell. Free viruses were removed, and at periodic intervals thereafter aliquots were withdrawn and subjected to plaque assay (*q.v.*). The number of plaques per aliquot remained constant for an initial period of time. Aliquots taken after this *latent* period showed a progressive increase in plaque numbers. During this time, infected cells were lysing and liberating infectious phage, each capable of producing a plaque. Once all cells had lysed, a plateau was reached, and so the curve describing plaque counts during the experiment showed a single step. The *eclipse* period refers to the time between viral attachment and the assembly of the first progeny phage. It is during this period that replication and assembly of the phages is occurring. Cells must be artificially lysed to determine when the earliest infectious particles appear. The latent period is longer than the eclipse period because the host cell does not normally lyse until many progeny have been assembled. *See* Appendix C, 1939, Ellis and Delbrück; burst size, plaque.

ONPG *o*-nitrophenyl galactoside, an unnatural substrate for beta galactosidase. It is cleaved by this enzyme (see illustration) into galactose and *o*-nitrophenol (a yellow compound, easily assayed spectrophotometrically). ONPG has been extensively used to determine enzyme activity associated with mutants of the *lac* operon (*q.v.*) in *E. coli*. Unlike IPTG (*q.v.*), ONPG is not an inducer of the operon, so these two substances are often used in combination.

ontogeny the development of the individual from fertilization to maturity.

Onychophora a phylum of about 70 species that are all topical or subtropical in distribution. They are commonly called peripatus or velvet worms. They are terrestrial and have between 14 and 43 pairs of unsegmented walking legs. Earlier forms were marine, and their fossils are found in rocks dating back to the Cambrian. Peripatus is sometimes called a living fossil (*q.v.*), and it shows a mixture of annelid and arthropod characters. Since it molts, it is placed in the Ecdysozoa (*q.v.*).

oocyte the cell that upon undergoing meiosis forms the ovum.

oogenesis the developmental process that results in the formation of the egg. Oogenesis involves a sequence of events, including mitotic proliferation of oogonial cells, meiotic divisions in the oocyte, vitellogenesis (*q.v.*) and oocyte growth, synthesis and localization of maternal products in the oocyte, specification of egg polarity, and formation of egg membranes. Most of these events entail interactions between the germ line (*q.v.*) and the surrounding soma (*q.v.*). *See* insect ovary types.

oogonium **1.** the female gametangium of algae and fungi. *Contrast with* antheridium. **2.** in animals, a mitotically active germ cell that serves as a source of oocytes. The stem cell shown on page 98 is an oogonium.

ookinete *See Plasmodium* life cycle.

oolemma the plasma membrane of the ovum.

ooplasm the cytoplasm of an oocyte.

ootid nucleus one of the four haploid nuclei formed by the meiotic divisions of a primary oocyte. Three of the nuclei are discarded as polar nuclei and the remaining one functions as the female pronucleus. *See* oriented meiotic division, polar body.

opal codon the mRNA stop codon UGA. *See* amber codon, ochre codon.

opaque-2 a mutant strain of corn that produces an increase in the lysine content of seeds. This was the first mutation shown to improve the amino acid balance in the proteins of an agriculturally important plant. Animal proteins, such as those in milk and beef, have a better balance of certain essential amino acids (like tryptophan and lysine) than do plant proteins. Mutants like *opaque 2* are of potential use in combating kwashiorkor (*q.v.*). *See* Appendix C, 1964, Mertz *et al.*, zein.

open population a population that is freely exposed to gene flow (*q.v.*).

open reading frame *See* reading frame.

operational definition a definition in terms of properties significant to a given experimental situation, without consideration of the more fundamental characteristics of the defined subject.

operator a chromosomal region capable of interacting with specific repressors, thereby controlling the function of adjacent cistrons. *See lac* operon, regulator gene.

operon a unit that consists of one or more cistrons that function coordinately under the control of an operator. The genome of the *E. coli* strain sequenced in 1997 contained about 2,200 operons. Of these, 73% had only one gene, 17% had two, 5% had three, and the rest had four or more. *See* Appendix C, 1961, Jacob and Monod; 1997, Blattner *et al.*; regulator gene.

operon network a collection of operons and their associated regulator genes that interact in the sense that the products of structural genes in one operon serve to suppress or activate another operon by acting as repressors or effectors.

opine a compound, specifically synthesized by crown gall plant cells, that can be used by agrobacteria as specific growth substances. Examples are nopaline [*N*-α-(1,3-dicarboxylpropyl)-L-arginine] and octopine [*N*-α-(D-1-carboxyethyl)-L-arginine]. *See Agrobacterium tumefaciens.*

opisthe the posterior daughter organism produced in a transverse division of a protozoan.

opisthokonta a monophyletic supergroup that contains animals and fungi. The conclusion that the Fungi are a sister group to Animalia and that fungi and plants belong to independent lineages is based on sequence data from SSU rRNAs and certain ubiquitous proteins. *See* Appendix A, Kingdoms 3 and 4; Appendix C, 1993, Baldauf and Palmer; 16S rRNA, translation elongation factors.

opportunism a theory that (1) all potential modes of existence will eventually be tried by some group and all potential niches will eventually become occupied, and (2) organisms evolve only as historical conditions permit and not according to what would theoretically be best.

opportunistic species a species specialized to exploit newly opened habitats because of its ability to disperse for long distances and to reproduce rapidly.

opsin the protein portion of a photosensitive molecule contained in the discs of the photoreceptors of the retina (*q.v.*). An opsin (see page 312) is a chain of amino acids, running from the amino-terminal end (N), exposed on the external aqueous surface of the disc, to a carboxyl terminal region (C), exposed to the internal aqueous surface of the disc. The chain has seven alpha helices that span the membrane. An opsin does not itself absorb light. Retinal (*q.v.*) is the chromophore that lies within the cluster of helices and undergoes a change in shape upon receiving a photon of light. *See* multiple transmembrane domain proteins.

opsonin any substance that promotes cellular phagocytosis. When antibodies bind to antigens by their Fab portions (*see* immunoglobulin), the shape of the molecule changes to expose the Fc region. Scavenger cells such as macrophages have Fc receptors on their surfaces. Thus, phagocytic cells can bind to and engulf antigen-antibody complexes. Neutrophils and macrophages have receptors for certain activated complement components. Thus, antigen-antibody-complement complexes also enhance phagocytosis through immune adherence. IgG antibodies are much more effective opsonins than IgM in the absence of complement, but IgM antibodies are more effective opsonins in the presence of complement.

optical antipodes enantiomers (*q.v.*).

optical density *See* Beer-Lambert law.

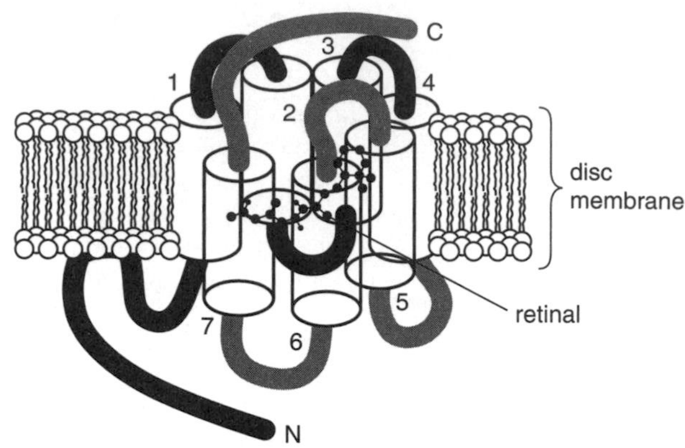

Opsin

optical isomers molecular isomers that in solution cause the rotation of the plane of a beam of plane-polanzed light passed through the solution. The rotation is due to the asymmetry of the molecule. Molecules with this property are given the prefix D or L depending on whether the plane is rotated to the right (dextro) or to the left (levo, laevo).

orange G an acidic dye often used in cytochemistry.

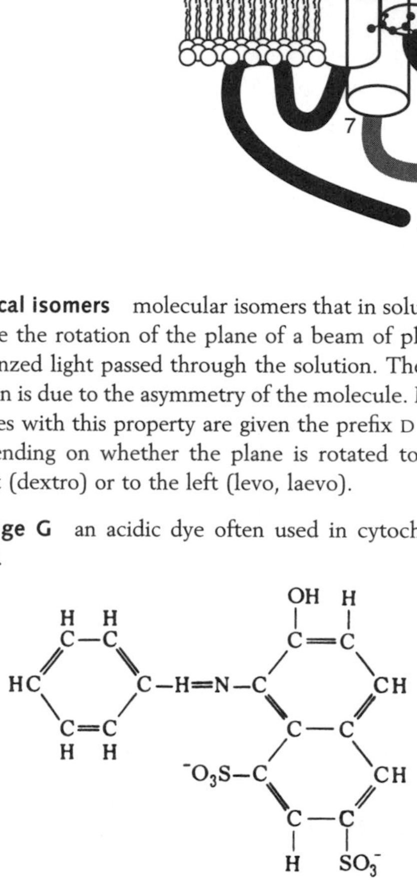

orangutan *See Pongo pygmaeus.*

orcein a dye used in cytology. *See* aceto-orcein.

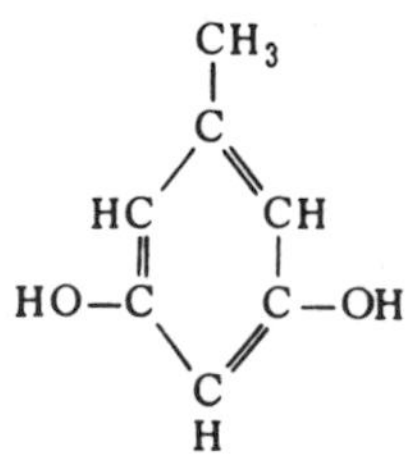

Orcein

ordered octad *See* ordered tetrad.

ordered tetrad a linear sequence of four haploid meiotic cells (or pairs of each of four haploid cells produced by a postmeiotic division) within a fungal ascus. This physical arrangement allows identification of chromatids participating in crossover events. Drawing A (page 313) illustrates that, in a tetrad heterozygous for alleles controlling ascospore pigmentation, single crossovers between these genes and the centromere will generate spores showing 2-2-2-2 and 2-4-2 segregation patterns. Drawing B illustrates that in *Neurospora* such patterns are observed, together with noncrossover asci showing 4-4 distributions.

Ordovician a period in the Paleozoic era during which marine invertebrates diversified. Brachiopods were the dominant species. The Cambrian genera of trilobites (*q.v.*) were replaced by new forms. The echinoderms bloomed with starfish, brittle stars, echinoids, and crinoids making their first appearances. Corals are found for the first time early in the Ordovician. Jawless fishes appeared and represent the first vertebrates. The Ordovician ended with a mass extinction during which the trilobites lost 50% of all families. *See* geologic time divisions.

ORF the symbol for *o*pen *r*eading *f*rame *See* reading frame, URF.

organ culture the maintenance or growth of organ primordia or the whole or parts of an organ *in vitro* in a way that may allow further differentiation or the preservation of architecture or function or both. *See also in vivo* culturing of imaginal discs.

organelle any complex structure that forms a component of cells and performs a characteristic function. Extensively studied organelles include: centrioles, chloroplasts, the endoplasmic reticulum, Golgi material, kinetosomes, lysosomes, microbodies, mitochondria, peroxisomes, proteosomes, quantasomes, ribosomes, and spindles (*all of which see*).

A

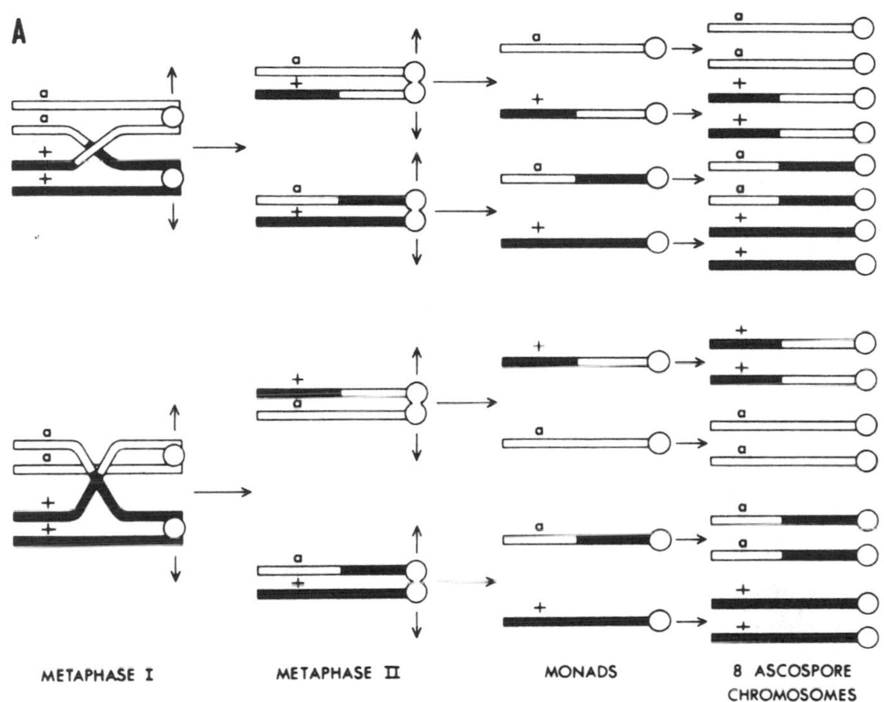

B

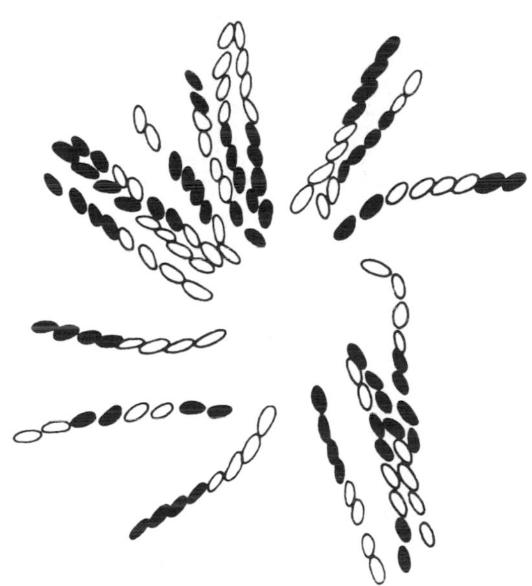

Ordered tetrad

organic 1. pertaining to organisms (dead or alive) or to the chemicals made by them. 2. chemical compounds based on carbon chains or rings. They may also contain oxygen, hydrogen, nitrogen, and various other elements.

organizer a living part of an embryo that exerts a morphogenetic stimulus upon another part, bringing about its determination and morphological differentiation. *See* *goosecoid*, Spemann-Mangold organizer.

organogenesis the formation of organs.

Oriental designating one of the six biogeographic realms (*q.v.*) of the globe, including the southern coast of Asia east of the Persian Gulf, the peninsula of India south of the Himalayas, eastern India, south China, Sumatra, Kalimantaro (Borneo), Java, Sulawesi (Celebes), and the Philippines. *See* Wallace's line.

oriented meiotic division an oocyte meiotic division, as in *Drosophila* where the spindles are oriented in single file with their long axes perpendicular to the egg surface. The nucleus farthest from the surface functions as the oocyte pronucleus. Aberrant chromosomes that are differentially distributed to the other nuclei are eliminated.

origin of replication *See* replication origin.

Origin of Species an abbreviated name for the most famous book by Charles Darwin that documented the phenomenon of evolution and elaborated a theory to explain its mechanism. The full title of the book was *On the Origin of Species by Means of Natural Selection, or the Preservation of Favoured Races in the Struggle for Life*. The first edition was published in 1859, and no biological treatise written before or since has produced an impact upon society equal to it. The 1,250 copies of the first edition were sold out the first day. *The Origin* went through six editions, the last in 1872.

***ori* site** a 422 base-pair segment of the *E. coli* chromosome where replication is initiated. *See* replicon.

ornithine cycle a cyclic series of reactions in which potentially toxic, nitrogenous products from protein catabolism are converted to urea that is innocuous. In the cycle diagrammed on page 315, ammonia is removed from the system and used in the conversion of ornithine to citrulline. Aspartic acid enters the cycle, and its amino group is incorporated into arginosuccinic acid before it can form ammonia. Arginosuccinic acid is converted to arginine, and the fumaric acid released enters the citric acid cycle (*q.v.*). Urea splits off arginine and regenerates ornithine. In humans, mutants are known that block the cycle at any one of its steps, as shown in the diagram. Blocking produces disorders that include: ornithine transcarbamylase deficiency, from blocking of step 1; citrullinuria (condensing enzyme deficiency), step 2; arginosuccinic aciduria (arginosuccinase deficiency), step 3; lysine intolerance (inhibition of arginase by excess lysine), step 4.

orphan drugs pharmaceuticals developed to treat diseases that afflict relatively few people.

orphans a name coined for previously undiscovered protein-coding ORFs, revealed by genome sequencing, that have no clear-cut homologs in any organisms. For example, 30% of the ORFs in *Saccharomyces cerevisiae* are orphans. *Orphans* and *URFs* are synonyms.

orphan viruses viruses found in the digestive and respiratory tracts of healthy people; hence they are nonpathogenic (orphan = without an associated disease). *See* reovirus.

orphons dispersed, single pseudogenes (*q.v.*) derived from tandemly repeated families or gene clusters, such as those for histones or hemoglohins. Orphons may serve as a reservoir of sequences that can evolve new functions, and have probably been important factors in the evolution of higher organisms. *See* hemoglobin genes.

ortet the single ancestral organism that produced a clone of genetically identical organisms (ramets) by budding. *See* modular organisms, ramets.

orthochromatic dye a dye that stains tissues a single color in contrast to a metachromatic dye (*q.v.*).

orthogenesis the concept of unidirectional change during the evolution of a group of related organisms. For example, the fossil record of the horse family (Equidae) shows a tendency toward an increase in the size of adults when more recent species are compared with ancestral ones. Trends of this sort were used in the past as evidence that evolution was driven toward a desired end by mystical forces. A diagram of orthogenic evolution through time shows a straight line with no side branches, since the ancestor evolves into a new species with no temporal overlap of ancestors and descendants. More detailed studies showed subsequently that the horse evolutionary tree contains dozens of side branches and that many new species coexisted in time with their immediate ancestors. *Contrast with* cladogenesis. *See* Appendix C, 1951, Simpson; *Hyracotherium*.

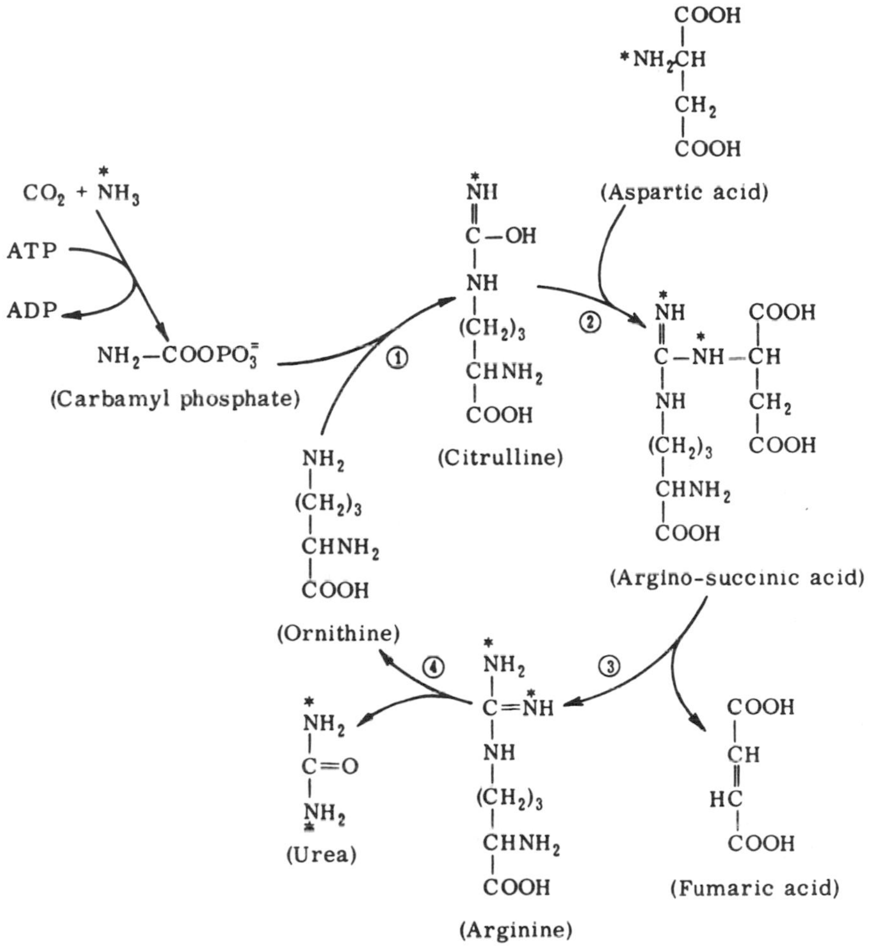

Ornithine cycle

orthologs genes or proteins found in different species that are so similar in their nucleotide or amino acid sequences that they are assumed to have originated from a single ancestral gene. The beta globin chain genes in humans and chimpanzees would be examples of orthologs. If one compares the genome of *Saccaromyces cerevisiae* (*q.v.*) and *Caenorhabditis elegans* (*q.v.*), most orthologs have "core functions." That is, they generate the proteins used in intermediary metabolism, DNA-, RNA-, and protein-metabolism, transport, secretion, and cytoskeletal structures. In contrast, the genes from C. *elegans* that function in intercellular signaling and gene regulation are not found in the yeast genome. *See* Appendix C, 1975, King and Wilson; hemoglobin genes, *Hox* genes, *Pan; contrast with* paralogs.

orthopteran an organism belonging to the Orthoptera, an order of the Hemimetabola containing cockroaches, locusts, grasshoppers, and similar insects.

orthoselection continuous selection on the members of a lineage over a long time, causing continued evolution in a given direction that may create an impression of "momentum" or "inertia" in evolutionary trends.

orthotopic transplantation the transplanation of grafts between identical sites in such a way that the graft maintains its normal orientation.

Oryctolagus cuniculus the rabbit, a mammal commonly reared in the laboratory and the subject of intensive genetic research. An extensive collection of mutations is available influencing a wide variety of morphological and physiological traits. The haploid chromosome number is 22, and about 60 genes have been distributed among 16 linkage groups. *See* Appendix A, Chordata, Mammalia, Lagomorpha; WHHL rabbit.

Oryza sativa rice. Together with wheat, corn, and potatoes, it is one of the world's four most important crops. The haploid chromosome number is 12. The rice genome (389 mbp) is the smallest of the cultivated species. For comparison, *Triticum aestivum* is 16 gbp. *See* Appendix A, Plantae, Angiospermae, Monocotyledonae, Graminales; Appendix E, Species Web Sites; Appendix F; helitron.

Oryzias latipes the medaka, or killifish, a freshwater fish common to Japan, Korea, and China. It is easily maintained in the laboratory and was the first fish in which Mendel's laws were shown to be valid. Y-linked inheritance was first demonstrated in the medaka and the guppy. N = 24; C = 720 mbp. *See* Appendix A, Chordata, Osteichythes, Neopterygii, Cypriniformes.

osmium tetroxide OsO_4, a compound often used as a fixative in electron microscopy.

osmosis diffusion of a solvent through a semipermeable membrane separating two solutions of unequal solute concentrations. The direction of solvent flow tends to equalize the solute concentrations.

otu a sex-linked female sterile gene in *Drosophila melanogaster* that is remarkable because different mutant alleles can produce quite different ovarian pathologies. One class of alleles produces ovarian tumors; hence the *otu* symbol. The tumors are composed of hundreds of single cells and clusters of two or three cells joined by ring canals (*q.v.*). These abnormalities preasumably arise from defective fusomes (*q.v.*). Another class of mutant alleles is characterized by germaria that either lack germ cells or contain germ cells that have undergone only one or two cell divisions. Mutants belonging to the third class can produce egg chambers, but the transport of nurse cell cytoplasm to the oocyte is inhibited. The nuclei of these abnormal nurse cells contain giant polytene chromosomes, and the largest have 8,000 times the haploid amount of DNA. Studies of the *otu* gene product have shown that this protein belongs to a highly conserved superfamily of cysteine proteases, many of which function to break down ubiquitin (*q.v.*). *See* cystocyte divisions, polyfusome, ubiquitin proteasome pathway.

Ouchterlony technique a gel-diffusion, antibody-antigen precipitation test that depends on horizontal diffusion from two or more opposite sources. An agar slab is prepared and two or more wells are cut in it. One (A) is filled with an aqueous suspension of antibody molecules, while each of the other wells (B and C in the illustration here) is filled with a different antigen preparation. The antigen and antibody molecules diffuse toward each other and eventually interact, forming curved precipitation lines. In example I, the antigens in wells B and C are different. In II, well B contains a single antigen and well C two antigens, one identical to that in B. *See* Appendix C, 1948, Ouchterlony.

Oudin technique a gel-diffusion, antibody-antigen precipitation test that depends on simple vertical diffusion in one dimension. A gel column is prepared containing a homogeneous distribution of antibody molecules. Above this is layered an aqueous suspension of antigen molecules. As these diffuse into the gel, a moving zone of antigen-antibody precipitate is formed. If several antigens and antibodies are present, separate zones of interaction will be seen. *See* Appendix C, 1946, Oudin.

outbreeding the crossing of genetically unrelated plants or animals; crossbreeding.

outcross *See* outbreeding.

outgroup a species or higher monophyletic taxon that is examined in the course of a phylogenetic study to determine which of two homologous character states may be inferred to be apomorphic. The most critical outgroup comparison involves the sister group of the taxon under study. *Compare with* sister group.

outlaw gene a gene favored by selection despite its disharmoneous effects on other genes in the same organism. *See* meiotic drive.

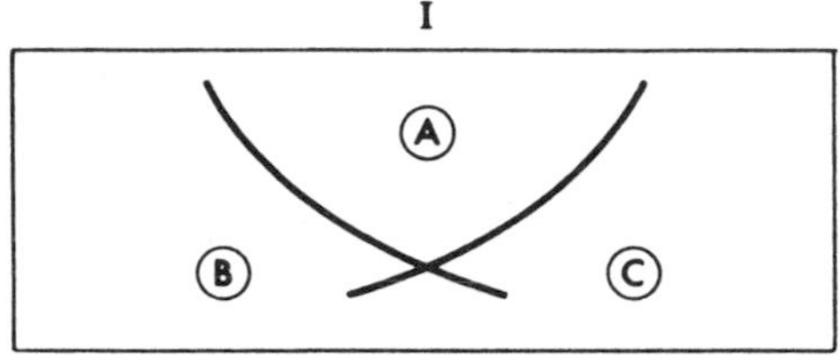

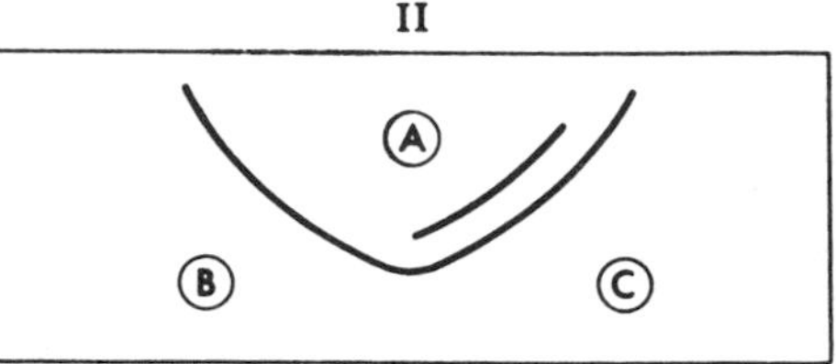

Ouchterlony technique

outron *See* trans-splicing.

ovariectomy the surgical removal of one or both ovaries.

ovariole one of several egg tubes constituting the ovary of most insects.

ovary the female gonad in animals or the ovule-containing region of the pistil of a flower.

overdominance the phenomenon of heterozygotes having a more extreme phenotype than either homozygote; monohybrid heterosis. Overdominance generally refers to the situation in which *AA′* individuals are more fit than *AA* or *A′A′* individuals.

overlapping code a hypothetical genetic code, first proposed by George Gamow, in which any given nucleotide is shared by two adjacent codons. The genetic code used in biological systems was later shown to be nonoverlapping. *See* overlapping genes for the few exceptions.

overlapping genes genes whose nucleotide sequences overlap to some degree. The overlap may involve regulatory sequences (e.g., tryptophan operator and promoter regions in *E. coli*) or structural genes (e.g., in bacteriophage phiX174, gene *E* lies entirely within gene *D*, but they are translated in different reading frames). *See* Appendix C, 1976, Burrell *et al.*; bidirectional genes.

overlapping inversion a compound chromosomal inversion caused by a second inversion that includes part of a previously inverted segment.

overwinding positive supercoiling of DNA, resulting in further tension in the direction of winding of the two strands of a duplex about each other.

ovicide any compound that destroys eggs, especially a compound that destroys insect eggs.

oviduct the tube carrying eggs from the ovary to the uterus.

ovine referring to members of the sheep family, especially the domestic sheep species *Ovis aries*.

oviparous laying eggs in which the embryo develops outside the mother's body and eventually hatches. *See* ovoviviparous, viviparous.

oviposition the laying of eggs by a female insect.

ovipositor an organ at the hind end of the abdomen in female insects, through which eggs are laid.

Ovis aries domestic sheep. The blood group genetics of this species has been intensively investigated. The haploid chromosome number is 27, and about 250 genes have been mapped. The haploid C value is 3.2×10^9 bp of DNA. *See* Appendix A, Chordata, Mammalia, Artiodactyla; Appendix F; sheep.

ovisorption the resorption of oocytes.

ovogenesis oogenesis (*q.v.*).

ovotestis the organ of some hermaphroditic animals that functions both as ovary and testis; the gonad of an animal that undergoes consecutive sexuality (*q.v.*).

ovoviviparous bringing forth young that develop from eggs retained within the maternal body, but separated from it by the egg membranes. Many fish, reptiles, molluscs, and insects are ovoviviparous. *See* oviparous, viviparous.

ovulation the release of a ripe egg from the mammalian ovarian follicle, frequently at the stimulus of a pituitary hormone.

ovule the structure found in seed plants which develops into a seed after the fertilization of an egg cell within it.

ovum an unfertilized egg cell.

oxidation classically defined as the combination of a molecule with oxygen or the removal of hydrogen from it. Since electrons are transferred to the oxidizing reagent, which becomes reduced, oxidation and reduction (*q.v.*) are always coupled.

oxidation-reduction reactions chemical reactions in which electrons are transferred from a reductant to an oxidant; as a consequence of the transfer, the reductant is oxidized and the oxidant is reduced.

oxidative phosphorylation the enzymatic phosphorylation of ADP to ATP, which is coupled to the electron transport chain (*q.v.*). Thus, respiratory energy is transformed into phosphate-bond energy. *See* adenosine phosphate, ATP synthase, chemiosmotic theory, mitochondrial proton transport.

oxidoreductases enzymes that transfer electrons. Catalase (*q.v.*) is an oxidoreductase.

oxygen the second most abundant of the biologically important elements. Atomic number 8; atomic weight 15.9994; valence 2^-; most common isotope ^{16}O.

oxyhemoglobin oxygenated hemoglobin.

oxytocin a polypeptide secreted by the hypothalamus and stored in the neurohypophysis. Oxytocin causes smooth-muscle contraction and may help terminate pregnancy.

Oxytricha *See* Stylonichia.

P

p 1. the smaller arm of a human chromosome. *See* symbols used in human cytogenetics. **2.** a symbol for protein; if followed by a number, that number indicates the relative molecular mass of the protein in kilodaltons. *See* p53. **3.** a symbol for plasmid; for example, pBR322 (*q.v.*).

P 1. probability. **2.** phosphorus. **3.** phosphate (when combined in an abbreviation such as ADP, ATP). **4.** symbol for panmictic index (*q.v.*).

P_1 the symbol denoting the immediate parents of the F_1 generation. The symbols P_2 and P_3 are used to designate the grandparental and great-grandparental generations, respectively, if one starts from the F_1 and works backward. This genetic symbolism was invented by W. Bateson. *See* Appendix C, 1900, Bateson.

^{32}P a radioactive isotope of phosphorus, widely used to label nucleic acids; it emits a strong beta particle and has a half-life of 14.3 days. ^{32}P was used in the famous Hershey-Chase experiment of 1952. *See* Appendix C.

P_i inorganic phosphate.

p34 (CDC2) a protein of relative molecular mass 34,000, hence the p34 designation. It functions in cell division control, hence the CDC acronym. This protein belongs to the Ser/Thr family of protein kinases (*q.v.*). CDC2 contains 297 amino acids, and phosphorylation at Thr-161 activates it, whereas phosphorylation at Thr-14 or Tyr-15 inactivates it. When combined with cyclin B, it forms the mitosis-promoting factor (MPF) (*q.v.*). For this reason CDC2 is sometimes called CDK2 (cyclin-dependant kinase-2). CDK2 is responsible for the separation of centrioles, the first step in centrosome duplication. In humans the gene for CDC2 contains 15,688 bases and has been mapped to 10q21.1. Homologous genes have been found in the mouse, *Drosophila*, *Xenopus*, *Caenorhabditis*, and *Schizosaccharomyces*.

p53 a protein of relative molecular mass 53,000, hence the designation p53. In 1993 *Science* magazine designated p53 as the "Molecule of the Year," and it has been referred to as the "cellular gatekeeper for growth and division." The protein is expressed in most normal tissues and has been highly conserved during evolution. The p53 protein normally plays a role in controlling the entry of cells into the S phase of the cell cycle, in transcriptional regulation, and in the induction of apoptosis (*q.v.*). The protein binds to DNA as a tetramer and activates the expression of adjacent genes that inhibit cell proliferation. A mutation in one allele will reduce the concentration of functionally active tetramers. If both alleles are affected, the inhibition will be reduced further. The gene that encodes the p53 protein is symbolized *TP53* (*q.v.*).

pachynema *See* meiosis.

pachytene stage *See* meiosis.

packing ratio the ratio of DNA length to the unit length of the fiber containing it.

paclitaxel a drug extracted from the bark of the Pacific yew and sold under the tradename Taxol by Bristol-Meyers Squibb. Paclitaxel binds to the tubulin of spindles and prevents them from breaking down after the chromosomes have separated. If spindles remain in place, the cell cannot complete cytokinesis. Taxol is the most broadly effective and commercially successful anticancer drug so far introduced. It is used primarily for treatment of cancers of the breast, ovary, and the most common type of lung cancer. *See* spindle poison.

PACs P1 artificial chromosomes (*q.v.*).

paedogenesis a type of precocious sexual maturity occurring in the larval stage of some animals. The eggs of paedogenetic females develop parthenogenetically.

paedomorphosis an evolutionary phenomenon in which adult descendants resemble the youthful stages of their ancestors; the opposite of "recapitulation," in which the early stages of descendants resemble the adult stages of their ancestors. Paedomorphosis may be produced either by the acceleration of sexual development (progenesis) or by retarded somatic development. *See* heterochrony.

Paeonia californica the California peony, a species with naturally occurring translocation complexes.

PAGE polyacrylamide gel electrophoresis. *See* electrophoresis, polyacrylamide gel.

pair bonding an intimate and long-lasting association between male and female animals of the same species, generally facilitating the cooperative rearing of their offspring.

paired a recessive lethal mutation on the left arm of chromosome 2 of *Drosophila melanogaster*. The *prd* gene belongs to the pair rule class of zygotic segmentation mutants (*q.v.*). It encodes a protein that contains DNA-binding segments. One, which is composed of amino acids 27 through 154, has been designated the "paired domain." A homeodomain comprises amino acids 213 through 272. Near the C terminus of the PRD protein is a region rich in histidines and prolines. This "PRD repeat" has been found in other *Drosophila* genes expressed in early embryogenesis. Genes that play an important role in embryogenesis of mice and men encode proteins with paired domains, and therefore these genes have been placed in the paired box (*Pax*) family. *See* Appendix C, 1986, Noll *et al.*; *Aniridia*, developmental control genes, DNA-binding motifs, *eyeless*, gene networking, homeobox, *Pax* genes, *Small eye*.

pairing synapsis.

pairing segments the segments of the X and Y chromosomes which synapse and cross over. The remaining segments, which do not synapse, are called the differential segments.

pair rule genes *See* zygotic segmentation mutants.

Palearctic designating one of the six biogeographic realms (*q.v.*) of the globe, including Eurasia except Iran, Afghanistan, the Himalayas, and the Nan-ling Range in China, Africa north of the Sahara, Iceland, Spitzbergen, and the islands north of Siberia.

Paleocene the most ancient of the Tertiary epochs. The placental mammals expanded at the expense of the marsupials. Early primates and grasses made their first appearance. The drifting of continents continued. *See* geological time divisions.

Paleogene a subdivision of the Tertiary period, incorporating the Oligocene, Eocene, and Paleocene epochs. *See* geologic time divisions.

Paleolithic that phase of human history prior to the cultivation of plants during which tools were manufactured, food was obtained by hunting, fishing, or collecting wild nuts and fruits. The Paleolithic culture lasted from about 500,000 years ago up to the beginning of the Neolithic stage about 10,000 years ago.

paleontology the study of extinct forms of life through their fossils, as opposed to neontology (*q.v.*).

paleospecies the successive species in a phyletic lineage that are given ancestor and descendant status according to the geological strata in which they appear. *See* anagenesis.

Paleozoic the earliest era of the Phanerozoic eon. Invertebrates flourished during this 320-million year interval. *See* geologic time divisions.

Paley's watch an argument developed by William Paley (1743–1805) for the existence of God based upon the commonsense notion that a watch is too complicated to have originated by accident; it presents its own evidence of having been purposely designed. This argument is commonly applied to living organisms by creationists.

palindrome a sequence of deoxyribonucleotide base pairs that reads the same (5′ to 3′) on complementary strands; tandem inverted repeats; example:

5′ AATGCGCATT 3′
3′ TTACGCGTAA 5′

Palindromes serve as recognition sites for restriction endonucleases, RNA polymerases, and other enzymes. In the human Y chromosome (*q.v.*) some palindromes are as long as 3 mb (*q.v.*). Gene conversions (*q.v.*) within these palindromes overwrite mutations that would lead to male sterility. *See* cruciform structure.

palynology the study of both living and fossil spores, pollen grains, and other microscopic propagules.

Pan the genus containing *P. troglodytes*, the common chimpanzee, and *P. paniscus*, the bonobo or pygmy chimpanzee. *Pan troglodytes* is the living primate genetically closest to man. *Pan* and *Homo* diverged from a common ancestor 5–6 million years ago. The haploid chromosome number for *P. troglodytes* is 24, and about 40 genes have been distributed among 19 linkage groups. *See* Appendix C, 1967, Sarich and Wilson; 1975, King and Wilson; 1984, Sibley and Alquist; 1988, Kazazian *et al.*; 2003, Anzai *et al.*; Appendix E; sequence similarity estimates.

Panagrellus redivivus a free-living nematode whose developmental genetics is under study, primarily for comparison with *Caenorhabditis elegans* (*q.v.*). It differs from *Caenorhabditis* in having XX females and XO males. Its cell lineage (*q.v.*) is known in part.

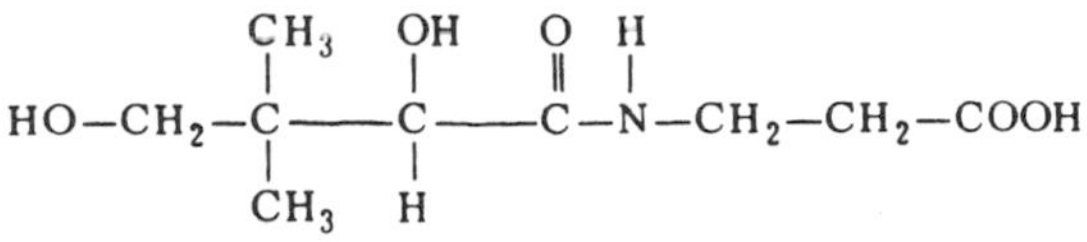

$$HO{-}CH_2{-}\underset{CH_3}{\overset{CH_3}{C}}{-}\underset{H}{\overset{OH}{C}}{-}\overset{O}{\overset{\|}{C}}{-}\overset{H}{N}{-}CH_2{-}CH_2{-}COOH$$

Pantothenic acid

pancreozymin a hormone secreted by the duodenum that causes secretion of pancreatic enzymes.

pandemic designating a disease simultaneously epidemic in human populations in many parts of the world.

panethnic referring to a hereditary disease that is found in a variety of ethnic groups.

pangenesis a defunct theory of development, popular in Darwin's time, proposing that small particles (pangenes) from various parts of the body distill into the gametes, resulting in a blending of parental characteristics in their offspring.

panhypopituitarism *See* pituitary dwarfism.

panicle *See* raceme.

panmictic index (P) a measure of the relative heterozygosity. $1 - P = F$, Wright's inbreeding coefficient (*q.v.*).

panmictic unit a local population in which mating is completely random.

panmixis random mating, as contrasted with assortative mating (*q.v.*).

panoistic ovary *See* insect ovary types.

pantothenic acid a water-soluble vitamin that functions as a subunit of coenzyme A (*q.v.*).

papain a proteolytic enzyme isolated from the latex of the papaya plant.

paper chromatography *See* chromatography.

Papilio glaucus the swallowtail butterfly, a species extensively studied in terms of isolating mechanisms (*q.v.*).

papillary pattern the pattern of dermal ridges on the fingertips and palms.

papilloma a benign cutaneous neoplasm; a wart. *See* human papilloma virus, Shope papilloma virus.

papovavirus a group of animal DNA viruses (including SV4O and polyoma) responsible for papillomas of the rabbit, dog, cattle, horse, and human.

parabiotic twins artificial "Siamese twins" produced by joining two animals surgically. Since their blood circulations will eventually anastomose, one can study the transmission of humoral agents from one "twin" to the other.

paracentric inversion an inversion (*q.v.*) that does not include the centromere.

Paracentrotus lividus a common sea urchin extensively used in studies of molecular developmental genetics. *See* echinoderm.

paracrine *See* autocrine.

paracrystalline aggregate a regular linear arrangement of stacked molecules.

paradigm (pronounced "paradime") a term with a variety of meanings in the scientific literature. In its weakest sense, it is used as a synonym for model, hypothesis, or theory. It is used most commonly to refer to a known example or incident that serves as a model or provides a pattern for a more general phenomenon. In a still more restricted sense, paradigm may refer to a ruling model that has replaced all others. Darwin's theory of evolution by natural selection is an example of such a paradigm. As time passes, the matching of new discoveries with the model may lead to a revision of the paradigm. An example of this would be the transformation of the one gene–one enzyme to the one gene–one polypeptide paradigm.

paraffin section a section of tissue cut by a microtome after embedding in a paraffin wax; the classical method of preparing tissues for microscopical study. *See* Appendix C, 1860, Klebs.

paragenetic referring to a chromosomal change that influences the expression of a gene rather than its structure. *See* position effect, Lyonization.

parallel evolution the occurrence of the same or a similar trend independently evolved in two or more lineages; the lineages are usually, although not necessarily, related to one another.

paralogs two or more different genes in the same species which are so similar in their nucleotide sequences that they are assumed to have originated following the duplication of a single ancestral gene. The human alpha and delta hemoglobin chain loci are examples of paralogous genes. The adjective *par-*

alogous is used when referring to nucleotide sequence comparisons of repetitive DNAs from a single species. For example, the rRNA genes from different chromosomes of species A might be homogeneous in sequence structure and those from species B might also be homogeneous. These would be paralogous comparisons. However, if rDNAs from species A and B were compared, the data might be more heterogeneous. This would be an *orthologous* comparison. *See* orthologs, unequal crossing over.

Paramecium a common genus of ciliate protozoa found in stagnant ponds. Common species are *P. aurelia, P. caudatum,* and *P. bursaria.* These are favorite species for the study of nuclear-cytoplasmic interactions. *See* Appendix C, 1976, Dippell; contractile vacuole, universal code theory.

Paramecium aurelia a cigar-shaped, heavily ciliated protoctist, 100–150 micrometers long, living in still or running fresh waters. *Paramecium aurelia* consists of a group of 14 syngens. Each syngen is genetically isolated from every other syngen and is also biochemically unique. However, all syngens are so similar morphologically that they have not been given individual species names. Each syngen has two mating types. Mating types were first discovered in syngen 1. *See* Appendix A, Protoctista, Ciliophora; Appendix C, 1937, Sonneborn; 1971, Kung; conjugation, killer paramecia.

parameter the value of some quantitative characteristic in an entire population. For example, the mean height of all males 20 years of age or older in a pygmy tribe. *Compare with* statistic.

parametric statistics *See* statistics.

paramutation a phenomenon discovered in corn where one allele influences the expression of another allele at the same locus when the two are combined in a heterozygote. The first allele is referred to as "paramutagenic," the second as "paramutable." The paramutable allele behaves like an unstable hypomorph when the paramutagenic allele is present.

paranemic spiral a spiral made up of two parallel threads coiled in opposite directions. The threads can be easily separated without uncoiling. *See* plectonemic spiral, relational coiling.

parapatric referring to populations or species that occupy adjacent areas with a narrow zone of overlap within which hybridization commonly occurs.

parapatric speciation a mode of gradual speciation in which new species arise from populations that maintain genetic contact during the entire process by a narrow zone of overlap. Gene flow between the populations is prevented by the rapid evolution of pre-mating isolation mechanisms. Also called *semigeographic speciation. Compare with* allopatric speciation, alloparapatric speciation, peripatric speciation.

paraphyletic 1. in classification, an incomplete clade, i.e., one from which one or more members of a halophyletic group have been omitted. A paraphyletic group is recognized by the absence of the homologs that define the excluded clades. For example, group AB in the illustration on p. 323 is paraphyletic because it is defined by the absence of feature 3 (features 1 and 2 are present in group CD as well). 2. in evolution, a monophyletic group that does not include all the groups descended from a single common ancestor. For example, here the AB group is paraphyletic, but group CD is excluded since it is not a direct descendant from common ancestor 1.

Parascaris equorum a nematode (commonly called the horse thread-worm) studied by early cytologists because it exhibited chromatin diminution (*q.v.*). *See* Appendix C, 1887, Boveri.

parasexuality any process that forms an offspring cell from more than a single parent, bypassing standard meiosis and fertilization. In fungi, for example, diploid nuclei arise from rare fusions of two genetically unlike nuclei in heterokaryons. Somatic crossing over occurs, and eventually haploid nuclei showing new combinations of genes are formed from the diploids (*see* Appendix C, 1952, Pontecorvo and Roper). In viruses, parasexual recombination can occur if genetically different mutant strains multiply in a host cell after it has been infected by viruses of both types (*see* Visconti-Delbrück hypothesis). In bacteria, there are three phenomena that can lead to parasexual recombination: conjugation, transduction, and transformation (*all of which see*). *Also see* sexduction, transfection.

parasitemia the presence of parasites at various developmental stages in circulating blood cells of the host. *See* glucose-6-phosphate dehydrogenase deficiency, malaria, merozoite.

Paranemic spiral

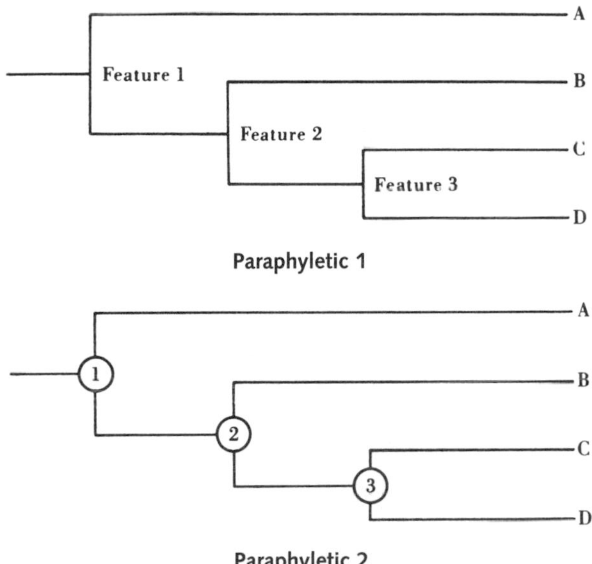

Paraphyletic 1

A
B
C
D
1
2
3

Paraphyletic 2

parasitic DNA *See* selfish DNA.

parasite theory of sex the proposal by W. D. Hamilton that sexual reproduction has evolved to provide a mechanism to generate the resistance of host species to rapidly multiplying parasites. The genetic recombination which accompanies meiosis (*q.v.*) rapidly produces new combinations of resistance factors in the host, and these overcome the susceptibility factors which arise more slowly in the parasites by mutation.

parasitism a symbiotic association that benefits one member (the parasite) but is harmful to the other (the host).

parathyroid hormone a hormone that controls calcium and phosphorus balance; it is synthesized by the cells of the parathyroid gland.

paratope the site within an immunoglobulin Fab that specifically interacts with an antigenic determinant (epitope).

Parazoa a subdivision of the animal kingdom containing organisms, like sponges, that lack tissues organized into organs, and have an indeterminate shape. *See* Appendix A.

parenchyma 1. plant tissues composed of thin-walled cells that fit loosely together, leaving intercellular spaces. 2. a reticulum of cells between the organs of an animal. 3. the cells performing the principal function of an organ.

parental ditype *See* tetrad segregation types.

parental imprinting the phenomenon whereby the degree to which a gene expresses itself depends upon the parent transmitting it. Huntington disease (*q.v.*) is an example. Individuals who receive this dominant gene show symptoms during adolescence if it is inherited from their father, but symptoms begin during middle age when the gene comes from their mother. The phenomenon of parental imprinting may result from differing patterns of DNA methylation (*q.v.*) which occur during gametogenesis in the two sexes. Thus, in oocytes the controlling elements of specific genes might be methylated, but demethylated in spermatocytes. Following fertilization, the cells of embryos would contain an inactive maternal and an active paternal allele. For such a system to maintain itself generation after generation, it would have to be reversible. Thus, when the offspring enters gametogenesis, it must demethylate all methylated alleles if it is a male and methylate all demethylated alleles if it is a female. Parental imprinting also occurs in flowering plants. The behavior of the *R* gene in maize provides an example. The gene controls the color of pigment grains in the aleurone (*q.v.*) of the kernel (*q.v.*). In the cross $rr \times RR$, the F_1 kernels are solid red when *RR* is the female parent but mottled if *RR* is the male parent. Extra paternal copies of *R* do not substitute for maternal *R* genes. So the maternal copy is preferentially expressed. *See* Appendix A, 1970, Kermicle; double fertilization, *H19*, Prader-Willi syndrome (PWS), *R* genes of maize.

parent-offspring conflict theory the proposal by David Haig that competition between the sexes occurred during the evolution of placental mammals. This resulted in a compromise between the robustness of the offspring during their period of uterine growth and the health of the mother. In the sperm, genes that increase the cost of the offspring are switched on. Such genes are active in the placenta and cause the embryo to grow rapidly and use up the mother's resources. In the egg, genes that decrease the cost of the offspring to the mother are activated. Such genes keep the unlimited growth of the embryo in check, so the mother survives the pregnancy. The genes occur in antagonistic pairs, and parental imprinting (*q.v.*) controls their functioning. *See* Appendix C, 1992, Haig.

parity the fact of having borne children. A woman of parity 0 has borne no live children although she may have been pregnant one or more times. A woman of parity 1 has given birth only once. However, the number of children produced at this birth may have exceeded one.

Parkinsonism a disease first described in 1817 as "shaking palsy" by the London physician James Parkinson. The disease arises from the death of neurons that synthesize DOPAMINE (*q.v.*), and these cells reside in a portion of the brain called the substantia nigra. The most common hereditary form of Parkinsonism is called early-onset, autosomal recessive Parkinsonism. It is caused by mutations in a gene now called *parkin* which resides at 6q25.2-q27. The gene product (PARKIN) is a protein of 465 amino acids which functions as a ubiquitin-conjugating enzyme. This observation suggests that Parkinsonism results from a defect in the system that normally removes potentially toxic proteins that accumulate in dopaminergic neurons. *See* proteasome, ubiquitin.

paroral cone a protuberance in the oral region of a conjugating ciliate that juts into the body of the partner. The haploid nucleus residing in the paroral cone survives and all others degenerate. *See* conjugation.

pars amorpha *See* nucleolus.

parsimony principle the principle that the simplest sufficient hypothesis is to be preferred, even if others are possible. Also called *Occam's razor.*

pars intercerebralis that medial region in the insect forebrain containing neurosecretory cells.

Parthenium argentatum the guayule, a desert plant, reproducing sexually or asexually depending upon its genotype and studied by geneticists accordingly.

parthenocarpy the natural or artificially induced formation of fruit without seeds because of either (1) lack of pollination, (2) lack of fertilization, or (3) death of the embryo at an early stage of development. *See* bananas.

parthenogenesis the development of an individual from an egg without fertilization. In most anisogamous eukaryotes, parthenogenesis is prevented by the inactivation of oocyte centrioles prior to fertilization. The centriole, which initiates cytokinesis, is brought in by the fertilizing sperm. *See* Appendix C, 1845, Dzierzon; arrhenotoky, artificial parthenogenesis, gynogenesis, heterogony, paedogenesis, sexuparous, thelytoky.

parthenote an organism produced by parthenogenesis (*q.v.*) and therefore having only genes of maternal origin.

partial denaturation an incomplete unwinding of the DNA double helix; GC-rich regions are more resistant to thermal disruption because three hydrogen bonds form between G and C, whereas only two form between A and T. *See* deoxyribonucleic acid.

partial diploid *See* merozygote.

partial dominance *See* incomplete dominance.

particle-mediated gene transfer a technique by which selected DNA molecules are quantitatively coated upon gold or tungsten spheres 1–3 micrometers in diameter and shot into plant or animal tissues with sufficient force to penetrate multiple layers of cells. The motive force to accelerate the particles comes from a high-pressure burst of helium gas stored in a tank connected to the gene gun. The targeted animal tissues are generally somatic (skin, liver, and pancreas), and a large number of reporter genes and genes with therapeutic potentials have been expressed in targeted cells. Gene guns have been used even more effectively in generating transgenic plants of economic importance.

particulate inheritance the Mendelian theory that genetic information is transmitted from one generation to another in the form of discrete units, so that the biological inheritance of offspring is not a solution in which the parental information is blended.

P1 artificial chromosomes (PACs) DNA cloning vectors containing regions of the P1 phage (*q.v.*) genome (*q.v.*) and capable of accepting large DNA inserts (i.e., up to 100 kilobases long). In this system, vector DNA and DNA fragments to be cloned are ligated and packaged *in vitro* into phage particles that can infect *Escherichia coli*. Once inside the bacterium, the packaged DNA is circularized and maintained as a plasmid (*q.v.*). A P1 plasmid replicon (*q.v.*) maintains the DNA at one copy per cell, and a P1 lytic replicon induces amplification of plasmid DNA under appropriate experimental conditions, allowing the isolation of large amounts of DNA for analysis. PACs are useful for cloning large genes, chromosome walking (*q.v.*), physical mapping, and shotgun sequencing (*q.v.*) of complex genomes. *Compare with* bacterial artificial chromosomes (BAC)s, yeast artificial chromosomes (YACs). *See* bacteriophage packaging, DNA vector, genomic library, kilobase, physical map, plasmid cloning vector.

parturition the act of giving birth to young.

Pascal's pyramid *See* binomial distribution.

PAS procedure periodic acid Schiff procedure (*q.v.*).

passage number the number of times a culture has been subcultured.

passenger in recombinant DNA research, a DNA segment of interest that will be spliced into a DNA vehicle for subsequent cloning.

passive equilibrium an unstable equilibrium resulting from selective neutrality of alleles at a genetic locus, as occurs in Hardy-Weinberg equilibrium; also called *neutral equilibrium*.

passive immunity the immunity against a given disease produced by injection into a host of serum containing antibodies formed by a donor organism that possesses active immunity to the disease. *See* active immunity.

Patau syndrome a well-defined set of congenital defects in humans caused by the presence of an extra chromosome 13. Also called trisomy 13 syndrome or D_1 trisomy syndrome. The median survival time is 3 days and the incidence is 1/10,000 live births. The condition was first described in 1960 by Klaus Patau and four colleagues. *See* human mitotic chromosomes.

patentability of genetically modified organisms the ability to obtain the exclusive legal rights to create genetically modified organisms. The United States Supreme Court has ruled that a microorganism which was genetically engineered to consume oil slicks can be patented. *See* Appendix C, 1980, Chakrabarty.

patent period the interval during an infection when the causative agent can be detected. *Compare with* latent period, prepatent period.

paternal effect gene a paternally expressed gene required for normal fertilization and early embryonic development. Males homozygous for recessive alleles of such genes produce sperm that cannot support normal development in the embryo, even if the egg is from a female that is homozygous for the wild-type allele, while the reciprocal cross (i.e., wild-type male × homozygous recessive female) produces viable embryos. The genotype of the father thus affects the phenotype of his offsprings. *Compare with* maternal effect gene, zygotic gene.

paternal inheritance hereditary traits that are governed by genes or self-reproducing organelles solely contributed to the offspring by the male parent. For example, in male and female animal embryos the centrioles are derived from a progenitor centriole carried by the fertilizing sperm. Y-linked genes functioning in sex determination and spermiogenesis are inherited by males from their fathers. *Contrast with* maternal inheritance. *See* Y chromosome.

paternal-X inactivation the method of dosage compensation found in marsupials where the paternal-X chromosome is inactivated in the somatic cells of females. *See* random-X inactivation.

path coefficient analysis a method invented by Sewall Wright for analyzing quantitatively the transmission of genes in regular and irregular breeding systems.

pathogenic producing disease or toxic symptoms.

pathovar a pathological variant of a bacterial species; symbolized *pv*. For example, the cause of citrus canker is *Xanthomonas campestris* pv. *citri*.

patroclinous designating an offspring that resembles the male more closely than the female parent. The sons of an attached-X female *Drosophila* are patroclinous in terms of their sex-linked genes. *See* matrocliny.

pattern specification the genetically controlled process which specifies the transformation of an apparently homogeneous population of cells into subpopulations of cells, each with a specific morphology and temporal behavior. *See* cell lineage mutations, compartmentalization, floral identity mutations, genetic networking, *Hox* genes, Positional Information Hypotheses, selector genes, zygotic segmentation mutations.

***Pax* genes** those genes that contain a *Paired* box, hence the abbreviation *Pax*. *Pax* genes encode a highly conserved family of transcription factors that play key roles in controlling the fates of cells, the development of patterns, and organogenesis. In *Drosophila* and various species of mammals, *Pax* genes specify the differentiation of the anterior brain, the cochlea, and the retina. However, *Pax* genes are also expressed in *Trichoplax adhaerans* (*q.v.*), a species which lacks both nervous tissues or sensory organs. *See* paired, *Pax-6* genes.

***Pax-6* genes** a subset of *Pax* genes (*q.v.*) that encode evolutionarily conserved transcription factors essential for eye development in insect and mammalian species. In *Drosophila*, the *Pax-6* genes (*q.v.*) are *eyeless (ey)* and *twin of eyeless (toy)*; in humans and rodents, the *Pax-6* genes are *Aniridia* (*q.v.*) and *Small eye* (*q.v.*), respectively.

P22 bacteriophage *See* P22 phage.

P blood group a human blood group identified by a glycolipid antigen specified by the dominant gene *P* on the long arm of chromosome 22. The P antigen binds to the fimbriae of certain strains of *E. coli*. Therefore, individuals homozygous for null *P* alleles are less susceptible to infection than P-positive individuals. *See* fimbria.

pBR322 a plasmid cloning vector that grows under relaxed control in *E. coli*. It contains ampicillin- and tetracycline-resistance genes and several convenient restriction endonuclease recognition sites. *See* Appendix C, 1979, Sutcliffe.

PCR polymerase chain reaction (*q.v.*).

PDGF platelet-derived growth factor (*q.v.*).

pea *See Pisum sativum*.

Peckhammian mimicry *See* mimicry.

Pecten irradians *See* Pelecypoda.

pectic acid a polymer made up of galacturonic acid subunits.

pectin a polysaccharide material within the cell wall and middle lamella. Pectin consists of pectic acid, many of whose COOH groups have been methylated.

pedigree a diagram setting forth the ancestral history or genealogical register. Symbols commonly used in such diagrams are illustrated on page 327 in the sample pedigree. Females are symbolized by circles and males by squares. Individuals showing the trait are drawn as solid figures. Offspring are presented beneath the parental symbols in order of birth from left to right. The arrow points to the propositus. The sex of individual II-3 is unknown. II-6 died at an early age and consequently her phenotype relative to the trait in question was unknown. II-7 and II-8 were dizygotic twins, whereas III-1 and III-2 were identical twins. Other symbols commonly encountered are also shown here. *See* first-degree relative, second-degree relative.

pedigree selection artificial selection of an individual to participate in mating based upon the merits of its parents or more distant ancestors.

pedogenesis *See* paedogenesis.

pelargonidin *See* anthocyanins.

Pelargonium zonale the geranium. Classical studies on the non-Mendelian inheritance of chloroplast mutants were performed on this species. *See* Appendix C, 1909, Correns and Bauer.

Pelecypoda the class containing the bivalve molluscs. The cytogenetics and quantitative genetics of certain species have been studied because of their economic importance. These include the American oyster (*Crassostrea virginica*), the clam (*Mercenaria mercenaria*), the mussel (*Mytilus edulis*), and the scallop (*Pecten jacobeus*).

P elements transposable elements in *Drosophila* that are responsible for one type of hybrid dysgene-

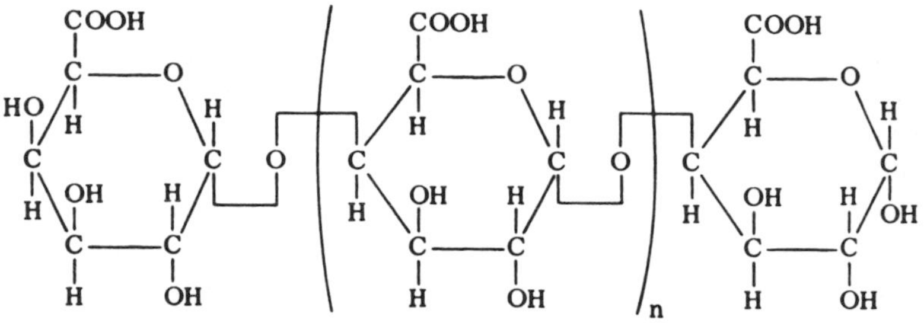

Pectic acid

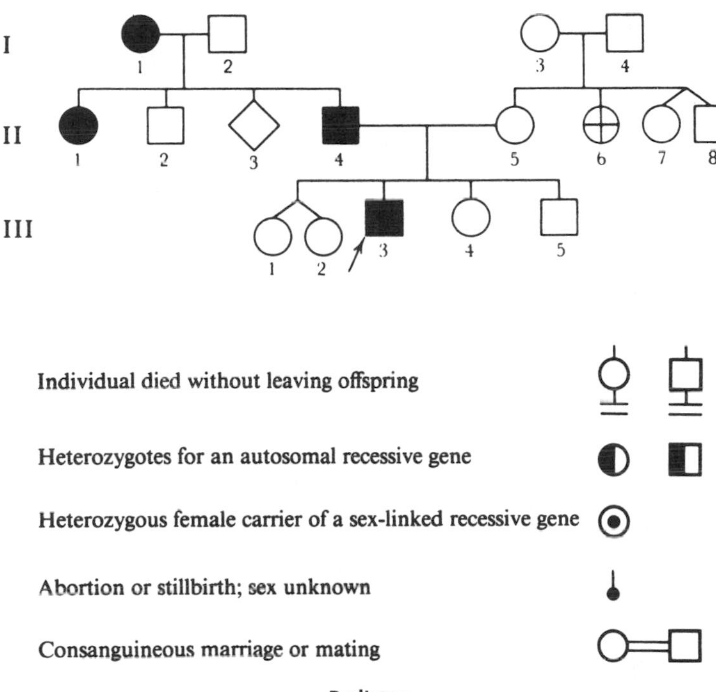

Pedigree

sis (*q.v.*). P elements have been cloned in *E. coli* plasmids. When DNA molecules carrying P elements are microinjected into *Drosophila* embryos, some P elements integrate into the germ-line chromosomes and are transmitted to the progeny of the injected flies. Active autonomous P elements are 2.9 kb in length and are flanked by inverted repeats that are 31 bp long. They have four exons (designated *ORF 0, 1, 2,* and *3*). All four exons encode an 87 kDa transposase, and the first three exons specify a 66 kDa repressor. P elements were discovered in *D. melanogaster,* but they are absent from other sister species. However, P elements are common in *D. willistoni* and species related to it. *D. melanogaster* is believed to have received its P elements in the 1950s from *D. willistoni.* Horizontal transfer may have been accomplished by ectoparasitic mites. *See* Appendix C, 1982, Bingham *et al.*, Spradling and Rubin; 1991, Houck *et al.*; 1994, Clark *et al.*; horizontal mobile elements (HMEs), promiscuous DNA, transposable elements, transposon tagging.

P element transformation the transfer of specific DNA segments into germ-line cells of *Drosophila* using the transposable P element to carry exogenous DNA fragments. *See* Appendix C, 1982, Spradling and Rubin.

Pelger-Huet anomaly *See* Pelger nuclear anomaly.

Pelger nuclear anomaly a hereditary abnormality in humans involving the nuclear morphology of the polymorphonuclear leukocytes. A similar syndrome occurs in the rabbit. Homozygotes show chondrodystrophy in the rabbit, but not in humans.

pellagra a disease caused by deficiency of niacin.

Pelomyxa a genus of amoebas. *P. carolinensis* is a favorite species for nuclear transplantation studies. *P. palustris* is of interest, since it lacks mitochondria and instead harbors anaerobic bacteria in a permanent symbiotic relationship. *See* Appendix A, Protoctista, Caryoblastea; serial symbiosis theory.

penetrance the proportion of individuals of a specified genotype that show the expected phenotype under a defined set of environmental conditions. For example, if all individuals carrying a dominant mutant gene show the mutant phenotype, the gene is said to show complete penetrance. *See* manifesting heterozygote.

penicillin any of a family of antibiotics derived from the mold *Penicillium notatum* and related species. It was once thought that the disruption of cell wall synthesis caused the cell wall to weaken and thus led to cell lysis. Evidence now exists that penicillin and almost all antibiotics indirectly activate a class of enzymes (autolysins) that are responsible for

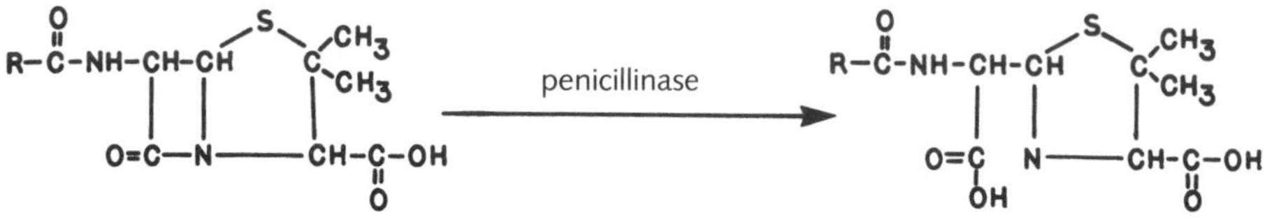

penicillinase

pencillin

penicilloic acid

cell lysis. Different penicillins differ only in the side chain symbolized by R in the above formula. In the parent molecule, R = $C_6H_5CH_2CO$–. Ampicillin is a penicillin derivative that is effective against a larger variety of Gram-negative bacteria than are most other penicillins. Here, R = $C_6H_5CH(NH_2)CO$–. Penicillin-resistant bacteria synthesize penicillinases. These enzymes attack the beta-lactam ring to produce penicilloic acid, which has no bacteriocidal activity. *See* Appendix C, 1929, Fleming; 1940, Florey *et al.*; 1949, Hodgkin *et al.*; *amp*[R], transformation.

penicillinases *See* penicillin.

penicillin enrichment technique *See* penicillin selection technique.

penicillin selection technique a method for isolating an auxotrophic mutant from a wild-type culture of bacteria by adding penicillin to minimal medium. Penicillin interferes with cell wall development, causing growing wild-type cells to rupture. However, nongrowing auxotrophic mutant cells are not killed. After one hour, about 99% of the wild-type cells have lysed, releasing their pool of metabolites into the medium. The culture must be filtered to remove these metabolites, because the auxotrophic mutants would use them for growth and be subject to penicillin-induced lysis; the filtering also removes the penicillin. Alternatively, the enzyme penicillinase can be used to destroy the penicillin. Surviving auxotrophs are then supplied with enriched medium. These cells produce the colonies that are harvested. *See* Appendix C, 1948, Lederberg and Zinder, Davis.

Penicillium notatum the ascomycote fungus that synthesizes penicillin (*q.v.*).

penicilloic acid *See* penicillin.

Pennsylvanian *See* Carboniferous.

pentabarbital Nembutal (*q.v.*).

pentosephosphate pathway a pathway of hexose oxidation that is an alternative to the glycolysis-citric acid cycle (*q.v.*). The pathway involves an interconversion of phosphates of 7-, 6-, 5-, 4-, and 3-carbon sugars, and of a sugar lactone and sugar acids in a cycle that effects the complete oxidation of glucose to CO_2 and H_2O with the formation of 36 molecules of ATP.

peplomers spikes that protrude from the envelopes of certain viruses. For example, the rabies virus (*q.v.*) has about 400 peplomers dispersed evenly over its surface, except for its planar end. Each spike consists of a trimer of virus-encoded glycoproteins. *See* enveloped viruses, virus.

pepsin a proteolytic enzyme from the gastric mucosa that functions at low pH.

peptidase an enzyme catalyzing the hydrolytic cleavage of peptide bonds (*q.v.*).

peptide a compound formed of two or more amino acids.

peptide bond a covalent bond between two amino acids formed when the amino group of one is bonded to the carboxyl group of the other and water is eliminated. *See* Appendix C, 1902, Hofmeister and Fischer; amino acid, translation.

$$-\underset{|}{\underset{H}{N}}-H + HO-\underset{\|}{\underset{O}{C}}- \longrightarrow -\underset{|}{\underset{H}{N}}-\underset{\|}{\underset{O}{C}}- + H_2O$$

peptidoglycan a heteropolymer present in the cell walls of most bacteria. It consists of linear polysaccharide chains cross linked by short peptides (containing ~8 amino acids). The structure that results forms a hollow net surrounding the bacterium, and it functions to strengthen the cell and to protect it from osmotic lysis. The cells of Archaebacteria (*q.v.*) do not contain peptidoglycans. *See* bacterial cell walls.

peptidyl transferase *See* translation.

peptidyl-RNA binding site *See* translation.

PER the protein encoded by the gene called *period* (*q.v.*).

perdurance in a concealed, yet durable state. In genetics literature, it refers to a situation in which the phenotypic expression of a gene remains unchanged, after it has been deleted or inactivated, because of the long-lived nature of its product.

perennation the survival of plants from growing season to growing season with a period of reduced activity in between.

perennial a plant that continues to grow from year to year.

perfect flower *See* flower.

perfusion the introduction of fluids into organs by injection into their arteries.

pericarp the wall of the ovary after it has matured into a fruit; it may be dry and hard (as in the case of a nut) or fleshy (as in the case of a berry). *See* kernel.

pericentric inversion an inversion that includes the centromere.

periclinal referring to a layer of cells running parallel to the surface of a plant part. *See* anticlinal.

periclinal chimera a plant made up of two genetically different tissues, one surrounding the other.

perikaryon that portion of the cell body of a neuron surrounding the nucleus as distinguished from the axon and dendrites.

perinuclear cisterna the fluid-filled reservoir enclosed by the inner and outer membranes of the nuclear envelope.

period the first gene shown to control a circadian clock (*q.v.*). Mutations at the *per* locus alter the rhythmicity of eclosion and of adult locomotor activity. This sex-linked gene has been cloned and sequenced in *D. melanogaster*, *pseudoobscura*, and *virilis*. Null mutants (per^0) show no rhythmicity of eclosion or locomotor activity. Missense mutations shorten (per^S) or lengthen (per^L) both rhythms. The per^+ gene is expressed in the normal adult nervous system, and it encodes a PER protein containing about 1,200 amino acids. The protein is present at extremely low levels, and it does not contain structural motifs of the sort that characterize DNA-binding domains, signal factors, or membrane-spanning domains. The gene also modulates the interpulse intervals during courtship "songs." These are generated by the extension and vibration of the males' wings, and the songs enhance the females' mating behavior. The songs are species-specific, and *per* has been shown to contain species-specific song instructions by introducing a cloned copy of *per* from *D. simulans* into the genome of a *D. melanogaster* per^0 mutant. *See* Appendix C, 1971, Konopka and Benzer; 1984, Bargiello and Young; 1991, Wheeler *et al.*; frequency.

periodic acid Schiff procedure a staining procedure for demonstrating polysaccharides. Abbreviated PAS. *See* Schiff's reagent.

periodicity in molecular genetics, the number of base pairs per turn of the DNA double helix.

periodic table an arrangement of the chemical elements (*q.v.*) in order of increasing atomic number (*see* page 330). Elements with similar properties are placed one under the other, yielding groups and families of elements. All living organisms consist of organic compounds derived from hydrogen (H), carbon (C), nitrogen (N), oxygen (O), phosphorus (P), and sulfur (S). The alkali metals, sodium (Na) and potassium (K), the alkaline earth metals magnesium (Mg) and calcium (Ca), and the halogens chlorine (Cl) and iodine (I) are also biologically important. Other elements are vital but occur in trace amounts. The dark boxes in the table show the relative abundance of the biologically important elements found in the human body. The values give the number of atoms in a total of 100,000. *See* Appendix C, 1869, Mendeleev; chemical elements.

H	60,562	S	130
O	25,670	Na	75
C	10,680	K	37
N	2,490	Cl	33
Ca	230	Mg	11
P	130	I, Mn, Fe, Co, Ni, Cu, Zn and Mo are each <1	

peripatric living in a region peripheral to that of the main body of the species.

peripatric speciation a model proposing that speciation occurs in small populations isolated on the periphery of the distribution of the parental population, as opposed to parapatric speciation (*q.v.*). The isolated populations may undergo shifts in their gene frequencies under the influence of genetic drift. This is most likely to occur if new populations arise from a few founder individuals and no gene flow occurs between the isolates and the main population. Consequently the most rapid evolutionary changes do not occur in widespread populous species, but in small founder populations. *See* Chronology C, 1954, Mayr; founder effect.

Peripatus *See* living fossil.

Periodic Table of the Elements

Key: atomic number / **Symbol** / standard atomic weight

1	2	3	4	5	6	7	8	9	10	11	12	13	14	15	16	17	18
1 **H** 1.008																	2 **He** 4.003
3 **Li** 6.941	4 **Be** 9.012											5 **B** 10.811	6 **C** 12.011	7 **N** 14.007	8 **O** 15.999	9 **F** 18.998	10 **Ne** 20.180
11 **Na** 22.990	12 **Mg** 24.305											13 **Al** 26.982	14 **Si** 28.086	15 **P** 30.974	16 **S** 32.065	17 **Cl** 35.453	18 **Ar** 39.948
19 **K** 39.098	20 **Ca** 40.078	21 **Sc** 44.956	22 **Ti** 47.867	23 **V** 50.942	24 **Cr** 51.996	25 **Mn** 54.938	26 **Fe** 55.845	27 **Co** 58.933	28 **Ni** 58.693	29 **Cu** 63.546	30 **Zn** 65.409	31 **Ga** 69.723	32 **Ge** 72.64	33 **As** 74.922	34 **Se** 78.96	35 **Br** 79.904	36 **Kr** 83.798
37 **Rb** 85.468	38 **Sr** 87.62	39 **Y** 88.906	40 **Zr** 91.224	41 **Nb** 92.906	42 **Mo** 95.94	43 **Tc** [98]	44 **Ru** 101.07	45 **Rh** 102.906	46 **Pd** 106.42	47 **Ag** 107.868	48 **Cd** 112.411	49 **In** 114.818	50 **Sn** 118.710	51 **Sb** 121.760	52 **Te** 127.60	53 **I** 126.904	54 **Xe** 131.293
55 **Cs** 132.905	56 **Ba** 137.327	57-71 *	72 **Hf** 178.49	73 **Ta** 180.948	74 **W** 183.84	75 **Re** 186.207	76 **Os** 190.23	77 **Ir** 192.217	78 **Pt** 195.078	79 **Au** 196.967	80 **Hg** 200.59	81 **Tl** 204.383	82 **Pb** 207.2	83 **Bi** 208.980	84 **Po** [209]	85 **At** [210]	86 **Rn** [222]
87 **Fr** [223]	88 **Ra** [226]	89-103 **	104 **Rf** [261]	105 **Db** [262]	106 **Sg** [266]	107 **Bh** [264]	108 **Hs** [277]	109 **Mt** [268]	110 **Ds** [271]	111 **Rg** [272]							

* Lanthanoids	57 **La** 138.906	58 **Ce** 140.116	59 **Pr** 140.908	60 **Nd** 144.24	61 **Pm** [145]	62 **Sm** 150.36	63 **Eu** 151.964	64 **Gd** 157.25	65 **Tb** 158.925	66 **Dy** 162.500	67 **Ho** 164.930	68 **Er** 167.259	69 **Tm** 168.934	70 **Yb** 173.04	71 **Lu** 174.967
** Actinoids	89 **Ac** [227]	90 **Th** 232.038	91 **Pa** 231.036	92 **U** 238.029	93 **Np** [237]	94 **Pu** [244]	95 **Am** [243]	96 **Cm** [247]	97 **Bk** [247]	98 **Cf** [251]	99 **Es** [252]	100 **Fm** [257]	101 **Md** [258]	102 **No** [259]	103 **Lr** [262]

Standard atomic weights (mean relative atomic masses) are here rounded up to 3 decimal places, where the accuracies of individual values make this possible. IUPAC 2001 standard atomic weights and corresponding uncertainties are reported in full in R. D. Loss, *Pure Appl. Chem.* **75**, 1107–1122 (2003). For elements that have no stable or long-lived nuclides, the mass number of the nuclide with the longest confirmed half-life is listed between square brackets.

Adapted from Periodic Table of IUPAC (International Union of Pure and Applied Chemistry: www.iupac.org), dated 4 Feb 2005. Reproduced with the permission of IUPAC.

peripheral protein a protein that is exposed on the outer or the inner surface of the plasma membrane and is connected to the membrane through ionic or covalent interactions with a membrane component. Peripheral proteins include many proteins from the membrane-supporting cytoskeleton (e.g., spectrin (*q.v.*)) or the extracellular matrix (e.g., fibronectin (*q.v.*)). *Compare with* integral protein. *See* lipid bilayer model.

peripherin *See* retinitis pigmentosum (RP).

peristalsis waves of muscular contractions that pass along tubular organs and serve to move the contents posteriorly.

perithecium the rounded or flask-shaped, fruiting body of certain ascomycete fungi and lichens. The mature fruiting body of *Neurospora*, for example, contains about 300 ascus sacs.

peritoneal sheath a network of anastomosing muscle fibers that holds together the ovarioles of an insect ovary.

peritrichous designating bacteria having flagella all over their surfaces.

permeability the extent to which molecules of a given kind can pass through a given membrane.

permease a membrane-bound protein in bacteria that is responsible for transport of a specific substance in or out of the cell; sometimes referred to as a *transport protein*. In *E. coli*, lactose permease actively transports lactose into the cell.

Permian the most recent of the Paleozoic penods. Reptiles flourished, including species with mammalian characteristics. Insects increased, while amphibians declined. Cycads and gingkos evolved and formed forests. The Permian ended with the most severe of all mass extinctions with over 95% of all species dying off. All trilobites became extinct. *See* continental drift, geologic time divisions.

permissible dose *See* maximum permissible dose.

permissive cells cells in which a particular virus may cause a productive infection (i.e., the production of progeny viruses). Cells in which infection is not productive are called *nonpermissive cells*. Some DNA tumor viruses may cause a productive infection in permissive cells of one species and a tumor in nonpermissive cells of another species.

permissive conditions environmental conditions under which a conditional lethal mutant (e.g., a temperature-sensitive mutant) can survive and produce wild-type phenotype.

permissive temperature *See* temperature-sensitive mutation.

Peromyscus a genus containing about 40 species of mice native to Central and North Amenca. The genetic data accumulated so far have been primarily related to the deermouse, *P. maniculatus*. More limited data exist for *P. boylei*, *P. leucopus*, and *P. polionotus*. Studies of biochemical variation and cytogenetics are presently the areas of most genetical research within the genus.

peroxisomes membrane-bound intracellular organelles containing at least four enzymes involved in the metabolism of hydrogen peroxide. They are thought to be important in purine degradation, photorespiration, and the glyoxylate cycle. Peroxisomes lack DNA. *See* microbody, protein sorting, sorting signals.

PEST sequence domains in proteins that are rich in proline, glutamic acid, serine, and threonine. The name is based on the one-letter codes for these amino acids. PEST sequences are found in short-lived proteins and are therefore thought to serve as degradation signals.

pet, PET *See* petites.

petites dwarf colonies of *Saccharomyces cerevisiae*. Such yeasts are slow growing because of mutations affecting mitochondria. *Segregational petites* contain mutated nuclear genes that result in mitochondrial defects. The wild-type allele of a segregational petite is designated *PET* and the mutant allele *pet*. The *PET* gene mutations fall into more than 215 complementation groups. In the case of *cytoplasmic* or *vegetative petites*, it is the mitochondrial genome that is affected. Wild-type strains are symbolized p^+. In most mutants (p^-), the mtDNA contains less than one-third of the normal genome. Some strains (p^0) have lost all the mtDNA. In such cases the yeast forms *promitochondria*. These have a normal outer membrane, but the inner membrane contains poorly developed cristae. In spite of the fact that the organelle cannot function in oxidative phosphorylation, it contains nuclear-encoded proteins such as DNA and RNA polymerases, all the enzymes of the citric acid cycle, and many inner membrane proteins. *See* Appendix C, 1949 Ephrussi *et al.*; *Podospora anserina*.

pfcrt **gene** a gene conferring resistance in mosquitoes to the insecticide chloroquine (*q.v.*). The acronym stands for "*P*lasmodium *f*alciparum *c*hloroquine-*r*esistance *t*ransporter," and the gene encodes a transmembrane protein of the food vacuole of the merozoite (*q.v.*). Mutations in this protein reduce uptake of the insecticide. *See* vacuoles.

petri dish a round, shallow glass or plastic dish with a loose-fitting cover in which microorganisms or dividing eukaryotic cells are cultured on a nutrient gel (usually agar). Richard Julius Petri, and assistant to Robert Koch, developed the technique around 1887. *See* plaque.

Pfu **DNA polymerase** a DNA polymerase from the bacterium *Pyrococcus furiosus* that possesses the lowest error rate of any commercially available thermostable DNA polymerase because its 3′ → 5′ exonuclease activity serves a proofreading function. This DNA polymerase generates amplicons with blunt ends, whereas *Taq* DNA polymerase (*q.v.*) adds an extra A residue to amplicons. *See* extremophiles.

pg picogram. *See* genome size.

p53 **gene** *See TP53*.

P granules polar granules (*q.v.*).

pH *See* hydrogen ion concentration.

phaeomelanin one of the pigments found in the coats of mammals. It is derived from the metabolism of tyrosine and is normally yellow in color. The amount of this pigment inserted into the hairs is quantitatively and qualitatively controlled by the *agouti* locus. *See* agouti, *MCIR* gene, melanin.

Phaeophyta a phylum that contains mostly marine species referred to as *brown seaweeds* (*e.g.*, kelps). The phylum is placed in the kingdom Protoctista (Appendix A) according to Margulis' Five Kindom scheme. Cavalier-Smith places phaeophytes in a sixth kingdom, the Chromista (*q.v.*).

phage an abbreviation of bacteriophage (*q.v.*), a virus that attacks bacteria. When referring to one or more virions of the same species, the word *phage* is used; thus, a cell may be infected by one or many lambda phage. *Phages* is used when referring to different species; thus, lambda and T4 are both phages.

phage conversion *See* prophage-mediated conversion.

phage cross a procedure requiring the multiple infection of a single bacterium with phages that differ at one or more genetic sites. Upon lysis of the host, recombinant progeny phages are recovered that carry genes derived from both parental phage types. *See* Visconti-Delbrück hypothesis.

phage induction the stimulation of prophage to enter the vegetative state, accomplished by exposing lysogenic cells to ultraviolet light. Hydrogen peroxide, x-rays, and nitrogen mustard (*q.v.*) also act as inducing agents. *See* zygotic induction.

phagocyte a cell that incorporates particles from its surroundings by phagocytosis.

phagocytosis the engulfment of solid particles by cells; the ingestion of microorganisms by leukocytes. *See* Appendix C, 1901, Mechnikov.

phagolysosome an organelle formed by the fusion of a phagosome (*q.v.*) and a lysosome (*q.v.*).

phagosome a membrane-bounded cytoplasmic particle produced by the budding off of localized invaginations of the plasmalemma. Recently phagocytosed particles are segregated within the cell in phagosomes.

phalloidin the most common phallotoxin produced by *Amanita phalloides*. It is a cyclic heptapeptide made up of (1) alanine, (2) D-threonine, (3) cysteine, (4) hydroxyproline, (5) alanine, (6) tryptophan, and (7) γ-δ-dihydroxyleucine.

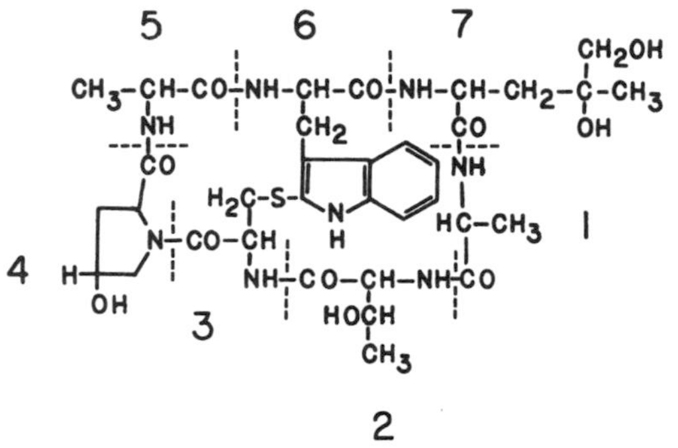

```
        5          6           7
                                          CH2OH
  CH3-CH-CO-NH-CH-CO-NH-CH-CH2-C-CH3
      NH        CH2         CO       OH
      CO                    NH
         H2C-S-     N                      1
      N-CO-CH       H       HC-CH3
4  H-                NH-CO-CH-NH-CO
   OH       3             HOCH
                           CH3
                            2
```

phallotoxins together with amatoxins (*q.v.*), the main toxic components produced by *Amanita phalloides* (*q.v.*). Phalloidin, one of the chief phallotoxins, forms tight complexes with filamentous actin. *See* rhodaminylphalloidin.

phanerogam an outmoded term referring to a plant belonging to the Spermatophyta (*q.v.*). *See* cryptogam.

Phanerozoic the geologic eon encompassing the Paleozoic, Mesozoic, and Cenozoic eras. During the 540-million-year interval an abundant fossil record was left in the rocks. *See* geologic time divisions.

phantom a volume of material approximating the density and effective atomic number of tissue. Radiation dose measurements are made within or on a phantom as a means of determining the radiation dose within or on a body under similar exposure conditions.

pharmacogenetics the area of biochemical genetics dealing with genetically controlled variations in responses to drugs.

pharming the genetic engineering of farm animals to produce pharmaceuticals.

phase contrast microscope light rays passing through an object of high refractive index will be retarded in comparison with light rays passing through a surrounding medium with a lower refractive index. The retardation or *phase change* for a given light ray is a function of the thickness and the index of refraction of the material through which it passes. Thus, in a given unstained specimen, transparent regions of different refractive indices retard the light rays passing through them to differing degrees. Such phase variations in the light focused on the image plane of the light microscope are not visible to the observer. The phase contrast microscope is an optical system that converts such phase variations into visible variations in light intensity or contrast. The phase microscope therefore allows cytologists to observe the behavior of living, dividing cells. *See* Appendix C, 1935, Zernicke.

phaseolin a glycoprotein that constitutes up to 50% of the storage protein in the cotyledons of the bean *Phaseolus vulgaris*. It is encoded by a family of about 10 genes. *See* Appendix C, 1981, Kemp and Hall.

Phaseolus the genus that includes *P. aureus*, the Mung bean; *P. limensis*, the lima bean; *P. vulgaris*, the red kidney bean. Most species are nodulating legumes (*q.v.*). *See* Plantae, Tracheophyta, Angiospermae, Dicotyledoneae, Leguminales; *Rhizobium*.

phene a phenotypic character controlled by genes.

phenetic taxonomy a system of classification based upon phenotypic characteristics without regard to phylogenetic relationships; also known as *numerical taxonomy*.

phenocopy the alteration of the phenotype, by nutritional factors or the exposure to environmental stress during development, to a form imitating that characteristically produced by a specific gene. Thus, rickets due to a lack of vitamin D would be a phenocopy of vitamin D-resistant rickets.

phenocritical period the period in the development of an organism during which an effect produced by a gene can most readily be influenced by externally applied factors.

phenogenetics developmental genetics.

phenogram a branching diagram linking taxons by estimates of overall similarity based on evidence from a sample of characters. Characters are not evaluated as to whether they are primitive or derived.

phenogroup any group of antigenically detectable factors of a blood group system that is inherited as a unit. Antigens in a phenogroup are encoded by alleles at a single locus. The B and C blood group systems in cattle are examples of two phenogroups. *See* Appendix C, 1951, Stormont *et al.*; haplotype.

phenome the sum of all of the manifest physical and behavioral characteristics of a cell or an organism.

phenomic lag phenotypic lag.

phenon a set of organisms grouped together by methods of numerical taxonomy (*q.v.*).

phenotype the observable characteristics of a cell or an organism, such as its size and shape, its metabolic functions, and its behavior. The genotype (*q.v.*) is the underlying basis of the phenotype, and the term is commonly used to describe the effect a particular gene produces, in comparison with its mutant alleles. Some genes control the behavior of the organism, which in turn generates an artefact outside the body. R. Dawkins uses the term *extended phenotype* to refer to the production of such an artefact (spider webs, bird nests, and beaver dams are examples). *See* Appendix C, 1909, Johannsen.

phenotypic lag the delay of the expression of a newly acquired character. Mutations may appear in a bacterial population several generations after the administration of a mutagen. Phenotypic lag may be due to any of the following reasons: (1) The mutagen may inactivate the gene at once, but this inactivity may not become apparent until the products previously made by the gene are diluted to a sufficient degree. A number of cell divisions occur before the concentration of these products falls below some critical level. (2) The "mutagen" may itself be inactive. It may undergo a series of reactions to yield a compound that is the true mutagen. The latent period would then be the time required for those reactions to take place. (3) The mutagen may cause the gene to become unstable. At a later time, it will return to a stable wild type or mutant state. (4) The microbe may be multinucleate, and a mutation may occur in but one nucleus. The latent period then would be the time required for nuclear segregation. *See* perdurance.

phenotypic mixing the production of a virus with a phenotype that does not match its genotype. During the assembly of a virus, nucleic acid and protein components are drawn randomly from two pools. In a host infected simultaneously by mutant and wild-type viruses, progeny phages are assembled without regard to matching the coat components to the genes in the nucleic acid core. Thus, discrepancies sometimes arise, and a virus is produced with coat proteins that are not specified by its genome. *See* pseudovirion, reassortant virus.

phenotypic plasticity a phenomenon in which a given genotype may develop different states for a character or group of characters in different environments; genotype-environment interaction (*q.v.*). *See* norm of reaction.

phenotypic sex determination control of gonad development by nongenetic stimuli. For example, the incubation temperature of fertilized eggs determines the type of sexual development in some turtle species.

phenotypic variance the total variance observed in a character. *See* genetic variance, variance.

phenylalanine *See* amino acid.

phenylketonuria a hereditary disorder of amino acid metabolism in humans, inherited as an autosomal recessive. Homozygotes cannot convert phenylalanine to tyrosine due to a lack of the liver enzyme phenylalanine hydroxylase. Brain dysfunction characteristic of the disease can be avoided by early dietary restriction of phenylalanine. Prevalence is 1/11,000. Abbreviated PKU. Phenylketonuria is sometimes called Følling disease after A. Følling, who discovered the underlying metabolic disorder in 1934. The *PKU* gene resides on the long arm of chromosome 12 between bands 22 and 24.1. It contains 13 exons and encodes a phenylalanine hydroxylase made up of 450 amino acids. The most frequent mutant allele in Caucasians is a base substitution that causes a skipping of exon 12 during mRNA splicing. *See* Appendix C, 1954, Bickel, Gerrard, and Hickmans; 1961, Guthrie; gene, Guthrie test, maternal PKU.

phenylthiocarbamide (PTC) a molecule belonging to a class of compounds called thioureas. Related compounds occur in broccoli, cabbage, kale, and turnips. PTC tastes bitter to some humans, but is tasteless to others. Human populations from various parts of the world show differences in the proportions that can taste PTC, ranging from 30%–90%. The ability to taste PTC is controlled by a taste receptor gene (*q.v.*). *See* Appendix C, 1931, Fox.

```
     H
     N———C—NH2
     |   ‖
     C   S
    // \
  HC   CH
   |   ‖
  HC   CH
    \\ /
     C
     H
```

phenylthiourea a synonym for phenylthiocarbamide (*q.v.*).

pheoplasts a brown plastid of brown algae, diatoms, and dinoflagellates. Also spelled *phaeoplast.*

pheromone a chemical signal released by one organism that, at low concentrations, can cause nonharmful responses in a second organism, often of the same species. Examples of pheromones are sex attractants, alarm substances, aggregation-promotion substances, territorial markers, and trail substances of insects. Pheromones functioning as sex attractants are also known in fungi and algae. *See* courtship ritual, mate choice, odorant, odorant receptor gene.

Philadelphia (Ph^1) chromosome an aberrant chromosome first observed by researchers at the University of Pennsylvania and named after the city where the discovery was made. The chromosome is often found in patients suffering from chronic myeloid leukemia, a disease in which several bone marrow–derived cell lineages proliferate uncontrollably. Ph^1 is generally a reciprocal translocation between human chromosomes 9 and 22, involving break points at 9q34 and 22q11. The translocation generates a fusion gene made up of regulatory elements from a gene on chromosome 22 and the majority of the ORF of a proto-oncogene (*q.v.*) from chromosome 9. *See* Appendix C, 1970, Nowell and Hungerford; 1971, O'Riordan *et al.*; 1972, Rowley; Abelson murine leukemia virus, *AML1* gene, Burkitt lymphoma, myeloproliferative disease.

philopatric descriptive of organisms that tend to remain in the location where they were born and raised.

phi X174 (ϕX174) virus a coliphage whose genome consists of a circular, positive sense, single stranded DNA molecule containing 5,286 nucleotides. Overlapping genes (*q.v.*) were discovered in this virus (*q.v.*). *See* Appendix C, 1959, Sinsheimer; 1967, Goulian *et al.*; 1977, Sanger *et al.*; Appendix F.

phloem the vascular tissue that conducts nutrient fluids in vascular plants.

phocomelia absence of the proximal portion of a limb or limbs, the hands or feet being attached to the trunk by a single bone. A genetic form occurs in humans that is inherited as an autosomal recessive. Phocomelia may also be caused by exposure of the developing embryo to the drug thalidomide (*q.v.*).

phosphate bond energy the energy liberated as one mole of a phosphorylated compound undergoes hydrolysis to form free phosphoric acid. *See* ATP.

phosphatidylinositol 3-kinase (PI 3-kinase) an enzyme that functions at DNA damage checkpoints (*q.v.*). *See RAD.*

phosphodiester any molecule containing the linkage

```
          O
          ‖
R—O—P—O—R′
          |
          O⁻
```

where R and R′ are carbon-containing groups, O is oxygen, and P is phosphorus. This type of covalent chemical bond involves the 5′ carbon of one pentose sugar (ribose or deoxyribose) and the 3′ carbon of an adjacent pentose sugar in RNA or DNA chains. *See* deoxyribonucleic acid.

phosphodiesterase I a 5′ exonuclease that removes, by hydrolysis, 5′ nucleotides from the 3′ hydroxy-terminus of oligonucleotides.

phospholipid a lipid containing phosphate esters of glycerol or sphingosine.

phosphorescence *See* luminescence.

phosphorus an element universally found in small amounts in tissues, a component of nucleic acids. Atomic number 15; atomic weight 30.4735; valence 5^+; most abundant isotope ^{31}P; radioisotope ^{32}P (*q.v.*).

phosphorylation the attachment of phosphate groups to appropriate target molecules. Biochemical phosphorylation reactions are used to trap energy and to form compounds that serve as intermediates during metabolic cycles. Phosphorylation is a major mechanism used by cells for the transduction of signals from protein to protein. *See* adenosine phosphate, cellular signal transduction, citric acid cycle, glycolysis, oxidative phosphorylation, phosphate bond energy, protein kinase.

phosvitin a phosphoprotein of relative molecular mass 35,000. Two molecules of phosvitin and one of lipovitellin comprise the basic subunit of the amphibian yolk platelet. *See* vitellogenin.

photoactivated crosslinking a technique for crosslinking a nucleic acid (e.g., a tRNA chain) to a polypeptide chain with which it is functionally associated (e.g., its cognate synthetase) by irradiating the synthetase-tRNA complex with ultraviolet light. The technique is used to locate the points of intimate contact between the two molecules. *See* RNase protection.

photoautotroph an organism that can produce all of its nutritonal and energy requirements using only inorganic compounds and light.

photoelectric effect a process in which a photon ejects an electron from an atom so that all the energy of the photon is absorbed in separating the electron and in imparting kinetic energy to it. *Contrast with* Compton effect.

photograph 51 Rosalind Franklin's x-ray diffraction photograph of the B form of deoxyribonucleic acid (*q.v.*). The photo provided Watson and Crick with certain data essential for constructing their three-dimensional model of DNA. It revealed the molecule to be a helix that undergoes a complete revolution every 3.4 nm with ten nucleotide pairs per turn. The photograph was taken May 1, 1952. *See* Appendix C, Bibliography, *Secret of Photo 51.*

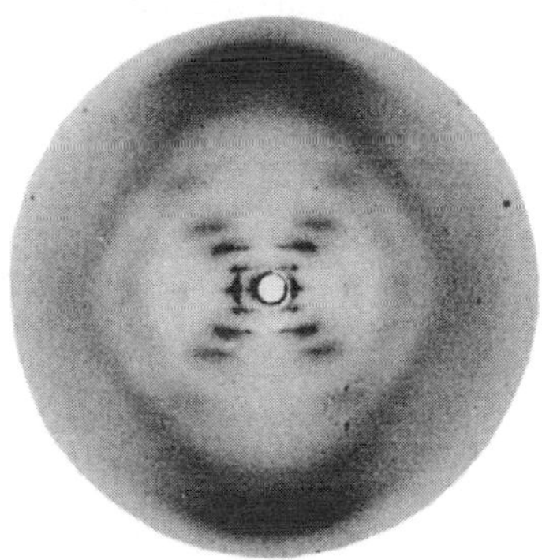

photographic rotation technique a technique used to establish the symmetry of a structure (such as a virus) observed in an electron micrograph. The micrograph is printed n times, the enlarging paper being rotated $360°/n$ between successive exposures. Structures with n-fold radial symmetry show reinforcement of detail, whereas the micrograph will show no reinforcement when tested for $n-1$ or $n+1$ symmetry.

photolyase *See* photoreactivating enzyme.

photolysis decomposition of compounds by radiant energy, especially light.

photomicrography the technique of making photographs through a light microscope.

photon a quantum of electromagnetic energy.

photoperiodism the response of organisms to varying periods of light. In plants, for example, the photoperiod controls flowering. *See* phytochrome.

photophosphorylation the addition of phosphate to AMP and ADP through the energy provided by light during photosynthesis.

photoreactivating enzyme an exonuclease catalyzing a photochemical reaction that removes UV-induced thymine dimers from DNA. This enzyme, called *photolyase,* is widely distributed in nature, from bacteria to marsupials, but is not found in placental mammals. Photolyase is very unusual in that it is the only enzyme (other than one involved in photosynthesis) that uses energy from light to drive its activities.

photoreactivation reversal of injury to cells caused by ultraviolet light, accomplished by postirradiation exposure to visible light waves. *See* Appendix C, 1949, Kelner; thymine dimer.

photoreceptor a biological light receptor. *See* cone, ommatidium, rod.

photosynthesis the enzymatic conversion of light energy into chemical energy in green plants, algae, and cyanobacteria that results in the formation of carbohydrates and oxygen from carbon dioxide and water. The net chemical reaction is $CO_2 + H_2O \rightarrow CH_2O + O_2$. Here CH_2O represents carbohydrate (*q.v.*) and chlorophyll (*q.v.*) functions as the light-harvesting molecule. The fixing of gaseous CO_2 into an organic product is catalyzed by ribulose-1, 5-bisphosphate carboxylase oxygenase (*q.v.*) *See* chloroplast.

phototrophs organisms that use light as the source of energy for metabolism and growth. Examples are cyanobacteria, algae and plants, all of which undergo photosynthesis (*q.v.*). *Contrast with* chemotrophs.

phototropism a growth movement induced by a light stimulus. Plant shoots grow toward a light source (are positively phototropic); roots grow away (show negative phototropism).

phragmoplast a differentiated region of the plant cell that forms during late anaphase or early telophase between the separating groups of chromosomes. The phragmoplast contains numerous microtubules that function to transport material used by the developing cell plate. Once the cell plate forms it divides the phragmoplast into two parts, and subsequently the cell plate is transformed into the middle lamella of the mature cell wall.

Phycomyces a genus belonging to the same family as the common bread molds *Mucor* and *Rhizopus.* Two species, *P. nitens* and *P. blakesleeanus,* are favorites for genetic studies. *See* Appendix A, Fungi, Zygomycota.

phyletic evolution the gradual transformation of one species into another without branching; anagenesis; vertical evolution.

phyletic speciation *See* phyletic evolution.

phylogenetic tree a diagram that portrays the hypothesized genealogical ties and sequence of historical ancestor/descendant relationships linking individual organisms, populations, or taxa. When species are considered they are represented as line segments, and points of branching correspond to subsequent speciation events. When possible, the lineage is presented in relation to a geological time scale. *See* Appendix C, 1936, Sturtevant and Dobzhansky; 1963, Margoliash; cladogram, PhyloCode.

phylogeny the relationships of groups of organisms as reflected by their evolutionary history.

phylum *See* classification.

Physarum polycephalum *See* Myxomycota.

physical map a map that shows the linear order of identifiable landmarks on chromosomal DNA, such as restriction enzyme cutting sites, sequence tagged sites, ORFs, and others. The distances between landmarks are always determined by methods other than genetic recombination (e.g., nucleotide sequencing, overlapping deletions in polytene chromosomes, electron micrographs of heteroduplex DNAs, etc.). The lowest resolution physical map would show the pattern of horizontal bands on each stained chromosome; the highest resolution map would give the complete nucleotide sequences for all chromosomes in the genome. Cloning vectors that can carry large DNA inserts are very useful in constructing physical maps of chromosomes or whole genomes. *Compare with* genetic map, linkage map. *See* bacterial artificial chromosomes (BACs), cosmid, P1 artificial chromosomes (PACs), yeast artificial chromosomes (YACs).

physiological saline an isotonic, aqueous solution of salts used for temporarily maintaining living cells.

physiology the study of the dynamic processes of living organisms.

phytochrome the molecule responsible for the photoperiodic control of flowering. During the day, a form of phytochrome that absorbs light at the far red end of the spectrum accumulates in plants. This form of the pigment inhibits flowering in short-day plants (*q.v.*) and stimulates flowering in long-day plants (*q.v.*). During darkness this compound reverts to a red-absorbing form that is stimulatory to the flowering of short-day plants and inhibitory to long-day plants. Whether a plant belongs to the short-or long-day class is genetically controlled.

phytohemagglutinin a lectin (*q.v.*) extracted from the red kidney bean, *Phaseolus vulgaris,* that agglutinates human erythrocytes and stimulates lymphocytes to undergo mitosis. *See* Appendix C, 1960, Nowell.

phytohormone a plant hormone. *See* auxin, cytokinin, gibberellin.

Phytophthora infestans the oomycete responsible for the highly destructive downy mildew or "late blight" of potatoes. This disease was the cause of Irish potato famine, 1843–1847, which was a principal reason for the waves of Irish immigration to the United States. *P. infestans* is incorrectly called a fungus in the earlier literature. *See* Appendix A, Protoctista, Oomycota.

phytotron a group of rooms used for growing plants under controlled, reproducible, environmental conditions.

pI isoelectric point (*q.v.*).

picogram 10^{-12} gram.

picornavirus a group of extremely small RNA viruses. The name is derived from the prefix *pico* (meaning small) + *RNA* + *virus.* The polio virus belongs to this group.

piebald designating an animal, especially a horse, having patches of black and white or of different colors. A pinto horse is piebald.

Pieris a genus of small butterflies extensively studied by ecological geneticists.

pig *See Sus scrofa.*

pigeon domestic breeds include the homing pigeon, carneaux, dragoon, white Maltese, white king, fantail, pouter, tumbler, roller, Jacobin, barb, carrier pigeon, and the ptarmigan. The species is *Columbia livia domestica*.

pigmentation *See* agouti, albinism, melanin, plumage pigmentation genes, *SLC 24A5*.

pilin *See* pilus.

pillotinas large symbiotic spirochaetes living in the hind gut of termites. These spirochaetes are of interest because they contain microtubules, and the presence of microtubules in spirochaetes supports the symbiotic theory of the origin of undulipodia. *See* motility symbiosis.

pilus (*plural* **pili**) a filamentous, hollow appendage extending from the surface of a conjugating bacterium. In *E. coli* pili serve as tubes through which DNA from a "male" (F^+ and Hfr) cell is transferred to a recipient, "female" (F^-) cell. Pili are composed of a glucophosphoprotein called *pilin*. Pilin molecules are arranged helically around a central canal about 2.5 nanometers in diameter. *See* androphages, F factor, fimbria.

pin the type of flower characterized by long styles and low anthers found among distylic species such as seen in the genus *Primula*. *See* thrum.

pinocytosis the engulfment of liquid droplets by a cell through the production of pinosomes (*q.v.*).

pinosome a membrane-bounded cytoplasmic vesicle produced by the budding off of localized invaginations of the plasma membrane. Recently pinocytosed fluids are segregated within the cell in pinosomes.

pioneer the first plant or animal species to become established in a previously uninhabited area.

PIR databases Protein Information Resource databases. *See* Appendix E.

pistil the female reproductive organ of the flower (*q.v.*), consisting of ovary, style, and stigma. *See* carpel.

pistillate designating a flower having one or more pistils and no stamens.

Pisum sativum the garden pea, Mendel's experimental organism. It has 7 pairs of chromosomes and a genome size of 4.1 gbp. *See* Appendix A, Plantae, Angiospermae, Dicotyledoneae, Leguminales, Appendix C, 1822–24, Knight, Gross, Seton; 1856, 1865, Mendel; 1990, Bhattcharyya *et al.*; Appendix E.

pitch the number of base pairs in one complete revolution of a DNA double helix. *See* twisting number.

pith parenchymatous tissue in the center of roots and stems.

Pithecanthropus erectus currently called *Homo erectus erectus* (*q.v.*).

pituitary dwarfism in humans, a form of ateliosis inherited as an autosomal recessive. There are two types: (1) *primordial dwarfism*, in which only growth hormone is deficient, and (2) *panhypopituitarism*, in which all hormones of the anterior pituitary are deficient. *See* ateliosis, hereditary growth hormone deficiencies, human growth hormone, Laron dwarfism, midget.

pituitary gland the master endocrine gland, which lies beneath the floor of the brain, within the skull, of vertebrates. *See* adenohypophysis, human growth hormone, hypothalamus, neurohypophysis.

pK dissociation constant (*q.v.*).

PKU phenylketonuria (*q.v.*).

placebo an inactive substance given to certain patients randomly chosen (without their knowledge) from a group. A new medicinal compound is administered to the other patients. The effectiveness of the compound is determined by comparing the progress of the treated patients with those receiving the placebo.

placenta an organ consisting of embryonic and maternal tissues in close union through which the embryo of a viviparous animal is nourished.

placoderms cartilagenous fishes representing the first vertebrates with jaws. Placoderm fossils appear in the upper Silurian and survive through the lower Permian.

plague a disease caused by *Yersinia pestis* (*q.v.*). Bubonic plague is transmitted to humans by the bites of fleas living on infected rats. Pneumonic plague is spread by inhaling respiratory droplets expelled by infected persons.

Plantae the plant kingdom. Its members are all eukaryotes made up of cells containing green plastids. *See* Appendix A; Eukaryotes.

plantain *See* Musaceae.

plant growth regulators small molecules synthesized by a variety of plant tissues that can act locally or at a distance to regulate plant development. The regulators include abscisic acid, auxins, cytokinins, ethylene, and gibberellins (*all of which see*).

plaque a clear, round area on an otherwise opaque layer of bacteria or tissue-cultured cells where the cells have been lysed by a virulent virus. In the example shown below, the petri dish contains a gel in which a nutrient broth is suspended. Covering the surface of the medium is a lawn of bacteria. These arose from a layer of 1×10^8 bacteria that was deposited on the agar surface. Subsequently, a small number of viruses was spread over these cells. The holes represent points at which a phage particle was present. The initial phage particle was adsorbed to a bacterium and after a short period the bacterium lysed, releasing new phage. These new phage in turn attacked neighboring bacteria and produced more phage. The process continued until holes visible to the naked eye appeared. Different phages can sometimes be recognized by the morphology of the plaques they produce. The numerous small plaques in the figure are from T6 bacteriophage, while the four large plaques are from phage T7. Animal viruses will also attack monolayer cultures of animal cells on petri plates, so it is possible to assay virus titer in the same way by plaque counts. *See* Appendix C, 1932, Ellis and Delbrück; 1952, Dulbecco.

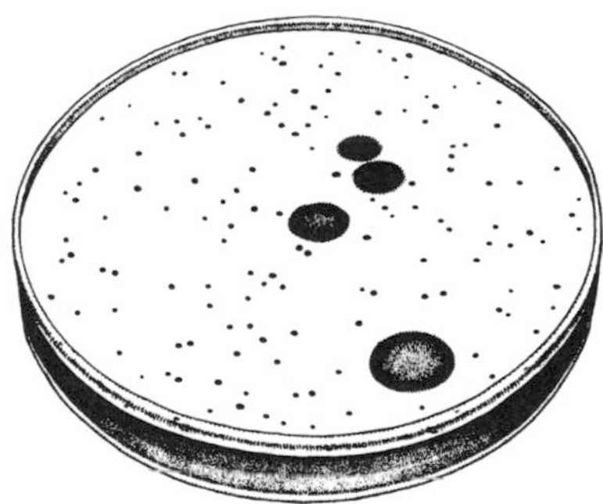

plaque assay a technique for counting the number of complete, infective phage in a culture by an appropriate dilution to ensure that no more than one phage can infect a given host cell, followed by counting the number of plaques that develop on a bacterial lawn.

plaque-forming cells in immunology, antibody-secreting cells that can cause a hemolytic plaque with the aid of complement on a lawn of erythrocytes. The term is also applied to cells in certain assay systems where the cell killing that creates the plaque is cell mediated, rather than antibody dependent.

plasma *See* blood plasma.

plasmablasts highly proliferative cells that are developmental intermediates between small B lymphocytes and immunoglobulin-secreting mature plasma cells.

plasma cell a terminally differentiated immunoglobulin-secreting cell of the B lymphocyte lineage.

plasmacytoma synonymous with myeloma (*q.v.*).

plasmagene a self-replicating cytoplasmic gene within an organelle or a symbiont of a eukaryotic cell or in a plasmid of a bacterial cell.

plasmalemma plasma membrane (*q.v.*).

plasma lipoproteins a multicomponent complex of proteins and lipids circulating in the plasma. Plasma lipoproteins are grouped by their densities into four classes: high-density lipoproteins (HDL), low-density lipoproteins (LDL), intermediate-density lipoproteins (IDL), and very-low-density lipoproteins (VLDL). LDLs are about 25% protein and 75% lipid, and about half of the lipid is cholesterol. LDLs serve as the major cholesterol transport system in human plasma. *See* Appendix C, 1975, Goldstein and Brown; familial hypercholesterolemia.

plasma membrane the membrane that surrounds the cell. It is made up of a bilayer of phospholipids that contain interspersed proteins. Cells have ways of confining membrane proteins to specific regions. For example, the cells that line the digestive tract have the proteins that function in transport localized at their apical surfaces. *See* cystic fibrosis, fluid mosaic concept, lipid bilayer model, unit membrane.

plasma membrane receptors *See* cell-surface receptors.

plasma protein any of the dissolved proteins of vertebrate blood plasma that are responsible for holding fluid in blood vessels by osmosis. *See* Appendix A, 1955, Smithies.

plasma thromboplastin component a protein (also called factor 9) that participates in the cascade of reactions that results in blood clotting (*q.v.*). A deficiency of factor 9 causes hemophilia B. The factor 9 gene, *HEMA* (*q.v.*), is X-linked. It encodes a protein of 461 amino acids, of which 46 are in a leader sequence peptide (*q.v.*), and the remaining 415 make up the mature clotting growth factor. This contains two EGF repeats. A catalytic domain gives it a serine protease activity. *See* epidermal growth factor, hemophilia, serine proteases.

plasma transferrins beta globins that bind and transport iron to the bone marrow and tissue-storage areas. Numerous heritable transferrin variants are known.

plasmid an extrachromosomal genetic element found in a variety of bacterial species that generally confers some evolutionary advantage to the host cell (i.e., resistance to antibiotics, production of colicins, etc.). Plasmids are double-stranded, closed DNA molecules ranging in size from 1 to 200 kilobases. Plasmids whose replication is coupled to that of the host so that only a few would be present per bacterium are said to be under *stringent control*. Under *relaxed control*, the number of plasmids per host cell may be from 10 to 100. *See* Appendix C, 1952, J. Lederberg; *Bacillus*, bacteriocins, *nif* (nitrogen-fixing) genes, pBR322, R (resistance) plasmid, sym-plasmid, Ti plasmid, *Vibrio cholerae*, virulence plasmid.

plasmid cloning vector a plasmid used in recombinant DNA experiments as an acceptor of foreign DNA. Plasmid cloning vectors are generally small and replicate in a relaxed fashion. They are marked with antibiotic resistance genes and contain recognition sites for restriction endonucleases in regions of the plasmid that are not essential for its replication. One widely used plasmid cloning vector is pBR322 (*q.v.*). *See* Appendix C, 1973, Cohen *et al.*; bacterial artificial chromosomes (BACs), P1 artificial chromosomes (PACs), Ti plasmid.

plasmid conduction the process whereby a conjugative plasmid can help a nonmobilizable plasmid to be transferred from a donor to a recipient cell. Nonmobilizable plasmids cannot prepare their DNA for transfer, but can become mobilized by recombination with a conjugative plasmid to form a single transferable DNA molecule. *See* plasmid donation, relaxation complex.

plasmid donation the process by which a nonconjugative plasmid is transferred from a donor to a recipient cell via the effective contact function provided by a conjugative plasmid. In *E. coli*, for example, the ColE 1 plasmid does not have genes for the establishment of effective contact, but an F plasmid in the same cell is conjugative and can provide this function.

plasmid engineering *See* recombinant DNA research, Ti plasmid.

plasmid fusion *See* replicon fusion.

plasmid rescue *See* transformation rescue.

plasmin an enzyme, cleaved from plasminogen, that hydrolyzes fibrin. *See* blood clotting.

plasminogen a blood proenzyme that is activated by cleavage of a single arg-val peptide bond to form the functional enzyme plasmin (*q.v.*).

plasmodesmata cytoplasmic threads that form delicate protoplasmic connections between adjacent plant cells.

plasmodium a multinucleate mass of cytoplasm lacking internal cell boundaries. The term is used to refer to the amoeboid stage in the life cycle of Sporozoa (*q.v.*) and to the vegetative stage of Myxomycota (*q.v.*).

Plasmodium the genus of protozoa that causes malaria in reptiles, birds, and mammals. The genus is estimated to have arisen 100–180 million years ago. Five species of medical importance are described below. Three of these species cause malaria characterized by a rhythmicity in the periodic fevers and chills experienced by the human host. These symptoms are associated with the release of merozoites (*q.v.*) from the synchronously rupturing red blood cells. In tertian and quartan malaria the cycles are 48 and 72 hours, respectively. In subtertian malaria the 24-hour period is frequently modified. *Plasmodium vivax* and *Plasmodium malariae* are species that cause benign tertian malaria and quartan malaria, respectively, both less serious diseases of humans. *P. vivax* is known to have 600–1,000 copies of genes that function to suppress the immune defenses of the host. These are clustered near the telomeres. *Plasmodium cynomolgi* and *P. knowlesi* are species causing malaria in monkeys. The liver stage in the malaria cycle was first observed in *P. cynomolgi* and the role of the circumsporozoite protein as an immunogenic decoy has been studied in *P. knowlesi*. *See* Appendix A, Protoctista, Protozoa, Apicomplexa, Sporozoa; Appendix C, 1948, Shortt and Garnham; *Plasmodium* life cycle.

Plasmodium falciparum the parasite that causes subtertian malaria, the most deadly form of the disease in humans. It is a sporozoan with a genome size of 22.8 mbp and about 5,300 genes spread among 14 chromosomes. The base composition is unusual in that 81% of the bases are As or Ts. The more common situation is seen in the yeast *Saccharomyces* where As and Ts make up only 62% of the total. About 60% of the proteins predicted to be encoded by the *Plasmodium* genome have no similarity to the proteins known in other organisms. Of particular interest are highly variable gene families that are clustered near the telomeres. Some of these genes encode proteins that are exported to the surface of the infected red blood cell where they bind to receptors in various tissues of the host. *P. falciparum* appears to have arisen within the past 6,000 years as the result of the lateral transfer between bird and human hosts. *See* Appendix C, 2002, Gardner *et al.*; Appendix E; Appendix F; apicoplast, artemisinin, *Plasmodium, Plasmodium* life cycle, *pfcrt* gene.

***Plasmodium* life cycle** the developmental changes which the malaria parasite, *Plasmodium* (*q.v.*) undergoes as it is transmitted from a vector to a host, and back to a vector. The life cycle of *P. falciparum*, the major parasite causing human malaria, is shown in the illustration. The saliva of an infected *Anopheles* (*q.v.*) female transmits the parasite in the sporozoite form when it bites a human being. Each sporozoite synthesizes a circumsporozoite (CS) protein (*q.v.*), which serves as an immune decoy. The sporozoites reach the liver, where they invade hepatocytes. They then undergo *schizogony*, a replicative process involving multiple rounds of rapid mitoses without cytokinesis (*q.v.*). Subsequently, the cytoplasm segments within the multinucleate mass to form hundreds of uninucleate merozoites. Following rupture of the hepatocyte, the merozoites enter the circulatory system and invade red blood cells.

Within the erythrocyte the parasite transforms into the trophozoic or feeding stage, ingesting massive mounts of hemoglobin (*q.v.*) from the cytoplasm of its host. The trophozoite hydrolyzes the globin into small peptides and releases the heme, which is converted in the food vacuole (*q.v.*) to a non-toxic crystalline form, called *hemozoin*. Shortly after its infection the erythrocyte forms thousands of projections on its surface, which contain adhesion molecules produced by the parasite. The infected cells can now adhere to the walls of the blood vessels and avoid being swept into and destroyed by the spleen. The trophocyte stage ends when the cells undergo schizogony. The host erythrocyte then ruptures, releasing merozoites which invade new erythrocytes. Erythrocytic schizogony is synchronized, causing cells to rupture in synchrony. This is the reason for the cyclic fever paroxysms experienced by humans suffering from malaria.

The merozoites in some of the erythrocytes differentiate into male and female gametocytes, and the erythrocytes containing these do not rupture. These gametocytes can be taken up by mosquitoes in the next blood meal, and it is in the mosquito that the sexual phase of the parasite life cycle occurs. Here male and female gametes fuse, resulting in the only diploid stage, the ookinete. This motile cell transverses the midgut epithelium of the mosquito and comes to rest between the epithelial cells and the basal lamina (an acellular membrane surrounding

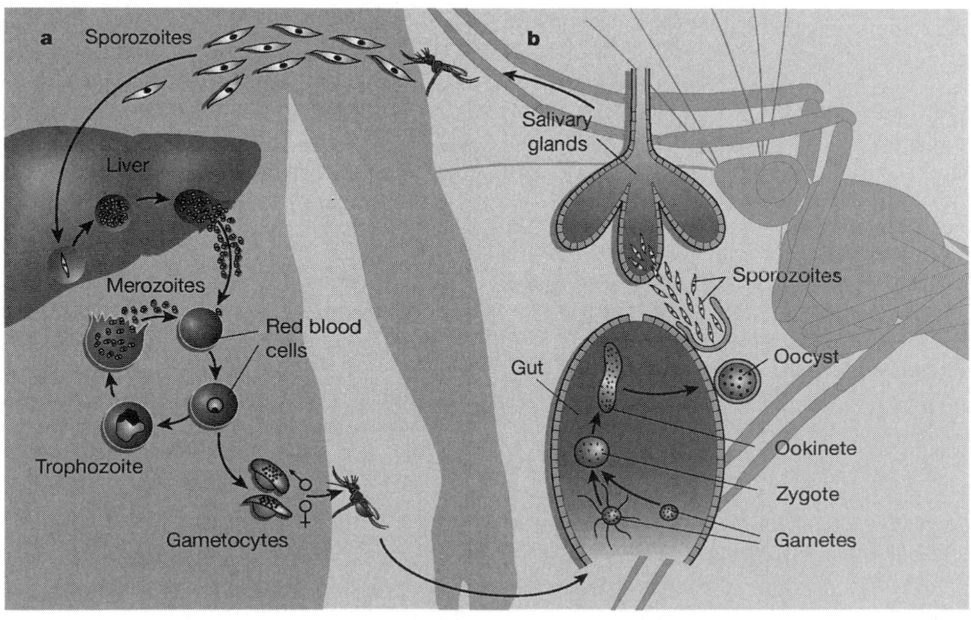

Plasmodium life cycle

the outside of the gut). The ookinete now completes meiosis to form haploid nuclei, which divide within a sporocyst to form thousands of sporozoites. When the mature sporocyst ruptures, the motile sporozoites penetrate the basal lamina and enter the hemocoel.

The sporozoites now migrate to the salivary glands, traverse the epithelial cells of the gland, and come to rest in the lumen. The life cycle is completed when the mosquito transfers sporozoites to a human during a blood meal. *See* Appendix C, 1880, Laveran; 1898, Ross; 1899, Grassi; 1948, Shortt and Garnham; hemoglobin C, sickle cell trait, vacuoles.

Plasmodium yoelii yoelii a species that causes malaria in wild African rats which can be readily bred in the laboratory. *P. yoelii* is used for comparative genomic and proteomic studies with *P. falciparum*. The *yoelii* genome has been sequenced and found to contain 23.1 mbp of DNA. There are about 5,800 genes spread among 14 chromosomes. The A+T content is 77%. The telomeric regions of chromosomes contain large numbers of genes that are homologs of those that function in *P. vivax* to evade the host's immune defenses.

plasmogamy the fusion of protoplasts of two haploid cells without the fusion of their nuclei, as in certain fungi.

plasmon all extrachromosomal hereditary agents considered collectively.

plasmosome a term in the older literature referring to the nucleolus (*q.v.*).

plasmotomy fission, unrelated to nuclear division, of a multinucleated protist into two or more multinucleated sibling cells.

plastid a self-replicating cytoplasmic organelle of algal and plant cells, such as a chloroplast, chromoplast, elaioplast, or leukoplast.

plastome a plastid genome.

plastome-genome incompatibility a form of genetic disfunction that affects plastid development. Examples are plastids inherited from one parent which fail to become fully pigmented in the nuclear background of species hybrids.

plastoquinone a group of quinones (*q.v.*) involved in the transport of electrons during photosynthesis in chloroplasts.

plate 1. a flat, round dish (petri plate) containing agar and nutrients for the culture of bacteria. 2. to spread or inoculate cells on the surface of semisolid medium in such a culture dish. 3. a geological plate. *See* plate tectonics.

platelet-derived growth factor (PDGF) a protein synthesized by platelets that is released into the serum during blood clotting. PDGF represents the major growth factor in human serum, and it is a potent mitogen for connective tissue and glial cells. There are extensive similarities between the amino acid sequences of PDGF and the product of the *v-sis* oncogene of the simian sarcoma virus (*q.v.*). This suggests that *v-sis* resulted from viral recombination with a host gene encoding PDGF. *See* Appendix C, 1983, Doolittle *et al.*; proto-oncogenes.

platelets anucleate, oval, colorless corpuscles present in blood. Platelets, which are one-third to one-half the size of erythrocytes, originate from projections pinched off the surface of megakaryocytes (*q.v.*), and function in blood clotting. *See* thrombocytopenia.

plate tectonics a theory that provides an explanation for the present-day global distribution of mountain building, volcanism, and earthquake activity along a series of linear belts. The theory postulates that the surface of the earth is a rigid outer shell, the lithosphere, which lies on a hotter semiplastic athenosphere. The brittle lithosphere is broken into a series of tectonic plates that move horizontally across the earth's surface. The plates are at once jostling and being constrained by neighboring plates, like an ever-shifting mosaic of tiles that change shape as they fill in intervening gaps. It is at plate boundaries that mountains are built and volcano and earthquake activities occur. Present-day continents are parts of tectonic plates, and they ride passively as magma comes to the surface and then moves laterally as the sea floor spreads. The driving mechanism of the convection currents is heat from the de cay of radioactive elements in the earth's core. Over long periods of time, rocks in the mantle deform like a fluid and move at rates in the order of centimeters a year. The movement of tectonic plates has been going on for at least 600 million years and sometimes can provide explanations for the biogeographic distribution of plants and animals. *See* Appendix C, 1968, Morgan, McKenzie, and Le Pinchon; biogeographic realms, continental drift, sea floor spreading, Sulawesi.

plating efficiency *See* absolute plating efficiency, relative plating efficiency.

platyrrhine referring to primates of the infraorder Platyrrhini that includes the New World monkeys, marmosets, and tamarins. These primates are characterized by nostrils that are far apart and face to the sides, and they have prehensile tails. *Compare with* catarrhine.

playback experiment an experiment designed to recover a DNA strand that has been saturated with RNA (*see* RNA-driven hybridization), and then using it in a further reassociation reaction to show that its $C_0t_{1/2}$ (*q.v.*) corresponds to that expected of nonrepetitive DNA.

plectonemic spiral a spiral in which two parallel threads coil in the same direction about one another and cannot be separated unless uncoiled. The component strands of a DNA duplex are plectonemically coiled. *See* paranemic spiral, relational coiling.

pleiomorphism the occurrence of variable phenotypes in a genetically uniform group of organisms. *See* phenotypic plasticity.

pleiotropy the phenomenon in which a single gene is responsible for a number of distinct and seemingly unrelated phenotypic effects.

Pleistocene the ice age, lasting from 10,000 B.C. to the beginning of the Pliocene. One of the two epochs of the Quaternary period. *Homo erectus* appeared, then *Homo sapiens*. *See* geologic time divisions.

pleomorphic having more than one form or shape.

plesiomorphic 1. in classification, referring to a character state that occurs in the group of organisms being considered, but also outside the group. Traits of this type cannot be used to define the group or to indicate that its members were derived from a common ancestor. 2. in evolution, an original primitive feature thought to have arisen in an ancestor of all the taxa being considered. *See* apomorphic, cladogram.

Pleurodeles salamanders of two species belonging to this genus, *P. waltlii* and *P. poireti*, have been studied both genetically and cytologically. Working maps of the oocyte lampbrush chromosomes of both species are available.

pleuropneumonia-like organisms a group of bacteria that do not form cell walls. PPLOs are included in the phylum Aphragmabacteria (*see* Appendix A). One PPLO, *Mycoplasma pneumoniae*, is the cause in humans of an atypical pneumonia.

Pliocene the most recent epoch in the Tertiary period, during which the first hominids appeared. North and South America joined together, and advanced placental mammals migrated into South America where they killed off many marsupials and primitive placental mammals. *See* Australopithecine, geologic time divisions.

-ploid a combining form used in cytology and genetics to designate a particular multiple of the chromosome set of the nucleus of an organism, as 16-ploid, 32-ploid, etc.

ploidy *See* polyploidy.

plumage pigmentation genes a group of genes controlling pigmentation of chicken feathers. Pigment will not be produced unless the gene C is present. A second gene *I*, which inhibits pigment formation, is located on a different chromosome. The White Leghorn breed of chickens has the genotype *IICC*, whereas the White Plymouth Rock is *iicc*. *See* poultry breeds.

pluripotent pertaining to any cell or early embryonic tissue that has a number of possible developmental fates but not all of the fates possessed by the zygote; also known as *multipotential*. *See* stem cells, totipotent.

plus and minus techniques *See* DNA sequencing techniques.

plus (+) and minus (–) viral strands 1. in a single-stranded RNA virus, a plus strand is one having the same polarity as viral mRNA and containing codon sequences that can be translated into viral protein. A minus strand is a noncoding strand that must be copied by an RNA-dependent RNA polymerase to produce a translatable mRNA. 2. in a single-stranded DNA virus, a plus strand is one contained in the virus particle or any strand having the same base sequence. A minus strand has a base sequence complementary to the plus strand; mRNA can be transcribed from the minus strand. *See* Baltimore classification of viruses, hairpin ribozyme.

Plectonemic spiral

No. of events observed	0	1	2	3	4	5	6	7
Probability	.3012	.3614	.2169	.0867	.0260	.0062	.0012	.0002

Poisson distribution

P-M hybrid dysgenesis *See* hybrid dysgenesis, M strain, P strain.

pneumococci bacteria that cause an inflammation of the lungs (classical lobar pneumonia). These bacteria belong to virulent strains of *Streptococcus pneumoniae* that can synthesize an external capsule which contains a complex polysaccharide. *See Streptococcus.*

Pneumococcus Transforming Principle (PTP) a substance isolated from heat-killed, virulent pneumococci which when added to cultures of living, non-virulent bacteria cause virulent pneumococci to appear. Thus bacteria of one phenotype are transformed to an alternative phenotype by a non-living material. The demonstration that the PTP is DNA rather than protein was one of the earliest lines of evidence that genes are made of DNA. *See* Appendix C, 1928, Griffith; 1944, Avery, MacLeod, and McCarty; transformation.

pod corn *Zea mays tunicata*, a primitive variety of corn characterized by kernels, each of which is enclosed in a husk.

podophyllotoxin an alkaloid isolated from the root of the mandrake, *Podophyllum peltatum*. Podophyllotoxin binds to tubulin (*q.v.*) and prevents it from polymerizing into microtubules. *See* colchicine, paclitaxel, spindle poison, vinca alkaloids.

Podospora anserina a filamentous fungus living on dung. Its genome size is 34 mbp distributed among 7 chromosomes. The fungus serves as a model for studies of the role of mitochondria in aging. *Podospora mycelia* undergo a time-dependent change during their culture. The pigmentation of the hyphae increases, they stop growing, and the apical cells die. During this "aging" process certain parts of the mitochondrial DNA become unstable. A mobile intron detaches from a specific gene, circularizes, and undergoes amplification. The gene involved normally functions to transcribe a subunit of cytochrome c oxidase. Therefore the energy production of the cell is compromized, and it dies. *See* Appendix A, Fungi, Ascomycota; cytochromes, gene amplification, petites.

point mutation **1.** in classical genetics, any mutation that is not associated with a cytologically detectable chromosomal aberration or one that has no effect on crossing over (and therefore is not an inversion) and complements nearby lethals (and therefore is not a deficiency). **2.** in molecular genetics, a mutation caused by the substitution of one nucleotide for another. *See* rotational base substitution.

Poisson distribution a function that assigns probabilities to the sequence of outcomes of observing no events of a specified type, one event, two events, and so on without limit. Events following a Poisson distribution are completely randomized. Suppose that within a defined spatial or temporal region one looks at the events that occur in nonoverlapping subregions. A Poisson distribution for the number of events observed in the total region implies that the events occurring in the subregions do not affect each other. A Poisson distribution will not be found if the events are correlated positively (in the case of clumping) or negatively (in the case of mutual repulsion). The Poisson is specified by the average number of events per observation, and its mean and variance are equal. The formula of the function is $P_i = (m^i e^{-m})/i!$ where m is the mean number of events; $i!$ is the factorial $i(i-1)(i-2) \ldots (2)(1)$; e is the base of natural logarithms, and i is the number for which the probability P_i is given. If $m = 1.2$, for example, the distribution is as shown in the table that appears above. Many natural distributions follow a Poisson, including the number of radioactive disintegrations of a radioisotope in a fixed period of time, or the number of larvae of a particular invertebrate species captured by towing a plankton net through a specified volume of seawater. The distribution was formulated by Simeon D. Poisson (1781–1840), a French mathematician and physicist.

pokeweed mitogen a lectin (*q.v.*) extracted from rhizomes of the pokeweed, *Phytolocca americana*, that stimulates the proliferation of lymphocytes, particularly mouse T and B lymphocytes.

poky the most famous of the mitochondrial mutants of *Neurospora*. These slow-growing, respiration-defective fungi have reduced numbers of ribosomes. This, in turn, is due to deletions in mtDNA that impair the synthesis of rRNA. *See* petites.

pol I, II, III *See* DNA polymerase.

polar referring to water-soluble chemical groups such as a hydrophilic side chain of an amino acid.

polar body the minute cell produced and discarded during the development of an oocyte. A polar body contains one of the nuclei derived from the first or second division of meiosis, but has practically no cytoplasm. *See* ootid nucleus.

polar fusion nucleus in plants, the product of the fusion of the two polar nuclei. This, after fusing with a male nucleus, gives rise to the tripoid endosperm nucleus. *See* double fertilization.

polar gene conversion a phenomenon in which a gradient of conversion frequencies exists from one end of a gene to the other; sites closer to one end of a gene usually have higher conversion frequencies than do those farther from that end.

polar granules electron-dense, membrane-less, RNA-protein complexes often associated with mitochondria, found in the pole plasm (*q.v.*) and subsequently incorporated into primordial germ cells (*q.v.*) in a variety of species. Also called *germinal granules* or *P granules*.

polarity gradient the quantitative effect of a polarity mutation in one gene on the expression of later genes in the operon. The effect is a function of the distance between the nonsense codon and the next chain-initiation signal.

polarity mutant 1. a mutant gene that is able to reduce the rate of synthesis of the proteins that normally would be produced by wild-type alleles of the genes lying beyond it on the chromosome. Such genes exert their effect during the translation of a polycistronic message (*q.v.*). *See* regulator gene, translation. 2. a mutant gene that influences polarized patterns of embryonic development. *See* bicoid, engrailed, hunchback, maternal polarity mutants, zygotic segmentation mutants.

polarization microscope a compound light microscope used for studying the anisotropic properties of objects and for rendering objects visible because of their optical anisotropy.

polar nuclei *See* ootid nucleus, pollen grain, polocyte.

polaron a chromosomal segment within which polarized genetic recombination takes place by gene conversion.

polar tubules microtubules of the spindle apparatus that originate at the centriolar or polar regions of the cell. *See* chromosomal tubules.

pole cell one of the cells that are precociously segregated into the posterior pole of the insect embryo before blastoderm formation. Among these cells are the progenitors of the germ cells. *See* Appendix C, 1866, Metchnikoff.

pole plasm in many vertebrate and invertebrate species, a specialized cytoplasmic region of the egg or the zygote that contains germ-cell determinants and other maternal products required for normal development in the early embryo. The cytoplasm located at the posterior pole of *Drosophila* and the vegetal pole of *Xenopus* embryos are examples of pole plasm. In *Drosophila*, several maternal effect genes involved in pole plasm formation have been identified and are known as *grandchildless* genes (*q.v.*). Females carrying mutations in these genes produce embryos that lack polar granules (*q.v.*) and show other developmental defects. *See* cytoplasmic determinants, cytoplasmic localization, maternal effect gene, maternal polarity mutants.

polio virus the cause of poliomyelitis (infantile paralysis). It is a positive-stranded RNA virus with a 6.1 kb genome. Enormous polysomes (*q.v.*) containing 60 or more ribosomes occur in infected cells. The entire genome is translated from a single initiation site to form a single polyprotein molecule. This is subsequently cleaved into both structural and nonstructural proteins. The structural proteins assemble to form the icosahedral capsule of the virus. The virus is remarkably stable, and it has been successfully grown from archaeological specimens centuries old. *See* icosahedron.

Polish wheat I *Triticum polonicum* (N = 14). *See* wheat.

pollen grain a microspore in flowering plants that germinates to form the male gametophyte (pollen grain plus pollen tube), which contains three haploid nuclei. One of these fertilizes the ovum, a second fuses with the two polar nuclei to form the 3N endosperm, and the third (the vegetative nucleus) degenerates once double fertilization (*q.v.*) has been accomplished.

pollen mother cell microsporocyte.

pollen-restoring gene a gene that permits normal microsporogenesis to occur in the presence of a cytoplasmic male sterility factor.

pollen tube the tube formed from a germinating pollen grain that carries male gametes to the ovum. *See* Appendix C, 1830, Amici.

pollination the transfer of pollen from anther to stigma. *See* Appendix C; 1694, Camerarius; pollen grain, self-pollination.

polocyte the small degenerate sister cell of the secondary oocyte. This cell generally divides into two polar bodies, which disintegrate. *See* polar body.

polyacrylamide gel a gel prepared by mixing a monomer (acrylamide) with a cross-linking agent (*N,N′*-methylenebisacrylamide) in the presence of a polymerizing agent. An insoluble three-dimensional network of monomer chains is formed. In water, the network becomes hydrated. Depending upon the relative proportions of the ingredients, it is possible to prepare gels with different pore sizes. The gels can then be used to separate biological molecules like proteins of a given range of sizes.

polyacrylamide gel electrophoresis *See* electrophoresis.

polyadenylation enzymatic addition of several adenine nucleotides to the 3′ end of mRNA molecules as part of the processing that primary RNA transcripts undergo prior to transport from the nucleus to the cytoplasm. The added segment is referred to as a "poly-A tail." Histone mRNAs lack poly-A tails. *See* Appendix C, 1971, Darnell *et al.*; posttranscriptional processing.

polyandry the state of having more than one male mate at one time.

poly-A tail *See* polyadenylation.

polycentric chromosome polycentromeric chromosome. *See* centromere.

polycentromeric chromosome *See* centromere.

polycistronic mRNA a messenger RNA that encodes two or more proteins. The messenger may later be cleaved into individual messages, each of which is translated into a single protein, or a giant polypeptide chain may be translated that is later cleaved to yield the individual proteins. Polycistronic mRNAs are common in prokaryotes. For example, the *lac* operon (*q.v.*) of *E. coli* generates a polycistronic mRNA. *Contrast with* monocistronic mRNA. *See* histone genes, polyprotein, retroviruses, transcription unit, trans-splicing, ubiquitin.

polyclonal an adjective applied to cells or molecules arising from more than one clone; e.g., an antigenic preparation (even a highly purified one) elicits the synthesis of various immunoglobulin molecules. These antibodies would react specifically with different components of the complex antigen molecule. Thus, the antibody preparation generated by such an antigen would be polyclonal in the sense that it would contain immunoglobulins synthesized by different clones of B lymphocytes.

polyclone *See* compartmentalization.

Polycomb (Pc) a *Drosophila* mutation that produces additional sex combs (*q.v.*) on the second and third pairs of legs in males. *Pc* is at 3-47.1 on the genetic map. Proteins encoded by the normal allele inhibit *Hox* genes. Binding sites of Pc proteins have been visualized by immunochemical staining of giant polytene chromosomes. The Pc protein and the heterochromatin-associated protein 1 (HP1) (*q.v.*) share a homologous domain 37 amino acids long near their N termini. The Pc proteins bind to histone 3 molecules (*q.v.*) that have been tagged by the addition of methyl groups to their tails. *See* Appendix C, 1989, Zink and Paro; histones, SUMO proteins.

polycomplex structures, observed in certain insects, within oocyte nuclei, formed by the fusion of components from synaptonemal complexes (*q.v.*) that have detached from the diplotene chromosomes.

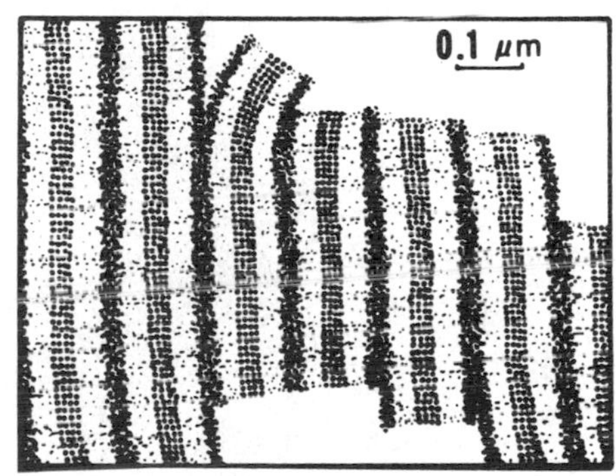

polycystic kidney disease one of the most common genetic diseases in humans with about 1 in 1,000 individuals affected. The major feature of PKD is the development of fluid-filled cysts in the kidneys that damage or destroy them. The disease is due to dominant mutations in a gene that maps to 13.3 on the short arm of chromosome 16. The gene (*PKD1*) spans 52 kilobase pairs and generates a transcript containing 14,148 nucleotides distributed among 46 exons. The predicted PKD1 protein, polycystin, is a glycoprotein with a carboxyl tail that contains about 225 amino acids and which protrudes into the cytoplasm. This is followed by about 1,500 amino acids containing transmembrane domains. The *N*-terminal extracellular portion of the protein contains about 2,500 amino acids, and these are subdivided into domains that are thought to bind to a variety of proteins and carbohydrates in the extracellular matrix. Polycystin is thought to function in cellular signal transduction (*q.v.*) mediated through its cytoplasmic tail. *See* Appendix C, 1995, Hughes *et al.*

polycythemia vera a disease in humans characterized by the overproduction of red blood cells. Erythroblasts in the bone marrow are hypersensitive to erythropoietin (*q.v.*). *See* Janis kinase 2.

polydactyly the occurrence of more than the usual number of fingers or toes.

polyembryony the formation of multiple embryos from a zygote by its fission at an early developmental stage. Monozygotic twins constitute the simplest example of polyembryony. Monozygotic quadruplets are commonly formed by armadillos. In certain parasitic wasps, as many as 2,000 embryos can be formed by polyembryony from a single zygote.

polyestrous mammal *See* estrous cycle.

polyethylene glycol a chemical used to promote the fusion of tissue-cultured cells, as in the production of a hybridoma (*q.v.*).

polyfusome a gelatinous mass assembled by the fusion of the adjacent fusomes (*q.v.*) formed at consecutive cystocyte divisions in *Drosophila*. The diagram illustrates a polyfusome in a cystocyte clone during the divison of 8 cells into 16. Cell 1 is obscured by the cells lying above it. In each of the other seven cells, a spindle and a ring canal (*q.v.*) can be seen. Pairs of centrioles lie at the spindle poles. The polyfusome protrudes through each ring canal and touches one pole of each spindle. As a result of this orientation, one cell of each dividing pair will retain all previously formed ring canals, while the other will receive none. These spindle-fusome alignments during the cycle of cystocyte divisions (*q.v.*) produce a branched chain of interconnected cells. There are always two central cells, each with four ring canals. In female sterile mutations characterized by ovarian tumors, polyfusomes often fail to form properly, and the pattern of germ cell divisions and their differentiation are abnormal. *See bag of marbles (bam)*, fusome, *hu-li tai shao (hts)*, *otu*, pro-oocyte.

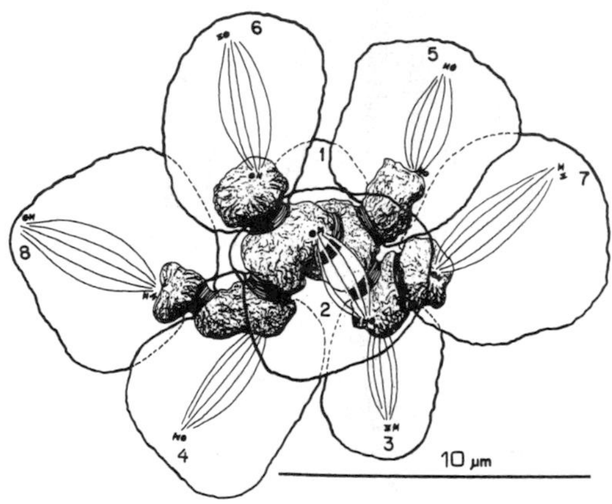

polygamy polandry and/or polygyny. *Compare with* monogamy.

polygene one of a group of genes that together control a quantative character. *See* Appendix C, 1941, Mather; oligogene, quantitative inheritance.

polygenic character a quantitatively variable phenotype dependent on the interaction of numerous genes.

polyglucosan a polymer such as glycogen made up of a chain of glucose units.

polygyny the mating of a male with more than one female during a single reproductive cycle. *Compare with* monogamy, polyandry.

polyhedrin *See* baculoviruses.

polylinker site a stretch of DNA engineered to have multiple sites for cleavage by specific restriction endonucleases (*q.v.*).

polymer a macromolecule composed of a covalently bonded collection of repeating subunits or monomers linked together during a repetitive series of similar chemical reactions. Each strand of DNA is a linear polymer of nucleotide monomers. A linear polypeptide chain is a polymer of amino acid monomers. *See* monomer, oligomer.

polymerase any enzyme that catalyzes the formation of DNA or RNA molecules from deoxyribonucleotides and ribonucleotides, respectively (e.g., DNA polymerase, RNA polymerase).

polymerase chain reaction (PCR) a highly sensitive technique for quickly amplifying a DNA segment. PCR involves three major steps. First, the reaction mixture containing the target DNA is heated to separate complementary DNA strands. Second, the mixture is cooled, and synthetic primers with nucleotide sequences complementary to each end of the DNA are allowed to anneal to the separated strands. Finally, the temperature is raised again, and a heat-stable DNA polymerase (*q.v.*) in the reaction mixture synthesizes new DNA strands by adding nucleotide bases to the primers. These steps are repeated for a series of replication cycles, each lasting a few minutes, in an automated cycler that controls the required temperature variations. The number of DNA strands doubles with each successive cycle, resulting in an exponential increase in the number of copies of the target DNA. Twenty cycles yield a millionfold amplification; 30 cycles yield an amplification factor of 1 billion. The ability of PCR to quickly and accurately generate billions of copies of the minutest amount of DNA has created a wealth of new practical applications in numerous areas, including DNA cloning and sequencing, screening for genetic disorders, detection of disease-causing organisms, DNA fingerprinting, and examination of species differences. *See* Appendix C, 1985, Saiki, Mullis *et al.*; 1993, Smith and Mullis; ligase chain reaction, reverse transcription-polymerase chain reaction, Taq DNA polymerase.

polymerization the formation of a polymer from a population of monomeric molecules.

polymerization start site the nucleotide in a DNA promoter sequence from which the first nucleotide of an RNA transcript is synthesized.

polymorphic locus a genetic locus, in a population, at which the most common allele has a frequency less than 0.95. *Compare with* monomorphic locus.

polymorphism the existence of two or more genetically different classes in the same interbreeding population (Rh-positive and Rh-negative humans, for example). The polymorphism may be transient, or the proportions of the different classes may remain the same for many generations. In the latter case, the phenomenon is referred to as *balanced polymorphism*. If the classes are located in different regions, *geographic polymorphism* exists. *See* Appendix C, 1954, Allison; 1966, Lewontin and Hubby.

polymorphonuclear leukocyte *See* granulocyte.

polyneme hypothesis the concept that a newly formed chromatid contains more than one DNA duplex. *Contrast with* unineme hypothesis.

polynucleotide a linear sequence of 20 or more joined nucleotides. *See* oligonucleotide.

polynucleotide kinase an enzyme that phosphorylates the 5′ hydroxyl termini produced by endonucleases (*q.v.*).

polynucleotide phosphorylase the first enzyme shown to catalyze the synthesis of polynucleotides. It was isolated from *Azotobacter vinelandii* in 1955, and it linked ribonucleotides together in a random fashion. Subsequently this enzyme was used to produce artificial messenger RNA molecules. *See* Appendix C, 1955, Grunberg-Manago and Ochoa; 1961, Nirenberg and Matthaei.

polyoma virus a virus that induces tumors in newborn mice, rats, and hamsters and can also transform cultured mouse or rat cells. The genome of the virus is a double-stranded, supercoiled, circular DNA molecule containing about 5,300 base pairs. *See* Appendix C, 1983, Rassoulzadegan *et al.*; oncogenic virus, transformation.

polyp 1. the sedentary form of a coelenterate. 2. a small stalked neoplasm projecting from a mucous surface (for example, an intestinal polyp).

polypeptide a polymer made up of less than 50 amino acids. *See* amino acid, peptide bond.

polyphasic lethal a mutation characterized by two or more lethal phases separated by developmental periods in which it produces no deaths.

polyphenism the occurrence of several phenotypes in a population that are not due to genetic differences between the individuals in question.

polypheny pleiotropy (*q.v.*).

polyphyletic group a group of species classified together, some members of which are descended from different ancestral populations. *Contrast with* monophyletic group.

polyploid designating a cell or an individual that has more than two sets of chromosomes.

polyploidy the situation where the number of chromosome sets is greater than two. If N is the value for one set of chromosomes, a somatic cell can be 2N (diploid), 3N (triploid), 4N (tetraploid), 5N (pentaploid), 6N (hexaploid), and so forth. When compared to diploids, polyploid cells are generally larger and metabolically more active. Most genes continue to be expressed at the same relative levels. However, a few genes seem to sense increasing gene dosage and raise or lower their levels of transcription appropriately. Polyploidy is a dominant factor in plant evolution, where rounds of large scale genomic duplication have been followed by selective gene loss. This conclusion arises from observations of annotated genomes where genes commonly occur in duplicate copies. The frequency of polyploidy varies across plant groups. It is rare in conifers, whereas 95% of fern species and 70% of angiosperms are polyploid. *See* Appendix A, Plantae; Appendix C, 1917, Winge; 1937, Blakeslee and Avery; 1999, Galitski *et al.*; allopolyploid, *Arabidopsis thaliana*, autopolyploid, bananas, colchicine, DNA chip, euploid, *Gossypium*, haploid or haploidy, *Nicotiana*, -ploid, *Raphanobrassica*, wheat.

polyprotein a cistronic product that is posttranslationally cleaved into several independent proteins. For example, an enkephalin precursor protein contains six copies of met-enkephalin and one copy of leu-enkephalin. *See* enkephalins, retroviruses.

polyribonucleotide phosphorylase *See* polynucleotide phosphorylase.

polyribosome polysome (*q.v.*).

polysaccharide a carbohydrate formed by the polymerization of many monosaccharide units. Starch, cellulose, and glycogen (*q.v.*) are examples of polysaccharides.

polysomaticism the phenomenon where an individual contains diploid and polyploid cells in the same tissue.

polysome a multiple structure containing a group of ribosomes held together by a molecule of messenger RNA. A contraction of *polyribosome*. *See* Appendix C, 1962, Warner *et al.*

polysomy the reduplication of some but not all of the chromosomes of a set beyond the normal diploid number. A metafemale *Drosophila* is polysomic (trisomic for the X).

polyspermy the penetration of more than one sperm into one ovum at the time of fertilization.

Polysphondylium pallidum *See* Acrasiomycota.

polytene chromosome a giant cable-like chromosome consisting of many identical chromatids lying in parallel. The chromatin is hypercoiled in localized regions, and since the chromatids are in register, a pattern of bands is produced vertical to the long axis of the chromosome. Polytene chromosomes are found within a limited number of organisms. They are present in the macronucleus anlage of some ciliates, in the synergids and antipodal cells of the ovules of certain angiosperms, and in various tissues of dipterans. The *Drosophila* salivary gland chromosomes (*q.v.*) have been studied most extensively. *See* Appendix C, 1881, Balbiani; 1912, Rambousek; 1934, Bauer; 1952, Beermann; 1959, Pelling; 1969, Ammermann; 1980, Gronemeyer and Pongs; *Anopheles*, Balbiani ring, *Calliphora erythrocephala, Chironomus, Culex pipiens, Glyptotendipes barbipes*, insulator DNAs, *otu* mutation, *Rhynchosciara, Sciara, Smittia*.

polytenization the continued replication of each interphase chromosome to produce giant chromosomes made up of multiple chromatids lying in parallel and forming a cable-like structure. *See* somatic pairing.

polythetic group a group of organisms that share a large number of features, no single one of which is either essential for group membership or is sufficient to make an organism a member of the group.

polytopic pertaining to the distribution of subspecies in two or more geographically discontinuous areas.

polytrophic meroistic ovary *See* insect ovary types.

polytypic species a species subdivided into a number of specialized races.

pome a fleshy, many-seeded fruit such as the apple or pear, in which the enlarged end of the flower stalk forms much of the flesh.

Pompe disease a hereditary glycogen storage disease in humans arising from a deficiency of the lysosomal enzyme α-1,4-glucosidase, due to a recessive gene on chromosome 17. Prevalence of the disease is 1/100,000.

Pongidae the family of primates containing all anthropoid apes.

Pongo pygmaeus the orangutan, a primate with a haploid chromosome number of 24. About 30 biochemical marker genes have been distributed among 20 linkage groups. *See* Hominoidea.

popcorn *See* corn, quantitative inheritance.

population a local (geographically defined) group of conspecific organisms sharing a common gene pool; also called a *deme*.

population biology the study of the patterns in which organisms are related in space and time. Such disciplines as ecology, taxonomy, ethology, population genetics, and others that deal primarily with the interactions of organisms or groups of organisms (demes, species, etc.) are included under this term.

population cage a special cage in which *Drosophila* populations can be reared for many generations. The cage is designed so that samples of the population can be conveniently withdrawn and food supplies can be replenished. *See* Appendix C, 1934, L'Héritier and Teissier.

population density 1. in ecology, the number of individuals of a population per unit of living space (e.g., per acre of land, per cubic meter of water, etc.). 2. in cell or tissue culture, the number of cells per unit area or volume of a culture vessel. *See* saturation density.

population doubling level in cell or tissue culture, the total number of population doublings of a cell line or strain since its initiation *in vitro*.

population doubling time *See* doubling time.

population genetics the study of the genetic composition of populations. Population geneticists try to estimate gene frequencies and detect the selective influences that determine them in natural populations. They also build mathematic models to elucidate the interaction of factors such as selection, population size, mutation, and migration upon the fixation and loss of linked and unlinked genes. *See* Appendix C, 1908, Hardy, Weinberg; 1930–32, Wright, Fisher, Haldane.

population structure the manner in which a population is subdivided into local breeding groups or demes, the sizes of such demes in terms of the number of breeding individuals, and the amount of migration or gene flow between demes.

Populus a genus containing trees such as the trembling aspen (*P. tremuloides*), the black cottonwood (*P. trichocarpa*), and the white poplar (*P. alba*.) Most species are dioecious (*q.v.*). The progeny of crosses between certain related species are typically grown on plantations. These hybrids are fertile and have exceptional growth and vigor. *Populus trichocarpa* is the first tree species selected for DNA sequencing because of its small genome size. It is ~550 mbp, which is only 4× larger than *Arabidopsis*, but 40× smaller than pine. *See* Appendix A, Plantae, Angiospermae, Dicotyledonae, Salicales.

porcine referring to members of the pig family, especially the domestic pig *Sus scrofa*.

porphyrias diseases caused by toxic accumulations of porphyrins (*q.v.*) and related compounds in tissues. Inherited porphyrias are due to mutations in genes that encode enzymes which catalyze steps in the biosynthesis of heme (*q.v.*). There are at least eight enzymes that control steps in the heme biosynthetic pathway. One is protoporphyrinogen oxidase (PPOX) which is encoded by a gene that maps to 1q22. This gene has 13 exons and spans about 8 kb. Mutations in the PPOX gene cause variegate porphyria (VP). The symptoms include photosensitivity, abdominal pain, and mental disturbances including hallucinations, depression, and paranoia. Attacks of VP are often triggered by drugs such as barbiturates. The disease has a very high prevalence in South African populations of Dutch descent, presumably because of a founder effect (*q.v.*).

porphyrin any of a class of organic compounds in which four pyrrole nuclei are connected in a ring structure usually associated with metals (like iron or magnesium). Porphyrins form parts of the hemoglobin, cytochrome, and chlorophyll molecules. *See* heme.

positional candidate approach in human genetics, a strategy for identifying the gene responsible for a disease by mapping the mutant gene to a specific chromosomal region and then looking for an appropriate candidate among the genes already localized in that region. Individuals suffering from the disease are then tested for mutations in the candidate gene. For example, a gene encoding a fibroblast growth factor receptor ($FGFR_3$) protein was discovered during a chromosome walk (*q.v.*) toward the Huntington disease gene. Next, the gene for achondroplasia (*q.v.*) was mapped to the same chromosomal region. Finally, the $FGFR_3$ genes of dwarfs were found to contain missense mutations, proving that mutations of the $FGFR_3$ gene were responsible for the retarded growth characterizing the disease.

positional cloning a strategy for identifying and cloning a gene based on a knowledge of its position in the genome (*q.v.*), with little or no information about the function or product of the gene at the outset. This strategy has been applied in a variety of plant and animal species. In humans, the chromosomal position of the gene of interest is usually determined by linkage analyses of families affected by a particular disease. A search is then made for genetically linked molecular markers, and the closest ones flanking the gene are used to start chromosome walking (*q.v.*) in order to identify additional markers with the closest possible linkage to the gene. The DNA defined by these molecular markers is then cloned and the gene residing between them identified by a variety of means, such as by searching databases for genes within the identified genomic region, by sequencing the gene-containing region and looking for an open reading frame(s), by comparing the suspected gene's sequence and expression patterns in mutant and wild-type individuals, and where possible, by the ability of the putative gene to rescue (*q.v.*) a mutant phenotype (*q.v.*). Once identified, the gene is cloned for further analysis. Human disease genes identified by positional cloning include cystic fibrosis, Duchenne muscular dystrophy, fragile X syndrome, and Huntington disease (*all of which*

See). *Also called* map-based cloning. *Compare with* functional cloning. *See* marker, open reading frame, positional candidate approach.

Positional Information Hypothesis a model developed by Louis Wolpert to explain pattern specification (*q.v.*) during development. His idea is that each population of cells in an embryonic structure lies in a field that contains a gradient of a chemical morphogen (*q.v.*). The position in the field determines the concentration of morphogen, and the cells are programmed to enter one of a number of developmental pathways depending on the concentration level of the morphogen to which they are exposed. *See* Appendix C, 1969, Wolpert.

position effects the change in the expression of a gene accompanying a change in the position of the gene with respect to neighboring genes. The change in position may result from crossing over or from a chromosomal aberration. Position effects are of two types: the *stable* (S) type and the *variegated* (V) type. S-type position effects are also called *cis-trans* position effects. S-type position effects involve cistrons that possess at least two mutated sites separable by intragenic recombination. In the *cis* configuration (m^1 m^2/++) a normal phenotype is observed, whereas in the *trans* configuration (m^1+/+m^2) a mutant phenotype is produced. A reasonable explanation for such an observation would be that the mRNA transcribed from a (++) chromatid would function normally, whereas the mRNAs transcribed from (m^1 m^2), (m^1+), or (+m^2) chromatids would not. V-type position effects generally involve the suppression of activity of a wild-type gene when it is placed in contact with heterochromatin because of a chromosome aberration. Under some conditions, the gene may escape suppression, and consequently the final phenotype may be variegated, with patches of normal and mutant tissues. *See* Appendix C, 1925, Sturtevant; 1936, Schultz; 1945, Lewis; heterochromatization, transvection.

positive assortative mating *See* assortative mating.

positive control control by a regulatory protein that must bind to an operator before translation can take place.

positive eugenics *See* eugenics.

positive feedback the enhancement or amplification of an effect by its own influence on the process that gives rise to it.

positive gene control enhancement of DNA transcription through binding of specific expressor molecules to promoter sites. For example, the binding of CAP-cAMP complexes to promoters of bacterial genes involved in catabolism of sugars other than glucose facilitates binding of RNA polymerase to these operons when glucose is absent. *See* glucose-sensitive operons. *Compare with* negative gene control.

positive interference the interaction between crossovers such that the occurrence of one exchange between homologous chromosomes reduces the likelihood of another in its vicinity. *Compare with* negative interference. *See* Appendix C, 1916, Muller.

positive sense ssDNA or RNA *See* plus (+) and minus (–) viral strands.

positive supercoiling *See* supercoiling.

positron a particle of the atomic nucleus equal in mass to the electron and having an equal but opposite (positive) charge.

postcoitum after mating.

postmating isolation mechanism *See* postzygotic isolation mechanism.

postmeiotic fusion a method for restoring diploidy in eggs produced by parthenogenesis, involving union of two identical haploid nuclei formed by a mitotic division of the egg nucleus.

postmeiotic segregation in ascomycete fungi such as *Neurospora*, the formation of heteroduplex regions (by meiotic crossing over) that results in aberrant 4 : 4 pattern of asci in which adjacent pairs of ascospores produced by mitotic division after meiosis have different genetic compositions. *See* tetrad segregation types.

postreductional disjunction referring to the separation of alleles at particular heterozygous loci during the first meiotic division. If the loci are represented by *A* and *A′*, in the case of postreductional disjunction the two chromatics that enter one sister nucleus have one *A* and one *A′* allele, whereas in the case of prereductional disjunction both have *A* alleles or both *A′* alleles.

postreplication repair repair to a DNA region after a replication fork has passed that region or in nonreplicating DNA.

posttranscriptional processing those modifications made to pre-mRNA molecules before they leave the nucleus; also called *nuclear processing*. A gene containing three exons (E_1, E_2, and E_3) and two introns (I_1 and I_2) is diagrammed (page 351). RNA polymerase II transcribes the 3′-5′ strand of the gene

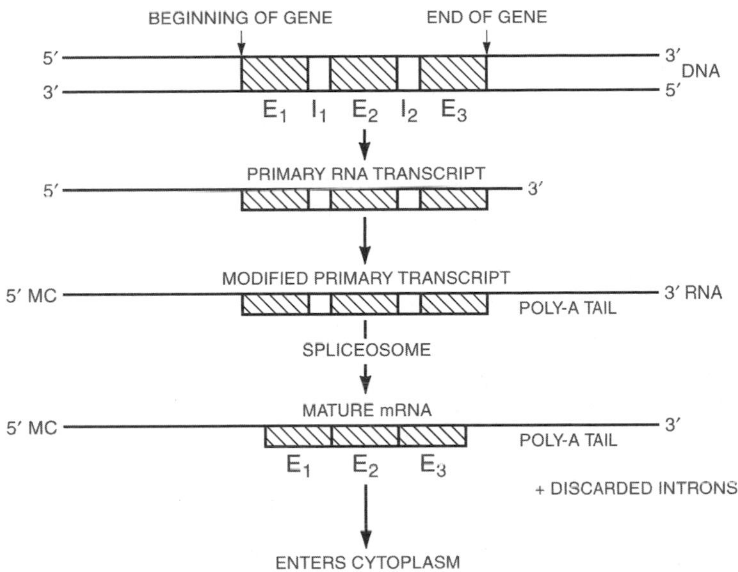

Posttransciptional processing

to form a 5′-3′ pre-mRNA molecule. Next, a methylated cap (MC) is added to the 5′ end of the primary transcript; a poly-A tail is added to the 3′ end. Finally, the introns are removed and the exons are spliced together during reactions that occur within a spliceosome, and the mature mRNA leaves the nucleus. *See* alternative splicing, Cajal body, cis-splicing, dystrophin, exon, hemoglobin genes, heterogeneous nuclear RNA (hnRNA), intron, methylated cap, polyadenylation, RNA editing, RNA splicing, small nuclear RNAs, snurposomes, spliceosome, transcriptosomes.

posttranslational processing alterations to polypeptide chains after they have been synthesized: e.g., removal of the formyl group from methionine in bacteria, acetylation, hydroxylation, phosphorylation, attachment of sugars or prosthetic groups, oxidation of cysteines to form disulfide bonds, cleavage of specific regions that convert proenzymes to enzymes, etc. *See* cystine, *N*-formylmethionine.

posttranslational sorting *See* protein sorting.

postzygotic isolation mechanism any factor that tends to reduce or prevent interbreeding between genetically divergent populations or species, but functioning after fertilization has occurred; includes hybrid inviability, hybrid sterility, and hybrid breakdown.

potassium an element universally found in small amounts in tissues. Atomic number 19; atomic weight 39.102; valence 1^+; most abundant isotope ^{39}K; radioisotope ^{42}K, half-life 12.4 hours, radiations emitted—beta particles and gamma rays.

potato *Solanum tuberosum*, a tetraploid with a genome size of 1.8 gigabases. Together with corn, wheat, and rice it is one of the four most valuable of the world's crops. Although it is called the *Irish potato*, *S. tuberosum* originated in South America. *Solanum* exists as two principal cultivated races designated as subspecies (ssp. *andigena* and ssp. *tuberosum*). Both subspecies arose in South America, but ssp. *andigena* was the first species introduced into Europe. This cultivated race was wiped out by the potato blights of the 1840s and was replaced by the American cultivar, which belonged to ssp. *tuberosum*. *See* Appendix A, Plantae, Angiospermae Dictotyledonae, Solanales; *Phytophthora infestans*.

potato virus Y a virus that causes diseases of commerically important crop plants, such as peppers, potatoes, and tomatoes. The virion is about 730 × 110 nm, and it contains a helically disposed ssRNA surrounded by protein subunits.

potency in developmental biology, the capacity of a cell or its descendants to give rise to differentiated structures (e.g., specific cell lineage(s), tissue(s), organ(s), or a whole organism), given a specific environment. *See* pluripotent, totipotent, unipotent.

Potorous tridactylus the rat kangaroo, a marsupial favorable for chromosomal studies because of the small number and individuality of its chromosomes. *See* Appendix A, Mammalia, Metatheria.

POU genes genes that encode related, DNA-binding proteins. The family is large, and it includes many genes expressed in the central nervous system. The DNA-binding domains consist of an upstream homeobox (*q.v.*) and a downstream POU-specific domain about 80 amino acids long. The name POU comes from the initials of the first genes included in the family (*P*it-1, *O*ct-1, and *U*nc-86.) The *Pit-1* gene is expressed in the pituitary gland of mouse embryos and *Oct-1* is expressed in many tissues of both developing and adult mice. The *unc-86* gene activates the differentiation of specific embryonic cells into neurons in *Caenorhabditis elegans*. *See* Appendix C, 1988, Herr *et al.*; selector genes.

poultry breeds Plymouth Rock, New Hampshire, White Leghorn, Blue Andalusian, Rhode Island Red, Rhode Island White, Australorp, and Orpington. *See* *Gallus domesticus*.

pox viruses viruses that belong to the family Poxviridae. These are the largest and most complex viruses known, with genomes made up of linear double-stranded DNAs. These molecules are 130–300 kbp and contain 200–300 genes. Pox viruses produce both specific and cross-reacting antibodies. For this reason it is possible to vaccinate against a deadly disease caused by one pox virus with a related species that causes a much milder disease. The classic example is vaccinating against smallpox (caused by the *Variola* virus) with the *Vaccinia* virus. Both *Variola* and *Vaccinia* viruses are believed to have evolved from the same rodent pox virus about 10,000 BC. *See* enveloped viruses, smallpox, smallpox vaccine, vaccine, virus.

pp inorganic pyrophosphate.

P particle *See* kappa.

P1 phage a temperate bacteriophage that is widely used in transduction experiments with *E. coli*. Its genome consists of a linear double-stranded DNA molecule of about 90 kilobases. The molecule is terminally redundant and cyclically permuted. *See* cyclically permuted sequences, P1 artificial chromosomes (PACs).

P22 phage a temperate bacteriophage that infects *Salmonella*. The prophage inserts at a specific site on the host chromosome (between *pro A* and *pro C*). Insertion is catalyzed by an integrase (*q.v.*) specified by the phage. Transduction (*q.v.*) was discovered in *Salmonella typhimurium* that carried the P22 prophage. *See* Appendix C, 1952, Zinder and Lederberg.

PPLO *p*leuro*p*neumonia-*l*ike *o*rganism (*q.v.*).

ppm parts per million.

ppt precipitate.

pp60c-src the 60-kilodalton protein kinase encoded by the *c-src* gene (*q.v.*) in normal cells. *See* pp60v-src.

pp60v-src the protein encoded by the oncogene of the Rous sarcoma virus. It is a 60-kilodalton phosphoprotein, hence the *pp60* in the name; the *v-src* indicates that it is encoded by viral gene *src*. The molecule is a protein kinase (*q.v.*) that phosphorylates tyrosine subunits in cellular proteins, particularly those that form the adhesion portions of the plasmalemma. *See* Appendix C, 1978, Collett and Erickson; pp60c-src.

Prader-Willi syndrome (PWS) a syndrome due to a genetic deletion of human chromosome 15 (q11–13). The condition was first described in 1956 by Andrea Prader and Heinrich Willi. PWS patients are generally mildly retarded and have insatiable appetites. They are obese from overeating. This condition is often discussed in conjunction with the Angelman syndrome (AS), which is also characterized by deletions in the same region of chromosome 15. Individuals with PWS have a very different phenotype than those with AS. In the case of PWS, the deleted chromosome 15 is usually of paternal origin, whereas in AS the deleted chromosome is maternally derived. Human chromosome 15 contains the genes SNRPN, IPW, and UBE3A lying in 1, 2, 3 order. Some deficiencies lack all three genes. Genes 1 and 2 are paternally imprinted, and gene 3 is maternally imprinted. In the diagram, the male and female signs mark the source of the chromosomes in the zygotes. The inactive genes are methylated, as shown by m's alongside the circles. The active genes produce products P_1, P_2, and P_3. The patient with Prader-Willi syndrome has lost the ability to produce the products of the genes that normally undergo paternal imprinting. The same deficiency (df) causes the Angelman syndrome (AS) because the maternally transmitted deficiency lacks UBE3A, which is maternally imprinted under normal circumstances. The UBE3A gene is active in localized regions of the female brain, where it specifies a ubiquitin-protein ligase. The PW syndrome may result from the loss of SNRPN and IPW or of other paternally expressed genes farther to the left. It is known that an imprinting center lies immediately to the left of SNRPN. This center contains CpG islands that are methylated on the maternal chromosome and unmethylated on the paternal chromosome. Patients who lack the 15q11-13 deficiency but have the PW or A syndromes often have mutations in the imprinting center. *See* DNA methylation, parental imprinting.

PRD domain, PRD repeat *See Paired*.

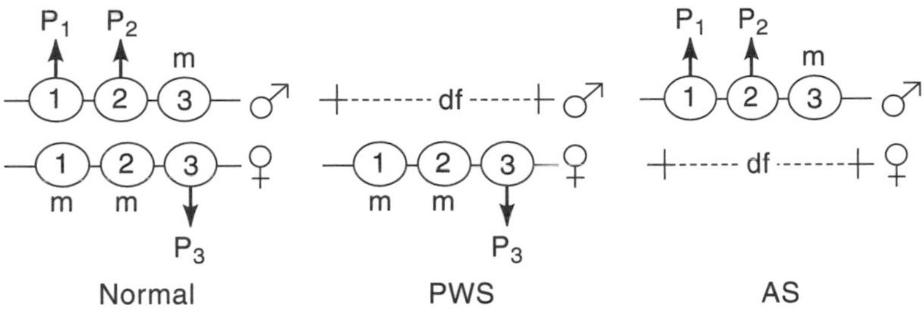

Prader-Willi syndrome

preadaptation *See* exaptation.

pre-adoptive parents *See* germinal choice.

Precambrian the eon between the Phanerozoic and the Hadean eons. The protists arose and evolved during this 3.2-billion-year interval. *See* geologic time divisions.

precursor ribosomal RNA *See* preribosomal RNA.

pre-embryo *See* embryo.

preferential association an immunological theory that specific viral antigens interact more strongly with certain allelic products of the major immunogene complex than with others. This preferential association may make the virus more immunogenic, and hosts with the strongly interacting allelic product would tend to be more immune to viral infection than would those with weakly interacting allelic products.

preformation *See* epigenesis.

prehensile adapted for grasping.

preimplantation genotyping the determination of the genotype of an *in vitro*-fertilized, human embryo prior to its implantation. The technique samples one blastomere from an eight cell embryo, and the selected templates are amplified by the polymerase chain reaction (*q.v.*). Tests are then run to see if the templates contain mutant copies of the gene under study. Embryos free of the defect are used to start the pregnancy.

premating isolation mechanism *See* prezygotic isolation mechanism.

premature initiation a second initiation of replication occurring before the first is completed; a phenomenon observed in bacteria grown in a complex nutrient broth or in some phage species that make replicas very rapidly.

prematurely condensed chromosomes interphase chromosomes that are experimentally forced to undergo rapid condensation to metaphase dimensions. This is done by fusing an interphase cell with a cell in mitosis. The interphase cell is induced to enter mitosis, and its chromosomes contract accordingly. *See* Appendix C, 1970, Johnson and Rao.

premessenger RNA the giant RNA molecule transcribed from a structural gene. It will undergo posttranscriptional processing (*q.v.*) before it leaves the nucleus.

prenatal genetic testing the sampling of cells from a fetus to determine whether or not it has a genetic disorder. Such testing is offered to mothers who will be older than 35 at the time of delivery, or to those who have had a previous child or pregnancy with a birth defect, or in situations where the family history indicates that the baby may be at risk of inheriting a genetic abnormality. *See* amniocentesis, chorionic villi sampling, Down syndrome, genetic counseling, informed consent, maternal contamination, pedigree.

prenylation the covalent binding of a protein to an isoprenoid lipid (*q.v.*), generally by its C-terminal cysteine. Prenylation increases the hydrophobicity of proteins and facilitates their interactions with membrane lipids.

prepatent period the interval between infection with a pathogen or a parasite and the time when the causative agent of the ensuing disease can be detected by conventional diagnostic techniques. *See* latent period, patent period.

prepattern a morphogenetic pattern superimposed upon a population of cells arranged in a two-dimensional array. Specific types of differentiation are stimulated in certain cells located in defined areas. An example of a gene that influences a prepattern is *engrailed* (*q.v.*). *See* compartmentalization.

preprimosome *See* primosome.

prepupal period the period between puparium formation and the eversion of the imaginal discs of the insect.

prereductional disjunction *See* postreductional disjunction.

preribosomal RNA the giant RNA molecule transcribed from a ribosomal RNA gene (*q.v.*). In *Drosophila*, it is 38S, in *Xenopus* it is 40S, and in HeLa cells it is 45S. After transcription, preribosomal-RNA is cleaved one or more times to generate the 5.8S, 18S, and 28S rRNAs that become components of ribosomes.

presenilins (PS1 and PS2) *See* Alzheimer disease.

presumptive in embryology, referring to the presumed fate of an embryonic tissue in a normal development. For example, if a tissue is "presumptive neural tube," this means that in the course of normal development it will become neural tube tissue.

prezygotic isolation mechanism any factor that tends to reduce or prevent interbreeding between members of genetically divergent populations or species and functioning before fertilization occurs; includes ecological, temporal, ethological, and other isolating factors. *See* courtship ritual, mate choice, sexual selection.

Pribnow box a segment upstream from the startpoint of prokaryotic structural genes to which the sigma subunit of the RNA polymerase binds. The segment is 6 base pairs long, and the nucleotides most commonly found are TATAAT. *See* Appendix C, 1975, Pribnow; canonical sequence, Hogness box, promoter.

primaquine-sensitivity *See* glucose-6-phosphate dehydrogenase deficiency, malaria.

primary culture a culture started from cells, tissues, or organs taken directly from the organism.

primary immune response *See* immune response.

primary ionization the ionization produced by the primary particles passing through matter as contrasted to the "total ionization," which includes the "secondary ionizations" of delta rays (*q.v.*).

primary nondisjunction sex chromosomal nondisjunction in diploid organisms with the XX, XY system of sex determination. In the homogametic sex, gametes are produced with two X chromosomes or none. In the heterogametic sex, primary nondisjunction during the first meiotic division produces gametes with no sex chromosome (O) or with an X and a Y. Primary nondisjunction during the second meiotic division produces XX and O or YY and O gametes.

primary sex ratio the ratio of male to female zygotes at conception.

primary sexual character an organ that functions in producing gametes; the ovaries and the testes.

primary speciation the splitting of one species into two, usually resulting from natural selection favoring different gene complexes in geographically isolated populations.

primary structure the specific sequence of monomeric subunits (amino acids or nucleotides) in a macromolecule (protein or nucleic acid, respectively). *See* protein structure.

primary transcript an RNA molecule as it was initially transcribed from DNA. In eukaryotic cells, a primary transcript usually contains introns (*q.v.*) that will be absent in the mature form of the RNA. *See* post-transcriptional modification.

primase in *E. coli*, the product of *dnaG* gene, responsible for initiation of precursor fragment synthesis in the lagging strand during discontinuous replication. Primase makes the RNA primer that is subsequently elongated by DNA polymerase III. The primase in *E. coli* consists of a single polypeptide of 60,000 daltons. Unlike RNA polymerase, primase is not inhibited by rifampicin (*q.v.*) and can polymerize deoxyribonucleotides as well as ribonucleotides *in vitro*. *See dna* mutations, DNA polymerase, replication of DNA, replicon, replisome.

primate a mammal belonging to the order Primates, which includes humans, the apes, and monkeys. *See* Appendix A.

primed in immunology, sensitization by contact of competent lymphocytes with antigens to which they are programmed to respond.

primed synthesis technique a method for nucleotide sequencing involving enzymatically controlled extension of a primer DNA strand. *See* DNA sequencing techniques.

primer DNA **1.** single-stranded DNA required for replication by DNA polymerase III in addition to primer RNA (*q.v.*). **2.** Oligonucleotides of single-stranded DNA synthesized by a gene machine (*q.v.*) for use in a polymerase chain reaction (*q.v.*).

primer RNA a short RNA sequence synthesized by a primase from a template strand of DNA and serving as a required primer onto which DNA polymerase III adds deoxyribonucleotides during DNA replication. Primers are later enzymatically removed and the gaps closed by DNA polymerase I, and the remaining nicks are sealed by ligase. *See* primase, replication of DNA.

primer walking a procedure that uses artificially synthesized primers about 18 bases long to bind to a unique DNA site. The primer is enzymatically extended by several hundred bases that are complementary to the target DNA. After sequencing the elongated primer, a sequence is selected near the far end to serve as a primer for the next "step" along the target DNA. A new 18 mers primer is then synthesized complementary to the far sequence from which the next round of extension can proceed.

primordial dwarfism *See* pituitary dwarfism.

primordial germ cells (PGCs) germ cell precursors that give rise to the germ line (*q.v.*). In *Drosophila* PGCs are known as pole cells. These cells, like PGCs in many other invertebrate and vertebrate species, arise during early zygotic divisions, contain the germ plasm (*q.v.*), and actively migrate to their final location in the somatic component of the gonad. *See* pole cell.

primordium the early cells that serve as the mitotic progenitors of an organ during development.

primosome a complex of proteins (including primase) required for the priming action that initiates synthesis of each Okazaki fragment in eukaryotic DNA replication. The complex minus primase is called a *preprimosome*. *See* replication of DNA.

Primula a genus of cowslips and primrose whose population genetics has been extensively studied.

prions infectious pathogens that cause neurodegenerative diseases such as the Creutzfeldt-Jakob disease of humans, scrapie of sheep, and bovine spongiform encephalopathy ("mad cow disease"). Prions are transmissible particles at least 100 times smaller than viruses and are composed exclusively of special proteins. The prion protein (symbolized *PrP*) is encoded by a chromosomal gene of the host. The normal cellular protein (PrP^C) is a component found in normal neurons and is folded into a conformation that is 40% alpha helixes and shows very few beta sheets. The modified protein from animals with scrapie (PrP^{Sc}) contains 30% alpha helixes and 45% beta sheets. Therefore, the disease protein represents a misfolded form of the normal PrP. The PrP^{Sc} proteins may act as templates upon which the PrP proteins are misfolded to magnify the production of pathogenic prions. The three-dimensional structure of prion proteins from a variety of mammalian species has been determined by nuclear magnetic resonance (NMR) spectroscopy (*q.v.*) in order to better understand the structural basis of prion transformation from the benign to the infectious form. Prion-like characteristics in a protein are not always detrimental; they can also be beneficial to a cell or organism. For example, the prion-like form of a neuronal CPEB protein (*q.v.*) in *Aplysia* is thought to play a role in maintaining synaptic changes associated with long-term memory storage. *See* Appendix C, 1982, 1997, Prusiner; 2003, Si *et al.*; chaperones, memory.

pro proline. *See* amino acid.

probability of an event the long-term frequency of an event relative to all alternative events, and usually expressed as a decimal fraction. Probabilities range between zero (if the event never occurs) and 1 (if the event always occurs and no alternative event ever occurs). In some cases we know a probability *a priori,* as in the case of a coin toss. In the long run, the coin will come up tails with a frequency of .5. More often, a probability must be estimated by averaging the results of many trials. *See* conditional probability, independent probabilities, significance of results.

proband propositus (*q.v.*).

proband method a method in human genetics for comparing the proportion in families of children in which a proband shows a specific trait with the proportion expected if the trait were inherited as a single gene. For example, if one considers a group of families, each with both parents heterozygous for a recessive gene and each with two children, the proportion of affected children is 57%, not 25%. This is because the families are chosen in the first place through an affected child, and all sibships in which just by chance no affected individuals occurred have been left out. Thus, there is an *ascertainment bias* that loads the results in favor of the trait. *See* Appendix C, 1910, Weinberg.

probe in molecular biology, any biochemical labeled with radioactive isotopes or tagged in other ways for ease in identification. A probe is used to identify or isolate a gene, a gene product, or a protein. For example, a radioactive mRNA hybridizing with a single strand of its DNA gene, a cDNA hybridizing with its complementary region in a chromosome, or a monoclonal antibody combining with a specific protein. *See* cDNA library, hybridoma, Southern blotting, strand-specific hybridization probes.

proboscipedia one of the homeotic mutations (*q.v.*) of *Drosophila* which belongs to the *Antennapedia* complex. The homeotic mutations figure on page 210 shows a normal fly head with its proboscis extending downward (A). The proboscis ends with a pair of labial palps which function in eating and taste. In the *pb* mutant (C) the proboscis has been transformed into a pair of legs. Therefore *pb* normally functions as a segment identity gene (*q.v.*). *See* *Antennapedia, Hox* genes.

procaryote *See* prokaryote.

Procaryotes *See* Prokaryotes.

processed gene an eukaryotic pseudogene (*q.v.*) lacking introns and containing a poly-A segment near the downstream end, suggesting that it arose by some kind of reverse copying from processed nuclear RNA into double-stranded DNA; also called *retrogene*.

processing **1.** posttranscriptional modifications of primary transcripts. **2.** antigen processing involves partial degradation by macrophages (and, in some cases, coupling with RNA) before the immunogenic units appear on the macrophage membrane in a condition that is stimulatory to cognate lymphocytes.

processive enzyme an enzyme that remains bound to a particular substrate during repetitions of the catalytic event.

Prochlorococcus marinus a marine cyanobacterium that is ubiquitous in the upper 100 m of oceans that occur in a latitudinal band from 40°N to 40°S. This is the most abundant species on earth, and over half of the total chlorophyll in the ocean surface is contained in these organisms. *See* chlorophyll, Cyanobacteria.

Prochloron a genus of cyanobacteria whose species contain the a and b forms of chlorophyll (*q.v.*) found in green plants. Therefore, *Prochloron* is sometimes called a living fossil (*q.v.*), a missing link in the evolution of the chloroplast (*q.v.*). *P. didemni*, the type species for this genus, lives in close association with marine ascidians. *See* Cyanobacteria, serial symbiosis theory.

proctodone a hormone, thought to be secreted by cells of the anterior intestine of insects, that terminates diapause (*q.v.*).

procumbent designating a plant stem that lies on the ground for all or most of its length (as in the case of vines). *See* runner.

productive infection viral infection of a cell that produces progeny via the vegetative or lytic cycle.

productivity fertility. In *Drosophila* the term is used specifically to refer to the number of progeny surviving to the adult stage among those produced per mated parental female in a specified time interval.

proenzyme a zymogen (*q.v.*).

proflavin an acridine dye (*q.v.*) that can function as an intercalating agent (*q.v.*). Treatment of T4 phage with proflavin resulted in *rII* mutants that had base additions or deletions. These were used to deduce the triplet nature of the genetic code. *See* Appendix C, 1961, Crick, Brenner *et al.*

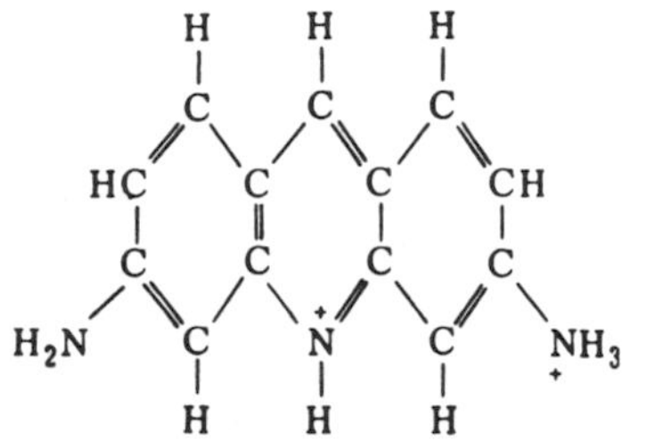

progenitor a person or organism from which a person, animal, or plant is descended or originates; an ancestor or parent.

progenote the hypothesized common ancestor of archaea, bacteria, and eukaryotes. *See* Sogin's first symbiont.

progeny the offspring from a given mating; members of the same biological family with the same mother and father; siblings.

progeny test the evaluation of the genotype of a parent by a study of its progeny under controlled conditions.

progeria a premature aging disease of humans. The hereditary form, Hutchinson-Gilford syndrome (*q.v.*) is inherited as an autosomal dominant and is caused by mutations in the lamin A gene (*LMNA*). Phenotypically old children usually die by age 13. Cytological studies of mutant lymphocytes show altered nuclear shapes and sizes, nuclear envelope interruptions, and chromatin extrusions. Cells have a reduced replicative life span and a reduced ability to repair damaged DNA. *See* lamins.

progesterone a steroid hormone secreted by the corpus luteum (*q.v.*) to prepare the uterine lining for implantation of an ovum; also later secreted by the placenta (*q.v.*); essential for the maintenance of pregnancy. The structure is drawn on page 357.

progestin *See* progestogens.

progestogens a group name for substances having progesteronelike activity; also termed *progestins*. *See* progesterone.

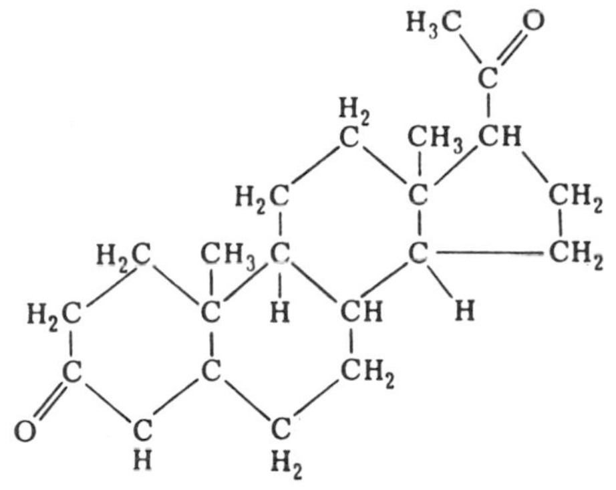

Progesterone

prognosis a forecast of the course and termination of a disease.

programmed cell death *See* apoptosis.

proinsulin a protein synthesized and processed by the beta cells of the pancreas. The molecule contains both the A and B peptides of insulin (*q.v.*) and an intervening C peptide containing 30 amino acids. Specific proteases cleave the precursor at two points, releasing the connecting peptide and the intact insulin molecule.

prokaryon synonymous with nucleoid (*q.v.*).

prokaryote member of the superkingdom Prokaryotes (*q.v.*).

Prokaryotes (*also* **Procaryotes**) the superkingdom containing all microorganisms that lack a membrane-bound nucleus containing chromosomes. Cell division involves binary fission. Centrioles, mitotic spindles, and mitochondria are absent. Aside from pillotinas (*q.v.*), prokaryotes also lack microtubules. The first cells, which are thought to have evolved about 3.9 billion years ago, were chemoautotrophic prokaryotes. Prokaryotes still make up the majority of the earth's biomass. Their total population (4–6 $\times 10^{30}$ cells) constitutes the largest living reservoir of the elements C, N, and P. The superkingdom Prokaryotes contains one kingdom, the Monera (*q.v.*). *See* Appendix A, Prokaryotes; Appendix C, 1937, Chatton; 1998, Whitman, Coleman, and Wiebe; biomass, genophore; *contrast with* Eukaryota.

prolactin *See* human growth hormone.

proline *See* amino acid.

promiscuous DNA DNA segments that have been transferred between organelles, such as mitochondria and chloroplasts, or from a mitochondrial genome to the nuclear genome of the host as a result of transpositional events happening millions of years ago. An example is a section of mitochondrial DNA present in the nuclear genome of *Strongylocentrotus purpuratus* (*q.v.*). The term is also used to refer to those plasmids that can transfer DNA horizontally between a wide variety of host species. Examples would be mariner elements and the Ti plasmid (*both of which see*). *See* Appendix C, 1983, Jacobs *et al.*

promitochondria aberrant mitochondria characteristically found in yeasts grown under anaerobic conditions. Promitochondria have incomplete inner membranes and lack certain cytochromes. *See* petites.

promoter 1. a region on a DNA molecule to which an RNA polymerase binds and initiates transcription. In an operon, the promoter is usually located at the operator end, adjacent but external to the operator. The nucleotide sequence of the promoter determines both the nature of the enzyme that attaches to it and the rate of RNA synthesis. *See* Appendix C, 1975, Pribnow; alcohol dehydrogenase, down promoter mutations, Hogness box, Pribnow box, regulator gene, up promoter mutations. 2. a chemical that, while not carcinogenic itself, enhances the production of malignant tumors in cells that have been exposed to a carcinogen.

promoter 35 S a promoter discovered in the Cauliflower Mosaic Virus. CaMV is naturally transmitted by aphids and is world wide in its distribution. The virus occurs in broccoli, cabbage, cauliflower, and turnips. Promoter 35 S has been used to activate the expression of foreign genes in genetically engineered plants, including corn, cotton, potato, rice, soybean, squash, sugar beets, and tomato. The Monsanto Company holds the patent rights to genetically modified plants and seeds that incorporate promoter 35 S. *See* Bt designer plants, GMO, Roundup, transgenic plants.

pronase an enzyme from *Streptomyces* that digests mucoproteins.

Prontosil a red dye used for treating leather. It was later found to successfully combat Streptococcal infections. Subsequently Prontosil was shown to breakdown *in vivo* into its component molecules, one of which was sulfanilamide. *See* Appendix C, 1938, Domagk; sulfa drugs.

pronucleus the haploid nucleus of an egg, sperm, or pollen grain. *See* Appendix A, 1877, Fol.

proofreading in molecular biology, any mechanism for correcting errors in replication, transcription, or translation that involves monitoring of individual units after they have been added to the chain;

also called *editing*. *See* dna mutations, DNA polymerase, RNA editing.

pro-oocyte one of the two cystocytes containing four ring canals that form synaptonemal complexes in *Drosophila melanogaster*. Upon entering the vitellarium, the anterior pro-oocyte switches to the nurse cell developmental pathway, leaving the posterior cell as the oocyte. *See* cystocyte divisions, polyfusome.

propagule usually referring to a vegative bud or shoot from a plant which, when separated, can produce a new individual and so propagate the species. More generally, any unicellular or multicellular reproductive body that can disseminate the species.

properdin pathway *See* complement.

prophage in lysogenic bacteria, the structure that carries genetic information necessary for the production of a given type of phage and confers specific hereditary properties on the host. *See* Appendix C, 1950, Lwoff and Gutman; cryptic prophage, lambda (λ) bacteriophage.

prophage attachment site either of the two attachment sites flanking an integrated prophage or the nucleotide sequences in a bacterial chromosome at which phage DNA can integrate to form a prophage.

prophage induction *See* induction.

prophage-mediated conversion the acquisition of new properties by a bacterium once it becomes lysogenized. A prophage, for example, confers upon its bacterial host an immunity to infection by related phages. Lysogenized bacteria also often show changes in their antigenic properties or in the toxins they produce. *See* Appendix C, 1951, Freeman; diphtheria toxin.

prophase *See* mitosis.

propositus (*female*, **proposita**) the clinically affected family member through whom attention is first drawn to a pedigree of particular interest to human genetics; also called *proband*.

prosimian a member of the most primitive primate suborder, the Prosimii, containing tree shrews and tarsiers.

Prosobranchiata one of the three subdivisions of the mollusc class Gastropoda. *See* Appendix A.

prospective significance the normal fate of any portion of an embryo at the beginning of development.

prostaglandin a group of naturally occurring, chemically related, long-chain fatty acids that exhibit a wide variety of physiological effects (contraction of smooth muscles, lower blood pressure, antagonism of certain hormones, etc.). The first prostaglandin was originally isolated from the prostate gland (hence the name), but they are now known to be produced by many tissues of the body.

prosthetic group that portion of a complex protein that is not a polypeptide. Usually the prosthetic group is the active site of such a protein. The heme groups of hemoglobin are examples of prosthetic groups.

protamines highly basic proteins that are bound to the DNA of sperm chromosomes. During spermiogenesis (*q.v.*) the histones of the nucleosomes break down and are replaced by protamines. These are shorter, simpler proteins that are very rich in arginine and have little or no lysine. Cysteine residues are distributed at relatively conserved positions along the molecules. Protamines form an alpha helix (*q.v.*) when bound to DNA. Protamine genes are turned on only in males and only in the testes. Protamines are translated from stored mRNA during a late spermatid stage.

protan *See* color blindness.

protandry **1.** the maturation of the pollen-bearing organs before the female organs on a monoecious plant. **2.** sequential hermaphroditism in animals, with the male stage preceding the female stage (*compare with* protogyny). **3.** the appearance of male animals earlier in the breeding season than females.

protanomaly *See* color blindness.

protanopia *See* color blindness.

protease an enzyme that digests proteins.

proteasome a cylindrical, multi-subunit protein complex that recognizes and degrades many intracellular proteins in a highly regulated, ATP-dependent manner. Proteasomes have been identified in prokaryotes and eukaryotes. In mammalian cells the proteasome is a *26S complex*, consisting of a *20S core complex* flanked by a *19S regulatory particle*, or "*cap*" on each end. The 19S caps serve to capture and unfold ubiquitin-conjugated proteins and guide them into the 20S core, where polypeptides are broken down into short peptides. Proteasome complexes from prokaryotes are simpler in form than those from mammals. *See* calnexin, cyclins, ubiquitin, ubiquitin-proteasome pathway (UPP).

protein a molecule composed of one or more polypeptide chains, each composed of a linear chain of amino acids covalently linked by peptide bonds. Most proteins have a mass between 10 and 100 kilodaltons. A protein is often symbolized by its mass in kDa. The p53 protein is an example. *See* Appendix C, 1838, Mulder, Berzelius; 1902, Hofmeister and Fisher; Appendix E, Individual Databases; amino acid, insulin, peptide bond, protein structure, translation.

protein clock hypothesis the postulation that amino acid substitutions occur at a constant rate for a given family of proteins (e.g., cytochromes, hemoglobins) and hence that the degree of divergence between two species in the amino acid sequences of the protein in question can be used to estimate the length of time that has elapsed since their divergence from a common ancestor.

protein databases *See* Appendix F.

protein engineering any biochemical technique by which novel protein molecules are produced. These techniques fall into three categories: (1) the *de novo* synthesis of a protein, (2) the assembly of functional units from different natural proteins, and (3) the introduction of small changes, such as the replacement of individual amino acids, into a natural protein. *See* Appendix C, 1965, Merrifield and Stewart.

protein kinase any member of a family of proteins that transfers phosphate groups from ATP to specific serine, threonine, or tyrosine molecules in proteins. Protein kinases are activated in response to specific chemical signals such as calcium ions, cyclic AMP, or mitogens. Phosphorylation of the protein substrate serves to amplify the signal inside the cell. The oncogenic protein synthesized by the Rous sarcoma virus is a protein tyrosine kinase. The chloride channels of epithelial cells are activated by reactions between protein kinases and the cystic fibrosis transmembrane regulator. *See* Appendix C, 1959, Krebs, Graves, and Fischer; 1978, Collett and Erickson; 1991, Knighton *et al.*; 1992, Krebs and Fischer; Abelson murine leukemia virus, Bruton tyrosine kinase, cellular signal transduction, cyclins, cystic fibrosis, epidermal growth factor (EGF), Janus kinase 2, maturation promoting factor, pp60v-src, protein kinase, *Src*, transforming growth factor-β (TGF-β).

proteinoid an amino acid polymer with a weight as high as 10,000 daltons formed under "pseudoprimeval conditions"' by heating to 70°C a dry mixture containing phosphoric acid and 18 amino acids. Such proteinoids are acted upon by proteolytic enzymes and have nutritive value for bacteria, but are nonantigenic.

protein sorting the sorting of newly synthesized proteins into correct compartments of the eukaryotic cell. In the case of *cotranslational sorting*, the ribosome is associated with the membrane of the endoplasmic reticulum via a signal recognition particle (*q.v.*). The protein enters the ER lumen as it is translated. It may be retained there, or it may be transferred via the Golgi apparatus (*q.v.*) to secretory vesicles, lysosomes, or the plasma membrane. In the case of *posttranslational sorting*, proteins begin their synthesis on ribosomes in the cytosol (*q.v.*). The proteins are then targeted to organelles such as mitochondria, chloroplasts, or peroxisomes, or they may enter the nucleus through nuclear pores. *See* endoplasmic reticulum, receptor-mediated translocation, sorting signals, translation.

protein splicing a phenomenon (known to occur in yeast, bacteria, and archaeons) during which a precursor protein has a segment excised from it and the N- and C-terminal fragments are subsequently spliced together. The excised segment is called an *intein* (*in*ternal pro*tein* sequence), and the spliced protein is composed of N- and C-*exteins* (*ex*ternal pro*tein* sequence). An intein cuts itself from its parent molecule and unites its former neighboring exteins with the usual peptide bond. Introns (*q.v.*) often encode a "homing endonuclease" (*q.v.*) that can excise a DNA segment, allowing it to move to a new genomic location. Analogously, many inteins contain a "homing endonuclease" segment in addition to a protein splicing region. This kind of intein can excise the DNA that encodes it out of a gene and allow the DNA to be transported elsewhere. A DNA polymerase in *Synechocystis* (*q.v.*) is encoded by two gene segments sandwiched between several other genes. Each segment terminates in half of an intein gene (a "split intein"). When their protein products make contact, the intein reassembles itself and splices the two polymerase segments together. *Compare with* fused protein, fusion gene. *See* Appendix C, 1990, Kane *et al.*; 1997, Klenk *et al.*; posttranslational processing.

protein structure The *primary* structure of a protein refers to the number of polypeptide chains in it, the amino acid sequence of each, and the position of inter-and intrachain disulfide bridges. The *secondary* structure refers to the type of helical configuration possessed by each polypeptide chain resulting from the formation of intramolecular hydrogen bonds along its length. The *tertiary* structure refers to the manner in which each chain folds upon itself. The *quaternary* structure refers to the way two or more of the component chains may interact. *See* Appendix C, 1951, Pauling and Corey; 1955, Sanger *et al.*; 1973, Anfinsen; alpha helix, beta pleated sheet.

protein tyrosine kinase *See* protein kinase.

proteolytic causing the digestion of proteins into simpler units.

proteome all of the proteins produced by a cell at any given time. Unlike the genome of a cell, which is normally invariant, the kinds or amounts of proteins produced by a cell may vary with such factors as stage of development, age, disease, drugs, and so forth. *See* metabolic control levels, serial analysis of gene expression (SAGE), transcriptome.

proteomics the large-scale study of all the expressed proteins, particularly their structures, functions, and interactions. Proteomics utilizes a diverse range of technologies, from genetic analysis and two-dimensional gel electrophoresis (*q.v.*) to x-ray chrystallography (*q.v.*), NMR spectroscopy (*q.v.*), and sequence alignment searches using advanced computer programs. *See* Appendix E, Individual Databases.

proter the anterior daughter organism produced by the transverse division of a protozoan.

Proterozoic the more recent of the two eras making up the Precambrian eon. Stromatolites (*q.v.*) occur in early Proterozoic strata, and by the end of the era animals as advanced as coelenterates and annelids were present. The origin of eukaryotes presumably occurred midway through the era. *See* Appendix C, 1954, Barghoorn and Tyler; geologic time divisions.

prothallus (prothalium) the independent gametophyte of a horsetail or fern. *See* Appendix A, Plantae, Tracheophyta.

prothetely an experimentally induced abnormality in which an organ appears in advance of the normal time because of a partially inhibited metamorphosis; for example, the formation of pupal antennae on a caterpillar.

prothoracic gland a gland located in the prothorax of insects that secretes ecdysone (*q.v.*). *See* ring gland.

prothoracicotropic hormone (PTTH) a peptide hormone produced by neurosecretory cells in the dorsum of the insect brain that stimulates the prothoracic gland (*q.v.*) to synthesize and secrete ecdysones.

prothrombin an inactive form of thrombin. *See* blood clotting.

protist an informal term used to refer to any single-celled (usually eukaryotic) organism.

protocooperation population or species interaction favorable to both, but not obligatory for either one.

Protoctista (pronounced "prototista") one of the five kingdoms of living organisms. It contains the eukaryotic microorganisms and their immediate descendants, i.e., the nucleated algae, flagellated water molds, slime molds, and protozoa. *See* Appendix A, Superkingdom Eukaryotes.

protogyny sequential hermaphroditism with the ovary functioning before the testis. *Compare with* protandry.

protomers single polypeptide chains (either identical or nonidentical) of a multimeric protein.

protomitochondria *See* petites.

proton an elementary particle of the atomic nucleus with a positive electric charge (equal numerically to the negative charge of the electron) and a mass of 1.0073 mass units.

proto-oncogene a cellular gene that functions in controlling the normal proliferation of cells and either (1) shares nucleotide sequences with any of the known viral *onc* genes, or (2) is thought to represent a potential cancer gene that may become carcinogenic by mutation, or by overactivity when coupled to a highly efficient promoter. Some proto-oncogenes (e.g., *c-src*) encode protein kinases that phosphorylate tyrosines in specific cellular proteins. Others (e.g., *c-ras*) encode proteins that bind to guanine nucleotides and possess GTPase activity. Still other oncogenes encode growth factors or growth factor receptors. *See* maturation promoting factor, Philadelphia (Ph^1) chromosome, platelet-derived growth factor.

protoplasm the substance within the plasma membrane of a cell; the nucleus and surrounding cytoplasm. *See* Appendix C, 1839, Purkinje.

protoplast the organized living unit of a plant or bacterial cell consisting of the nucleus (or nucleoid), cytoplasm, and surrounding plasma membrane, but with the cell wall left out of consideration. Protoplasts can be generated experimentally; e.g., the walls of *E. coli* cells can be removed by lysozyme treatment. Aphragmabacteria (*see Mycoplasma*) lack cell walls and in this sense are protoplasts.

protoplast fusion a mechanism for achieving genetic transformation by joining two protoplasts or joining a protoplast with any of the components of another cell.

Protostomia one of the two major subdivisions of the Bilateria. It contains the annelids, molluscs, and several smaller phyla. The protostome egg undergoes spiral cleavage (*q.v.*), and each of the cells produced is determined to serve as the progenitor of a specific type of tissue. The blastopore (*q.v.*) becomes the adult mouth, and the anus forms anew at the end of the gastrula sac. *Compare with* Deuterostomia. *See* Appendix A.

prototroph 1. an organism that is able to subsist on a carbon source and inorganic compounds. For most bacteria, the carbon source could be a sugar; green plants use carbon dioxide. **2.** a microbial strain that is capable of growing on a defined minimal medium; wild-type strains are usually regarded as prototrophs.

Protozoa a kingdom erected in Cavalier-Smith's classification to contain the majority of unicellular heterotrophic eukaryotes. Protozoa contain 80 S ribosomes, they lack chloroplasts, and their undulipodia lack mastigonemes (*q.v.*). *See* Chromista.

provirus 1. a virus that is integrated into a host cell chromosome and is transmitted from one cell generation to another without causing lysis of the host. 2. more specifically, a duplex DNA sequence in an eukaryotic chromosome (corresponding to the genome of an RNA retrovirus) that is transmitted from one cell generation to another without causing lysis of the host. Such proviruses are often associated with transformation of cells to the cancerous state. *See* mouse mammary tumor virus.

proximal toward or nearer to the place of attachment (of an organ or appendage). In the case of a chromosome, the part closest to the centromere.

Prunus the genus that includes *P. amygdalus*, the almond; *P. armeniaca*, the apricot; *P. avium*, the cherry; *P. domestica*, the plum; *P. persica*, the peach.

Przewalski horse (pronounced *she-val-ski)* a horse that once roamed the vast grasslands of central Asia, but now is found only in zoological parks. *See Equus przewalskii.*

pseudoalleles genes that behave as alleles in the *cis-trans* test (*q.v.*) but can be separated by crossing over. *See* Appendix A, 1949, Green and Green.

pseudoautosomal genes *See* human pseudoautosomal region.

Pseudocoelomata a subdivision of the Protostomia containing animals having a body cavity that is not lined with peritoneum. The space is formed by dispersion of mesenchyme. *See* Appendix A.

pseudocopulation the mode of pollination in certain orchids in which structures of the flower closely resemble a female insect, and the male insects attempting copulation serve to transfer pollen from one flower to another.

pseudodiploid a condition in which the chromosome number of a cell is the diploid number characteristic of the organism but, as a consequence of chromosomal rearrangements, the karyotype is abnormal and linkage relationships may be disrupted.

pseudodominance the phenotypic expression of a recessive allele on one chromosome as a consequence of deletion of the dominant allele from the homolog.

pseudoextinction disappearance of a taxon by virtue of its being evolved by anagenesis into another taxon.

pseudogamy the parthenogenetic development of an ovum following stimulation (but not fertilization) by a male gamete or gametophyte; synonymous with *gynogenesis*.

pseudogene a gene bearing close resemblance to a known gene at a different locus, but rendered nonfunctional by additions or deletions in its structure that prevent normal transcription and/or translation. Pseudogenes are usually flanked by direct repeats of 10 to 20 nucleotides; such direct repeats are considered to be a hallmark of DNA insertion. Two classes of pseudogenes exist: (1) *Traditional pseudogenes* (as exemplified in the globin gene families) appear to have originated by gene duplication and been subsequently silenced by point mutations, small insertions, and deletions; they are usually adjacent to functional copies and show evidence of being under some form of selective constraint for several millions of years after their formation. (2) *Processed pseudogenes* lack introns, possess a remnant of a poly-A tail, are often flanked by short direct repeats, and are usually unassociated with functional copies; all of which suggests their formation by the integration into germ-line DNA of a reverse-transcribed processed RNA. Processed pseudogenes are rare in yeast and *Drosophila*, but common in mammals. For example, in humans there are 20 pseudogenes that are believed to have arisen from actin and beta tubulin mRNAs. *See* Appendix C, 1977, Jacq *et al.*; hemoglobin genes, leprosy bacterium, orphons, processed gene.

pseudohermaphroditism a condition in which an individual has gonads of one sex and secondary sexual characters of the other sex or of both sexes. Pseudohermaphrodites are designated as male or female with reference to their sex chromosome constitution or the type of gonadal tissue present.

Pseudomonas a genus of Gram-negative, motile bacteria that grow as free living organisms in soil, river water, marshes, and coastal marine, habitats and as pathogens of plants and animals. Geneticists often study strains of *P. aeruginosa* which are resistant to antibiotics and disinfectants and are responsible for many infections in humans. This species is the predominant cause of mortality in patients with cystic fibrosis (*q.v.*). The bacterium is characterized by a single polar flagellum. Its genome contains 6.3 mbp of DNA and about 5,570 ORFs have been identified. Lysogeny (*q.v.*) is common in *P. aeruginosa.* The 6.2 mbp genome of *P. putida* has also been sequenced and found to contain 5,420 ORFs. *P. putida* is a species with diverse metabolic and transport systems, which colonizes soil and water habitats, as well as the roots of crop plants. It has unusual abilities in breaking down aromatic and other toxic compounds, and it can tolerate heavy metals. *See* Appendix A, Bacteria, Proteobacteria; bacteriocins, bioremediation.

pseudotumor an aggregation of blackened cells in *Drosophila* larvae, pupae, and adults of certain genotypes. Such "tumors" result from encapsulation during the larval stage of certain tissues by hemocytes and subsequent melanization of these masses.

pseudouridine *See* rare bases.

pseudovirion a synthetic virus consisting of the protein coat from one virus and the DNA from a foreign source. *See* phenotypic mixing, reassortment virus.

pseudo-wild type the wild phenotype of a mutant, produced by a second (suppressor) mutation.

psilophytes early vascular plants that were transitional between algae and true plants. They had branches but no leaves.

P site *See* translation.

psoralens photosensitive cross-linking reagents that act on specific base-paired regions of nucleic acids. *See* trimethylpsoralen.

P strain the paternally contributing strain of *Drosophila* in a P-M hybrid dysgenesis cross. P strains differ genetically from M strains in that they contain multiple P factors in their genomes. *See* hybrid dysgenesis, M strain, P elements.

^{32}P suicide inactivation of phages due to the decay of radiophosphorus molecules incorporated into their DNA.

psychosis a generic term covering any behavioral disorder of a far-reaching and prolonged nature.

PTC abbreviation for *p*henyl*t*hio*c*arbamide or *p*lasma *t*hromboplastin *c*omponent (*both of which see*).

PTK protein tyrosine kinase. *See Src.*

pteridine *See* Drosophila eye pigments.

pteridophytes the ferns, horsetails, club mosses, and other vascular spore-bearing plants.

pteroylglutamic acid folic acid (*q.v.*).

pterygote an insect belonging to a division that includes all winged species. Some pterygotes (e.g., fleas) are wingless, but they are believed to have been derived from winged ancestors. *See* apterygotes, Appendix A.

PTTH *pro*thoracico*t*ropic *h*ormone (*q.v.*).

Pu abbreviation for any purine (e.g., adenine or guanine). *See* R3. *Compare with* Py.

puff *See* chromosomal puff.

pufferfish *See Takifugu rubripes* and *Tetraodon nigroviridis.*

pulse-chase experiment an experimental technique in which cells are given a very brief exposure (the pulse) to a radioactively labeled precursor of some macromolecule, and then the metabolic fate of the label is followed during subsequent incubation in a medium containing only the nonlabeled precursor (the chase).

pulsed-field gradient gel electrophoresis a technique for separating DNA molecules by subjecting them to alternately pulsed, perpendicularly oriented electrical fields. The technique has allowed separation of the yeast genome into a series of molecules that ranged in weight between 40 and 1800 kilobases and represent intact chromosomes. *See* Appendix C, 1984, Schwartz and Cantor.

pulvillus the last segment of the foot in an insect. It has a pad with a claw on either side.

punctuated equilibrium a term describing a pattern seen in the fossil record of relatively brief episodes of speciation followed by long periods of species stability. Although this pattern conflicts with the pattern of gradualism (*q.v.*), no special developmental, genetic, or ecological mechanisms are required to explain it. Both the gradual and the punctuated shifting equilibrium pattern can be simulated from mathematical equations that include only terms for random mutation, natural selection, and population size. *See* Appendix C, 1972, Eldredge and Gould; 1985, Newman *et al.*

Punnett square the checkerboard method commonly used to determine the types of zygotes produced by a fusion of gametes from the parents. The results allow the computation of genotypic and phenotypic ratios. This matrix was first shown in a textbook by R. C. Punnett titled *Mendelism* and published in 1911.

pupal period the developmental period between pupation and eclosion.

puparium formation the formation of a pupal case by the tanning of the skin molted from the last-instar larval insect.

pupation that stage in the metamorphosis of the insect signaled by the eversion of the imaginal discs.

purebred derived from a line subjected to inbreeding (*q.v.*).

pure culture a culture that contains only one species of microorganism. *See* Appendix C, 1881, Koch.

pure line a strain of an organism that is homozygous because of continued inbreeding.

purine *See* bases of nucleic acids.

puromycin an antibiotic that, because of its structural resemblance to the terminal aminoacylated adenosine group of aminoacyl tRNA, becomes incorporated into the growing polypeptide chain and causes the release of incompleted polypeptide chains (which are terminated with a puromycin residue) from the ribosome.

P value probability value. A decimal fraction showing the number of times an event will occur in a given number of trials. *See* probability of an event.

Py abbreviation for any pyrimidine (e.g., thymine, cytosine, uracil). *See* Y. *Compare with* Pu.

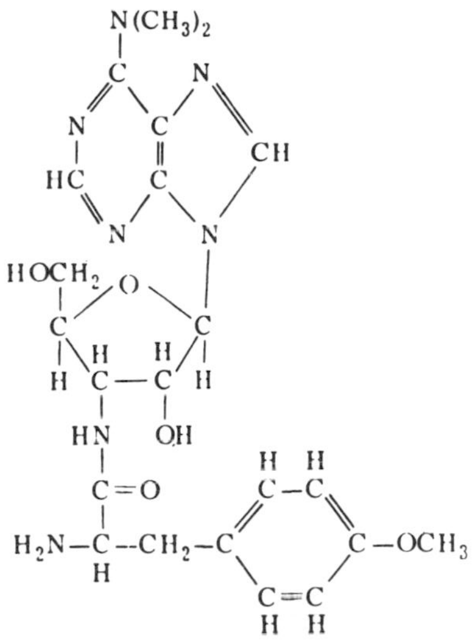

Puromycin

pycnosis the contraction of the nucleus into a compact, strongly staining mass, taking place as the cell dies.

pycocin *See* bacteriocin.

pyloric stenosis the constriction of the valve between the stomach and intestine, a congenital disorder of high heritability.

pyrenoid a small, round protein granule surrounded by a starch sheath found embedded in the chloroplasts of certain algae and liverworts.

pyrethrins diterpene insecticides found in plant tissues. These compounds were first extracted from pyrethrum (chrysanthemum) flowers.

pyridoxal phosphate the coenzyme of both amino acid decarboxylating enzymes and transaminating enzymes.

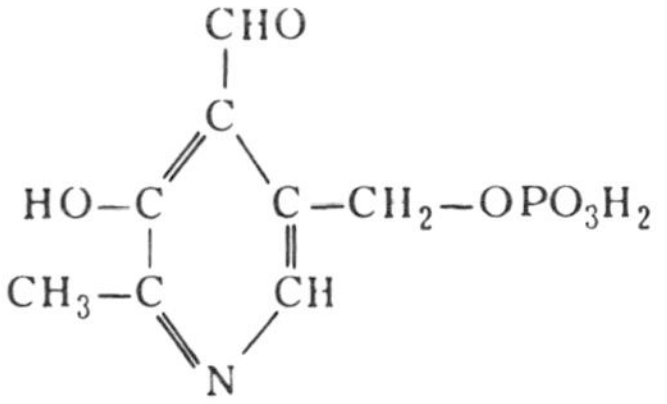

pyridoxine vitamin B_6.

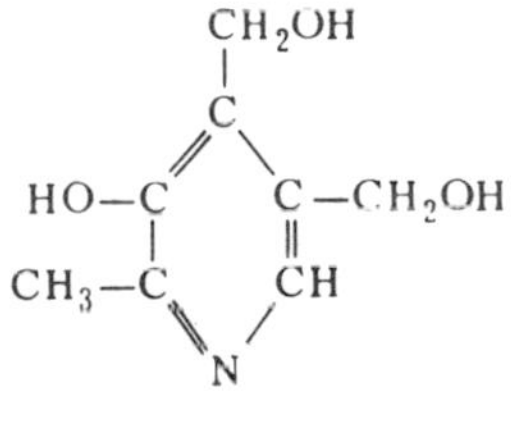

pyrimidine *See* bases of nucleic acids.

pyrimidine dimer the compound formed by UV irradiation of DNA whereby two thymine residues, or two cytosine residues, or one thymine and one cytosine residue occupying adjacent positions in the polynucleotide strand become covalently joined. *See* thymine dimer.

pyronin Y a basic dye often used in cytochemistry. In 2 M magnesium chloride at *p*H 5.7, pyronin Y stains only undegraded RNA. *See* methyl green.

Pyronin Y

pyrrole molecules ring-shaped compounds containing one nitrogen and four carbon atoms that are components of porphyrin (*q.v.*) molecules.

Q

q *See* symbols used in human cytogenetics.

Q_{10} temperature coefficient; the increase in a reaction or other process (expressed as a multiple of the initial rate) produced by raising the temperature 10°C.

Qa in the mouse, a series of loci located very close to the major histocompatibility complex (H-2) whose products are expressed on the surfaces of some lymphocyte classes and subclasses.

Q bands *See* chromosome banding techniques.

Q beta (Qβ) phage an RNA virus that infects *E. coli*. Its genome consists of a circular, positive sense single-stranded RNA molecule. This strand acts both as a template for the replication of a complementary strand and as an mRNA molecule that directs the translation of viral proteins. Q beta phage is one of the smallest known viruses, measuring 24 nm in diameter. Its icosahedral capsid is composed of 180 copies of a single coat protein. *See* Appendix C, 1965, 1967, Spiegelman *et al.*; 1973, Mills *et al.*; 1983, Miele *et al.*; Appendix F; androphages, bacteriophage, *in vitro* evolution, Q beta replicase, virus.

Q beta replicase the enzyme that catalyzes the replication of Qβ phage. *See* RNA-dependent RNA polymerase.

QTLs quantitative trait loci; genes that control the expression of traits (such as height or skin color in humans, pesticide resistance in insects, and ear length in corn) that show quantitative inheritance (*q.v.*).

quadrivalent a meiotic association of four homologous chromosomes; synonymous with *tetravalent*.

quadruplex *See* autotetraploidy.

quail *See Coturnix coturnix japonica.*

quantasome a photosynthetically active particle found in the grana of chloroplasts. Each quantasome is an oblate ellipsoid with axes of about 100 and 200 Ångstroms. Chlorophyll (*q.v.*) is localized within quantasomes.

quantitative character a character showing quantitative inheritance (beef and milk production in cattle, egg production in hens, DDT resistance in *Drosophila*, stature, weight, and skin pigmentation in humans).

quantitative inheritance phenotypes that are quantitative in nature and continuous in distribution are referred to as quantitative characters (*q.v.*). During their genetic transmission, there is an absence of clear-cut segregation into readily recognizable classes showing typical Mendelian ratios. An often-used example is ear length in maize, as illustrated by the histograms on page 366. When crosses are made between individuals from lines showing large quantitative differences in ear length, the offspring are intermediate. When F_1 individuals are crossed, the F_2 population has a mean that is very similar to the F_1 mean, but some individuals produce ears as long or as short as the grandparents. Such results are explained by the "multiple factor hypothesis," which assumes that the quantitative character depends on the cumulative action of multiple genes (or polygenes), each on a separate chromosome, and each producing a unitary effect. In the corn example, a simple model would employ three genes, each existing in two allelic forms. Each capital-letter gene might be responsible for three units of "growth potential," and each small-letter gene, for one unit. Thus the capital-letter genes are all interchangeable in the sense that each produces the same phenotypic effect, and the same is true for the small-letter genes. The long- and short-eared parental individuals would be *AABBCC* and *aabbcc*, respectively, and their offspring would be *AaBbCc*. These would show little variability, because all plants would be genetically identical. The segregation of the alleles in the F_2 population would produce 27 different genotypic classes, and the cumulative action of the genes would generate 7 phenotypic classes. The most common genotype (making up one-eighth of the total population) would be *AaBbCc*, genetically identical to the F_1 plants. But there would also be plants of genotype *AABBCC* and *aabbcc* (each making up one-sixty-fourth of the population) and these would be phenotypically and genetically identical to the grandparents. There would also be individuals with various intermediate ear lengths of genotypes (*AABBCc*,

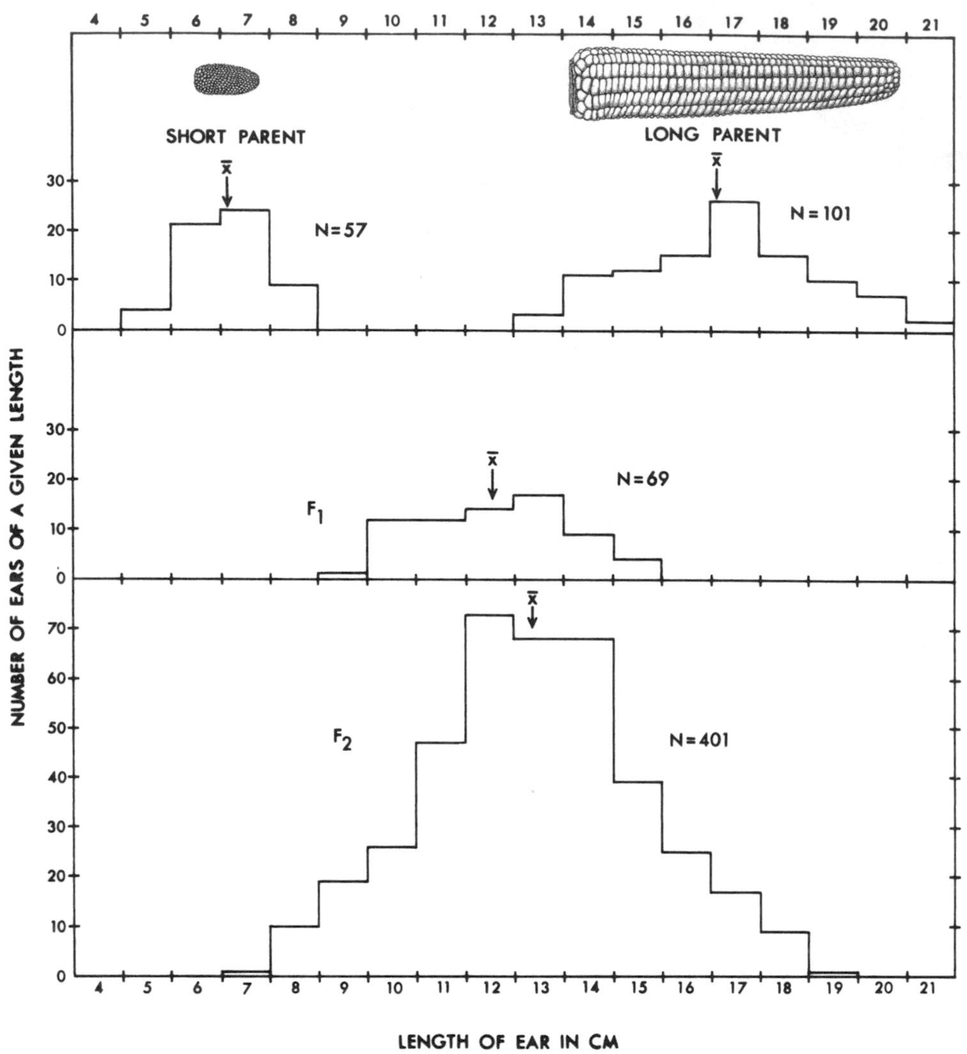

Distribution of ear lengths in parents, F_1 and F_2 generations of a cross between Tom Thumb popcorn and Black Mexican sweet corn.

Quantitative inheritance in maize

*aabbc*C, *AAbb*CC, etc.) and the result would be an F_2 population with a mean equivalent to the F_1, but with a distribution whose width depended on the number of segregating alleles. By comparing the variances of the F_1 and F_2 populations, one can estimate the number of segregating gene pairs responsible for the trait. *See* Appendix C, 1889, Galton; 1909, Nilsson Ehle; 1913, Emerson and East; Wright's polygenic estimate.

quantum according to the quantum theory, energy is radiated in discrete quantities of definite magnitude called quanta and absorbed in a like manner.

quantum speciation the rapid evolution of new species, usually within small, peripheral isolates, with founder effects and genetic drift playing important roles. *See* evolution.

quartet a group of four nuclei or of four cells arising from the two meiotic divisions.

Quaternary the most recent of the two geologic periods making up the Cenozoic era. *See* geologic time divisions.

quaternary protein structure *See* protein structure.

Quercus the genus of oaks including: Q. *alba*, the white oak; Q. *coccinea*, the scarlet oak; Q. *palustris*, the pin oak; Q. *suber*, the cork oak.

quick-stop mutants of *E. coli* that immediately cease replication when the temperature is increased to 42°C.

quinacrine an acridine derivative used in the treatment of certain types of cancer and malaria. It is also used as a fluorochrome in chromosome cytology. *See* Appendix C, 1970, Caspersson *et al.*; 1971, O'Riordan *et al.*

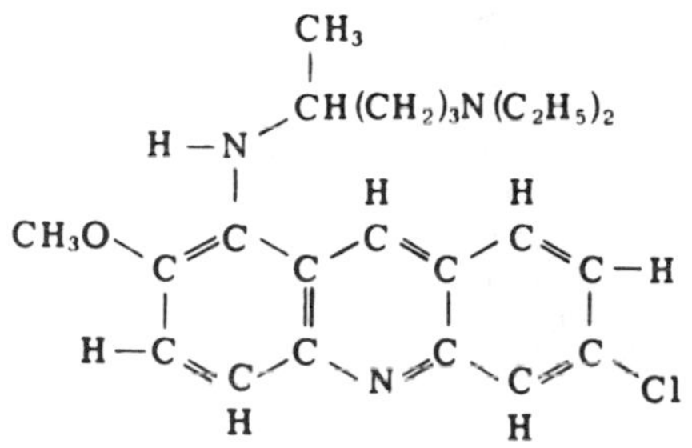

quinone a compound belonging to a class of molecules that function in biological oxidation-reduction systems.

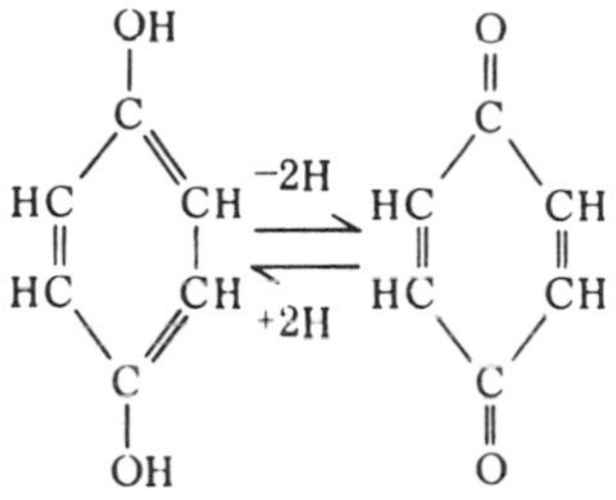

q.v. which see. An abbreviation for Latin, *quod vide*.

r 1. reproductive potential. 2. ring chromosome; *see* symbols used in human genetics. 3. roentgen. 4. correlation coefficient; *see* correlation.

R 1. a chemical radical. Used to show the position of an unspecified radical in a generalized structural formula of a group of organic compounds. 2. a drug-resistant plasmid conferring resistance to one or more antibiotics on bacteria in which it resides. 3. the single-letter symbol for purine. *Compare with* Y.

rII a segment on the chromosome of the T4 bacteriophage that was the first to be subjected to fine structure mapping. Mutants at the rII locus failed to produce plaques or produced abnormal plaques, depending on the strain of *E. coli* used as hosts. Benzer mapped over 1,600 of these mutations. In the diagram each is represented by a box and they reside in adjacent genes (cistrons A and B). The smallest unit of recombination (the recon) corresponded to a distance of two nucleotide pairs. The mutational sites (mutons) were the equivalent of one or two nucleotides. Note that each cistron contains a mutational hot spot. *See* Appendix C, 1955, 1961, Benzer; 1978, Coulondre *et al.*; beads on a string, proflavin.

rabbit *See Oryctolagus cuniculus.*

rabies virus a virus (*q.v.*) belonging to the rhabdoviridae (*q.v.*). It can multiply in many species of mammals, and it induces aggressive-biting behavior by the infected host which maximizes the chances of spreading the viral infection.

Rabl orientation a chromosome orientation sometimes observed in interphase nuclei where centromeres are grouped near one pole and the telomeres all point toward the opposite pole. The arrangement has been interpreted to mean that centromeres and telomeres attach to opposite sides of the nuclear lamina. The orientation is named after C. Rabl, who first described the phenomenon in 1885. *See Drosophila* salivary gland chromosomes, lamins.

race a phenotypically and/or geographically distinctive subspecific group, composed of individuals inhabiting a defined geographical and/or ecological

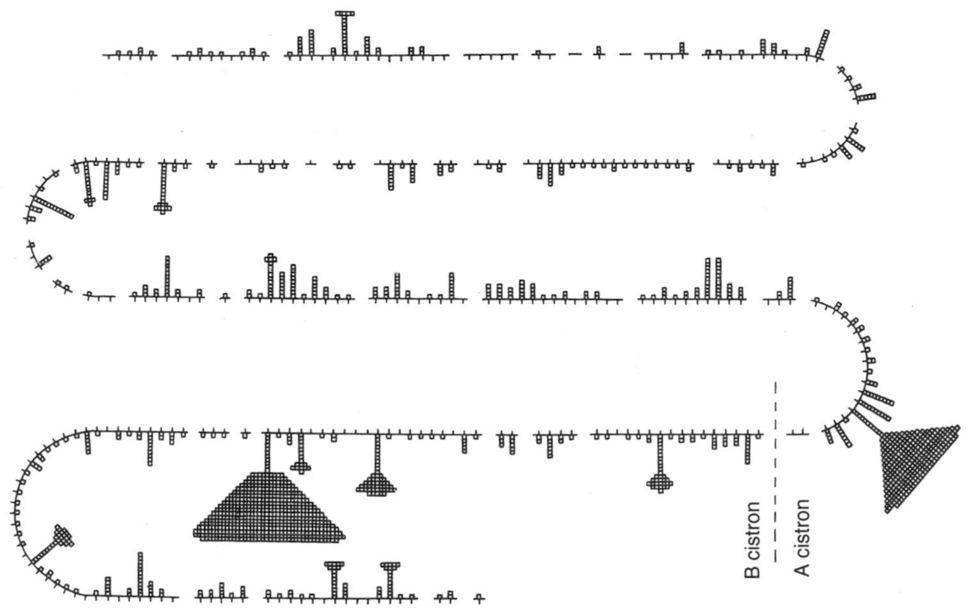

rII

region, and possessing characteristic phenotypic and gene frequencies that distinguish it from other such groups. The number of racial groups that one wishes to recognize within a species is usually arbitrary but suitable for the purposes under investigation. *See* ecotype, subspecies.

raceme an inflorescence as in the hyacinth, in which the flowers are borne on pedicels arising from the rachis. A branched raceme, such as may be seen in oats, rice, wheat, and rye, is called a panicle. *See* spike.

rachet a tool with teeth on the rim of a wheel which is prevented from moving backward by a rotating pivot. A rachet is sometimes used as a metaphor for evolution moving relentlessly forward. *See* Dollo law, Muller rachet.

rad abbreviation of *r*adiation *a*bsorbed *d*ose. A unit defining that energy absorbed from a dose of ionizing radiation equal to 0.01 joule per kilogram. 1 rad = 0.01 gray.

RAD the symbol for a gene that repairs radiation damage to DNA. In *Schizosaccharomyces pombe* there are four *Rad* genes that are involved at DNA damage checkpoints (*q.v.*). The wild-type allele of *rad3* encodes a protein called a phosphatidylinositol 3-kinase (Pl 3-kinase). This enzyme transduces signals from the environment that detect damage to DNA. The normal allele of a gene in humans that is responsible (when mutated) for the hereditary disease *ataxia-telangiectasia* encodes a homologous Pl 3-kinase. *See* ATM kinase.

radial cleavage a pattern of cell divisions seen in the developing embryos of deuterostomes, such as echinoderms and amphibians. The first two cleavages are vertical and the third is horizontal. As a result, each of the blastomeres in the upper tier of four cells lies directly over the corresponding blastomeres in the lower tier. *Compare with* spiral cleavage.

Radiata a subdivision of the Eumetazoa containing animals, such as jelly fish and coral polyps, characterized by radial symmetry. *See* Appendix A.

radiation the emission and propagation of energy through space or a medium in the form of waves. When unqualified, radiation usually refers to electromagnetic radiations (radio waves, infrared, visible light, ultraviolet, x-rays, and gamma rays), and, by extension, ionizing particles. *See* microbeam irradiation, radiation units, recoil energy.

radiation absorbed dose *See* rad.

radiation chimera an experimentally produced animal containing hemopoeitic cells of a genotype different from that of the rest of the organism. Recipients receive a single dose of radiation that kills the stem cells of the bone marrow and much of the differentiated hemopoeitic tissue. Very shortly thereafter, they receive an intravarious inoculation of bone marrow or fetal liver cells from nonirradiated donors. The injected stem cells home to the recipient's bone marrow sites and begin repopulating them, and ultimately they replace the recipient's hemopoeitic tissues.

radiation dosage *See* dose, phantom.

radiation genetics the scientific study of the effects of radiation on genes and chromosomes. The science began with the demonstration for *Drosophila* and corn that x-rays produced deleterious mutations. *See* Appendix C, 1927, Muller; 1928, Stadler.

radiation-induced chromosomal aberration a chromosomal aberration (*q.v.*) induced through breakage caused by ionizing radiation. In the figure on page 370 are shown the origin and mitotic behavior of a variety of radiation-induced aberrations. Original break positions are indicated by short diagonal lines.

radiation sickness a syndrome characterized by nausea, vomiting, diarrhea, psychic depression, and death following exposure to lethal doses of ionizing radiation. The median lethal radiation dose for humans is between 400 and 500 roentgens. Such a dose leaves only about 0.5% of the body's reproducing cells still able to undergo continued mitosis. Since each cell continues to function normally in the physiological sense, death is not immediate. Damage shows up first in tissues with a high mitotic rate (the blood cell–forming tissues of the bone marrow, for example). Death occurs when the surviving cells are unable to restore by mitosis the needed numbers in time to maintain the physiological functioning of the various vital tissues.

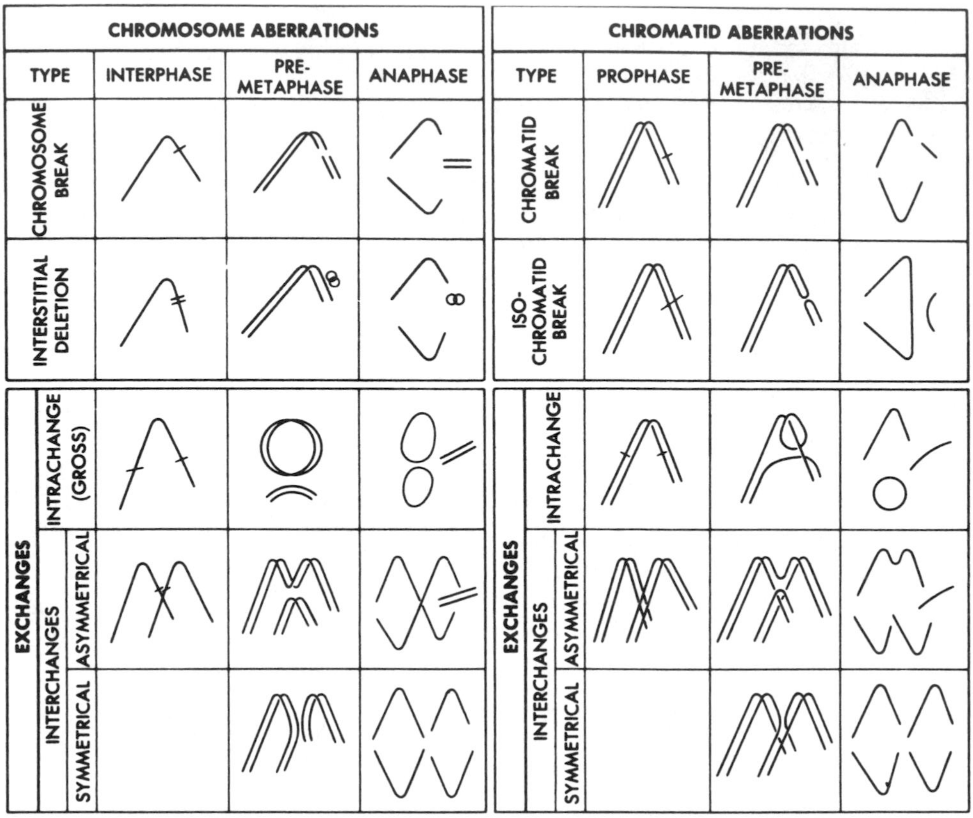

CHROMOSOME ABERRATIONS			
TYPE	INTERPHASE	PRE-METAPHASE	ANAPHASE
CHROMOSOME BREAK			
INTERSTITIAL DELETION			
EXCHANGES — INTRACHANGE (GROSS)			
EXCHANGES — INTERCHANGES — ASYMMETRICAL			
EXCHANGES — INTERCHANGES — SYMMETRICAL			

CHROMATID ABERRATIONS			
TYPE	PROPHASE	PRE-METAPHASE	ANAPHASE
CHROMATID BREAK			
ISO-CHROMATID BREAK			
EXCHANGES — INTRACHANGE			
EXCHANGES — INTERCHANGES — ASYMMETRICAL			
EXCHANGES — INTERCHANGES — SYMMETRICAL			

Radiation-induced chromosomal aberration

radiation units *See* Gray (Gy), rad, rem, Roentgen (R), roentgen equivalent physical (rep), Sievert (Sv).

radical scavenger a molecule with a high affinity for free radicals. If a radical scavenger is added to a biological system prior to irradiation, it may act as a protective agent.

radioactive decay the disintegration of the nucleus of an unstable nuclide accompanied by the spontaneous emission of charged particles and/or photons.

radioactive isotope an isotope with an unstable nucleus that stabilizes itself by emitting ionizing radiations. The use of radioisotopes in biology dates back to 1943 when the X-10 reactor at the Oak Ridge Laboratory in Tennessee started their commercial production. *See* autoradiography, labeled compound, radioimmunoassay, tritium.

radioactive series a succession of nuclides, each of which transforms by radioactive disintegration into the next until a stable nuclide results.

radioactivity the spontaneous disintegration of certain nuclides accompanied by the emission of one or more types of radiation, such as alpha particles, beta particles, and gamma photons.

radioautograph autoradiograph (*q.v.*).

radioautographic efficiency autoradiographic efficiency (*q.v.*).

radioautography autoradiography (*q.v.*).

radiobiology a branch of biology that deals with the effects of radiation on biological systems. It includes radiation genetics (*q.v.*).

radiogenic element an element derived from another element by atomic disintegration.

radiograph a shadow image made on photographic emulsion by the action of ionizing radiation. The image is the result of the differential attenuation of the radiation during its passage through the object being radiographed. A chest x-ray negative is a radiograph.

radioimmunoassay a highly sensitive technique for the quantitative determination of antigenically active substances that are present in very small amounts, such as hormones. The concentration of an unknown, unlabeled antigen is determined by comparing its inhibitory effect on the binding of radioactively labeled antigen to specific antibody with the inhibitory effect of known standards. Symbolized RIA. *See* Appendix C, 1957, Berson and Yalow; iodine.

radiological survey the evaluation of the radiation hazards incident to the production, use, or existence of radioactive materials or other sources of radiation under a specified set of conditions.

radiomimetic chemical a chemical that mimics ionizing radiations in terms of damage to nucleic acids. Radiomimetic compounds include sulfur mustards, nitrogen mustards, and epoxides (*all of which see*).

radioresistance the relative resistance of cells, tissues, organs, or organisms to the injurious action of radiation. Ultraviolet-resistant bacteria, for example, can excise ultraviolet-induced thymine dimers from their DNA. *See Deinococcus radiodurans.*

radiotracer *See* labeled compound, radioactive isotope.

radon the name used to refer to the many isotopes of element 86. Radon is an inert gas that is readily soluble in water. All its isotopes are radioactive with short half-lives, and all decay with the emission of densely ionizing alpha particles (*q.v.*). While such particles are too weak to penetrate the skin, they are very dangerous when radon is ingested or inhaled. Radon is found in nature because it is continuously formed by the radioactive decay of the longer-lived, precursor elements uranium and thorium, which occur in certain minerals. The most common radon isotope in human environment is ^{222}Rn, which has a half-life of 3.8 days. Radiation from radon is responsible for over half of the average exposure to humans from ionizing radiation.

RAG-1* and *RAG-2 genes that synergistically activate V(D)J recombination (*q.v.*). In the gnathostomes studies so far, these two genes are adjacent and are coordinately transcribed but only in lymphatic tissue (*q.v.*). In most cases their ORFs do not contain introns. In humans, *RAG-1* and *RAG-2* are on the short arm of chromosome 11. Coordinate expression of *RAG-1* and *RAG-2* is regulated by genetic elements on the 5′ side of the *RAG-2* gene. The proteins encoded by *RAG-1* and *RAG-2* coexist in a complex that resides in the periphery of the lymphocyte nucleus. The complex recognizes and binds to recombination signals that flank the V, D, and J segments of the *Ig* and *TCR* genes. Each recombination signal consists of a row of seven specific bases (CACAGTC), then a spacer, and then a row of nine bases (ACAAAAACC). The spacer contains either 12 or 23 bases, but these show no consistent ordering of specific nucleotides. The RAG-1 protein contains a homeobox (*q.v.*) by which it binds to the recombination signal. The *RAG* transposon excises a donor DNA segment and inserts it into the maturing fusion product of the *Ig* gene. In gnathosomes, *Ig* and *TCR* genes must be assembled before they can be expressed, and *RAG-1* and *RAG-2* are transcribed only in B and T lymphocytes (*q.v.*). Agnathans and invertebrates lack these molecules and cannot form antigen-specific lymphocytes. This suggests that soon after the divergence of jawed and jawless vertebrates a transposon inserted itself into the germ line of the gnathostome ancestor, and this transposon was the source of *RAG-1* and *RAG-2*. *See* Appendix A, Chordata, Craniata, Agantha, Gnathostoma; Appendix C, 1990, Oettinger *et al.*; 1996, Spanopoulou *et al.*; immunoglobulin genes, *recombination activating genes (RAGs)*, somatic recombination, T cell receptor genes, Tc1/mariner element.

ramets buds that can detach from a plant or animal and result in the asexual production of offspring genetically identical to each other and the parent. Ramets can also refer to the specific offspring produced by asexual budding from a single ancestral organism (the ortet). *See* modular organisms.

Rana frogs of this genus have been used widely in research. The leopard frog, *R. pipiens*, is the most common species bred in the laboratory, and many of its mutations have been recovered and analyzed. This was the species in which the first experiments were performed involving transfer of somatic diploid nuclei into enucleated eggs. Mutant strains are also available for *R. sylvatica*, *R. esculenta*, and *R. temporaria*. *R. esculenta* is the only anuran for which working maps of the lamp-brush chromosomes are available. *See* Appendix C, 1952, Briggs and King.

r and K selection theory a theory in population ecology that attempts to establish whether environmental conditions favor the maximization of r (the intrinsic rate of natural increase) or of K (the carrying capacity of the environment). When populations can expand without food reserves limiting their growth, then r selection is in control. When food reserves limit population size, K selection takes over, and increase in one genotype must be at the expense of another. Whereas r selection operates in ecological situations where food reserves fluctuate drastically, and species are favored that reproduce rapidly and produce large numbers of offspring. K selection operates in populations that are close to the environmental carrying capacity, and species are favored that reproduce slowly and generate a few offspring that are well adapted to a relatively stable environment.

random assortment *See* assortment.

random cloning *synonymous with* shotgun cloning (*q.v.*).

random drift genetic drift (*q.v.*).

random mating a population mating system in which every male gamete has an equal opportunity to join in fertilization with every female gamete, including those gametes derived from the same individuals (if the species is monoecious or hermaphroditic); panmixis. The Hardy-Weinberg law (*q.v.*) assumes random mating. *Contrast with* mate choice.

random primers randomly generated oligodeoxyribonucleotides, some of which anneal to complementary sequences in the template nucleic acid and serve as primers in reactions involving reverse transcriptase.

random sample a sample of a population selected so that all items in the population are equally likely to be included in the sample.

random sampling error *See* experimental error, sampling error.

random sequencing *synonymous with* shotgun sequencing (*q.v.*).

random-X inactivation the method of dosage compensation (*q.v.*) found in eutherian mammals. *See* paternal-X inactivation.

Raphanobrassica the classic example of a fertile allotetraploid, obtained from hybrids between the radish and cabbage. *See* Appendix C, 1927, Karpechenko.

rapidly reannealing DNA, rapidly reassociating DNA repetitious DNA (*q.v.*).

rapid-lysing (*r*) mutants mutants of T-even phage that enhance the rate at which *E. coli* host cells are lysed; on a bacterial lawn, r-plaques are larger than wild-type plaques (r^+). *See* plaque.

rare bases purines (other than adenine and guanine) and pyrimidines (other than cytosine and uracil) found in transfer RNA (*q.v.*). Formulas are shown on page 373. *See* bases of nucleic acids.

rare earth any of the series of very similar metals ranging in atomic number from 57 to 71. *See* periodic table.

Rassenkreis *See* circular overlap, polytypic species.

rat *See Rattus.*

rat kangaroo *See Potorous tridactylus.*

Rattus the genus of rats including *R. norvegicus*, the brown rat, and *R. rattus*, the black rat. The white laboratory rat is an albino form of *R. norvegicus*. The laboratory rat has 21 chromosome pairs including the sex chromosomes X and Y. The genome size is 2.75 gbp, and the estimated number of structural genes is 25,000. *See* Appendix A, Mammalia, Rodentia; Appendix E.

Rauscher leukemia virus a virus (*q.v.*) isolated from the plasma cells of leukemic mice by Frank Rauscher. This virus was the source of one of the first reverse transcriptases to be isolated. *See* Appendix C, 1970, Baltimore; retrovirus, RNA-dependent DNA polymerase.

R bands *See* chromosome banding techniques.

RB the symbol for the gene causing retinoblastoma (*q.v.*).

RBC red blood cell.

***rbc* genes** genes that encode the components of the enzyme ribulose-1,5-bisphosphate carboxylase-oxygenase (*q.v.*). The large subunits of the enzyme are encoded by *rbcL* genes; the small subunits by *rbcS*. In prokaryotes the *rbcL* and *rbcS* genes are part of the same operon. In most photosynthetic eukaryotes, *rbcL* genes are in the chloroplast genome and *rbsS* genes are in the nuclear genome. *See* photosynthesis.

RBE relative biological effectiveness (*q.v.*).

rDNA 1. in general, any DNA regions that code for ribosomal RNA components. 2. specifically, a tandem cluster of eukaryotic rRNA genes with a sufficiently atypical base composition to allow its isolation directly from sheared genomic DNA. In recent literature, rDNA is also used to refer to hybrid molecules formed by uniting two or more heterologous DNA molecules. To avoid confusion, the symbol rtDNA should be used for such recombinant DNA molecules and rDNA should be reserved for ribosomal DNA. *See* Appendix C, 1967, Birnstiel.

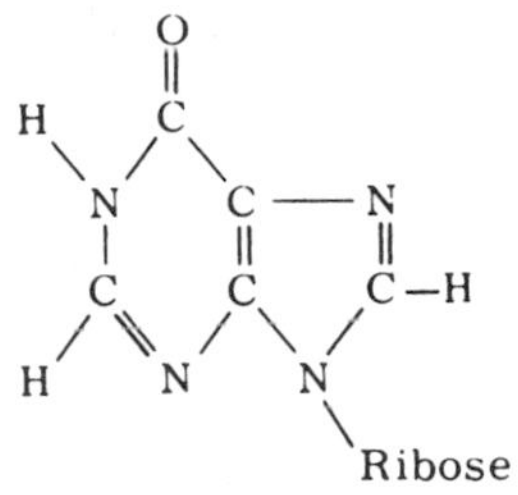

inosine (I)

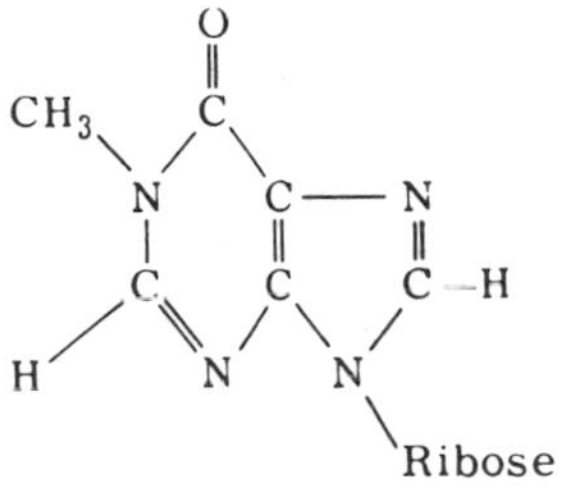

1-methylinosine (I^m)

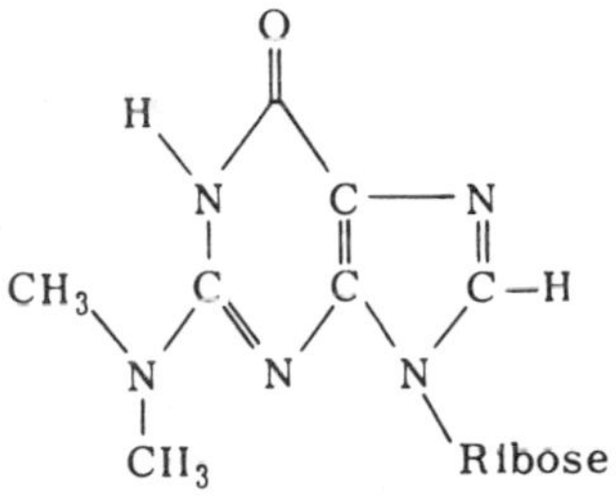

N^2-dimethylguanosine (G^d)

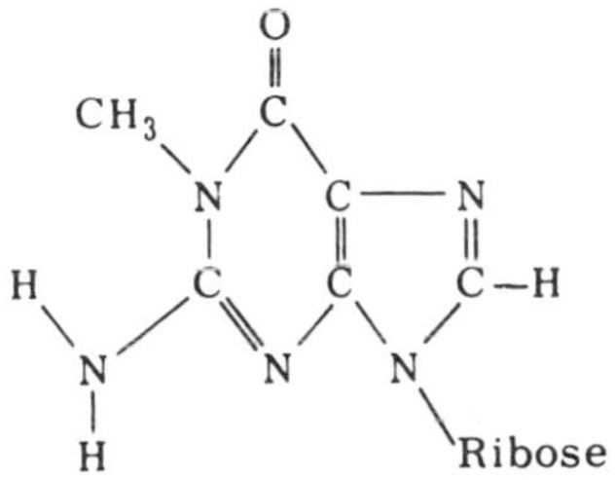

1-methylguanosine (G^m)

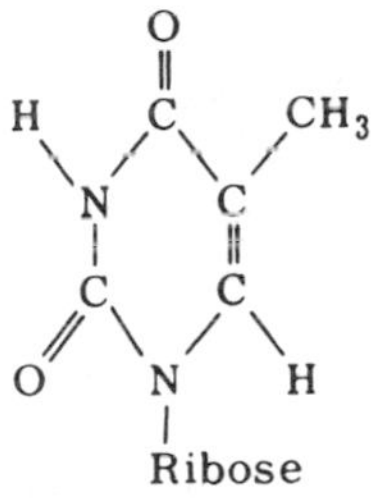

ribothymidine (T^r)

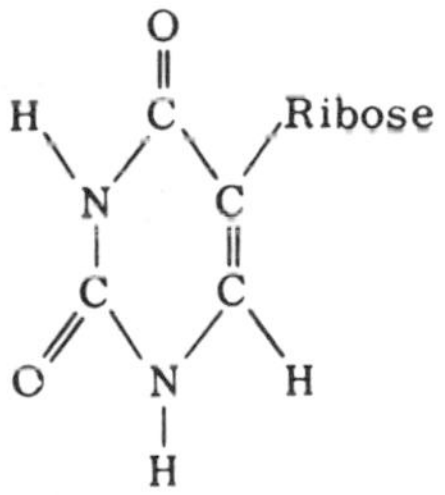

pseudouridine (Ψ)

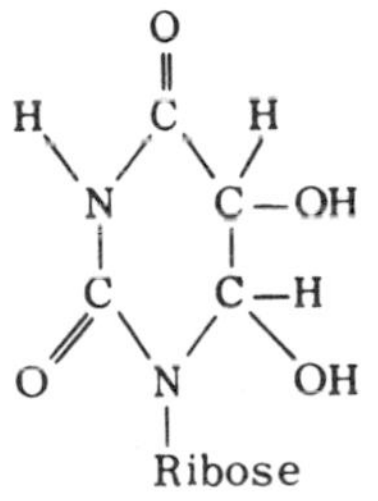

5,6 dihydrouridine (U^d)

Rare bases

rDNA amplification The genes for rRNA are preferentially replicated during oogenesis in amphibia. In *Xenopus laevis,* for example, there are 2,000 rDNA repeats integrated into the chromosomes of the oocyte. However, there are 2 million DNA repeats distributed among about 1,000 extrachromosomal nucleoli that lie near the periphery of the nucleus of each diplotene oocyte. These amplified genes arose from single copies of the chromosomal rDNA, and during pachynema they replicated extrachromosomally by a rolling circle (*q.v.*) mechanism. These extrachromosomal nucleoli function to transcribe the rRNAs stored in the growing oocyte. Amplification of rDNA also occurs commonly in insects with panoistic ovaries and in the macronuclei of protozoa, such as *Tetrahymena. See* Appendix C, 1968, Gall, Brown and Dawid; insect ovary types, Miller trees, nucleolus, nucleolus organizer.

reading the unidirectional process by which mRNA sequences are decoded (translated) into amino acid sequences (polypeptide chains).

reading frame a nucleotide sequence that starts with an initiation codon, partitions the subsequent nucleotides into amino acid–encoding triplets, and ends with a termination codon. The interval between the start and stop codons is called the *open reading frame* (ORF). If a stop codon occurs soon after the initiation codon, the reading frame is said to be *blocked.*

reading frame shift Certain mutagens (acridine dyes, for example) intercalate themselves between the strands of a DNA double helix. During subsequent replication, the newly formed complementary strands may have a nucleotide added or subtracted. A cistron containing an additional base or missing a base will transcribe a messenger RNA with a *reading frame shift*. That is, during translation the message will be read properly up to the point of loss or addition. Thereafter, since the message will continue to be read in triplets, all subsequent codons will specify the wrong amino acids (and some may signal chain termination). *Contrast with* in-frame mutation. *See* acridine orange, acriflavin, amino acid, nonsense mutation, proflavin, translation.

reading mistake the incorrect placement of an amino acid in a polypeptide chain during protein synthesis.

reads overlapping base sequences generated during shotgun sequencing (*q.v.*) from which stretches of contiguous sequences, or *contigs* are assembled.

readthrough 1. transcription beyond a normal terminator sequence in DNA, due to occasional failure of RNA polymerase to recognize the termination signal or due to the temporary dissociation of a termination factor (such as rho in bacteria) from the terminator sequence. 2. translation beyond the chain-terminator (stop) codon of an mRNA, as occurs by a nonsense suppressor (*q.v.*) tRNA. An example of readthrough is found in the tobacco mosaic virus (*q.v.*). Here a 183 kd protein is formed that contains amino acids specified by ORFs 1 and 2.

reannealing in molecular genetics, the pairing of single-stranded DNA molecules that have complementary base sequences to form duplex molecules. Reannealing and annealing (*q.v.*) differ in that the DNA molecules in the first case are from the same source and in the second case from different sources. *See* Appendix C, 1960, Doty *et al.*; Alu family, mouse satellite DNA, reassociation kinetics, repetitious genes.

reassociation reannealing (*q.v.*).

reassociation kinetics a technique that measures the rate of reassociation of complementary strands of DNA derived from a single source. The DNA under study is fragmented into pieces several hundred base pairs in length and then disassociated into single strands by heating. Subsequently, the temperature is lowered and the rate of reannealing (*q.v.*) is monitored. Reassociation of DNA is followed in the form of a cot curve, which plots the fraction of molecules that have reannealed against the log of cot. Cot values are defined as $C_0 \times t$, where C_0 is the initial concentration of single-stranded DNA in moles of nucleotides per liter and t is the reannealing time in seconds. Typical cot curves are shown on page 375. DNAs reannealing at low cot values (10^{-4}–10^{-1}) are composed of highly repetitive sequences, DNAs reannealing at cot values between 10^0 and 10^2 are moderately repetitive, and DNAs reannealing at higher cot values are nonrepetitive. *See* Appendix C, 1968, Britten and Kohne; Alu family, delta T50H, mouse satellite DNA, repetitious DNA.

reassortant virus a virion consisting of DNA from one virus and protein from another viral species; e.g., through genetic engineering, a hybrid virus has been made containing genes from the human influenza virus and capsid proteins that provoke immunity, but also containing avian influenza virus genes that slow the rate of viral replication. *See* phenotypic mixing, pseudovirion.

recapitulation the theory first put forth by Ernst Haeckel that an individual during its development passes through stages resembling the adult forms of its successive ancestors. The concept is often stated "ontogeny recapitulates phylogeny" and is sometimes referred to as the *biogenetic law*.

RecA protein the product of the *RecA* locus of *E. coli*. The protein is of great antiquity, since it occurs in virtually all bacteria. The RecA monomer contains 352 amino acids. The monomers are packed to form a continuous right-handed spiral with six monomers per turn of the helix. The spiral filament contains a deep groove that can accommodate up to three strands of DNA. The RecA protein is a DNA-dependent ATPase, and ATP is hydrolyzed during genetic recombination processes. *See* Appendix C, 1965, Clark; 1992, Story, Weber, and Steitz.

receptor element *See* controlling elements.

receptor-mediated endocytosis endocytosis that involves the binding of a ligand, such as vitellogenin (*q.v.*) to a plasma membrane receptor followed by the lateral movement of the ligand-receptor complex through the membrane toward a coated pit. The cytoskeleton of each coated pit is a basketlike network of hexagons and pentagons formed by the assembly of three-legged protein complexes called *triskelions*. Each triskelion is composed of three molecules of *clathrin*, a 185-kilodalton protein, and three smaller polypeptides. Once a clathrin-coated pit contains a large number of ligand-receptor complexes, it invaginates further into the cytoplasm and eventually a small vesicle is pinched off the pit. This endocytotic vesicle is called a *receptosome*. Once the ligands have been internalized in a receptosome, the

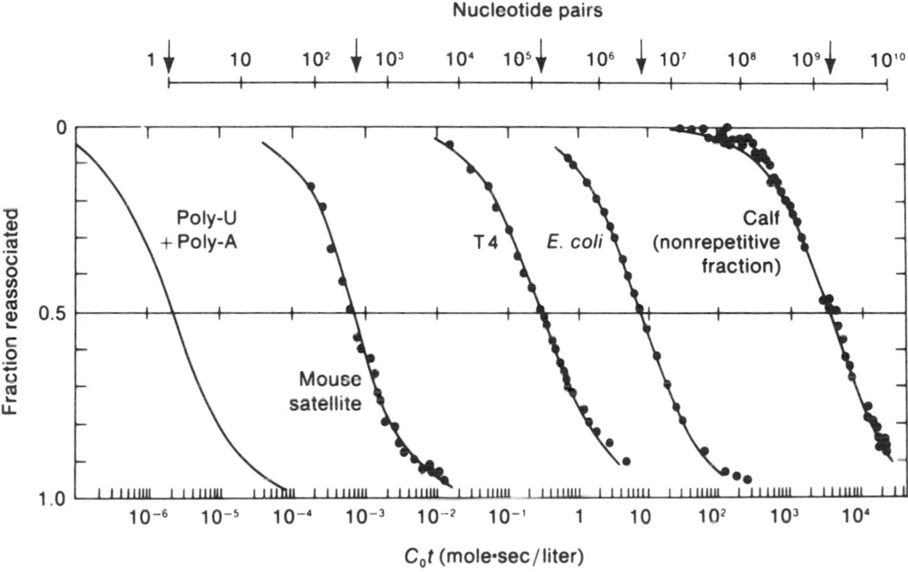

Reassociation kinetics

For each of the DNA samples tested, the number of base pairs in the genome is indicated by an arrow on the logarithmic scale at the top of the graph. The poly-U + poly-A sample is a double helix of RNA, with one strand containing only A and the other strand only U. The mouse satellite DNA is a fraction of nuclear DNA in mouse cells that differs in its physical properties from the bulk of the DNA. The calf DNA represents only those sequences that are present in single copies per haploid genome. The denatured DNA samples were fragmented by mechanical shear to chain lengths of about 400 nucleotides and incubated at a temperature near 60°C. The fraction reassociated was measured by the decrease in UV absorption as double strands formed.

receptor molecules are returned intact to the plasma membrane.

receptor-mediated translocation a hypothesis concerning the translocation of nascent polypeptides across the endoplasmic reticulum membrane. As shown in the diagram on page 376, soon after the signal sequence peptide of the nascent chain emerges from the ribosome it is recognized by a specific receptor called the signal recognition particle (SRP). The second component of the translocation process is the docking protein. It is bound to the surface of the ER membrane, and it serves as a receptor for the SRP. Since the SRP binds to both the docking protein and the signal sequence of the protein being translated, it serves to bring the ribosome into the vicinity of the ER membrane. Subsequently, the ribosome binds to a ribosome receptor on the ER and the nascent polypeptide is threaded through a pore in the membrane and into the ER lumen. A peptidase then removes the signal peptide from the newly synthesized protein molecule. *See* Appendix C, 1975, Blobel and Dobberstein; 1991, Simon and Blobel; leader sequence peptide, signal hypothesis, signal recognition particle, translation, translocon.

receptors *See* cell-surface receptors, cellular signal transduction.

receptosome *See* receptor-mediated endocytosis.

recessive complementarity *See* complementary genes.

recessive gene in diploid organisms, a gene that is phenotypically manifest in the homozygous state but is masked in the presence of its dominant allele. Usually the dominant gene produces a functional product, while its recessive allele does not. Therefore, the normal phenotype is produced if the dominant allele is present (in one or two doses per nucleus), and the mutant phenotype appears only in the absence of the normal allele (i.e., when the recessive gene is homozygous). By extension, the terms dominant and recessive are used in the same sense for heterokaryons and merozygotes.

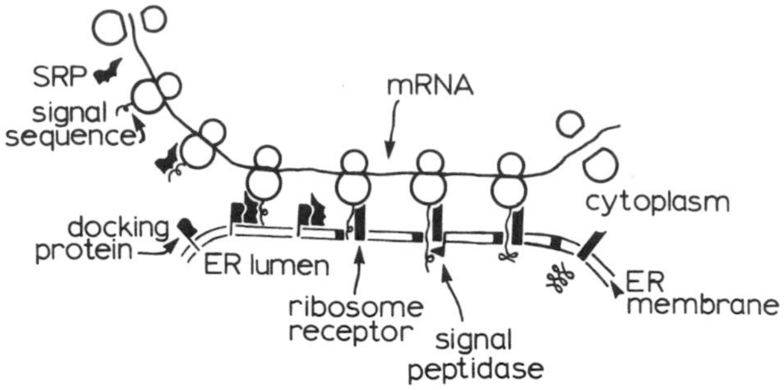

Receptor-mediated translocation

The mRNA is moving from left to right in this diagram.

recessive lethal an allele that kills the cell or organism that is homozygous or hemizygous for it. *See* lethal mutation.

reciprocal crosses crosses of the forms A ♀ × B ♂ and B ♀ × A ♂, where the individuals symbolized by A and B differ in genotype or phenotype or both. Reciprocal crosses are employed to detect sex linkage, maternal inheritance, or cytoplasmic inheritance (*all of which see*).

reciprocal genes complementary genes (*q.v.*).

reciprocal hybrids hybrid offspring derived from reciprocal crosses of parents from different species.

reciprocal recombination in the gametes of dihybrids, the production of new linkage arrangements that are different from those of the maternal and paternal homologs. For example, if the nonallelic mutants *a* and *b* were present in the coupling configuration *AB/ab*, crossovers would generate the reciprocal recombinant gametes *Ab* and *aB* in equal numbers.

reciprocal translocation *See* translocation.

***rec*⁻ mutant** a class of mutations characterized by defective recombination. Such mutants are also radiation-sensitive, which suggests that enzymes functioning during the naturally occurring breakage and rejoining characterizing meiotic crossing over may also repair damage caused by mutagens.

recognition protein *See* cyclins.

recoil energy the energy imparted to the positively charged ion formed during the radioactive transmutation of an atom. A high-energy beta particle is emitted concurrently.

recombinant **1.** the new individuals or cells arising as the result of recombination. **2.** recombinant DNA or a clone containing recombinant DNA.

recombinant DNA a composite DNA molecule created *in vitro* by joining a foreign DNA with a vector molecule.

recombinant DNA technology techniques for joining DNA molecules *in vitro* and introducing them into living cells where they replicate. These techniques make possible (1) the isolation of specific DNA segments from almost any organism and their amplification in order to obtain large quantities for molecular analysis, (2) the synthesis in a host organism of large amounts of specific gene products that may be useful for medicine or industry, and (3) the study of gene structure-function relationships by *in vitro* mutagenesis of cloned DNAs. *See* Appendix C, 1972, Jackson, Symons, and Berg; 1973, Cohen *et al.*; 1974, Murray and Murray; 1975, Asilomar Conference, Benton and Davis; 1976, Efstratiadis *et al.*, Kan *et al.*; 1977, Collins and Holm, Gilbert; 1978, Maniatis *et al.*; 1979, Goeddel *et al.*; 1980, Chakrabarty, Berg *et al.*; 1981, Wagner, Kemp and Hall; 1982, Eli Lilly; 1985, Smithies *et al.*; 1994, Whitham *et al.*; expression vectors, gene cloning.

recombinant inbred (RI) lines inbred lines, each derived independently from an F_2 generation produced from crossing two unrelated, inbred, progenitor lines. Each RI line has a characteristic combination of genes with a different pattern of alternative alleles at multiple loci. This technique has been used in mice to fix chance recombinants in a homozygous state in a group of strains derived from two unrelated but highly inbred progenitor strains.

recombinant joint the edge of a heteroduplex region where two recombining DNA molecules are connected.

recombinant RNA technology techniques that unite foreign RNA molecules or splice different RNAs from the same species. For example, a heterologous RNA sequence can be constructed by ligation of two or more different RNA molecules with T4 RNA ligase. *See* Appendix C, 1983, Miele, Mills and Kramer.

recombination *See* genetic recombination.

recombination activating genes (RAGs) in humans the *RAG-1* and *RAG-2* genes are about 8 kb apart, and they have been mapped to 11p13. Missense mutations in both *RAGs* have been shown to be the cause of hereditary immunodeficiency syndromes. *See Rag-1* and *RAG-2*.

recombination frequency the number of recombinants divided by the total number of progeny. This frequency is used as a guide in assessing the relative distances between loci on a genetic map. *See* Morgan unit.

recombination hotspot special regions on chromosomes where the frequency of meiotic crossing over is elevated. Such hotspots are likely to be initiation sites for recombination. In the human genome, recombination hot spots occur at 200 kb intervals, usually between genes.

recombination mapping *See* linkage map.

recombination nodules (RNs) electron dense structures seen in electron micrographs of synaptonemal complexes (*q.v.*). In *Drosophila*, since the number of RNs in mid pachytene oocytes is about the same as the number of chiasmata observed at diplonena, it is assumed that RNs participate in the exchange process. The *mei-W68* mutation abolishes meiotic exchange in the oocytes of homozygotes, and it also suppresses the formation of RNs. The *SPO11* gene of *Saccharomyces cerevisiae* is an ortholog of *mei W68*. *SPO11* encodes topoisomerase II, an enzyme that produces double-strand breaks in DNA. *See* Appendix C, 1975, Carpenter; centromeric coupling, topoisomerase.

recombination repair formation of a normal DNA molecule by exchanging correct for incorrect segments between two damaged molecules.

recombination suppression *See* crossover suppressor.

recombinators any sequences of nucleotides that promote genetic recombination in their neighborhood. An example would be the chi sequence (*q.v.*) in the *E. coli* chromosome.

recon the smallest unit of DNA capable of recombination (corresponding to an adjacent pair of nucleotides in *cis* position). *See* Appendix C, 1955, Benzer.

record of performance a record of an animal with respect to certain economically important characteristics. Such data are used by livestock breeders in the artificial selection and development of improved breeds.

recurrence risk the risk that a genetic defect that has appeared once in a family will appear in a child born subsequently.

recurrent parent backcross parent.

red blood cell erythrocyte. *See* hemoglobin, sickle-cell anemia.

Red Queen hypothesis one of two major mathematical models concerning the likely evolutionary state of communities under conditions of constancy in the physical environment. *Stationary models* predict that evolution would grind to a halt. The Red Queen hypothesis predicts that evolution would continue because (1) the most important component of the species environment is other species in the community, and (2) not all species will be at their local adaptive peaks, and hence are capable of further evolution even though the physical environment has stabilized. Any evolutionary advance made by one species will, through a close network of interactions, represent a deterioration in the biotic environment of all other species in that same community. Consequently, these other species become subject to selective pressures to achieve evolutionary advances of their own, simply to catch up. The name for this hypothesis is derived from the Red Queen in *Through the Looking Glass*, who said: "Now here, you see, it takes all the running you can do to keep in the same place." *See* lag load, zero sum assumption.

reductase an enzyme responsible for reduction in an oxidation reduction reaction.

reduction classically defined as the addition of hydrogen or electrons. Most biological reductions involve hydrogenations, and hydrogen transfer reactions are usually mediated by NADPH. In cases involving electron transfer, cytochromes (*q.v.*) are reduced. *See* nicotinamide-adenine-dinucleotide phosphate, oxidation.

reduction divisions the division that halves the zygotic chromosome number. *See* Appendix C, 1883, van Beneden; 1887, Weismann; meiosis.

reductionism a philosophy that each phenomenon in the natural world can be understood from a knowledge of its component parts. *See* mechanistic philosophy.

reductive evolution a downsizing of the genome which often occurs in obligate intercellular parasites. The deletion of a subset of genes or their conversion into pseudogenes is tolerated, because the host now supplies the products normally controlled by the dispensable genes. *See* leprosy bacterium, regressive evolution, *Rickettsia prowazeki.*

redundant cistrons cistrons frequently repeated on a chromosome. Examples are the cistrons in the nucleolus organizer coding for the ribosomal RNA molecules.

redundant code *See* degenerate code.

redundant DNA *See* repetitious DNA.

refractive index the ratio of the velocity of light in a vacuum to its velocity in a given substance. *See* phase contrast microscope.

regeneration the process whereby a whole animal or part(s) of an animal is(are) reformed after being lost or damaged in a fully developed individual. The phenomenon of regeneration is typical of most plants but is restricted in animals primarily to less complex forms such as certain flatworms (planarians) or polyploid cnidarians (e.g., hydroids). Among the more complex animals, regeneration of whole limbs or other body parts is much less common in a few groups such as salamanders (*see* axolotl), which can regrow limbs, tails, heart muscle, jaws, spinal cord, and so on. Crabs can regrow lost claws. Some lizards can regrow lost tails. In humans, regeneration is mainly limited to superficial wound healing, although a few human tissues can regenerate (e.g., blood, liver). *See* dedifferentiation. *Compare with* metamorphosis.

regression coefficient the rate change of the dependent variable with respect to the independent variable. The change in mutation frequency per unit change in radiation dose, for example, would be determined by the regression coefficient of the regression line (*q.v.*).

regression line a line that defines how much an increase or decrease in one factor may be expected from a unit increase in another. *See* line of best fit, scatter diagram.

regressive evolution the reduction in morphological complexity as the result of the loss of unneeded structures or biochemical pathways. An example would be the loss of eyes and pigments by animals such as fish and crustaceans living in caves. *See* mitosomes, reductive evolution.

regulation the power of an embryo to continue normal or approximately normal development or regeneration in spite of experimental interference by ablation, implantation, transplantation, etc.

regulative development embryonic development in which the fates of all parts of the embryo are not fixed before fertilization. In such development an ablated part can be repaired, or even separated blastomeres can form identical twins. *See* mosaic development.

regulator element *See* controlling elements.

regulator gene a gene whose primary function is to control the rate of synthesis of the products of other distant genes. The regulator gene ($_rG$) controls the synthesis of a protein repressor (R), which inhibits the action of an operator gene ($_oG$) and thus turns off the operon it controls. In the illustration below, the horizontal line represents a chromosome upon which four genes reside. The left gene can be distant from the other three closely linked genes (in fact, $_rG$ can be on a different chromosome). Genes $_sG_1$ and $_sG_2$ are structural genes or cistrons of the conventional sort that produce specific proteins P_1 and P_2, respectively, through the formation of specific messenger RNA molecules. The repressor is present in exceedingly small amounts. It possesses two sites, one of which can attach to the operator and one of which can bind an effector (E) molecule. Once bound to E, however, the repressor changes shape and cannot attach to the operator. The effector mol-

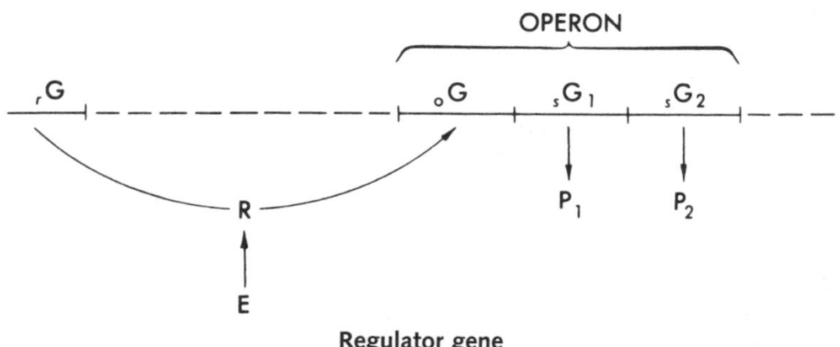

Regulator gene

ecule is generally a substrate of an enzyme produced by ${}_sG_1$ or ${}_sG_2$. Thus the system is an inducible one, since synthesis of P_1 and P_2 proceeds only in the presence of E. *See* allosteric effect, constitutive mutations, *cro* repressor, derepression, inducible system, *lac* operon, lambda repressor, operon, repressible system, selector genes.

regulatory sequence a DNA sequence involved in regulating the expression of the structural gene(s) in the common operon. Examples include attenuators, operators, and promoters. *See* gene, regulator gene.

regulons a group of operons that are under the control of the same regulatory protein. An example is the SOS regulon that is induced when there is DNA damage. The SOS regulon is under the control of a repressor protein encoded by the *LexA* gene. The LexA protein represses four operons (specifying RecA, UvrA, UvrB, and UvrC), all of which are involved in recombination and DNA repair. *See* RecA protein, SOS boxes.

reiterated genes genes that are present in multiple copies that are clustered together on specific chromosomes. Ribosomal RNA genes, transfer RNA genes, and histone genes are examples of such tandem multigene families.

rejection in immunology, destruction of a cell or tissue graft by the immune system of the recipient, directed against antigens on the graft that are foreign to the recipient. *See* Appendix C, 1914, Little; 1927, Bauer; 1948, Snell; histocompatibility molecules.

relational coiling the loose plectonemic coiling of two chromatids about one another. The two chromatids cannot separate until uncoiling is completed. *See* plectonemic spiral.

relative biological effectiveness the ratio of the doses of different ionizing radiations required to produce the same biological effect.

relative molecular mass (Mr) the mass of a molecule relative to the dalton (*q.v.*). Mr has no units and has replaced the term *molecular weight* in the recent chemical literature.

relative plating efficiency the percentage of inoculated cells that give rise to colonies, relative to a control where the absolute plating efficiency is arbitrarily set as 100. *See* absolute plating efficiency.

relaxation complex a group of three proteins tightly bound to some *E. coli* supercoiled plasmids that convert supercoiled DNA to a nicked open circle. When heated or treated with alkali, proteolytic enzymes, or detergents, one of these proteins nicks one strand at a specific site, thereby relaxing the supercoil to a nicked open circular form. During relaxation, the two smaller proteins are released, but the largest protein becomes covalently attached to the 5′-P end of the nick. Nicking plays a role in transfer of the plasmid during conjugation. The site of the nick establishes the *transfer origin*.

relaxation of selection cessation of selection in an experimental situation.

relaxation proteins *See* helix-destabilizing proteins.

relaxed control *See* plasmid.

release factors specific proteins that read termination codons and cause the release of the finished polypeptide.

releaser in ethology (*q.v.*), the particular physical attributes or behavior of one animal that stimulates another animal to perform a specific response.

relic coil the relaxed spirals often seen in prophase chromosomes. These coils are believed to be left over from the tightly coiled condition that the chromosome maintained at the previous metaphase.

relict surviving beyond others of a kind, as a species that persists in one region after becoming extinct elsewhere or a surviving species of a group of which others are extinct.

relief in photography, etc., to produce the effect of a third dimension in a two-dimensional image, as by the use of shadows. *See* shadow casting.

rem *r*oentgen *e*quivalent *m*an, the dose of ionizing radiation that has the same biological effect as one *rad* of x-rays. It is equal to *rads* × *rbe* (relative biological effectiveness). *See* Sievert (Sv).

renaturation the return of a protein or nucleic acid from a denatured state to its native three-dimensional configuration.

Renner complex a group of chromosomes (and the genes within them) that are distributed as a unit generation after generation. Such complexes are found in species belonging to the genera *Oenothera* and *Rhoeo*, for example, where whole sets of chromosomes are involved in a series of interchanges. At first meiotic metaphase, one sees rings of bivalents, rather than independent tetrads.

reovirus a virus whose name is derived from *r*espiratory and *e*nteric *o*rphan virus. Reoviruses have been found in humans, but their relation to any disease is uncertain. The term "orphan" is used for viruses "without a disease." They are one of the few viruses that contain double-stranded RNA.

rep *r*oentgen *e*quivalent *p*hysical (*q.v.*).

repair *See* DNA repair.

repair synthesis enzymatic excision and replacement of regions of damaged DNA as when UV-induced thymine dimers are removed. *See* Appendix C, 1964, Setlow and Carrier, Boyce and Howard-Flanders; 1965, Clark and Margulies; cut-and-patch repair.

rep DNA repetitious DNA (*q.v.*).

repeated epitope *See* sporozoite.

repeated gene families synonymous with multigene families (*q.v.*).

repeat-induced point mutation (RIP) a mechanism, first detected in *Neurospora crassa*, by which repetitive DNA sequences are inactivated through mutation. During the sexual reproductive cycle, CG to AT mutations are induced in duplicated sequences, and this creates targets in the DNA of vegetative cells that are subsequently methylated. DNA methylation (*q.v.*) prevents gene transcription, and therefore repeat-induced point mutation causes the epigenetic silencing of repetitive DNA sequences. RIP is a process unique to fungi, and it may have evolved as a defense against transposable elements (*q.v.*).

repeating unit the length of a nucleotide sequence that is repeated in a tandem cluster.

repeats small tandem duplications (*q.v.*).

repetition frequency the number of copies of a given DNA sequence present in the haploid genome.

repetitious DNA nucleotide sequences that occur repeatedly in chromosomal DNA. For example, in the kangaroo rat, more than 50% of its genome contains three sequences—AAG, TTAGGG, and ACACAGCGGG—each repeated 1–2 billion times. Sequences of this sort are said to be *highly repetitive*. Sometimes such tandemly arranged, highly repetitive sequences are localized. For example, *Drosophila nasutoides* has such repetitive sequences restricted to one of its four chromosomes. Highly repetitive sequences are also localized in constitutive heterochromatin (*q.v.*). There are two classes of dispersed, highly repetitive DNA. They are abbreviated SINEs and LINEs, for *S*hort and *L*ong *IN*terspersed *E*lements. SINEs are usually shorter than 500 base pairs and occur in 10^5–10^6 copies. In humans, the Alu family (*q.v.*) is a SINE. LINEs are DNAs longer than 5 kilobases and present in at least 10^4 copies per genome. The human genome contains one family of LINEs (LI). Retrotransposons (*q.v.*) are found in this family. *Middle repetitive* DNA consists of segments 100–500 base pairs in length repeated 100 to 10,000 times each. This class of rep DNA contains the genes transcribed into rRNAs and tRNAs. *See* Appendix C, 1966, Waring and Britten; 1968, Britten and Kohne; 1970, Pardue and Gall; 1978, Finnegan *et al.*; 1988, Kazazian *et al.*; C value paradox, *Dipodomys ordii*, X-chromosome inactivation.

repetitive DNA repetitious DNA (*q.v.*).

repetitive genes *See* multigene families.

replacement sites positions within a gene at which point mutations alter the amino acid specification.

replacement vector *See* lambda cloning vehicle.

replica plating a technique used to produce identical patterns of bacterial colonies on a series of petri plates. A petri plate containing bacterial colonies is inverted, and its surface is pressed against a cylindrical block covered with velveteen. In this way, 10 to 20% of the bacteria are transferred to the fabric. Subsequently, bacteria-free plates are inverted and pressed against the velveteen disc to pick up samples of the colonies. About eight replicas may be printed from a single pad in this way. If the medium contained in each of the secondary plates differs in its selective properties, then the hundreds of different bacterial clones transferred may be scored simultaneously for their responses to a given agent. *See* Appendix C, 1952, Lederberg and Lederberg.

replicase a nucleotide polymerase that replicates DNA or RNA molecules. In nature, DNA replicases require an RNA primer. RNA replicases do not require primers, and this is one reason for speculating that the first replicating molecule was RNA, not DNA. *See* DNA polymerase, polymerase chain reaction, primer RNA, replication of DNA, RNA-dependent RNA polymerase.

replication a duplication process requiring copying from a template.

replication-defective virus a virus defective in one or more genes essential for completing the infective cycle.

replication eye (bubble) the appearance under the electron microscope of a portion of a DNA molecule that is undergoing replication. The eye-shaped region is formed by bidirectional replication forks

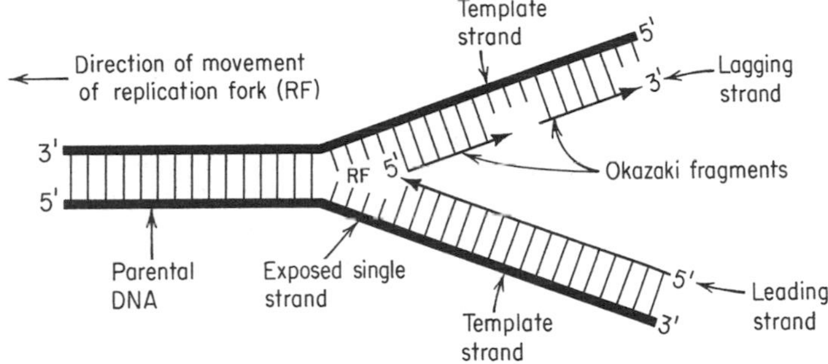

Replication of DNA

growing away from the origin. *See* D loop, replication of DNA.

replication fork *See* replication of DNA.

replication of DNA during DNA replication the two strands of the duplex molecule separate to form a *replication fork.* DNA polymerase (*q.v.*) then adds complementary nucleotides starting at the 3′ end. The strand that is continuously replicated in this way is referred to as the *leading strand.* The other strand is replicated discontinuously in short pieces. These Okazaki fragments are 1,000–2,000 nucleotides long, and each is built by the extension of a priming RNA molecule about 10 bases long. The extension proceeds from 5′ to 3′. The segments are synthesized one after the other, so that the left Okazaki fragment in the diagram is the one that was synthesized last. Subsequently, the RNA primers are removed, the gaps are filled, and nicks are sealed by DNA ligases (*q.v.*). Eventually, the lagging strand is extended until it reaches the 3′ end of the template strand. When the last RNA primer is removed, there will be a 3′ terminal overhang. It follows that genetic information will be lost during each replication cycle. This is prevented by having the chromosomes capped with repetitive noncoding DNA. *See* Appendix C, 1968, Okazaki *et al.*; 1970, Smos and Inman; primase, replicon, replisome, telomerase, telomere, zygotene DNA.

replication origin a nucleotide sequence at which DNA synthesis begins; termed an *ori* site. Circular bacterial genophores have a single *ori* site, whereas there are many *ori* sites on each eukaryotic chromosome. *See* replicon.

replication rate the speed at which deoxyribonucleotides are polymerized at a replication fork by DNA polymerases. In *E. coli,* with a single replication fork, the replication rate is approximately 50,000 base pairs per minute or about 833 per second. *Compare with* transcription rate, translation rate.

replicative forms double-stranded nucleic acid molecules seen at the time of the replication of single-stranded DNA and RNA viruses.

replicator a DNA segment that contains a replication origin (*q.v.*).

replicon a genetic element that behaves as an autonomous unit during DNA replication. In bacteria, the chromosome functions as a single replicon, whereas eukaryotic chromosomes contain hundreds of replicons in series. Each replicon contains a segment to which a specific RNA polymerase binds and a replicator locus at which DNA replication commences. The polymerase makes an RNA primer called an initiator. *See* Appendix C, 1963, Jacob and Brenner; 1968, Huberman and Riggs; autonomously replicating sequences, DNA fiber autoradiography, *ori* site, primase.

replicon fusion the joining of two complete replicating systems, mediated by a transposon. If two plasmids are joined (one of which carries a transposon), the process is called *plasmid fusion.*

replisome a ribosome-sized particle that contains a complex of proteins which carry out the replication of DNA. In *E. coli* it contains DNA polymerase III, a primosome (*q.v.*), a helicase (*q.v.*) that unwinds the DNA ahead of the polymerase, a gyrase (*q.v.*) that relieves supercoiling (*q.v.*) ahead of the replication fork, and single-stranded DNA-binding proteins that prevent the strands from reannealing. *See* DNA polymerase, replication of DNA.

reporter gene a gene (*R*) that (1) can generate a product which is easy to visualize and (2) can respond to a regulatory signal meant for a second gene. It is this gene (*X*) that is of primary interest, but there may be no convenient method of recording its activity. The tissue-specific expression of gene *X* can be monitored by letting the *R* gene intercept a regulatory signal meant for *X*. *R* responds by synthesizing its product and thus "reports" the time and place of activation of *X*. The *lac Z* gene of *E. coli* has been used extensively as a reporter. Its product is beta galactosidase. This enzyme normally breaks down lactose, but under the conditions of the experiment an analog of lactose is provided. When this compound (5-bromo-4-chloro-indolyl-β-D-galactoside) is cleaved by beta galactosidase, it yields 5-bromo-4-chloro-indigo, a blue dye. The production of this blue pigment in a particular tissue during development reports the activation of the gene under study. *See* enhancer trap, *lac* operon.

rep protein a helicase (*q.v.*) identified in the *rep* mutant strain of *E. coli* that hydrolyzes ATP while forcing the strands of the DNA helix apart.

representation *See* abundance.

representational difference analysis (RDA) a technique used to detect DNA segments in one sample (the *tester pool*) that are not in another (the *driver pool*). DNAs from each pool are fragmented by digestion with restriction endonucleases (*q.v.*), and the fragments are amplified by the polymerase chain reaction (*q.v.*). The amplicons from one pool are then allowed to hybridize with those from the other, and the hybrid molecules are removed. This subtractive hybridization leads to the enrichment of fragments that are unique to the tester pool. Such a technique could be used to isolate cDNAs from genes specifically activated by certain hormones.

repressible enzyme an enzyme whose rate of production is decreased when the intracellular concentration of certain metabolities is increased.

repressible system a regulatory system in which the product of a regulator gene (the repressor) blocks transcription of the operon only if it first reacts with an effector molecule (called the repressing metabolite). Thus, mRNA synthesis occurs only in the absence of the effector. *See* regulator gene. *Compare with* inducible system.

repressing metabolite *See* repressible system.

repression **1.** the inhibition of transcription or translation when a repressor protein binds to an operator locus on DNA or to a specific site on a mRNA. **2.** the cessation of synthesis of one or more enzymes when the products of the reactions they catalyze reach a critical concentration.

repressor a protein (synthesized by a regulator gene) that binds to an operator locus and blocks transcription of that operon. *See cro* repressor, *lac* repressor, lambda repressor, regulator gene.

reproduction probability the average number of children of patients with a specific hereditary disease in relation to the average number of children of comparable individuals who do not have the hereditary disease. The reproduction probability is a measure of the selective disadvantage of a hereditary disease. *See* fitness.

reproductive death the suppression of the proliferative ability of a cell that otherwise would divide indefinitely. *See* genetic death.

reproductive isolation the absence of interbreeding between members of different species. *See* isolating mechanisms.

reproductive potential the theoretical logarithmic rate of population growth when unimpeded by environmental limitations; also known as the biotic potential; symbolized *r*. *See* r and K selection theory.

reproductive success for a given individual, the number of offspring that survive to reproduce. *See* fitness.

repulsion *See* coupling, repulsion configurations.

ReqQ DNA helicase *See* helicase.

RER *r*ough *e*ndoplasmic *r*eticulum. (*q.v.*).

RES *r*eticulo*e*ndothelial *s*ystem (*q.v.*).

rescue the restoration of a defective cell or tissue to normal development through such means as cytoplasmic transplantation or transformation (*q.v.*) with a normal gene.

residual genotype background genotype (*q.v.*).

residual homology in species hybrids, the homology remaining between those chromosomes that have been derived from a common ancestral chromosome and have since diverged through the accumulation of mutations. *See* homoeologous chromosomes.

residue in biochemistry, referring to a small subunit that forms a compoment of a larger molecule. Thus, a protein digested by a protease yields amino acid residues, and nucleases release nucleotide residues from nucleic acids.

resistance factor a class of episomes that confer antibiotic resistance to the recipient bacterium. *See* R plasmid.

resistance genes genes in plants that confer resistance to specific strains of pathogens. The first such gene to be cloned was the *R* gene in the tomato, which encodes a protein kinase (*q.v.*). This produces a resistance to certain species of *Pseudomonas* (*q.v.*).

resistance transfer factor (RTF) *See* R plasmid.

resolvase an enzyme catalyzing the site-specific recombination (*q.v.*) between two transposons present as direct repeats in a *cointegrate structure* (*q.v.*). The term also refers to a family of site-specific recombinases. *See* site-specific recombinase.

resolving power the ability of any magnifying system to reveal fine detail. This ability is often measured as the minimum distance between two lines or points at which they are resolved as two rather than as a single blurred object. The maximum resolving power of the light microscope is about 0.2 micrometer; that of the electron microscope, about 0.5 nanometer.

resource tracking a hypothesis involving host-parasite coevolution according to which ectoparasites track a particular resource, such as a type of skin, hair, or feathers. If in addition there is opportunity for a given species of parasite to disperse to unrelated host species, then there will be no direct parallel relationship between the taxonomy of the hosts and that of their parasites (*contrast with* Fahrenholz's rule). In birds, for example, the same species of mite may be found on birds from different orders, and a single species of bird may be parasitized by different lice species. Thus, birds and their parasites show little phylogenetic parallelism.

respiration the aerobic, oxidative breakdown and release of energy from fuel molecules.

respiratory pigment a substance that combines reversibly with oxygen, thus acting as a carrier of it (hemoglobin, for example). The color of respiratory pigments is due to their prosthetic groups. In red proteins like hemoglobin or cytochrome c, the prosthetic group is heme (*q.v.*).

responder in immunology, an animal capable of mounting an immunological response to a particular antigen.

resting cell (nucleus) any cell (nucleus) not undergoing division. The cell (nucleus) is nevertheless very active metabolically. *See* cell cycle.

restitution the spontaneous rejoining of experimentally induced, broken chromsomes to produce the original configuration.

restitution nucleus **1.** a nucleus containing double the expected number of chromosomes owing to a failure of the mitotic apparatus to function properly. **2.** an unreduced product of meiosis. A diploid nucleus, resulting from the failure of the first or second meiotic division. *See* meiosis.

restricted transduction *See* transduction.

restriction ability of a bacteriophage to infect bacteria belonging to certain strains but not others. *See* DNA restriction enzyme.

restriction allele *See* DNA restriction enzyme.

restriction and modification model a theory proposed by W. Arber to explain host-controlled restriction of bacteriophage growth. According to this model, the DNA of the bacterium contains specific nucleotide sequences that are recognized and cleaved by the restriction endonucleases carried by that cell. The bacterium also contains methylases that methylate these sequences. This chemical modification thus protects the DNA of the bacterium from its own endonucleases. However, these serve to degrade foreign DNA introduced by phages. *See* Appendix C, 1962, Arber; 1972, Kuhnlein and Arber.

restriction endonuclease any one of many enzymes that cleave foreign DNA molecules at specific recognition sites. Restriction endonucleases are coded for by genes called restriction alleles. The enzymes are named by a symbol that indicates the bacterial species from which they were isolated, followed by a Roman numeral that gives the chronological order of discovery when more than one enzyme came from the same source. Some restriction endonucleases, the organisms from which they were isolated, and their target nucleotide sequences are illustrated on page 384. The arrows indicate the cleavage sites. Note that *Bam*HI and *Eco*RI cleave the strands of DNA at specific sites four nucleotides apart. Such staggered cleavage yields DNA fragments with protruding 5′ termini. Such ends are said to be "sticky" or "cohesive" because they will hydrogen bond to complementary 3′ ends. As a result, the end of any DNA fragment produced by an enzyme, such as *Eco*RI, can anneal with any other fragment produced by that enzyme. This property allows splicing of foreign genes into *E. coli* plasmids. Enzymes like *Hin*dII produce flush or blunt-ended fragments. Restriction endonucleases are used extensively to map DNA regions of interest. *See* Appendix C, 1962, Arber; 1968, Smith *et al.*; 1970, Smith and Wilcox; 1971, Danna and Nathans; 1972, Mertz and Davis, Hedgpeth *et al.*; Appendix E, Individual Databases; Alu family, polylinker site.

*Bam*HI *Bacillus amyloliquefaciens* ↓
5′ . . . G-G-A-T-C-C . . . 3′
3′ . . . C-C-T-A-G-G . . . 5′
↑

*Eco*RI *Escherichia coli RY13* ↓*
5′ . . . G-A-A-T-T-C . . . 3′
3′ . . . C-T-T-A-A-G . . . 5′
↑

*Hind*II *Hemophilus influenzae Rd* C ↓ A *
5′ . . . G-T-(T)-(G)-A-C-3′
3′ . . . C-A-(A)-(C)-T-G-5′
G ↑ T

$\overset{*}{\text{A}}$ is N^6-methyladenine;
C A
(T)-(G) signifies that either base can occupy that position.

Restriction endonucleases

restriction fragment a fragment of a longer DNA molecule digested by a restriction endonuclease.

restriction fragment length polymorphisms (RFLPs) variations occurring within a species in the length of DNA fragments generated by a specific endonuclease. Such variations are generated by mutations that create or abolish recognition sites for these enzymes. For example, restriction endonuclease mapping of human structural genes for beta hemoglobin chains has shown that patients with the sickle cell mutation produce abnormal restriction fragments. Since restriction enzyme analyses can be performed on DNA from amniotic fluid cells, RFLPs are now used in the prenatal diagnosis of genetic defects. A human mutant gene can be mapped, even when the nature of its product is unknown, by searching through DNA samples in a pedigree that contains the mutation, and looking for RFLPs that also segregate with the mutation. *See* Appendix C, 1978, Kan and Dozy; 1980, Botstein *et al.*; alphoid sequences, DNA fingerprint technique, variable numbers of tandem repeats locus.

restriction map a diagram portraying a linear array of sites on a DNA segment at which one or more restriction endonucleases (*q.v.*) cleave the molecule.

restriction site a deoxyribonucleotide sequence at which a specific restriction endonuclease cleaves the molecule.

restrictive conditions any environmental condition (e.g., temperature or type of host) under which a conditional mutation either cannot grow or expresses the mutant phenotype. *See* temperature sensitive mutation.

restrictive transduction *See* transduction.

reticulate evolution the netlike lineage relation seen for a series of related allopolyploid species. The cross-links represent places where hybridization has occurred and allotetraploid species have arisen. Reticulate evolution is common in plants. *See* dendritic evolution.

reticulocyte an immature erythrocyte at an active stage in the synthesis of hemoglobin.

reticuloendothelial system a network of phagocytic cells residing in the bone marrow, spleen, and liver of vertebrates, where they free the blood or lymph of foreign particles.

retina a delicate, multilayered, light-sensitive membrane that lines the inner eyeball and is connected to the brain by the optic nerve. The deepest layer of retinal cells contains melanin (*q.v.*). This pigment functions to absorb light that might scatter among the photoreceptors that lie above it and create confusing neural signals. Photoreceptor cells are contained in the visual epithelium, which overlies the pigmented layer. The photoreceptor cells are modified cilia that contain, in their outer segments, hundreds of thin membranes arranged like stacks of coins (see page 385). The photosensitive pigments are attached to these membranes. There are two morphologically distinct types of photoreceptors. The rod cells make up 95% of all photoreceptors and are responsible for perceiving light, particularly when it is at low intensity. Cone cells respond to different wavelengths of light and are responsible for color vision. The supply of photopigments and the outer segment discs themselves are continually replenished. As new discs are generated at the proximal end of the outer segment, distal worn-out discs are phagocytosed by cells of the pigmented layer. *See*

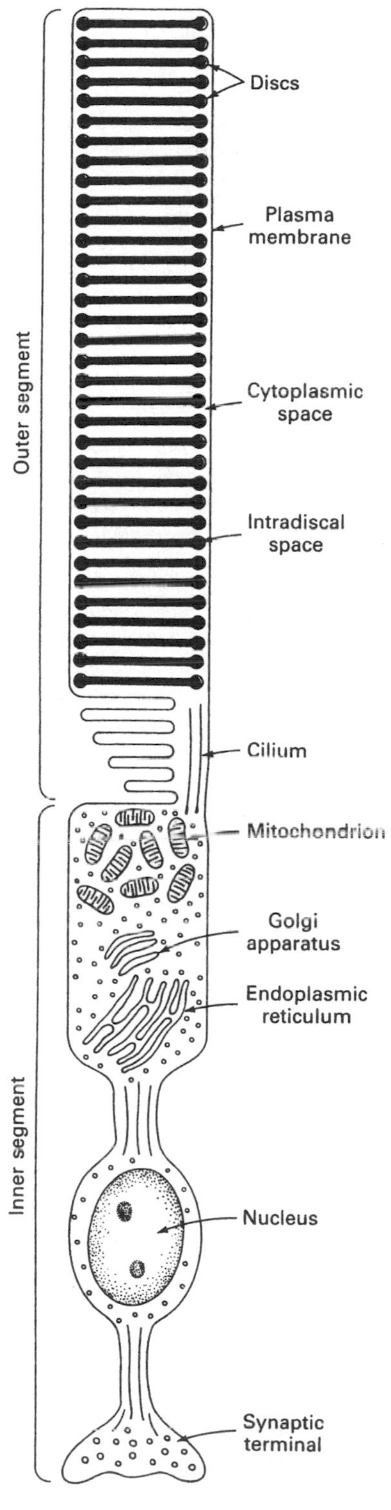

Retina

color blindness, cone pigment genes, ocular albinism, opsin, retinitis pigmentosum (RP), rhodopsin (RHO).

retinal a light-absorbing, carotenoid pigment that is derived from vitamin A (*q.v.*). Upon absorbing a photon of light, this chromophore (*q.v.*) undergoes a change in shape that involves the rotation of the starred carbon atom and the four carbon chain attached to it. This is the primary event in visual excitation that eventually leads to a nerve impulse. *See* opsin.

```
CH3   CH3
   \ /
    C            CH3             CH3   H
   / \           |       ★       |     |
H2C   C-CH=CH-C=CH-CH=CH-C=CH-C=O
 |    ||
H2C   C-CH3
   \ /
   CH2
```

retinitis pigmentosum (RP) a pathological condition where the retina, when viewed through an ophthalmoscope, is seen dotted with jet-black pigment spicules. These represent dying photoreceptor cells. There are at least seven autosomal dominant, one autosomal recessive, and three X-linked recessive forms of the disease. One mutant autosomal gene for RP maps at the site (on 3p) of the gene that encodes rhodopsin (*q.v.*). Another gene (at 6p) encodes a glycoprotein called *peripherin*. This is an adhesion molecule localized in the rim of each disc. It joins adjacent discs and maintains their integrity as they lie stacked together in the outer segment of the photoreceptor cells. The most common RP mutation in American patients is *P23H*, a missense mutation in the rhodopsin gene. The mutated rhodopsin forms high molecular weight aggregates which impair the ubiquitin-proteasome pathway (*q.v.*) of protein degradation. *See* retina, rhabdomere.

retinoblastoma a malignant neoplasm composed of primitive retinal cells usually occurring in children less than three years old. Retinoblastoma occurs in hereditary and non-hereditary forms. In the case of hereditary retinoblastoma, patients are generally affected bilaterally. The gene involved (*RB*) is located on the long arm of chromosome 13 at band 14. The normal gene contains 27 exons. The *RB* gene product is a nuclear phosphoprotein (pRB). The phosphates are usually attached to serine and threonine residues. The DNA binding of pRB depends on its degree of phosphorylation (only the underphosphorylated molecule binds). Mutant alleles of the *RB* gene have been found in other types of human cancers besides retinoblastomas. Such mutations often involve deletions in exons 13–17. Introduction of the wild-type *RB* gene into these tumor cells suppresses their uncontrolled growth. *See* Appendix C, 1971, Knudson; 1989, Hong *et al.*; 1990, Bookstein *et al.*; anti-oncogenes, Knudson model, retina.

retinoic acid a morphologically active compound that exists in a concentration gradient along the limb bud of tetrapods and stimulates the development of limbs and digits. Retinoic acid is vitamin A (*q.v.*) with the terminal CH_2OH replaced by a COOH group.

retinol synonymous with vitamin A (*q.v.*).

retrodiction the act of predicting the yet undiscovered results of past events. Useful theories of evolution should allow retrodictions that can be validated by looking in the right places within the fossil record.

retrogene *See* processed gene.

retrogression evolution toward a less complex state; characteristic of some parasitic groups. For example, tapeworms, having no digestive system, are thought to have evolved from free-living flatworms with a digestive system.

retroposons transposable elements (*q.v.*) that mobilize via an RNA intermediate. Each DNA segment in the host chromosome is transcribed into RNA and then reverse-transcribed via a reverse transcriptase (*q.v.*) into a DNA segment. This is reinserted into the host genome, usually at a new site. Retroposons are the most abundant transposons in plants. These elements account for 70% of the nuclear DNA in maize. The best-understood retroposons are retroviruses (*q.v.*) such as HIV (*q.v.*). *Retroposon* is a shortened form of *retrotransposon*, which also appears in the literature. *See* Appendix C, 1985, Boeke *et al.*; 1990, Biessmann *et al.*; centromere, copia elements, repetitious DNA, telomere, Ty elements.

retroregulation the ability of downstream DNA sequences to regulate translation of an mRNA.

retrotransposon *See* retroposon.

retroviruses RNA viruses that utilize reverse transcriptase (*q.v.*) during their life cycle. This enzyme allows the viral genome to be transcribed into DNA. The name *retro*virus alludes to this "backward" transcription. The transcribed viral DNA is integrated into the genome of the host cell where it replicates in unison with the genes of the host chromosome. Therefore retroviruses violate the central dogma (*q.v.*) during their replication. Most retrovirions carry two copies of a linear, positive sense, ssRNA held together by hydrogen bonds. Most retroviruses carry the genes *gag, pol,* and *env*. The *gag* gene encodes the structural proteins of the capsid, and *env* codes for the glycoproteins that form spikes on the envelope of the viral membrane. The *pol* region encodes a protease, a reverse transcriptase, and an integrase (*q.v.*). The transcription of the retroviral genome results in the production of a polycistronic mRNA. This is translated to yield a polyprotein that is subsequently cleaved by the viral protease into the functional subunits. The retroviruses are assigned to three groups: the Lentivirinae, the Oncovirinea, and the Spumavirinae. HIV viruses I and II (*q.v.*) belong to the Lentivirinae. The Oncovirinae contain viruses that attack birds, such as the Rous sarcoma virus, rodents (the Friend, Moloney, and Rauscher leukemia viruses and the mammary tumor virus), carnivores (the feline leukemia and sarcoma viruses), and primates (the simian sarcoma virus) (*all of which see*). *Also see* enveloped viruses, reverse transcription, virus.

reversals situations in phylogenetic analysis in which a derived character state changes (reverses) to a preexisting character state. For example, among the vertebrates, limbs are considered to be derived from a limbless state; but snakes are thought to have evolved from tetrapod (four-legged) ancestors and in the process have lost their limbs.

reverse genetics an experimental approach for determining gene function in which the sequence of a gene of unknown function is deliberately altered *in vitro* (*q.v.*) and introduced back into the cell or the organism, or the expression of the gene is disrupted in order to determine its phenotypic effect. Gene modification or knockout (*q.v.*) is achieved by such means as chemical and transposon-induced mutagenesis, gene targeting (*q.v.*), and RNA interference (RNAi) (*q.v.*). With the availability of tools for genome manipulation and whole-genome sequence data, reverse genetics has gained prominence in genetic analysis in a variety of organisms, including bacteria, yeast, the nematode, the fruit fly, the mouse, and many plant species. This approach is the *reverse* of the regular or *forward genetics* approach, in which one begins with a mutant allele or phenotype and then identifies the wild-type gene and its biological effect. *See* functional cloning, gene targeting, knockout, positional cloning, transgenic animals.

reverse loop pairing *See* inversion.

reverse mutation a change in a mutant gene which restores its ability to produce a functional protein. *Compare with* forward mutation.

reverse selection selection in an experimental situation for a trait opposite to the one selected earlier (e.g., selection first for increased numbers of thoracic bristles in *Drosophila*, then for decreased numbers).

reverse transcriptase RNA-dependent DNA polymerase (*q.v.*).

reverse transcription DNA synthesis from an RNA template, mediated by reverse transcriptase. Since retroviruses undergo reverse transcription, they have two genomic forms: ssRNA in their virions and dsDNA in their prophages. *See* RNA-dependent DNA polymerase.

reverse transcription-polymerase chain reaction (RT-PCR) The most sensitive technique for mRNA detection and quantitation, involving amplification of a reverse transcription product. In this technique, an RNA template is reverse transcribed using an RNA-dependent DNA polymerase (*q.v.*). The resulting cDNA (*q.v.*) is then amplified using the polymerase chain reaction (*q.v.*). cDNA corresponding to the rarest of transcripts can be detected by this method. RT-PCR products are quantitated using any of a variety of methods, providing information about the expression levels of a particular gene or relative transcription of multiple RNAs.

reversion reverse mutation (*q.v.*).

revertant **1.** an allele that undergoes reverse mutation. **2.** an organism bearing such an allele.

reverted Bar *See Bar.*

R_f in paper chromatography, a ratio given by the distance traveled by the solute divided by the distance traveled by the solvent. For a given solute molecule the R_f varies with the solvent, and therefore the solvent must be specified for any R_f value.

RF **1.** replicative form; **2.** recombination frequency.

R factor resistance factor (*q.v.*).

RFLP pronounced "rif lip." *See* restriction fragment length polymorphisms.

***R* genes of maize** a family of genes responsible for determining the temporal and spatial pattern of anthocyanin (*q.v.*) pigmentation in the corn plant. The gene products are proteins that bind to DNA and activate the transcription of the structural genes that function directly in anthocyanin synthesis.

rhabdomere a specialized organelle found in the photoreceptor cells of an ommatidium (*q.v.*). In *Drosophila* each rhabdomere is composed of ~60,000 tightly packed microvilli which expand the membrane surface to accommodate hundreds of millions of molecules of rhodopsin (*q.v.*). Each rhabdomere functions like the discs in the outer segments of the photoreceptor cells of the vertebrate retina (*q.v.*). *See* microvillus.

rhabdoviridae a family of enveloped viruses (*q.v.*) with negative-strand RNA genomes. The rabies virus (*q.v.*) belongs to this family. It binds to a receptor on peripheral nerves and is then transmitted to cells of the central nervous system and salivary glands.

rhesus factor *See* Rh factor.

rhesus monkey *See Macaca mulatta.*

rheumatoid factor a distinctive gamma globulin commonly present in the serum of patients with rheumatoid arthritis.

Rh factor an antigen occurring on the erythrocytes of certain human beings. The Rh system actually contains several antigens. The most important one was first found in the Rhesus monkey—hence the name. Persons of genotype *r/r* produce no antigen and are classified as Rh-negative. *R/R* and *R/r* individuals produce the antigen. A pregnant mother who is Rh-negative but is carrying an Rh-positive child may produce antibodies against the child *in utero*, causing the child to develop a hemolytic disease called erythroblastosis fetalis. However, this occurs only if the mother has been previously exposed to the Rh antigen from cells entering her circulation during the birth of a previous Rh-positive baby. The Rh antigens are encoded by two closely linked and highly homologous genes (*RHD* and *RHCE*) located at 1p34-36. The most clinically important antigen, D, is encoded by *RHD*, and the Rh-negative phenotype results from deletions in the *RHD* gene. The highest frequency of Rh in the world is seen in Basques (*q.v.*). *See* Appendix C, 1939, Levine and Stetson; 1947, Mourant; RhoGAM.

Rhizobium a genus of nitrogen-fixing bacteria that live as symbionts in the root nodules of leguminous plants. The most studied species are *R. meliloti* and *R. trifoli.* The genes involved in host specificity, nodulation, and nitrogen fixation are carried on large plasmids, and restriction maps have been constructed of those regions carrying symbiotically important genes. *See* Appendix A, Bacteria, Proteobacteria; Appendix C, 1981, Hombrecher *et al.*; Appendix E; bacteroids, *nif* genes, plasmid, restriction map, sym-plasmid.

Rhizopoda the phylum of protoctists containing singled-celled amoebas. *See Amoeba proteus,* Appendix A.

rhodoplast the red plastid of red algae.

rhodopsin (RHO) the light-sensitive chromoprotein found in the rod cells of the retina. Rhodopsin consists of the protein opsin (*q.v.*) in combination with retinal (*q.v.*). In the primary event of visual excitation, rhodopsin (also called *visual purple*) is bleached to a yellow compound. In humans, the gene encoding RHO resides at 3q 21.3–24. RHO is a *multiple transmembrane domain protein* (*q.v.*). The cone pigment genes (*q.v.*) show considerable sequence similarities to RHO. *See* color blindness, retinitis pigmentosum (RP), rhabdomere.

rhodamine B a fluorochrome commonly used to tag compounds that bind to specific cell components. *See* rhodaminylphalloidin.

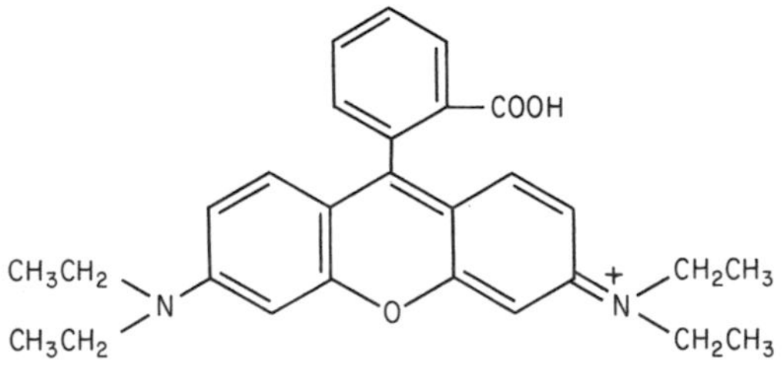

COOH
CH_3CH_2 N CH_3CH_2 O + N CH_2CH_3 CH_2CH_3

rhodaminylphalloidin a fluorescent derivative of phalloidin used to specifically stain actin filaments in whole mounts of cells. *See* fluorescence microscopy, phallotoxins.

rho factor an oligomeric protein in *E. coli* that attaches to certain sites on its DNA to assist in termination of transcription.

RhoGAM the trade name for anti-Rh gamma globulin used in the prevention of Rh hemolytic disease. *See* Appendix C, 1964, Gorman *et al.*; Rh factor.

rho(ρ)-independent terminators *E. coli* DNA sequences recognized by RNA polymerase as transcription termination signals in the absence of rho (ρ) factors.

Rhynchosciara a genus of fungus gnats belonging to the Sciaridae. Two species, *R. angelae* and *R. hollaenderi*, can be reared in the laboratory and have been extensively studied because of the gigantic polytene chromosomes found in various larval tissues (such as the salivary glands, Malpighian tubules, and anterior mid gut). RNA puffs occur often, and they are tissue and developmental stage specific. The DNA puffs are rare and were first discovered in *Rhynchosciara*. They occur in several regions of the polytene chromosomes of the salivary glands of fourth instar larvae. The DNA in each puff undergoes a 16-fold increase, and the mRNAs transcribed from the amplified genes are translated into salivary proteins. *See* Appendix A, Arthropoda, Insecta, Diptera; gene amplification, *Sciara*.

RIA radioimmunoassay (*q.v.*).

riboflavin vitamin B_2, a subunit of flavin adenine dinucleotide and flavin mononucleotide. The structure appears below.

riboflavin phosphate flavin mononucleotide (*q.v.*).

ribonuclease the enzyme that hydrolyzes ribonucleic acid.

ribonuclease A an enzyme made up of 124 amino acids, first isolated from bovine pancreas. Ribonuclease A hydrolyzes pyrimidines at the 3′ phosphate group and cleaves the 5′ phosphate linkage to the adjacent nucleotide. The end products of digestion are pyrimidine 3′ phosphates and oligonucleotides with pyrimidine 3′ phosphate termini. Ribonuclease A was the first protein to be subjected to reversible chemical modifications that demonstrated that the linear sequence of amino acids determined the unique three-dimensional structure of the protein. *See* Appendix C, 1961, Anfinsen *et al.*

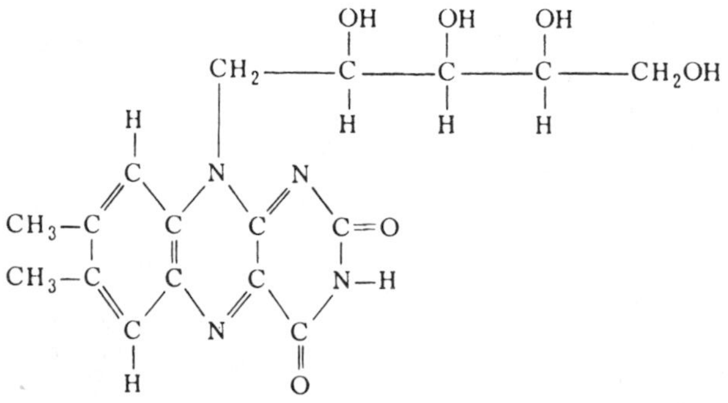

OH OH OH
CH_2—C—C—C—CH_2OH
H H H
H C N N
CH_3—C C C C=O
CH_3—C C C N—H
C N C
H O

Riboflavin

ribonuclease H an endonuclease that specifically degrades the RNA strands of RNA-DNA hybrid molecules. However, the enzyme does not hydrolyze RNAs complexed with Morpholinos (*q.v.*).

ribonuclease P a universally occurring enzyme which, by removing the 5′ ends from precursor tRNA molecules, converts them into their functional states. RNase P is a heterodimer made up of one large RNA and one small protein molecule. The protein alone has no catalytic activity. The RNA alone can catalyze the cleavage, although less efficiently than the intact enzyme. RNase P was the first RNA found to exhibit the characteristics of a true enzyme. *See* Appendix C, 1983, Guerrier-Takada *et al.*; 1989, Cech and Altmann; ribozyme.

ribonuclease T1 an endoribonuclease that specifically attacks the 3′ phosphate groups of guanosine nucleotides and cleaves the 5′ phosphate linkage to the adjacent nucleotide. The end products of digestion are guanosine 3′ phosphates and oligonucleotides with guanosine 3′ phosphate termini.

ribonucleic acid (RNA) any of a family of polynucleotides characterized by their component sugar (ribose) and one of their pyrimidines (uracil). RNA molecules are single stranded and have lower molecular weights than DNAs. There are three classes of RNAs: (1) messenger RNA (*q.v.*), (2) ribosomal RNA (*see* ribosome), and (3) transfer RNA (*q.v.*). *See* Appendix C, 1941, Brachet and Caspersson; hydrogen bonding, plus (+) and minus (–) viral strands.

ribonucleoprotein a complex macromolecule containing both RNA and protein and symbolized by RNP.

ribonucleotide an organic compound that consists of a purine or pyrimidine base bounded to ribose, which in turn is esterified with a phosphate group.

ribose a five-carbon sugar. *See* ribonucleic acid (RNA).

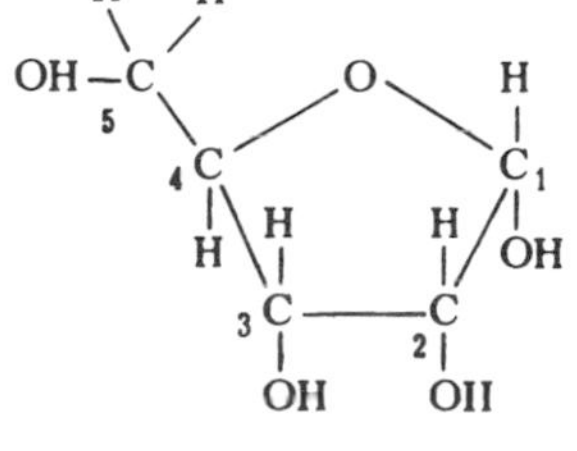

ribosomal binding site *See* Shine-Dalgarno sequence.

ribosomal DNA *See* rDNA.

ribosomal DNA amplification *See* rDNA amplification.

ribosomal precursor RNA *See* preribosomal RNA.

ribosomal protein *See* ribosome.

ribosomal RNA (rRNA) *See* ribonucleic acid (RNA), ribosome.

ribosomal RNA genes rRNA genes reside as tandem repeating units in the nucleolus organizer regions of eukaryotic chromosomes. Each unit is separated from the next by a nontranscribed spacer. Each unit contains three cistrons coding for the 28S, 18S, and 5.8S rRNAs. The transcriptional polarity of the unit is 5′-18S-5.8S-28S-3′. Ribosomal RNA genes are often symbolized by rDNA. *See* Miller trees, preribosomal RNA, rDNA amplification, RNA polymerase.

ribosomal stalling *See* attenuation.

ribosome one of the ribonucleoprotein particles, 10–20 nanometers in diameter, that are the sites of translation. Ribosomes consist of two unequal subunits bound together by magnesium ions. Each subunit is made up of roughly equal parts of RNA and protein. Each ribosomal subunit is assembled from one molecule of ribosomal RNA that is noncovalently bonded to 20 to 30 smaller protein molecules to form a compact, tightly coiled particle. In eukaryotes, the rRNAs of cytoplasmic ribosomes are formed by cistrons localized in the nucleolus organizer region (*q.v.*) of chromosomes. Animal ribosomes also contain a 5.8S rRNA, which is hydrogen bonded to the 28S rRNA and is derived from the same intermediate precursor as the 28S rRNA (see table on page 390). *See* Appendix C, 1956, Palade and Siekevitz; 1959, McQuillen *et al.*; 1961, Jacob and Monod, Waller and Harris, Littauer; 1964, Brown and Gurdon; 1965, Sabatini *et al.*; 1980, Woese *et al.*; 2001, Yusupov *et al.*; cyclohexamide, preribosomal RNA, receptor-mediated translocation, ribosomes of organelles, 16S rRNA, translation.

ribosomes of organelles ribosomes of chloroplasts and mitochondria show a variety of sedimentation constants. Chloroplast ribosomes are 70S, mitochondrial ribosomes from plant cells are 78S, mitochondrial ribosomes from fungal cells are 73S, and mitochondrial ribosomes from mammalian cells are 60S. Both mitochondria and chloroplasts are believed to have arisen from free-living prokaryotes that fused with primitive nucleated cells. The fact that translation starts with *N*-formylmethionine on

Sedimentation Characteristics of Prokaryotic and Eukaryotic Ribosomes

Ribosomal Source	Ribosome	Large Subunit	Small Subunit	RNAs of Large Subunit	RNAs of Small Subunit
Bacteria	70S	50S	30S	23S (2,904 nt) 5S (120 nt)	16S (1,542 nt)
Mammalian cells	80S	60S	40S	28S (4,718 nt) 5.8S (160 nt) 5S (120 nt)	18S (1,874 nt)

Ribosome

the ribosomes of both mitochondria and chloroplasts supports the theory of their endosymbiotic orgin. *See* initiator tRNA, serial symbiosis theory, 5S rRNA.

riboswitches in prokaryotes and some eukaryotes, genetic control elements found in untranslated regions of some messenger RNAs (mRNAs), which directly and selectively bind target metabolites and regulate the transcription or translation of the bound mRNAs. The binding of the metabolite to the RNA usually induces a structural change in the mRNA, which causes either the termination of transcription or inhibition of translation initiation. Riboswitches may also regulate gene expression at the level of pre-mRNA splicing, or mRNA processing or stability. A number of distinct riboswitches have been identified, which regulate a variety of metabolic pathways. These genetic elements do not require proteins for ligand recognition and regulatory functions and are thought to be of ancient origin, from a time period prior to the emergence of proteins.

5-ribosyluracil pseudouridine. *See* rare bases.

ribothymidine *See* rare bases.

ribotype the RNA complement of a cell.

ribozyme an RNA molecule with catalytic activity. An example is the self-splicing rRNA of *Tetrahymena thermophila*. The gene for the 26S rRNA contains an intron 413 base pairs long. The precursor rRNA molecule transcribed from this gene includes a copy of this intervening sequence, which must be deleted by RNA splicing. If the pre-rRNAs are incubated *in vitro* with certain cations and guanosine, the intervening sequences are excised and the mature sequences are ligated. Ribonuclease P (*q.v.*) is another example of a ribozyme. Perhaps in the primordial biosphere ribozymes functioned as the first enzymes, and as life evolved proteins began to fine-tune the catalysis and eventually replaced ribozymes as enzymes. *In vitro* selection experiments have generated ribozymes that can catalyze self-alkylation reactions. This finding suggests that RNA may be capable of a broad range of catalytic activities. *See* Appendix C, 1981, Cech *et al.*; 1989, Cech and Altman; 1995, Wilson and Szostak; hairpin ribozyme, hammerhead ribozyme, *in vitro* evolution.

ribulose-1,5-bisphosphate carboxylase-oxygenase (RuBisCO) an enzyme responsible for virtually all photosynthetic CO_2 fixation. It is located in the stromal surface of the thylakoid membranes of chloroplasts (*q.v.*). RuBisCO is probably the most abundant protein in the biosphere, since it makes up about 40% of the protein in green leaves. RuBisCO catalyzes the reaction:

$$\text{D-ribulose-1,5-bisphosphate} + CO_2 \rightarrow \text{2,3 phospho-D-glycerate.}$$

The enzyme consists of eight large subunits (Mr 56,000 each) and eight small subunits (Mr 14,000 each). *See* cyanelles, Cyanobacteria, *rbc* genes.

rice *Oryza sativa* (*q.v.*).

Ricinus communis the castor bean. The genetics of sex determination has been extensively studied in this species.

rickets a deficiency disease of growing bone due to insufficient vitamin D in the diet. *See* vitamin D–resistant rickets.

Rickettsia a genus of Gram-negative, oval to rod-shaped, nonmotile, obligatory intracellular prokaryotes, placed in the Proteobacteria. All are spread by arthropod vectors (lice, fleas, mites, and ticks). *See* Appendix A, Prokaryotes, Proteobacteria.

Rickettsia prowazeki the bacterium responsible for louse-borne epidemic typhus. Its genome contains 1,111,523 base pairs and only 834 ORFs, compared to the 4,300 ORFs of the free-living bacterium *E. coli*. Because of its parasitic habit, this rickettsia has undergone reductive evolution (*q.v.*). For example, it has lost genes that encode enzymes that metabolize sugars and synthesize amino acids and nucleotides. Based on the sequences of nucleotides in the rRNA of the small subunit of the ribosomes, these bacteria are closest to the ancestor of mitochondria. *See* Appendix A, Bacteria, Proteobacteria; Appendix C, 1998, Anderson *et al.*; mitochondrion, serial symbiosis theory.

rifampicin the most commonly used of the rifamycins (*q.v.*).

rifamycins a group of antibiotic molecules produced by *Streptomyces mediterranei* that interfere with the beta subunit of prokaryotic RNA polymerases and thereby inhibit initiation of transcription. *See* RNA polymerase.

rift a place where the earth's surface is cracking apart because of tectonic activity. An example of an undersea rift is the Galapagos rift, about 380 miles north of the Galapagos Islands, where undersea vent communities (*q.v.*) were first discovered. *See* Appendix C, 1977, Corliss and Ballard.

right splicing junction the boundary between the right (3′) end of an intron and the left (5′) end of an adjacent exon in mRNA; also called the *acceptor splicing site*.

ring canals canals connecting sister cystocytes in the *Drosophila* egg chamber. The rim of each canal is formed by a partially closed contractile ring (*q.v.*). The canal system is initially plugged by a polyfusome (*q.v.*). Once this dissolves, the ring canals allow a stream of cytoplasm to flow from the nurse cells (*q.v.*) to the oocyte. As each ring matures, it develops a coating of actin filaments and increases in diameter and thickness. During the growth of the egg chamber, proteins encoded first by the gene *hts* and later by *kel* serve to organize the actin filaments by cross-linking them in various ways. This process requires phosphorylation reactions catalyzed by *Src* protein tyrosine kinases. *See* cystocyte divisions, *hu-li tai shao (hts)*, *kelch (kel)*, and *Src.*

ring chromosome 1. an aberrant chromosome with no ends. 2. a ring-shaped chromosomal association seen during diakinesis in normal metacentric tetrads with two terminal chiasmata.

Ringer solution a physiological saline containing sodium, potassium, and calcium chlorides used in physiological experiments for temporarily maintaining cells or organs alive *in vitro*. Ringer solution is sometimes simply designated as "ringer."

RING finger a protein motif that helps add ubiquitin (*q.v.*) to proteins, thereby marking them for destruction. The RING finger is an evolutionarily conserved structure found in more than 200 proteins, in which two loops of amino acids are pulled together at their base by eight cysteine or histidine residues that bind two zinc ions. *See* metaloprotein.

ring gland a gland lying above the hemispheres (h) of the brain of the larval *Drosophila*. Its lateral extremities encircle the aorta (a) like a ring, hence its name. The gland contains three endocrine tissues: the corpus allatum (ca), the prothoracic gland (pg), and the corpus cardiacum (cc). The diagrams here show the ring gland viewed from the side (A) and from above (B). Other symbols are as follows: n_1 = afferent nerve to the corpus cardiacum from the brain; n_2 = efferent nerve from the corpus cardiacum; n_3 = nerve from corpus cardiacum to the corpus allatum; o = oesophagus; vg = ventral ganglion. *See* allatum hormones, ecdysone.

RIP repeat-induced point mutation (*q.v.*).

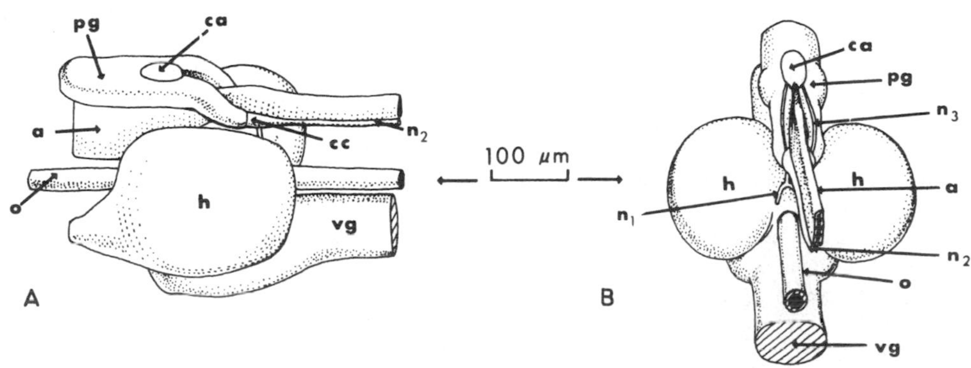

Ring gland

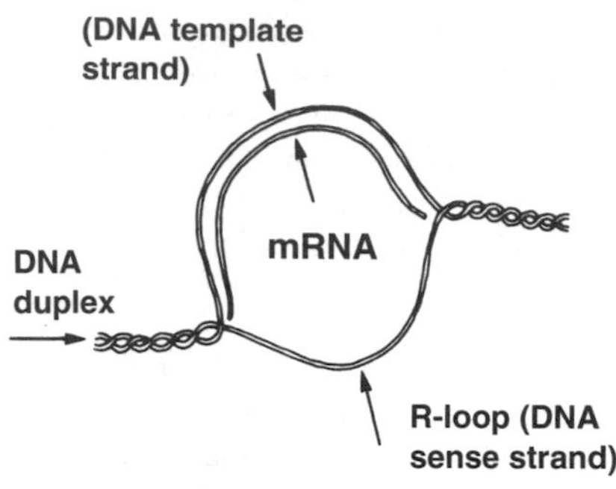

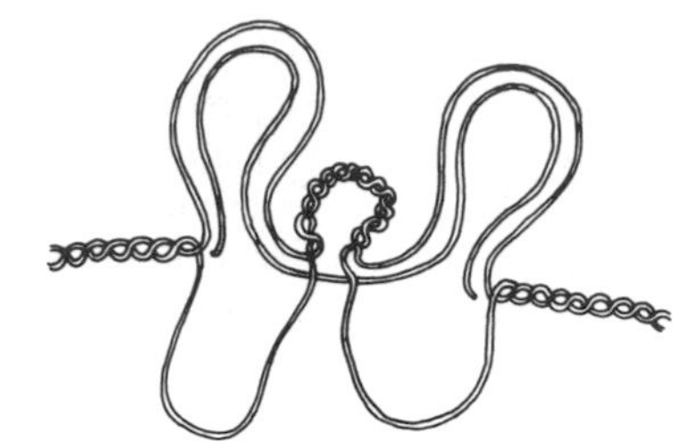

R-loop mapping

RISC RNA-induced silencing complex. *See* RNA interference (RNAi).

RK a symbol used in *Drosophila* studies to indicate *rank* or valuation of a given mutant. For example, RK1 indicates the best and most used mutants, with sharp classification, excellent viability, and accurate genetic localization. RK5 mutants show poor penetrance, low viability, and their chromosomal loci are not accurately determined.

R loop during molecular hybridization, the single-stranded sense strand of DNA that is prevented from reannealing because its complementary template strand is base-paired with an mRNA exon as a heteroduplex.

R-loop mapping a technique for visualizing under the electron microscope the complementary regions shared by a specific eukaryotic RNA and a segment of one strand of a DNA duplex. The RNA-DNA hybrid segment displaces one of the DNA strands, causing it to form a *loop*; hence the name of the technique. Double-stranded regions appear thicker than single-stranded regions in electron micrographs. Introns cannot hybridize with mature mRNA (from which introns have been removed); thus, one intron results in two R loops, two introns yield three R loops, etc. *See* Appendix C, 1977, Chow and Berget.

RMRP a gene located on human chromosome 9 at p13. It encodes an *R*NA component of a *M*itochondrial *R*NA *P*rocessing endoribonuclease (hence the abbreviation). Therefore the RNAs transcribed from the nuclear gene are transferred to mitochondria where they form enzyme components. Mutations of *RMRP* result in a hereditary disease called cartilage-hair hypoplasia (CHH) (*q.v.*). *See* Appendix C, 2001, Ridanpaa *et al.*

RNA ribonucleic acid (*q.v.*).

RNA amplification a technique that allows the synthesis by T7 RNA polymerase of antisense molecules from cDNAs containing a T7 promoter. *See* Appendix C, 1990, Van Gelder *et al.*

RNAase *See* RNase.

RNA coding triplets *See* amino acid, start codon, stop codon.

RNA-dependent DNA polymerase an enzyme that synthesizes a single strand of DNA from deoxyribonucleoside triphosphates, using RNA molecules as templates. Such enzymes occur in oncogenic RNA viruses. This class of enzymes, also known as *reverse transcriptases*, can be used experimentally to make complementary DNA (cDNA) from purified RNA. The functioning of these polymerases contradicts the central dogma (*q.v.*) in the sense that the direction of information exchange between DNA and RNA is reversed. *See* Appendix C, 1970, Baltimore, Temin and Mizutani; Rous sarcoma virus (RSV), telomerase.

RNA-dependent RNA polymerase an enzyme that uses an RNA template to synthesize a complementary RNA molecule. All RNA viruses use such polymerases to replicate their genomes and to transcribe mRNAs from their minus strands. *See* plus (+) and minus (–) viral strands, replicase, RNA replicase.

RNA-driven hybridization an *in vitro* technique that uses an excess of RNA molecules to ensure that all complementary sequences in single-stranded DNA undergo molecular hybridization. *See* DNA-driven hybridization.

RNA editing a mechanism for modifying the nucleotide compositions of previously formed mRNAs by adding or deleting uridine molecules at precise sites within the coding regions of mRNAs. The phenomenon was discovered in the mitochondria of trypanosomes, and the edited molecules allowed translation of functional proteins in situations where the unedited mRNAs were defective. In trypanosomes the mtDNA is made up of 25–50 maxicircles (each 20–40 kilobases) and hundreds of minicircles (each 1–3 kilobases). The maxicircles contain the mitochondrial genes, while the minicircles encode guide RNAs, each about 40 nucleotides long. The gRNAs bind to specific sites on the mRNAs and subsequently insert or delete uridines. *See* Appendix C, 1986, Benne *et al.*; 1990, Blum, Bakalara and Simpson; mitochondrial DNA, proofreading, *Trypanosoma*.

RNA gene a DNA segment coding for one of the various types of nonmessenger RNA (rRNA, 5S RNA, or tRNA).

RNA interference (RNAi) a cellular process in which endogenous or exogenous double-stranded RNA (dsRNA) (*q.v.*) molecules induce gene silencing by mediating the degradation of target messenger RNAs (mRNAs), or by other regulatory means. In this process, dsRNA precursor molecules are processed into double-stranded small interfering RNAs (siRNAs) (*q.v.*), approximately 22 nucleotides long, by the enzyme Dicer (*q.v.*). The siRNA duplex then associates with an endonuclease-containing, multiprotein complex, known as RNA-induced silencing complex (RISC) and unwinds in an ATP-dependent manner. The activated RISC and its associated siRNA subsequently bind complementary sequences in the target mRNA, and the mRNA is cleaved and destroyed. RNAi has some intriguing features: gene silencing can spread from one tissue to the entire organism, and it can also be transmitted through the germ line for some generations. RNAi occurs naturally in a variety of organisms and is thought to play a role in regulating gene expression and in protecting organisms from viruses and transposable elements. It can also be induced experimentally as a tool for creating loss of function phenotypes (e.g., by transfecting cells or embryos with chemically synthesized siRNAs). In *Caenorhabditis elegans* RNAi has been used to analyze the function of all genes residing on a specific chromosome. *See* Appendix C, 1998, Fire *et al.*; 2000, Fraser *et al.*; RNA transfection.

RNA ligase an enzyme, such as T4 RNA ligase (*q.v.*), that can join RNA molecules together. *See* Appendix C, 1972, Silber *et al.*

RNA-P I, II, III *See* RNA polymerase.

RNA phage an RNA bacteriophage such as MS2 and Q beta.

RNA polymerase (RNA-P) an enzyme that transcribes an RNA molecule from the template strand of a DNA molecule. Two kinds of RNA polymerases are known in prokaryotes: one produces the RNA primer required for DNA replication; the other transcribes all three types of RNA (mRNA, tRNA, rRNA). In eukaryotes, each type of RNA is transcribed by a different RNA polymerase. RNA polymerase I (RNA-PI) resides in the nucleolus and catalyzes the synthesis of rRNA. RNA-PII is localized in the nucleoplasm, where it catalyzes the synthesis of mRNA. RNA-PII is specifically inhibited by alpha amanitin (*q.v.*). RNA-PIII makes tRNA, 5S RNA, and other small RNA molecules. The structure of the RNA polymerase of *E. coli* is known in great detail. It has the general formula $\omega\alpha\alpha\sigma\beta'\beta$. The component proteins, their relative molecular masses, the encoding genes, and their locations are shown here.

Protein	*Mr*	*Gene*	*Map Positions (minutes)*
omega	11,000	?	?
alpha	36,500	*rpo A*	72
sigma	70,000	*rpo D*	66.5
beta′	151,000	*rpo* C	89.5
beta	155,000	*rpo B*	89.5

The catalytic site for RNA polymerization is thought to reside in the beta protein, and this is also the place where rifampicin (*q.v.*) binds. The sigma protein functions in promoter recognition and the initiation of RNA synthesis. Eukaryotic RNA polymerases are still more complex, since they contain about 10 rather than five types of subunit. *See* Appendix C, 1961, Weiss and Nakamoto; 1969, Roeder and Rutter, Burgess *et al.*; amatoxins, Cajal body, Miller trees, muscular dystrophy, nucleolus, nucleolus organizer, polynucleotide, phosphorylase, ribosomal RNA genes, strand terminologies, TATA box binding protein, terminator, T7 RNA polymerase, transcription unit.

RNA primer *See* primer RNA.

RNA processing *See* posttranscriptional processing.

RNA puff *See* chromosomal puff.

RNA replicase an RNA-dependent RNA polymerase (*q.v.*). *See* MS2 and Q beta.

RNase any enzyme hydrolyzing RNA.

RNase protection a technique for locating the points of effective contact between a nucleic acid chain and a cognate polypeptide chain; the complex (e.g., tRNA and its cognate aminoacyl-tRNA synthetase) is treated with a group of RNases that digest all of the RNA except those regions in contact with the synthetase. *See* photoactivated cross-linking.

RNA splicing the removal of noncoding regions from a large precursor RNA molecule, and the nucleotide sequences transcribed from nonadjacent DNA segments are then joined together to produce a smaller mature RNA. *See* alternative splicing, intron, posttranscriptional processing, splice junctions, spliceosome.

RNA transfection the experimental introduction of exogenous single- or double-stranded RNA (ds RNA) molecules into a cell or an embryo, which results in a phenotypic change that is transient in nature (i.e., is not stably inherited). Transfection of cells with a specific messenger RNA (mRNA) (*q.v.*) can be used to express a particular protein, whereas transfection with certain single- or ds RNA molecules can produce RNA interference (*q.v.*). RNA transfection can be achieved through a number of methods, including microinjection (*q.v.*), electroporation (*q.v.*), the use of viral vectors, and the use of liposomes (*q.v.*). Furthermore, in *C. elegans*, RNA transfection (and resulting RNA interference) can be achieved by soaking the animals with ds RNA or feeding them *E. coli* cells which express ds RNA.

RNA vector a retrovirus used to transfer a provirus carrying a beneficial foreign gene into the chromosome of a host. An example would be using a genetically engineered HIV which contains a functioning beta globin gene to infect a mouse genetically engineered to have a beta globin deficiency. *See* Appendix C, 2001, Pawliuk *et al.*

RNP ribonucleoprotein (*q.v.*).

rNTP ribonucleoside 5′-triphosphates.

Robertsonian translocation *See* centric fusion.

rod one of the elongate, unicellular photoreceptors in the vertebrate retina (*q.v.*), involved with vision in dim light. Rods do not discriminate color differences. *See* rhodopsin.

Roentgen (R) the quantity of ionizing radiation that liberates 2.083×10^9 ion pairs in a cubic centimeter of air (at 0°C and at a pressure of 760 millimeters of mercury) or approximately two ion pairs per cubic micron of a substance such as protein (which has a density of 1.35). A gram of tissue exposed to 1 roentgen of gamma rays absorbs about 93 ergs. The unit is named in honor of W. C. Roentgen, the discoverer of x-rays, who received the Nobel Prize in physics in 1901 for this work.

roentgen equivalent physical (rep) the amount of ionizing radiation that will result in the absorption in tissue of 93 ergs per gram.

rogue a variation from the standard variety, usually inferior.

rolling circle a model mechanism for the replication of DNA molecules, so named because the growing point can be imagined as rolling around a circular template strand. The circular DNA is shown here in A. In B, a nick opens one strand, and the free 3′-OH end is extended by DNA polymerase. The newly synthesized strand displaces the original parental strand as it grows (C, D). By E, the polymerase has completed one revolution, and by F, two revolutions. The result is a molecule containing three unit genomes, one old and two new. The displaced strand can then serve as a template for a complementary strand. This mechanism is used to generate concatemeric duplex molecules (e.g., phage lambda, amplified rDNA in amphibian oocytes, etc.). This type of DNA replication is sometimes called *sigma replication* because the structure produced by the rolling circle resembles the Greek lower case sigma (σ). The rolling circle mechanism also occurs in viroids (*q.v.*). Here the infectious (+) viroid RNA serves as a template for an RNA polymerase, which generates a concatomeric, complementary (–) strand. In turn, this serves as template for synthesis of a (+) concatomeric RNA, which is later cleaved into genomic units that are then ligated to form circles. *See* Appendix C, 1968, Gilbert and Dressler; hammerhead ribosome, plus (+) and minus (–) viral strands, theta replication.

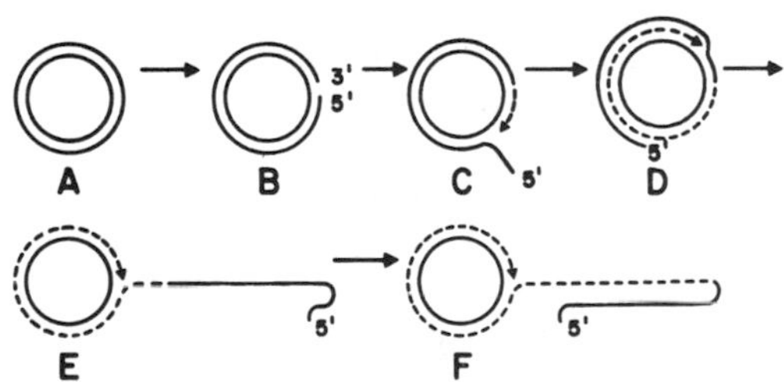

rolling-circle transposon a helitron (*q.v.*).

Romalea microptera the lubber grasshopper. Meiosis has been extensively studied in this species.

Romanov the ruling family of Russia for almost 300 years. The last Tsar, Nicholas II, and his wife, Alexandra, had five children of which the first four were girls (Olga, Tatiana, Maria, and Anastasia). Then came Alexei, the male heir to the throne. Unfortunately he inherited hemophilia (*q.v.*) from his grandmother, Victoria, Queen of England. As a result he was in poor health during his short life. After the Revolution the Romanovs were imprisoned in the Ipatiev House at Yekaterinberg, Siberia. On July 16, 1918, the entire family, their doctor, and three servants were murdered by a Bolshevik firing squad. In 1991 a grave was discovered containing the fragmented remains of nine skeletons. Tests of mtDNA showed that five of the bodies were related. Subsequent matches with the mtDNA of surviving relatives identified Nicholas and Alexandra. The Tsar's mtDNA was found to be heteroplasmic (*q.v.*). The bodies of 14-year-old Alexei and his sister Maria were never found. *See* Appendix C, 1994, Gill *et al.*; mitochondrial DNA (mtNDA).

root cap a cap of cells covering the apex of the growing point of a root and protecting it as it is forced through soil.

root hair a tubular outgrowth of an epidermal cell of a root which functions to absorb water and nutrients from the soil.

root nodules a small swelling on roots of legumes (*q.v.*) produced as a result of infection by symbiotic nitrogen-fixing bacteria. *See Rhizobium.*

Rosa the genus that includes the rose species, which have been extensively hybridized. Commercially grown species include *R. centifolia, R. damascena*, and *R. multiflora*.

Rose chamber a closed culture vessel permitting long-term observation of explanted cells under phase microscopy. The fluid culture medium may be periodically renewed without disturbing the growing cells.

R_0t the product of RNA concentration and the time of incubation in an RNA-driven hybridization; the analog of C_0t values used to describe DNA-driven hybridization reactions.

rotational base substitution a break is induced by radiation at a corresponding point in both complementary strands of a DNA molecule. The bond broken is between the base and the sugar molecule to which it is attached. Thus two complementary bases (held together by hydrogen bonds) are detached from their backbones. If the pair rotates before it is reinserted in the molecule, the resulting transversional mutation would be termed a rotational base substitution. Many radiation-induced point mutations may result in this way.

rotation technique *See* photographic rotation technique.

Roundup the trade name of a glycophosphate herbicide (produced by the Monsanto Company) that kills most plants. When sprayed on weeds, the glycophosphate is absorbed by the leaves, enters the cells, and inactivates a critical enzyme. Plants cannot synthesize certain essential amino acids without this enzyme. The target enzyme does not occur in animals, so Roundup is not toxic to humans. The herbicide is rapidly degraded in the soil. Monsanto next produced crops that were genetically engineered to carry resistance genes for glycophosphate. Seeds for these crops are sold under the trade names Roundup Ready corn, Roundup Ready cotton, Roundup Ready potatoes, Roundup Ready soy beans, etc. *See* GMO, transgenic plants.

rough endoplasmic reticulum *See* endoplasmic reticulum.

Rous sarcoma virus (RSV) the first oncogenic virus to be discovered. It is an RNA virus that induces tumors in chickens. The RSV was one of the first retroviruses shown to produce a reverse transcriptase (*q.v.*). The genes *gag, pol*, and *env*, which characterize all retroviruses, were first identified in this virus. The RSV genome also contains *src*, an oncogene, so named because it induces *sarcomas*. The *src* gene codes for a protein kinase, pp60v-src (*q.v.*), which is localized in the plasmalemma. Vertebrate cells contain a gene homologous to the *src* gene. To distinguish the two, the viral gene is abbreviated *v-src* and the cellular gene *c-src*. The two genes differ in that *v-src* has an uninterrupted coding sequence, whereas *c-src* contains seven exons separated by six introns. The *c-src* gene is a proto-oncogene (*q.v.*). *See* Appendix C, 1910, Rous; 1970, Temin and Mizutani; 1975, Wang *et al.*; 1978, Collett and Erickson; 1981, Parker, Varmus, and Bishop; 1989, Bishop and Varmus; oncogene hypothesis, retroviruses.

royal hemophilia classical hemophilia (*q.v.*) transmitted by a defective X chromosome first carried by Queen Victoria of Great Britain and passed on to plague three generations of European royalty.

R17 phage a small RNA androphage (*q.v.*). *See* bacteriophage.

rpo *See* RNA polymerase.

R (resistance) plasmid an extrachromosomal DNA molecule that confers on bacteria resistance to one or more antibiotics. It consists of two components: the resistance transfer factor (RTF) required for transfer of the plasmid between bacteria, and the r-determinants (genes conferring antibiotic resistance). R plasmids were first isolated from strains of *Shigella dysenteriae* that showed resistance to multiple antibiotics. *See* Appendix C, 1960, Watanabe and Fukusawa.

rRNA ribosomal RNA. *See* ribosome.

rRNA transcription unit *See* Miller trees.

r strategy a type of life cycle exploiting high reproductive rate to achieve survival. *See* r and K selection theory.

RSV Rous sarcoma virus (*q.v.*).

rtDNA *See* rDNA.

RTF resistance transfer factor. *See* R plasmid.

rTU rRNA transcription unit. *See* Miller trees.

RuBisCO an abbreviation for the enzyme *ri*b*u*lose-1, 5-*bis*phosphate *c*arboxylase-*o*xygenase (*q.v.*).

ruffled edges *See* lamellipodia.

ruminant mammals even-toed hoofed mammals of a type that chews a cud. They include cattle, sheep, and goats and their relatives.

runner a procumbent shoot that takes root, forming a new plant that eventually is freed from connection with the parent by decay of the runner. The runner serves as a vegetative propagule. *See Frageria*, modular organisms.

runt (run) a pair rule gene of *Drosophila* residing at 1-65. The *run* gene interacts with many other genes that regulate embryonic development (*bicoid, fushi tarazu*, and *hunchback* are examples). *Runt* is expressed in subsets of neuroblasts in each neuromere at early stages of neurogenesis. The *run* gene encodes a transcription factor containing 509 amino acids that is localized in the nucleus. This Runt protein contains a DNA-binding motif 128 amino acids long. This Runt domain is also found in the protein encoded by *lozenge* (*q.v.*). Many other transcriptional regulators produced by both invertebrates and vertebrates contain Runt domains. In humans the symbol RUNX (RUNT-related) is given to these proteins, and one of them, RUNX1 (also called AML1) controls the differentiation of leucocytes. *See acute myeloid leukemia 1* gene, zygotic segmentation mutants.

runting disease a pathological condition seen in young experimental animals inoculated with allogeneic immunocompetent cells that produce a graft-versus-host reaction (*q.v.*).

rut the period of sexual activity; estrus.

S

s 1. selection coefficient (*q.v.*) 2. standard deviation (*q.v.*) 3. sedimentation coefficient (*q.v.*) 4. second.

S 1. Svedberg unit (*q.v.*) 2. Silurian. 3. sulfur. 4. DNA synthesis phase of the cell cycle (*q.v.*).

^{35}S a beta-emitting radioactive isotope of sulfur with a half-life of 87.1 days; commonly used to label proteins via their sulfur-containing amino acids cysteine and methionine. ^{35}S was used in the famous Hershey-Chase experiment of 1952. *See* Appendix C.

S_1, S_2, S_3, etc. the representation for continued selfing (self-fertilization) of plants. S_1 designates the generation obtained by selfing the parent plant; S_2, the generation obtained by selfing the S_1 plant, etc.

Saccharomyces cerevisiae the species of budding yeast used by brewers and bakers. Strains of brewer's and baker's yeast have special properties. Baker's yeast can raise bread five times faster than brewer's yeast, but baker's yeast gives beer an undesirable yeasty flavor and settles out poorly. When nutrients are plentiful, wild type strains of *S. cerevisiae* proliferate as diploid cells. If starved, they go through meiosis to form haploid spores. These can later germinate and proliferate as haploid cells, or they can fuse to reform diploids. The demonstration that yeast DNA has a high content of AT relative to GC disproved the tetranucleotide hypothesis (*q.v.*). *S. cerevisiae* was the first eukaryote to have its entire genome sequenced. It contains 12,068,000 base pairs, divided among 17 chromosomes. Its 5,885 ORFs make up about 70% of the genome. Only about 4% of the genes contain introns (*q.v.*). This is unusual, since 40% of the genes of *Schizosaccharomyces pombe* (*q.v.*) contain introns. In *S. cerevisae* there are 140 rRNA genes in a large tandem array on chromosome 12. The 40 genes for snRNAs and the 275 genes for tRNAs are dispersed among all the chromosomes. *S. cerevisiae* is a favorite species for studying the genetic regulation of progression through the cell cycle. *See* Appendix A, Fungi, Ascomycota; Appendix C, 1949, Ephrussi *et al.*; 1965, Holley *et al.*; 1970, Khorana *et al.*; 1973, Hartwell *et al.*; 1974, Dujon *et al.*; 1979, Cameron *et al.*; 1980, Clark and Carbon; 1985, Boeke *et al.*; 1989, Kaback, Steensma, and DeJonge; 1992, Oliver *et al.*; 1996, Goffeau *et al.*; 1999, Galitsky *et al.*; 2000, Rubin *et al.*; Appendix E, Individual Databases; bud, cassettes, centromere, Chargaff rule, genetic code, mitochondrial DNA, omnipotent suppressors, orphan, petites, *Plasmodium*, Ty elements, universal code theory, yeast artificial chromosomes (YACs), yeast two-hybrid system.

S-adenosylmethionine *See* 5-methylcytosine.

SAGE *See* serial analysis of gene expression (SAGE).

Saint-Hilaire hypothesis the proposal by Etienne Geoffroy Saint-Hilaire that arthropods and vertebrates have a common body plan. However, the plan in the two groups is inverted, since the nervous system of an insect is ventral, while that of a mammal is dorsal. Recent studies on genes that control the dorsoventral differentiation of embryonic cells support the idea that the pattern for the body axis turned upside down during the early evolution of chordates. *See* Appendix C, 1822, Saint-Hilaire; 1994, Arendt and Nubler-Jung.

salivary gland chromosomes polytene chromosomes found in the interphase nuclei of the salivary gland cells in larval diptera. These chromosomes undergo complete somatic pairing; consequently, the mature salivary gland chromosome consists of two homologous polytene chromosomes fused side by side. *See* Appendix C, 1881, Balbiani; 1912, Rambousek; 1933, Painter; 1934, Bauer; 1935, Bridges; *Drosophila* salivary gland chromosomes.

salivary gland squash preparation 1. a rapid method of preparing insect polytene chromosomes for microscopic investigation without sectioning them. The organ is simply squashed in a drop of stain placed between slide and coverslip. *See* aceto-orcein. 2. a method of preparing giant polytene chromosomes for localization of specific DNA sequences via *in situ* hybridization. Larval salivary glands are quickly squashed in an appropriate fixative between slide and coverslip, frozen, and the coverslip removed. The squashed specimens are then dehydrated and prepared for hybridization with labeled nucleic acid probes. *See in situ* hybridization.

***S* allele** *See* self-sterility genes.

Salmo the genus containing various fish species of economic importance; especially *S. salar*, the Atlantic salmon and *S. gairdneri*, the rainbow trout. *See* Appendix A, Chordata, Osteichythes, Neopterygii, Salmoniformes.

Salmonella together with *Escherichia* and *Shigella*, one of the genera of enteric bacteria containing favorite species for genetic study. *Salmonella typhimurium*, the mouse typhoid bacillus, has been the most studied, but considerable work has also been done with *S. abony*, *S. flexneri*, *S. minnesota*, *S. montevideo*, *S. pullorum*, and *S. typhosa*. The first method for assaying bacteriophage titer was worked out for *S. typhimurium*, and the tranduction phenomenon was also discovered in this species. *See* Appendix A, Bacteria, Proteobacteria; Appendix C, 1917, d'Herelle; 1952, Zinder and Lederberg; histidine operon, virulence plasmid.

saltation 1. a theory that new species originate suddenly from one or more mutations with large phenotypic effects ("macromutations"); referred to by R. Goldschmidt as "hopeful monsters." 2. quantum speciation (*q.v.*). *See* evolution.

saltatory replication lateral amplification of a chromosome segment to produce a large number of copies of a specific DNA sequence. *See* gene amplification, rDNA amplification.

salt linkage *See* electrostatic bond.

salvage pathways metabolic pathways for the synthesis of nucleosides and nucleotides utilizing preformed purines and pyrimidines or nucleosides. Examples include the conversion of a base to a nucleoside, interconversion of bases and nucleosides, conversion of nucleosides to nucleotides, and interconversions by base alterations. *Compare with de novo* pathway.

Salvarsan dihydroxy diamino arsenobenzene dihydrochloride, a drug used for 30 years in the treatment of syphilis until the advent of penicillin. *See* Appendix C, 1909, Ehrlich; 1940, Florey *et al.*

samesense mutation a point mutation (usually in the third position of a codon) that does not change the amino acid specificity of the codon so altered; a "silent" mutation. *See* degenerate code.

sampling error variability due to the limited size of the samples.

Sandhoff disease a lysosomal storage disease (*q.v.*) with symptoms similar to the Tay-Sachs disease (*q.v.*). K. Sandhoff and three colleagues published the first description of the condition in 1968. The Sandhoff and Tay-Sachs diseases are due to mutations involving the same enzyme, hexosaminidase (*q.v.*). The catalytic form of the enzyme is a heterodimer, and each gene encodes a different monomer.

Sanger-Coulson method *See* DNA sequencing techniques.

saprobe an organism, often a fungus, that lives on and derives its nutrition from dead organic matter.

saprophyte a plant saprobe (*q.v.*).

sarcoma a cancer of mesodermal origin (e.g., connective tissue). *See* Rous sarcoma virus, simian sarcoma virus.

sarcomere the repeating unit, about 2.5 micrometers long, within striated muscle fibers, containing a set of interacting actin and myosin filaments.

sarcosomes the mitochondria of the flight muscles of insects.

Sargasso Sea an elliptical region of the northwest Atlantic Ocean covered with huge mats of floating sargassum seaweed. The central area is usually calm with little wind or currents, and it rotates clockwise from the West Indies past Bermuda to the Azores.

SARS severe acute respiratory syndrome. *See* coronavirus.

sat DNA or RNA satellite DNA or satellite RNA (*q.v.*).

satellite a distal chromosomal segment separated from the rest of the chromosome by a thin chromatic filament or stalk called the *secondary constriction.*

satellite (sat) DNA any fraction of the DNA of an eukaryotic species that differs sufficiently in its base composition from that of the majority of the DNA fragments to separate as one or more bands distinct from the bands containing the majority of the DNA during isopycnic CsCl gradient centrifugation. Satellite DNAs obtained from chromosomes are either lighter (A + T rich) or heavier (G + C rich) than the majority of the DNA. Satellite DNAs are usually highly repetitious. *See* Alu family, mouse satellite DNA, reassociation kinetics, repetitious DNA.

satellite (sat) RNA a small, linear, single-stranded RNA molecule that is encapsulated within certain specific plant RNA viruses. An example is the sat RNA of the cucumber mosaic virus, which in its monomeric form is 335 nucleotides long. The satellite RNA depends on the supporting helper virus to provide a protective coat protein and some of the enzymes necessary for its replication. The satellite RNA is not necessary for the replication of the helper virus. The presence of the sat RNAs may attenuate the symptoms of the helper virus in the infected plant. *See* hairpin ribozyme, hammerhead ribozyme, viroid, virusoid.

saturation density the maximum cell number attainable under specified culture conditions in a culture vessel. This term is usually expressed as the number of cells per square centimeter in an anchorage-dependent culture, or per cubic centimeters in a suspension culture.

saturation hybridization an *in vitro* reaction in which one polynucleotide component is in great excess, causing all complementary sequences in the other polynucleotide component to enter a duplex form. *See* DNA-driven hybridization reaction, RNA-driven hybridization.

scaffold an ordered series of contigs (*q.v.*) separated by gaps whose approximate lengths are known. *See* chromosome scaffold, shotgun sequencing.

scanning electron microscope *See* electron microscope.

scanning hypothesis a theory to explain initiation of translation in eukaryotes according to which a 40S ribosomal subunit attaches at or near the mRNA cap and then drifts in the 3′ direction until it encounters an AUG start codon; at this point, an initiation complex is formed and the reading frame becomes established.

scape a leafless flower-bearing stem arising from ground level (as in the dandelion and daffodil).

scarce mRNA *See* complex mRNA.

scatter diagram a diagram in which observations are plotted as points on a grid of x and y coordinates to see if there is any correlation (*q.v.*). For example, one might plot for a number of species the LD 50 (*q.v.*) against the average DNA content per somatic cell nucleus. Finding a correlation suggests that the variables may be interrelated. No correlation suggests that the variables chosen have no bearing on one another.

scattering the change of direction of particles or waves as a result of a collision or interaction. Scattering of electrons by the specimen, for example, is responsible for the electron microscopic image.

SCE sister chromatid exchange (*q.v.*).

Schiff reagent a compound developed about 1866 by Hugo Schiff. The reagent attaches to and colors aldehyde-containing compounds. Used in the PAS and Feulgen procedures (*q.v.*).

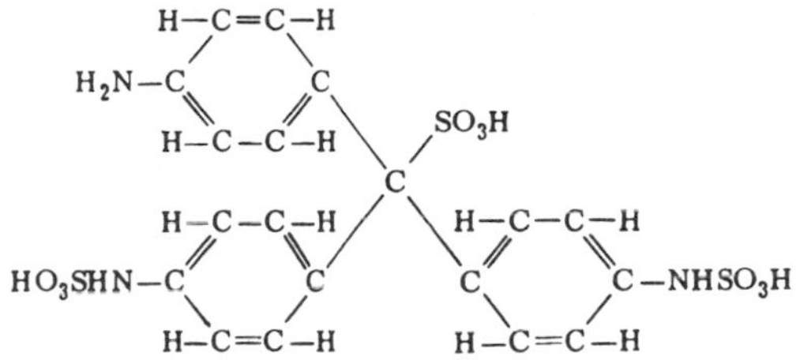

```
              H—C=C—H
               /     \
        H2N—C         C
             \\      // \
              H—C—C—H    \    SO3H
                          \  /
                           C
              H—C—C—H    /   \    H—C—C—H
              //     \\ /     \   //     \\
   HO3SHN—C          C         C          C—NHSO3H
             \       /          \        /
              H—C=C—H            H—C=C—H
```

Schiff reagent

Schistocerca gregaria a locust whose population dynamics has been extensively studied.

schistosomiasis also called *bilharziasis* or *snail fever;* a helminthic infection of humans involving 200 million persons in Africa, the Middle East, the Orient, and South America. The causative agent is the blood fluke *Schistosoma mansoni*, and its snail vector is *Biomphalaria glabrata*. The genetic control of susceptibility of *B. glabrata* to *S. mansoni* is a subject of active investigation.

schizogony a series of rapid mitoses without increase in cell size which gives rise to schizonts.

schizont a sporozoan spore arising from schizogony.

schizophrenia the name given to a constellation of symptoms including hallucinations and delusions, disorders of thinking and concentration, and erratic behavior. Schizophrenia appears to be a family of diseases of high heritability that together afflict about 1 percent of all humans. Sequence variations in the neuregulin gene, *Nrg1*, have been associated with schizophrenia in Icelandic and Scottish populations, and *Nrg1* mutations in mice produce behavioral patterns consistent with mouse models of this disease. *See* neuregulins (NRGs).

Schizophyllum a genus of fungi that form fan-shaped, bracket-type fruiting bodies. *S. commune* has been the subject of considerable genetic research, and two linkage groups have been mapped. *See* Appendix A, Fungi, Basidiomycota.

Schizosaccharomyces pombe a fungus that can vary in shape from single globose to cylindrical cells that reproduce by septal fission (*q.v.*). A mitotic spindle forms during G2, and the chromosomes condense. However, the nuclear envelope remains intact. *S. pombe* can also form true hyphae (*q.v.*). Sexual reproduction results when haploid cells fuse to produce a zygote. This undergoes meiosis to form an ascus that contains four haploid ascospores. These develop into vegetative cells. Its genome contains 12.5×10^6 bp of DNA. Comparison of the genes of *Saccharomyces cerevisiae* (*q.v.*) and *Schizosaccharo-*

myces pombe has revealed that genes of the latter are far more likely to contain introns than those from *Saccharomyces*. *S. pombe* is a favorite for the study of the genetic regulation of the progression through the cell cycle. *See* Appendix A, Fungi, Ascomycota; Appendix C, 1976, Nurse, Thuriaux and Nasmyth; 1994, Chikashige *et al.*

Schwann cell the cell that enfolds a myelinated nerve fiber and first described by the German histologist Theodor Schwann. *See* myelin sheath.

Sciara a genus of fungus gnats extensively studied in terms of the cytogenetics of chromosome diminution (*q.v.*). The giant polytene chromosomes of the larval salivary glands of *S. coprophila* have been mapped, and certain DNA puffs have been extensively studied. *See* chromosomal puff, *Rhynchosciara*.

scintillation counter *See* liquid scintillation counter.

scission a severance of both strands of a DNA molecule; a cut (*q.v.*). *Compare with* nick.

scleroproteins the very stable fibrous proteins, present mainly as surface coverings of animals. Keratin and collagen are examples of scleroproteins.

scrapie *See* prion.

scurvy *See* ascorbic acid.

scutellum **1.** the single cotyledon of a grass embryo. **2.** a shield-shaped metathoracic tergite of *Drosophila*.

SD **1.** standard deviation. **2.** segregation distortion (*q.v.*).

SD sequence *See* Shine-Dalgarno sequence.

SDS sodium dodecyl sulfate (*q.v.*).

SDS-PAGE technique *See* electrophoresis.

SE, S.E. standard error (*q.v.*).

Se *Secretor* gene (*q.v.*).

sea floor spreading the horizontal spreading of the sea floor as tectonic plates separate. A good example is the floor of the Atlantic Ocean where the mid-Atlantic ridge marks the line of separation between the Eurasian plate, which is moving east, and the North American plate, which is moving west. The rending of the crust is accompanied by outpouring of the volcanic materials that produce the ridge, while the void between the separating plates is filled with molten rock that rises from below and solidifies. Thus new sea floor is added to the trailing edges of the separating plates as they move laterally, and consequently the sea floor widens. *See* Appendix C, 1960, Hess; 1963, Vine and Matthews; continental drift, plate tectonics.

Searle translocation a reciprocal X-autosome translocation in the mouse which exhibits paternal X inactivation in the somatic cells of female heterozygotes. All other X-autosome translocations studied in the mouse show random inactivation of either normal or rearranged chromosome. *See* Lyon hypothesis.

seasonal isolation a type of ecological reproductive isolation in which different species become reproductively active at different times; temporal isolation.

sea squirt *See Ciona intestinalis.*

sea urchin *See* echinoderm.

secondary constriction a thin chromatic filament connecting a chromosomal satellite to the rest of the chromosome.

secondary DNA *See* skeletal DNA hypothesis.

secondary gametocytes *See* meiosis.

secondary immune response *See* immune response.

secondary nondisjunction sex chromosomal nondisjunction in an XXY individual resulting in gametes containing either two X chromosomes, one X, one Y, or an X and a Y.

secondary protein structure *See* protein structure.

secondary sex ratio the ratio of males to females at birth; in contrast to the primary sex ratio at conception.

secondary sexual character a characteristic of animals other than the organs producing gametes that differs between the two sexes (e.g., mammary glands, antlers, and external genitalia). *See* primary sexual character, sex combs.

secondary speciation the fusion through hybridization of two species that were formerly geographically isolated, followed by establishment of a new adaptive norm through natural selection.

second cousin *See* cousin.

second degree relative when referring to a specific individual in a pedigree (*q.v.*), any individual who is two meioses away from that individual (a grandparent, a grandchild, an uncle, an aunt, a niece, a nephew, or a half-sibling). Any relative with whom one quarter of one's genes are shared. *Contrast with* first-degree relative.

second division segregation ascus pattern in ascomycetes, a 2-2-2-2 or 2-4-2 linear order of spore phenotypes within an ascus. These patterns indicate that a pair of alleles (e.g., those controlling spore pigmentation) separated in the second meiotic division because crossing over occurred between that locus and the centromere. *See* ordered tetrad.

second law of thermodynamics *See* thermodynamics.

second messenger small molecules or ions generated in the cytoplasm in response to binding of a signal molecule to its receptor on the outer surface of the cell membrane. Two major classes of second messengers are known: one involves cyclic adenosine monophosphate and the other employs a combination of calcium ions and either inositol triphosphate or diacylglycerol. *See* G proteins.

second site mutation *See* suppressor mutation.

secretin a hormone that stimulates the secretion of pancreatic juice. The epithelial cells of the duodenum release secretin when activated by the acidic contents of the stomach.

secretion the passage out of a cell or gland of compounds synthesized within it.

***Secretor* gene** a dominant autosomal gene in humans that permits the secretion of the water-soluble forms of the A and B blood-group antigens into saliva and other body fluids. The *Se* gene is not linked to the *I* locus. *See* A, B antigens.

secretory vesicle a vesicle that contains a secretory product. Secretory vesicles in the acinar cells of the pancreas, for example, contain precursors of digestive enzymes in a highly concentrated form (200 times as high as their concentration in the ER). Proteins destined for secretion contain sorting signals (*q.v.*) that target them to secretory vesicles. *See* Golgi apparatus.

sedimentation coefficient (*S*) the rate at which a given solute molecule suspended in a less dense solvent sediments in a field of centrifugal force. The sedimentation coefficient is a rate per unit centrifugal field. The *S* values for most proteins range between 1×10^{-13} sec and 2×10^{-11} sec. A sedimentation coefficient of 1×10^{-13} sec is defined as one Svedberg unit (S). Thus a value of 2×10^{-11} sec would be denoted by 200S. For a given solvent and temperature, *S* is determined by the weight, shape, and degree of hydration of the molecule. *See* Appendix C, 1923, Svedberg.

seed a mature ovule containing an embryo in an arrested state of development, generally with a food reserve. *See* kernel.

seeding efficiency in cell culture, the percentage of the cells in an inoculum that attach to the culture vessel within a specific length of time; synonymous with *attachment efficiency*.

segmental alloploid *See* allosyndesis.

segmental interchange a translocation.

segmentation *See* metamerism.

segmented genome a viral genome fragmented into two or more nucleic acid molecules. For example, the alfalfa mosaic virus has four different RNA segments, each packaged in a different virion. Successful infection requires that at least one RNA of each type enters the cell. Such a virus is said to be *heterocapsidic*. If all fragments of a segmented genome are present in the same virion (e.g., influenza virus), the virus is said to be *isocapsidic*.

segment identity genes genes that determine the type of differentiation the cells in a specific *Drosophila* segment will undergo. These genes express themselves later in development than the zygotic segmentation genes. While mutations in these genes cause the deletion of certain body parts and are generally lethal, mutations of segment identity genes allow the mutants to survive but with inappropriate structures developing in specific segments. For example, a bizarre four-winged fly results if the mutation converts halteres (*q.v.*) to wings. The segment identity genes are located in two clusters on the right arm of chromosome 3 (the *Antp* complex and the *bx* complex). *See* *Antennapedia*, *bithorax*, homeotic mutations, Hox genes, metamerism, pattern specification, zygotic segmentation mutations.

segment polarity genes *See* zygotic segmentation mutants.

segregational lag delayed phenotypic expression of an induced mutation in one nucleoid of a multinucleoid bacterium. The lag period is the time required for the fission of the parent to produce a cell containing only the mutant chromosome.

segregational load the genetic disability sustained by a population due to genes segregating from advantageous heterozygotes to less fit homozygotes.

segregational petites *See* petites.

segregation distortion a distortion of expected Mendelian ratios in a cross due to dysfunction or lethality in gametes bearing certain alleles. This form of meiotic drive (*q.v.*) is represented in *Drosophila melanogaster* by the segregation distorter (*SD*) mutation on chromosome 2. SD/sd^+ heterozygous males produce both *SD* and sd^+ spermatozoa, but only those carrying *SD* are functional. *SD* cannot achieve fixation, however, because it is lethal when homozygous.

segregation of chromosomes *See* disjunction.

segregation of genes *See* Mendel's laws.

segregation ratio distortion the distortion of the 1 : 1 segregation ratio produced by a heterozygote. Such distortions may arise because of abnormalities of meiosis that result in an *Aa* individual's producing unequal numbers of *A*- and *a*-bearing gametes, or it may arise from *A*- and *a*-bearing gametes, being unequally effective in producing zygotes.

selectins a family of structurally related lectins (*q.v.*), so named because they mediate the selective contact between cells. Selectins are glycoproteins that form an integral part of the membrane of certain cells. At the *N*-terminal end of each molecule is a lectin domain followed by an EGF domain. Next comes a short consensus sequence, which is repeated several times, and finally a transmembrane domain. Selectins are distinguished by capital letters that indicate the specific cells synthesizing the molecules (E,L,P; endothelial cells, lymphocytes, platelets). *See* epidermal growth factor.

selection the process determining the relative share allotted individuals of different genotypes in the propagation of a population. The selective effect of a gene can be defined by the probability that carriers of the gene will reproduce. *See* alloprocoptic selection, artificial selection, balanced selection, directional selection, disruptive selection, frequency-dependent selection, group selection, indirect selection, kin selection, normalizing selection, r and K selection, sexual selection, stabilizing selection theory.

selection coefficient (*s*) the proportionate reduction in the average gametic contribution to the next generation made by individuals of one genotype relative to those of another genotype (usually the most fit). For example, if the best adapted genotypes are *AA* and *Aa*, and they are not being selected against, $s = 0$, and their fitness $= (1 - s) = 1$. If individuals of genotype *aa* leave only on the average 80% as many progeny (proportionate to their numbers in the population) as the other genotypes, then the selection coefficient against *aa* individuals is 0.2 or 20%, and their fitness is $(1 - 0.2) = 80\%$.

selection differential the difference between the average value of a quantitative character in the whole population and the average value of those selected to be parents of the next generation. *See* record of performance.

selection pressure the effectiveness of natural selection in altering the genetic composition of a population over a series of generations.

selective advance the increment in the average value (measured for a quantitative character being selected in a population) from one generation to the next, usually a fraction of the selection differential (*q.v.*).

selective medium a medium designed to allow growth of only those cells of a specific genotype. *Compare with* nonselective medium.

selective neutrality a situation in which the phenotypic manifestations of certain mutant alleles are equivalent to that of the wild-type allele in terms of their fitness values. *See* neutral gene theory, silent mutations.

selective plating a method for selectively isolating recombinants. Two different auxotropic mutants are plated upon a minimal medium. Only the recombinant class receiving the normal allele of each mutant can multiply under these conditions.

selective silencing any mechanism that consistently eliminates plasmagenes of one parent from the zygote, such as the destruction of chloroplasts or chloroplast DNA in some algae and plants or of sperm mitochondria in some animals.

selective system any experimental technique that aids in the detection and isolation of a specific (usually rare) genotype. *See* penicillin selection technique.

selective variant in microbial genetics, a mutation that confers upon the organism the ability to exist under conditions that kill off all organisms not possessing the mutation. Examples of selective variants are mutations conferring resistance to antibacterial agents or the ability to synthesize some essential metabolite lacking in the medium.

selector genes a subset of regulatory genes that control choices between alternative developmental pathways. For example, a cell may continue along pathway A, unless it receives a signal to enter pathway B. The signal is often a protein encoded by a selector gene that binds to specific genomic sites, activating one or more of the genes necessary for the new pattern of development. *See Antennapedia, apetala-2, bicoid, bithorax, caudal,* compartmentalization, *decapentaplegic,* developmental control genes, *dorsal,* downstream genes, *engrailed, eyeless, fushi tarazu, goosecoid, gurken, hedgehog, Hox* genes, *hunchback,* metamerism, *nanos, oskar, paired, Polycomb, sevenless, SRY.*

self to undergo self-pollination (*q.v.*) or self-fertilization (*q.v.*).

self-assembly the spontaneous aggregation of multimeric biological structures involving formation of weak chemical bonds between surfaces with complementary shapes. For example, most of the components of the phage T4 capsid (head, tail, base plate, and tail fibers) are self-assembled.

self-compatible said of a plant that can be self-fertilized.

self-fertilization the fusion of male and female gametes from the same individual.

self-incompatibility self-sterility (*q.v.*).

selfish DNA (*also called* **junk DNA** or **parasitic DNA**) **1.** functionless segments of DNA that are replicated along with the rest of the chromosomal regions that serve vital functions. Examples would be pseudogenes (*q.v.*) and tandemly repeated and dispersed repetitive DNA segments that appear to serve no function, yet accumulate by unequal crossing over (*q.v.*). **2.** the term is also used to refer to a parasitic DNA that has the ability to engineer its host genetically so that the host cell is better able to survive in nature. Examples would be R (resistance) plasmids and Ti plasmids (*both of which see*). *See* Appendix C, 1980, Doolittle and Sapienza, Orgel and Crick; 1997, Yoder, Walsh, and Bestor; C value paradox, DNA methylation, repetitious DNA.

selfish operon a model to explain the origin in bacteria of clusters of genes that have similar functions and are conditionally expressed. The idea is that genes whose products are used only during conditions that happen infrequently are continually mutating to inactive alleles. When conditions that require the missing gene products reoccur, the bacteria die off unless horizontal transfer of active alleles can take place. But only small segments of DNA are transferred by episomes. Therefore, when two or more genes are needed simultaneously, selection will favor genes that are near one another. Also, cotranscribed genes are more likely to function adaptively in a new host, since the host needs to be able to recognize only one promoter. Thus the transmission of DNA by horizontal mobile elements (*q.v.*) tends to select the stepwise formation of gene clusters that function as operons. *See* Appendix C, 1996, Lawrence and Roth; sympatric speciation.

self-pollination the transfer of pollen to the stigmas of the same plant.

self-splicing rRNA *See* ribozyme.

self-sterility the inability of some hermaphrodites to form viable offspring by self-fertilization.

self-sterility genes genes that prevent the deleterious effects of inbreeding in monoecious plants by controlling the rate of growth of the pollen tube down the style. Self-incompatibility is controlled by a highly polymorphic *S* locus. Growth of a pollen tube in the style is arrested when the *S* allele carried by the pollen matches one of the two *S* alleles carried by the pistil. *S* genes have been cloned and shown to encode glycoproteins with RNase activity. In self-pollinations, the RNase is internalized by receptors on the pollen tube surface. Once inside the pollen tube, the enzyme degrades RNAs essential to its further growth.

SEM scanning electron microscope. *See* electron microscope.

semelparity reproduction that occurs only once in the life of an individual (e.g., annual plants, Pacific salmon). *Compare with* iteroparity.

semen a biochemically complex nutrient fluid containing spermatozoa which is transferred to the female during copulation.

semiconservative replication the method of replication of DNA in which the molecule divides longitudinally, each half being conserved and acting as a template for the formation of a new strand. *See* Appendix C, 1953, Watson and Crick; 1957, Taylor *et al.*; 1958, Meselson and Stahl; 1963, Cairns; 1964, Luck and Reich. *Compare with* conservative replication.

semidiscontinuous replication a mode of DNA replication in which one new strand is synthesized continuously, while the other is synthesized discontinuously as Okazaki fragments. *See* replication of DNA.

semidominance the production of an intermediate phenotype in individuals heterozygous for the gene concerned; also known as *partial dominance. See* incomplete dominance.

semidwarf a term used to distinguish mutant strains of wheat that are of agricultural importance from the extremely short dwarfs of purely genetic interest. Semidwarfs grow from half to two-thirds the height of and have greater yields than those of standard varieties.

semigeographic speciation *See* parapatric speciation.

semilethal mutation a mutation causing death of more than 50%, but not of all individuals, of mutant genotype.

seminiferous tubule dysgenesis Klinefelter syndrome (*q.v.*).

semiochemistry the study of the chemical signals that mediate interactions between members of different species. *See* pheromone.

semipermeable membrane any membrane that permits passage of molecules selectively.

semispecies incipient species.

semisterility a situation in which half or more of all zygotes are inviable (as in *Oenothera* crosses that maintain only heterozygotes). *See* balanced lethal system.

semisynthetic antibiotic a natural antibiotic that has been chemically modified in the laboratory to enhance its stability.

Sendai virus a virus, first isolated in Japan, that causes an important and widespread infection of laboratory mice; it belongs to the Paramyxoviridae. The virus is widely used in cell fusion studies. The viruses so modify the surfaces of infected cells that they tend to fuse. Even UV-killed viruses adsorb on host cells and promote their fusion. *See* Appendix C, 1965, Harris and Watkins; enveloped viruses.

senescence the process of aging (*q.v.*).

sense codon any of the 61 triplet codons in mRNA that specify an amino acid.

sense strand *See* strand terminologies.

sensitive developmental period a period during development when there is an enhanced chance that genetic malfunction will bring development to a standstill. In *Drosophila* these sensitive periods correspond to the onset of embryonic, larval, pupal, or adult development, and it is during such periods that many new systems are differentiated and put to immediate test. Gastrulation (*q.v.*) is a sensitive period for amphibians.

sensitive volume **1.** that portion of an ionization chamber that responds to radiation passing through it. **2.** that biological volume in which an ionization must occur to produce a given effect (such as a mutation). *See* target theory.

sensitizing agent an agent which, when added to a biological system, increases the amount of damage done by a subsequent dose of radiation.

sepal *See* floral organ primordia.

separase a cysteine protease related to the caspases (*q.v.*) that cleaves cohesin complexes and allows sister chromatids to separate and be drawn to opposite poles during the anaphase stage of mitosis (*q.v.*). During meiosis (*q.v.*) separase can only cleave those cohesin complexes that are not protected by Sgo (*q.v.*). *See* cohesin, sister chromatid cohesion.

sepiapterin *See* *Drosophila* eye pigments.

septal fission cell division that characterizes yeasts like *Schizosaccharomyces pombe* (*q.v.*), where mitosis is followed by the formation of a septum that segregates the sibling nuclei into two cells of roughly equal size (*contrast with* bud).

sequence the order in which amino-acid or nucleotide residues are arranged in a protein, DNA, and so forth; to ascertain the sequence of amino-acid or nucleotide residues in (a protein, DNA, etc.). *See* DNA sequencing techniques, protein structure.

sequencer an apparatus for determining the sequence of amino acids or other monomers in a biological polymer.

sequence similarity estimates conclusions made from nucleotide by nucleotide comparison of DNA segments from species that have diverged only recently. For example, when the class 1 segments of the major histocompatibility complexes (MHC) (*q.v.*) of man and chimpanzee were compared, nucleotide substitutions were found less often than insertions and deletions (indels). This was unexpected, since it was always assumed that indels were more difficult to generate than base substitutions. *See* Appendix C, 2003, Anzai *et al.*

sequence tagged site (STS) a short DNA sequence, readily located and amplified by the polymerase chain reaction, that uniquely identifies a physical genomic location. Expressed sequence tags (ESTs) are STSs obtained from cDNAs (*q.v.*). *See* physical map.

ser serine. *See* amino acid.

SER *s*mooth *e*ndoplasmic *r*eticulum.

serial analysis of gene expression (SAGE) a method that allows the quantitative and simultaneous analysis of a large number of transcripts. Short diagnostic sequence tags can be isolated from a tissue (e.g., pancreas), concatenated, and cloned. Sequencing of many (perhaps 1,000) tags reveals a gene expression pattern characteristic of the target tissue's function. *See* proteome, transcriptome.

serial homology the resemblance between different members of a single, linearly arranged series of structures within an organism (the vertebrae are an example).

serial symbiosis theory the theory that eukaryotic cells evolved from bacterial ancestors by a series of symbiotic associations. In its most modern form, it suggests that the mitochondria and microtubule organizing systems of present-day eukaryotes evolved from bacteria and spirochaetes that lived as symbionts in a line of single-celled eukaryotes that were the ancestors of both fungi and animals. A subline of these protoctists subsequently entered an endosymbiosis with cyanobacteria. These evolved into chloroplasts, and the algae and plant lineages developed from this group. *See* Appendix C, 1978, Schwartz and Dayhoff; 1981, Margulis; 1986, Shih *et al.*; cryptomonads, cyanelles, endosymbiont theory, *Pelomyxa*, ribosomes of organelles, *Rickettsia prowazeki*, symbiogenesis.

sericins a group of proteins found in silk (*q.v.*).

sericulture the culture of *Bombyx mori* (*q.v.*) for the purpose of silk production.

serine *See* amino acid.

serine proteases a family of homologous enzymes which require the amino acid serine in their active site and appear to use the same mechanism for catalysis. Members include enzymes involved in digestion (trypsin, chymotrypsin, elastase), blood coagulation (thrombin), clot dissolution (plasmin), complement fixation (Cl protease), pain sensing (kallikrein), and fertilization (acrosomal enzymes).

serology the study of the nature, production, and interactions of antibodies and antigens.

serotonin a cyclic organic compound (5-hydroxytryptamine) that causes certain smooth muscles to contract rapidly and increases capillary permeability. The symptoms of anaphylaxis are due in large part to serotonin released from platelets that accumulate in the capillary bed of the lung. Serotonin also plays an important role in the metabolism of the central nervous system.

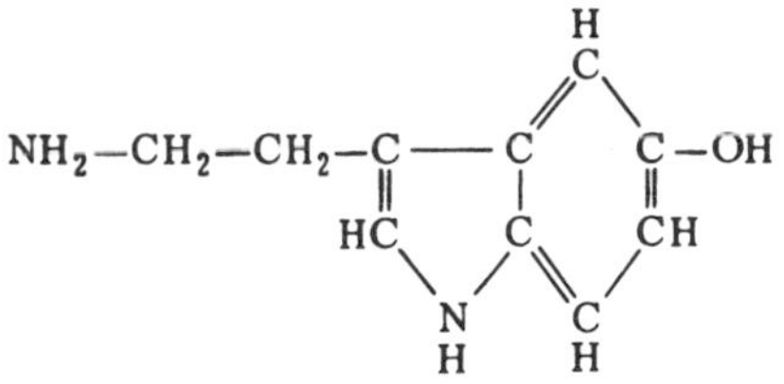

```
                                H
                                C
                              //  \
NH2—CH2—CH2—C ——— C        C—OH
            ||       |        ||
           HC        C       CH
             \     /  \\    /
               N        C
               H        H
```

serotype an antigenic property of a cell (bacterial cell, red blood cell, etc.) identified by serological methods.

serotype transformation *See* antigenic conversion.

serum the fluid remaining after the coagulation of blood.

seta *See* chaeta.

sevenless a gene in *Drosophila melanogaster* that controls the development of R7, the seventh photoreceptor cell within an ommatidium (*q.v.*). In the absence of the wild-type allele of *sevenless*, R7 develops as a cone cell. It appears that the protein encoded by this cell fate gene is a membrane-bound receptor that transmits positional information that controls the type of differention the R7 cell undergoes. *See* Appendix C, 1986, Tomlinson and Ready; developmental control genes, ommatidium, selector genes.

severe acute respiratory syndrome (SARS) *See* coronavirus.

Sewall Wright effect the concept advanced by S. Wright in 1955 that alleles may be fixed or lost, especially from small populations, because of random sampling errors and without regard to their adaptive values.

sex 1. in its broadest sense, sex is any process that recombines in a single organism genes derived from more than a single source. In prokaryotes, sex may involve genetic recombination between two autopoietic cells, or between an autopoietic cell (like *E. coli*) and a nonautopoietic episome (like phage lambda). Eukaryotic sex always involves two autopoietic organisms and leads to the alternating generation of haploid and diploid cells. Meiosis results in the formation of haploid gametes that unite in the process of fertilization to restore the diploid condition. Prokaryotic sex probably evolved more than 3 billion years ago, in the Archean era, while meiotic sex evolved among the protoctists late in the Proterozoic era, about 1 billion years ago. **2.** a classification of organisms or parts of organisms according to the kind of gamete produced; larger, nutrient-rich gametes are female; smaller, nutrient-poor gametes are male. Meiosis in some organisms produces morphologically indistinguishable *isogametes,* in which case the sexes are arbitrarily designated "plus" and "minus." An individual that produces both male and female gametes is *monoecious* (plant) or *hermaphroditic* (animal).

sex cell gamete (*q.v.*).

sex chromatin a condensed mass of chromatin representing an inactivated X chromosome. Each X chromosome in excess of one forms a sex chromatin body in the mammalian nucleus. *See* Barr body, late-replicating X chromosome.

sex chromosomes the homologous chromosomes that are dissimilar in the heterogametic sex. *See* X chromosome, W, Z chromosomes, Y chromosomes.

sex comb a row of bristles arranged like the teeth of a comb on the forelegs of male *Drosophila.* Toward the end of the courtship ritual (*q.v.*) the male drums these tibial bristles on the dorsal surface of the female's abdomen. Sex combs play a crucial role in determining acceptance or rejection by the female during courtship.

sex-conditioned character a phenotype that is conditioned by the sex of the individual. For example, a sex-conditioned, autosomal gene may behave as a dominant in males and as a recessive in females. Furthermore, in the homozygous female the condition may be expressed to a minor degree. Human baldness is an example of a sex-conditioned character. Also called a sex-influenced character.

sex determination the genetic or environmental process by which sexual identity is established in an organism, beginning with the initial commitment by embryonic cells to a particular sexual fate and ending with sex-specific terminal differentiation. The sex of most dioecious species is established by genotypic sex determination (*q.v.*). In mammals sex is specified by the nature of the sperm that fertilizes the egg. Y-bearing sperm produce male zygotes; X-bearing sperm, female zygotes. The Y chromosome gene, *SRY* (*q.v.*) specifies testis formation in the embryonic gonad, and feminization occurs when it is absent. in *Drosophila* and C. *elegans* sex determination is initiated by an assessment of the X:A ratio (*q.v.*). This ratio is communicated to master switch gene(s), which in turn influence downstream genes that effect sex-specific differentiation. In hymenopteran insects sex is determined by haplodiploidy (*q.v.*). Some dioecious species (e.g., many reptiles) employ environmental sex determination (*q.v.*). *See* Appendix C, 1902, McClung; 1925, Bridges; 1958, Page *et al.*; androgen receptor gene, genic balance, germ line sex determination, Klinefelter syndrome, somatic sex determination, Turner syndrome.

sexduction the process whereby a fragment of genetic material from one bacterium is carried with the sex factor *F* to a second bacterium.

sex factor *See* fertility factor.

sex gene pool theory of speciation a theory of speciation (*q.v.*) applicable to all higher sexual organisms and based on evidence for the rapid evolution of sexual traits. This theory partitions the species gene pool (*q.v.*) into (1) those genes that primarily affect mating and reproduction and (2) those that are essential for other aspects of the life of the organism (its development, metabolism, viability, etc.). According to this theory, genes in pools 1 and 2 are selected differentially, with sex-related genes changing preferentially during the early stages of speciation. *See* Appendix C, 2000, Singh and Kulathinal; mate choice, sexual selection.

sex hormone any hormone produced by or influencing the activity of gonads: e.g., gonadotropins, estrogens, androgens. Sex hormones are responsible for development of certain secondary sexual characteristics (e.g., growth of facial hair and muscular development in men).

sex index in *Drosophila,* the ratio of X chromosomes to autosome sets (A) (e.g., a male has a sex index of 0.5, a female is 1.0, a metamale 0.33, and a metafemale 1.5).

sex-influenced character sex-conditioned character (*q.v.*).

sex-limited character a phenotype expressed in only one sex, although it may be due to a sex-linked or autosomal gene. Examples: the recessive, female sterile genes of *Drosophila;* the genes influencing milk and egg production in farm animals.

sex linkage a special case of linkage occurring when a gene that produces a certain phenotypic trait (often unrelated to primary or secondary sexual characters) is located on the X chromosome. The result of this situation is that in certain crosses the phenotypic trait in question may be observed only in individuals of the heterogametic sex, differences between reciprocal crosses (*q.v.*) may also be observed, and the trait will be observed much less frequently among members of the homogametic sex. Genes residing on the Y chromosome will influence only the heterogametic sex. *See* Appendix C, 1820, Nasse; 1910, 1911, Morgan; sex chromosomes.

sex pilus *See* F pilus.

sex ratio the relative proportion of males and females of a specified age distribution in a population.

sex ratio organisms spiroplasmas (*q.v.*) responsible for male-specific lethality in certain *Drosophila* species. SROs are transmitted in the ooplasm, and they kill male embryos. The Y chromosome is not associated with SRO-induced male lethality, and therefore only embryos with two X chromosomes can withstand the infection. Sex ratio spiroplasmas are referred to as sex ratio spirochaetes in the earlier literature. *Spiroplasma poulsonii,* a sex ratio spiroplasma of *Drosophila willistoni,* has been successfully cultivated on a complex, cell-free medium. Its genome contains 2.04×10^6 bp of DNA.

sex reversal the change from functioning as one sex to functioning as the other sex. The change may be a normal occurrence (*see* consecutive sexuality) or it may be experimentally or environmentally induced.

sexual differentiation the process by which sexual determination is phenotypically expressed through the proper development of sexual organs and characteristics.

sexual isolation ethological isolation (*q.v.*).

sexually dimorphic trait a morphological trait seen only in one sex of a species. Examples would be the antlers of elk, the manes of lions, the brilliant tail plumage of peacocks, and the sex combs of fruit flies. All these characters are found in males, not females. *See* mate choice.

sexually antagonistic genes genes that generate phenotypes beneficial to one sex but deleterious to the other. For example, a gene that produces a colorful pattern on the sides of a fish might be useful for males that participate in mating displays to attract females. It would be deleterious to females if it made them more vunerable to predation. The accumulation of such genes on the Y chromosome would ensure that they are expressed only in males, provided mechanisms evolve that prevent recombination between the X and Y chromosomes in males.

sexual reproduction reproduction involving fusion of haploid gamete nuclei, which result from meiosis. *See* parasite theory of sex.

sexual selection selection at mating; the concept first proposed by Charles Darwin that in certain species there occurs a struggle between males for mates and that characteristics enhancing the success of those bearing them would have value and be perpetuated irrespective of their general value in the struggle for existence. In the current literature, sexual selection is generally subdivided into *intrasexual* and *epigamic* selection. Intrasexual selection is what Darwin had in mind and involves the power to conquer other males in battle. In epigamic selection, the female is the active selective agent, making a choice from among a field of genetically variable males of her own species. The results of epigamic selection may be seen in the elaborate sexual displays of certain male birds and insects. Sexual selection for particular types of males may be the first stage in speciation. Rapid speciation driven by sexual selection of this type has been shown to occur in the drosophilids of the Hawaiian Islands and the cichlid fishes of Lake Victoria, Africa. *See* Appendix C, 1871, Darwin; 1981, Lande; 1985, Carson; Hawaiian Drosophilidae, mate choice, prezygotic isolation mechanism, sex gene pool theory of speciation.

sexuparous producing offspring by sexual reproduction. The term is used to describe the sexual phase in a species that alternates sexual reproduction with parthenogenesis. This phenomenon occurs in aphids, for example.

Sgo the symbol for a gene and the protein it encodes that ensures the cohesion of centromeres during the first meiotic division. The protein binds to centromeric cohesins and prevents the separase enzyme from cleaving them. *Sgo* is the acronym for Sugoshin—"guardian spirit" in Japanese. *Sgo* is degraded during telophase I. Therefore the centromeric cohesin molecules can be cleaved by separases during the second meiotic division, and sister chromatids can then separate. *See* cohesin, meiosis, separase.

SGSI a gene in yeast that is similar in base sequence to the human gene that causes Werner syndrome (*q.v.*). In both species, mutations at this locus speed up aging. In mutant yeast cells there is an excessive production of small circles of ribosomal DNA. Eventually the rDNA exceeds the amounts of DNA that comprise the total genome in the normal yeast cell. *See* gene amplification.

shadow casting the use of a vacuum evaporator (*q.v.*) to deposit a coating of a heavy metal on a submicroscopic particle. The coating is cast at an angle so that metal will build up on one side of the particle and cover the specimen support except in the shadow of the particle. The length and shape of the shadow allows calculation of the dimensions of the particle.

SH domains *See Src.*

shearing in molecular biology, the process whereby a DNA sample is broken into pieces of fairly uniform size (e.g., by subjecting them to shearing forces in a Waring blender).

sheep a mammal belonging to the genus *Ovis* and domesticated for its meat, milk, wool, and hide. The domestic sheep, *Ovis aries* (*q.v.*), is distributed worldwide, and common breeds are Merino, Rambouillet, Finn Dorset, Debouillet, Lincoln, Leicester, Cotswold, Romney, Corridale, Columbia, Romeldale, Panama, Montadale, Polwarth, Targee, Hampshire, Shropshire, Southdown, Suffolk, Cheviot, Oxford, Tunis, Ryeland, and Blackface Highland. The most famous member of this species was Dolly (*q.v.*).

shift *See* translocation.

shifting balance theory a theory, proposed by Sewell Wright, that maintains that biological evolution proceeds most rapidly when subpopulations of a species remain isolated for a time sufficient to acquire distinctive adaptations, followed by reestablishment of gene flow, the broadening of genetic diversity, and the enhancement of evolutionary flexibility. *See* Appendix C, Wright.

Shine-Dalgarno (S-D) sequence in *E. coli* the mRNA consensus sequence 5′ AGGAGGU 3′ that is between 6 and 8 bases upstream from the AUG translation initiation codon. The S-D sequence forms complementary base pairs with a consensus sequence found at the 3′ end of the 16S rRNA molecule (*q.v.*) in the 30S subunit of the ribosome. The S-D sequence thus serves as the binding site for bacterial mRNA molecules on ribosomes. *See* Appendix C, 1974, Shine and Dalgarno; translation.

Shope papilloma virus an icosahedral particle about 53 nanometers in diameter containing 8 kbp of DNA. The virus (*q.v.*) was discovered in 1933 by R. E. Shope. The virus produces papillomas in wild, cottontail, North American rabbits.

short-day plant a plant with a flowering period that is accelerated by daily exposures to light of less than 12 hours. *See* photoperiodism.

short-period interspersion a genomic pattern in which moderately repetitive DNA sequences (each about 300 base pairs in length) alternate with nonrepetitive sequences of about 1,000 base pairs.

short tandem repeat (STR) *See* microsatellites, STR analysis.

short term memory consolidation *See* CREBs, spaced training.

shotgun cloning a technique in which the genomic DNA of an organism is randomly broken into smaller fragments by mechanical shearing or partial digestion with restriction endonucleases (*q.v.*) and cloned in appropriate vectors to produce a genomic library (*q.v.*). The clones are then used in shotgun sequencing (*q.v.*).

shotgun sequencing a method for determining the nucleotide sequence of a large stretch of DNA by randomly sequencing a genomic library (*q.v.*) derived from it containing clones with overlapping DNA fragments and using powerful computers to assemble the sequence data into a continuum. The application of this method to determine the nucleotide sequence of a whole genome is called whole-genome sequencing (WGS) or assembly. In this approach randomly-selected inserts, whose physical location in the genome need not have been mapped, are sequenced using a primer from within the cloning vector. A more recent approach is to prepare more than one such library with small and large insert sizes and to sequence the ends of randomly selected inserts. When a sufficiently large number of such fragments has been sequenced, they theoretically contain enough overlaps such that subsequent assembly by computational methods yields a continuous genomic sequence. In practice, the preliminary sequence assembled in this way contains gaps of unknown sequence. These gaps are filled in a variety of ways, such as by further sequencing using primers derived from DNA sequences flanking the gaps, or by identifying small-insert library clones that span the gaps. The final sequence is then analyzed, annotated, and stored in publicly accessible databases. The first free-living organism whose genome was sequenced by WGS methodology was *Haemophilus influenzae*. Subsequent advances in the development of large-insert vectors, sequencing, and computational technology have allowed WGS of a large number of genomes, including those of *Drosophila melanogaster*, *Anopheles gambiae*, *Takifugu rubripes*, *Homo sapiens*, and microbial populations from specific habitats. *See* Appendix C, 1995, Fleischmann *et al.*; 2000, Adams *et al.*; 2001, Venter *et al.*; 2002, Aparicio *et al.*, Holt *et al.*; 2004, Venter *et al.*, Tyson *et al.*; Appendix E, Individual Databases, Species Web Site Addresses; Appendix F; community genome sequencing, DNA vector, genomic annotation, primer DNA, shotgun cloning.

shuttle vector a cloning vector able to replicate in two different organisms—e.g., in *E. coli* and yeast. These DNA molecules can therefore shuttle between the different hosts. Also called *bifunctional vectors*.

sib a shortened form for sibling.

siblings brothers and/or sisters; the offspring of the same parents.

sibling species species that are almost identical morphologically but are reproductively isolated; also called *cryptic species*.

sibmating a brother-sister mating.

sibship all the siblings in a family.

sickle cell anemia a generally fatal form of hemolytic anemia seen in individuals homozygous for an autosomal, codominant gene H^S. The erythrocytes of such individuals contain an abnormal hemoglobin, Hb^S (*q.v.*). These cells undergo a reversible alteration in shape when the oxygen tension of the plasma falls slightly, and they assume elongate, filamentous, and sickle-like forms. Such red cells show a greatly shortened life span, since they tend to clump together and are rapidly destroyed. About 0.2% of babies of African descent born in the United States suffer from sickle-cell anemia. *See* Appendix C, 1949, Neel, Pauling *et al.*; 1957, Ingram; 1978, Kan and Dozy; gene, hemoglobin C.

sickle-cell hemoglobin *See* hemoglobin S.

sickle-cell trait the benign condition shown by individuals carrying both the normal gene, H^A, and the sickle-cell gene, H^S. The erythrocytes of such individuals produce both Hb^A and Hb^S. Such heterozygotes are healthy, and their erythrocytes can be caused to sickle only under conditions where the oxygen concentration is drastically reduced. H^A/H^S individuals suffer far less severely from *Plasmodium falciparum* infections than do H^A/H^A individuals. This malaria plasmodium enters the erythrocyte and lives by engulfing cytoplasm into its food vacuole. Parasites are unable to feed efficiently upon H^A/H^S cells because Hb^S molecules are insoluble compared to Hb^A and cause a great increase in the viscosity of the cytoplasm. About 9% of the African-American population of the United States shows the sickle-cell trait. *See* Appendix C, 1954, Allison; hemoglobin, hemoglobin S, malaria, *Plasmodium* life cycle.

siderophilin plasma transferrin (*q.v.*).

Sievert (Sv) the amount of ionizing radiation that liberates one joule of energy per kilogram of tissue. One Sv equals 100 rem. The millesievert (mSv) is often used in estimates of annual dose rates to human populations. The average U.S. citizen receives approximately 4 mSv per year, with radon (*q.v.*) accounting for 55% and dental and medical x-rays about 15%. Doses of more than five Sieverts are usually fatal to humans. *See* radiation units.

Sigma (Σ) the summation of all quantities following the symbol.

sigma (σ) factor a polypeptide subunit of the RNA polymerase of *E. coli*. This molecule by itself has no catalytic function, but it serves to recognize specific binding sites on DNA molecules for the initiation of RNA transcription. *See* Appendix C, 1969, Burgess *et al.*; Pribnow box, RNA polymerase.

sigma replication *See* rolling circle.

sigma virus a virus that confers CO_2 sensitivity upon *Drosophila melanogaster*.

signal hypothesis the notion that the *N*-terminal amino acid sequence of a secreted polypeptide is critical for attaching the nascent polypeptide to membranes. *See* Appendix C, 1975, Blobel and Dobberstein; receptor-mediated translocation, signal peptide.

signal peptide a sequence of about 20 amino acids at or near the amino terminus of a polypeptide chain that attaches the nascent polypeptide and its ribosome to the endoplasmic reticulum. Polypeptides thus anchored are "flagged" to be processed in the Golgi apparatus prior to release from the cell. *See* Appendix C, 1999, Blobel.

signal recognition particle (SRP) a nucleoprotein particle that functions during receptor-mediated translocation (*q.v.*). The particle contains a 7S RNA molecule and six different proteins. More than 75% of the total 7S RNA in animal cells is present in SRPs. There is a high degree of homology between the 7S RNAs of different animal species. All cells analyzed thus far have SRPs that select proteins destined to be secreted or integrated into the plasma membrane and target them to the endoplasmic reticulum (*q.v.*) in eukaryotic cells or to the plasma membrane in prokaryotic cells. In *E. coli,* the SRP consists of a single protein and a 4.5S RNA molecule. Other cells have SRPs that contain additional proteins and larger RNA molecules, but all share the same evolutionarily conserved core of nucleoprotein. *See* Appendix E, Individual Databases; protein sorting, viroid.

signal sequence synonym for leader sequence (*q.v.*).

signal transduction *See* cellular signal transduction.

significance of results if the probability values are equal to or less than .05 but greater than .01, the results are said to be *significantly different;* probability values ≤ .01 but > .001 are called *highly significant,* and those ≤ .001 are called *very highly significant* by convention. *See* P value.

silent allele an allele that has no detectable product. *See* null allele.

silent mutation a gene mutation that has no consequence at the phenotypic level; i.e., the protein product of the mutant gene functions just as well as that of the wild-type gene. Functionally equivalent amino acids may sometimes substitute for one another (e.g., leucine might be replaced by another nonpolar amino acid such as isoleucine). *See* neutral mutation, samesense mutation, synonymous codons.

silk the cocoon filament spun by the fifth-instar larva of *Bombyx mori.* Each cocoon filament contains two cylinders of fibroin, each surrounded by three layers of sericin. Fibroin is secreted by the cells of the posterior portion of the silk gland. These cells undergo 18 to 19 cycles of endomitotic DNA replication before they begin transcribing fibroin mRNAs. The fibroin gene is present in only one copy per haploid genome. It resides on chromosome 23 and is about 18 kilobases long. The fibroin gene is fundamentally an extensive array of 18 base-pair repeats coding for Gly-Ala-Gly-Ala-Gly-Ser. The similarity of the fibroin gene to a satellite DNA suggests that the gene grew to its current size and continues to evolve by unequal crossing over. The sericin proteins receive their name because of the abundance of serine, which makes up over 30% of the total amino acids. There are at least three sericins, and one of these is encoded by a gene on chromosome 11. All sericins are secreted by the cells from the middle region of the silk gland. Several mutations are known that influence silk production. Some (*Fib* and *Src-2*) represent mutations in the cistrons coding for fibroin and sericin molecules. Others (*Nd, Nd-s,* and *flc*) seem to have defects in the intracellular transport and secretion of fibroin. *See* Appendix C, 1972, Suzuki and Brown; 1979, Perdix-Gillot; beta pleated sheet, endopolyploidy.

silkworm the larva of *Bombyx mori* (*q.v.*).

Silurian the Paleozoic period during which the first life appeared on land. It consisted of psilophytes (*q.v.*) and millipedes. In the oceans, agnathans diversified and placoderms arose. *See* geologic time divisions.

silversword alliance a group of 30 species of plants belonging to three genera endemic to the Hawaiian islands and most closely related to California tarweeds in the "Madia" lineage. They get their common name from the visually impressive silversword, *Argyroxiphium sandwicense,* which lives in dry alpine habitats on the Haleakala and Mauna Kea volcanoes. Silverswords are members of the sunflower family (Asteraceae) and represent one of the world's most spectacular examples of adaptive radiation (*q.v.*) on islands. They populate a variety of habitats, show a diverse range of leaf and flower morphologies, and grow as shrubs, trees, climbing vines, or herbaceous mats. Most species can be crossed and some even produce fully fertile hybrids. The entire set of taxa originated from seeds presumably caught on the feathers of migrating sea birds and transferred from North America to Kauai soon after it rose above the

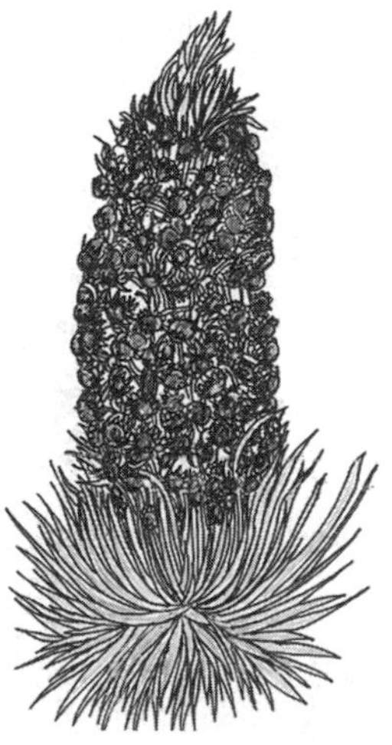

The **Hawaiian silversword** (*Argyroxiphium sandwicense*) is an impressive sight framed by the lava background of its volcanic habitat. When in flower the plant may be 3 meters tall and 75 cm wide. As many as 800 compound flower heads are carried on the inflorescense, and the individual flowers are pink or wine red.

sea. Studies of chloroplast and nuclear gene sequences show that the ancestral silversword arose as an allotetraploid (*q.v.*) from an interspecific cross between two diploid species of tarweed. As new islands emerged from the Hawaiian islands hotspot and were rafted to the northwest, the silversword dispersed southeastward to successively younger islands. In these new habitats they evolved into new species. *See* Appendix A, Plantae, Trachaeophyta, Angiospermae, Asterales; Appendix C, 1991, Baldwin *et al.*; hotspot archipelago.

simian immunodeficiency viruses (SIVs) viruses that are genetically similar to the HIVs but occur in wild populations of monkeys and apes. The SIV most closely related to HIV-1 was found in the chimpanzees of Gabon. Perhaps the progenitor of HIV-1 was a zoonotic virus (*q.v.*) which infected natives that butchered chimpanzees for food. The date of origin of HIV-1 has been estimated to have been about 1931. *See* Appendix C, 2000, Korber *et al.*.

simian sarcoma virus retrovirus first found in the woolly monkey (*q.v.*). This oncogenic virus carries the *v-sis* oncogene, which encodes a transforming protein $p28^{sis}$. Nucleotide sequences related to *v-sis* have been located on human chromosome 22 at a site subsequently called *c-sis*. *See* Appendix C, 1983, Doolittle *et al.*; platelet-derived growth factor, protooncogene, Rous sarcoma virus.

simian virus any of a group of viruses that attack nonhuman primates.

simian virus 40 a DNA virus that readily infects cultured primate cells. SV40 is lytic in monkey cells, but temperate in mouse cells, causing occasional neoplastic transformations. The virus replicates in the nuclei of host cells and may become stably integrated into the host genome. The virus genome consists of a circular DNA molecule containing 5,227 base pairs. The complete nucleotide sequence for this virus has been determined, and it contains both conventional genes, overlapping genes, and split genes. *See* Appendix C, 1971, Dana and Nathans; 1978, Reddy *et al.*; Appendix F; enhancer, oncogenic virus, transformation.

simple-sequence DNA satellite DNA (*q.v.*).

simplex *See* autotetraploidy.

Sinanthropus pekinensis a group of extinct hominids, originally found near Beijing (formerly spelled Peking), China, but no longer considered a distinct species; now included in the species *Homo erectus*.

SINEs *See* repetitious DNA.

single-copy plasmids plasmids maintained in bacterial cells in a ratio of one plasmid for each host chromosome.

single-event curve a dose-response curve in radiobiology which gives a linear relation when the log of survival is plotted against radiation dose. *See* multiple-event curve, target theory.

single-nucleotide polymorphisms (SNPs) small variations in DNA sequence in which at any given position a single nucleotide (*q.v.*) is replaced by one of the other three nucleotides. Those SNPs (pronounced "snips") found in protein-coding regions are designated *cSNPs*. Most of the genetic variation in man is thought to be in the form of SNPs, and the human genome (*q.v.*) contains at least 10 million of them, or one at every 300 base pair interval. Human SNPs tend to be inherited stably as haplotype blocks and can serve as useful landmarks when scanning the genome for specific mutations. The SNPs known to date, together with their exact locations in the human genome, are catalogued in a public database called *dbSNP*. *See* Appendix E, Individual Databases; haplotype, sequence similarity estimates.

single-strand assimilation the process whereby a single strand of DNA displaces its homologous strand in a duplex, forming a D loop (*q.v.*). The reaction is mediated by RecA protein (*q.v.*) and is involved in recombination and heteroduplex formation.

single-stranded DNA binding protein in *E. coli*, a tetrameric protein of 74,000 daltons that binds to the single-stranded DNA generated when a helicase (*q.v.*) opens the double helix. This stabilizes the single-stranded molecule and prevents reannealing or the formation of intrastrand hydrogen bonds. *Compare with* helix-destabilizing proteins. *See* replisome.

single-strand exchange pairing of one strand of duplex DNA with a complementary strand in another DNA molecule, displacing its homolog in the other duplex. *See* 5-bromodeoxyuridine.

sire the male parent in animal breeding. *Compare with* dam.

siRNAs small interfering RNAs (*q.v.*).

sister chromatid cohesion cohesion between the two chromatids formed from each chromosome during the S phase of the cell cycle (*q.v.*). The evolutionarily conserved multiprotein complex cohesin (*q.v.*) is the primary effector of sister chromatid cohesion in all eukaryotes. This cohesion ensures that the two chromatids attach to microtubules with opposite orientations, which is a precondition for each being drawn subsequently toward the opposite pole of the dividing cell. *See* chromatid, microtubules, separase.

sister chromatid exchange the exchange of DNA sequences between sister chromatids due to breakage at apparently homologous sites followed, after switching partners, by reunion of the broken ends. *See* crossing over.

sister chromatids identical nucleoprotein molecules joined by a centromere. *See* chromatid.

sister group a species or higher monophyletic taxon that is hypothesized to be the closest genealogical relative of a given taxon. Sister taxa are derived from an ancestral species not shared by any other taxon. Graphically, sister groups appear on a cladogram (*q.v.*) as lineages that arise as branches from a single divergence node. *Compare with* outgroup.

site the position occupied by a mutation within a cistron.

site-specified mutagenesis a technique that introduces nucleotide alterations of known composition and location into a gene under study. *See* oligonucleotide directed mutagenesis.

site-specific recombinase an enzyme which brings together two short stretches of similar DNA sequences and catalyzes a reciprocal exchange between them, called site-specific recombination (*q.v.*). Sequence comparisons have identified two major families of site-specific recombinases, the *integrase* family and the *resolvase* family, which differ from one another in protein sequence and structure, and in the mechanism of recombination. The integrase family is named after the lambda bacteriophage integrase (*q.v.*), and members of this family are found in bacteriophage, bacteria, and yeast. Integrases from the human immunodeficiency virus (HIV) and avian sarcoma virus (ASV) have been widely studied. The resolvase family is named after the recombinase encoded by the Tn3 and γσ transposons, which is involved in resolution of a cointegrate structure (*q.v.*). Members of this family are found in prokaryotes. *See* FLP/FRT recombination.

site-specific recombination a type of recombination catalyzed by a site-specific recombinase (*q.v.*), in which there is exchange of genetic material between two short, defined DNA segments containing only a small region of homology. This process involves breakage and reformation of specific phosphodiester bonds within the DNA targets, and does not require DNA synthesis or energy from a nucleotide cofactor. Recombination between inverted repeats results in inversion of a DNA segment, between repeat sequences on two DNA molecules in their cointegration, and between repeats on one DNA molecule in the excision of a DNA segment. The integration of lambda phage into the bacterial genome and its excision, and FLP/FRT recombination (*q.v.*), are examples of processes involving site-specific recombination. *Compare with* homologous recombination.

skeletal DNA hypothesis a concept developed to explain the observation that over 90% of the genomes of most eukaryotes are made up of noncoding DNAs. According to Cavalier-Smith, this "secondary" DNA functions as a skeleton that increases the volume of the nucleus and so gives the coding genes more space in which to operate. The skeletal DNA determines the size of the nucleus, and this in turn controls cell volume, which is optimized by selection. *See* Appendix C, 1999, Beaton and Cavalier-Smith; cryptomonads, C value paradox, junk DNA, selfish DNA.

SLC 24A5 a gene at 15q21.1 in humans that encodes a protein which functions as a cation exchanger in melanosomes. The gene exists in two allelic forms varying at one codon. The ancestral form (*A*) is present in most African populations, while the derived form (*a*) is present in most Europeans. *AA* individuals are dark skinned, *aa* are light skinned, and *Aa* are intermediate. This polymorphism accounts for most of the skin pigmentation differences between human races. *See* Appendix C, 2005, Lamason *et al.*; melanin, melanocytes.

sleeping sickness the common name for African human trypanosomiasis, a disease which is caused by infection with the parasite *Trypanosoma brucei* and which affects populations in sub-Saharan Africa, as well as in South America and Asia. Human infection results from the bite of a tsetse fly infected with the parasite. The disease is characterized by fever in the early stages, followed by heart, kidney, and neurological impairment, and eventually, death. *Also called* African sleeping sickness. *See* Appendix C, 2005, Berriman *et al.*; Glossina, *Trypanosoma*.

sliding filament model a theoretical mechanism that explains muscle contraction by the making and breaking of cross bridges between adjacent thick (myosin) and thin (actin) filaments. A similar model is postulated to explain the lengthening of the microtubules of the spindle apparatus, whereby separation of chromosomes or chromatics is affected during cell division.

slime molds *See* Acrasiomycota, Myxomycota.

slippage strand mispairing *See* microsatellites.

slow component in a reassociation reaction, the last component to reassociate, usually consisting of nonrepetitive (unique) DNA sequences.

slow stop mutants temperature-sensitive *dna* mutants of *E. coli* that complete the current round of DNA replication when placed at the restrictive temperature, but cannot initiate another round of replication.

Slp sex-limited protein; a serum protein that is normally found only in male mice and encoded by a gene in the major histocompatibility complex (H-2).

small angle x-ray diffraction the technique used in the analysis of widely spaced repetitions, such as the groups of atoms that form monomeric subunits of a polymer. *See* large angle x-ray diffraction, x-ray crystallography.

small cell lung carcinoma *See* p53.

small cytoplasmic RNAs the cytoplasmic counterparts of small nuclear RNAs (*q.v.*); found in small ribonucleoprotein particles (*q.v.*) in their native state.

***Small eye* (*Sey*)** a gene in mice and rats that controls eye development. Animals heterozygous for a mutant allele have underdeveloped eyes. Homozygotes lack eyes altogether. *Sey*, *Pax-6*, and *ey* are homologous genes. *See* *Aniridia, eyeless, paired.*

small interfering RNAs (siRNAs) noncoding RNAs, approximately 22 nucleotides long, which silence gene expression by binding with perfect homology to complementary sequences in target messenger RNAs in the cell and mediating their destruction. siRNAs have also been shown to silence genes at the transcriptional level. siRNAs are derived from endogenous or exogenous, double-stranded RNA (*q.v.*) precursors that fold into hairpin structures and are processed by the enzyme Dicer (*q.v.*), such that both strands of the dsRNA give rise to different siRNAs. siRNAs have been used as an experimental tool to inhibit gene expression (see RNA interference (RNAi) (*q.v.*)). *Compare with* small temporal RNAs (stRNAs).

small nuclear RNAs a family of small RNA molecules that bind specifically with a small number of proteins to form small nuclear ribonucleoprotein particles. These snRNPs (pronounced "snurps") play a role in the posttranscriptional modification of RNA molecules. *See* Appendix C, 1979, Lerner and Steitz; posttranscriptional processing, snurposomes, spliceosome, transcriptosomes, Usn RNAs.

small nucleolar RNAs molecules (abbreviated *snoRNAs*) that are characterized by "boxes" that contain four to six nucleotides. These conserved sequence elements are distant in the primary sequence, but they are brought into proximity in the folded RNAs as a result of base pairing of complementary sequences that flank the boxes. The loop that results is called a *stem-box structure*, and it functions to target the molecule to the nucleolus (*q.v.*). There are at least 150 different snoRNAs, and these function in the cleavage of pre rRNA molecules and the subsequent modifications of the products. For example, there is a large group of snoRNAs that play a role in the methylation of ribose-2-hydroxyl groups at conserved positions in rRNAs. *See* Miller trees, nucleolin, preribosomal RNA, ribose, ribosome, 16S RNA, transcriptosomes.

smallpox vaccine the antigenic preparation used to elicit an active immunity to smallpox. Commercial vaccines contain freeze-dried Vaccinia virus. *See* immunity, pox viruses, vaccine, *Variola* virus.

small temporal RNAs (stRNAs) in *Caenorhabditis elegans* and presumably in other organisms, noncoding RNAs, approximately 22 nucleotides long, which regulate the timing of developmental events by binding to partially complementary sequences in the 3′ untranslated regions of protein-coding messenger RNAs and inhibiting their translation. MicroRNAs (*q.v.*) encoded by the *lin-4* and *let-7* genes of C. *elegans* are examples of stRNAs. These genes produce approximately 70 nucleotides-long, double-stranded, stem loop precursor RNAs, which are processed by the enzyme Dicer (*q.v.*), such that only one strand of the dsRNA gives rise to the mature stRNA. *Compare with* small interfering RNAs (siRNAs).

Smittia a genus of chironomid possessing giant polytene chromosomes (*q.v.*) and hence subjected to cytological study. *See* Chironomus.

smooth endoplasmic reticulum (SER) endoplasmic reticulum that lacks ribosomes; sometimes called the *agranular reticulum*. SER is common in cells that are actively synthesizing compounds other than proteins, that is, carbohydrates, lipids, or steroids. This intricate network of tubes and sacs is seen in cells such as oil gland cells in the epidermis, cells from glands that synthesize steroid hormones, and cells that line the small intestine. *See* endoplasmic reticulum.

smut 1. a fungus disease of cereals characterized by black masses of spores. 2. any basidiomycete fungus of the order Ustilagnales that causes smut disease.

snapdragon *See Antirrhinum majus.*

sneak synthesis *See* background constitutive synthesis.

snoRNAs small nucleolar RNAs (*q.v.*).

SNPs *See* single-nucleotide polymorphisms.

snRNA a small nuclear RNA (*q.v.*).

snRNPs small nuclear ribonucleoproteins. *See* small nuclear RNAs.

S1 nuclease an endonuclease from *Aspergillus oryzae* that selectively degrades single-stranded DNA to yield 5′ phosphoryl mono- or oligonucleotides.

snurposomes organelles found in amphibian germinal vesicles, equivalent to the speckles or clusters of interchromatin granules of other cell types. Snurposomes are composed of densely packed particles, each 20–30 nm in diameter. These contain splicing snRNPs and other factors involved in splicing pre-mRNAs. Snurposomes are 1–4 μm in diameter, and they are observed in the matrix of Cajal bodies (*q.v.*) or attached to their surfaces. Snurposomes also float free in the nucleoplasm. They were given their name because they stain strongly with antibodies that localize splicing snRNPs (snurps). *See* posttranscriptional processing, transcriptions.

snurps *See* small nuclear RNAs.

social Darwinism a theory originated by the British philosopher Herbert Spencer, proposing that most of the "progress" in human societies has been brought about by competition (economic, military) and the "survival of the fittest." Spencer believed that human progress required a struggle and competition, not only between individuals but also between social classes, nations, states, and races, and he ranked human races and cultures according to their assumed levels of evolutionary attainment.

social evolution a continued increase in the complexity of human society resulting from the selection, transmission, and utilization of the useful information gained in each generation.

sociobiology the study of animal behavior from a genetic perspective.

SOD *See* superoxide dismutase.

sodium an element universally found in small amounts in tissues. Atomic number 11; atomic weight 22.9898; valence 1^+; most abundant isotope ^{23}Na; radioisotopes: ^{24}Na, half-life 15 hours, radiations emitted—beta particles and gamma rays; ^{22}Na, half-life 2.6 years, radiations emitted—positrons and gamma rays.

sodium dodecyl sulfate $CH_3(CH_2)_{11}SO_3Na$ (sodium lauryl sulfate), an anionic detergent used in the SDS-PAGE method of protein fractionation. *See* electrophoresis.

Sogin's first symbiont the hypothesized ancestor of eukaryotes. As outlined in the diagram (page 415), this organism arose by fusion of two prokaryotes with complementing metabolic capabilities. The first had a fragmented RNA-based genome that encoded the RNAs functioning in translation and the elaboration of cytoskeletal proteins. Once equipped with a cytoskeleton, the prokaryote could engulf other microorganisms. The second organism was a primitive archeon with a relatively unfragmented DNA genome that encoded metabolically active proteins. The first organism engulfed the second, which eventually functioned as the nucleus in the chimera. *See* Appendix C, 1991, Sogin; serial symbiosis theory.

Solanum tuberosum *See* potato.

solenoid structure a supercoiled DNA produced during the condensation of chromosomes in the nuclei of eukaryotes. This chromosome shortening is achieved by folding the linear array of nucleosomes into a helical fiber with six nucleosomes per turn. The molecules of histone 1 aggregate into a helical polymer along the center of the solenoid and stabilize it. A top view of the solenoid is shown. *See* Appendix C, 1976, Finch and Klug; chromatin fibers, histones.

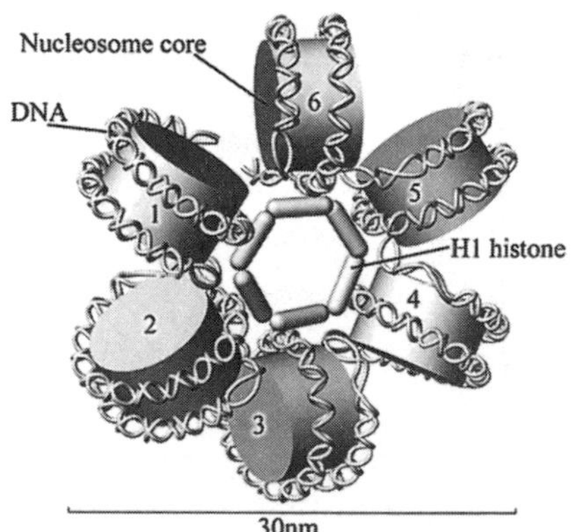

solution hybridization liquid hybridization (*q.v.*).

soma the somatic cells of a multicellular organism in contrast to the germ cells.

somatic cell any cell of the eukaryotic body other than those destined to become sex cells. In diploid organisms, most somatic cells contain the 2N num-

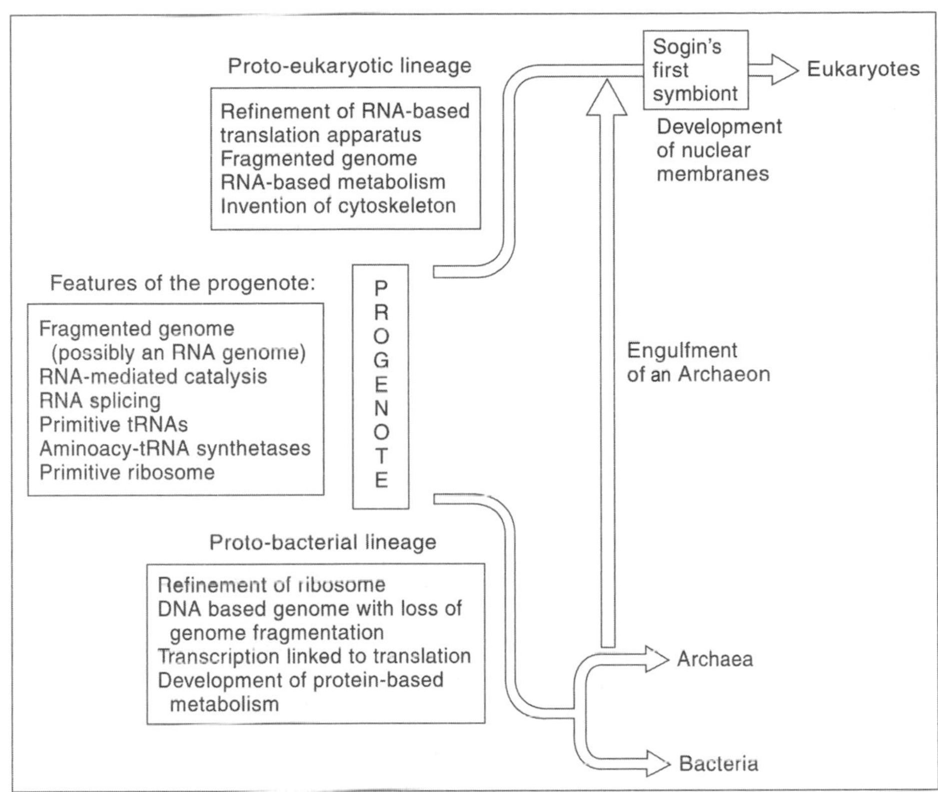

Sogin's first symbiont

ber of chromosomes; in tetraploid organisms, somatic cells contain the 4N number, etc.

somatic cell genetic engineering correction of genetic defects in somatic cells by genetic engineering: e.g., insertion of genes for insulin production into defective pancreatic cells. Such correction would not be hereditary.

somatic cell genetics the genetic study of asexually reproducing body cells, utilizing cell fusion techniques, somatic assortment, and somatic crossing over. *See* Appendix C, 1964, Littlefield; 1965, Harris and Watkins; 1967, Weiss and Green; 1969, Boon and Ruddle; 1985, Smithies *et al.*

somatic cell hybrid a hybrid cell resulting from cell fusion (*q.v.*).

somatic cell nuclear transfer therapeutic cloning (*q.v.*).

somatic crossing over the exchange of DNA between non-sister chromatids in a somatic cell. *Also called* mitotic crossing over. *See* mitotic recombination, site-specific recombination.

somatic doubling the doubling of the diploid chromosome set. Such doubling may be induced experimentally by applying the alkaloid colchicine in a lanolin paste to somatic tissues that are undergoing mitosis.

somatic mutation a mutation occurring in any cell that is not destined to become a germ cell. If the mutant cell continues to divide, the individual will come to contain a patch of tissue of genotype different from the cells of the rest of the body. *Compare with* gametic mutation.

somatic pairing the conjoining of the homologous chromosomes in somatic cells, a phenomenon seen in dipterans. The fact that the polytene chromosomes of *Drosophila* undergo somatic pairing makes possible the identification of chromosomal rearrangements, the mapping of deficiencies, and, as a result, the cytological localization of genes. *See* Diptera, transvection.

somatic recombination genetic recombination that does not involve germ cells but rather somatic cells, usually of a specific type and at a particular developmental stage. For example, somatic recombination occurs in developing B lymphocytes (*q.v.*). V (D) J recombination (*q.v.*) results in the joining of any one of many variable *Ig* gene segments to one of a few constant segments. The arrangement that results is different in the cells that produce the antibody from all other somatic cells and germ cells. *Compare with* somatic crossing over.

somatic sex determination the genetic and developmental process that specifies sexual identity and sex-specific development of the somatic cells of an organism. *Compare with* germ line sex determination. *See* sex determination.

somatoclonal variation the appearance of new traits in plants that regenerate from a callus in tissue culture. Some of the variations represent single nucleotide changes; others involve chromosomal translocations, losses, or duplications. Much of the variation occurs during tissue culture, rather than as a result of unmasking the variation present in the parent plant. *See* gametoclonal variation.

somatocrinin growth hormone releasing hormone. *See* human growth hormone.

somatomammotropin *See* human growth hormone.

somatostatin a polypeptide hormone that stimulates the release of growth hormone by the pituitary and of insulin and glucagon by the pancreas. The gene for this 14 amino acid peptide was chemically synthesized, spliced into a plasmid, and cloned in *E. coli*. The transformed bacteria secreted somatostatin, and this led to the first commercial production of a synthetic human protein. *See* Appendix C, 1977, Itakura *et al.*; human growth hormone.

somatotropin *See* human growth hormone.

sonicate subject (a biological sample) to ultrasonic vibration so as to fragment the cells, macromolecules, and membranes. A biological sample that has been subjected to such treatment.

Sonic hedgehog (Shh) the vertebrate homolog of the *Drosophila* gene *hedgehog* (*q.v.*). In humans *Shh* has been mapped to 7q36. *Shh* encodes a signal protein that controls the patterning of the ventral neural tube, the anterior-posterior limb axis, and the ventral somites.

Sordaria fimicola an ascomycete fungus often used in studies of gene conversion (*q.v.*).

sorting *See* protein sorting.

sorting signals segments several amino acids long in proteins that target them to their final destinations. For example, the *nuclear-targeting signal* is four to eight amino acids in length, and it contains several positively charged residues and usually one or more prolines. The targets can occur at a variety of places in different nuclear proteins. The *peroxisomal-targeting signal* is usually located near the carboxy terminus of the protein, and it consists of three amino acids (serine, lysine, and leucine). There are also targeting signals that cause specific proteins to be retained in the ER or Golgi apparatus or to be targeted to *lysosomes*. In animals like *Caenorhabditis* or *Drosophila*, about 5% of the proteins contain sorting signals that direct them to mitochondria. However, in plants, like *Arabidopsis*, nearly 25% of the nuclear genes direct encoded proteins to either chloroplasts or mitochondria. *See* protein sorting.

SOS boxes the operator sequences in *E. coli* DNA that are recognized by a repressor called the LexA protein. This protein represses several loci involved in DNA repair functions. *See* regulon, SOS response.

SOS response a cellular response to extensive DNA damage in which certain genes, called SOS genes, are sequentially activated in order to repair the damaged DNA. In *E. coli* about 20 such genes have been identified, including *lexA*, whose product normally represses the SOS genes. The remainder include genes such as *uvrA*, *uvrB*, *recA*, *sulA*, and *umuC*. Among the functions assigned to these genes are recombinational repair, nucleotide excision repair, inhibition of cell division, and error-prone repair. Normally, SOS genes are repressed by the LexA protein, which binds to operator sequences, called SOS boxes (*q.v.*), upstream of each of these genes. When DNA is damaged, single-stranded regions become exposed, and these interact with the RecA protein (*q.v.*) to form a complex (RecA*), which acquires protease activity and facilitates the cleavage of the LexA repressor (*q.v.*). The cleaved LexA protein is unable to bind DNA, thus allowing the SOS genes to be de-repressed. When the DNA has been repaired, RecA becomes inactivated, LexA is no longer cleaved and accumulates in the cell, and the SOS genes are shut down. In addition to accurate, error-free repair, the SOS response also induces DNA repair that leads to mutagenesis, i.e., error-prone or mutagenic repair, in which the DNA template is read with reduced fidelity. Thus, in the presence of extensive DNA damage the cells survive, albeit at the cost of introducing some errors in their DNA. The acronym, SOS, is derived from "Save Our Souls," the Morse code signal given by ships in danger and conveys that this is an emergency response in cells that are in danger of dying. *See* Appendix C, 1967, Witkins.

South African clawed frog *See Xenopus.*

Southern blotting a technique, developed by E. M. Southern, for transferring electrophoretically resolved DNA segments from an agarose gel to a nitrocellulose filter paper sheet via capillary action. Subsequently, the DNA segment of interest is probed with a radioactive, complementary nucleic acid, and its position is determined by autoradiography. A similar technique, referred to as *northern blotting*, is used to identify RNAs. For example, an electropherogram containing a multitude of different mRNAs could be probed with a radioactive cloned gene. In cases where proteins have been separated electrophoretically, a specific protein on an electropherogram can be identified by the *western blotting* procedure. In this case, the probe is a radioactively labeled antibody raised against the protein in question. *See* Appendix C, 1975, Southern; 1977, Alwine *et al.*; probe.

soybean *See Glycine max.*

spaced training referring to experiments on memory during which repeated training sessions are given with short rest intervals between the sessions. The term *massed training* refers to repeated training sessions with no such rest intervals. Comparisons of the results from both types of experiments have shown that the memory generated immediately after training is short lived and disruptable. During a rest period such short-term memory (STM) is consolidated into a longer-lasting, more stable, long-term memory (LTM). For the consolidation of STM into LTM, the syntheses of the protein products of specific genes are required. *See* CREBs.

spacer DNA untranscribed segments of eukaryotic and some viral genomes flanking functional genetic regions (cistrons). Spacer segments usually contain repetitive DNA. The function of spacer DNA is not presently known, but it may be important for synapsis. *See* transcribed spacer.

special creation a nonscientific philosophy asserting that each species has originated through a separate act of divine creation by processes that are not now in operation in the natural world.

specialized 1. an organism having a narrow range of tolerance for one or more ecological conditions. 2. a species having a relatively low potential for further evolutionary change; the opposite of generalized.

specialized transduction *See* transduction.

speciation 1. the splitting of an ancestral species into daughter species that coexist in time; horizontal evolution or speciation; cladogenesis. 2. the gradual transformation of one species into another without an increase in species number at any time within the lineage; vertical evolution or speciation; phyletic evolution or speciation. *See* Appendix C, 1954, Mayr; 1975, King and Wilson; 1985, Carson; 2000, Singh and Kulathinal; alloparapatric speciation, allopatric speciation, cichlid fishes, evolution, founder effect, Hawaiian Drosophilidae, parapatric speciation, peripatric speciation, punctuated equilibrium, selector genes, sex gene pool theory of speciation, sexual selection, silversword alliance, sympatric speciation.

species 1. biological (genetic) species: reproductively isolated systems of breeding populations. 2. paleospecies (successional species): distinctly different appearing assemblages of organisms as a consequence of species transformation (*q.v.*). 3. taxonomic (morphological; phenetic) species: phenotypically distinctive groups of coexisting organisms. 4. microspecies (agamospecies): asexually reproducing organisms (mainly bacteria) sharing a common morphology and physiology (biochemistry). 5. biosystematic species (ecospecies) populations that are isolated by ecological factors rather than ethological isolation (*q.v.*).

species group superspecies (*q.v.*).

species selection a form of group selection (*q.v.*) in which certain species (produced by cladogenesis) continue the cladogenic process and others become extinct.

species transformation the transformation of a species (A) into another (species B) during the passage of time. Species transformation does not increase the number of species, since species A and B do not coexist in time. *See* anagenesis, speciation, vertical evolution.

specific activity the ratio of radioactive to nonradioactive atoms or molecules of the same kind. Sometimes given as the number of atoms of radioisotope per million atoms of stable element. Also expressed in curies per mole.

specific immune suppression an immune response in which the initial exposure to a particular antigen results in the loss of the ability of the organism to respond to subsequent exposures of that antigen, but not to different antigens. *See* immunological tolerance.

specific ionization the number of ion pairs per unit length of path of the ionizing radiation in a given medium (per micron of tissue, for example).

specificity selective reactivity between substances: e.g., between an enzyme and its substrate, between a hormone and its cell-surface receptor, or between an antigen and its corresponding antibody.

specificity factors proteins that temporarily associate with the core component of RNA polymerase and determine to which promoters the enzyme will bind (e.g., the sigma factor, *q.v.*). *See* antispecificity factor.

specimen screen the support for sections to be viewed under the electron microscope consisting of a disc made of copper or gold mesh.

spectrin a protein that is a major component of the plasma membranes of animal cells. It is composed of two different polypeptide chains, alpha and beta, which form heterodimers. Each polypeptide contains tandemly repeated sequences that can fold upon themselves and so give the spectrin filament great flexibility. In the cell membrane, spectrin filaments form a pentagonal network in which their ends attach to junctions made of actin and other proteins. Spectrin has been identified as one of the molecular components of the spectrosome (*q.v.*) and the fusome (*q.v.*) in *Drosophila*. *See* peripheral protein.

spectrophotometer an optical system used in biology to compare the intensity of a beam of light of specified wave length before and after it passes through a light-absorbing medium. *See* microspectrophotometer.

spectrosome a prominent spectrin-rich, spherical mass found in the cytoplasm of germ line stem cells (*q.v.*) and cytoblasts (*q.v.*) in the *Drosophila* ovary. The spectrosome is rich in cytoskeletal proteins such as actin (*q.v.*), α- and β-spectrin, the adducin-like Hts protein, and ankyrin (*q.v.*). This organelle is though to be a precursor of the fusome (*q.v.*) and to anchor the mitotic spindle during germ line stem cell and cystoblast divisions. *hts* and *α-spectrin* mutations eliminate spectrosome and fusome formation and result in aberrant mitotic spindle orientation during germ line stem cell and cystoblast/cystocyte divisions. *See* adducin, ankyrin, cystocyte divisions, *hu-li tai shao (hts)*, spectrin.

speech-language disorder 1 an extremely rare condition, showing autosomal dominant inheritance, that affects a British family (the KE family) and causes a severe language disorder. The afflicted individuals are unable to learn certain rules of grammar and tense, and they cannot enunciate certain verbal patterns. The gene involved is *FOXP2*, located at 7q31, and it spans approximately 600 kb of DNA. Two functional copies of *FOXP2* are required for the acquisition of normal spoken language. The gene contains over 20 exons, and it spans approximately 600 kb of DNA. There are at least four mRNAs transcribed from *FOXP2* by alternative splicing (*q.v.*), and these transcripts are plentiful in the fetal brain. The most common splice form encodes a protein 715 amino acids long. A segment of this protein contains a DNA-binding site, suggesting that it functions as a regulator of transcription. Homologs of *FOXP2* have been identified in the chimpanzee, the gorilla, and the orangutan. The *FOXP2* proteins of the apes are all identical to each other, but the human protein has different amino acids at two sites. Perhaps these changes gave new properties to the protein which influenced neural systems in ways that eventually led to the acquisition of speech. *See* Appendix C, 2001, Lai *et al.*

spelt *Triticum spelta* (N = 21), the oldest of the cultivated hexaploid wheats, grown since the latter days of Roman Empire. *See* wheat.

Spemann-Mangold organizer named after Hans Spemann and Hilde Mangold who published the details of their tissue implantation experiments in 1924. They showed that tissue from the dorsal blastopore lip of the amphibian gastrula can induce a secondary body axis in another embryo. The secondary brain and spinal cord did not arise from the transplanted cells, but from the presumptive ventral epidermis of the host. They concluded that the implanted material contained diffusible "organizing factors" that determined the future differentiation of the adjacent host tissues. The Spemann-Mangold organizer is now known to play a vital role during development in all members of the Chordata (*q.v.*). *See* blastoporal lip, chordamesoderm, gastrulation, *goosecoid*, *Triton*, xenograft.

S period *See* cell cycle.

sperm a single male gamete or spermatozoon. Sperm can also refer to multiple male gametes or spermatozoa.

spermateleosis spermiogenesis (*q.v.*).

spermatheca the organ in a female or a hermaphrodite which receives and stores the spermatozoa donated by the mate.

spermatid one of four haploid cells formed during meiosis in the male. Spermatids without further division transform into spermatozoa, a process known as spermiogenesis (*q.v.*).

spermatocyte a diploid cell that undergoes meiosis and forms four spermatids. A primary spermatocyte undergoes the first of the two meiotic divisions and gives rise to two secondary spermatocytes. Each of these divides to produce two haploid spermatids.

spermatogenesis the developmental process that results in the formation of mature sperm in an organism. Spermatogenesis involves a series of events, including mitotic divisions in spermatogonia (*q.v.*), meiotic divisions in spermatocytes, and morphological changes in spermatids that lead to the formation of mature spermatozoa. Many of these events require interactions between the germ line (*q.v.*) and the surrounding soma (*q.v.*). *See* spermatocyte, spermatid, spermatozoon, spermiogenesis.

spermatogonia mitotically active cells in the gonads of male animals that are the progenitors of primary spermatocytes.

Spermatophyta in older taxonomies the division of the plant kingdom containing the contemporary dominant flora. Spermatophytes are characterized by the production of pollen tubes and seeds. All angiosperms and gymnosperms are included in the Spermatophyta. *See* Appendix A.

spermatozoon a single male gamete or sperm (plural, *spermatozoa*). When the word is used as an advective the spelling is *spermatozoan*.

sperm bank a depository where samples of human semen are stored in liquid nitrogen at −196°C; when needed, perhaps years later, a sample can be thawed and used in artificial insemination.

spermiogenesis the series of morphological and chemical changes that transform the spermatids resulting from the meiotic divisions of a spermatocyte into functional spermatozoa. In most animals, excess cytoplasm is expelled from the spermatid, and the acrosome (*q.v.*) and the flagellum (*q.v.*) are formed. An interesting exception is found in the Nematoda which have amoeboid sperm.

sperm polymorphism the production of normal and aberrant sperm during spermatogenesis. The normal sperm are called *eupyrene*, those containing subnormal numbers of chromosomes are *oligopyrene*, and those lacking a nucleus altogether are *apyrene*. Apyrene and oligopyrene sperm are formed by certain snails (*Viviparus malleatus* is an example) and moths (*Bombyx mori*), but the function of these abnormal gametes is unknown.

sperm sharing a phenomenon occurring in Brazilian freshwater snails of the genus *Biouphalaria* in which a simultaneous hermaphrodite (acting mechanically as a male) transfers sperm to its partner that was collected when it functioned as a female in a previous mating. Sperm sharing may occur both within and between species. The term *sperm commerce* refers to the transfer of a sperm donor's own sperm along with exogenous sperm from a previous mating. *See* hermaphrodite.

Sphaerocarpus donellii a species of liverwort used in the classic mutagenesis experiments which showed that the wave length of UV specifically absorbed by DNA, not protein, was the most effective in producing mutations. *See* Appendix A, Plantae, Bryophyta, Hapaticae; Appendix C, 1939, Knapp *et al.*; ultraviolet radiation.

S phase *See* cell cycle.

sphenophytes horse tails, a group of plants that originated during the Devonian and is represented today by the rush *Equisetum*. In Carboniferous forests, sphenophytes grew to heights of 15 meters.

spheroplast a protoplast (*q.v.*) to which some cell wall remnants are attached. For example, a rod-shaped bacterium treated with lysozyme becomes spherical because the enzyme removes peptidoglycan components that give rigidity to the cell wall.

sphingomyelin a molecule belonging to a family of compounds that occur in the myelin sheath of nerves. All sphingomyelins contain sphingosine, phosphorylcholine, and a fatty acid.

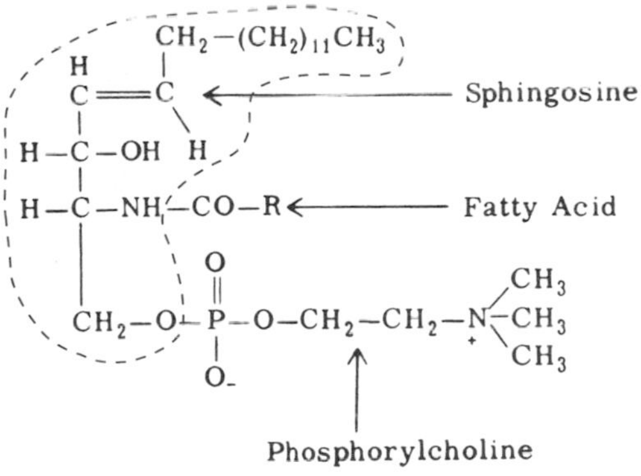

$CH_2-(CH_2)_{11}CH_3$
H–C=C–H ← Sphingosine
$H-C-OH$
$H-C-NH-CO-R$ ← Fatty Acid
$CH_2-O-P(=O)(O^-)-O-CH_2-CH_2-N^+(CH_3)_3$
↑ Phosphorylcholine

sphingosine an amino dialcohol component of the sphingolipids, which are abundant in the brain.

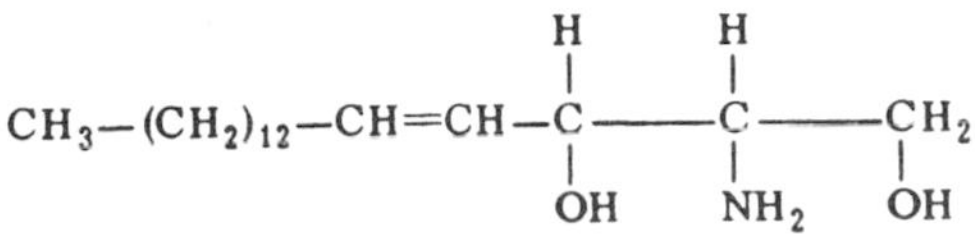

$$CH_3-(CH_2)_{12}-CH{=}CH-\underset{OH}{\overset{H}{C}}-\underset{NH_2}{\overset{H}{C}}-\underset{OH}{CH_2}$$

spike an inflorescence, such as the catkin of the pussy willow, in which the flowers arise directly from a central axis, the rachis. *See* raceme.

spikelet in grasses, a secondary spike bearing few flowers.

spinal bulbar muscular atrophy one of several neurological diseases due to an unstable *trinucleotide repeat* (*q.v.*). The X-linked gene involved encodes an androgen receptor, and the trinucleotide repeat is located in the coding portion of the gene.

spindle a collection of microtubules responsible for the movement of eukaryotic chromosomes subsequent to their replication. *See* centromere, centrosome, Fungi, meiosis, microtubule organizing centers (MTOCs), mitosis, spindle pole body.

spindle attachment region (*also* spindle fiber attachment, spindle fiber locus) centromere (*q.v.*).

spindle checkpoint a checkpoint (*q.v.*) that protects the integrity of the genome by initiating a delay in the cell cycle if all the chromosomes are not properly attached to the spindle. *See MAD* mutations.

spindle fiber one of the microtubular filaments of a spindle.

spindle poison any compound that binds to certain molecular components of spindles and causes them to malfunction. A subset of the spindle poisons, notably paclitaxel, vinblastine, and vincristine (*all of which see*), have turned out to be potent anticancer drugs. Others like colchicine and podophyllin (*both of which see*) have been clinical failures. *See* tubulin.

spindle pole body in yeast, the organelle that organizes nuclear and cytoplasmic microtubules into a mitotic spindle. The spindle pole body in fungi is the functional equivalent of the centrosome (*q.v.*) in animals. During mitosis in yeast, the nuclear envelope does not break down, and the spindle pole body remains embedded in the nucleus. *See* tubulin.

spineless-aristapedia one of the homeotic mutations (*q.v.*) of *Drosophila* located at 3-58.5. The distal portions of the antennae are transformed into leglike structures with claws. The homeotic mutations illustration on page 210 allows a comparison of a normal pair of antennae, each bearing a bristle-like arista (Fig. A) with a mutant antenna which lacks an arista and has distal claws (Fig. B). Mutations of the *ssa* gene demonstrate that legs and antennae are homologous structures and suggest that antennae of arthropods evolved from an anterior pair of legs. *See* metamerism.

spiral cleavage a type of embryonic development seen in invertebrates such as annelids and molluscs. The first and second divisions of the zygote are in vertical planes but at right angles to one another, producing a quartet of blastomeres. The next divisions are horizontal, cutting off successive quartets. However, each quartet is slightly displaced from the one above, giving a spiral appearance to the embryo. The direction of the spiral is genetically determined.

spirochete (*also* spirochaete) bacteria that are nonflagellated, spiral, and move by flexions of the body. *See* Appendix A, Eubacteria, Spirochaetae; *Treponema pallidum*.

spiroplasmas helical, motile bacteria that resemble spirochaetes. Unlike spirochaetes, spiroplasmas lack a cell wall, and they are therefore included in the Aphragmabacteria (*see* Appendix A). Spiroplasmas are responsible for certain plant diseases and cause male-specific lethality among the progeny of female *Drosophila* carrying them. *See* sex ratio organisms.

splice junctions segments containing a few nucleotides that reside at the ends of introns and function in excision and splicing reactions during the processing of transcripts from split genes. The sequence at the 5′ end of any intron transcript is called the *donor junction* and the sequence at the 3′ end the *acceptor junction*. U1 RNA (*q.v.*) contains a segment adjacent to its 5′ cap that exhibits complementarity to the sequences at the donor and acceptor splice junctions of introns. U1 binds to such segments, causing introns to loop into a lariat (*q.v.*) that allows intron excision and exon splicing. *See* Usn RNAs.

spliceosome the intranuclear organelle in which the excision and splicing reactions that remove introns from premessenger RNAs occur. *See* alternative splicing, Cajal body, exon, intron, posttranslational processing, RNA splicing, small nuclear RNAs, splice junctions, Usn RNAs.

splicing 1. RNA splicing: the removal of introns and the joining of exons from eukaryotic primary RNA transcripts to create mature RNA molecules of the cytoplasm. 2. DNA splicing. *See* recombinant DNA research.

splicing homeostasis a phenomenon in which a maturase (*q.v.*) helps to catalyze the excision of an intron from its own primary RNA transcript. In so doing, the maturase destroys its own mRNA and thereby limits its own level of activity.

split genes genes containing coding regions (exons) that are interrupted by noncoding regions (introns). This type of genetic organization is typical of most eukaryotic genes and some animal viral genomes, but introns are not found in prokaryotic organisms. *See* Appendix C, 1977, Roberts and Sharp; adenovirus, R-loop mapping.

sp. n. new species.

SPO 11 a gene which encodes a type 2 topoisomerase (*q.v.*) in *Saccharomyces cerevisiae.* It was isolated as a sporulation mutant, hence the *spo* symbol. No meiotic recombination was observed in the mutant, although normal synaptonemal complexes (*q.v.*) appeared during meiotic prophase. The SPO 11 topoisomerase catalyzes meiosis-specific DNA double-strand breaks. The *Drosophila* mutant *mei-W68* is a homolog of *SPO 11*. *See* Appendix C, 1997, Keeney, Giroux, and Kleckner; double-strand break (DSB) formation, recombination nodules (RNs).

sponge body a membrane-less, cytoplasmic structure with a sponge-like appearance, found in female germ line cells of *Drosophila* and thought to function in the assembly and transport of materials required for RNA localization in the oocyte (*q.v.*). Sponge bodies generally consist of endoplasmic reticulum-like cisternae and vesicles embedded in an electron-dense matrix that is devoid of ribosomes. They contain RNA and protein, and are often surrounded by mitochondria. They are first observed during early oogenesis (*q.v.*) near the nurse cell nuclear membrane, change in morphology as development progresses, migrate through the ring canals (*q.v.*), and dissociate toward later oogenesis into smaller particles that are incorporated into the ooplasm (*q.v.*). Sponge bodies share morphological and functional characteristics with Balbiani bodies and mitochondrial clouds. *See* Balbiani body, cytoplasmic localization, mitochondrial cloud, nurse cells.

spontaneous generation the origin of a living system from nonliving material. *See* Appendix C, 1668, Redi; 1769, Spallanzani; 1864, Pasteur.

spontaneous mutation a naturally occurring mutation.

spontaneous reaction exergonic reaction (*q.v.*).

sporangium a structure housing asexual spores.

spore **1.** sexual spores of plants and fungi are haploid cells produced by meiosis. **2.** asexual spores of fungi are somatic cells that become detached from the parent and can either germinate into new haploid individuals or can act as gametes. **3.** certain bacteria respond to adverse growth conditions by entering a spore stage until more favorable growth conditions return. Such spores are metabolically inert and exhibit a marked resistance to the lethal effects of heat, drying, freezing, deleterious chemicals, and radiation.

spore mother cell a diploid cell that by meiosis gives rise to four haploid spores.

sporogenesis the production of spores.

sporophyte the spore-producing, 2N individual. In the higher plants the sporophyte is the conspicuous plant. In lower plants like mosses, the gametophyte is the dominant and conspicuous generation. *See* alternation of generations.

Sporozoa a class of parasitic protoctists in the phylum Apicomplexa that reproduce sexually with an alternation of generations (*q.v.*). Both haploids and diploids undergo schizogony (*q.v.*) to produce small infective spores. All species of *Plasmodium* belong to the Sporozoa. *See* Appendix A, malaria.

sporozoite the stage in the life cycle of the malaria parasite that infects humans. Lance-shaped sporozoites reside in the salivary gland of the *Anopheles* mosquito and are delivered to the bloodstream of the victim when the mosquito takes a meal. The major surface antigen of the sporozoite is the circumsporozoite (CS) protein. In *Plasmodium knowlesi*, the CS protein contains a 12-amino-acid epitope that is repeated 12 times. When host antibodies bind to the CS protein, it sloughs off and is renewed. Thus the CS protein serves as an immune decoy. The nucleotide sequence of the gene encoding the entire CS protein has been determined. Unlike most eukaryotic genes, it is not interrupted by introns. *See* Appendix C, 1983, Godson *et al.*; malaria, *Plasmodium* life cycle.

sporulation **1.** the generation of a bacterial spore. **2.** production of meiospores by fungi and many other eukaryotic organisms.

spreading position effect the situation in which a number of genes in the vicinity of a translocation or inversion seem to be simultaneously inactivated. *See* Appendix C, 1963, Russell.

38, 40, 45S preribosomal RNAs *See* preribosomal RNA.

src the oncogene of the Rous sarcoma virus (*q.v.*). The human *SRC* gene lies at 20 q12-q13, while the *Drosophila Src* gene is on the third chromosome at 64B. *Src* genes encode proteins that function as protein tyrosine kinases and are characterized by SH2 and SH3 domains. These are important for intra- and intermolecular interactions that regulate both the catalytic activity of the molecules and their recruitment of substrates. SH2 domain is about 100 amino acids long, while SH3 domain is about 50 amino acids long and is rich in proline. *Src* PTCs regulate the actin cytoskeleton, and they play a role in the morphogenesis of ring canals (*q.v.*) during oogenesis. *See* actin, *c-src*, domain, *v-src*.

src tyrosine kinase *See* pp60v-src.

30S, 40S, 50S, 60S ribosomal subunits *See* ribosome.

60S, 70S, 73S, 78S, 80S ribosomes *See* ribosome, ribosomes of organelles.

4S RNA transfer RNA (tRNA) (*q.v.*).

7S RNA *See* signal recognition particle.

5S rRNA a small RNA molecule that is a component of most ribosomes. The 5S rRNA molecule shown in the illustration is from *E. coli*. 5S rRNA occurs in the large ribosomal subunit in the cytosol of all prokaryotes and eukaryotes. While the molecule stabilizes the structure of the large ribosomal subunit, 5S rRNA does not contribute directly to any of the active sites in the subunit. 5S rRNA occurs in the ribosomes of the mitochondria of plants and in the ribosomes of their chloroplasts. However, the ribosomes of the mitochondria of fungi and animals lack 5S rRNAs. In humans, the 5S rRNA locus is near the telomere of the short arm of chromosome 1. In *Drosophila melanogaster*, it is on 2R at 56 E-F. *See* Appendix C, 1963, Rosset and Monier; 1970, Wimber and Steffensen; 1973, Ford and Southern; 1985, Miller, McLachlan, and Klug; *Drosophila* salivary gland chromosomes, ribosomal RNA genes, ribosome, RNA polymerase, *Xenopus*.

5.8S rRNA a component of the large ribosomal RNA molecule that is transcribed in the nucleolus. 5.8S rRNA is the structural equivalent of the 5′-terminal 160 nucleotides of prokaryotic 23S rRNAs. Thus, in eukaryotes, the 5.8S and 28S coding sequences are separated by an internal transcribed spacer that is absent from the rDNA unit that is transcribed into the RNA of the large subunit of prokaryotic ribosomes. The 5.8S and 28S molecules are eventually separated by posttranscriptional excision of the spacer. However, these molecules remain associated by intermolecular base pairing interactions as the large subunit of the ribosome matures. *See* Miller trees, ribosomal RNA genes, ribosome.

16S rRNA the RNA molecule found in the small ribosomal subunits of prokaryotes. This RNA is often abbreviated SSU rRNA (small subunit rRNA). The secondary structure of the 16S rRNA of *E. coli* is shown on page 423. This 30S subunit also contains 20 specific proteins. The folding pattern results from hydrogen bonding of C to G and A to U molecules. The nucleotides are numbered starting with 1 at the 5′ end and ending with 1,542 at the 3′ end. Comparison of the nucleotide sequence of 16S rRNAs from widely diverse species has allowed the construction of a "universal tree of life" (*q.v.*). *See* Appendix C, 1977, Woese and Fox; 1980, Woese et al.; ribosome, Shine-Dalgarno (S-D) sequence.

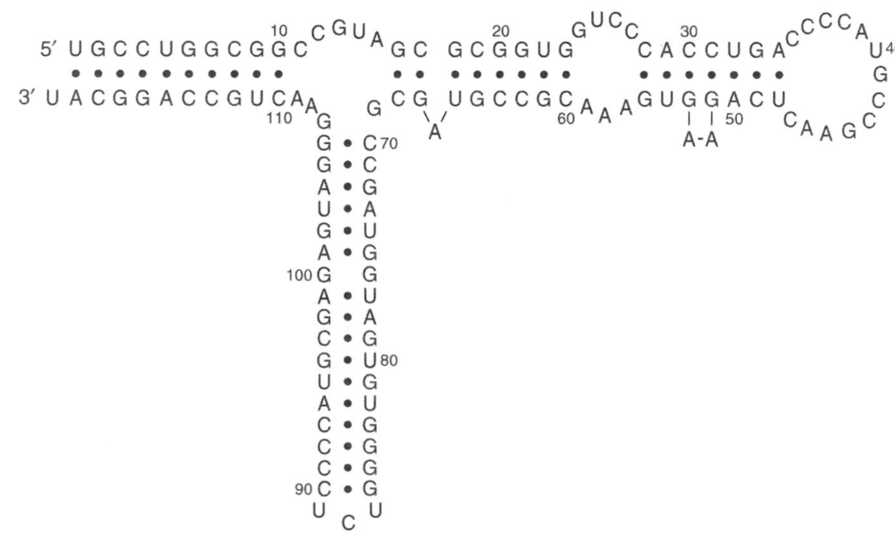

5S rRNA

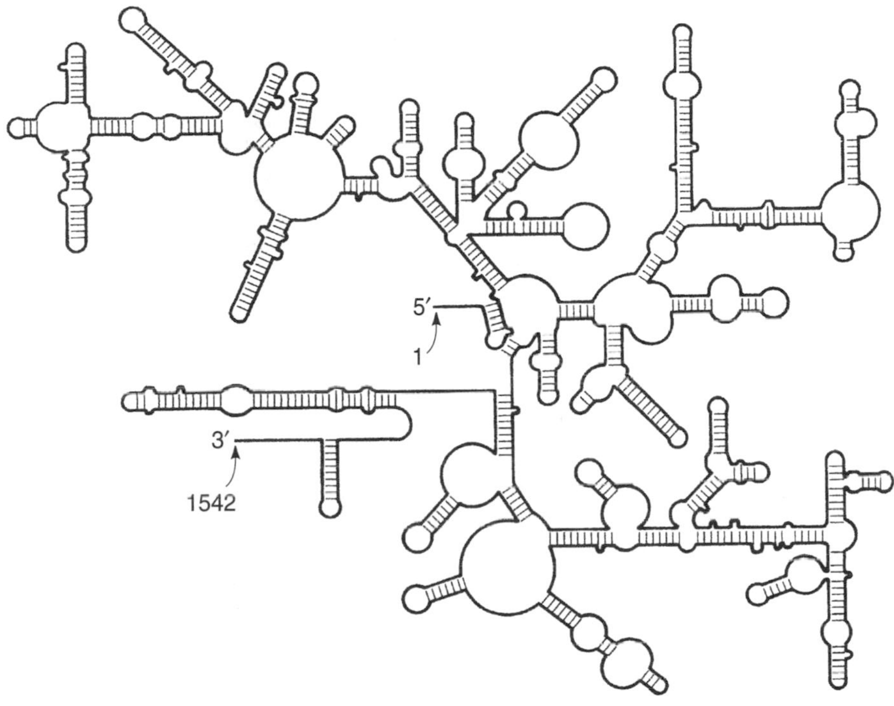

16S rRNA

16S, 18S, 23S, 28S rRNAs the RNA molecules that reside in the subunits of ribosomes. Prokaryotes have 16S and 23S RNAs in their small and large subunits, respectively. Eukaryotes have 18S and 28S RNAs in their small and large units, respectively.

5S rRNA genes genes that are transcribed into 5S rRNAs. Such genes occur in tandemly linked clusters in all eukaryotes. In *Xenopus laevis*, 5S rRNA genes account for 0.5% of the entire genome. There are three separate 5S rRNA multigene families. Two of these, the *major oocyte* and *trace oocyte* families, are expressed only in oocytes, while a third, *somatic* 5S rDNA, is expressed in all types of somatic cells. The major oocyte, trace oocyte, and somatic 5S rDNAs are present in 20,000, 1,300, and 400 copies, respectively, per haploid genome.

SRY *s*ex-determining *r*egion *Y*, the gene at p11.3 on the Y chromosome that is both required and sufficient to initiate testis development in human embryos. *SRY* is an intronless gene that spans 3.8 kb. It encodes a 204 amino acid protein which regulates the transcription of the genes that function in sexual differentiation by binding to target sequences in their DNAs. XY individuals with loss-of-function mutations in the *SRY* gene are phenotypically female, but with rudimentary ovaries. *See* Appendix C, 1987, Page *et al.*; 2003, Skaletsky *et al.*; human Y chromosome, selector genes, sex determination, Y chromosome.

SSC sister-strand crossover. *See* sister chromatid exchange.

ssDNA single-stranded DNA.

SSU rRNA small subunit rRNA. *See* 16S rRNA.

stabilizing selection normalizing selection (*q.v.*).

stable equilibrium an equilibrium state of alleles at a genetic locus to which the population returns following temporary disturbances of the equilibrium frequencies. For example, a locus with overdominance should form a stable equilibrium as long as selection favoring heterozygotes remains constant.

stable isotope a nonradioactive isotope of an element.

stacking 1. the planar alignment of adjacent flattish nitrogen bases in a DNA double helix. 2. stacking of dye molecules on RNA to yield metachromasy (*q.v.*).

staggered cuts the result of breaking two strands of duplex DNA at different positions near one another, as occurs by action of many restriction endonucleases (*q.v.*).

stamen the pollen-bearing organ of the angiosperm flower. It consists of a filament bearing a terminal anther. *See* flower.

standard deviation (s) a measure of the variability in a population of items. The standard deviation of a sample is given by the equation

$$s = \sqrt{\Sigma(x - \bar{x})^2/N - 1}$$

where N is the number of items in the sample and $\Sigma(x - \bar{x})^2$ is the sum of the squared deviations of each measurement from the mean ($\bar{x}$).

standard error (SE) a measure of variation of a population of means.

$$SE = \frac{s}{\sqrt{N - 1}}$$

where N = the number of items in the population and s = standard deviation.

standard type the most common form of an organism.

Stanford-Binet test used to gauge intelligence, it consists of a series of questions and problems grouped for applicability to ages up to 16 years. Some questions require verbal recognition and others recognition of form and manual skills. The subject's performance is expressed in terms of his mental age. *See* intelligence quotient.

Staphylococcus a genus of spherical, Gram-positive bacteria, belonging to the family Staphylococcaceae. Of the 19 species identified, only two—*S. aureus* and *S. epidermis*—are considered relevant to human health. *S. aureus* is found predominantly in the nasal passages and *S. epidermis* on the skin of normal humans. *S. aureus* has a genome size of 2.80 mbp and an estimated gene number of 2,600. These microbes cause disease or damage tissue when they move away from their normal habitats, particularly in individuals with weakened immune systems. As a human pathogen, *S. aureus* produces toxins that can cause a wide array of infections and toxic effects, such as boils, pneumonia, meningitis, urinary tract infections, bone infections, food poisoning, and toxic shock syndrome (*q.v.*). Pathogenesis by *S. epidermis* is relatively less understood. Ninety percent of *Staphylococcus* strains are resistant to penicillin and penicillin-derived antibiotics, presenting a challenge for doctors to treat *Staphylococcus*-derived ailments. Many genes encoding virulence factors have been characterized, proteins involved in pathogenesis identified, and factors associated with drug resistance detected. *See* Appendix A, Bacteria, Deinocci; Appendix E, Species Web Sites; Gram-staining procedure.

starch the storage polysaccharide of most plants. It is a polymer made up of α-D-glucose molecules. See formula below.

start codon a group of three adjacent ribonucleotides (AUG) in an mRNA coding for the methionine in eukaryotes (formylated methionine in bacteria) that initiates polypeptide formation; also called an *initiation codon*. *See* genetic code, initiator tRNA.

start kinase *See* cyclins.

startpoint in molecular genetics, the base pair on DNA that corresponds to the first nucleotide incorporated into the primary RNA transcript by RNA polymerase.

startsite synonym for startpoint (*q.v.*).

stasigenesis referring to a period during the paleontological history of a lineage during which little or no significant evolutionary change occurred.

stasipatric speciation speciation resulting from the dispersion of a favorable chromosomal rearrangement that yields homozygotes that are adap-

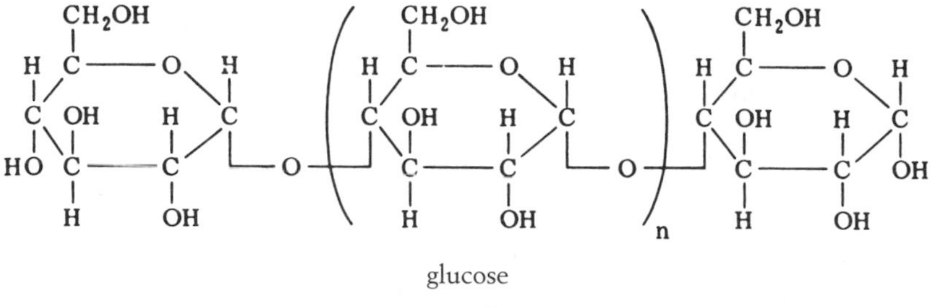

Starch

tively superior in a particular part of the geographical range of the ancestral species.

stasis in evolutionary studies, the persistence of a species over a span of geological time without significant change.

stationary phase a period of little or no growth that follows the exponential growth phase (*q.v.*) in a culture of microorganisms or in a tissue culture.

statistic the value of some quantitative characteristic in a sample from a population. *Compare with* parameter.

statistical errors a "type one" statistical error occurs when a purely random fluctuation is taken as evidence for a positive effect. The risk of making a "false positive" error of this sort is symbolized by the Greek letter *alpha* (α). A type 2 statistical error results when we fail to detect an effect when there is one. The risk of making a "false negative" error of this sort is symbolized by the Greek letter *beta* (β). Often a false negative error may be extremely costly, and so α is set at a very low value, but this increases the risk of type 2 errors. *See* confidence limits, null hypothesis method, significance of results.

statistics the scientific discipline concerned with the collection, analysis, and presentation of data. The analysis of such data depends on the application of probability theory. Statistical inference involves the selection of one conclusion from a number of alternatives according to the result of a calculation based on observations. *Parametric* methods in statistical analysis assume that the data follow a defined probability distribution (e.g., a *normal, binomial,* or *Poisson distribution, all of which see*), and the results of the calculations are valid only if the data are so distributed. The Student's *t* test (*q.v.*) is an example of a parametric procedure. *Nonparametric* methods in statistical inference are free from assumptions as to the shape of the underlying probability distribution. The Mann-Whitney rank sum test and the sign test are examples of nonparametric procedures. *See* analysis of variance, chi-square test, Gaussian curve, null hypothesis, Student's *t* test.

status quo hormones synonym for allatum hormones (*q.v.*).

steady-state system a system whose components seem unchanging because material is entering and leaving the system at identical rates.

stem cells undifferentiated or partially differentiated animal or plant cells that can proliferate and are pluripotent (*q.v.*) or totipotent (*q.v.*) in nature. Stem cells are generally divided into the following two classes. (1) *Embryonic stem cells* (ESCs) are undifferentiated cells from the early embryo that can proliferate and are pluripotent or totipotent, i.e., during normal development or upon transfer into an appropriate host or environment, they have the potential to differentiate into every adult cell type or to produce a whole animal from a single cell. The term usually refers to stem cells from the mammalian embryo. The mammalian zygote (*q.v.*) and cells produced by early zygotic divisions up to around the blastocyst (*q.v.*) stage are examples of ESCs. ESCs derived from cultured mouse blastocyst cells differentiate into teratocarcinomas when injected into immunodeficient mice, produce pure lines of pluripotent cells under appropriate experimental conditions, and when injected into a host blastocyst, form nearly all the tissues of the chimeric adult animal. (2) *Adult tissue stem cells* (TSCs) are partially differentiated, post-embryonic or postnatal plant or animal cells that have the potential to proliferate, self-renew, and produce one or more types of differentiated progeny. Through *in vivo* (*q.v.*) and *in vitro* (*q.v.*) manipulations, TSCs have been identified in a variety of tissues (e.g., bone marrow, central nervous system, the epidermis, intestinal epithelium, skeletal muscle, the germ line, and shoot and root apical meristems), where they serve to replace cells that die, are lost due to injury, or are continually depleted during the life of the organism. TSCs differ from ESCs in that during normal development TSCs are more limited in their *in vivo* developmental potential, and their cell division gives rise to one daughter cell that acts as a stem cell and another that produces differentiated progeny. The stem cell state and the developmental capacity of the daughter cells is influenced by signals from the surrounding environment, and TSCs show plasticity in choosing their course of differentiation when their microenvironment is altered. Mutations that affect stem cell fate have been identified in both plants and animals. *See* chimera, cystocyte divisions, teratocarcinoma.

stem structure in molecular biology, the base-paired (unlooped) segment of a single-stranded RNA or DNA hairpin (*q.v.*). Also known as a *stem and loop structure.*

stereochemical structure the three-dimensional arrangement of the atoms in molecules.

stereoisomers molecules that have the same structural formula, but that differ in the spatial arrangement of dissimilar groups bonded to a common atom.

steric relating to stereochemical structure (*q.v.*).

sterile 1. unable to reproduce. 2. free from living microorganisms; axenic.

sterile male technique a technique used in controlling noxious insects. Large numbers of artificially reared males are given nonlethal but sterilizing doses of ionizing radiation and then released in nature. The natural populations are so overwhelmed by these males that females are almost always fertilized by them. As a result, the fertilized eggs produced are rendered inviable, and a new generation cannot be produced.

sterilization 1. elimination of the ability to reproduce. 2. the process of killing or removing all living microorganisms from a sample.

steroid a lipid belonging to a family of saturated hydrocarbons containing 17 carbon atoms arranged in a system of four fused rings. The hormones of the gonads and adrenal cortex, the bile acids, vitamin D, digitalis, and certain carcinogens are steroids.

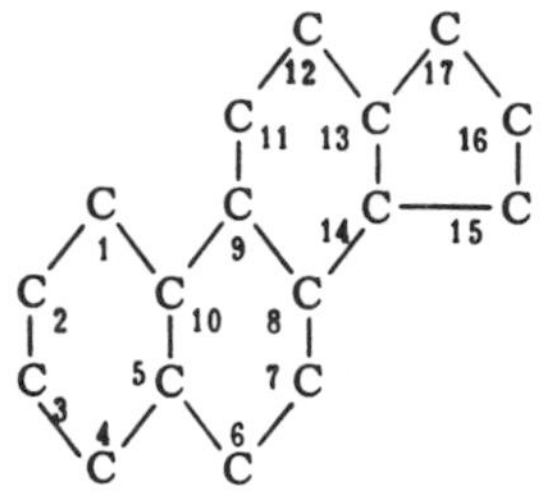

steroid receptor a cytoplasmic receptor protein that can bind to a specific steroid hormone. The receptor-hormone complex then moves into the nucleus and binds to a specific DNA site to regulate gene activity.

***steroid sulfatase (STS)* gene** a pseudoautosomal gene in the mouse. *See* human pseudoautosomal region.

sterol a compound with the general chemical ring structure of a steroid, but with a long side chain and an alcohol group. Cholesterol (*q.v.*) is an example of a sterol.

sticky ends complementary single-stranded projections from opposite ends of a DNA duplex or from different duplex molecules that are terminally redundant. Sticky ends allow the splicing of hybrid molecules in recombinant DNA experiments. Many restriction endonucleases (*q.v.*) create sticky ends by making staggered cuts (*q.v.*) in a palindromic restriction site. Also called *cohesive ends*. *See* Appendix C, 1970, Smith and Wilcox.

stigma the receptive surface usually at the apex of the style of a flower on which compatible pollen grains germinate.

stillbirth the birth of a dead fetus.

stochastic process a process that can be visualized as consisting of a series of steps, at each of which the movement made is random in direction.

stock 1. that part of a plant, usually consisting of the root system together with part of the stem, onto which is grafted a scion. 2. an artificial mating group, as, for example, a laboratory stock of mutant *Drosophila*. *See* strain.

stoloniferous referring to a plant having a creeping horizontal stem that takes root at several points to produce new plants. *See* modular organism.

stop codon a ribonucleotide triplet signaling the termination of the translation of a protein chain (UGA, UAG, UAA). *See* Appendix C, 1965, Brenner *et al. Compare with* start codon.

strain an intraspecific group of organisms possessing only one or a few distinctive traits, usually genetically homozygous (pure-breeding) for those traits, and maintained as an artificial breeding group by humans for domestication (e.g., agriculture) or for genetic experimentation. There is no clear distinction between the terms *strain* and *variety*, but the latter is generally applied when the differences between such intraspecific groups is substantial. *See* cultivar, pathovar, stock.

STR analysis STR stands for *short tandem repeat*, and the method relies on the variability in the STRs that are scattered along the human chromosomes to distinguish the DNAs of different individuals. The FBI uses a standard set of probes that bind to 13 specific STR regions to generate DNA profiles. The odds that any two humans (except identical twins) will have a match at all 13 loci is about one in a billion. *See* CODIS, DNA fingerprint technique, microsatellites, repetitious DNA.

strand displacement a replication mechanism, used by certain viruses, in which one DNA strand is displaced as a new strand is being synthesized.

strand-specific hybridization probes specifically designed RNA transcripts used for blot or *in situ* hybridization experiments. A special plasmid vector is synthesized that contains a promoter for a phage RNA polymerase and an adjacent polylinker site (*q.v.*) which allows insertion of a DNA fragment in a specific direction. The vector is then cleaved with an appropriate restriction enzyme, and the gene fragment to be analyzed is ligated into the vector and propagated in *E. coli.* After purification, the plasmid DNA is used as a template for transcription by the specific phage RNA polymerase. By using appropriately labeled ribonucleoside triphosphates, radioactive transcripts of high specific activity are produced. These have two advantages over DNA probes obtained by nick translation (*q.v.*). (1) Since the RNA is strand specific, one strand of DNA can be analyzed at a time. (2) The sensitivity of hybridization is increased, since the RNA will not self-anneal. DNA probes, on the other hand, compete with their own complementary strands.

strand terminologies names given to distinguish the two strands of a DNA molecule. Each strand of a DNA molecule has a 5′ end and a 3′ end. The 5′ end has a PO_4 molecule connected to the number 5 carbon of the first sugar. The 3′ end has an OH group connected to the number 3 carbon of the last sugar. The two strands of each DNA molecule are aligned in an antiparallel configuration, that is, they point in opposite directions. Terminologies for these strands depend on conventions adopted for messenger RNA. Since it represents a set of instructions, mRNA is considered to be a "sense" molecule, and therefore a synthetic RNA molecule with a complementary nucleic acid sequence has been named an *antisense RNA* (*q.v.*). Here the prefix *anti* signifies *opposite to* or *lying alongside*. When the nucleotide sequence of an mRNA is printed in a scientific publication, the 5′ end is always drawn above (as in the diagram below) or to the left. The direction of transcription is then down or from left to right. When mRNA is translated in a ribosome, the amino end of the new protein is the first and the carboxyl end the last to be formed. The DNA strand that serves as the template for mRNA is called the *template strand*. The other strand will contain segments that are identical in sequence to the codons in the mRNA, assuming one substitutes Ts for Us. For this reason, the DNA strand complementary to the template strand is labeled "sense." It is the sense strand that is drawn when a "gene sequence" is presented in the literature. *Upstream* refers to the 5′ direction and *downstream* to the 3′ direction on the sense strand. For example, the promotor sequence will be upstream (to the left) of the first exon, and the polyadenylation site will be downstream (to the right) of the last exon. Other terms such as coding strand,

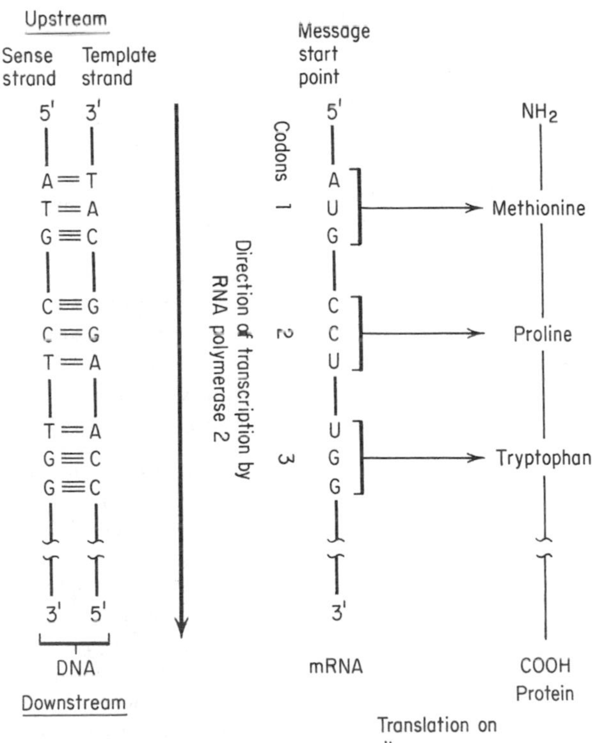

Strand terminologies

anticoding strand, and antisense strand are found in the literature, but since they are used inconsistently, they should be avoided in the future. *See* deoxyribonucleic acid, leader sequence, plus (+) and minus (–) viral strands, polyadenylation, posttranscriptional processing, trailer sequence, transcription unit.

stratigraphic time divisions geologic time divisions (*q.v.*).

strawberry *See Frageria.*

streak plating a technique of spreading microorganisms over the surface of a solidified medium for the purpose of isolating pure cultures.

streptavidin a biotin-binding protein synthesized by *Streptomyces avidinii*. *See* biotinylated DNA.

Streptocarpus the genus containing the Cape primroses. The inheritance of flower pigmentation has been thoroughly studied in various species in this genus. *See* anthocyanins.

Streptococcus a genus of Gram-positive bacteria that occur as parasites and pathogens, particularly in the lungs and intestines of various animal species. Two species of immense medical importance are described below. *Streptococcus pyogenes* is responsible for more human diseases than other bacterial species. These include impetigo, rheumatic fever, scarlet fever, septicemia, "strep" throat, and toxic shock syndrome (*q.v.*). The genome is a circular DNA molecule containing 1,852,442 bp. Ten percent of the 1,752 ORFs are located within resident prophages. *S. pyogenes* can produce at least 40 different virulence factors. *Streptococcus pneumoniae* is the cause of bacterial pneumonia, and the Pneumoccus Transforming Principle (PTP) (*q.v.*) was isolated from the organism. The genome consists of a single circular chromosome composed of 2,160,837 bp of DNA. There are 2,236 genes and biological roles have been assigned to 64% of the proteins they are predicted to encode. The genome of *S. pneumococcus* is rich in insertion sequences (*q.v.*), but most of these are nonfunctional because of insertions, deletions, and point mutations. The virulence of this pneumococcus is associated with its ability to synthesize a polysaccharide capsule. A 13 gene cluster has been identified that is likely to be involved in the biosynthesis and secretion of this structure. *See* Appendix A, Bacteria, Endospora; Appendix C, 1928, Griffith; 1944, Avery, MacLeod, and McCarty; 1964, Fox and Allen; 2001, Ferretti *et al.*, Tettelin *et al.*

streptolydigins a group of antibiotics that, when bound to the beta subunit of bacterial RNA polymerase, prevent transcriptional elongation.

Streptomyces a genus of soil-inhabiting bacteria containing over 500 species. Some of these are notable for their synthesis of many useful compounds, including the majority of the antibiotics used in human and veterinary medicine, immunosuppressants, and herbicides. *Streptomyces*-derived antibiotics include streptomycin, streptonigrin, neomycin, chloramphenicol, and tetracyclines (*all of which See*). Streptomyces are also of interest for use in bioremediation (*q.v.*), since they are able to break down a diverse range of molecules, including aromatic compounds, organic acids, sugars, and alcohols. Of all the species, *S. coelicolor* is the most widely studied and has become the model organism for genetic analysis. It has a linear chromosome containing the largest number of predicted genes (7,825) for any prokaryote. Its genome contains an unprecedented number of regulatory genes. *See* Appendix A, Bacteria, Actinobacteria; Appendix E, Species Web Sites; Appendix F; antibiotic, streptavidin.

streptomycin an antibiotic produced by *Streptomyces griseus* that binds to the 30S subunit of the bacterial ribosome and leads to faulty translation of the advancing messenger tape. *See* ribosome, translation.

streptomycin suppression seen in bacterial mutants with an altered ribosomal protein (S12). This enables them to initiate polypeptide synthesis in the presence of streptomycin, and it also reduces the extent of misreading induced by that antibiotic. Such cells are converted from streptomycin-sensitive to streptomycin-resistant.

streptonigrin an antibiotic produced by *Streptomyces flocculus* that causes extensive chromosomal breakage.

stress fibers bundles of parallel-aligned, actin-containing microfilaments underlying the plasma membrane of cultured eukaryotic cells. Stress fibers permit cells to attach to the substratum and generate the stress or tension that causes them to assume a flattened shape. *See* fibronectin.

stringency the condition with regard to temperature, ionic strength, and the presence of certain organic solvents such as formamide (*q.v.*), under which nucleic acid hybridizations are carried out. With conditions of high stringency, pairing will occur only between nucleic acid fragments that have a high frequency of complementary base sequences. Conditions of weaker stringency must be used if the nucleic acids come from organisms that are genetically diverse. Thus, if one were trying to isolate an alcohol dehydrogenase gene from a silkworm genomic library using a cloned gene from *Drosophila mel-*

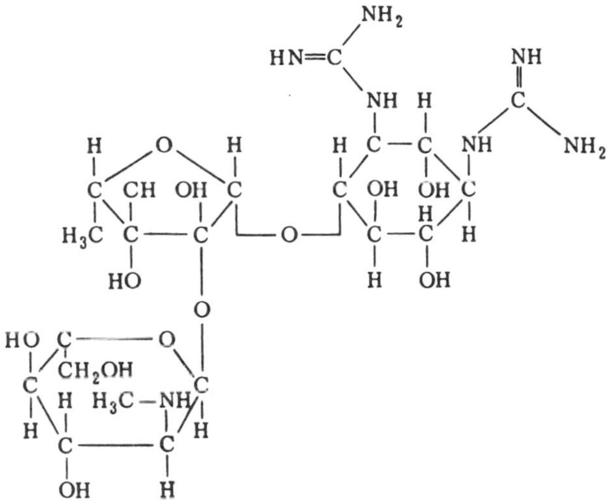

Streptomycin

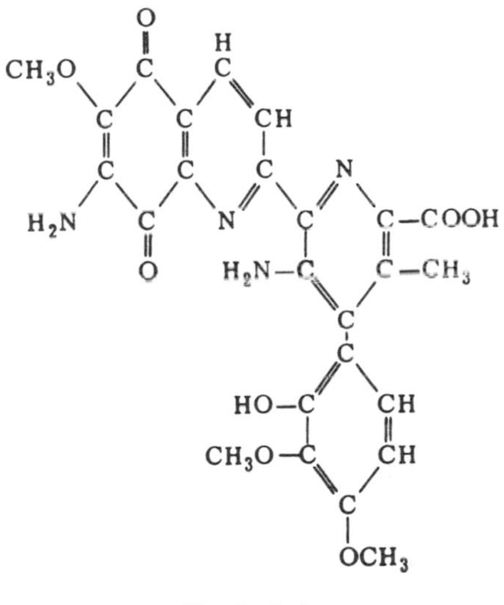

Streptonigrin

anogaster as a probe, less stringent conditions would be used than if the library came from *D. virilis*.

stringent control *See* plasmid.

stringent response the cessation of tRNA and ribosome synthesis by bacteria under poor growth conditions.

stRNAs small temporal RNAs (*q.v.*).

stroma the protein background matrix of a chloroplast or mitochondrion.

stromatolites living or fossil microbial mats dominated by cyanobacteria and fine sediment (usually calcium carbonate) trapped by these photosynthetic microbial communities. The oldest stromatolites are more than 3 billion years old and are among the oldest known fossils. *See* Appendix C, 1980, Lowe.

Strongylocentrotus purpuratus a common sea urchin used in studies of molecular developmental genetics. During oogenesis and egg maturation, large reservoirs of histone mRNAs are produced by females. Histone genes (*q.v.*) were first isolated from this species. Its estimated genome size is 845,000 kilobases. *See* echinoderm.

strontium^{90} a radioisotope of strontium with a half-life of 28 years generated during the explosion of nuclear weapons. ^{90}Sr is one of the major sources of radiation due to fallout.

structural change chromosomal aberration (*q.v.*).

structural gene a DNA segment whose own structure (nucleotide sequence) determines the structure (amino acid sequence) of a specific polypeptide. *See* gene, *lac* operon.

structural heterozygote a cell or an individual multicellular organism characterized by a pair of homologous chromosomes, one normal and the other containing an aberration, such as an inversion or a deficiency.

structural protein any protein that substantially contributes to shape and structure of cells and tissues: e.g., the actin and myosin components of muscle filaments, the proteins of the cytoskeleton, collagen, etc.

struggle for existence the phrase used by Darwin to describe the competition between animals for environmental resources such as food or a place to live, hide, or breed. Darwin wrote in *On the Origin of Species* "I use the term *struggle for existence* in a large and metaphorical sense . . . including (which is more important) not only the life of the individual, but success in leaving progeny."

STS sequence tagged site (*q.v.*).

Student *t* test a statistical method used to determine the significance of the difference between two sample means. The method was developed by the British statistician W. S. Gosset, who used the pseudonym "Student" in his publications. *See* page 74 for *t* distribution.

style a slender column of tissue arising from the top of the ovary and through which the pollen tube grows.

Stylonychia a genus of ciliates in which the macronuclear anlage undergo endomitotic DNA replication to form giant, banded, polytene chromosomes. Subsequently, the macronucleus undergoes a major reorganization of its DNA. The polytene chromosomes are destroyed, and over 90% of the DNA is eliminated. The remaining DNA molecules are present as gene-sized pieces, and these undergo a series of replications as the macronucleus matures. Therefore, the macronucleus comes to contain multiple copies of a subset of the genes found in the micronucleus. A similar sort of chromatin elimination occurs in ciliates of the related genus *Oxytricha*. In *Stylonychia lemnae*, UAA and UAG encode the amino acid glutamine rather than serving as termination codons. *See* Appendix A, Protoctista, Ciliophora; Appendix C, 1969, Ammermann; genetic code, nuclear dimorphism.

subculture a culture made from a sample of a stock culture of an organism transferred into a fresh medium.

subdioecy a sexual state of certain plants in which some unisexual individuals show imperfect sexual differentiation.

sublethal gene *See* subvital mutation.

submetacentric a chromosome that appears J-shaped at anaphase because the centromere is nearer one end than the other.

subpopulations breeding groups within a larger population or species, between which migration is restricted to a significant degree.

subspecies 1. a taxonomically recognized subdivision of a species. 2. geographically and/or ecologically defined subdivisions of a species with distinctive characteristics. *See* race.

substitutional load the cost to a population in genetic deaths of replacing an allele by another in the course of evolutionary change. *See* genetic load.

substitution vector *See* lambda cloning vector.

substrain a population of cells derived from a cell strain by isolating a single cell or groups of cells having properties or markers not shared by all cells of the strain.

substrate 1. the specific compound acted upon by an enzyme. 2. substratum.

substrate-dependent cells *See* anchorage-dependent cells.

substrate race a local race of organisms selected by nature to agree in coloration with that of the substratum.

substratum the ground or other surface upon which organisms walk, crawl, or are attached.

subtertian malaria *See* malaria.

subtractive hybridization *See* representational difference analysis (RDA).

subvital mutation a gene that significantly lowers viability, but causes the death before maturity of less than 50% of those individuals carrying it. *Contrast with* semilethal mutation.

sucrose the sugar of commerce, a disaccharide composed of glucose and fructose.

sucrose gradient centrifugation *See* centrifugation separation.

Sudan black B a commonly used lysochrome.

***sue* mutations** *See* suppressor-enhancing mutations.

sugar *See* carbohydrate, glucose.

suicide genes genes whose products facilitate apoptosis (*q.v.*).

Sulawesi a peculiarly shaped island east of Borneo. Sulawesi straddles the equator, with the Celebes Sea to the north and the Molucca Sea to the east. In the middle Miocene, about 15 million years ago, the Australian plate, bounded on the north by New Guinea, collided with the Eurasian plate. Sulawesi received land from both plates. This explains why this island has animals, some of whose ancestors came from Asia and others from Australia. *See* biogeographic realms, plate tectonics, Wallace line.

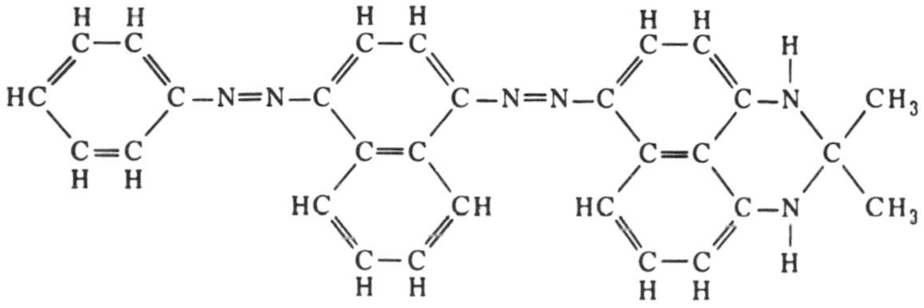

Sudan black B

sulfa drugs a group of compounds also called *sulfonamides*. The simplest is sulfanilamide, and it bears a close resemblance to p-aminobenzoic acid.

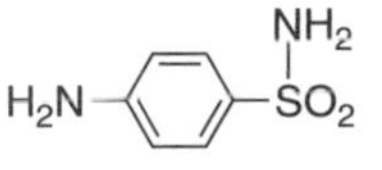

Sulfanilamide

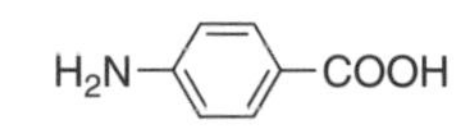

p-Aminobenzoic Acid

Sulfanilamide and p-aminobenzoic molecules compete during the enzymatic synthesis of folic acid (*q.v.*). The bacterial enzyme dihydropteroate synthetase is tricked into making a dihydropteroate containing sulfanilamide. This product cannot be converted to folate, and the bacteria are starved of the folate they require to divide, and die. Humans get the folate they need in their diet and therefore are not harmed by doses of sulfanilamide that kill bacteria. *See* Prontosil.

sulfatide lipidosis an autosomal-recessive disease in man due to a defect in the production of the lysosomal enzyme arylsulfatase A. The clinical symptoms are paralysis, blindness, and dementia, leading to death during childhood.

sulfonamides *See* sulfa drugs.

sulfur an element universally found in small amounts in tissues. Atomic number 16; atomic weight 32.064; valence 2^-, 4^+, 6^+; most abundant isotope ^{32}S; radioisotope ^{35}S (*q.v.*).

sulfur-containing amino acids cysteine, cystine, methionine. *See* amino acids.

sulfur-dependent thermophiles a group of prokaryotes that generally live in sulfur-rich hot springs and generate energy by metabolizing sulfur. They are placed in the Crenarcheota of the Archaea (*see* Appendix A). Members of one subgroup, called *eocytes*, are thought to be immediate relatives of the first eukaryotes. Some eocyte genera are *Acidianus*, *Desulfurococcus*, *Pyrodictium*, *Sulfolobus*, and *Thermodiscus*.

sulfur mustard mustard gas, the first chemical mutagen to be discovered. *See* Appendix C, 1941, Auerbach and Robson; nitrogen mustard.

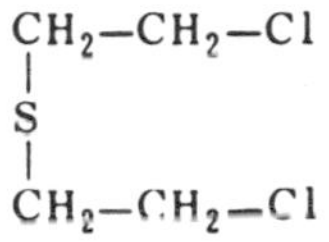

$$\begin{array}{l} CH_2{-}CH_2{-}Cl \\ | \\ S \\ | \\ CH_2{-}CH_2{-}Cl \end{array}$$

SUMO proteins a conserved family of small *u*biquitin-like *mo*difier proteins which become covalently conjugated to target proteins and modify the properties of these proteins. A SUMO protein is structurally related to ubiquitin (*q.v.*), and like ubiquitin, binds by its C terminus (*q.v.*) end to a lysine residue in the target protein. However, protein modification by SUMO does not lead to protein degradation; rather, the modified protein plays a role in regulating a diverse range of functions, such as nucleocytoplasmic transport, gene transcription, chromosome separation, DNA repair, and protein stability. SUMO proteins, like those of the Polycomb group, sometimes also silence genes by inducing the condensation of localized chromosomal regions. The posttranslational modification of a substrate protein by SUMO ligation is called *sumoylation*. Sumoylation is a reversible, dynamic process, and many enzymes involved in SUMO activation, conjugation, and deconjugation have been identified. SUMO proteins are found in animals, fungi, and plants. In humans there are at least four SUMO isoforms, with distinct functions and subcellular localization. *See Polycomb (Pc)*, ubiquitin-proteasome pathway (UPP).

supercoiling the coiling of a covalently closed circular duplex DNA molecule upon itself so that it crosses its own axis. A supercoil is also referred to as a *superhelix*. The B form of DNA is a right-handed double helix. Winding of the DNA duplex in the same direction as that of the turns of the double helix is called *positive supercoiling*. Twisting of a duplex DNA molecule in a direction opposite to the turns of the strands of the double helix is called *negative supercoiling*. *See* DNA gyrase, replisome.

superdominant overdominant (*q.v.*).

superfemale metafemale (*q.v.*).

supergene a chromosomal segment protected from crossing over and so transmitted from generation to generation as if it were a single recon.

supergene family *See* gene superfamily.

superhelix *See* supercoiling.

superinfection the introduction of such a large number of viruses into a bacterial culture that each bacterium is attacked by several phages.

supermale metamale (*q.v.*).

Superman *See* cadastral genes.

supernatant the fluid lying above a precipitate in a centrifuge, following the centrifugation of a suspension.

supernumerary chromosome a chromosome present, often in varying numbers, in addition to the characteristic invariable complement of chromosomes. *See* Appendix C, 1928, Randolph; B chromosomes.

superovulation the simultaneous release of more than the normal number of eggs from an ovary. This can be induced artificially by hormone treatment in cattle and other livestock for embryo transfer (*q.v.*) to surrogate mothers.

superoxide anion a highly reactive and destructive radical generated by the one electron reduction of O_2. The reaction may be symbolized:

$$O_2 + e^- \rightarrow O_2^-$$

See free radical, superoxide dismutase (SOD).

superoxide dismutase (SOD) an antioxidant enzyme (*q.v.*). The most common SOD in eukaryotes is a homodimeric enzyme containing one copper and one zinc atom per monomer. Prokaryotes like *E. coli* have SODs that contain manganese or iron. SODs remove superoxide anions from cells by catalyzing the conversion of two of these radicals to hydrogen peroxide and molecular oxygen:

$$O_2^- + O_2^- \xrightarrow[2H^+]{SOD} H_2O_2 + O_2$$

In eukaryotes, superoxide dismutase is symbolized SOD1 to distinguish it from the SOD2 of mitochondria. *See* familial amyotrophic lateral sclerosis, free radical hypothesis of aging, indolephenoloxidase, superoxide anion.

superrepression an uninducible state for a gene usually attributed to (1) a defective operator locus to which a functional repressor protein cannot bind, or (2) a mutant regulatory gene whose repressor protein product is insensitive to the inducer substance; a phenomenon that causes a gene to be permanently "turned off."

superspecies a complex of related allopatric species (often called a species group). Such species are grouped together because of their morphological similarities. That the superspecies constitutes a natural grouping is demonstrated by finding in the genus *Drosophila* that whenever species hybrids are produced under laboratory conditions the parents are from the same species group.

supersuppressor a mutation that can suppress the expression of specific alleles of mutations at a number of different chromosomal sites; usually a nonsense suppressor.

supervital mutation a mutation that increases the viability of individuals bearing it above the wild-type level.

suppression **1.** the restoration of a lost or aberrant genetic function (*see* suppressor mutation). **2.** in immunology, a specific or nonspecific state of induced nonresponsiveness in the immune system. *See* immunological suppression, suppressor T cell.

suppressor-enhancing mutations genetic alterations that boost the activity of both temperature-sensitive as well as temperature-insensitive suppressors.

suppressor mutation a mutation that compensates for some other mutation, resulting in a normal or near-normal phenotype in the double mutant. Two main types of suppressor mutations occur: intergenic and intragenic. An *intergenic suppressor mutation* suppresses the effect of a mutation in another gene. Some intergenic suppressors change the physiological conditions so that the proteins encoded by the suppressed mutants can function. Other intergenic suppressors actually change the amino acid sequence of the mutant protein. For example, the intergenic suppressor may cause a base substitution in a tRNA gene. This results in an anticodon that reads a mutated codon of the mRNA of the suppressed mutant so as to insert a functionally acceptable amino acid in the protein responsible for the phenotype. An *intragenic suppressor mutation* suppresses the effect of a mutation in the same gene in which it is located. Some intragenic suppressors restore the original reading frame after a frameshift. Other suppressor mutations produce new amino acid substitutions at different sites from those produced by the first mutation. However, the second changed amino acid compensates functionally for the first. Intragenic suppressor mutations are also called *second site mutations*.

***Suppressor of variegation 205* [*Su(var)205*]** *See* heterochromatin protein 1 (HP1).

suppressor T cell a subpopulation of T lymphocytes (designated Ts) whose function is to suppress the response of other lymphocytes to a particular antigen.

surface-dependent cells *See* anchorage-dependent cells.

surrogate mother a woman (or other female mammal) that receives an embryo transplant from another donor.

survival of the fittest the corollary of Darwin's theory of natural selection; namely, that as a result of the elimination by natural selection of those individuals least adapted to the environment, those that ultimately remain are the fittest.

survival value the degree of effectiveness of a given phenotype in promoting the ability of that organism to contribute offspring to the future populations.

suspension culture a type of *in vitro* culture in which the cells multiply while suspended in liquid medium. *See* anchorage-dependent cells.

Sus scrofa the pig. Domesticated pigs are generally given the subspecies name *domestica*. The haploid chromosome number is 19, and about 350 genes have been mapped. Because of its anatomical and physiological similarities with humans and the ease with which it can be bred in large numbers, the domesticated pig is the most likely source of organs for replacement of those incapable of continued function in humans. Unfortunately, pig organs transplanted into humans are rapidly rejected by the recipient's immune system. The generation of genetically engineered pigs may eventually overcome this rejection problem. *See* Appendix A, Chordata, Mammalia, Artiodactyla; Appendix E; swine, transgenic animals, xenoplastic transplantation.

SV 40 simian virus 40 (*q.v.*).

Svedberg *See* sedimentation coefficient.

sweepstakes route a potential migration pathway along which species disperse with difficulty. Chance events play a large role in colonization of new areas along this route. For example, birds blown far out to sea by a storm may accidentally land on an oceanic island and colonize it, but it is unlikely that this will happen a second time for that species.

sweet corn *See* corn.

swine any of a number of domesticated breeds of the species *Sus scrofa* (*q.v.*). Popular breeds include Berkshire, Chester White, Duroc, Hampshire, Hereford, Ohio Improved Chester, Poland China, Spotted Poland China, Tamworth, Yorkshire.

switchback evolution the recurrent reversals between alternative patterns of development during the evolution of particular groups of plants or animals. Insect ovary types (*q.v.*) provide an example. Reversions from the polytrophic to the panoistic type of oogenesis have occurred within the Mecoptera and the Neuroptera. Reversions from the telotrophic to the panoistic type have occurred within the Megaloptera and the Thysanoptera. *See* Appendix A, Eumetazoa, Bilateria, Coelomata, Arthropoda, Insecta.

switch gene a gene that causes the epigenotype to switch to a different developmental pathway.

switching sites break points at which gene segments combine in gene rearrangements.

swivelase *See* gyrase, topoisomerase.

symbiogenesis the evolutionary process by which bacterial symbionts were transformed into certain of the organelles found in eukaryotic cells, such as mitochondria and chloroplasts. *See* Appendix C, 1910, Mereschkowsky; apicoplast, serial symbiosis theory.

symbiont an organism living in a mutually beneficial relationship with another organism from a different species: e.g., the coexistence of algae and fungi in lichens.

symbiosis any interactive association between two or more species living together. *See* commensalism, lichen, mutualism, parasitism, serial symbiosis theory.

symbiotic theory of the origin of undulipodia the theory proposed by L. Margulis that the ancestral eukaryote acquired undulipodia (*q.v.*) as the result of a motility symbiosis with spirochaetes.

symbols used in human cytogenetics A–G, the chromosome groups; 1–22, the autosome numbers; X, Y, the sex chromosomes; p, the short arm of a chromosome; q, the long arm of a chromosome; ace, acentric; cen, centromere; dic, dicentric; inv, inversion; r, ring chromosome; t, translocation; a plus (+) or minus (–) when placed before the autosome number or group letter designation indicates that the particular chromosome is extra or missing; when placed after a chromosome arm, a plus or minus designation indicates that the arm is longer or shorter than usual; a diagonal (/) separates cell lines when describing mosaicism. Examples: 45,XX, –C = 45 chromosomes, XX sex chromosomes, a missing chromosome from the C group; 46, XY, t (Bp–; Dq+) = a reciprocal translocation in a male between the short arm of a B and the long arm of a D group chromosome; inv (Dp+, q–) = a pericentric inversion involving a D chromosome; 2p+ = an increase in the length of the short arm of a chromosome 2; 46,XX, r = a female with one ring X chromosome; 45,X/46, XY = a mosaic of two cell types, one with 45 chromosomes and a single X, one with 46 chromosomes and XY sex chromosomes. *See* human mitotic chromosomes.

symmetrical replication bidirectional replication (*q.v.*).

sympatric speciation in multicellular eukaryotes, an uncommon process where populations inhabiting (at least in part) the same geographic range become reproductively isolated. However, sympatric bacterial speciation, defined as the origin of new bacterial species that occupy definable ecological niches, is relatively common. It occurs as a result of incorporation of horizontal mobile elements (*q.v.*) that carry adaptive genes. It has been estimated that *E. coli* has received 31 kb of DNA per million years from HMEs. *See* Appendix C, 1997, Lawrence and Ochman; introgressive hybridization.

sympatric species species whose areas of distribution coincide or overlap.

sympatry living in the same geographic location. *Compare with* allopatry.

sym-plasmid a plasmid found in the symbiotic bacteria that inhabit the root nodules of legumes. One such plasmid NRG234, has had its DNA analysed. It is a 536,165 bp circle, containing 416 ORFs. Among these are symbiotic determinants, which include nodulation and nitrogen fixation genes.

symplesiomorphic character an ancestral or plesiomorphic character shared by two or more different taxa. *See* cladogram.

synapomorphic character a derived or apomorphic (*q.v.*) character shared by two or more different taxa. *See* cladogram.

synapsis the pairing of homologous chromosomes during the zygotene state of meiosis that results from the construction of a synaptonemal complex (*q.v.*). *See* Appendix C, 1901, Montgomery.

synapsis-dependent allelic complementation *See* transvection.

synaptonemal complex a tripartite ribbon consisting of parallel, dense, lateral elements surrounding a medial complex. See illustration. The lateral elements lie in the central axes of the paired homologous chromosomes of a pachytene bivalent. The medial complex contains a system of interdigitating protein filaments that are oriented perpendicularly to the lateral elements and serve to maintain their parallel configuration during meiotic synapsis. *See* Appendix C, 1956, Moses and Fawcett; *Gowen crossover suppressor,* meiosis.

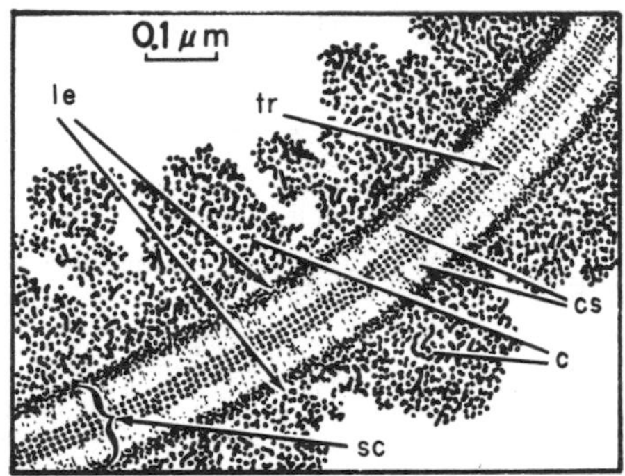

Synaptonemal complex

A drawing of a segment of a bivalent as seen under the electron microscope. (c) chromatin; (cs) central space; (le) lateral element; (sc) synaptonemal complex; (tr) transverse rods of the medial complex.

syncaryon synkaryon (*q.v.*).

syncytial blastoderm the stage during insect embryogenesis in which the cleavage nuclei lie at the surface of the egg in a common cytoplasm. Subsequently the cellularization of the blastoderm occurs. In *Drosophila* there are 13 rapid synchronous divisions resulting in about 8,000 nuclei which migrate to the cortical ooplasm and then cellularize. The transition from the syncytial blastoderm to the cellular blastoderm corresponds to the time the control of embryogenesis by maternal genes switches to control by the zygotic genome.

syncytium a multinucleate tissue whose constitutent cells have only partial cell boundaries and are connected through a common cytoplasm. A syncytium arises when nucleated cells fail to completely separate from one another during meiosis or mitosis (as in sperm development) or when nucleated cells fuse with one another (as in the fetal-maternal interface called a *syncytiotrophoblast*).

syndactyl having webbed digits either as a normal aspect of the species or, in man, pathologically.

syndesis meiotic chromosomal synapsis.

syndrome a group of symptoms that occur together, characterizing a disease.

Synechocystis a genus of bacteria that belongs to the Cyanobacteria (*q.v.*). The cells are coccoid and divide by binary fission in two or three planes to produce clusters of cells. One species from this genus (referred to as *sp. strain PCC 6803*) has had its genome completely sequenced. It is made up of 3.57 mb of DNA and contains 3,168 ORFs. Genes homologous to 45 of the ORFs from *Synechocystis* have been found in the chloroplasts of a wide variety of photosynthetic protoctists and land plants. *See* Appendix C, 1996, Kaneko *et al.*; Appendix F; *Arabidopsis thaliana*, chloroplast, serial symbiosis theory.

synergid one of two haploid cells that lie beside the ovum in the embryo sac (*q.v.*). Synergids of angiosperms are the source of chemical cues that guide pollen tubes to ovules. *See* double fertilization.

synergism the phenomenon in which the action of two agents used in combination is more effective than the sum of their individual actions.

synezis the clumping of chromosomes into a dense knot that adheres to one side of the nucleus. Synezis is a common occurrence during leptonema in microsporocytes.

syngamy the union of the nuclei of two gametes following fertilization to produce a zygote nucleus; karyogamy.

syngen *See Paramecium aurelia.*

syngeneic pertaining to genetically identical organisms such as identical twins or the members of a highly inbred strain. Because syngeneic animals have the same antigens on their tissues, they can exchange skin or organ grafts successfully. *Compare with* allogeneic, congenic strain.

syngraft a graft wherein the recipient receives a graft of tissue from a genetically identical donor (e.g., from an identical twin or from a member of the same highly inbred line). *Contrast with* allograft, autograft, xenoplastic transplantation.

synkaryon 1. the zygote nucleus resulting from the fusion of two gametic nuclei. 2. the product of nuclear fusion in somatic cell genetic experiments.

synomone *See* allomone.

synonym in taxonomy, a different name for the same species or variety.

synonymous codons same-sense codons. For example, UUU and UUC code for the same amino acid-phenylalanine. *See* degenerate code, genetic code.

syntenic genes genes that are orthologs (*q.v.*) and reside on the same chromosome in two species. For example, when *Drosophila melanogaster* and *D. pseudoobscura* are compared, the vast majority of their genes are found to be syntenic. However, the sequences of the genes in each chromosome arm have been extensively shuffled. *See* Appendix C, 2005, Richards *et al.*

synthetase an enzyme catalyzing the synthesis of a molecule from two components, with the coupled breakdown of ATP or some other nucleoside triphosphate.

synthetic lethal a lethal chromosome derived from normally viable chromosomes by crossing over.

synthetic linkers short, chemically synthesized DNA duplexes containing sites for one or more restriction endonucleases. Synthetic linkers are used most commonly in the cloning of blunt-ended DNA molecules.

synthetic polyribonucleotides RNA molecules made without a nucleic acid template, either by enzyme action or chemical synthesis. *See* Appendix C, 1961, Nirenberg and Matthaei; 1967, Khorana; polynucleotide phosphorylase.

syphilis a sexually transmitted disease caused by the spirochaete *Treponema pallidum* (*q.v.*).

systematics the study of classification; taxonomy based on evolutionary relationships.

T

t *See* symbols used in human cytogenetics.

t the Student's *t* statistic that is used for testing the difference between the means of two samples. *See* Student's *t* test, and page 65.

T thymine or thymidine.

tachyauxesis *See* heterauxesis.

tachytelic evolution *See* evolutionary rate.

Takifugu rubripes in the earlier literature this marine pufferfish is called *Fugu rubripes*. Pufferfish have the most compact genomes seen in vertebrates. The value (400 Mb) is 7.5 times smaller than the human genome. The number of ORFs is similar in both species, but the pufferfish has genes with smaller introns, the intergenic sequences are smaller, and there is less repetitive DNA. The compaction of the genome appears to result from the selection of deletions during the evolution of the fish. Duplicate genes are common in the pufferfish genome, and a phylogenetic analysis suggests the ancestral genome was duplicated early during the evolution of the Osteichthyes. *See* Appendix A, Chordata, Osteichthyes, Neopterygii, Tetraodontiformes; Appendix C, 2002, Aparicio *et al.*; Appendix E, Species Web Sites; Huntington disease (HD), Linnean system of binomial nomenclature, shotgun sequencing, *Tetraodon nigroviridis*.

TAOS1 a gene on the short arm of chromosome 11 which is amplified and overexpressed in human oral squamous cell carcinomas. The name is an acronym for *t*umor *a*mplified and *o*verexpressed *s*equence *1*.

tandem duplication an aberration in which two identical chromosomal segments lie one behind the other. The order of the genes in each segment is the same.

tandem repeat *See* microsatellites, tandem duplication.

T antigen a "tumor" antigen found in the nuclei of cells infected or transformed by certain oncogenic viruses such as polyoma. The antigen is thought to be a protein coded for by a virus cistron.

Taq DNA polymerase a DNA polymerase synthesized by the thermophilic bacterium *Thermus aquaticus*. This enzyme, which is stable up to 95°C, is used in the polymerase chain reaction (*q.v.*). *See* ligase chain reaction.

target number *See* extrapolation number.

target organ the receptor organ upon which a hormone has its effect.

target theory a theory developed to explain some biological effects of radiation on the basis of ionization occurring in a very small sensitive region within the cell. One or more "hits," that is, ionizing events, within the sensitive volume are postulated to be necessary to bring about the effect. *See* Appendix C, 1936, Timofeyeff-Ressovsky and Delbrück; extrapolation number.

target tissue 1. the tissue against which antibodies are formed. 2. the tissue responding specifically to a given hormone.

tassel the staminate inflorescence of corn.

taste receptor gene a gene in humans which resides at 7q35-q36. It is symbolized *TAS2R38*, and human polymorphisms in the ability to taste phenylcarbamide (*q.v.*) are due to base substitutions at various sites within the gene. The gene contains a single exon 1,002 base pairs long, and it encodes a G protein-coupled receptor (*q.v.*).

TATA box Hogness box (*q.v.*); pronounced "tahtah."

TATA box-binding protein (TBP) an essential transcription factor for RNA polymerases I, II, and III of eukaryotes. TBP does not occur in eubacteria, but archaebacteria (*q.v.*) contain a TBP that has amino acid sequence similarities to the eukaryotic TBP. These and other data suggest that archaebacteria and eukaryotes are more closely related than either is to eubacteria.

tautomeric shift a reversible change in the location of a hydrogen atom in a molecule that alters it from one to another isomer. Thymine and guanine are normally in *keto* forms, but when in the rare *enol* forms (see diagram on page 437) they can join by three hydrogen bonds with keto forms of guanine or thymine, respectively. Likewise, cytosine and adenine are normally in amino forms, but when in the rare *imino* forms they can join by two hydrogen bonds with amino forms of adenine or cytosine, respectively. Tautomeric shifts that modify the pairing of nucleotides can result in base substitutions and, as a result, mutations.

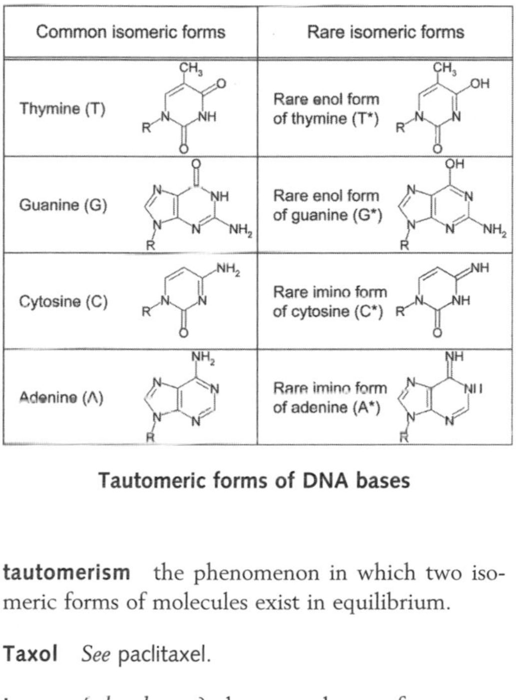

Common isomeric forms	Rare isomeric forms
Thymine (T)	Rare enol form of thymine (T*)
Guanine (G)	Rare enol form of guanine (G*)
Cytosine (C)	Rare imino form of cytosine (C*)
Adenine (A)	Rare imino form of adenine (A*)

Tautomeric forms of DNA bases

tautomerism the phenomenon in which two isomeric forms of molecules exist in equilibrium.

Taxol *See* paclitaxel.

taxon (*plural* taxa) the general term for a taxonomic group whatever its rank.

taxonomic category the rank of a taxon in the hierarchy of classification. *See* classification.

taxonomic congruence the degree to which different classifications of the same organisms postulate the same groupings. When the classifications compared are based on different sources of information (independent sets of data), congruence provides a measure of the degree to which the classifications remain stable as various lines of evidence are considered.

taxonomic extinction nonsurvival of a taxon, either by extinction or by pseudoextinction (*q.v.*).

taxonomist a specialist in taxonomy (*q.v.*).

taxonomy the study of the classification of living things. Classically, taxonomy is concerned with description, naming, and classification on the basis of morphology. More recently, taxonomists have been concerned with the analyses of patterns of variation in order to discover how they evolved, with the identification of evolutionary units, and with the determination by experiment of the genetic interrelationships between such units. *See* Appendix C, 1735, Linné; PhyloCode.

Taxus brevifolia the Pacific yew, a small, slow-growing evergreen tree native to the northwestern United States. The spindle poison paclitaxel (*q.v.*) is extracted from its bark.

Tay-Sachs disease a lethal hereditary disease due to a deficiency of hexosaminidase A. This deficiency results in storage of its major substrate (Gm2 ganglioside). Progressive accumulation of this compound causes developmental retardation, followed by paralysis, mental deterioration, and blindness. Most patients die by the age of three. The alpha chain of hexosaminidase A is encoded by a gene, *HEXA*, on the long arm of chromosome 15 between bands 22 and 25. About 2% of all Ashkenazi Jews are heterozygous for a defective *HEXA* allele. There are two distinct common mutations, as well as other rare ones. The beta chain of hexosaminidase A is encoded by a gene, *HEXB*, on autosome 5. Humans homozygous for mutant alleles of *HEXB* suffer from storage of gangliosides. The condition, Sandhoff disease, has symptoms similar to Tay-Sachs disease. Both *HEXA* and *HEXB* contain 14 exons and are believed to have arisen from a single ancestral gene. The disease gets its name from Warren Tay and Bernard Sachs, who published accounts of its pathologies in 1881 and 1887, respectively. *See* Appendix C, 1935, Klenk; Ashkenazi, ganglioside, hexosaminidase, lysosomal storage diseases.

T bacteriophages *See* T phages.

T box genes any gene containing a conserved motif encoding a DNA-binding site. In mice, amphibians, and fishes, these genes encode proteins necessary for the development of mesodermal structures. In humans the T box gene (*TBX1*) maps to 22q11.2, and the protein it encodes shares a 98% amino acid identity with the mouse TBX. In *Drosophila*, T box genes are expressed during eye development. *See*

Appendix C, 1990, Hermann *et al.*; 1994, Bollag *et al.*; brachyury, DNA-binding motifs, T complex.

TBP TATA box-binding protein (*q.v.*).

TCA trichloroacetic acid (*q.v.*).

TCA cycle an abbreviation for tricarboxylic acid cycle (*q.v.*).

TψC loop the hairpin loop nearest the 3′ end of tRNA molecules, containing the modified base pseudouridine (ψ). This loop is thought to interact with ribosomal RNA. *See* transfer RNA (tRNA).

T cell T lymphocyte. *See* lymphocyte.

T cell receptor (TCR) a heteromeric protein on the surface of T lymphocytes (*q.v.*) that specifically recognizes histocompatibility molecules (*q.v.*). T cell receptors are made up of two different polypeptide chains that are joined by disulfide bonds and are embedded in the plasmalemma with their carboxyl ends extending into the cytoplasm and their amino ends reaching outside the cell. The membrane portion of the T cell receptor is associated with a collection of CD3 proteins that transmit, from the outside of the cell to the inside, information as to whether or not the T cell receptor is occupied. The receptor recognizes as nonself the histocompatibility molecules on foreign cells, and it can also recognize antigenic sites on smaller molecules, provided these are presented in association with self-histocompatibility molecules. *See* immunoglobulin domain superfamily.

T cell receptor genes genes that encode the component polypeptides of T cell receptors (*q.v.*). There are two types of receptors: those containing an alpha and a beta chain, and those containing a delta and a gamma chain. In humans, both the alpha and the gamma chains are encoded by genes on the long arm of chromosome 14. The beta chain gene is located on the long arm of chromosome 7, and the gamma chain gene resides on the short arm of chromosome 7. As in the case of the immunoglobulins, the T cell receptor polypeptide chains are encoded by gene segments that are reshuffled during the differentiation of the precursor cells. The rearrangement of segments occurs in thymocytes before the genes encoding the polypeptides are expressed. As a result, T cell receptors have more than 10^7 different amino acid sequences. *See* Appendix C, 1984, Davis and Mak; V(D)J recombination.

T4, T8 cells classes of helper and suppressor T lymphocytes, respectively, characterized by antigenic markers that react with monoclonal antibodies designated anti-T4 and anti-T8, respectively. *See* lymphocyte.

Tc1/mariner element transposable elements that are 1,300–2,400 bp in length and contain a single gene that encodes a transposase (*q.v.*). The DNA segment is characterized by terminal inverted repeats. The transposon family is named after its two best-studied members, the *Tc1* transposon of *Caenorhabditis elegans* and the *mariner* transposon of *Drosophila mauritiana*. *See* mariner elements, transposons.

T complex a region on chromosome 17 of the mouse; it contains genes that affect tail length. Heterozygous mice with only one functional gene have shortened or missing tails. Homozygotes (T^-/T^-) die as embryos with defects in mesoderm-derived tissues. *See* Appendix C, 1990, Hermann *et al.*; *brachyury*, T box genes.

T-DNA a group of seven genes (collectively referred to as transferred DNA) of the Ti plasmid (*q.v.*) that integrates into the nuclear DNA of the host plant during tumor induction. T-DNA is always present in crown gall cells of plants. *See Agrobacterium tumefaciens.*

T4 DNA ligase an enzyme encoded by *E. coli* phage T4 that not only seals nicks in double-stranded DNA but also has the unique ability to join two DNA molecules that have completely base-paired (blunt) ends. This latter property is useful in forming recombinant DNA molecules.

T4 DNA polymerase an enzyme encoded by coliphage T4 that catalyzes the synthesis of DNA in the 5′ to 3′ direction and also has 3′ to 5′ exonuclease activity. If DNA is incubated with T4 DNA polymerase in the absence of deoxyribonucleoside triphosphates, the DNA will be partially degraded by the exonuclease. If the four dNTPs are now added, the degraded strand will be resynthesized by the polymerase. Thus, if the alpha phosphates of the added nucleotides are ^{32}P-labeled, a highly radioactive product can be obtained. The technique serves as an alternative to nick translation (*q.v.*).

tectonic plates *See* plate tectonics.

tektins a class of proteins found attached to the peripheral microtubules in sperm tails. Tectin filaments are 2 nanometers in diameter, 50 nanometers long, and are positioned longitudinally along the walls of the outer doublet, where the A and B subfibers join. *See* axoneme, Y chromosome.

teleology the explanation of a phenomenon such as evolution by the purposes or goals it serves. Teleological explanations usually invoke supernatural powers and are therefore nonscientific.

teleonomy the doctrine that the existence in an organism of a structure or function implies that it has conferred an advantage on its possessor during evolution.

telestability destabilization of a DNA double helix at a site distant from the site of binding of a protein. For example, binding of the cAMP-CAP complex to the *lac* operon of *E. coli* facilitates the distal formation of an open promoter site in which RNA polymerase can initiate transcription. *See* catabolite activating protein.

telocentric chromosome a chromosome with a terminal centromere.

telolecithal egg one in which the yolk spheres are accumulated in one hemisphere. *See* centrolecithal egg, isolecithal egg, vegetal hemisphere.

telomerase a reverse transcriptase containing an RNA molecule that functions as the template for the telomeric repeat. The first telomerase was isolated from *Tetrahymena* (*q.v.*). It is a large ribonucleoprotein complex weighing about 500 kilodaltons. The RNA of the *Tetrahymena* telomerase contains 159 nucleotides, and its secondary structure is shown in the drawing. The nine specific nucleotides form the *templating domain*, which is complementary to the G-rich strand of the telomere (*q.v.*). The functioning of telomerases seems to be activated in dividing embryonic cells and gametocytes. Telomerase function is repressed in differentiated somatic cells but reactivated in cancer cells. In human telomerase, the templating domain is 5′-CUAACCCUAAC-3′ and the telomeric repeat is (TTAGGG) n. Antisense RNAs designed to bind with telomerases cause HeLa cells (*q.v.*) to die after 23 to 26 doublings. *See* Appendix C, 1985, Greider and Blackburn; 1994, Kim *et al.*; 1995, Feng *et al.*; RNA-dependent DNA polymerase.

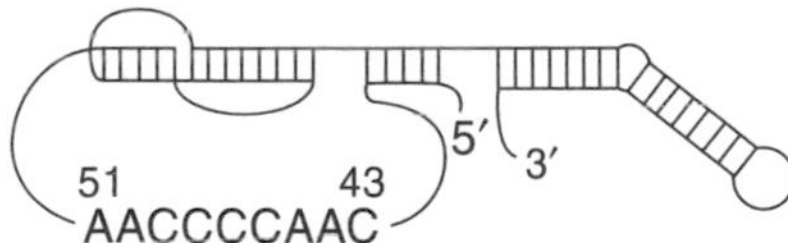

telomere a specialized DNA sequence found at the ends of eukaryotic chromosomes. The first telomeres to be sequenced belonged to *Tetrahymena thermophila*. They contained an A_2C_4 segment in one DNA strand and a T_2G_4 segment in the other, repeated in tandem about 60 times. The telomeres from all species subsequently studied showed the same pattern: a short DNA sequence, one strand G-rich and one C-rich, that is tandemly repeated many times. These telomere-specific repeats make it possible to identify chromosomes that have arisen by telomere-telomere fusions. Chromosomes lose nucleotides from their ends each cell division, and the shortening of telomeres may provide cells with a mitotic clock. Telomeric sequences can be added back to the chromosome ends, one base at a time by a telomerase (*q.v.*). The maintenance of telomeres is required for cells to escape from replicative senescence and to continue to multiply. In human leucocytes, telomeres shorten at a rate of 1,000 base pairs per year during the first 4 years of life. Then for about 20 years telomeres stay at lengths of about 12 kb. Thereafter there is a gradual loss (700 bp/yr) through old age. At the end of chromosomes, the 3′ overhang of the DNA duplex folds upon itself to form a telomeric loop (t-loop). The sequestered ends within t-loops are protected from enzymes that might degrade them. *Drosophila* chromosomes do not have conventional telomeres. Instead, telomere-specific retrotransposons are present in multiple copies on normal chromosome ends, and these retroposons can transpose to heal terminally deleted chromosomes. Bacterial chromosomes are generally circular. The spirochaete *Borrelia burgdorferi*, however has a linear major chromosome, and some of its plastids are also linear. Telomeric DNA forms covalently closed hairpin structures. *See* Appendix C, 1938, Muller, McClintock; 1971, Olonikov; 1972, Watson; 1978, Blackburn and Gall; 1990, Biessman *et al.*; 1991, Ijdo *et al.*; 1998, Frenck, Blackburn, and Shannon; Appendix E; bouquet configuration, centric fusion, guanine-quartet model, Hayflick limit, lamins, marginotomy, replication of DNA, sheep, telomere-led chromosome movement, *Tetrahymena*, tissue culture.

telomere-led chromosome movement the movement of chromosomes seen during meiotic prophase when all chromosomes associate by their telomeres, and these assume the leading position as the homologous chromosomes synapse and undergo crossing over. *See* Appendix C, 1994, Chikashige *et al.*; nucleoporins (Nups).

telomeric fusions end-to-end attachments of chromosomes during the evolution of a species which results in a reduction in the chromosome number from the ancestral value. *See Muntiacus*.

telomeric fusion site a segment on human chromosome 2 that contains nucleotide sequences which once resided at or near telomeres. This region (2q13-2q14.1) is where two nonhomologous autosomes fused end-to-end to produce a single V-shaped chromosome. This telomere fusion occurred early in human evolution, and it explains why the human diploid chromosome number is 46, not 48—the value for chimpanzees, gorillas, and orangutans. *See* Appendix C, 1991, Ijdo *et al.*.

telomeric repeat-binding factor 2 (TRF2) a protein that binds to TTAGG repeats and controls telomere length by inhibiting telomerases (*q.v.*). In humans the gene that encodes it is located at 16q22.1. TRF2 protects human telomeres from end-to-end fusions.

telomeric silencing the repression by telomeres of transcription by genes in adjacent DNA domains. Telomeres also appear to reduce the accessibility of subtelomeric chromatin to modification by DNA methylases. *See* DNA methylation.

telophase *See* mitosis.

telotrophic meroistic ovary *See* insect ovary types.

telson the most posterior arthropod somite in which the posterior opening of the alimentary canal is located. *See* maternal polarity mutants.

TEM transmission electron microscope. *See* electron microscope.

temperate phage a nonvirulent bacterial virus that infects but rarely causes lysis. It can become a prophage and thereby lysogenize the host cell.

temperature-sensitive mutation a mutation that is manifest in only a limited temperature range. The product of such a gene generally functions normally, but is unstable above a certain temperature. Thus, the mutant when reared at the lower (permissive) temperature is normal, but when placed at the higher (restrictive) temperature shows the mutant phenotype. *See* Appendix C, 1951, Horowitz and Leupold; 1971, Suzuki *et al.*; albinism, Himalayan mutant.

template the macromolecular mold for the synthesis of a negative antitemplate macromolecule. The antitemplate then serves as a mold for the template. Thus the duplication of the template requires two steps. A single strand of DNA serves as a template for the complementary strand of DNA or mRNA.

template strand the strand of a DNA segment that is transcribed into mRNA. *See* strand terminologies.

template switching in *E. coli*, a bizarre *in vitro* reaction often accompanying strand displacement in which DNA polymerase I shifts from the original template strand to the displaced strand.

temporal isolation *See* seasonal isolation.

teosinte various Mexican wild grasses that are interfertile with corn. The wild ancestor of corn has been identified as *Zea mays* ssp. *parviglumis*, an annual teosinte growing in the Balsas river valley of southern Mexico. The initial domestication occurred about 9,000 years ago. *See* Appendix C, 1939, Beadle; 2002, Matsuoka *et al.*; consilience, *Zea mays* spp *mays*.

teratocarcinoma embryonal tumors originating in the yolk sac or gonads of amniotes and capable of differentiating into a variety of cell types. These tumors are used to study the regulatory mechanisms involved in embryological development. *See* Appendix C, 1975, Mintz and Illmensee.

teratogen any agent that raises the incidence of congenital malformations.

teratoma a tumor composed of an unorganized aggregation of different tissue types.

terminal chiasmata the end-to-end association of homologous chromosome arms resulting from terminalization (*q.v.*).

terminal deletion *See* deletion.

terminal inverted repeats (TIR) sequences of nucleotides found at each end of a transposon (*q.v.*), but in reverse order. Each transposon family is defined by the fact that its members share the same TIRs. In maize, Ac and Ds have the same 11 bp TIR; whereas the Suppressor-mutator (Spm) transposable element has a 13 bp TIR. A Spm transposase does not recognize an Ac TIR and *vice versa*.

terminalization in cytology, the progressive shift of chiasmata from their original to more distal positions as meiosis proceeds through diplonema and diakinesis. *See* Appendix C, 1931, Darlington.

terminal redundancy referring to the repetition of the same sequence of nucleotides at both ends of a DNA molecule.

terminal taxa the groups that occur at the ends of branches in a cladogram.

terminal transferase a deoxyribonucleotidyltransferase that is used by molecular biologists to add a homopolymer tail, e.g., polydeoxyadenylate, to each end of a vehicle DNA. The enzyme is then used to add poly T tails to a passenger DNA. The passenger and vehicle are then annealed via their complementary termini, ligated, and cloned. *See* Appendix C, 1972, Lobban and Kaiser.

termination codon a codon that signals the termination of a growing polypeptide chain. *See* Appendix C, 1965, Brenner *et al.*; amber mutation, ochre mutation, opal codon, stop codon, universal code theory.

termination factors *See* release factors.

termination hairpin, termination sequence *See* terminators.

terminators nucleotide sequences in DNA that function to stop transcription; not to be confused with terminator codons that serve as stop signals for

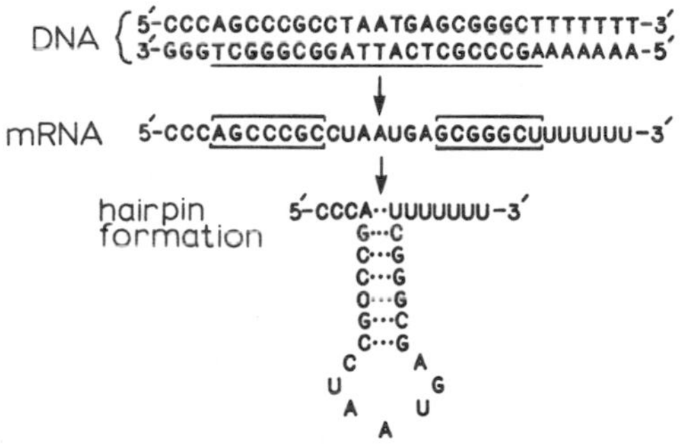

```
DNA  { 5′-CCCAGCCCGCCTAATGAGCGGGCTTTTTTT-3′
     { 3′-GGGTCGGGCGGATTACTCGCCCGAAAAAAA-5′
                        ↓
mRNA   5′-CCCAGCCCGCCUAAUGAGCGGGCUUUUUUU-3′
                        ↓
hairpin         5′-CCCA··UUUUUUU-3′
formation            G···C
                     C···G
                     C···G
                     O···G
                     G···C
                     C···G
                    C     A
                   U       G
                    A     U
                       A
```

translation. In the illustration here, the lower DNA strand is being transcribed from left to right. The RNA segment transcribed from the underlined DNA forms a hairpin-shaped loop because the two blocks of nucleotides have complementing base sequences. This tends to force the adjacent region of the DNA/RNA hybrid to open up. Since it consists of polyribo-U and polydeoxy-A regions that bind weakly, the mRNA molecule will detach at this point. *See* attenuator, exon.

territoriality the defense by an animal or group of an area against members of the same species.

territory an area of the habitat occupied by an individual or group. If members belonging to the same species enter the territory, they are attacked as trespassers.

Tertiary the older of the two geologic periods making up the Cenozoic era. *See* geologic time divisions.

tertiary base pairs the specific base pairs of a tRNA molecule responsible for its three-dimensional folding. Most of these base pairs are evolutionarily conserved in all tRNA molecules.

tertiary nucleic acid structure the three-dimensional conformation of a nucleic acid strand (chain) formed by folding of the strand and formation of intrastrand complementary base pairing (e.g., transfer RNA, *q.v.*).

tertiary protein structure *See* protein structure.

tesserae functionally different patches of endoplasmic reticulum, each bearing a characteristic set of enzymes.

test cross a mating between an individual of unknown genotype, but showing the dominant phenotype for one or more genes, with a tester individual known to carry only the recessive alleles of the genes in question. The test cross reveals the genotype of the tested parent. For example, an individual showing the A and B phenotypes is crossed to an *aabb* tester. If the F_1 contains individuals of AB, Ab, aB, and ab phenotypes in a 1 : 1 : 1 : 1 ratio, this reveals that (1) the tested parent produced gametes with genotypes identical to the F_1 phenotypes and in the same proportions, and (2) the tested parent was AaBb. The first test crosses were made in 1862 by Gregor Mendel.

tester strain a multiply recessive strain that provides the genotypically known mate used in a test cross.

testicular feminization *See* androgen insensitivity syndrome.

testis (*plural* **testes**) the gamete-producing organ of a male animal.

testosterone a masculinizing, steroid hormone secreted by interstitial cells of the testis.

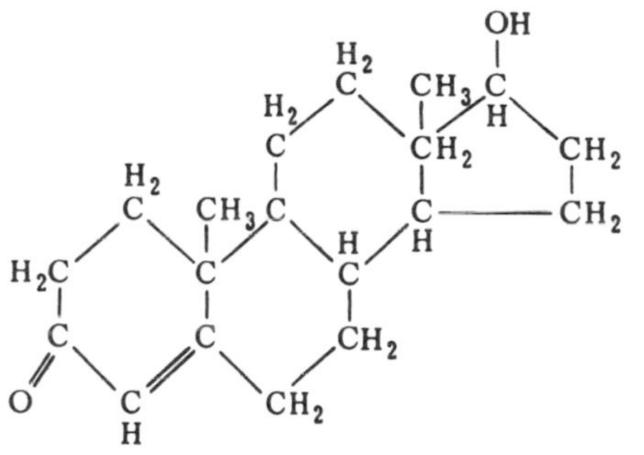

test-tube baby the production of a child by *in vitro* fertilization, followed by embryo transplantation to complete gestation in a normal uterus. This may be provided by the biological or surrogate mother.

tetra-allelic referring to a polyploid in which four different alleles exist at a given locus. In a tetraploid $A_1A_2A_3A_4$ would be an example.

tetracyclines a family of antibiotics obtained from various species of *Streptomyces*. Tetracyclines bind to the 30S subunit of prokaryotic ribosomes and prevent the normal binding of aminoacyl-tRNA at the A site. The structure of a typical tetracycline appears below. *See* cyclohexamide, ribosome, ribosomes of organelles, translation.

```
       OH      O      OH           O
       |       ||     |     OH     ||
       C       C      C      |     C       CONH2
     //  \   /   \  //  \    |   /   \   /
  HC       C       C      \  C  /      C
  |        ||      |         |         ||
  HC       C       C         C         CH
    \\   /   \   / H \    /  H  \    /
       C       C       C           C
       H      / \      H2          H2
            OH   CH3
```

tetrad 1. four homologous chromatids (two in each chromosome of a bivalent) synapsed during first meiotic prophase and metaphase. *See* meiosis. 2. four haploid products of a single meiotic cycle.

tetrad analysis the analysis of crossing over by the study of all the tetrads arising from the meiotic divisions of a single primary gametocyte. To perform such an analysis, one must use an organism in which the meiotic products are held together, as for example, in the case of meiospores confined in an ascus sac. Genera suitable for such analyses include *Ascobolus, Aspergillus, Bombardia, Neurospora, Podospora, Saccharomyces, Schizosaccharomyces, Sordaria,* and *Sphaerocarpus.*

tetrad segregation types For a bivalent containing the genes *A* and *B* on one homolog and *a* and *b* on the other, three patterns of chromatid segregation are possible: *AB, AB, ab, ab* (referred to as the parental ditype); *AB, Ab, aB, ab,* where two chromatids are recombinant (the tetratype); and *Ab, Ab, aB, aB,* where all chromatids are recombinant (the nonparental ditype).

tetrahydrofolate *See* folic acid.

Tetrahymena a genus containing *T. pyriformis,* the species for which the most genetic information is available, and *T. thermophila,* the species in which UAA and UAG were shown to encode glutamine rather than serving as stop codons. The nuclear reorganization that takes place following conjugation (*q.v.*) in these ciliates makes them a rich source of telomeres and the enzymes that work on them. This is because during the regeneration of a new macronucleus, the DNA of the micronucleus is split at specific sites into hundreds of thousands of pieces. New telomeres are synthesized at each new end, and each chromosome fragment undergoes many cycles of replication. The *Tetrahymena* macronucleus contains 20,000 to 40,000 telomeres! In *T. thermophila,* each macronucleus carries about 45 copies of each expressed gene, and it is responsible for the phenotype of the cell. The micronucleus, which is transcriptionally inactive, contains five pairs of metacentric chromosomes. *See* Appendix A, Protoctista, Ciliophora; Appendix E; genetic code, nuclear dimorphism, telomerase, telomere.

tetramer a structure resulting from an association of four subunits. If the subunits are all identical, they form a homotetramer; if the subunits are not all identical, they form a heterotetramer.

tetranucleotide hypothesis the proposal that DNA is a linear, single-stranded polynucleotide consisting of four repeating bases (adenine, thymine, guanine, and cytosine) linked to each other by a deoxyribose phosphate ester backbone. *See* Appendix C, 1929, Levene and London; Chargaff rule.

Tetraodon nigroviridis the green spotted pufferfish. A species, which like *Takifugu rubripes* (*q.v.*), possesses a very small genome. However, it has the advantage of being a popular aquarium fish that is easily reared in tap water.

tetraparental mouse a mouse developed by artificial fusion of embryonic cells from two genetically different blastulas.

tetraploid having four haploid sets of chromosomes in the nucleus. *See* allotetraploid, autotetraploid, polyploidy.

tetrasomic having one chromosome in the complement represented four times in each nucleus.

tetratype *See* tetrad segregation types.

tetravalent *See* quadrivalent.

thalassemias a group of human anemias due to imbalance in the ratio of alpha and/or beta hemoglobin subunits. Since there are four alpha genes per genome, deletions (commonly produced by unequal crossing over) can result in an individual having any number of alpha genes from zero to four. The complete absence of alpha genes produces *hydrops fetalis* (*q.v.*). With only one alpha gene, excess beta chains form a tetramer (β_4), resulting in hemoglobin H disease. Individuals with two or three alpha genes are almost indistinguishable from normal. Epidemiological studies have shown that individuals with alpha thalassemia trait (--/aa or -a/-a) are more resistant to malaria than aa/aa individuals. Incomplete beta chains can be produced by nonsense codons. Deletions in beta genes are commonly produced by unequal crossing over, as are the hybrid chains containing δ and β segments (Hb Lepore) or A_γ and β segments (Hb Kenya). Beta thalassemia (also called Cooley anemia) is a hemoglobinopathy in which few functional beta globin chains are made. A point mutation, within an intron that alters the cutting and splicing signal, causes an extra piece of intron RNA to be present in processed mRNA; the extra piece shifts the reading frame and causes translation to stop prematurely, yielding a truncated and nonfunctional beta globin molecule. *See* Appendix C, 1976, Kan *et al.*; 1986, Costantini *et al.*; Desferal, hemoglobin fusion genes, hemoglobin homotetramers. http://www.thalassemia.org

thelytoky a type of parthenogenesis in which diploid females are produced from unfertilized eggs and males are absent or rare. There are two types of thelytoky, meiotic (automictic) and ameiotic (apomictic). In automictic thelytoky, meiosis takes place, but the reduction in chromosome number is compensated for later in the life cycle. The most widespread method of doing this is to have a haploid polar body nucleus fuse with a haploid egg nucleus (autofertilization). In apomictic thelytoky, the maturation division in the egg is equational and therefore the egg nucleus remains diploid.

Theobroma cacao the cacao tree, source of chocolate.

theobromine a mutagenically active purine analog. It is the main alkaloid stimulant in chocolate. The enzyme caffeine synthase catalyzes the conversion of theobromine into caffeine (*q.v.*). *See* alkaloid, bases of nucleic acids.

```
      O     CH3
      ‖      |
      C      N
     / \   /   \
    N    C       C—H
    |    ‖       ‖
  O=C    C ——— N
     \  /
      N
      |
     CH3
```

therapeutic cloning the proposal (sometimes also called *somatic cell nuclear transfer*) to generate embryonic stem cells (*q.v.*) that are genetically matched to a donor organism. The purpose is to later induce them to differentiate into a specific tissue to provide grafts. The procedure would involve transplanting the nucleus from a somatic cell of an adult individual to an enucleated egg cell. The diploid egg would be stimulated to undergo early embryonic development *in vitro*. The cultured cells would be genetically identical to the individual that provided the transplanted nucleus. If by appropriate chemical treatment the cells could then be stimulated to differentiate into a specific tissue, it would be a perfect match for the nuclear donor. Therefore diseased or damaged tissues could be replaced by the cloned cells without risk of graft rejection (*q.v.*). *See* nuclear transfer.

thermal denaturization profile *See* melting profile.

thermal neutron a fast neutron from uranium fission that has been slowed down by elastic collision with a moderator such as graphite to energies equivalent to those of gas molecules at room temperature (approximately 0.025 electron volts). The biological effect of thermal neutrons is attributable to the summation of capture and decay radiations. In biological material, the reactions ^{1}H (n, γ) ^{2}H and ^{14}N (n, p) ^{14}C are the most important sources of tissue ionization. The relative importance of these reactions depends on the size of the organism. Protons from nitrogen capture are the major cause of the biological effects of thermal neutrons in an organism the size of *Drosophila*.

thermoacidophiles bacteria that live in extremely acidic hot springs. Species belonging to the genus *Thermoplasma* are examples. They are placed in the archaebacteria (*q.v.*) on the basis of the nucleotide sequences of their 16S rRNAs.

thermophilic heat loving. Said of bacteria that grow at temperatures between 45°C and 65°C (found in fermenting manure and hot springs). *See* hyperthermophile.

Thermotoga maritima a thermophilic bacterium that lives in geothermal marine sediments. It has an optimum growth temperature of 80°C. Its genome is a circular DNA molecule that contains 1,860,725 base pairs. There are 1,877 ORFs, each with an average size of 947 bps. These coding sequences cover 95% of the chromosome. The largest gene family encodes ABC transporters (*q.v.*). The organization of the majority of the genes of *T. maritima* places it in the bacteria. However, a quarter of the genome is archaeal in nature. The mosaic nature of the *T. maritima* genome suggests that extensive lateral gene transfer as occurred during the evolution of this species. *See* Appendix A, Bacteria, Thermotogae; Appendix C, 1999, Nelson *et al.*; Appendix E; horizontal transmission, hyperthermophile.

Thermus aquaticus an aerobic, Gram-negative, heterotrophic, thermophilic bacterium discovered in the natural hot springs of Yellowstone National

Park, Wyoming. This bacterium became the source of the Taq DNA polymerase used in the polymerase chain reaction (*q.v.*) and the Taq DNA ligase used in the ligase chain reaction (*q.v.*). *T. aquaticus* is also the source of the ribosomes that were used in the crystallographic studies which showed the 3D structure of 70S ribosomes complexed with tRNA and mRNA molecules. *See* Appendix A, Bacteria, Deinococci; Appendix C, 2001, Yusupov *et al.*

theta replication a bidirectional mode of replication of circular DNA molecules from a replication origin in which, midway through the replication cycle, the structure produced resembles the Greek letter θ. Also called a *Cairns molecule*. *See* Appendix C, 1963, Cairns; D loop, sigma replication.

thiamine vitamin B_1, the anti-beriberi factor.

```
        NH2            CH3
         |              |
         C              C=====C-CH2-CH2OH
       // \            /      |
      N    C-CH2-N           |
      |    ||     +\         |
 CH3-C     C-H      CH-----S
      \\  /
        N
```

thiamine pyrophosphate a coenzyme of the carboxylases and aldehyde transferases.

thin layer chromatography *See* chromatography.

thioglycolic acid treatment a procedure used to rupture disulphide bridges linking adjacent protein chains.

Thio-tepa trade name for triethylenethiophosphoramide, a mutagenic, alkylating agent.

```
                 S
                 ||
   H2C \         ||        / CH2
    |   > N  -   P  -  N <   |
   H2C /         |         \ CH2
                 N
                / \
             CH2---CH2
```

third cousin *See* cousin.

thirty-seven percent survival dose the radiation dose at which the number of hits equals the number of targets. The dose at which there is an average of one hit per target. *See* target theory.

thr threonine. *See* amino acid.

three-point cross a series of crosses designed to determine the order of three, nonallelic, linked genes upon a single chromosome on the basis of their crossover behavior.

three-strand double exchange *See* inversion.

threonine *See* amino acid.

threshold dose the dose of radiation below which the radiation produces no detectable effect.

threshold effect hypothesis the notion that certain traits with a polygenic basis develop only if the additive effects of contributory alleles exceed a critical value. This hypothesis is often used to explain many all-or-none phenomena with a polygenic mode of inheritance (e.g., susceptibility vs. resistance to certain diseases).

thrifty gene hypothesis the proposal that there exist in human populations genes that facilitate the efficient utilization of food and its conversion into stored fat. The bearers of such genes gain weight during periods of plenty, and as a result are more likely to survive periods of famine. Since populations in early stages of human evolution were alternately exposed to unpredictable periods of feast and starvation, thrifty genes were selected and retained in the gene pool. However, in contemporary societies food is usually available in unlimited amounts and life is sedentary; now thrifty genes become deleterious, since they increase susceptibility to the diabetes type 2 family of diseases. The observation that the symptoms of diabetes disappear under conditions of starvation supports this hypothesis. *See* Appendix C, 1962, Neel; diabetes mellitus, obese.

thrombin *See* blood clotting.

thrombocyte blood platelet. *See* platelets.

thrombocytopenia the condition in humans where the circulation has an abnormally low concentration of platelets (*q.v.*).

thrum the type of flower characterized by short styles and high anthers found among distylic species such as seen in the genus *Primula*. *See* pin.

Thy-1 antigen an antigen on the plasma membrane of thymocytes (*q.v.*) that can be used to distinguish them from other lymphocyte groups.

thylakoid *See* chloroplast.

thymidine the deoxyriboside of thymine. *See* nucleoside.

thymidine kinase an enzyme catalyzing the phosphorylation of thymidine to thymidine monophosphate.

thymidylate kinase an enzyme catalyzing the phosphorylation of thymidine monophosphate and thymidine diphosphate to thymidine diphosphate and thymidine triphosphate, respectively.

thymidylic acid *See* nucleotide.

thymine *See* bases of nucleic acids.

thymine dimer two thymine molecules joined by bonds between their number 5 and 6 carbons, as shown in the accompanying illustration. The reaction forming such dimers occurs when ultraviolet radiations interact with DNA. These dimers, which block future DNA replication, are removed during cut-and-patch repair (*q.v.*). *See* Appendix C, 1961, Wacker, Dellweg, and Lodemann; error-prone repair, xeroderma pigmentosum.

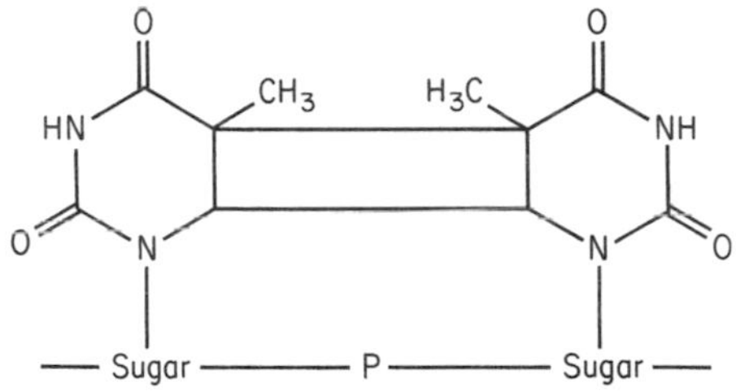

thymocyte a thymus-derived lymphocyte or T lymphocyte.

thymonucleic acid *See* nucleic acid.

thymus an organ lying in the chest of mammals, formed embryologically from gill pouches. This organ functions to populate the body with lymphoid cells. It reaches a maximum size about the time of sexual maturity and then atrophies.

thyroglobulin *See* thyroid hormones.

thyroid hormones thyroxine and triiodothyronine are synthesized by the thyroid gland in response to thyrotropin, a hormone from the anterior pituitary gland. The synthesis of thyroid hormones begins with the selective accumulation of inorganic iodide by the epithelial cells of the gland. The trapped iodide is then oxidized to iodine, and tyrosine residues of the glycoprotein thyroglobulin are then iodinated, converting them to monoiodotyrosine residues (a). The iodotyrosyl residues are then coupled with the elimination of the alanine side chain to form triiodothyronine (b) and thyroxine (c). The structures of these compounds are shown in the accompanying drawings. *See* cretinism.

thyroid-stimulating hormone (TSH) a glycoprotein hormone stimulating secretion by the thyroid. TSH is produced by the adenohypophysis.

thyrotropin *See* thyroid hormones.

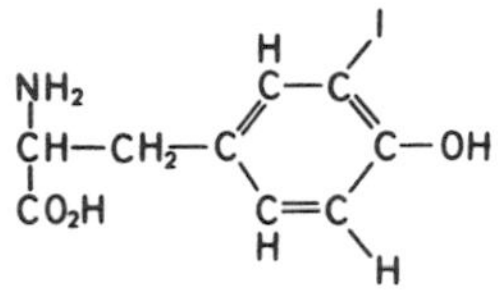

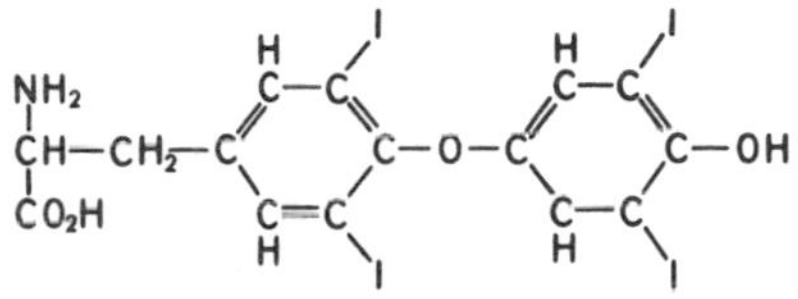

Thyroid hormones

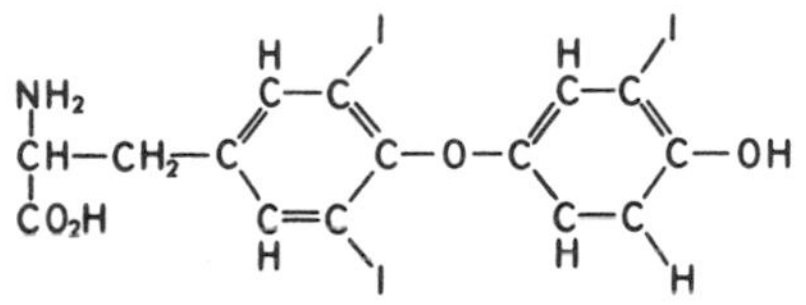

thyroxine *See* thyroid hormones.

TIGR the abbreviation for *T*he *I*nstitute for *G*enomic *R*esearch, a not-for-profit research institute whose primary concern is the comparative analysis of the genomes of prokaryotes and eukaryotes of genetic interest. TIGR was founded in 1992 by J. Craig Venter, and it became the world's largest DNA sequencing institute of that time. The first two bacterial genomes were worked out at TIGR. However, in 1998 Venter left TIGR to set up an even larger facility at Celera (*q.v.*), and his wife, Claire M. Fraser, took the reins at TIGR. *See* Appendix C, 1995, Fleischmann, Venter, and Fraser; Appendix E, Master Web Sites; *Haemophilis influenzae, Mycoplasma genetalium.*

tiller a grass side shoot produced at the base of a stem.

TIM *See* mitochondrial translocase.

timber line the line in high latitudes and in high elevations in all latitudes beyond which trees do not grow.

time-lapse photomicrography a technique in which living cells are photographed using phase contrast, Nomarski, or fluorescence microscopy with a camera that records images at selected intervals (e.g., one frame per minute), and the sequence of images subsequently displayed at a more rapid speed (e.g., 24 frames per second). Time is thus speeded up for a clearer understanding of cellular dynamics.

Ti plasmid a *t*umor-*i*nducing (hence the acronym) plasmid found in the bacterium *Agrobacterium tumefaciens* (*q.v.*) that is responsible for crown gall disease of dicotyledonous plants. When Ti DNA is integrated into the DNA of the host plant, cytokinins and auxins (*both of which see*) are produced. It is these hormones that cause gall formation. Only a small part of the plasmid actually enters the plant; the rest stays in the bacterium, where it has other functions. The wild-type plasmid produces tumor cells, but it can be modified so that it can carry foreign genes into cells without making the recipient cells tumorous. During tumor induction, a specific segment of the Ti plasmid, called the T-DNA (transferred DNA), integrates into the host plant nuclear DNA. Ti-mediated tumorigenesis is the first case of a horizontal mobile element (*q.v.*) that transfers DNA between cells that belong to separate kingdoms. *See* Appendix C, 1974, Zaenen *et al.*; 1981, Kemp and Hall; genetic engineering, promiscuous DNA, selfish DNA.

Tiselius apparatus electrophoresis apparatus.

tissue a population consisting of cells of the same kind performing the same function.

tissue culture the maintenance or growth of tissue cells *in vitro* in a way that may allow further differentiation and preservation of cell architecture or function, or both. *Primary cells* are those taken directly from an organism. Treating a tissue with the proteolytic enzyme trypsin dissociates it into individual primary cells that grow well when seeded onto culture plates at high densities. Cell cultures arising from multiplication of primary cells in tissue culture are called *secondary cell cultures*. Most secondary cells divide a finite number of times and then die. A few secondary cells may pass through this "crisis period" and be able to multiply indefinitely to form a continuous *cell line*. Cell lines have extra chromosomes and are usually abnormal in other respects as well. The immortality of these cells is a feature shared in common with cancer cells. *See* Appendix C, 1907, Harrison; 1965, Hayflick; telomerase.

tissue typing identification of the major histocompatibility antigens of transplant donors and potential recipients, usually by serological tests. Donor and recipient pairs should be of identical ABO blood group, and in addition should be matched as closely as possible for H antigens in order to minimize the likelihood of allograft rejection. *See* histocompatibility molecules.

titer the amount of a standard reagent necessary to produce a certain result in a titration (*q.v.*).

titin the largest protein known, consisting of a continuous chain of 27,000 amino acids. Each molecule spans a distance greater than one micrometer (from the Z discs to the M discs in the striated muscles of vertebrates). Titins act as springs, pulling the muscle fiber back into shape after it is stretched. Titins come in a variety of isoforms generated by alternative splicing (*q.v.*). The human titin gene (*TTN*) is located on the long arm of chromosome 2 and contains 80,780 base pairs. It is subdivided into 178 exons, the largest of which contains 17,106 bases. The *Drosophila* titin gene is located at the distal end of 3L. *Drosophila* titin is a component of muscle sarcomeres, but it also is localized in chromosomes. Here it presumably organizes higher-order chromosome structure and provides elasticity. *See* Appendix C, 1995, Labeit and Kolmer.

TLC thin-layer chromatography. *See* chromatography.

T lymphocyte the lymphocyte responsible for cell-mediated immunological reactions, such as graft rejection, and characterized by the possession of T cell receptors (*q.v.*). T lymphocytes differentiate within the microenvironment of the thymus gland. Mature T cells can be divided into two groups (CD4 and CD8) on the basis of their ability to recognize certain classes of histocompatibility molecules (HCMs). $CD4^+$ cells, which recognize class II HCMs, function as helper T lymphocytes (*q.v.*). $CD8^+$ cells, which recognize class I HCMs, function as cytotoxic T lymphocytes (*q.v.*). *See* histocompatibility molecules, V(D)J recombination. *Compare* B lymphocyte.

T_m the temperature at which a population of double-stranded nucleic acid molecules becomes half-dissociated into single strands; referred to as the *melting temperature* for that system.

TMV tobacco mosaic virus.

TNFs tumor necrosis factors (*q.v.*).

Tn 5 a bacterial transposon that is favorite for studies of the three-dimensional interactions of its transposase proteins with the DNA target sites at the ends of the element during its excision from one host and transfer to another.

Tn 5 transposition intermediate a hairpin-shaped complex between the Tn 5 transposon and two transposases, each bound to recognition sequences at the ends of the Tn 5 element, after its release from the host chromosome. The intermediate can now attach to target sequences in a new host chromosome where it can catalyze the reintegration of the sequence. *See* Appendix C, 2000, Davies *et al.*

tobacco *See Nicotiana.*

tobacco mosaic virus a virus that produces lesions on the leaves of tobacco plants. An infected cell contains about 1×10^7 particles, and a kg of infected tobacco leaves can yield several grams of TMV. This was the first virus to be discovered, the first to be seen with the electron microscope, the first to be purified, the first shown to be built up of regularly arranged subunits, the first from which an infectious nucleic acid was obtained, and the first to have its method of assembly worked out in detail. The TMV particle is a rigid cylinder 18 nm in diameter and 300 nm long. Its genetic material is a single, positive sense strand of RNA that contains 6,400 nucleotides. There are four ORFs. The first two function in replication, the third facilitates movement of the virus from cell to cell, and the fourth encodes the 158 amino acids of the coat protein subunit. The protein making up each capsomere has sites that bind to the proteins of adjacent capsomeres and to the central RNA molecule. This forms a right-handed helix of 8 nm diameter. There are about 17 protein subunits per turn of the helix, and each is attached to three nucleotides. The shell of a completed virus contains 2,130 subunits. *See* Appendix C, 1892, Ivanovski; 1898, Beijerinck; 1935, Stanley; 1937, Bawden and Pirie; 1939, Kausche, Pfankuch, and Ruska; 1955, Fraenkel-Conrat and Williams; 1956, Gierer, Schramm, and Fraenkel-Conrat; 1959, Franklin, Caspar, and Klug; 1960, Tsugita *et al.*; 1982, Goelet *et al.*; Appendix F; *Nicotiana*, readthrough, virus.

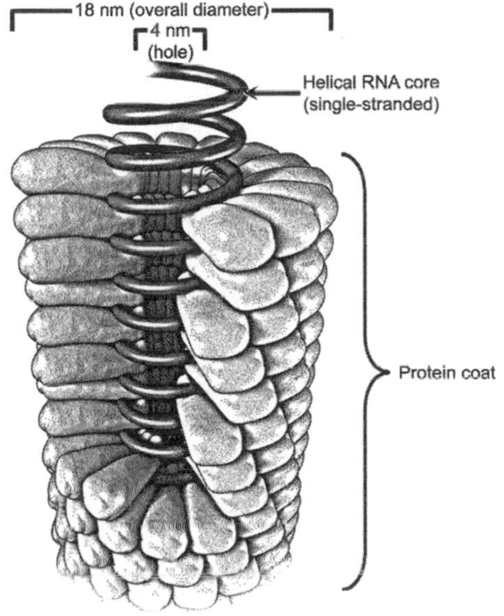

tolerance *See* immunological tolerance.

toluidine blue a metachromatic basic dye used in cytochemistry.

```
                      H                 H
                      C                 C
                   //   \             /   \\
 H3C       HC            C—N=C             C—CH3
    \       |            ‖     |           |
     N—C                 C—S—C             C—NH2
    /     \\           /       \\       / +
 H3C         C                   C
             H                   H
```

TOM *See* mitochondrial translocase.

tomato *See Lycopersicon esculentum.*

T24 oncogene a gene isolated from a line of human bladder carcinoma cells. This oncogene appears to be homologous to the oncogene carried by the Harvey murine sarcoma virus (*q.v.*). The change that leads to the activation of the T24 oncogene is due to a single base substitution. *See* Appendix C, 1982, Reddy *et al.*

tonofilaments synonymous with keratin filaments. *See* intermediate filaments, keratin.

topoisomerase an enzyme that can interconvert topological isomers of DNA. These enzymes alter DNA topology by changing the linking number of circular duplex DNAs or by interconverting knotted and catenated forms. Topoisomerase has replaced a number of earlier terms (DNA relaxing enzyme, swivelase, untwisting enzyme, nick-closing enzymes). Topoisomerases are divided into two classes: Type 1 enzymes make a transient break in one strand of the duplex, whereas type 2 enzymes introduce transient double-strand breaks. During the relaxation of DNA by type 1 topoisomerases, an intact strand of the helix is passed through the break in its complementary strand. *See mei-W68, SPO 11.*

topoisomerase poisons compounds such as Adriamycin (*q.v.*) that inhibit DNA polymerase II and cause cells to arrest at the DNA damage checkpoint (*q.v.*). Poisoned cells subsequently undergo apoptosis (*q.v.*).

topological isomers *See* linking number.

topology the study of the properties of geometrical figures that are subjected to deformations such as bending or twisting.

tormogen cell *See* trichogen cell.

tortoiseshell cat a cat showing patches of orange and more darkly pigmented fur in its coat. The sex-linked gene O is responsible for the conversion of eumelanin to phaeomelanin, which gives the fur an orange coloration. The O gene is epistatic to those autosomal genes, which give the coat a black or agouti color. Since the X chromosome is randomly inactivated in somatic cells during development, females heterozygous for O will show the tortoiseshell phenotype. The term *calico* is sometimes applied to tortoiseshell females that also have patches of white fur. Such females also contain the dominant spotting gene *S*.

totipotent descriptive of cells that have the capacity to differentiate into all of the cells of the adult organism. A zygote is normally totipotent, but most cells of the embryo become progressively restricted in this capacity as development progresses. *See* pluripotent, stem cells.

toxic shock syndrome (TSS) a rare illness caused by toxins produced by certain strains of *Staphylococcus aureus* and characterized by a sudden onset of fever, vomiting, diarrhea, sore throat, muscle aches, rash, and a drop in blood pressure. Other symptoms may appear as the illness progresses, leading to shock and, possibly, death (5% of all cases are fatal). An alternate form, called *Streptococcal toxic shock syndrome (STSS)* is caused by *Streptococcus pyogenes* and appears most often following a skin infection caused by these bacteria. *See Staphylococcus, Streptococcus.*

toxin a poisonous substance elaborated by a microorganism, as well as some fungal, plant, and animal species. An example is alpha amanatin (*q.v.*).

toxoid a poisonous protein that has been detoxified without harm to its antigenic properties.

TP53 the symbol for the tumor suppressor gene that encodes the p53 protein. *TP53* contains 11 exons and is located at 17p13.1. Two promoters have been identified at the 5′ end of *TP53*. The first is 100–250 bp upstream of exon 1, and the second is in intron 1. The gene was first isolated in 1983 from mouse cells and subsequently from human, rat, rhesus monkey, cat, sheep, horse, rabbit, hamster, chicken, and *Xenopus*. Mutations of *TP53* occur in about half of all human cancers. All of the small-cell lung carcinomas and the cancers found in patients with Li-Fraumeni syndrome (*q.v.*) have *TP53* mutations. When introduced into cultured cells from human cancers, the *TP53* genes suppress their proliferation. *See* Appendix C, 1983, Oren and Levine; 1990, Baker *et al.*; anti-oncogenes.

T phages virulent viruses that attack *Escherichia coli* and other enterobacteria. The chromosome is contained in a large capsule of protein, and it is injected into the host through a hollow, tubular tail. The T-even viruses (T2, T4, and T6) have heads 80 × 110 nm, while the T-odd phages (T1, T3, T5, and T7) have isometric heads about 60 nm in diameter. T2, T4, and T6 phages differ in the cell wall receptors to which they bind. Their dsDNAs are linear, cyclically permuted, and terminally redundant. The DNA contains 5-hydroxymethylcytosine (*q.v.*), rather than cytosine. T2 was the first phage to be observed under the electron microscope (1942, Luria and Anderson), and it was utilized in the famous Hershey-Chase experiment (1952). T4, with a genome of 166 kbp, is the best-known of all the T phages. About 300 of its genes have been characterized. Of the 43 phage-encoded proteins, 16 are used in constructing the head and 27 for the tail. T4 was the subject of the classic studies by Benzer (1955), Crick, Brenner *et al.* (1961), Brenner, Stretton, and Kaplan (1965), and Edgar and Wood (1966). In the T7 phage DNA replication involves a concatenation (*q.v.*) of multiple head-to-tail copies of the genome. *See* Appendix C, 1949, Hershey and Rotman; 1961, Rubinstein, Thomas, and Hershey; Appendix F; bacteriophages, cyclically permuted sequences, rII, triplet code T4 RNA ligase, T7 RNA polymerase, viruses.

tracer *See* labeled compound, radioactive isotope.

traction fiber one of the fibers connecting the centromeres of the various chromosomes to either centriole. *See* mitotic apparatus.

trailer sequence a nontranslated segment at the 3′ end of mRNA following the signal that terminates translation, but exclusive of the poly-A tail. The trailer contains the binding site for the polyadenylating enzyme. Some mRNAs contain blocks of nucleotides in their trailers that bind to receptor molecules that are localized within specific regions of cells. *See* Appendix C, 1988, Macdonald and Stuhl; bicoid, exon, leader sequence, polyadenylation.

trait *See* character.

trans *See cis-trans* configuration.

***trans*-acting locus** a genetic element, such as a regulator gene (*q.v.*) that encodes a diffusible product that can influence the activity of other genes. *Trans*-acting genes can be on different DNA molecules from the genes they control. *Contrast with cis*-acting locus.

transcribed spacer that part of a primary rRNA transcript that is discarded during the formation of functional RNAs of the ribosome (*q.v.*).

transcript a length of RNA or DNA that has been transcribed respectively from a DNA or RNA template.

transcriptase RNA polymerase (*q.v.*). *Contrast with* reverse transcriptase.

transcription the formation of an RNA molecule upon a DNA template by complementary base pairing; mediated by RNA polymerase (*q.v.*). Transcription can proceed in opposite directions along different genes within the same chromosome. *See* Appendix C, Taylor, Hradecna, and Szybalski.

transcription factors proteins characterized by DNA-binding segments that enable them to attach to chromosomes and regulate the transcription of specific genes. *Arabidopsis* has over three times more transcription factors than have been identified in *Drosophila* or *Caenorhabditis*, and of the 29 classes of transcription factors that have evolved in this species, 16 are unique to plants. Transcription factors that contain zinc atoms are abundant in animals and fungi. For example, in *Drosophila*, *Caenorhabditis*, and *Saccharomyces* zinc-coordinating proteins make up 51%, 64%, and 56%, respectively, of their transcription factors. However, over 80% of *Arabidopsis* transcription factors lack zinc. *See* zinc finger proteins.

transcription rate the speed at which ribonucleotides are polymerized into RNA chains by RNA polymerases. The transcription rate for mRNA molecules in bacteria at 37°C is about 2,500 nucleotides per minute or about 14 codons per second. This transcription rate corresponds closely to the translation rate (*q.v.*) in bacteria. *Compare with* replication rate. *See* gene expression.

transcription unit the segment of DNA between the sites of initiation and termination of transcription by RNA polymerase; more than one gene may reside in a transcription unit. A polycistronic message (*q.v.*) may be translated as such and the translational product may be enzymatically cleaved later into two or more functional polypeptide chains. RNA is transcribed in a 5′ to 3′ direction from the template strand of the gene. However, when describing the nucleotide sequence of a specific gene, the convention has been adopted to give it the same nucleotide sequence as the RNA transcript, except that each uridine is replaced by thymidine. Any element to the left of the initiation site is said to be "5′ to" or "upstream of" the gene. Any element to the right is "3′ to" or "downstream of" the gene. Nucleotides are numbered starting at the initiation site and receive positive values to the right and negative values to the left. Thus, in a specific gene the binding site for RNA polymerase II might include nucleotides –80 to –5, and the first intron might contain nucleotides +154 to +688. *See* Appendix C, 1967, Taylor *et al.*; coding strand, Miller trees, polyprotein, RNA polymerase, strand terminologies.

transcriptome the totality of mRNAs being produced by a cell at any given time. *See* metabolic control levels.

transcripton a unit of genetic transcription.

transdetermination change in developmental fate of a cell or group of cells. *See in vivo* culturing of imaginal discs.

transduced element the chromosomal fragment transferred during transduction.

transductant a cell that has been transduced. *See* transduction.

transduction the transfer of bacterial genetic material from one bacterium to another using a phage as a vector. In the case of *restrictive* or *specialized transduction* only a few bacterial genes are transferred. This is because the phage has a specific site of integration on the host chromosome, and only bacterial genes close to this site are transferred. In the case of *generalized transduction* the phage can integrate at almost any position on the host chromosome, and therefore almost any host gene can be transferred with the virus to a second bacterium. Transducing phage are usually defective in one or more normal phage functions, and may not be able to replicate in a new host cell unless aided by a normal "helper" phage. *See* Appendix C, 1952, Zinder and Lederberg; abortive transduction.

trans face *See* Golgi apparatus.

transfection a term which is a hybrid between *transformation* and *infection* and refers to the experimental introduction of exogenous DNA or RNA into a cell or embryo, resulting in either a hereditary or a transient change in the affected cells. The term generally denotes one of the following. (1) The transformation (*q.v.*) of bacterial cells with purified viral nucleic acids, resulting in the production of the complete virus in the cells. (2) The transformation of cultured animal cells with purified DNA and incorporation of this DNA into the cell's genome (*q.v.*). In this case the term *transfection* has been adopted rather than *transformation* because the latter term is used in another sense in studies involving cultured animal cells (i.e., the conversion of normal cells to a state of unregulated growth by oncogenic viruses). (3) The transformation of cells or embryos with single- or double-stranded RNA molecules, resulting in the expression of specific proteins or in the silencing of specific genes. Transfection with RNA

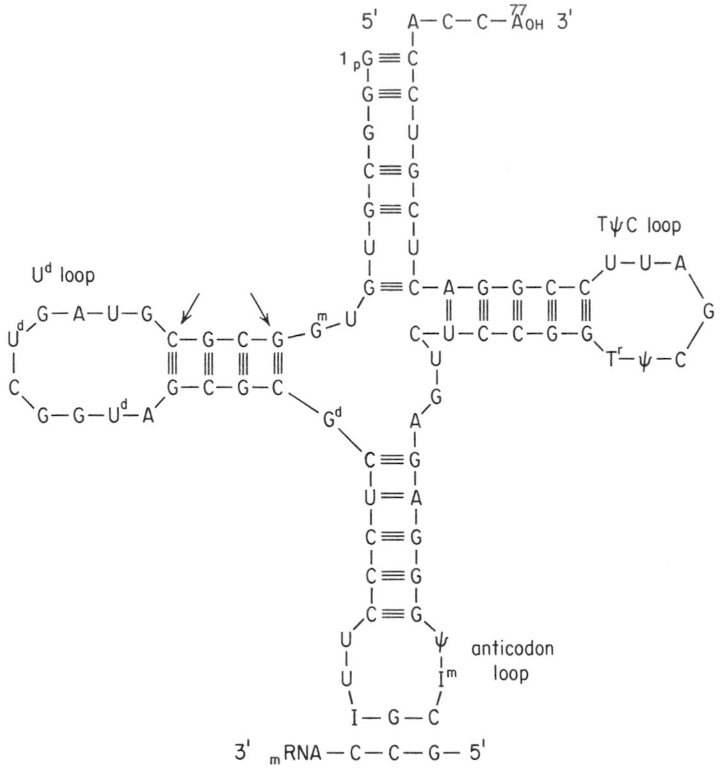

Alanine tRNA

molecules produces changes that are not permanently transmissible. *See* Appendix C, 1972, Jackson, Symons and Berg; 1985, Smithies *et al.*; gene therapy, RNA interference, RNA transfection.

transfectoma a myeloma cell into which immunoglobulin genes, either wild type or altered *in vitro*, have been transfected and expressed. Novel chimeric immunoglobulin molecules can be produced by this technique, including unique combinations of heavy and light chains, or combinations of variable regions with different constant regions (both within and between species). *Contrast with* hybridoma. *See* immunoglobulin.

transferases enzymes that catalyze the transfer of functional groups between donor and acceptor molecules. The most common molecules transferred are amino, acyl, phosphate, and glycosyl groups.

transfer factor a dialyzable extract (lymphokine) from sensitized T lymphocytes that can transfer some types of cell-mediated immunity from one individual to another.

transferred immunity *See* adoptive transfer.

transferrins *See* plasma transferrins.

transfer RNA (tRNA) an RNA molecule that transfers an amino acid to a growing polypeptide chain during translation (*q.v.*). Transfer RNA molecules are among the smallest biologically active nucleic acids known. For example, an alanine transfer RNA isolated from yeast (shown above) contains 77 nucleotides and is folded back upon itself and kept in a "clover leaf" configuration by the characteristic pairings of the bases G to C and A to U. All transfer RNAs attach to their amino acids by the 3′ end, which contains a terminal adenylic preceded by two cytidylic acids. The 5′ end always carries a terminal guanylic acid. The P (near 1) and the OH (near 77) show the positions of the phosphoric acid and hydroxyl groups and the 5′ and 3′ ends of the molecule. Transfer RNAs contain several purines and pyrimidines not generally encountered in other RNA molecules. These *rare bases* (*q.v.*) are formed following transcription, since nuclei are known to contain enzymes that are capable of modifying certain bases on preformed RNA. The site recognized by tRNA synthetase is believed to be located in the neck region adjacent to the dihydrouridine loop (see arrows).

The anticodon occupies positions 36–38. In the drawing, A, U, C, and G have their usual meanings. The rare bases are symbolized as follows: ψ, pseudouridylic acid; T^r, ribothymidylic acid; U^d, dihydrouridylic acid; G^m, methylguanylic acid; G^d, dimethylguanylic acid; I, inosinic acid; and I^m, methylinosinic acid. *See* Appendix C, 1958, Crick, Zamecnik; 1965, Holley *et al.*; 1969, 1971, Dudock *et al.*; 1970, Khorana *el al.*; 1973, Kim *et al.*; codon bias, initiator tRNA, isoacceptor RNA, rare bases, ribonucleic acid (RNA), tRNA genes.

trans-filter induction *in vitro* inductions using organizer tissues separated from reactive cells by a Millipore filter (*q.v.*). The trans-filter induction system allows one to interrupt induction at any time by removing the inducing cells from the surface of the filter.

transformants cells or multicellular organisms that show an inherited modification after exposure to a transforming principle or after incorporating exogenous DNA. *See* transformation.

transformation 1. in microbial genetics, the phenomenon by which genes are transmitted from one bacterial strain to another in the form of soluble fragments of DNA. These may originate from live or dead cells. The DNA fragments dissolved in the external medium can penetrate cells only if they have receptor sites for the DNA on their surfaces. Once inside, a fragment usually replaces, by recombination, a short section of the DNA of the receptor cell that contains a zone of homology. *Also called* bacterial transformation. See Appendix C, 1928, Griffith; 1944, Avery *et al.*; 1964, Fox and Allen; 1970, Mandel and Higa; 1972, Cohen *et al.*; Pneumoccus Transforming Principle (PTP). 2. in genetic engineering of bacterial and eukaryotic cells, the acquisition of a new genetic characteristic through the uptake of exogenous DNA from the surrounding medium, DNA microinjection directly into the cell, or other means. For example, ampicillin-sensitive *E. coli* cells that take up a plasmid (*q.v.*) containing the ampicillin-resistant gene amp^R (*q.v.*) are transformed to ampicillin resistance. *See* competence, plasmid cloning vector. 3. in multicellular organisms, the acquisition of a new genetic makeup through experimental transfer and incorporation of exogenous DNA into the genome (*q.v.*). For example, microinjection of purified recombinant DNA (*q.v.*) into a *Drosophila* embryo produces a transformed adult fly which contains the exogenous DNA in its cellular DNA. The insertion of exogenous DNA into the germ cell DNA of the transformant (*q.v.*) is called *germ line transformation*, which ensures that the transforming DNA is transmitted to the next generation. *See* transgenic animals, P element, P element transformation. 4. in cultured cells, the conversion of normal animal cells to a state of unregulated growth by oncogenic viruses (*q.v.*) or by mutations in anti-oncogenes (*q.v.*). Such transformations are generally accompanied by alterations in cell shape, changed antigenic properties, and loss of contact inhibition (*q.v.*). *Also called* cellular transformation or transfection. *See* Appendix C, 1980, Capecchi.

transformation rescue the suppression of a mutant phenotype by introducing a specific wild-type nucleotide sequence into an embryo. A transposable element is generally used as a vector. The introduced wild-type gene sequences are said to "rescue" the mutant phenotype to normality. Transformation rescue is commonly used in germ line transformation experiments to identify functional components of a gene or to identify genomic sequences that correspond to a genetic locus. *Also called* plasmid rescue. *See* Appendix C, 1982, Spradling and Rubin; P elements.

transformation series the various expressions of a character ordered in a hypothesized sequence from the most primitive, plesiomorphic state to the most derived, apomorphic state. This sequence may be linear (e.g., $A^0 \rightarrow A^1 \rightarrow A^2$), or it may be branched.

transforming growth factor β (TGF-β) a large family of intercellular signaling molecules. These proteins are serine/threonine kinases, and they act by binding receptors on cell surfaces. The factors act as dimers, and their receptors are also dimers. TGF-βs play a role in embryonic development, cellular differentiation, hormone secretion, and immune suppression. *See* activin, cellular signal transduction, *decapentaplegic*.

transgene a foreign gene that is introduced into an organism by injecting the gene into newly fertilized eggs. Some of the animals that develop from the injected eggs will carry the foreign gene in their genomes and will transmit it to their progeny.

transgenic animals animals into which cloned genetic material has been experimentally transferred. In the case of laboratory mice, one-celled embryos have been injected with plasmid solutions, and some of the transferred sequences were retained throughout embryonic development. Some sequences became integrated into the host genome and were transmitted through the germ line to succeeding generations. A subset of these foreign genes expressed themselves in the offspring. *See* Appendix C,

1980, Gordon *et al.*; 1986, Costantini *et al.*; 1987, Keuhn; *Sus scrofa.*

transgenic plants genetically engineered plants that contain useful genes from other species. *See* Appendix C, 1981, Kemp and Hall; Bt designer plants, promoter 35S, Roundup.

transgressive variation progeny phenotypes outside the range of that which occurs in the parents; usually attributed to polygene segregation.

transient diploid a relatively short stage in the life cycles of predominantly haploid fungi or algae during which meiosis occurs.

transient polymorphism polymorphism existing in a population during the period when an allele is being replaced by a superior one.

transit cells in descriptions of stem cell differentiation, transitional, proliferating cells that are descendants of stem cells and that give rise to one or more mature, differentiated cell lineage(s). Transit cells have properties intermediate between stem cells (*q.v.*) and differentiated cells, as well as the plasticity to behave like one or the other of these cell types in response to specific environmental signals. *Also called* transit amplifying cells.

transition *See* base-pair substitution.

translation the formation of a protein directed by a specific messenger RNA (mRNA) molecule (see illustration on page 453). Translation occurs in a ribosome (*q.v.*). A ribosome begins protein synthesis once the 5′ end of an mRNA tape is inserted into it. As the mRNA molecule moves through the ribosome, much like a tape through the head of a tape recorder, a lengthening polypeptide chain is produced. Once the leading (5′) end of the messenger tape emerges from the first ribosome, it can attach to a second ribosome, and so a second identical polypeptide can start to form. When the 3′ end of the mRNA molecule has moved through the first ribosome, the newly formed protein is released and the vacant ribosome is available for a new set of taped instructions. The assembly of amino acids into a peptide starts at the amino end (*N* terminus) and finishes at the carboxyl end (C terminus). There are two binding sites for transfer RNA (tRNA) in the ribosome. The P site (peptidyl-tRNA binding site) holds the tRNA molecule that is attached to the growing end of the nascent polypeptide. The A site (aminoacyl-tRNA binding site) holds the incoming tRNA molecule charged with the next amino acid. The tRNAs are held so that their anticodons form base pairs with adjacent complementary codons of the mRNA moving through the ribosome. In the diagram above, an mRNA molecule is shown progressing through a ribosome. At time T_0, codon 5 of the messenger tape occupies the P site and codon 6 the A site. About half a second later, the mRNA has advanced to the left by one codon. The bond between tRNA I and the 5th amino acid of the nascent polypeptide has been split, and it has been linked to amino acid 6. The atoms involved in the rearrangements are in boldface type. This reaction is catalyzed by *peptidyl transferase*, an enzyme that is bound tightly to the ribosome. Transfer RNA molecule I is discharged from the ribosome, tRNA II has entered the P site and is now attached to a polypeptide chain six amino acids long. A new tRNA (III) carrying an appropriate anticodon enters the A site. Note that, in the diagram, letters represent atoms, whereas circled letters represent molecules (i.e., nucleotides or amino acid residues). *See* Appendix C, 1959, McQuillen *et al.*; 1961, Dintzis; 1963, Okamoto and Takanami, Noll *et al.*; 1964, Gilbert; 1974, Shine and Dalgarno; 1976, Pelham and Jackson; amino acid, elongation factors, initiation factors, leader sequence peptide, *N*-formylmethionine, peptide bond, receptor-mediated translocation, start codon.

translational amplification a mechanism for producing large amounts of a polypeptide based upon prolonged mRNA lifetime. Since there is only a single copy of the ovalbumin gene per genome, large numbers of ovalbumin molecules produced by the cells of the chicken oviduct are generated by translational amplification.

translational control the regulation of gene expression through determining the rate at which a specific RNA message is translated.

translation elongation factors ubiquitous proteins that transport aminoacyl tRNAs to the ribosomes and participate in their selection by the ribosomes. Translation elongation factors are symbolized *Ef-Tu* in prokaryotes and *Ef-1* in eukaryotes. Comparisons of the amino acid sequences in specific segments of Efs from a variety of prokaryotes and eukaryotes have shown that these molecules are the slowest evolving proteins discovered so far. Therefore, certain sequences from Efs have been used to construct phylogenetic trees to determine critical steps in the early evolution of life. *See* endokaryotic hypothesis, opisthokonta.

translation rate the speed at which amino acids are polymerized into polypeptide chains on ribosomes. In bacteria at 37°C, the translation rate is approximately 15 amino acids per second. The transla-

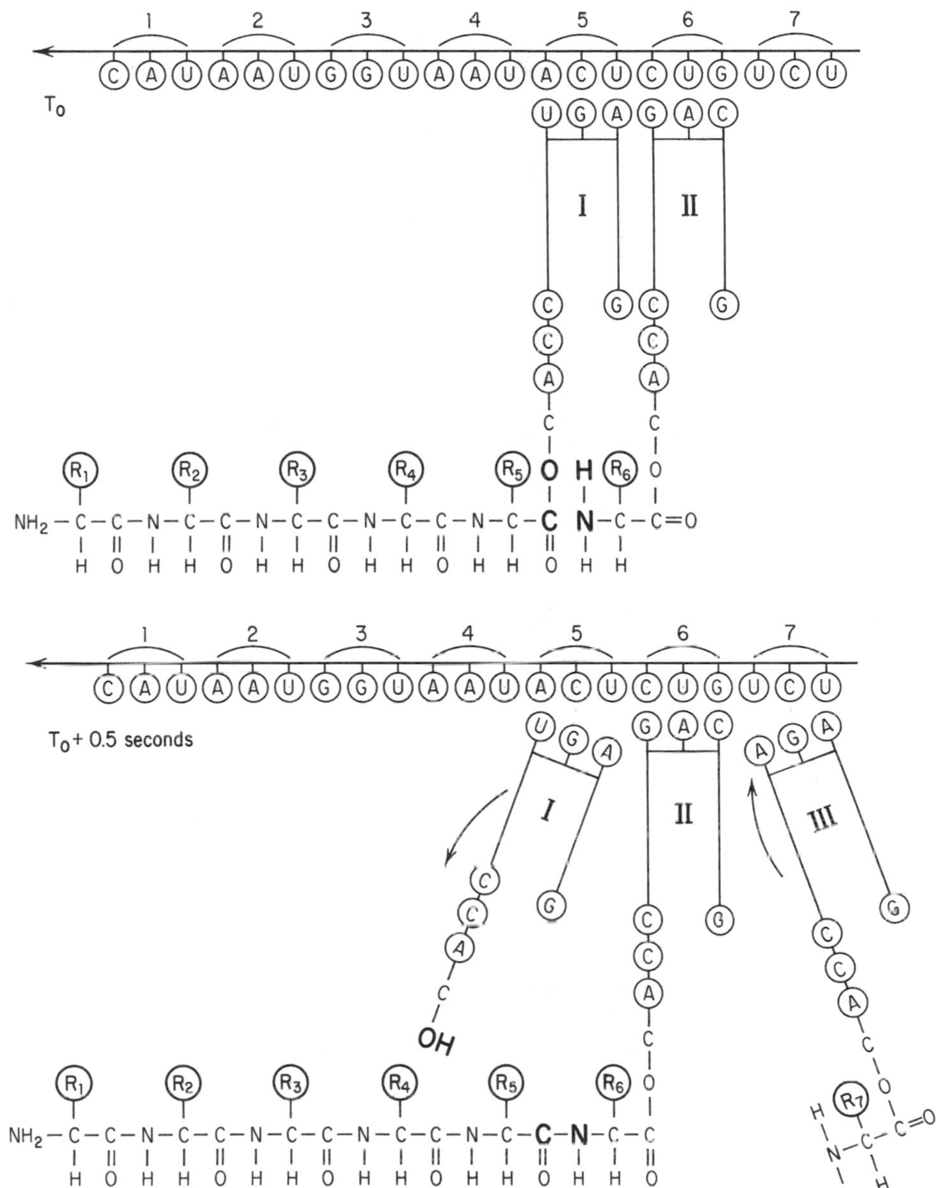

Translation

tion rate for eukaryotic cells is much slower (e.g., two amino acids per second in red blood cells *in vitro*). *Compare* replication rate, transcription rate.

translocase a protein that forms a complex with GTP and the ribosome. Translocation of the charged tRNA from the A site to the P site is coupled with the hydrolysis of GTP to GDP and with the release of the translocase. *See* elongation factors, mitochondrial pore-forming translocases, translation.

translocation 1. the movement of mRNA through a ribosome during translation (*q.v.*) Each translocation exposes an mRNA codon in the A site for base pairing with a tRNA anticodon. **2.** a chromosome aberration which results in a change in position of a chromosomal segment within the genome, but does not change the total number of genes present. Various types of translocations are illustrated on page 454. An *intrachromosomal translocation* is a three-break aberration that results in the transposition of a chromosomal segment to another region of the same chromosome. Such an aberration is often called a *shift*. *Interchromosomal translocations* involve interchanges between nonhomologous chromosomes. A

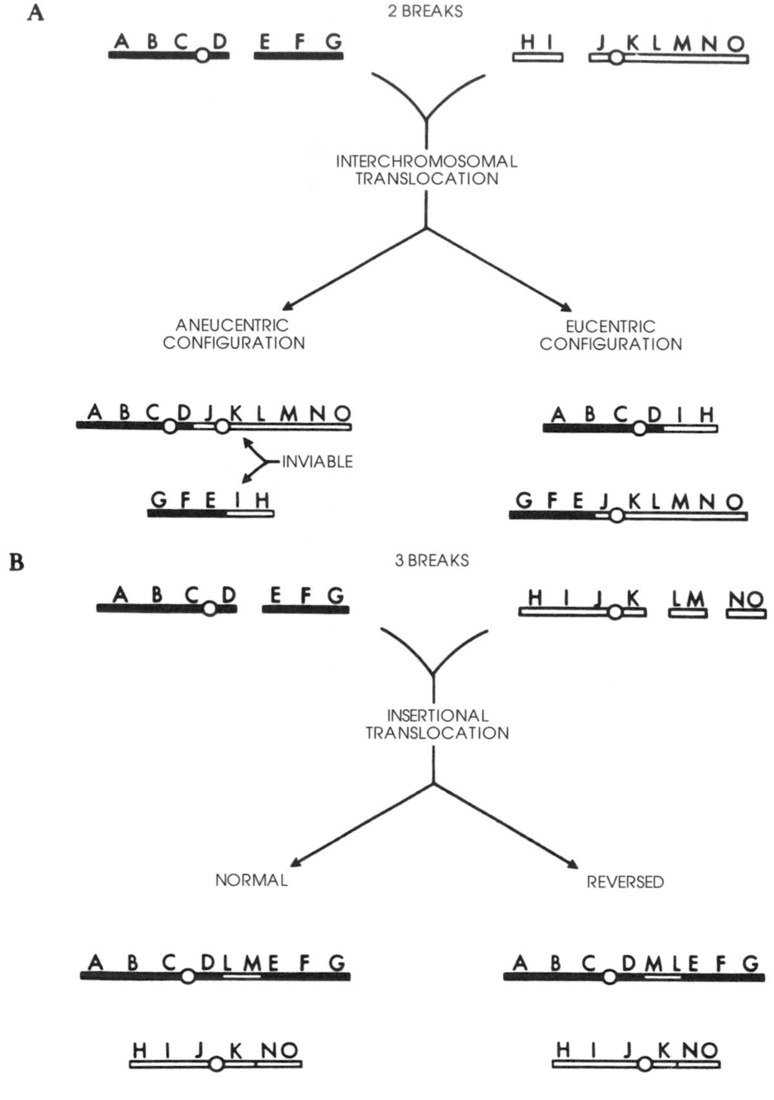

Translocation

reciprocal or eucentric translocation (Figure A, right) is a two-break aberration that results in an exact interchange of chromosomal segments between two nonhomologous chromosomes and produces two monocentric translocated chromosomes. A *nonreciprocal or aneucentric translocation* (Figure A, left) is a two-break aberration that results in a dicentric and an acentric translocated chromosome. A three-break interchromosomal translocation (Figure B) can generate a deficient chromosome and a recipient chromosome containing an intercalated segment of the other nonhomologous chromosome. This is called an *insertional translocation*. *See* Appendix C, 1923, Bridges; Burkitt lymphoma, Cattanach translocation, centric fusion, Philadelphia (Ph[1]) chromosome.

translocation Down syndrome familial Down syndrome, caused by having three copies of chromosome 21—two as separate chromosomes and one translocated to another chromosome, usually number 14. In families where one parent is a translocation heterozygote, the probability of having a child with Down syndrome is .33. *See* Appendix C, 1960, Polani *et al.*

translocation heterozygote an individual or cell in which two pairs of homologous chromosomes have reciprocally exchanged nonhomologous segments between one member of each pair. Each chromosome pair contains both homologous and nonhomologous segments, i.e., one normal (untranslocated chromosome) and one translocated chromosome. A translocation heterozygote (illustrated below) forms a quadrivalent chromosomal association during pachynema (*see* meiosis), and the subsequent segregation of the four chromosomes is determined by their centromere orientations. In the case of *adjacent disjunction*, (or *adjacent segregation*), each daughter cell receives a normal and a translocated chromosome. The gametes produced from such cells are inviable because they contain certain genes in duplicate and are deficient for others. Two kinds of adjacent disjunction are recognized. In *adjacent-1 segregation*, homologous centromeres go to opposite anaphase poles. In *adjacent-2 segregation*, homologous centromeres go to the same anaphase pole. In the case of *alternate disjunction*, both translocated chromosomes go to one cell and both normal ones to the other. The gametes that result are viable because they contain all the genetic material.

translocation of proteins *See* receptor-mediated translocation.

translocation mapping gene mapping by the use of translocation chromosomes as markers. The semisterility that is usually associated with structural translocation heterozygotes is a phenotypic marker that can be used to locate the position of the break point of the translocation relative to other more conventional gene markers.

translocon an organelle in the endoplasmic reticulum (*q.v.*) that allows a protein manufactured on a ribosome to pass from the cytoplasm to the ER lumen. The mammalian translocons are composed of a small number of specific proteins, and some of these are homologous to proteins that were identified in early genetic screens for secretion mutants in *Saccharomyces*. When bound to a ribosome, the aqueous pore that spans the translocon is estimated to be about 5 nm in diameter. The translocon does not disassemble when the ribosome leaves. However, it does change its conformation, and the pore diameter is reduced to about 2 nm. *See* Appendix C, 1975, Blobel and Dobberstein; 1991, Simon and Blobel; receptor-mediated translocation.

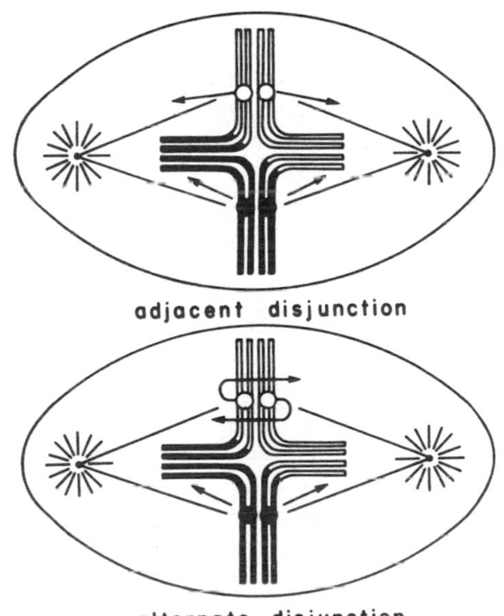

Translocation heterozygote

7 transmembrane domain (7TM) receptor an integral protein (*q.v.*) belonging to a large superfamily of transmembrane receptors that contain seven membrane-spanning domains. Most of the members of 7 TM receptors are G protein-coupled receptors (*q.v.*), each of which binds a signaling molecule on the extracellular side of the membrane and transduces a signal on the cytoplasmic side to initiate or inhibit biochemical reactions within the cell. Examples of 7 TM receptors are sensory and neurotransmitter receptors.

transmission electron microscope *See* electron microscope.

transmission genetics that part of genetics concerning the mechanisms involved in the transfer of genes from parents to offspring.

transmutation the transformation of one element into another accompanying radioactive decay (*q.v.*).

transplantation transfer of a part of an organism to another organism or to another position in the same organism. In zoology, the term is used interchangeably with graft (*q.v.*). In botany, graft is used in the above sense, and transplantation is used in the sense of planting again in a different place. *See* rejection.

transplantation antigen a protein coded by genes in the major (e.g., human HLA, mouse H-2) or minor histocompatibility loci and present on most vertebrate cells. Transplantation antigens are targets for T lymphocytes if the graft bears antigens different from that of the host. *See* histocompatibility molecules.

transposable elements DNA sequences that move from one chromosomal site to another. They were discovered in maize by McClintock during her analysis of the *Activator-Dissociation* system (*q.v.*). Transposable elements were subsequently detected in bacteria as insertion sequences (symbolized *IS1*, *IS2*, etc.). Later bacterial transposable elements were discovered that carried genes that conferred antibiotic resistance in addition to the segments that functioned in transposition. These transposons were labeled *Tn1*, *Tn2*, and so on. Next the transposable elements discovered in yeasts were called *Ty elements* (*q.v.*). The P elements (*q.v.*) of *Drosophila melanogaster* proved to be transposons (*q.v.*), and mariner elements (*q.v.*) were shown to be present in many other insects. At least half of all *Drosophila* mutations are now known to result from the insertion of transposable elements. The first transposable element of *Caenorhabditis elegans* was called *Tc1*. Homologs of Tc1 and mariner are thought to have a monophyletic origin and are therefore placed in a superfamily that constitutes the largest group of DNA transposons in nature. Transposons from the Tc1/mariner superfamily occur in fungi, plants, ciliates, and animals, including nematodes, arthropods, fish, frogs, and humans. In any given population most transposons are inactive. This is because most are defective. For example, in maize lines there will be a few hundred *Ds* elements for each *Ac* element. *See* Appendix C, 1950, McClintock; 1969, Shapiro; 1974, Hedges and Jacob; 1979, Cameron *et al.*; 1982, Bingham *et al.*; 1983, Bender *et al.*; 1984, Pohlman *et al.*; 1985, Boeke *et al.*; 1986, Semeshin *et al.*; 1988, Kazazian *et al.*; 1994, Whitham *et al.*; *Arabidopsis thaliana*, copa elements, C value paradox, *Dotted*, lysogenic conversion, Minos element, R elements, retroposon, site-specific recombination, terminal inverted repeats (TIRs), Tn 5 transposition intermediate, transposase.

transposase an enzyme encoded by a segment of a transposable element (*q.v.*). The transposase mediates both the excision of a transposable element from the chromosome of a host and its subsequent reintegration into another site. Each transposase generally contains a DNA-binding domain near its N-terminal end, a domain that serves as a nuclear recognitional signal, and a catalytic domain. The nuclear recognition signal allows the transposase to take advantage of the receptor-mediated transport machinery of host cells to enter the host nucleus. The catalytic domain is responsible for the DNA cleavage and joining reactions that let the transposon excise from one site and reintegrate into another site in the host genome. *See* *RAG-1* and *RAG-2*, Tn 5 transposition intermediate.

transposition 1. the process of insertion of a replica of a transposable element at a second site; another replica remains at the original insertion site. This type of transposition is a major source of gene expansion during evolution. This is because the old transposon remains at its original site, while its replica subsequently occupies a new position in the host genome. Any adjacent genes excised along with the transposon are also duplicated during the reintegration process. *See* evolutionary mechanisms, Tn 5. 2. movement of a chromosomal segment to a new location within the same chromosome, or to a different chromosome, without reciprocal exchange. *See* translocation.

transposons one kind of transposable element in both prokaryotes and eukaryotes that is immediately flanked by inverted repeat sequences, which in turn are immediately flanked by direct repeat sequences. Transposons usually possess genes in addition to those needed for their insertion (e.g., genes for resistance to antibiotics, sugar fermentation, etc.). With the passage of time *transposon* has come to be used as a synonym for any *transposable element*. *See* Appendix C, 1974, Hedges and Jacob; integron, retroposon.

transposon tagging a technique used to isolate mutations induced by the insertion of a transposable element into genes of interest. The mutations are identified by the phenotypic changes that result from their inactivation. The chromosomal locations of the mutations can then be identified by hybridization of the inserted sequences with polynucleotide probes complementary to the transposon. P elements (*q.v.*) have been used to tag *Drosophila* genes, and *Activator* and other maize transposons have served to tag genes from maize and heterologous species. *See* Appendix C, 1994, Whitman *et al.*; *Activator-Dissociation* system, tobacco mosaic virus.

trans-splicing joining of an exon from one gene to an exon of a different gene. In trypanosomes, trans-splicing involves the addition of a short segment that consists of 39 nucleotides to the 5′ ends of all mRNAs. The intron-like sequence at the 5′ end of pre-mRNAs destined for trans-splicing has been named an *outron*. With the exception of platyhelminths and nematodes, in multicellular eukaryotes from plants to humans all pre-mRNA splicing involves *cis*-splice sites. Nearly all *Caenorhabditis elegans* genes contain introns, and these are removed by conventional *cis*-splicing. Polycistronic messages are processed by the addition of a trans-spliced donor to an outron that occurs at the beginning of each cistron. In this way the multigene messages are processed into single gene units. The gene clusters are regulated and transcribed from promoters upstream of the most 5′ gene. *Compare with* cis-splicing. *See* exon shuffling, operon, *Trypanosoma*.

transvection a type of position effect (*q.v.*) in *Drosophila* where the ability of one allele of a gene to influence the activity of another on an opposite homolog requires that these two loci be synapsed. For example, a^1/a^1 and a^2/a^2 individuals both show mutant phenotypes, whereas the a^1/a^2 hybrids appear to be wild types. However, if a rearrangement is introduced that prevents the pairing of a^1 and a^2, the mutant phenotype is now expressed. For this reason, the phenomenon has also been called *synapsis-dependent allelic complementation*. The model that explains transvection assumes that each gene consists of a regulatory element (RE) and a nearby transcription unit (TU). In a^1 the regulatory element is functional, while the transcription unit is defective. In a^2 the reverse is true. In the hybrid (diagrammed here), the normal activity of the gene is restored because the a^1 regulatory element can enhance the activity of the a^2 transcription unit, provided that the synapsis of the homologs keeps RE close to TU. *See* Appendix C, 1945, Lewis; *cis-trans* configurations, somatic pairing.

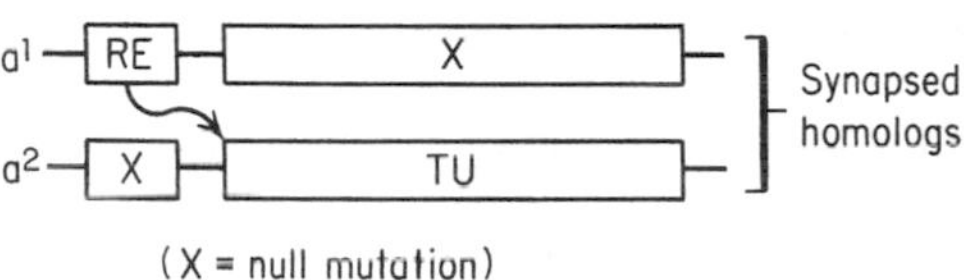

trend an apparently directional evolutionary change in a character within a lineage, a chronocline. For example, many mammalian lineages show a tendency to increase in size over part of their evolutionary history. *See* orthogenesis.

Treponema pallidum the bacterium that causes the sexually transmitted disease syphilis. It is an obligate parasite of humans. Successful chemotherapy of syphilis began with Salvarsan (*q.v.*). After 30 years, it was replaced by penicillin. The complete genome sequence of *T. pallidum* has been recently completed and shown to contain 1.14×10^6 nucleotide base pairs within which are 1,041 ORFs. *See* Appendix A, Prokaryotae, Bacteria, Spirochaetae; Appendix C, 1909, Ehrlich *et al.*.

TRF2 the abbreviation for telomeric repeat-binding factors 2 (*q.v.*).

triallelic referring to a polyploid in which three different alleles exist at a given locus. In a tetraploid, $A_1A_2A_3A_3$ and $A_1A_2A_2A_3$ would be examples.

Triassic the most ancient of the Mesozoic periods, during which the first dinosaurs and mammals arose. Gymnosperms and ferns were the dominant plants. At the end of the Triassic, a mass extinction wiped out about 25% of all animal families. Pangaea began to break apart. *See* continental drift, geologic time divisions.

Tribolium a genus of flour beetles containing *T. castaneum* and *T. confusum*, genetically well-known species. At present about 125 mutations are known for *castaneum* and 75 for *confusum*. *See* Appendix A, Animalia, Insecta, Coleoptera; *Hox* genes.

tricarboxylic acid cycle a synonym for citric acid cycle (*q.v.*).

trichloroacetic acid a compound commonly used to precipitate proteins during biochemical extractions.

```
     Cl
     |
Cl—C—COOH
     |
     Cl
```

trichocyst the minute protrusible, spindle-shaped organelles in the ectoplasm of many ciliates.

trichogen cell a large cell that secretes the long, tapering hair of the insect bristle. A smaller tormogen cell forms the circular chitinous socket around the base of the bristle.

trichogyne a receptive hypha extending from the ascogenous mycelium of a fungus such as *Neurospora*.

Trichoplax adhaerans a marine species which represents the simplest animal phylum, the Placozoa. In culture these animals are a few millimeters in diameter and about 10 μm thick. There are four cell types: (a) the flat covering cells of the dorsal epithelium, each with a single cilium, (b) ventral cylindrical, ciliated epithelial cells, (c) ventral, non-ciliated gland cells that secrete digestive enzymes, and (d) cells containing contractile fibers in their cytoplasm which reside in the space between the dorsal and ventral epithelia. The ventral ciliated cells are adhesive and capable of resorbing materials that have been digested from the substrate. *Trichoplax* has the smallest genome of all animals surveyed to date, 40 mb, and its haploid chromosome number is 6. A *Pax* gene has been isolated and shown to be expressed in the margin of *Trichoplax*. The gene product may specify the differentiation of the fiber cells that are regarded as protomesodermal cells. *See* *Pax* genes.

triethylenethiophosphoramide *See* Thio-tepa.

triiodothyronine *See* thyroid hormones.

trilobites the earliest of the arthropods, a group found throughout the Paleozoic era. A dorsal view of a Cambrian trilobite, *Cedaria,* is shown here. *See* ecdysis.

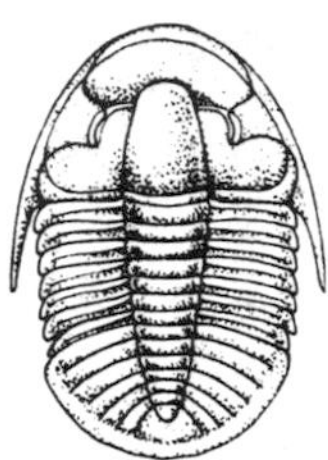

trimesters traditional terms used in medicine which divide the 9-month period of normal pregnancy in women into three stages, each lasting 3 months. The differentiation of cells and tissues in the embryo take place during the first trimester. The mother feels fetal movements early in the second trimester, and there is a rapid growth and maturation of the organ systems. During the third trimester the organ systems are completed, and fat accumulates under the skin of the fetus. It moves into position for birth which brings the third trimester to an end.

trimethylpsoralen a low-molecular-weight, planar molecule that undergoes photochemical reactions with pyrimidines. Trimethylpsoralen molecules intercalate into double-stranded DNA. Upon exposure to ultraviolet light, the molecules attach covalently to pyrimidines forming both monoadducts and interstrand cross links. Trimethylpsoralen has no effects on proteins.

trinucleotide repeats unstable DNA sequences found in several human genes. Normally the triplets are repeated in tandem 5–50 times. When the number rises above the normal range, mutant disease syndromes appear. The triplets subject to expansion start with C and end with G. Examples are given in the table below. *Also called* triplet repeats. *See* microsatellites.

trioecy a sexual trimorphism in plants that can exist as either male, female, or bisexual individuals.

triparental recombinant a progeny phage containing marker genes derived from each of three different phages that simultaneously infected the host cell. The observation of triparental recombinants demonstrates that repeated recombinational events occur in the infected cell between replicating nucleic acid molecules derived from all parent viruses. *See* Visconti-Delbrück hypothesis.

tripartite ribbons synaptonemal complexes (*q.v.*).

triplet a unit of three successive bases in DNA or RNA that codes for a specific amino acid. *See* amino acid, genetic code, translation.

triplet code a code in which a given amino acid is specified by a set of three nucleotides. *See* Appendix C, 1961, Crick, Brenner *et al.*; amino acid, genetic code, proflavin, reading frame, translation.

triplet repeats *See* trinucleotide repeats.

triplex *See* autotetraploid.

triploid an organism having three haploid sets of chromosomes in each nucleus. In those situations where triploid tissues have two paternal and one maternal complement, the triploid is said to be android, whereas gynoid triploids have two maternal and one paternal set of chromosomes.

triskelion *See* receptor-mediated endocytosis.

trisomic an organism that is diploid but contains one extra chromosome, homologous with one of the existing pairs, so that one kind of chromosome is present in triplicate. *See* Appendix C, 1920, Blakeslee *et al.*; Down syndrome, Edwards syndrome, metafemale, Patau syndrome.

Trinucleotide repeats

CHROMOSOME LOCUS	TRINUCLEOTIDE REPEATED	NORMAL RANGE OF REPEATS	RANGE PRODUCING DISEASE	HEREDITARY DISEASE
Xq11-12	CAG	13–30	30–62	Spinal bulbar muscular atrophy
Xq27.3	CGG	6–54	50–1500	Fragile-X syndrome
4p16.3	CAG	9–37	37–121	Huntington disease
19q13.3	CTG	3–37	44–3000	Myotonic dystrophy

trisomy 13 syndrome Patau syndrome (*q.v.*).

trisomy 18 syndrome Edwards syndrome (*q.v.*).

trisomy 21 syndrome Down syndrome (*q.v.*).

tritanomaly *See* color blindness.

tritanopia *See* color blindness.

tritiated thymidine, uridine *See* tritium.

Triticum the genus containing various species of wheat. Wheat species are generally grouped together on the basis of their chromosome numbers. The einkorn wheats are diploids (N = 7), the emmer wheats are tetraploids (N = 14), and the vulgare wheats are hexaploids (N = 21). The hexaploids, of which *T. aestivum* is the most common example, contain three genomes, A, B, and D, each of which is composed of a set of seven pairs of nonhomologous chromosomes. The A and D genomes were derived from *T. monococcum* and *T. taushii*, respectively. The source of the B genome is unknown, and the donor may be extinct. The genome size for *T. aestivum* is 16 gigabase pairs. *See Aegilops*, durum, einkorn, emmer, spelt.

tritium the radioactive isotope of hydrogen, with a half-life of 12.46 years. Tritium-labeled thymidine and uridine are used to tag newly synthesized DNA and RNA, respectively. Tritium is the radioisotope of choice in radioautography because it emits an extremely weak beta particle when it undergoes radioactive decay. In a medium of unit density, the average tritium beta particle will penetrate only 1 micrometer. Therefore, in autoradiographs of tritium-labeled cells, silver grains of the photographic emulsion will be localized within 1 micrometer of the decaying atoms. Symbolized ^{3}**H.** Tritium was first used in autoradiography in the classic experiments of Taylor, Woods, and Hughes. *See* Appendix C, 1957.

Triton a genus containing several species of small aquatic salamanders. *See* Spemann-Mangold organizer.

Triturus a genus of salamanders that are studied for their lampbrush chromosomes (*q.v.*). Species for which working chromosome maps are available include *T. alpestris*, *T. cristatus*, *T. helveticus*, *T. italicus*, *T. marmoratus*, and *T. vulgaris*. Another favorite, *T. viridescens*, has been renamed *Notophthalmus viridescens* (*q.v.*). *See* Appendix A, Chordata, Amphibia, Urodela.

trivalent an association of three homologous chromosomes in meiosis.

tRNA transfer RNA (*q.v.*).

tRNA genes genes that are transcribed into tRNAs. Most tRNA genes are present in multiple copies. For example, the average repetition frequency for each tRNA gene species in yeast, *Drosophila*, humans, and *Xenopus*, has been estimated to be 5, 10, 15, and 200, respectively, per haploid genome. The mitochondrial genome of the malaria parasite lacks tRNA genes, and therefore these mitochondria must import all their tRNAs.

tRNA isoacceptors those tRNAs that accept the same amino acid, yet differ in primary sequence, either in the anticodon or in other regions of the molecule, or both. Such tRNA isoacceptors that differ in nucleotide sequences are generally encoded by different genes.

T4 RNA ligase an enzyme from bacteriophage T4 that catalyzes an ATP-dependent covalent joining of 5′ phosphate and 3′ hydroxyl termini of oligoribonucleotides.

tRNA-modifying enzymes enzymes that function to alter the primary tRNA transcript, generally by adding a single methyl group to a base of the 2′-hydroxyl group of ribose. There are at least 45 different tRNA-modifying enzymes in *E. coli*, and about 1% of its genome is devoted to encoding these enzymes. *See* rare bases, transfer RNA.

T7 RNA polymerase an enzyme, encoded by a gene in bacteriophage T7, that consists of a single protein. It carries out all the reactions accomplished by the transcriptases of prokaryotes and eukaryotes which are composed of multiple protein subunits (see RNA polymerase (RNA P). Because of its structural simplicity, T7 RNA-P is a favorite for the study of the organization of the transcription process, and it has practical applications in the synthesis of RNA probes. *See* RNA amplification, T phages.

tRNA suppressor *See* nonsense suppressor.

tRNA synthetase recognition site the site on the tRNA molecule that is bound to the aminoacetyl synthetase. In the case of yeast phenylalanine tRNA, this region is located adjacent to the dihydrouridine loop and consists of the four nucleotide pairs bracketed by arrows on the tRNA on page 397. *See* Appendix C, 1971, Dudock *et al.*; amino acid activation, transfer RNA.

trophoblast the nonembryonic part of the blastocyst, that attaches to the uterine wall and later develops into the fetal portion of the placenta.

trophocyte nurse cell (*q.v.*).

trophozoite *See Plasmodium* life cycle.

tropomyosin a protein dimer made up of subunits of Mr 35,000 found associated with actin in striated muscles. Tropomyosin is involved in the regulation of muscle contraction by Ca^{++} ions. Tropomyosins exist as a number of isoforms (*q.v.*). The differences in amino acid composition between isoforms generally involve a small segment of the molecule. In *Drosophila*, tropomyosin isoforms are known that result from alternative splicing (*q.v.*).

trp tryptophan. *See* amino acid.

true breeding line a group of genetically identical homozygous individuals that, when intercrossed, produce only offspring that are identical to their parents. *See* pure line.

truncation selection a method employed by breeders in which members of a population are chosen for mating, saving only those whose phenotypic merit (on a quantitative scale) is either above or below a certain value (the truncation point).

Trypanosoma a genus of eukaryotic flagellate parasites that pursue a life cycle in two different hosts, a mammal and an insect. *Trypanosoma brucei* causes sleeping sickness (*q.v.*) in humans. Its 26 megabase genome has been sequenced and predicted to contain 9,068 genes, distributed among 11 chromosomes. Approximately 900 of these are pseudogenes and 1,700 are *T. brucei*-specific. *T. brucei* also has a variable number of microchromosomes (totaling about 10 mb) that contain largely repetitive sequences. *T. cruzi* causes Chagas disease (*q.v.*) in humans. Its 60.4 mb genome contains about 12,000 genes, distributed among approximately 28 chromosomes (the exact chromosome number is not known). *See* Appendix A, Protoctista, Zoomastigina; Appendix C, 2005, Berriman *et al.*, El-Sayed *et al.*, Ivens *et al.*; Appendix F; antigenic variation, *Glossina*, kinetoplast, RNA editing, trans-splicing, *Leishmania*.

trypsin a proteolytic enzyme that cleaves peptide chains on the carboxyl side of lysine and arginine residues only; initially secreted from the pancreas in the inactive form of trypsinogen, and activated to trypsin by the enzyme enterokinase in intestinal juice. *See* Appendix C, 1876, Kühne.

tryptophan *See* amino acid.

tryptophan synthetase the enzyme that catalyzes the union of indole and serine to form tryptophan. In *E. coli*, tryptophan synthetase is a tetrameric aggregate of two alpha chains and two beta chains. *See* Appendix C, 1948, Mitchell and Lein.

TSH thyroid-stimulating hormone (*q.v.*).

***ts* mutation** temperature sensitive mutation (*q.v.*).

***TSPY* gene** a Y-linked gene which is the most spectacular example of a tandemly amplified gene in mammals. It encodes a testis-specific protein that is concentrated in the cytoplasm of spermatogonia (*q.v.*). Humans contain 20–60 copies of this gene, bulls up to 200. *See* human Y chromosome, Y chromosome, Y linkage.

T suppressor cell *See* suppressor T cell.

***t* test** Student's *t* test (*q.v.*).

tube nucleus the vegetative nucleus that resides in a growing pollen tube.

tuberculosis bacterium *See Mycobacterium tuberculosis.*

tubulin the principal protein component of microtubules. Tubulin is a dimer composed of α and β subunits, each of Mr 55,000. Microtubules are polymerized from $\alpha\beta$ dimers. The drug colchicine (*q.v.*) binds to $\alpha\beta$ dimers and prevents the addition of subunits to elongating microtubules. There are isoforms of both α and β tubulins, and at least some of these are products of different genes. In *Drosophila*, for example, four different genes have been identified that encode α tubulins; another four encode β tubulins. A very minor species of tubulin, γ-tubulin, also exists. The γ-tubulin molecules form an open ring structure that serves as a nucleation site upon which α and β-tubulin molecules assemble to form microtubules. γ-Tubulin ring complexes are located in the pericentriolar material. In *Saccharomyces cerevisiae*, the gene for gamma tubulin has been localized to chromosome 12. *See* Appendix C, 1995, Zheng *et al.*; Moritz *et al.*; centriole, paclitaxel, *Vinca* alkaloids.

tumor a clump of cells (usually disfunctional) due to abnormal proliferation. Benign tumors are not life-threatening (e.g., warts); malignant tumors are potentially lethal cancers.

tumor necrosis factor receptor *See* tumor necrosis factors (TNFs).

tumor necrosis factors (TNFs) cytokines (*q.v.*) that have cytotoxic effects on tumor cells but not on normal cells. One group (TNFα) is secreted by macrophages (*q.v.*); a second (TNFβ) by helper T lymphocytes (*q.v.*) and cytotoxic T lymphocytes (*q.v.*). The tumor necrosis factor receptor (TNFR) is a transmembrane protein. Its extracellular domain binds to the TNF, while its intracellular portion combines with various adaptor proteins to form a complex that activates upstream procaspases. Thus it is the binding of a death ligand like the TNF to its receptor that signals the cell to activate its caspases (*q.v.*) and to undergo apoptosis (*q.v.*).

tumor-specific transplantation antigen an antigen found on a tumor cell but not on the normal cells or tissues of the individual in which it arose, and that can lead to immune rejection of the tumor *in situ* or if transplanted.

tumor suppressor gene synonymous with antioncogene (*q.v.*).

tumor virus *See* oncogenic virus.

tunicate *See Ciona intestinalis.*

tunicin a cellulose-like fiber that makes up the matrix of the tunic that surrounds sea squirts. The genome of *Ciona intestinalis* (*q.v.*) includes a group of genes engaged in cellulose synthesis, and these evolutionary innovations may have arisen by horizontal transmission (*q.v.*) of genes from bacterial symbionts.

Turbatrix aceti a free-living nematode whose developmental genetics is under study, primarily for comparison with *Caenorhabditis elegans* (*q.v.*). Its cell lineage is known in part.

turbid plaque *See* plaque.

Turner syndrome a group of abnormalities in humans due to monosomy for the X chromosome. Such individuals are female in phenotype, but are sterile. The ovaries are rudimentary or missing. The syndrome is named after Henry H. Turner, the physician who described the clinical features of the condition in 1938, but the underlying chromosomal abnormality went unnoticed for 21 years. Girls with Turner syndrome have normal intelligence but often have behavioral problems. These seem to be related to the parental origin of the X chromosome. Girls who inherit the X from their fathers are significantly better adjusted socially than those who inherit the X from their mothers. This finding suggests that there are imprinted genes on the X that somehow enhance social adjustment and only the paternal alleles are switched on. *See* Appendix C, 1959, Ford *et al.*; 1997, Skuse *et al.*; parental imprinting. http://www.turner-syndrome-us.org

turnover the dynamic replacement of atoms in a tissue or organism without any net change in the total number of atoms. *See* Appendix C, 1942, Schoenheimer.

turnover number the number of molecules of a substrate transformed per minute by a single enzyme molecule under optimal conditions.

twins pairs of individuals produced at one birth. Monozygotic (MZ) or identical twins have identical sets of nuclear genes. MZ twins result from the separation of blastomeres and thus represent a form of asexual or clonal reproduction. Dizygotic (DZ) or fraternal twins arise when two eggs are released and fertilized. DZ twins, therefore, are no more similar genetically than are siblings that share only half of their genes, on the average. In humans, the frequency of monozygotic twinning is relatively constant (1/240 births). The frequency of dizygotic twinning varies in different races. In Caucasian populations, it is about double the frequency of MZ twinning. *See* Appendix C, 1869, Galton; 1874, Dareste; 1875, Galton; 1927, Bauer.

twin spots paired patches of tissue genetically different from each other and from the background tissue, produced by mitotic crossing over in an individual of heterozygous genotype during development of the organism. In the *Drosophila* example diagrammed on page 462, twin spots occur on a female of genotype y +/+ *sn*. Most of the thorax and abdomen show wild-type pigment and long, black bristles, but there are adjacent patches of yellow and singed tissue. These twin spots are clones of cells derived from the reciprocal products of an exchange between the singed gene and the centromere.

twin studies the use of data collected from twins (*q.v.*) to evaluate the relative roles of heredity and environment in determining various human physical and behavioral characteristics. One method compares monozygotic and dizygotic twins reared together and another compares identical twins who were adopted by different families and reared apart. *See* Appendix C, 1875, Galton; concordance.

twisting number the number of base pairs in duplex DNA divided by the number of base pairs per turn of the helix. *See* pitch.

twofold rotational symmetry the situation in which the two DNA strands of a duplex have the same base sequence when read with the same polarity. For example, the *Eco*RI (*q.v.*) endonuclease recognizes a DNA segment six base pairs long that has twofold rotational symmetry. *See* palindrome.

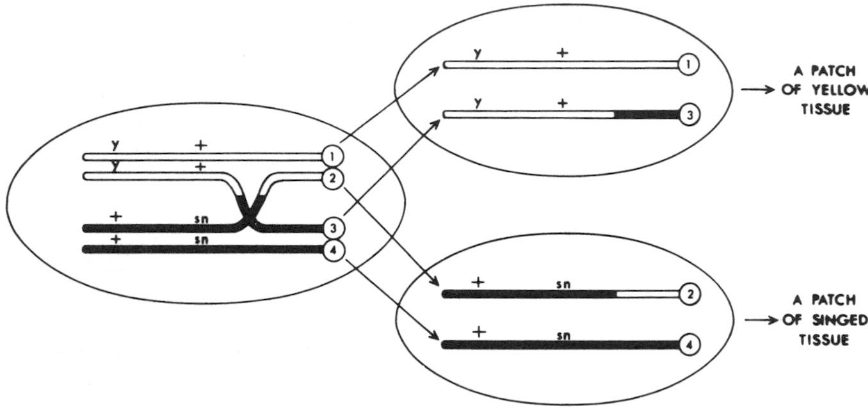

Twin spots

two genes–one polypeptide chain participation of more than one gene in the coding of a given polypeptide chain, as occurs in the production of immunoglobulins. *See* immunoglobulin genes, one gene–one polypeptide hypothesis.

two-point cross a genetic recombination experiment involving two linked genes.

two-strand double exchange *See* inversion.

Ty elements retroposons (*q.v.*) of the yeast *Saccharomyces cerevisiae*. The symbol Ty comes from *t*ransposon-*y*east. There are 52 Ty elements per haploid yeast genome, and each consists of a central region of about 5.6 kilobase pairs of DNA flanked by direct repeats about 330 base pairs long. *See* Appendix C, 1979, Cameron *et al.*; 1985, Boeke *et al.*

type specimen the individual chosen by the taxonomist to serve as the basis for naming and describing a new species.

typological thinking the consideration of the members of a population as replicas of or deviations from a hypothetical type.

tyr tyrosine. *See* amino acid.

tyrosinase an enzyme that catalyzes the hydroxylation of tyrosine to DOPA (*q.v.*). This spontaneously oxidizes to DOPA quinone, and these molecules then polymerize to form melanin. Human tyrosinase is made up of 529 amino acids, and it contains a signal peptide, two copper binding sites, and a hydrophobic transmembrane region. Any mutation that inhibits the ability of the protein to bind copper deactivates the enzyme. *See* albinism.

tyrosine *See* amino acid.

tyrosine-protein kinase *See* protein kinase.

tyrosinemia a hereditary disease in humans arising from a deficiency of the enzyme p-hydroxyphenylpyruvic oxidase.

U

U 1. uracil or uridine. 2. uranium.

U20 *See* nucleolin.

ubiquinone coenzyme Q (*q.v.*).

ubiquitin a highly conserved, 76 amino acid protein *ubiquit*ous in all eukaryotes, which regulates the selective, intracellular degradation of many proteins by the proteasome (*q.v.*), or in some cases, by the lysosome (*q.v.*) or vacuole (*q.v.*). Ubiquitin is a globular protein with a functionally important "tail" at the C-terminus (*q.v.*). The C-terminus glycine forms a covalent bond with a lysine residue or a lysine side chain on the substrate protein. A polyubiquitin chain usually forms after a ubiquitin molecule conjugates with the target protein, in which the C-terminus of each ubiquitin is linked to a lysine residue of the previous molecule. Ubiquitin-tagged proteins are then recognized and degraded by the proteasome or the lysosome/vacuole. Ubiquitination of proteins also serves as a signal for endocytosis (*q.v.*) and intracellular transport of proteins. In addition, ubiquitin plays a role in chromatin structure, since about 10% of the H2A histone molecules in nucleosomes are ubiquitinated. Ubiquitin is encoded by a family of genes whose translation products are fusion proteins consisting of either a single ubiquitin molecule fused to a ribosomal protein, or multiple ubiquitins fused head to tail. In the former case ubiquitin-conjugated ribosomal proteins are targeted to the ribosome (*q.v.*), and ubiquitin is released following their incorporation into the ribosomal-protein complex. In the latter case ubiquitin monomers are formed post-translationally through the action of deubiquitinating enzymes (*q.v.*) that cleave the fusion protein. *See* histone, nucleosome, *otu*, polycistronic mRNA, ubiquitin-proteasome pathway (UPP).

ubiquitin ligase *See* cyclins.

ubiquitin-proteasome pathway (UPP) a highly specific and regulated intracellular process that functions to recognize, tag, and break down many proteins in eukaryotic cells. In this process a protein is tagged for degradation through covalent ligation to ubiquitin (*q.v.*), followed by degradation by the 26S proteasome (*q.v.*). The conjugation of ubiquitin to the substrate protein (ubiquitination) occurs in multiple steps. First, the C-terminus (*q.v.*) glycine residue of ubiquitin is activated by the enzyme E1 in an ATP-requiring step. Activated ubiquitin is then transferred to one of several E2 enzymes (ubiquitin-carrier proteins or ubiquitin-conjugating enzymes). Ubiquitin is further transferred from E2 to an E3 enzyme belonging to the ubiquitin-protein ligase family. Finally, the E3 enzyme catalyzes the covalent linkage of ubiquitin to the substrate protein. The specificity of ubiquitin-protein conjugation is dependent on the recognition of particular signals on the substrate protein by the appropriate E3 enzyme. The attachment of a ubiquitin molecule to the target protein is usually followed by the formation of a polyubiquitin chain, which is then recognized by the 19S regulatory particle of the proteasome. The protein is unfolded and translocated to the central cavity of the proteasome, the 20S core complex, where it undergoes ATP-dependent degradation. The resulting peptides, 8-9 amino acids long, leave the proteasome and enter the cytosol where peptidases break them down further. Ubiquitin monomers are also released for future use by the activity of deubiquitinating enzymes (*q.v.*). Ubiquitin-mediated protein degradation is important for eliminating defective proteins. Furthermore, through the highly selective and regulated targeting of a wide range of substrate proteins, UPP controls numerous cellular functions, including cell-cycle progression, cellular signal transduction (*q.v.*), and regulation of transcription (*q.v.*). UPP has also been linked to apoptosis (*q.v.*) and to the processing of many MHC class 1 antigens. Many pathological conditions are associated with abnormalities in ubiquitin-mediated processes. *See* Appendix C, 1980, Hershko *et al.*; *otu*, ritinitis pigmentosum (RP), SUMO proteins.

UDPG uridine diphosphate glucose (*q.v.*).

UH2A *See* ubiquitin.

ultracentrifuge a powerful centrifuge that can attain speeds as high as 60,000 rpm and generate sedimenting forces 500,000 times that of gravity. The instrument is used to sediment macromolecules. *See* Appendix A, 1923 Svedberg; centrifugation separation techniques, centrifuge, sedimentation constant.

ultramicrotome an instrument that cuts ultrathin sections (50–100 nanometers thick) of plastic embedded tissues using knives made from polished diamonds or from triangles of broken plate glass. *See* Appendix C, 1950, Latta and Hartmann; 1953, Porter-Blum and Sjöstrand; knife breaker.

ultrasonic pertaining to ultrasound.

ultrasound sound waves of frequency higher than the human audible limit of about 20,000 vibrations per second.

ultrastructure fine structure, especially within a cell, that can be seen only with the high magnification obtainable with an electron microscope.

ultraviolet absorption curve the curve showing the relation between the relative amount of ultraviolet radiation absorbed by a solution of molecules and the wave length of the incident light.

ultraviolet microscope an optical system utilizing ultraviolet radiation. Since glass filters out UV, quartz transmitting or glass reflecting lenses must be used in the UV microscope. Such a microscope has double the resolving power of the light microscope. Furthermore, if monochromatic UV of a wavelength absorbed by nucleic acids (260 nanometers) is used, nucleic acid–rich structures may be photographed in unstained cells. In combination with a spectrophotometer, the UV microscope provides a method for the quantitative estimation of nucleic acids in cells.

ultraviolet radiation that part of the invisible electromagnetic spectrum just beyond the violet with wavelengths between 1,000 and 4,000 Ångstroms. Wavelengths around 260 nanometers are absorbed by DNA. *See* Appendix C, 1939, Knapp *et al.*; melanin, thymine dimer, UV reactivation.

uncharged tRNA a tRNA molecule to which no amino acid is attached.

underdominance the unusual situation where a heterozygote shows an attribute, such as viability or fertility, that is lower than either homozygote. For example, the New Zealand Black (NZB) strain of mouse spontaneously develops a disease that resembles lupus erythematosis in humans (*q.v.*). New Zealand White (NZW) mice are normal in this regard. The hybrid offspring from crossing these inbred strains of mice (NZB × NZW) develop a more severe disease than that of the NZB strain.

undersea vent communities chemosynthetic organisms that live at great depths and in the absence of sunlight along tectonically active rifts where lava erupts from ocean floor. The first community of this type was discovered around the vents of sulfide hot springs at a depth of 8,000 feet about 380 miles north of the Galapagos Islands. *See* Appendix C, 1977, Corliss and Ballard; hyperthermophile.

underwinding coiling of a DNA molecule in a left-handed direction, i.e., opposite to that of the double helix; negative supercoiling.

undulipodium any cellular projection surrounding a cylindrical shaft containing a bundle of eleven microtubules, nine of which form a circle around the central pair, while the cortical microtubules are doublets. Cilia and flagellae (*q.v.*) are examples. Mastigote protoctists possess undulipodia; amastigote protoctists do not. *See* axoneme, Rhizopoda.

unequal crossing over a recombinational event that involves recombining sites that are misalinged, and nonreciprocal recombinant chromosomes are formed as a result. The phenomenon was discovered in *Drosophila* at the *Bar* (*q.v.*) locus. Here improper pairing of a duplicated chromosomal segment was followed by unequal crossing over. The result was one crossover chromatid with one copy of the segment and another with three copies. Unequal crossing over is responsible for generating duplications and deletions at many sites in human chromosomes where clusters of duplicated DNA segments occur. *See* Appendix C, 1925, Sturtevant; cone pigment genes (CPGs), hemoglobin Lepore, silk, thalassemias.

ungulate a hoofed mammal. Perissodactyla plus Artiodactyla. *See* Appendix A.

uniformitarianism a geological theory that "the present is the key to the past." In other words, the phenomena of volcanism, crustal movements, erosion, glaciation, etc., that can be seen today have been operating through billions of years of earth history, and they are the primary forces that have made the earth what it is today. *Contrast with* catastrophism.

uniform resource locator (URL) the symbolic representation of an address used on the World Wide Web. *See* Appendix E for examples of URLs.

unineme hypothesis the concept that a newly formed chromatid contains only one DNA duplex extending from one end to the other. *Contrast with* polyneme hypothesis. *See* Appendix C, 1973, Kavenoff and Zimm.

uniovular twins monozygotic twins (*q.v.*).

uniparental disomy *See* disomy.

uniparental inheritance a phenomenon in which all offspring of a given mating seem to have received certain phenotypes from only one of the parents (usually the female) regardless of the genotype or phenotype of the other parent; such inheritance is usually the result of macromolecules or organelles stored in the cytoplasm. *See* holandric.

unipotent an adjective pertaining to a cell that has the potential to give rise to only one type of differentiated progeny. *Compare with* pluripotent, totipotent.

unique DNA a class of DNA determined by C_0t analysis to represent sequences that are present only once in the genome. Most structural genes and their introns are unique DNAs. *See* reassociation kinetics.

unisexual flower a flower having only stamens or only carpels. A plant can bear either one or both kinds of unisexual flowers. *See* flower.

unit character a term used by early geneticists for traits that segregate according to Mendel's laws.

unit evolutionary period the time in millions of years during which a divergence of 1% occurs between the initially identical nucleotide sequence in two branches of a lineage under study. The UEP for the globin gene family is 10.4.

unit membrane the trilamellar membrane seen when the sectioned plasma membrane is viewed under the electron microscope. The membrane, which has a total thickness of about 75 Ångstroms, appears as two strata each about 20 Ångstroms thick separated by a light interzone about 35 Ångstroms wide. *See* fluid mosaic concept.

univalent a single chromosome seen during meiosis when bivalents are also present. A univalent has no synaptic mate. An example of a univalent would be the sex chromosome of an XO male.

universal code theory the assumption that the genetic code is used exclusively by all forms of life. This is true with a few exceptions. In yeast mitochondria, CUA codes for threonine instead of leucine; and in mammalian mitochondria, AUA codes for methionine instead of isoleucine, and codons AGA and AGG signal termination instead of coding for arginine. UGA codes for tryptophan instead of signaling termination in mitochondria from both sources. In four ciliates, *Tetrahymean thermophila, Stylonychia lemnae, Paramecium primaurelia,* and *P. tetraurelia,* UAA and UAG encode the amino acid glutamine rather than serving as termination codons. In *Mycoplasma capricolum,* as in mitochondria, UGA encodes tryptophan rather than serving as a termination codon. *See* Appendix C, 1961, von Ehrenstein and Lipmann; 1979, Barrell *et al.*; 1985, Horowitz and Gorowsky, Yamao; genetic code.

universal donor an individual with type O blood, who is able to donate red blood cells to O, A, B, or AB recipients. *See* blood group.

universal recipient an individual with type AB blood, who can receive red blood cells from AB, A, B, or O donors. *See* blood group.

universal tree of life this phylogenetic tree (see illustration on page 466) is based on comparisons of the nucleotide sequences of the RNAs in the small ribosomal subunits from various prokaryotes and eukaryotes. The vertical length of each branch corresponds quantitatively to the number of base sequence changes that have occured in that lineage since its divergence from its nearest neighbor. Most of the tree is made up of microorganisms. The animal, plant, and fungal kingdoms are but 3 of the 23 lineages. All living organisms fall into three groups: Bacteria, Archaea, and Eukaryotes. The eukaryotes are more closely related to the archaeons than to the bacteria. The common ancestor at the base of the tree lies between the Archaea and the Bacteria. This prokaryote was probably a hyperthermophile (*q.v.*). *See* Appendix A, Prokaryotes, Eukaryotes; Appendix C, 1977, Woese and Fox; 1980, Woese *et al.*; ribosome, 16S rRNA.

univoltine *See* voltinism.

unordered tetrad *See* nonlinear tetrad.

unscheduled DNA synthesis DNA synthesis that occurs at some stage in the cell cycle other than the S period, generally to repair damaged DNA. *See* interphase cycle, repair synthesis.

unstable equilibrium the situation where the equilibrium value for an allele in a population fluctuates because of temporary environmental changes. For example, these may lead to a sudden selection for or against the allele, or a drastic reduction in the size of the population may result in genetic drift away from the former equilibrium value.

unstable mutation a mutation with a high frequency of reversion. The original mutation may be caused by the insertion of a controlling element (*q.v.*), and its exit produces a reversion.

untwisting enzyme *See* topoisomerase.

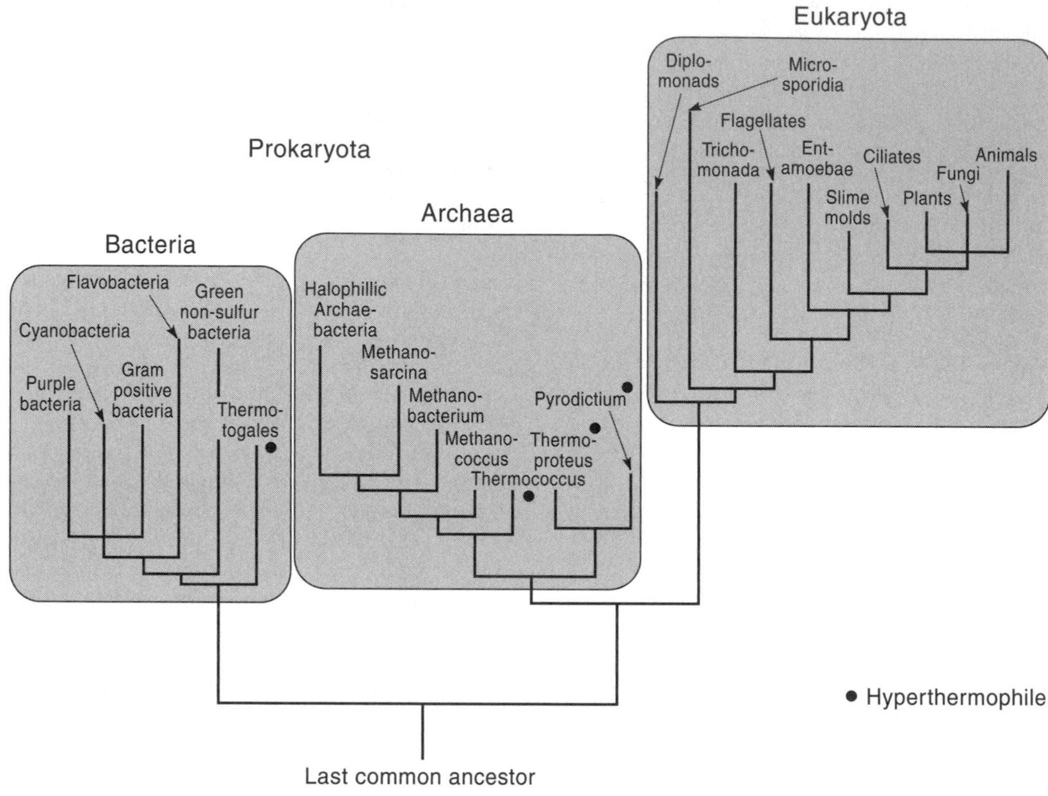

Universal tree of life

unwinding proteins proteins that bind to, destabilize, and unwind the DNA helix ahead of the replicating fork. *See* Appendix C, 1970, Alberts and Frey; gene 32 protein.

u orientation *See* n orientation.

up promoter mutations mutations in promoter sites that increase the rate of initiation of transcription; promoters with this property are called "high level or strong promoters."

upstream *See* strand terminologies, transcription unit. *Compare with* downstream.

uracil *See* bases of nucleic acids.

uracil fragments during DNA replication in *E. coli*, polymerases I and III occasionally make mistakes and incorporate dUTP instead of TTP. Several enzymes remove these uracils from both leading and lagging strands, creating "uracil fragments."

urea *See* ornithine cycle.

urease a nickel-dependent metaloenzyme that catalyzes the hydrolysis of urea to form ammonia and carbon dioxide. Urease was the first enzyme to be isolated in pure, crystalline form. The source was seeds from the jack bean plant (*Canavlia ensiformis*). *See* Appendix C, 1926, Summer.

urethane a carcinogen that induces tumorous nodules in the lungs of mammals.

$$NH_2{-}\underset{\underset{O}{\|}}{C}{-}O{-}CH_2{-}CH_3$$

URF unidentified reading frame. The open reading frames (ORFs) uncovered in the DNA of a species of interest are matched against those in large databases that contain all known ORFs from all kinds of organisms. A substantial number of the new ORFs are left over that do not resemble genes that encode any known proteins. These ORFs are therefore also URFs. Such genes are sometimes called *orphans*. (*q.v.*).

uric acid the end product of nucleic acid catabolism in mammals; the main nitrogenous constituent of the urine of reptiles and birds. See structure on page 467.

uridine *See* nucleoside.

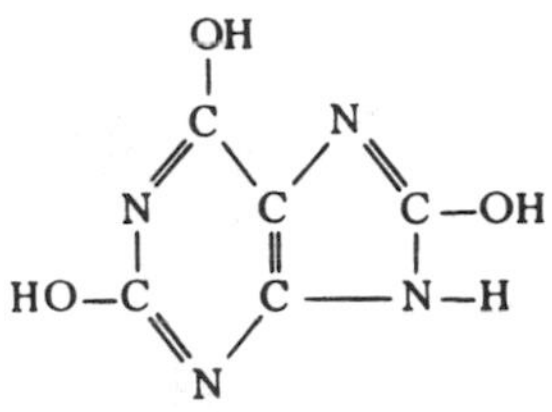

Uric acid

uridine diphosphate galactose *See* uridine diphosphate glucose.

uridine diphosphate glucose (UDPG) a compound (shown below) that acts as a coenzyme and substrate to various enzymes. UDPG may be converted to uridine diphosphate galactose by the enzyme epimerase. These coenzymes play essential roles in carbohydrate metabolism.

uridylic acid *See* nucleotide.

Urkingdom a kingdom that some authorities have suggested be erected to house the Archaebacteria (*q.v.*).

URL uniform resource locator (*q.v.*).

Usn RNAs the U class of small nuclear RNAs (*q.v.*). These molecules range in size from 60 to 216 nucleotides and are rich in uridine. Five of Usn RNAs (U_1, U_2, U_4, U_5, and U_6) have been studied in the most detail. The first four all have a trimethylguanosine cap structure at the 5′ end. Most Usn RNAs are associated with seven proteins, some of which are common to all five RNAs, while others are specific for U_1 or U_2. These snRNPs make up about a third of the mass of the spliceosomes, and they function in the excision and splicing reactions that take place in this organelle. *See* cajal body, exon, intron, lupus erythematosus, posttranscriptional modification, splice junctions.

Ustilago a genus in the *Basidiomycota* (*see* Appendix A). These are yeastlike smut fungi, and two species, *U. maydis* and *U. violacea*, have been subjects of genetic research, especially in terms of recombination-defective and radiation-sensitive mutations.

uteroglobin a protein synthesized by the cells of the rabbit endometrium and present in the uterine fluids. Uteroglobin is composed of two identical subunits of 70 amino acids, and the subunits are held together by two disulphide bridges. Reduced uteroglobin binds steroids, such as progesterone. There is one uteroglobin gene per genome. It is 3 kilobases long and contains two intervening sequences and three exons. Some authors have proposed that uteroglobin exerts a stimulatory effect on the blastocyst (hence its other name, *blastokinin*).

utrophin a gene which maps to 6q24 and encodes DRP, a dystrophin-related protein. DRP has an amino acid sequence with an 80% identity to dystrophin (*q.v.*), and it is synthesized in fetal muscle cells. Since the urotrophin protein functions like dystrophin in early development, upregulation of this gene has been suggested as a possible therapeutic approach for the treatment of Duchenne muscular dystrophy (*q.v.*).

UV ultraviolet radiation (*q.v.*).

UV-induced dimers *See* thymine dimers.

UV reactivation a phenomenon in which survival of an ultraviolet-irradiated lambda phage is greater on an irradiated host than on an unirradiated host. The repair mechanism involved in UV-reactivation utilizes an error-prone replication system of the host such as that of the SOS response (*q.v.*).

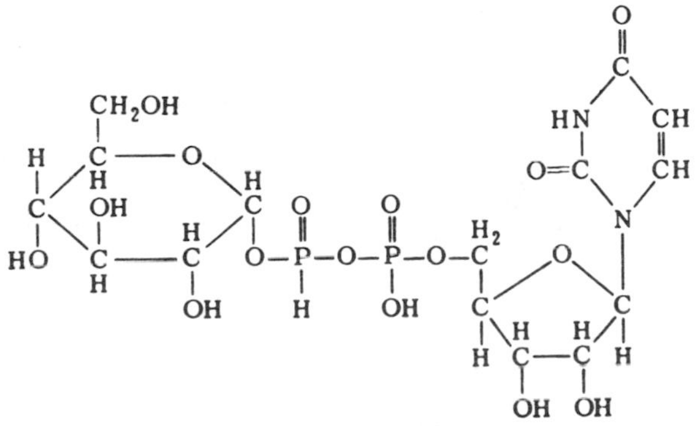

Uridine diphosphate glucose (UDPG)

V

vaccine any substance that, when injected into an animal's body, stimulates the development of active immunity against a specific infectious agent (a bacterium, virus, protozoan parasite, etc.) or against a harmful product of such an organism (e.g., a toxin produced by a bacterium). Most vaccines consist of antigenic material, either whole organisms (containing many different antigens) or specific antigenic parts of an organism (such as a viral coat protein). Vaccines may be developed from the whole organisms that are live, dead or inactivated, or attenuated. These kinds of vaccines induce an immune response (production of antibodies and activated lymphocytes) that directly attacks the immunizing antigens as well as any identical (or, in some cases, closely related) antigens from organisms that may subsequently infect the animal host. An exception to this generalization occurs with DNA vaccines. Here a plasmid (*q.v.*) may be genetically engineered to contain one or more specific genes from a potential pathogenic organism. Upon artificial introduction into host cells, such inserted genes may become activated and produce their corresponding protein products. When these proteins are combined with the distinctive "self proteins" of the host cell (e.g., class II HLA antigens of humans) and displayed on the cell's surface, they can be recognized as foreign antigens by killer T-lymphocytes. These attack and kill such infected cells and prevent further replication of the pathogenic organism. Thus the immune response to DNA vaccines is not directed at the vaccine itself but rather against the foreign protein products of the plasmid genes. *See* Appendix C, 1798, Jenner; 1881, Pasteur; antibody, antigen, diphtheria toxin, histocompatibility molecules, immune response, immunoglobulin, lymphocyte, poxviruses, smallpox vaccine.

vaccinia virus (VV) a double-stranded DNA virus with a genome size of 192 kbp. *See* pox viruses, smallpox vaccine.

vacuoles fluid-filled, membrane-enclosed vesicles that are multipurpose organelles in all plant cells. Some vacuoles function like lysosomes (*q.v.*), since they contain a variety of hydrolytic enzymes and can serve as compartments for storing waste products

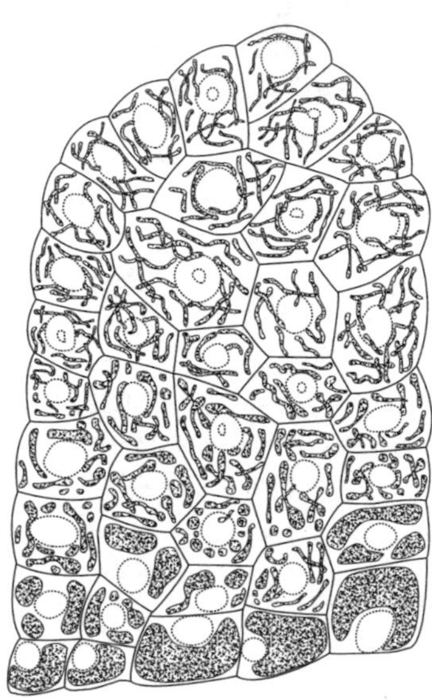

of intracellular digestion. Vacuoles can make up as much as 90% of the volume of cells, and they control the osmotic pressure that presses cells outward and keeps the plant from wilting. The cells of seeds can store proteins for years. Once the seeds germinate, the proteins are hydrolyzed by the vacuole to supply the developing embryo with amino acids. The pigments that give color to flowers and fruits are also stored in vacuoles. The drawing shows a population of cells in a rose petal. Filamentous vacuoles that contain red anthocyanin pigment (stippling) swell, anastomose, and eventually form a single vacuole in the mature cells (at the bottom). Cell nuclei are shown as dashed circles. Vacuoles also occur in the cells of fungi and in protozoans. In the malaria parasite, ingested hemoglobin is degraded in the food vacuole. *See* anthocyanins, chloroquine, contractile vacuole, trophozoite.

vacuum evaporator the vacuum chamber containing a set of electrodes through which a current can be passed to heat a metal foil placed between the electrodes. The heated metal evaporates, and atoms from it coat the specimen lying below the electrodes. The coating is cast at an angle, and as a consequence the specimen appears in relief when viewed under the electron microscope. A vacuum evaporator equipped with graphite electrodes is also employed to prepare the carbon films that sometimes are used to support ultrathin sections on specimen screens. *See* shadow casting.

vagility the capacity of an organism or species to disperse in a given environment.

val valine. *See* amino acid.

valence 1. in chemistry, a number representing the combining or displacing power of an atom; the number of electrons lost, gained, or shared by an atom in a compound; the number of hydrogen atoms with which an atom will combine, or which it will displace. 2. in serology, the number of antigen-combining sites on an antibody molecule.

valine *See* amino acids.

valinomycin *See* ionophores.

van der Waals forces the relatively weak, short-range forces of attraction existing between atoms or molecules, caused by the interaction of varying dipoles.

van Leeuwenhoek on the Web an impressive 78-page Website on the father of microbiology. http://www.euronet.nl/users/warnar/leeuwenhoek/html.

var variety.

variable any organismal attribute that may have different values in various situations (e.g., between members of different species, between members of the same species, within an individual at different times, etc.).

variable domain that domain of an immunoglobulin light or heavy chain that has variable amino acid sequences within an individual.

variable number of tandem repeats locus (VNTR locus) any gene whose alleles contain different numbers of tandemly repeated oligonucleotide sequences. Such alleles when cleaved by a specific restriction endonuclease will produce fragments that differ in length. Such restriction-length polymorphisms (*q.v.*) serve as convenient markers in linkage studies. *See* DNA fingerprint technique.

variable region the *N*-terminal portion of an immunoglobulin chain that binds to antigen.

variance when all values in a population are expressed as plus and minus deviations from the population mean, the variance is the mean of the squared deviations.

variant an individual that is different from an arbitrary standard type (usually wild type) for that species. Variants are not necessarily mutants; for example, many birth defects are simply developmental accidents or environmentally induced. *See* phenocopy.

variate a specific quantitative value of a variable (*q.v.*).

variation divergence among individuals of a group, specifically a difference of an individual from others of the same species that cannot be ascribed to a difference in age, sex, or position in the life cycle. The variations of evolutionary significance are gene-controlled phenotypic differences of adaptive significance.

variegated position effect *See* position effects.

variegation 1. irregularity in the pigmentation of plant tissues due to a variety of causes (virus infection, segregation of normal and mutant plastids, bridge-breakage-fusion-bridge cycles, transposable elements, etc.). 2. irregularity in the pigmentation of animal tissues or their products (hair, feathers, etc.) due to a variety of causes (X chromosome inactivation, defective embryonic migration of melanocytes, localized physiological conditions, mitotic recombination, etc.). *See* tortoiseshell cat.

variety *See* strain.

Variola virus (VAR) the virus (*q.v.*) that causes smallpox, a disease responsible for more deaths throughout human history than all armed conflicts. Worldwide use of the smallpox vaccine (*q.v.*) led to the eradication of the disease. The remaining stocks of the virus are stored under constant surveillance at two laboratories, one at the Centers for Disease Control in Atlanta, Georgia, USA, and the other at the Research Institute for Viral Preparations in Moscow, Russia. *See* Appendix F.

vasopressin a peptide hormone secreted by the hypothalamus and stored in the neurohypophysis, which constricts arterioles and promotes resorption of water by the kidney tubules. Also called the antidiuretic hormone.

VAST *v*ector *a*lignment *s*earch *t*ool, an algorithm which allows the direct comparison of protein chains in three dimensions.

V(D)J recombination the process by which most vertebrates assemble immunoglobin (Ig) and T cell receptor (TCR) genes during the development of lymphoid cells. In the germ cells, the genes that encode the variable portions of the Ig and T cell receptor heterodimers are split into V (variable), J (joining), and sometimes D (diversity) segments. In immature lymphoid cells, segments of each type are joined together to make a V-J or a V-D-J fusion product. V(D)J recombination involves DNA cleavage catalyzed by an element that behaves like a transposase (*q.v.*). The element is the product of two genes, *RAG-1* and *RAG-2* (*q.v.*). *See* Appendix C, 1990, Oettinger *et al.*; allelic exclusion, B lymphocyte, immunoglobulin chains, immunoglobulin genes, somatic recombination, T cell receptor genes, T lymphocyte.

VDR vitamin D receptor (*q.v.*).

vector an organism (such as the malaria mosquito) that transfers a parasite from one host to another. *See* DNA vector, plasmid cloning vector, RNA vector, shuttle vector.

vegetal hemisphere the surface of the amphibian egg farthest from the nucleus, the yolk-rich hemisphere of the egg.

vegetative designating a stage or form of growth, especially in a plant, distinguished from that connected with reproduction.

vegetative cell an actively growing cell, as opposed to one forming spores.

vegetative nucleus 1. the macronucleus of a ciliate. 2. the tube-nucleus of a pollen grain.

vegetative petites *See* petites.

vegetative reproduction in plants, the formation of a new individual from a group of cells, without the production of an embryo or seed. More generally, asexual reproduction. *See* agamospermy, apomixis.

vegetative state the noninfective state during which a phage genome multiplies actively and controls the synthesis by the host of the materials necessary for the production of infective particles.

vehicle a plasmid or bacteriophage possessing a functional replicator site, and containing a genetic marker to facilitate its selective recognition, used to transport foreign genes into recipient cells during recombinant DNA experiments; also called a *vector.*

***vermilion* (*v*)** a sex-linked, recessive eye-color mutation in *Drosophila melanogaster.* This was the first *Drosophila* mutation to be understood biochemically. The *vermilion* gene encodes tryptophan oxygenase, an enzyme that converts tryptophan to formylkynurenine (*q.v.*). This is the first step in the reaction chain that leads to xanthomatin, the brown pigment in the *Drosophila* eye. If formylkynurenine is supplied in the diet of larvae containing mutant alleles of *v*, the adults that develop show normal eye color. The first *vermilion* mutation to be isolated resulted from a transposon (*q.v.*) insertion. *See* Appendix C, 1935, Beadle and Ephrussi; *Drosophila* eye pigments.

vermilion plus substance formylkynurenine; so called because the synthesis of this compound in *Drosophila melanogaster* is controlled by the plus or wild-type allele of the gene *vermilion* (*q.v.*).

vernalization the treatment of germinating seeds with low temperatures to effect their flowering. Winter varieties of certain cereals, if vernalized, can be sown in the spring and harvested in the summer.

Veronica a large genus of hardy herbs belonging to the family Scrophulariaceae. Classic studies on the genetic control of self-sterility were performed on this species.

vertical classification a system of classification that recognizes taxa corresponding to clades and groups transitional forms with their descendants rather than with their ancestors; the opposite of *horizontal classification* (*q.v.*).

vertical evolution the process whereby an ancestral species changes through time (without splitting) to become distinctively different, and therefore recognized as a new species; phyletic evolution. *See* anagenesis, speciation.

vertical transmission 1. passage of genetic information from one cell or individual organism to its progeny by conventional heredity mechanisms (mitosis, meiosis), in contrast to horizontal transmission (*q.v.*). 2. transmission of a parasite from parent to offspring via the egg or *in utero.*

vervet monkey another name for the African green monkey, *Cercopithecus aethiops* (*q.v.*).

V gene one of many (perhaps hundreds) of genes coding for the variable (*N*-terminus) region of an immunoglobulin chain.

viability a measure of the number of individuals surviving in one phenotypic class relative to another class, taken as standard, under specified environmental conditions.

Vibrio cholerae the comma-shaped bacterium that is the cause of cholera (*q.v.*). The *V. cholerae* genome consists of two circular chromosomes. The larger contains 2.96 mbp, the smaller 1.07 mbp. Together they have 3,885 ORFs. The smaller chromosome contains many genes of unknown function, and it may represent a captured plasmid (*q.v.*). The large chromosome has the majority of genes for essential cell functions and pathogenicity. The cholera toxin is encoded in the genome of a virus that has been integrated into the large chromosome. *See* Appendix A, Bacteria, Proteobacteria; Appendix C, 1883, Koch; 2000, Heidelberg *et al.*; Appendix E.

vicariance distribution a discontinuous biogeographical distribution of organisms that previously inhabited a continuous range. The current gaps in the distribution were caused by some extrinsic factor (geologic or climatic).

Vicia faba the broad bean; also called the fava bean, the horse bean, the Windsor bean. A plant often used in cytogenetics because its cells contain a small number ($N = 6$) of large chromosomes. The semiconservative nature of DNA replication was first demonstrated by analyzing autoradiographs of ^{3}H thymidine-labeled chromosomes from cells of *Vicia* root tips. *See* Appendix C, 1957, Taylor *et al.*; favism.

villus a finger-like projection extending from an epithelium. Such a villus is composed of many cells. *Contrast with* microvillus.

vimentin a 55,000-dalton, cytoskeletal protein commonly found in fibroblasts. In glial cells it is copolymerized with an acidic protein of 50,000 daltons, while in muscle cells it is combined with desmin (*q.v.*).

vinblastine a vinca alkaloid (*q.v.*).

***Vinca* alkaloid** any one of a number of anticancer drugs isolated from *Vinca rosea* (*q.v.*). The drugs block the division of cancer cells by acting as spindle poisons (*q.v.*). Like colchicine (*q.v.*) the *Vinca* alkaloids bind to tubulin (*q.v.*) and interfere with the assembly of the microtubules of mitotic spindles.

Vinca rosea the Madagascar periwinkle, source of the vinca alkaloids.

vincristine a vinca alkaloid (*q.v.*).

vinculin a fibrous protein responsible for anchorage of actin filaments to the inner side of the cell membrane. Vinculin is located in patches called *adhesion plaques* on the cell membrane that are thought to be responsible for intercellular adhesion. Cells infected with Rous sarcoma virus (*q.v.*) produce a kinase that phosphorylates the tyrosine residues of vinculin. It is hypothesized that the phosphorylation of vinculin both destabilizes actin linkages (allowing transformed cells to become rounded) and weakens intercellular adhesion (allowing metastasis).

viral-specific enzyme any enzyme produced in the host cell after viral infection and encoded by a viral gene.

viral transformation *See* transformation.

virion a completed virus particle consisting of a nucleic acid core and a protein coat.

viroid a disease-causing agent of plants consisting of a circular, single-stranded RNA molecule typically 270–380 nucleotides long and therefore thousands of times smaller than the most diminutive virus. Since its RNA does not encode proteins, the viroid must rely on host enzymes for its replication. This occurs in the nucleus of the host cell where the viroid RNA is localized within the nucleolus. Replication occurs by a rolling circle mechanism that generates an oligomeric concatomer that is several times the length of the original viroid. Hammerhead ribozymes (*q.v.*) cleave the concatomer into genomic units, and these are subsequently circularized. Viroids are not encapsulated in a protein coat, and their genomes are not integrated into the host genome. The pathogenic effects of viroids on their host results from the fact that the RNA of the viroid contains segments that are complementary to the 7S RNAs of the signal recognition particles (*q.v.*) of their hosts. Thus, the viroid behaves like an antisense RNA (*q.v.*) and blocks the formation and functioning of signal recognition particles. Typical viroids are the apical stunt and planta macho viroids of tomatoes and the cadang cadang viroid of coconut palms. *See* Appendix C, 1967, Diener and Raymer; rolling circle, virusoid.

virulence the relative ability of an organism to produce disease.

virulence plasmids genetically related plasmids found in various *Salmonella* species and in certain *E. coli* strains that give the bacteria the ability to invade intestinal cells.

virulent phage a phage that causes lysis of the host bacterium. *Contrast with* temperate phage.

virus an ultramicroscopic, obligate, intracellular parasite incapable of autonomous replication. Viruses can reproduce only by entering a host cell and using its translational system. Viruses are generally classified according to the type of nucleic acid they contain and the morphology of the nucleocapsid

Virus

Nucleic Acid Types		Virus Families	Examples	Shapes	Hosts
ds DNA	l	Myoviridae	T4 phage	tailed phage	bacteria
	l	Siphoviridae	lambda phage	tailed phage	bacteria
	c	Papovaviridae	SV40	icosahedral	vertebrates
	l	Adenoviridae	HAdv-2	icosadedral	vertebrates
	l	Herpesviridae	EBV, HCMV, HHV	icosahedral	vertebrates
	l	Poxviridae	VAR, VV	brick-shaped	vertebrates
ss DNA (+)	c	Microviridae	phi X174	icosahedral	bacteria
ds RNA	ls	Reoviridae	human reovirus	icosahedral	animals, plants
ss RNA (+)	l	Leviviridae	MS2, Q beta	icosahedral	bacteria
	l	Tobamoviridae	TMV	elongated rod	plants
	l	Coronaviridae	SARS	spherical	birds, mammals
ss RNA (–)	ls	Rhabdoviridae	rabies virus	bullet-shaped	animal
	l	Paramyxoviridae	Sendai virus	pleomorphic	animal
	ls	Orthomyxoviridae		spherical	vertebrates
ssRNA (+) reverse transcribed	ls	Retroviridae	RSV, ALV, MMTV, MoMLV, HIV	spherical	vertebrates

(*q.v.*). The table above compares properties of some of the viruses mentioned in the dictionary. Viruses have been isolated which parasitize organisms that belong to the Eukaryotes, the Bacteria, and the Archaea. Certain double-stranded DNA viruses from each group have major coat proteins called double-barrel trimers that are architecturally similar. This means that these viruses all evolved from a common ancestor that existed before cellular life evolved on earth. In the intervening 3 billion years mutations have erased all similarities in nucleotide sequences in the genomes of these viruses. Viral nucleic acid molecules are linear (l), circular (c), or linear, but in two or more segments (ls). Virus acronyms: ALV (avian leukosis virus), EBV (Epstein-Barr virus), HAdv-2 (human adenovirus–2), HCMV (human cytomegalovirus), HHV (human herpes virus), HIV (human immunodeficiency virus), MMTV (mouse mammary tumor virus), MoMLV (Moloney murine leukemia virus), RSV (Rous sarcoma virus), SARS (severe acute respiratory syndrome), SV40 (simian virus 40), TMV (tobacco mosaic virus), VAR (variola virus), VV (vaccinia virus). *See* Appendix A, Prokaryotes; Appendix C, 1971, Baltimore; 2004, Rice *et al.*; Appendix E, Individual Databases; Appendix F; bacteriophages, Baltimore classification of viruses, enveloped viruses, herpes virus, oncogenic viruses, peplomers, plus (+) and minus (–) viral strands.

virusoid molecules that, like viroids (*q.v.*), are single-stranded, circular RNAs. Unlike viroids, they are encapsulated within the virion of several plant viruses. The velvet tobacco mosaic virus is an example.

virus receptors sites on the cell membrane to which viruses attach. Such sites contain neuraminic acid (*q.v.*).

viscoelastic molecular weight determination a method using a viscometer that allows the determination of the molecular weights of the largest molecules present in a solution. The technique is very useful in determining the molecular weights of very long DNA molecules, since a fraction of these are fragmented during the isolation procedure. *See* Appendix C, 1973, Kavenoff and Zimm.

Visconti-Delbrück hypothesis according to this proposal, bacteriophages multiply upon entering a host, and the replicating units so formed mate repeatedly. Mating occurs in pairs and is at random with respect to the pairing partner. During any given mating cycle, a segment of genetic material from one parent can exchange with that from a second parent phage, yielding recombinant units. *See* Appendix C, 1953, Visconti and Delbrück.

visibles referring to phenotypically observable mutants, as opposed to lethals, which are scored by the absence of an expected class of individuals in a cross designed to detect induced mutants.

visual pigments molecules that participate in reactions that occur after the absorption of a photon

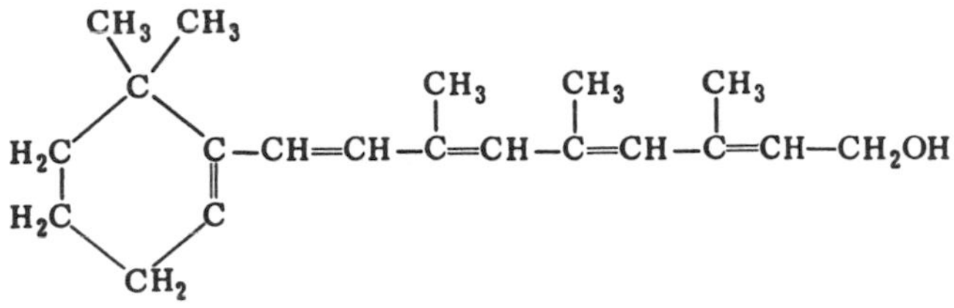

Vitamin A

of light. *See* chromatophore, colorblindness, cone pigment genes (CPGs), opsin, rhodopsin.

visual purple *See* rhodopsin.

vitalism a philosophy holding that the phenomena exhibited in living organisms are the result of special forces distinct from chemical and physical ones. *See* mechanistic philosophy.

vital stain a dye used to stain living cells (Janus green, methylene blue, trypan blue, etc.).

vitamin an organic compound (often functioning as a coenzyme) that is required in relatively minute amounts in the diet for the normal growth of a given organism.

vitamin A a fat-soluble vitamin functioning as a precursor to retinal (*q.v.*) and retinoic acid (*q.v.*). Its structure is shown at the top of the page. Vitamin A is generated by the splitting in two of a molecule of beta carotene. *See* carotenoids.

vitamin B complex a family of water-soluble vitamins, including thiamin (B_1), riboflavin (B_2), nicotinic acid, pantothenic acid, pyridoxin (B_6), and cobalamin (B_{12}).

vitamin C ascorbic acid, an important regulator of the oxidation-reduction state of protoplasm.

vitamin D calciferol; a fat-soluble vitamin required in man for the prevention of rickets. Its structure is shown below. Vitamin D mediates the absorption

Vitamin C

of calcium and phosphorus from the intestine and promotes the mineralization of bone. In humans 7-dehydrocholesterol is secreted at the surface of the skin, where it is converted by ultraviolet radiation from sunlight into vitamin D. A 10 minute exposure per week of the upper body to sunlight will provide sufficient vitamin D. Most naturally occurring foods lack vitamin D, but saltwater fish are an exception. *See* cholesterol.

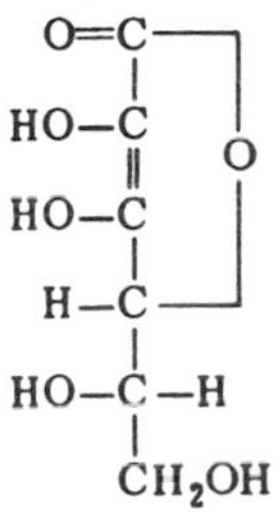

vitamin D receptor (VDR) a protein encoded by a gene on the long arm of human chromosome 12. The gene contains 9 exons, and the binding to vitamin D is due to the protein segment encoded by exons 7, 8, and 9. Exons 2 and 3 each encode zinc-finger domains, and mutations in these regions abolish the DNA-binding function of the protein. Thus, VDR belongs to the family of steroid, nuclear hormone receptors. *See* androgen receptor, zinc-finger protein.

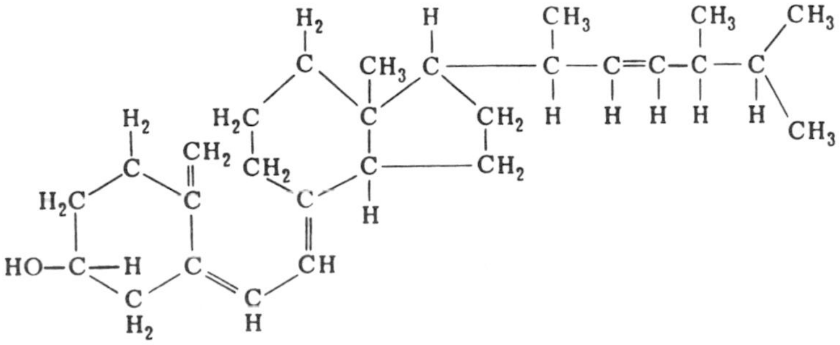

Vitamin D

vitamin D-resistant rickets a group of hereditary diseases in which patients show a reduction in the levels of calcium and phosphorous in their blood and skeletal changes characteristic of rickets, although they have adequate dietary vitamin D. An autosomal recessive form of the disease has been shown to result from mutations in a gene that encodes a vitamin D receptor (*q.v.*).

vitamin E alpha tocopherol, a vitamin functioning as an antioxidant.

vitamin H biotin (*q.v.*).

vitellarium the portion of the insect ovariole posterior to the germarium. Egg chambers complete development within the vitellarium.

vitelline membrane a membrane that surrounds the ovum. In *Drosophila* the term is used specifically for the membrane that immediately surrounds the oolemma and is formed by the fusion of deposits in the intercellular space between the oocyte and the columnar follicle cells that invest it.

vitellogenesis the formation of yolk.

vitellogenic hormone *See* allatum hormones.

vitellogenin a protein synthesized by vitellogenic females and incorporated into the yolk spheres of the developing oocyte. In *Xenopus laevis*, vitellogenins are synthesized by the liver. In *Drosophila melanogaster*, vitellogenins are synthesized by abdominal and thoracic fat bodies and by the columnar follicle cells surrounding the oocyte. *See* lipovitellin, phosvitin.

viviparous **1.** producing living young rather than eggs. Embryogenesis occurs within the mother's body, as with most mammals. **2.** bearing seeds that germinate within the fruit, as in the mangrove.

Viviparus malleatus a prosobranchiate snail showing a bizarre type of spermatogenesis with the production of oligopyrene sperm. *See* sperm polymorphism.

v-myc *See myc.*

VNTR locus *See* variable number of tandem repeats locus.

voltinism a polymorphism in terms of whether or not the embryos produced by an insect enter diapause (*q.v.*). For example, in *Bombyx mori* univoltine strains produce only diapause embryos. Bivoltine strains produce a nonhibernating brood, then diapause embryos.

von Gierke disease a hereditary glycogen storage disease in humans arising from a deficiency of the enzyme glucose-6-phosphatase. Inherited as an autosomal recessive. Prevalence 1/100,000.

von Willebrand disease the most common blood clotting disorder of humans. It is due to a deficiency of the von Willebrand factor (vWF), which is synthesized by endothelial cells and megakaryocytes. The vWF consists of multimers made up of monomers, each containing 2,050 amino acids. The vWF is first synthesized as a precursor molecule containing 2,813 amino acids. The antihemophilic factor (AHF) (*q.v.*) is extremely susceptible to proteolytic degradation. In the plasma, vWF combines with AHF and stabilizes it. If vWF is reduced because of severe damage to the *vWF* gene, there is also a marked reduction in AHF. The vonWillebrand factor is encoded by the *VWF* gene which is located at the end of the short arm of human chromosome 12, which encodes the vWF. The gene is 178 kilobases in length and contains 52 exons. The first 17 exons encode the signal peptide and the propolypeptide. The remaining 35 exons encode the mature subunit and the 3′ noncoding region of the pre-mRNA. Most von Willebrand patients have one normal *vWF* gene and one gene carrying a missense or nonsense mutation in the ORF. They make about half the usual amount of vWF, and their disease is mild. Such heterozygotes occur at a frequency of about 8 per 1,000 individuals. Homozygotes are very rare (less than one in a million), and they bleed uncontrollably when injured. The syndrome is named after Eric vonWillebrand, a Danish physician who first described it in 1931. *See* blood clotting, hemophilia.

***v-sis* gene** *See* simian sarcoma virus.

***v-src* gene** *See* Rous sarcoma virus.

vulgare wheats *See* wheat.

Vulpes vulpes the red fox, a species bred on a large scale on ranches for its pelt. Numerous mutations influencing fur color are known.

W

Wallace effect the hypothesis put forth by Alfred Russel Wallace that natural selection favors the evolution of mechanisms that ensure the reproductive isolation of sexual populations that have reached the level of elementary biological species. Reproductive isolation prevents the production of sterile hybrids which compete for food reserves. *See* isolating mechanism.

Wallace line a zone of contact between two entirely distinct terrestrial biotas, those now called the Oriental and the Australian biogeographic realms. The Wallace linke follows a deepwater zone with the Philippines and Borneo on the west and Sulawesi and the Moluccas on the east. The line cuts through Indonesia between Bali and Lombok. *See* Appendix C, 1859, 1869, Wallace; biogeographic realms, Linnean Society of London, plate tectonics, Sulawesi.

waltzer one of the many neurological mutants in the laboratory mouse. Homozygotes are deaf and characteristically show a circling and head shaking behavior.

Waring blender an electric kitchen appliance used to homogenize mixtures, but used in the laboratory to generate the shearing forces required to detach conjugating bacteria, to strip bacteriophages or their "ghosts" from host cell surfaces, to homogenize tissue samples, etc. *See* interrupted mating experiment, shearing.

warning coloration conspicuous colors or markings on an animal that is poisonous, distasteful, or similarly defended against predators. Such coloration is presumed to facilitate learning, on the part of predators, to avoid the possessor of such markings.

Watson-Crick model *See* deoxyribonucleic acid.

wax any esters of fatty acids and long-chain monohydroxyalcohols.

weak interactions forces between atoms, such as ionic bonds, hydrogen bonds, and van der Waals forces, which are weak relative to covalent bonds (*q.v.*).

weed killer *See* Roundup.

weighted mean the mean obtained when different classes of observations or quantities are given different weights (are multiplied by different factors) in the calculation.

Weismannism the generally accepted concept proposed by August Weismann that acquired characters are not inherited and that only changes in the germ plasm are transmitted from generation to generation. *See* Appendix C, 1883, Weismann.

Werner syndrome a genetic disease that causes premature aging of people in their twenties, beginning with graying of hair and development of wrinkles, followed by diseases of old age (e.g., atherosclerosis, cataracts, osteoporosis, diabetes). Few people afflicted with this syndrome survive to age 50. The Werner syndrome gets its name from Otto Werner, who gave the first description of the disease in 1905. When grown in culture, cells from Werner syndrome patients undergo only about a third the number of doublings characteristic of normal cells. The gene involved (*WRN*) is at 8p12, and it encodes a protein of 1,432 amino acids. The protein is a helicase (*q.v.*) that unwinds DNA and resides in the nucleus and the nucleolus. The enzyme also shows exonuclease activity, is ATP dependent, and contains two transcription initiation sites. Homologous genes have been found in the mouse and the rat. *See* Hayflick limit.

western blotting *See* Southern blotting.

wheat the world's most important grain crop, produced by species of the genus *Triticum* (*q.v.*).

WHHL rabbit the *W*atanabe-*h*eritable *h*yper*l*ipidemic rabbit, which has an LDL receptor activity in its liver cells that is less than 5% that of normal rabbits. WHHL rabbits are a model for humans homozygous for familial hypercholesterolemia (FH) (*q.v.*).

White Leghorn *See* plumage pigmentation genes.

White Plymouth Rock *See* plumage pigmentation genes.

whole-arm fusion, whole-arm transfer *See* centric fusion.

whole genome shotgun (WGS) assembly or sequencing *See* shotgun sequencing.

wild type the most frequently observed phenotype, or the one arbitrarily designated as "normal." Often symbolized by "+" or "wt."

wild-type gene the allele commonly found in nature or arbitrarily designated as "normal."

Wilms tumor a malignant kidney tumor of children often associated with deletions in the short arm of chromosome 11. The *WT1* gene, located at 11p13, encodes specific proteins synthesized during the development of the genitourinary tract. Four different proteins are produced by alternative splicing (*q.v.*) of *WT1* mRNA. These proteins contain zinc fingers which bind to and prevent transcription of certain genes that encode growth factors. The kidney tumors are named after Max Wilms, who described them in a monograph on tumor pathology he published in 1899. *See* anti-oncogenes, zinc finger proteins.

Wilson disease (WD) a hereditary human disease described in 1912 by the neurologist S.A.K. Wilson. WD is caused by mutations in the *ATP7B* gene (*q.v.*), and over 200 have been identified. Most mutations are of the missense variety. The gene product is an enzyme that transports copper through the plasma membranes of liver and brain cells. The disease is characterized by a reduction both in the biliary excretion of copper and in the incorporation of copper into ceruloplasmin (*q.v.*). Liver and brain damage result from the copper overload. The prevalence of WD is 1 per 75,000 live births. *See* missense mutant.

winter variety a variety of cereal which must be sown in the autumn of the year preceding that in which the plants should flower. If sown in the spring, they will not flower in the same growing season. *See* vernalization.

Wnt the symbol for a group of homologous genes which encode proteins that regulate cell-to-cell interactions during development. The *Drosophila* gene *wingless (wg)* controls the segmental pattern of the embryo, and the mouse was later shown to have a homologous gene called *integration (int)*. Once it became clear that these genes were conserved in both invertebrates and vertebrates, the symbol *Wnt* was chosen, *w* from *wg* and *nt* from *int*, to reflect the widespread evolutionary distribution of this family. The proteins encoded by *Wnt* genes are cysteine-rich, secreted glycoproteins which interact with specific cell-surface receptors that in turn initiate a signaling cascade which turns on specific genes in target cells. In the mouse embryo the timing of *Wnt* expression in the central nervous system indicates that *Wnt* genes regulate the differentiation of forebrain and spinal cord. In *Xenopus* embryos the ectopic expression of *Wnt* can cause a second neural axis to develop, and a two-headed larva will be produced. *See* cellular signal transduction.

wobble hypothesis a hypothesis developed to explain how one tRNA may recognize two codons. The anticodon in each tRNA is a base triplet. The first two bases of the mRNA codon pair according to the base pairing rules. The third base in the anticodon, however, has a certain amount of play or wobble that permits it to pair with any one of a variety of bases occupying the third position of different codons. Thus, U in the third position would recognize A or G, for example, and transfer RNA with a CUU anticodon would bind to either of two codons (GAA or GAG). *See* Appendix C, 1966, Crick.

Wolbachia a genus of Gram-negative bacteria that resemble *Rickettsia* (*q.v.*) and live as endosymbionts in many invertebrates (including nematodes, mites, spiders, crustaceans, and insects). Wolbachia are inherited maternally by transovarial transmission, and they often manipulate the reproductive behavior of their hosts. For example, in some insects the infecting Wolbachia secrete toxins that kill Y-bearing sperm, and female-biased sex ratios result. In a strain of the beetle *Callosobruchus chinensis* the X chromosome has been shown to contain a Wolbachia DNA fragment. This is about 11 kbp and contains 12 ORFs, somewhat more than 1% of the genome of the Wolbachia. The structure of the transferred segment is highly preserved which suggests that the transfer was recent. *Wolbachia pipientis* (wMel) is an obligate intracellular parasite of *Drosophila melanogaster*. A recent survey of cultures kept at the Bloomington *Drosophila* Stock Center at Indiana University found that 30% of the strains were infected with Wolbachia. The wMel genome consists of a 1,267,782 bp circle of DNA. It contains very high levels of repetitive DNA and mobile DNA elements. However, there is no evidence for recent lateral gene transfer between wMel and its *Drosophila* host. *See* Appendix A, Prokaryotae, Bacteria, Protobacteria; Appendix C, 2002, Kondo *et al.*; horizontal transmission.

wolf the gray wolf, *Canis lupus*, is the ancestor of the domestic dog. In Europe, what had been considered pure wolf populations have turned out to be hybrids with domestic or feral dogs. *See Canis familiaris;* introgressive hybridization.

Wolman disease a lysosomal storage disease frist described by M. Wolman and three colleagues in 1961. The condition is inherited as an autosomal recessive, and it is due to a deficiency of an acid lipase which leads to progressive accumulation of triglycerides and cholesterol esters in lysosomes of homozygotes. The mutations responsible are in a gene symbolized *LIPA* (for lipase A), and it is located between bands 24 and 25 on the long arm of chromosome 10.

working hypothesis a hypothesis that serves as the basis for future experimentation.

woolly monkey *Lagothrix lagothricha*, a species inhabiting the rain forests of the Amazon basin. The source of the simian sarcoma virus (*q.v.*).

Wright's equilibrium law an expression of the zygotic proportions expected in a population experiencing a certain amount of inbreeding. For a pair of alleles *A* and *a* with frequencies *p* and *q*, respectively, the zygotic proportions are expected to be $AA = p^2 + Fpq$: $Aa = 2pq\ (1 - F)$: $aa = q^2 + Fpq$, where *F* is Wright's inbreeding coefficient (*q.v.*). The Hardy-Weinberg Law (*q.v.*) is a special case of Wright's equilibrium law in which $F = 0$.

Wright's inbreeding coefficient (*F*) the probability that two allelic genes united in a zygote are both descended from a gene found in an ancestor common to both parents. Also, the proportion of loci at which an individual is homozygous. *See* Appendix C, 1968, Wright.

Wright's polygene estimate in the case of quantitative inheritance (*q.v.*), the number of segregating pairs of polygenes (*n*) can be estimated from the variances computed for the F_1 and F_2 populations. If these are symbolized s^2_{F1} and s^2_{F2}, respectively, and the means for the high and low P_1 strains are X_h and X_l, respectively, then

$$n = \frac{(X_h - X_l)}{8(s^2_{F2} - s^2_{F1})}$$

See Appendix C, 1968, Wright; quantitative inheritance.

writhing number the number of times the axis of a DNA molecule crosses itself by supercoiling.

wt wild type (*q.v.*).

W, Z chromosomes the sex chromosomes of an animal in which the female is the heterogametic sex (*Bombyx mori*, for example). In such cases the W chromosome is female-determining and the male is ZZ. *See* Bkm sequences.

x *See* basic number.

X "crossed with" or "mated to" (as in A ♀ × B ♂).

X_2 the offspring of an F_1 test cross.

X:A ratio the ratio of the number of X chromosomes to the number of autosomal sets. A diploid organism with two X chromosomes (i.e., 2X:2A) has an X:A ratio of 1:1, or 1. With one X chromosome (i.e., 1X:2A) this ratio is 1:2, or 0.50. *See* genic balance, sex determination.

xantha any of many chloroplast mutations in various cereal species. The *xantha* 3 mutant of barley, for example, is characterized by chloroplasts that accumulate excessive numbers of pigment granules and never develop orderly arrays of grana.

xanthommatin *See Drosophila* eye pigments.

X chromosome the sex chromosome found in double dose in the homogametic sex and in single dose in the heterogametic sex. *See* human X chromosome.

X-chromosome inactivation in mammalian development, the repression of one of the two X chromosomes in the somatic cells of females as a method of dosage compensation. At an early embryonic stage in the normal female, one of the two X chromosomes undergoes inactivation, apparently at random. From this point on, all descendant cells will be clonal in that they will have the same X chromosome inactivated as the cell from which they arose. Thus, the mammalian female is a mosaic composed of two types of cells—one that expresses only the paternal X chromosome, and another that expresses only the maternal X chromosome. In some cells and tissues, the inactivated X chromosome can be seen as a dense body in the nucleus (referred to as a Barr body or sex chromatin). In abnormal cases where more than two X chromosomes are present, only one X remains active and the others are inactivated. In marsupials, the paternal X is selectively inactivated during female development. In the somatic cells of human females the "inactivated" X chromosome is not completely silenced. Fifteen percent of all X-linked genes escape inactivation, and another 10% vary in their expression. Therefore over 15% of the genes on the X are transcribed at higher levels in women than in men. *See* Appendix C, 1949, Barr and Bertram; 1961, Lyon and Russell; 1962, Beutler *et al.*; 1963, Russell; 1998, Lyon; Cattanach translocation, human X chromosome, *XG, XIST.*

xenia referring to the situation in which the genotype of the pollen influences the developing embryo or the maternal tissue of the fruit so as to produce a phenotypically demonstratable effect upon the seed.

xenogeneic transplantation xenoplastic transplantation (*q.v.*).

xenograft a tissue from an animal that is transferred to another species. The classical experiments of Spemann and Mangold involved xenografts between gastrulas of one newt species and another. The embryos of *Triton taenitus* were pigmented, while those of *T. cristatus* had little or no pigment. Therefore the source of the cells in the induced secondary structures in the graft hybrid could be ascertained. *See* Appendix C, 1924, Spemann and Mangold.

xenoplastic transplantation the transplantation between individuals of different genera or widely distant species. *See Sus scrofa.*

Xenopus a genus of aquatic anurans found in sub-Saharan Africa. They are commonly called *South African clawed frogs*. The 16 species in the genus have genome sizes that range from 3.5×10^9 bp to 1.6×10^{10} bp. These size differences are probably the result of a series of chromosome doublings that occurred within the past 40 million years. The ancestral chromosome number for the genus appears to have been 18, but there are now species with 36, 72, and 108 chromosomes. *X. laevis* and *X. borealis* are favorites for research in molecular genetics. Studies on the nucleolar mutants of *X. laevis* have shown that the nucleolus contains about 450 rRNA genes. This frog has one class of 5S rRNA genes transcribed in the oocyte and another in somatic cells. In *Xenopus* lampbrush chromosomes there are about 20,000 copies of the oocyte 5S rRNA genes per haploid genome, and these are distributed among large chromomeres that terminate the long arms of 15 of the 18 bivalents. There are also 1,300 copies of the somatic 5S rRNA genes, and these are distributed at nonterminal sites along the chromosomes. Since this species has a diploid chromosome number of 36, it can be designated a tetraploid from an evolutionary standpoint. *See* Appendix A, Chordata, Amphibia, Anura; Appendix C, 1966, Wallace and Birnsteil; 1967, Birnsteil; 1968, Davidson, Crippa, and Mirsky; 1973, Ford and Southern; Appendix E, Species Web Site Addresses; Cajal body, concerted evolution, polyploidy, ribosomal RNA genes.

xeroderma pigmentosum a group of hereditary diseases inherited as autosomal recessives in which the skin is extremely sensitive to sunlight or ultraviolet light, and death is usually due to skin cancer. Normal skin cells can repair UV damage to DNA by cut-and-patch repair (*q.v.*). Skin cells from patients with xeroderma pigmentosum contain mutations in genes functioning in this process. For example, the *XPA* gene at 9q encodes a DNA-binding protein that recognizes damaged regions. Helicases encoded by *XPB* (2q) and *XPD* (19q) unwind the double helix and expose these damaged segments. Endonucleases encoded by *XPG* (13q) then cut out the defective segments. *See* Appendix C, 1968, Cleaver. http://www.xps.org.

XG the first X-linked gene in humans shown to escape X-chromosome inactivation. *XG* resides on Xp between *MIC2* (*q.v.*) and the centromere. It spans the pseudoautosomal boundary and has its first three exons in the pseudoautosomal region and the rest in the X-specific region. The Xg blood group antigen is encoded by *XG*. The gene product is a protein that shows a 50% sequence identity to CD99, the product of the *MIC2* gene. It therefore appears that the two genes share a common ancestor.

X-inactivation *See* X chromosome inactivation.

Xiphophorus maculatus the platyfish, and *X. helleri*, the swordtail. Pigment cell genetics and the genetics of sex determination have been intensively studied in laboratory strains of these freshwater species. *See* Appendix A, Chordata, Osteichythes, Neopterygii, Antheriniformes.

XIST a gene located in the X chromosome inactivation center of humans at Xq13. The homologous gene in the mouse is symbolized *Xist*. *XIST* is the acronym for X-Inactive Specific Transcript, and the major transcript is an RNA molecule about 19 kb long. Shorter transcripts are also generated by alternative splicing (*q.v.*). The *XIST* gene is 232,103 base pairs long. The euchromatic part of the X is six times longer than that of the Y. *XIST* RNA is transcribed in the nuclei of female somatic cells, where it coats the X chromosome and causes its inactivation. The transcribing gene is on the chromosome that is being silenced. Chromosomal regions subject to inactivation are enriched with DNA LINE-1 elements. These L1 elements may serve as targets to which *XIST* RNAs bind. *See* Appendix C, 1996, Penny *et al.*; 1998, Lyon; repetitious DNA, X-chromosome inactivation.

X linkage the presence of a gene located on the X chromosome; usually termed "sex linkage" (*q.v.*).

XO the symbolic designation of the situation in some heterogametic organisms in which the X chromosome is present and the Y chromosome is absent.

XO monosomy *See* Turner syndrome.

x radiation radiations produced when high-speed electrons strike a metallic target. X-rays have wavelengths in the range between ultraviolet and gamma radiation and are ionizing radiations.

x-ray crystallography the use of the diffraction patterns produced by x-ray scattering from crystals to determine the three-dimensional structure of the atoms or molecules in the crystal. X-ray crystallography and nuclear magnetic resonance spectroscopy (*q.v.*) are the only techniques capable of determining the 3D structures of proteins and nucleic acids at atomic resolution. *See* Appendix C, 1913, Bragg and Bragg; 1949, Hodgkin *et al.*; 1951, Wilkins and Gosling; 1952, Franklin and Gosling; 1953, Watson and Crick; 1958, Kendrew *et al.*; 1959, Franklin, Caspar, and Klug; 1960, Perutz *et al.*; 1976, Finch and Klug; isomorphous replacement, large angle x-ray diffraction, photograph 51, small angle x-ray diffraction.

XXY trisomy *See* Klinefelter syndrome.

XYY trisomy a human karyotype observed in about 1 in 1,000 male births. Most adult XYY males are over six feet tall. A few are sterile, and some are mentally retarded or have behavioral disorders. XYY individuals make up a greater-than-average proportion of the patients in mental-penal institutions. This may be accounted for in part by their diminished intelligence, which may make it easier for them to be apprehended.

Y

Y the single-letter symbol for pyrimidine. *See* R3.

YAC yeast artificial chromosome (*q.v.*).

Y chromosomal DNA lineages paternal lines of evolutionary descent traced using molecular markers on the Y chromosome. Studies of 1,000 men from 22 geographical areas traced the ancestral Y chromosome to an African male who lived about 59,000 years ago. In the popular press this male was called *Y chromosome Adam*. The observation that this man lived thousands of years later than *mitochondrial Eve* simply means that at the time Eve's mtDNA had become fixed in *Homo sapiens*, males still had a variety of Y chromosomal DNAs. It took tens of thousands of years more before one particular version of the Y became fixed in our species. *See* Appendix C, 2001, Underhill *et al.*; mitochondrial DNA lineages.

Y chromosome the sex chromosome found only in the heterogametic sex. In *Drosophila melanogaster* (*q.v.*) the Y chromosome is composed almost entirely of heterochromatin (*q.v.*). In many *Drosophila* species, the Y chromosome develops prominent lampbrush loops in the nuclei of spermatocytes. The loops consist of a DNA axis to which fibers of transcribed RNA are attached. This RNA is associated with large amounts of protein. The loop-associated transcripts lack open reading frames. Proteins such as tektins (*q.v.*) are bound to specific loops. However, tektins are not encoded by Y-linked genes. Therefore, the RNAs associated with the Y loops may bind specific exogenous proteins that are destined for assembly into sperm axonemes. In humans the Y is rich in heterochromatin and only contains about 1% as many genes as the X. One of these, *SRY* (*q.v.*), is responsible for switching development into the male pathway. *See* Appendix C, 1968, Hess and Meyer; 1987, Page *et al.*; 1993, Pisano *et al.*; human gene maps, human Y chromosome, Jews, *TSPY* gene.

yeast any fungus that generally exists as single cells and usually reproduces by budding. When "yeast" is mentioned in genetics, *Saccharomyces cerevisiae* (*q.v.*), the yeast used by bakers and brewers, is usually the species being referred to. *Schizosaccharomyces pombe* (*q.v.*), a fission yeast, is also mentioned in the more recent literature.

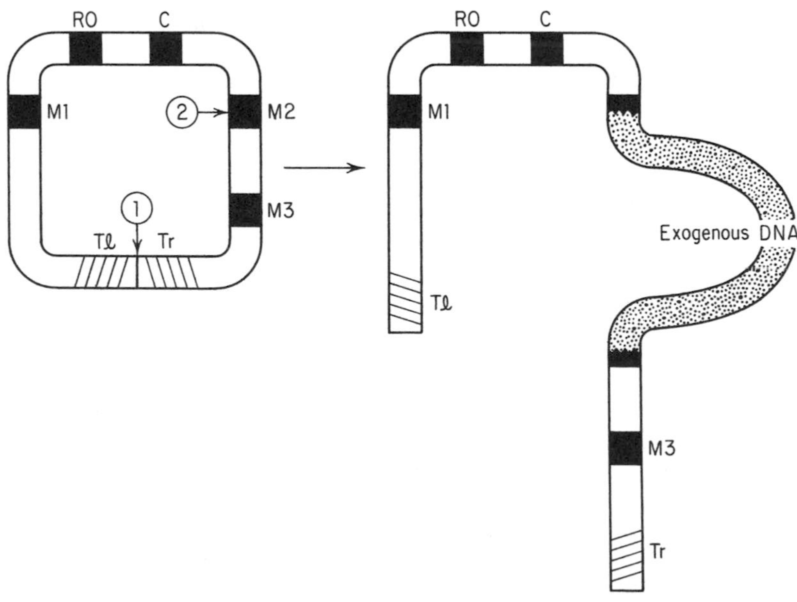

Yeast artificial chromosomes

yeast artificial chromosomes (YACs) genetically engineered circular chromosomes that contain elements from chromosomes contributed by *Saccharomyces* and segments of foreign DNAs that can be much larger than those accepted by conventional cloning vectors (*q.v.*). As shown in the diagram on page 480, YACs are generated from synthetic minichromosomes that contain a yeast centromere (C), a replication orgin (RO), and fused telomeres (Tℓ and Tr). In addition, the circular chromosome contains three marker genes (*M1*, *M2*, and *M3*), which when expressed, allow selection of the cells carrying the plasmid and sites 1 and 2, which allow specific restriction endonucleases (*q.v.*), to break the molecule. Cleavage at 1 opens the ring, while cleavage at 2 generates centric and acentric fragments with ends that will accept foreign DNA fragments. Once these are ligated, an artificial chromosome is generated with a short and a long arm. This contains the spliced segment of foreign DNA to be cloned. Such artificial chromosomes are distributed normally during subsequent yeast divisions, and so colonies containing YACs are generated. In cells possessing the insert, the *M1* and *M3* markers are expressed, but the damaged *M2* is not. So religated YACs can be distinguished from unbroken plasmids. YACs can accept DNA inserts up to 1,000 kilobases long. *Compare with* bacterial artificial chromosomes (BACs) and P1 artificial chromosomes (PACs). *See* Appendix C, 1987, Burke, Carle, and Olson; DNA vector, kilobase, plasmid cloning vectors.

yeast nucleic acid *See* nucleic acid.

yeast two-hybrid system an *in vivo* (*q.v.*) method for identifying protein-protein interactions, based on the properties of a transcriptional activator protein. The simplest version of this system is based on the yeast protein, GAL4 (*q.v.*), whose DNA-binding domain (BD) binds with an upstream activator sequence (UAS) and the activation domain (AD) interacts with the transcription complex to stimulate transcription of a downstream gene (see illustration A). In this two-hybrid scheme, two plasmids (*q.v.*) encoding two hybrid proteins are constructed and introduced into yeast cells. One hybrid contains the GAL4 BD fused to a known protein (protein X), and the second hybrid is a fusion between the GAL4 AD and a second protein (protein Y). These hybrids are coexpressed in a yeast strain lacking GAL4 activity and containing a reporter gene (*q.v.*), such as the bacterial *lac Z* gene, with a binding site for GAL4. Either hybrid by itself is incapable of inducing transcription (illustration B). An interaction between proteins X and Y, however, brings the BD and AD in close proximity, and GAL4 activity is reconstituted (illustration C). This leads to transcriptional activation of the reporter gene and allows X-Y interaction to be monitored by β-galactosidase (*q.v.*) activity. This approach has been modified to screen protein sequences in libraries. In this case, the second hybrid is a fusion between the GAL4 AD and proteins encoded by genomic (*q.v.*) or cDNA library (*q.v.*) sequences. Interaction between the known protein and a protein encoded by one of the library plasmids is detected by expression of the reporter gene. *See* Appendix C, 1989, Field and Song; GAL4, lac operon.

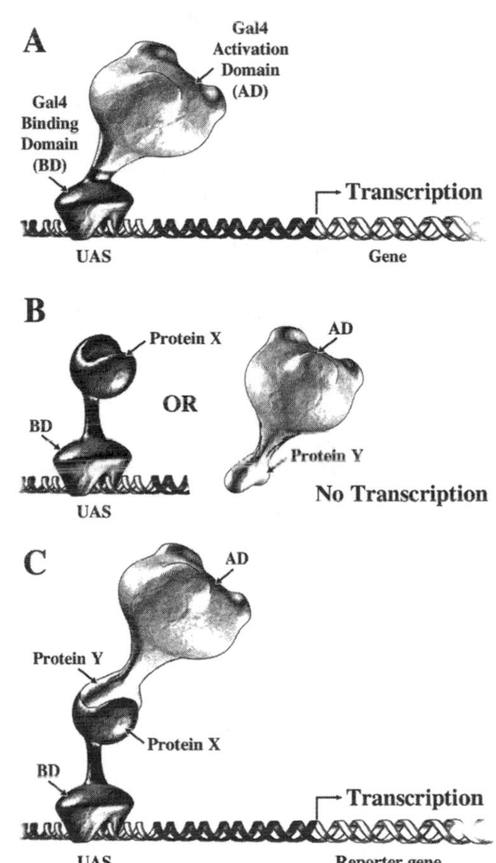

yeast two-micron plasmid *See* FLP/FRT recombination.

Yersinia pestis a Gram-negative bacterium that is the cause of plague. The genus gets its name from Alexandre Yersin, a colleague of Pasteur, who isolated the bacterium in 1894. Bubonic plague is maintained in rat populations and transmitted to humans by the bites of rat fleas. Pneumonic plague occurs under crowded conditions where infected persons spread the bacteria in respiratory droplets which are directly inhaled by nearby people. *Y. pestis* appears to have been a relatively harmless gut pathogen until about 1,500 years ago when it picked up genes that allowed it to colonize fleas and to multiply in the bloodstream of humans. Its genome has been sequenced and found to consist of a 4.65 Mb chromosome and three plasmids of 96.2 kb, 70.3 kb, and 9.6 kb. *See* Appendix A, Bacteria, Proteobacteria.

Y fork the point at which a DNA molecule is being replicated; the two template strands of the parental molecule separate, forming the arms of a Y-shaped structure. The unreplicated double-stranded DNA distal to the arms forms the base of the Y. *See* replication of DNA.

Y linkage genes located on the Y sex chromosome, exhibiting holandric (*q.v.*) inheritance. *See Oryzias latipes.*

yolk the complex collection of macromolecules and smaller nutrient molecules with which the oocyte is preloaded prior to fertilization. *See* lipovitellin, phosvitin, vitellogenin.

Y-suppressed lethal a sex-linked, recessive lethal that causes death of XO *Drosophila melanogaster* but allows survival of normal males.

Z

Z atomic number, the number of protons in the nucleus of the neutral atom.

Z chromosome the sex chromosome found in both heterogametic females and homogametic males. *See* W, Z chromosomes.

Z DNA *See* deoxyribonucleic acid.

Zea mays* spp *mays maize or Indian corn, one of the world's most important food sources. The haploid chromosome number is 10, and cytological maps are available for the pachytene chromosomes. The genome contains 2.4 gbp of DNA, of which half is located in retrotransposons (*q.v.*). *See* Appendix A, Plantae, Angiospermae, Monocotyledoneae, Graminales; Appendix C, 1909, Shull; 1913, East and Emerson; 1928, Stadler; 1931, 1933, 1934, 1938, McClintock; 1938, Rhoades; 1950, McClintock; 1964, Mertz *et al.*; 1984, Pohlman *et al.*, corn, kwashiorkor, *opaque-2, R* genes of maize, teosinte.

zeatin *See* cytokinins.

zebra fish a popular, easy-to-rear aquarium fish. Its scientific name is *Danio rerio* (*q.v.*). In the earlier literature its genus is sometimes given as *Branchydanio*.

zebras *See* Equidae.

Zebu the Brahman (*q.v.*) breed of cattle.

zein a group of alcohol soluble proteins that function as storage proteins in maize kernels. The proteins are encoded by a multigene family, are synthesized in the developing endosperm, and account for more than 50% of the protein in mature seeds. Unfortunately zein is practically devoid of lysine. *See* kwashiorkor, *opaque-2, Zea mays.*

zero-order kinetics the progression of an enzymatic reaction in which the formation of product proceeds at a linear rate with the time. This rate is not increased if additional substrate is added. *See* first-order kinetics.

zero sum assumption an aspect of the Red Queen hypothesis (*q.v.*) proposing that the beneficial effect enjoyed by a species in evolutionary advance is precisely matched by the sum of the negative effects experienced by all other species in the community. The zero sum assumption derives in part from the notion that the total resources available in the system is constant and the rate of evolution will also be constant.

zero time binding DNA strands of DNA containing intramolecular repeats that form duplexes at the start of a reassociation reaction.

Zimmermann cell fusion a technique developed by Ulrich Zimmermann in which cells are exposed to a low-level, high-frequency electric field that orients them into chains. A direct current pulse is then used to open micropores in adjoining cell membranes. These micropores allow mixing of the cytoplasms, and the cells may eventually fuse. The Zimmermann technique may also alter the permeability of the plasmalemma so that DNA fragments the size of genes can enter the cell.

zinc a biological trace element. Atomic number 30; atomic weight 65.37; valence 2^+; most abundant isotope ^{64}Zn, radioisotope ^{65}Zn, half-life 250 days, radiation emitted—positrons.

Zn zinc.

zinc finger proteins proteins possessing tandemly repeating segments that bind zinc atoms. Each segment contains two closely spaced cysteine molecules followed by two histidines. Each segment folds upon itself to form a fingerlike projection. The zinc atom is linked to the cysteines and histidines at the base of each loop as shown here, where C circles represent cysteine molecules, H circles represent histidine molecules, and unlabeled circles represent the other amino acids of the polypeptide finger. The zinc fingers serve in some way to enable the proteins to bind to DNA molecules, where they regulate transcription. *See* Appendix C, 1985, Miller *et al.*; 1987, Page *et al.*; androgen receptor (AR), motifs, transcription factors, vitamin D receptor (VDR), Wilms tumor.

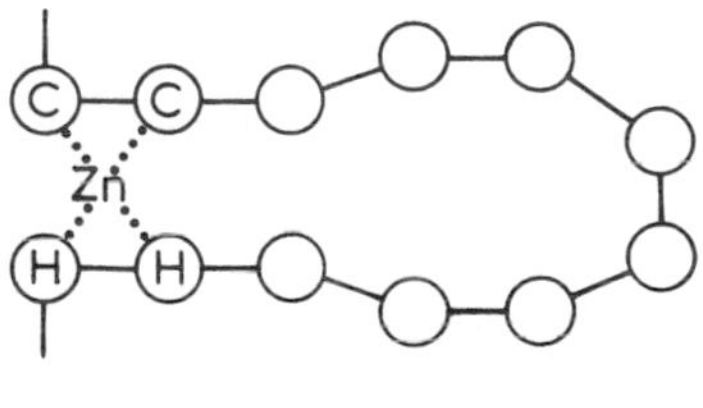

Zip 1 a gene in *Saccharomyces cerevisiae* that is a homolog of the *Gowen crossover suppressor* (*q.v.*) of *Drosophila melanogaster*. *Zip 1* encodes a protein that functions both in centromeric coupling (*q.v.*) and the construction of synaptomemal complexes (*q.v.*).

zonal electrophoresis a technique that makes possible the separation of charged macromolecules and the characterization of each molecule in terms of its electrophoretic mobility. *See* electrophoresis.

zona pellucida one of the envelopes surrounding the mammalian ovum that produces various substances that attract homologous sperm, prevent entry of foreign sperm, and prevent polyspermy (*q.v.*)

zoogeographic realms the divisions of the land masses of the world according to their distinctive faunas. *See* biogeographic realms.

zoogeography the study of the geographical distribution of animals.

zoonotic viruses viruses that can undergo transformations which allow them to cross various species boundaries to infect humans. For example, the influenza A virus left its original host, the duck, infected pigs, and finally humans.

ZPG zero population growth; a population status in which birth and death rates are equivalent.

ZR515 a synthetic juvenile hormone analog that mimics the effects of JH and is more resistant to breakdown by the esterases normally found in insect hemolymph. *See* allatum hormone.

Z, W chromosomes *See* W, Z chromosomes.

zwitterion a dipolar ion. For example, amino acids in solution at neutral pH are in dipolar form, with the amino group protonated ($-NH_3^+$) and the carboxyl group dissociated ($-COO^-$).

zygosity testing the testing of individuals born from a multiple gestation (twins, triplets, etc.) to see if they are homozygotic or dizygotic. The process generally involves comparisons of DNA profiles. The purpose may be to identify a suitable donor for organ transplantation or to determine the risks of other siblings developing a genetic disease, if one is diagnosed with the condition. *See* DNA forensics, twins.

zygotene (zyg) DNA DNA replicated during the zygotene stage of meiosis. During the premeiotic S phase, only about 99.7% of the DNA is replicated. The remainder replicates during zygotene and is intimately involved in synapsis. *See* Appendix C, 1971, Hotta and Stern.

zygonema *See* meiosis.

zygote the diploid cell resulting from the union of the haploid male and female gametes.

zygotene stage *See* meiosis.

zygotic gene a gene expressed in the early embryo. Embryos with mutations in zygotic genes are phenotypically abnormal, and this phenotype is dependent on genetic contributions from both parents, i.e., upon the genotype of the zygote, rather than the genotype of the mother or the father alone. *Also called* zygotic effect gene. *Compare with* maternal effect gene, paternal effect gene. *See* Appendix C, 1987, Nüsslein-Volhard *et al.*; 1988, 1989, Driever and Nüsslein-Volhard; 1995, Lewis *et al.*; *bicoid, hunchback*, zygotic segmentation mutants.

zygotic induction the induction of vegetative replication in a prophage that is transferred during conjugation to a nonlysogenic F^- bacterium.

zygotic lethal in *Drosophila*, a lethal gene whose effect is apparent in the embryo, larva, or adult, but that does not render inviable any gamete that carries it.

zygotic meiosis *See* meiosis.

zygotic segmentation mutants mutations in *Drosophila melanogaster* that are zygotically expressed and control the spatial pattern of development of the embryo. The mutations fall into three classes that are defined by the pattern of cuticular defects they produce. The *gap genes* are active in contiguous domains along the anteroposterior axis of the embryo and regulate segmentation within each domain. The *pair rule genes* are expressed in stripes along the blastoderm with a periodicity that corresponds to every other segment. *The segment polarity genes* regulate the spatial pattern within each segment. *See* Appendix C, 1980, Nüsslein-Volhard and Wieschaus; 1989, Driever and Nüsslein-Volhard; *fushi tarazu, Hox* genes, *hunchback*, maternal polarity mutants, metamerism, selector genes.

zymogen the enzymatically inactive precursor of a proteolytic enzyme. Zymogens usually become activated by posttranslational modifications. For example, the zymogen pepsinogen is converted to the digestive enzyme pepsin by cleavage in a particular peptide sequence.

zymogen granules enzyme-containing particles elaborated by the cells of the pancreas.

Appendix A: Classification

Classification: the subdivision of organisms into an evolutionary hierarchy of groups. The formal hierarchy proceeding from the largest to the smallest group is kingdom, phylum, class, order, family, genus, and species. To allow further subdivisions, the names *grade* or *division* are sometimes placed between kingdom and phylum, the name *branch* is placed between phylum and class, the name *cohort* between class and order, and the name *tribe* between family and genus. In addition, the prefixes *super-* and *sub-* may be added to any group name.

CLASSIFICATION OF LIVING ORGANISMS

superkingdom Prokaryotes (*q.v.*)
  kingdom 1 Prokaryotae
    subkingdom Archaea (formerly Archaebacteria)
      phylum Euryarchaeota (methanogens and halophiles) *Archaeoglobus, Halobacterium, Methanococcus, Pyrococcus, Thermoplasma*
      phylum Crenarchaeota (sulfur-dependent thermophiles) *Desulfurococcus, Sulfolobus, Thermoproteus*
    subkingdom Bacteria (formerly Eubacteria)
      phylum Proteobacteria (*Agrobacterium, Azotobacter, Escherichia, Haemophilus, Helicobacter, Pseudomonas, Rhizobium, Rickettsia, Salmonella, Serratia, Treponema, Yersinia*)
      phylum Spirochaetae (*Borrelia, Spirochaeta, Treponema*)
      phylum Cyanobacteria (*Anabaena, Prochlorococcus, Prochloron, Synechocystis*)
      phylum Saprospirae (*Saprospira*)
      phylum Chloroflexa (*Chloroflexus*)
      phylum Chlorobia (*Chlorobium*)
      phylum Aphragmabacteria (*Mycoplasma, Spiroplasma*)
      phylum Endospora (*Bacillus, Clostridium, Streptococcus*)
      phylum Pirellulae (*Chlamydia*)
      phylum Actinobacteria (*Actinomyces, Cornebacterium, Mycobacterium, Streptomyces*)
      phylum Deinococci (*Deinococcus, Staphylococcus, Thermus*)
      phylum Thermotogae (*Thermotoga*)
superkingdom Eukaryotes (*q.v.*)
  kingdom 2 Protoctista (*q.v.*)
      phylum Archaeprotista (*Barbulanympha, Giardia*)
      phylum Microspora (*Nosema*)
      phylum Glaucocystophyta (*Glaucocystis, Cyanospora*)
      phylum Caryoblastea (*Pelomyxa*)
      phylum Dinoflagellata (*Gonoyaulax*)
      phylum Rhizopoda (*Amoeba*)

phylum Chrysophyta (*Ochromonas*)
phylum Haptophyta (*Prymnesium*)
phylum Euglenida (*Euglena*)
phylum Cryptomonada (*Chroomonas, Cryptomonas, Rhodomonas*)
phylum Zoomastigina (animal flagellates, *Leishmania, Trypanosoma*)
phylum Xanthophyta (*Vaucheria*)
phylum Eustigmatophyta (*Vischeria*)
phylum Bacillariophyta (diatoms, *Diatoma*)
phylum Phaeophyta (brown algae, *Fucus, Macrocystis*)
phylum Rhodophyta (red algae, *Cyanidioschyzon, Polysiphonia*)
phylum Gamophyta (conjugating green algae and desmids, *Spirogyra* and *Micrasterias*)
phylum Chlorophyta (green algae forming flagellated gametes, *Acetabularia, Chlamydomonas, Chlorella, Mesostigma, Volvox*)
phylum Actinopoda (radiolarians and heliozoans, *Acanthocystis, Sticholonche*)
phylum Foraminera (*Fusulina, Globigerina*)
phylum Ciliophora (ciliates, *Paramecium, Stylonychia, Tetrahymena*)
phylum Apicomplexa (*Cryptosporidium, Plasmodium*)
phylum Labyrinthulomycota (slime nets, *Labryrinthula*)
phylum Acrasiomycota (cellular slime molds, *Dictyostelium, Polysphondylium*)
phylum Myxomycota (plasmodial slime molds, *Echinostelium, Physarum*)
phylum Plasmodiophoromycota (*Plasmidiophora*)
phylum Hypochytriomycota (*Hypochytrium*)
phylum Chytridiomycota (*Blastocladiella*)
phylum Oomycota (*Phytophthora, Saprolegnia*)

kingdom 3 Fungi (*q.v.*)
phylum Zygomycota (conjugating fungi, *Mucor, Phycomyces, Pilobolus, Rhizopus*)
phylum Basidiomycota (club fungi, *Agaricus, Amanita, Puccinia, Schizophyllum, Ustilago*)
phylum Ascomycota (sac fungi, *Aspergillus, Neurospora, Penicillium, Podospora, Saccharomyces, Schizosaccharomyces, Sordaria*)

kingdom 4 Animalia (*q.v.*)
subkingdom Parazoa (*q.v.*)
phylum Placozoa (*Trichoplax*)
phylum Porifera (sponges, *Euplectella*)
subkingdom Mesozoa (*q.v.*)
phylum Mesozoa (*Dicyema*)
subkingdom Eumetazoa (*q.v.*)
grade Radiata (*q.v.*)
phylum Cnidaria (Coelenterates)
class Hydrozoa (hydroids, *Hydra*)
class Scyphozoa (true jelly fish, *Physalia*)
class Anthozoa (corals and sea anemones, *Metridium*)
phylum Ctenophora (comb jellies, *Mnemiopsis*)
grade Bilateria (*q.v.*)
subgrade Protostomia (*q.v.*)
superphylum Acoelomata (*q.v.*)
phylum Platyhelminthes (flatworms)
class Turbellaria (planarians)
class Trematoda (flukes, *Schistosoma*)
class Cestoda (tapeworms)

phylum Nemertina (Rhynchocoela) (ribbon worms)
phylum Gnathostomulida
superphylum Pseudocoelomata (*q.v.*)
phylum Acanthocephala (spiny-headed worms)
phylum Entoprocta (entoprocts)
phylum Aschelminthes (*q.v.*)
subphylum Rotifera (rotifers)
subphylum Gastrotricha (gastrotrichs)
phylum Loricifera (loriciferans)
subphylum Kinorhyncha (kinorhynchs)
subphylum Priapulida (priapulids)
subphylum Nematoda (round worms, *Ascaris*, *Caenorhabditis*, *Parascaris*)
subphylum Nematomorpha (Gordiacea) (horsehair worms)
superphylum Coelomata (*q.v.*)
division Tentaculata
phylum Phoronida (phoronids)
phylum Ectoprocta (bryozoa)
phylum Brachiopoda (brachiopods)
division Inarticulata (*q.v.*)
phylum Sipunculoidea (sipunculids)
phylum Mollusca (molluscs)
class Amphineura (chitins)
class Scaphopoda (tooth shells)
class Gastropoda (snails, Aplysia, *Cepaea*, *Limnaea*)
class Pelecypoda (bivalves, *Crassostrea*, *Mytilus*)
class Cephalopoda (squids, octopuses, *Nautilus*)
division Articulata (*q.v.*)
phylum Echiuroidea (echiuroids)
phylum Pentastomida (tongue worms) (parasites)
phylum Tardigrada (water bears)
phylum Annelida (segmented worms)
class Polychaeta (marine worms)
class Oligochaeta (earthworms)
class Hirudinea (leeches)
phylum Onychophora (*Peripatus*)
phylum Arthropoda
branch Chelicerata
class Merostomata (king crabs)
class Pycnogonida (sea spiders)
class Arachnida (scorpions, opilionids, mites, spiders)
branch Mandibulata
class Crustacea
order Branchiopoda (shrimps)
order Ostracoda (ostracods)
order Copepoda (copepods)
order Cirripedia (barnacles)
order Malacostraca (lobsters and crabs)
class Myriapoda
order Diplopoda (millipedes)
order Chilopoda (centipedes)
class Hexapoda
subclass Entognatha
order Collembola (springtails)
order Protura (proturans)

order Diplura (campodeans)
subclass Insecta
cohort Apterogota (primitively wingless insects)
order Archaeognatha (machilids)
order Zygentoma (silverfish)
cohort Pterygota (winged insects)
subcohort Paleoptera (extended wing)
order Ephemeroptera (Mayflies)
order Odonata (dragonflies)
subcohort Neoptera (hinged wing)
superorder Hemimetabola (no pupal stage)
order Embioptera (embiids)
order Phasmida (stick insects)
order Orthoptera (grasshoppers)
order Grylloblatteria (grylloblatids)
order Dictyoptera (roaches, *Blattella*)
order Isoptera (termites)
order Dermaptera (earwigs)
order Psocoptera (booklice)
order Phthiraptera (sucking lice)
order Hemiptera (true bugs)
order Plecoptera (stone flies)
order Thysanoptera (thrips)
superorder Holometabola
order Coleoptera (beetles, *Tribolium*)
order Raphidioptera (snakeflies)
order Megaloptera (alder flies)
order Neuroptera (lacewings)
order Hymenoptera (*Apis, Microbracon, Mormoniella*)
order Trichoptera (caddis flies)
order Lepidoptera (moths, *Biston, Bombyx, Ephestia, Lymantria*)
order Mecoptera (scorpion flies)
order Siphonaptera (fleas)
order Diptera (*Aedes, Anopheles, Chironomus, Culex, Drosophila, Glyptotendipes, Lucilia, Musca, Rhynchosciara, Sciara*)
subgrade Deuterostomia (*q.v.*)
phylum Echinodermata
class Crinoidea (sea lilies)
class Asteroidea (starfish)
class Ophiuroidea (brittle stars)
class Echinoidea (sea urchins, *Strongylocentrotus*)
class Holothuroidea (sea cucumbers)
phylum Chaetognatha (arrow worms)
phylum Pogonophora (beard worms)
phylum Chordata (notochord-bearing animals)
subphylum Acraniata (*q.v.*)
branch Hemichordata (acorn worms and pterobranchs)
branch Urochordata (tunicates)
branch Cephalochordata (lancelets, *Branchiostoma*)
subphylum Craniata (*q.v.*)
branch Agnatha (jawless vertebrates)
class Cyclostomata (lampreys)

branch Gnathostoma (jawed vertebrates)
- **class** Chondrichthyes (elasmobranchs)
- **class** Osteichthyes (bony fish)
  - **subclass** Palaeopterygii (ancient fishes)
    - **order** Acipenseriformes (sturgeons)
    - **order** Semionotiformes (gars)
  - **subclass** Neopterygii (modern fishes)
    - **order** Salmoniformes (*Salmo*)
    - **order** Anguilliformes (eels)
    - **order** Cypriniformes (*Carassius, Oryzias*)
    - **order** Cyprinidontiformes (*Danio, Fundulus, Lebistes*)
    - **order** Perciformes (*Tilapia, cichlids*)
    - **order** Siluriformes (catfishes)
    - **order** Elopiformes (tarpons)
    - **order** Antheriniformes (*Xiphophorus*)
    - **order** Clupeiformes (herrings)
    - **order** Gasterosteiformes (sea horses)
    - **order** Pleuronectiformes (flounders)
    - **order** Tetraodontiformes (*Tukifugu, Tetraodon*)
  - **subclass** Crossopterygii (lobe-finned fishes)
    - **order** Dipnoi (lungfishes, *Protopterus*)
    - **order** Actinista (coelacanths, *Latimeria*)
- **class** Amphibia
    - **order** Apoda (caecilians)
    - **order** Urodela (salamanders, *Ambystoma, Notophthalmus, Pleurodeles, Triton, Triturus*)
    - **order** Anura (frogs and toads, *Rana, Xenopus*)
- **class** Reptilia (turtles, alligators, lizards, snakes)
- **class** Aves (birds)
  - **subclass** Palaeognathae (flightless birds, ostriches, emus)
  - **subclass** Neognathae (modern birds)
    - **order** Anseriformes (ducks, *Anas*)
    - **order** Galliformes (quail, turkeys, *Coturnix, Gallus*)
    - **order** Columbiformes (pigeons and doves, *Columba*)
    - **order** Psittaciformes (parrots)
    - **order** Passeriformes (song birds)
- **class** Mammalia
  - **subclass** Protheria (egg-laying mammals)
  - **subclass** Metatheria (marsupials, *Monodelphis, Potorous*)
  - **subclass** Eutheria (placental mammals)
    - **order** Insectivora (moles, shrews)
    - **order** Chiroptera (bats)
    - **order** Edentata (sloths)
    - **order** Carnivora (carnivores, *Canis, Felis, Vulpes, Mustela*)
    - **order** Cetacea (whales)
    - **order** Proboscidea (elephants)
    - **order** Pinnipedia (seals)
    - **order** Perissodactyla (odd-toed ungulates, *Equus*)

order Artiodactyla (even-toed ungulates, *Bos*, *Camelus*, *Ovis*, *Sus*)
order Rodentia (rodents, *Cavia*, *Chaetodipus*, *Chinchilla*, *Dipodomys*, *Mesocricetus*, *Mus*, *Peromyscus*, *Rattus*)
order Lagomorpha (rabbits, *Oryctolagus*)
order Scandentia (tree shrews)
order Dermoptera (flying lemurs)
order Primates (lemurs, tarsiers, monkeys, apes, humans, *Cercopithecus*, *Gorilla*, *Homo*, *Macacca*, *Pan*)

kingdom 5 Plantae
phylum Bryophyta (*q.v.*)
subphylum Hepaticae (liverworts, *Sphaerocarpos*)
subphylum Anthocerotae (hornworts)
subphylum Musci (mosses)
phylum Tracheophyta (vascular plants)
subphylum Psilophyta (*Psilotum*)
subphylum Lycopodophyta (clubmosses and quillworts)
subphylum Sphenopsida (horsetails)
subphylum Pteropsida (ferns and seed plants)
superclass Filicinae (ferns)
superclass Gymnospermae (cone-bearing, seed plants)
class Pteridospermophyta (seed ferns)
class Cycadophyta (cycads)
class Ginkgophyta (ginkgos)
class Coniferophyta (conifers)
class Gnetophyta (gnetophytes)
superclass Angiospermae (flowering plants)
class Dicotyledoneae
order Magnoliales (magnolia, avocado)
order Rosales (rose, apple, plum, strawberry, raspberry, currant, hemp)
order Leguminales (*Glycine*, *Pisum*, *Phaseolus*)
order Salicales (willows, *Populus)*
order Fagles (beech, oak, birch)
order Geraniales (Nasturtium, *Pelargonium)*
order Cactales (cactuses)
order Scrophulariales (*Antirrhinum*, *Collinsia*, privet)
order Ranales (buttercups, water lillies, lotus, *Ranunculus*)
order Myrtales (myrtle, eucalyptus, *Oenothera*)
order Cruciales (cabbage, turnip, *Arabidopsis*, radish)
order Cucurbitales (*Cucurbita*, *Cucumis*)
order Caryophyllales (carnations, *Mirabilis*)
order Gentianales (gentians, olives, lilacs, *Vinca*)
order Primulales (*Primula*)
order Malyales (*Gossypium*, linden, elm, *Theobroma*, jute)
order Rubiales (coffee, quinine)
order Papaverales (poppy)

order Hamamelidales (sycamore)
order Urticales (nettle, elm, mulberry)
order Jugandales (walnut, hickory)
order Linales (flax)
order Cornales (dogwood, mangrove)
order Proteales (macadamia nut)
order Sarranceniales (pitcher plant, sundew)
order Theales (tea)
order Umbellales (carrot, parsnip, celery, ginseng)
order Solanales (*Datura*, *Lycopersicon*, morning glory, sweet potato, *Nicotiana*, *Solanum*)
order Rhamnales (grapes, ivy)
order Boraginales (heliotropes)
order Lamiales (lavender, mint, verbenas)
order Asterales (composites, silverswords, daisies, sunflowers, *Haplopappus*)

class Monocotyledoneae
order Palmales (palms)
order Graminales (grasses, *Hordeum*, *Oryzea*, *Triticum*, *Zea*)
order Liliales (lilies, tulips, amaryllis, iris, *Colchicum*)
order Commelinales (*Tradescantia*)
order Arales (callas, taros)
order Zingiberales (bananas)
order Orchidales (orchids)
order Bromeliales (pineapple)

For further information on the classification and evolution of life on earth consult:

Futuyma, D. J. 1998 *Evolutionary Biology*. Third edition, Sinauer Associates Publishers, Sunderland, MA.

Margulis, L. 1974 The Classification and Evolution of Prokaryotes and Eukaryotes. Chapter 1, pages 1–41. In *The Handbook of Genetics*, R. C. King, editor, vol. 1, *Bacteria, Bacteriophages, and Fungi*, Plenum Press, New York.

Nielsen, C. 2001 *Animal Evolution: Interrelationships of Living Phyla*. Second edition, Oxford University Press, Oxford, England.

Strickberger, M. L. 2000 *Evolution*. Third edition, Jones and Bartlett Publishers, Boston.

Tudge, C. 2000 *The Variety of Life. A Survey and a Celebration of all the Creatures that Have Ever Lived*. Oxford University Press, Oxford, England.

Appendix B: Domesticated Species

Domesticated species, organisms that have been trained to live with or be of service to humans, include agricultural plants, livestock, household pets, laboratory animals, and others. Below is a listing of the common and scientific names of a variety of economically important domesticated organisms. Species that are given their own entries in the dictionary (i.e., barley, chicken, and corn) are omitted from this list.

alfalfa *Medicago sativa*
almond *Prunus amygdalus*
alpaca *Lama pacos*
amaryllis *Amaryllis belladonna*
anise *Pimpinella anisum*
apple *Pyrus malus*
apricot *Prunus armeniaca*
artichoke *Cynara scolymus* (globe)
ash *Fraxinus americana* (white)
asparagus *Asparagus officinalis*
aster *Aster novaeangliae* (New England)
avocado *Persea americana*
balsam fir *Abies balsamea*
bamboo *Bambusa vulgaris*
banana *Musa* species
bass *Micropterus salmoides* (large mouth)
bean *Vicia faba* (broad): *Ricinus communis* (castor); *Phaseolus limensis* (lima); *P. aureus* (Mung); *P. vulgaris* (string)
beech *Fagus grandifolia* (American); *F. sylvatica* (European)
beet *Beta vulgaris*
begonia *Begonia rex*
birch, paper *Betula papyrifera*
blackberry Cultivated blackberries are derived chiefly from three species of *Rubus: R. argutus, R. alleghaniensis,* and *R. frondosus.*
blueberry *Vaccinium corymbosum*
bluegrass *Poa pratensis*
Brazil nut *Bertholletia excelsa*
breadfruit *Artocarpus communis*
broccoli *Brassica oleracea italica*
broomcorn *Sorghum vulgare technicum*
Brussels sprouts *Brassica oleracea gemmifera*
buckwheat *Fagopyrum sagittatum*
cabbage *Brassica oleracea capitata*
calabash *Lagenaria siceraria*
camellia *Camellia japonica*
canary *Serinus canaria*
cantaloupe *Cucumis melo cantalupensis*
cardamom *Elettaria cardamomum*
carnation *Dianthus caryophyllus*
carp *Cyprinus carpio*
carrot *Daucus carota sativa*
cashew nut *Anacardium occidentale*
cassava *Manihot esculenta*
cauliflower *Brassica oleracea botrytis*
cedar, eastern red *Juniperus virginiana*
celery *Apium graveolens*
cherry *Prunus cerasus* (sour), *P. avium* (sweet)
chestnut (European) *Castenea sativa*
chick pea *Cicer arietinum*
chili pepper *Capsicum annuum*
Chinese cabbage *Brassica rapa*
chive *Allium schoenoprasum*
chrysanthemum *Chrysanthemum morifolium*
cinnamon *Cinnamomum zeylanicum*
clove *Syzygium aromaticum*
clover *Trifolium pratense* (red), *T. repens* (white)
cocoa *Theobroma cacao*
coconut *Cocos nucifera*
coffee *Coffea arabica*
coriander *Coriandrum sativum*
cowpea *Vigna sinensis*
crabapple *Pyrus ioensis*
cranberry *Vaccinium macrocarpon*
crocus *Crocus susianus* (cloth of gold)
currant *Rhibes sativum*
daffodil *Narcissus pseudo-narcissus*
date palm *Phoenix dactilifera*

dill *Anethum graveolens*
Douglas fir *Pseudotsuga menziesii*
duck *Anas platyrhynchos* (mallard)
ebony *Diospyros ebenum*
eggplant *Solanum melongena esculentum*
elephant *Elephas maximus* (Indian)
elm *Ulmus americana*
endive *Cichorium endivia*
ermine *Mustela erminea*
fig *Ficus carica*
foxglove *Digitalis purpurea*
geranium *Pelargonium graveolens*
gerbil *Merinoes unguiculatis* (Mongolian)
ginger *Zingiber officinale*
gladiolus *Gladiolus communis* (one of hundreds of species)
goat *Capra hircus*
goose *Cygnopsis cygnoid* (Chinese)
grape *Vitis vinifera* (common wine)
grapefruit *Citrus paradisii*
guava *Psidium guajava*
hazelnut *Corylus americana*
hemlock *Tsuga heterophylla* (western)
hemp *Cannabis sativa* (marijuana), *Agave sisalana* (sisal)
hickory *Carya ovata* (shagbark)
holly *Ibex opaca* (American), *I. aquifolium* (European)
hollyhock *Althea rosea*
honey locust *Gleditsia triacanthos*
hop *Humulus lupulus*
huckleberry *Gaylussacia baccata* (black)
hyacinth *Hyacinthus orientalis*
iris *Iris versicolor grandiflorum* (blue flag)
ivy *Hedera helix* (English)
jasmine *Jasminum officinale*
juniper *Juniperus communis*
kapok *Ceiba pentandra*
laburnum *Laburnum anagyroidies*
lavender *Lavandula officinalis*
lemon *Citrus limon*
lentil *Lens culinaris*
lettuce *Lactuca sativa*
licorice *Glycyrrhiza glabra*
lime *Citrus aurantifolia*
lingonberry *Vaccinium vitis-idaea* (also called mountain cranberry or lowbush cranberry)
llama *Lama glama*
lotus *Nelumbo lutea* (yellow)
lychee *Litchi chinensis*
macadamia *Macademia integrifolia*
maguey *Agava cantala* (Manila), *A. atrovirens* (pulque)
mahogany *Swietenia mahagoni* (West Indian)
mango *Mangifera indica*
marigold *Tagetes erecta* (garden)
marten *Martes americana*
millet *Pennisetum glaucum*
morning-glory *Ipomoea purpurea*
mullberry *Morus rubra*
mushrooms
  common edible *Agaricus bisporus*
  Chinese *Volvariella volvacea*
  shiitake *Lentinus edodes*
musk ox *Ovibos moschatus*
mustard *Brassica hirta* (white)
narcissus *Narcissus poeticus*
nasturtium *Tropaeolum majus*
oak *Quercus suber* (cork), *Q. alba* (white)
okra *Hibiscus esculentus*
olive *Olea europaea*
orange *Citrus aurantium* (sour): *C. sinensis* (sweet); the naval orange is a cultivar of this species.
oyster *Crassostraea virginica*
Pak choy (Chinese cabbage) *Brassica campestris*
pansy *Viola tricolor*
papaya *Carica papaya*
parakeet *Melopsittacus undulatus*
parsnip *Pastinaca sativa*
passionfruit *Passiflora edulis*
peach *Prunus persica*
peanut *Arachis hypogaea*
pear *Pyrus communis*
pecan *Carya illinoensis*
peony *Paeonia officinalis*
pepper *Capsicum frutescens* (red), *Piper nigrum* (black)
peppermint *Mentha piperita*
perch *Perca flavescens* (yellow)
persimmon *Diospyros kaki* (Japanese)
philodendron *Philodendron cordatum*
phlox *Phlox drummondii*
pickerel *Esox niger* (chain)
pike *Esox lucius* (northern)
pine *Pinus lambertiana* (sugar)
pineapple *Ananas comosus*
pistachio nut *Pistacia vera*
plum *Prunus domestica*
poinsettia *Euphorbia pulcherrima*
poppy *Papaver somniferum* (opium)
quince *Cydonia oblonga*
radish *Raphanus sativus*
raspberry *Rubus occidentalis* (black); *R. idaeus* (red)
redbud *Cercis canadensis*
reindeer *Rangifer tarandus*
rhubarb *Rheum officinale*
rose Many species of *Rosa* are grown commercially. The most common are *R. centifolia*, *R. damascena*, and *R. multiflora*

rubber *Hevea brasiliensis*
rutabaga *Brassica napus*
rye *Secale cereale*
sandalwood *Santalum album*
sequoia *Sequoia gigantia* (big tree), *S. sempervirens* (coastal redwood)
sesame *Sesamum indicum*
sorghum *Sorghum bicolor*
spearmint *Mentha spicata*
spinach *Spinacia oleracea*
spruce *Picea pungens* (Colorado blue)
strawflower *Helichrysum bracteatum*
sugar cane *Saccharum officinarum*
sunflower *Helianthus annuus*
sweet potato *Ipomoea batatas*
Swiss chard *Beta vulgaris cicla*
sycamore *Platanus occidentalis*
tangerine *Citrus reticulata*
taro *Colocasia esculenta*
tea *Camellia sinensis* (*Thea sinensis*)
teak *Tectona grandis*
thyme *Thymus vulgaris*
timothy *Phleum pratense*
trout *Salvelinus frontinalis* (brook), *S. namaycush* (lake), *Salmo gairdneri* (rainbow)
tulip *Tulipa gesneriana*
tulip tree *Liriodendron tulipifera*
turkey *Meleagris gallopavo*
turnip *Brassica campestris*
vanilla *Vanilla plantifolia*
vicuña *Vicugna vicugna*
violet *Saintpaulia ionantha* (African)
walnut *Juglans regia* (English)
water buffalo *Bubalus babalis*
watermelon *Citrullus vulgaris*
wild rice *Zizania aquatica*
willow *Salix babylonica* (weeping)
yak *Bos grunniens*
yam *Dioscorea alata*
yucca *Yucca brevifolia* (Joshua tree)
zinnia *Zinnia elegans*

Appendix C: Chronology

Genetics, cytology, and evolutionary biology have received stimulation from both related and quite independent sciences. In many cases, the development of a particular physical instrument or technique has led to a golden age of discovery. Often research in various areas has advanced in nonsynchronous spurts, and consequently it is difficult to develop courses in genetics, cytology, and evolutionary biology from a strictly historical standpoint. The student, however, should have some idea of the chronological order in which certain events having a bearing on these sciences took place. The following chronology will fill this need, even though many experts will complain about the inclusion of some events and the omission of others. Furthermore, a decade from now some of the recent discoveries may be relegated to less prominent positions. The student should keep the following thought in mind when perusing this catalog. In science, a great unifying concept generally does not spring full-blown from the mind of a single individual. Rather, when the time is ripe, perhaps a dozen authorities may grope about for an explanation, and all may be on the verge of the answer. However, often one scientist may first express the unifying concept in a clear fashion, and as a matter of convenience he or she is the one listed as the progenitor of the idea.

Recently, more and more papers, reporting the work of international teams of scientists, have appeared. For example, the entries between 1989 and 1996 that describe the cloning of different human genes refer to 14 reports with an average of 25 authors each. The entries listed between 1995 and 2004 that describe the genomes of 25 different prokaryotes have an average of 54 authors each. The consortium that sequenced the *Caenorhabdites* genome in 1998 contained 407 scientists! Space limitations prevent us from listing more than three authors in each new entry, and therefore we regret that a large fraction of the scientists participating in such team projects fail to get the credit they deserve.

Chronology

1590 Z. and H. Janssen combine two double convex lenses in a tube and produce the first compound microscope.

1651 W. Harvey puts forward the concept that all living things (including man) originate from eggs.

1657 R. de Graaf discovers follicles in the human ovary, but interprets them incorrectly as eggs.

1665 R. Hooke publishes *Micrographia,* in which he gives the first description of cells.

1668 F. Redi disproves the theory of spontaneous generation of maggots.

1673 A. van Leeuwenhoek sends the first of a series of letters to the Royal Society of London describing his observations with homemade microscopes. This correspondence continues for 50 years. He is the first to see "animalcules" (bacteria and protozoa) and red blood cells of humans and other mammals. Later he described human spermatozoa and those from arthropods, molluscs, fishes, birds, and other mammals.

1694 J. R. Camerarius conducts early pollination experiments and reports the existence of sex in flowering plants.

1735 C. V. Linné publishes the first edition of the *Systema Naturae*. Sixteen editions of this taxonomic work are completed during his lifetime. It is the 10th edition of this book (published in 1753) that serves as the starting point for the modern scientific naming of animals, as his *Species Plantarum* is for plants. Linné originated the "Linnean" system of binary nomenclature used to this day. His insistence on the constancy and objective classification of species posed the problem of the method of the origin of species. Linné invented the names that are used to this day for about 7,700 plant and 4,400 animal species, including *Homo sapiens*. C. Linné was granted nobility in 1741, becoming Carl von Linné (C. V. Linné), and he also used the Latinized form of his name (Carolus Linnaeus).

1761–67 J. G. Kölreuter carries out crosses between various species of *Nicotiana* and finds that the hybrids are quantitatively intermediate between their parents in appearance. The hybrids from reciprocal crosses are indistinguishable. He concludes that each parent contributes equally to the characteristics of the offspring.

1769 L. Spallanzani demonstrates that the "spontaneous generation" of microorganisms in a nutrient medium can be prevented, provided the vessel is sealed and subjected to the temperature of boiling water for 30 minutes or more. In 1780 he performs artificial insemination experiments with amphibians and demonstrates that physical contact of the egg with spermatic fluid is necessary for fertilization and development.

1798 T. R. Malthus publishes anonymously *An Essay on the Principle of Population*. This essay subsequently influences the thinking of both Charles Darwin and Alfred Russel Wallace and leads them to the concept of natural selection.

Edward Jenner publishes *An Inquiry into the Causes and Effects of the Variolae Vaccinae, a Disease Discovered in Some of the Western Countries of England. Particularly Gloucestershire and Known by the Name of Cow Pox*. In it he gave the first account of vaccination with cowpox virus to prevent smallpox. He thus establishes the principle of active immunization and initiates the science of immunology.

1809 J. B. de Monet Lamarck puts forward the view that species can change gradually into new species through a constant strengthening and perfecting of adaptive characteristics, and that these acquired characteristics are transmitted to the offspring.

1818 W. C. Wells suggests that human populations in Africa have been selected for their relative resistance to local diseases. He is thus the first to enunciate the principle of natural selection.

1820 C. F. Nasse points out that hemophilia occurs only in males, but is passed on by females, who themselves are not bleeders. The phenomenon is now called sex linkage.

1822 E. G. Saint-Hilaire suggests that the chordate body plan arose by inverting the dorsoventral axis of an ancestor whose body plan resembled that of arthropods.

1822–24 T. A. Knight, J. Goss, and A. Seton all independently perform crosses with the pea and observe dominance in the F_1 and segregation of various hereditary characters in the F_2. However, they do not study later generations or determine the numerical ratios in which the characters are transmitted.

1825 F. V. Raspail founds the science of histochemistry by using the iodine reaction for starch.

J. E. Purkinje discovers the germinal vesicle in the avian egg.

1827 K. E. von Baer gives the first accurate description of the human egg.

1830 G. B. Amici shows that the pollen tube grows down the style and into the ovule of the flower.

1831 R. Brown coins the name *nucleus* for a structure he observes within each of the cells of an orchid.

On December 27, H.M.S. *Beagle* sets sail from Plymouth for a voyage around the world. It carries as naturalist the 22-year-old Charles Darwin.

1835 The *Beagle* reaches the Galapagos Islands, on September 15, and Darwin spends five weeks surveying the plant and animal life.

1837 Darwin realizes, after he and other experts go over the collections from the Galapagos, that many species are unique to various islands. This suggests that the islands were colonized by a few species from the mainland, and from these evolved new species specialized to live on each of the many new environments provided. These conclusions stimulated Darwin to start accumulating data to support a theory of evolution through natural selection. He makes a diagram of an evolutionary tree in his *First Notebook on Transmutation of Species.*

Hugo von Mohl provides the first description of chloroplasts (chlorophyllkörnen) as discrete bodies within the cells of green plants.

1838 The word *protein* first appears in the chemical literature in a paper by G. J. Mulder. The term, however, was invented by J. J. Berzelius.

1838–39 M. J. Schleiden and T. Schwann develop the cell theory. Schleiden notes nucleoli within nuclei.

1839 J. E. Purkinje coins the word *protoplasm.*

1841 A. Kölliker shows spermatozoa to be sex cells that arise by transformation of cells in the testis.

1845 J. Dzierzon reports that drones hatch from unfertilized eggs, worker and queen bees from fertilized eggs.

R. Remak concludes that cell rearrangements early in vertebrate gastrulation lead to the formation of three specific tissue layers, which he names *ectoderm, mesoderm,* and *endoderm.*

1855 A. R. Wallace accumulates evidence in favor of geographical speciation during his studies of the fauna of the Malay Archipelago. He is led to doubt the dogma of the constancy of species and begins to develop a theory of evolution identical to Darwin's.

R. Virchow states the principle that new cells come into being only by division of previously existing cells.

1856 Gregor Mendel, a monk at the Augustinian monastery of St. Thomas in Brünn, Austria (now Brno, Czechoslovakia), begins breeding experiments with the garden pea, *Pisum sativum.*

1858 Essays by Darwin and Wallace, which contain the theory of evolution based upon natural selection, are communicated to the Linnean Society of London at its June meeting and published in the next issue of the society's journal.

1859 Darwin publishes *On the Origin of Species.*

A. R. Wallace, after studying the zoogeography of the Malay peninsula, points out that a line can be drawn defining the zone of contact between two distinct terrestrial biotas. To the east of the Wallace line the fauna is Australian, and to the west the fauna is Oriental.

1860 T. A. E. Klebs introduces paraffin embedding.

1861 L. Pasteur discovers that certain microorganisms can flourish in the absence of oxygen and that some are even poisoned by it. He postulates subsequently that fermentation is the method used by microbes like yeast to derive energy from sugar under anaerobic conditions.

1864 L. Pasteur demonstrates that air-borne microscopic bacteria and mold spores can infect sterile nutrient cultures. His essay that disproves "spontaneous generation" inaugurates a new epoch in bacteriology.

1865 Gregor Mendel presents the results and interpretations of his genetic studies on the garden pea at the Brünn Society for the Study of Natural Science at their monthly meetings held February 8 and March 8.

1866 G. Mendel's *Versuche über Pflanzenhybriden* (*Experiments on Plant Hybridization*) is published and ignored.

E. Metchnikoff proposes that the pole cells of the insect embryo are the precursor cells of the oocyte and nurse cells.

1867 W. Waldeyer publishes the first histologic study of a developing human cancer. He describes how breast cancer develops as a localized proliferation of glandular cells. They then invade adjacent tissue and reach nearby blood vessels. From here they spread throughout the body via the lymphatic and circulatory systems.

1868 T. H. Huxley concludes from a study of the first specimen of *Archaeopteryx* that it represents the most reptilian bird, a link between flightless feathered birds and the dinosaurs.

1869 D. I. Mendeleev constructs the earliest periodic table of elements.

F. Galton publishes *Hereditary Genius*. In it he describes a scientific study of human pedigrees from which he concludes that intelligence has a genetic basis.

A. R. Wallace publishes *The Malay Archipelago, the Land of the Orangutan and the Bird of Paradise*. In it he reveals that the animals on the Maylay Archipelago are divided into two groups; the western species are similar to the animals found in India, and the eastern species resemble the animals of Australia.

1870 W. His invents the microtome.

1871 F. Miescher publishes a technique for the isolation of nuclei and reports the discovery of nuclein (now known to be a mixture of nucleic acids and proteins).

Charles Darwin publishes *The Descent of Man and Selection in Relation to Sex*. In it, the theory of sexual selection (i.e., selection at mating) is first enunciated.

1872 J. T. Gulick describes variations in shell coloration among natural populations of land snails living in valleys of Oahu. He suggests that geographical isolation of small populations of such animals may be a necessary prerequisite to the formation of new species.

1873 A. Schneider gives the first account of mitosis.

1874 C. Dareste draws attention to the distinction between monozygotic and dizygotic twins.

1875 F. Galton demonstrates the usefulness of twin studies for elucidating the relative influence of nature (heredity) and nurture (environment) upon behavioral traits.

O. Hertwig concludes from a study of the reproduction of the sea urchin that fertilization in both animals and plants consists of the physical union of the two nuclei contributed by the male and female parents.

E. Strasburger describes cell division in plants, and later (1884) coins the terms *prophase*, *metaphase*, and *anaphase*.

1876 O. Bütschli describes nuclear dimorphism in ciliates.

A. R. Wallace publishes *The Geographical Distribution of Animals* in which he describes the distinctive fauna which characterize the six zoogeographic realms of the earth.

W. F. Kühne discovers trypsin in pancreatic juice and coins the term *enzyme*.

1877 H. Fol reports watching the spermatozoan of a starfish penetrate the egg. He was able to see the transfer of the intact nucleus of the sperm into the egg, where it became the male pronucleus. He coins the term *aster*.

E. Abbe begins to publish important contributions to the theory of microscopic optics.

R. Koch develops the method used to this day for obtaining pure cultures of bacteria.

L. Pasteur and his coworkers perform experiments (between 1877 and 1888) that prove that specific bacteria and viruses are the primary causes of various diseases that plague both domesticated animals and humans.

1879 W. Flemming studies mitosis in the epithelium of the tail fin of salamanders. He shows that nuclear division involves a longitudinal splitting of the chromosome and a migration of the sister chromatids to the future daughter nuclei. He also coins the term *chromatin*.

1880 A. Laveran observes the malaria parasite in the red blood cells of a soldier who is suffering from malaria. This is the first case of a disease shown to be caused by a single-celled eukaryote.

1881 E. G. Balbiani discovers "cross-striped threads" within the salivary gland cells of *Chironomus* larvae. However, he does not realize that these are polytene chromosomes.

L. Pasteur develops the first artificially produced vaccine, against anthrax, a deadly disease of cattle, sheep, and humans.

1882 W. Flemming discovers lampbrush chromosomes and coins the term *mitosis*.

R. Koch describes the microscopic structure of *Mycobacterium tuberculosis*, the bacterium he proves to be the cause of human tuberculosis.

1883 E. van Beneden studies meiosis in a species of the round worm, *Ascaris*, which (fortunately) has a diploid chromosome number of only four. He shows that the gametes contain half as many chromosomes as the somatic tissues and that the characteristic somatic number is reestablished at fertilization. He also describes fertilization in mammals.

R. Koch isolates *Vibrio cholerae*, the bacterium that causes cholera, and describes its microscopic anatomy.

W. Roux suggests that the filaments within the nucleus that stain with basic dyes are the bearers of the hereditary factors.

A. Weismann points out the distinction in animals between the somatic cell line and the germ cells, stressing that only changes in germ cells are transmitted to further generations.

A. F. W. Schimper proposes that chloroplasts are capable of division and that green plants owe their origin to a symbiotic relationship between chlorophyll-containing and colorless organisms.

1884 A. Kossel isolates basic proteins from the nuclei of goose erythrocytes and names them *histones*.

H. C. Gram devises a diagnostic staining procedure that allows bacteria to be assigned to Gram-negative or Gram-positive categories.

1887 A. Weismann postulates that a periodic reduction in chromosome number must occur in all sexual organisms.

T. Boveri discovers chromatin diminution during the embryogenesis of parasitic nematodes and shows that the process is linked to the differentiation between germ cells and somatic cells.

1888 W. Waldeyer coins the word *chromosome* for the filaments referred to by Roux (1883).

T. Boveri describes the centriole.

E. Roux and A. E. J. Yersin demonstrate that the diphtheria bacillus produces a toxin.

1889 F. Galton publishes *Natural Inheritance*. In it he describes the quantitative measurement of metric traits in populations. He thus founds biometry and the statistical study of variation.

1890 R. Altmann reports the presence of "bioblasts" within cells and concludes that they are "elementary organisms" that live as intracellular symbionts and carry out processes vital to their hosts. Later (1898), C. Benda names these organelles *mitochondria*.

E. von Behring shows that blood serum from previously immunized animals contains factors that are specifically lethal to the organisms used for the immunization. These factors are now called antibodies.

1891 J. W. Tutt describes industrial melanism in the British peppered moth and suggests that the phenomenon results from natural selection of those moths which carry color variations that make them inconspicuous to predators.

1892 D. I. Ivanovski demonstrates that the tobacco mosaic disease is caused by an agent that can pass through filters with pores too small for bacteria and can neither be seen with the light microscope nor grown upon bacteriological media.

1896 E. B. Wilson publishes the first edition of the *Cell in Development and Heredity*. This treatise distills the information gained concerning cytology in the half century since Schleiden and Schwann put forth the cell theory. The influential third edition published 30 years later is three times the size of the first.

1898 R. C. Ross shows from experiments done in India that *Anopheles* mosquitoes can transmit malaria to birds.

M. W. Beijerinck shows that the tobacco mosaic disease agent multiplies only in host cells. He proposes that the agent is a molecule dissolved in the juice of the infected plant and endowed with the power of replication, but only with the help of mechanisms provided by the host cell.

C. Golgi develops histological procedures that selectively stain axons and dendrites. During observations of nervous tissues, he discovers the organelle we now call the *Golgi apparatus* in his honor.

S. G. Navashin discovers the double fertilization that occurs in seed-bearing plants and concludes that endosperm cells are triploid.

1899 G. Grassi shows from experiments done in Italy that the malaria vector in humans is the *Anopheles* mosquito.

1900 H. de Vries in Amsterdam, Holland, and C. Correns in Tübingen, Germany publish the results of breeding experiments that paralleled Mendel's studies. They used *Pisum sativum* (and other plant species as well) and got the same F1 and F2 ratios. When de Vries published his results, he never mentioned Mendel. However, in his paper published later in 1900, Correns gave Mendel credit for his analysis done 35 years earlier. Correns was the first to observe the 9:3:3:1 ratio when analyzing the results of dihybrid crosses. In Great Britain, W. Bateson became Mendel's chief apostle. Over the next decade he had Mendel's paper translated into English, and he invented the terms *allelomorph*, *epistasis*, *genetics*, *heterozygote*, *homozygote*, and *Mendelism*. He also introduced the notation used in diagramming the generations in breeding experiments (P1, P2, P3, F1, F2, F3, etc.).

J. Loeb demonstrates that frog and sea urchin eggs can begin parthenogenic development after mechanical stimulation.

K. Pearson develops the chi-square test.

K. Landsteiner discovers the blood-agglutination phenomenon in humans.

P. Ehrlich proposes that antigens and antibodies bind together because they have structural complementarity.

1901 H. de Vries adopts the term *mutation* to describe sudden, spontaneous, drastic alterations in the hereditary material of *Oenothera*.

T. H. Montgomery studies spermatogenesis in various species of Hemiptera. He concludes that maternal chromosomes only pair with paternal chromosomes during meiosis.

K. Landsteiner demonstrates that humans can be divided into three blood groups: A, B, and C. The designation of the third group was later changed to O.

I. I. Mechnikov reports observing white blood cells engulfing bacteria and thus founds the study of cellular immunity.

E. von Behring wins the Nobel Prize for his studies on antiserum therapy.

1902 C. E. McClung notes that in various insect species, equal numbers of two types of spermatozoa are formed; one type contains an "accessory chromosome" and the other does not. He suggests that the extra chromosome is a sex determinant, and he next argues that sex must be determined at the time of fertilization, not just in insects, but perhaps in other species (including humans).

T. Boveri studies the development of haploid, diploid, and aneuploid sea urchin embryos. He finds that in order to develop normally, the organism must have a full set of chromosomes, and he concludes that the individual chromosomes must carry *different* essential hereditary determinants.

W. S. Sutton advances the chromosome theory of heredity, which proposes that the independent assortment of gene pairs stems from the behavior of the synapsed chromosomes during meiosis. Since the direction of segregation of the homologs in a given bivalent is independent of those belonging to any other bivalent, the genes they contain will also be distributed independently.

F. Hofmeister and E. Fischer propose that all proteins are formed by the condensation of amino acids bound through regularly recurrent peptide linkages.

R. C. Ross receives the Nobel Prize in Medicine for his studies of the transmission of the malaria parasite by mosquitoes.

1903 W. Waldeyer defines centromeres as chromosome regions with which the spindle fibers become attached during mitosis.

1904 A. F. Blakeslee discovers heteromixis in fungi.

A. Laveran publishes *Les Trypanosomes et Trypanosomiasis*, a great monograph which summarizes the current status of knowledge concerning diseases caused by flagellated protozoa.

1905 L. Cuénot performs crosses between mice carrying a gene that gives them yellow fur. Since they always produce yellow furred and agouti offspring in a 2 : 1 ratio, he concludes they are heterozygous. W. E. Castle and C. C. Little show in 1910 that yellow homozygotes die *in utero*. This dominant allele in the agouti series (A^y) is thus the first gene shown to behave as a homozygous lethal.

R. Koch receives the Nobel Prize in Physiology and Medicine for his researches on the etiology of tuberculosis.

1906 W. Bateson and R. C. Punnett report the first case of linkage (in the sweet pea).

1907 R. G. Harrison cultures fragments of the central nervous systems of frogs in hemolymph and observes the outgrowth of nerve fibers. In so doing, he invents tissue culture.

E. F. Smith and C. O. Townsend show that a specific bacterium, *Agrobacterium tumefaciens*, is responsible for crown gall disease.

A. Laveran receives the Nobel Prize in Medicine for his contributions to the understanding of protozoan parasites that cause diseases, especially the sporozoans and trypanosomes responsible for malaria and African sleeping sickness, respectively.

1908 G. H. Hardy and W. Weinberg, working independently, formulate the so-called Hardy-Weinberg law of population genetics.

I. I. Mechnikov and P. Ehrlich share the Nobel Prize in Physiology and Medicine for their contributions to immunology.

1909 G. H. Shull advocates the use of self-fertilized lines in production of commercial seed corn. The hybrid corn program that resulted created an abundance of foodstuffs worth billions of dollars.

A. E. Garrod publishes *Inborn Errors of Metabolism*, the earliest discussion of the biochemical genetics of humans (or any other species). He concludes that the metabolic disease alkaptonuria is a "rare recessive character in the Mendelian sense."

F. A. Janssens suggests that exchanges between nonsister chromatids produce chiasmata.

C. C. Little initiates a breeding program that produces the first inbred strain of mice (the strain now called DBA).

W. Johannsen's studies of the inheritance of seed size in self-fertilized lines of beans leads him to realize the necessity of distinguishing between the appearance of an organism and its genetic constitution. He invents the terms *phenotype* and *genotype* to serve this purpose, and he also coins the word *gene*.

C. Correns and E. Bauer study the inheritance of chloroplast defects in variegated plants, such as *Mirabilis jalapa* and *Pelargonium zonale*. They find that the inability to form healthy chloroplasts is in some cases inherited in a non-Mendelian fashion.

H. Nilsson Ehle puts forward the multiple-factor hypothesis to explain the quantitative inheritance of seed-coat color in wheat.

P. Ehrlich leads a team of organic chemists who synthesize Salvarsan, the first drug to successfully treat spirochaete infections.

1910 T. H. Morgan discovers white eye and consequently sex linkage in *Drosophila*. *Drosophila* genetics begins.

W. Weinberg develops the methods used for correcting expectations for Mendelian segregation from human pedigree data under different kinds of ascertainment applied to data from small families.

C. Mereschkowsky proposes that certain of the organelles found in cells began as symbionts, and he coins the term *symbiogenesis* to describe this process.

P. Rous shows that injection of a cell-free filtrate from chicken sarcomas induces new sarcomas in recipient chickens.

1911 T. H. Morgan proposes that the genes for white eyes, yellow body, and miniature wings in *Drosophila* are linked together on the X chromosome.

W. R. B. Robertson points out that a metacentric chromosome in one orthopteran species may correspond to two acrocentrics in another and concludes that during evolution metacentrics may arise by the fusion of acrocentrics. Whole-arm fusions are called *Robertsonian translocations* in his honor.

1912 A. Wegener proposes the continental drift concept.

F. Rambousek suggests that the "cross-striped threads" within the salivary gland cells of fly maggots are chromosomes.

T. H. Morgan demonstrates that crossing over does not take place in the male of *Drosophila melanogaster.* He also discovers the first sex-linked lethal.

1913 Y. Tanaka reports that crossing over does not take place in the female of *Bombyx mori.* In this species, the female is the heterogametic sex.

E. M. East and R. A. Emerson use a multiple factor hypothesis to explain the inheritance of ear length in *Zea mays.*

W. H. Bragg and W. L. Bragg demonstrate that the analysis of x-ray diffraction patterns can be used to determine the three-dimensional atomic structure of crystals.

A. H. Sturtevant provides the experimental basis for the linkage concept in *Drosophila* and produces the first genetic map.

1914 C. B. Bridges discovers meiotic nondisjunction in *Drosophila.*

C. C. Little postulates that the acceptance or rejection of transplanted tumors in mice has a genetic basis.

T. Boveri suggests that cancers arise as the result of the proliferation of a single cell that has a genetic imbalance due to errors in the number of chromosomes it received during mitosis.

1915 F. W. Twort isolates the first filterable bacterial virus.

R. B. Goldschmidt coins the term *intersex* to describe the aberrant sexual types arising from crosses between certain different races of the gypsy moth, *Lymantria dispar.*

J. B. S. Haldane, A. D. Sprunt, and N. M. Haldane describe the first example of linkage in vertebrates (mice).

T. H. Morgan, A. H. Sturtevant, H. J. Muller, and C. B. Bridges publish *The Mechanism of Mendelian Heredity*, which summarizes the early work on *Drosophila.*

C. B. Bridges discovers *bithorax*, the first homeotic mutation of *Drosophila.*

W. H. Bragg (father) and W. L. Bragg (son) share the Nobel Prize in physics for initiating the study of x-ray crystallography.

1916 H. J. Muller discovers interference in *Drosophila.*

1917 F. d'Herelle discovers viruses that attack *Salmonella typhimurium.* He coins the term *bacteriophage* and develops methods for assaying virus titre.

O. Winge calls attention to the important role of polyploidy in the evolution of angiosperms.

C. B. Bridges discovers the first chromosome deficiency in *Drosophila.*

1918 H. J. Muller discovers the balanced lethal phenomenon in *Drosophila.*

1919 T. H. Morgan calls attention to the equality in *Drosophila melanogaster* between the number of linkage groups and the haploid number of chromosomes.

C. B. Bridges discovers chromosomal duplications in *Drosophila.*

1920 A. F. Blakeslee, J. Belling, and M. E. Farnham describe trisomics in the Jimson weed, *Datura stramonium.*

1921 F. G. Banting and C. H. Best isolate insulin and study its physiological properties.

H. J. Muller calls attention to the similarities between bacterial viruses and genes, and he predicts that phage studies will provide insights into the molecular nature of genes.

C. B. Bridges reports the first monosomic (haplo-4) in *Drosophila*.

1922 L. V. Morgan discovers attached-X chromosomes in *Drosophila*.

J. B. S. Haldane points out that the members of the heterogametic sex are often absent, rare, or sterile among the offspring of species hybrids.

A. F. Blakeslee, and three colleagues discover a haploid *Datura*.

1923 C. B. Bridges discovers chromosomal translocations in *Drosophila*.

T. Svedberg builds the first ultracentrifuge.

A. E. Boycott and C. Diver describe "delayed" Mendelian inheritance that controls the direction of the coiling of the shell in the snail *Limnaea peregra*. A. H. Sturtevant suggests that the direction of coiling of the *Limnaea* shell is determined by the character of the ooplasm, which is in turn controlled by the mother's genotype.

The XX-XY type of sex determination is demonstrated for certain dioecious plants: for *Elodea* by J. K. Santos, for *Rumex* by H. Kihara and T. Ono, and for *Humulus* by O. Winge.

1924 R. Feulgen and H. Rossenbeck describe the cytochemical test that is used for DNA localization and the determination of C values.

H. Spemann and H. Mangold demonstrate that a living part of an embryo can exert a morphogenetic stimulus upon another part, bringing about its morphological differentiation (embryonic induction). They thus discover and name the *organizer*.

1925 C. B. Bridges completes his cytogenetic analysis of the aneuploid offspring of triploid *Drosophila*, and he defines the ratios between the sex chromosomes and autosomes that control sexual phenotype.

A. H. Sturtevant proves the high reversion rate at the *Bar* locus in *Drosophila* is due to unequal crossing over. His analysis also uncovers the position effect phenomenon.

F. Bernstein suggests that the A B O blood groups are determined by a series of allelic genes.

T. H. Goodspeed and R. E. Clausen produce an amphidiploid in *Nicotiana*.

1926 E. G. Anderson establishes that the centromere of the X chromosome of *Drosophila* is at the end opposite the locus of *yellow*.

S. S. Chetverikov initiates the genetic analysis of wild populations of *Drosophila*.

J. B. Sumner isolates the first enzyme in crystalline form and shows it (urease) to be a protein.

A. H. Sturtevant finds the first inversion in *Drosophila*.

R. E. Clausen and T. H. Goodspeed describe the first analysis of monosomics in a plant (*Nicotiana*).

N. I. Vavilov publishes *Origin and Geography of Cultivated Plants* in which the *center of origin hypothesis* is developed.

1927 K. M. Bauer reports that the rejection of skin grafts does not occur when skin is transplanted from one monozygotic twin to another.

A. L. Du Toit concludes from the patterns of geological similarities between the east coast of South America and the west coast of South Africa that the two continents were once juxtaposed. These results provide the earliest evidence of continental drift.

J. Belling proposes that interchanges between nonhomologous chromosomes result in ring formations at meiosis.

J. B. S. Haldane suggests that the genes known to control certain coat colors in various rodents and carnivores may be evolutionarily homologous.

J. Belling introduces the acetocarmine technique for staining chromosome squashes.

B. O. Dodge initiates genetic studies on *Neurospora*.

G. D. Karpechenko generates *Raphanobrassica*, an allotetraploid hybrid of the radish, *Raphanus sativa*, and the cabbage, *Brassica oleracea*.

H. J. Muller reports the artificial induction of mutations in *Drosophila* by x-rays.

1928 L. J. Stadler reports the artificial induction of mutations in maize, and demonstrates that the dose-frequency curve is linear.

F. Griffith shows that a transformation to virulence can be induced in pneumococci by mixing heat-killed bacteria from virulent strains with living bacteria from non-virulent strains. This lays the foundation for the work of Avery, MacLeod, and McCarthy (1944).

L. F. Randolph distinguishes supernumerary chromosomes from the normal chromosomes of the plant cell. He calls the normal ones "A chromosomes" and the supernumerary ones "B chromosomes."

E. Heitz coins the terms *euchromatin* and *heterochromatin*.

1929 A. Fleming reports that a mold of genus *Penicillium* secretes a substance that prevents the growth of certain bacteria. He names this antibacterial substance *penicillin*.

Karl Kohmann discovers ATP.

P. A. Levene and E. S. London propose a tetranucleotide hypothesis for the structure of thymonucleic acid (DNA).

C. D. Darlington suggests that chiasmata function to hold homologs together at meiotic metaphase I and so ensure that they pass to opposite poles at anaphase I.

R. C. Tryon demonstrates successful selection for rate of maze learning in the rat.

1930 From 1930 to 1932, a group of books and papers is published that constitutes the mathematical foundation of population genetics (1930, by R. A. Fisher; 1931, S. Wright; 1932, J. B. S. Haldane).

R. E. Cleland and A. F. Blakeslee demonstrate that the peculiar patterns of the transmission of groups of genes in various *Oenothera* races result from a system of balanced lethal and reciprocal translocation complexes.

K. Landsteiner is awarded the Nobel Prize in physiology or medicine for his contributions to immunology. His elucidation of the ABO blood group system made successful blood transfusions possible.

1931 C. Stern and, independently, H. B. Creighton and B. McClintock provide the cytological proof of crossing over.

C. D. Darlington suggests that chiasmata can move to the ends of bivalents without breakage of the chromosomes. This process of *terminalization,* as he called it, is now known to occur in some species but not in others.

B. McClintock shows in maize that if a segment of a chromosome has become inverted, individuals heterozygous for such an inversion often show reversed pairing at pachynema.

A. L. Fox synthesizes phenylthiocarbamide and observes that it tastes bitter to some people and others cannot taste it at all.

1932 M. Knoll and E. Ruska invent the prototype of the modern electron microscope.

1933 T. S. Painter initiates cytogenetic studies on the salivary gland chromosomes of *Drosophila*.

H. Hashimoto works out the chromosomal control of sex determination for *Bombyx mori*.

A. W. K. Tiselius reports the invention of an apparatus that permits the separation of charged molecules by electrophoresis.

B. McClintock demonstrates in maize that a single exchange within the inversion loop of a paracentric inversion heterozygote generates an acentric and a dicentric chromatid.

T. H. Morgan receives a Nobel Prize for his development of the theory of the gene.

1934 M. Schlesinger reports that certain bacteriophages are composed of DNA and protein.

P. L'Héritier and G. Teissier experimentally demonstrate the disappearance of a deleterious gene from populations of *Drosophila melanogaster* maintained in population cages for many generations.

A. Følling discovers phenylketonuria, the first hereditary metabolic disorder shown to be responsible for mental retardation.

H. Bauer postulates that the giant chromosomes of the salivary gland cells of fly larvae are polytene.

A. F. Blakeslee utilizes *Datura* aneuploids to expose the morphogenic effects of genes residing on specific chromosomes.

B. McClintock shows that the nucleolus organizer of *Zea mays* can be split by a translocation and that each piece is capable of organizing a separate nucleolus. She thus sets the stage for the later demonstration (1965) that the genes for rRNA are present in multiple copies.

1935 J. B. S. Haldane is the first to calculate the spontaneous mutation frequency of a human gene.

E. Klenk identifies the glycolipid that accumulates in the brain of patients with the Tay-Sachs disease as a ganglioside.

F. Zernicke describes the principle of the phase microscope.

G. W. Beadle and B. Ephrussi and A. Kuhn and A. Butenandt work out the biochemical genetics of eye-pigment synthesis in *Drosophila* and *Ephestia* respectively.

W. M. Stanley succeeds in isolating and crystallizing the tobacco mosaic virus. He mistakenly believes it is pure protein.

C. B. Bridges publishes the salivary gland chromosome maps for *Drosophila melanogaster.*

H. Spemann receives a Nobel Prize for his studies on embryonic induction.

1936 J. Schultz notes the relation of the mosaic expression of a gene in *Drosophila* to its position relative to heterochromatin.

N. V. Timofeyeff-Ressovsky and M. Delbrück make estimates of gene volumes from target theory calculations.

T. Caspersson uses cytospectrophotometric methods to investigate the quantitative chemical composition of cells.

J. J. Bittner shows that mammary carcinomas in mice can be caused by a virus-like factor transmitted through the mother's milk.

A. H. Sturtevant and T. Dobzhansky publish the first account of the use of inversions in constructing a chromosomal phylogenetic tree.

C. Stern discovers somatic crossing over in *Drosophila.*

R. Scott-Moncrieff reviews the inheritance of plant pigments. The major part of this work was done by a group of English geneticists at the John Innes Horticultural Institute, and these early workers established that gene substitutions resulted in chemical changes in certain flavanoid and carotenoid pigments.

1937 T. Dobzhansky publishes *Genetics and the Origin of Species*, a milestone in evolutionary genetics.

A. F. Blakeslee and A. G. Avery report that colchicine induces polyploidy.

T. M. Sonneborn discovers mating types in *Paramecium.*

F. C. Bawden and N. W. Pirie show that the tobacco mosaic virus, although being made mostly of protein, also contains a small amount (about 5%) of RNA.

P. A. Gorer discovers the first histocompatibility antigens in the laboratory mouse.

H. A. Krebs discovers the citric acid cycle.

H. Karström points out that the synthesis of certain bacterial enzymes is stimulated when the substrates these enzymes attack are added to the medium. He coins the term "adaptive enzymes" for these and differentiates them from the "constitutive enzymes" that are always formed irrespective of the composition of the medium.

P. L'Héritier and G. Teissier demonstrate frequency-dependent selection of mutants in laboratory populations of *Drosophila melanogaster.*

E. Chatton stresses the fundamental differences between the group of organisms comprising the bacteria and blue-green algae, which he named *prokaryotes,* and all other living organisms, which he called *eukaryotes.*

1938 H. J. Muller defines a *telomere* as a functional gene that seals the chromosome at each end. He points out that a chromosome cannot persist indefinitely without having its ends sealed by these specialized chromomeres.

B. McClintock shows that a maize chromosome which lacks a telomere will fuse with any nearby chromosome which also lacks a telomere. The dicentric chromosome that results will generate a bridge at the next mitosis, and a breakage-fusion-bridge cycle will follow. Therefore telomeres play a capping function vital to the stability of each chromosome.

T. M. Sonneborn discovers the killer factor of *Paramecium*.

M. M. Rhoades describes the mutator gene *Dt* in maize.

H. Slizynska makes a cytological analysis of several overlapping *Notch* deficiencies in the salivary gland X chromosomes of *Drosophila melanogaster* and determines the band locations of the *w* and *N* genes.

1939 E. L. Ellis and M. Delbrück perform studies on coliphage growth that mark the beginning of modern phage work. They devise the "one-step growth" experiment, which demonstrates that after the phage adsorbs onto the bacterium, it replicates within the bacterium during the "latent period," and finally the progeny are released in a "burst."

G. A. Kausche, E. Pfankuch, and H. Ruska publish the first electron microscope photographs of a virus (tobacco mosaic).

A. A. Prokofyeva-Belgovskaya discovers the heterochromatization phenomenon.

P. Levine and R. E. Stetson discover maternal immunization by a fetus carrying a new blood group antigen inherited from the father. This antigen is subsequently identified with the human Rh blood group system as the cause of erythroblastosis fetalis.

A. W. K. Tiselius and E. A. Kabat demonstrate that antibodies belong to the gamma class of serum globulins.

E. Knapp and three colleagues demonstrate that the effectiveness of ultraviolet light in inducing mutations in *Sphaerocarpus donnelli* corresponds to the absorption spectrum of nucleic acid.

G. W. Beadle proposes that corn is a domesticated form of teosinte and suggests that mutations in as few as five genes can account for the morphological transformation of teosinte into corn.

G. Domagk receives the Nobel Prize for his discovery of Prontosil. This and other sulfa drugs revolutionize the treatment of infectious diseases.

1940 W. Earle establishes the strain L permanent cell line from a C3H mouse.

H. W. Florey, E. Chain, and five colleagues successfully extract and purify penicillin. They show in experiments with mice that it is by far the most powerful chemotherapeutic agent then known against bacterial infections.

1941 G. W. Beadle and E. L. Tatum publish their classic study on the biochemical genetics of *Neurospora* and promulgate the one gene–one enzyme theory.

J. Brachet and T. Caspersson independently reach the conclusion that RNA is localized in the nucleoli and cytoplasm and that a cell's content of RNA is directly linked with its protein synthesizing capacity.

C. Auerbach and J. M. Robson use Muller's *C1B* technique to prove that mustard gas induces mutations in *Drosophila*. Because of secrecy imposed during World War II on research with poison gas, the results, which opened the field of chemical mutagenesis, were not published until 1946.

A. J. P. Martin and R. L. M. Synge develop the technique of partition chromatography and use it to determine the amino acids in protein hydrolyzates.

A. H. Coons, H. J. Creech, and R. N. Jones develop immunofluorescence techniques to demonstrate the presence of antibody-reactive sites on specific cells.

K. Mather coins the term *polygenes* and describes polygenic traits in various organisms.

1942 R. Schoenheimer publishes *The Dynamic State of Body Constituents* and describes the use of isotopically tagged compounds in metabolic studies. He introduces the concepts of "metabolic pools" and the "turnover" of the organic compounds in cells.

S. E. Luria and T. F. Anderson publish the first electron micrographs of bacterial viruses. T2 has a polyhedral body and a tail!

G. D. Snell sets out to develop highly inbred strains of mice to study the genes responsible for graft rejection.

1943 A. Claude isolates and names the microsome fraction and shows that it contains the majority of the RNA of cells.

J. Hammerling makes grafts between species of giant algae belonging to the genus *Acetabularia*. He shows that the basal nucleus of each cell controls the morphological development of the cytoplasm at the apex of the stalk centimeters away.

S. E. Luria and M. Delbrück initiate the field of bacterial genetics when they demonstrate unambiguously that bacteria undergo spontaneous mutation.

1944 O. T. Avery, C. M. MacLeod, and M. McCarty purify and chemically characterize the pneumococcus transforming principle. It is 99.9% DNA, and they later show that deoxyribonuclease inactivates it. They conclude that genetic information is carried by DNA molecules.

T. Dobzhansky describes the phylogeny of the gene arrangements in the third chromosome of *Drosophila pseudoobscura* and *D. persimilis*.

E. L. Tatum, D. Bonner, and G. W. Beadle use mutant strains of *Neurospora crassa* to work out the intermediate steps in the synthesis of tryptophan.

1945 R. R. Humphrey demonstrates that the female is the heterogametic sex in urodeles.

M. J. D. White publishes *Animal Cytology and Evolution*, the first monograph to review progress in the study of the evolutionary cytogenetics of animals.

S. E. Luria demonstrates that mutations occur in bacterial viruses.

E. B. Lewis describes the stable position effect phenomenon in *Drosophila*.

R. D. Owen reports that in cattle dizygotic twins are born with, and often retain throughout life, a stable mixture of each other's red blood cells. This chimerism, which results from vascular anastomoses within the chorions of the fetuses, provides the first example of immune tolerance.

A. Fleming, E. B. Chain, and H. W. Florey receive the Nobel Prize for the discovery, purification, and chemical characterization of penicillin.

1946 A. Claude introduces cell fractionation techniques based upon differential centrifugation and works out methods for characterizing the fractions biochemically.

Genetic recombination in bacteriophage is demonstrated by M. Delbrück and W. T. Bailey and by A. D. Hershey.

J. Lederberg and E. L. Tatum demonstrate genetic recombination in bacteria.

Mutations are chemically induced in *Drosophila* by J. A. Rapoport using formaldehyde.

J. Oudin develops the gel-diffusion, antigen-antibody precipitation test that bears his name.

Nobel Prizes are awarded to H. J. Muller for his contributions to radiation genetics; to J. B. Sumner for crystallizing enzymes; and to W. M. Stanley for his studies on the purification and chemical characterization of viruses.

1947 A. M. Mourant suggests that earliest European populations were Rh^- and that modern-day Basques have the highest Rh^- frequencies known because through isolation they have maintained the genetic characteristics of their paleolithic ancestors.

David Lack publishes *Darwin's Finches*. In this monograph he describes how the finches of the Galapagos archipelago are transformed by evolution from competing to non-competing species by "fine-tuning" their habitat preferences and beak morphologies for different foods.

1948 A. Boivin, R. Vendrely, and C. Vendrely analyze the DNA of nuclear suspensions of several beef tissues. They observe that the amount of DNA per somatic nucleus is the same and twice the amount in sperm nuclei. They predict that the amount of DNA per cell is constant and distinct for each species.

H. E. Shortt and P. C. C. Garnham discover the stage in the life cycle of the malaria parasite specific to hepatocytes in primates.

H. K. Mitchell and J. Lein show that tryptophan synthetase is missing in certain mutant strains of *Neurospora*. This finding constitutes the first direct evidence for the one gene-one enzyme theory.

P. A. Gorer, S. Lyman, and G. D. Snell discover the major histocompatibility locus in the mouse. It resides on chromosome 17 and is named H2.

O. Ouchterlony develops the double-diffusion antigen-antibody precipitation test that bears his name.

H. J. Muller coins the term *dosage compensation*.

J. Lederberg and N. Zinder, and, independently, B. D. Davis develop the penicillin selection technique for isolating biochemically deficient bacterial mutants.

J. Clausen, D. D. Keck, and W. M. Hiesey describe the genetic structure of ecotypes among species of herbaceous plants along an altitudinal transect in the Sierra Nevada mountains of California.

G. D. Snell introduces the term *histocompatibility gene* and formulates the laws of transplantation acceptance and rejection.

1949 B. Ephrussi, H. Hottinger, and A. M. Chimenes discover *petite* cytoplasmic mutations in yeast.

A. D. Hershey and R. Rotman generate a genetic map for T2 bacteriophage that contains three linkage groups. In subsequent studies these linkage groups are found to merge into a linear molecule that has cyclically permuted sequences.

D. M. C. Hodgkin and three colleagues use x-ray crystallography to determine the structure of the penicillin molecule at atomic resolution.

A. Kelner discovers photoreactivation of potential UV damage by visible light in *Saccharomyces*.

J. V. Neel provides genetic evidence that the sickle-cell disease is inherited as a simple Mendelian autosomal recessive.

L. Pauling and three colleagues show that the H^S gene produces an abnormal hemoglobin.

M. L. Barr and E. G. Bertram demonstrate that the sex chromatin is morphologically different in the neurons of male and female cats.

1950 B. McClintock discovers the *Ac, Ds* system of transposable elements in maize.

W. Hennig publishes *Grundzüge einer Theorie der phylogenetischen Systematik*. It sets forth the criteria by which genealogic relationships between groups of organisms can be uncovered. An extensively revised English translation titled *Phylogenetic Systematics* will be published in 1966. This serves as an introduction to cladistics for scientists in English.

H. Swift uses Feulgen microspectrophotometry to determine the DNA contents of individual nuclei for a variety of tissues from mice, frogs, and insects. In accordance with the 1948 prediction of Boivin, Vendrely, and Vendrely, the DNA contents fall into classes that are integer multiples of a haploid value, defined as C.

E. Chargaff lays the foundations for nucleic acid structural studies by his analytical work. He demonstrates for DNA that the numbers of adenine and thymine groups are always equal, and so are the numbers of guanine and cytosine groups. These findings later suggest to Watson that DNA consists of two polynucleotide strands joined by hydrogen bonding between A and T and between G and C. The finding that the DNAs from yeast and from the tubercle bacillus are relatively rich in AT and GC, respectively, disproved the tetranucleotide hypothesis.

A. Lwoff and A. Gutman study a lysogenic strain of *Bacillus megatherium* and demonstrate that each bacterium harbors a noninfectious form of a virus that gives the host the capacity to generate new phage without the intervention of exogenous bacteriophages. They propose the term *prophage* for this noninfectious phase. Together with L. Siminovitch and N. Kjeldgaard, Lwoff shows that prophage can be induced by ultraviolet light to produce infective virus.

H. Latta and J. F. Hartmann introduce glass knives for ultramicrotomy.

E. M. Lederberg discovers lambda, the first viral episome of *E. coli*.

1951 G. Gey establishes the human HeLa permanent cell line.

J. Mohr is the first to demonstrate autosomal linkage in humans (between the genes specifying the Lewis and Lutheran blood groups).

C. Stormont, R. D. Owen, and M. R. Irwin describe serological cross-reactions in the multiple allelic B and C blood group systems of cattle.

Y. Chiba demonstrates the presence of DNA in chloroplasts using the Feulgen-staining cytochemical technique.

N. H. Horowitz and U. Leupold generate large populations of temperature-sensitive mutations to determine what percentage of the genes in *E. coli* and

N. crassa perform functions that are indispensable. The values obtained were 23% and 46%, respectively.

C. Petit reports the existence of a minority-genotype advantage in populations of *Drosophila melanogaster* and points out that this phenomenon can lead to frequency-dependent selection and stable polymorphism.

G. G. Simpson publishes a phylogeny of the horse family that shows that the evolutionary tree has multiple side branches. Earlier paleontologists mistakenly used the evolution of the horse as an example of orthogenesis.

L. Pauling and R. B. Corey propose that most proteins form one or both of two secondary structures: the alpha helix and the beta sheet.

V. J. Freeman reports that a specific bacteriophage confers upon its host, *Corynebacterium diphtheriae*, the ability to produce the diphtheria toxin.

M. H. F. Wilkins and R. Gosling prepare DNA for x-ray crystallographic analysis and take the first x-ray photograph that shows a meaningful diffraction pattern.

1952 R. E. Franklin and R. Gosling discover that, under conditions of high humidity, DNA undergoes a transformation from an A to a B configuration. Franklin subsequently proposes that the B form is a helical molecule with the phosphate groups on the inside.

F. H. C. Crick concludes from Franklin's crystallographic data that the DNA chains in the helix are aligned in an antiparallel configuration.

D. M. Brown and A. Todd demonstrate that DNA and RNA are 3′-5′ linked polynucleotides.

G. E. Palade publishes the first high-resolution electron micrographs of mitochondria. The organelle has an outer membrane and an inner one, which has shelf-like invaginations. He calls these *cristae mitochondriales*.

R. Dulbecco adapts the techniques of bacterial virology to study animal viruses. He counts plaques made by western equine encephalomyelitis virus on monolayers of cells obtained from chick embryos.

D. Mazia and K. Dan isolate the sea urchin mitotic apparatus and start work on its biochemical characterization.

N. D. Zinder and J. Lederberg describe transduction in *Salmonella typhimurium*. The transducing phage was P22.

J. Lederberg discovers and names plasmids.

J. Lederberg and E. M. Lederberg invent the replica plating technique.

J. T. Patterson and W. S. Stone publish *Evolution in the Genus Drosophila*, which summarizes an encyclopedic body of information dealing with the chromosomal evolution of this most studied genus of flies.

A. H. Bradshaw reports that certain populations of grasses living near mine entrances in Great Britain can tolerate high concentrations of heavy metals (copper, lead, zinc). This is evidence for the recent natural selection of tolerant genotypes.

W. Beermann observes stage and tissue specificities in the puffing patterns of polytene chromosomes and suggests that these are the phenotypic reflections of differential gene activities.

A. D. Hershey and M. Chase demonstrate that the DNA of phage enters the host, whereas most of the protein remains behind.

G. Pontecorvo and J. A. Roper describe the parasexual cycle in *Aspergillus nidulans*.

R. Briggs and T. S. King develop a method for taking a nucleus of a living cell from a *Rana pipiens* blastula and implanting it into an oocyte whose own nucleus had been removed. Many of the oocytes carrying these implanted diploid somatic nuclei underwent cleavage and some eventually developed into tadpoles. However, when nuclei from gastrulae were used, most of the implanted embryos died. This showed that the nuclei of embryonic cells acquire developmental restrictions concomitant with cell specialization.

A. J. P. Martin and R. L. M. Synge receive the Nobel Prize in chemistry for their invention of chromatographic separation techniques.

1953 J. D. Watson and F. H. C. Crick publish the double-helix model of DNA. It proposes that each molecule is composed of two antiparallel, helically intertwined chains held together by hydrogen bonds between purines and pyrimidines. The supporting crystallographic data appear in the same issue of *Nature* in a second paper by Wilkins, A. R. Stokes, and H. R. Wilson, and a third paper by Franklin and Gosling. They conclude that the B configuration is the one found in living cells. In a subsequent issue of *Nature,* Watson and Crick propose the semiconservative mechanism for the replication of DNA molecules.

C. C. Patterson uses a radioactive dating procedure (the uranium-lead clock) to determine the age of the earth (4.6 billion years).

K. R. Porter takes electron micrographs of tissue-cultured animal cells. He describes and names the endoplasmic reticulum and identifies it as the source of cytoplasmic basophilia.

C. C. Lindegren discovers gene conversion in *Saccharomyces*.

A. Howard and S. R. Pelc demonstrate by autoradiography that during the cell-division cycle of plants there exists a period following mitosis during which DNA synthesis does not take place (G_1), a subsequent period of DNA synthesis during which the DNA content of the interphase nucleus is doubled (S), a second growth period (G_2), and then mitosis.

W. Hayes discovers polarized behavior in bacterial recombinations. He isolates the Hfr H strain of *E. coli* and shows that certain genes are readily transferred from Hfr to F^- bacteria, whereas others are not.

R. E. Billingham, L. Brent, and P. B. Medawar show that immunological tolerance can be produced experimentally.

Porter-Blum and Sjöstrand ultramicrotomes become commercially available.

J. B. Finean, F. S. Sjöstrand, and E. Steinmann publish the first electron micrographs of sectioned chloroplasts.

G. D. Snell finds that the major histocompatibility complex of the mouse (H-2) is composed of multiple loci.

N. Visconti and M. Delbrück put forth a hypothesis to explain genetic recombination in bacteriophages.

1954 A. J. Dalton and M. D. Felix provide the first detailed description of the ultrastructure of the Golgi complex.

A. C. Allison provides evidence that individuals heterozygous for the sickle-cell gene are protected against subtertian malaria infection. This is the first case of genetic balanced polymorphism described in a human population.

H. Bickel, J. Gerrard, and E. M. Hickmans report that babies with phenylketonuria show great improvement in mental development and behavioral performance after being fed a synthetic diet low in phenylalanine.

J. Dausset observes that some patients who had received multiple blood transfusions produced antibodies against antigens found on the white blood cells of other individuals but not against those of their own cells. These antibodies defined the first HLA antigens and led to the definition of the human histocompatibility system.

E. S. Barghoorn and S. A. Tyler report finding fossils of filamentous and coccoid microorganisms in sedimentary rocks over 2 billion years old. This discovery demonstrates that life existed in the Proterozoic era.

E. Mayr advances the peripatric speciation concept.

L. Pauling receives the Nobel Prize in chemistry for his research into the nature of the chemical bond and the structure of complex molecules including proteins.

1955 M. B. Hoagland obtains cell-free preparations that synthesize protein.

F. Sanger and five colleagues are the first to work out the primary structure for a protein. They show that insulin contains two polypeptide chains held together by disulfide bridges.

S. Benzer works out the fine structure of the *r* II region of phage T4 of *E. coli*, and coins the terms *cistron*, *recon*, and *muton*.

H. Fraenkel-Conrat and R. C. Williams reconstitute "hybrid" tobacco mosaic virus from nucleic acid and protein components arising from different sources. The reconstituted particles are normal in appearance and infectivity. This is the first example of the self-assembly of an active biological structure.

O. Smithies uses starch-gel electrophoresis to identify plasma protein polymorphisms.

N. K. Jerne puts forth the natural-selection theory of antibody formation. According to this proposal, antibody molecules are already present in the host, having developed during fetal life. An invading foreign antigen selects the antibody molecule that provides the best fit and binds to it. The formation of this complex stimulates the further production of the selected antibody. These concepts are incorporated into later clonal selection theories.

M. Grunberg-Manago and S. Ochoa isolate polynucleotide phosphorylase, the first enzyme involved in the synthesis of a nucleic acid.

N. E. Morton develops the lod score method for estimating linkage from pedigree data.

R. H. Pritchard studies the linear arrangement of a series of allelic adenine-requiring mutants of *Aspergillus*. He concludes that crossing over can occur between different alleles of the same gene, provided they are characterized by mutations at different subsites.

C. de Duve and four colleagues describe intracellular vesicles that contain hydrolytic enzymes and name them *lysosomes*.

F. Jacob and E. L. Wollman experimentally interrupt the mating process in *E. coli* and show that a piece of DNA is inserted from the donor bacterium into the recipient.

P. Grabar and C. A. Williams devise the technique of immunoelectrophoresis to analyze complex mixtures of antigenic molecules.

H. H. Flor puts forth the "gene-for-gene" hypothesis after studying the genetics of a parasite and its host simultaneously (the *Melampsora-Linum* system).

1956 J. H. Tjio and A. Levan demonstrate that the diploid chromosome number for humans is 46.

C. O. Miller and four colleagues obtain from hydrolyzed DNA of herring sperm a substance that promotes cell division in tissue cultured plant cells. They determine the structure of the molecule and name it kinetin.

B. C. Heezen and M. Ewing discover the Mid Oceanic Ridge, a formation of underwater mountains and rifts that girdles the globe.

T. T. Puck, S. J. Cieciura, and P. I. Marcus succeed in growing clones of human cells *in vitro*.

G. E. Palade and P. Siekevitz isolate ribosomes.

A. Kornberg, I. R. Lehman, and S. E. Simms isolate and purify DNA polymerase 1 from *E. coli*. This is the first enzyme shown to take directions from a template molecule.

M. J. Moses and D. Fawcett independently observe synaptonemal complexes in spermatocytes.

W. L. Brown and E. O. Wilson define character displacement and give examples of the phenomenon in insects, fish, amphibians, and birds.

A. Gierer and G. Schramm and H. Fraenkel-Conrat demonstrate independently that a chemically pure nucleic acid, namely tobacco mosaic virus RNA, is infectious and genetically competent.

E. Volkin and L. Astrachan infect *E. coli* with T2 bacteriophage and demonstrate the immediate synthesis of small amounts of a labile RNA. Its base composition is similar to the genome of the virus, not the bacterium. They don't realize it, but they have identified the first mRNA.

1957 S. A. Berson and R. S. Yalow report the first use of the radioimmunoassay procedure for the detection of insulin antibodies developed by patients in response to the administration of exogenous insulin.

J. H. Taylor, P. S. Woods, and W. L. Hughes are the first to use tritiated thymidine for high-resolution autoradiography in experiments that demonstrate the semiconservative distribution of label during chromosome replication in *Vicia faba*.

E. W. Sutherland and T. W. Rall isolate cyclic AMP and show that it is an adenine ribonucleotide.

V. M. Ingram reports that normal and sickle-cell hemoglobin differ by a single amino acid substitution. This is the first proof that a gene mutation results in an abnormal amino acid sequence in a protein.

A. Todd receives the Nobel Prize for his studies on the structure of nucleosides and nucleotides.

1958 F. Jacob and E. L. Wollman demonstrate that the single linkage group of *E. coli* is circular and suggest that the different linkage groups found in different Hfr strains result from the insertion at different points of a factor in the circular linkage group that determines the rupture of the circle.

F. H. C. Crick suggests that during protein formation the amino acid is carried to the template by an adaptor molecule containing nucleotides and that the adaptor is the part that actually fits on the RNA template. Crick thus predicts the discovery of transfer RNA.

P. C. Zamecnik and his colleagues characterize amino acid-transfer RNA complexes.

H. G. Callan and H. G. MacGregor demonstrate that the linear integrity of chromatids of amphibian lampbrush chromosomes is maintained by DNA, not protein.

H. B. Kettlewell organizes surveys to determine the carbonaria frequencies in peppered moth populations throughout the United Kingdom. These show that the highest frequencies of melanic forms are located in industrial areas.

M. Okamoto and, independently, R. Riley and V. Chapman discover genes that control the pairing of homoeologous chromosomes in wheat.

F. C. Steward, M. O. Mapes, and K. Mears succeed in rearing sexually mature plants from single diploid cells derived from the secondary phloem of roots of the wild carrot, *Daucus carota*. They conclude that each cell of the multicellular organism has all the ingredients necessary for the formation of the complete organism.

M. Meselson and F. W. Stahl use the density gradient equilibrium centrifugation technique to demonstrate the semiconservative distribution of density label during DNA replication in *E. coli*.

J. C. Kendrew and five colleagues produce a three-dimensional model of the myoglobin protein molecule obtained by x-ray analysis. They show that the molecule is composed almost entirely of alpha helices.

Nobel Prizes are awarded to G. W. Beadle, E. L. Tatum, and J. Lederberg for their contributions to genetics and to F. Sanger for his contributions to protein chemistry.

1959 J. Lejeune, M. Gautier, and R. Turpin show that Down syndrome is a chromosomal aberration involving trisomy of a small telocentric chromosome.

C. E. Ford and four colleagues discover that females suffering from Turner syndrome are XO.

P. A. Jacobs and J. A. Strong demonstrate that males suffering from Klinefelter syndrome are XXY.

S. J. Singer conjugates ferritin with immunoglobulin to produce a labeled antibody that is readily recognized under the electron microscope.

R. L. Sinsheimer demonstrates that bacteriophage phiX174 of *E. coli* contains a single stranded DNA molecule.

E. G. Krebs, D. J. Graves, and E. H. Fischer isolate and purify the first protein kinase.

F. M. Burnet improves Jerne's selective theory of antibody formation by suggesting that the antigen stimulates the proliferation of only those cells that are genetically programmed to synthesize the complementary antibodies.

G. M. Edelman resolves immunoglobulin G into heavy and light chains.

R. E. Franklin, D. L. D. Caspar, and A. Klug complete an x-ray crystallographic analysis that shows the architecture of the tobacco mosaic virus in three dimensions.

A. Lima-de-Faria demonstrates by autoradiography that heterochromatin replicates later than euchromatin.

M. Chèvremont, S. Chèvremont-Comhaire, and E. Baeckeland demonstrate DNA in mitochondria using a combination of autoradiographic and Feulgen-staining techniques.

K. McQuillen, R. B. Roberts, and R. J. Britten demonstrate in *E. coli* that ribosomes are the sites where protein synthesis takes place.

E. Freese proposes that mutation can occur as the result of single, base-pair changes in DNA. He coins the terms *transitions* and *transversions*.

C. Pelling finds selective labeling of puffed regions of polytene chromosomes after they are incubated in a nutrient solution containing ^{3}H uridine.

R. H. Whittaker suggests the grouping of organisms into five kingdoms: the bacteria, the eukaryotic microorganisms, animals, plants, and fungi.

S. Brenner and R. W. Horne develop the negative staining procedure for electron microscopy of subcellular particles.

S. Ochoa and A. Kornberg receive Nobel Prizes for their studies on the *in vitro* synthesis of nucleic acids.

1960 P. Nowell discovers phytohemagglutinin and demonstrates its use in stimulating mitoses in human leukocyte cultures.

A. Tsugita and five colleagues determine the complete sequence of the 158 amino acids in the coat protein subunit of the tobacco mosaic virus.

P. Siekevitz and G. E. Palade describe the synthesis of secretory proteins on membrane-bound ribosomes.

P. Doty and three colleagues demonstrate that complementary strands of DNA molecules can be separated and recombined.

P. E. Polani and four colleagues document the first case of Down syndrome caused by a Robertsonian translocation.

T. Watanabe and T. Fukusawa describe the mechanism of antibiotic resistance that results from the transfer of R plasmids during bacterial conjugation.

G. Barski, S. Sorieul, and F. Cornefert report the first successful *in vitro* hybridization of mammalian cells.

U. Clever and P. Karlson experimentally induce specific puffing patterns in polytene chromosomes by injecting *Chironomus* larvae with ecdysone.

H. H. Hess proposes the theory of sea floor spreading.

M. F. Perutz and five colleagues determine the three-dimensional structure of hemoglobin at 5.5 Å resolution.

P. B. Medawar and F. M. Burnet receive a Nobel Prize for their studies on immunological tolerance.

1961 F. Jacob and J. Monod publish "Genetic regulatory mechanisms in the synthesis of proteins," a paper in which the theory of the operon is developed.

F. Jacob and J. Monod suggest that ribosomes do not contain the template responsible for the orderly assembly of amino acids. They propose that instead each DNA cistron causes synthesis of an RNA molecule of limited life span that harbors the amino acid sequence information in its nucleotide sequence. This molecule subsequently enters into temporary association with a ribosome and so confers upon it the ability to synthesize a given protein. Within months S. Brenner, F. Jacob, and M. Meselson identify a specific mRNA that is synthesized in *E. coli* after phage infection.

M. F. Lyon and L. B. Russell independently provide evidence suggesting that in mammals one X chromosome is inactivated in some embryonic cells and their descendants, that the other is inactivated in the rest, and that mammalian females are consequently X-chromosome mosaics.

S. Benzer discovers two sites in the rII region of the chromosome of phage T4 of *E. coli* that show exceptionally high rates of spontaneous mutation. He calls these *hot spots*.

J. Josse, A. D. Kaiser, and A. Kornberg demonstrate that there is a difference in polarity between the complementary strands of the DNA helix, so that the sugars of one strand are oriented in a direction opposite to those in the other strand. This result confirms Crick's antiparallel model of 1952.

V. M. Ingram presents a theory explaining the evolution of the four known kinds of hemoglobin chains from a single primitive myoglobinlike heme protein by gene duplication and translocation.

B. D. Hall and S. Spiegelman demonstrate that hybrid molecules can be formed containing one single-stranded DNA and one RNA molecule which are complementary in base sequence. Their technique opens the way to the isolation and characterization of messenger RNAs.

S. B. Weiss and T. Nakamoto isolate an RNA polymerase from *E. coli* that uses DNA as a template.

F. H. C. Crick, S. Brenner, and two colleagues show that the genetic language is made up of three-letter words.

G. von Ehrenstein and F. Lipmann combine messenger RNA and ribosomes from rabbit reticulocytes with amino acid-transfer RNA complexes derived from *E. coli*. Since this cell-free system synthesized a protein similar to rabbit hemoglobin, they conclude that the genetic code is universal.

W. Beermann demonstrates that a puffing locus on a *Chironomus* polytene chromosome is inherited in a Mendelian fashion.

A. Wacker, H. Dellweg, and E. Lodemann show that thymine dimers are formed when DNA is irradiated with ultraviolet light.

M. W. Nirenberg and J. H. Matthaei develop a cell-free system from *E. coli* that incorporates amino acids into protein when supplied with template RNA preparations. They show that the synthetic polynucleotide, polyuridylic acid, directs the synthesis of a protein resembling polyphenylalanine.

M. Meselson and J. J. Weigle demonstrate in phage lambda that recombination involves breakage and reunion (but not replication) of the chromosome.

H. Dintzis shows that the direction of synthesis of the hemoglobin molecule is from amino to carboxyl termini.

H. Moor and three colleagues develop the first freeze-fracture procedure that permits ultrastructural observations, which are impossible with conventional sectioning methods.

U. Z. Littauer shows that ribosomes contain only two high-molecular-weight species of RNA, with sedimentation values of 16S and 23S in bacteria and 18S and about 28S in animals.

I. Rubenstein, C. A. Thomas, and A. D. Hershey demonstrate that the DNA in the head of the T2 bacteriophage constitutes a single chromosome.

C. B. Anfinsen and three colleagues show for ribonuclease A that the linear sequence of amino acids determines the unique three-dimensional structure of the protein.

P. D. Mitchell proposes the theory of chemiosmosis to explain how ATP synthesis is accomplished.

R. Guthrie develops a blood screening test for phenylketonuria. Screening of every newborn infant in Massachusetts for PKU will begin a year later.

G. P. Georgiev and V. L. Mantieva discover and characterize heterogeneous nuclear RNA.

L. Hayflick and P. S. Moorehead discover that the *in vitro* life span of human diploid cells in tissue culture is limited to about 50 doubling cycles.

J. P. Waller and J. I. Harris find that bacterial ribosomes contain a large number of different proteins.

1962 H. Ris and W. Plaut show by electron microscopy that chloroplasts contain DNA.

E. Zuckerkandl and L. Pauling calculate the approximate times of derivation of different hemoglobin chains from their common ancestors during eukaryotic evolution.

F. M. Ritossa reports that the salivary gland chromosomes of *Drosophila buskii* respond to heat shocks by puffing.

E. Beutler, M. Yeh, and V. E. Fairbanks study the erythrocytes of women heterozygous for glucose-6-phosphate dehydrogenase deficiency. They find a mixed population of normal and G6PD-deficient cells and conclude that the adult human female is a mosaic of cells containing an inactivated X chromosome, either maternal or paternal.

S. Cohen isolates the epidermal growth factor from the salivary glands of mice.

J. F. A. P. Miller, R. A. Good *et al.*, and N. L. Warner *et al.* experimentally demonstrate the distinction between T and B lymphocytes.

R. R. Porter uses enzymes to cleave immunoglobulin molecules. He demonstrates that each molecule has two antigen-binding portions (F_{ab}) and a crystallizable segment (F_c) that does not bind antigen. He shows that the heavy and light chains are present in 1 : 1 ratio and suggests the four-chain model.

D. A. Rodgers and G. E. McClearn discover differences between mouse strains in alcohol preference.

J. V. Neel suggests that diabetes mellitus represents a "thrifty" genotype rendered detrimental by "progress."

U. Henning and C. Yanofsky show that amino acid replacements can arise from crossing over within triplets.

J. B. Gurdon reports that a normal fertile frog can develop from an enucleated egg injected with a nucleus from an intestinal cell. This experiment demonstrates that somatic and germinal nuclei are qualitatively equivalent.

Polyribosomes are discovered independently in three laboratories (by A. Gierer, by J. R. Warner, A. Rich, and C. E. Hall, and by T. Staehelin and H. Noll).

A. M. Campbell proposes that episomes become integrated into host chromosomes by a crossover event resembling the exchanges that were previously reported between synapsed ring- and rod-shaped chromosomes in eukaryotes.

W. Arber puts forward the restriction and modification model that predicts the presence of restriction endonucleases.

Nobel Prizes in Physiology or Medicine are shared by J. D. Watson, F. H. C. Crick, and M. H. F. Wilkins for their studies on the three-dimensional structure of DNA. The Prize in Chemistry goes to M. F. Perutz and J. C. Kendrew for their x-ray crystallographic studies of hemoglobin and myoglobin.

1963 B. B. Levine, A. Ojida, and B. Benacerraf publish the first paper on the immune-response genes of guinea pigs.

R. Rosset and R. Monier discover 5S rRNA and conclude that it is a constituent of the ribosome. They demonstrate subsequently that it is a component of the large ribosomal subunit.

T. Okamoto and M. Takanami show that mRNA binds to the small ribosomal subunit.

H. Noll, T. Staehelin, and F. O. Wettstein demonstrate the tape mechanism of protein synthesis.

J. G. Gall produces evidence that the lampbrush chromatid contains a single DNA double helix.

B. J. McCarthy and E. T. Bolton use their DNA-agar technique to measure genetic relatedness between diverse species of organisms.

E. Hadorn demonstrates allotypic differentiation in cultured imaginal discs of *Drosophila*.

F. Jacob and S. Brenner publish the replicon model.

R. Sager and M. R. Ishida isolate chloroplast DNA from *Chlamydomonas*.

K. R. Porter and M. A. Bonneville publish a collection of electron micrographs and associated legends that provides the first atlas of the ultrastructure of different cell types.

J. T. Wilson proposes the hot spot archipelago hypothesis to explain the origin of the Hawaiian islands.

F. J. Vine and D. H. Matthews show that the rate of sea floor spreading can be calculated by dating parallel belts of rocks on the sea floor that show differences in the directions of their magnetic fields.

M. J. Schlesinger and C. Levinthal demonstrate that *in vitro* complementation results from the formation of a hybrid protein. The hybrid proteins studied were alkaline phosphatase molecules of *E. coli*. They were dimers that contained monomers with mutants at different sites.

J. Cairns demonstrates by autoradiography that the genophore of *Escherichia coli* is circular and that during its semiconservative replication Y-shaped, replicating forks proceed in opposite directions from a starting point and generate two circular offspring genophores.

E. Margoliash determines the amino acid sequences for cytochrome c derived from a wide variety of species and generates the first phylogenetic tree for a specific gene product.

L. B. Russell shows in the mouse that, when an X chromosome containing a translocated autosomal segment undergoes inactivation in somatic cells, the autosomal genes closest to the breakpoint are also inactivated. Thus the X inactivation spreads into the attached autosomal segment.

E. Mayr publishes *Animal Species and Evolution*. This volume provides a synthesis of modern ideas concerning the mechanism of speciation, and it has a profound influence on scientists working in this area.

I. R. Gibbons first isolates dynein from the arms on the microtubules of ciliary axonemes.

1964 R. B. Setlow and W. L. Carrier and, independently, R. P. Boyce and P. Howard-Flanders describe the mechanism of excision repair in bacteria.

A. S. Sarabhai and three colleagues establish the colinearity of gene and protein product in the case of the protein coating the head of virus T4 of *E. coli*.

C. Yanofsky and four colleagues establish the colinearity of gene and protein product in the case of tryptophan synthetase for *E. coli*.

M. Meselsen shows for lambda (λ) bacteriophage that genetic recombination occurs by breakage and rejoining of double stranded DNA molecules.

M. S. Fox and M. K. Allen show that transformation in *Streptococcus pneumoniae* involves incorporation of segments of single-stranded donor DNA into the DNA of the recipient.

E. T. Mertz, L. S. Bates, and O. E. Nelson show that the *opaque-2* mutation modifies the amino acid composition of the mature endosperm, resulting in a striking improvement in the nutritional quality of maize seed.

J. G. Gorman, V. J. Freda, and W. Pollack demonstrate that the sensitization of Rh-negative mothers can be prevented by administration of Rh antibody immediately after delivery of their first Rh-positive baby.

D. J. L. Luck and E. Reich isolate mitochondrial DNA from *Neurospora*. They demonstrate subsequently (1966) that this DNA replicates by the classical semiconservative mechanism.

G. Marbaix and A. Burny isolate a 9S RNA from mouse reticulocytes and suggest that it may be mRNA.

R. Holliday puts forth a model that defines a sequence of breakage and reunion events which must occur during crossing over between the DNA molecules of homologous chromosomes.

J. W. Littlefield develops a method for selecting somatic cell hybrids utilizing HGPRT$^-$ and TK$^-$ fibroblasts cultured on HAT medium.

D. D. Brown and J. B. Gurdon show that no synthesis of the 18S and 28S rRNAs occurs in *Xenopus* tadpoles homozygous for a deficiency covering the nucleolus organizer.

W. D. Hamilton puts forth the genetical theory of social behavior.

W. Gilbert finds that nascent proteins bind to the large ribosomal subunit, as do the tRNAs.

D. M. C. Hodgkin receives the Nobel Prize in chemistry for her pioneering studies in x-ray crystallography. It was in her laboratory at Oxford University that the three-dimensional structures of cholesterol, penicillin, cephalosporin, cobalamin, and insulin were elucidated.

1965 R. B. Merrifield and J. Stewart develop an automated method for synthesizing polypeptides on a solid supporting polymeric matrix. Some of the same automation principles will later be adopted for automated nucleic acid synthesis by instruments called "gene machines."

D. D. Sabatini, Y. Tashiro, and G. E. Palade show that the large subunit of ribosome attaches to the ER membrane.

R. W. Holley and his colleagues determine the complete sequence of alanine transfer RNA isolated from yeast.

N. Hilschmann and L. Craig report that immunoglobulin molecules are made up of carboxyl-terminal segments that are constant in their amino acid composition and amino-terminal segments that are variable. This finding poses the problem of how a gene can code for those portions of the protein that vary in their amino acid compositions.

P. Karlson and four colleagues determine the complete structural configuration of ecdysone.

D. H. Carr publishes cytological studies of spontaneous abortuses from a Canadian population. Over 20% show chromosomal abnormalities.

S. Spiegelman and four colleagues succeed in the *in vitro* synthesis of a self-propagating infectious RNA (bacteriophage Q beta of *E. coli*) using a purified enzyme (Q beta replicase).

S. Brenner, A. O. W. Stretton, and S. Kaplan deduce that UAG and UAA are the codons that signal the termination of a growing polypeptide.

F. M. Ritossa and S. Spiegelman demonstrate that multiple transcription units producing the ribosomal RNAs of *Drosophila* reside in the nucleolus organizer regions of each X and Y chromosome.

H. Harris and J. F. Watkins use the Sendai virus to fuse somatic cells derived from man and mouse and produce artificial interspecific heterokaryons.

A. J. Clark identifies in *E. coli* the first gene (*recA*) whose product functions in genetic crossing over.

F. Sanger, G. G. Brownlee, and B. G. Barrell describe a method for fingerprinting oligonucleotides from partially hydrolyzed RNA preparations.

R. Rothman demonstrates that lambda phage has a specific attachment site on the *E. coli* chromosome.

W. J. Dreyer and J. C. Bennett propose that antibody light chains are encoded by two distinct DNA sequences, one for the variable region and the other for the constant region. They suggest that there is only one constant region, but that the variable region contains hundreds of different minigenes.

F. Jacob, J. Monod, and A. Lwoff receive a Nobel Prize for their contributions to microbial genetics.

1966 B. Weiss and C. C. Richardson isolate DNA ligase.

M. M. K. Nass reports that mitochondrial DNA is a circular double-stranded molecule.

F. H. C. Crick puts forward the wobble hypothesis to explain the general pattern of degeneracy found in the genetic code.

J. Adams and M. Cappecchi show that *N*-formylmethionyl-tRNA functions as the initiator of the polypeptide chain forming on a bacterial ribosome.

W. Gilbert and B. Müller-Hill demonstrate that the lactose repressor of *E. coli* is a protein.

M. Ptashne shows that the phage lambda repressor is a protein and that it binds directly to the lambda DNA molecule.

F. M. Ritossa, K. C. Atwood, and S. Spiegelman show that the *bobbed* mutants of *Drosophila* are partial deficiencies of ribosomal DNA.

H. Röller and three colleagues determine the structural formula for the juvenile hormone of *Hyalophora cecropia*.

M. Waring and R. J. Britten demonstrate that vertebrate DNAs contain repetitious nucleotide sequences.

R. S. Edgar and W. B. Wood analyze the genetically controlled steps in the assembly of the T4 bacteriophage.

E. Terzaghi and five colleagues confirm that the genetic code is translated by the sequential reading of triplets of bases starting at a defined point in the mRNA for phage T4 lysozyme.

V. A. McKusick publishes the one-volume *Mendelian Inheritance in Man*, a catalog that lists 1,487 genetic disorders in *Homo sapiens*. By the time of publication of the three-volume 12th edition in 1998, the number of recognized hereditary diseases had risen to 8,600.

H. Wallace and M. L. Birnstiel demonstrate that an anucleolate deletion in *Xenopus laevis* removes more than 99% of the rDNA.

R. C. Lewontin and J. L. Hubby use electrophoretic methods to survey gene-controlled protein variants in natural populations of *Drosophila pseudoobscura*. They demonstrate that between 8 and 15% of all loci in the average individual genome are in the heterozygous condition. Using similar techniques, H. Harris demonstrates the existence of extensive enzyme polymorphisms in human populations.

R. R. Ernst and W. A. Anderson show that the sensitivity of nuclear magnetic resonance spectroscopy can be greatly increased by replacing the slow radiofrequency sweep of the specimen by short, intense RF pulses.

P. Rous receives the Nobel Prize for his studies on oncogenic viruses.

1967 S. Spiegelman, D. R. Mills, and R. L. Peterson report the results of an *in vitro* evolution experiment that generates the smallest self-duplicating molecule known to science.

H. G. Khorana and his co-workers use polynucleotides with known repeating di- and trinucleotide sequences to solve the genetic code.

K. Taylor, Z. Hradecna, and W. Szybalski show that transcription in phage lambda can proceed in opposite directions in different genes on the same chromosome. Therefore, mRNA can originate from transcription units residing in the + and in the – strand of the same double helix.

B. Mintz uses allophenic mice to demonstrate that melanocytes that provide color to the fur of the mouse are derived from 34 cells that have been determined at an early stage in embryogenesis.

J. B. Gurdon transplants somatic nuclei into frog eggs at different developmental states. The synthesis of RNA and DNA of transplanted nuclei changes to the kind of synthesis characteristic of the host cell nucleus.

L. Goldstein and D. M. Prescott perform nuclear transplantations in *Amoeba*. These show there are specific proteins that move from the cytoplasm to the nucleus, and these presumably control the nucleic acid metabolism of the nuclei they enter.

C. B. Jacobson and R. H. Barter report the use of amniocentesis for intrauterine diagnosis and management of genetic defects.

C. C. F. Blake and four colleagues publish the three-dimensional structure of lysozyme at 2 Å resolution. This gives the first indication as to how an enzyme molecule is shaped to accommodate its substrate.

M. Goulian, A. Kornberg, and R. L. Sinsheimer report the successful *in vitro* synthesis of biologically active DNA. The template they presented to the purified *E. coli* DNA polymerase was single-stranded DNA from phiX 174.

M. L. Birnstiel reports the isolation of pure rDNA from *Xenopus laevis*.

T. O. Diener and W. B. Raymer show that the potato spindle tuber disease is caused by a viroid.

M. C. Weiss and H. Green use the HAT selection procedure to localize the gene for thymidine kinase. This was the first use of somatic cell genetics to localize human genes.

W. M. Fitch and E. Margoliash develop a method for constructing phylogenetic trees based on mutation distances separating homologous proteins. They use data provided by mutations identified in the cytochrome C genes from 20 different species ranging from yeasts to humans.

S. Ohno publishes *Sex Chromosomes and Sex-linked Genes*. In this book he proposes that the genes on the X chromosomes of all eutherian mammals are conserved as a single linkage group because of the constraints imposed by X chromosome inactivation.

E. M. Witkin discovers error-prone repair in *E. coli*. Her subsequent work on the molecular mechanisms for repairing damaged DNA uncovers the SOS response.

V. M. Sarich and A. C. Wilson contrast the immunological properties of protein albumen between chimpanzees, gorillas, and humans. They conclude that the African apes and mankind shared a common ancestor 4 to 6 million years ago.

1968 R. T. Okazaki and four colleagues report that newly synthesized DNA contains many fragments. These represent short lengths of DNA that are replicated in a discontinuous manner and then spliced together.

W. Gilbert and D. Dressler put forth the "rolling circle" model of DNA replication.

W. J. Morgan, D. P. McKenzie, and X. Le Pinchon develop the concept of plate tectonics to explain continental drift.

Amplification of ribosomal DNA during amphibian oogenesis is reported by J. G. Gall and by D. D. Brown and I. B. Dawid.

M. Kimura proposes the neutral gene theory of molecular evolution.

H. O. Smith, K. W. Wilcox, and T. J. Kelley isolate and characterize the first specific restriction endonuclease (*Hind*II).

D. Y. Thomas and D. Wilkie demonstrate recombination of yeast mitochondrial genes.

R. P. Donahue and three colleagues assign the Duffy blood group locus to chromosome 1 in humans. This is the first gene localized in a specific autosome.

S. M. Gartler reports that HeLa cells contaminate many cell lines derived from other sources. Subsequent studies have shown that such contaminations are widespread and often invalidate the conclusions drawn from research on these contaminated cultures.

J. A. Huberman and A. D. Riggs demonstrate that mammalian chromosomes contain serially arranged replicons each about 30 micrometers long.

S. Wright publishes volume 1 of his four-volume series *Evolution and the Genetics of Populations*. The final volume will be completed 10 years later.

E. H. Davidson, M. Crippa, and A. E. Mirsky show that more than 60% of the RNA labeled during oogenesis in *Xenopus laevis* is synthesized during the lampbrush stage and stored during the remaining months of oocyte maturation. This RNA is presumably a long-lived mRNA stored for use in early embryogenesis.

O. Hess and G. Meyer report extensive studies on structural modifications of the Y chromosome in various *Drosophila* species showing that the Y chromosome contains segments that generate lampbrush loops in spermatocytes. These loops are necessary for stage-specific steps spermiogenesis.

S. A. Henderson and R. G. Edwards demonstrate that the number of chiasmata per oocyte declines with increasing maternal age in the mouse and that the number of univalents increases with age. If the same occurs in human females, one would expect (as has been demonstrated) an increase in aneuploid offspring with advancing maternal age.

J. E. Cleaver shows that the repair replication of DNA is defective in patients with xeroderma pigmentosum.

R. J. Britten and D. E. Kohne demonstrate that Cot curves can be used to determine the relative abundances of repetitive and nonrepetitive DNA sequences in the genomes of different species.

R. W. Davis and N. Davidson visualize deletion mutations of bacteriophage lambda utilizing experimentally produced heteroduplex DNA molecules.

R. W. Holley, H. G. Khorana, and M. W. Nirenberg receive Nobel Prizes for discoveries concerning the interpretation of the genetic code and its function in protein synthesis.

1969 J. Abelson and six colleagues provide proof of the proposed mechanism of nonsense suppression by determining the actual nucleotide sequence of mutant tyrosine transfer ribonucleic acids.

L. Wolpert proposes the Positional Information Hypothesis to explain the mechanism of pattern formation during embryogenesis.

The *in situ* hybridization techniques for the cytological localization of specific nucleotide sequences are developed by J. G. Gall and M. L. Pardue and by H. John, M. L. Birnstiel, and K. W. Jones.

B. C. Westmoreland, W. Szybalski, and H. Ris develop an electron microscopic technique for physically mapping genes in lambda phage. They photograph heteroduplex DNA molecules obtained by annealing of the – strand of one parent and the + strand of a second parent that has deletions, insertions, substitutions, or inversions.

B. Dudock and three colleagues show that tRNA is folded into a three-dimensional cloverleaf.

R. Burgess and three colleagues isolate and identify the sigma factor from RNA polymerase.

H. Harris and four colleagues demonstrate by cell fusion experiments the existence of anti-oncogenes.

O. L. Miller and B. R. Beatty publish electron micrographs showing amphibian genes in the process of transcribing RNA molecules.

J. R. Beckwith (with five associates) reports the isolation of pure *lac* operon DNA from *E. coli*.

G. M. Edelman (with five associates) publishes the first complete amino acid sequence for human gamma G_1 immunoglobulin.

Y. Hotta and S. Benzer and W. L. Pak and J. Grossfield independently induce and physiologically characterize neurological mutants in *Drosophila*.

C. Boon and F. Ruddle correlate the loss of particular chromosomes from a somatic hybrid cell line containing both human and mouse chromosomes with the loss of specific phenotypic characters. This approach permits assignment of specific loci to certain human chromosomes.

R. E. Lockard and J. B. Lingrel purify the 9S RNA fraction obtained from polysomes of mouse reticulocytes and show that it directs the synthesis of mouse hemoglobin beta chains. They thus confirm the suggestion of Marbaix and Burny (1964).

A. Ammermann reports that in the hypotrichous ciliate *Stylonychia mytilus* macronuclear anlage undergo endomitotic DNA replication to form polytene chromosomes. Subsequently, the major portions of these are destroyed, and over 90% of the macronuclear DNA is degraded and excreted into the medium.

R. C. Roeder and W. R. Rutter demonstrate that the RNA polymerases of eukaryotes belong to three major classes.

R. Huebner and G. Todaro put forth the oncogene hypothesis. This suggests that all cancers arise from the expression of genes that are implanted in the germlines of host organisms eons earlier by oncogenic viruses. These oncogenes lie silent until activated subsequently by carcinogens of various types

H. A. Lubs describes a fragile site on the human X chromosome and shows that it is present in mentally retarded males. Subsequent studies show that this locus (Xq27) is associated with a common form of X-linked mental retardation.

J. A. Shapiro detects mutations of the galactose operon of *E. coli* caused by insertion sequences.

M. Delbrück, S. E. Luria, and A. D. Hershey receive a Nobel Prize for their contributions to viral genetics.

1970 B. M. Alberts and L. Frey isolate the protein product of gene 32 of phage T4 and demonstrate that this protein binds cooperatively to single-stranded DNA. They suggest that the gene 32 protein functions to initiate unwinding of the DNA molecule so that replication can begin.

H. G. Khorana and 12 colleagues report the total synthesis of the gene for an alanine tRNA from yeast.

M. Mandel and A. Higa develop a general method for introducing DNA into *E. coli*. They demonstrate that placing the cells in a cold calcium chloride solution renders them permeable to nucleic acid fragments.

J. Yourno, T. Kohno, and J. R. Roth succeed in fusing two bacterial enzymes into one large protein molecule that combines the functions of both. The enzyme fusion was accomplished by fusing the *his D* and *his* C genes in the histidine operon of *Salmonella*, using a pair of frame shift mutations.

D. Baltimore reports the existence of an RNA-dependent DNA polymerase in an oncogenic RNA virus (the Rauscher mouse leukemia virus).

H. M. Temin and S. Mizutani discover a reverse transcriptase in another retrovirus (the Rous sarcoma virus).

P. Nowell and D. Hungerford report that patients suffering from chronic granulocytic leukemia often carry an abnormal minute chromosome in their cells. This chromosome is later nicknamed the *Philadelphia chromosome*.

J. L. Kermicle demonstrates that the *R* gene in maize shows parental imprinting.

M. L. Pardue and J. G. Gall demonstrate that pericentric heterochromatin is rich in repetitious DNA.

D. E. Wimber and D. M. Steffensen localize the 5S RNA cistrons on the right arm of chromosome 2 of *Drosophila melanogaster*.

T. Caspersson, L. Zech, and C. Johansson use quinacrine dyes in chromosomal cytology and demonstrate specific fluorescent banding patterns in human chromosomes.

R. Sager and Z. Ramanis publish the first genetic map of non-Mendelian genes. This group of eight genes resides on a chloroplast chromosome of *Chlamydomonas*.

R. T. Johnson and P. N. Rao induce premature chromosome condensation by fusing mitotically active cells with interphase cells *in vitro*.

M. Smös and R. B. Inman use the denaturation mapping technique to study chromosome replication in bacteriophage lambda. They show that replication begins at a unique origin and that both forks are growing points that progress in opposite directions around the circular molecule.

H. O. Smith and K. W. Wilcox discover that certain restriction endonucleases can generate DNA termini, in one step, that have projecting single-stranded ends.

M. Rodbell and L. Birmbaumer find that GTP is required for the hormonal activation of adenylcyclase. This suggests that a GTP-binding protein participates in signal transmission.

1971 A. M. Olonikov suggests that marginotomy of DNA limits the mitotic potential of clones of somatic cells.

D. Baltimore proposes a classification that arranges viruses within six groups that depend upon the relationship between their genomes and the mechanisms used to generate their mRNAs.

Y. Masui and L. D. Smith demonstrate independently the existence of a *maturation promoting factor* in the cytoplasm of activated *Xenopus* oocytes.

M. L. O'Riordan and three colleagues report that all 22 pairs of human autosomes can be identified visually after staining with quinacrine hydrochloride.

They demonstrate that the Philadelphia chromosome is an aberrant chromosome 22.

Y. Hotta and H. Stern characterize the DNA that is synthesized during meiotic prophase in the lily. Synthesis during zygonema represents the delayed replication of a small fraction of the DNA that failed to replicate during the previous S phase. The DNA synthesized during pachynema has the characteristics of repair replication.

S. H. Howell and H. Stern demonstrate that an endonuclease present in lily microspores reaches its highest concentration early in pachynema, the stage when crossing over is thought to occur.

C. A. Thomas coins the term C *value paradox*.

A. G. Knudson proposes that the normal retinoblastoma locus functions as a dominant anti-oncogene.

J. E. Darnell and three colleagues suggest that during the posttranscriptional processing of premessenger RNA, a polyadenylic acid segment is added and that this poly-A tail somehow stabilises the mRNA.

S. Altman and J. D. Smith isolate RNase P from *E. coli* and show it is a ribonucleoprotein that removes the 5′ leader from tRNA precursor molecules.

H. Klenow isolates and characterizes the fragment of DNA polymerase that now bears his name.

B. Dudock and three colleagues present evidence for the phenylalanyl tRNA synthetase recognition site being adjacent to the dihydrouridine loop.

C. R. Merril, M. R. Geier, and J. C. Petricciani infect fibroblasts cultured from a patient suffering from galactosemia with transducing lambda phage carrying the galactose operon. The cells then make the missing transferase and survive longer in culture than uninfected galactosemic cells.

R. J. Konopka and S. Benzer report recovery of induced clock mutants of *Drosophila*.

D. T. Suzuki, T. Grigliatti, and R. Williamson isolate a temperature-sensitive paralytic mutant of *Drosophila*.

J. E. Manning and four colleagues detect circular DNA molecules in lysates of *Euglena* chloroplasts.

C. Kung induces and isolates behavioral mutants of *Paramecium aurelia* and shows that many mutants have electrophysiological defects in their plasma membranes.

K. Danna and D. Nathans use restriction endonucleases to cleave the circular DNA of simian virus 40 into a series of fragments and then deduce their physical order.

E. W. Sutherland receives a Nobel Prize for discovering cyclic AMP and the enzyme that creates it, adenylcyclase.

1972 P. Lobban and A. D. Kaiser develop a general method for joining any two DNA molecules, employing terminal transferase to add complementary homopolymer tails to passenger and vehicular DNA molecules.

A. F. Zakharov and N. A. Egolina develop the BUDR labeling technique to produce harlequin chromosomes.

J. D. Watson points out that, during a series of replication cycles, DNA molecules should shorten until they become inviable. This is because DNA molecules are not copied all the way to their 5′ ends during each replication cycle. The clarification of the structure and functioning of telomeres (1978) and telomerases (1983) provided the solution to Watson's dilemma.

G. H. Pigott and N. G. Carr show that ribosomal RNAs from cyanobacteria hybridize with DNA from the chloroplasts of *Euglena gracilis*. This genetic homology provides strong support to the theory that chloroplasts are the descendants of endosymbiotic cyanobacteria.

S. J. Singer and G. L. Nicholson put forth the fluid mosaic model of the structure of cell membranes.

B. Benacerraf and H. O. McDevitt show for the mouse that *Ir* genes are linked with the H-2 complex.

N. Eldredge and S. J. Gould propose a punctuated equilibrium model for the evolution of species.

Y. Suzuki and D. D. Brown isolate and identify the mRNA for silk fibroin from *Bombyx mori*, and Suzuki, L. P. Gage, and Brown characterize the fibroin gene.

D. D. Brown, P. C. Wensink, and E. Jordon describe the concerted evolution of rRNA gene families in two species of *Xenopus*.

R. Silber, V. G. Malathi, and J. Hurwitz discover RNA ligase.

D. A. Jackson, R. H. Symons, and P. Berg report splicing the DNA of SV40 virus into the DNA of the lambda virus of *E. coli*. They are thus the first to join the DNAs of two different organisms *in vitro*.

M. L. Pardue and three colleagues locate the histone genes on chromosome 2 of *Drosophila melanogaster*.

J. Hedgpeth, H. M. Goodman, and H. W. Boyer identify the nucleotide sequence in the DNA of lambda (λ) bacteriophage that is recognized by a specific endonuclease.

S. N. Cohen, A. C. Y. Chang, and L. Hsu show that *E. coli* can take up circular plasmid DNA molecules and that transformants in the bacterial population can be identified and selected utilizing antibiotic resistance genes carried by the plasmids.

U. Kuhnlein and W. Arber report the isolation of recognition-site mutations in coliphages. This confirms Arber's 1962 restriction and modification proposal to explain the host-controlled restriction of virus growth.

J. Mertz and R. W. Davis show that cleavage of DNA by R1 restriction endonuclease generates cohesive ends.

D. E. Kohne, J. A. Chisson, and B. H. Hoyer use DNA-DNA hybridization data to study the evolution of primates. They conclude that man's closest living relative is the chimpanzee.

B. Daneholt shows that Balbiani ring BR2 of *Chironomus* generates a 75S RNA which serves as a message for translating giant salivary polypeptides.

J. Rowley shows that the *Philadelphia chromosome* is the result of a reciprocal translation between chromosomes 22 and 9.

G. M. Edelman and R. R. Porter receive a Nobel Prize for their studies on the chemical structure of antibodies.

1973 P. D. Boyer proposes that there are three catalytic sites in ATP synthase that interact and interconvert in a cyclic fashion.

D. R. Mills, F. R. Kramer, and S. Spiegelman publish the sequence for the 218 nucleotides in the shortest replicating RNA. The molecule (MDV-1) is a variant derived from the RNA genome of Q beta phage that underwent *in vitro* evolution.

S. H. Kim and seven colleagues propose a three-dimensional structure for yeast phenylalanine transfer RNA.

J. T. Finch and three colleagues show that the long fibers that distort the erythrocytes of patients with sickle-cell anemia are tubes about 17 nm in diameter. These are hollow cables made up of six strands that consist of strings of molecules of deoxygenated hemoglobin S.

L. H. Hartwell and three collagues define 32 genes in *Saccharomyces cerevisiae* that encode products essential for specific consecutive steps in the cell division cycle.

R. Kavenoff and B. H. Zimm use a newly developed viscoelastic method for measuring the molecular weights of DNA molecules isolated from cells from different *Drosophila* species. They conclude that a chromosome contains one long molecule of DNA and that it is not interrupted in the centromere region.

P. Debergh and C. Nitsch succeed in culturing haploid tomato plants directly from microspores.

W. G. Hunt and R. K. Selander analyze a zone of hybridization between two subspecies of the house mouse, using gel electrophoresis to trace the boundary.

P. J. Ford and E. M. Southern show for *Xenopus laevis* that different 5S RNA genes are transcribed in the oocyte than in somatic cells.

W. Fiers and three colleagues are the first to sequence a gene coding for a protein (the coat protein of the RNA coliphage MS2).

B. E. Roberts and B. M. Patterson report the preparation of a wheat germ cell-free system for the *in vitro* translation of experimentally supplied mRNAs.

A. Garcia-Bellido, P. Ripoll, and G. Morata report the developmental compartmentalization of the wing disc of *Drosophila*.

S. N. Cohen and three colleagues construct the first biologically functional, hybrid bacterial plasmids by *in vitro* joining of restriction fragments from different plasmids.

J. M. Rosenberg and five colleagues produce the first x-ray crystallographic structure with atomic resolution for a segment of a double helix. It proves that the pairing of purines to pyrimidines is of the type Watson and Crick proposed in their model for DNA published 20 years earlier.

C. B. Anfinsen, in his Nobel Prize address, summarizes the evidence for concluding that the three-dimensional conformation of a protein is dictated by its amino acid sequence.

1974 J. Shine and L. Dalgarno show that the 3′ terminus of *E. coli* 16S rRNA contains a stretch of nucleotides that is complementary to ribosome binding sites of various coliphage mRNAs. They suggest that this region of the 16S rRNA may play a base-pairing role in the termination and initiation of protein synthesis on mRNA.

I. Zaenen and four colleagues discover the tumor-inducing plasmid of the crown gall bacterium.

K. M. Murray and N. E. Murray manipulate the recognition sites for restriction endonucleases in lambda (λ) bacteriophage so that its chromosome can be used as a receptor site for restriction fragments from foreign DNAs. Lambda thus becomes a cloning vehicle.

A. Tissieres, H. K. Mitchell, and U. M. Tracy find that heat shocks result in the synthesis of six new proteins in *Drosophila*. These are also synthesized by tissues that do not have polytene chromosomes.

B. Dujon, P. P. Slonimski, and L. Weill propose a model for recombination and segregation of mitochondrial genomes in *Saccharomyces cerevisiae*. According to it, mtDNA molecules are present in the zygote cell in multiple copies. These pair at random, and during any mating cycle a segment from one parent can exchange with that from a second parent mtDNA yielding recombinant units.

R. D. Kornberg proposes that chromatin is built up of repeated structural units of 200 base pairs of DNA and two each of the histones H2A, H2B, H3, and H4. These structures, which are later called nucleosomes, are isolated by M. Noll. A. L. Olins, and D. E. Olins publish the first electron micrographs of chromatin spreads from nuclei that show nucleosomes.

B. Ames develops a rapid screening test for detecting mutagenic and possibly carcinogenic compounds.

S. Brenner describes methods for inducing, isolating, and mapping mutations in the nematode *Caenorhabditis elegans*.

R. W. Hedges and A. E. Jacob discover in *E. coli* that ampicillin-resistance genes can be transferred between plasmids that show no DNA homology. The agent responsible is a mobile DNA sequence, which they named a *transposon*.

J. Ott invents the first computer program (LIPED) for the efficient computation of lod scores.

C. A. Hutchison and three colleagues demonstrate the maternal inheritance of mitochondrial DNA in horse-donkey hybrids.

G. L. Stebbins publishes *Flowering Plants: Evolution Above the Species Level*, which brings a modern framework of cytogenetics and systematics to the study of plant evolution.

A. Claude, C. de Duve, and G. Palade receive Nobel Prizes for their contributions to cell biology.

1975 G. Köhler and C. Milstein perform experiments with mouse cells that show that somatic cell hybridization can be used to generate a continuous "hybridoma" cell line producing a monoclonal antibody.

Molecular biologists from around the world meet at Asilomar, California, to write a historic set of rules to guide research in recombinant DNA experiments.

The NIH Recombinant DNA Committee issues guidelines aimed at eliminating or minimizing the potential risks of recombinant DNA research.

L. L. Goldstein and M. S. Brown demonstrate that normal fibroblasts have binding sites for low-density lipoproteins, whereas fibroblasts from humans homozygous for the hypercholesterolemia gene lack these receptors.

M. Grunstein and D. S. Hogness develop the colony hybridization method for the isolation of cloned DNAs containing specific DNA segments or genes.

A. T. C. Carpenter identifies recombination nodules in *Drosophila melanogaster*. She points out the correlation between the number of RNs and the number of meiotic exchanges.

D. Pribnow determines the nucleotide sequences of two independent bacteriophage T7 promoters, and compares these and other known promoter sequences to form a model for promoter structure and function.

E. M. Southern describes a method for transferring DNA fragments from agarose gels to nitrocellulose filters. The filters are subsequently hybridized to radioactive RNA and the hybrids detected by autoradiography.

W. D. Benton and R. W. Davis describe a rapid and direct method for screening plaques of recombinant lambda bacteriophages that involves transfer of phage DNA to a nitrocellulose filter and detection of specific DNA sequences by hybridization to complementary labeled nucleic acids.

F. Sanger and A. R. Coulson develop the "plus and minus" method for determining the nucleotide sequences in DNA by primed synthesis with DNA polymerase.

M. C. King and A. C. Wilson point out that 99% of the proteins that have been studied in humans and chimpanzees have identical amino acid sequences. They conclude that the biological differences between these two species must be largely the result of mutations that involve regulatory rather than structural genes.

G. Morata and P. A. Lawrence show in *Drosophila* that the *engrailed* mutation allows cells of the posterior wing compartment to mix with those of the anterior compartment. Therefore, the normal allele of this gene functions to define the boundary conditions between the sister compartments of the developing wing.

B. Mintz and K. Illmensee inject XY diploid cells from a malignant mouse teratocarcinoma into mouse blastocysts that then are transferred to foster mothers. Cells derived from the carcinoma appear in both somatic and germ cells of some F_1 males. When these are mated, some F_2 mice contain marker genes from the carcinoma. The experiments demonstrate that the nuclei of teratocarcinoma cells remain developmentally totipotent, even after hundreds of transplant generations during which they functioned in malignant cancers.

S. L. McKenzie, S. Henikoff, and M. Meselson isolate mRNAs for heat-shock proteins and show that they hybridize to specific puff sites on the *Drosophila* polytene chromosomes.

L. H. Wang and three colleagues locate within the RNA genome of the Rous sarcoma virus the segment responsible for its oncogenic activity.

G. Blobel and B. Dobberstein put forth the signal hypothesis.

R. Dulbecco, H. Temin, and D. Baltimore receive Nobel Prizes for their studies on oncogenic viruses.

1976 H. R. B. Pelham and R. J. Jackson describe a simple and efficient mRNA-dependent *in vitro* translation system using rabbit reticulocyte lysates.

W. Fiers and 11 colleagues complete their analysis of MS2 RNA. This is the first virus to have its genome sequenced from beginning to end.

R. V. Dippell shows in *Paramecium* that kinetosomes contain RNA (not DNA) and that RNA (not DNA) synthesis accompanies kinetosome reproduction.

N. Hozumi and S. Tonegawa demonstrate that the DNA segments coding for the variable and constant regions of an immunoglobulin chain are distant from one another in the chromosomes isolated from mouse embryos, but the segments are adjacent in chromosomes isolated from mouse plasmacytomas. They conclude that somatic recombination during the differentiation of B lymphocytes moves the constant and variable gene segments closer together.

W. Y. Kan, M. S. Golbus, and A. M. Dozy are the first to use recombinant DNA technology in a clinical setting. They develop a prenatal test for alpha thalassemia utilizing molecular hybridization techniques.

P. M. Nurse, P. Thuriaux, and K. Nasmyth elucidate the genetic control of the cell division cycle in *Schizosaccharomyces pombe*. Among the genes that control mitosis are those that encode cyclin-dependent kinases.

M. F. Gellert and three colleagues discover DNA gyrase to be the enzyme that converts a relaxed, closed, circular DNA molecule into a negatively supercoiled form.

W. Y. Chooi shows that ferritin-labeled antibodies raised against proteins (isolated from rat ribosomes) bind to the terminal knobs of fibers extending from Miller trees (isolated from the ovarian nurse cells of *Drosophila*). This observation proves that Miller trees are rRNA transcription units and shows that at least some ribosomal proteins attach to a precursor rRNA molecule before its transcription is completed.

B. G. Burrell, G. M. Air, and C. A. Hutchison report that phage phiX174 contains overlapping genes.

Formal guidelines regulating research involving recombinant DNA are issued by the National Institutes of Health in the United States.

H. Boyer and R. Swanson found Genentech, an event which marks the beginning of the biotech industrial revolution.

A. Efstratiadis and three colleagues are the first to enzymatically generate eukaryotic gene segments *in vitro*. They synthesize double-stranded DNA molecules that contain the sequences transcribed into the mRNAs for the alpha and beta chains of rabbit hemoglobin.

J. T. Finch and A. Klug propose that the 300 Å threads seen in electron micrographs of fragmented chromatin are formed by the folding of DNA-nucleosome filaments into solenoids.

L. H. Miller and three colleagues conclude that the Duffy blood group antigens (Fy^a and Fy^b) serve as receptors for the merozoites of *Plasmodium vivax* and that individuals of blood group Fy^-/Fy^- are resistant to *P. vivax* infections because their red cells lack these receptors.

1977 A. Knoll and E. S. Barghoorn find microfossils which they interpret as undergoing cell division in rocks 3,400 million years old. This discovery pushes back the age of life on earth to the lower Archean eon.

J. B. Corliss and R. D. Ballard aboard *Alvin*, a deep-diving minisubmarine, discover communities of hyperthermophilic bacteria, tube worms, clams, and other organisms, living in the Galapagos rift.

E. M. Ross and A. G. Gilman show that adenylcyclase is regulated by a protein that binds GTP. This G protein is purified three years later and shown to be a heterotrimer.

K. Itakura and six colleagues chemically synthesize a gene for human somatostatin and express it in *E. coli.* This leads to the commercial production of the first artificial human protein, somatostatin.

S. M. Tilghman and eight colleagues clone the first protein-encoding gene (mouse beta-hemoglobin) using bacteriophage lambda as a vector.

C. Jacq, J. R. Miller, and G. G. Brownlee describe the presence of "pseudogenes" within the 5S DNA cluster of *Xenopus laevis* oocytes.

J. C. Alwine, D. J. Kemp, and G. R. Stark prepare diazobenzyloxymethyl (DBM) paper and describe methods for transferring electrophoretically separated bands of RNA from an agarose gel to the DBM paper. Specific RNA bands are then detected by hybridization with radioactive DNA probes, followed by autoradiography. Since this method is the reverse of that described by Southern (1975) in that RNA rather than DNA is transferred to a solid support, it has come to be known as "northern blotting."

F. Sanger and eight colleagues report the complete nucleotide sequence for the DNA genome of bacteriophage phiX174.

E. W. Silverton, M. A. Navia, and D. R. Davies determine the three-dimensional structure of the human immunoglobulin molecule.

M. Leffak, R. Grainger, and H. Weintraub show that "old" histone octamers remain intact during DNA replication and that "new" octamers consist entirely of proteins synthesized immediately before replication.

C. Woese and G. E. Fox conclude from their studies of the nucleotide sequences of the 16S rRNAs of certain newly discovered microorganisms that they should be placed in a domain separate from the other bacteria (the Archaea).

W. Gilbert induces bacteria to synthesize useful nonbacterial proteins (insulin and interferon).

A. M. Maxam and W. Gilbert publish the "chemical method" of DNA sequencing.

R. J. Roberts and P. A. Sharp lead groups that discover split genes in adenovirus 2. R-loop mapping by L. Chow and S. Berget shows the position of intron loops. Intervening noncoding segments are then described for genes that encode animal proteins, namely, the rabbit beta-globin gene (A. Jeffreys and R. A. Flavell) and the chicken ovalbumin gene (R. Breathnach, J. L. Mandel, and P. Chambon).

J. Weber, W. Jelinek, and J. E. Darnell report that alternative splicing of nonconsecutive DNA segments in the adenovirus-2 genome can produce multiple mRNAs.

J. F. Pardon and five colleagues use neutron contrast matching techniques to demonstrate that in nucleosomes the DNA segment that attaches to the histone octamer is on the outside of the particle.

J. Sulston and H. R. Horvitz work out the postembryonic cell lineages for *Caenorhabditis elegans*.

J. Collins and B. Holm develop cosmids for cloning large DNA fragments.

F. Lee and C. Yanofsky explain the mechanism of attenuation that takes place in the tryptophan operon of *E. coli.*

R. S. Yalow receives a Nobel Prize for developing the radioimmunoassay procedure.

1978 R. M. Schwartz and M. O. Dayhoff compare sequence data for a variety of proteins and nucleic acids from an evolutionarily diverse assemblage of prokaryotes, eukaryotes, mitochondria, and chloroplasts. Their computer-generated evolutionary trees identify the times during evolution when protoeukaryotic organisms entered into symbiosis with mitochondria and chloroplasts (about 2 and 1 billion years ago, respectively).

W. Gilbert coins the terms *intron* and *exon*.

T. Maniatis and seven colleagues develop a procedure for gene isolation, which involves construction of cloned libraries of eukaryotic DNA and screening these libraries for individual sequences by hybridization to specific nucleic acid probes.

M. S. Collett and R. L. Erickson report that the product of the *src* gene of the Rous sarcoma virus is a protein kinase.

W. Bender, R. Spierer, and D. Hogness describe a method for sequencing genes they call *chromosome walking*.

E. B. Lewis concludes that the component genes in the *bithorax* complex have related functions in *Drosophila* segmentation and that they evolved from a smaller number of ancestral genes by their duplication and subsequent specialization.

C. Coulondre and three colleagues show that sites in the DNA of *E. coli* identified as mutational hot spots contain the modified pyrimidine, 5-methylcytosine.

V. B. Reddy and eight colleagues publish the complete nucleotide sequence for simian virus 40 and correlate the sequence with the known genes and mRNAs of the virus.

Y. W. Kan and A. M. Dozy demonstrate the value of using restriction-fragment-length polymorphisms as linked markers for the prenatal diagnosis of sickle-cell anemia.

C. A. Hutchison and five colleagues demonstrate that it is possible to introduce specific mutations at specific sites in a DNA molecule.

E. H. Blackburn and J. G. Gall demonstrate that telomeres from *Tetrahymena pyriformis* consist of short DNA sequences (one strand containing AACCCC, the other TTGGGG) repeated tandemly 30 to 70 times.

R. T. Schimke and three colleagues show that cultured mouse cells exposed to methotrexate develop resistance by amplifying the genes that encode the enzymes that serve as the target for the drug.

W. Arber, H. O. Smith, and D. Nathans share the Nobel Prize in Physiology or Medicine for the development of techniques utilizing restriction endonucleases to study the organization of genetic systems.

P. D. Mitchell receives the Nobel Prize in Chemistry for his contribution to the understanding of biological energy transfer through the formulation of the chemiosmotic theory.

1979 J. G. Sutcliffe determines the complete 4,362 nucleotide pair sequence of the plasmid cloning vector pBR322.

J. C. Avise, R. A. Lansman, and R. O. Shade successfully use restriction endonucleases to measure mitochondrial DNA sequence relatedness in natural populations.

The National Institutes of Health relax guidelines on recombinant DNA to allow viral DNA to be studied.

S. Perdix-Gillot reports giant nuclei that are over 1 million–ploid in the cells of the silk gland of *Bombyx mori*.

B. G. Barrell, A. T. Bankier, and J. Drouin report that the genetic code of human mitochondria has some unique, nonuniversal features.

E. F. Fritsch, R. M. Lawn, and T. Maniatis determine the chromosomal arrangement and structure of human globin genes utilizing recombinant DNA technology.

J. R. Cameron, E. Y. Loh, and R. W. Davis discover transposable elements in yeast.

N. Wexler and a group of Venezuelan colleagues begin a study of natives who are living in three fishing villages on the shore of Lake Maracaibo. Huntington disease is prevalent in this population, and eventually an eight-generation pedigree is constructed that contains over 11,000 people. Analyses of DNA samples of this group lead to the localization and eventual sequencing of the HD gene by MacDonald and coworkers in 1993.

D. V. Goeddel and nine colleagues construct a gene that encodes human growth hormone (HGH) using recombinant DNA technology. The synthesized gene is expressed in *E. coli* under the control of the *lac* promoter and a polypeptide having the properties of HGH is synthesized.

M. R. Lerner and J. A. Steitz report the discovery of small nuclear ribonucleoproteins (snurps).

1980 L. Olsson and H. S. Kaplan produce the first human hybridomas that manufacture a pure antibody in laboratory culture.

A. Hershko, A. Ciechanover, I. A. Rose, and three other colleagues demonstrate that proteins destined to be destroyed undergo ATP-dependent conjugation to a specific protein (APF-1). They later show that APF-1 is ubiquitin and that it delivers doomed proteins to the 26S proteasome where they are broken down into short peptides for reuse.

The United States Supreme Court rules that genetically modified microorganisms can be patented. General Electric, on behalf of A. Chakrabarty, obtains a patent for a genetically engineered microorganism capable of consuming oil slicks.

D. Lowe describes stromatolites from the Archean of Western Australia. They contain 3.8 billion-year-old fossils that resemble cyanobacteria.

J. W. Gordon and four colleagues produce the first transgenic mice by direct injection of cloned DNA into the pronucleus of a fertilized egg.

M. R. Capecchi describes a technique for efficient transformation of cultured mammalian cells by direct microinjection of DNA into cells with glass micropipettes.

C. Woese and 10 colleagues publish the secondary structure for 16S ribosomal RNA.

D. Botstein and three colleagues describe the method of using restriction fragment length polymorphisms to construct genetic linkage maps of the human genome.

W. F. Doolittle and C. Sapienza and, independently, L. E. Orgel and F. H. C. Crick point out that the genomes of all species are littered with DNA segments that contribute nothing to the fitness of the species and persist only because they are efficient replicators. These authors name this collection of DNA segments *selfish DNA* and suggest that these DNAs represent the ultimate parasites.

H. Gronemeyer and O. Pongs demonstrate that, in *Drosophila melanogaster* salivary glands, beta ecdysone binds directly to sites on polytene chromosomes where ecdysone-inducible puffs occur.

C. Nüsslein-Volhard and E. Wieschaus describe the isolation and characterization of zygotic segmentation mutations of *Drosophila melanogaster.*

L. Clark and J. A. Carbon clone the gene that corresponds to the centromere of yeast chromosome 3.

A. R. Templeton provides a new theoretical framework for speciation by the founder effect.

Nobel Prizes in Physiology and Medicine go to G. D. Snell, J. Dausset, and B. Benacerraf for their contributions to immunogenetics.

P. Berg, W. Gilbert, and F. Sanger receive Nobel Prizes in Chemistry for their contributions to the experimental manipulation of DNA.

1981 R. C. Parker, H. E. Varmus, and J. M. Bishop demonstrate that the tumorigenic properties of the Rous sarcoma virus are due to a protein encoded by the *v-src* gene. Cells from various vertebrates contain a homologous gene, *c-src.* The two genes differ in that *v-src* has an uninterrupted coding sequence, whereas *c-src* contains seven exons separated by six introns.

L. Margulis publishes *Symbiosis in Cell Evolution.* Here she summarizes the evidence for the theory that organelles such as mitochondria, chloroplasts, and kinetosomes evolved from prokaryotes that lived as endosymbionts in the ancestors of modern-day eukaryotes.

R. Lande proposes a new model of speciation based on sexual selection on polygenic traits. This model results in a revival of interest in sexual selection.

J. D. Kemp and T. H. Hall transfer the gene of a major seed storage protein (phaseolin) from beans to the sunflower via a plasmid of the crown gall bacterium *Agrobacterium tumefaciens*, creating a "sunbean."

T. R. Cech, A. J. Zaug, and P. J. Grabowski report the discovery of a self-splicing rRNA in *Tetrahymena thermophila.* This is the first demonstration that a macromolecule other than a protein can act as a biological catalyst.

W. F. Anderson and three colleagues determine the three-dimensional structure of the *cro* repressor at 2.8 Å resolution.

G. Hombrecher, N. J. Brewin, and A. W. B. Johnson demonstrate that the ability of *Rhizobium* bacteria to nodulate legumes and fix atmospheric nitrogen is due to plasmid-linked genes.

P. R. Langer, A. A. Waldrop, and D. C. Ward develop a procedure for synthesizing biotinylated DNA probes that hybridize normally with complementary DNA, providing an anchor for streptavidin-linked, color-generating systems.

S. Anderson and 13 colleagues work out the complete nucleotide sequence and genetic organization of the human mitochondrial genome.

H. Sakano and three colleagues discover two segments in the heavy chain immunoglobulin gene of the mouse, which serve as recognition sites for a somatic DNA recombinase.

M. E. Harper and G. F. Saunders demonstrate that single-copy genes can be mapped on human mitotic chromosomes utilizing an improved *in situ* hybridization technique.

J. Banerji, S. Rusconi, and S. Schaffner show that the transcription of the beta-globin gene is enhanced hundreds of times when this gene is linked with certain SV40 nucleotide sequences that they name "enhancer sequences."

J. G. Gall and four colleagues localize histone mRNAs that are being transcribed on the lampbrush chromosomes of salamander oocytes.

M. Chalfie and J. Sulston identify among the touch-insensitive mutants of *Caenorhabditis elegans* five genes that affect a specific set of six sensory neurons.

K. E. Steinbeck and three colleagues demonstrate that the resistance of a weed, *Amaranthus hybridus*, to triazine herbicides is controlled by a chloroplast gene that encodes a polypeptide to which the herbicide binds. Resistant strains of the weed produce a modified gene product that fails to bind triazine.

J. D. Walker sequences the eight genes of the *E. coli atp* operon. These encode the protein subunits of ATP synthase.

1982 Eli Lilly and Company markets a Genentech-licensed, recombinant, human insulin. This is the first product generated by this new technology.

E. P. Reddy and three colleagues report that the genetic change that leads to the activation of an oncogene carried by a line of human bladder carcinoma cells is due to a single base substitution in this gene. The result is the incorporation of valine instead of lysine in the 12th amino acid of the protein encoded by the oncogene.

P. Goelet and five colleagues determine the complete nucleotide sequence for the RNA genome of the tobacco mosaic virus.

P. M. Bingham, M. G. Kidwell, and G. M. Rubin show that P strains of *Drosophila* contain 30 to 50 copies per genome of a transposable P element. This is the cause of hybrid dysgenesis. Then A. C. Spradling and Rubin demonstrate that cloned P elements, when microinjected into *Drosophila* embryos, become integrated into germ-line chromosomes and that P elements can be used as vectors to carry DNA fragments of interest into the *Drosophila* germ line.

E. R. Kandel and J. G. Schwartz utilize the gill-withdrawal reflex in *Aplysia* to study the molecular control of memory formation. They eventually show that the long-term facilitation of sensory neurons requires the activation of cAMP-responsive memory genes.

S. B. Prusiner shows that the infectious agent that causes scrapie is a protein, which he calls a prion.

A. Klug receives the Nobel Prize for his contributions to the analysis of crystalline structures of biological importance, especially virus particles, tRNA, and nucleosomes.

1983 E. A. Miele, D. R. Mills, and F. R. Kramer construct the first recombinant RNA molecule by inserting a synthetic foreign deca-adenylic acid into a variant of the RNA genome of phage Q beta via the action of the Q beta replicase.

H. J. Jacobs and six colleagues report the presence of promiscuous DNA in the sea urchin.

T. Hunt and four colleagues demonstrate that sea urchin eggs contain a maternal mRNA which encodes a protein that is synthesized after fertilization and is cyclically destroyed and resynthesized during cleavage divisions. They name the protein *cyclin*.

I. S. Greenwald, P. W. Sternberg, and H. R. Horvitz demonstrate that the *lin-12* mutant of *Caenorhabditis* functions as a developmental control gene.

M. Oren and A. J. Levine isolate and identify a cDNA of the p53 mRNA from a SV-40-transformed mouse cell line.

S. D. Gillies and three colleagues show that a tissue-specific enhancer is located in the first intron of the heavy-chain immunoglobulin gene.

W. Bender and seven colleagues sequence genes in the *bithorax complex* of *Drosophila* and show that spontaneous mutations by *bx*, *Ubx*, and *bxd* are associated with insertions of transposable elements.

M. P. Scott and six colleagues sequence another group of segment identity genes and work out the organization of the *Antennapedia* locus.

G. N. Godson and four colleagues clone the gene that encodes the circumsporozoite protein of *Plasmodium knowlesi*, and they show that the protein contains a repetitive epitope that serves as a decoy to the host immune system.

C. Guerrier-Takada and four colleagues show that ribonuclease P consists of one protein and one RNA subunit and that the latter is the catalytic subunit.

L. Montagnier in France and R. Gallo in the United States lead teams that independently publish accounts of the discovery of the virus that causes AIDS.

M. Kimura and T. Ohta estimate 1.8×10^9 years as the time of divergence of eukaryotes and prokaryotes through comparative studies of the nucleotide sequences of 5S rRNAs from humans, yeasts, and bacteria.

M. Rassoulzadegan and six colleagues isolate a recombinant DNA clone from the polyoma virus that immortalizes cultured fibroblast cells from rat embryos. They also show that only the amino-terminal portion of the protein encoded by the viral gene carries the immortalizing function.

R. F. Doolittle and six colleagues demonstrate that the simian sarcoma virus oncogene, *v-sis*, is derived from the gene encoding a platelet-derived growth factor.

E. Hafen, M. Levine, and W. J. Gehring work out a technique for *in situ* hybridization of labeled DNA probes to RNA transcripts in frozen tissue sections. They succeed in localizing transcripts of homeotic genes to specific regions of developing *Drosophila* embryos.

R. Mann, R. C. Mulligan, and D. Baltimore genetically engineer the Moloney murine leukemia virus so that it can be used safely as a vector in gene transfer experiments with mammalian hosts.

B. McClintock receives the Nobel Prize for her discovery of transposable genetic elements.

1984 D. C. Schwartz and C. R. Cantor show that pulsed field gradient electrophoresis can be used to separate DNA fragments as large as 2,000 kbp. This method overcomes the limitation of agarose gel electrophoresis, which can only separate molecules of much smaller sizes (50 kbp or less).

J. Gitschier and eight colleagues report the cloning of the gene encoding the antihemophilic factor in humans.

C. G. Sibley and J. E. Ahlquist show from DNA-DNA hybridization data that humans are more closely related to chimpanzees than to any other hominoid and estimate that the species diverged 5 or 6 million years ago.

R. F. Pohlman, N. V. Fedoroff, and J. Messing determine the nucleotide sequence of the maize transposable element *Activator*.

F. S. Collins and four colleagues identify mutations upstream of the gamma-globin gene that cause it to be switched on in adults.

W. McGinnis and six colleagues discover and name the conserved *homeobox* sequence in *Drosophila* homeotic genes, and they find that the mouse also contains genes that influence segmentation and possess homeoboxes.

J. C. W. Shepherd and four colleagues show that yeast mating-type regulatory proteins contain homeoboxes.

T. A. Bargiello and M. W. Young clone and sequence *period*, the first gene known to control a biological clock.

M. Davis and T. Mak identify and clone the genes for the T cell receptor.

N. K. Jerne, G. Köhler, and C. Milstein receive the Nobel Prize in Medicine for their contributions to immunology.

R. B. Merrifield is awarded a Nobel Prize in Chemistry for his work in automated peptide synthesis.

1985 J. R. Miller, A. D. McLachlan, and A. Klug report the isolation and characterization of a zinc finger protein from *Xenopus* oocytes. This protein binds to the 5S RNA gene and controls its transcription.

M. P. Williamson, T. F. Havel, and K. Wüthrich publish the first atomic resolution structure of a protein, a bull seminal proteinase inhibitor, using nuclear magnetic resonance spectroscopy.

The universal code theory has to be amended because codons that serve as termination signals according to the "universal" genetic code are found to encode amino acids in certain ciliates and bacteria. For example, in *Stylonychia lemnae* UAA and UGA encode glutamine (S. Horowitz and M. A. Gorowsky) and in *Mycoplasma capricolum* UGA encodes tryptophan (F. Yamao).

C. M. Newman, J. E. Cohen, and C. Kipnis demonstrate mathematically that the punctuated shifting equilibrium patterns of species formation seen in the fossil record are to be expected on traditional grounds and do not require special mechanisms to explain them.

C. W. Greider and E. H. Blackburn isolate a telomerase from *Tetrahymena pyriformis*.

O. Smithies and four colleagues report the successful insertion of DNA sequences into human tissue culture cells by homologous recombination at the beta-globin locus. This is an early example of transfection.

J. D. Boeke and three colleagues discover the first retroposons in *Saccharomyces*.

S. M. Mount and G. M. Rubin determine the complete nucleotide sequence of a *Drosophilia* copia element and conclude that it is a retroposon.

A. J. Jeffries, V. Wilson, and S. L. Thien develop the DNA fingerprint technique and point out its potential use in forensic science.

R. K. Saiki, K. B. Mullis, and five colleagues report the use of the polymerase chain reaction to allow enzymatic amplification *in vitro* of specific beta-hemoglobin gene fragments.

H. L. Carson concludes from a study of the ecological genetics of the Hawaiian Drosophilidae that the evolution of this group is driven by sexual selection which ensures the choice of mates with the greatest Darwinian fitness.

M. S. Brown and J. L. Goldstein receive the Nobel Prize for identifying the low-density lipoprotein receptor pathway and for demonstrating that familial hypercholesterolemia is a genetic defect in this pathway.

1986 M.-C. Shih, G. Lazar, and H. M. Goodman show that the nuclear genes that encode chloroplast glyceraldehyde-3-phosphate dehydrogenase of higher plants are direct descendants of the genes from the symbionts that gave rise to the chloroplast. Later during evolution, these genes were transferred from the chloroplast to the nuclear genome.

L. E. Hood and three colleagues invent the first automated DNA sequencer. As these machines are perfected, the sequencing of genomes becomes thousands of times faster, and the Human Genome Project is greatly facilitated.

T. C. James and S. C. R. Elgin identify HP1 (heterochromatin protein 1) in *Drosophila melanogaster*.

A. Tomlinson and D. F. Ready report the discovery of *sevenless*, a mutation in *Drosophila* that controls the developmental fate of a specific cell in the ommatidium.

A. G. Amit and three colleagues determine the three-dimensional structure of an antigen-antibody complex at a resolution of 2.8 Å.

F. Costantini, K. Chada, and J. Magram demonstrate that cloned normal beta-hemoglobin genes can be experimentally substituted for defective thalassemia genes in the mouse. They inject cloned normal genes into the fertilized thalassemic eggs. The mice that develop possess red blood cells that can synthesize normal beta-hemoglobin chains. These transgenic mice transmit this ability to their offspring.

J. Nathans, D. Thomas, and D. S. Hogness isolate and characterize the human visual pigment genes.

M. Noll and four colleagues identify a gene (*paired*) that encodes a protein with a DNA-binding site (the *paired* domain). This domain is later identified in mammalian regulatory proteins. Noll's group shows that regulatory genes often contain multiple conserved domains and suggests that genes which share one or more of these domains form networking families that program the early development of multicellular organisms.

R. Benne and five colleagues discover RNA editing in trypanosomes.

H. M. Ellis and H. R. Horvitz isolate genes in *Caenorhabditis elegans* that cause the programmed death of specific cells.

The complete nucleotide sequence and gene organization of the chromosomes from chloroplasts is determined for two plant species. In the case of the liverwort, *Marchantia polymorpha*, the genome contains 121 kilobase pairs (K. Ohyama and 12 colleagues), while the genome of tobacco, *Nicotiana tabacum*, contains 155 kilobase pairs (K. Shinozaki and 22 colleagues). Some chloroplast genes are found to contain introns.

E. U. Selker and three colleagues characterize the phenomenon of repeat-induced point mutation (RIP) in *Neurospora*.

V. F. Semeshin and five colleagues observe new bands and interbands at the site where a transposable element had inserted into a *Drosophila* polytene chromosome.

R. Levi-Montalcini and S. Cohen receive the Nobel Prize in Physiology for their studies on growth factors.

E. Ruska receives the Nobel Prize in Physics for designing the first electron microscope.

1987 M. R. Kuehn and four colleagues introduce a human gene into the mouse to allow its study in a convenient laboratory rodent. They employ a mutant allele of the gene encoding HPRT and use a retrovirus as a vector to insert it into cultured mouse embryonic germ cells. These are then implanted into mouse embryos to form chimeras. Strains of mice carrying the human gene are obtained from these chimeras.

C. Nüsslein-Volhard, H. G. Frohnhöfer, and R. Lehmann show that a small group of maternal effect genes exist in *Drosophila* that determine the polarized pattern of development of the embryo.

E. P. Hoffman, R. H. Brown, and L. M. Kunkel isolate dystrophin, the protein encoded by the musculardystrophy locus.

D. C. Wiley and five colleagues determine the three-dimensional structure of HLA-A2, a human class I histocompatibility molecule.

D. C. Page and eight colleagues clone a segment of the human Y chromosome that contains a gene which encodes a factor influencing testis differentiation. Within the Y chromosome fragment is a 1.2 kb ORF that appears to encode a zinc finger protein.

R. L. Cann, M. Stoneking and A. C. Wilson compare the extent of sequence divergence in the mtDNA of individuals belonging to geographically distinct human populations. They erect a genealogical tree that suggests that all mtDNAs can be traced back to a common African maternal ancestor.

C. J. O'Kane and W. J. Gehring successfully utilize enhancer traps to identify the positions in *Drosophila* embryos of elements that are functioning to activate the transcription of specific genes.

D. T. Burke, G. F. Carle, and M. V. Olson describe a technique for cloning large segments of exogenous DNA by means of yeast artificial chromosomes.

R. E. Dewey, D. H. Timothy, and C. S. Levings show that cytoplasmic male sterility in maize is due to a protein encoded by the mitochondrial genome.

K. H. Wolf, W. H. Li, and P. M. Sharp report for various plant species that the rates of nucleotide substitutions in chDNAs are on average five times slower than the rates shown for nuclear genes.

J. E. Anderson, M. Ptashne, and S. C. Harrison describe the three-dimensional structure of the lambda (λ) bacteriophage repressor-operator complex.

S. Tonegawa wins the Nobel Prize for his elucidation of the genetic mechanism that generates antibody diversity.

1988 W. Driever and C. Nüsslein-Volhard demonstrate that the *bicoid* gene encodes a protein that is distributed in an exponential concentration gradient along the anteroposterior axis of the embryo.

P. M. Macdonald and G. Struhl show that a 625-nucleotide segment in the trailer of a message encoded by the maternal polarity gene *bicoid* is responsible for the anterior localization of this mRNA in the *Drosophila* oocyte.

W. H. Landschulz, P. F. Johnson, and S. L. McKnight discover the leucine zipper and propose that it functions as a DNA binding site.

W. Herr and 10 colleagues discover a new DNA-binding domain (POU) encoded by a family of homeotic genes. Many POU genes are expressed only in the nervous system.

R. R. Brown and seven colleagues clone the human androgen receptor gene and show that mutations within it cause the hereditary androgen insensitivity syndrome.

D. C. Wallace and seven colleagues report that a human, maternally inherited disease, Leber's hereditary optic neuropathy, is caused by a mutation in mitochondrial DNA.

H. H. Kazazian and five colleagues discover two cases of hemophilia A due to insertions of truncated transposable elements. They subsequently isolate a complete transposable element that is a likely progenitor of one of these insertions. They show the transposon resides on chromosome 22 and that homologous elements occur in chimpanzee and gorilla at the same genomic location. This finding suggests that the element has been occupying the same chromosomal site since the evolutionary divergence of humans, chimpanzees, and gorillas 7 million years ago.

V. Sorsa publishes a two-volume monograph that reviews the encyclopedic literature concerning polytene chromosomes and presents electron microscope maps of *Drosophila* salivary-gland chromosomes.

The first U.S. patent is issued for a genetically altered animal. Harvard University receives the patent for "oncomice," developed by P. Leder and T. Stewart.

S. L. Mansour, K. R. Thomas, and M. R. Capecchi describe a general strategy for gene targeting in the laboratory mouse.

1989 W. Driever and C. Nüsslein-Volhard show that in *Drosophila* the protein encoded by the *bicoid* gene acts by switching on the *hunchback* segmentation gene.

B. Zink and R. Paro show by immunostaining that a protein encoded by the *Polycomb (Pc)* gene binds to a limited number of discrete sites along the *Drosophila* polytene chromosomes. The sites include the *Antennapedia* complex and the *bithorax* complex, which contain genes known to be repressed by *Pc*.

S. Field and O. Song develop the yeast two-hybrid system for identifying protein-protein interactions, based on the properties of the GAL4 protein of *S. cerevisiae*. This system is later modified by various lab groups as a screen to identify protein sequences encoded by genomic or cDNA libraries which interact with a known protein.

J. J. Brown and three colleagues determine the structure of the "Dotted" transposon of maize.

L. H. Hartwell and T. A. Weinert introduce the concept of controls called "checkpoints" that ensure the order of events in the cycle of cell division.

L.-C. Tsui and 24 colleagues identify the cystic fibrosis gene, predict the amino acid sequence of the protein it encodes, and determine the nature of its most common mutant allele.

M. Srivastava and four colleagues clone and sequence the cDNA for human nucleolin.

J. R. Williamson, M. K. Raghuraman, and T. R. Cech present the guanine quartet model of telomere structure.

D. B. Kaback, H. Y. Steensma, and P. De Jonge show that crossing over on the shortest chromosome of yeast is two times higher than the average for the whole genome. They conclude that this ensures that at least one crossover will occur in every bivalent, a necessity for proper segregation of the homlogs during the first meiotic division.

Y. Q. Qian and five colleagues show that the *Antennapedia* homeobox protein binds to DNA through a helix-turn-helix motif.

F. D. Hong and seven colleagues determine the structure of the retinoblastoma gene. The RB transcript is encoded in 27 exons dispersed over about 200 kilobase pairs of genomic DNA.

M. Horowitz and five colleagues determine the structure of the human glucocerebrosidase gene. They also sequence a nearby pseudogene. Mutations in the functional gene are the cause of Gaucher disease.

J. M. Bishop and H. E. Varmus receive the Nobel Prize in Medicine for their studies on the oncogenes of retroviruses.

T. R. Cech and S. Altman receive the Nobel Prize in Chemistry for their demonstration that certain RNAs have enzymatic functions.

1990 W. French Anderson describes the first successful example of human gene therapy. Lymphocytes from a four-year-old girl suffering from adenosine deaminase deficiency are grown in culture and later incubated with a retroviral vector carrying a normal gene that encodes the missing enzyme. The transformed cells are reinjected into the patient, where they multiply and correct the disease.

M. K. Bhattcharyya and four colleagues show that one of the mutations (*Wrinkled seed*) used by G. Mendel in his classic experiments is due to the insertion of a transposon in a gene encoding an enzyme that controls the starch content of pea embryos.

S. J. Baker and four colleagues show that the introduction of wild-type *p53* genes suppress the proliferation of human cancer cells.

R. Bookstein and four colleagues show that cells from some human prostate cancers contain mutated retinoblastoma genes and that the uncontrolled growth of these cells is suppressed when wild-type *RB* alleles are introduced into them.

B. Blum, N. Bakalara, and L. Simpson propose that RNA editing is performed by guide RNA molecules.

R. N. Van Gelder and five colleagues devise a method for amplifying RNA utilizing an RNA polymerase from bacteriophage T7.

B. G. Herrmann and four colleagues clone the *T* complex, which is required for the formation of mesoderm in the mouse.

F. Yamamoto and four colleagues work out the molecular basis of the ABO blood group system.

J. Malicki, K. Schughart, and W. McGinnis introduce a homeobox gene from the mouse into *Drosophila* embryos and observe that it can induce homeotic transformations similar to those produced by the *Antennapedia* gene. Therefore, genes from animals that have been evolving independently for hundreds of millions of years generate products that function interchangeably.

D. Malkin and 10 colleagues show that the defects underlying the Li-Fraumeni syndrome are mutations in the *p53* gene. Subsequent studies reveal that *p53* mutations are present in about half of all human cancers.

F. Barany invents the ligase chain reaction. This provides a rapid screening procedure for identifying mutations in selected DNA sequences.

X. Fang and three colleagues clone the gene in *Plasmodium vivax* that encodes its Duffy receptor.

H. Biessmann and six colleagues show that a specific retrotransposon can transpose to the broken ends of *Drosophila* chromosomes and "heal" them.

M. A. Oettinger and three colleagues identify *RAG-1* and *RAG-2*, genes whose products catalyze V(D)J recombination.

P. M. Kane and five colleagues discover protein splicing in yeast.

1991 S. M. Simon and G. Blobel demonstrate that translocons in the endoplasmic reticulum contain aqueous pores through which proteins manufactured on ribosomes pass from the cytoplasm to the ER lumen.

G. M. Preston and P. Agre isolate the cDNA for *aquaporin-1*.

B. G. Baldwin and three colleagues compare the chloroplast DNAs of Hawaiian silverswords with those of North American tarweeds and conclude that species from two genera of California tarweeds are the closest living relatives of the Hawaiian silverswords.

L. Buck and R. Axel report the cloning and characterization of 18 different genes from a multigene family of about a thousand genes which encode odorant receptors in the rat. This is the first report in which odorant receptors are described and molecularly characterized in any species.

M. L. Sogin proposes that the ancestor of eukaryotes was a chimera formed by the fusion of prokaryotes with complementing metabolic capabilities.

M. A. Houck and three colleagues suggest that mites may transfer P elements between *Drosophila* species.

D. A. Wheeler, J. C. Hall, and five colleagues succeed in introducing cloned *Drosophila simulans period* genes into the genomes of *D. melanogaster* carrying inactive *per* alleles. Transduced males "sing" the *simulans*' song.

J. W. Ijdo and four colleagues identify specific nucleotide sequences in band q13 of human chromosome 2, which mark the site of telomere-telomere fusions that converted two rod-shaped ancestral chromosomes into the V-shaped chromosome 2 of modern humans. This resulted in a reduction in the number of chromosome pairs to 23 from the 24 pairs characteristic of chimpanzees, gorillas, and orangutans.

A. J. M. H. Verkerk and 20 colleagues identify the *FMR-1* gene at the fragile site of the human X chromosome and demonstrate that the gene contains an expanded CGG triplet in patients suffering from fragile X–associated mental retardation.

D. R. Knighton and six colleagues determine the three-dimensional structure of the catalytic core shared by all known eukaryotic protein kinases.

R. R. Ernst is awarded the Nobel Prize in Chemistry for his contributions to the development of high-resolution nuclear magnetic resonance (NMR) spectroscopy.

1992 G. G. Oliver and 146 colleagues from a consortium of 35 European laboratories publish the first complete nucleotide sequence for a eukaryotic chromosome. Chromosome III of *Saccharomyces cerevisiae* is the third smallest. It is 315,357 bp long and contains 182 ORFs, of which 117 (80%) show no significant homology to any previously sequenced yeast genes.

R. M. Story, I. T. Weber, and T. A. Steitz determine the three-dimensional structure of RecA, a protein playing a central role in crossing over and DNA repair of *E. coli.*

M. C. Rivera and J. A. Lake make phylogenetic studies of the translation elongation factors isolated from various prokaryotes and eukaryotes. Comparisons of amino acid sequences identify a subgroup of archaeons as the immediate relatives of eukaryotes and therefore the source of the nucleus.

D. Haig proposes the theory of parent–offspring conflict to explain the evolution of parental imprinting.

E. G. Krebs and E. H. Fischer receive the Nobel Prize for discovering protein kinases and elucidating their roles in signal transduction.

1993 M. C. Mullins and C. Nüsslein-Volhard generate hundreds of developmental mutants in the zebra fish, opening a new era in study of the genetic control of vertebrate development.

D. R. Rosen, T. Siddique, and 32 colleagues identify 11 different ALS mutations in 13 families. The mutations are all in the gene that encodes the SOD enzyme.

R. Hallick and seven colleagues determine the complete nucleotide sequence for the DNA of chloroplasts from *Euglena gracilis*. Some chloroplast genes contain twintrons.

A. Chaudhuri and five colleagues clone the gene for the Duffy blood group factor. It encodes a 338 amino acid protein that is attached to the erythrocyte plasma membrane and is required for the invasion of certain malaria parasites.

S. L. Baldauf and J. D. Palmer conclude from a phylogenetic study of combined sequence data from certain ubiquitous proteins that animals and fungi are each other's closest relatives. Therefore both animals and fungi have been placed in a monophyletic supergroup, the Opisthokonta.

G. Maroni publishes the first atlas of the comparative morphology of the genes of a specific eukaryote. The monograph illustrates 90 *Drosophila* genes that transcribe mRNAs ranging in size from 319 to 4,749 base pairs.

C. Pisano, S. Bonaccorsi, and M. Gatti report that a protein which is not encoded by Y-linked genes binds to a specific, giant, lampbrush loop on the Y chromosome in *Drosophila* spermatocytes. This protein is a component of the sperm tail. They suggest that the Y loops in spermatocytes bind exogenous specific proteins and facilitate their assembly into axonemes.

L. Pereira and six colleagues determine the organization of the *FBNI* gene. This encodes fibrillin, and mutations in it cause Marfan syndrome.

M. E. MacDonald and 56 colleagues belonging to the Huntington's Disease Research Group clone and sequence the Huntington disease gene and show that an unstable trinucleotide repeat is expanded in victims of the disease.

J. A. Tabcharani and six colleagues demonstrate that the cystic fibrosis transmembrane conductance regulator functions as a channel capable of conducting multiple anions. They show that positively charged amino acids in the sixth membrane helix of the transmembrane domains of the CFTR protein are required for halide transport.

R. J. Roberts and P. A. Sharp receive the Nobel Prize in Medicine for discovering split genes.

M. Smith and K. B. Mullis receive the Nobel Prize in Chemistry for inventing the site-directed mutagenesis technique and the polymerase chain reaction, respectively.

1994 N. Morral and 30 colleagues from 19 European laboratories study the microsatellites associated with the ΔF508 mutation in the cystic fibrosis genes of CF families from various parts of Europe. They conclude that the mutation originated in southwestern Europe at least 50,000 years ago.

D. E. Nilsson and S. Pelger show from a computer simulation that an organ similar in complexity to a fish eye can evolve from a patch of skin containing photosensitive cells in a relatively short time (~400,000 generations).

M. Chalfie and three colleagues demonstrate that green fluorescent proteins can be used to visualize sites in cells where specific genes are being expressed.

P. Gill and eight colleagues identify the remains of the Romanov family by DNA analysis.

S. E. Gabriel and four colleagues find a positive correlation between the amount of cystic fibrosis conductance regulator protein in intestinal cells and the amount of fluid secretion induced by cholera toxin. They propose that cystic fibrosis heterozygotes are resistant to cholera, and this selective advantage is responsible for the high frequency of the gene in human populations.

W. C. Orr and R. S. Sohal construct transgenic lines of *Drosophila* bearing extra copies of catalase and superoxide dismutase genes. The aging process is slowed in these flies.

N. W. Kim and nine colleagues develop a sensitive assay for telomerase activity. Using it they show that human somatic cells from differentiated tissues lack telomerase activity, whereas cells from a variety of cancers contain active telomerases. Normal ovaries and testes also were positive for telomerase activity.

Y. Chikashige and six colleagues observe the movement of chromosomes during meiotic prophase in *Schizosaccharomyces pombe* by fluorescence microscopy. They report that the telomeres group together and assume a leading position during movement of the chromosomes.

T. Tully and eight colleagues isolate genes that control the formation of memory in *Drosophila*.

Y. Zhang and five colleagues clone the *obese* gene of the mouse and determine its structure. The product appears to be a secretory protein that controls the size of the body fat depot.

R. J. Bollag and five colleagues demonstrate that the *T* genes of the mouse encode a protein motif (the T box), which binds to DNA. This T box also occurs in genes with critical roles in the development of amphibians, fishes, and insects.

S. Whitham and five colleagues use the maize *Activator* transposable element to tag and clone a disease-resistance gene in tobacco.

Y. Miki and 44 colleagues identify *BRCA1*, a human anti-oncogene that, when mutated, confers susceptibility to breast and ovarian cancer.

D. Arendt and K. Nübler-Jung provide support for the Saint-Hilaire hypothesis of 1822. Their comparative studies of the expression of homologous genes that control the dorsoventral patterning of embryonic cells show that in flies and mice these genes have opposite effects. Dorsalization genes in *Drosophila* cause ventralization in *Mus*, whereas genes that cause ventralization in flies specify dorsal patterns in mice.

J. B. Clark, W. P. Maddison, and M. G. Kidwell report phylogenetic studies that show horizontal transfer of P elements has occurred at least twice in the genus *Drosophila*.

M. Rodbell and A. G. Gilman receive a Nobel Prize for discovering G proteins and elucidating their role in cellular signal transduction.

1995 G. Halder, P. Callaerts, and W. J. Gehring demonstrate in *Drosophila melanogaster* that the gene *eyeless* is a master control gene for eye morphogenesis.

M. Schena and three colleagues used DNA microarray technology to simultaneously monitor the expression of 45 different genes in *Arabidopsis*. The microarrays were prepared by high-speed robotic printing of cDNAs on glass.

C. Wilson and J. W. Szostak report *in vitro* evolution experiments which generate RNAs that can catalyze self-alkylation reactions.

J. Hughes and seven colleagues publish the sequence of the 4,320 amino acids in polycystin, the product of the *PKD1* gene. Mutations in this gene cause polycystic kidney disease in humans.

J. Feng and 15 colleagues induce senescence in HeLa cells by adding an antisense RNA that contains a message opposite to the templating domain of human telomerase.

S. Baxendale and 10 colleagues compare human and puffer fish Huntington disease genes and show that the human gene is over seven times larger because its introns are larger, not its exons.

R. Wooster and 40 colleagues identify *BRCA2*.

R. D. Fleischmann, J. C. Venter, and 38 colleagues publish the first complete nucleotide sequence of a free-living organism (*Haemophilus influenzae*). A few months later, C. M. Fraser, J. C. Venter, and 27 colleagues publish the complete nucleotide sequence of *Mycoplasma genitalium*.

R. Sherrington, P. H. St. George-Hyslop, and 31 colleagues isolate and characterize a gene on chromosome 14 which is responsible for 80% of the cases of early-onset, familial Alzheimer's disease. Two months later, G. D. Shellenberg and 21 colleagues report locating a gene on chromosome 1 that encodes a protein showing great similarities in amino acid sequence to the product of the *AD* gene on chromosome 14. Mutations in the *AD* gene on chromosome 1 are responsible for the other 20% of the cases of early-onset, familial *AD*. The products of these genes are called *presenilin 1* and *2*.

S. Labeit and B. Kolmer clone the cDNA for cardiac titin. This is the largest protein known, some 50 times the size of average proteins.

K. Zhao, C. M. Hart, and U. K. Laemmli purify a protein from *Drosophila* that binds to insulator DNAs and demonstrate by immunostaining that this protein attaches to hundreds of interbands and many puff boundaries on polytene chromosomes.

S. Horai and four colleagues compare the nucleotide sequences for the entire mitochondrial genomes of three individual women (Japanese, European, and African) and females belonging to four species of apes. The analysis supports the theory that all human mtDNA molecules are derived from a woman who lived in Africa about 140,000 years ago.

L. A. Tartaglia and 18 colleagues identify a gene, *OB-R*, that encodes a leptin receptor and show that the mRNA for this membrane-bound protein is transcribed in the hypothalamus.

M. Moritz, Y. Zheng, B. Alberts, and five colleagues identify gamma-tubulin-containing ring complexes in centrosomes and show that they function as microtubule nucleating sites.

E. B. Lewis, E. Wieschaus, and C. Nüsslein-Volhard receive the Nobel Prize in Medicine for their analyses of the genetic mechanisms that control cell differentiation during embryogenesis and metamorphosis in *Drosophila*.

1996 G. D. Penny and four colleagues use gene targeting to demonstrate that in order for an X chromosome to undergo inactivation, the *Xist* gene on that X must be transcriptionally active.

B. Lemaitre and four colleagues elucidate the genetic control in *Drosophila* of the synthesis of different antimicrobial peptides in response to bacterial or fungal infections.

C. Bult and 39 colleagues show that most of the genes making up the genome of the archaeon *Methancoccus jannaschii* have no equivalent in other organisms.

J. Dubnau and G. Struhl, as well as R. Rivera-Pomar and four colleagues, show that a homeobox protein can control translation by binding to discrete target sequences on specific mRNAs.

J. G. Lawrence and J. R. Roth propose the selfish operon model to explain the evolution of gene clusters in bacteria.

M. Lewis and seven colleagues determine the crystalline structure of lactose operon repressor proteins complexed with operator DNA or inducer.

A. Goffeau and 15 colleagues publish "Life with 6,000 Genes," a review of the structure of the yeast genome. The complete nucleotide sequences for all 16 chromosomes of *Saccharomyces cerevisiae* took the combined labors of 600 scientists in North America, Europe, and Japan and is the first genome available for a eukaryote.

E. Spanopoulou and five colleagues show that the protein encoded by the *RAG-1* gene contains a homeobox through which it binds to lymphocyte DNA during V(D)J recombination. They point out that *RAG-1/RAG-2* complexes behave like the transposases of nematodes.

G. Burger and three colleagues conclude from a study of the comparative structure of mitochondrial ribosomal proteins that the mitochondria in all eukaryotes have a monophyletic origin.

T. Kaneko and 23 colleagues completely sequence the genome of the cyanobacterium *Synechocystis* and determine the position of over 3,000 ORFs. Many of these genes are later identified in the chloroplasts of photosynthetic protoctists and land plants.

B. A. Krizek and E. M. Meyerowitz present a model that explains the transformations brought about by homeotic mutations during the development of the *Arabidopsis* flower.

1997 F. R. Blattner and 16 colleagues sequence the genome of *Escherichia coli* and begin to assign functions to its genetic elements.

F. Kunst and 150 collaborators publish the complete nucleotide sequence for the genome of *Bacillus subtilis* and describe its genetic organization.

H-P. Klenk and 50 colleagues determine the genome structure of *Archaeoglobus fulgidus* and compare it to *Methanococcus jannashii*, the other archaeon for which sequence data are available. There are surprising qualitative differences. Abundant genes that allow protein splicing occur in *Methanococcus*, but none occur in *Archaeoglobus*.

I. Wilmut and four colleagues report the successful cloning of a mammal. The sheep Dolly has chromosomes derived from a cell of the udder of a pregnant female. Once mature, Dolly gave birth to a healthy lamb. This offspring, however, was the result of a normal mating and gestation.

M. Krings and five colleagues are able to isolate and sequence segments of mtDNA from bones of Neandertal fossils and compare them to homologous segments from modern humans. They conclude that Neandertals constitute a species distinct from *Homo sapiens*.

J. Summerton and D. Weller describe the design, preparation, properties, and potential uses of Morpholinos, synthetic polymers with antisense characteristics.

C. F. Fraser and 34 colleagues determine the genomic sequence of the Lyme disease spirochaete, *Borrelia burgdorferi*. It has a main chromosome that is linear, as are some of its plasmids.

F. Yang and three colleagues use chromosome-specific paint probes to demonstrate that the Indian muntjak (2N = 6) has a reduced chromosome number because of the end-to-end fusion of different chromosomes. For example, its chromosome 3 is an assemblage of seven chromosomes present in the Chinese muntjak (2N = 46).

S. Keeney, C. N. Giroux, and N. Kleckner show Spo11 to be a DNA double-strand-break-producing topoisomerase that is responsible for meiotic crossing-over in yeast.

J. C. Lawrence and H. Ochman suggest that most bacteria contain mosaic genomes and show that in a species like *E. coli*, 15–30% of the genome is made up of DNA sequences contributed from other species by horizontal mobile elements.

J. A. Yoder, C. P. Walsh, and T. H. Bestor suggest that DNA methylation is a mechanism that evolved to suppress the effects of selfish DNA.

D. H. Skuse and nine colleagues present evidence from girls with Turner syndrome that the X chromosome contains imprinted genes that affect their social behaviors.

P. D. Boyer and J. E. Walker share the Nobel Prize in Chemistry for their contributions to the enzymology of ATP synthase.

S. B. Prusiner is awarded the Nobel Prize in medicine for his elucidation of the molecular structure of prions.

1998 S. T. Cole and 41 coworkers sequence the DNA of *Mycobacterium tuberculosis* and work out its genetic structure.

S. G. Anderson and nine colleagues determine the nucleotide sequence of the *Rickettsia prowazeki* genome and conclude that this parasitic bacterium has undergone reductive evolution. They also stress the similarities between the 16S RNAs of mitochondria and these bacteria.

The C. *elegans* Sequencing Consortium, made up of 407 scientists associated with the Sanger Centre in Cambridge, England, and the Washington University of Medicine in St. Louis, Missouri, USA, determine the nucleotide sequence and gene organization for the first multicellular species, the nematode *Caenorhabiditis*.

R. W. Frenck, E. H. Blackburn, and K. M. Shannon show for peripheral leukocytes that telomeres shorten as humans age. However, the rate of loss of telomeric repeats is most rapid during the first four years of life and more gradual during the period from 25 to 80.

Y. J. Lin, L. Serounde, and S. Benzer isolate *methuselah*, a gene that extends the life span of *Drosophila*, and they subsequently determine the properties of the protein it encodes.

M. Lyon proposes that DNA sequences on mammalian X chromosomes, called *LINE-1 elements*, may interact with *XIST* RNA to facilitate gene silencing by helping this RNA spread along the chromosome.

A. Fire and five colleagues show that the injection of double-stranded RNA into *Caenorhabditis* silences specific genes. They call this phenomenon "RNA interference."

R. S. Stephens and 11 colleagues sequence the genome of *Chlamydia trachomatis* and investigate its genetic structure. They conclude that the species contains many genes with phylogenetic origins from eukaryotes.

E. S. Belyaeva and five colleagues discover a gene that controls the underreplication of heterochromatin in polytene chromosomes of *Drosophila*.

J. G. Gall and C. Murphy show that demembranated *Xenopus* sperm heads, when injected into oocyte nuclei from *Xenopus* or *Notophthalmus*, swell and liberate their chromosomes, which then take on a transcriptionally active lampbrush morphology.

W. B. Whitman, D. C. Coleman, and W. J. Wiebe estimate the total number of prokaryotes on earth to be $4\text{–}6 \times 10^{30}$ cells. Their calculations show that prokaryotes are the largest living reservoir of C, N, and P.

1999 K. Petren, B. R. Grant, and P. R. Grant work out the phylogeny of Darwin's finches based on microsatellite DNA length variations among the related Galapagos species.

T. Galitski and four colleagues demonstrate that most genes in *Saccharomyces cerevisae* continue to be expressed at the same relative levels in yeast strains of different ploidies (1N, 2N, 3N, 4N). However, a small subset of genes exists whose transcription rates are dramatically induced or repressed as ploidy levels rise.

J. D. Evans and D. E. Wheeler demonstrate that, during the larval differentiation of genetically identical female honeybees into worker or queen castes, different sets of specific genes are switched on or off.

K. E. Nelson and 28 colleagues determine the genome sequence of *Thermotoga maritima*. They conclude that although this hyperthermophile belongs to the bacteria, it has acquired a significant portion of its genome by horizontal transmission of genes from archaeons.

R. M. Andrews and five colleagues resequence human mtDNAs. They analyze the original DNA sample used by Anderson *et al.* (1981) and also mtDNA from HeLa cells. They uncover several errors and suggest some simple revisions to correct the Cambridge reference sequence and clarify its position in mtDNA lineages.

G. P. Copenhaver and 13 colleagues analyze the centromeres of *Arabidopsis thaliana* at the nucleotide level and show that they contain genes capable of transcription.

M. J. Beaton and T. Cavalier-Smith show for a group of cryptomonad species that differ in cell volumes by a 10-fold factor that nuclear and nucleomorph genome sizes obey different scaling laws. Nuclei in larger cells have more DNA, but nucleomorphs do not. This finding supports the hypothesis that noncoding DNA has a skeletal function in eukaryotic nuclei.

O. White and 31 colleagues determine the genome sequence and the genetic organization of the radioresistant bacterium *Deinococcus radiodurans* R1.

I. Dunham and 216 colleagues are the first to sequence a human chromosome. They show that the smallest chromosome (number 22) contains 545 genes spread along a 33.4 mb molecule of DNA.

The MHC Sequencing Consortium (consisting of 28 contributors from eight international centers) publishes a map of the gene loci in the human major histocompatibility complex.

J. G. Gall and three colleagues demonstrate that many proteins and RNAs that function in the synthesis and posttranscriptional processing of RNAs are assembled in *Cajal bodies.*

G. Blobel receives the Nobel Prize for deciphering the method used by cells to target newly synthesized proteins to the endoplasmic reticulum or other organelles.

2000 P. Underhill and an international group of 20 colleagues publish a study of paternal lines of evolutionary descent that they traced by following markers in the DNA of Y chromosomes from humans belonging to ethnic groups from different parts of the world.

W. V. Ng and 42 colleagues determine the genome sequence and genetic organization of the archaeon *Halobacterium species NRC1.*

A collaborative, international group of 152 scientists that called itself the *Arabidopsis Genome Initiative* publishes a genome sequence for the first plant, *Arabidopsis thaliana.* They conclude that about 70% of its 25,500 genes are duplicated and that the actual number of different genes is less than 15,000.

C. Lemieux, C. Otis, and M. Turmel sequence the genome of the chloroplast of the green alga *Mesostigma viride.* They conclude that this chDNA shows an organization that predates the split between chlorophytes and green plants about 800 million years ago.

A. C. Bell and G. Felsenfeld show that CTCF, a DNA-binding protein, serves to insulate an imprintable gene (*Igf2*) from its enhancer.

M. Hattori and 63 colleagues determine the nucleotide sequence of human chromosome 21 and show that it contains only 40% as many genes as chromosome 22, which is similar in size.

F. Catteruccia and six colleagues develop a method for introducing foreign genes into malaria mosquitoes that utilizes the Minos transposon.

D. R. Davies and three colleagues report the three-dimensional structure of Tn5 transposase complexed with the recognition sequences that terminate the transposon, and they propose a mechanism for transposition that involves a synaptic complex, transposition intermediate that is hairpin shaped.

M. D. Adams and 189 colleagues publish the genome sequence of *Drosophila melanogaster* euchromatin. They estimate that this genome contains 13,600 structural genes.

G. M. Rubin and 54 colleagues publish an analysis of the comparative genomics of yeast, worm, fly, and human that is encyclopedic in scope. They find, for example, that at least 30% of the structural genes of *Drosophila* have orthologs in *Caenorhabiditis*. Of 289 genes associated with human diseases, 61% have an ortholog in *Drosophila*.

N. G. Jablonski and G. Chaplin propose a comprehensive theory to explain the variation in human skin color.

A. G. Fraser and five colleagues use RNA interference to assign phenotypes to 90% of the genes on chromosome 1 of *Caenorhabditis elegans*. This technique enables them to increase the number of sequenced genes with known phenotypes from 70 to 378.

J. F. Heidelberg and 31 colleagues determine the DNA sequences and the organization of genes within the two chromosomes of the cholera bacterium, *Vibrio cholerae*.

M. F. Hammer and eleven colleagues show that Jewish and Middle Eastern non-Jewish populations share a common pool of Y-chromosome markers.

B. Korber and nine colleagues present evidence that HIV-1 emerged from chimpanzees about 70 years ago.

R. S. Singh and R. J. Kulathinal propose a sex gene pool theory of speciation applicable to all higher sexual organisms. R. S. Singh then links it to a speciation model that also includes asexual organisms and plants.

E. Kandel receives the Nobel Prize for his contributions to the molecular definition of long-term memory.

2001 M. M. Yusupov and six colleagues determine the three dimensional structure of the bacterial 70S ribosome at 5.5Å resolution. The ribosomes are from a species of thermophilic bacteria.

J. J. Ferretti and 22 colleagues sequence the genome of *Streptococcus pyogenes* and illustrate its genetic structure. They determine the location of the 40 different genes associated with virulence.

H. Tettelin and 38 colleagues sequence the genome of *Streptococcus pneumoniae* and assign functions to many of the coding sequences contained in its DNA molecule.

R. Pawliuk and 13 colleagues use a genetically engineered lentivirus as an RNA vector to integrate a normal human beta hemoglobin gene into the chromosome of a host. This is a mouse genetically engineered to have defects in beta-chain synthesis. Normal genes are integrated into hematopoetic stem cell chromosomes, and the transgenic mouse synthesizes normal levels of hemoglobin.

D. W. Wood, E. W. Nester, and 49 colleagues describe the organization of the genome of *Agrobacterium tumefaciens* C58.

C. S. L. Lai and four colleagues isolate and characterize *FOXP2*, a gene required for the development of language in children.

M. Ridanpaa and twelve colleagues show that a hereditary disease of humans *cartilage-hair hypoplasia (CHH)* is caused by mutations in the *RMRP* gene. This is the first untranslated nuclear gene that upon mutation is found to cause a human disease.

V. V. Kapitonov and J. Jura use an *in silico* analysis to identify rolling-circle transposons.

O. Masden and nine colleagues and W. J. Murphy and five colleagues use gene sequence data to construct phylogenies for placental mammals. The resulting phylogenetic trees differ substantially from the traditional ones developed from comparative anatomy and fossil data.

The International Human Genome Sequencing Consortium, led by F. S. Collins and consisting of hundreds of scientists from around the world and J. C. Venter and colleagues at Celera Genomics independently determine the draft sequence of the euchromatic portion of the human genome. Annotated maps of the human genome are published in *Nature* vol. 409 (Feb. 15 issue) and *Science* vol. 291 (Feb. 16 issue).

L. H. Hartwell, R. T. Hunt, and P. M. Nurse receive the Nobel Prize in Physiology or Medicine for their discoveries of key chemicals that regulate the cell division cycle.

2002 M. J. Gardner and 44 other members of an international consortium successfully sequence the genome of *Plasmodium falciparum*, the protozoan causing subtertian malaria, the most dangerous form of the disease.

R. H. Holt and 122 other members of an international consortium successfully sequence the genome of *Anopheles gambiae*, a principal vector of malaria, and analyze the functions of many of the genes uncovered.

K. Kondo and four colleagues document the first case of horizontal gene transfer between a specific prokaryote (the bacterium *Wolbachia*) and a specific eukaryote (the beetle *Callosobruchus*).

P. Dehal and 86 colleagues generate a draft sequence of the genome of the ascidian *Ciona intestinalis*. This was the first Urochordate to have its genome sequenced.

X. Huang and three colleagues develop a technique for identifying amplified and overexpressed genes in the chromosomes of cultured human cancer cells. The first gene identified by this technique is *TAOS*.

Y. Matsuoka and five colleagues identify the oldest surviving teosinte ancestor of corn and conclude that highland farmers started its domestication in southern Mexico about 9,000 years ago.

S. Aparicio and 40 colleagues present the draft sequence and initial analysis of the genome of *Takifugu rubripes*.

The Nobel Prize in Medicine is awarded jointly to Sydney Brenner, H. Robert Horvitz, and John E. Sulston for their work concerning genetic regulation of organ development and programmed cell death.

K. Wüthrich shares the Nobel Prize in Chemistry with J. B. Fenn and K. Tanaka for his contributions to the development of nuclear magnetic resonance spectroscopy as a tool for determining the three-dimensional structure of biological molecules.

2003 H. T. Skaletsky and 39 colleagues sequence the male-specific region of the human Y chromosome and describe the organization of its heterochromatic and euchromatic elements.

K. Si, S. Lindquist, and E. R. Kandel discover that a neuronal CPEB protein from *Aplysia*, which regulates protein synthesis at activated synapses, alters its form and behaves like a prion in its biologically active state. They propose that conversion to the prion-like state plays a role in the maintenance of synaptic changes that allow long-term memory storage.

M. W. Nachman, H. E. Hoekstra, and S. L. D'Agostino elucidate the molecular genetic mechanisms for adaptive melanism in a desert mouse *Chaetodipus intermedius*. Mutations in a gene that controls the syntheses of yellow or black melanins produce coat colors that have been selected to provide camouflage for the mice that live in dark- *vs* light-colored natural environments.

J. E. Galagan and 76 colleagues publish a draft sequence for the genome of *Neurospora* and analyze the structure and functioning of the genetic system revealed.

T. Anzai and 21 colleagues compare the nucleotide sequences in homologous 1.75 mbp stretches of DNA from humans and chimpanzees. The segments contain the major histocompatibility complexes. Unexpectedly, the majority of the evolutionary sequence divergence between the two primates is found to be due not to single base substitutions, but to insertions and deletions (indels).

The International Human Genome Sequencing Consortium announces the successful completion of the Human Genome Project, nearly 2 years ahead of schedule. The euchromatic portion of the human genome is completely sequenced with an error rate of less than 1 per 10,000 bases.

P. Agre receives the Nobel Prize in chemistry for discovering aquaporins, the channels that facilitate the movement of water molecules through cell membranes.

2004 G. W. Tyson and nine colleagues use community genome sequencing to determine the genomes and metabolic interrelations of archaean species flourishing in a toxic acid pool at the bottom of a mine shaft.

J. C. Venter and 22 colleagues use community genome sequencing on the prokaryotes filtered out of a 1,500 liter sample of surface water from the Sargasso Sea. They sequence 1.05 billion base pairs of DNA representing 1.2 million genes from an unknown number of marine species. Approximately 800 of the genes encode light sensitive proteins.

M. Matsuzaki and 41 colleagues sequence the genome of *Cyanidioschyzon merolae* and determine its functional morphology. This red alga has the smallest genome of all photosynthetic eukaryotes.

G. Rice and eight colleagues study the tertiary and quaternary structures of the coat proteins of certain dsDNA viruses that attack species of Archaea, Bacteria and Eukaryotes. They demonstrate conformational similarities in these proteins and conclude that these viruses evolved from a common ancestor that lived prior to the formation of the three domains of cellular life.

E. Birney, M. Clamp, and R. Durbin publish the algorithms GeneWise and Genomewise which are widely used in studies of comparative evolutionary genomics.

S. Simonsson and J. G. Gurdon show that DNA demethylation is necessary for the nuclear reprogramming of *Xenopus* somatic cells.

R. Axel and L. Buck share the Nobel Prize in Medicine for their discoveries of odorant receptors and the molecular basis for odor recognition.

The Nobel Prize in Chemistry is awarded to A. Ciechanover, A. Hershko, and I. Rose for their discovery of ubiquitin-mediated protein degradation.

2005 M. T. Ross and 284 colleagues (representing 21 institutions in 6 countries) publish the DNA sequence of the human X chromosome and a discussion of its organization and evolution.

L. Eichinger together with an international group of 95 scientists sequence the genome of the slime mold *Dictyostelium discoideum*, analyze its repertoire of genes and proteins, and determine its position in the evolutionary tree of eukaryotes.

S. Richards and 51 colleagues sequence the genome of *Drosophila pseudoobscura* and compare the order of its genes to that of *Drosophila melanogaster*. They identify a core set of about 10,000 genes that have been conserved since the divergence of the two species 25–50 million years ago.

Three international teams of scientists sequence and compare the genomes of three trypanosomatid parasites, which cause major diseases in millions of people in many regions of the developing world. M. Berriman *et al.* sequence and analyze the genome of *Trypanosoma brucei*, N. M. El-Sayed *et al.* of *Trypanosoma cruzi*, and A. C. Ivens *et al.* of *Leishmania major*. These studies provide insights into the biology of these parasites and information crucial for the development of new therapeutic drugs.

R. L. Lamason and 24 colleagues show that *golden*, a pigmentation mutation first observed in the zebrafish, identifies *SLC 24A5*, the gene responsible for the skin color differences between African and European human populations.

Scientists Listed in the Chronology

Here we arrange alphabetically the names of scientists credited with various discoveries and the years key papers by them were published. In situations where a research contribution made at a later time is referred to in an earlier entry, both years are given. Thus, Benda, 1890 (1898) indicates that his 1898 publication is referred to in an 1890 entry (on Altmann's bioblasts). If two or more publications by the same author are cited for the same year, the number of entries appears in parentheses, for example, Jacob, F., . . . 1961(3). The list also contains an entry for Nobel Prizes. Here one can find the years when awards were given to geneticists and molecular biologists.

Beaton, M. J., 1999
Beatty, B. R., 1969
Beckwith, J. R., 1969
Beermann, W., 1952, 1961
Beijerinck, M. W., 1898
Bell, A. C., 2000
Belling, J., 1920, 1927
Belyaeva, E. S., 1998
Benacerraf, B., 1963, 1972, 1980
Benda, C., 1890 (1898)
Bender, W., 1983
Benne, R., 1986
Bennett, J. C., 1965
Benton, W. D., 1975
Benzer, S., 1955, 1961, 1969, 1971, 1998
Berg, P., 1972, 1980
Berget, S., 1977
Bernstein, F., 1925
Berriman, M., 2005
Berson, S. A., 1957
Bertram, E. G., 1949
Berzelius, J. J., 1838
Best, C. H., 1921
Bestor, T. H., 1997
Beutler, E., 1962
Bhattacharyya, M. K., 1990
Bickel, H., 1954
Biessmann, H., 1990
Billingham, R. E., 1953
Bingham, P. M., 1982
Birnbaumer, L., 1970
Birney, E., 2004
Birnstiel, M. L., 1966, 1967, 1969
Bishop, J. M., 1981, 1989
Bittner, J. J., 1936
Blackburn, E. H., 1978, 1985, 1998
Blake, C. C. F., 1967
Blakeslee, A. F., 1904, 1920, 1922, 1930, 1934, 1937
Blattner, F. R., 1997
Blobel, G., 1975, 1991, 1999
Blum, B., 1990
Boeke, J. D., 1985
Boivin, A., 1948
Bollag, R. J., 1994
Bolton, E. T., 1963
Bonaccorsi, S., 1993
Bonner, D., 1944
Bonneville, M. A., 1963
Bookstein, R., 1990
Boon, C., 1969
Botstein, D., 1980
Boveri, T., 1887, 1888, 1902, 1914
Boyce, R. P., 1964
Boycott, A. E., 1923
Boyer, H. W., 1972, 1976
Boyer, P. D., 1973, 1997
Brachet, J., 1941
Bradshaw, A. H., 1952
Bragg, W. H., 1913, 1915
Bragg, W. L., 1913, 1915
Breathnach, R., 1977
Brenner, S., 1959, 1961(2), 1963, 1965, 1974, 2002
Brent, L., 1953
Brewin, N. J., 1981
Bridges, C. B., 1914, 1915, 1917, 1921, 1923, 1925, 1935
Briggs, R., 1952
Britten, R. J., 1959, 1966, 1968
Brown, D. D., 1964, 1968, 1972(2)
Brown, D. M., 1952
Brown, J. J., 1989
Brown, M. S., 1975, 1985
Brown, R., 1831
Brown, R. H., 1987
Brown, T. R., 1988
Brown, W. L., 1956
Brownlee, G. G., 1965, 1977
Buck, L., 1991, 2004
Bult, C., 1996
Burger, G., 1996
Burgess, R., 1969
Burke, D. T., 1987
Burnet, F. M., 1959, 1960
Burny, A., 1964 (1969)
Burrell, B. G., 1976
Butenandt, A., 1935
Bütschli, O., 1876

Cairns, J., 1963
Callaerts, P., 1995
Callan, H. G., 1958
Camerarius, J. R., 1694
Cameron, J. R., 1979
Campbell, A. M., 1962
Cann, R. L., 1987
Cantor, C. R., 1984
Capecchi, M. R., 1980, 1988
Cappecchi, M., 1966
Carbon, J. A., 1980
Carle, G. F., 1987
Carpenter, A. T. C., 1975
Carr, D. H., 1965
Carr, N. G., 1972
Carrier, W. L., 1964
Carson, H. L., 1985
Caspar, D. L. D., 1959
Caspersson, T., 1936, 1941, 1970
Castle, W. E., 1905 (1910)
Catteruccia, F., 2000
Cavalier-Smith, T., 1999
Cech, T. R., 1981, 1989(2)
C. *elegans* Sequencing Consortium, 1998
Chada, K., 1986
Chain, E., 1940, 1945
Chakrabarty, A., 1980
Chalfie, M., 1981, 1994
Chambon, P., 1977
Chang, A. C. Y., 1972
Chaplin, G., 2000

Grunberg-Manago, M., 1955
Grunstein, M., 1975
Guerrier-Takada, C., 1983
Gulick, J. J., 1872
Gurdon, J. B., 1962, 1964, 1967, 2004
Guthrie, R., 1961
Gutman, A., 1950

Hadorn, E., 1963
Hafen, E., 198
Haig, D., 1992
Haldane, J. B. S., 1915, 1922, 1927, 1930 (1932), 1935
Haldane, N. M., 1915
Halder, G., 1995
Hall, B. D., 1961
Hall, C. E., 1962
Hall, J. C., 1991
Hall, T. H., 1981
Hallick, R., 1993
Hamilton, W. D., 1964
Hammer, M. F., 2000
Hammerling, J., 1943
Hardy, G. H., 1908
Harper, M. E., 1981
Harris, H., 1965, 1966, 1969
Hams, J. I., 1961
Harrison, R. G., 1907
Harrison, S. C., 1987
Hart, C. M., 1995
Hartmann, J. F., 1950
Hartwell, L. H., 1973, 1989, 2001
Harvey, W., 1651
Hashimoto, H., 1933
Hattori, M., 2000
Havel, T. F., 1985
Hayes, W., 1953
Hayflick, L., 1961
Hedges, R. W., 1974
Hedgpeth, J., 1972
Heezen, B. C., 1956
Heidelberg, J. F., 2000
Heitz, E., 1928
Henderson, S. A., 1968
Henikoff, S., 1975
Hennig, W., 1950
Henning, U., 1962
Herr, W., 1988
Herrmann, B. G., 1990
Hershey, A. D., 1946, 1949, 1952, 1961, 1969
Hershko, A., 1980, 2004
Hertwig, O., 1875
Hess, H. H., 1960
Hess, O., 1968
Hiaga, A., 1970
Hickmans, E. H., 1954
Hilschmann, N., 1965
His, W., 1870
Hoagland, M. B., 1955
Hodgkin, D. M. C., 1949, 1964
Hoekstra, H. E., 2003
Hoffman, E. P., 1987
Hofmeister, F., 1902
Hogness, D. S., 1975, 1978, 1986
Holley, R. W., 1965, 1968
Holliday, R., 1964
Holm, B., 1977
Holt, R. A., 2002
Hombrecher, G., 1981
Hong, F. D., 1989
Hood, L. E., 1986
Hooke, R., 1665
Horai, S., 1995
Horne, R. W., 1959
Horowitz, M., 1989
Horowitz, N. H., 1951
Horowitz, S., 1985
Horvitz, H. R., 1977, 1983, 1984, 1986, 2002
Hotta, Y., 1969, 1971
Hottinger, H., 1949
Houck, M. A., 1991
Howard, A., 1953
Howard-Flanders, P., 1964
Howell, S. H., 1971
Hoyer, B. H., 1972
Hozumi, N., 1976
Hradecna, Z., 1967
Hsu, L., 1972
Huang, X., 2002
Hubby, J. L., 1966
Huberman, J. A., 1968
Huebner, R. I., 1969
Hughes, J., 1995
Hughes, W. L., 1957
Humphrey, R. R., 1945
Hungerford, D., 1970
Hunt, R. T., 1983, 2001
Hunt, W. G., 1973
Hurwitz, J., 1972
Hutchison, C. A., 1974, 1976, 1978
Huxley, T. H., 1868

Ijdo, J. W., 1991
Illmensee, K., 1975
Ingram, V. M., 1957, 1961
Inman, R. B., 1970
International Human Genome Sequencing Consortium, 2001, 2003
Irwin, M. R., 1951
Ishida, M. R., 1963
Itakura, K., 1977
Ivanovsky, D. I., 1892
Ivens, A. C., 2005

Jablonski, N. G., 2000
Jackson, D. A., 1972
Jackson, R. J., 1976
Jacob, A. E., 1974
Jacob, F., 1955, 1958, 1961(3), 1963, 1965
Jacobs, H. J., 1983

Penny, G. D., 1996
Perdix-Gillot, S., 1979
Pereira, L., 1993
Perutz, M. F., 1960, 1962
Peterson, R. L., 1967
Petit, C., 1951
Petren, K., 1999
Petricciani, J. C., 1971
Pfankuch, E., 1939
Pigott, G. H., 1972
Pirie, N. W., 1937
Pisano, C., 1993
Plaut, W., 1962
Pohlman, R. F., 1984
Polani, P. E., 1960
Pollack, W., 1964
Pongs, O., 1980
Pontecorvo, G., 1952
Porter, K. R., 1953, 1963
Porter, R. R., 1962, 1972
Prescott, D. M., 1967
Preston, G. M., 1991
Pribnow, D., 1975
Pritchard, R. H., 1955
Prokofyeva-Belgovskaya, A. A., 1939
Prusiner, S. B., 1982, 1997
Ptashne, M., 1966, 1987
Puck, T. T., 1956
Punnett, R. C., 1906
Purkinje, J. E., 1825, 1839

Qian, Y. Q., 1989

Ragkuraman, M. K., 1989
Rall, T. W., 1957
Ramanis, Z., 1970
Rambousek, F., 1912
Randolph, L. F., 1928
Rao, P. N., 1970
Rapoport, J. A., 1946
Raspail, F. V., 1825
Rassoulzadegan, M., 1983
Raymer, W. B., 1967
Ready, D. F., 1986
Reddy, E. P., 1982
Reddy, V. B., 1978
Redi, F., 1668
Reich, E., 1964
Remak, R., 1845
Rhoades, M. M., 1938
Rice, G., 2004
Rich, A., 1962
Richards, S., 2005
Richardson, C. C., 1966
Ridanpaa, M., 2001
Riggs, A. D., 1968
Riley, R., 1958
Ripoll, P., 1973
Ris, H., 1962, 1969
Ritossa, F. M., 1962, 1965, 1966
Rivera, M. C., 1992
Rivera-Pomar, R., 1996
Roberts, B. E., 1973
Roberts, R. B., 1959
Roberts, R. J., 1977, 1993
Robertson, W. R. B., 1911
Robson, J. M., 1946
Rodbell, M., 1970, 1994
Rodgers, D. A., 1962
Roeder, R. C., 1969
Röller, H., 1966
Roper, J. A., 1952
Rose, I., 1980, 2004
Rosen, D. R., 1993
Rosenberg, J. M., 1973
Ross, E. M., 1977
Ross, M. T., 2005
Ross, R. C., 1898, 1902
Rossenbeck, H., 1924
Rosset, R., 1963
Roth, J. R., 1970, 1996
Rothman, R., 1965
Rotman, R., 1949
Rous, P., 1910, 1966
Roux, E., 1888
Roux, W., 1883
Rowley, J., 1972
Rubin, G. M., 1982, 1985, 2000
Rubinstein, I., 1961
Ruddle, F. H., 1969
Rusconi, S., 1981
Ruska, E., 1932, 1986
Ruska, H., 1939
Russell, L. B., 1961, 1963
Rutter, W. R., 1969

Sabatini, D. D., 1965
Sager, R., 1963, 1970
Saiki, R. K., 1985
Saint-Hilaire, E. G., 1822 (1944)
Sakano, H., 1981
Sanger, F., 1955, 1958, 1965, 1975, 1977, 1980
Santos, J. K., 1923
Sapienza, C., 1980
Sarabhai, A. S., 1964
Sarich, V. M., 1967
Saunders, G. F., 1981
Schaffner, S., 1981
Schellenberg, G. D., 1995
Schena, M., 1995
Schimke, R. T., 1978
Schimper, A. F. W., 1883
Schleiden, M. J., 1838–39
Schlesinger, M., 1934, 1963
Schneider, A., 1873
Schoenheimer, R., 1942
Schramm, G., 1956
Schughart, K., 1990
Schultz, J., 1936
Schwann, T., 1838–39

Bibliography

Here one can find references to some of the epoch-making books listed in the chronology, as well as appreciations of great geneticists, collections of important scientific papers, and histories of certain scientific breakthroughs. Systematic readings from this bibliography will put flesh upon the skeleton provided by the chronology.

Adelberg, E. A., editor 1960 *Papers on Bacterial Genetics*. Little, Brown, Boston.

Allen, G. E. 1978 *Thomas Hunt Morgan: The Man and His Science*. Princeton University Press, Princeton, NJ.

Angier, N. 1988 *Natural Obsessions: The Search for the Oncogene*. Houghton Mifflin, Boston.

Babcock, E. B. 1950 *The Development of Fundamental Concepts in the Science of Genetics*. American Genetics Assoc., Washington, D.C.

Barlow, W. G. 2000 *The Cichlid Fishes: Nature's Grand Experiment*. Perseus Publishing Co., Cambridge, MA.

Bearn, A. G. 1993 *Archibald Garrod and the Individuality of Man*. Oxford University Press, Oxford, England.

Berg, P. and M. Singer 2003 *George Beadle, An Uncommon Farmer. The Emergence of Genetics in the 20th Century*. Cold Spring Harbor Press. Cold Spring Harbor, New York.

Blunt, W. 1971 *The Compleat Naturalist: A Life of Linnaeus*. Viking Press, New York.

Boveri, T. 1914 *Zur Frage der Enstehung Maligner Tumoren*. Verlag der Gustav Fischer. Jena; translated into English (*The Origin of Malignant Tumors*) by Marcella O'Grady Boveri, 1929, Williams and Wilkins, Baltimore.

Bowler, P. J. 1984 *Evolution: The History of an Idea*. Harvard University Press, Cambridge, MA.

Boyer, S. H., editor 1963 *Papers in Human Genetics*. Prentice Hall, Englewood Cliffs, NJ.

Brenner, S. 2001 *My Life in Science*. As told to L. Wolpert. An edited interview with additions by E. C. Friedberg and E. Lawrence. BioMed Central, London.

Brock, T., editor 1961 *Milestones in Microbiology*. Prentice Hall, Englewood Cliffs, NJ.

Brosseau, G. E. 1967 *Evolution: A Book of Readings*. W. C. Brown Publishers, Dubuque, IA. (This collection contains articles by Lamarck, Malthus, Wallace, Allison, Kettlewell, and others.)

Browne, J. 1995 *Charles Darwin: Voyaging. Volume 1 of a Biography*. Knopf, New York.

Browne, J. 2002 *Charles Darwin: The Power of Place: Volume 2 of a Biography*. Knopf, New York.

Butler, J. M. 2001 *Forensic DNA Typing: Biology and Technology Behind STR Markers*. Academic Press, London.

Cairns, J., G. Stent, and J. D. Watson 1966 *Phage and the Origins of Molecular Biology*. Cold Spring Harbor Laboratory Press, Plainview, NY.

Carlson, E. A. 1966 *The Gene: A Critical History*. W. B. Saunders, Philadelphia.

Carter, G. S. 1957 *A Hundred Years of Evolution*. Macmillan, New York.

deChadarevian, S. 2002 *Designs for Life: Molecular Biology after World War II*. Cambridge University Press, Cambridge, UK.

Charles, D. 2001 *Lords of the Harvest: Biotech, Big Money and the Future of Food*. Perseus Publishers, Cambridge, MA.

Clark, R. W. 1984 *The Survival of Charles Darwin: A Biography of a Man and an Idea*. Random House, New York.

Conversations in Genetics 2004 Collections of 10 videos of interviews of prominent geneticists including Seymour Benzer, Leland Hartwell, François Jacob, Edward Lewis, and Janet Rowley. Directed by R. E. Esposito under the auspices of the Genetics Society of America. https://genetics.faseb.org/gsa-dvd/genestory.htm.

Cook-Deegan, R. 1994 *The Gene Wars: Science, Politics, and the Human Genome.* W. W. Norton, New York.

Corwin, H. O. and J. B. Jenkins, editors 1976 *Conceptual Foundations of Genetics: Selected Readings.* Houghton Mifflin, Boston.

Creager, A. N. H. 2001 *The Life of a Virus: Tobacco Mosaic Virus as an Experimental Model, 1930–1965.* University of Chicago Press, Chicago.

Crick, F. 1988 *What Mad Pursuit: A Personal View of Scientific Discovery.* Weidenfeld and Nicholson, New York.

Crotty, S. 2003 *Ahead of the Curve: David Baltimore's Life in Science.* University of California Press, Berkeley, CA.

Dampier, W. C. 1943 *A History of Science.* Cambridge University Press, Cambridge, England.

Darwin, C. 1839 *The Voyage of the Beagle.* Edited with an extensive historical introduction by Janet Browne and Michael Neve. Penguin Books, Ltd., London, 1989.

Darwin, C. 1859 *The Origin of Species by Means of Natural Selection or the Preservation of Favoured Races in the Struggle for Life.* The 1998 edition published by Oxford University Press contains an introduction, notes, and a glossary by Gillian Beer.

Darwin, C. 1871 *The Descent of Man, and Selection in Relation to Sex.* The edition published by Princeton University Press in 1981 contains introductions by J. T. Bonner and R. E. May.

Darwin, C., and A. R. Wallace 1958 *Evolution by Natural Selection.* Cambridge University Press, Cambridge, England. This collection of essays includes the papers by Darwin and Wallace that were presented *in absentia* at the July 1, 1858 meeting of the Linnaean Society in London.

Davies, K. 2001 *Cracking the Genome: Inside the Race to Unlock Human DNA.* The Free Press: Simon & Schuster, Inc., New York.

Dawes, B. 1952 *A Hundred Years of Biology.* Duckworth, London.

Dawkins, R. 1976 *The Selfish Gene.* Oxford University Press, New York. The 1989 edition contains two new chapters, endnotes, and an updated bibliography.

Dawkins, R. 1982 *The Extended Phenotype: The Long Reach of the Gene.* Oxford University Press, New York.

Dawkins, R. 1986 *The Blindwatchmaker.* W. W. Norton, New York.

Dawkins, R. 1997 *Climbing Mount Improbable.* W. W. Morton, New York.

Dawkins, R. 1998 *Unweavering the Rainbow: Science, Delusion, and the Appetite for Wonder.* Houghton Mifflin, Boston.

de Beer, G. 1964 *Charles Darwin.* Doubleday, New York.

de Kruif, P. 1926 *Microbe Hunters.* Harcourt Brace Jovanovich, New York.

Dennett, D. C. 1995 *Darwin's Dangerous Idea: Evolution and the Meanings of Life.* Simon & Schuster, New York.

Desmond, A., and J. Moore 1991 *Darwin.* Michael Joseph Publishers, London.

Dobell, C. 1932 *Antony van Leeuwenhoek and His "Little Animals" Being Some Account of the Father of Protozoology and Bacteriology and His Multifarious Discoveries in These Disciplines.* Staples Press Ltd., London.

Dobzhansky, T. 1937 *Genetics and the Origin of Species.* Columbia University Press, New York.

Dobzhansky, T. 1970 *Genetics and the Evolutionary Process.* Columbia University Press, New York.

Dubos, R. J. 1950 *Louis Pasteur, Free Lance of Science.* Little, Brown, Boston.

Dunn, L. C., editor 1951 *Genetics in the 20th Century.* Macmillan, New York.

Dunn, L. C. 1965 *A Short History of Genetics.* McGraw-Hill, New York.

Fedoroff, N. and N. M. Brown 2004 *Mendel in the Kitchen: A Scientist's View of Genetically Modified Foods.* Joseph Henry Press, Washington, D.C.

Fischer, E. P., and C. Lipson 1988 *Thinking About Science: Max Delbrück and the Origins of Molecular Biology.* W. W. Norton, New York.

Fisher, R. A. 1930 *The Genetical Theory of Natural Selection.* Oxford University Press, Oxford, England.

Fruton, J. S. 1999 *Molecules and Life. Historical Essays on the Interplay of Chemistry and Biology.* Second edition, Wiley-Interscience, New York.

Futuyma, D. J. 1995 *Science on Trial: The Case for Evolution.* Second edition, Sinauer, Sunderland, MA.

Gabriel, M. L., and S. Fogel, editors 1955 *Great Experiments in Biology.* Prentice Hall, Englewood Cliffs, NJ.

Gall, J. G., K. R. Porter, and P. Siekevitz 1981 Discovery in Cell Biology. *J. Cell Biol.* **91**(3), Part 2.

Galton, F. 1869 *Hereditary Genius.* Macmillan, London.

Gardner, E. J. 1960 *History of Life Science.* Burgess, Minneapolis, MN.

Garrod, A. E. 1909 *Inborn Errors of Metabolism.* Reprinted by H. Harris (see below).

Gehring, W. J. 1999 *Master Control Genes in Development and Evolution: The Homeobox Story*. Yale University Press, New Haven, CT.

Gillham, N. W. 2001 *A Life of Sir Francis Galton: From African Exploration to the Birth of Eugenics*. Oxford University Press, New York.

Glass, B., O. Temkin, and W. L. Straus, Jr. 1959 *Forerunners of Darwin, 1745–1859*. Johns Hopkins University Press, Baltimore, MD.

Goldschmidt, R. B. 1956 *Portraits from Memory*. University of Washington Press, Seattle.

Goldstein, L. 1966 *Cell Biology*. W. C. Brown, Dubuque, IA. (A collection of classic papers on cytology.)

Grant, P. R. 1999 *Ecology and Evolution of Darwin's Finches*. Second edition, Princeton University Press, Princeton, NJ.

Grant, V. 1956 The Development of a Theory of Heredity. *Am. Sci.* **44**:158–179.

Haldane, J. B. S. 1932 *The Causes of Evolution*. Harper and Row, New York.

Hall, S. S. 1987 *Invisible Frontiers: The Race to Synthesize a Human Gene*. Atlantic Monthly Press, New York.

Hamburger, V. 1988 *The Heritage of Experimental Embryology: Hans Spemann and the Organizer*. Oxford University Press, New York.

Harper, P. S., editor, 2004 *Landmarks in Medical Genetics. Classic Papers with Commentaries*. Oxford University Press, New York.

Harris, H. 1963 *Garrod's Inborn Errors of Metabolism*. Oxford University Press, London.

Henig, R. M., 2000 *The Monk in the Garden: The Lost and Found Genius of Gregor Mendel, the Father of Genetics*. Houghton Mifflin Company, Boston.

Hennig, W. 1966 *Phylogenetic Systematics*. University of Illinois Press, Urbana. This is an English translation by D. D. Davis and R. Zangerl of a book first published in German in 1950.

Hodge, J., and G. Radick, editors 2003 *The Cambridge Companion to Darwin*. Cambridge University Press, U.K.

Hooper, J. 2000 *Of Moths and Men: An Evolutionary Tale. The Untold Story of Science and the Peppered Moth*. W. W. Norton & Company, New York.

Hsu, T. C. 1979 *Human and Mammalian Cytogenetics: An Historical Perspective*. Springer-Verlag, New York.

Hughes, A. 1959 *A History of Cytology*. Abelard-Schuman, New York.

Ingram, V. M. 1963 *The Hemoglobins in Genetics and Evolution*. Columbia University Press, New York.

Irvine, W. 1955 *Apes, Angels, and Victorians*. McGraw-Hill, New York.

Judson, H. F. 1996 *The Eighth Day of Creation: Makers of the Revolution in Biology*. Expanded edition, Cold Spring Harbor Laboratory Press, Plainview, NY.

Keller, E. F. 1983 *A Feeling for the Organism: The Life and Work of Barbara McClintock*. W. H. Freeman, San Francisco.

Kelves, D. J. 1985 *In the Name of Eugenics: Genetics and the Uses of Human Heredity*. Harvard University Press, Cambridge, MA.

Kimura, M. 1983 *The Neutral Theory of Molecular Evolution*. Cambridge University Press, Cambridge, England.

Kittredge, M. 1991 *Barbara McClintock*. Chelsea House Publishers, New York.

Kobilinsky, L., T. Liotti, and J. Oeser-Sweat 2004 *DNA: Forensic and Legal Applications*. Wiley Publishers, New York.

Koerner, L. 1999 *Linnaeus: Nature and Nation*. Harvard University Press, Cambridge, MA.

Kohn, D., editor 1986 *The Darwinian Heritage*. Princeton University Press, Princeton, NJ.

Lack, D. 1947 *Darwin's Finches*. Cambridge University Press, Cambridge, England.

Larson, E. J. 1997 *Summer for the Gods: The Scopes Trial and America's Continuing Debate over Science and Religion*. Harvard University Press, Cambridge, MA.

Larson, E. J. 2001 *Evolution's Workshop: God and Science on the Galapagos Islands*. Basic Books, New York.

Lawrence, P. A., 1992 *The Making of a Fly: The Genetics of Animal Design*. Blackwell Scientific Publications, Cambridge, MA.

Levine, L., editor 1971 *Papers on Genetics: A Book of Readings*. C. V. Mosby, St. Louis, MO.

Lewis, E. B., editor 1961 *Selected Papers of A. H. Sturtevant on Genetics and Evolution*. W. H. Freeman, San Francisco.

Lwoff, A., and A. Ullman 1979 *Origins of Molecular Biology: A Tribute to Jacques Monod*. Academic Press, New York.

MacFadden, B. J. 1992 *Fossil Horses: Systematics, Paleobiology, and Evolution of the Family Equidae*. Cambridge University Press, Cambridge, England.

Maddox, B. 2002 *Rosalind Franklin: The Dark Lady of DNA*. HarperCollins, New York.

Manger, L. N. 1979 *A History of the Life Sciences*. Marcel Dekker, New York.

Margulis, L. 1981 *Symbiosis in Cell Evolution*. W. H. Freeman, San Francisco.

Margulis, L., K. V. Schwartz, and M. Dolan 1999 *Diversity of Life*. Second edition, Jones and Bartlett Publishers, Sudbury, MA.

Maroni, G. 1993 *An Atlas of Drosophila Genes: Sequences and Molecular Features*. Oxford University Press, New York.

Massie, R. K. 1995 *The Romanovs: The Final Chapter*. Random House, New York.

Maynard Smith, J. 1993 *The Theory of Evolution*. Canto edition, Cambridge University Press, Cambridge, U.K.

Mayr, E. 1963 *Animal Species and Evolution*. Harvard University Press, Cambridge, MA.

Mayr, E. 1982 *The Growth of Biological Thought: Diversity, Evolution and Inheritance*. Harvard University Press, Cambridge, MA.

Mayr, E. 2001 *What Evolution Is*. Basic Books, New York.

McCarty, M. 1985 *The Transforming Principle: Discovering That Genes Are Made of DNA*. W. W. Norton, New York.

McElheny, V. K. 2003 *Watson and DNA: Making a Scientific Revolution*. Perseus Publishing, Cambridge, MA.

McKusick, V. A. 1998 *Mendelian Inheritance in Man*. Twelfth edition, 3 vols., Johns Hopkins University Press, Baltimore, MD.

McMurray, E. J. 1995 *Notable Twentieth-Century Scientists*. Gale Research Inc., New York.

Medvedev, Z. A. 1969 *The Rise and Fall of T. D. Lysenko*. Translated by I. Michael Lerner, Columbia University Press, New York.

Moore, J. A., editor 1972 *Readings in Heredity and Development*. Oxford University Press, New York.

Moore, J. A., editor 1987 *Genes, Cells and Organisms: Great Books in Experimental Biology*, vol. 17. The collected papers of Barbara McClintock. The Discovery and Characterization of Transposable Elements. Garland, New York.

Moore, R. 1961 *The Coil of Life. The Story of the Great Discoveries in Life Sciences*. Knopf, New York.

Morange, M. 1998 *A History of Molecular Biology*. Translated by M. Cobb, Harvard University Press, Cambridge, MA.

Morgan, T. H. 1926 *The Theory of the Gene*. Yale University Press, New Haven, CT.

Morgan, T. H., A. H. Sturtevant, H. J. Muller, and C. B. Bridges 1915 *The Mechanism of Mendelian Heredity*. Henry Holt and Company, New York.

Moses, K., editor 2002 *Drosophila Eye Development*. Springer-Verlag, New York.

Muller, H. J. 1962 *Studies in Genetics*. Indiana University Press, Bloomington, IN.

Müller-Hill, B. 1996 *The Lac Operon: A Short History of a Genetic Paradigm*. Walter de Gruyter, Berlin.

Nathan, D. G. 1995 *Genes, Blood, and Courage: A Boy Called Immortal Sword*. Harvard University Press, Cambridge, MA.

National Academy of Sciences of the USA: *Biographical Memoirs*. Columbia University Press, New York. Published at periodic intervals (vol. 57 appeared in 1987).

Neel, J. V. 1994 *Physician to the Gene Pool: Genetic Lessons and Other Stories*. Wiley, New York.

Nobel Lectures (Including Presentation Speeches and Laureates' Biographies). Physiology or Medicine. Jan Lindsten, editor, vol. 1 (1971–1980), vol. 2 (1981–1990), World Scientific, River Edge, NJ.

Nordenskjöld, E. 1928 *The History of Biology. A Survey*. Knopf, New York.

Ohno, S. 1967 *Sex Chromosomes and Sex-linked Genes*. Springer-Verlag, Berlin.

Olby, R. 1994 *The Path to the Double Helix: The Discovery of DNA*. Dover Publications, New York.

Olson, S. 2002 *Mapping Human History. Discovering the Past Through Our Genes*. Houghton Mifflin Company, Boston.

Oppenheimer, J. 1963 Theodor Boveri: The Cell Biologists' Embryologist. *Quart. Rev. Biol.* **38**: 245–249.

Orel, V. 1984 *Mendel*. Oxford University Press, Oxford, England.

Patterson, J. T., and W. S. Stone 1952 *Evolution in the Genus* Drosophila. Macmillan, New York.

Perutz, M. F. 1998 *I Wish I'd Made You Angry Earlier: Essays on Science, Scientists, and Humanity*. Cold Spring Harbor Laboratory Press, Plainview, NY. (This collection gives useful insights into the personalities of Jacob, Pauling, Delbrück, Luria, Crick, and Watson.)

Peters, J. A., editor 1961 *Classical Papers in Genetics*. Prentice Hall, Englewood Cliffs, NJ.

Porter, K., and M. A. Bonneville 1963 *An Introduction to the Fine Structure of Cells and Tissues*. Lea and Febiger, Philadelphia.

Portugal, F. H., and J. S. Cohen 1977 *A Century of DNA: A History of the Discovery of the Structure and Function of the Genetic Substance*. MIT Press, Cambridge, MA.

Provine, W. B. 1971 *The Origins of Theoretical Population Genetics*. University of Chicago Press, Chicago.

Provine, W. B. 1986 *Sewall Wright and Evolutionary Biology*. University of Chicago Press, Chicago.

Punnett, R. C. 1950 The Early Days of Genetics. *Heredity* **4**:1–10.

Ridley, M. 1994 *The Red Queen: Sex and the Evolution of Human Nature.* Macmillan Publishing Company, New York.

Royal Society *Biographical Memoirs of Fellows of the Royal Society.* Burlington House, Picadilly, London. Published at periodic intervals (vol. 41 appeared in 1995).

Schoenheimer, R. 1942 *The Dynamic State of Body Constituents.* Harvard University Press, Cambridge, MA.

Schopf, J. W. 1999 *Cradle of Life: The Discovery of Earth's Earliest Fossils.* Princeton University Press, Princeton, NJ.

Secret of Photo 51 A video portraying the contribution of Rosalind Franklin to the discovery of the structure of DNA. Narrated by Sigourney Weaver. http://www.pbs.org/wgbh/nova/photo51.

Shipman, P. 1998 *Taking Wing:* Archaeopteryx *and the Evolution of Bird Flight.* Simon & Schuster, New York.

Simpson, G. G. 1951 *Horses: The Story of the Horse Family in the Modern World and Through Sixty Million Years of History.* Oxford University Press, New York.

Singer, C. 1959 *A History of Biology.* Abelard-Schuman, New York.

Solnica-Krezel, L., editor 2002 *Pattern Formation in Zebrafish.* Springer-Verlag, New York.

Sorsa, V. 1988 *Chromosome Maps of Drosophila.* Vols. 1 and 2. CRC Press, Boca Raton, FL.

Srinivasan, P. R., J. S. Fruton, and J. T. Edsall, editors 1979 The Origin of Modern Biochemistry. *Ann N.Y. Acad. Sci.* **325**:1–375.

Stansfield, W. D. 2000 *Death of a Rat: Understandings and Appreciations of Science.* Prometheus Books, Amherst, NY.

Stebbins, G. L. 1974 *Flowering Plants, Evolution above the Species Level.* Harvard University Press, Cambridge, MA.

Stent, G. S., editor 1960 *Papers on Bacterial Viruses.* Little, Brown, Boston.

Stent, G. S., editor 1981 *The Double Helix: A Personal Account of the Discovery of the Structure of DNA by James D. Watson. Text, Commentary, Reviews and Original Papers.* W. W. Norton, New York.

Stern, C., and E. R. Sherwood 1966 *The Origin of Genetics, A Mendel Source Book.* W. H. Freeman, San Francisco. (Contains a translation of Mendel's paper.)

Stubbe, H. 1972 *History of Genetics from Prehistoric times to the Rediscovery of Mendel's Laws.* Translated from the Revised Second German edition of 1965 by T. R. W. Walters. MIT Press, Cambridge, MA.

Sturtevant, A. H. 1965 *A History of Genetics.* Harper and Row, New York.

Sulston, J. and G. Ferry 2001 *The Common Thread: A Story of Science, Politics, Ethics, and the Human Genome.* Joseph Henry Press, Washington, D.C.

Taylor, J. H., editor 1965 *Selected Papers on Molecular Genetics.* Academic Press, New York.

Terzaghi, E. A., A. S. Wilkins, and D. Penny, editors 1984 *Molecular Evolution: An Annotated Reader.* Jones and Bartlett, Boston.

Thomas, L. 1974 *The Lives of a Cell: Notes of a Biology Watcher.* Viking, New York.

Thomas, L. 1979 *The Medusa and the Snail.* Viking, New York.

Van Oosterzee, R. 1997 *Where Worlds Collide: The Wallace Line.* Cornell University Press, Ithaca, NY.

Vavilov, N. I. 1992 *Origin and Geography of Cultivated Plants.* Translated from Russian to English by D. Löve. Cambridge University Press, Cambridge, England.

Voeller, B. R., editor 1968 *The Chromosome Theory of Inheritance: Classic Papers in Development and Heredity.* Appleton-Century-Crofts, New York.

Wallace, A. R. 1876 *The Geographical Distribution of Animals (with a Study of the Relations of Living and Extinct Faunas as Elucidating the Past Changes on the Earth's Surface).* 2 volumes, Macmillan and Co., London.

Ward, H. 1927 *Charles Darwin: The Man and His Welfare.* Bobbs-Merrill, Indianapolis, IN.

Waterson, A. P., and L. Wilkinson 1978 *An Introduction to the History of Virology.* Cambridge University Press, Cambridge, England.

Watson, J. D., and J. Tooze 1981 *The DNA Story: A Documentary History of Gene Cloning.* W. H. Freeman, San Francisco.

Weiner, J. 1994 *The Beak of the Finch: A Story of Evolution in Our Time.* Knopf, New York.

Weiner, J. 1999 *Time, Love, Memory: A Great Biologist and His Quest for the Origins of Behavior.* Alfred A. Knopf, New York. (A biography of Seymour Benzer.)

Welxer, A. 1995 *Mapping Fate: A Memoir of Family, Risk, and Genetic Research.* Random House, New York. (A family battles Huntington disease.)

White, M. J. D. 1973 *Animal Cytology and Evolution.* Third edition, Cambridge University Press, Cambridge, England.

Wilkins, M. 2004 *The Third Man of the Double Helix: The Autobiography of Maurice Wilkins*. Oxford University Press, Oxford.

Williams, T. I., editor 1982 *A Biographical Dictionary of Scientists*. Third edition, Wiley, New York.

Williams-Ellis, A. 1966 *Darwin's Moon: A Biography of Alfred Russel Wallace*. Blackie & Son Ltd., Glasgow.

Willier, B. J., and J. M. Oppenheimer, editors 1964 *Foundations of Experimental Embryology*. Prentice Hall, Englewood Cliffs, NJ.

Willis, C. 1991 *Exons, Introns and Talking Genes: The Science behind the Human Genome Project*. Basic Books, New York.

Wilson, E. B. 1925 *The Cell in Development and Heredity*. Third edition, Macmillan, New York.

Wright, S. 1931 Evolution in Mendelian Populations. *Genetics* **16**:97–159.

Wright, S. 1968 *Evolution and the Genetics of Populations*. University of Chicago Press, Chicago. Volumes 2, 3, and 4 were published in 1969, 1977, and 1978, respectively.

Zubay, G. L., and J. Marmur, editors 1973 *Papers in Biochemical Genetics*. Second edition, Holt, Rinehart and Winston, New York.

Appendix D: Periodicals Covering Genetics, Cell Biology, and Evolutionary Studies

A listing of the titles of current periodicals and the addresses or URLs of their publishers. The names and addresses of publishers of multiple journals are placed at the end of this list.

Acta Virologica
Academic Press
Advanced Drug Delivery Reviews
Elsevier Science
Advances in Agronomy
Academic Press
Advances in Applied Microbiology
Academic Press
Advances in Biological and Medical Physics
Academic Press
Advances in Biophysics
Elsevier Science
Advances in Cell and Molecular Biology
Academic Press
Advances in Cell Biology
Appleton-Century-Crofts
Advances in Clinical Chemistry
Academic Press
Advances in Enzymology
John Wiley and Sons
Advances in Experimental Medicine and Biology
Plenum
Advances in Genetics
Academic Press
Advances in Genome Biology
JAI Press
55 Old Post Rd, No. 2
Greenwich, Connecticut 06836
Advances in Human Genetics
Plenum
Advances in Immunology
Academic Press
Advances in Metabolic Disorders
Academic Press
Advances in Microbial Physiology
Academic Press
Advances in Morphogenesis
Academic Press
Advances in Pediatrics
Year Book Medical Publishers
Advances in Protein Chemistry
Academic Press
Advances in Radiation Biology
Academic Press
Advances in Teratology
Academic Press
Advances in Viral Oncology
Raven Press
Advances in Virus Research
Academic Press
Agri Hortique Genetica
Plant Breeding Institute
Weibullsholm
Landskrona, Sweden
Agronomy Abstracts
American Society of Agronomy
Agronomy Journal
American Society of Agronomy
AIDS Research and Human Retroviruses
Mary Ann Liebert Inc. Publishers
American Journal of Diseases of Children
American Medical Association
American Journal of Epidemiology
Annual Reviews
American Journal of Human Genetics
http://www.journals.uchicago.edu/AJHG/journal/
American Journal of Medical Genetics
Wiley Interscience
American Journal of Mental Deficiency
Boyd Printing
49 Sheridan Avenue
Albany, New York 12210
American Naturalist
University of Chicago Press

Analytical Biochemistry
Methods in the Biological Sciences
Academic Press
Animal Breeding Abstracts
Commonwealth Agricultural Bureaux
Animal Cell Biotechnology
Academic Press
Animal Genetics
Blackwell Publishing
Animal Production
Longman
Annales de Génétique
Elsevier Science
Annales de Génétique et da Sélection Animal
Institut National de la Recherche Agronomique
Annales de l'Institut Pasteur
Elsevier Science
Annals of Human Biology
Taylor and Francis Group
Annals of Human Genetics
Cambridge University Press
Annals of Medicine
Taylor & Francis Group
Annals of Science
Taylor & Francis Group
Annual Review of Biochemistry
Annual Reviews
Annual Review of Biophysics and Biomolecular Structure
Annual Reviews
Annual Review of Cell and Developmental Biology
Annual Reviews
Annual Review of Ecology and Systematics
Annual Reviews
Annual Review of Entomology
Annual Reviews
Annual Review of Genetics
Annual Reviews
Annual Review of Genomics and Human Genetics
Annual Reviews
Annual Review of Immunology
Annual Reviews
Annual Review of Medicine
Annual Reviews
Annual Review of Microbiology
Annual Reviews
Annual Review of Neuroscience
Annual Reviews
Annual Review of Plant Physiology and Plant Molecular Biology
Annual Reviews
Antisense & Nucleic Acid Drug Development
Mary Ann Liebert, Inc., Publishers
Archives d'Anatomie Microscopique et de Morphologie Experimentale
Masson et Cie
Archiv für Mikroskopische Anatomie und Entwicklungsmechanik
Springer-Verlag
Archives of Biochemistry and Biophysics
Academic Press
Archives of Disease in Childhood
British Medical Association
Archives of Insect Biochemistry and Physiology
Wiley Interscience
Archives of Microbiology
Springer-Verlag
Archives of Virology
Springer-Verlag
Australian Journal of Agricultural Research
Commonwealth Scientific and Industrial Research Organization
Australian Journal of Biological Sciences
Commonwealth Scientific and Industrial Research Organization
Bacteriological Proceedings
American Society for Microbiology
Bacteriological Reviews
American Society for Microbiology
Behavior Genetics
Plenum
Bibliographia Genetica
Martinus Nijhoff
Biken Journal
Osaka University
Research Institute for Microbial Diseases
3 Dojima Nishimachi
Kita-ku
Osaka, Japan
Biochemical and Biophysical Research Communications
Academic Press
Biochemical Genetics
Plenum
Biochemical Medicine and Metabolic Biology
Academic Press
Biochemistry and Cell Biology
National Research Council of Canada
Biochimica et Biophysica Acta
Elsevier Science
Biocytologia
Masson et Cie.
BioEssays
Wiley Interscience
Biological Bulletin
Marine Biological Laboratory
Woods Hole, MA 02543
Biological Reviews of the Cambridge Philosophical Society
Cambridge University Press
Biology of the Cell
Elsevier, Paris
Biomedical Ethics Reviews
Humana Press
Biometrics
Blackwell Publishing
Biophysical Journal
Rockefeller University Press

Biophysik
Springer-Verlag
Biopolymers
John Wiley and Sons, Inc.
BioSystems
Elsevier Science
Biotechniques and Histochemistry
Taylor & Francis Group
Biotechnology and Bioengineering
John Wiley and Sons
Biotechnology and Genetic Engineering Reviews
intercept@andover.co.uk
Blood Journal
bloodsubs@hematology.org
Blut
Springer-Verlag
BMG Genetics
http://www.biomedcentral.com/bmcgenet/
BMC Genomics
http://www.biomedcentral.com/bmcgenomics/
BMC Medical Genetics
http://www.biomedcentral.com/bmcmedgenet/
Botanical Review
New York Botanical Garden
Brain, Behavior and Evolution
S. Karger AG
British Journal of Cancer
Nature Publishing Group
British Journal of Haematology
Blackwell Scientific
British Medical Bulletin
Longman Group
British Poultry Science
Longman Group
Brookhaven Symposia in Biology
Brookhaven National Laboratory
Biology Department
Upton, NY 11973
Canadian Journal of Microbiology
National Research Council of Canada
Cancer Gene Therapy
http://www.nature.com/cgt/
Cancer Genetics and Cytogenetics
Elsevier Science
Carcinogenesis
IRL Press
Carlsburg Research Communications
Springer-Verlag
Cell
Cell Press
Cell & chromosome
http://www.cellandchromosome.com/home
Cell and Tissue Kinetics
Blackwell Publishing
Cell and Tissue Research
Springer-Verlag
Cell Biology International Reports
Academic Press
Cell Death and Differentiation
Nature Publishing Group
Cell Stress & Chaperones
Churchill Livingstone
Cellular and Molecular Biology
Pergamon Press
Cellular Immunology
Academic Press
Cellular Microbiology
Blackwell Publishing
Chromosome Research
Kluwer Academic Publishers
Chromosomes Today
Plenum
Chromosoma
Springer-Verlag
Clinical Genetics
http://www.blackwellpublishing.com
Clinical Immunology and Immunopathology
Academic Press
Clinical Pediatrics
J. B. Lippincott
Cloning and Stem Cells
Mary Ann Liebert Inc. Publishers
Cold Spring Harbor Symposia on Quantitative Biology
Cold Spring Harbor Laboratory
Comparative and Functional Genomics
Wiley Interscience
Comparative Biochemistry and Physiology
Pergamon Press
CRC *Critical Reviews in Biochemistry and Molecular Biology*
CRC *Critical Reviews in Biotechnology*
CRC *Critical Reviews in Immunology*
CRC *Critical Reviews in Microbiology*
Chemical Rubber Company Press
2000 NW Corporate Blvd.
Boca Raton, FL 33431
Critical Reviews in Biochemistry and Molecular Biology
Taylor & Francis Group
Critical Reviews in Oncology/Hematology
Elsevier Science
Crop Science
Crop Science Society of America
http://crop.scijournals.org
Current Advances in Cell and Developmental Biology
Elsevier Science
Current Advances in Genetics and Molecular Biology
Elsevier Science
Current Biology
Cell Press
Current Gene Therapy
Bentham Science
Current Genetics
Springer-Verlag
Current Genomics
Bentham Science
Current Issues in Molecular Biology
http://www.horizonpress.com/cimb/

Current Molecular Medicine
Bentham Science
Current Opinion in Biotechnology
Elsevier Science
Current Opinion in Cell Biology
Elsevier Science
Current Opinion in Genetics and Development
Elsevier Science
Current Opinion in Microbiology
Elsevier Science
Current Pharmacogenomics
Bentham Science
Current Problems in Pediatrics
Year Book Medical Publishers
Currents in Modern Biology
Elsevier Science
Current Topics in Cellular Regulation
Academic Press
Current Topics in Developmental Biology
Academic Press
Current Topics in Microbiology and Immunology
Springer-Verlag
Current Topics in Radiation Research
Elsevier Science
Cytogenetics and Genome Research
S. Karger AG
Cytopathology
Blackwell Publishing
Development
Company of Biologists Ltd.
Developmental and Comparative Immunology
Pergamon Press
Developmental and Cell Biology
Cambridge University Press
Developmental Biology
Academic Press
Developmental Dynamics
Wiley Interscience
Developmental Genetics
Wiley-Liss
Development Genes and Evolution
Springer-Verlag
Development, Growth and Differentiation
Blackwell Publishing
Differentiation
Springer-Verlag
Disease Markers
Wiley Interscience
Dissertation Abstracts Online
DNA—A Journal of Molecular and Cellular Biology
Mary Ann Liebert Inc. Publishers
DNA and Cell Biology
Mary Ann Liebert, Inc., Publishers
DNA Repair Reports
Elsevier Science
DNA Sequence: The Journals of DNA Sequencing and Mapping
Harwood Academic
Drosophila Information Service
c/o James N. Thompson, Jr.
Department of Zoology
730 Van Vleet Oval
University of Oklahoma
Norman, Oklahoma 73019 U.S.A.
Drug Resistance Updates
Churchill Livingstone
Dysmorphology and Clinical Genetics
Wiley Interscience
Electron Microscopy Reviews
Pergamon Press
EMBO Journal
Nature Publishing Group
EMBO Reports
Nature Publishing Group
Endeavor
Pergamon Press
Endocrine Journal
Macmillan Journals
Environmental and Molecular Mutagenesis
Wiley Interscience
Environmental Mutagen Society Newsletter
(online version)
Enzyme
S. Karger AG
Ethology
Blackwell Publishing
Eukaryotic Cell
American Society for Microbiology
Euphytica
Kluwer online
European Journal of Biochemistry
Springer-Verlag
European Journal of Cell Biology
Wissenschaftliche Verlagsgesellschaft
European Journal of Human Genetics
Nature Publishing Group
European Journal of Immunology
Academic Press
Evolution and Development
Blackwell Publishing
Evolutionary Anthropology
John Wiley and Sons
Evolution: International Journal of Organic Evolution
evol.allenpress.com
Evolutionary Biology
Appleton-Century-Crofts
Evolutionary Ecology
Kluwer online
Experimental and Molecular Pathology
Academic Press
Experimental Cell Biology
S. Karger AG
Experimental Cell Research
Academic Press
Extremophiles: Life under Extreme Conditions
Springer

FASEB Journal
Federation of American Societies for Experimental Biology
9650 Rockville Pike
Bethesda, MD 20814

Federation of European Biochemical Societies, Symposia
Elsevier Science

Functional & Integrative Genomics
Springer-Verlag

Fungal Genetics and Biology
Academic Press

Gene
Elsevier Science

Gene Analysis Techniques
Elsevier Science

Gene Function & Disease
Wiley Interscience

Genen en Phaenen
Journal of the Dutch Genetical Society
Institute of Genetics
State University of Utrecht
Opaalweg 20
Utrecht, Holland

Genes and Development
http://www.genesdev.org/

Genes and Function
Blackwell Publishing

Genes and Immunity
http://www.nature.com/gene

Genes, Brain and Behavior
Blackwell Publishing

Genes, Chromosomes and Cancer
Wiley Interscience

GeneScreen: An International Journal of Medical Genomes
Blackwell Publishing

Genesis: The Journal of Genetics and Development
Wiley Interscience

Genes to Cells
Blackwell Publishing

Gene Therapy
http://www.nature.com/gt

Genetica
Kluwer Academic

Genetical Research
Cambridge University Press

Genetic Engineering
Academic Press

Genetic Engineering: Principles and Methods
Plenum Press

Genetic Epidemiology
Wiley Interscience

Genetic Resources and Crop Evolution
Kluwer Academic Publishers

Genetics
http://www.genetics.org/

Genetics and Molecular Biology
gmb.editor@sbg.org.br

Genetics, Selection, Evolution
www.edpsciences-usa.org

Genetic Testing
Mary Ann Liebert, Inc., Publishers

Gene Therapy
Stockton Press

Genetik
Gustav Fischer Verlag

Genetika
Mezhdunarodnaya Kniga

Genome
National Research Council of Canada

Genome Biology
http://genomebiology.com/home/

Genome Research
Cold Spring Harbor Laboratory Press

Genomics
Academic Press

Glia
Wiley-Liss

Growth Factors
Harwood Academic Publishers

Harvey Lectures
Academic Press

Hemoglobin
Taylor & Francis Group

Héreditas
Blackwell Publishing

Heredity
Nature Publishing Group

Histochemical Journal
Kluwer Academic Publishers
England

Histochemistry
Springer-Verlag

Human Biology
Wayne State University Press
5959 Woodward Ave.
Detroit, MI 48202

Human Gene Therapy
Mary Ann Leibert Inc. Publishers

Human Genetics
Springer-Verlag

Human Heredity
S. Karger AG

Human Molecular Genetics
Oxford University Press

Human Mutation
Wiley Interscience

Hybridoma
Mary Ann Liebert Inc. Publishers

Immunity
Cell Press

Immunobiology
Gustav Fischer Verlag

Immunochemistry
Pergamon Press

Immunogenetics
Springer-Verlag

Immunological Reviews
Munksgaard Förlag

Immunology
Blackwell Publishing
Immunology Letters
Elsevier Science
Immunology Today
Elsevier Science
Infection and Immunity
American Society for Microbiology
Insect Molecular Biology
Blackwell Publishing
International Journal of Biochemistry and Cell Biology
Elsevier Science
International Journal of Insect Morphology and Embryology
Pergamon Press
International Journal of Invertebrate Reproduction and Development
Elsevier Science
International Journal of Peptide and Protein Research
Munksgaard Förlag
International Journal of Radiation Biology
Taylor & Francis Group
International Review of Cytology
Academic Press
International Society for Cell Biology, Symposia
Academic Press
Intervirology
S. Karger AG
In Vitro
Williams and Wilkins
Isis (contains articles on the history of science and medicine)
University of Chicago Press
Journal of Agricultural Genomics
CABI Publishing
Journal of Agricultural Science
Cambridge University Press
Journal of Animal Breeding and Genetics
Blackwell Publishing
Journal of Applied Bacteriology
Academic Press
Journal of Bacteriology
American Society for Microbiology
Journal of Biochemistry
JB-online
Journal of Biochemistry, Molecular Biology and Biophysics
Harwood Academic Publishers
Journal of Biological Chemistry
Williams and Wilkins
Journal of Biomolecular NMR
Kluwer Academic Publishers
Journal of Biosocial Science
Blackwell Scientific
Journal of Cell Biology
Rockefeller University Press
Journal of Cell Science
Company of Biologists Ltd.
Journal of Cellular Physiology
Wiley Interscience
Journal of Cereal Science
Academic Press
Journal of Chemical Technology and Biotechnology
Elsevier Science
Journal of Chronic Diseases
Pergamon Press
Journal of Dairy Research
Cambridge University Press
Journal of Dairy Science
www.adsa.org./jds/
Journal of Experimental Biology
Company of Biologists
Journal of Evolutionary Biochemistry and Physiology
Kluwer Academic Publishers
Journal of Evolutionary Biology
Blackwell Publishing
Journal of Experimental Zoology
Wiley Interscience
Journal of Gene Medicine
Wiley Interscience
Journal of General and Applied Microbiology
Journal Press
Journal of General Microbiology
Cambridge University Press
Journal of General Virology
Cambridge University Press
Journal of Genetic Counseling
Plenum
Journal of Genetic Psychology
Journal Press
Journal of Genetics
http://www.ias.ac.in/jgenet
Journal of Heredity
Oxford University Press
Journal of Histochemistry and Cytochemistry
Elsevier Science
Journal of Horitcultural Science and Biotechnology
http://www.jhortscib.com
Journal of Human Evolution
Academic Press
Journal of Human Genetics
Springer-Verlag
Journal of Immunogenetics
Blackwell Publishing
Journal of Immunological Methods
Elsevier Science
Journal of Immunology
Williams and Wilkins
Journal of Inherited Metabolic Disease
Kluwer Academic Publishers
Journal of Insect Physiology
Pergamon Press
Journal of Mammalian Evolution
Plenum
Journal of Medical Genetics
http://jmg.bmjjournals.com/
Journal of Medical Primatology
Wiley Interscience
Journal of Medical Virology
Wiley Interscience

Journal of Mental Deficiency Research
Blackwell Scientific
Journal of Molecular and Applied Genetics
Raven Press
Journal of Molecular Biology
Academic Press
Journal of Molecular and Cellular Immunology
Springer-Verlag
Journal of Molecular Evolution
Springer-Verlag
Journal of Molecular Medicine
Springer-Verlag
Journal of Molecular Microbiology and Biotechnology
S. Karger AG
Journal of Molecular Modeling
Springer-Verlag
Journal of Morphology
Wiley Interscience
Journal of Neurogenetics
Harwood Academic
Journal of Obstetrics and Gynaecology
Taylor & Francis
Journal of Pediatrics
C. V. Mosby
Journal of Submicroscopic Cytology
S. Karger AG
Journal of the National Cancer Institute
http://jncicancerspectrum.oupjournals.org
Journal of Natural History
Taylor & Francis Group
Journal of Theoretical Biology
Academic Press
Journal of Ultrastructure and Molecular Structure Research
Academic Press
Journal of Virology
American Society for Microbiology
Lancet: the journal
http://www.thelancet.com
Life Sciences
Pergamon Press
The Linnean
Linnean Society of London (LSL)
Burlington House
Piccadilly, London, W1J OBF
Also publishes: Biological Journal of the LSL, Botanical Journal of the LSL, Transactions of the LSL (3rd series), and Zoological Journal of the LSL.
Mammalian Genome
Springer-Verlag
Mechanisms of Development
Elsevier Science
Metabolism
Grune and Stratton
Methods in Cell Biology
Academic Press
Methods in Cell Physiology
Academic Press
Methods in Enzymology
Academic Press
Methods in Immunology and Immunochemistry
Academic Press
Methods in Medical Research
Year Book Medical Publishers
Methods in Molecular and Cellular Biology
Wiley Interscience
Methods in Virology
Academic Press
Microbiological Reviews
American Society for Microbiology
Microbiological Sciences
Blackwell Publishing
Microscopy Research and Technique
Wiley Interscience
Mikrobiologiya
Mezhdunarodnaya Kniga
Modern Cell Biology
Wiley-Liss
Molecular and Cellular Biochemistry
Kluwer Academic
Molecular and Cellular Biology
American Society for Microbiology
Molecular and Cellular Neuroscience
Academic Press
Molecular and Cellular Probes
Academic Press
Molecular and Cellular Proteomics
www.mcponline.org
Molecular and Developmental Evolution
Wiley Interscience
Molecular and General Genetics
Springer-Verlag
Molecular Biology and Evolution
University of Chicago Press
Molecular Biology and Medicine
Academic Press
Molecular Biology Reports
Kluwer Academic Publishers
Molecular Biology SSSR
Mezhdunarodnaya Kniga
Molecular Breeding
Kluwer Academic Publishers
Molecular Carcinogenesis
Wiley Interscience
Molecular Cell
Cell Press
Molecular Diagnostics
Churchill Livingstone
Molecular Ecology
Blackwell Publishing
Molecular Genetic Medicine
Academic Press
Molecular Genetics and Metabolism
Elsevier Science
Molecular Human Reproduction
Oxford University Press

Molecular Immunology
Pergamon Press
Molecular Medicine
Blackwell Publishing
Molecular Microbiology
Blackwell Publishing
Molecular Microbiology and Medicine
Academic Press
Molecular Pharmacology
Academic Press
Molecular Phylogenetics and Evolution
Elsevier Science
Molecular Psychiatry
Stockton Press
Molecular Reproduction and Development
Wiley Interscience
Molecular Therapy
Elsevier Science
Monatshefte für Chemie
Springer-Verlag
Monatsschrift für Kinderheilkunde
Springer-Verlag
Monographs in Human Genetics
S. Karger AG
Mouse Genome
Oxford University Press
Mutagenesis
Oxford University Press
Mutation Research
Elsevier Science
Mycologia
New York Botanical Garden
National Institute of Genetics (Mishima), Annual Report
http://www.nig.ac.jp/section/annual.html
Nature, Nature Biotechnology, Nature Cell Biology, Nature Genetics, Nature Immunology, Nature Medicine, Nature Neuroscience, Nature Structural Biology
Nature Publishing Group (Journals)
Nature Reviews Cancer, Nature Reviews Genetics, Nature Reviews Immunology, Nature Reviews Molecular Cell Biology, Nature Reviews Neurosciences
Nature Publishing Group (Reviews)
Naturwissenschaften
Springer-Verlag
NCI Monographs
http://www3.oup.co.uk/jncmon/etoc.html
Neurogenetics
Springer-Verlag
Neuron
Cell Press
New Genetics and Society
Taylor & Francis Group
Nucleic Acids Research
IRL Press at Oxford University Press
Nucleosides & Nucleotides
Marcel Dekker, Inc.
P.O. Box 5005
185 Cimarron Road
Monticello, NY 12701-5185
The Nucleus
Cytogenetics Laboratory
Department of Botany
University of Calcutta
35, Ballygunge Circular Rd.
Calcutta 19, India
Oak Ridge National Laboratory Symposia
(supplements to the *Journal of Cell Physiology*)
Wistar Institute Press
Oat Newsletter
Dr. J. A. Browning
Department of Botany
Iowa State University
Ames, IA 50010
Obstetrics and Gynecology
Elsevier Science
Oncogene and Oncogene Reviews
Nature Publishing Group
Oncogene Research
Harwood Academic
Ophthalmic Genetics
Taylor & Francis Group
Origins of Life and the Evolution of the Biosphere
Kluwer Academic Publishers
Oxford Surveys in Evolutionary Biology
Oxford University Press
Oxford Surveys on Eukaryotic Genes
Oxford University Press
Paleobiology
Harwood Academic
Pasteur Institute
(see *Annales de l'Institut Pasteur*)
Pathologia et Microbiologia
S. Karger AG
Pediatric Research
Williams and Wilkins
Pediatrics
http://pediatrics.aappublications.org/
Perspectives in Biology and Medicine
Johns Hopkins University Press
Pharmacogenomics Journal
http://www.nature.com/tpj/
Philosophical Transactions of the Royal Society of London, Series B, Biological Sciences
http://www.journals.royalsoc.ac.uk/app/home/
Photochemistry and Photobiology
Pergamon Press
Physiological Genomics
www.physiologicalgenomics.org
Phytochemistry
Pergamon Press
Planta
Springer-Verlag
Plant Biology
Georg Thieme Verlag KG Stuttgart
Plant Breeding Abstracts
Commonwealth Agricultural Bureaux
Plant Genetic Resources: Characterization and Utilization
CABI Publishing

Plant Molecular Biology
Kluwer Academic
Plasmid
Academic Press
Poultry Science
http://www.poultryscience.org/ps/
Prenatal Diagnosis
Wiley Interscience
Proceedings of the National Academy of Sciences of the United States of America
http://www.pnas.org/
Proceedings of the Royal Society of Edinburgh, Section B (Biological Sciences)
Royal Society of Edinburgh
22 George St.
University of Edinburgh
Edinburgh EH 2–2 PQ, Scotland
Proceedings of the Royal Society of London, Series B, Biological Sciences
Royal Society of London
6 Carlton House Terrace
London SW1Y SAG, England
Progress in Biophysics and Molecular Biology
Elsevier Science
Progress in Histochemistry and Cytochemistry
Gustav Fischer Verlag
Progress in Medical Genetics
Grune and Stratton
Progress in Medical Virology
S. Karger AG
Progress in Nucleic Acid Research and Molecular Biology
Academic Press
Progress in Theoretical Biology
Academic Press
Protein Science
Cambridge University Press
Proteins; Structure, Function and Genetics
Wiley-Liss
Protoplasma
Springer-Verlag
Quarterly Review of Biology
http://www.journals.uchicago.edu/QRB/journal/
Quarterly Reviews of Biophysics
Cambridge University Press
Radiation Botany
Pergamon Press
Radiation Research
Academic Press
Recent Progress in Hormone Research
Academic Press
Resumptio Genetica
Martinus Nijhoff
Revue Suisse de Zoologie
Revue Suisse de Zoologie
Muséum d'Histoire Naturelle
Geneva, Switzerland
RNA
Cambridge University Press
Science
http://www.sciencemag.org/
Scientific American
http://www.sciamdigital.com/
Seminars in Cell and Developmental Biology
Academic Press
Sequence: The Journal of DNA Mapping and Sequencing
Harwood Academic
Sexual Plant Reproduction
Springer-Verlag
Silvae Genetica
J. P. Sauerlander's Verlag
Finkenhofstrasse 21
D-60322 Frankfurt a. M.
Social Biology
University of Chicago Press
Society for Developmental Biology, Symposia
http://www.sdbonline.org
Society for Experimental Biology, Symposia
Cambridge University Press
Society for General Microbiology, Symposia
Cambridge University Press
Somatic Cell and Molecular Genetics
Plenum
Soviet Genetics
(an English translation of *Genetika*)
Plenum
Stain Technology
Williams and Wilkins
Stem Cells and Development
Mary Ann Liebert, Inc. Publishers
Studies in Drosophila Genetics
See *University of Texas Publications.*
Sub-Cellular Biochemistry
Plenum
Teratology
Wiley Interscience
Theoretical and Applied Genetics
Springer-Verlag
Theoretical Population Biology
Academic Press
Tissue and Cell
Longmans Group
Tissue Antigens
Munksgaard Förlag
Traffic (intracellular transport)
http://www.traffic.dk
Transactions of the British Mycological Society
Cambridge University Press
Transactions of the New York Academy of Sciences
New York Academy of Sciences
Transgenic Research
Chapman and Hall
Transgenics
Harwood Academic
Transplantation
Williams and Wilkins

Transplantation Proceedings
Grune and Stratton
Transplantation Reviews
Williams and Wilkins
Trends in Biochemical Sciences
Elsevier Trends Journals
Trends in Biotechnology
Elsevier Trends Journals
Trends in Cell Biology
Elsevier Trends Journals
Trends in Ecology and Evolution
Elsevier Trends Journals
Trends in Genetics
Elsevier Trends Journals
Trends in Microbiology
Elsevier Trends Journals
Trends in Molecular Medicine
Elsevier Trends Journals
Trends in Neuroscience
Elsevier Trends Journals
Trends in Plant Science
Elsevier Trends Journals
Tribolium Information Bulletin
Dr. A. Sokoloff
Natural Sciences Division
California State College
San Bernadino, CA 92407
Tumor Targeting
Stockton Press
Twin Research
Stockton Press
UCLA Symposia on Molecular and Cellular Biology
Wiley-Liss
Ultrastructure in Biological Systems
Academic Press
University of Texas, M. D. Anderson Hospital and Tumor Institute—Symposia on Fundamental Cancer Research
Texas Medical Center
Houston, TX 77025
Virology
Academic Press
Virus Genes
Kluwer Academic Publishers
Virus Research
Elsevier Science
World's Poultry Science Journal
CABI Publishing
Yearbook of Obstetrics and Gynecology
Year Book Medical Publishers
Yearbook of Pediatrics
Year Book Medical Publishers
Yeast
John Wiley and Sons
Zebrafish
Mary Ann Liebert, Inc. Publishers
Zeitschrift für Immunitätsforschung
Gustav Fischer Verlag
Zygote
Cambridge University Press

MULTIJOURNAL PUBLISHERS

Academic Press, 1250 Sixth Avenue, San Diego, CA 92101

American Medical Association, 535 North Dearborn Street, Chicago, IL 60610

American Society of Agronomy, 677 South Segoe Road, Madison, WI 53711

American Society for Microbiology, 1752 N Street NW, Washington, DC 20036-2904

Annual Reviews, Inc., 4139 El Camino Way, Palo Alto, CA 94303-0139

Appleton-Century-Crofts, 440 Park Avenue South, New York, NY 10016

Bentham Science, www.bentham.org

Blackwell Scientific Publications Inc., 238 Main Street, Cambridge, MA 02142

British Medical Association, Tavistock Square, London WC1H 9JR, England

CABI Publishing, http://www.cabi-publishing.org/

Cambridge University Press, 40 West 20th Street, New York, NY 10011

Cambridge University Press (Journals on Line), http://www.journals.cup.org

Cell Press, 1100 Massachusetts Avenue, Cambridge, MA 02138

Chapman & Hall, Subscription Department RSP, 400 Market Street, Suite 750, Philadelphia, PA 19106

Churchill Livingstone, 1–3 Baxter's Place, Leith Walk, Edinburgh, EH1 3AF, URL: rsh.pearson-pro.com

Cold Spring Harbor Laboratory, P.O. Box 100, Cold Spring Harbor, NY 11724

Commonwealth Agricultural Bureaux, Farnham Royal, Bucks, England

Commonwealth Scientific and Industrial Research Organization, 314 East Albert Street, East Melbourne, Victoria 3002, Australia

Company of Biologists Ltd., Department of Zoology, University of Cambridge, Downing Street, Cambridge CB23EJ, England

C. V. Mosby Co., 11830 Westline Industrial Drive, St. Louis, MO 63141

Elsevier Science, 655 Ave. of the Americas, New York, NY 10010-5107

Elsevier Trends Journals, P.O. Box 882, Madison Square Station, NY 10159

Grune and Stratton Inc., 111 Fifth Avenue, New York, NY 10003

Gustav Fischer Verlag, P.O. Box 7-20143, Stuttgart, Germany

Harper and Row Publishers, 2350 Virginia Avenue, Hagerstown, MD 21740

Harwood Academic Publishers, P.O. Box 786, Cooper Station, New York, NY 10276, http://www.gbhap.com/
Humana Press, http://humanapress.com
Institut National de la Recherche Agronomique, Service des Publication, Route de Saint Cyr, 78-Versailles, France
IRL Press at Oxford University Press, 198 Madison Avenue, New York, NY 10016
Japan Publications Trading Co. Ltd., 175 Fifth Avenue, New York, NY 10010
Lippincott Williams & Wilkins, 530 Walnut Street, Philadelphia, PA 19106
John Wiley and Sons, 605 Third Avenue, New York, NY 10158
Journal Press, 2 Commercial Street, Provincetown, MA 02657
Kluwer Academic Publishers has merged with Springer-Verlag
Longman Group Ltd., 43–45 Annandale Street, Edinburgh, Scotland
Macmillan Journals Ltd., 4 Little Essex Street, London, WC2R 3LF, England
Marcel Dekker, Inc., http://www.dekker.com
Martinus Nijhoff, P.O. Box 269, The Hague, Holland
Mary Ann Liebert Inc., Publishers, 2 Madison Avenue, Larchmont, NY 10538
Masson et Cie., 120 Boulevard Saint Germain, F75280, Paris 6e, France
Mezhdunarodnaya Kniga, 39 Dimitrova Ulitza, 113095 Moscow, Russia
Munksgaard Förlag, 35 Norre Sogade, DK 1016, Copenhagen K, Denmark
National Research Council of Canada, Research Journals Publishing Dept., Ottawa 2, Ontario, Canada K1OR6
Nature Publishing Group, the trading name of Nature America, Inc., 345 Park Avenue South, New York, New York 10010-1707
New York Botanical Garden, Bronx, NY 10458
Oliver and Boyd Ltd., Tweeddale Court, 14 High Street, Edinburgh EH1 1YL, Scotland
Oxford University Press, 198 Madison Avenue, New York, NY 10016
Oxford University Press (Journals on Line), URL: www.hmg.oupjournals.org
Plenum Publishing Co., 233 Spring Street, New York, NY 10013
Rockefeller University Press, 1230 York Avenue, New York, NY 10021
S. Karger AG, 26 West Avon Road, P.O. Box 529, Farmington, CT 06085
Springer-Verlag, 175 Fifth Avenue, New York, NY 10010
Stockton Press, http://www.stockton-press.co.uk
Taylor & Francis Group, http://www.tandf.co.uk/journals/
University of Chicago Press, 5720 South Woodlawn Avenue, Chicago, IL 60637
Wiley Interscience (Journals on Line), www.interscience.wiley.com
Wissenschaftliche Verlagsgesellschaft GMBH, P.O. Box 40, D-7000 Stuttgart 1, Germany
Wistar Institute Press, 3631 Spruce Street, Philadelphia, PA 19104
Year Book Medical Publishers, Inc., 35 East Wacker Drive, Chicago, IL 60601

FOREIGN WORDS COMMONLY FOUND IN SCIENTIFIC TITLES

D = Dutch; *F* = French; G = German; *I* = Italian; *J* = Japanese; *L* = Latin; *R* = Russian; *Sp* = Spanish; *Sw* = Swedish

Abbildung (G) figure
Abhandlung (G) dissertation, transaction, treatise, paper
Abstammungslehre (G) theory of descent, origin of species
Abteil, Abteilung (G) division
Acta (L) chronicle
allgemein (G) general
angewandt (G) applied
Annalen (G) annals
Anzeiger (G) informer
Arbeiten (G) work
Atti (I) proceedings
Band (G) volume
Beiheft (G) supplement
Bericht (G) report
Bokhandel (Sw) bookstore
Boktryckeri (Sw) press
Bunko (J) library
Bunkyo (J) education
Buchbesprechung (G) book review
Comptes Rendus (F) proceedings
Daigaku (J) university
Doklady (R) proceedings
Entwicklungsmechanik (G) embryology
Ergänzungshefte (G) supplement
Ergebnis (G) conclusion
Folia (L) leaflet, pamphlet, journal
Förlag (Sw) publisher
Forschung (G) research
Fortbildung (G) construction
Forstgenetik (G) forestry genetics

Fortschritt (G) advance, progress
gesamt (G) general
Gesellschaft (G) association, society
hebdomadaire (F) weekly
Hefte (G) number (of a periodical)
Helvetica (L) Swiss
Hoja (Sp) paper, pamphlet, record, journal
Iberica (L) referring to Spain and Portugal
Idengaku (J) genetics
Inhalt (G) contents
Jahrbuch (G) yearbook, annual
Kenkyusho (J) research institute
Kniga (R) book
Kunde (G) science
Lebensmittel (G) nutrition
Lehrbuch (G) textbook
Mezhdunarodnaya (R) international
Monatsblätter (G) monthly journal
Nachrichten (G) news
Naturwissenschaft (G) natural science
Nauk (R) science
Österreich (G) Austria
Planches (F) plates
real (Sp) royal
Recueil (F) collection
Rendiconti (I) account
Resumptio (D) review
Revista (Sp) review
Rundschau (G) overview, survey
Sammlung (G) collection
Säugetier (G) mammal
Schriften (G) publication
Schweizerische (G) Swiss
Scripta (I) writing
Séance (F) session, meeting
Seibutsugaku (J) biology
Seiken (J) biological institute
Shokubutsugaku (J) botany
Silvae (L) forest
sperimentale (I) experimental
Shuppan (J) publication
Teil (G) part
Tierärtliche Medizin (G) veterinary medicine
Tijdschrift (D) magazine, periodical
Tome (F) volume
Toyo (J) East, Orient
Travaux (F) work
Trudy (R) works, reports
Untersuchungen (G) research
Vererbungslehre (G) genetics
vergleichen (G) comparative
Verhandlung (G) proceeding, transaction
Verlag (G) publishing house
Verslag (D) report, account
Vorbericht (G) preliminary report
Wissenschaft (G) science
Wochenschrift (G) weekly publication
Zasshi (J) magazine
Zeitschrift (G) periodical, journal, magazine
Zeitung (G) newspaper
Zellforschung (G) cytology
Zentralblatt (G) overview or survey
Ziho (J) journal
Züchtung (G) breeding, culturing, rearing

Appendix E: Internet Sites

The World Wide Web contains a vast amount of information of potential use to researchers, educators, biotechnicians, students, and the general public. Any list of addresses (Uniform Resource Locators or URLs) from this storehouse of knowledge cannot help but be incomplete. Many websites are updated periodically, and some may even be eliminated by the time any list can be printed and disseminated. This list was compiled in July 2001. The first listings constitute a sample set of "master" websites that were chosen because they contain a relatively large number of links to various subjects, mainly in the fields of genetics, molecular and cell biology, evolution, and biotechnology. A brief description of some of the linked contents of each site is provided, but most sites contain far more subjects than those listed here. The second listing contains a sampling of individual websites of more specific content and may or may not contain links to other useful sites. The third listing gives the URLs for many of the species listed in this dictionary.

MASTER WEB SITES

All the virology on the WWW electron micrographs and macromolecular images, immunology, laboratory techniques, resources and databases, taxonomy and phylogeny, genome sequence data, emerging viruses information and research, specific virus servers and information, AIDS information and research, plant viruses servers and information, viral diseases, vaccines and treatments, organizations and groups of interest to virologists, educational resources, general virology information and news, and related internet resources for virologists.
http://www.tulane.edu/~dmsander/garryfavweb.html

Animal Diversity Web an online illustrated encyclopedia of the natural history of hundreds of animal species, together with descriptions of classifications at the levels of phylum and class.

CEPH-Généthon Centre d'Etude du Polymorphisms Humaine A center for research in gene therapy.
http://www.genethon.fr

DEAMBULUM: Genomes and Organisms databanks sequences, genomes, structures, dictionaries, model organisms, mitochondria, metabolism, comparative genome databases, phylogeny and taxonomy, microbiology, virology, parasitology, linkage and genetic mapping, BLAST comparison with genomes.
http://www.infobiogen.fr/services/deambulum/english/genomes1.html

Genetics Virtual Library (Oak Ridge National Laboratory, Department of Energy) organism index (transgenic, C. *elegans*, cattle, slime mold, *Drosophila*, fish, fungi, horse, microbes, mosquito, parasites, plant, poultry, rodents, sheep, swine, yeast); human genome project (U.S. research sites, international research sites, human chromosome-specific sites).
http://www.ornl.gov/TechResources/Human_Genome/v1.html
http://public.ornl.gov/hgmis/external/new.cfm?organism-name

Genome Sites Index (Ernest Orlando Lawrence Berkeley National Laboratory) databases, genome information sources (The Sanger Center, Whitehead

Institute/MIT Genome Center, Stanford Human Genome Center, Biotech Life Sciences Directory).
http://www-gsd.lbl.gov/GenSites.htm

Human Genome Central offers links to a variety of Websites that deal with human genomics and proteomics, genetic diseases, ethical dilemmas, etc.
http://www.ensembl.org/genome/central

Human Genome Epidemiology Network (HuGENet) issues reviews from the Centers for Disease Control and Prevention and maintains a database on literature dealing with the allelic variants of genes that influence disease susceptibility and which occur at different frequencies in various human populations.
http://www.cdc.gov./genomics/hugenet

National Biotechnology Information Facility, New Mexico State University storehouse of some 7,000 biotechnology links to gene and protein sequence databases, atlas of protein side-chain interactions, databases of telomere papers, patent databases, and much more.
www.nbif.org

National Center for Biotechnology Information (NCBI), National Library of Medicine, National Institutes of Health cancer genome anatomy project, clusters of orthologous groups, electronic PCR, genes and disease, human genome resources, human/mouse homology maps, malaria genetics and genomics, ORF finder, retrovirus resources, serial analysis of gene expression, recently sequenced genomes for organelles, viruses, prokaryotes, and eukaryotes.
http://www.ncbi.nlm.nih.gov/index.html

Nature Magazine Genome Gateway contains links to human and non-human genome projects and resources, institutes, and companies devoted to genome research, genome-related publications, and bioethics. There are links to Albert Einstein Genome Center, Cardio Genomics PGA, Centre for Law and Genetics, Compugen, Double Twist, Euchromatin Network, Functional Genomics, Genemap 99, Genome Database, Genoplante, Human Genome Central, Incyte Genomics, National Human Genome Research Institute, National Laboratory for Genetics of Israeli Populations, Mosquito Genomics WWW Server, NCBI Human Genome Resources, Parasite Genome, RATMAP, UmanGenomics, Whitehead Institute, Wormbase, and many more.
http://www.nature.com/genomics/links/

Nucleic Acids on the Internet major nucleic acid sequence databases, homology search engines, journals online.
http://bmbsgill.leeds.ac.uk/bmbknd/DNA/DNA.html

Rockefeller University Computing Services; Miscellaneous Scientific Servers List
http://cs.rockefeller.edu/index.php3?page=toolkit

Science Magazine Functional Genomics Web Site reports on advances in genomics and biotechnology found in *Science* and other journals. Lists new Web resources on model organisms, genome map advances, upcoming meetings, sites dealing with ethics and society, emerging lab techniques, etc.
http://www.sciencegenomes.org

TIGR (The Institute for Genomic Research) CMR (Comprehensive Microbial Resource) Rockville, MD
http://www.tigr.org/tigr-scripts/CMR2/CMRHomePage.spl

Weizmann Institute of Science, Israel general list of Science Web sites.
http://bioinformatics.weizmann.ac.il/hotmolecbase/webscience.htm

WWW Resources for Molecular Genetics (Washington University in St. Louis) databases (literature, sequence/structure, phylogenetic, annotated protein family, sequence searching, protein motif/domain searching), genome project information (national, international), ACEDB databases for model organisms.
http://www.genetics.wustl.edu/bio5491/bio5491.html

INDIVIDUAL DATABASES

Cancer Genome Anatomy Project (CGAP) funded by the National Cancer Institute (NIH).
www.ncbi.nlm.nih.gov/ncicgap

Celera Genomics.
www.celera.com

CMS Molecular Biology Resource (C. M. Smith, curator).
http://www.sdsc.edu/ResTools/cmshp.html

Dendrome a collection of forest tree genome databases and other forest genetic information resources.
http://dendrome.ucdavis.edu/

DNA Data Bank of Japan (DDBJ)
http://www.ddbj.nig.ac.jp/

Drosophila sites
Berkeley Drosophila Genome Project (BDGP).
http://www.fruitfly.org/

The Bloomington Drosophila Stock Center.
http://flystocks.bio.indiana.edu/
Drosophila Chromosome 4 Site.
http://www.biology.ualberta.ca/dros4.hp/4th_main.htm
Drosophila DNA Microarray Homepage.
http://quantgen.med.yale.edu
The Drosophila Virtual Library.
http://www.ceolas.org/fly/index.html
European Drosophila Genome Project (EDGP).
http://edgp-dev.ebi.ac.uk/
Fly base. http://flybase.bio.indiana.edu/
FlyTrap. http://gravitaxis.bioc.rice.edu/flytrap/
Fly View, A Drosophila image database.
http://pbio07.uni-muenster.de/
Functional Genomics for UK Drosophilists.
http://www.gen.cam.ac.uk/%7Eflychip/
Interactive Fly, a cyberspace guide to Drosophila genes and their roles in development.
http://sdb.bio.purdue.edu/fly/aimain/laahome.htm
The Szeged *Drosophila melanogaster* P Insertion Mutant Stock Centre.
http://gen.bio.u-szeged.hu/stock/

EcoCyc Encyclopedia of *E. coli* genes and metabolism.
http://ecocyc.pangeasystems.com

EMBL Nucleotide Sequence Database European Bioinformatics Institute, an EMBL outstation.
http://www.ebi.ac.uk

Ensembl Human Genome Central.
http://www.ensembl.org/genome/central

ENZYME enzyme nomenclature database.
http://ca.expasy.org

ERGO A curated database of public and proprietary genomic DNA, with connected similarities, functions, pathways, functional models, clusters and more. The system presents these data interconnected with WWW links, but it also allows searches and comparisons. Users may annotate and comment genes and pathways, but cannot currently edit sequences. It is probably the most extensive such integration available and is being actively developed by Integrated Genomics Inc., Chicago, USA.
http://wit.integratedgenomics.com/IGwit/
http://wit.integratedgenomics.com/ERGO/CGI/genomes.cgi?page_type=MODELS_PAGE

GDB Human Genome Database
http://gdbwww.gdb.org

GenBank National Center for Biotechnology Information http://www.ncbi.nlm.nih.gov
http://www.ncbi.nlm.nih.gov/Genbank/index.html

GenProtEC genes and proteins of *E. coli* K12.
http://genprotec.mbl.edu/

Hereditary Disease Foundation
http://www.hdfoundation.org/index.html
http://www.hdfoundation.org/links.htm

Human Gene Mutation Database at the Institute of Medical Genetics in Cardiff, Wales.
http://archive.uwcm.ac.uk/uwcm/mg/hgmd0.html

Human Genome Project Information U.S. Department of Energy.
http://www.ornl.gov/hgmis

Human Genome Resources National Center for Biotechnology Information (NCBI).
http://www.ncbi.nlm.nih.gov/genome/guide

Image Library of Biological Macromolecules Institute für Molekulare Biotechnologie, Jena, Germany.
http://www.imb-jena.de/IMAGE.html

Kyoto Encyclopedia of Genes and Genomes (KEGG) linking genes to biochemical pathways.
http://www.genome.ad.jp/kegg/

Little People of America
http://www.lpaonline.org

March of Dimes Birth Defects Foundation
http://www.modimes.org

Mendelian Inheritence in Man: OMIM Online Mendelian Inheritance in Man: a catalog of inherited human diseases.
http://www3.ncbi.nlm.nih.gov/omim

MIPS (Munich Information Center for Protein Sequences) database for protein sequences, homology data, yeast genome information.
http://www.mips.embnet.org

miRNA database http://www.sanger.ac.uk/software/Rfam/miRNA/

MITOMAP A compendium of polymorphisms and mutations of the human mitochondrial DNA.
http://www.gen.emory.edu/mitomap.html

Molecular Modeling DataBase (MMDB) a NCBI database containing a subset of 3-D structures of biological molecules contained in the Protein Data Bank (PDB), as well as tools for analyzing them.
http://www.ncbi.nlm.gov/Structure/MMDB/mmdb.shtml

Molecular Probe Data Base (MPDB) Information on ca. 4,300 synthetic oligonucleotides with a sequence of up to 100 nucleotides.
http://www.biotech.ist.unige.it/interlab/mpdb.html

Mouse Databases
http://www.informatics.jax.org/
http://www.rodentia.com/wmc/

National Center for Biotechnology Information
http://www.ncbi.nlm.nih.gov/

National Human Genome Research Institute
http://www.nhgri.nih.gov

National Newborn Screening and Genetics Resource Center (NNSGRC)
http://genes-r-us.uthscsa.edu/

Online Mendelian Inheritance in Animals (OMIA)
http://www.angis.org.au/omia/

Plants website (United States Department of Agriculture) databases with plant nomenclature, classification, distribution maps, links, and other information.
http://plants.usda.gov/index.html

Protein Data Bank (PDB) a database for the collection and dissemination of three-dimensional structures of biological molecules.
http://www.rcsb.org/pdb/

Protein families database network
http://www.proweb.org

Protein Information Resource (PIR) and the PIR-International Protein Sequence Database
http://pir.georgetown.edu

ProtoMap Automatically generated hierarchical classification of all SWISSPROT and TrEMBL proteins
http://protomap.cornell.edu/

Radiation Hybrid Database
http://www.ebi.ac.uk/RHdb

Rare Human Diseases, National Institutes of Health
http://rarediseases.info.nih.gov/ord/index.html

Rebase restriction enzymes and methylases.
http://www.neb.com/rebase

Ribosomal Database Project (RDP)
http://rdp.cme.msu.edu/html/

Sanger Centre
http://www.sanger.ac.uk/Projects

SBASE The SBASE protein domain library sequences over 200,000 annotated structural, functional, ligand-binding and topogenic segments of proteins, cross-referenced to all major sequence databases and sequence pattern collections.
http://www3.icgeb.trieste.it/~sbasesrv/
http://sbase.abc.hu/sbase/

Signal Recognition Particle Database (SRPDB)
http://psyche.uthct.edu/dbs/SRPDB/SRPDB.html

Single Nucleotide Polymorphisms (SNPs)
http://www.ncbi.nlm.nih/gov/SNP/

siRNA resource
www.rockefeller.edu/labheads/tuschl/sirna.html

SWISS-PROT protein sequence data bank.
http://www.expasy.ch/

Telomere Database Washington University School of Medicine, St. Louis.
http://www.genlink.wustl.edu/teldb/index.html

TreeBase a database designed to manage and explore information on phylogenetic relationships. All types of phylogenetic data (e.g., trees of species, trees of populations, trees of genes) are available. The database can be used as an interactive tool for browsing published evolutionary trees representing all biotic taxa.
http://treebase.org

TRRD transcriptional regulatory regions database.
http://www.bionet.nsc.ru/trrd

Universal Virus Database Australian National University, Canberra. The official catalog of the International Committee on Taxonomy of Viruses (ICTV).
http://life.anu.edu.au/viruses/welcome.htm

Yeast Protein Database (YPD) complete proteome of *Saccharomyces cervesiae*. YPD also provides proteomes for *S. pombe* and *Caenorhabditis*.
http://www.proteome.com/YPDhome.html

SPECIES WEB SITE ADDRESSES

Agrobacterium tumefaciens http://helios.bto.ed.ac.uk/bto/microbes/crown.htm
Anopheles gambiae http://klab.agsci.colostate.edu/
Arabidopsis thaliana http://www.arabidopsis.org/
Archaeoglobus fulgidus http://www.tigr.org/tigr-scripts/CMR2/GenomePage3.spl?database=gaf
Bacillus subtilis http://genolist.pasteur.fr/SubtiList/genome.cgi?gene_list
Bos taurus http://www.tigr.org/tdb/btgi/
Caenorhabditis elegans http://www.sanger.ac.uk/Projects/C_elegans/

Canis familiaris http://mendel.berkley.edu/dog.html

Chlamydomonas reinhardi http://www.botany.duke.edu/dcmb/chlamy.htm

Danio rerio http://zebra.biol.sci.edu/

Deinococcus radiodurans http://www.tigr.org/tigr-scripts/CMR2/GenomePage3.spl?database=gdr

Dictyostelium discoideum http://dictybase.org/dicty.html

Drosophila melanogaster http://flybase.bio.indiana.edu

Equus caballus http://www.vgl.ucdavis.edu/~lvmillon/

Escherichia coli http://www.cgsc.biology.yale.edu/top.html

Felis catus http://www.ri.bbsrc.ac.uk/cgi-bin/map

Gallus domesticus http://poultry.mph.msu.edu/

Glycine max http://www.ncbi.nlm.nih.gov/cgi-bin/Entrez/map_search?chr=soybean.inf

Gossypium species http://plants.usda.gov/cgi_bin/plant_profile.cgi?symbol=GOSSY#distribution

Haemophilus influenzae http://www.tigr.org/tigr-scripts/CMR2/GenomePage3.spl?database=ghi

Helicobacter pylori http://www.tigr.org/tigr-scripts/CMR2/GenomePage3.spl?database=ghp

Homo sapiens *http://www.ncbi.nlm.nih.gov/Omim/*

Methanococcus jannaschii http://www.tigr.org/tigr-scripts/CMR2/GenomePage3.spl?databases=arg

Mus musculus http://www.ncbi.nlm.nih.gov/genome/guide/mouse/
http://www.informatics.jax.org/mgihome/GXD/aboutGXD.shtml

Mycobacterium leprae http://www.biochem.kth.se/MycDB.html

Mycobacterium tuberculosis http://www.tigr.org/tigr-scripts/CMR2/GenomePage3.spl?database=gmt

Mycoplasma genitalium http://www.tigr.org/tigr-scripts/CMR2/GenomePage3.spl?database=gmg

Mycoplasma pneumoniae http://www.zmbh.uni-heidelberg.de/M_pneumoniae/MP_Home.html

Neurospora crassa http://www.genome.ou.edu/fungal.html

Ovis aries http://www.ri.bbsrc.ac.uk/sheepmap/

Oryza sativa http://rgp.dna.affrc.go.jp/

Oryzia latipes Medakafish Home Page

Pan troglodytes http://sayer.lab.nig.ac.jp/~silver/

Pisum sativum http://pisum.bionet.nsc.ru/

Plasmodium falciparum http://PlasmoDB.org

Rattus norvegicus http://www.ncbi.nlm.nih.gov/genome/guide/rat/

Rhizobium http://www.rhizobium.umn.edu

Saccharomyces cerevisiae http://genome-www.stanford.edu/Saccharomyces/

Solanum tuberosum http://www.tigr.org/tdb/stgi/

Staphylococcus aureus http://www.sanger.ac.uk/Projects/S_aureus/

Streptococcus pyogenes http://dna/chem.ou.edu/strep.html

Streptomyces coelicolor http://www.sanger.ac.uk/Projects/S_coelicolor/

Sus scrofa http://www.ri.bbsrc.ac.uk/pigmap/pigbase.html

Synechocystis PCC6803 http://www.kazusa.or.jp/cyano/cyano.html

Takifugu rubripes *See* Fugu Genome Browser Gateway.

Tetrahymena pyriformis http://www.lifesci.ucsb.edu/~genome/Tetrahymena/

Thermotoga maritima http://www.tigr.org/tigr-scripts/CMR2/GenomePage3.spl?database=btm

Triticum aestivum http://www.genoscope.cns.fr/externe/English/Projets/Projet_BG/BG.html

Vibrio cholerae http://www.tigr.org/tigr-scripts/CMR2/GenomePage3.spl?databases=gvc

Xenopus http://www.xenbase.org

Zea mays http://www.agron.missouri.edu/main.html

Appendix F: Genome Sizes and Gene Numbers

	Organisms	Genome sizes	Total ORFs	Descriptions
1	MS2	3.6×10^3	4	RNA bacteriophage
2	Q beta (Q β) phage	4.2×10^3	3	RNA bacteriophage
3	SV40	5.2×10^3	8	Simian Virus 40
4	phi X174 (ϕ X174) virus	5.4×10^3	9	ssDNA bacteriophage
5	TMV	6.4×10^3	4	Tobacco Mosaic Virus (RNA)
6	HIV	9.3×10^3	10	AIDS virus (RNA)
7	mitochondrion	16.6×10^3	37	Human mitochondrion
8	HAdV-2	35.9×10^3	11	Human adenovirus 2
9	lambda (λ) phage	48.5×10^3	50	lysogenic bacteriophage
10	chloroplast	118×10^3	135	from *Mesostigma viride*
11	T4 phage	169×10^3	300	DNA virus of *E. coli*
12	Var	185×10^3	206	Variola virus
13	human cytomegalovirus	230×10^6	208	a herpes virus
14	*Mycoplasma genitalium*	0.58×10^6	470	Bacteria, Aphragmabacteria
15	*Chlamydia trachomatis*	1.04×10^6	894	Bacteria, Pirellulae
16	*Rickettsia prowazeki*	1.11×10^6	834	Bacteria, Proteobacteria
17	*Methanococcus jannashii*	1.64×10^6	1,682	Archaea, Euryarchaeota
18	*Haemophilus influenzae*	1.83×10^6	1,752	Bacteria, Proteobacteria
19	*Thermotoga maritima*	1.86×10^6	1,877	Bacteria, Thermotogae
20	*Streptococcus pyogenes*	1.85×10^6	1,752	Bacteria, Proteobacteria
21	*Streptococcus pneumoniae*	2.16×10^6	2,236	pneumococcus
22	*Archaeoglobus fulgidus*	2.18×10^6	2,436	Archaea, Crenarchaeota
23	*Halobacterium NRC-1*	2.57×10^6	2,682	Archaea, Crenarchaeota
24	*Encephalitozoon cuniculi*	2.71×10^6	1,997	Protoctista, Microspora
25	*Deinococcus radiourans*	3.28×10^6	3,187	Bacteria, Deinococci
26	*Synechocystis PCC6803*	3.57×10^6	3,168	Bacteria, Cyanobacteria
27	*Vibrio cholerae*	4.03×10^6	3,885	Bacteria, Proteobacteria
28	*Mycobacterium tuberculosis*	4.41×10^6	3,924	Bacteria, Actinobacteria
29	*Bacillus subtilis*	4.21×10^6	4,100	Bacteria, Endospora
30	*Escherichia coli*	4.64×10^6	4,288	Bacteria, Proteobacteria
31	*Pseudomonas aeruginosa*	6.26×10^6	5,570	Bacteria, Proteobacteria
32	*Streptomyces coelicolor*	8.26×10^6	7,825	Bacteria, Actinobacteria
33	*Saccharomyces cerevisiae*	12.07×10^6	5,885	Fungi, Ascomycota
34	*Schizosaccharomyces pombe*	12.5×10^6	5,400	fission yeast
35	*Cyanidioschyzon merolae*	16.5×10^6	5,331	Protoctista, Rhodophyta
36	*Plasmodium flaciparum*	22.8×10^6	5,268	Protoctista, Apicomplexa
37	*Trypanosoma brucei*	26×10^6	9,068	Protoctista, Zoomastigina
38	*Neurospora crassa*	38.6×10^6	10,082	filamentous fungus
39	*Caenorhabditis elegans*	100.3×10^6	19,100	Animalia, Nematoda
40	*Ciona intestinalis*	160×10^6	15,582	Deuterostomia, Urochordata
41	*Arabidopsis thaliana*	120×10^6	25,500	Plantae, Angiospermae
42	*Oryza sativa*	389×10^6	37,544	Plantae, Angiospermae

	Organisms	Genome sizes	Total ORFs	Descriptions
43	*Drosophila melanogaster*	176×10^6	13,000	Arthropoda, Insecta, Diptera
44	*Anopheles gambia*	278×10^6	13,683	Arthropoda, Insecta, Diptera
45	*Bombyx mori*	430×10^6	18,510	Arthropoda, Insecta, Lepidoptera
46	*Takifugu rubripes*	0.4×10^9	31,000	Osteichthyes, pufferfish
47	*Gallus domesticus*	1×10^9	23,000	Aves, Galliformes
48	*Mus musculus*	2.5×10^9	30,000	laboratory mouse
49	*Homo sapiens*	3.2×10^9	31,000	Mammalia, Primates

The range in genome sizes and gene numbers for organisms or cell organelles defined in this dictionary. In all but five organisms, the genomes are given in nucleotide base pairs, since their genomes are double-stranded DNA molecules. The exceptions are viruses containing RNA (1,2,5,6) or ssDNA (4). The first 13 organisms are obligate symbionts or parasites, organisms 14–32 are prokaryotes, and the remaining 17 are eukaryotes.

Illustration Credits

Page 26 After Blakeslee, 1934. Drawings of *Datura* seed capsules by E. John Pfiffner.

Page 44 From Beermann, W., 1952, Chromosomenkonstanz und spezifische Modifikationen der Chromosomenstruktur in der Entwicklung und Organdifferenzierung von *Chironomus tentans*. *Chromosoma* 5: 139–198. © 1952, Springer-Verlag. Reproduced with permission of Springer Science and Business Media.

Page 50 Redrawn from *General Zoology*, 4th ed., by T. I. Storer and R. L. Usinger. © 1965 by McGraw-Hill, Inc. Reproduced with permission of the estate of Robert L. Usinger and the Archives of the California Academy of Sciences, San Francisco.

Page 52 Reprinted by permission of Kettlewell, H. B., 1956, Further selection experiments on industrial melanism in the Lepidoptera. *Heredity* 10: 287–301. © 1956, Macmillan Publishers Ltd. Drawings of peppered moths by E. John Pfiffner.

Page 74 Redrawn from J. F. Crow, *Genetics Notes*, Burgess Publishing Company (1983). Reproduced with permission of J. F. Crow.

Page 85 Drawing of citric acid cycle by Vikram K. Mulligan.

Page 98 Redrawn from G. H. Beale, *The Genetics of Paramecium*, 1954, Fig. 1, p. 27, Cambridge University Press, UK. Reproduced with permission of Cambridge University Press, UK.

Page 115 From *Principles of Human Genetics*, 2nd ed., by Curt Stern. © 1960, 1973, 1980 by W. H. Freeman and Company. Used with permission.

Page 130 From *General Genetics* by A. M. Serb and R. D. Owen. © 1952 by W. BW/BBY Serral Gent, W. H. Freeman and Company. Used with permission.

Page 134 Reprinted with permission from M. D. Adams et al., 2000, The genome sequence of *Drosophila melanogaster*. *Science* 287 (5461): 2185–2195. © 2000 American Association for the Advancement of Science.

Page 144 Drawings of enveloped viruses by Robert S. King.

Page 181 Adapted from Harold L. Levin, *The Earth Through Time*, 7th ed., Fig. 1–15. © 2003 by John Wiley & Sons, Inc. Reprinted with permission of John Wiley & Sons, Inc.

Page 183 From E. Schnepf, W. Koch, and G. Deichgraber, "Zur Cytologie und taxonomischen Einordnung von Glaucocytis." *Archiv fur Mikrobiologie* 55 (1966): 151. Reproduced with permission of Springer Science and Business Media.

Page 184 Drawing of glucose molecule by Vikram K. Mulligan.

Page 186 Drawing of glycolysis by Vikram K. Mulligan.

Page 192 Reprinted with permission from *Nature* (A. C. Bell and G. Felsenfeld, 2000, Methylation of a CTCF-dependent boundary controls imprinted expression of the Igf2 gene. *Nature* 405: 482–485). © 2000 Macmillan Magazines Limited, London, UK.

Page 195 From D. S. Falconer, 1960, *An Introduction to Quantitative Genetics*. The Ronald Press Co., New York.

Page 209 Adapted from H. Potter and D. Dressler, 1976, On the mechanism of genetic recombination: electron microscopic observation of recombination intermediates. *Proc. Natl. Acad. Sci. USA* 73: 3000–3004, Fig. 6. Reproduced with permission of H. Potter, University of South Florida College of Medicine, Tampa, FA. D. Dressler is at Oxford University, UK.

Page 220 Adapted from T. E. Creighton, 1999, *Encyclopedia of Molecular Biology*, Vol. 2, Fig. 1, p. 1167. © 1999 by John Wiley & Sons. Reprinted with permission of John Wiley & Sons, Inc.

Page 227 From Stansfield, W. D., *Serology and Immunology*, 1981, Fig. 2.11, page 45. Reproduced with permission of W. D. Stansfield.

Page 248 Image courtesy of Joseph G. Gall, Carnegie Institution of Washington, Baltimore, MD. Reproduced with permission of J. G. Gall.

Page 286 Illustration by Jon D. Moulton, Gene Tools, LLC, Philomath, OR. Reproduced with permission of J. D. Moulton.

Page 298 Drawing by Dan-E. Nilsson, Lund University, Lund, Sweden. Reproduced with permission of D.-E. Nilsson.

Page 304 From L. W. Sharp, *Fundamentals of Cytology*, page 85. © 1943, McGraw Hill, NY.

Page 312 Adapted from K. R. Gegenfurtner and L. T. Sharpe eds., *Color Vision: From Genes to Perception*, 1999, Fig. 1.2, p. 6, Cambridge University Press, UK. Reproduced with permission of Cambridge University Press, UK.

Page 335 Reprinted by permission from *Nature*. R. E. Franklin and R. E. Gosling, Molecular configuration in sodium thymonucleate. *Nature* 171 (April 25, 1953): 740–741. © 1953 Macmillan Publishers Ltd.

Page 340 From D. F. Wirth, The parasite genome: biological revelations. *Nature* **419** (3 October 2002): 495–496.

Page 346 From P. D. Storto and R. C. King, 1989, The role of polyfusomes in generating branched chains of cystocytes during *Drosophila* oogenesis. *Dev. Genet.* 10: 70–86, Fig. 8. © 1989 by John Wiley & Sons, Inc. Reprinted with permission of Wiley-Liss, Inc., a subsidiary of John Wiley & Sons, Inc.

Page 366 Adapted from R. A. Emerson and E. M. East, The inheritance of quantitative characters in maize. *Nebraska Agr. Exp. Sta. Res. Bull.* 2 (1913).

Page 368 From S. Benzer, 1955, Fine structure of a genetic region in bacteriophage. *Proc. Natl. Acad. Sci.* 41: 344–354. Reproduced with permission of S. Benzer.

Page 370 From A. Hollaender, ed., *Radiation Biology*, Vol. 1 (1954), page 716, McGraw-Hill, NY.

Page 375 Reprinted with permission from R. J. Britten and D. E. Kohne, 1968, Repeated sequences in DNA. *Science* 161: 529–540. © 1968 American Association for the Advancement of Science.

Page 385 From *Biochemistry*, 3rd ed., by Lubert Stryer. © 1975, 1981, 1988 by Lubert Stryer. Used with permission of W. H. Freeman and Company.

Page 406 Scanning electron microscope image by F. R. Turner, courtesy of FlyBase.

Page 410 Drawing of the silversword *Argyroxiphium sandwicense* by R. S. King.

Page 414 Drawing of solenoid structure by R. S. King.

Page 415 Adapted from *Current Opinion in Genetics and Development* Vol. 1, M. L. Sogin, Early evolution and the origin of eukaryotes, pp. 457–463. © 1991, reproduced with permission from Elsevier and from M. L. Sogin.

Page 422 From P. B. Moore, Chapter 10, Fig. 3C, in *Ribosomal RNA: Structure, Evolution, Processing and Function in Protein Biosynthesis*, R. A. Zimmermann and A. E. Dahlberg (eds.), 1996. CRC Press, Florida.

Page 423 From *Biochemistry*, 3rd ed., by Lubert Stryer. © 1975, 1981, 1988 by Lubert Stryer. Used with permission of W. H. Freeman and Company.

Page 434 Reprinted from *International Review of Cytology*, Vol. 28, R. C. King, The meiotic behavior of the *Drosophila* oocyte, Fig. 4A, page 136. © 1970, reproduced with permission from Elsevier.

Page 437 Chart of tautomeric shifts by Vikram K. Mulligan.

Page 439 From D. Gilley and E. H. Blackburn, 1996, Specific RNA residue interactions required for enzymatic functions of *Tetrahymena* telomerase. *Mol. Cell. Biol.* 16: 66–75, Fig. 1A. Reproduced with permission of American Society for Microbiology and of D. Gilley.

Page 447 Drawing of tobacco mosaic virus by Vikram K. Mulligan.

Page 466 From André Brack, ed., *The Molecular Origins of Life*, 1998, Fig. 16, page 338, Cambridge University Press, NY.

Page 468 Reproduced with permission from *Plant Cell Vacuoles: An Introduction* by D. N. De (2000), Fig. 1.1, page 4. Published by CSIRO Publishing, Melbourne, Australia (http://www.publish.csiro.au/pld/2318.htm).

Page 481 Illustration of yeast two-hybrid system by Vikram K. Mulligan.